Handbook of Research on Computational Grid Technologies for Life Sciences, Biomedicine, and Healthcare

Mario Cannataro
University Magna Graecia of Catanzaro, Italy

Volume I

Medical Information Science REFERENCE
MEDICAL INFORMATION SCIENCE REFERENCE
Hershey · New York

Director of Editorial Content:	Kristin Klinger
Senior Managing Editor:	Jamie Snavely
Managing Editor:	Jeff Ash
Assistant Managing Editor:	Carole Coulson
Typesetter:	Jeff Ash
Cover Design:	Lisa Tosheff
Printed at:	Yurchak Printing Inc.

Published in the United States of America by
Information Science Reference (an imprint of IGI Global)
701 E. Chocolate Avenue
Hershey PA 17033
Tel: 717-533-8845
Fax: 717-533-8661
E-mail: cust@igi-global.com
Web site: http://www.igi-global.com/reference

and in the United Kingdom by
Information Science Reference (an imprint of IGI Global)
3 Henrietta Street
Covent Garden
London WC2E 8LU
Tel: 44 20 7240 0856
Fax: 44 20 7379 0609
Web site: http://www.eurospanbookstore.com

Library of Congress Cataloging-in-Publication Data

Handbook of research on computational grid technologies for life sciences, biomedicine, and healthcare / Mario Cannataro, editor.
p. cm.

Summary: "This book provides methodologies and developments of grid technologies applied in different fields of life sciences"--Provided by publisher.

Includes bibliographical references and index.
ISBN 978-1-60566-374-6 (hardcover) -- ISBN 978-1-60566-375-3 (ebook) 1. Bioinformatics. 2. Computational biology. 3. Problem solving--Data processing. I. Cannataro, Mario, 1964-

QH324.2.H3576 2009
570.285--dc22

2009001527

British Cataloguing in Publication Data
A Cataloguing in Publication record for this book is available from the British Library.

To Francesco and Matteo

Editorial Advisory Board

List of Contributors

Table of Contents

Detailed Table of Contents

Grid technology, one of the key technologies for the "European Research Area", offers rapid computation, large scale data storage and flexible collaboration by harnessing together the power of large numbers of computers, from end-users' desktops to powerful workstations and clusters of more powerful machines. However, a major challenge is to take the technology out of the laboratory to the citizen. The application of Grid technology to biomedical and healthcare informatics, in short HealthGrid, presents some difficult challenges. The chapter presents the results of the SHARE project (http://www.eu-share.org) that identified the key developments, i.e. technical advances, social actions, economic investments and ethical or legal initiatives, needed to achieve wide adoption and deployment of HealthGrids throughout Europe. The project analyses several case studies and discusses technical, ethical, legal, social and economic issues which may impede early deployment of HealthGrids.

The emerging technology of HealthGrids holds the promise to successfully integrate health information systems and various healthcare entities onto a common, globally shared and easily accessible platform. This chapter presents a taxonomy of different types of HealthGrid resources and proposes some solutions for the resource discovery problem, an emerging challenge in HealthGrids. A discussion on discovering and integrating data resources is also provided.

Data provenance is key for understanding and interpreting the results of scientific experiments. This chapter introduces and characterises data provenance in scientific workflows using illustrative examples taken from real-world workflows. Scientific workflows and related provenance data are first defined. Then the chapter proposes a taxonomy that characterizes provenance in scientific workflows. Such taxonomy is used for comparing and analysing provenance capabilities supplied by existing scientific workflow systems.

Provenance tracking is an indispensable element of each e-Science infrastructure for conducting in silico experiments. The chapter proposes an ontology-based provenance model which captures the execution

Section II
Grids for Genomics and Proteomics

This section discusses the use of the Grid for the management and analysis of genomics and proteomics data, the basic data at the biological level. After describing main issues for the parallel and distributed implementation of BLAST, a cornerstone of all genomics analysis, the Section discusses some significant Grid-based implementations of genomics applications. Then, the Section introduces proteomics with a special focus on mass spectrometry-based proteomics and different Grid-based proteomics applications, ranging from biomarker discovery to protein identification and protein classification, are discussed.

Vincent Breton, IN2P3, CNRS, Clermont-Ferrand, France
Eddy Caron, Université de Lyon, LIP, CNRS-ENS-Lyon-UCBL-INRIA, France
Frédéric Desprez, INRIA, Université de Lyon, LIP, CNRS-ENS-Lyon-UCBL-INRIA, France
Gaël Le Mahec, CNRS, IN2P3, UBP Clermont-Ferrand, France

As grids become more and more attractive for solving complex problems with high computational and storage requirements, bioinformatics algorithms start to be ported on large scale platforms. The BLAST kernel was one the first application ported on such platform. However, although simple parallelization was enough for the first proof of concept, its use in production platform needed more optimized algorithms. The chapter reviews existing parallelization and "gridification" of the BLAST algorithm as well as related issues such as data management and replication. A case study using the DIET middleware over the Grid'5000 experimental platform is also presented.

Luciano Milanesi, Istituto di Tecnologie Biomediche – Consiglio Nazionale delle Ricerche, Italy
Ivan Merelli, Istituto di Tecnologie Biomediche – Consiglio Nazionale delle Ricerche, Italy
Gabriele Trombetti, Istituto di Tecnologie Biomediche – Consiglio Nazionale delle Ricerche, Italy
Paolo Cozzi, Istituto di Tecnologie Biomediche – Consiglio Nazionale delle Ricerche, Italy
Alessandro Orro, Istituto di Tecnologie Biomediche – Consiglio Nazionale delle Ricerche, Italy

A common ongoing task for Functional Genomics is to compare full organisms' genome with those of related species, to search in huge database for functional annotation of novel sequences and to identify specific patterns of them, such as ESTs, genes, and microRNA. The prediction of these patterns has a relevant computational cost, while public genome archives exceed one billion sequence traces from over 1,000 organisms and this number is increasing rapidly. Thus Functional Genomics applications require significant computational infrastructures, where reusable tools and resources can be accessed.

The chapter describes issues and challenges in porting main Functional Genomics applications, from gene prediction, to sequence alignment, and phylogenetic applications on the Grid. The implementation and evaluation of a software environment for the management of distributed bioinformatics computations over the Grid is also presented.

This chapter describes PheGee@Home, a grid-based comparative genomics tool that finds candidate genes responsible for a given phenotype, where a phenotype is intended as the physical manifestation of the interplay of genetic, epigenetic and environmental factors. The tool facilitates the discovery and prioritization of candidate genes controlling or contributing to the genetically determined portion of a specified phenotype. Due to prohibitively long runtimes the system architecture is based on a desktop grid environment and commodity graphics hardware to significantly accelerate PheGee. Authors validate this approach by showing the comparison of microbial genomes on a grid testbed.

The number of fully sequenced genomes increases daily, producing an exponential explosion of the sequence, annotation and metadata databases. Data analysis on a genome-wide level has become a data- and computation-intensive task. However, most genomics applications can be partitioned into many independent tasks that can be scheduled on high performance computers or Grids. The problem addressed by the chapter is the Grid-based analysis of Gene Ontology data and its associations to gene products of any kind of organism to find gene products with similar functionalities. The chapter presents a system to partition the computation of the full search into a large number of jobs and to submit these jobs to the Grid infrastructure as long as all jobs are processed successfully, guaranteeing an analysis of the data without missing any information.

Recent advances in high throughput technologies such as mass spectrometry enabled the researchers to collect a huge amount of data when analysing biological samples. Computational Proteomics regards the computational methods for analyzing spectra data in qualitative (i.e. peptide/protein identification in tandem mass spectrometry), and quantitative proteomics (i.e. protein expression measurement in samples), as well as in biomarker discovery (i.e. the identification of a molecular signature of a disease directly from spectra). This chapter presents main standards, tools, and technologies for building scalable, reusable, and portable applications in this field. The chapter surveys available solutions for computational proteomics and describes MS-Analyzer, a Grid-based software platform for the integrated management and analysis of spectra data. MS-Analyzer provides efficient spectra management through a specialized spectra database, and supports the semantic composition of pre-processing and data mining services to analyze spectra on the Grid.

Proteomics is currently one of the most promising fields in bioinformatics as it provides important insights into the function of proteins of organisms. Mass spectrometry is one of the techniques to study the proteome, and several software tools exist for this purpose. The chapter first introduces the protein identification problem, an important application in proteomics, then it presents a meta-computing approach for solving such problem by combining different existing predictors. The chapter presents an extendable software platform called swissPIT that combines different existing tools and exploits Grid infrastructures to speed up the data analysis process for the proteomics pipeline.

The goal of this chapter is the presentation of Data Mining techniques for knowledge extraction in proteomics, taking into account both the particular features of most proteomics issues (such as data retrieval and system complexity), and the opportunities and constraints found in a Grid environment. The chapter discusses how new and potentially useful knowledge can be extracted from proteomics data utilizing Grid resources in a transparent way. As a case study, the problem of protein classification is considered. After introducing an overview of Data Mining algorithms with emphasis on the specific needs of protein classification, the chapter presents a unified methodology for complex Data Mining processes on the Grid, highlighting the different application types and the benefits and drawbacks in each case. Finally, the methodology is validated through real-world case studies, deployed over the EGEE grid environment.

The chapter presents BioSimGrid, a distributed biomolecular simulation database. It is a general-purpose database for trajectories from molecular dynamics simulations. The presentation of BioSimGrid explains how to install the system, and how to deposit, query, and analyze trajectories with real Python code examples for each step. The chapter also presents the underlying concepts in the implementation of BioSimGrid: relational database, distributed computing, and the input/output (deposit and analysis) modules. The chapter concludes by discussing emerging trends in biomolecular simulations and concerns in the further development of BioSimGrid and similar biological databases.

The New York State Grid (NYS Grid) is an integrated computational and data grid that provides access to users from around the world to a wide variety of resources. This Grid can be accessed via a Web portal, where the users have access to their data sets and applications, but do not need to be made aware of the details of the data storage or computational devices that are specifically employed in solving their problems. The chapter presents a Grid-enabled version of the SnB and BnP programs which respectively implement the Shake-and-Bake method of molecular structure (SnB) and substructure (BnP) determination. The programs were run on the NYS Grid and their performance evaluation is presented. In particular, SnB has been run simultaneously through the Grid Portal on all the computational resources of the NYS Grid as well as on more than 1100 of the over 3000 processors available through the Open Science Grid.

Section IV
Grids for Medical Informatics

This Section focuses on Grid-based medical informatics applications. Since a core aspect of medical applications regards the acquisition and analysis of biomedical images, the Section first introduces different aspects of biomedical images visualization on the Grid. Then, as significant applications in the field, Grid analysis of radiological data and 3D electron microscopy reconstruction are discussed. Finally, as an example of biomedical instrumentation made available on the Grid, a hybrid mock circulatory system to test cardiovascular prostheses is presented.

Visualization, the art and science of representing data visually, is now recognized as an equal partner in the conduct of science via the simulation and modeling paradigm. Different aspects of visualization, such as image preprocessing/analysis, and some computational geometric and computational topological processes are amenable to deployment over compute Grids, but there has been equal focus on the collaborative aspect of Grid computing which is driving collaboration-based visualization systems. The chapter introduces some aspects of visualization and the grid. The chapter first surveys some of the roles of visualization as they relate to the role of Grid computing within a biomedical context. Then it examines certain scheduling strategies that are believed to have value in terms of the distribution of visualization tasks over Grid fabrics.

Volume II

Grid technologies and infrastructures can contribute to harnessing the full power of computer-aided image analysis into clinical research and practice. A main challenge in medical images analysis is to fill the gap between the grid middleware and the requirements of clinical applications, that can be summarized in large volumes of data, sensitivity of medical information, and complexity of medical datasets. This chapter reports on the goals, achievements and lessons learned from the AGIR (Grid Analysis of Radiological Data) project. AGIR faces these challenges by providing some core grid medical services (data management, responsiveness, compression, and workflows) supporting medical data processing, and by grid-enabling a panel of applications ranging from algorithmic research to clinical use cases.

Three-dimensional electron microscopy allows scientists to study biological specimens and to understand how they behave and interact with each other depending on their structural conformation. Electron microscopy projections of the specimens are taken from different angles and are processed to obtain a virtual three-dimensional reconstruction for further studies. Nevertheless, the whole reconstruction process is neither straightforward nor cheap in terms of computational costs, thus different computing paradigms have been applied in order to overcome such high costs. The chapter explores the main tasks present in a typical three-dimensional electron microscopy reconstruction process showing parallelization and gridification issues. In addition, important aspects like fault-tolerance are widely covered; given that the distributed nature of a grid infrastructure makes it inherently unstable and difficult to predict. Issues on Automatic Grid Job Management and Automatic Files Replication on the EGEE Grid are also described.

In the last decades cardiovascular diseases greatly increased worldwide and bioengineering provided new technologies and cardiovascular prostheses to medical doctors and surgeons. The design of active and passive devices aroused notable interests becoming more and more challenging as well as crucial. In this framework, it is important to faithfully reproduce the interaction between the prostheses and the cardiovascular system when in-vitro experiments are performed. For this reason, new and improved kind of test benches become necessary. Purely hydraulic mock circulatory systems showed low flexibility to allow tests of different cardiovascular devices and low precision when a reference mathematical model must be reproduced. In this chapter a new bench is described. It combines the computer model of the cardiovascular system and its real-time interaction with the device to be tested. A possible architecture to deploy the solution adopted in a Grid environment to allow remote experimentation is presented.

Section V
Collaborative Grids for Healthcare and Clinical Applications

Healthcare applications at the population level as well as applications used in the clinical context are the focus of this Section. These applications are the most difficult to port on the Grid, due to the high number of patients involved, the partitioning of data among different health centers, the privacy and security problems and the need of collaboration among various operators. The use of the Grid as a support for collaboration in epidemiology, as well as the integration and analysis of epidemiology data are fully described. Two important HealthGrids for paediatrics and brain injury are presented showing how the Grid is a mature technology also for the clinical setting.

Epidemiology constitutes one relevant use case for the adoption of grids for health. It combines challenges that have been traditionally addressed by grid technologies, such as managing large amounts of distributed and heterogeneous data, large scale computing and the need for integration and collaboration tools, but introduces new challenges traditionally addressed from the e-health area. Although grid technology has been applied to epidemiology, e.g. for data federation and for evaluating statistical epidemiological models, however, epidemiology presents important constraints that are not solved. The chapter presents the most important problems of epidemiology, such as the semantic integration of data, the effective management of security and privacy, the lack of exploitation models for the use of infrastructures, the instability of Quality of Service and the seamless integration of the technology on the epidemiology environment. Then it presents an analysis of how these issues are being considered in state-of-the-art research.

The chapter presents IntegraEPI, a Grid-based system for space-time visualization, monitoring, modeling and analysis of epidemic data. The system integrates data from heterogeneous epidemic databases and provides analytical and computational methods to increase the predicting capability of the public health system when dealing with epidemic outbreak and prevention. By using IntegraEPI, Health authorities will be able to decide about a set of possible actions that will be previously tested in a virtual population interacting in an urban infrastructure, considering its environmental factors, and finally compare the simulated data to consolidated data of real epidemic dynamics.

The Health-e-Child project is developing a Grid-based healthcare platform for European paediatrics and providing seamless integration of traditional and emerging sources of biomedical information. It aims to provide data integration across heterogeneous biomedical information in order to facilitate improved

Research environments for modern, cross-disciplinary scientific endeavors have to unite multiple users, with varying levels of expertise and roles, along with multitudes of data sources and processing units. The high level of required integration contrasts with the loosely-coupled nature of environments which are appropriate for research. A main problem is to support integration of dynamic service-based infrastructures with data sources, tools and users in a way that conserves ubiquity, extensibility and usability. This chapter presents the basic concepts of semantics-based collaborative environments, including semantic data, semantic metadata, annotations, and services integration. Then the authors demonstrate that using semantics as an integration mechanism, i.e. combining formal representations of domain knowledge with techniques like data integration, semantic annotations and shared vocabularies, enables the development of systems for modern e-Science (collaborative laboratories named "collaboratories"). As case study, the way how several semantically-augmented experiments are modeled in the ViroLab virtual laboratory for virology is presented.

Marian Bubak, Institute of Computer Science AGH, Poland & University of Amsterdam, The Netherlands
Maciej Malawski, Institute of Computer Science AGH, Poland
Tomasz Gubala, ACC CYFRONET AGH, Poland & University of Amsterdam, The Netherlands
Marek Kasztelnik, ACC CYFRONET AGH, Poland
Piotr Nowakowski, ACC CYFRONET AGH, Poland
Daniel Harezlak, ACC CYFRONET AGH, Poland
Tomasz Bartynski, University of Amsterdam, The Netherlands & ACC CYFRONET AGH, Poland
Joanna Kocot, ACC CYFRONET AGH, Poland
Eryk Ciepiela, ACC CYFRONET AGH, Poland
Wlodzimierz Funika, Institute of Computer Science AGH, Poland
Dariusz Krol, ACC CYFRONET AGH, Poland
Bartosz Balis, Institute of Computer Science AGH, Poland
Matthias Assel, University of Stuttgart, Germany
Alfredo Tirado-Ramos, University of Amsterdam, The Netherlands

Advanced research in Life Sciences requires new information technology solutions to support complex computer simulations, collaborative result analysis and annotation, as well as software reuse. This chapter presents the ViroLab virtual laboratory, which is an integrated system of dedicated tools and services, providing a common space for planning, building, improving and performing in-silico experiments by different groups of users. Within the virtual laboratory, collaborative applications are built as experiment plans, using a notation based on the Ruby scripting language. During experiment execution, provenance data is created and stored. The virtual laboratory enables access to distributed, computational and data resources, in Grid systems, clusters and standalone computers. The process of application development as well as the architecture and functionality of the virtual laboratory are demonstrated using a real-life example from the HIV treatment domain.

This chapter provides an overview of Grid middleware and applications related to biomedical and Life Sciences disciplines. Various technologies, including web-based solutions, are presented. One of the solutions, the UNICORE framework, in its recent version implements key grid standards and specifications. The system architecture and capabilities, such as security, workflow and data management are described. Special attention is given to the idea of a 'gridbean', which expands the UNICORE use for different applications. Examples of gridbeans are provided and the capabilities of UNICORE are illustrated through specific examples built using this grid middleware. In particular, the Chemomentum workbench and its use for in-silico design and modeling in chemistry and Life Sciences are both described.

Nowadays many biomedicine studies are dealing with large, distributed, and heterogeneous repositories as well as with computationally demanding analyses, and complex integration techniques are more often required to handle this complexity. This chapter describes the Bio Med Portal, a Grid oriented platform that aims to promote collaboration and cooperation among scientists and healthcare research groups, enabling the remote use of resources integrated in complex software platform services forming a virtual laboratory. The Bio Med Portal is designed to host several medical services and it is able to deploy several analysis algorithms. The scope of this chapter is both to present a Grid application with its own medical use case and to emphasize the benefit that a new design paradigm based on Grid could provide to research groups spread in geographically distributed sites.

Repositories are digital stores that manage data and metadata providing their access to users, offering an easy-to-use service and a powerful system to handle digital assets. This chapter presents the gLibrary/DRI (Digital Repositories Infrastructure) platform, a Grid-based digital repository that takes advantage of the Grid features such as VO authentication, file catalogues, and metadata services. The main goal of the platform is to reduce the cost in terms of time and effort that a repository provider spends to get its repository deployed. This is achieved by providing a common infrastructure and a set of mechanisms that repository providers use to define the data model, the access to the content and the storage model. Two use cases are also presented: a mammograms repository example that provides clinicians with a tool that eases diagnostics process and an algorithmic repository based on the Poincare Surface Section.

Section VIII
Selected Readings

This Section is a short collection of suggested readings of different authors, aiming to enrich this book with others knowledge, experience, thought and insight. After introducing Cloud Computing and describing the porting of applications to Grids and Clouds, the Section describes a bio-inspired approach for the construction of a self-organizing Grid information system, and concludes discussing Grid-based implementation of phylogenetic analysis.

Cloud Computing is an emerging style of computing in which dynamically scalable resources are provided as a service. The chapter describes main stages of implementing applications on Grid and Cloud infrastructures. As a case study, the chapter presents the Distributed European Infrastructure for Supercomputing Applications (DEISA) and describes the DEISA Extreme Computing Initiative (DECI) for porting and running scientific grand challenge applications on the DEISA Grid. The chapter concludes proposing the top ten rules of building a sustainable Grid.

The chapter describes a bio-inspired approach for the construction of a self-organizing Grid information system, with the aim to foster the use of swarm intelligence, multi-agent and bio-inspired paradigms in the field of distributed computing. The chapter also describes the SO-Grid Portal, a simulation portal through which registered users can simulate and analyze the ant-based protocols and perform "parameter sweep" studies.

Phylogenetic data analysis represents an extremely compute-intensive area of Bioinformatics and thus requires high-performance technologies. In particular, the chapter focuses on hostparasite co-phylogenetic analysis, another compute- and memory-intensive problem. After introducing some tools for conducting such co-phylogenetic studies, the chapter describes their enhanced Grid-based implementations. Since the computational core of the problem is embarrassingly parallel, the chapter shows how the parallel implementation fits well to a computational Grid and reduces the response time of large scale analyses.

Foreword

"Empirical science may have to be considered, not merely as a scientific theory – that is, a system of asserted statements arranged according to certain rules – but rather as a complex consisting partly of such statements and partly of human activities." Alfred Tarski, Introduction to Logic xi-xii.

"We don't really have a health care delivery system in this country. We have an expensive plethora of uncoordinated, unlinked, economically segregated, operationally limited micro systems, each performing in ways that too often create suboptimal performance both for the overall health care infrastructure and for individual patients." George Halverson, Health Care Reform Now!: A Prescription for Change, 2007.

The development of the grid since the early 1990s has led from a system of methods for sharing computation and information resources, to a system of active virtual communities: from computation through sharing, to sharing through computation. Our ability to extend and exploit both perspectives is of extraordinary relevance to biology and medicine. It gives us the opportunity to organize increasingly sophisticated, flexible and powerful systems of logic, data, rules, computation and human activities within social networks in which time and space constraints are reduced to a minimum. In so doing, we may hope to create the infrastructure, methods, and (virtual) organizations needed to deliver the benefits of evidence-based and personalized medicine to the seven billion inhabitants of planet Earth.

What are we sharing and distributing when we apply the grid to healthcare and biomedical research? Computation, software, and information, certainly – but also to a great extent we are enabled to share computed results, which given an appropriate infrastructure can be passed from process to process (workflows), stored and retrieved, layered in mutual support, interpreted, counterposed and resolved. The Grid now enables us to construct a connected system of computational "collaboratories" within virtual organizations (VOs) that embrace different specialties and disciplines in biomedicine and healthcare, as well as engineering and software, and that must ultimately also extend to healthcare providers, payers, and of course patiens. In this way, we create *an ecosystem of actions, tools and information*. The Grid as VO is a system of radical enabling technologies for this scientific knowledge ecosystem.

Virtual cooperation and complex layering of technologies and results are now available on a global scale, allowing interpretations and conclusions to be extracted from masses of data obtained in ultra high-throughput experiments. The results of one group's experiment, or data from huge numbers of clinical observations, can now be data mined, consumed by workflow processes and used as input to further interpretations and conclusions, with great rapidity and agility. These results can then drive further experiments, inform the design of, and participation in, clinical trials, and ultimately ensure that better care is delivered to more patients, faster than before. At the core of this knowledge ecosystem are grid-enabled virtual organizations, made possible through organized and uniform service models around computational, communications and storage resources, with shared commitments to common semantics and models of component behavior.

In perhaps no other area of endeavor does the emergence of virtual organizations present such complex issues – and striking opportunities – as in biomedical research and healthcare. In the U.S. alone, health care costs surpassed $2 trillion in 2006[1] with approximately 6% of these expenditures going to biomedical research[2]. It is widely accepted that this "nonsystem," while outstanding in its technical capabilities, is far from effective in delivering the right care to the right people at the right time. Grids and VOs promise to contribute significantly to productive organization of service delivery, improvement of quality of care, and more rapid translation of basic medical research into advances in clinical care. Yet realizing this vision involves a host of challenges, technical and social.

Technical challenges include achieving the uniformity of infrastructures and services, and network, computational and communication resources, required to make sharing feasible; balancing local versus global policies and requirements (access, security, trust, load, service provisioning); and developing computable semantic representations of data, metadata, policies and services. Social challenges include adequate funding of the research required to build out the grids and VOs that will support a healthcare revolution; balancing research with product development, and the quest for the perfect with the deployment of the good; the agile capture, mapping and translation of user requirements into working systems features with adequate fidelity; overcoming the substantial disincentives that currently hinder sharing across the healthcare nonsystem; and building methods and systems that can scale not only to many patients and providers, but also to many countries, societies, and economies.

This book provides a comprehensive and focused view of the state of research in the application of grids and virtual organizations to these and other challenges. Reporting in particular on the results of the European Union's substantial HealthGrid program, the book includes descriptions of both cutting-edge research and substantial deployments. We believe it will make a significant contribution to meeting the challenges that currently face healthcare, and will be of great importance in broadening the application of an extremely powerful system of technology and social organization in biomedicine.

ENDNOTES

1 U.S. Centers for Medicare and Medicaid Services, Office of the Actuary, National Health Statistics Group, 2006 National Health Care Expenditures Data, January 2008. Retrieved from http://www.cms.hhs.gov/nationalhealthcareexpenddata/01_overview.asp?

2 Moses, H. 3rd,, Dorsey, E.R., Matheson, D.H., Their, S.O.,(2005, September 21). Financial anatomy of biomedical research. *JAMA 294*(11), 1333-42.

Tim Clark & Ian Foster
March 23, 2009

__Tim Clark__ is director of Informatics at the MassGeneral Institute for Neurodegenerative Disease and an Instructor in Neurology at Harvard Medical School. Tim was formerly Vice President of Informatics at Millennium Pharmaceuticals, where his team built one of the first integrated bio- and chemi-informatics software platforms in the pharmaceutical industry. Tim is also a Founding Editorial Board member of the journal Briefings in Bioinformatics. His research group's program is focused on applications of semantic web and social computing to biomedicine and scientific publishing.

***Ian Foster** is director of the Computation Institute, a joint institute of the University of Chicago and Argonne National Laboratory, where he is also the Arthur Holly Compton Distinguished Service professor of Computer Science and an Argonne Distinguished Fellow. He received a BSc (Hons I) degree from the University of Canterbury, New Zealand, and a PhD from Imperial College, United Kingdom, both in computer science. His research deals with distributed, parallel, and data-intensive computing technologies, and innovative applications of those technologies to scientific problems. Methods and software he has developed underpin many large national and international cyberinfrastructures. Dr. Foster is a fellow of the American Association for the Advancement of Science and the British Computer Society. His awards include the British Computer Society's award for technical innovation, the Global Information Infrastructure (GII) Next Generation award, the British Computer Society's Lovelace Medal, R&D Magazine's Innovator of the Year, and an honorary doctorate from the University of Canterbury.*

Preface

Grid technologies emerged from specific needs in basic sciences demanding high-computing power and large store repositories like physics, and from the availability of high-speed high-bandwidth networks and high-performance computers. In the last few years the term Grid has evolved toward a concept of ubiquitous and transparent computing that extends the network paradigm offered by the Internet and offers a new way to think and use computing resources and databases. As this evolution continued new terms like Data Grid, Knowledge Grid, and Semantic Grid emerged, including new functionalities and capabilities regarding, respectively, the efficient management of data stores, the discovery and sharing of knowledge, and the use of semantic-rich services and workflows.

Presently Grids start to be available as a commodity and have reached a maturity that are now available for different application domains, as it happened with the Internet in recent years. In particular, Grids are currently used in many sectors of Life Sciences, from basic sciences such as genomics, proteomics, and bioinformatics, to biomedical research and applications such as drug discovery, biomedical imaging, and biomedicine, and to clinical research and healthcare. So, the application of Grid technologies to Life Sciences, yielded new terms like Bioinformatics Grid, Biomedical Grid, and Health Grid.

The term "HealthGrid" is used to describe the application of Grid technology in the field of biomedical and healthcare informatics. The scientific and industrial community recognized the importance of governing and coordinating technical, standardization and organization activities of HealthGrids by establishing a set of international initiatives.

The international HealthGrid Association (http://www.healthgrid.org/) is an organization based in Europe that promotes the application of Grid technologies in the biomedical sector. With the HealthGrid White Paper (http://whitepaper.healthgrid.org), it established an agenda and defined the concepts, the opportunities and the likely benefits of such an approach. During the last few years it has been working to develop a strong relationship with clinicians and researchers to gain feedback from the use of Grid in the biomedical and healthcare sectors. The HealthGrid.US Alliance (http://www.healthgrid.us), an affiliate of the international HealthGrid Association, is a partnership of scientific, medical and technology professionals whose goal is to promote the application of advanced information technology to solve cutting-edge problems in biomedical science and Healthcare.

Starting from the HealthGrid vision, the European SHARE project (http://www.eu-share.org) identified the key developments – technical advances, social actions, economic investments and ethical or legal initiatives – needed to achieve wide adoption and deployment of HealthGrids throughout Europe. The full road map (http://roadmap.healthgrid.org) analyses several case studies and discusses technical, ethical, legal, social and economic issues which may impede early deployment of HealthGrids.

Relevant projects are addressing these challenges worldwide and include, among others, sharing datasets for cancer cure (caBIG – https://cabig.nci.nih.gov/, ACGT – http://www.eu-acgt.org/), science portals for better visualization of brain morphology (BIRN – http://www.nbirn.net/), collaborative sharing of mammograms (MammoGrid), 3D spatio-temporal visualization of the mammalian cells (Virtual Cell – http://www.visiblecell.com/), accurate multi-scale computational models of the heart and of cancer tumours (Integrative Biology – http://www.integrativebiology.ox.ac.uk/), "virtual" research repository of clinical, laboratory and genetic data sets spread within hospitals and research organisations (MMIM – http://www.mmim.org.au/). These and other projects have begun to demonstrate the power and potential of the Grid approach in biomedicine.

Among the others, HealthGrid is currently attempting to face the following questions:

- How to address the needs of the multiple laboratories collecting genomics and post-genomics data and willing to analyse them in an up-to-date and competitive environment?
- How to solve the need for large computation power needed in medical informatics?
- How to address the need for easy accessible data and computing power for both scientist and medical staff?
- How to make information on all levels, from molecule to population, accessible and understandable to the large variety of people which could benefit from such a knowledge?

As will be explained in the following, the book aims to give a contribution to those and other questions regarding the methods, techniques, benefits and problems of using Grid technology in the biological, biomedical and healthcare domains.

The book surveys state of the art research, methodologies and applications of Grid technology in different fields of Life Sciences, from basic science to clinical and healthcare applications. According to an agreed taxonomy, the fields where life science research takes place can be classified as molecule, cell, tissue, individual, and population levels.

Considering data and computational models, bioinformatics is related to the molecule and cell; medical informatics is related to tissue/organ and individual, while public health informatics (in short healthcare informatics) is related to population.

According to that taxonomy, the book discusses the application of Grid technology to each of the information levels where life science research takes place, by focusing on three main areas: bioinformatics for the study of molecules and cells, medical informatics for the study and simulation of tissues and organs, and healthcare informatics related to population.

On the other hand, the Grid is a valuable tool for collaboration and cooperation, enabling the remote use of medical instruments in the so called virtual laboratory. Thus, the use of the Grid to improve collaboration among scientists and health centers and to enhance the healthcare provision to the final users is another important topic that has been addressed by the book.

Finally, the application of Grid technology to Life Sciences poses news challenges about security and privacy of patient's data, provenance of data, protection and reliability of information, and collaborative analysis of biomedical data. The book introduces these requirements and discusses new methodologies and techniques, such as the Virtual Laboratory concept, used to realize HealthGrids and BioGrids. So, one of the main questions the book attempts to answer is "How current Grids have to be extended to support Life Sciences and become HealthGrids?".

Guided by these considerations, the book's contents are organized in a systematic way by considering different orthogonal dimensions: the complexity level of information being considered (molecule, cell, tissue/organ, individual, population), the support to collaboration and knowledge sharing (e.g. virtual laboratory), as well as the technology, infrastructures and middleware currently available for planning and building HealthGrids.

The chapters of the book, organized accordingly to those dimensions of analysis, are divided into the following eight sections:

- Section I Infrastructures and Services for HealthGrids and Biogrids
- Section II Grids for Genomics and Proteomics
- Section III Grid-Based Bioinformatics Environments
- Section IV Grids for Medical Informatics
- Section V Collaborative Grids for Healthcare and Clinical Applications
- Section VI Grid-Based Virtual Laboratories for Bioinformatics and e-Science
- Section VII Building and Deploying HealthGrids
- Section VIII Selected Readings

Section I introduces the main concepts and defines technical, methodological and organizational challenges of HealthGrids and BioGrids. After discussing the roadmap toward future HealthGrids (Chapter I), the Section defines basic resources and their discovery in HealthGrids (Chapter II). Then, taking into account the large use of scientific workflows in BioGrids and HealthGrids, the Section introduces the data provenance concept (Chapter III) and discusses ways to query and exploit provenance data (Chapter IV). Data protection and security, an important aspect that must be faced when considering biomedical data, is also discussed (Chapter V). Finally, a review of distributed data mining and knowledge discovery systems useful for implementing the analysis layer of HealthGrids is also presented (Chapter VI).

At the molecular level, common trends are the increasing size and complexity of biological data and the increasing number and heterogeneity of collected databases. In fact, biological data is being more and more produced by different high-throughput technologies, like micro array and mass spectrometry, and such novel data must be correlated to well known sequences and structures databases, like SwissProt and PDB. Such requirements can be faced by Grids that provide the storage and the computing power needed to store, analyse and correlate such data.

In particular, two current trends in developing BioGrids are reported in the book (Sections II and III): (a) well established bioinformatics tools like BLAST are being ported, eventually through parallelization, on the Grid, allowing to face the analysis issues of the overwhelming biological data; and (b) complete software environments allowing to combine basic bioinformatics tools through workflow of Web/Grid services are deployed on the Grid, supporting the complete life cycle of complex "in silico" experiments.

Section II discusses the use of the Grid for the management and analysis of genomics and proteomics data, the basic data at the biological level. After describing main issues for the parallel and distributed implementation of BLAST (Chapter VII), a cornerstone of all genomics analysis, the Section discusses some significant Grid-based implementations of genomics applications (Chapters VIII to X). Then, the Section introduces proteomics with a special focus on mass spectrometry-based proteomics (Chapter XI), and different Grid-based proteomics applications, ranging from biomarker discovery to protein identification (Chapter XII) and protein classification (Chapter XIII), are discussed.

Section III discusses Grid-based software environments for bioinformatics. Grid-based Problem Solving Environments specifically devoted for bioinformatics (Chapter XIV), as well as Grid-based implementations of relevant bioinformatics applications, such as docking (Chapter XV), bio-molecular simulation (Chapter XVI) and molecule structure determinations (Chapter XVII), are presented.

Section IV focuses on Grid-based medical informatics applications. Since a core aspect of medical applications regards the acquisition and analysis of biomedical images, the Section first introduces different aspects of biomedical images visualization on the Grid (Chapter XVIII). Then, as significant applications in the field, Grid analysis of radiological data (Chapter XIX) and 3D electron microscopy reconstruction (Chapter XX) are discussed. Finally, as an example of biomedical instrumentation made available on the Grid, a hybrid mock circulatory system to test cardiovascular prostheses is presented (Chapter XXI).

Considering the research at individual and population level, the Grid is used both as a technology to enable compute intensive applications such as simulation of an organ or organism, and also as a mean to enhance collaboration among individuals and to enable interoperability among applications and databases. As an example, personalized medicine will be based on the integration of genotype, phenotype and environment information (e.g. clinical, genomic, proteomic, and drug toxicity data).

Healthcare applications at the population level as well as applications used in the clinical context are the focus of Section V. These applications are the most difficult to port on the Grid, due to the high number of patients involved, the partitioning of data among different health centers, the privacy and security problems and the need of collaboration among various operators. The use of the Grid as a support for collaboration in epidemiology (Chapter XXII), as well as the integration and analysis of epidemiology data (Chapter XXIII) are fully described. Two important HealthGrids for paediatrics (Chapter XXIV) and brain injury (XXV) are presented showing how the Grid is a mature technology also for the clinical setting.

Section VI discusses concepts and properties of Virtual Laboratories, an abstraction for cooperative data analysis and distributed collaboration among scientists, and their applications in Life Sciences. After describing the foundations of modern virtual laboratories (Chapter XXVI), such as formalisms for representing domain knowledge, data integration, semantic annotations and shared vocabularies, the Section describes some emergent virtual laboratories that focus on distributed collaboration and use of provenance data (Chapter XXVII), transparent use of the Grid (Chapter XXVIII) and support for specific domains like bioinformatics (Chapter XXIX).

Section VII describes main infrastructures, middleware, and tools for building and deploying HealthGrids. After introducing the main requirements posed on Grid middleware by biomedical applications, and how these are satisfied by a well known Grid middleware (Chapter XXX), the Section describes a Grid-based portal that aims to promote collaboration and cooperation among scientists and healthcare research groups on the Grid (Chapter XXXI). The management and deployment of Grid repositories for biomedical digital contents such as mammograms is also discussed (Chapter XXXII).

Finally, Section VIII is a short collection of suggested readings of different authors, aiming to enrich this book with others knowledge, experience, thought and insight. Chapter XXXIII introduces Cloud Computing, an emerging style of computing in which dynamically scalable resources are provided as a service, and describes the porting of applications to Grids and Clouds. Chapter XXXIV describes a bio-inspired approach for the construction of a self-organizing Grid information system. Chapter XXXV discusses Grid-based implementation of phylogenetic analysis.

In summary, the book records the main experiences and best practices on using Grids in bioinformatics, genomics, proteomics, medical informatics, and healthcare informatics, allowing a better understanding of benefits and added values in concrete biomedical and clinical applications.

Readers of the book can explore the power and flexibility of Grid solutions with respect to either the complexity level of life exploration or the level of cooperation among scientists.

In particular, the book reports the main methodologies currently used for exploiting Grid's features in Life Sciences, and it may be an important aid for engineers, computer scientists, and clinicians working on biomedicine and healthcare.

The book contains also a comprehensive description of tools, infrastructures and organizational issues currently available for planning and building HealthGrids and BioGrids. Thus, healthcare practitioners may find several interesting cases and best practices useful to plan and build HealthGrids in real situations, as well as arguments to evaluate the impact on organization.

Many open research issues remain in these fields and new challenges emerge continuously. I hope that this book will serve to introduce readers to the potential developments and benefits of Grid technology for Life Sciences and to the challenges posed by biomedical and healthcare applications to current Grid.

Mario Cannataro
University Magna Graecia of Catanzaro, Italy
January 2009

Acknowledgment

I would like to acknowledge the help of all involved in the preparation, collation and review process of the handbook, without whose support the project could not have been satisfactorily completed.

Deep appreciation and gratitude is due to the members of the Editorial Advisory Board, who shared with the editor the main decisions about chapter selection and handbook organization and who also contributed in the review process. Special thanks go to Tim Clark and Ian Foster who kindly wrote the Foreword to this handbook.

Most of the authors of chapters included in this handbook also served as referees for chapters written by other authors. Thanks go to all of them who provided constructive and comprehensive reviews. Moreover, special thanks go to the following additional reviewers: Eugenio Cesario, Marcelo Lobosco, Gregorio Mercurio and Renato Pajarola.

Thanks also to University of Catanzaro and ICAR-CNR, Italy, who provided the WWW resources for the handbook web site. Deep appreciation is due to Pierangelo Veltri and all the members of the Bioinformatics Laboratory, University of Catanzaro, Italy, for their work on bioinformatics and biomedical applications. Special thanks go to Valeria Farrelli, who helped me in the recruitment of prospective authors by organizing the invitation emails, to Pietro Hiram Guzzi who developed the handbook web site pages, and to Antonio Oliverio who solved some connectivity problems.

Special thanks also go to the publishing team at IGI Global, whose contributions throughout the whole process from inception of the initial idea to final publication have been invaluable. In particular to Julia Mosemann and Beth Ardner, who continuously prodded via e-mail for keeping the project on schedule, to Lindsay Johnston, who first invited me to this exciting project, and to Jessica Thompson, Kristin M. Klinger, and Jan Travers who helped me in starting this editorial activity.

Special thanks go to Domenico Talia, University of Calabria, Italy, and to Giovanni Cuda, University of Catanzaro, Italy, who introduced me to the exciting fields treated in this handbook, respectively Grids and Life Sciences.

In closing, I wish to thank all of the authors for their insights and excellent contributions to this handbook.

Mario Cannataro
University Magna Graecia of Catanzaro, Italy
January 2009

Section I
Infrastructures and Services for HealthGrids and BioGrids

Chapter I
SHARE:
A European Healthgrid Roadmap

Mark Olive
University of the West of England, UK

Hanene Boussi Rahmouni
University of the West of England, UK

Tony Solomonides
University of the West of England, UK

Vincent Breton
IN2P3, CNRS, Clermont-Ferrand, France

Nicolas Jacq
*HealthGrid (International)**

Yannick Legré
*HealthGrid (International) & IN2P3, CNRS, Clermont-Ferrand, France**

Ignacio Blanquer
Universidad Politécnica de Valencia, Spain

Vicente Hernandez
Universidad Politécnica de Valencia, Spain

Isabelle Andoulsi
*Universitaires Notre-Dame de la Paix, Belgium**

Jean Herveg
Universitaires Notre-Dame de la Paix, Belgium

Celine Van Doosselaere
*European Health Management Association (International)**

Petra Wilson
*European Health Management Association (International)**

Alexander Dobrev
Empirica GmbH, Germany

Karl Stroetmann
Empirica GmbH, Germany

Veli Stroetmann
Empirica GmbH, Germany

ABSTRACT

The principal goal of this chapter is to elucidate the future requirements of healthgrids if they are to become the infrastructure of choice for biomedical research and healthcare. These requirements take many forms, technical, organizational and economic, with initiatives required in the domains of ethical and legal regulation. Thus, particular objectives of the chapter are to explore and analyse each of these domains to a sufficient depth to be able to make sense of the overall picture.

INTRODUCTION

Grid technology, one of the key technologies for the 'European Research Area', offers rapid computation, large scale data storage and flexible collaboration by harnessing together the power of large numbers of computers, from end-users' desktops to powerful workstations and clusters of more powerful machines. The grid was devised for use in scientific fields, such as particle physics and bioinformatics, in which large volumes of data, or very rapid processing, or both, are necessary. The impact of this concept has already reached beyond eScience, to eBusiness, eGovernment and eHealth. However, a major challenge is to take the technology out of the laboratory to the citizen. The term 'healthgrid' is used to describe the application of this technology to biomedical and healthcare informatics. This domain of application presents some difficult challenges.

The SHARE project (http://www.eu-share.org) was asked to identify the key developments needed to achieve wide adoption and deployment of healthgrids throughout Europe. The project was asked to organise these as milestones on a road map, so that all technical advances, social actions, economic investments and ethical or legal initiatives necessary for healthgrids would be seen together in a single coherent document. The full road map (http://eu-share.org/about-share/deliverables-and-documents.html) includes an extensive analysis of several case studies exploring their technical requirements, full discussion of the ethical, legal, social and economic issues which may impede early deployment, and concludes with an attempt to reconcile the tensions between technological developments and regulatory frameworks.

Grid technology (Foster *et al*, 2004), one of the key technologies for the 'European Research Area', offers rapid computation, large scale data storage and flexible collaboration by harnessing together the power of a large number of commodity computers or clusters of other basic machines. The grid was devised for use in scientific fields, such as particle physics and bioinformatics, in which large volumes of data, or very rapid processing, or both, are necessary. The impact of this concept is expected to reach far beyond eScience, to eBusiness, eGovernment and eHealth. However, a major challenge is to take the technology out of the laboratory to the citizen.

The concept of grids for health was born in Europe in 2002 and has been carried forward through the HealthGrid initiative. This European collaboration has edited a White Paper (Breton *et al*, 2005) setting out the concept and benefits of emerging grid technologies in different applications in healthcare.

The White Paper defines the concept of a healthgrid as follows:

HealthGrids are grid infrastructures comprising applications, services or middleware components that deal with the specific problems arising in the processing of biomedical data. Resources in healthgrids are databases, computing power, medical expertise and even medical devices.

The EU-funded SHARE project has identified some important challenges towards wide deployment and adoption of healthgrids in Europe. The project has devised a strategy to address the issues identified in the Action Plan for European e-Health and has devised roadmaps for technological developments, legal and ethical initiatives and socio-economic investments needed for successful uptake of healthgrids in the next ten years.

The SHARE roadmaps express certain measurable goals and objectives for the HealthGrid community, provide an analysis of the technical gaps to be bridged in order to achieve a number of staged technical objectives, explore the ethical, legal, social and economic (ELSE) conditions of such developments, analysing the extent to which technology and its environment will need to be reconciled, and articulate a strategy for the concurrent achievement of these goals and objectives subject to realistic contextual conditions.

These roadmaps have been developed from three major inputs:

- An analysis of user requirements in a carefully triangulated set of domains through current projects and scripted use-cases;
- A technical road map which sets out the key objectives for a viable 'knowledge healthgrid' to be achieved in a span of 10-15 years;
- A conceptual map of ELSE conditions, constraints and requirements which must be addressed before a knowledge healthgrid can be deployed in a real healthcare setting.

The conceptual map of ethical, legal, social and economic issues considered the regulatory challenges that any real healthgrid must meet, including: legal challenges concerning rights to privacy and confidentiality, 'right to know' and duty of care; ethical challenges concerning primary and secondary use of data whether individual or aggregated; legal and ethical challenges concerning provenance and quality of information; legal, ethical and economic challenges to the use of healthcare data in commercial and public research, including questions of ownership of data; legal and ethical challenges in the communication of genetic information and the resultant 'lateral leakage' of information; legal and ethical challenges to the communication of medical data across borders; social and legal challenges concerning the formal professional competencies of different healthcare actors; and legal, ethical and socio-economic challenges of 'exceptional cases', such as assisted reproduction, organ donation and transplantation.

After some brief background information on the HealthGrid initiative, this chapter will discuss the benefits of healthgrids from a variety of perspectives (users, as well as patients and wider benefits to society). This is followed by a summary of the use case scenarios developed during the project, organized by grid paradigm; note the addition of the 'collaboration' paradigm, an emerging, central theme of many e-health projects. Then there will be a quick run through of the main technical, ethical and socio-economic issues covered, and the requirements derived from the various use case scenarios. Finally, the chapter concludes with recommendations and a summary of the roadmaps produced, with some concrete technical, legal and socio-economic recommendations.

BACKGROUND

Early grid projects, while encompassing potential applications to the life sciences, did not address the specificities of an e-infrastructure for health, such as the deployment of grid nodes in clinical centres and in healthcare administrations, the connection of individual physicians to the grid and the strict regulations ruling access to personal data. However, a community of researchers did emerge with an awareness of these issues and an interest in tackling them. The context in which this initiative was realized is nicely captured in the *Resource Book of IST Projects* (EU, 2002) which evidences a shift from a pull to a push economy: the world of healthcare had to be made aware of the potential of technology. Apart from the projects funded by the EU, a number of others funded, for example, by the Istituto Nazionale di Fisica Nucleare (INFN) in Italy and the UK e-Science Programme served to reinforce the point.

Once these pioneering projects in the application of grid technologies to the health area had been completed, and the technology to address high level requirements in a grid environment had been under development and making good progress, it was necessary to consider how to consolidate. Because these projects had a finite lifetime and the healthgrid paradigm required a sustained effort over a much longer period, and besides because there was an obvious need for these projects to cross-fertilize, the '*HealthGrid* initiative', represented by the *HealthGrid* association (http://www.HealthGrid.org), was initiated to provide the necessary long-term continuity. Its goal is to encourage and support collaboration between autonomous projects in such a way as to ensure that requirements really are met and that the wheel, so to speak, is not re-invented repeatedly at the expense of other necessary work.

Apart from research, where the value of grid computing is well established, a healthgrid may be deployed to support the full range of healthcare activities, from screening through diagnosis, treatment planning to epidemiology and public health. The results of several major studies of the interface between bioinformatics and medical informatics have been published with a remarkable promise of synergy between the two disciplines, leading to what had already begun to be referred to as 'personalized medicine'. From the point of view of *healthgrid*, this made clear the need to unify the field and to put its various elements in perspective: how would they – improved evidence bases, imaging, genetic information, pharmacology, epidemiology – fit together? What was their relative importance in the unfolding programme of work?

The European community working on applications of grid computing to health and biomedicine joined forces in *HealthGrid* and defined a vision of grids as the infrastructure of choice for biomedical research in the first place and healthcare delivery in the longer term. In the light of the HealthGrid White Paper and its impact, the EC funded a 'specific support action' project, SHARE, to explore exactly what it would mean to realize the vision of the White Paper, investigate the issues that arise and define a roadmap for research and technology which would lead to wide deployment and adoption of healthgrids in the next ten years.

BENEFITS OF HEALTHGRIDS

For the Healthcare Professional/ Biomedical Researcher

Healthcare systems both in developed and in developing countries face major economic and capacity challenges to maintain quality of care in the face of the growing demands of ageing populations and the increasingly sophisticated treatments available. Add to this the desire to improve access to new care methods, and the challenge of delivering care becomes significant. In an attempt to meet these demands, health systems have increasingly looked at deploying information technology to scale resources, to reduce queues, to avoid errors and to provide modern treatments into remote communities, for example.

From the individualised care point of view, in order for clinicians to make the best diagnosis and decide on treatment all the relevant health information of the patient needs to be available and transparently accessible to them regardless of the location where it is stored. Moreover, computer-aided tools are now essential for interpreting patient-specific data in order to determine the most suitable therapy from the diagnosis.

To store and process medical images, genetic information and other patient data, a large amount of computing power is needed. Large computing resources are also needed for keeping statistics of patient records, for knowledge extraction using data mining, and for the simulation of organisms and diseases using complex biomedical models. Grid technology has undoubtedly much to offer medical professionals, as illustrated by the following examples.

The delivery of medical information and certain services through the internet is familiar. In healthgrid computing, we seek an extension of the concept to consider how to provide large scale services to the user on demand. Some examples will serve to illustrate:

- Consider a radiologist who needs to manipulate an image: we want to provide a set of services, some of which may require heavy processing, making them available on her desktop 'transparently', as if they were programs simply running on her computer.
- Consider a public health service which monitors certain infectious diseases and has to trigger an alert in case of a suspected epidemic. The identification of unusual patterns would in many cases be the critical step to halting the problem.
- Consider a surgical simulation prior to maxillofacial surgery, to determine how the patient's face may appear after one manoeuvre versus another, the presence of sufficient tissue to allow the operation or to demand transplantation, and even to involve the patient in the decision.
- Consider a 'neglected disease' like malaria. Malaria is neglected by the pharmaceutical industry because there is no prospect of profit in it. Relatively little progress has been made towards the eradication of this well understood disease, notwithstanding substantial investments of public funds in research projects. *In silico* lead generation may possibly be coupled with investment in plant by the poorer nations that suffer from it to lead to a locally sustainable solution.
- Consider the possibility of linking genomic information to imaging in diseases like juvenile idiopathic arthritis. The genome will indicate susceptibility long before the disease is expressed, but equally, signs picked up from imaging may obviate the need for genetic screening, thus avoiding some of the most acute problems associated with it.
- Consider more abstractly the nature of evidence-based practice, the volume of scientific literature that provides the evidence base and the accumulation of evidence from practice that occurs as a matter of routine healthcare. How can these be integrated? How can they be used without violating any ethical restrictions on use of data, confidentiality, privacy, security? How can they be shared without violating any data protection laws?

These are simple examples of foreseeable beneficial advances within the next generation of developments. For the radiologist or the maxillofacial surgeon, the services described would be important new tools, while for the provider of such services the underlying grid would be an ideal e-marketplace. For the public health service concerned with infectious diseases or the academic unit concerned with neglected diseases, healthgrids would provide power and flexibility beyond what could be achieved using traditional approaches. For a group of physicians seeking to improve treatment through research, as described in the fifth scenario above, the possibility of collaboration and concerted scientific work through a grid offers a completely new concept. Knowledge management, our concluding example, is an issue of increasing concern in the medical domain; a solution of the kind described here would be of immense benefit, even without the additional service of automated policy bridging and compliance.

Of course, there are challenging problems even among these scenarios. Standards have not stabilised in the grid world any more than in current healthcare systems, so data exchanges will present problems straight away. Codes and coding languages are also still not universally adopted, while the application provider will wish to protect investments in software licence rights. These are none the less tractable problems, ones in which the very nature of grids would make easier to address. It is arguably the broader modernisation process that faces the more complex challenges:

- Connecting and understanding patient records across organisation structures and national borders.

- Ensuring that information is secured and is accessed in compliance with a multilayered regulatory framework.
- Discovering trustworthy sources of information.
- Handling the enormous volume of data involved, for example, in genetic medicine.
- Incorporation of 'legacy' traditional information networks and technology in healthgrids.
- In view of these questions, it is essential to understand also the socio-economic advantages of grids.

Socio-Economic Benefits

Modern healthcare services are expected to be available around the clock, seven days a week, so that systems with pervasive access and near-absolute fault tolerance are indispensable. However, it is difficult for these applications to run non-stop with a high quality of service. Grids could help by providing a platform of collaboration, allowing the linking centres which co-operate to achieve better continuity and quality of service. Medical staff will then be able to share experience, knowledge and 'second opinion' with other internal and external staff. The distributed architecture of grids with the availability of high-bandwidth networks responds well to the requirements of healthcare provision. There are also optimistic stakeholders' views towards medical research, healthcare and computing capabilities combined to better satisfy the patient.

HealthGrids promise many benefits to mobile patients as well as citizens. It could help a travelling individual to receive the right treatment in an emergency situation, thanks to the ability of the grid to facilitate communication between the local hospital of the patient and the admitting hospital abroad in order to exchange necessary heath related information.

In addition, healthgrids enable the mobility of a patient within EU states and allow them to receive medical treatment in a country of their choice. This could help solve problems of long waiting lists in states with busy hospitals and lack of medical staff. In economic terms, the grid could provide an optimal solution for healthcare. It allows a better use of resources and maintenance of tasks, an improved global IT organisation, scalable costs, and a large and consolidated IT business within the healthcare organisation.

Health tourism is a growing concept which can enrich the economy of countries where modern medical treatments (plastic surgery, dental surgery, reproductive medicine, laser surgery for vision correction, etc.) are evolving and having higher success rates than others. This domain can benefit from healthgrid technology as it facilitates the exchange of patient health records and the communication between foreign hospitals and health insurance companies to facilitate the referral and payment process.

Transferring medical images for the purpose of a second opinion to another hospital requires high bandwidth connections between hospitals. healthgrid technology can provide automated workflows that could be considered a better alternative to manual workflows, such as agreement over the phone and fax transmission of data. These manual workflows are still used by clinicians at present, but are labour-intensive and can cause errors.

USE CASE SCENARIOS

Grid Paradigms

Grids are often differentiated into computational, data and collaboration grids. The ideal grid, envisaged as a servant of a new paradigm of scientific research called 'e-science', must provide transparent processing power, storage capacity and communication channels for scientists who may from time to time join the grid, do some work and then leave, so that the alliances they form in their scientific endeavours might be described as 'virtual organisations' (VOs). Different sciences have different needs, and the grid concept has become differentiated: particle physics generates enormous amounts of data which must be quickly stored, but not necessarily instantly processed; on the other hand, data in bioinformatics is not large by comparison but requires intensive processing. In extending the application of grids to e-health, collaboration becomes pre-eminently necessary.

Computing Grids: An Example from Innovative Medicine

Drug discovery is the long term, multi-stage and high cost process by which drugs are discovered and/or designed. The drug discovery goal is to find new molecules that bind with specific macromolecules known to play a key role in a disease process, in a manner that changes their function, either to increase resistance to or to reduce the virulence of some pathogen. Reducing the research time in the discovery stage and having enhanced information about substances affecting the selected target ('leads') are key priorities for pharmaceutical companies worldwide.

In silico drug discovery, including analysis of the gene expression data, target function prediction and target three-dimensional (3D) structure prediction, is one of the most promising strategies to speed up the drug discovery process, avoiding time consuming and costly *in vitro* and *in vivo* tests. It is estimated that the efficiency gains of this approach could result in 35% cost savings, or about US$300 million, and a 15% time reduction, or two years of development time per drug (WISDOM, 2006).

In silico drug discovery requires advances in data integration (including data format standardisation, dataflow definition in a distributed system, services for data and meta-data registration, and development and sharing of ontologies and knowledge representations), workflow enactment to ease data management and data mining and to assist the scientist and the decision-maker in organising their work in a flexible manner, access to computing and data resources (computing 1 million docking probabilities or modelling 1000 compounds on one target protein requires a few TFlops for one day), and collaboration between public and private partners, involving the concrete sharing of data and knowledge, software and workflow, and infrastructures such as computing, storage and networks. Security and the effective protection of intellectual property and sensitive information are key challenges for pharmaceutical industries, but also for academic institutes in most cases.

Data Grids: Examples from Epidemiology and Public Health Informatics

Epidemiology

The epidemiology use case is defined as a system able to link the information from distributed and heterogeneous databases, identify patients, complete episodes and automatically improving quality without interrupting clinical practice. With this data, complex epidemiological models are fed and simulated producing aggregated prospective results, in a reliable way. This use case is representative of different applications and systems including ontological information systems, infectious surveillance networks, pharma-epidemiology analysis of efficiency and cost, and study of propagation models for diseases.

The main users (from the highest-concept level to lowest) are public health authorities, epidemiologists and pharmaceutical companies. The data is normally owned by clinical care (both public and private). In order to achieve this use case the user needs to go through different steps including automatic data gathering, data quality improvement, processing of the data, and presentation of results. An early project that foreshadowed some of these potential applications was the EURIS project (Almeida, 2003).

A Surveillance Network for Avian Flu

A use case in the field of public health informatics is a surveillance network of data repositories offering services to the research communities working on avian flu and issuing an alert in case of pandemic risk. Indeed, the ability of the international community to respond efficiently to the possible emergence of a human-to-human transmissible avian influenza virus depends on its capacity to quickly assess any evolution of the disease. Many countries have set up very efficient national networks for collecting data and monitoring outbreaks. However, there is presently no international surveillance network that allows the sharing of data collected at a national level.

The starting point would be to set up a data grid collecting public and private information on avian flu. Public data would be made available to all registered users while access to private data would be strictly controlled through grid authorisation and authentication mechanisms. The repositories would share a common model allowing distributed queries.

Collaboration Grids: Examples from Tertiary Healthcare and VPH

Virtual Physiological Human (VPH)

The EuroPhysiome initiative has led to the concept of the Virtual Physiological Human (VPH) (STEP Consortium, 2002). This concept goes beyond the idea of a model to an altogether more ambitious methodological and technological framework to enable the investigation of the human body as a single complex system. Just like the biological human, the VPH will be made up of interacting subsystems which need to be understood at a hierarchy of levels, from the molecular to the cellular, through organs and systems to the whole individual. These are often associated with a different discipline of scientific knowledge and have been served by a different specialism within informatics. The metholodological aspect of VPH as a project is to integrate these into an appropriate framework within which observations made in laboratories, hospitals, and in the field all over the world can be collected, catalogued, organised, shared and combined in a variety of ways. It should also allow experts collaboratively to analyse observations and develop systemic hypotheses that involve the knowledge and insights from multiple scientific disciplines. The ultimate scientific goal would be to interconnect predictive models defined at different scales, with different methods, and with different levels of detail, into systemic networks that allow complex hypotheses to be formulated as models which can be tested through simulation and experiment. This may be illustrated by the study of tumour growth in cancer, in which molecular and cellular processes may be studied together to understand first what happens in the individual cell and how these proliferate to form a malignant colony.

There is a great deal of similarity in the conception and ambition of the VPH and of healthgrids. The vast scope and integrative approach of the VPH call for the kind of resource sharing mechanisms provided by a grid infrastructure. The need for a medium of collaboration for a large but heterogeneous community of scientists and the need for coherent multi-scale modelling likewise argue in favour of a grid approach.

Tertiary and Specialist Healthcare

The Health-e-Child project dealing with paediatrics, and the MammoGrid and e-DiaMoND projects dealing with breast cancer, are examples of projects that have made use of grid technologies to benefit healthcare professionals and patients in a tertiary healthcare setting.

Paediatrics is an area where not only the disease in question is changing, but the child as well, as he or she grows. The Health-e-Child project addressed specific diseases within certain medical domains; cardiology, rheumatology and neurology. Through consultation with these communities, a number of requirements for the project were identified. The primary requirements from cardiology are related to imaging and integrated disease modelling. Requirements from rheumatology include the construction of homogenous juvenile inflammatory arthritis subtypes (which will require vertically integrated data), and models for predictive disease outcome, and for progressive organ damage. The project determined that a variety of models using integrated data would be useful for neurologists, including models for surgical decision-making, post surgery treatment, and models supporting automatic tumour detection and change quantification. The potential for individualised brain models has also been recognised, which will require deforming a generic brain atlas in order for it to match the geometry of a patient.

Breast cancer as a medical condition, and mammograms as images, exhibit many dimensions of variability across a population. Likewise, the way diagnostic systems are used and maintained by clinicians varies between imaging centres and breast screening programmes, as does the appearance of the mammograms generated. A distributed database that reflects the spread of pathologies across a broad population is an invaluable tool for the epidemiologist; understanding the variation in image acquisition protocols is essential to a radiologist in a screening programme. Exploiting emerging grid technology, the aim of the MammoGrid project was to develop a potentially EU-wide prototype database of mammograms to be used to investigate a set of important healthcare applications and to explore the potential of the grid to support effective collaboration between healthcare professionals (Warren, 2007a, 2007b) In particular, the project aimed to prove that grid infrastructures can be practically used for collaborative medical image analysis. MammoGrid and e-DiaMoND effectively demonstrated the viability of the grid by harnessing its power to enable radiologists from geographically dispersed hospitals to share standardised mammograms, to compare diagnoses (with and without computer aided detection of masses and microcalcifications) and to perform sophisticated epidemiological studies across national boundaries.

Knowledge Grids: An Example from General Healthcare

For any given domain, a distinction is often drawn between declarative knowledge ('know what') and procedural or operational knowledge ('know how'). In the domain of healthcare, both kinds occur. What is often referred to as 'the scientific basis' of medicine – the scientific results that constitute the evidence for 'evidence-based practice' – is present in various research publications to which different standards of credibility are attributed. For example, research results based on a randomised, double-blind, controlled clinical trial are held to be the gold standard, provided they were also submitted to adequate peer review. Evidence based on one physician's own practice, although not negligible, would be considerably less reliable. On the other hand, 'best ways' of treating patients – in a particular context – are often described in integrated care pathways (ICPs). It is not unreasonable to claim that declarative knowledge in medicine tends to be disseminated through peer-reviewed publication and operational knowledge through such things as guidelines and care pathways. In a healthgrid environment, these are brought together for the better treatment of patients and at the same time to improve research; indeed, the interplay between healthcare and research, e.g. through appropriately controlled 'secondary use', would be an important element in a full healthgrid environment.

Chronic obstructive pulmonary disease (COPD) refers to an airway obstruction caused by chronic inflammation. It is usually progressive, not fully reversible, and often occurs as a result of smoking but other factors such as air pollution can also contribute to the development of COPD.

In the UK almost 900,000 people have been diagnosed with the disease, and the true number of people suffering from the condition is estimated to be around 1.5 million.

According to NHS guidance, the management of the disease should be tailored to the individual, with adjustments being made based on responses to treatment. The guidance includes a large number of drugs, including some off-label drugs. In the UK, the National Institute for Health and Clinical Excellence (NICE) has issued national guidelines for the treatment of the disease, but these are frequently modified to account for local variations and priorities. As a result, the procedure for assessing and treating the disease will vary even within a single country, let alone between countries. The evidence on which this guidance is based comes from a variety of sources, such as national studies by NICE and systematic reviews with an international scope.

Two main concerns exist for general healthcare; supporting the travelling patient, such as migrating elderly populations, and enabling decision support systems that can account for local variations in best practice and clinical evidence.

TECHNICAL, ETHICAL, LEGAL AND SOCIO-ECONOMIC ISSUES

Grid Technology

A 'grid' – not *the* grid – is now understood to mean an Internet-like infrastructure which extends the concept of the Internet in several significant ways:

- Like the Internet, a grid would provide access to information services but in addition would provide pooled storage, processing power and collaboration in so-called 'virtual organisations' (VOs);
- Use of a grid will be reciprocal – while a user subscribes and takes advantage of services provided by a grid, the user's resources are pooled and are available to all grid subscribers;
- The process is transparent – the grid allocates resources and provides an interface to services which give the appearance that the user is accessing just one powerful machine.

Major IT companies have agreed to develop web services as the technology to enable the deployment of services on the Internet. It has been also adopted by the Open Grid Forum which is the acknowledged body to propose and develop standards for grid technology.

Moreover, web service technology provides the bridge between the grid world and the Semantic Web which is about common formats for interchange of data and about language for recording how the data relates to real world objects. Although many current grid infrastructures do not offer a web service interface to their services, we will concentrate our 'state of the art' on web services because it is the relevant technology for the future. We will then go on to discuss the status of existing grid infrastructures, the technologies they use and the services they offer.

The initial idea behind web services was to enable the World Wide Web increasingly to support real applications and a means of communication among them. Thus, a set of standards and protocols have been proposed to allow interaction between distant machines over a network. These interactions are made possible through the use of standardised interfaces which describe the available operations in a service, the nature and form of messages exchanged (requests and responses), and the physical location of the service on the network. A language, Web Service Description Language (WSDL) has been devised to describe such interfaces (W3C, 2001).

One of the subtle advantages of a so called Service Oriented Architecture (SOA) is that it leads to loosely coupled components that may be substituted by better services so long as they comply with the interface specification. Thus in a grid, services can be offered in a way that approximates an ideal marketplace, a market with perfect information.

Technical Issues

Basic technical requirements include the adequacy of network infrastructures for the effective operation of a grid, the usability of grid interfaces and simplicity of administration of grid nodes. Other healthgrid-specific issues can be divided into standardisation, communication and security.

Standardisation Issues

Standards are essential for the deployment of services which integrate data in bioinformatics and medical informatics, but are also vital for data coming from different medical disciplines or from different countries in Europe. These standards are needed for building data models, producing ontologies and the development of knowledge management services. The adoption of standards for the exchange of biological and medical information is still limited to a few specific fields. Moreover, they need to be compatible with grid standards so as to allow their implementation on healthgrids. Of course, many of these issues arise in any data integration programme, but they appear all the more sharply in grids where the heterogeneity of the underlying system is expected to be transparent, or at least not to impact on the user.

Communication Issues

Lack of information about grids and grid technologies is frequently identified as one of the key reasons why there has been very little interest in them from the field of medical research. It is essential that all relevant actors to be kept well informed by the HealthGrid community of the potential benefits of the technology to them. Success stories demonstrating the impact of grids for medical research will be vital for convincing medical researchers of these benefits. As a result, there is a need for a suitable demonstration environment, offering very easy access to the grid for non experts and providing services that will help convince the medical research community. On this dissemination environment, dedicated efforts to promote the technology can then be developed.

Security Issues

Deployment of a data grid for medical research will only be possible when the middleware can provide all the necessary guarantees in terms of management of personal data. We perceive the specific technical requirements related to the handling of medical data on the grid to be as follows.

- Manipulation of personal data on the grid must obey strict regulations. These regulations vary between European member states.
- Services for the anonymisation and pseudonymisation of medical data must be provided.
- Medical data is the property of the patient.[1] A mechanism must be set up to allow individuals to access their data on the grid.
- For healthcare purposes, the authentication of healthcare professionals on the grid cannot be handled by requesting all of them to get a grid certificate. A mechanism must be set up so that professional cards can be used to provide authentication on the grid.

Ethical Issues

Ethical issues in healthgrids correspond to well known ethical principles, *autonomy*, *beneficence* and *justice*. Each has individual value, but the three must be taken as a whole offering a system of ethics in which the needs of the individual are balanced against those of society.

Autonomy and HealthGrids

Most common belief systems give a special place to the autonomy of the individual - the right of the individual to control his or her own person. Autonomy is intimately tied up with the legal duties of consent and confidentiality, both of which could prove difficult in the context of healthgrids. Thus, in healthgrids, one of the key ethical issues will be in the possible compromise of the patient's autonomy that will arise from sharing his or her data with people who are yet to be identified. It is worth noting that it has been argued, notably by the European Article 29 Data Protection Working Party[2], that consent may have only a very limited place as a justification of the sharing of health related data in the electronic age. The limitation is based on the argument that to exercise autonomy one must be able to make decisions unfettered by coercion. If it is accepted that sharing health data allows doctors to provide better care, then a patient who refuses permission to share such information will be *de facto* opting for a lower quality of healthcare, arguable therefore he or she is not able to withhold consent freely and is therefore not able to act autonomously. It is argued therefore that robust system of security of information and ethical practice should be adopted in which patients will be able to trust, notwithstanding that their information is shared, and providing for special opt-out possibilities when the nature of the information is especially sensitive.

Beneficence and Non-Malfeasance

The ethical duty of Beneficence and Non-Malfeasance is the duty to aim to do good, or in the words of the Hippocratic Oath at least to do no harm. This ethical duty is frequently used to justify the adoption of specific health technologies to improve treatment. The argument with respect to healthgrids is that in order to act ethically, a healthcare professional is obliged to use suitable grid applications if they are available. A healthcare professional refusing to use standard medical technology or refusing to prescribe antibiotics would be considered in breach of his or her duty of beneficence, thus, as the sophistication of grid aided diagnosis develops we will one day arrive at a time when a practitioner not linked to the appropriate grid networks will be in breach of his or her duty.

However, until we have reached a time when grid applications are stable, well 'fed' with data and fully integrated into the evidence base of good clinical practice such arguments will not apply. At present, in the more experimental stages of the healthgrid it will be important to ensure that the use of the applications does no harm, but perhaps most importantly to ensure that the patient is aware of any possible medical and social damage.

Justice

The ethical principle of justice concerned with the duty to achieve a fair distribution of resources as well as the need to develop an overall just medical system in which the greatest health benefit of the greatest number is achieved is the principle of justice. In most legal systems this ethical principle is used to support social systems of distributive justice which provide for tools as taxation to distribute wealth on such a way that all may be afforded an acceptable minimum of social care. The developments of applications such as MammoGrid have established that the sharing of a very large number of mammogram images across a wide network that allows radiologists to test suspect images against a known and tested database of cases significantly contributes to the early detection of breast cancer. The healthgrid in this case not only acts to the benefit of the known patient whose suspect image is submitted to the tool, but to the overall health of the population.

Legal Issues

The legal issues presented here (and analysed more extensively in several SHARE deliverables) were chosen for their relevance to healthgrid technologies. Other legal concerns such as competition issues are of relevance for grid technology, but in order for a full and complete analysis to be made, these were intentionally omitted. Regarding competition issues in relation to eHealth, we refer to the reports and analysis from the *Legally eHealth* project[3].

Data Protection

The ethical principle of autonomy is legally underpinned by the duty of data protection. This is taken very seriously at EU level; as well as the inclusion of privacy protection in the European charter of fundamental rights, the EU has developed the robust directive on data protection to promote privacy.

However, the current legislation is not adequate to support most of the longer running research initiatives around which healthgrids are based. As the current EU level legislation stands, member states can enact specific legislation covering specific tools such as healthgrids in order to exempt scientists and medical practitioners using healthgrids from some of the more onerous duties of the directive.

No member state has addressed legislation to this particular issue and so healthgrids are burdened with onerous data protection requirements which could deter scientists from adopting healthgrid technology and using its enhanced computational and data acquisition power.

Liability

Using grids blurs the liability issues in terms of medical practice. While the EU has a range of legislation designed to protect citizens from harm resulting from goods offered on the market, the construction of healthgrids makes it difficult to ascertain at which EU level legislation would apply to each part of the system.

This is particularly the case with the law on medical devices, which is unclear with respect to healthgrids. In September 2007, the European Parliament and the Council adopted Directive 2007/47/EC of 5 September 2007 amending Council Directive 90/385/EEC on the approximation of the law of the Member States relating to active implantable medical devices, Council Directive 93/42/EEC concerning medical devices and Directive 98/8/EC concerning the placing of biocidal products on the market (OJ, L 247/21, 21.09.2007). In particular, it is stated in Recital 6 of this directive that *"it is necessary to clarify that software in its own right, when specifically intended by the manufacturers to be used for one or more of the medical purposes set out in the definition of a medical device, is a medical device. Software for general purposes when used in healthcare setting is not a medical device."* An amended definition of what a medical device is also to be found in the directive.

Intellectual Property

In the EU, legislation dictates that the owner of copyrighted software running a healthgrid has exclusive rights to reproduce his work, produce derivative works, distribute copies to the public, and perform and/or display the work publicly. Under these circumstances any natural or legal person would have to pay to use the computer programs that constitute one of the most important components of healthgrids. Conversely, an open standards approach to software co-development could help in the development and implementation of HealthGrids. The open source licensing model actually uses copyright and contract principles to retain control of the work while enabling its use effectively for free, and could thus encourage use and development.

Organisational, Social and Cultural Issues

Both at the individual and the societal level, issues like universality of availability of full healthcare services to all citizens, equal access to healthcare, and equally high quality of services are key issues. Geographical factors relate mainly to equal access to quality care independent of location. ICT-based systems pose new problems like access to EHRs by insurance companies or employers, and even police and prosecutors. The opinions and attitudes of patient and citizen associations and lobbying groups, often magnified by the media, could have a strong impact through public (policy) discussions of these topics on the implementation and diffusion of healthgrids.

The organisational level is always complex. Perspectives, confirmed by two recent research studies, include:

- Changing care pathways that need new information, skills, knowledge and processes for healthcare providers
- Changing roles of healthcare professionals, teams and healthcare organisations
- The transfer of roles between healthcare professionals, teams and healthcare organisations
- Increased collaborative working and exchange of information between providers
- New relationships between citizens and healthcare professionals and organisations
- New strategic partnerships for third party financers and healthcare providers.

Cultural issues are a key factor in health services, including the great diversity of attitudes, behaviour and knowledge exchange among professional and non-professional staff involved in healthcare, and the impact this has on the quality, efficiency and organisation of services. Education and training, professional standards and bodies, rules and regulations, attitudes and behaviour all have an influence here.

REQUIREMENTS FROM USE CASES

An analysis of user requirements from the use case scenarios outlined in section 3, with reference to the technical, ethical, legal and socio-economic issues identified in section 4, lead to the formulation of the following challenges and requirements for each use case.

Innovative Medicine

The European Federation of Pharmaceutical Industries and Associations (EFPIA) identified pre-competitive barriers to innovation. The objective for the future would be to identify as soon as possible in the pre-clinical phase the reasons for lack of efficacy, despite promising pre-clinical data, and the potential for adverse drug reactions and pre-clinical toxicity. The identified key bottlenecks in the R&D process are:

- Predictive pharmacology at the discovery research stage;
- Predictive toxicology at the preclinical development stage;
- Identification of biomarkers at the translational medicine stage;
- Patient recruitment and validation of biomarkers at the clinical development stage;
- Risk assessment with regulatory authorities at the pharmacovigilance stage.

The knowledge management area is identified as key to leveraging the potential of new technologies such as genomics and proteomics and to analyse the huge quantity and diversity of information in an integrated way, such as the capture, analysis and interpretation of knowledge generated for one potential drug candidate from discovery, non-clinical and clinical development all the way to lifecycle management. One of the major bottlenecks is the lack of availability of databases across R&D that might facilitate data integration.

Epidemiology

The main challenges to this use case include the general problem of access to distributed, critically sensitive and heterogeneous data, resulting in costly computing processes. Patient-centric analyses normally deal with smaller amounts of data and require a pre-existing knowledge of models of healthy and diseased organs or tissues. Population-level analyses normally deal with the integration of larger, poorer-quality data. Semantics are especially relevant in those approaches.

The research requirements for the epidemiology use case can be summarised as follows:

- *Effective semantic annotation of data.* Data is poorly coded and interoperability of coding is not trivial. Extracting knowledge from medical data, however, is a main objective.
- *Effective integration of distributed and heterogeneous data.* Integrating distributed resources requires exchange protocols, secure mechanisms, patient de- and re-identification, and automatic data analysis services.
- *Availability of efficient infrastructures and usage policies.* Applications will require sufficient resources and a reliable infrastructure to work on under a clear Quality of Service (QoS) promise.
- *Robust security mechanisms* insuring that the different components of the infrastructure behave in way that conforms to the European norms controlling the handling of personal data.
- *Usability of applications and services.* The tools should be available through protocols and interfaces similar to those used in the users' normal research. Not only must the applications be as compliant as possible with current systems and interfaces, but so must the technologies.
- *Ensuring that the research is done in a secure, ethically and legally compliant framework.*
- *Reliability, scalability and pervasiveness.* All the previous services must be robust and should be scalable without reducing performance.

A Surveillance Network for Avian Flu

The two main concerns in this use case are the evolution of the virus and its capacity for human to human transmission.

- *Services for molecular epidemiology.* Once a new virus stream has been sequenced, its comparison to the previously identified streams allows the evolution of the virus to be measured. Genome sequences can also be used to guide the search for vaccines and drugs. The data must be properly integrated and its exploitation must build upon expert skills in epidemiology.
- *Mathematical modelling and computer based simulations for the testing of outbreak hypotheses.* Disease epidemiology requires measuring the evolution of the disease in time and space using GIS (geographic information systems) and environment data as well as understanding the disease transmission, reservoir, immunity and treatment.
- *The reliability, relevance, and completeness of the data stored.* Reliability depends on the mechanism to collect data. Relevance depends on the mechanism to update data while completeness is influenced by the capacity to collect data from multiple scientific disciplines and multiple countries.
- *The usability of the environment.* Scientists are so busy they will not take time to contribute information and operate services if these services are not easy to use.
- *The integration of several types of data for disease monitoring*: medical / epidemiologi-

cal data on human and animal cases, geographical data for each outbreak (location, number of cases, geographic distribution of the casualties, environment, and population density), virus genome sequences, phylogenetic trees, and proteomics.

- *Text mining services.* In order to extract information from this literature on demand, advanced text mining services will be required.

Virtual Physiological Human (VPH)

With respect to grid computing, the following challenges and requirements have been identified:

- *Knowledge management.* The accumulative nature of VPH requires that everything is organised within solid knowledge management models. This should make it possible to keep even very large information spaces organised and usable.
- *Fostering grid adoption.* To foster grid adoption in the VPH community, it is highly recommended to identify a few VPH CPU intensive applications which could benefit immediately from deployment on existing grid infrastructures like EGEE or DEISA. The deployment of these applications will allow the identification of missing services on the existing infrastructures, and will raise the awareness of grids in the VPH community.[4]
- *Access to resources.* Researchers require access to all available resources in a uniform way, from those provided by their own department to specialised HPC resources. Access to these should be as seamless as possible, with simulations at different scales being automatically migrated and appropriate resources being used as required.
- *The usability of interfaces.* Current grid portals require the user to specify parameters such as memory to be allocated at execution time; this would not be appropriate for VPH users.
- *Grid usage models.* The nature of VPH simulations means that processing is time critical, and current models of HPC use would not be appropriate. Instead, models which permit a large number of grid nodes to be used for a relatively short time ('burst mode') with little or no waiting time should be established. Resource co-reservation will also be required, particularly where multi-scale simulations that run over multiple sites are concerned.
- *Shared storage for large data/model repositories.* The imaging datasets concerned can be several hundred megabytes in size, but after pre-processing to generate a predictive model, the modelling and simulation data can be as much as a hundred gigabytes in size. With potentially thousands of these, there is a clear need for multi-terabyte storage facilities, connected to distributed HPC resources via high speed networks. These must incorporate the required security and confidentiality measures.
- *Methods to solve multiple predicted models, in a coupled way.* As the coupling of predictive models at different scales is central to VPH's description of human physiology, coupled methods to solve multiple models will be required. This is considerably complicated by the fact that the models concerned may be very different both in conceptual nature and mathematical nature. Even relatively simple VPH problems can be considerably complicated by variations between individual subjects and treatment procedures.
- *Direct prediction from medical images.* Methods for transforming a medical imaging dataset into a subject specific predictive model that do not require the costly pre-processing phase are being developed, such as the Boltzmann Lattice in haemo-

dynamics and voxel meshes for hard tissue simulations. However, these are enormously computationally intensive, requiring fifty or more teraflops of computational power to solve in less than a day.

Tertiary Healthcare: The MammoGrid, e-DiaMoND and Health-e-Child Projects

Several technical issues were identified by MammoGrid and e-DiaMoND, including the standardisation of mammograms, design of an appropriate clinical workstation and distribution of data, images and clinician queries across a grid-based database while respecting patient confidentiality and security protocols.

- *Image quality.* While clinically significant signs are subtle, many parameters also affect the appearance of an image. For mammograms, these include image acquisition parameters, such as degree of breast compression, tube voltage and beam intensity, and anatomical and physiological data, which show marked variation across the population, at different times in the menstrual cycle and throughout the course of a woman's life. The way diagnostic imaging systems are used and maintained by clinicians also varies between imaging centres and breast screening programmes. In order to study the epidemiology of breast cancer, it is necessary to understand this variability. This is also a prerequisite for the integration of computer-aided detection (CADe[5]) tools and quality control in the process, as well as for comparison between images and for radiological training programmes.
- *Image standardisation.* There are a large number of standards that are sometimes conflicting or overlapping with worldwide accepted standards like DICOM and the defacto SMF standard for the standardisation of mammograms. These must be disambiguated and adhered to for a production grid to be deployed even at a national level.
- *Storage of medical images.* Images have to be stored unaltered and uncompressed for legal reasons. Also, several copies of the mammogram need to be stored which increases the storage requirements. The e-DiaMoND project decided to use expensive high speed disks for storing the currently active data (those currently undergoing treatment), and cheaper lower performance disks for storing data that is being migrated to or from long term slower storage. For the MammoGrid project only metadata about the images were stored in the shared database.

The following additional requirements were identified by the Health-e-Child project:

- *Vertical integration of clinical data.* Practitioners within different medical disciplines require information at different levels and with different granularity. Integrating genomic, imaging and proteomic data could provide medical practitioners and researchers with a unified, coherent view of the patient's current and past health.
- *Storage and sharing of biomedical information.* Biomedical information should be accessible from geographically distributed sources, including information from the local hospital intranet and databases such as gene databanks on the internet. Access to data must be secure, with appropriate levels of encryption, and access rights must be enforced. Additionally, data sources must have sufficient availability and responsiveness to assure an appropriate quality of service.
- *Biomedical query processing.* Clinicians from different disciplines will require different views of the integrated record provided by the project. Given the vertically integrated nature of the patient records, these queries

will be considerably complex, but must be executed sufficiently fast for use by clinicians during consultations or treatment.

- *Integrated disease modelling.* Disease modelling is required at all levels addressed by the project, from molecular modelling such as searching for gene defects, to *in silico* physiological models of the whole body at a high level. These models should aid clinicians when attempting to determine if a patient has or is likely to develop a certain disease, and what the best treatment might be.
- *Tools to support image annotation.* Image annotation tools should support the creation and modification of annotations by cardiologists, rheumatologists and radiologists, and should be stored as metadata. For second readings by radiologists, standard ways of annotating images are needed for better collaboration between experts.
- *Queries to find records of similar cases.* Based on similarity criteria specified by the user, a mechanism should exist to find and display similar cases, with any accompanying annotations, metadata and records of clinical decisions taken.
- *Macroscopic computational models for key organs and diseases.* Generic models should be constructed, which can then be adjusted according to patient-specific clinical data in order to produce individualised models.

General Healthcare: Chronic Obstructive Pulmonary Disease (COPD)

Evidence from national studies may not be available to a doctor in a different geographical location. In order to continue treating the patient concerned, the doctor (or decision support system) must be aware of the evidence and guideline/pathway that informed the plan of care for that patient, and any deviations from that plan that have occurred to date. This may not a trivial matter of simply retaining a link to the relevant material, as there may be language barriers, and local reasons why the guidance followed in one country would not be appropriate in another.

The guidance also mentions drugs that are not certified for the treatment of COPD (off-label) despite this evidence coming from high quality systematic reviews. Different drugs will be certified for the treatment of the condition in different countries, complicating the process of following a single guideline or pathway regardless of travel. In fact, the patient concerned may be travelling for the express purpose of receiving different or less costly treatment in another country.

Knowledge of prior history of exacerbations and smoking is essential for the proper treatment of the disease, and therefore the doctor concerned must be able to access, comprehend and update the patient's record. This requires standardisation of electronic health records (EHRs) and electronic integrated care pathways (eICPs). When it comes to decision support, a standard interface format, such as the proposed HL7 VMR (virtual medical record), will also be a necessity.

RECOMMENDATIONS AND ROADMAPS

In order to translate the broad range of issues highlighted in our discussion above into a manageable programme of necessary developments, we have adopted certain broadly acceptable assumptions: first, that the fundamental architecture will be service-oriented; second, that it will be necessary to distinguish and layer, from the bottom up, core infrastructure services, generic healthgrid services and specific biomedical or healthcare application services; and third, that the typology of grids will be reflected in the nature of the challenges. Moreover, there are interdependencies between challenges. Thus challenges have been organised according to a number of key criteria:

- Is the proposed healthgrid essentially a computational grid, a data grid or a collaboration grid? Could it potentially develop into a knowledge grid?
- Is the necessary development to achieve any given stage likely to be delivered by generic grid research or is it particular to healthgrids?
- Is some prerequisite standard or other agreed framework necessary for the achievement of any particular milestone?

Proposed Roadmaps

In this section, we present a technical roadmap for the adoption of the grid technology for healthcare. In the previous section, for three families of grids, computing, data and knowledge grids, we have identified a number of research challenges which have been characterised by key words. The diagram overleaf represents how research challenges address different layers of services from core infrastructure to applications.

Reading the diagram from left to right, we note two pillars of challenges for computational and data grids, with knowledge grids supported on these two pillars. For simplicity, the challenges of collaboration grids have been subsumed under those of computational and data grids. The image conveys the sense that knowledge grids will be built at a higher level (and, necessarily, at a later stage) than computational and data grids.

Reading the diagram from the bottom up, we have depicted challenges according to their closeness to the underlying infrastructure, or in terms of our criteria above, whether they are likely to be addressed by generic grid services, by healthgrid service or by specific applications.

The following comments can be made from the picture:

- Interoperability as well as improved distributed data management must be core functionalities of the infrastructure
- Quality of service is required from both core and HealthGrid services for successful healthcare / biomedical applications
- healthgrid services should be accessible on demand, with a high degree of usability.

Figure 1. Research challenges and their relationship to HealthGrid service layers and grid family

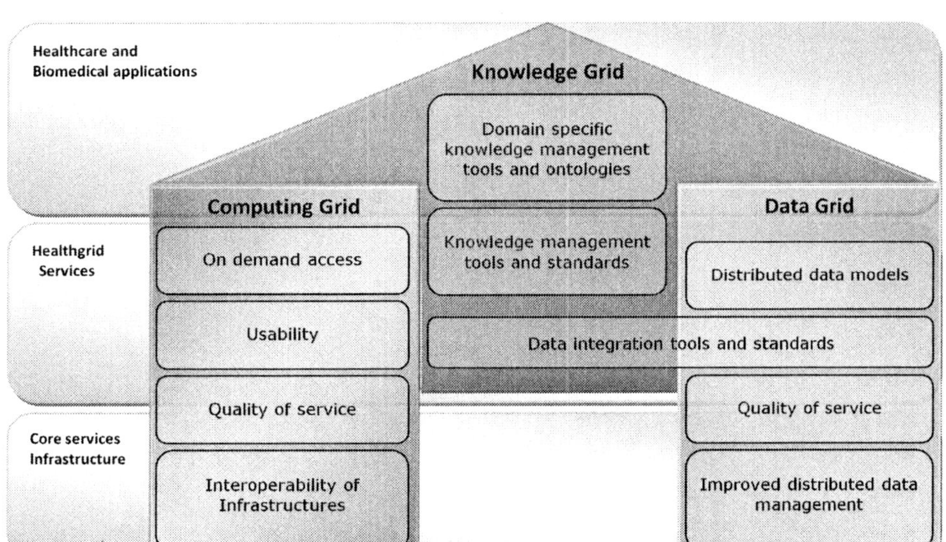

Distributed data models must also be provided.

- Data Integration Tools and Standards are healthgrid services which stand at the interface between data and knowledge grids.
- Knowledge Management Tools and Standards require the availability of proper job and data management tools. They stand at the interface between generic healthgrid services and the application specific developments.
- Domain Specific Knowledge Management and Ontologies are the responsibility of individual research communities. Their interface to the knowledge grid is achieved using Knowledge Management Tools and Standards.

On the basis of this analysis, we have represented in Figure 2 the research challenges according to their complexity and an estimated time when they should be overcome. The figure inspired from The Innovative Medicine Case Study[6], also indicates the level of adoption by the research communities. As can be seen clearly from the picture, we identify several distinct roadmaps:

- Research and development for computing grids should allow offering the quality of services needed for biomedical research and healthcare at a 5-year horizon.
- Data grids are expected to reach maturity at a 10-year horizon as the core technology is not yet mature.
- Collaboration grids are achievable with different levels of sophistication at different stages.
- Knowledge grids depend on the quality of services for distributed data integration and the capacity of the research communities to agree on standards and ontologies. As a consequence, their maturity is not expected before 15 years.

This model arises in the field of innovation studies, and distinguishes between: *visionaries*, pioneers both in research and applications; *early adopters*, who recognise the potential for rapid benefits and take up the technology quickly, often introducing further innovation; *early majority* who incur relatively little risk in adopting the technology; and *late majority*, those who take virtually no risk and finally adopt a technology because there is no other option.

Figure 2. Research challenges as a function of time and complexity

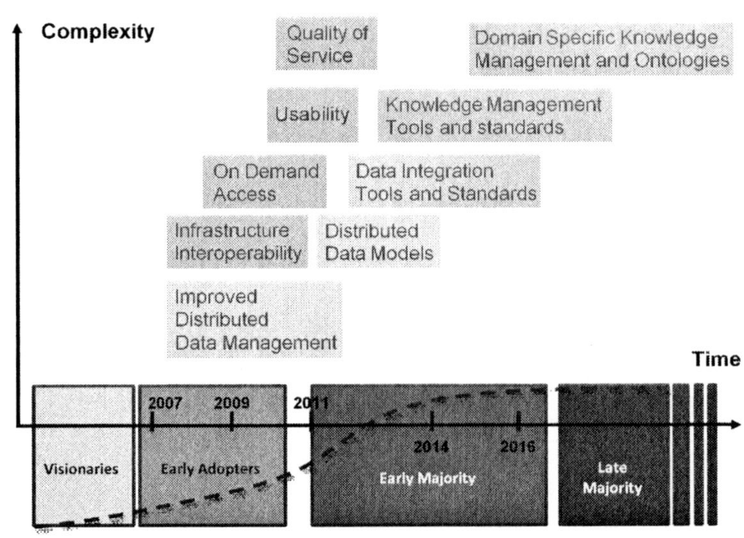

The diagram depicts priorities: even for early adopters, infrastructure interoperability and distributed data management are already necessary; on demand access, usability and quality of service are at the first point of inflection, before rapid expansion, while with sophisticated AI tools in the later stages, a second inflection occurs and the technologies become routinely accepted.

Mapping Ethical, Legal, Social and Economic (ELSE) Requirements

- *Liability Issues.* The current state of EU legislation does not cover liability issues that are specific to healthgrids. The following tasks could help minimise liability concerns for healthgrid usage. Effort need to take place at the organisational level to incorporate strategies and guidelines in order to minimise liability risks.
- *Trust and Acceptance.* Pilot projects and prototype applications are essential in order to demonstrate the usability of grid services in clinical and research workflows
- *Data Protection.* Adopting evolving privacy enhancing technologies in order to insure the enforcement of data protection regulation within data management processes and clinical workflows.
- *Sustainability Guarantees.* The development and deployment of data grids will benefit from more focused prospective assessments of the socio-economic impact in order to identify existing and potential barriers.
- *Education and New Skills Requirements.* Training and educational programs to increase users' confidence in the use of healthgrid products and services.
- *Intellectual Property.* There is a contradiction between intellectual property rights and the needs of grid technology, which will require that access to databases, knowledge and software is free of rights. Contract law and agreements could be an option to regulate the IP issues related to knowledge integration, ontologies and software reuse.
- *Governance and Delegation.* Some analysis need to be performed to explore the requirement of data controller for data stored and processed on a European grid environment.
- *Policies and Codes of Conduct.* Discussions should take place between different healthgrid stakeholders to decide on the importance and benefit of applying for new legislation to address healthgrid related legal and ethical issues.
- *Dissemination and Publicity.* Dissemination and publicity programmes need to precede the deployment of knowledge grids. This includes workshops, conferences, and magazines to attract the user community and build awareness of healthgrid facilities.

CONCRETE RECOMMENDATIONS

Technical Recommendations

It is important that technical research and development be conducted in close collaboration with user communities. At certain stages it must be driven and validated by user groups, although there is always scope for innovators to introduce unforeseen possibilities to users. The research communities involved in the definition of the roadmap expressed their interest and support for the deployment of prototypes and test cases on existing grid infrastructures. We recommend that these infrastructures and tools continue to be made available to applications requiring computing services and data management.

Indeed, some projects are already using the DEISA and EGEE infrastructures for scientific production in the fields of epidemiology, medical imaging and drug discovery. However, these initiatives come from pioneers and are not sufficient to achieve a wider adoption in these research communities. We recommend that:

- More attention be paid to such initiatives so that they may influence the evolution of the technology to make it better fit the needs of the community;
- Two projects within the framework of the EuroPhysiome initiative be identified that could directly benefit from the computing and data management resources of the EGEE and DEISA infrastructures; these should be deployed in parallel on the two infrastructures in order to investigate interoperability issues and identify bottlenecks.

In terms of encouraging biomedical applications to fully exploit grids, we recommend:

- Linking certain advanced health domains to an e-science infrastructure;
- The adaptation of epidemiology data sources to grid models and grid-enabled gateways to epidemiological data, using medical informatics-related connectors such as HL7, DICOM, ENV13606, or similar.

In the same spirit, in order to foster the uptake of grids in the biomedical research and healthcare communities, we recommend:

- The release of open-source components for medical data interfacing;
- Building a core reference database of validated experimental and clinical research data extracted from the literature in innovative medicine and to explore whether a grid infrastructure could support this activity;
- The creation of disease-specific European imaging networks towards the establishment of standards, validation of imaging biomarkers, development of regional centres of excellence in innovative medicine and exploration of grid infrastructures to support such activity.

We recognise that there are a number of concerns (for example: security and standards) in which problems exist irrespective of the use of grids. It is important to understand the nature of these problems and the extent to which the use of grids complicates them. Results could be concrete implementation recommendations (for example: security improvement) and a suggested list of health applications requiring security which may be able to be deployed on a grid. In the field of standards, we believe that the HealthGrid initiative provides the right framework to coordinate the development of the different standards in collaboration with the OGF and the various medical informatics standardisation bodies. We recommend:

- The active pursuit of standards for the sharing of medical images and electronic health records on the grid within the already existing medical informatics standardisation bodies;
- The active pursuit of ontology matching and development for healthgrids;

We believe that technology transfer between EC projects should receive more prominent and active encouragement. In particular, we recommend:

- The commission implements collaboration measures in the funding mechanism for projects;
- Targeted capacity building so that projects may access grid resources on demand, without previous agreement or request; European grid infrastructures should be freely accessible to European projects;
- Porting of one or two biomedical grid applications, already successfully deployed on grid infrastructures, to e-science en-

vironments using OGSA-compliant grid toolkits.

Finally, to return to a frequent theme in this analysis, we recommend:

- The encouragement of cross community interaction, in order to build meaningful dialogue between grid developers and health researchers.

Legal Recommendations

Liability in a healthgrid System. Using grids blurs the liability issues in terms of medical practice. A stepwise approach should therefore be taken to develop the liability framework, distributing legal responsibility appropriately across healthgrid users. Such an approach would help to favour the reliance on the system while providing legal certainty for all stakeholders, including patients. Moreover, the European Commission should consider supporting the adoption of EU level guidelines that would identify the various parties involved in delivering healthgrid services and annex services and establish the various liabilities that each party must accept. Such guidelines should be widely disseminated in order to develop users' confidence in the use of healthgrids in general. In particular it should be investigated whether specific guidelines on those specific services could be drafted under the provisions for a code of conduct established in directive 2000/31 on eCommerce.

Product safety. In the framework of the European level legislation applicable to product safety, national authorities have been established to monitor product safety and to take appropriate measures to protect consumers. Under these circumstances, an information system has been put in place that imposes collaboration between distributors, producers and the national authorities but also between member states and the European Commission (RAPEX).

At present, this system is not used at all for products used in the composition of grid systems. The European Commission should thus adopt policy tools encouraging the use of the RAPEX system for such products.

HealthGrid as a medical device. The law on medical devices is very unclear with respect to healthgrids. While it may be argued that a healthgrid could fall within the ambit of the current medical devices directive in that it is a software tool that impacts on a medical act, the whole construction of the directive is based upon physical goods (which might have a software component) that are placed on the market for purchase or lease. In this situation, many of the currently available monitoring devices are covered only by general product liability, but not by specific liability provision.

In this framework, special guidelines should be issued in order to clarify the application of medical devices legislation to specific tools used in healthgrids.

Patient consent. In February 2007, the European working party on data protection, established under article 29 of the directive issued a working paper looking at the applicability of data protection legislation to Electronic Health Record (EHR) systems. In its report, the working party noted in particular the limitation of the use of consent to permit the processing of heath data. The working party notes that if processing health data in an EHR system is the primary way of processing health data in a given health system, then a patient's care may be compromised if he or she opts-out of such a system by not giving his or her consent to the creation of an EHR. Accordingly, consent should not be used as it cannot be said to be truly and freely given.

The remaining provisions setting aside the general prohibition on article 8 of the directive 95/46/CE can also be said to pose some problems – notably the idea that a patient ought to know the full finality of the use of data before his or her data may reasonably be used. But, as noted by the data protection working party there are some problems in using consent as a valid basis for processing data in eHealth applications. Indeed, if the creation of, for example, electronic medical records is a necessary and unavoidable consequence of the medical situation, withholding consent may be to the patient's detriment.

Specified and explicit purposes. According to the data protection directive, data may only be collected for specified and explicit purposes. If healthgrids can be used for risk detection, disease monitoring and preventive care, legal guidelines should be established that clarify the circumstances in which professionals can make further use of personal data related to health in the interests of public health. Such guidelines should allow for secondary uses even where such uses could not have been foreseen at the time of data collection.

Technical and organisational security measures. Efforts should be made to harmonise national standards on the technical and organisational measures of data security. While the data protection directive calls for such standards to be adopted, little has been done at a regulatory level to harmonise guidelines across the EU.

Intellectual property rights. It might be desirable for the commission to develop guidelines for the use of open licensing and open standards, which could address the tension between the intellectual property rights of developers and the needs of the grid technology. Such an open standards software approach could then be a solution to help the development and implementation of healthgrids.

On the other hand, the use of healthgrids in the drug discovery sector raises the issue of the ownership of both methods used to discover the medicines and the results achieved. Indeed, all the grid nodes that contribute resources to compute the docking probabilities could claim some ownership of the results and the designers of the software used in the process would certainly be in position to claim ownership of the method. In this context, one may ask whether it is important to know, say, which grid node was the one to identify a particular candidate molecule.

In this context, it is of essential interest, notably in patents, to determine guidelines that would determine, in case of collaboration in the research, what every actor is entitled to according to his contribution to the system.

Privacy policies and codes of conduct. A directive or code of conduct on privacy and health information infrastructure should be developed within the context of directive 95/46/EC and could take the form of either a dedicated directive or could be an EU-level code of conduct to be approved by the European working party on data protection set up under article 29 of the directive. This could help to solve the problem of data processing legitimacy. In particular, it could provide possible bases of legitimacy other than the data subject's consent. It could also provide the following solutions:

- Appropriate safeguards to allow for the further processing of personal data (and especially of medical data) for substantial public interests (without requiring the data subject's consent) like scientific research. An example of appropriate safeguard would be a first coding by the initial data controller and a second coding by a trusted third party gathering all the data from the data controllers before sending them to the researchers,
- Appropriate safeguards to allow keeping the data for longer periods for scientific use; terms under which identification numbers or other identifiers may be used; terms under which (coded) personal data may be transferred to third countries for scientific research.

Socio-Economic Recommendations

Trust and Acceptance. Trust is a very important element in any interaction between the different members of a society. In the market context, trust is crucial for successful business to business collaborations. Similarly, in a healthgrid domain a good collaboration will not be achieved unless a trust relationship exists between the different users and stakeholders. Pilot projects and prototype applications, which are an inherent part of the technology roadmap, need to be future oriented in the sense that the ultimate routine operation users have to be persuaded both of their value and their applicability, i.e. their ability to fit into real clinical or research workflows. This has to be taken seriously from the very beginning, even in proof-of-technology demonstrators: the goal should always be to give users, especially clinicians, tools that they would consider using with patients in real healthcare situations. Trust and acceptance can be greatly enhanced by the establishment of appropriate ethics committee structures to advise on the observance of ethical principles.

Estimation of Costs and Benefits. Ex-ante analyses over time, based on initial pilot experience, have to focus on ensuring acceptance, technical and regulatory certainty, and sufficient private incentives in the steps to follow. An inherent part of such assessments should be to estimate potential net benefits (i.e. expected benefits less expected costs over time), accounting for different risks and for optimism bias in estimations. Such studies will facilitate access to initial funding, but can also be beneficial in the necessary dissemination work among the health sector.

Sustainability Guarantees. Work towards achieving the next milestone in complexity – data grids – will benefit from more focused prospective assessments of socio-economic impact in order to a) identify already existing, as well as potential barriers, and b) build convincing business cases ensuring sustainability. The analysis of alternative resource allocation options from a societal perspective, but also on organisational level, becomes necessary.

An organisational milestone can be defined here in the move from technology science towards service provision. By that stage, a notable amount of legal and regulatory certainty has to be achieved, so that private incentives can be assessed and adjusted (including via government intervention) if necessary.

Cross-Organisational Interoperability. The effective deployment of knowledge grids will crucially depend on collaboration between institutions, meaning more than "simple" access to each others' data and computing resources. This collaboration requires the utilisation of human resources and in some cases a significant strategic re-orientation and re-organisation of working processes and even management structures. As the health sector, including clinical research and public health, is (and should be) highly regulated, policy makers on regional, national, and EU level should review the existing regulatory framework against the requirements arising from the exploitation of knowledge grids. Particular attention should be given to flexibility of government regulated budgets and reimbursement schemes. The latter should encourage cross-organisational collaboration, including such beyond national borders, by means of using knowledge grids.

CONCLUSION

Although the term 'grid' is used throughout, it should also be noted that this area of research is continually moving, with 'clouds' and other forms of so called 'cyber-infrastructures' emerging; regardless of the label used, the recommendations and roadmaps defined here – as well as the user-driven approach to requirements – are still applicable.

This chapter has presented an abbreviated version of the SHARE vision. This began with the White Paper and resulted, after technical and ethical, legal, social and economic analyses, and consideration of several important use cases, in the road map. The reader is invited to explore the full version of this roadmap available, as are all

SHARE documents, from the SHARE website which covers the issues presented here in far greater detail.

ACKNOWLEDGMENT

This chapter is based on a deliverable from the SHARE Specific Support Action, funded by the European Communities under the Framework Programme 6. It is reproduced here with permission of the EC as the copyright holder of the original document. Co-funding by partner institutions is also acknowledged.

REFERENCES

Almeida, J., et al (2003). EURISWEB – Web-based epidemiological surveillance of antibiotic-resistant pneumococci in Day Care Centers. *BMC Medical Informatics and Decision Making, 3,* (9).

Berman, F., Fox, G. &, Hey, A.J.G. (2003). *Grid Computing: Making the Global Infrastructure a Reality.* Hoboken, NJ: John Wiley.

Breton, V., Dean, K. & Solomonides, T. (2005). The HealthGrid White Paper. In *Proceedings of Third HealthGrid Conference: Studies in Health Technology and Informatics,* Vol. 112. Amsterdam, The Netherlands: IOS Press (pp 249–321). Retrieved from http://www.healthgrid.org

European Union (2002). *Resource Book of IST Projects Relating to Health (1998-2002).* Retrieved from http://cordis.europa.eu/ist/ka1/health/projectbooklet/index.htm.

Foster, I., Kesselman, C. & Tuecke, S. (1999). *The Grid.* San Francisco: Morgan Kaufmann Publishers.

Foster, I., Kesselman, C. & Tuecke, S. (2004). *The Grid 2.* San Francisco: Morgan Kaufmann Publishers.

STEP (2002). *Seeding the EuroPhysiome: A Roadmap to the Virtual Physiological Human.* Retrieved from http://www.biomedtown.org/biomed_town/STEP/Reception/step_presentations/RoadMap/vph_roadmap_printed_3.pdf

W3C (2001). *Web Services Description Language (WSDL).* Retrieved from http://www.w3.org/TR/wsdl. Warren, R. et al (2007a). A Prototype Distributed Mammographic Database for Europe 2007. *Clinical Radiology 62*(11) pp. 1044-51. Amsterdam, The Netherlands: Elsevier.

Warren, R. et al (2007b). A Comparison of Some Anthropometric Parameters Between an Italian and a UK Population: "proof of principle" of a European project using MammoGrid. *Clinical Radiology 62*(11), pp. 1052-60. Amsterdam, The Netherlands: Elsevier

WISDOM (2006). *Wide In-Silico Docking On Malaria.* Retrieved from http://wisdom.eu-egee.fr/

ADDITIONAL READINGS

Grid computing was introduced in the first edition of *The Grid* (Foster *et al,* 1999) and amply developed in its second edition, essentially a new book, (Foster *et al,* 2004) and in (Berman *et al,* 2003).

The *HealthGrid* initiative and organization have organized an annual conference since 2003, the proceedings of which have been published by IOS Press in the Netherlands since 2005. The first volume, in 2005, includes the most complete version of the seminal White Paper (Breton *et al,* 2005) which gave rise to the SHARE project.

All SHARE public documents are freely available for download from the official project website, http://eu-share.org/about-share/deliverables-and-documents.html, and the interested reader is encouraged to explore the detailed documents and road maps relating to technologies, to ethical, legal, social and economic (ELSE) issues and to the use case scenarios. These documents also offer extensive references.

A further source is the HealthGrids Knowledge Base, a growing database of information about projects, individuals, technologies, organizations and references at http://kb.HealthGrid.org/.

KEY TERMS AND DEFINITIONS

The following definitions are useful for understanding the document content.

Authentication: Verifying and confirming the identity of a grid user.

Authorisation: Restricting access to resources based on what a user has been granted access to.

Data: Any and all complex data entities from observations, experiments, simulations, models, and higher order assemblies, along with the associated documentation needed to describe and interpret them.

Data Controller: The person or organisation responsible for the manner in which any personal data is processed.

Data Mining: Automatically searching large volumes of data for patterns or associations.

Data Model: A model that describes in an abstract way how data is represented in an information system. A data model can be a part of ontology, which is a description of how data is represented in an entire domain.

Data Processor: Any person who processes data on behalf of a data controller.

Data Subject: An individual who is the subject of personal data.

Grid: A fully distributed, dynamically reconfigurable, scalable and autonomous infrastructure to provide location independent, pervasive, reliable, secure and efficient access to a coordinated set of services encapsulating and virtualising resources.

Informed Consent: A legal term referring to a situation where a person can be said to have given their consent based upon an appreciation and understanding of the facts and implications of an action.

Metadata: May be regarded as a subset of data, and are data about data. Metadata summarise data content, context, structure, inter-relationships, and provenance (information on history and origins). They add relevance and purpose to data, and enable the identification of similar data in different data collections.

Middleware: A software stack composed of security, resource management, data access, accounting, and other services required for applications, users, and resource providers to operate effectively in a grid environment.

Ontology: The systematic description of a given phenomenon, which often includes a controlled vocabulary and relationships, captures nuances in meaning and enables knowledge sharing and reuse. Typically, ontology defines data entities, data attributes, relations and possible functions and operations.

Processing: Obtaining, recording or holding the data, or carrying out any operation on the data, including organising, adapting or altering it. Retrieval, consultation or use of the data, disclosure of the data, and alignment, combination, blocking, erasure or destruction of the data are all legally classed as processing.

Roadmapping: An extended look at the future of a chosen field of inquiry, leading to an outline or map of how and by what means to achieve certain goals.

SOAP: A protocol for exchanging XML messages over a network. It defines the structure of the XML messages (the SOAP envelope), and a framework that defines how these messages should be processed by software.

The Article 29 Data Protection Working Party: A working party established by article 29 of directive 95/46/EC. It is the independent EU advisory body on data protection and privacy. Its tasks are laid down in article 30 of directive 95/46/EC and in article 14 of directive 97/66/EC.

Virtual Organisation: A group of grid users with similar interests and requirements working collaboratively and/or sharing resources regardless of geographical location.

Web Service: A software system designed to allow inter-computer interaction over a network to perform a task. Other computers interact with a web service, in a manner prescribed by its interface, using messages which are enclosed in a SOAP envelope and are often conveyed by HTTP. Software applications can use web services to exchange data over a network.

Workflow: A set of components and relations between them, used to define a complex process from simple building blocks. Relations may be in the form of data links which allow the output of one component to be used as the input of another, or control links which state some conditions on the execution of a component.

XML: An annotation technology used to describe structured data within a document using mark-ups and tags, similar to HTML. The main difference between the two is that the elements in XML can be given a definition depending on their usage which may be semantic rather than presentational. XML is a text format and can be read easily either by humans or machines.

XML Schema: A definition of the structure of an XML document. A schema contains a set of rules that dictate how an XML document must look like in order to be an instance of this schema. The relationship between a schema and an XML document implementing it can be compared with a class definition and an instance in object-oriented programming.

ENDNOTES

* Affiliation at the time of the project

1 This is arguably the majority moral position, but is in fact not the legal reality in every Member State. For example, in the UK medical records formally belong to the Secretary of State for Health.

2 The website of the WG is: http://ec.europa.eu/justice_home/fsj/privacy/workinggroup/index_en.htm.

3 E.g. http://ec.europa.eu/information_society/activities/health/docs/studies/legally-ehealth-report.pdf

4 Some FP6 related projects, such as VIROLAB, Health-e-child and @neurist, may serve as models.

5 CADe is used to differentiate computer-aided detection from computer-aided diagnosis, a significantly more difficult goal. In the literature both are sometimes indifferently referred to as 'CAD'.

6 SHARE deliverable D5.2a.

ABBREVIATION LIST

DICOM	The Digital Imaging and Communications in Medicine Standard
EC	European Commission
EHR	Electronic Health Record
ELSE	Ethical, Legal and Socio-Economic
EPAL	Enterprise Privacy Authorization Language
EU	European Union
HL7	The Health Level 7 Standard
HPC	High Performance Computing
IPR	Intellectual Property Rights
OGSA	Open Grid Services Architecture
QoS	Quality of Service
SHARE	Supporting and structuring HealthGrid Activities and Research in Europe
SOKU	Service Oriented Knowledge Utility
VPH	Virtual Physiological Human
W3C	World Wide Web Consortium
WPx	Work Package x
WP3	SHARE Technology and Security Activity
WP4	SHARE Health Policy, Legal, Social and Economics Activity
WP5	SHARE Applications Activity
WP6	SHARE Roadmap Synthesis and Validation Activity

Chapter II
Types of Resources and their Discovery in HealthGrids

Aisha Naseer
Brunel University, UK

Lampros K. Stergioulas
Brunel University, UK

ABSTRACT

Adoption of cutting edge technologies in order to facilitate various healthcare operations and tasks is significant. There is a need for health information systems to be fully integrated with each other and provide interoperability across various organizational domains for ubiquitous access and sharing. The emerging technology of HealthGrids holds the promise to successfully integrate health information systems and various healthcare entities onto a common, globally shared and easily accessible platform. This chapter presents a systematic taxonomy of different types of HealthGrid resources, where the specialized resources can be categorised into three major types; namely, Data or Information or Files (DIF); Applications & Peripherals (AP); and Services. Resource discovery in HealthGrids is an emerging challenge comprising many technical issues encapsulating performance, consistency, compatibility, heterogeneity, integrity, aggregation and security of life-critical data. To address these challenges, a systematic search strategy could be devised and adopted, as the discovered resource should be valid, refined and relevant to the query. Standards could be implemented on domain-specific metadata. This chapter proposes potential solutions for the discovery of different types of HealthGrid resources and reflects on discovering and integrating data resources.

INTRODUCTION

Advanced communication channels and networks are needed as Healthcare is currently going through a series of technological advancements and modifications. As the two communities of Computer Science and Biotechnology join forces to create new technologies for the advancement of medical science and improvement of medical service delivery (Stewart, 2004), this means that more people will be able to lead normal, healthy lives.

It is widely recognised today that the healthcare industry requires customized solutions with respect to information integration. The information sharing techniques currently available are not sufficient to meet the requirements of an integrated healthcare system. The state of electronic information integration in healthcare lags noticeably behind other business domains such as banking, insurance and electronic commerce (Bilykh et al., 2003). There is a need for health information systems to be fully integrated with each other and provide interoperability across various organizational domains for ubiquitous access and sharing. Moreover, due to rapid progress of biotechnology, an increasing number of life science databases are becoming available that are being operated and managed individually (Tohsato et al., 2005). Many existing solutions still do not offer the desired levels of utility/functionality or sophistication that a health information system demands.

The emerging technology of HealthGrids (HGs) holds the promise to successfully integrate health information systems and various healthcare entities onto a common (global) platform that would be shared and easily accessible. These entities include both humans and machines, such as scientists, scientific tools, medical instruments, physicians, patients and all types of healthcare data, etc. In such a scenario, each health information system is composed of various distinct components, which are integrated in a way that each component has its well-defined semantics and ontology and is well-aware of all other components.

The chapter starts by reflecting on the practical application of Grid technology to healthcare, and then presents a systematic taxonomy of the HealthGrid (HG) resources, outlining their characteristic features. The proposed taxonomy of the HG resources is based on their functionality, purpose, and application area. This chapter also discusses the issue of resource discovery in HealthGrids, which is an emerging challenge comprising many technical issues, such as performance, consistency of data/information, efficient retrieval of resources, compatibility of platforms, integrity of medical data, aggregation of storage resources and security of life-critical data, etc. Considering the challenge of resource discovery, it examines the problem of heterogeneity, issues of medical coding and terminology, and the role of semantic technologies. It also proposes potential solutions for the discovery of different types of HG resources. Finally, it reflects on discovering and integrating data resources and the future of HealthGrids and draws some conclusions.

CONTEXT: HEALTHCARE AND GRID TECHNOLOGY

HealthGrid is a Grid infrastructure dedicated to the management of healthcare resources that encompasses and integrates the various Grid components and healthcare components with consistent, compatible and meaningful coordination among them, to facilitate provision of the healthcare services (Naseer and Stergioulas, 2006c)

Thus, HealthGrid is a type of Grid that is used in the context of healthcare. HealthGrids are expected to possess enhanced, customized capabilities and features, such as:

1. Remote access services
2. Common distributed databases for healthcare
3. Information sharing
4. Integration of heterogeneous information from disparate sources
5. Common/standardized storage mechanisms
6. Efficient computation & data retrieval
7. Large-scale data processing
8. Shared access to computing resources
9. Social healthcare services
10. Provision of secure access to:
    a. Patient's medical history
    b. Medical Images (e.g. mammograms)
    c. Standard formats of files & information for comparison
    d. Library of examples for training & diagnosis
    e. Health support services
    f. Drug details & clinical trials
    g. Health information systems

Many Grid projects related to healthcare are currently running across Europe, such as CrossGrid (CrossGrid.org, 2008) that targeted parallel (MPI) computing and interactivity, DataTAGrid (DataTag, 2008) which focuses on interoperability, and DataGrid (Breton et al., 2003) which is a prototype of the BioMedicalGrid and supports better medical record management and improved diagnosis. Other projects such as the MammoGrid (MammoGrid, 2008), GEMSS (Jones et al., 2008) and e-Diamond (Brady et al., 2003) for UK, the NDMA (Emmen, 2008) for US and MEDIGrid (Boccia et al., 2005) for France are also in their completion phase.

The case for the use of Grid technology in healthcare arises mainly from the need to improve, safeguard and effectively exploit the available *life-significant medical information*, the need to protect the privacy of *personal, life-sensitive health information*, and the need to provide *integrated healthcare services* and have in place effective, global *channels of collaboration*.

To exploit effectively the wealth of medical information, there is an urgent need to *integrate, manipulate, process, and analyse* huge heterogeneous datasets from disparate sources. More systematic use of Grid technology in healthcare will not only help meet the current needs for data processing, but will ensure that future demand for even more capacity to deal with far larger volumes of data can be met.

Estrella et al. (2007) discuss that Grid computing holds the promise of harnessing extensive computing resources located at geographically dispersed locations that can be used by a dynamically configured group of collaborating institutions. It defines a suitable platform on which distributed medical informatics applications could be based. Particularly, Grids can address issues relevant to medical domains such as data distribution, heterogeneity, data processing and analysis, security and confidentiality, standardization and compliance, etc. HealthGrids are a good way to address these needs and provide reasonable solutions the challenges of modern healthcare. Piggott et al. (2004) explores the potential use of Grid technology in Healthcare, such as integration of heterogeneous data sets from multiple diverse sources systems. Thus, if successfully implemented, the HealthGrid will have a high impact towards lower costs and greater benefits for healthcare in the long run. In this respect, the HealthGrid could be driver of the next generation of healthcare IT.

Moreover, HealthGrids can prove to be an *effective channel for international collaborations* where the world's scientific minds can collectively work, such as to conduct a group-wise analysis, and might produce solutions that would effectively address complex medical problems (for instance, a disease or remedy). There are many *other reasons* why the healthcare domain needs Grid technology such as extensive computational power, ready availability of the various network resources, elimination of idle resources, efficient problem solving, concurrent and ubiquitous access to Grid resources, availability of massive data storage required in healthcare, making healthcare solutions/systems more efficient, etc.

All of the above and many more emerging issues demand to be addressed in a sophisticated manner by an advanced and reliable solution. HealthGrids are a good way to address these needs and provide reasonable solutions the challenges of modern healthcare. Thus, if successfully implemented, the HealthGrid will have a high impact towards lower costs and greater benefits for healthcare in the long run. In this respect, the HealthGrid could be the driver of the next generation of healthcare IT.

TYPES OF RESOURCES ON HEALTHGRIDS

HealthGrids vie to bring the e-Health concept into reality, by helping to provide personalized and quality healthcare at a cost-effective price. In addition to possessing characteristics of generic Grid resources (Naseer and Stergioulas, 2006b), the HealthGrid resources bear specialized features, which are adapted to the needs of healthcare domain. For example, the Data Resource (Naseer and Stergioulas, 2006b) can encapsulate specialized health-specific data such as:

- Patients' records (patients' complete health history, etc.)
- Doctors' data (doctor's expertise, doctor's patients list: treated or under treatment, etc.)
- Data files containing simulation or test results
- Digital images or DICOM files (mammograms etc.)
- Medical communications
- Data related to pharmaceutical companies (drug design, drug development, clinical trials, effectiveness trials, etc.)
- Diseases and other information, etc.

With a view to offer a unified taxonomy of the specialised HealthGrid resources, Figure 1 proposes and illustrates a new hierarchy of the various resources available on a HealthGrid. The specialized HealthGrid resources can be Data, Information or Files (DIF), Applications & Peripherals (AP), and Services. The DIF and AP resources are on Level 1 of the hierarchy whereas the Service resource is on Level 2. This is because Services are built on top of the DIF and AP resources. Each of these resources is discussed in detail in the following sections.

Figure 1. A hierarchy of HealthGrid resources

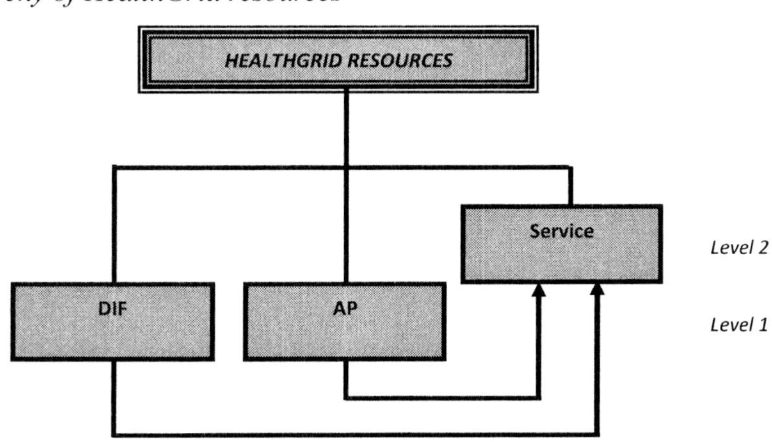

Data, Information or File (DIF) Resources

Information can be extracted from data which resides in computer files, therefore data, information or files are all considered as a single type of HealthGrid resource. This kind of resource comprises all sorts of medical records and healthcare information such as Patient Records; Healthcare Staff Details; Drug (medication) Details; and Diseases & Other Biomedical/Scientific Information.

Figure 2 depicts the breakdown of DIF resources in the form of a tree structure. Each of the end nodes of this tree can be further broken down into sub-categories, such as shown in Figure 3 to Figure 6.

The DIF resources encapsulate all sorts of medical and health-related data or information which is contained in electronic files, such as patient records (including their test results e.g. simulation results, radiotherapy reports, medical images or mammograms, etc.), healthcare staff details, drug details (including clinical trials and drug effects, etc.), diseases & other biomedical scientific information.

Figure 3 depicts the structure of a patient record. It includes the patient's personal details and his/her complete health (medical) history, the time he/she first needed a health treatment (patient's first entry to the clinic), the illness or disease diagnosed, the healthcare staff that attended the patient, observations made about patient's health, various tests and their results such as mammograms or simulation results, the type of treatment or medicines prescribed or recommendations made with regards to the various test results of the patients, after-treatment effects, patient's current health status, etc. In fact, the patient record is the most sensitive data resource and is of crucial importance since it encompasses many critical elements related to personalized healthcare. A study in (Whiddett et al., 2006) suggests that patient's information could be distributed with their consultation. Mostly, patients are happy to consider sharing their information with health professionals if they are first consulted.

Figure 4 depicts the structure of a Healthcare staff record which contains details about their area of expertise, their specializations or specialized domains, patients under their treatment or treated and other comments about the staff's details and/or expertise. The healthcare staffs include health specialists (doctors, physicians, consultants/surgeons, healthcare practitioners, etc.), medical lab technicians, pharmacists (drug developers, analyzers, etc.), health analysts, etc.

The Drug Detail is one of the nodes of the DIF Resource tree (Figure 2). This includes data related to a particular drug, not only about drug specifications but also about its effects, etc.

Figure 5 depicts the structure of a drug record which contains details about a particular drug or medicine such as drug features, drug specifications, details about various clinical trials for a particular drug, drug evolution or drug development,

Figure 2. DIF resource tree

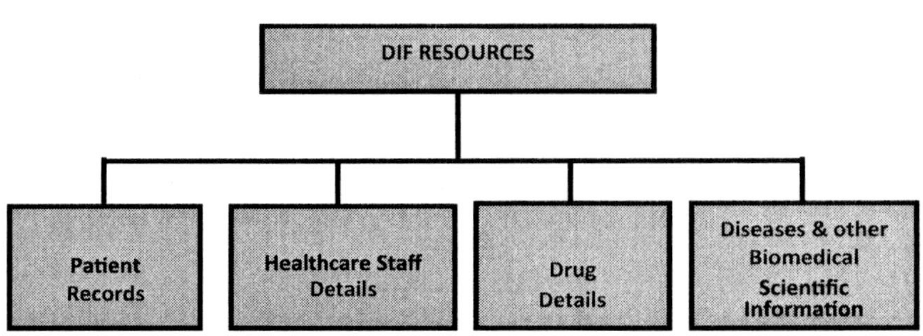

Figure 3. A typical structure of a patient record

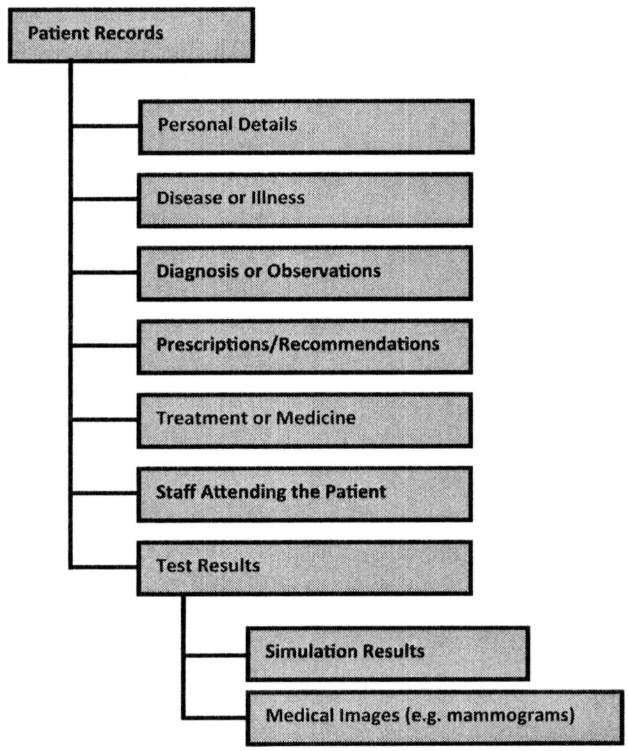

Figure 4. Healthcare staff details

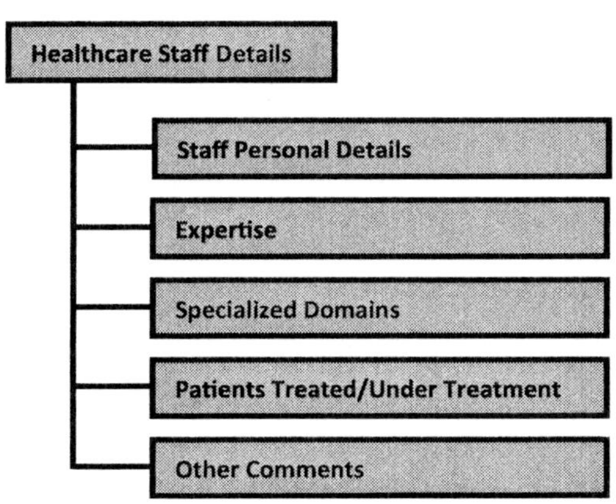

Figure 5. Drug details

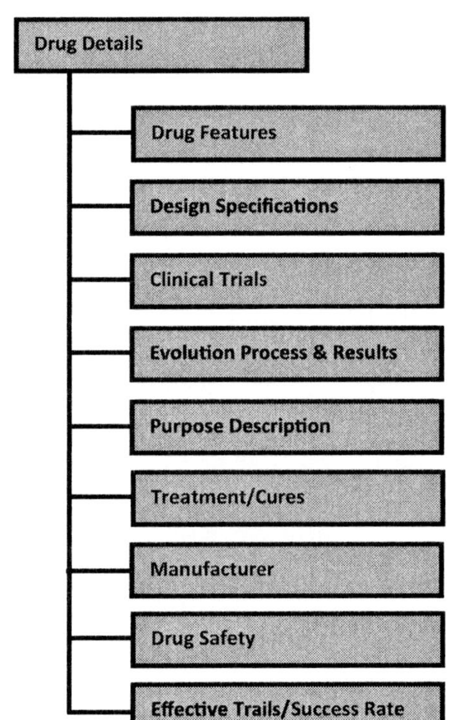

purpose of using this drug, the disease it cures (for what treatment it is used and/or recommended), manufacturers/pharmaceutical industries, drug safety (such as implications, etc.), effectiveness of the drug, and its success rate.

Figure 6 depicts a detailed breakdown of "Diseases & Other Biomedical Scientific Information" such as disease description (including disease nature and occurrence, etc.), disease type or category, disease symptoms, level of infection, severity level (such as high/moderate/low), causes, cure, and specialized treatment, etc.

It can be seen through all of the above mentioned nodes (Figure 3 to Figure 6) of the DIF resource that there exist enormous amounts of health-related data with different hierarchies. The healthcare data is produced every second in the real-world and needs to be handled and catered in a systematic manner. This data if made available on a globally shared channel such as a HealthGrid could prove to be very useful for many healthcare stakeholders such as healthcare specialists and health analysts, medical lab technicians, pharmacists, medical equipment providers, healthcare organizations and even patients or the general public. All of them need a globally shared channel for their collaborative work for solving the various healthcare problems, and in one way or another will be positively influenced by the successful deployment of HealthGrids.

Figure 6. Diseases and other biomedical scientific information

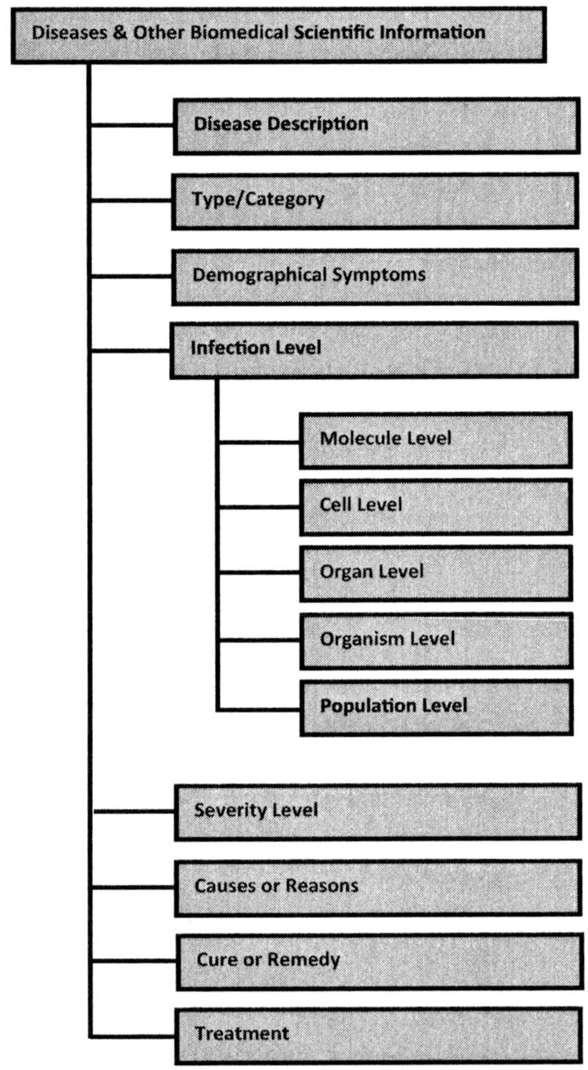

Application and Peripheral (AP) Resources

Applications such as device drivers or programs are installed onto computerized peripheral devices or machines such as image scanners, etc. Since software applications need hardware to work and cannot run alone, therefore both applications and peripherals are considered as a single HealthGrid resource. Such a resource comprises of dedicated and specialized healthcare applications and automated peripheral devices attached to the HealthGrid, such as image-scan processing applications installed on computerized organ scanners and image processors, MRI software and scanning machines, magnetic strip scanners (immuno-chromatography reader) for analysis, comparison and management of images, pattern classifiers, PACS (Picture Archiving & Communication Systems), simulation software and devices supporting simulation of body parts for various purposes.

Figure 7 depicts the breakdown of Application and Peripheral resources dedicated to healthcare.

These two subcategories merge at the bottom of the tree to provide automated and/or programmed systems that can carry out various computational health-related tasks/operations.

These resources can be deployed and/or managed through a dedicated HealthGrid, for example an *InstruGrid* (Instrument Grid), which could be designed specifically to access, retrieve and manage the scientific instruments used for the medical purposes. These instruments containing the logical resources (software applications) and physical resources (hardware machinery or peripherals) of the HealthGrid can be utilized in order to carry out specialized healthcare processes or medical tasks such as remote image scanning, distributed simulation or automatic activation of the programmed machines (e.g. personalized healthcare monitoring systems).

Service Resources

Service resources (also referred to as service-type resources or simply services) on a HealthGrid are compositions of basic healthcare-specific services, which are built using "standard" Grid services to provide high-quality customized healthcare. These services can be categorized according to their applicability into two levels: namely the Operational Level services and the Management Level services (Naseer and Stergioulas, 2006a).

Almost always, the service-type resources access/employ DIF and AP resources. For example, a Grid-enabled data Comparison & Analysis application can be used to carry out analytical studies of images or scans from any part of the world. Thus, a purpose-built data-access service can provide access to the Comparison & Analysis application (AP resource) as well as concurrent access to these image files (DIF resource). Figure 8 depicts a classification of the various types of service resources.

The first class of dedicated HealthGrid services provides *concurrent and ubiquitous access to or retrieval of* various other HealthGrid resources such as data, files, information, applications and peripherals, etc. The *on-demand registration & scheduling* services are responsible: (a) for authenticating the user registration and user authorization, groups or companies to access and use HealthGrid resources; and (b) for job scheduling

Figure 7. AP resource tree

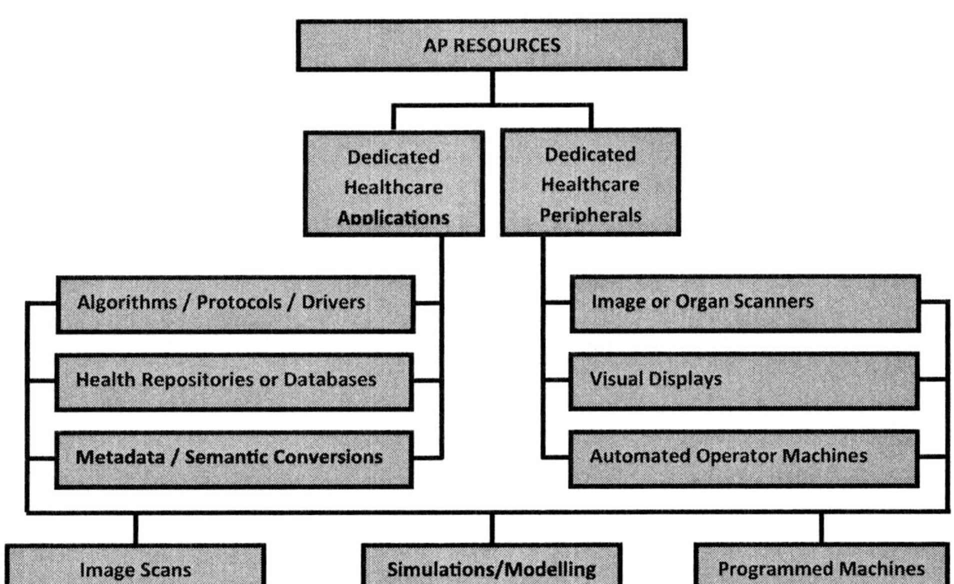

Figure 8. HealthGrid service resource tree

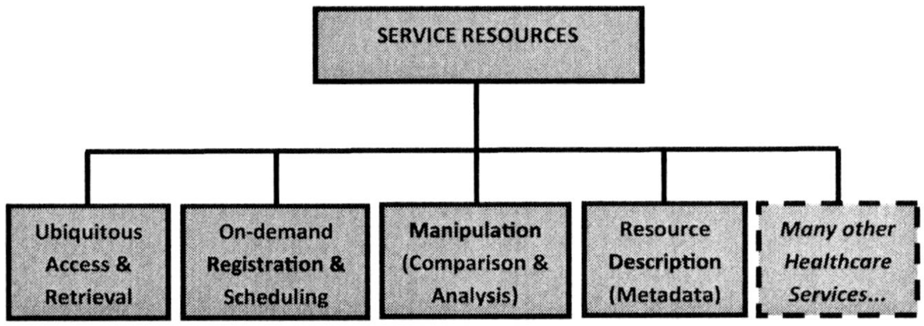

and prioritizing healthcare processing tasks. The third class of services is responsible for performing *manipulation* healthcare operations *(e.g. data integration, analysis, comparison and reviewing)* by recruiting other HealthGrid resources. Some service-type resources are responsible for supporting online group-based data sharing in different parts of the world, a task made feasible by the so-called *resource defining services* (using metadata or semantic descriptions of the resources). There is a diverse range of many other healthcare services (Naseer and Stergioulas, 2006a), which can be provided on HealthGrids.

RESOURCE DISCOVERY IN HEALTHGRIDS

The Resource Discovery Challenge

The successful adoption and integration of Grid technology in the healthcare industry has not yet been achieved fully due to reasons such as lack of awareness, reliability, trust, and security. Probably these are also some of the reasons why the healthcare community is reluctant to embrace an automation revolution as witnessed in other industries (e.g. banking, finance, marketing, stock exchange, etc.). For example, human consultants are still given preference on personalized healthcare systems.

Grid-enabled sharing of data, experience and expertise will facilitate the archiving, curation, retrieval and analysis of imaging data from multiple sites and enable large-scale clinical studies in neurology. To achieve this goal, the NeuroGrid (Geddes et al., 2005) seems to be built upon existing Grid technologies and tools (developed within the UK e-Science programme), aiming to integrate image acquisition, storage and analysis, and to support collaborative working within and between neuro-imaging medical centres. Moreover, the Biomedical Information Research Network (BIRN) (Ellisman and Peltier, 2004) is devoted to neurology and is exploring the use of Virtual Data Grid (VDG) to support multiscale brain mapping. BIRN currently participates in three testbed projects; namely Function BIRN, Morphometry BIRN and Mouse BIRN (Stewart, 2004).

In HealthGrids the problem of data management and discovery is magnified due to the highly sensitive nature of health-related data. The information contained on a HealthGrid has to be handled (stored, retrieved, shared) with care so as to meet the patient's confidentiality & privacy, as much as the other data security constraints. There are a number of issues that are needed to be addressed in order to achieve successful data management & discovery: for example, encoding of medical terms, file format compatibility issues, ontologies matching issues, heterogeneity issues, data archiving & distributed image analysis issues, etc.

In the case of health-related data, confidentiality and security are crucial factors to be considered before the data is exchanged across various domains, organizations or among various individuals. As mentioned earlier, Grid resources are highly heterogeneous with respect to their dynamic status and their dispersion across different geographical regions or different platforms, and therefore are difficult to manage. In an interconnected environment of healthcare, the successful implementation of *resource discovery* techniques need to cope with issues such as scalability, autonomy and heterogeneity, security and privacy, maintainability and resolvability, reliability and robustness, active and autonomous coordination, mobility and ubiquity, etc (Bilykh et al., 2003). Moreover, differences in medical terminologies, organizational heterogeneity and ubiquitous management of resources can hinder consistent and accurate resource discovery on HealthGrids.

Heterogeneity and Coding Issues

In general, the level of data integrity is based on the level of heterogeneity of the various data resources used and includes syntactic heterogeneity (i.e. differences in the data models and data types, which can be easily resolved) and semantic heterogeneity (i.e. differences in the underlying meaning of data, which play an essential role during heterogeneous resource integration and interoperation) (Verschelde et. al, 2004). Many geographically distributed healthcare organizations have data stored in different formats, different database management systems, for different types of medical tests, and in different descriptions. Even the medical terminologies used across various organization boundaries vary widely, thus making resource discovery a highly complex process, since there is a possibility that the information is misinterpreted by the user (individual or organization), which may lead to serious medical errors. Moreover, the remote processing of a dataset is also an issue as regards to obtaining consistent (error-free) results every time.

Various healthcare terminologies and classification systems (Coiera, 2003) have been developed such as the International Classification Diseases (ICD-10), the Systematized Nomenclature of Medicine (SNOMED), the Medical Subject Headings (MeSH), the Unified Medical Language System (UMLS), the Medical Dictionary for Regulatory Activities (MedDRA) and the Health Level Seven (HL7), etc. The UMLS is the latest system with the potential of wide acceptance, since it provides a mapping between 100 different terminology systems and incorporates a number of other systems such as ICD-9, ICD-10, MeSH, ICPC-93, WHO Adverse Drug Reaction Terminology, SNOMED II, SNOMED III and UK Clinical Terms. The knowledge sources of UMLS are the metathesaurus, the semantic network and the lexicon. UMLS is used in (Slaughter et al., 2006) to identify the underlying semantics of health consumers' questions and physicians' answers in order to analyse the semantic patterns within their texts. Semantic relationships are manually identified within the question-answer pairs from Ask-the-Doctor Web sites. Identification of the semantic relationship instances within the texts is based on the relationship classes and structure of the UMLS Semantic Network since its relationship classes are hierarchical.

None of the healthcare terminologies and classification systems has so far succeeded in resolving the issue of heterogeneity in medical terminologies. However, the coding systems should be compared on specific tasks and results should cautiously be generalized to other tasks and populations. Similarly, the poor performance of the coding systems on tasks outside their design should not reflect badly on their assessment in terms of performance capability (which should be always assessed within the intended scope of application). Moreover, as one of the requirements for the next generation Grids, they need to posses features such as the heterogeneous capabilities and performances of access networks and devices (Cannataro, 2006). Therefore there is a well-recognised need to have a flexible and detailed medical terminology and classification system to integrate medical or health-related data over the HealthGrids, which are suitable channels for ubiquitous access to, sharing of, and processing of masses of health-related data. There are many other technical issues which make resource discovery a major challenge and need to be addressed for the successful implementation of HealthGrids.

Resource Discovery Problems and Solutions

Resource Discovery in HealthGrids is an important and timely issue that relates to a number of "information integration" problems in health informatics. The integration of medical information electronically on a national and international level in a way conformant to all the organizational policies and Grid constraints is a big challenge. Moreover, coding & terminology are far from unified across the sector, and medical coding systems are not ready to incorporate and manage the emerging *genetic information* (Breton et al., 2005). Thus it is important to maintain standardized metadata and standardized translation between the various medical terminologies used in different countries. Resources on the Grids need to be managed in a sophisticated manner in order to optimize their discovery. One P2P-based approach (Mastroianni et al., 2004) manages global queries on multiple Index Services, where the Metadata models are studied using an XML-based approach for heterogeneous resource representation and management in Grid-based data mining applications (Knowledge Grid).

Emerging Grid technologies combine Web Services with Grid infrastructures and employ Semantic Web technologies for solving problems of Grid deployment and management. Such *semantic description schemes* can provide the glue for coordinating different sorts of resources, services and application on a HealthGrid and between HealthGrids (Stevens et al., 2004), (MammoGrid, 2007).

Grid services are a type of Grid resource (service resource) with special capabilities and features to get tasks successfully accomplished on Grid networks. Grid services are used to access other types of Grid resources. Technically defined as lines of code that are packaged into a component to perform a dedicated functionality. In terms of their function, services implemented on a network infrastructure could resemble thread-like neuron structures or impulses in a human body. Thus, the output of one service can serve as an input to another service. Web services are considered as the most efficient and reliable communicators of messages from one place to another regardless of geographical or technological heterogeneity. The emerging Web services provide a framework for an application-to-application interaction that grants access to business-to-business, e-science, and e-government services over the Internet (Naseer and Stergioulas, 2007).

However, the state-of-the-art is that Grid services are no longer considered to be separate from the Web services. In fact, according to the Open Grid Services Infrastructure (OGSI) version 1.0 specification (Tuecke et al., 2003), a Grid service is considered to be a Web service that conforms to a set of conventions (interfaces and behaviours) which define how a client interacts with a Grid service for such purposes as service lifetime management, inspection, and notification of service state changes (Foster et al., 2005).

Currently, a lot of work is being devoted to Grid resource discovery using web services. Each of these endeavours constitutes a step towards the integration of Grid services into Web services, which might provide a viable solution to the problem of resource discovery in Grids. This integration is necessary since Web services cannot directly be used as Grid services due to intrinsic limitations of Web services such as statelessness, etc. (whereas a Grid service must have a state since it is prone to dynamic changes and has more complex functionality than an ordinary Web service).

To serve the purpose of resource discovery, Grid services could be glued to Web services using specialized object-oriented techniques such as encapsulation or inheritance, etc. For the e-Health dream to become reality, successful implementation of these concepts is crucial (Naseer and Stergioulas, 2007). Since Grid technology has been developed as a way to support large-scale distributed resource manipulation, using shared heterogeneous pooled resources across administrative domains. With the growing success of Web Services, the opportunity naturally arises for a confluence between these two increasingly mature areas. A new class of distributed applications could even emerge as a result - in fact, there's clear evidence that this is happening. Certain applications, called pervasive management support systems, combine sensor network and Grid technologies to support the ongoing management of evolving physical systems.

Since HealthGrids contain life-sensitive data, there is a need for secured access mechanisms. Therefore, it is necessary for web services, which enable remote access to medical data, to be secured in an appropriate fashion (Power et al., 2006). There is also a need to have *enhanced investigation mechanisms* in healthcare information systems so as to facilitate the process of universal automation and ubiquitous sharing and to achieve successful resource discovery on HealthGrids.

One of the mainstream Grid technologies, OGSA-DAI (Open Grid Service Architecture Data Integration and Access) (Antonioletti et al., 2005) is a powerful technology that possesses strong features for providing interfaces to heterogeneous data sources on Grids in order to integrate them. OGSA-DAI acts as an interface for each data source that is available on the Grids. It is an open-source middleware designed to facilitate controlled access, management and integration of distributed heterogeneous data resources and provides a ready-made framework that promotes locality and product transparency to connect data resources to the Grid environment (Crompton et al., 2006).

OGSA-DAI has been used quite intensively for the Semantic-based approach in order to facilitate successful data (resource) discovery. A mediator-wrapper architecture and ontology based semantic information has been proposed (Tan et al., 2007) to wrap the heterogeneous data source. It employs a mediator structure to supply accessing interface for the data sources, and it builds virtual data source (VDS) to support standard OGSA-DAI query interface. A MIP-Grid (Grid-enabled Medical Image Processing Application System) is presented (Huang et al., 2006), that is based on OGSA-DAI middleware. It aims at providing high performance medical image process services in a large distributed grid computing environment. OGSA-DAI allows uniform access to and integration of data held in heterogeneous data resources.

Each of the HealthGrid resources described in this chapter has its own discovery problems, some of which along with their possible solutions are discussed in the Table 1.

CONCLUSION

HealthGrids can be used to support many kinds of healthcare operations and tasks. In this chapter a unified taxonomy of HealthGrid resources has been proposed and the issue of resource discovery in HealthGrids has also been discussed. A refined hierarchy of the various types of HG resources is presented, where the specialized resources can be categorised into three major types; namely, Data or Information or Files (DIF); Applications & Peripherals (AP); and Services.

To address the challenge of resource discovery in HealthGrids, a systematic search strategy should be devised and adopted, as the discovered resource should be valid, refined and relevant to the query. Standards should be implemented on

Table 1. Resource discovery problems and possible solutions

Resource Type	*Sources of Problems in Resource Discovery*	*Possible Solutions*
DIF	Encoding medical terms Compatibility of file formats Matching ontologies Semantic descriptions Data/information consistency Ubiquitous & easy availability Heterogeneous Databases/Languages Data archiving & distributed image analysis	Devising and using interoperable standards for healthcare terminologies and classification systems. For example, this can be based on any of the following systems: SNOMEDor UMLS (Coiera, 2003), or MedDRA (Medical Dictionary for Regulatory Activities, 2008). Also designing generic protocols for ubiquitous and easy availability of data and standardization of data formats. In a similar vein of work, the HL7 community (Health Level Seven Inc., 2008), (Dolin et al., 2001) is creating standards for the exchange, management and integration of electronic healthcare information.
AP	Reliability Compatibility • inter & intra applications • inter & intra peripherals • inter & intra Grid infrastructures Grid-enabled applications & peripherals Consistency	Designing dedicated Grid-enabled applications and peripherals having platform and organizational compatibility. Also compatibility with inter- and intra- Grid infrastructures to support ubiquitous performance and exchange of information.
SERVICE	Accuracy Integrity Cost effectiveness User friendliness Efficiency and performance Security and user authentication [Privacy, Integrity, Confidentiality]	Designing specialized protocols to enable controlled access and performance monitoring to ensure high performance at a cost-effective price. Also providing easy to use flexible IDE (integrated development environment) for dedicated HealthGrid interfaces. Moreover, implementing security measures for user privacy and confidentiality.

domain-specific metadata. Moreover, a critical question arises as to how metadata can support the integration of two or more heterogeneous objects. There is also a need to have a semantic integration of various resources that are geographically or organisationally spread, so that they can be shared and utilized globally on a HealthGrid. The emergent semantic networks ensure the integrity of meaning between different concepts and can play an important role in solving this complex integration problem. Moreover, one approach could be using the mainstream or contemporary Grid technologies such as OGSA-DAI (as a candidate solution) to resolve the data federation problem. In summary, the chapter has addressed the challenge of resource discovery; discussed the problem of heterogeneity, issues of medical coding and terminology, and the role of semantic technologies; and proposed potential solutions for different types of resources.

As a final remark, HealthGrids offer accurate and reliable sources of health information that can be accessed any time from any place thus serving as an effective channel for international collaborations. They can and should become a major driver in the race towards successful e-Health and an important ingredient of the next generation of healthcare IT. A successfully implemented HealthGrid infrastructure could support all the facets of healthcare sector, and help realise the vision of personalized healthcare. Successful implementation of HealthGrids will have a high impact towards lower costs and greater benefits for healthcare in the long term.

REFERENCES

Antonioletti, M., Atkinson, M., Baxter, R., Borley, A., Hong, N.P.C., Collins, B., et al. (2005). The design and implementation of grid database services in OGSA-DAI. *Concurrency Computation Practice and Experience*, 17(2-4), 357-376.

Bilykh, I., Bychkov, Y., Dahlem, D., Jahnke, J.H., McCallum, G., Obry, C., et al. (2003). Can grid services provide answers to the challenges of national health information sharing? *In Proceedings of the 2003 conference of the Centre for Advanced Studies on Collaborative research (CASCON '03)* (pp. 39-53). Armonk, NY: IBM Press.

Boccia, V., Guarracino, M.R., D'Amore, L. & Laccetti, G. (2005). A grid enabled PSE for medical imaging: experiences on MedIGrid. In *Proceedings of the 18th IEEE Symposium on computer-based medical systems* (pp. 529-536). Washington, DC: IEEE Computer Society.

Brady, J. M., Gavaghan, D. J., Simpson, A. C., Parada, M. M. & Highnam, R. P. (2003). eDiaMoND: A grid-enabled federated database of annotated mammograms. In Berman, F., Fox, G. C. & Hey, A. J. C. (Eds.). *Grid computing: making the global infrastructure a reality,* (3rd Ed.) (pp. 923-943). Hoboken, NJ: Wiley.

Breton, V., Dean, K. & Solomonides, T. (2005). The HealthGrid white paper. *Studies in Health Technology and Informatics, 112,* 249-321.

Cannataro, M. (2006). Next-generation Grids: requirements and knowledge-based services: Research Articles, *Concurrency and Computation: Practice & Experience, 18*(8), 887-898.

Coiera E. (2003). *Guide to health informatics.* London: Hodder Arnold.

Crompton, S. Y., Matthews, B. M., Gray, W. A., Jones, A. C., White, R. J. & Pahwa, J. S. (2006). OGSA-DAI and bioinformatics grids: Challenges, experience and strategies. In *Proceedings of the 6th IEEE International Symposium on Cluster Computing and the Grid (CCGRID 2006),* (pp. 193-200). Washington, DC: IEEE Computer Society.

CrossGrid.org. (2008). *CrossGrid project homepage.* Retrieved March 8th, 2008, from http://www.crossgrid.org/main.html

DataTag (2008). *DataTag project description.* Retrieved March 8th 2008, from http://datatag.web.cern.ch/datatag/project.html

Dolin, R. H., Alschuler, L., Beebe, C., Biron, P. V., Boyer, S. L., Essin, D., et al. (2001). The HL7 Clinical Document Architecture. *Journal of the American Medical Informatics Association, 8*(6), 552-569.

Ellisman, M. & Peltier, S. (2004). Medical data federation: the biomedical informatics research network. In Foster, I. & Kesselman, C. (Eds.), *The Grid: Blueprint for a New Computing Infrastructure* (2nd Ed.). San Francisco, CA: Morgan Kaufmann.

Estrella, F., Hauer, T., McClatchey, R., Odeh, M., Rogulin, D. & Solomonides, T. (2007). Experiences of engineering Grid-based medical software. *International Journal of Medical Informatics, 76*(8), 621-32.

Foster, I., Czajkowski, K., Ferguson, D. E., Frey, J., Graham, S., & Maguire, T., et al. (2005). Modeling and managing state in distributed systems: The role of OGSI and WSRF. *Proceedings of the IEEE, 93*(3), 604-612.

Geddes, J., Lloyd, S., Simpson, A., Rossor, M., Fox, N., Hill, D., Et al. (2005). NeuroGrid: using grid technology to advance neuroscience. In *Proceedings of the 18th IEEE symposium on computer-based medical systems,* (pp. 570-572). Washington, DC: IEEE Computer Society.

Health Level Seven Inc. (2008). *Health Level Seven homepage.* Retrieved March 8, 2008 from http://www.hl7.org/

Huang, X., Huang, L. & Li, M. (2006). Grid-enabled medical image processing application system based on OGSA-DAI techniques: In Proceedings of the International workshops on advanced Web, network technologies and Applications, pp. 460-464. In Shen, H. T., Li, J., Li, M., Ni, J. & Wang, W. (Eds.), *Lecture notes in*

Computer Science, (Vol. 3842). Aachen, Germany: Springer-Verlag.

Jones, D. M., Fenner, J. W., Berti, G., Kruggel, F., Mehrem, R. A., Backfrieder, W., Moore, R. and Geltmeier, A. (2004). The GEMSS Grid: An evolving HPC environment for medical applications. In *Proceedings of HealthGrid 2004.* Retrieved March 8, 2008, from http://www.ccrl-nece.de/gemss/Reports/Jones-healthgrid2004.pdf

MammoGrid (2008). *European federated mammogram database implemented on a grid structure.* Retrieved March 8, 2008 from http://mammogrid.vitamib.com

Mastroianni, C., Talia, D. & Trunfio, P. (2004). Metadata for managing grid resources in data mining applications. *Journal of Grid Computing, 2*(1), 85-102.

Medical Dictionary for Regulatory Activities (2008). Retrieved March 8, 2008, from http://www.meddramsso.com/MSSOWeb/index.htm

Naseer, A. and Stergioulas, L. K. (2006a). Discovering HealthGrid services. In *Proceedings of the IEEE International Conference on Services Computing (SCC'06), Chicago, IL,* (pp.301-306).

Naseer, A. & Stergioulas, L. K. (2006b). Resource discovery in grids and other distributed environments: States of the art. *Multiagent and Grid Systems - An International Journal, 2*(2), 163-182.

Naseer, A. & Stergioulas, L. K. (2006c). "A Taxonomy of HealthGrids and HealthGrid Resources". *11th International Symposium on Health Information Management Research* (iSHIMR2006), Halifax, Canada (pp. 93-99).

Naseer, A. and Stergioulas, L. K. (2007). Combining web services and grid services: practical approaches and implications to resource discovery. In Periorellis, P. (Ed.), *Securing Web Services: Practical Usage of Standards and Specifications,* (pp. 298-317). Hershey, PA: IGI Global..

Emmen, A.. (2008). National Digital Mammography Archive announces installation at University of Pennsylvania Health System. *EnterTheGrid - PrimeurMonthly* Retrieved March 8, 2008 from *http://www.hoise.com/primeur/05/articles/monthly/AE-PR-02-05-1.html* .

Piggott, D., Teljeur, C. & Kelly, A. (2004). Exploring the potential for using the grid to support health impact assessment modelling. *Parallel Computing, 30*(9-10), 1073-1091.

Power, D. J., Politou, E. A., Slaymaker, M. A. & Simpson, A. C. (2006). Securing web services for deployment in HealthGrids. *Future Generation Computer Systems, 22*(5), 547-570.

Slaughter, L. A., Soergel, D. & Rindflesch, T. C. (2006). Semantic representation of consumer questions and physician answers. *International Journal of Medical Informatics, 75*(7), 513-529.

Stevens, R., McEntire, R., Goble, C., Greenwood, M., Zhao, J., Wipat, A. & Li, P. (2004). myGrid and the drug discovery process. *Drug Discovery Today: BIOSILICO, 2*(4), 140-148.

Stewart, C. A. (2004). Introduction to the edition. *Communications of the ACM, 47*(11), 30-33.

Tan, H., Chen, X. & Gu, J. (2007). Semanteme-based processing mechanism under data grid environment. *Journal of Huazhong University of Science and Technology (Natural Science Edition), 35*(4), 22-25.

Tohsato, Y., Kosaka, T., Date, S., Shimojo, S. & Matsuda, H. (2005). Heterogeneous database federation using grid technology for drug discovery process,. In *Lecture notes in Computer Science,* (Vol. 3370) (pp. 43-52). Aachen,Germany: Springer-Verlag.

Tuecke, S., Czajkowski, K., Foster, I., Frey, J., Graham, S., & Kesselman, C., et al. (2003). Open grid services infrastructure (OGSI) (No. Ver. 1.0)., *Global Grid Forum (GGF).* Retrieved March 8, 2008 from http://www.globus.org/alliance/publications/papers.php#GSSpec

Verschelde, J.L., Santos, M.C.D., Deray, T., Smith, B., & Ceusters, W. (2004) Ontology-Assisted Database Integration to Support Natural Language Processing and Biomedical Data-mining. *Journal of Integrative Bio-informatics*.

Whiddett, R., Hunter, I., Engelbrecht, J. & Handy, J. (2006). Patients' attitudes towards sharing their health information. *International Journal of Medical Informatics, 75*(7), 530-541.

KEY TERMS AND DEFINITIONS

Grid Resource: A Grid resource can be any real or conceptual object that is needed to be accessed by other entities, such as human users of the system or programmes that generate requests for accessing particular resources.

Grid Services: A type of Grid resource (service resource) with special capabilities and features to get tasks successfully accomplished on Grid networks. These are used to access other types of Grid resources.

HealthGrid: A type of Grid. It is a Grid infrastructure dedicated to the management of healthcare resources that encompasses and integrates the various Grid components and healthcare components with consistent, compatible and meaningful coordination among them, to facilitate provision of the healthcare services.

Heterogeneity: A difference in format, platform, semantics or ontology.

Integration or Interoperability: A link or connection that enables communication among two or more resource types.

OGSA-DAI: Open Grid Service Architecture Data Integration and Access, it is a powerful technology that possesses strong features for providing interfaces to heterogeneous data sources on Grids in order to integrate them.

Resource Discovery: The operation of tracking, accessing, matching, selecting and eventually requesting the right or the most accurately suitable resource for the successful accomplishment of the desired job.

Taxonomy: A classification.

Web Services: Considered as the most efficient and reliable communicators of messages from one place to another regardless of geographical or technological heterogeneity.

Chapter III
Data Provenance in Scientific Workflows

Khalid Belhajjame
University of Manchester, UK

Paolo Missier
University of Manchester, UK

Carole A. Goble
University of Manchester, UK

ABSTRACT

Data provenance is key to understanding and interpreting the results of scientific experiments. This chapter introduces and characterises data provenance in scientific workflows using illustrative examples taken from real-world workflows. The characterisation takes the form of a taxonomy that is used for comparing and analysing provenance capabilities supplied by existing scientific workflow systems.

INTRODUCTION

Workflow technology is increasingly adopted in modern science as a means for modelling and automating the enactment of experiments (Ludäscher *et al.* (006) and Taylor, I. (006). An experiment is specified as a series of analysis operations that are connected using data links that define the flow of data between the outputs and the inputs of the constituent operations, and control links that define temporal dependencies between operations' executions. The analysis of the results obtained by the enactment of an experiment allows the scientist to verify, amongst others, the validity of a hypothesis or confirm a known fact. To draw such conclusions however, the results returned by the experiment enactment are by no mean sufficient, rather, more information that describes how and where the returned results were obtained from is needed. This information is known in the literature as data provenance or data lineage (Buneman *et al.* (006). Generally speaking, data provenance can be thought of as the process by which the data instances and the analysis operations that were used to derive a given data instance are identified.

While data provenance in databases, notably in datawarehouses, is a fairly understood problem (Ludäscher *et al.* (006), it is not comprehended yet in workflows. It was largely recognised by the researches in this area at the second provenance challenge workshop[2] that existing workflow systems provide little support for managing and querying data provenance. A step in this direction, in this chapter we study and propose a taxonomy that thoroughly characterises data provenance in scientific workflows. It is probably worth mentioning that different from the survey on data provenance in eScience by Simmhan *et al.* (2005), in which the authors attempt to review data provenance in general, be it in databases, workflows or other contexts, in this chapter, we focus on data provenance in scientific workflows and identify their peculiar characteristics. The proposed taxonomy in used as an evaluation and comparison grid for existing workflow provenance systems, thereby highlighting their similarities, differences and limitations. We also discuss open issues that need further attention and investigation from the researchers in the area.

The chapter is organised as follows. We begin by motivating the need for supporting data provenance in scientific workflows using examples of queries a scientist may need an answer for and which can be automatically answered if provenance information is collected. We then formally define scientific workflows and present a real-world workflow to motivate the need for data provenance. We go on to present the taxonomy that characterises data provenance in scientific workflows. The taxonomy is composed of a set of dimensions and associated values. A dimension describes a specific aspect of data provenance and a value represents a possible implementation of a dimension. The taxonomy presented is used for characterising and comparing provenance-related functionalities provided by existing scientific workflow systems, such as Taverna (Oinn *et al.* (2006)) and Vistrails (Callahan *et al.* (2006)), highlighting their similarities, differences. It is also used to point out open issues that need further investigation from the provenance community.

BACKGROUND

In this section, we formally define what a scientific workflow is. We then go on to present example of workflow provenance queries taken from the domain of bioinformatics.

Scientific Workflow

A scientific workflow is a collection of interrelated tasks, also called activities or processors, which models a scientific experiment[3]. Different objectives may be aimed from the modelled experiment, e.g., test the validity of a given scientific workflow; verify a known fact; compare the results obtained by a given scientist with the results obtained by other scientists worldwide. A scientific workflow can be decomposed into sub-processes, which contribute to achieve the goal of the global scientific experiment. The enactment of the tasks that compose a scientific workflow is typically constrained by dependency rules among tasks. These dependency rules can be classified into two classes: data-flow dependencies and controlflow dependencies. Data flow dependencies specify the flow of data between the outputs and inputs of the tasks, whereas controlflow dependencies specify temporal dependencies between them. For the purpose of this chapter, we define a scientific workflow by the tuple:

$$wf = \langle OP,\ PARAM,\ PL,\ DL,\ CF \rangle$$

where *OP* is the set of operations from which the workflow is composed, *PARAM* is the set of inputs and outputs of the operations in *OP*, *PL* is a set of links that associate each operation in *OP* with its inputs and outputs, *DL* is the set of data links connecting the operations in *OP*, and *CF* is a set of controlflow dependencies.

Operation An operation represents the unit of functionality supplied by a service, and it represents a task within the scientific workflow. It is defined as:

$$op = \langle nameOp, loc, desc \rangle$$

where *nameOp* is the operation identifier, *loc* is the end point of the service to which the operation belongs (it also serves as the service identifier), and *desc* is a textual description of the action performed by the operation.

Parameter A service operation is associated with input and output parameters. A parameter is defined by the pair:

$$\langle nameP,\ typeP \rangle$$

Where *nameP* is the identifier of the parameter, which must be unique within the operation to which the parameter belongs, and *typeP* is the data type of the parameter.

The Parameters in *PARAM* are associated to their operations in *OP* using the parameter links in *PL*. Therefore, *PL* can formally be defined as:

$$PL \subseteq (PARAM \times OP) \cup (OP \times PARAM)$$

Specifically, let *op* be an operation and *p* a parameter.

- $<p,op> \in PL$: means that p is an input of *op*.
- $<op,p> \in PL$: means that *p* is an output of *op*.

In the following we denote the inputs and the outputs of a given operation by *inputs(op)* and *outputs(op)*, respectively.

Data links A data link describes a data flow between the output of one operation and the input of another. Let *IN* be the set of all input parameters of all operations present in the workflow *wf* and *OUT* the set of all their output parameters (see Box 1).

The set of data links connecting the operations in *wf* must then satisfy:

$$DL \subseteq OUT \times IN$$

Note that this condition does not specify that every possible connection between output and input is a data link, rather, it specify that data links must be connection that associate output parameters to input parameters.

Controlflow dependencies are defined using ordering operators. Given the operations $(op_1, op_2, \ldots op_n)$, the following controlflow dependencies can be specified.

- op_1 *sequence* op_2: means that the operation op_2 is enacted once the execution of op_1 terminates.
- op_1 *and-split* $(op_2, \ldots, op_n)$: all the operations in $(op_2, \ldots, op_n)$ are enacted once the execution of op_1 terminates.
- op_1 *or-split* $(op_2, \ldots, op_n)$: one operation in $(op_2, \ldots, op_n)$ is enacted once the execution of op_1 terminates.
- $(op_1, \ldots, op_{n-1})$ *and-join* op_n: the operation op_n is enacted when the execution of all the operations in $(op_1, \ldots, op_{n-1})$ terminates.
- $(op_1, \ldots, op_{n-1})$ *or-join* op_n: the operation op_n is enacted when the execution of an operation in $(op_1, \ldots, op_{n-1})$ terminates.

Formally, we define controlflow dependencies as:

Box 1.

$$IN = \bigcup_{op \in wf.OP} inputs(op) \qquad OUT = \bigcup_{op \in wf.OP} outputs(op)$$

$CF \subseteq (A \times OPERATOR) \cup (OPERATOR \times A)$

A here denotes the set of operations that compose the workflow *and OPERATOR* is the set of operators used for defining controlflow dependencies between the activities. Specifically, see Box 2.

Notice that the controlflow between operations is defined independently of the parameters of the operations in question.

Workflow instance An instance of a workflow represents its execution. In our context, a workflow instance is used primarily to keep track of the data used as inputs and delivered as outputs of the operation executions. We define an instance wf_i of the workflow $wf = \langle nameWf, OP, DL, CF \rangle$ by the graph:

$$wf_i = \langle N_{OP}, N_{PARAM}, E_{PL}, E_{DF}, E_{CF} \rangle$$

$N_{OP} \cup N_{PARAM}$ are the nodes of the graph. Specifically; N_{OP} represents the instances of the workflow operations *OP*, and N_{PARAM} represents the instances of the parameters used as inputs or delivered as outputs of the operation instances in N_{PARAM}.

$E_{PL} \cup E_{DF} \cup E_{CF}$ are the edges that connect the nodes in $N_{OP} \cup N_{PARAM}$. E_{PL} are instances of parameter links, E_{CF} are instances of the controlflow edges in *wf.CF*, and E_{DF} are instances of the dataflow edges in *wf.DF*.

In the remainder of the chapter, we assume the existence of the function *instancesOf()*, which given a workflow, operation, a parameter, a parameter link, a data link or control link, return the associated instances. We also consider the functions *paths()*, with the signature presented below, to retrieve the paths that relates the parameters and the operations using the parameter links, data links and control links of workflows (see Box 3).

OP, *PARAMETER* and *PATHS* denote the domains of operations, parameters and paths, respectively.

Workflow Provenance

The traces of execution of scientific workflows are stored in what we call *provenance logs*. Provenance queries are issued over provenance logs and are used to trace back the origin of the results of the workflows, i.e., where the results come from, and the processing they underwent, i.e., the operations that have been applied to data.

As an example, consider the scientific workflow shown in Figure 1 to which we refer in the rest by *wf1*. It is used for constructing the phylogenetic tree for proteins. Given a protein accession specified by the user, a homology search is first performed to retrieve its homologous from the Uniprot[4] database using the operation *BlastP*. Returned homologous undergo a multiple alignment analysis, using the *ClustalW* operation. The output of this operation is then used to feed the last operation in the workflow, *Phylip*, which constructs a phylogenetic tree specifying the relationships between the protein input by the user and its homologues.

Box 2.

$$OPERATOR = \{Sequence, AND-split, AND-join, OR-split, OR-join\}.$$

Box 3.

$$paths: (OP \cup PARAMETER) \times (OP \cup PARAMETER) \rightarrow PATHS$$

Figure 1. wf1, a scientific workflow for constructing the phylogenetic tree of proteins

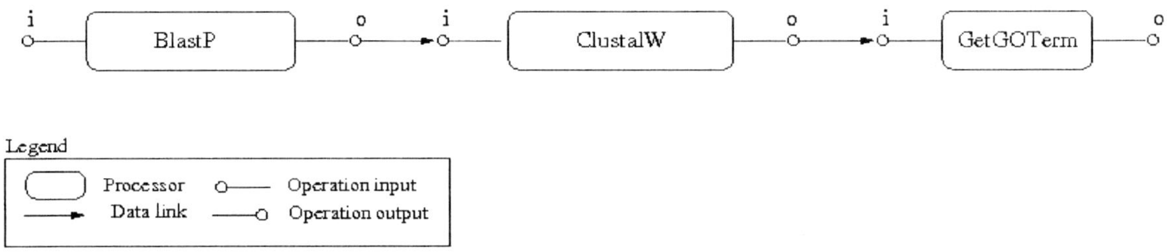

Examples of provenance queries that may help the scientists in analysing the results of executions of this workflow are as follows:

- **Q_1**: *Which proteins contributed to the construction of a given phylogenetic tree, pyt?*
- **Q_2**: *What are the operations that delivered the phylogenetic tree pyt?*
- **Q_3**: *What are the operations that contributed to the production of the phylogenetic tree pyt?*

The above queries can be answered by analysing the provenance logs collected when enacting the workflow, i.e., the workflow instances.

The result of the first query Q_1 is the set of instances of the input of the operation *BlastP* that are connected to the phylogenetic tree *pyt* with a path. Specifically, see Box 4.

The result of the second query is the set of operations that produced the phylogenetic tree *pyt*, which we denote by *deliver(pyt)* (see Box 5).

The result of the query Q_3 is the set composed of the operations that delivered the phylogenetic tree *pyt*, *deliver(pyt)*, and any operation connected to those operations in the workflow *wf1*. Specifically, see Box 6.

Notice that the query Q_1 allows retrieving data-related provenance, in the sense that the results are the data products that contributed to the value of *pyt*, whereas the queries Q_2 and Q_3 allow retrieving process-related provenance since they returns the operations the execution of which contributed to the value of *pyt*.

Box 4.

$$\{p, \ (p \in instancesOf(i)) \ and \ ((i, BlastP) \in wf1.PL) \ and \ (paths(p, pyt) \neq \phi)\}$$

Box 5.

$$deliver(pyt) = \{op \in wf1.OP, \ ((op, o) \in wf1.PL) \ and \ (pyt \in instancesOf(o))\}$$

Box 6.

$$\{op \in wf1.OP, (op \in deliver(pyt) \ or \ (\exists \ op' \in deliver(pyt)), paths(op, op') \neq \phi)\}$$

TAXONOMY FOR CHARACTERISING PROVENANCE IN SCIENTIFIC WORKFLOWS

This section presents the dimensions that we identified for characterising provenance in scientific workflows. Specifically, the following aspects are covered by the taxonomy. *Dataflow vs. controlflow based provenance tracking*: there are two kinds of dependencies that connect the analysis operations in a scientific workflow, namely dataflow and controlflow. The first dimension shows how these dependencies are exploited for tracing back a provenance of workflow results by identifying the data instances that contributed to the value of the result. *Black box analysis operations*: in databases, data is generally derived using queries that have a well defined semantics. Therefore, identifying the tuples that were used to derive a given tuple is generally a feasible operation (Cui, Y. & Widom, J. (2003)). In a workflow context, we do not have this luxury; the analysis operations that compose the workflows are black boxes supplied by third party providers. Because of this, the operational semantics of the analysis operations is not known making it harder to compute the provenance of the workflow results. In view of that, we show how the semantic annotations of the web services that supplies the analysis operations may, in certain cases, help overcome this issue and computing data provenance (Belhajjame *et al.* (2008)). *Fine grained/coarse grained provenance*: data provenance in workflows can have different granularity, for instance in certain cases, the provenance of certain data results can be identified as the database or a data set used by a given analysis operation, in other cases it can be finer by identifying only the data items that were actually used in deriving the results. The ability to support fine grained provenance of workflow results depends partly on the information the provenance systems has on the internal operational semantics of the constituent operations of the workflow. *Instance identification*: It is not always trivial to know what the scientist use for identifying data products. In certain cases, s/he may use the value of the data product as the identifier; in other cases, s/he may rely on keys, such as LSID[5]. *Scope*: The scope of provenance can be either local or global. It is local if the provenance of a data product is computed with respect to a given service operation, subworkflow or workflow, and it is global if it is computed considering the entire provenance store.

Dataflow vs. Controlflow

Two kinds of provenance queries can be issued over the instances of workflows for analysing and interpreting the provenance of collected data products: *dataflow* queries and *controlflow* queries.

In dataflow-oriented analysis of workflow provenance, the origin of the workflow results is computed based on the dataflow between the operations that compose the workflow. Consider for example, the workflow presented earlier in Figure 1. The operations that compose this workflow are connected using exclusively data flow dependencies. Therefore, the provenance queries that may be posed against the instances of this workflow are data-flow based.

Let $<N_{OP}, N_{PARAM}, E_{PL}, E_{DF}, E_{CF}>$ be a workflow instance, and *dp1* and *dp2* two data products in N_{PARAM}. And, let $PATH_{df}$ be the set of paths that connects the nodes in $N_{OP}\ U N_{PARAM}$ using the links in $E_{PL}\ U\ E_{DF}$. The data products that contributed to a given instance *ins* in N_{PARAM} with respect to the workflow dataflow is a subset of N_{PARAM}, N', such that for each element *ins'* in N' there exists a path in $PATH_{df}$ that connects *ins'* to *ins*.

In controlflow-oriented analysis of workflow provenance, the origin of the workflow results are computed based on the controlflow dependencies defined between the operations that compose the workflow. To illustrate this, consider a version of the workflow presented earlier in which some of the operations are also connected using controlflow dependencies (see Figure 2). This workflow contains two additional operations *IdentifyUser* and *OpenSession*. These operations are typically needed when the analysis operations are provided by grid services. From controlflow perspective, the input *usedId* provided as input to the operation *IdentifyUser* belong to the provenance of the results of the operation *Phylip*.

Let $<N_{OP}, N_{PARAM}, E_{PL}, E_{DF}, E_{CF}>$ be a workflow instance, and *dp1* and *dp2* two data products in N_{PARAM}. And, let $PATH_{cf}$ be the set of paths that connects the nodes in N_{OP} U N_{PARAM} using the links in E_{PL} U E_{CF}. The data products that contributed to a given instance *ins* in N_{PARAM} with respect to the control flow is a subset of N_{PARAM}, *N'*, such that for each element *ins'* in *N'* there exists a path in $PATH_{cf}$ that connects *ins'* to *ins*.

A hybrid approach, in which both the dataflow and the controlflow dependencies are used in computing the origin of results, is also possible. In this case, all the paths that connect the nodes in N_{OP} U N_{PARAM} using the links in E_{PL} U E_{DF} U E_{CF} are considered when tracing back the origin of a data product.

As we shall see later on, most of provenance systems in the literature consider only dataflow analysis of workflow provenance.

Service Operation: Black Box vs. White Box

In databases, data is generally derived using operators that have a well defined semantics. Therefore, identifying the tuples that were used to derive a given tuple is often a feasible operation (Cui, Y. & Widom, J. (2003)). As an example, Figure 3-(a) illustrates a relational query issued against two protein databases to retrieve the accession and sequence of certain proteins. The query makes use of three relational operators, *selection*, *projection* and *join*. The operational semantics of these operators is well defined and allows tracing back the tuples that were used to derive the results of the query and to derive the association that relates the inputs and outputs of the query.

Consider now the scientific workflow illustrated in Figure 3-(b). The workflow is composed of two analysis operations: *GetEntry* and *ExtractAC*. Because these operations are black boxes that are provided by third party services, users are unable to understand the transformation carried out by these operations and to understand whether the outputs of those operations are actually derived and depend on the values of their inputs.

If on the other hand, the operational semantics of the analysis operations that compose the work-

Figure 2. Example of scientific workflow containing controlflow dependencies

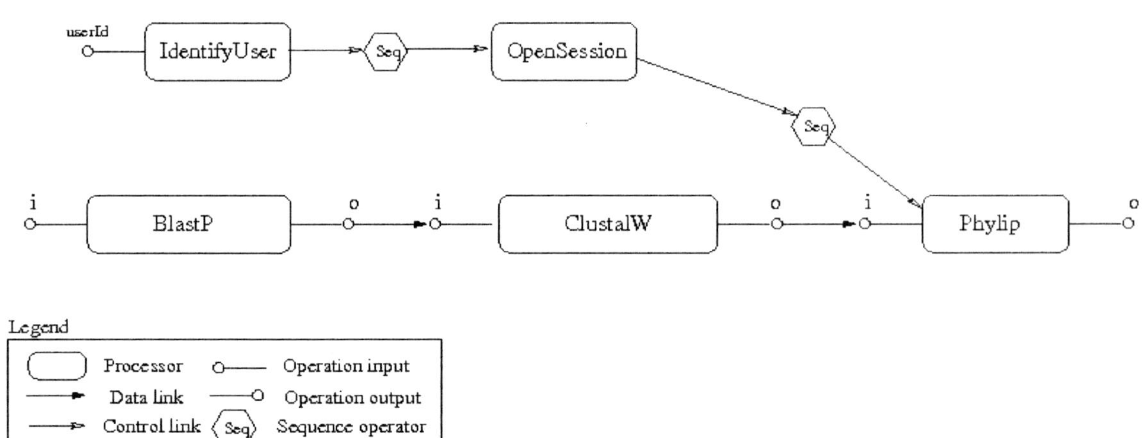

Figure 3. Example of data transformation

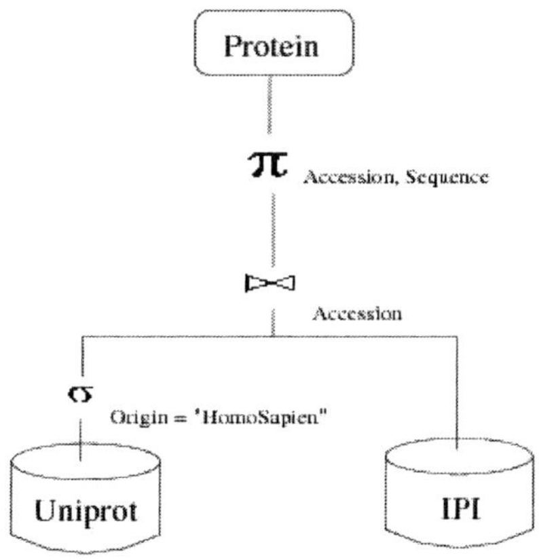

(a) A query over two protein databases

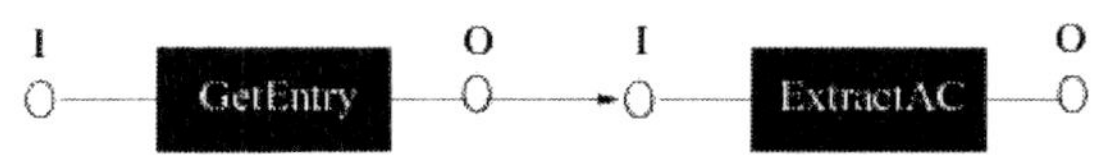

(b) A scientific workflow

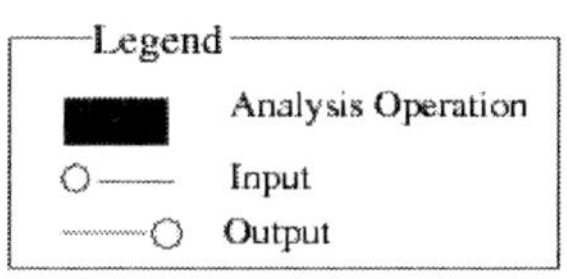

flow is accessible, then it is possible to understand the transformation that the data inputs undergo in order to produce the data obtained as a result of the operation execution. Analysis operations accompanied with operational semantics are called *white box* operations as opposed to the *black box* operations.

Fine Grained Provenance vs. Coarse Grained Provenance

If the analysis operations that compose workflows are black boxes then the dependencies between the operation inputs and outputs are not known. Because of this, when processing data provenance queries, it is often (if not always) assumed that every operation output is dependent on every input of that operation. Consider for instance that the provenance query has to do with computing the data products that we used to contribute to the value of a given data instance. Because of the above assumption, the results of this query may include some false positives, i.e., data products that did not contribute to the value of the data instance.

Assume now that the dependencies between the inputs and outputs of the operations that compose the workflow are known: these dependencies can be provided for instance by the operation semantics of the operations, if available (McIlraith *et al.* (2001) and Patil *et al.* (2004)). Consider again the above provenance query. In this case, the results of the provenance query will be free from false positive: that is it will include only the data products that truly contributed to the value of the data instance in question. We call the results of this provenance query *fine-grained provenance results*, as opposed to results that contain false positives, which we call *coarse-grained provenance results*.

Instance Identification

It is not always trivial to know what the scientist use for identifying data products. Practice shows that the following identification schemes are adopted:

- *Value-based identification*: In this case, data products are identified based on their values. Take the example of the scientific workflow presented earlier in Figure 1. A user may be interested in locating the proteins homologous to the protein identified by the value *P15435*: this is a Uniprot accession number.

- *Key-based identification*: In this case, data items are identified using keys. For instance, one of the common scheme used for identifying bioinformatics data products are LSID (Clark, T. (2003)), a specific kind of URIs.
- *Context-based identification*: This is perhaps the most commonly used scheme. The user identifies a data product using a combination of the value of the data product and contextual information specifying, for instance, the workflow/operation that outputs the data product and/or the time of its delivery. For example, the user may issue the following provenance query. *Return the homologous of the protein having the value 'P17116' and that was produced by the analysis operation 'GetProtein'.*

Scope

The scope of provenance can be either local or global.

- *Local-provenance*: In this case, the provenance of a given data products include the data instances input and output with respect to a given analysis operation, workflow or sub-workflow, specified by the user.
- *Global-provenance*: In this case, the scope of the provenance expands the provenance store. In other words, when processing provenance queries, all the data stored within the provenance store are considered to be potential candidates.

Using the first or the second scheme is usually driven by the specific requirements of the user.

Table 1 summarises the dimensions we identified for characterising provenance in scientific workflows. Each dimension is associated with the set of possible values.

PROVENANCE LANDSCAPE

In this section, we analyse a selection of popular provenance systems that were proposed for supporting scientists in the analysis and interpretation of the results of scientific workflows, and use, for this purpose, the taxonomy presented in the previous section.

Taverna

Taverna is a workbench that allows users to specify and enact data-driven scientific workflows (Oinn *et al.* (2004)). The workbench proved to be very popular among bioinformaticians where the focus is on modelling *in silico* experiments. The Taverna system comes with a family of plug-ins, e.g., the Feta service registry, which allows users to discover web services using semantic information

Table 1. Taxonomy for characterising provenance in scientific workflows

	Dimension	Values
1	Dataflow	*(i) true, (ii) false*
2	Controlflow	*(i) true, (ii) false*
3	Precision	*(i) fine-grained, (ii) coarse-grained*
4	Instance identification	*(i) key, (ii) value*
5	Scope	*(i) local, (ii) global*

that describe the task performed by the services as well as the semantic domain of their input and output parameters. Taverna also comes with a plug-in that allows collecting and browsing data provenance. Provenance in Taverna, as within most scientific workflow systems, is collected automatically when enacting workflows. Stored provenance is then explored using a graphical interface. With regard to the taxonomy presented in this chapter, Table 2 provides the value of each of the dimension identified.

It shows that Taverna supports a data-flow-based lineage of provenance information and that it caters for a coarse-grained provenance tracking. The data collected by provenance can be identified by key, using LSIDs and/or by value, and the scope of provenance queries can be both local and global.

Vistrails

Vistrails is a workflow system that is used for specifying and enacting scientific workflows (Callahan et al. (2006)). Particularly, Vistrails is suitable for visualisation pipelines, in other words, workflows that manipulate and produce images. It supports workflow versioning and caching of the results of workflow execution to enhance the performance of the workflow enactor. From provenance perspective, Table 3 characterises Vistrails using the dimensions of the taxonomy presented.

Karma

Different from Taverna and Vistrails, Karma is not a workflow workbench; rather it is a framework that allows collecting and browsing provenance (Simmhan *et al.* (2006)). Data provenance is collected based on a loosely coupled publish-subscribe mechanism. Table 4 characterises Karma in terms of the taxonomy presented. The dependencies between collected data products are drawn based on the dataflow of the workflow. Therefore, the relationships between collected data products are data-flow based and coarse-grained. In Karma, data products are identified using an ad-hoc identification schema that the authors refer to using the term GUID, for globally unique identifier.

Table 2. Characteristics of the provenance plug-in supplied by Taverna workbench

	Dimension	Values
1	Dataflow	*(i) true*
2	Controlflow	*(ii) false*
3	Precision	*(ii) coarse-grained*
4	Instance identification	*(i) key, (ii) value*
5	Scope	*(i) local, (ii) global*

Table 3. Characteristics of the provenance capabilities provided by Vistrails

	Dimension	Values
1	Dataflow	*(i) true*
2	Controlflow	*(ii) false*
3	Precision	*(ii) coarse-grained*
4	Instance identification	*(ii) value*
5	Scope	*(i) local*

Table 4. Characteristics of the Karma framework

	Dimension	Values
1	Dataflow	*(i) true*
2	Controlflow	*(ii) false*
3	Precision	*(ii) coarse-grained*
4	Instance identification	*(i) key*
5	Scope	*(i) local*

Table 5. Characteristics of the Zoom system

	Dimension	Values
1	Dataflow	*(i) true*
2	Controlflow	*(ii) false*
3	Precision	*(ii) coarse-grained*
4	Instance identification	*(i) key*
5	Scope	*(i) local*

Zoom

Zoom is a system that allows users to explore workflow provenance in a customised manner (Biton *et al.* (2007) and Biton *et al.* (2008)). Given a workflow specification, users can group the constituent operations into modules. The obtained abstract workflow is then used as a base model for browsing data provenance. Table 5 characterises the Zoom system in terms of the dimensions of the taxonomy presented. As for Karma, the dependencies between data products are inferred purely from the dataflow of the workflow and, therefore, are coarse-grained. Data products are identified by keys, and the scope of data provenance queries is limited within the workflow that delivers the data products.

PASOA

The objective of the PASOA (Provenance Aware Service Oriented Architecture) project is to investigate the accuracy of collected provenance data. The authors of this project developed a platform-independent framework for validating the results of workflow executions (Miles *et al.* (2007) and Wong *et al.* (2005)). Table 6 characterises this framework in terms of the dimensions of the taxonomy presented. The workflows from which data is collected are data-driven workflows, and, therefore, the dependencies derived between collected data provenance are data-flow based.

To assess the accuracy of collected data provenance, reasoning is performed over the documented provenance of experiment results and semantic descriptions of services that compose workflows. Thanks to the use of these descriptions, the analysis of data provenance can be fine-grained in certain cases: that is where the semantic annotations of web services allow to draw in a precise manner the dependencies between the inputs and outputs of services (Miles *et al.* (2007)). The scope

Table 6. Characteristics of the PASOA framework

	Dimension	Values
1	Dataflow	*(i) true*
2	Controlflow	*(ii) false*
3	Precision	*(i) fine-grained*
4	Instance identification	*(i) key*
5	Scope	*(i) local*

of collected data products is within the workflow from which data products are collected, in other words, it is local.

FUTURE TRENDS

The analysis presented in this paper confirmed the limitations mentioned in the introduction regarding current provenance systems. Notably, it showed that most of analysed systems with the exception of PASOA, do not support a fine grained interpretation of the provenance of workflow data products. The absence of a fine-grained provenance capture means a more difficult task for the user (scientist) in analysing the results of execution of workflows and verifying the scientific hypothesis they were enacted for their verification. It is worth noting that in this regard, a work in progress is being conducted by the Taverna team (see Missier, P., Belhajjame, K, Zhao, J. & Goble, C.A. (2008)) to capture fine-grained provenance. In doing so, they rely on annotations that describes the relationship between the inputs and outputs of the analysis operations that compose the workflow.

The analysis also showed that the scope of provenance is in most cases local. In other words, the origin of a data product can be traced back up to the inputs of the workflow that produced that piece of data. In practice, however, the scientists may want to trace back the origin of a piece of data with respect to several workflow not only one, thereby, allowing him/her to discover the relationships between data products that were produced in the context of multiple workflows. We are not aware of any provenance system in which the scope of the captured provenance is global.

It is worth noting that we attempted to be selective when designing the taxonomy to capture the dimensions we believe are important in capturing data provenance in scientific workflow (including grid workflows). Therefore, we do not claim that the taxonomy is exhaustive. Indeed, it can be extended by considering other issues, such as the query language for querying workflow provenance.

CONCLUSION

Data provenance is key to understanding and interpreting the results of scientific experiments. This chapter introduced data provenance in scientific workflows and characterised existing work in the area using a taxonomy proposed for this purpose. This analysis allowed drawing the commonalities between existing provenance systems. For example, it showed that most of these systems are used for collecting provenance of data-driven workflows. In reality however, controlflow information may be of relevance when analysing data provenance. For example, if during the workflow execution the data produced by an analysis operation was checked, e.g., by the workflow user, before being used to feed the execution of the succeeding analysis operation in the workflow, then this information can be of value when analysing the provenance, as it provides information about the quality of a data product (Missier, P., Embury, S.M. & Stapenhurst, R (2008)).

In addition to analysing existing workflow provenance systems, the taxonomy presented in this chapter provides valuable guidelines that can be adopted by developers and designer of provenance systems, e.g., to ensure that their system implements the dimensions of interest to their target users.

REFERENCES

Belhajjame, K., Embury, S.M., Paton, N. (2008). Automatic annotations of web services based on workflow definitions. *ACM Transactions on the Web, 2* (2), 1-34.

Biton, O., Cohen-Boulakia, S. & Davidson, S. (2007). Zoom* userviews: Querying relevant provenance in workflow systems. In *VLDB* (pp. 1366–1369).

Biton, O., Cohen-Boulakia, S., Davidson, S. & Hara, C. (2008) Querying and managing provenance through user-views in scientific workflows. In *ICDE*. Los Alamitos, CA: IEEE.

Buneman, P., Chapman, A. & Cheney, J. (2006) Provenance management in curated databases. In Chaudhuri, et al.

Callahan, S.P., Freire, J., Santos, E., Eduardo Scheidegger, C., Silva, C.T. & Vo, H.T. (2006) Vistrails: visualization meets data management. In Chaudhuri, et al. (pp. 745–747).

Chaudhuri, S., Hristidis, V. & Polyzotis, N. (2006) *Proceedings of the ACM SIGMOD International Conference on Management of Data, Chicago, Illinois, USA, June 27-29, 2006*. New York: ACM.

Clark, T. (2003). Editorial: Identity and interoperability in bioinformatics. *Briefings in Bioinformatics*, *4*(1),4–6.

Cui, Y. & Widom, J. (2003). Lineage tracing for general data warehouse transformations. *VLDB J*, *12*(1),41–58.

Ludäscher, B., Altintas, I., Berkley, C., Higgins, D., Jaeger, E., Jones, M., et al. (2006). Scientific workflow management and the kepler system. *Concurrency and Computation: Practice and Experience*, *18*(10),1039–1065.

McIlraith, S.A., Son, T.C. & Zeng, H. (2001) Semantic web services. *IEEE Intelligent Systems*, *16*(2), (46–53.

Miles, S., Wong, S.C., Fang, W., Groth, P.T., Zauner, K.P. & Moreau, L. (2007) Provenance-based validation of e-science experiments. *J. Web Sem.*, *5*(1), 28–38.

Missier, P., Belhajjame, K, Zhao, J. & Goble, C.A. (2008) Data lineage model for Taverna workflows with lightweight annotation requirements. *IPAW 2008*. New York: Springer.

Missier, P., Embury, S.M. & Stapenhurst, R. (2008) Exploiting provenance to make sense of automated data acceptance decisions in scientific workflows. *IPAW 2008*. Springer.

Oinn, T.M., Addis, M., Ferris, J., Marvin, D., Senger, M., Greenwood, R.M., et al. (2004). Taverna: a tool for the composition and enactment of bioinformatics workflows. *Bioinformatics*, *20*(17), 3045–3054.

Oinn, T.M, Greenwood, R.M., Addis, M., Alpdemir, N., Ferris, J., Glover, K., et al. (2006). Taverna: lessons in creating a workflow environment for the life sciences. *Concurrency and Computation: Practice and Experience*, *18*(10) 1067–1100.

Patil, A.A., Oundhakar, S.A., Sheth, A.P. & Verma; K. (2004) Meteor-s web service annotation framework. In S. I. Feldman, M. Uretsky, M. Najork, & C. E. Wills (Eds.), *WWW*, (pp. 553–562). New York; ACM.

Simmhan, Y., Plale, B. & Gannon, D. (2005) A survey of data provenance in e-science. *SIGMOD Record*, *34*(3) 31–36.

Simmhan, Y., Plale, B., Gannon, D. & Marru, S. (2006) Performance evaluation of the karma provenance framework for scientific workflows. In L. Moreau and I. T. Foster (Eds.), *IPAW* (pp. 222–236). New York: Springer.

Taylor, I. (2006) Triana generations. In *e-Science*, (pp. 143). Los Alamitos, CA: IEEE Computer Society.

Wong, S.C., Miles, S., Fang, W., Groth, P.T. & Moreau, L. (2005) Provenance-based validation of e-science experiments. In Y. Gil, E. Motta, V. R. Benjamins, & M. A. Musen (Eds.), *Interna-*

tional Semantic Web Conference: Lecture Notes in Computer Science, (Vol. 3729), (pp. 801–815). New York: Springer.

KEY TERMS AND DEFINITIONS

Data Provenance: data provenance can be thought of as the process by which the data instances and the analysis operations that were used to derive a given data instance are identified.

***In silico* Experiment:** An *in silico* experiment is a routine that employs computational analysis tools to verify, amongst other things, the validity of a scientific hypothesis or demonstrate a known fact.

Semantic Annotations: Semantic annotations are specifications that define the meaning and the form of an object. They are often encoded in the form of relationships that links the object subject to description to concepts from ontologies.

Semantic Annotation of Web Services: They are semantic annotations that describe the task performed by a web service and the forms and the domains of the parameters they take as input and the result they delivers as a result of a service execution.

Web Service: A web service can be defined as a software program that provides an API whereby it can be invoked over the internet using XML-based standard protocols.

Workflow: A workflow is the computerised representation of a process (e.g., software construction, registration process at the university). It specifies the various activities of the process that have to be executed in some order, the flow of data between activities and the multiple collaborating agents that execute activities to fulfil a common objective.

Workflow System: A workflow system, a.k.a. workflow management system, is a software responsible for enacting workflows. It instantiates the activities that compose a workflow, assigns their execution to agents and coordinates these executions.

ENDNOTES

* Corresponding author, e-Mail: Khalid.Belhajjame@manchester.ac.uk

2 http://twiki.pasoa.ecs.soton.ac.uk/bin/view/Challenge/SecondWorkshopAgenda

3 In the following, we use the terms *experiment* and *scientific workflow* interchangeably.

4 http://www.expasy.uniprot.org

5 http://lsids.sourceforge.net

Chapter IV
Provenance Tracking and End-User Oriented Query Construction

Bartosz Balis
Institute of Computer Science AGH, Poland

Marian Bubak
Institute of Computer Science AGH, Poland & University of Amsterdam, The Netherlands

Michal Pelczar
ACC CYFRONET AGH, Poland

Jakub Wach
ACC CYFRONET AGH, Poland

ABSTRACT

Provenance tracking is an indispensable element of each e-Science infrastructure for conducting in silico experiments. However, enabling end-users who are non-IT experts to query provenance and experiment data in a meaningful way is equally important. The authors propose an ontology-based provenance model which captures the execution of in silico experiments, as well as domain-specific semantics of data and computations used in those experiments. They demonstrate how ontologies can serve as inter-lingua for end-users, provenance tracking system, and query tools. Query Translation Tools (QUaTRO), enabling end-user oriented, ontology-guided visual querying over provenance records and experiment data, are also presented. In those tools, they also show how the ontology models enable semantic information integration of provenance metadata and experiment data, enabling queries capable of exploring the structure of provenance and associated experiment data. Their approach is demonstrated on a Drug Resistance application deployed in the ViroLab Project.

INTRODUCTION

The term 'e-Science' (Hey 2002) was coined to denote a new type of scientific research based on the collaboration within a number of scientific areas, enabled by a next generation infrastructure, wherein people, computing resources, data and instruments are brought together to bring a new quality to the everyday work of researchers. The infrastructure in question is usually identified with Grid systems which offer at least two benefits important for loosely-coupled cross-institution research and collaboration: virtualization and sharing of resources, and building of virtual organizations. Recently, the increasing importance of semantics and knowledge for a future e-Science infrastructure has been emphasized (De Roure 2005). The term *Semantic Grid* has been used to denote a Grid infrastructure wherein information services are enhanced with well-defined meaning enabling better cooperation between computers and people (Goble, De Roure 2004). They key enabler to achieve the vision of Semantic Grid are Semantic Web technologies, such as ontologies which are a standard, highly sharable, and machine-processable way to represent vocabularies and semantic relationships in a given domain (Goble, Corcho 2006).

Scientific experiment results are neither reliable nor reusable without their *provenance*, i.e. the information on the origin and history of these results (Groth 2006). Though provenance in computer science originated in the database community, the difference between database provenance and provenance in e-Science has been pointed out (Tan 2007). Both types of provenance describe the origin of a piece of data, i.e. they answer the question what other pieces of data contributed to a given result. However, while in the case of a database, the piece of data is a result of a 'whitebox' database query, in e-Science, it is a result of a 'blackbox' process. Hence different provenance models and different methods to compute provenance are needed in the two communities (Ibid.).

The importance of provenance tracking in e-Science environments has been pointed out many times and numerous provenance approaches have been proposed. However, providing an adequate end-user support for provenance querying is also a challenge. It has been recognized that the need for provenance queries goes beyond the lineage of a single data item and searching or mining over many provenance records might be useful (Moreau 2007). Nevertheless, there are technical and conceptual barriers preventing or making it difficult for end-users of e-Science environments, such as domain researchers and specialists, to construct complex queries using query languages such as XQuery, SQL or SPARQL. Therefore, there is a need for provenance query support which would enable end-users to construct powerful queries in an easy way.

The goal of this chapter is to propose an ontology-based provenance model which not only can capture the execution of in *silico experiments* including the domain semantics of data and computations used in an experiment, but also enable end-user oriented visual querying over provenance records. We demonstrate how ontologies are useful as a common language for: (1) *end-users* who use ontologies as a query language; (2) *query tools* using them to represent and evaluate queries, and (3) *provenance repository* in which the ontologies serve as the data model. Moreover, we show how the proposed ontology model enables *semantic information integration* – recognized as one of the key requirements for a *Semantic Grid* infrastructure for e-Science (De Roure 2005). Though many semantic data integration approaches have been proposed, we show how ontologies represented in OWL/RDF are used to integrate not only experiment data from various heterogeneous sources, but also data and provenance metadata, where the latter describes the process wherein the former has been used. In this approach, end users are given a unified interface to construct queries that integrate experiment data and provenance metadata.

The remainder of this chapter is structured as follows. We begin by presenting the related work. Next, our provenance model is presented, followed by the presentation of our provenance tracking & querying approach. Then, we demonstrate this approach on a case study Drug Resistance experiment. Finally, we analyze the querying capabilities enabled by the presented provenance model, and discuss some implementation issues. We conclude with a summary of the current achievements and a discussion of future work.

BACKGROUND

The essential provenance querying capabilities concentrate on obtaining derivation paths of experiment results and this is the focus of many provenance models. For example, provenance approach presented in (Bowers 2006) is based on capturing only three types of events – *read, write* and *state-reset* – for each actor in a workflow. In another approach, realized in the context of the PROVENANCE Project, a provenance model based on a handful of so-called *p-assertions* is introduced (Groth 2006). It has been pointed out that those low-level approaches, while a powerful basis for high-level models, in itself limit the provenance querying capabilities (Zhao 2006).

Some provenance approaches are based on high-level, structured relational models, such as the KARMA provenance model (Simmhan 2007) and the virtual data provenance model (Zhao 2006). These models typically capture the workflow computational model and record information about workflows, invocations and data sets. However, relational models are inherently less flexible than ontology models. For example, extending the provenance model, e.g. adding a new aspect of information (such as resource usage), or some new attributes to existing tables, could easily break the operation of existing tools that rely on the database schema. Furthermore, KARMA does not support annotation-based querying, while the virtual data provenance model supports semantics only as key-value annotations.

A complex RDF/ontology provenance model is presented in (Zhao 2004). This model introduces four interconnected ontologies that describe provenance records: (i) *process ontology* which describes actual run of a workflow; (ii) *data ontology* which model data used as inputs and outputs of the experiment; (iii) *organizational ontology* in which the experimenter and its organization are described; (iv) *knowledge ontology* which allows to associate user-defined tags or annotations from external taxonomies with data products. In this way, the semantics of the data products used in the experiment and relationships between data products are captured. However, the model of the workflow execution record is semantics-free and is based on a syntactic graph model. The described ontologies are used to annotate those records with domain semantics as RDF metadata. In addition, the semantics of computations used in the experiment is not captured.

Many ontology-based information integration approaches have been proposed (Wache 2001). A representative example from the bioinformatics domain is the TAMBIS project which aims at integration of diverse bioinformatics data sources and provides visual tools that enable transparent querying over those sources, creating an illusion of a single query language and a unified data model (Stevens 2003). The basis for the integration is the TaO ontology which provides a common vocabulary for biological concepts, and mappings from the ontology terms to external data sources.

In contrast to this and similar projects, our main focus is not on information integration across heterogeneous data sources, but rather integration which is centered on provenance, whose purpose is to enable querying for *data with certain features that has been used in an experiment with specified properties* (or vice versa, etc.). In other words, our approach is oriented towards *semantic information integration of data and metadata* in a unified querying framework. Consequently, we view our tools as complimentary to those that concentrate on ontology-based information integration across multiple data sources. Moreover, no provenance model known to us attempts to achieve semantic information integration between experiment data and provenance metadata.

Overall, though many provenance models have been shown to support complex provenance queries (Zhao 2007) (Simmhan 2007) (Kim 2007) defined by the First Provenance Challenge (Moreau 2007), the aspect of visual querying of provenance in an end-user-oriented fashion is not addressed by the mentioned works.

The approach to provenance tracking and querying presented in this chapter is evaluated in the context of the ViroLab Project[1] (Sloot 2006) which provides a virtual laboratory for conducting *in silico* experiments related to diagnosis of infectious diseases, especially the HIV virus[2] (Gubala 2008).

PROVENANCE MODEL

We adopt provenance model based on ontologies modeling the execution of *in silico* experiments as a workflow of execution stages associated with input and output data. A partial ontology graph of the provenance model with domain ontologies from a drug resistance domain is shown in Fig. 1. The model comprises of a number of interconnected ontologies:

- The *in silico experiment ontology* lies in the center of the model providing the general structure of a provenance record. It contains generic concepts, such as *Experiment*, *Execution Stage*, etc. Notably, an execution stage can either be a computation or a data access operation, so that data access operations can be explicitly captured as part of the experiment execution. Data items are associated with the execution stages as their inputs and outputs. Information about memory and CPU usage, as well as the duration of a computation, enables performance-related uses of the execution records.
- A specific *ViroLab computation ontology* describes the model of computation resources in ViroLab, so called Grid Objects. Alternatively, models of other computation resources, such as web services, can easily be added.
- *Data domain ontology* describes the semantics of data items. The domain-specific data entities are associated with the generic concept *DataItem* via a subclassing relationship. The figure depicts a few examples of a data ontology related to a drug ranking application.
- *Computation domain ontology* models the semantics of computations in a similar way as data ontologies do for data items. The figure depicts a few concepts, subclasses of a generic *Computation* class, also related to the drug ranking scenario.
- *Semantic relationships* can also be added to the model, notably those connecting computations with data, such as a *DrugRanking* data item being a *resultRanking* from a *NewDrugRanking* computation.

The described ontology model serves as an information model for representing executions of *in silico* experiments. Those execution records are represented as a graph of ontology individuals compliant with the ontology models described above. The individuals are created in the process of provenance tracking as an aggregation of low-level execution events. This process is described in detail in the next section.

PROVENANCE TRACKING AND QUERYING

Provenance tracking is based on monitoring the experiment execution. Monitoring captures low-

Figure 1. Ontology models as a base for provnance. Generic ontology of an in silico experiment captures the execution stages, inputs, and outputs of an experiment. Domain ontologies describe the semantics of data and computations used in the experiment. In this case parts of domain ontologies from a drug resistance domain are shown.

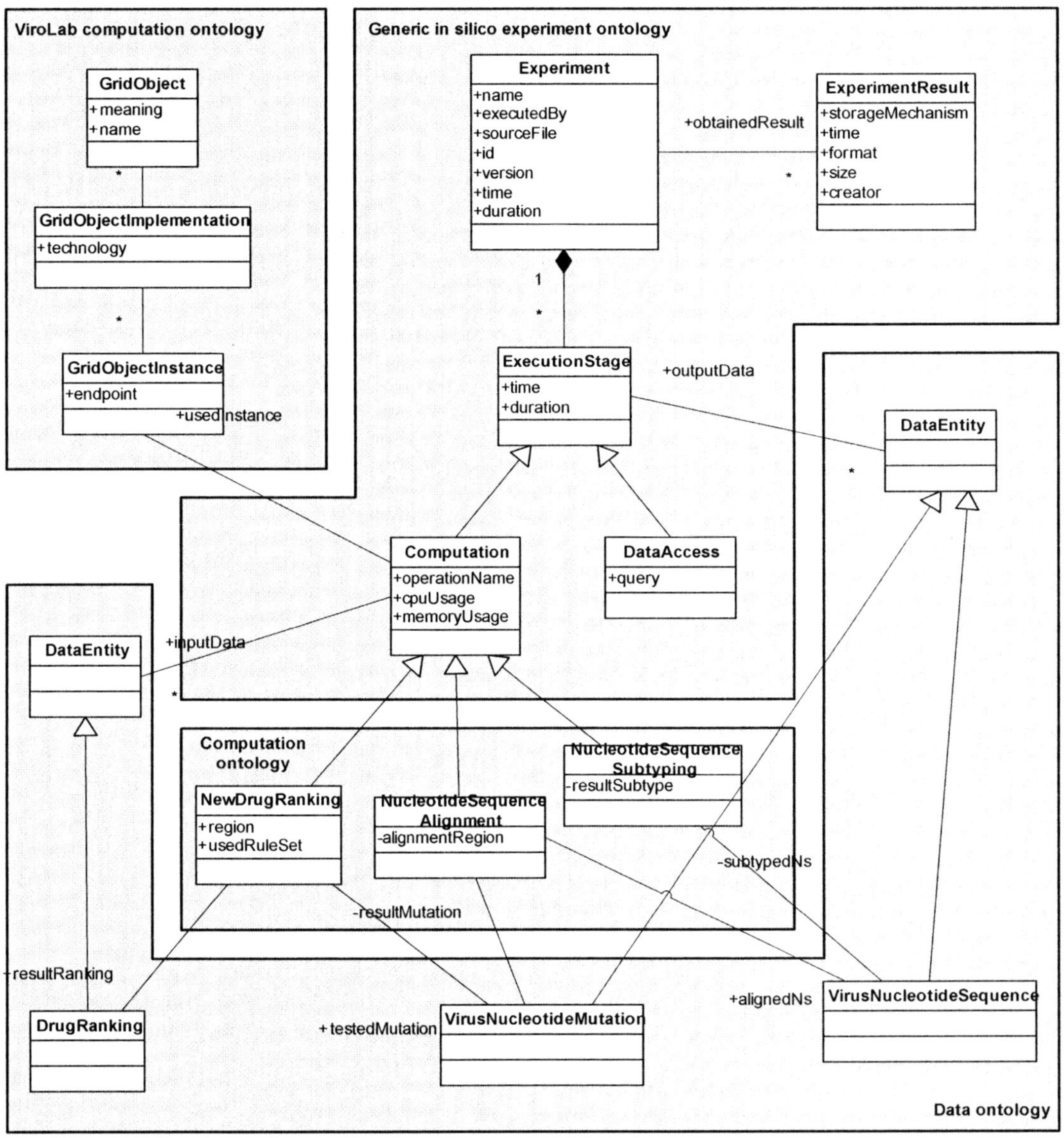

level events which are aggregated, processed, and translated into high-level ontology individuals which are stored in a Provenance Tracking System (PROToS) which maintains the provenance repository (Balis 2008). A general view of the provenance tracking and querying process is presented in Fig. 2. The enactment engine – *GSEngine* – is executing *in silico* experiments according to some *experiment plan*. In ViroLab, the experiment plan is written in Ruby scripts.

Translation of Low-Level Events to Ontology Individuals

In the course of the experiment's execution, events are generated from various locations, including the GSEngine, the experiment activities (instances of Grid Objects), as well as other sources (not shown). Those events are collected by a Monitoring System, and passed to a Semantic Aggregator which, according to a set of predefined rules, aggregates the low-level events and translates them into high-level ontology individuals. The aggregation rules specify which events are needed to produce a given piece of information, and how this piece of information should be created from those events. Basically, the following rules need to be specified:

- When new individuals should be instantiated,
- How to compute attributes for existing individuals,
- Which relationships should be established between individuals.

The basic prerequisite for aggregation is a correct correlation of events, i.e., the identification of events that belong together. To this end, we use hierarchical correlation identifiers (so called Experiment Correlation Identifiers – ECID) which uniquely identify the application's execution unit (e.g. *Exp0/Activity1/Task1* for *Task1* of a parallel application invoked from activity *Activity1* of experiment *Exp0*). The simplest rule which describes how to create a new Experiment individual states that two events are required for this: *ExperimentStarted* and *ExperimentFinished*, and that they should have the same ECID at the first level of hierarchy (i.e. the experiment identifiers should match). Similarly, a new experiment stage is created out of two other events: *GridOperationInvoking* and *GridOperationInvoked* (domain-specific specializations of generic events *ProcessStarted* and *ProcessFinished*), only this time ECIDs must match up to the second level.

Attributes (including object relationships between individuals) are computed according to so called derivation rules. A simple rule, for example, specifies that the duration of an experiment – *Experiment/duration* property – should be computed from two attributes of two events: *ExperimentStarted/time* and *ExperimentFinished/time*. One can specify an arbitrary piece of code (a Java class) which should be invoked in order to compute a given attribute. In a more complex case, an attribute can be obtained, for example, with a complex query to a database.

Finally, let us discuss how domain-specific individuals are created as opposed to generic ones. The hierarchy of basic events reflects the hierarchy within the ontology models. For example, the enactment engine can either generate a generic event *Computation*, or – if the semantics of the computation is known – a domain event, e.g. *NewDrugRanking*. The latter event is a specialization of the former, thus it inherits all its attributes plus it adds some domain-specific ones. The Aggregator instantiates a proper individual depending on whether domain events have arrived or not.

Query Translation Tools

The provenance records can be queried visually using *Query Translation Tools* (QUaTRO) which are in turn using the PROToS query interface to access the underlying provenance store. QUaTRO tools can actually query the provenance repository, experiment data repositories, or both at the same time, depending on the constructed query. Effectively, *semantic information integration* between data sources as well as provenance metadata, is achieved in QUaTRO. An end-user uses the QUaTRO web interface to construct queries, which are translated by the QUaTRO query engine into XQuery or SQL queries to underlying provenance and experiment data repositories.

It is interesting to note that several actors use the same provenance ontology model which effectively plays the role of inter-lingua between tools and end-users:

Figure 2. Provenance tracking and querying system architecture. Low-level events describing the execution of an experiment are captured by means of monitoring of the experiment execution. Those events are collected, aggregated by a Semantic Aggregator and translated into high-level ontology representation which is published in PROToS. QUaTRO tools enable construction of queries over both provenance repository and experiment data repository, using PROToS and a Data Access Service, respectively.

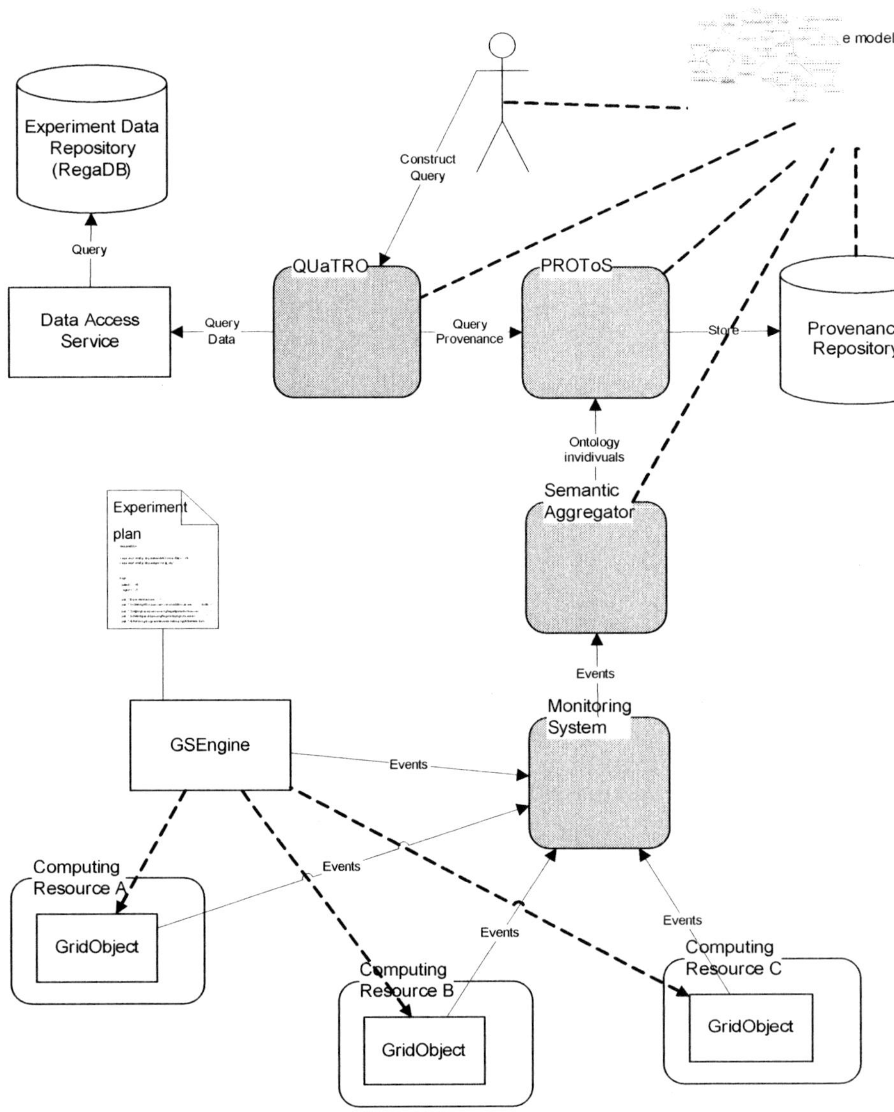

- The end-user uses the ontology model as a query language;
- In PROToS, the ontology model serves as a data model to store provenance records;
- For QUaTRO, the ontology model is a structure to represent and evaluate queries;
- The Semantic Aggregator uses ontologies as a data model to which the output records should conform.

The next section explores the visual querying in QUaTRO in more detail.

END-USER ORIENTED QUERYING OVER PROVENANCE AND EXPERIMENT DATA

Drug Resistance Example

This section is based on a case study *drug resistance* scenario and some example queries based on provenance tracking thereof. The scenario in question is based on the *Drug Ranking System* (DRS) application deployed in the ViroLab project. The goal of this application is to determine best combinations of drugs for a particular HIV virus obtained from a patient's blood. A typical version of this scenario consists of the following stages:

- *Stage 1*: Obtain the HIV virus nucleotide sequences from a database;
- *Stage 2a*: Align input nucleotide sequence. Compare it to reference strains. Result: mutations of this particular virus mutant;
- *Stage 2b*: Determine the subtype of the HIV virus;
- *Stage 3*: Apply selected *rule sets* to mutations to determine drug rankings.

Domain events for this scenario generated in the course of provenance tracking describe the experiment execution stages and their respective input and output data. These are summarized in Table 1.

Visual Query Construction

QUaTRO enables construction of powerful queries to underlying repositories of experiment data and provenance by means of a wizard-based easy-to-use graphical user interface. Query construction results in building a query tree which is started by selecting an initial concept, e.g. *NewDrugRanking, Experiment, NucleotideSequence*. The query form automatically expands and allows to follow properties of the starting concept which can lead to *other concepts* (in which case the tree can be further expanded), as well as *literal values*. In addition, some concepts can be *mapped to a database*, meaning that the instances of those concepts are stored as rows in a database table. In such a case, a mapping can be defined which associates the given class with a table in a database. A set of data sources can also be defined which store the records of this type. When the user selects a concept mapped to a database, the tools dynamically retrieve the list of columns of the proper database table and allow the user to further constrain the query by setting some values for those attributes. Let us note, however, that whether a concept is mapped to a database or not is completely transparent to the end user which just sees concepts, relationships and attributes. At any time, the user can also select a logical operator (at the moment only logical 'and' is implemented) in which case a new tree branch is created.

When the construction of the query tree is complete, the tree is analyzed in a bottom-up manner. Subtrees represent subqueries which can refer to different repositories of data or provenance. The results of subqueries are merged in upper nodes until the tree root is achieved. Let us examine the following DRS query example:

Select all computations of drug rankings performed by 'John Doe' for blood samples taken

Table 1. Domain events for Drug Ranking System scenario

Execution stage	***Input data***	***Output data***
NucleotideSequenceAlignment	*VirusNucleotideSequence*	*MutationList*
VirusSubtyping	*VirusNucleotideSequence*	*HIVVirusSubtype*
NewDrugRanking	*MutationList, RuleSet*	*DrugRanks*

on Jun-28-2007, and in which rule set 'HIVDB' in version 4.2.8 was used.

The query constructed in a QUaTRO tool web form is presented in Figure 3. The form is dynamically composed based on the concepts and relationships chosen by the user. The available relationships and attributes of an entity are automatically and dynamically presented as selection lists. In this case, the 'RuleSet' entity is actually mapped to a database, and its attributes 'name' and 'version' are actually database column names. Literals are either entered manually by the user into a text field, or are automatically chosen out of the available range of values. In the case of a database attribute this range of values can be dynamically determined by examining the actual values which occur in the database table.

Query Representation and Evaluation

This query is internally represented as a tree depicted in Figure 4a. The nodes of the query tree denote either classes (*NewDrugRanking*, *RuleSet*), literals (*JohnDoe*, 2007 – 06–28, HIV DB, 4.2.8), or logical operators. The tree edges are one of the follows:

- Object properties which link to another concept (*usedRuleSet*);
- Datatype properties which link to a literal of a given datatype (*executedBy*, *dateOfBloodSample*);
- Database properties which denote that a literal they link to is actually stored in a database (table *RuleSets*, columns *name* and *version*).

This representation is neutral with respect to query languages used by the underlying data repositories. The evaluation of a query amounts to translating the internal representation to appropriate query languages and joining the results. The query tree is analyzed bottom-up, starting from its leaves and following the edges up to the tree root. Subqueries represented in particular subtrees are evaluated and joined in upper nodes, if feasible. Currently the records are transparently joined in a way semantically similar to the relational inner join – the records are joined according to ontological identifier – object property resource pair, analogous to primary – foreign keys in relational models. The query is evaluated in 10 steps as shown in Figure 4b:

- *Steps 1-3*: Identifiers of rule sets with name 'HIVDB' and version '4.2.8' are retrieved from a database using an SQL query.
- *Step 4*: Those identifiers are converted to our internal 'database ID' format. The database IDs link ontology records with database records.
- *Step 5*: The rule-sets individuals with the obtained database IDs are retrieved from the experiment information store, using XQuery.

Figure 3. Sample DRS query in QUaTRO web form

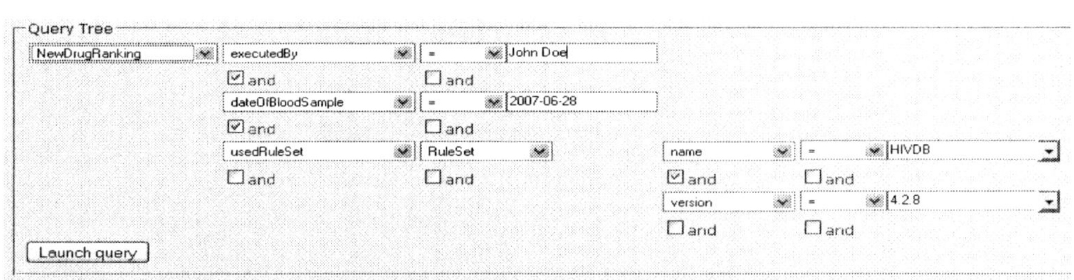

- *Step 6*: Next, the identifiers of the retrieved rule sets are extracted.
- *Steps 7-10*: XQuery requests are issued to retrieve from the experiment information those *NewDrugRanking* individuals which (i) used the rule sets with identifiers extracted in the previous step, (ii) have the date of blood sample 2007 – 06 – 28, and (iii) were executed by *JohnDoe*.

QUERYING CAPABILITIES

Mixed Data - Metadata Queries

The presented provenance model and querying approach enables integration of not only data gathered in multiple data sources, but also of data and metadata, where the latter is describing the process wherein the former has been used. This approach enables not only querying for provenance of experiment results, but also to is-

Figure 4. Query (a) representation, (b) evaluation in QUaTRO

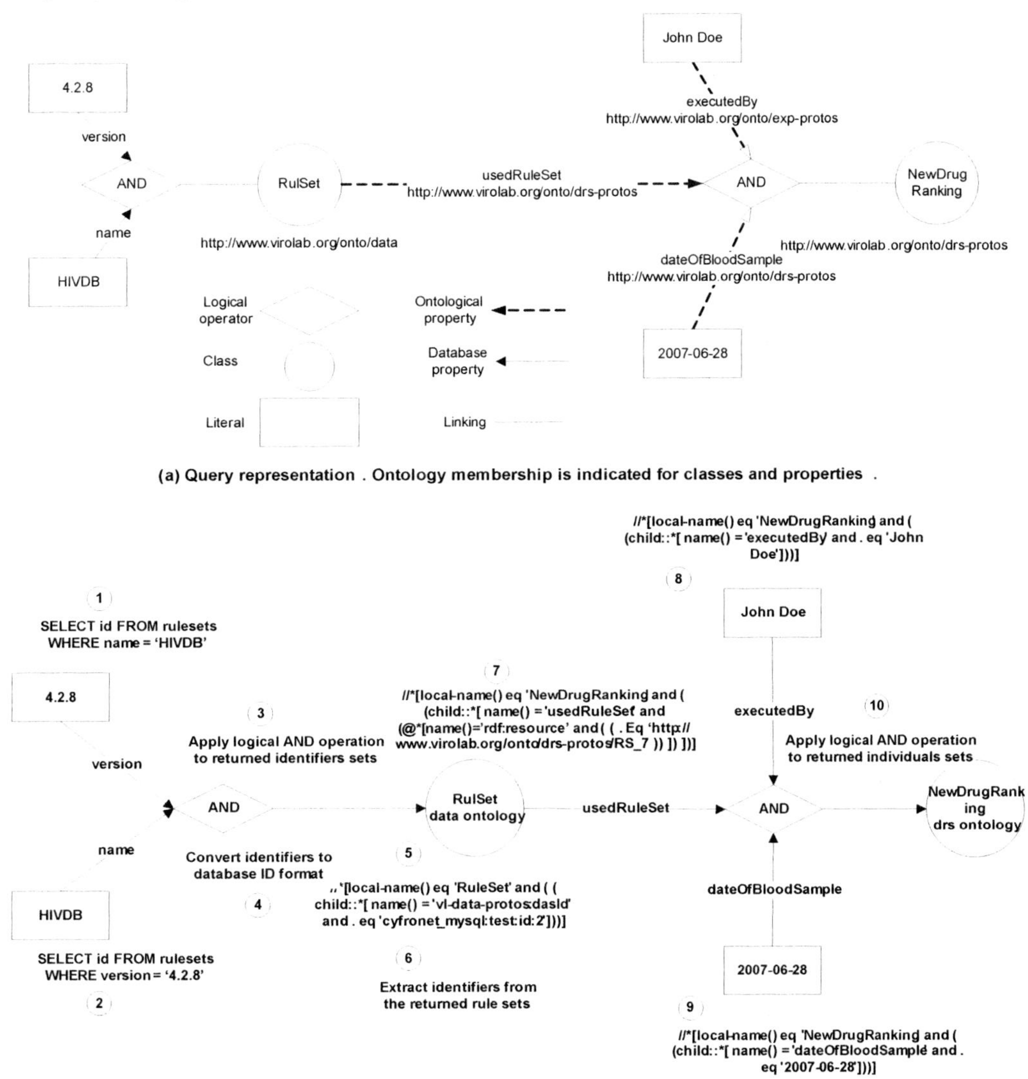

(a) Query representation . Ontology membership is indicated for classes and properties .

(b) Query evaluation .

sue queries such as *Select all experiments of type 'Drug ranking' whose results contained recommendation for 'Drug X'*. Querying provenance alone would enable to obtain all experiments of type 'Drug ranking' which resulted in *some* drug recommendations. However, the actual results containing detailed descriptions of what drugs were recommended, are not part of the experiment execution record, but of actual experiment data which is stored in a database. Only a combined request to the experiment information store and a data repository will allow constructing queries of this kind.

Obviously, the possibility of such queries is dependant on whether the experiment data is stored in a structured way in repositories that allow querying for data schemas. This does not always have to be the case. For example, no data model might exist for experiment results which may be stored as files or unstructured records in simple key-value repositories.

Obtaining Provenance Data Dependency Graphs

Provenance of a data result can be computed thanks to unique identifiers of data items. Furthermore, if the unique identifier is preserved from execution to execution (e.g. a result previously stored is used as an input for another experiment), provenance spanning multiple experiments, not only a single one as in most approaches, can be computed. However, it must be stressed, that this capability is dependent on the capability of the underlying execution and storage services to assign and preserve unique identifiers of the data entities. A provenance query is a recursive one, and is not directly supported by the QUaTRO visual query construction tool. However, a provenance query has a well defined pattern, therefore it has been implemented as a pre-existing query template which recursively builds the provenance graph based on the relationship established by the same identifiers of data items. In addition, we recognize that the data derivation relationship can well be explicitly captured in the provenance model as a semantic relationship connecting two data items. We plan to add this relationship in the future. It would be computed in the recording phase which would enable not only faster queries for provenance, but also browsing the derivation relationships by the end users.

Querying with Partial Semantic Information

Collecting semantics about data and computations used in the experiment may not be trivial. In some cases, only partial semantic information may be available, e.g., one related to data, but not computations or vice versa. Our model is flexible with respect to the availability of semantic information. This is achieved thanks to using *subClass* and *subProperty* OWL relationships whereby generic concepts are connected to domain-specific concepts. Figure 5 presents four possible records of a drug ranking computation: one with full semantics, one with data semantics only, one with computation semantics only, and one with no domain semantics.

In the presence of full semantics (Figure 5a), complex queries are supported including those which refer to domain attributes or data base attributes (as long as data base mappings are available). The user can construct a query directly using domain-specific concepts and relationships, for example, '*NewDrugRankings whose usedRuleset was a Ruleset ...*'.

If only data semantics is available (Figure 5d), the data items are connected to the generic *Computation* individual via generic relationships *inputData* and *outputData*. The user can now query for all '*Computations whose inputData was a Ruleset ...*'.

When only the computation semantics is available (Figure 5c), a domain-specific individual

NewDrugRanking is connected to generic ViroLabDataEntity individuals via either generic relationships, or domain-specific relationships such as *usedRuleSet* or *testedNS*, as long as information required to establish those relationships is available. In this case the user can, for example, query for all '*NewDrugRankings performed by …*'.

Let us note that in the absence of any domain semantic information (Figure 5b), still useful queries are feasible, namely those related to attributes modeled at the generic level. The user can still query, for example, for provenance data dependency graph, or for all computations performed by a certain user.

Reasoning Over the subClass Property

Thanks to using the subClass and subProperty OWL properties, generic queries are possible which are transparently extended to all subclasses of a given class. For example, when the user queries for all computations with certain properties, all existing subclasses of class *Computation* can be implicitly taken into account. As a result, the user will obtain all domain-specific computations which satisfy the query's constraints. At the moment we implement the reasoning related to this capability in our query engine. Namely, we examine the ontology models in search for all

Figure 5. Representation of a drug ranking computation: (a) with full domain semantics, (b) with no domain semantics, (c) with only computation semantics, (d) with only data semantics.

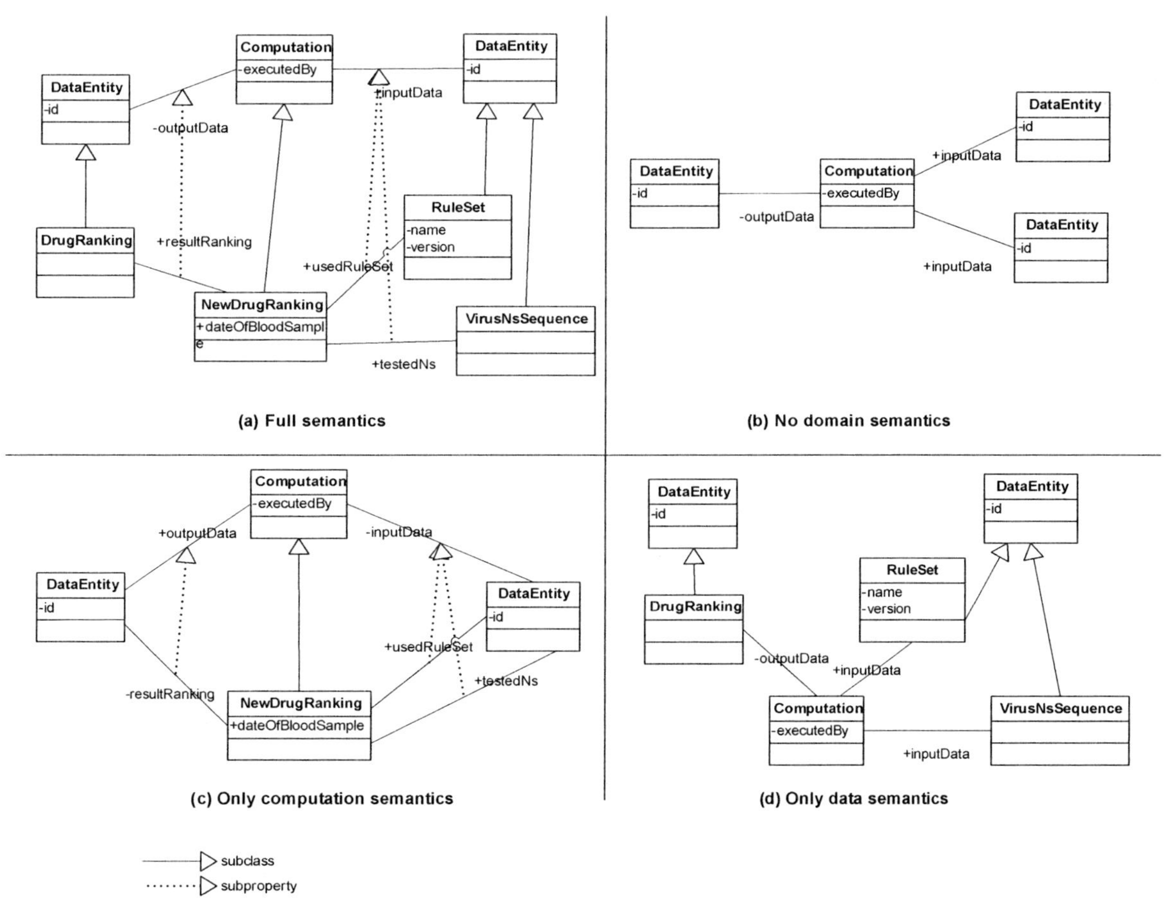

subclasses and we replace the original superclass with a list of those subclasses (a logical alternative in an XQuery expression). An alternative approach would be to use Jena reasoning capabilities or the SPARQL language. However, those approaches are inefficient, as we point out in the next section. In contrast, our approach applies reasoning only at the level of the class model and is thus much more efficient.

IMPLEMENTATION ISSUES

In the current implementation of the querying engine, XQuery is used to query ontology records. However, there are specialized query languages for ontologies, such as SPARQL, which enable, for example, reasoning based on specific relationships, such as transitiveness or subclassing. Despite this fact, it has been decided to implement the current version of the query engine based on the XQuery language. The reasons for this are purely practical. In the current ontology frameworks, taking benefit of full reasoning capabilities is very expensive. For example, in Jena, all ontology models and individuals need to be loaded into memory in order to enable reasoning. However, a system for storing execution records, especially in the Grid context, needs to be capable of scaling to a huge volume of stored records. Also, persistent storage of OWL ontologies with native SPARQL support is immature. Jena supports storing ontologies in a RDBMS, but this should be rather treated as proof of concept, not a production-quality feature. On the other hand, XQuery is becoming standard in XML querying. Finally, our system has not abandoned a typical ontology support. Storage nodes of the storage system are running the Jena framework, and do currently support native SPARQL querying. Because of performance reasons mentioned earlier, it is disabled by default. However, we are planning to migrate to native ontology support as the technologies become mature. A careful separation of the query engine from the underlying query languages makes the system well-prepared for this.

FUTURE TRENDS

Grid and Semantic Web technologies are viewed as two pillars of a future cyber-infrastructure for e-Science (Stein 2008). They are combined in the vision of the Semantic Grid, where seamless access to underlying resources and semantic information integration are key features of the infrastructure. Provenance tracking and querying services – essential for researchers – need to be part of the Grid middleware. On the other hand, an information model for provenance is needed whereby information integration and tool interoperability is enabled, for which Semantic Web technologies are well suited. The presented approach extends this vision by emphasizing end user-oriented visual querying over provenance.

Future work concerns both storage and querying of provenance. We plan to support a distributed storage for provenance repository in order to ensure efficient and scalable querying, essential in a large-scale Grid. We also work on implementation of the query engine on the basis of more ontology-oriented technologies, such as the Jena reasoning engine or the SPARQL query language. Moreover, we plan several enhancements as to visual querying capabilities, such as support of reverse relationships in the ontologies so that many-to-one queries would be possible. The rendering of query results, including the provenance graphs, is also a work in progress.

CONCLUSION

We have presented an approach to tracking and querying provenance and experiment data which enables end-users who are non-IT experts to compose complex queries in a simple way. The ontology models exploited in our approach proved to be an adequate inter-lingua between: (1) end-users who use ontology concepts as a query language while interacting with a graphical user interface; (2) query tools in which ontologies are used to represent and evaluate queries; (3) provenance repository where ontologies serve as the data model.

We have learned that researchers often need to build complex queries to retrieve interesting input data for *in silico* experiments and the queries themselves are important piece of knowledge. End-user oriented querying removes the burden of learning database query languages and relational or XML data models from end-users, thus enabling a wider community to explore data and provenance repositories.

In presenting the end-user oriented approach to querying provenance, we have also demonstrated how the ontology model of executions can contribute to information integration, not only across different data sources but also between data and metadata, where metadata describes the processes wherein the data was used. As has been shown, this integration enables useful querying capabilities, otherwise not possible.

Currently, prototypes of PROToS and QUaTRO are implemented and verified for the ViroLab Drug Resistance scenario.

ACKNOWLEDGMENT

This work was partially funded by the European Commission, Project ViroLab IST-027446, the related Polish grant SPUB-M, and AGH grant 11.11.120.777, and ACC Cyfronet AGH grant 500-08.

REFERENCES

Balis, B., Bubak, M., & Wach, J. (2008). Provenance Tracking in the ViroLab Virtual Laboratory. In *Proceedings of PPAM 2007, Seventh International Conference on Parallel Processing and Applied Mathematics, Lecture Notes in Computer Science*, (Vol. 4967), (pp. 381-390). New York: Springer.

Bowers, S., McPhillips, T. M., Ludascher, B., Cohen, S., & Davidson, S. B. (2006). A Model for User-Oriented Data Provenance in Pipelined Scientific Workflows. In L. Moreau & I. T. Foster, (Eds.), *Provenance and Annotation of Data, International Provenance and Annotation Workshop, IPAW 2006, Lecture Notes in Computer Science*, (Vol. 4145) (pp. 133-147). Chicago, IL: Springer.

Goble, C., Corcho, O., Alper, P. & De Roure, D. (2006). E-science and the semantic web: A symbiotic relationship. In *Discovery Science 2006, Barcelona, Spain, LNAI* (Vol. 4265)(pp. 1-12). New York: Springer.

Goble, C. A., & De Roure, D. (2004). The Semantic Grid: Myth Busting and Bridge Building. In *Proceedings of the 16th Eureopean Conference on Artificial Intelligence, ECAI'2004, including Prestigious Applicants of Intelligent Systems, PAIS 2004, Valencia, Spain, August 22-27, 2004* (pp. 1129-1135). Amsterdam, the Netherlands: IOS Press.

Groth, P., Jiang, S. Miles, S. Munroe, S., Tan, V., Tsasakou, S., & Moreau, L. (2006). *An Architecture for Provenance Systems* (Tech. Rep.). Southampton, UK: University of Southampton..

Gubala, T., Balis, B., Malawski, M., Kasztelnik, M, Nowakowski, P., Assel, M., et al. (2008). ViroLab Virtual Laboratory. In *Proceedings of Cracow Grid Workshop '07, ACC CYFRONET AGH, Krakow.*

Hey, A. J. G., & Trefethen, A. E. (2002). The UK e-Science Core Programme and the Grid. *Future Generation Computer Systems 18*(8), 1017-1031.

L. Moreau, et al. (2007). The first provenance challenge. *Concurrency and Computation: Practice and Experience, 20*(5), 409-418. San Francisco, CA: Wiley InterScience.

De Roure, D., Jennings, N.R., & Shadbolt, N.R. (2005). The Semantic Grid: Past, Present, and Future. In *Proceedings of the IEEE, 93*(3), 669-681.

Kim, J., Deelman, E., Gil, Y., Mehta, G., & Ratnakar, V. (2007). Provenance trails in the Wings/Pegasus system. In *Concurrency and Computation: Practice and Experience*, San Francisco, CA: Wiley InterScience, *20*(5), 587-597.

Simmhan, Y. L., Plale, B., & Gannon, D. (2005). A survey of data provenance in e-science. *SIGMOD Rec. 34* (3), 31-36.

Simmhan, Y. L., & Plale, B., & Gannon, D. (2007). Query capabilities of the Karma provenance framework. In *Concurrency and Computation: Practice and Experience, 20*(5) 441-451. San Francisco, CA:Wiley InterScience

Sloot, P. M., Tirado-Ramos, A., Altintas, I., Bubak, M., & Boucher, C. (2006). From Molecule to Man: Decision Support in Individualized E-Health. *Computer, 39*(11), 40–46.

Stein, L. D. (2008). Towards a cyber-infrastructure for the biological sciences: progress, visions and challenges. *Nature Rev. Genet., 9*, 678-688.

Stevens, R., Goble, C., Paton, N.W., Bechhofer, S., Ng, G., Baker, P., and Brass A. (2003. May). Complex Query Formulation Over Diverse Information Sources in TAMBIS. In Z. Lacroix & T. Critchlow, (Eds.), *Bioinformatics: Managing Scientific Data*. San Fransisco, CA : Morgan Kaufmann.

Tan, W. C. (2007). Provenance in Databases: Past, Current, and Future. In *IEEE Data Eng. Bull. 30*(4),3-12.

Wache, H., Vogele, T., Visser, U., Stuckenschmidt, H., Schuster, G., Neumann, H., & Hiibner, S., (2001). Ontology-based integration of information--a survey of existing approaches. In *Proceedings of the International Workshop on Ontologies and Information Sharing*, pp. 108-117.

Zhao, J., Wroe, C., Goble, C. A., Stevens, R., Quan, D., Greenwood, R. M. (2004). Using Semantic Web Technologies for Representing E-science Provenance. In *International Semantic Web Conference, LNCS* (Vol. 3298), (pp. 92–106). New York: Springer.

Zhao, J., Goble, C., Stevens, R., & Turi, D. (2007). Mining Taverna's Semantic Web of Provenance. In *Concurrency and Computation: Practice and Experience, 20*(5), 463-472.

Zhao, Y., Wilde, M., & Foster, I. T. (2006). Applying the virtual data provenance model. In L. Moreau and I. T. Foster (Eds.) *Provenance and Annotation of Data, International Provenance and Annotation Workshop, IPAW 2006, Chicago, IL, USA, May 3-5, 2006, Revised Selected Papers, Lecture Notes in Computer Science,* (Vol. 4145) (pp. 148–161). New York: Springer.

KEY TERMS AND DEFINITIONS

End-user oriented querying is a query construction methodology which enables non-IT experts to construct powerful queries to repositories of experiment data and provenance using domain-specific terms instead of those of underlying data models and query languages.

e-Science is a new type of scientific research based on the collaboration within a number of scientific areas, enabled by a next generation infrastructure, wherein people, computing resources, data and instruments are brought together to bring a new quality to the everyday work of researchers.

Experiment is a process that combines together data with a set of activities that act on that data in order to yield experiment results.

Experiment plan is a *recipe* that describes the process of certain experiment execution in the environment of the virtual laboratory.

Ontology is a specification of a conceptualization which provides a shared vocabulary for a given domain, as well as relationships between and constraints imposed upon concepts.

Provenance of a piece of data is metadata which describes the derivation history of this piece of data (Simmhan 2005). It can be expressed either as a process that led to that piece of data or its data dependency graph. Provenance has several important uses in e-Science including estimation of quality and reliability of a scientific result, a means to audit a piece of data, or a prerequisite to repeat an experiment.

Semantic Grid is and extension of the Grid "in which information and services are given well-defined meaning, better enabling computers and people to work in cooperation" (Goble 2004).

Virtual laboratory is a set of integrated components that, used together, form a distributed and collaborative space for science.

ENDNOTES

1. ViroLab Project: www.virolab.org
2. ViroLab Virtual Laboratory: virolab.cyfronet.pl

Chapter V
Data Protection and Data Security Regarding Grid Computing in Biomedical Research

Yassene Mohammed
Georg-August-University, Germany

Fred Viezens
Georg-August-University, Germany

Frank Dickmann
Georg-August-University, Germany

Jürgen Falkner
Fraunhofer Institute for Industrial Engineering IAO, Germany

Thomas Lingner
Georg-August-University, Germany

Dagmar Krefting
University Medicine Berlin, Germany

Ulrich Sax
Georg-August-University, Germany

ABSTRACT

This chapter describes security and privacy issues within the scope of biomedical Grid Computing. Grid Computing is of rising interest for life sciences (Konagaya, 2006) and has been used since many years in sciences like high energy physics. Anyhow, medical applications on the grid require a special focus on data security and data protection issues. Based on general security and privacy rules, the authors describe the current state of the art of grid security. Then they describe which additional security measures have to be established in different biomedical grid scenarios. Legal aspects have to be taken into account as well as the current possibilities and flaws of grid security technology. Describing the enhanced security concept in MediGRID (MediGRID, 2005) they outline how medical Grid Computing could fulfill privacy regulations used in more demanding environments.

INTRODUCTION

Following the power grid paradigm, Grid computing implies, that a grid user has easy and transparent access to compute power, storage resources, data, and applications (I. Foster & Kesselmann, 1999). In practice there are still some hurdles in doing so. Nevertheless Grid Computing will become a key infrastructural component on the way to a broadly applied computational medicine (EU-SHARE, 2006; Iakovidis, 2005). In this context HealthGrid contains three aspects: the computing aspect for highly intense calculation tasks, the data aspect for distributed access to a large amount of data and the knowledge aspect for a sophisticated utilization of data (Figure 1).

Within biomedical Grid projects, a grid infrastructure for medical and biomedical research has been established mainly using Computing Grid resources. It enhances interdisciplinary and widely location-independent collaboration of researchers. The interoperability between different medical and biomedical grids provides a broader range of applications and datasets for the biomedical applications and provides opportunities for new collaborations (Montagnat et al., 2008).

Concerning security matters of medical and biomedical applications, different types of grid usage need different security levels according to legal and ethical (Iavindrasana et al., 2008)

Figure 1. Three aspects of grid-functions in healthcare (adapted from Iakovidis, 2005)

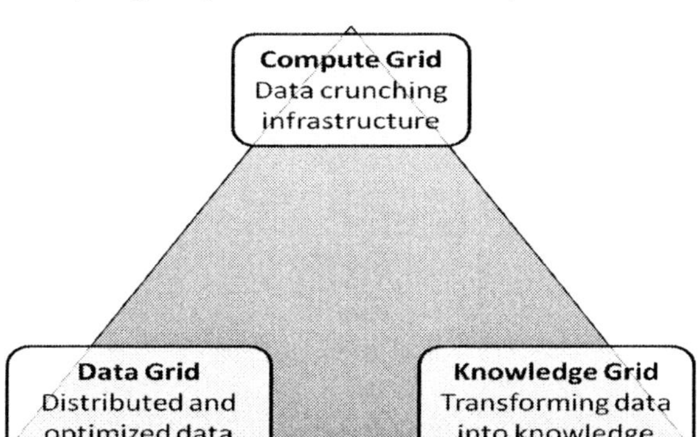

requirements as well as Good Clinical Practice GCP (Mohammed, 2006). These requirements include privacy as well as intellectual property issues. Whereas privacy addresses confidentiality of identifying data of patients and intellectual property is directed to protect clinical and research information from unauthorized access. Trust in privacy upholds patients trust in health services. Without trust patients might avoid healthcare services which can lead to severe problems, for the patient but nonetheless for the whole healthcare system of a society. Also beyond doubt scientists and healthcare professionals need to have full confidence in the offered grid infrastructure in order to store their precious data onto the Data Grid. Thus technical implementations must ensure a reliable confidentiality of any information transfer and storage. (Garcia, Dikaiakos, Kyprianou, Bilas, & Marazakis, 2008)

While Computing Grid applications have to decide whether a user may use one resource or not, Data Grid applications can only be operated securely with fine grained access control with respect to data elements. Security solutions for Knowledge Grids still have to be outlined, as liability requirements have to be met additionally.

Regarding access control for the Computing Grid scenario, most authentication and authorizations challenges are solved due to the usage of grid proxy certificates and virtual organization solutions (Pommerening, 2006). Fine grained access control solutions - the right to access one document within a data store, or even accessing different sections within the same document (e.g. medical structured documents) - are still under development (Mohammed, Sax, Viezens, & Rienhoff, 2007; Schiffers, 2008; Witzig, 2008) .

For Data Grid applications in patient care and medical research the usage of international standardized ontologies, terminologies and nomenclatures is essential (Blanquer, Hernandez, & Segrelles, 2006). The problem of how to establish data repositories using standardized terminologies could not be solved by the grid projects yet, although for example caBIG and caGRID (Oster et al., 2007; Welch et al., 2003) try to tackle those issues. Knowledge Grid applications are not to be found yet in biomedical Grid Computing (Konagaya, 2006).

BACKGROUND

General Requirements in Distributed Systems

Information systems (compute, networks and application systems, thus hard- and software), which have a use for the real world, must be reliable. That means that they must fulfill the following four basic requirements:

- **Accessibility:** In particular information must be available to the right time at the right place and prepared to secure the systems in case of error situations.
- **Authenticity:** In particular, is the identity of persons and processes/services free of doubts?
- **Accountability:** In particular the responsibility for critical actions
- **Privacy:** Secrets should be protected and access rights to private information should be regulated.

In a distributed system persons communicate with objects (services, processes, partial jobs) and objects communicate among them. Therefore each data access or communication process between two entities requires the entities to mutually identify, authenticate and authorize themselves:

- **Identification:** Who are you?
- **Authentication:** Prove it! (e.g. with a second factor, a token or a directory entry)
- **Authorization:** Are you allowed to perform this action?

Reliability of information security in complex distributed systems such as a grid will only be achieved, if those issues are addressed with professional IT security know-how and thorough safety analysis. This includes a careful design of applications and use of strong mechanisms at all levels. This is especially applicable in proving the interaction of the system components and the interaction of various mechanisms of possible vulnerabilities.

Special Requirements in Biomedical Applications

Unlike common Grids, data processing in the biomedical community has to face particular challenges. Beyond the problem of finding the relevant data sets via metadata description using the adequate searching tools, access control enforcement is of paramount importance, as the owners of the data are foremost patients.

Due to the heterogeneity of the data we need an ontology process to homogenize the data. In contrast to data in classical grid applications, high dimensional data are typical for medical applications. Given semantic interoperability the researcher can correlate and analyze the data using suitable biomedical informatics methods. Biomedical data are not only heterogeneous; they contain numerous types of information and have different levels of privacy. They vary from aggregated data describing population and diseases (epidemiology, clinical practice, clinical trials), going more granular to patient data and pathological descriptions (health record, clinical history, physical exams), ending with cellular and molecular data (histology, genetic test results and genomic data) (Butte & Kohane, 2006). Having these data online with the suitable tools to correlate, merge and analyze creates new challenges for data protection and data security. Data Protection and Data Security are special issues in clinical research projects and must be explicitly considered. Implementation of enhanced security measures is absolutely necessary. Yet no middleware would approximately fulfill the GCP (Switula, 2000) nor FDA (FDA, 2004) regulations concerning clinical studies. Those requirements do not only apply to the biomedical community, other communities like engineering deal with similar privacy problems concerning intellectual property rights.

Maybe the most advanced usage of Grid Computing in Biomedicine can be found in the pharmaceutical industry. The pharmaceutical industry is a data and knowledge-driven industry. The number of high performance computing applications in pharmacy is rather limited (when compared to engineering simulation applications in e.g. automotive industry). Thus, Grid-enabled solutions for the pharmaceutical industry do not necessarily have to deal with number crunching, although the most prominent application of grid technology in pharmacy is still the structure-based virtual screening application as published by Novartis or as done in the course of the WISDOM project. However, complex information extraction processing workflows ("text mining") in scientific literature including patents seem to open a new field for grid-enabled problem solving in pharmacy including new security challenges concerning intellectual property rights.

Example: MediGRID

The main goal of MediGRID (MediGRID, 2005) was to set up generic grid application secenarios on life sciences. Those scenarios are as heterogeneous as the life science community itself. Therefore we distinguish several application classes with different security requirements (Falkner & Sax, 2008; Krefting et al., 2008). The project launched with applications of low or no security requirements. For these cases, security issues like identification and authorization are mainly determined by the demands of the resource providers. Table 1 shows the identified classes of processed data and their use and users regarding security requirements.

In the further progress of MediGRID the privacy concepts grew more complex due the usage of human data. Every computational medical application involving any human data have to meet regulatory requirements, encompassing data protection, data safety and reliability. The principles of confidentiality and privacy have to

Table 1. Security levels and application classes in biomedical grid environments (Krefting et al., 2008)

Processed data	Security Level	Application Classes	User
Non-human data	low	- basic research - knowledge bases - demo versions	- researcher - all - all
Anonymized human data, no risk of re-identification	low	- basic research - clinical research - demo versions	- researcher - res./physician - all
Anonymized human data with risk of re-	medium	- basic research - clinical research	- researcher - res./physician
Pseudonymized human data	medium or high	- clinical research - clinical application	- res./physician - physician
Patient data	high	- clinical application - telemedicine	- physician - physician/ patient

be respected at all times within the grid processing. Fine grained access control with personalized authentication and authorization is required. The patient - as owner of his data - has the right to be informed why, where and how long his data is processed and stored. Therefore, medical grid applications must be equipped with a comprehensible audit track in order to fulfill this requirement (a-posteriori). Furthermore it has to guarantee the patient, that his data will only be stored and processed in a trustworthy environment (Tracking, a-priori). This is a challenge in Grid Computing, as every grid node has to be assessed concerning the trustworthiness using trust metrics. Current grid middleware cannot fulfill all these requirements, as standard security methods do not scale in heterogeneous, distributed environments.

The MediGRID project is based on the core D-Grid (D-Grid, 2008) infrastructure: The different community grids can choose between Globus Toolkit (Ian Foster, 2006), gLite and Unicore as basic grid middleware. The D-Grid supports several technologies on top of the middleware stack, such as OGSA-DAI for distributed database access, dCache for distributed data management, the GridSphere Portal Framework for the setup of grid portals, the Grid Resource Registry Service (GRRS) and the virtual organization (VO) Membership Registration Service (VOMRS) for resource and user management, respectively. These technologies also include grid-wide monitoring services and (so far) low level accounting. It also provides concepts for authentication and authorization as well as for the setup and management of firewall rules. D-Grid supports a public key infrastructure (PKI). MediGRID focuses on fine-grained user management, using the provided VO and SubVO structure, and the development of strictly portal-based graphical user-interfaces. Access is secured by using X.509 certificates instead of username and password, but SRB has a separate user management and is not connected to Globus mapfiles by default.

PKI based authentication and authorization provides all legal features for fully secured grid usage on a basic level. Therefore D-Grid PKI and VOMRS infrastructure is used to register to MediGRID. The user needs to request a PKI certificate at a trusted certification authority (CA). The next step is the login at the VOMRS. In case the validity check of the user certificate is passed, the user is identified by the system and can request membership in different virtual organizations, among them MediGRID. In MediGRID, the facility of a more fine-grained differentiation into sub-VOs (so called groups) is enabled by the user management for a simple modeling of roles as a first step towards role based access control within the grid. During the registration process the applicant has to accept the grid usage policies of the respective VO, so a certain legal basis for the provision and use of grid resources is given. As the information from the VOMRS can be retrieved by trusted services, all grid resources are kept up-to-date with respect to the user registrations; and the necessary local accounts, role mappings and authorization rules are implemented automatically. Within the MediGRID project, the GridSphere Portal Framework has been extended in terms of user management functionality by linking it to the VOMRS. Furthermore, an extension for certificate-based login reduces the portal login and registration effort and lowers usage barriers. Once the user possesses this grid certificate he or she will face the second large barrier on the way to use grid applications, services and resources. It is possible for the user to log on to the portal, as the primary user interface for MediGRID, but this does not mean that grid applications, services and resources can be instantly used. For authentication and authorization between the middleware components, they need access to a complete certificate pair (public and private key) of the user, which is usually solved by issuing a grid credential, generated from intermediary proxy certificates stored on the MyProxy server.

Grid Computing Authorization Technologies

In order to analyze the possibilities to fulfill the privacy needs we examined the Grid Security Infrastructure (GSI) in the Globus Toolkit version 4 (GT4), which could be considered as the standard grid middleware for biomedical research. The security tools in GT4 (The Globus Security Team, 2005) deals with authentication, communication security, authorization, and supporting functions for managing user credentials and maintaining group membership information.

The Grid Security Infrastructure builds the core for security in the GT4 middleware. The use of this security infrastructure in combination with the job submission, the data management, and the execution management tools contrives the secure Grid infrastructure. Whereas these tools are part of GT4 or 3rd party GT4-plugins like Storage Resource Broker (SRB) (Rajasekar et al., 2003.) – a Data Grid management system – and Data Access and Integration Services (OGSA-DAI) (Antonioletti et al., 2005; Karasavvas et al., 2005; *Open Grid Services Architecture - Data Access and Integration Services*).

Table 2 summarizes some of the efforts regarding authorization in the Grid Computing environment.

Table 2. Grid Computing authorization technologies. Mostly uses we find the grid-mapfile- a file-based Access Control List (ACL) mechanism. It maps a certificate subject to a local user's login name, authorizing them to access and use GSI-enabled services. It is up to this administrator to verify that the certificate subject belongs to the authorized user and is mapped to the appropriate local user ID. All executed tasks by this user are running with this local ID (Alfieri et al., 2004; D. Chadwick & Otenko, 2003; D. W. Chadwick & Otenko, 2002; Chakrabarti, 2007; Pearlman, 2002; Seitz, Montagnat, Pierson, Oriol, & Lingrand, 2005; The Globus Security Team, 2005).

	Project	Middleware	Authorization
grid-mapfile	Globus Alliance	In all GT versions	- Access Control List (ACL)
GSI	Globus Alliance	In WS GT (GT4)	- Implement SAML and grid-mapfile (as Web Services)
GridShib	Globus Alliance	Plug-in for GT4	- Federated authorization infrastructure (Shibboleth) with Grid technology (GT4). - Use SAML and X.509
CAS	Globus Alliance	In WS GT (GT4)	- Implement GSI and SAML - VO based: server stores all authorizations concerning the VO members
VOMS	EGEE	gLite(GT2.4)	- Implement GSI and can use SAML - Attribute Certificates in newer VOMS versions - VO based: server only stores the group/role memberships, where the local resources interface their access control lists to these groups/roles
Sygn	MEDIGRID (France)	μGrid	- Implement own "Authorization Certificates" - No standard / Proof of concept
PERMIS	EC project independent from Grid activities.	Independent Plug-in for GT4 and other projects	- Implement x.509 Attribute Certificate - Implement Privilege Management Infrastructure (PMI) parallel to the Public Key Infrastructure (PKI)

Shortcomings of Current Grid Middlewares

Beyond anonymization and pseudonymization, which are to be accomplished before uploading sensitive data into the grid, the current grid middleware technologies fulfill the security needs to a good degree. Nevertheless, we still need to enforce who did what when and why, namely to permit accountability on the grid in order to fulfill the legal requirements. Our analysis shows the need for supplemental "Enhanced Security" elements for a Healthgrid:

1. **Auditing (a posteriori):** An audit trail consisting of log files and activity protocols. This requirement is one of the FDA regulations for clinical trials which cannot be fulfilled currently within a grid environment.
2. **Trackability (a priori):** Trackability requires knowledge about where transfers, transactions, calculations and storage of person related data take place. Auditing and tracking possibilities cover the accountability requirement. This function is partially covered by some grid workflow management systems, but not part of the middleware services.
3. **Access control:** Fine grained access control with respect to access rights within medical documents and within medical databases are needed. Current access control on file level (e.g. grid-mapfiles (Oster et al., 2007)) do not suffice, as structured medical documents(Dolin et al., 2006; HL7, 2005) provide different sections with a different degree of confidentiality. Fine grained, role based access control fulfill the requirements of releasing only necessary data and retain the doctor-patient confidentiality (see Figure 2).
4. **Trust and trust-delegation:** Trust relations and delegation as well as trust hierarchies have to be set up electronically. Using the data of a minor or a person with dementia requires that an authorized person signs electronically on behalf of those persons (eConsent). These workflows are described

Figure 2. the lack of a mechanism to transfer the security policy information from the Policy Information Point (PIP) to the resources side and enforce it there by connecting this information correspondingly to the resources Policy Decision Point (PDP) and Policy Enforcing Point (PEP). Currently this is solved inadequately by manual mapping being carried out by the resources administrators.

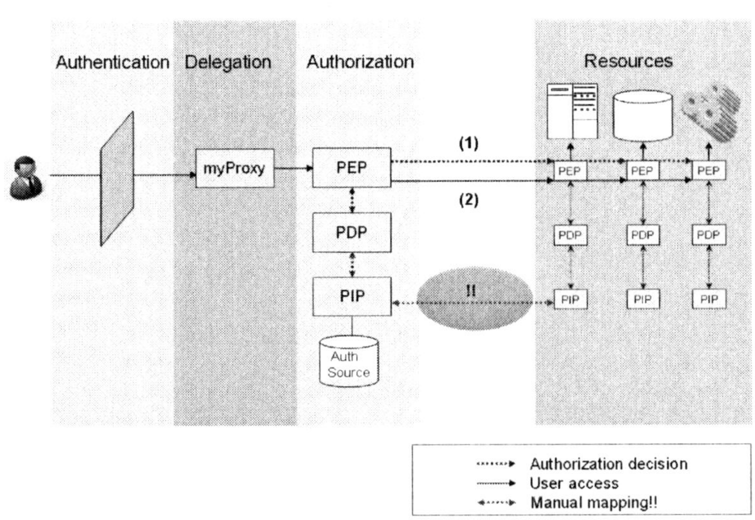

in some projects (Bergmann, Bott, Hoffmann, & Pretschner, 2005; TMF, 2006), but have to be adapted for grid usage.

5. **Safety:** Security of data in dynamic grid environments requires policies for data storage and policies for data management. Safety reflects the need to develop and adopt suitable_policies for the use and storage of data; a complementary safeguard principle when intending to use sensitive data considering the accessibility in time (long term archiving) and place (replicas).
6. **Validation of the service components:** Each service in the grid makes use of different software components. Part of these components is delivered with the middleware. The other part is developed for a community specific application or goal. The problem is that we still do not have a validation procedure neither for grid middlewares nor for grid-based applications. Compared with operating systems, there is no similar validation for grid middlewares like Common Criteria (United States. Dept. of Defense., 1999). Nevertheless, the middlewares themselves are still under intensive development, which means it is difficult to obtain a long term stable release in order to be validated.

ADVANCED DATA SECURITY AND DATA PROTECTION TECHNIQUES

The Need of "Privacy Enhanced Security" (PET)

The use of information and communication technologies (ICT) for the sake of the protection of human privacy is widely known as the privacy-enhancing technologies (PET). PETs shall protect privacy by eliminating or reducing personal data or by preventing unnecessary or undesired processing of personal data, all without losing the functionality of the data system (Borking, 2005). John Broking was the first to coin the term PET when he developed the Identity Protector (Hes & Borking, 1998). Nevertheless, the PET technologies exist since the 1980s (Federrath, 2005).

In their revised report (Hes & Borking, 1998), Hes and Borking argue that "It is only the application of privacy-enhancing technologies that is lacking, not the technologies it self." PETs are primarily developed for the exchange of information. In this context, in Grid Computing it is not only exchange of information, it is also computing power, storage capacity and application sharing.

PETs are indeed helpful, but they do not meet all the requirements for a secure Grid environment. PET techniques are more concerned about de-identification, which is not sufficient for data security and data privacy in the grid. Especially when it comes to sensitive data (patient data), which is the case in medical applications. In medical applications, the security and data protection technologies in the grid should not meet only the legal framework, they should also fulfill the Good Clinical Practice requirements. In this context and as the Grid is a distributed environment, all new regulatory and safety requirements should be taken into account.

Legislation and Realisation

A legal framework for the protection, security and transport of personal data as well as patient data is introduced in different EU directives (95/46/EC processing of personal data; 97/66/EC protection of privacy in the telecommunications sector; 99/93/EC a framework for electronic signatures; 2002/58/EC privacy and electronic communications). Country specific implementations vary among the EU countries.

The legal framework implies special requirements regarding data security and data protection in most grid middlewares: (1) Confidentiality (2) Integrity and authenticity, (3) Accessibility, and (4) Accountability.

Additional data protection requirements arise dealing with person related data especially in the healthcare sector:

- **Data necessity principle:** Disclose all person related data of a patient, but not more than the necessary data for the treatment.
- **Context of treatment:** Person related data of a patient should be disclosed only to the personnel participating in his treatment.
- **Patient consent:** The patient should formally agree on the handling of his person related data
- **The guarantee of patient rights:** The possibility of rectification, blocking, deletion of his personal data should be offered.

Needed Legal Actions for the Future

The elements of the Enhanced Security consider the current requirements of data protection and data security intending to make grid technologies better suitable for the biomedicine community. In the future, Enhanced Security also should be flexible to fulfill future legal requirements and new developments in the medical area, e.g. genome wide association studies (see Figure 3).

The very promising aspects of virtualization and virtual organizations were being discussed since the mid-1990s as a technical solution. But until now there is no legal framework for virtualization. Therefore, it is necessary to study data protection aspects in virtual environments and to develop generic solutions for it. The current technology should be conformed according to the current legal framework (Allaert, 2002; Herveg & Poullet, 2003). Anyhow, as long as there is no clear legal definitions of virtualization concepts, it remains difficult, to conform the grid technology to a legal framework, as virtualization is a very important concept in Grid Computing. A kind converging between the legal and the technical developments regarding data protection and data security is necessary.

FUTURE TRENDS

Grid technology has not reached healthcare yet. The reasons for this situation reach from immature and non-reliable technology to quality of service and privacy concerns.

For example the development of most grid projects is being performed on open research networks, whereas the healthcare sector usually

Figure 3. The limits of using Grid Computing for medical applications. ID: identifying data, PSD: pseudonymized data. The re-identification of the object (patient) becomes easier with additional data types. The legal framework allows the use of Grid Computing only with non-identifying data as well as pseudonymized data (Mohammed et al., 2007; Sax, Mohammed, Viezens, & Rienhoff, 2006).

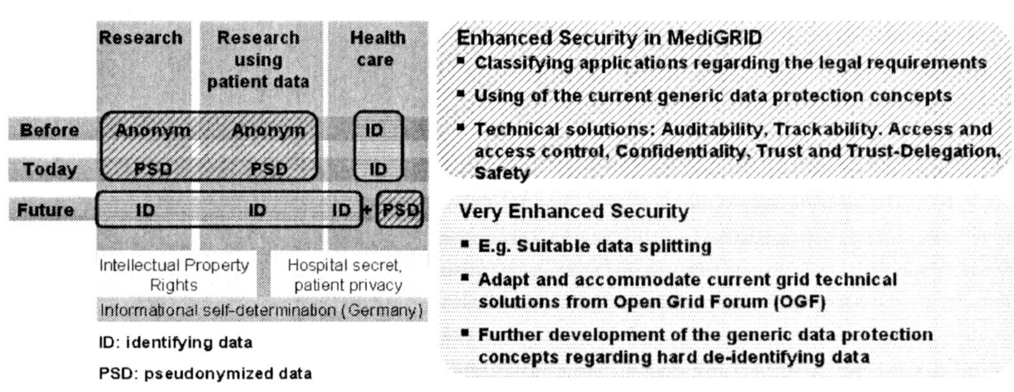

works on secure, private networks for clinical routine use. A further potential obstacle is the legal framework in the EC member states, which has to evolve to allow the transfer of medical data within a holistic European Healthgrid.

Finally we have to take into account that grid technology needs to be disseminated and accepted by the potential users in order to successfully set up business cases that allow a sustainable operation of this infrastructure in the healthcare environment. Assessing dissemination, acceptance and usage of grid technology in life sciences the potential users can be clustered in the application classes (1) Computing Grid, (2) Data Grid and (3) Knowledge Grid (Iakovidis, 2005; Konagaya, 2006).

Most current Healthgrid applications are using Computing Grids (for instance: GATE radiotherapy (De Beenhouwer, Staelens, Vandenberghe, & Lemahieu, 2008) on EGEE grid, OpenMolGrid (Maran et al., 2007) on Unicore, Globus MEDICUS (Erberich et al., 2007) on Globus). Having a look at bioinformatics for examples we find some Data Grid application (de la Calle, Maojo, & Benito, 2006). So future Healthgrid projects (Freund et al., 2006; Tsiknakis et al., 2006) will mainly address Knowledge Grid applications – which are not to be found yet (Legré, Y., Viezens, Rienhoff, & Sax, 2008).

CONCLUSION

Mostly, all current Healthgrid projects are facing similar challenges concerning standardization, reliability, user friendliness, security, acceptance and finally sustainability. The technology seems to be maturing, nevertheless solutions addressing privacy and intellectual property are not addressed yet or still under construction. This is be reflected by the fact, that most Healthgrid projects run in an R&D environment, Grid Computing has not reached healthcare yet. In the expert's opinion, the timeframe for real world applications depends on factors like (1) Interoperability, (2) legislation and security, (3) acceptance and (4) sustainable cost models. Therefore, the most urgent security problems should be solved in simple projects in Computing and Data Grids involving the corresponding community. Due to semantic and interoperability problems, Knowledge Grids need more time, and in turn the security problems in such grids are to be discovered, analyzed and solved later. Nevertheless, this does not mean that we should not being preparing ourselves for them (Very Enhanced Security, see Figure 3).

ACKNOWLEDGMENT

This work was supported by the e-Science research program of the German Federal Ministry of Education and Research (Bundesministerium für Bildung und Forschung (BMBF), 2008), promotion number: 01AK803A-H and (BMBF, 2008, Services@MediGRID), promotion number: 01IG07015A

REFERENCES

Alfieri, R., Cecchini, R., Ciaschini, V., dell'Agnello, L., Frohner, A., Gianoli, A., et al. (2004). VOMS, an Authorization System for Virtual Organizations. In F. Fernández Rivera (Ed.), *Grid computing: first European Across Grids Conference, Santiago de Compostela, Spain, February 13-14, 2003 : revised papers. Lecture notes in computer science, 2970* (pp. xi, 328 p.). New York: Springer.

Allaert, F. A. (2002). *Security standards for healthcare information systems: a perspective from the EU ISIS MEDSEC Project.* Amsterdam, the Netherlands: IOS Press.

Antonioletti, M., Atkinson, M., Baxter, R., Borley, A., Hong, N. P. C., Collins, B., et al. (2005). The design and implementation of Grid database services in OGSA-DAI. *Concurrency and Computation-Practice & Experience, 17*(2-4), 357-376.

Bergmann, J., Bott, O. J., Hoffmann, I., & Pretschner, D. P. (2005). An eConsent-based System Architecture Supporting Cooperation in Integrated Healthcare Networks. *Stud Health Technol Inform, 116*, 961-966.

Blanquer, I., Hernandez, V., & Segrelles, D. (2006). TRENCADIS – A Grid Architecture for Creating Virtual Repositories of DICOM Objects in an OGSA-Based Ontological Framework. In *Biological and Medical Data Analysis* (pp. 183-194).

Borking, J. J. (2005, October 3, 2005). *Privacy Standards for Trust.* Paper presented at the 27th International Conference on Privacy and Personal Data Protection, Montreux, Switzerland.

Bundesministerium für Bildung und Forschung (BMBF). (2008). *3. D-Grid call: BMBF-Förderbekanntmachung „Grid-Dienste für Wirtschaft und Wissenschaft"*. Retrieved May 26, 2008, from http://www.bmbf.de/foerderungen/12378.php

Butte, A. J., & Kohane, I. S. (2006). Creation and implications of a phenome-genome network. *Nat Biotechnol, 24*(1), 55-62.

Chadwick, D., & Otenko, O. (2003). *A Comparison of the Akenti and PERMIS Authorization Infrastructures in Ensuring Security in IT Infrastructures.* Paper presented at the Proceedings of the ITI First International Conference on Information and Communications Technology (ICICT 2003) Cairo University, Egypt.

Chadwick, D. W., & Otenko, A. (2002). The PERMIS X.509 role based privilege management infrastructure. *Proceedings of the seventh ACM symposium on Access control models and technologies.* Monterey, CA: ACM.

Chakrabarti, A. (2007). *Grid computing security.* New York: Springer.

D-Grid. (2008). *D-Grid.* Retrieved July 21, 2008, from www.d-grid.de

De Beenhouwer, J., Staelens, S., Vandenberghe, S., & Lemahieu, I. (2008). Acceleration of GATE SPECT simulations. *Med Phys, 35*(4), 1476-1485.

de la Calle, G., Maojo, V., & Benito, M. (2006). The INFOBIOMED Network of Excellence: facilitating training and mobility in biomedical informatics in Europe. *Stud Health Technol Inform, 124*, 893-898.

Dolin, R. H., Alschuler, L., Boyer, S., Beebe, C., Behlen, F. M., Biron, P. V., et al. (2006). HL7 Clinical Document Architecture, Release 2. *J Am Med Inform Assoc, 13*(1), 30-39.

Erberich, S. G., Silverstein, J. C., Chervenak, A., Schuler, R., Nelson, M. D., & Kesselman, C. (2007). Globus MEDICUS - federation of DICOM medical imaging devices into healthcare Grids. *Stud Health Technol Inform, 126*, 269-278.

EU-SHARE. (2006). *SHARE - Roadmap for a European Health Grid.* Retrieved 28.06.2006, from http://www.eu-share.org/

Falkner, J., & Sax, U. (2008). *Service Certificates and Portal Security - Longterm requirements towards security mechanisms (German).* Paper presented at the D-Grid Security Workshop 2008, Göttingen.

FDA. (2004, April 1). *FDA 21 CFR Part 11.* Retrieved September 21, 2005, from http://www.accessdata.fda.gov/scripts/cdrh/cfdocs/cfcfr/CFRsearch.cfm?CFRPart=11

Federrath, H. (2005, 22-26 August). *Privacy Enhanced Technologies: Methods – Markets – Misuse* Paper presented at the Trust, Privacy and Security in Digital Business Copenhagen, Denmark.

Foster, I. (2006). *Globus Toolkit Version 4: Software for Service-Oriented Systems.* Paper presented at the IFIP International Conference on Network and Parallel Computing.

Foster, I., & Kesselmann, C. (1999). *The Grid: Blueprint for a New Computing Infrastructure*. San Francisco, CA: Morgan Kaufmann Publishers Inc.

Freund, J., Comaniciu, D., Ioannis, Y., Liu, P., McClatchey, R., Morley-Fletcher, E., et al. (2006). Health-e-child: an integrated biomedical platform for grid-based paediatric applications. *Stud Health Technol Inform, 120*, 259-270.

Garcia, J. L., Dikaiakos, M. D., Kyprianou, T., Bilas, A., & Marazakis, M. (2008). Data Privacy considerations in Intesive Care Grids. *HealthGrid Conference 2008*. Chicago, IL.

Herveg, J. A. M., & Poullet, Y. (2003,January 16-172003). *Directive 95/46 and the use of GRID technologies in the heathcare sector: selected legal issues*. Paper presented at the Healthgrid 2003, Lyon, France.

Hes, R., & Borking, J. J. (1998). *Privacy-enhancing technologies: the path to anonymity* (Rev. Rd.). The Hague, the Netherlands: Registratiekamer.

HL7. (2005). HL7 Receives ANSI Approval of Three Version 3 Specifications Including CDA, Release 2.

Iakovidis, I. (2005). *HealthGrid – 3 sided concept: ICT for Health, ISTAG WG*. Retrieved August 30, 2007, from http://www.who.int/classifications/terminology/iakovidis.pdf

Iavindrasana, J., Loiacono, L., Müller, H., Periz, I., Summers, P., & Wright, J. (2008). Access to Clinical Information Systems for Research in the Life Sciences - Security and Privacy Considerations, *HealthGrid Conference 2008*. Chicago, IL.

Karasavvas, K., Antonioletti, M., Atkinson, M., Hong, N. C., Sugden, T., Hume, A., et al. (2005). Introduction to OGSA-DAI Services. In *Scientific Applications of Grid Computing, First International Workshop* (Vol. 3458)(pp. 1-12). Beijing: Springer Berlin / Heidelberg.

Konagaya, A. (2006). Trends in life science grid: from computing grid to knowledge grid. *BMC Bioinformatics, 7 Suppl 5*, S10.

Krefting, D., Bart, J., Beronov, K., Falkner, J., Hartung, M., Hoheisel, A., et al. (2008). MediGRID: A Userfriendly Secured Grid Infrastructure. *Future Generation Computer Systems Journal - International Journal of Grid Computing*.

Legré, Y., Y., M., Viezens, F., Rienhoff, O., & Sax, U. (2008). HealthGRID/ SHARE: Grids for Health - International Interoperability. In *eHealth Conference 2007 - Experts' Special Interest Sessions* (Vol. 58) (pp. 81-99). Bonn, Germany: nanos Verlag oHG.

Maran, U., Sild, S., Mazzatorta, P., Casalegno, M., Benfenati, E., & Romberg, M. (2007). Grid Computing for the Estimation of Toxicity: Acute Toxicity on Fathead Minnow (Pimephales promelas). In *Distributed, High-Performance and Grid Computing in Computational Biology* (pp. 60-74).

MediGRID. (2005). *www.medigrid.de*. Retrieved March 14, 2007.

Mohammed, Y. (2006). Erweiterte Sicherheit und Datenschutz Techniken. In U. Sax, Y. Mohammed, F. Viezens & O. Rienhoff (Eds.), *Grid-Computing in der Biomedizinischen Forschung – Datenschutz und Datensicherheit* (Vol. 90, pp. 70-77). München, Germany: Urban &Vogel.

Mohammed, Y., Sax, U., Viezens, F., & Rienhoff, O. (2007). Shortcomings of Current Grid Middlewares Regarding Privacy in HealthGrids. *Stud Health Technol Inform, 126*, 322-329.

Montagnat, J., Frohner, Á., Jouvenot, D., Pera, C., Kunszt, P., Koblitz, B., et al. (2008). A Secure Grid Medical Data Manager Interfaced to the gLite Middleware. *Journal of Grid Computing, 6*(1), 45-59.

Open Grid Services Architecture - Data Access and Integration Services. Retrieved March 13, Accessed 2006, from www.ogsadai.org.uk/

Oster, S., Langella, S., Hastings, S., Ervin, D., Madduri, R., Phillips, J., et al. (2007). caGrid 1.0: An Enterprise Grid Infrastructure for Biomedical Research. *J Am Med Inform Assoc*, M2522.

Pearlman, L. W., V. Foster, I. Kesselman, C. Tuecke, S. (2002). A Community Authorization Service for Group Collaboration. In IEEE Computer Society, TC on Distributed Processing, Naval Postgraduate School (U.S.) & United States, Office of Naval Research. (Eds.), *Third International Workshop on Policies for Distributed Systems and Networks, June 5-7, 2002, Monterey, California, USA: proceedings* (pp. 50-59). Los Alamitos, CA.: IEEE Computer Society.

Pommerening, K. (2006). Grundlagen des sicheren Grid-Computing. In *Grid-Computing in der biomedizinischen Forschung* (Vol. 90, pp. 63-69). München, Germany: Urban & Vogel.

Rajasekar, A., Wan, M., Moore, R., Schroeder, W., Kremenek, G., Jagatheesan, A., et al. (2003.). Storage Resource Broker - Managing Distributed Data in a Grid. *Computer Society of India Journal, 33*(4), 42-54

Sax, U., Mohammed, Y., Viezens, F., & Rienhoff, O. (2006). Grid-Computing in der biomedizinischen Forschung. In K. Überla, O. Rienhoff & N. Victor (Eds.), *Medizinische Informatik, Biometrie und Epidemiologie* (Vol. 90). München, Germany: Urban & Vogel.

Schiffers, M. (2008). VO Management, *IVOM Workshop*. Regionales Rechenzentrum für Niedersachen, Hannover, Germany.

Seitz, L., Montagnat, J., Pierson, J. M., Oriol, D., & Lingrand, D. (2005). Authentication and Authorisation Prototype on the µgrid for Medical Data Management. In T. Solomonides, R. McClatchey, V. Breton, Y. Legré & S. Nørager (Eds.), *From Grid to Healthgrid - Proceedings of Healthgrid 2005* (pp. 222 - 233). Washington, DC: IOS Press.

Switula, D. (2000). Principles of good clinical practice (GCP) in clinical research. *Sci Eng Ethics, 6*(1), 71-77.

The Globus Security Team. (2005). *Globus Toolkit Version 4 Grid Security Infrastructure: A Standards Perspective*.

TMF. (2006). *AG Datenschutz*. Retrieved March 16, 2006, from www.tmf-ev.de/site/DE/int/AG/DS/container_ag_ds.php

Tsiknakis, M., Kafetzopoulos, D., Potamias, G., Analyti, A., Marias, K., & Manganas, A. (2006). Building a European biomedical grid on cancer: the ACGT Integrated Project. *Stud Health Technol Inform, 120*, 247-258.

United States. Dept. of Defense. (1999). *Common criteria & methodology for information technology security evaluation*

Welch, V., Siebenlist, F., Foster, I., Bresnahan, J., Czajkowski, K., Gawor, J., et al. (2003). Security for Grid services. *Proceedings of the 12th IEEE International Symposium on High Performance Distributed Computing, 2003*, (pp. 48-57).

Witzig, C. (2008). Interoperability Shibboleth - gLite, *IVOM Workshop*. Regionales Rechenzentrum für Niedersachen, Hannover, Germany.

KEY TERMS AND DEFINITIONS

Accountability: Actions of a person, especially modifications that she performs on data can be traced.

Anonymization: The changing of personal data in such a way that the re-identification of the person/subject is no longer possible (absolute Ananymization) or only with large amount of time, cost and manpower possible (de facto anonymized, scientific use files).

Authenticity/Authentication: Ensure that the person is the one she claimed to be. In regard to authentic document, authenticity is defined as integrity plus freshness.

Authorization: The act of providing and checking the authority of the user on a specific resource, informally speaking "who can do what".

Confidentiality: The assurance that data are not made available or disclosed to unauthorized person.

Integrity: Ensure data cannot be changed/deleted/altered by unauthorized party/person.

PET: Privacy-enhancing technologies; the use of information and communication technologies for the sake of the protection of human privacy.

Privacy: The ability and/or right to protect the personal secrets; it extends to the ability and/or right to prevent invasions of the personal space (the exact definition varies from one country to another). Privacy is applicable to natural persons (human beings) but not to legal persons (such as corporations).

Pseudonymization: Before releasing the data (for example for a specific research) the identifying data will be replaced with pseudonym. Only the trusted party who perform the pseudonymization can match between the identifying data and the pseudonym.

Chapter VI
Parallel, Distributed, and Grid-Based Data Mining:
Algorithms, Systems, and Applications

Moez Ben HajHmida
Faculty of Sciences of Tunis, Tunisia

Antonio Congiusta
University of Calabria, Italy & University of Salerno, Italy

ABSTRACT

Knowledge discovery has become a necessary task in scientific, life sciences, and business fields, both for the growing amount of data being collected and for the complexity of the analysis that need to be performed on it. Classic data mining techniques, developed for centralized sites, often reveal themselves inadequate, due to some unique characteristics of today's data sources. In such cases, sequential approaches to data mining cannot provide for scalability, in terms of the data dimensionality, size, and runtime performance. Moreover, the increasing trend towards decentralized business organizations, distribution of users, software, and hardware systems magnifies the need for more advanced and flexible approaches and solutions. Life science is one of the application areas that best resemble such scenario. This chapter presents the state of the art about the major data mining techniques, systems and approaches. A detailed taxonomy is drawn by analyzing and comparing parallel, distributed and Grid-based data mining methods, with a particular focus on the exploitation of large and remotely dispersed datasets and/or high-performance computers.

INTRODUCTION

Data mining aims at extracting hidden information from large data repositories (e.g., databases, file archives, digital libraries) for building valuable knowledge patterns and predictive models. The main data mining tasks are association rules discovery, classification and clustering.

Data mining is a massive computing task that deals with memory resident data. With the huge amount of stored data in centralized or distributed systems, traditional data mining techniques encounter limitations and shortcomings that often lead to inefficiencies. The need for parallel and distributed computing becomes inevitable to deal with large-scale data mining (Freitas, 1998; Kargupta, 2000), and for addressing complex needs and scenarios encountered in business as well as research organizations.

Parallel Data Mining (PDM) is targeted to tightly-coupled systems, like shared or distributed memory machines, and clusters based on fast networks. Distributed Data Mining (DDM) deals with loosely-coupled systems: clusters with average-fast or slow networks and geographically distributed computing nodes. The main differences between PDM and DDM are the number of involved computing nodes, the communication costs, and the degree of data distribution.

The advances in network technologies have produced huge amount of data stored on geographically distributed databases and repositories. When these amounts of data are owned by different organizations and hosted by non-dedicated computing resources, parallel and distributed data mining techniques start showing their limits. The Grid is the computing architecture that provides means for utilizing geographically distributed resources as a single meta-system. The emergence of such new infrastructure is highly beneficial to large-scale and compute-intensive data mining, as it offers new opportunities to optimize and speed-up mining processes.

Unlike previous parallel and distributed data mining surveys (Kargupta, 2000; Zaki, 2000), this chapter differentiates between:

- PDM, where learning methods are often platform dependent, databases are centralized (for example in a cluster or a supercomputer), and network connections are reliable and fast;
- DDM, in which learning methods are based on sharing nothing machines with a much slower network and databases are naturally distributed;
- and, in addition, the Grid Data Mining (GDM) category of algorithms and systems, which is deeply explored.

Although GDM shares many commonalities with PDM and DDM, there are platform peculiarities and requirements implying that efforts and obtained results in such area cannot be compared (in a homogeneous way) with those achieved by PDM and DDM.

The remainder of this chapter is organized as follows: Section 2 contains a background on classical data mining techniques; Section 3 presents parallel systems and the related programming paradigms, it details the techniques used in parallel data mining and classify them on the basis of the employed method; Section 4 presents the main differences between parallel and distributed data mining techniques, then it discusses the major distribution methods and their transition from parallel to distributed systems; Section 5 contains a description of the knowledge discovery process in Grid environments, it focuses on Grid infrastructures and frameworks designed for such purpose; finally Section 6 draws conclusions and highlights some future trends.

BACKGROUND

Association rules discovery aims at finding all the itemsets (set of attributes) in a database that frequently occur together, the so called *frequent itemsets*, and the derived association rules. The main algorithm for this data mining task is *Apriori* (Agrawal, 1996), which is an iterative algorithm that needs multiple scans of the database. If n is the number of items (attributes), the complexity of Apriori algorithm is exponential, i.e. $O(2^n)$. To speed up the algorithm, the amount of generated candidates and the number of database scans have to be optimized.

Classification aims at assigning data items to one of n predefined categorical classes. Since the category being predicted is pre-labeled, classification is also known as supervised induction. There are several classification techniques like *decision trees* (Quinlan, 1993), *rule induction* (Cohen, 1995), *neural networks* (Lippmann, 1987), *bayesian networks* (Guan, 1991), *support vector machines* (Boser, 1992), and *evolutionary algorithms* (Freitas, 2002).

Clustering is used to partition database records into subsets or clusters, such that elements of the same cluster share common properties that distinguish them from other clusters. The goal is to maximize intra-cluster similarity and minimize inter-cluster similarity. Unlike classification which has predefined categories, clustering must in essence autonomously come up with the categories. For this reason clustering is also called unsupervised induction. There are five main methods in clustering: *partitioning*, *hierarchical*, *density-based*, *grid-based* and *model-based*. One of the main known algorithms for clustering is *k-means*.

PARALLEL DATA MINING

Parallel Data Mining (PDM) has emerged for exploiting tightly-coupled, parallel systems in increasing performances of the mining algorithms. Typically a PDM algorithm shares the entire database and constructs a global model. Data is communicated via high speed buses, shared file systems, or dedicated networks, and is available from centralized data-stores.

Parallel systems are divided into three categories:

- Shared-memory,
- Distributed-memory, and
- Hierarchical systems.

In shared-memory systems, all processors share all the system memory. Each processor has direct access to the entire set of data. Access to the same data needs synchronization and sequential memory access. Within a distributed-memory system each processor has its own system memory that cannot be accessed by other processors. Shared data is transferred usually by message passing. Hierarchical systems combines shared and distributed systems approaches. They are composed by multiprocessor nodes in which memory is shared by intra-node processors and distributed over inter-node processors. Typically hierarchical systems use fast networks and may have shared disks.

Exploitation of parallelism in data mining algorithms follows two main paradigms: data parallelism and task parallelism (Freitas, 1998). In *data parallelism*, the database is partitioned among all processors and each processor executes locally the same task on its own data partition. In *task parallelism*, different tasks are executed concurrently on different partitions but processors need to have access to the entire database. In this case, the database is explicitly replicated (distributed-memory) or shared by the different processors (shared-memory). *Hybrid parallelism* is the combination of data and task parallelism it is well suited for hierarchical systems. In large-scale databases, when the entire database does not fit in memory, it is preferable to use data parallelism. Moreover, the distribution of the database among the processors makes the data mining algorithm more scalable (Zaki, 2000). When the search space is large, task parallelism is preferred. In this case the search space can be divided by partitioning attributes or candidate patterns (Freitas, 1998).

Data mining techniques are widely used in genomics and computational biology. One of the major challenges was the analysis of the high-dimension DNA or protein sequences. Parallel data mining algorithms proved their efficiency when applied to genomic problems (Jourdan, 2004, Guarracino, 2006).

Table 1 summarizes the major parallel data mining techniques. For each technique, this table enumerates the used paradigms, the target platform, and an estimation of the communication cost as reported by the literature on parallel algorithms.

Parallel Association Rules

Parallel association rules mining algorithms can be classified according to three main parallelization approaches:

- **Task parallelism**, in which candidate sets are distributed among the processors;
- **Data parallelism**, where the database is partitioned among the processors;
- **Hybrid parallelism**, which is the combination of the two above paradigms.

The main approach for task parallelism is *Data Distribution* (DD). DD is an example of Apriori parallelization where candidate itemsets are partitioned among processors. Each processor counts its candidate itemsets for all the database fragments. To compute the global support, each processor needs to scan the entire database at each iteration. Thus, DD needs fragments exchange between processors, which produce a communication overhead. The *Intelligent Data Distribution* algorithm (IDD) solves this problem through a ring-based, all-to-all broadcast mechanism. The entire database is still communicated, but a transaction might not be processed if it does not contain relevant items (Zaki, 1999).

The *Count Distribution* (CD) algorithm is another Apriori parallel formulation based on the data parallel paradigm. In the CD algorithm, the database is partitioned over multiple processors. Each processor executes locally the Apriori algorithm, and runs a global reduction of partial candidate counts to obtain the global counts. At each pass *k*, the CD algorithm requires synchronization, with a communication overhead of $O(|Ck|*N)$,

Table 1. Parallel data mining techniques

	Association Rules	Classification	Clustering
Paradigm	*Data parallelism* *Task parallelism* *Hybrid parallelism*	*Data parallelism* *Task parallelism* *Hybrid parallelism*	*Data parallelism* *Task parallelism*
Platform	*Distributed memory* *Hierarchical*	*Distributed memory* *Shared memory*	*Distributed memory*
Communication cost	*Exponential*	*Linear to exponential*	*Linear*

where $|Ck|$ is the cardinality of the k-candidate set (candidate set of length k), and N is the number of processors. CD presents a bottleneck with large number of processors (Zaki, 1999). The *Fast Parallel Mining* (FPM) algorithm outperforms the CD algorithm thanks to local and global pruning techniques. Moreover, FPM needs only one round of messages exchange, local support information is broadcasted to all processors thus reducing the communication overhead (Zaki, 1999).

For distributed memory machines, Zaki (1999) proposed the *Common Candidate Partitioned Database* (CCPD) algorithm. CCPD parallelizes candidate generation, each processor generates a disjoint candidate subset. The global candidate set is built in parallel and shared by all processors. CCPD obtains reasonable speedup, but the serial I/O is detrimental to performance.

Asynchronous Parallel Mining (APM) is another algorithm for shared memory machines, which is based on the DIC algorithm (Brin, 1997). APM logically divides the database into many small, equally-sized virtual partitions and applies DIC in parallel. Experiments indicate that APM outperforms CD by a factor ranging from four to five (Zaki, 1999).

The *Hybrid Distribution* (HD) algorithm combines CD and IDD algorithms. P processors are split in G groups, the database is partitioned among the G groups, and the candidates are partitioned among the P/G processors of a group. Each group executes the CD algorithm, while the IDD algorithm is executed within each group. The advantage of the HD algorithm is that it reduces database communication costs by $1/G$. Experiments showed that HD has the same performance as CD, but can handle much larger databases (Zaki, 2000).

It is possible to observe that data-parallel algorithms require less communication overhead with respect to task-parallel ones, essentially because the first ones exchange or share itemsets, while the second ones exchange or share database partitions. The hybrid approach reduces the communication overhead and offers better load balancing. But all these approaches have been proposed for parallel machines with shared memory or distributed memory and fast networks. In case of slow networks, a large number of frequent candidates will produce a high communication cost. The use of maximal frequent itemsets can resolve this problem. Two hierarchical parallel algorithms have been proposed to such purpose: ParMaxEclat and ParMaxClique (Zaki, 2000). These algorithms exploit the capabilities of a hierarchical platform (N nodes of P processors) through a three-phased computation. In the first phase, the master node generates prefix equivalence classes then distributes them among the processors. Instead of candidate itemsets, transaction "id" lists (*tidlist*) are exchanged. A tidlist is a list of all transaction *ids* containing a given item X. In the second phase, each processor generates independently all frequent itemsets from its classes. Eventually, final results are aggregated, inducing a communication cost of $O(t*N)$, where t is the number of database transactions. Experiments showed that in spite of the low speedup of ParMaxClique, it outperforms CD (Zaki, 2000).

Table 2 presents a summary of the major characteristics of the parallel rule-discovery algorithms reviewed thus far. The table shows that most of the parallel algorithms are based on Apriori, thus their communication costs are function of the generated candidates. ParMaxEclat and ParMaxClique are derived from their sequential versions and APM is based on the DIC algorithm. These algorithms outperform the Apriori-based ones. The communication complexity presented in this table, is the cost of a pass k. The predominant factor in the communication complexity is the number of generated candidates, which do not change by increasing the number of nodes. Thus, increasing the number of nodes produces an overcrowding of the network traffic. Despite ParMaxEclat and ParMaxClique do not generate candidate itemsets, transmitting tidlist among processors incurs heavy communication cost. For

Table 2. Parallel rule discovery algorithms

Algorithm	Paradigm	Platform	Communication cost
CD	*Data parallelism*	*Distributed memory*	*O(|Ck|*N)*
CCPD	*Data parallelism*	*Shared memory*	*No communication, but I/O overhead*
FPM	*Data parallelism*	*Distributed memory*	*O(|Cp|*N)*
DD **IDD**	*Task parallelism*	*Distributed memory*	*Full database exchange per iteration*
APM	*Task parallelism*	*Shared memory*	*No communications, but I/O overhead*
HD	*Hybrid parallelism*	*Distributed memory*	*Database exchange reduced by 1/G w.r.t. CD*
ParMaxEclat **ParMaxClique**	*Task parallelism*	*Hierarchical*	*O(t*N)*

N = number of processors
|Cp| = cardinality of the potential candidate itemsets
t = number of database transactions
|Ck| = cardinality of k-candidate set
|C| = cardinality of the candidate itemsets
G = number of processor groups

these reasons, the scalability of parallel association rule discovery is limited to a reduced number of nodes (no more than 32 nodes).

Parallel Classification

A wide variety of methods has been proposed for this data mining task. Parallelizing a classification algorithm implies considering the method characteristics that permit to find the best parallelism. In the following, parallel classification algorithms are presented for each specific method of classification.

Decision Trees

The initial approaches to parallelizing decision trees were not scalable on large databases and massively parallel machines. These limitations were resolved by the SPRINT algorithm (Shafer, 1996). SPRINT reduces the multiple sorting passes on the data to only one pass; it partitions horizontally the data among all processors. The decision tree is replicated on all processors. Each processor computes locally the best split point then exchanges its statistics to determine the global best split point. SPRINT scalability is limited because the hash table used to split the attributes generates a communication overhead of $O(N)$ per processor, where N is the number of examples.

The ScalParc algorithm (Joshi, 1998) is an extension of SPRINT that uses a distributed hash table to reduce the communication overhead to $O(log(N))$. Srivastava (1999) proposed a dynamic hybrid approach that migrates gradually from data parallelism to task parallelism in order to reduce the load imbalance and the communication overhead incurred by the decision tree when it becomes bushy. This approach distributes the dataset among P processors. All processors are assigned to a single partition. Each processor computes synchronously with the other processors the same decision tree node. The local information is exchanged by global reduction. This exchange overhead grows with the number of leaf nodes. When the communication becomes costly, the hybrid algorithm splits these nodes equally among two processor partitions. On each processor partition, the tree nodes are synchronously computed then split among two processor partitions.

Rule Induction

Most of parallel rule induction algorithms are based on decision tree induction (Freitas, 1998). Decision trees are learned in parallel from disjoint data-set fragments (Provost, 1996). The learned decision trees are converted to rules and the rules are combined into a single rule set. For conflicting rules, rule conflict resolution techniques are proposed (Hall, 1998).

Artificial Neural Networks (ANN)

There are three main approaches in parallel neural networks:

- The first approach parallelizes the computation by partitioning the nodes of the network and their weights among processors.
- In the second approach, namely "pattern partition", network nodes and weights are replicated on each processor and the patterns are partitioned among the processors. In this case, weights are communicated to update the network. This approach is preferred when the number of patterns is large.
- The two approaches can be combined into a hybrid one (Freitas, 1998).

While the second approach is regarded as a coarse-grained parallelism, the first neural network approach is considered a fine-grained parallelism.

Neural networks performances are sensitive to initial conditions and network topology. Thus, neural networks learning can be seen as an optimization problem that can be resolved by Genetic algorithms to find an optimal topology or to find an initial set of good weights (Abraham, 2000).

Support Vector Machines (SVM)

Compared to the other classification algorithms SVM provide good models that do not overfit the training examples, but it cannot deal with very large datasets. To support a huge amount of data, SVM algorithms should be parallelized. An incremental learning approach has been proposed by Tveit (2003). Data is distributed among nodes on a tree topology. Each node computes an update, called "increment", and sends it to its parent node. The nodes are continuously updated from increment data in a synchronous mode.

Collobert (2002) proposed an algorithm based on a mixture of classification models, called "expert mixture". The training set is divided into M random partitions of size N/M, where N is the number of examples in the training set. Each node is trained on its local partition to produce an "expert". Then, a neural network algorithm is trained on the whole training set, minimizing a cost function called *gater*. Experts are sorted in descending order according to the weights assigned by the *gater*. Each example is assigned to the best expert and a mixture of experts is obtained through a weighted combination. This algorithm reduces the computation complexity from $O(N^2)$, in case of sequential SVM, to $O(N)$.

Evolutionary Algorithms (EA)

As mentioned by Freitas (2002), there are two schemes of parallel Genetic Algorithms (GA) in data mining: coarse-grained parallelism (island model) and fine-grained parallelism (cellular model). In the island model, the population is divided into several equal subpopulations. Each processor runs a sequential GA independently and in parallel on its local subpopulation.

Occasionally, most accurate individuals of the population migrate randomly from one sub-population to another. In a fine-grained parallel GA, usually one individual is assigned to each processor. The individuals communicate only with their neighbors. Parallel GAs seem to be complex in the determination of migration rate, communication topology, and optimal number of sub-populations. They also imply heavy communication overhead.

Table 3 shows that parallel algorithms are essentially proposed for decision trees, ANN, GA and SVM techniques. These parallelization efforts are aimed at reducing the computation time and the search space.

Parallel Clustering

Dhillon (2000) proposed a parallelization of *k-means* algorithm on distributed memory machines. This parallelization is based on the observation that computation of distance metrics is inherently data parallel. Thus, they can be executed synchronously and in parallel for each data point. The parallelization strategy is to divide the N data points among P processors. To compute the distance measures in parallel, the centroids (barycenters of the clusters) must be available to each processor. To enable this potential parallelism, a local copy of the centroids is maintained for

Table 3. Parallel classification algorithms

Algorithm	Paradigm	Method	Remarks
SPRINT **ScalParc**	*Data parallelism*	*Decision tree*	*High communication*
Dynamic hybrid approach	*Hybrid parallelism*	*Decision tree*	*High communication*
Fine-grained ANN	*Task parallelism*	*ANN*	*High communication*
Coarse-grained ANN	*Data parallelism*	*ANN*	*Lower communication* *Replicates nodes and weights*
Cellular model	*Task parallelism*	*GA*	*Communicates with neighbors* *Restricted communication*
Island model	*Data parallelism*	*GA*	*Complex parameters setting* *High communication*
Incremental SVM	*Data parallelism*	*SVM*	*Needs synchronizations*
Expert mixture SVM	*Data parallelism*	*SVM*	*Reduces computation complexity*

each processor. Each iteration of this parallel algorithm consists of an asynchronous computation phase followed by a synchronous communication phase. The parallelization is based on the Message Passing Interface (MPI), such that at each iteration the algorithm communicates the mean square error using a broadcast operation, that produces a communication overhead of $O(N)$, where N is the number of data points. Furthermore, this approach suffers from the general shortcoming of k-means, i.e. the number of clusters has to be defined by the user and is not automatically computed.

A parallel subspace clustering algorithm MAFIA (Merging of Adaptative Finite Intervals) has been proposed for a shared-nothing architecture (Nagesh, 1999). This approach distributes the dataset among P processors. Each processor builds a local histogram. The MPI reduce operation is used to distribute the global histogram on each processor. Then, each processor identifies the local counts of the dense units; finally, the global counts are obtained through another reduce communication. Experiments with 16 processors showed that the parallelization achieves near linear speed-up. This is due to the data parallelism and the negligible communication overhead, if compared to computation time.

A parallel implementation of the AutoClass (Cheeseman, 1996) algorithm, called PAutoClass has been proposed by Foti (2000). The PAutoClass algorithm divides the clustering task among the processors of a parallel machine. Processors work on their own partition and exchange their intermediate results. Each processor computes locally the weights of the different items, which are summed to produce the class weights. A reduce operation is then executed to compute the global classes weights.

Table 4 presents a summary of the parallel clustering algorithms reviewed thus far. Compared to parallel k-means and PAutoClass, MAFIA communication is negligible. MAFIA algorithm is designed to work in parallel, on the contrary to the two others which are parallel versions of sequential algorithms.

Comparison and Discussion

Parallel data mining algorithms can be classified by the target architecture and the used paradigm. This classification can be expressed by the taxonomy of Figure 1. This figure shows that parallelization efforts are concentrated more on association rule discovery than other data mining techniques.

A great part of parallel association-rule techniques is based on the Apriori algorithm. The developed implementations perform extensive I/O, incurring exponential communication overhead. Major parallel classification algorithms are essentially proposed for decision trees and neural networks on parallel and distributed memory machines. But for parallel clustering, only few algorithms are proposed. These parallelization techniques are motivated by the need of reducing the computation time and the search space. Although the computation time is considerably reduced, the communication overhead increases

Table 4. Parallel clustering algorithms

Algorithm	Paradigm	Method	Remarks
Parallel k-means	*Data parallelism*	*Partitioning*	*High communication*
MAFIA	*Hybrid parallelism*	*Grid-based*	*Negligible communication overhead*
PAutoClass	*Data parallelism*	*Model-based*	*Exchanges weights*

with the growing of the number of processors. From these considerations a general observation can be derived: the scalability of parallel data mining techniques is limited to a reduced number of nodes (generally no more than 32 nodes).

For very large datasets, like genomic data, data parallelism approach incurs less communication overhead than task parallelism, which needs to access the entire database. The hybrid approach combines both of them and offer better load balancing.

DISTRIBUTED DATA MINING

The objective of Distributed Data Mining (DDM) is to perform the data mining operations using the peculiarities and the availability of distributed computational resources (Kargupta, 2000). It has also been influenced by the PDM literature because of the possibility to support PDM locally. DDM can be seen as a gradual transition from tightly-coupled, fine-grained parallel machines to loosely-coupled, medium-grained LAN of workstations, and finally very coarse-grained WAN (Zaki, 2000).

The main differences between PDM and DDM are the scale of the environment, communication costs, and the level of data distribution. Distributed Data Mining (DDM) deals with loosely-coupled systems: generally share-nothing machines interconnected by high-latency networks and geographically distributed (Zaki, 2000). In these systems data is commonly shared through message passing.

Typically a DDM algorithm partitions the database at several sites and mines a local model for each partition, then merges them to build a global model. DDM algorithms try to learn from the distributed data generally without exchanging the raw data.

Bioinformatics data is being massively stored on distributed datasets. Distributed data mining methods are needed to process this huge amount

Figure 1. Taxonomy of parallel association rule, classification, and clustering

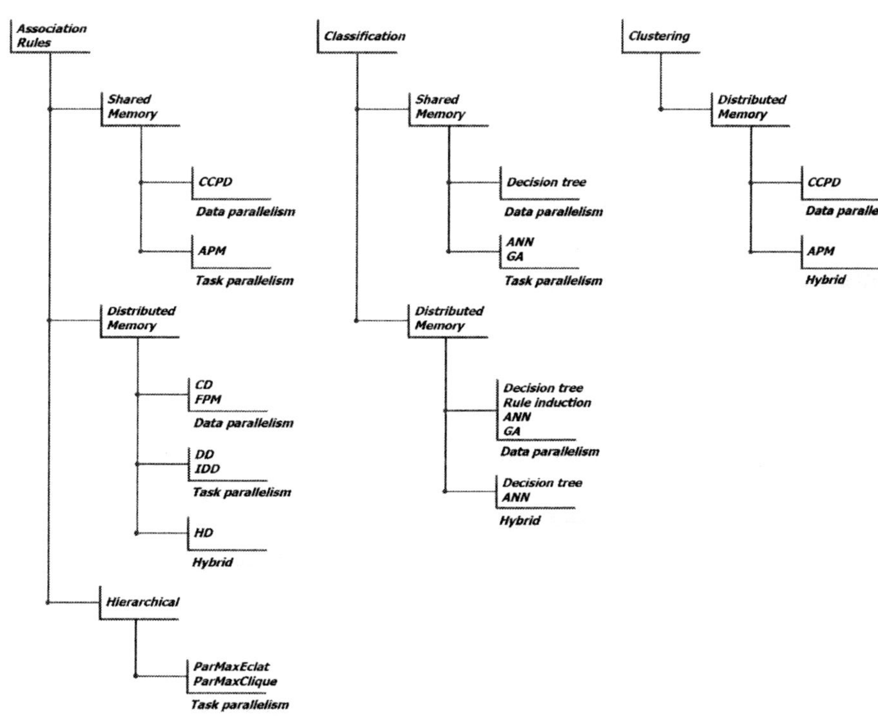

Table 5. Distributed data mining techniques

	Association rules	Classification	Clustering
Paradigm	*Data parallelism*	*Data parallelism*	*Data parallelism*
Platform	*Share-nothing*	*Share-nothing*	*Share-nothing*
Communication cost	*Exponential*	*Linear*	*Linear*

of data. Also in biomedicine, distributed data mining is applied to find treatments for diseases by analysing more than one million proteins that regulate body functions. More generally, life sciences applications build distributed systems based on distributed data mining methods such as the *Human Genome* project[1] and the *LSDIS* lab[2].

Table 5 summarizes the major distributed data mining techniques. For each technique, this table enumerates the used paradigms, the target platform, and an estimation of the average communication costs.

Distributed Association Rules

The design of distributed association rule discovery algorithms is directly derived from parallel algorithms by taking into consideration the different environmental factors, like the possibility of slow networks. Thus the basic method to build upon the distributed algorithm should have a low communication cost, like CD algorithm.

The *Fast Distributed Mining* (FDM) algorithm is a distributed version of the CD algorithm (Cheung, 1996). It is designed for a distributed database environment and it assumes that the database is horizontally partitioned. FDM is based on the observation that a globally frequent itemset must be locally frequent at a site. To reduce the number of candidates, a site generates candidates from itemsets that are globally frequent and locally frequent. For example, if the set of frequent itemsets is *{A, B, C, D, E}* and the set of locally frequent itemsets at a site is *{A, B, C}*, the generated candidate set is *{AB, AC, BC}*. FDM uses also a count polling technique to optimize the communication cost, reduced to $O(|Cp|*N)$, where $|Cp|$ is the cardinality of the potential candidate itemsets and N the number of nodes. This technique requires two rounds of message exchanges at each iteration. First, each node requests the local support from other nodes, to compute the global support of its candidates. In the second round, each node broadcasts the computed global supports to other nodes. The main problem of FDM is that Cp quickly increases to 1 as N increases, especially when the database is heterogeneous (Schuster, 2001).

The *Distributed Decision Miner* (DDMiner) algorithm reduces the communication overhead to $O(Pr_{above} *|C|*N)$, where Pr_{above} is the probability that a candidate itemset has a support greater than the minimal support (Schuster, 2001). DDMiner verifies that an itemset is frequent before collecting its support counts from the other nodes. For FDM a locally large itemset is considered as globally large, while for DDMiner it is not identified as large until verified by messages exchange. In DDMiner, each node computes local and global count hypothesis. Message exchange is done only if the two hypotheses disagree. These hypotheses are usually updated by messages received from other nodes.

To avoid the multiple database scans of Apriori based algorithms, Schuster (2003) proposed a parallelization of the Sampling algorithm (Toivonen,

1996) for geographically distributed databases. This version of the Sampling algorithm is called *D-Sampling*, it scans the database only once and is based on a modified version of the DDM algorithm. It begins by loading a sample of the database into memory, as to establish the candidate set. Each partition of the database is scanned in parallel to compute the candidate counts. The modified version of DDMiner is executed a second time to identify the globally frequent itemsets. D-Sampling produces the same communication cost per candidate of DDMiner, but reduces the total communication cost by producing fewer candidates. Experiments showed that D-Sampling outperforms both DDMiner and FDM.

Table 6 presents a summary of the characteristics of distributed rule discovery algorithms reviewed thus far. Most of them is based on the Apriori algorithm, thus their communication cost is function of the generated candidates. D-sampling is based on its sequential version: the Sampling algorithm. It is also possible to observe that there are more efforts devoted to parallelize algorithms (see Table 2) than to distribute them. The reason is that rule discovery needs high volumes of communications at each step, becoming not scalable for distributed environments where the network speed is low. The communication complexity presented in the table is the cost of a pass *k*.

Distributed Classification

Most of the DDM classification approaches are based on the *ensemble learning* method. Ensemble (also called committee) methods are learning algorithms that construct a set of classifiers (base classifiers) and then classify new data points by taking a (weighted or unweighted) vote of their predictions (Dietterich, 2000). Typically, an ensemble learning method runs the learning algorithm several times, each time with a different subset of the training examples. The major implementations of such an approach are stacking, bagging, and boosting:

- *Stacking* is a method in which the output pattern of an ensemble of trained classifiers serves as input to a second-level classifier (Wolpert, 1992).
- *Bagging* (bootstrap aggregating) produces replications of the training set by using sampling and replacement: each replication of the training set has the same size of the original set, but some examples (sampling) can appear more than once, while others don't appear at all (replacement). From each replication, a classifier is generated. The generated classifiers are used to classify each sample from the test-set using a voting scheme. Bagging improves the accuracy of unstable algorithms (ANN, decision tree) but

Table 6. Distributed rule discovery algorithms

Algorithm	Paradigm	Communication cost	Remarks
FDM	*Data parallelism*	$O(\lvert Cp\rvert * N)$	*Distributed version of CD*
DDMiner	*Data parallelism*	$O(Pr_{above} * \lvert C\rvert * N)$	*Apriori based*
D-Sampling	*Data parallelism*	*Less than DDMiner*	*Sampling based*

N = number of processors
$\lvert C\rvert$ = cardinality of the candidate itemsets
Pr_{above} = probability that a candidate itemset has a support greater than the minimal support

degrades the accuracy of stable algorithms (k-nearest neighbors) (Dietterich, 2000).

- *Boosting*, like bagging, manipulates the training examples to generate multiple hypotheses (Dietterich, 2000). Boosting is a family of methods for accelerating a learning algorithm: AdaBoost (two-class problem), AdaBoost.M1 and AdaBoost.M2 (multiple-class problems) and AdaBoostR (regression) are among the most known (Freund, 1997). Boosting maintains a set of weights over the training examples. For each iteration, the learning algorithm is invoked to minimize the error on the training set. Weights of misclassified examples are increased and weights of correctly classified examples are decreased. The final classifier is constructed through a weighted vote of the individual classifiers.

The ensemble learning approach, in which the same program runs several times on different data, can be applied to a distributed environment, where data is stored on different sites, thus producing different local models to be aggregated into a global model. Based on this observation, several methods have been proposed such as:

- Meta-learning,
- Knowledge probing, and
- Collective learning.

The *meta-learning* method was proposed for homogeneous, distributed databases (Kargupta, 2000). First, a base classifier is trained from the local training-set at each site. The produced classifiers are collected to a central site where their predictions are generated on a separate validation-set to form meta-level data. Finally, the global classifier (meta-classifier) is trained from the meta-level data. The meta-classifier algorithm can be an *arbiter* or a *combiner* of the different predictions composing the meta-level data.

An example of meta-learning based system is JAM (Java Agents for Meta-Learning) (Kargupta, 2000). By combining different learning systems, meta-learning improves the predictive performance and exploit task parallelism. It also reduces the communication cost when it exchanges the learned models instead of training examples. Meta-learning is scalable with respect to the data size when it learns from small subsets that fit in memory. For the scalability on the number of sites, meta-learning has been enriched with a hierarchical approach, yielding however admittedly poor results: reduced predictive performance and increased communication overhead with large number of subsets (Kargupta, 2000).

Knowledge probing is another variant of meta-learning (Kargupta, 2000). It was proposed as a technique to probe descriptive knowledge from a black box model, such as neural network. In meta-learning, one or more classifiers are trained to arbitrate or to combine the output of the base models. These models together with the base models are used to give predictions on unseen data items. In knowledge-probing, the second learning phase is performed on the probing-set, where class values are the combination of predictions from the base models. The result is one descriptive model at the base level rather than the meta-level. Knowledge-probing has the disadvantage that the final model is not intelligible, because ANN network connections and weights are not expressive. It does not describe how the data are correlated with the assigned classes. In addition, it does not scale-up with large numbers of classes and computing nodes.

The *collective data mining* (CDM) methodology proposes a solution to learn from heterogeneous sites using regression or decision tree models. Foundation of CDM is the observation that any function can be represented in a distributed fashion using an appropriate set of basic functions (Kargupta, 2000). CDM guarantees the correctness of the learned model by the use of an orthonormal representation (Fourier transformation).

It first chooses an orthonormal representation that is appropriate for the type of data model to be constructed. Then, it generates approximate orthonormal basis coefficients at each local site. If necessary, it moves an appropriately chosen sample of the datasets from each site to a central site and generates the approximate basis coefficients, corresponding to non-linear cross terms. Finally, it combines the local models, transforms the model into the user-described canonical representation, and outputs the model.

A distributed version of the boosting algorithm is proposed in Fan (1999), where each classifier is trained using a fraction of the training set. In this distributed version, the classifiers are trained from disjoint partitions of the data set (D-sampling). In a *D-sampling* partition the weighted training-set is partitioned on different sites. At each boosting round the classifiers are first learned on disjoint datasets and next exchanged among the sites. Experiments showed that the distributed boosting approach has worse performance compared to the traditional boosting algorithm.

Another distributed version of the boosting algorithm is proposed in Lazarevic (2002). This method uses the AdaBoost.M2 procedure. At each boosting round, the classifiers are first learned on disjoint datasets. The produced models are exchanged among the sites to form an ensemble, which takes the role of the hypothesis. Afterwards, the exchanged models are combined and their weighted voting-ensemble is constructed on each disjoint data set. In this way, each distributed site maintains a local version of the global distribution without exchanging the complete weight vector. Experimental results showed that the proposed algorithm achieves classification accuracy comparable or even slightly better than that of boosting applied to centralized data (Lazarevic, 2002).

Tsoumakas (2002) proposed a new approach called *average stacking* to make scalable the stacking approach that suffers from the large size of the meta-level training-set. The proposed approach reduces the amount of the communicated data to combine the learned classifiers. The predictions of these classifiers (probability that classifier C_i assigns class k for instance e) compose the meta-level training set. In this approach, the training examples are built by averaging the classifiers predictions. Thus, the size of the training set becomes independent from the number of the base classifiers. Experiments showed that the execution time of average stacking outperforms stacking, still maintaining similar performance in terms of predictive accuracy. Compared to meta-learning, average stacking has worse predictive and run-time performance, the reason being that this approach reduces the execution time of the stacking method but communicates the same amount of data.

A *fragmentation* approach to rule discovery has been described by Cho (2002). From each data site (or fragment), only a single rule is generated for each class. Some of these individual rules are selected to form a rule-set for each class. These rule subsets represent adequately the hidden knowledge of the entire database. The selection criteria are the product of confidence and support, while relative deviation is used to resolve conflicting rules.

Ant colonies is a powerful EA approach that can be used for induction rules. Abraham (2006) proposed a possible use of the AntMiner+ approach in a dynamic, distributed environment.

Table 7 summarizes the characteristics of distributed classification. Efforts done in parallelizing classification algorithms (Table 3) are comparable to those done in distributing them. Except rule induction, all the other distributed algorithms are inspired by the ensemble-learning method. The communicated data is constituted by local models, which can be in different formats, e.g. orthonormal representation in CDM. CDM is the most accurate algorithm, thanks to its mathematical foundations. Rule fragmentation has the lowest communication overhead, but the highest error rate.

Table 7. Distributed classification algorithms

Algorithm	Paradigm	Method	Remarks
Rule fragmentation	*Data parallelism*	*Rule induction*	*Communicates rules* *Reduced accuracy*
JAM	*Data parallelism* *Task parallelism*	*Meta-learning*	*Exchange models* *Bad scale-up*
Knowledge Probing	*Data parallelism*	*Knowledge probing*	*Exchange models* *Bad scale-up*
CDM	*Data parallelism*	*Collective learning*	*Orthonormal representation* *Correctness of learned model*
Distributed boosting	*Data parallelism*	*Ensemble learning*	*Exchange models*
Average stacking	*Data parallelism*	*Ensemble learning*	*Worse than meta-learning (JAM in particular)*

Distributed Clustering

Most of distributed clustering algorithms are designed for heterogeneous distributed databases. The *Collective Hierarchical Clustering* (CHC) algorithm, for vertically distributed data sets, generates local models that are combined to obtain the global model. Each local site applies the chosen hierarchical clustering algorithm to generate a local *dendogram* (Zaki, 2000). CHC gathers local dendograms to a central site (facilitator site) to produce a global dendogram. The CHC algorithm is based on the nearest-neighbor clustering algorithm. The facilitator site uses dissimilarities metrics to produce the global dendogram. The communication cost of this algorithm is $O(n)$, where n is the number of elements in the data set.

The *RACHET* (Recursive Agglomeration of Clustering Hierarchies by Encircling Tactic) algorithm also exchanges dendograms with the same communication cost (Samatova, 2002). RACHET focuses on the reduction of the communication cost by using traditional hierarchical-clustering algorithms for massive, distributed data sets. Lazarevic (2000) proposed a method for learning from heterogeneous spatial databases. In each local site, similar regions are identified by independently applying a spatial clustering algorithm. For each cluster on the local data set, a *convex hull* is constructed. Each site receives the different convex hulls from all other sites. Using these hulls, the correspondences between the clusters from all spatial data sets are identified. For situations in which the exact matching cannot be determined, averaging of the local regression models, built-up associating neighbor clusters with high connectivity, is used for overlapping regions.

The *Density Based Distributed Clustering* (DBDC) algorithm is based on representative elements. On each site, it performs the DBSCAN clustering algorithm (Januzaj, 2004). The best representative elements for each locally found cluster are identified to constitute the local model.

Table 8. Distributed clustering algorithms

Algorithm	Paradigm	Method	Remarks
CHC	*Data parallelism*	*Hierarchical*	*Exchanges dendograms*
RACHET	*Data parallelism*	*Hierarchical*	*Exchanges dendograms*
DBDC	*Data parallelism*	*Density-based*	*Exchanges models*
Convex hulls	*Data parallelism*	*Density-based*	*Exchanges hulls and models*

Each local model is transferred to a central site, where the local models are merged in order to form a global model. The global model is created by a new clustering of the representatives with the DBSCAN algorithm. To each local representative a global cluster-identifier is assigned and this resulting global clustering is sent to all sites. This approach reduces the communication cost.

Table 8 presents a summary of the major algorithms for distributed clustering. Hierarchical algorithms are directly applicable to the distributed scenario. Thanks to dendograms, exchanged data is compact, thus produces less communication overhead. In density-based algorithms, the density estimation can be computed on different subsets of data then aggregated by density approximation when models are exchanged. It can be observed that hierarchical and density-based methods are more suitable for distributed environments.

Comparison and Discussion

Distributed data mining algorithms can be classified on the basis of the target architecture and the used paradigm. This classification can be expressed by the taxonomy of Figure 2. This figure shows that efforts in designing distributed algorithms are more concentrated on classification than other data mining techniques, and all of them are restricted to data parallelism approach. The majority of distributed association-rules algorithms are directly derived from their parallel versions. This kind of task needs high volumes of communication at each step: this makes them not scalable on distributed environments where the network speed is normally low. On the other hand, distributed classification approaches are based on ensemble methods. This approach minimizes considerably the communication cost, but decreases the predictive performance and yields reduced efficiency.

The communicated data is constituted by local models that can be in different formats. Most of distributed clustering algorithms is designed for heterogeneous, distributed databases and based on hierarchical or density methods. The exchanged data is constituted by local dendograms or models. The most critical phase in a distributed classification or clustering algorithm is the aggregation of local models. This phase is decisive for the quality of the final model.

GRID DATA MINING

Grid computing has emerged as an important new branch of distributed computing focused on large-scale resource sharing and showing also high-performance orientation (Reed, 2003). In many sectors, like bioinformatics (Jithesh, 2005) and biomedical applications (Sinnott, 2004), it is necessary to perform the analysis of very large data sets. Data repositories are often large, geographically distributed, and their complexity is increasing (Berman, 2001).

Figure 2. Taxonomy of distributed association rules, classification, and clustering algorithms

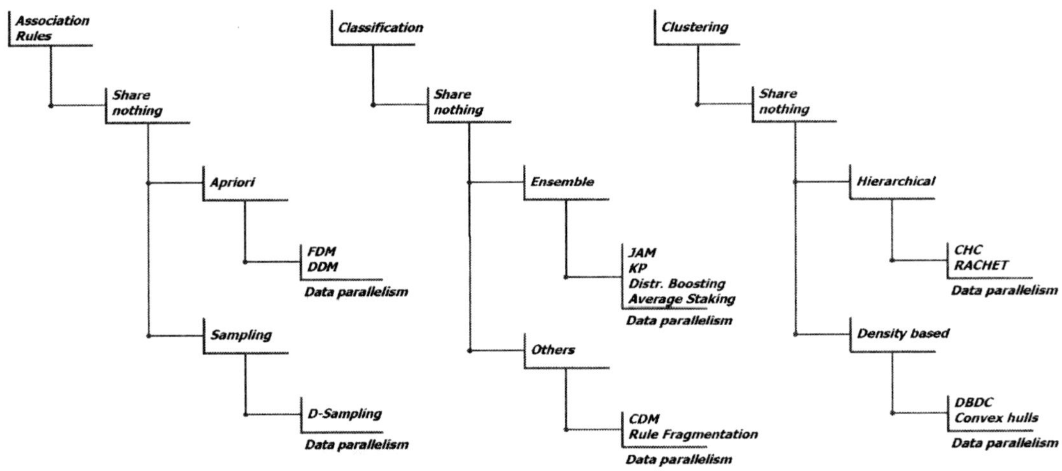

The Grid provides protocols and services to share computing resources, data, and software managed by multi-institutional virtual organizations. Grid services are provided by middleware systems such as Globus, Legion, gLite, and Unicore. These middleware environments are meta-systems that run on top of existing operating systems and networks. Globus is the most used middleware; it aims at building Grid settings spanning several administrative domains and is designed as a toolkit for supporting grid applications (Foster, 1999). However, services for data access and management provided by today's middleware are not sufficient to easily build Grid-based data mining applications. The development of Grid systems supporting data mining applications is needed to analyze and extract knowledge from huge amount of data stored within such environments. Efforts for Grid-enabling data mining applications produced a variety of systems, algorithms, and other tools that will be presented in the following sections.

General Purpose GDM Systems

Several projects such as TeraGrid and InfoGrid focused on the creation of Grid infrastructure providing tools for management and manipulation of distributed data sources. Other projects like ADaM on NASA's Information Power Grid, Discovery Net, Grid Miner, GATES, the KnowledgeGrid, DataMiningGrid, and KDDML-G have focused on the creation of systems for distributed knowledge discovery over Grid infrastructures.

The *TeraGrid* project[3] aims to provide a Grid infrastructure (Catlett, 2007) built over resources available at four main sites (San Diego Supercomputing Center, National Center for Supercomputing Applications, Caltech and Argonne National Lab). The deployment of TeraGrid involved a Grid infrastructure that enables access to more than 250 teraflops of computing power and more than 30 petabytes of computing storage that are connected via a 40 Gbps network. The architecture of the TeraGrid adopts the existing Grid software technologies for building a "virtual system".

TeraGrid offers a set of application services, called *TeraGrid Application Services*, implemented upon basic software components, called *Grid Services*. The *Grid Services* consist of three distinct layers:

- The *Basic layer* (called resource layer) provides authentication, resource allocation, data access, and resource information services;

- The *Core layer* (called collective layer) consists of advanced data movement, job scheduling, and monitoring services;
- The *Advanced layer* provides services such as super-scheduling, resource discovery services, repositories, etc.

One of the most challenging applications for the TeraGrid is the synthesis of knowledge from very large data sets. The first application is the creation of the Biomedical Informatics Research Group to allow researchers at geographically different locations to share and access brain image data and extract useful patterns and models from it. This enables TeraGrid to act as a knowledge Grid in the biomedical informatics domain (Berman, 2001).

InfoGrid is a data integration middleware designed to operate on Grids. Its main objective is providing information access and analysis means for knowledge discovery on scientific datasets (Giannadakis, 2003). Instead of providing a universal query system, where everything is abstracted to the user, InfoGrid gives an individualized view of the resources for each scientific domain. The main entity of the InfoGrid architecture is the *Wrapper*. Wrappers are distributed entities on a Grid; each node publishes a directory of the wrappers it owns. A wrapper may wrap executables and information sources or even be composed by other published wrappers (*Composite Wrappers*). A wrapper provides a set of query construction interfaces that can be used to query a resource in its native language. It also provides a set of administration interfaces that can be used to configure its properties (access control properties, linkage to other resources, configuration files).

The *ADaM* (Algorithm Development and Mining) project[4] is an agent-based platform for knowledge discovery researched at Alabama University (Rushing, 2005). It consists of over 100 data mining and image processing components and is primarily designed for scientific and remote-sensing data. The ADaM system has been Grid enabled by making use of Globus and Condor-G. It offers a set of data mining operations like classification, clustering, association rule mining, etc. While it provides this great number of components and a composition mechanism to design complex mining processes, ADaM lacks a proper workflow support and a graphical user interface.

The Discovery Net project[5] provides a set of services for knowledge discovery through the creation of high-level, re-usable, and distributed application workflows. The infrastructure is being used by scientists from three different application domains including Life Sciences, Environmental Monitoring and Geo-hazard Modeling. Discovery Net is a client/server system using the low-level features of Globus and defining also ad-hoc protocols and components, such as the Discovery Process Mark-up Language (DPML), for data flow management. The knowledge discovery process is based on a pipeline model. Discovery Net comprises two levels of services: core services and high-level services. Core services include data, execution, and component services on top of which are built composition and deployment services. Data services include distributed and local data management and also integration functionalities provided by Infogrid wrappers. Execution services allow for component composition and also the mapping from DPML to an execution plan on the Grid. Composition services allow users to design the graph of the discovery process, in which nodes represent operations and edges represent data flows. Deployment services make possible the publication as new services of knowledge processes.

The *GridMiner* (Brezany, 2003) system[6] is designed to allow the use of data mining and On-Line Analytical Processing (OLAP) techniques on top of Grid infrastructures. Built on top of the Open Grid Service Infrastructure (OGSA) (Reed, 2003), GridMiner-Core is an extensible framework providing a set of services like job scheduling, data distribution and heterogeneity handling, and data pre-processing and analysis support. GridMiner-Core services provide workload management, information and logging, repository access and manipulation, and replica management to upper-level services. The repository access and manipulation component is composed by *Data Access* and *Data Mediation* services. Data Access implements the data access to databases and other Grid data repositories, and Data Mediation provides a view of distributed data sources by integrating them logically into *virtual data* (VDS). The upper layer, called *Knowledge Discovery Services*, comprises data preprocessing and data mining tools. An optional component, called *Orchestration Service*, is implemented on top of this infrastructure to allow for aggregating a sequence of data mining operations into a job. It acts as a workflow engine that executes the steps involved in the complete data mining task (either sequentially or in parallel). Distributed Grid-enabled Induction of Decision Trees (DIGIDT) (Hofer, 2004) is proposed as a pilot application for GridMiner. It extends the concepts proposed in SPRINT (Shafer, 1996) but has a modified data partitioning scheme and workload management strategy.

Grid-based AdapTive Execution on Streams (GATES) (Chen, 2004) is a middleware[7] for processing distributed data streams based on OGSA and Globus 3. It provides self-adaptation mechanisms to face real-time stream processing constraints and automatic resource discovery mechanisms. GATES is a three-layered system including: design layer, deployment layer and execution layer. The first layer distinguishes between the developer and the user of an application. The developer divides an application into stages, chooses adjustment parameters, implements the processing at each stage and finally permits the user to start the application by sending him the URL link to the configuration file. The user activities are only starting and stopping an application. In the second layer a *Launcher* gets the configuration file passed by the user, analyzes it then sends it to the *Deployer*. The *Deployer* is responsible for initiating the different stages of the computation at the available resources. The third layer is composed by the instances of GATES grid services which make network connections with each other and execute the *stage* functionalities to perform the desired computations. The execution schema is thus a pipeline of stages where each stage is composed by one or more nodes executing a GATES Grid service.

The *Knowledge Grid* (Cannataro, 2003) is a high-level, service-oriented framework[8] providing for Grid-based data mining tools and services. The Knowledge Grid system uses the low-level features of Globus allowing for the creation of geographically distributed data mining applications and related tasks, such as data management and knowledge representation. The system architecture is organized into two service layers built on top of generic Grid services:

- The *Core K-grid Layer* supports definition, composition, execution, and metadata management of data mining computations over the Grid;
- The *High-level K-grid Layer* provides services and tools for data access, tools and algorithm access, execution plan management, and results presentation.

Such services have been implemented through a software prototype, named *VEGA* (Visual Environment for Grid Applications), which is a visual environment for composing complex grid applications. Recent developments of the Knowledge Grid (Congiusta, 2007) have accomplished a re-design and re-implementation of all the Knowledge Grid services as WSRF-compliant Grid services. In such latter version, the Knowledge Grid exposes two main set of services: Resource Management Services and Execution Management Services. Through such services it is possible to design and execute knowledge discovery applications. In particular, the design phase is followed by an application-management phase, in which the application model is further processed and causes interactions and information flows between services at the different levels of the architecture. This is aimed at refining the specification, generating *execution plans*, and submitting the application for execution, according to specific requirements and invocation patterns mandated and managed by the environment.

Very similar to GridMiner and the Knowledge Grid, the *DataMiningGrid* (DMG) system[9] allows for the composition of complex workflows of existing data mining applications and their execution in Grid environments (Stankovski, 2007). Built on top of the Globus Toolkit and other Grid standards, the DataMiningGrid system comprises two layers: *DMG high-level Services* and *DMG applications*. The three high-level services of the DMG system are represented by:

- *DMG Resource Broker* that provides resources discovery and selection and job management (submission, monitoring, etc.);
- *Information Services* which discover, characterize and monitor resources, services and computation;
- *Data Services* which are based on OGSA-DAI for data management.

The two DMG applications are the workflow editor and manager, based on Triana (Churches, 2005), and the Web client that is intended for end users with limited knowledge of data mining and Grid technologies. It also permits the user to add new data mining applications to the framework. The main advantage of the DMG system is the support for fine-grained parallelism and its efficient resource broker.

KDDML-G (Romei, 2007) is a system for knowledge discovery on the Grid facilitating the analysis of immovable and confidential distributed data. KDDML-G is built by porting the *KDDML (KDD Markup Language)* system (Romei, 2006)

Table 9. General purpose Grid-based systems

System	Standard	Middleware	Workflow management	Application domain
TeraGrid	*WSRF*	*Globus 4*	*BPEL*	*Life Sciences*
InfoGrid	*OGSA*	*Globus 4*	*ad-hoc*	*Data integration*
ADaM	*SOAP*	*SOAP::Lite*	*ad-hoc*	*Scientific imaging*
Discovery Net	*OGSA*	*Unicore*	*DPML*	*Life Sciences*
GridMiner	*OGSA*	*Globus 4*	*ad-hoc*	*Independent*
GATES	*OGSA*	*Globus 3*	*pipeline*	*Data streams*
Knowledge Grid	*WSRF*	*Globus 4*	*ad-hoc*	*Independent*
DataMiningGrid	*WSRF*	*Globus 4*	*Triana*	*Independent*
KDDML-G	*none*	*none*	*Query composition*	*Independent*

onto the Grid. The support of the Grid environment is accomplished by extending the original KDDML architecture with suitable components and functionalities, such as query specification, execution coordination, result visualization, and *KPQ (KDDMLProcessingQuery)* which provides an interface between the GUI layer and the KDDML core layer. The core-layer components are:

- The Resource Discoverer,
- The Constraint Compiler, and
- The Execution Coordinator.

The *Resource Discoverer* is responsible for providing XML documents containing Grid resource information (i.e. location, size, mobility of the data sets). The *Constraint Compiler* transforms the constraints generated by KPQ to be subsequently used by the ASSIST programming environment (Romei, 2006). The *Execution Coordinator* interprets the mapping of the query and coordinates the parallel and distributed execution of sub-queries. A user interface assists the user during the query composition and the results visualization phases. Although the system is intended for running in Grid environments, it has only been tested on a cluster machine.

Table 9 presents a summary of the major general-purpose systems available for Grid-based data mining, the supported standards, the used middleware, the workflow support, and the application domains. Most of the reviewed systems is built on top of Globus, provides an ad-hoc workflow manager, and follows the OGSA architecture.

Algorithms Ported to Grid

The Grid-based Distributed Max-Miner (GridDMM) (Luo, 2006) is an algorithm for mining maximal frequent itemsets from databases on a data-Grid system. To deal with the Grid environment, it is designed to have low communication and synchronization overhead. GridDMM implements a local mining phase and a global mining phase. During the local mining phase, each node mines the local database to discover the local maximal frequent itemsets. In the global mining phase they form a set of maximal candidate itemsets for a top-down search. The global candidate itemsets are stored on a prefix-tree. Counts are exchanged between neighbor nodes using the Grid-enabled Message Passing Interface (MPICH-G2). Experiments on GridDMM have been done using a cluster of workstations and the Globus Toolkit; they show that GridDMM algorithm has better performance than the CD algorithm. Its performance is scalable in terms of the database size and the number of nodes.

The *DisDaMin* project (DIStributed DAta MINing) (Fiolet, 2006) proposes new exploitable algorithms for Grids. First DisDaMin fragments the data using clustering methods then uses asynchronous collaborative techniques adapted to the specificities of Grid execution. The use of this fragmentation method makes it possible to carry out optimal local processing on each node, with a minimum amount of communications. Simulations were performed on the French national Grid GRID5000 (built on top of Xtremweb middleware) showing the efficiency of DisDAMin.

A method for integrating the Apriori algorithm in distributed databases with the Globus Toolkit is proposed in Aflori (2007). Yang (2005) suggested an architecture for decision trees in Grid environments. This architecture has been experimented with a modified version of the SPRINT algorithm. A study of different parallel implementations of Self-Organizing Maps for clustering analysis in a dynamic Grid environment is proposed in Luengo (2004). Experimentations made by these works show the flexibility provided by Grid services and the time overhead they incur.

Table 10 presents a summary of the major algorithms ported to Grid according to the supported standards, the used middleware, and the target application. The dominant middleware is Globus, whereas, often communications are based on MPI.

Table 10. Algorithms ported to Grid

Algorithm	Standard	Middleware	Target application
GridDMM	*MPICH-G2*	*Globus 4*	*Maximal association rules*
DisDaMin	*MPICH-G2*	*Java RMI*	*Association rules and clustering*
Apriori	*SOAP*	*Globus 3*	*Association rules*
Decision Tree	*MPICH-G2*	*Globus 2*	*Classification*
Self-Organizing Maps	*MPICH*	*MPICH-P4*	*Clustering*

Data Mining Suites Ported to Grid

Research has also been done to adapt *Weka* (Witten, 2005), a widely used java based data mining tool, to the Grid infrastructure. Weka provides a large collection of machine learning algorithms written in Java for data pre-processing, classification, clustering, association rules, and visualization, which can be invoked through a common graphical user interface. Some of the projects working on this objective include: *GridWeka*, *WekaG*, *Weka4WS* and Federated Analysis Environment for Heterogeneous Intelligent Mining (*FAEHIM*).

GridWeka (Khoussainov, 2004) is an ongoing work at the University of Dublin, which distributes classification algorithms using the cross-validation method over computers in an ad-hoc Grid. The system includes two main components: *Weka Server* and *Weka Client*. Each participating machine runs the original Weka as server component. The Weka client allows users to specify the target algorithms and datasets, and also to specify resources and tasks constraints. Even if GridWeka provides a way to use multiple resources to execute distributed data mining tasks, there are some criticisms about its design that make the system able to work only within an ad-hoc environment, which does not constitute a Grid per se. Also interoperability is greatly limited.

Another system using the client-server architecture is *WekaG* (Pérez, 2005). The WekaG server implements data mining algorithms while the client side instantiates grid services. WekaG is an implementation of a more general architecture called *Data Mining Grid Architecture* (*DMGA*) based on Weka and Globus Toolkit 3. WekaG provides the same user interface as Weka and can support parallel algorithms.

Weka4WS (Talia, 2005) is a framework that extends the open-source Weka toolkit (Weka) to support distributed data mining in Grid environments. Weka4WS also extends the Weka GUI to enable the invocation of the data mining algorithms that are exposed as Web Services on remote machines. The architecture of Weka4WS includes three kinds of nodes: *storage nodes*, which contain the datasets to be mined; *compute nodes*, on which remote data mining algorithms are run; *user nodes*, which are the users' local machines. Remote execution is managed using basic WSRF mechanisms (state management, notifications, etc.), while the Globus Toolkit 4 services are used for standard Grid functionalities, such as security and file transfer.

Very similar is the Federated Analysis Environment for Heterogeneous Intelligent Mining (*FAEHIM*)) project (Taylor, 2005), aimed at supporting data mining and the Grid services approach. The main data mining tasks being supported include:

classification, clustering, association rule mining, and visualization. The algorithms implementations are derived from the Weka library and converted into a set of Grid services. Complex mining processes are designed and enacted using the Triana workflow engine.

Table 11 presents a summary of the major works aimed at porting the Weka suite to the Grid. Details are provided on the system architecture, the supported standards, the used middleware, and the workflow support. Is it possible to observe that most of the reviewed systems are built on top of Globus, use existing workflow standards, and are based on Web/Grid services.

Comparison and Discussion

This section discusses and compares the reviewed projects by taking into considerations the following criteria:

- *Workflow support*, means employed to provide complex tasks composition and management,
- *Grid standards*, the Grid standards supported by the system, and
- *Middleware*, Grid middleware on top of which the system is built.

Teragrid is the largest storage and computing infrastructure including managed services to build scientific applications. But it only offers the possibility to use specific tools, it is not possible not to deploy or integrate custom solutions. By contrast, Infogrid provides an integration mechanism based on Wrappers. Infogrid is more flexible, it permits the user to have individualized views of the resources and dynamic data integration by providing a universal query system and a wrapper interface. Teragrid uses Globus 4 as Grid middleware, OGSA-DAI for data management and implements WSRF-compliant services, Infogrid makes only use of OGSA-DAI.

The benefits of Infogrid have been recognized by the Discovery Net project, which has adopted its functionalities for supporting data integration. The main advantage of Discovery Net is the use of the DPML language for workflow representation. ADaM has its proper composition mechanism, GridMiner provides a workflow engine for sequential or parallel executions, GATES uses a pipeline model, the Knowledge Grid provides a graphical interface (VEGA) for designing complex applications, DataMiningGrid is based on the Triana engine, while KDDML-G provides a graphical interface for query composition.

The WSRF standard is supported by the Knowledge Grid, DataMiningGrid, and TeraGrid systems whereas Discovery Net, Grid Miner, and GATES are only based on the general OGSA architecture. ADaM and KDDML-G architecture do not adopt any Grid standard.

Table 11. Data mining suites ported to Grid

System	Architecture	Standard	Middleware	Workflow
GridWeka	Client/Server	none	none	none
WekaG	Client/Server	OGSA/OGSI	Globus 3	ad-hoc
Weka4WS	Grid Services	WSRF	Globus 4	Based on Weka workflow engine
FAEHIM	Web Services	SOAP	Apache Axis	Triana

Figure 3. Taxonomy of Grid-based data mining systems and algorithms

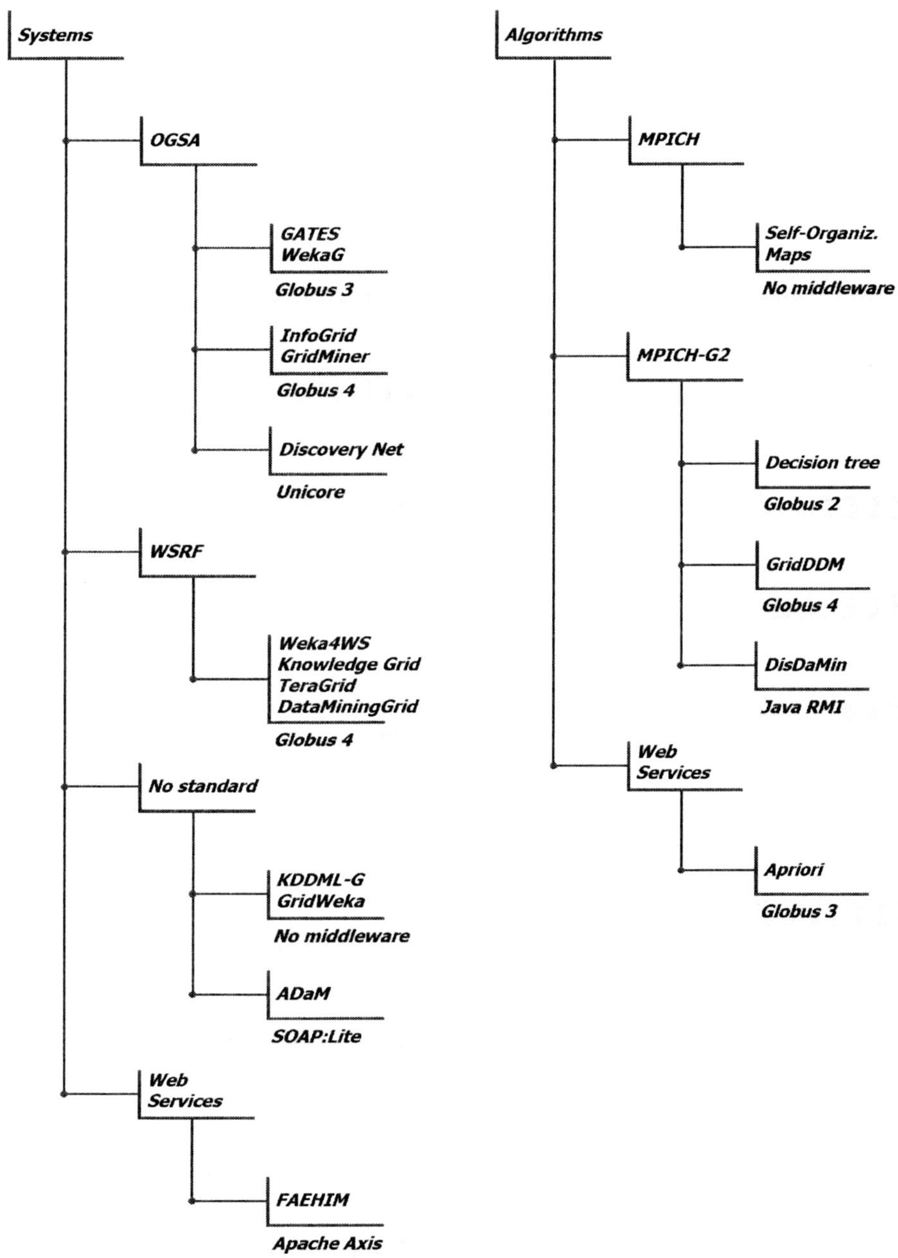

Most of the reviewed systems are based on the Globus middleware. GridMiner, the Knowledge Grid, TeraGrid, InfoGrid, and DataMiningGrid are based on Globus 4; GATES is based on Globus 3. Discovery Net uses Unicore and KDDML-G was experimented on a Beowulf cluster.

ADaM can be considered as a toolkit rather than a framework. GATES could be used to build a more complete framework but it seems to be abandoned since 2006. KDDML is a promising middleware language and system that fills the main requirements of a knowledge grid system.

The Knowledge Grid, GridMiner and DataMiningGrid are the most complete ongoing projects embracing the emerging trends and providing an almost complete set of functionalities.

GridWeka and WekaG tried to adapt Weka to a Grid infrastructure using the client-server architecture. But this architecture is not so flexible and powerful to support large scale requirements such as security, resource management etc. By contrast, Weka4WS and FAEHIM are based on a services approach. The former provides the design of complex discovery processes by porting the Weka KnowledgeFlow to Web services, and the latter uses Triana.

While most of such systems are based on Grid technologies and provide remote and concurrent execution, they do not support parallel data mining algorithms described in the first section, neither complex distributed scenarios described in the second section of this chapter; they only support ensemble learning and meta-learning techniques. Some initiatives for the design of new algorithms fitting the constraints of Grid environments were also reported. One of the most convincing is GridDMM, which demonstrated good performance and scalability on a cluster of workstations using Globus.

Grid-based data mining systems and algorithms can be classified by the supported standard and the used middleware. This classification can be summarized by the taxonomy of Figure 3.

FUTURE TRENDS

The availability of new computing platforms and paradigms has for sure driven the majority of the novelties registered in the last years in information technology; in particular Grid computing has represented one of the most demanding, challenging, and promising of such platforms. Based on such evolution, also new visions and scenarios can be portrayed for data mining: they propose the emergence of complex long-lived distributed knowledge discovery applications involving a large number of world-wide computing nodes, each hosting datasets and/or mining algorithms, and in which all the complexities due to the management of heterogeneities, applications, and the infrastructure are handled by appropriate high level services and programming facilities.

The subject of data mining in life science, while relatively young compared to data mining in other application fields, such as finance and marketing, is already benefitting from classical and more recent mining techniques. On the basis of the outlined evolutions and improvements that have involved mining algorithms and systems, it is foreseeable an ever more tight coupling of data mining and life sciences, especially in the distributed and Grid fields. It is worth observing, in fact, that life science is often characterized as a community, in which cooperation, interchanges, and interoperability play a major role; while, at the same time, many Grid systems are today adopting service orientation, which is an enabling approach for cooperative applications. Along this direction, other further, possible key points are for sure the SaaS (software as a service) and the virtual organization paradigms.

CONCLUSION

Data mining techniques are key elements for knowledge discovery applications aimed at extracting valuable knowledge, implementing business intelligence strategies, and improve company competitiveness.

Life science is an important application domain in which continuous advances in data generation devices, data acquisition, and storage technology have enabled biomedical research and healthcare organizations to accumulate vast amounts of heterogeneous data that can be exploited for achieving important new discoveries or therapeutic improvements. Extracting useful knowledge from such data has proven extremely challenging however. Traditional data analysis and mining tools and techniques are often inefficient because of the massive size of the involved data sets and the non-traditional nature of the biomedical data.

In this chapter a significant set of data mining techniques has been presented, in particular those addressing the exploitation of large and remotely dispersed datasets and/or high-performance computers, and those proposing interesting new approaches suitable to address an ever more extensive set of challenging applications. They have been categorized into parallel, distributed, and Grid-based data mining.

Comparisons and evaluations on the presented techniques and approaches have been discussed throughout the different sections of the chapter, providing insights on their main peculiarities and applications, as well as discussions on the fundamental properties that permitted to categorize them.

ACKNOWLEDGMENT

Moez Ben HajHmida wants to acknowledge Yahya Slimani for the precious time, the guidance, and the help provided. Antonio Congiusta is thankful to Domenico Talia for having shared his advises and suggestions on this work.

REFERENCES

Abraham, A., & Nath, B. (2000). Hybrid heuristics for optimal design of artificial neural networks. In R. John & R. Birkenhead (Eds.), *Advances in Soft Computing Techniques and Applications* (pp. 15-22). Berlin, Germany: Springer-Verlag.

Abraham, A., Grosan, C., & Ramos, V. (Eds.). (2006). *Swarm Intelligence in Data Mining, Studies in Computational Intelligence*. Berlin, Germany: Springer-Verlag.

Aflori, C., & Craus, M. (2007). Grid implementation of the Apriori algorithm. *Advanced Engineering Software*, *38*(5), 295-300.

Agrawal, R., Mannila, H., Srikant, R., Toivonen, H., & Verkamo, A. I. (1996). Fast discovery of association rules. In , U. Fayyad et al. (Eds.), *Advances in Knowledge Discovery and Data Mining* (pp. 307-328). Menlo Park, CA: AAAI Press.

Ali, A. S., & Taylor, I. J. (2005). Web services composition for distributed data mining. In *The 2005 IEEE International Conference on Parallel Processing Workshops* (pp. 11-18).

Alsairafi, S., Emmanouil, F. S., Ghanem, M., Giannadakis, N., Guo, Y., Kalaitzopoulos, D., et al. (2003). The design of discoverynet: Towards open grid services for knowledge discovery. *International Journal of High Performance Computing and Applications*, *17*(3), 297-315.

Berman, F. (2001). Viewpoint: From teragrid to knowledge grid. *Commun. ACM*, *44*(11), 27-28.

Boser, B. E., Guyon, I. M., & Vapnik, V. N. (1992). A training algorithm for optimal margin classifiers. *In Proceedings of the fifth annual workshop on Computational learning theory* (pp. 144-152). New York: ACM.

Brezany, P., Hofer, J., Tjoa, A., & Wohrer, A. (2003). Gridminer: An infrastructure for data mining on computational grids. *In Data Mining on Computational Grids APAC'03*.

Brin, S., Motwani, R., Ullman, J. D., & Tsur, S. (1997). Dynamic itemset counting and implication rules for market basket data. In J. Peckham (Ed.), *International Conference on Management of Data* (pp. 255-264). New York: ACM Press.

Cannataro, M. & Talia, D. (2003). The knowledge grid: An architecture for distributed knowledge discovery. *Commun. ACM*, *46*(1), 89-93.

Catlett, C. et al (2007). TeraGrid: Analysis of Organization, System Architecture, and Middleware Enabling New Types of Applications. In L. Grandinetti (Ed.), *Advances in Parallel Computing*. Amsterdam, the Netherlands: IOS Press.

Cheeseman, P., & Stutz, J. (1996). Bayesian classification (autoclass): Theory and results. *In Advances in Knowledge Discovery and Data Mining* (pp. 153-180).

Chen, L., Reddy, K., & Agrawal, G. (2004). Gates: A grid-based middleware for processing distributed data streams. *In the 13th International Symposium on High Performance Distributed Computing* (pp. 192-201). New York: IEEE Computer Society.

Cheung, D. & Xiao, Y. (1998). Effect of data skewness in parallel mining of association rules. *In the Second Pacific-Asia Conference on Research and Development in Knowledge Discovery and Data Mining* (pp. 48–60). Berlin: Springer-Verlag.

Cho, V., & Wüthrich, B. (2002). Distributed mining of classification rules. *Knowledge Information Systems*, *4*(1), 1-30.

Churches, D., Gombas, G., Harrison, A., Maassen, J., Robinson, C., Shields, M., Taylor, I., & Wang, I. (2005). Programming scientific and distributed workflow with triana services. *Concurrency and Computation: Practice and Experience*, *18*(10), 1021-1037.

Cohen, W. W. (1995). Fast effective rule induction. In A. Prieditis and S. Russell (Eds.), *12th International Conference on Machine Learning*, (pp 115-123). San Francisco, CA: Morgan Kaufmann.

Collobert, R. , Bengio, S., & Bengio, Y. (2002). A parallel mixture of svms for very large scale problems. In *Advances in Neural Information Processing Systems*. Cambridge, MA: MIT Press.

Congiusta, A., Talia, D., & Trunfio, P. (2007). Distributed data mining services leveraging WSRF. *Future Generation Computing Systems*, *23*(1), 34-41.

Dhillon, I. S., & Modha, D. S. (2000). A data-clustering algorithm on distributed memory multiprocessors. In *Large-Scale Parallel Data Mining* (pp. 245-260). Lecture Notes in Artificial Intelligence.

Dietterich, T. G. (2000). An experimental comparison of three methods for constructing ensembles of decision trees: Bagging, boosting, and randomization. *Machine Learning*, *40*, 139-157.

Fan, W., Stolfo, S. J., & Zhang, J. (1999). The application of adaboost for distributed, scalable and on-line learning. In *The fifth ACM SIGKDD international conference on Knowledge discovery and data mining* (pp. 362-366). New York: ACM.

Fiolet, V., Olejnik, R., Lefait, G., & Toursel, B. (2006). Optimal grid exploitation algorithms for data mining. *In the Fifth International Symposium on Parallel and Distributed Computing* (pp. 246-252). New York: IEEE Computer Society.

Foster, I., & Kesselman, K. (Eds.). (1999). *The Grid: Blueprint for a New Computing Infrastructure*. San Francisco, CA: Morgan Kaufmann.

Foti, D., Lipari, D., Pizzuti, C., & Talia, D. (2000). Scalable parallel clustering for data mining on multicomputers. *In the 15th Workshops on Parallel and Distributed Processing* (pp. 390-398). Berlin, Germany: Springer-Verlag.

Freitas, A. A. (1998). A survey of parallel data mining. In *2nd International Conference on the Practical Applications of Knowledge Discovery and Data Mining* (pp. 287-300). London: The Practical Application Company.

Freitas, A. A. (Ed.). (2002). *Data Mining & Knowledge Discovery with Evolutionary Algorithms*. Berlin, Germany: Springer-Verlag.

Freund, Y. & Schapire, R. E. (1997). A decision-theoretic generalization of on-line learning and an application to boosting. *Journal of Computing System Science*, *55*(1), 119-139.

Giannadakis, N., Rowe, A., Ghanem, M., & Guo, Y. (2003). Infogrid: providing information integration for knowledge discovery. *Information Sciences 155*, 199-226.

Guan, J. W., & Bell, D. A. (1991). *Evidence Theory and Its Application*. Amsterdam, the Netherlands: Elsevier Science.

Guarracino, M. R., Cifarelli, C., Seref, O., & Pardalos, P. M. (2006). A Parallel Classification Method for Genomic and Proteomic Problems. *In 20th international Conference on Advanced information Networking and Applications* (Vol. 02) (pp. 588-592). AINA. Washington, DC: IEEE Computer Society,.

Hall, L. O., Chawla, N., & Bowyer, K. W. (1998). Combining decision trees learned in parallel. In *Working Notes of the KDD-97 Workshop on Distributed Data Mining* (pp. 10-15).

Hofer, J. (2004). *Distributed decision tree induction within the grid data mining framework gridminer-core* (Tech. Rep. No. 2004-04). Vienna, Austria: University of Vienna, Institute for Software Science.

Januzaj, E., Kriegel, H. P., & Pfeifle, M. (2004). Scalable density-based distributed clustering. *In the 8th European Conference on Principles and Practice of Knowledge Discovery in Databases* (pp. 231-244). Berlin, Germany: Springer-Verlag.

Jithesh, P. V., Kelly, N., Wasnik, S., Donachy, P., Harmer, T., Perrott, R., et al. (2005). Bioinformatics application integration in genegrid. *In the UK e-Science All Hands Meeting.*

Joshi, A., Karypis, G., & Kumar, V. (1998). ScalparC: A new scalable and efficient parallel classification algorithm for mining large datasets. In *11th Inter national Parallel Processing Symposium.* New York: IEEE Computer Society Press.

Jourdan L., Dhaenens C., Talbi E.-G. (2004). A Parallel adaptive GA for linkage disequilibrium in genomics. NIDISC'2004 Workshop. *IEEE International Parallel & Distributed processing symposium (IPDPS 2004),* Santa Fe, NM.

Kargupta, H., & Chan, P. (Eds.) (2000). *Advances in Distributed and Parallel Knowledge Discovery.* Cambridge, MA: MIT Press.

Khoussainov, R., Zuo, X., & Kushmerick, N. (2004). Grid-enabled weka: A toolkit for machine learning on the grid. *ERCIM News, 59.*

Lazarevic, A., & Obradovic, Z. (2002). Boosting algorithms for parallel and distributed learning. *Distributed and Parallel Databases, 11*(2), 203-229.

Lazarevic, A., Pokrajac, D., & Obradovic, Z. (2000). Distributed clustering and local regression for knowledge discovery in multiple spatial databases. *In 8th European Symposium on Artificial Neural Networks* (pp. 129-134).

Lippmann, R. P. (1987). An introduction to computing with neural nets. *IEEE ASSP Magazine, 4*(22), 7-25.

Luengo, F., Cofino, A. S., & Gutierrez, J. M. (2004). Grid oriented implementation of self-organizing maps for data mining in meteorology. *Lecture Notes in Computer Science, 2970*, 163-171.

Luo, C., Pereira, A. L., & Chung, S. M. (2006). Distributed mining of maximal frequent itemsets on a data grid system. *Journal of Supercomputing, 37*(1), 71-90.

Nagesh, H., Goil, S., & Choudhary, A. (1999). *Mafia: Efficient and scalable subspace clustering for very large data sets* (Tech. Rep. No. 9906-010), Center for Parallel and Distributed Computing.

Pérez, M. S., Sànchez, A., Herrero, P., Robles, V., & Pena, J. M. (2005). Adapting the weka data mining toolkit to a grid based environment. *In 3rd Atlantic Web Intelligence Conference* (pp. 492-497).

Provost, F. P., & Hennessy, D. N. (1996). Scaling up: Distributed machine learning with cooperation. *In the Thirteenth National Conference on Artificial Intelligence* (74-79). Menlo Park, CA: AAAI Press.

Quinlan, J. R. (Ed.). (1993). *C4.5: Programs for Machine Learning.* San Francisco, CA: Morgan Kaufmann.

Reed, D. A., Mendes, C. L., Lu, C., Foster, I., & Kesselman, C. (Eds.). (2003). *The Grid 2: Blueprint for a New Computing Infrastructure - Application Tuning and Adaptation.* San Francisco, CA: Morgan Kaufman.

Romei, A., Ruggieri, S., & Turini, F. (2006). Kddml: a middleware language and system for knowledge discovery in databases. *Data Knowledge Engineering, 57*(2), 179-220.

Romei, A., Sciolla, M., Turini, F., & Valentini, M. (2007). Kddml-G: a grid-enabled knowledge discovery system. *Concurr. Comput. : Pract. Exper., 19*(13), 1785-1809.

Rushing, J., Ramachandran, R., Nair, U., Graves, S., Welch, R., & Lin, H. (2005). ADaM: a data mining toolkit for scientists and engineers. *Computers and Geosciences, 31*, 607-618.

Samatova, N. F., Ostrouchov, G., Geist, A., & Melechko, A. (2002). RACHET: An Efficient Cover-Based Merging of Clustering Hierarchies from Distributed Datasets. *Distributed and Parallel Databases, 11*(2), 157-180.

Schuster, A., & Wolff, R. (2001). Communication-efficient distributed mining of association rules. *In ACM SIGMOD international conference on Management of data* (pp. 473-484). New York: ACM.

Schuster, A.,Wolff, R., & Trock, D. (2003). A High-Performance Distributed Algorithm for Mining Association Rules. In *Third IEEE International Conference on Data Mining.*

Shafer, J. C., Agrawal, R., & Mehta, M. (1996). SPRINT: A scalable parallel classifier for data mining. In T. M. Vijayaraman, A. P. Buchmann, C. Mohan, & N. L. Sarda (Eds.), *22nd Int. Conf. Very Large Databases* (pp. 544-555). San Francisco, CA: Morgan Kaufmann.

Sinnott, R., Atkinson, M., Bayer, M., Berry, D., Dominiczak, A., Ferrier, M., Gilbert, D., Hanlon, N., Houghton, D., Hunt, E., & White, D. (2004). Grid services supporting the usage of secure federated, distributed biomedical data. *In the UK e-Science All Hands Conference. Engineering and Physical Sciences Research Council.*

Skillicorn, D. B. (2002). The case for datacentric grids. *In the 16th International Parallel and Distributed Processing Symposium* (p. 313). New York: IEEE Computer Society.

Srivastava, A., Han, E., Kumar, V., & Singh, V., (1999). Parallel formulations of decision-tree classification algorithms. *Data Mining and Knowledge Discovery, 3,* 237-261.

Stankovski, V., Swain, M., Kravtsov, V., Niessen, T., Wegener, D., Kindermann, J., & Dubitzky, J. (2007). Grid-enabling data mining applications with dataminininggrid: An architectural perspective. *Future Generation Computing Systems, 24*(4), 259-279.

Talia, D., Trunfio, P., & Verta, O. (2005). Weka4WS: A WSRF-Enabled Weka Toolkit for Distributed Data Mining on Grids. *In the 9th European Conference on Principles and Practice of Knowledge Discovery in Databases, 3721,* 309-320.

Ting, K. M., & Witten, I. H. (1997). Stacked generalizations: When does it work? *In International Joint Conference on Artificial Intelligence* (pp. 866-873).

Toivonen, H. (1996). Sampling large databases for association rules. In T. M. Vijayaraman, A. P. Buchmann, C. Mohan, and N. L. Sarda (Eds.),

International Conference on Very Large Data Bases. San Francisco, CA: Morgan Kaufman.

Tsoumakas, G. & Vlahavas, I. (2002). Distributed data mining of large classifier ensembles. *In the Second Hellenic Conference on Artificial Intelligence* (pp. 249-256).

Tveit, A., & Engum, H. (2003). Parallelization of the Incremental Proximal Support Vector Machine Classifier using a Heap-based Tree Topology. In *Parallel & Distributed computing for Machine Learning*.

Witten, I. H., & Frank, E. (Eds.). (2005). *Data mining: Practical machine learning tools and techniques*. San Francisco, CA: Morgan Kaufmann.

Wolpert, D. H. (1992). Stacked generalization. *Neural Networks*, *5*, 241-259.

Wüthrich, B. (1997). Discovering probabilistic decision rules. *International Journal of Intelligent Systems in Accounting, Finance and Management*, *6*(4), 269-277.

Yang, C. T., Tsai, S. T., & Li, K. C. (2005). Decision tree construction for data mining on grid computing environments. *In the 19th International Conference on Advanced Information Networking and Applications* (pp. 421-42). New York: IEEE Computer Society.

Zaki, M. J. (1999). Parallel and Distributed Association Mining: A Survey. *IEEE Concurrency*.

Zaki, M. J., & Ho, C. T. (Eds.). (2000). *Large-Scale Parallel Data Mining*. Lecture Notes in Computer Science. Berlin, Germany: Springer.

KEY TERMS AND DEFINITIONS

Association Rules: Describe frequent co-occurrences in sets.

Classification: Technique for learning a function from training data labeled by class membership.

Clustering: A way to form clusters of patterns without a-priori knowledge.

Distributed Computing: Method of computer processing in which different parts of a program run on two or more computers communicating with each other over a network.

Data Mining: A subset process of knowledge discovery. It is concerned with the application of mining algorithms to data.

Grid Computing: Based on a parallel and distributed system that enables the sharing, selection, and aggregation of geographically distributed autonomous resources dynamically and at runtime, depending on their availability, capability, performance, cost, and users' quality-of-service requirements.

Knowledge Discovery: Process of automatically finding novel, interesting, and useful patterns in large volumes of data.

Parallel Computing: Form of computing in which many instructions are carried out simultaneously.

ENDNOTES

1 http://genomics.energy.gov
2 http://lsdis.cs.uga.edu
3 http://www.teragrid.org
4 http://datamining.itsc.uah.edu/adam
5 http://www.lesc.ic.ac.uk/projects/dnet.html
6 http://www.gridminer.org
7 http://www.cse.ohio-state.edu/~chenlia/gates/gates.html
8 http://grid.deis.unical.it/kgrid
9 http://www.dataminninggrid.org

Section II

Grids for Genomics and Proteomics

Chapter VII
High Performance BLAST Over the Grid

Vincent Breton
IN2P3, CNRS, Clermont-Ferrand, France

Eddy Caron
Université de Lyon, LIP, CNRS-ENS-Lyon-UCBL-INRIA, France

Frédéric Desprez
INRIA, Université de Lyon, LIP, CNRS-ENS-Lyon-UCBL-INRIA, France

Gaël Le Mahec
CNRS, IN2P3, UBP Clermont-Ferrand, France

ABSTRACT

As grids become more and more attractive for solving complex problems with high computational and storage requirements, bioinformatics starts to be ported on large scale platforms. The BLAST kernel, one of the main cornerstone of high performance genomics, was one the first application ported on such platform. However, if a simple parallelization was enough for the first proof of concept, its use in production platform needed more optimized algorithms. In this chapter, we review existing parallelization and "gridification" approaches as well as related issues such as data management and replication, and a case study using the DIET middleware over the Grid'5000 experimental platform.

INTRODUCTION

Genomics acquiring programs, such as full genomes sequencing projects, are producing larger and larger amounts of data. The analysis of these raw biological data requires very large computing resources. Functional sites and signatures of proteins are very useful for analyzing these data or for correlating different kinds of existing biological data. These methods are applied, for example, to the identification and characterization of the potential functions of new sequenced proteins, and to the clusterization into protein families of the sequences contained in international databanks. The sites and signatures of proteins can be expressed by using the syntax defined by the PROSITE databank, and written as a "protein regular expression". Searching one such site in a sequence can be done with the criterion of the identity between the searched and the found patterns. Most of the time, this kind of analysis is quite fast. However, in order to identify non perfectly matching but biologically relevant sites, the user can accept a certain level of error between the searched and the matching patterns. Such an analysis can be very resource consuming. Parallel computing (Zomaya, 2006) and Grids (Krishnan, 2004) are then obvious responses to performance issues.

Genomics involving BLAST (Basic Local Alignment Search Tool) applications (Altschul, Gish, Miller, Myers & Lipman, 1990) (Korf, Yandell, & Bedell, 2003) are the common tool for searching sequences similarities within protein and DNA databases. The BLAST is used to search homologies between amino-acids or nucleotides sequences. It quantifies similarities using some sensibility parameters and a similarity score matrix. Sequences are known as similar if they have enough common characters and not too many differences, which can be substitutions, deletions or insertions. The search for similarities between sequences can be done at a global or local level. At the global level, we search for the best alignment between two sequences permitting large different areas inside the sequences (Needleman, & Wunsch, 1970). At the local level, we search subsequences alignments, not using the subsequences without any alignment for the total similarity score computing (Goad, & Kanehisa, 1982) (Smith, & Waterman, 1981).

However, as the size of available databases increases (now up to hundreds of Gigabytes) (NCBI. Ncbi website. http://www.ncbi.nih.gov/), the search of many sequences is time consuming. Many researches have been done to get faster algorithms. Their goal is to reduce the search scope while looking at high scoring alignments. Heuristics have been produced and among them the BLAST is the most popular. Still, as databases size and number still increase, the use of parallel computing becomes mandatory. Several approaches exist from parallelizing the BLAST kernel itself (and distributing the databases) to splitting the input sequences to different servers and replicating the databases. All these approaches can be combined to get the best performance.

The rest of this chapter is organized as follow. The next section will be devoted to a short presentation of Bioinformatics Grids around the world. Then, in the third section, we present the BLAST kernel and different parallelization levels. In the fourth section, we present some algorithms and heuristics used for the joint replication and scheduling problem which is used for the gridification of the BLAST kernel. The next section presents middleware used for bioinformatics application over the Grid. Finally, before a conclusion, we describe a test case of the deployment of a BLAST application over DIET, a middleware implementing the GridRPC API of the Open Grid Forum (OGF).

BACKGROUND

Several projects around the world aim at setting Bioinformatics Grids. These projects are built upon de facto standards like Globus (Globus. http://www.globus.org) or Condor (Frey, Tannenbaum, Livny, Foster, & Tuecke, 2002).

The main challenge for data analysis in life sciences is to offer to the molecular biologists an integrated and up-to-date view of an exponentially growing volume of data in a multiplicity of formats. Data providers put a huge amount of effort into providing data resources that are comprehensive, user-friendly and cross-linked to other databases; but different data providers use different methods. This means that a researcher might have to search ten or more different databases to find all the information pertaining to a particular set of candidate genes. If they are doing these kinds of searches on a regular basis, they will want their own local copies of all those databases. Maintaining up-to-date and fully functioning versions of all those databases and associated tools to search and exploit them is a huge and complex task.

To address these challenges, projects such as the Canadian BioGrid (Stark, Breitkreutz, Reguly, Boucher, Breitkreutz, & Tyers, 2006), the EuroGrid BioGrid (Bala, Pytlinski, Nazaruk, Alessandrini, Girou, Erwin, Mallmann, MacLaren, Brooke, & Myklebust, 2002), the Asia Pacific BioGrid (Yang, Luo, & Lai, 2005), GenoGrid (Lavenier, Leroy, Mac Wing, Andonov, Hurfin, Raipin-Parvedy, Mouchard, & Guinand, 2003), the cancer biomedical informatics Grid (Buetow, 2003) (Covitz, Hartel, Schaefer, De Coronado, Fragoso, Sahni, Gustafson, & Buetow 2003) (Saltz, Oster, Hastings, Langella, Kurc, Sanchez, Kher, Manisundaram, Shanbhag, & Covitz, 2006), the Simdat project (Angele, & Weiten, 2005) and the Biomedical Informatics Research Network (Grethe, Baru, Gupta, James, Ludaescher, Martone, Papadopoulos, Peltier, Rajasekar, & Santini, 2005) share computing and storage resources with dedicated web interfaces to provide access to public databases and analysis tools which are integrated to the environment. In Europe, EGEE (Jacq, Blanchet, Combet, Cornillot, Duret, Kurata, Nakamura, Silvestre, & Breton, 2004) aims at deploying a secured Grid infrastructure offering resources to multiple research communities including life sciences that other projects like BioinfoGRID (Milanesi, & Merelli, 2006) exploit for data production or analysis in relation to molecular biology.

Grids should take away the need for bioinformaticians to maintain local copies of databases and software, and should allow them to fully concentrate on homogeneous, bioscientist-friendly interfaces. As a consequence, bioscientists could shift emphasis from technical database and software maintenance issues to scientific data and information integration issues. Efforts like MyGrid (Stevens, Robinson, & Goble, 2003), Proteus (Cannataro, Comito, Lo Schiavo, & Veltri, 2004), and ProGenGrid (Aloisio, Cafaro, Fiore, & Mirto, 2005) develop in silico workbenches to enable distributed data integration, workflow management and environment personalization. They are buit upon semantic web technologies, which are deployed to manipulate the biological objects and concepts while hiding all the underlying technicalities.

The development of these high level environments is eased by the development of standard interfaces to the bioinformatics tools and databases. The Embrace network of excellence (Embrace. Embrace website. http://www.embracegrid.info) explores the integration of molecular biology data through standardized access to bioinformatics resources throughout Europe and beyond. Standardization should allow turning the relationship between user and provider on its head by enabling data providers to provide well defined interfaces to their databases that will conform to the same standards, essentially creating a Grid that will allow users to make the most of dispersed data resources.

BLAST PARALLELIZATION AND GRIDIFICATION

In this section, we will survey available approaches for the parallelization of the BLAST kernel and existing results on porting this kernel over the Grid. Several levels of parallelization can be applied for the BLAST kernel used for multiple sequences alignment (De Carvalho,

& Lifschitz, 2003) (Pedretti, Casavant, Braun, Scheetz, Birkett, & Roberts, 1999) (Wu, & Tseng, 2006). As we have as an input a set of sequences to be compared against a huge database, we can first have a fine grain parallelization with all the alignments of the comparison executed in parallel. Then, a medium grain approach consists in partitioning the database and performing multiple sequences examination in parallel. Finally, multiple sequences can be processed in batch mode on a replicated database.

Global Alignments

A global alignment search consists in searching the best alignment of a sequence with another one using the entire sequences. Instead of a local alignment search, the entire sequence is used to calculate the "alignment score". The Needleman-Wunsh algorithm (Needleman, & Wunsch, 1970) is a dynamic programming algorithm which performs a global alignment between two amino-acids sequences. This algorithm searches for the optimal alignment between two complete sequences by using a similarity score matrix and a linear gap penalty to evaluate a global alignment score. An example of such a substitution matrix is given in Figure 1.

Then, the best alignment score is obtained by constructing an alignment matrix using a dynamic program which takes into account the gap penalty each time a gap is detected. For example, with the substitution matrix given in Figure 1, a gap penalty of -2 and the two sequences AAGCTAAT and AGTTAT, we obtain the alignment shown in the table below.

For the biologists, the global alignment of two sequences is generally less expressive than the local alignments, which detect homologies after mutations or lineage characterizations more efficiently.

Figure 1. A simple substitution matrix for DNA sequences alignment

$$\begin{array}{c} \\ A \\ C \\ G \\ T \end{array} \begin{array}{c} \begin{array}{cccc} A & C & G & T \end{array} \\ \begin{pmatrix} 2 & -1 & -1 & -1 \\ -1 & 2 & -1 & -1 \\ -1 & -1 & 2 & -1 \\ -1 & -1 & -1 & 2 \end{pmatrix} \end{array}$$

Local Alignments

A local alignment between two amino-acids sequences detects similarity region in larger sequences. By detecting such similarities, the biologists can obtain clues about the function of the target protein/gene even if the two sequences are very different. By using a substitution matrix taking into account the specificities of the biological sequences and the mutation probabilities (for example the BLOSUM matrices (Henikoff, & Henikoff, 1992)), a local alignment can be very expressive about the sequence function in a protein. For these reasons, the local alignment searches are the most widely used by the biologists.

The Smith-Waterman algorithm (Smith, & Waterman, 1981) is one of the first algorithms used to search such local alignments. Like the Needleman-Wunsh algorithm, it uses a dynamic programming method to evaluate the best alignment between two sequences regarding to a substitution matrix and a gap penalty. To accelerate these time-consuming searches, two heuristics methods have been proposed: FASTA (Pearson, & Lipman, 1988) and BLAST (Altschul, & al. 1990)

	A	A	G	C	T	A	A	T	
	A	-	G	-	T	T	A	T	
With a score of	2	-2	+2	-2	+2	-1	+2	+2	= 5

which is probably the most widely used alignment search application today. BLAST approximates the results obtained by the previously mentioned algorithms an order of magnitude faster than using them for a comparable sensitivity. On a first step, BLAST compiles a list of high scoring words (a short sequence which produces a score over a threshold when aligned with a subsequence of the request). On a second step BLAST searches the hits of these words in the database. During the last step, the algorithm tries to extend the hits using some heuristics methods.

BLAST Kernel Parallelization

The first approach parallelizes the BLAST kernel itself. It involves accelerating the comparison of a single pair of sequences by performing the alignments in a fully parallel way. This can be done by using the parallelism available inside one processor or by a SMP machine. This approach is currently exploited in the NCBI BLAST itself which allows to choose how many processors can be used to proceed to the alignments search.

Parallelization on a SMP Machine

The NCBI BLAST application uses threads to parallelize the alignments search process. Usually, researchers submit sequences to be aligned on every sequence of a biological database. Then, the parallelization of BLAST on SMP machines simply consists in using the available processors to search alignments on different sequences of the database simultaneously. Figure 2 presents such a parallelization. In (Kim, & Han, 2006), the authors analyze this parallelization on a 8 processors SMP machine. When the query sequences are long enough, the SMP parallelization gives a speedup linear in number of processors. Another approach is to use specific hardware to perform the alignment search like IBM Blue/

Figure 2. BLAST SMP parallelization

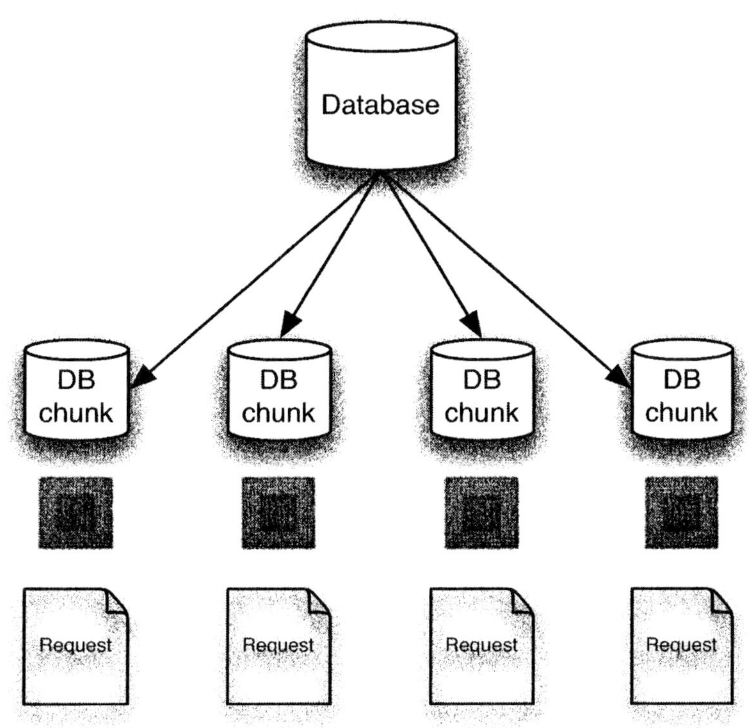

Gene (Rangwala, Lantz, Musselman, Pinnow, Smith, & Wallenfelt, 2005). Using this specific hardware, the researchers can obtain very good performance, with a speedup close to the linear one. With more than 2000 processor units, the time to search an alignment on a long query request is greatly reduced. More recently, researchers proposed a Smith-Waterman algorithm implementation using the recent GPUs parallelism capacities (Liu, Schmidt, Voss, Schroder, & Muller-Wittig, 2006). The alignment search algorithms fit to such a parallelization and the authors showed that a well designed implementation gives very good results.

Parallelization on a Distributed Memory Machine

The second approach consists in distributing the database itself (Braun, Pedretti, Casavant, Scheetz, Birkett, & Roberts, 2001) (De Carvalho, & Lifschitz, 2003) (Pedretti, & al., 1999) (Carey, Darling, & Chun Feng, 2003). Each request is distributed on different parts of the database on which it applies. However, BLAST associates to a result an expected value, which represents the number of hits with a sufficient score that would be "expected" by chance and this value is dependent on the size of the database. So, this partitioning needs to merge the results and to correct the expected values obtained after the process. All the current implementations of this principle partition the databases in equal parts. For an heterogeneous platform, it could be possible to choose a better partitioning system taking into account the storage and computation capacities of the different sites. mpiBLAST (Carey, & al., 2003) implements such a database partitioning system for a BLAST execution on cluster using MPI. By distributing tasks on several CPU, mpiBLAST performs the same parallelization than whose performed inside the NCBI BLAST described above. Moreover, mpiBLAST divides the databases into chunks. If there are enough nodes, these chunks can fit entirely in memory avoiding costly IO.

We tested the performance of mpiBLAST on large cluster by increasing number of nodes used and increasing request sizes. The measured performance show that mpiBLAST is very efficient on medium-size cluster but does not seem to be scalable to the Grid scale. Figure 3 shows the results of the tests we performed. With more than about 60 nodes, performance that were improving start to decrease. With less than 4 nodes,

Figure 3. BLAST databases partitioning with mpiBLAST (Proteins database of 150MB)

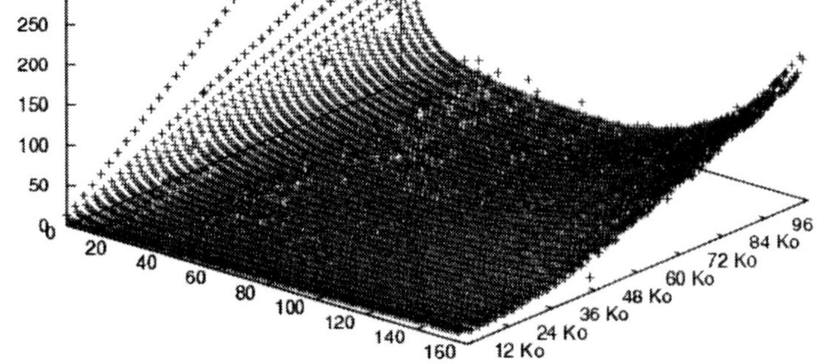

Figure 4. Requests distribution

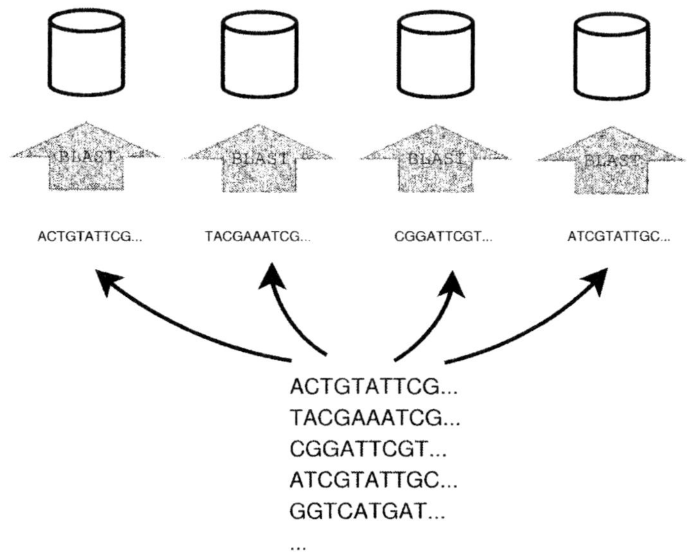

mpiBLAST is inefficient and it is more efficient to use the standard BLAST application.

Multiple Sequences Division

The last way of improving the performance of large scale BLAST over the Grid is to divide the multiple sequences. We consider a request as a set of sequences to compare to a database. Each sequence in the request can be treated on the different sites storing the database, using the different processors (Pedretti, & al., 1999). Figure 4 presents the principle of this task distribution over the different computing nodes. As for the databases partitioning, it is possible to choose the "size" of the sub-request or the number of requests inside a sub-request regarding to the nodes capacities. The results have then to be merged into a single result file to obtain the same complete response than for a local BLAST execution.

Different Levels Combination

The development of a portal for sequence comparison involves several parallelism levels on several different databases using parallel clusters accessible over a Grid. The choice of one strategy depends on the platform characteristics, the size of the input data and the size and number of databases.

To combine different levels of parallelism, we have to take into account several parameters:

- The number of sequences inside the request.
- The number of sequences inside the database.
- The number of available nodes.
- The database availability on each node.
- The number of processors on each node.
- The memory available on the nodes.

If the number of sequences in a request is low and the database very large, it is useful to divide the database and probably less efficient to distribute the sub-requests. If a node has more CPU than the number of sequences in a portion of database, some CPU are not used etc.

In a previous, we have seen that mpiBLAST can largely improve the performance of an alignment search using BLAST. However, mpiBLAST does not seem to be scalable at a Grid scale. These results have to be taken into account for the combination of the different parallelism levels.

Combining the kernel parallelization using threads on SMP computer with any other parallelization level is transparent and do not need special requirements. However, when choosing where a request should be scheduled, it is important to take into account the number of CPU available on the target cluster.

When combining database partitioning and sub-request distribution, a complex analysis is needed. The time needed to complete a request on a node depends on the size of the database partition and the number of sub-requests sent to it. The easiest parameter to control is the number of sequences in a sub-request. Indeed, the database partition process needs to reformat the database and to distribute the chunks over the nodes. However, when a request does not contain enough sequences to be distributed over the different nodes, the only way to improve the performance is to divide the database. Table 1 shows the constraints and advantages of these two parallelism levels. These two parallelization levels imply some replication needs:

- **Databases partitioning:** The databases chunks have to be replicated on each site we want to perform the search. It could be possible to adapt the chunk size to the node memory, computing, and storage capacities.
- **Requests partitioning:** The entire databases have to be replicated on each site on which we want to perform the search. The sub-requests can have different sizes to be scheduled on the nodes with different computing capacities.
- **Databases and requests partitioning:** Databases have to be replicated on the different sites. We can replicate the entire databases or their chunks. The scheduling policy has to take into account if a request is performed on a complete database or on

Table 1. Different BLAST parallelizations

Database partitioning	Request partitioning
✔ The databases chunks can fit in memory avoiding IO access to disk. ✔ Performance increases even with requests containing few sequences. ✔ Only databases chunks have to be replicated avoiding to replicate the entire databases on the nodes. ✘ Not easily scalable to the grid scale. ✘ Needs the correction of the results *e-values*. ✘ The databases chunk have to be distributed over the nodes.	✔ A very large number of sequences can be treated simultaneously. ✔ Highly and easily scalable. ✘ Small sets of sequences treatments don't take profit of all the available nodes. ✘ Needs IO disk access if the databases cannot fit in memory. ✘ The entire databases have to be replicated on the nodes.
Database & Request partitioning	
✔ Take benefit of all the available nodes. ✔ Scalable to the grid scale. ✔ The databases chunks can fit in memory avoiding IO access. ✘ Needs complex scheduling and replication strategies.	

a chunk to merge the results and to choose where to perform the operation.

What emerges from this analysis is that the more we combine parallelism levels, the more the scheduling and data replication are complex and important.

JOINT SCHEDULING AND REPLICATION ALGORITHMS

In this section, we will describe algorithms and heuristics used to schedule BLAST requests and to replicate databases over the Grid. We also present our new approach using request frequencies variation detection to correct databases distribution and job scheduling along the submission process.

Problem Statement

Life science applications usually manage large databases that should be replicated to let applications scale. Grid middleware are then asked to manage data and schedule computation tasks at the same time. These two important operations have to be tightly coupled to get the best results. In existing Grid computing environments, data replication and scheduling are usually two independent tasks. In some cases, replication managers are used to find the best replicas in terms of access costs. However the choice of the best replication scheme should be done at the same time as the scheduling of computation requests.

Replication Algorithms

In computation grids, some works are focused on replication (Lamehamedi, Szymanski, Zujun Shentu, & Deelman, 2002); an example is Datagrid project from the CERN (DATAGRID. The European DataGrid Project. http://www.eu-datagrid.org). OptorSim (Carvajal-Schiaffino, Cameron, Millar, Nicholson, Stockinger, & Zini 2003), built within this project, allows one to simulate data replication algorithms over a Grid. A modified version of this tool was used in our experiments.

In (Bell, Cameron, Capozza, Millar, Stockinger, & Zini, 2002), several strategies are simulated including unconditional replication (oldest file deleted, LRU) and an economic approach. The target application is data management for the Datagrid physics application. Simulation shows that the economic model is as fast as classical algorithms. In (Kosar, & Livny, 2004), the authors describe Stork, a scheduler for data mapping in Grid environments. Data are considered as resources that have to be managed as computation resources. This environment is mainly used to map data close to computations during the scheduling of task graphs in Condor.

Join Replication and Scheduling

To get the best performance, the scheduling process and replication scheme have to be tightly joint. The scheduling of tasks depends on the data distribution and the data distribution depends on the database usage. When the platform storage capacity is large enough to replicate all the databases on all nodes, we just have to choose a good scheduling strategy. However, even in such a case, it is generally better to save the storage capacity anticipating the growth of the databases (or the execution of other applications).

In (Mohamed, & Epema, 2004), the scheduling of computations is linked to a previous data mapping. Tasks are scheduled on the least loaded processors close to sites where data are stored. Replication is also used to improve performance. Some researches seek to schedule computation requests and data mapping on remote sites at the same time. In (Ranganathan, & Foster, 2003) several strategies are evaluated to manage data and computation scheduling. These strategies are either strongly related to the scheduling of computation or completely disconnected. In

(Chakrabarti, Dheepak, & Sengupta 2004), the authors present an algorithm in which performance is iteratively improved by working alternatively on the data mapping and the task mapping. However, the approach still separates replication and scheduling, but in a connected fashion. In (Desprez, & Vernois, 2006), authors present a static algorithm that provides simultaneous data management and scheduling using a steady-state approach. Using a model of the platform, the number of requests as well as their distribution, and the number and size of databases, a linear program is defined to satisfy the constraints at every level of the platform in steady-state. The solution of this linear program gives a mapping for the databases on the servers as well as providing, for each kind of job, the server on which they should be executed. However, even if this algorithm is very efficient when the requests frequencies and the load of the platform do not change over time, it reaches its limits rapidly on a highly dynamic environment. It was then improved by adding heuristics that allow some data to be deleted or moved on some specific computing servers and adapt scheduling policies (Desprez, & Vernois, 2007).

Another approach consists in re-evaluating the data distribution and job scheduling when detecting a request frequency variation. This method avoids the use of complex Grid monitoring services: The only dynamic information needed by the scheduler is collected by the scheduler itself. We have the following problem formulation: m available servers to store data and to achieve the submitted tasks, n databases on which the tasks will be applied, and p linear-time complexity algorithms using the databases.

Users submit their job to a centralized scheduler. The jobs have a negligible communication cost regarding to their execution time. Large sets of jobs are submitted along the process. Table 2 presents the problem parameters.

In (Desprez, & Vernois, 2006), the authors analyzed the execution trace of a bioinformatics cluster. They noted that by choosing a large enough time interval, frequencies of each type of request are constant. They denoted $f_{k,j}$ the frequency of the jobs of type $R_{k,j}$. The following linear program resolution gives a data mapping on the sites and how many jobs of each type should be executed on each node to maximize the platform throughput for a given set of frequencies $f_{k,}$ (see Box 1).

Table 2. Problem formulation

Computation servers	Databases	Algorithms
$\{S_i\}_{i\in[1..m]}$: Servers set	$\{d_j\}_{j\in[1..n]}$: Data set	$\{a_k\}_{k\in[1..p]}$: Algorithms set
Comput. power of S_i: ω_i	Size of d_j: $size_j$	Linear complexity of a_k: $\alpha_k \times size_j + c_k$
Storage capacity of S_i: m_i	Request $R_{k,j}$: application of a_k on d_j	
Database distribution matrix $\delta_i^j = \begin{cases} 1 & \text{if } d_j \text{ stored on } S_i \\ 0 & \text{otherwise} \end{cases}$		
$n_i(k,j)$: The number of request $R_{k,j}$ to compute on S_i		

Box 1.

$$\begin{cases} \sum_{i=1}^{m} \delta_i^j \geq 1 & \forall j \in [1..n] & \text{There is at least one replica of each data.} \\ \sum_{j=1}^{n} \delta_i^j \cdot size_j \leq m_i & \forall i \in [1..m] & \text{The server storage capacity is limited.} \\ n_i(k,j) \leq v_{k,j}.\delta_i^j.\frac{w_i}{\alpha_k \cdot size_j + c_k} & \forall i \in [1..m] & \text{Each } n_i(k,j) \text{ is restricted by the server capacity.} \\ \sum_{k=1}^{p} \sum_{j=1}^{n} n_i(k,j)(\alpha_k \cdot size_j + c_k) \leq w_i & \forall i \in [1..m] & \text{A server cannot compute more than its capacity.} \\ \sum_{i=1}^{m} n_i(k,j) = f_{k,j} \cdot TP & \forall k \in [1..p], \forall j \in [1..n] & \text{Each } n_i(k,j) \text{ is restricted by the ratio } f_{k,j}. \\ \delta_i^j \in 0,1 & & \\ \text{Maximize } TP & & \end{cases}$$

Due to the variety of users on a Grid and how they use it, the time interval to obtain constant frequencies can be very large. Two solutions are possible to correct this problem:

- By detecting a node CPU overload or underload, we can choose to redirect the next tasks on the less used nodes and avoid to add them to the most used ones. In (Desprez, & Vernois, 2007), different databases redistribution and task scheduling algorithms are evaluated using such a detection system.
- Another approach consists in detecting frequencies variations. Then, when a significant difference is detected, the databases distribution mapping and the tasks scheduling schema are re-evaluated.

Semi Static Approach

In (Desprez, & Vernois, 2007), the authors present an evaluation of different algorithms combination to choose:

- An overloaded server:
    - The most loaded server in computation volume.
    - The most loaded server in computation time.
    - The most over from median server in computation.
- A data to be replicated:
    - The most costly data in computation.
    - The data having the most different effective usage than expected.
- An underloaded server:
    - The less loaded server in computation volume.
    - The less loaded server in computation time.
    - The most under from median server in computation.

Then, the chosen data is replicated on the chosen underloaded server to balance the most loaded server load on it. The evaluation of each algorithm possible combination shows that the most effective combinations are based on the choice of the most costly data in computation to be replicated and the sharing of the load between the most loaded and the less loaded servers.

Frequencies Variation Detection

By detecting a task type frequency variation from the frequency used to compute the SRA linear program solution, the algorithm re-evaluates the optimal data distribution and job scheduling schema.

For our experiments, we used the OptorSim Grid simulator (Carvajal-Schiaffino, & al., 2003). We simulated a Grid platform with 9 sites which have from 90 to 680 worker nodes of different computation power. The size of the databases

Figure 5. Greedy scheduling

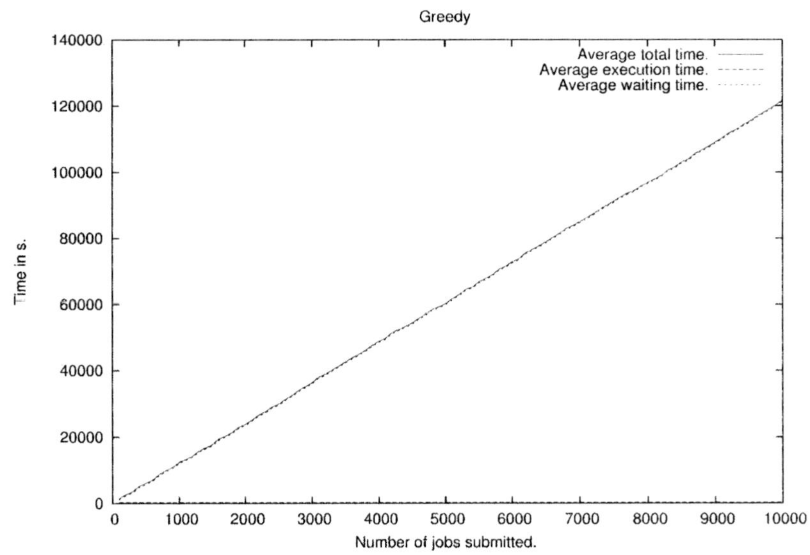

Figure 6. MCT scheduling

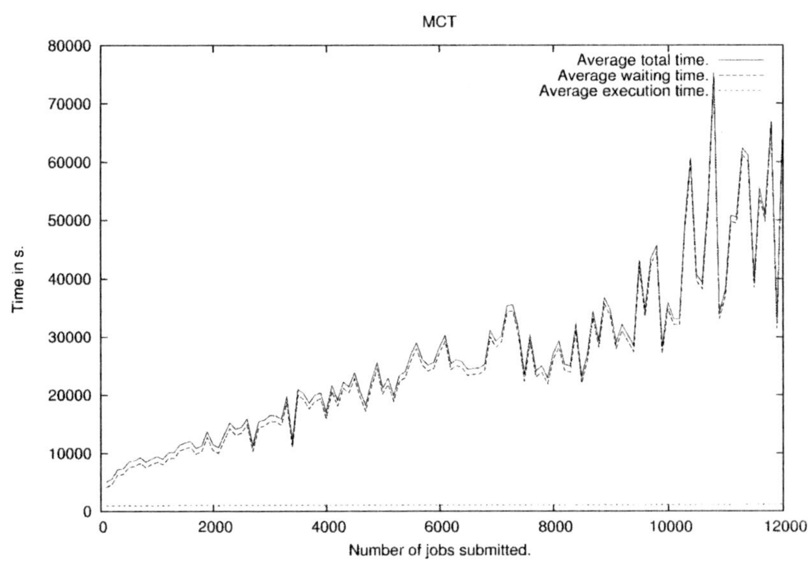

varies from 1GB to 5GB and the jobs are submitted with a rate of 1 job per 3.5 seconds. In a first step we evaluated the performance of the classical SRA algorithm. Each kind of job is submitted according to a frequency randomly fixed before the beginning of the process. These frequencies are constant along the experiment. We proceed to the evaluation of the following algorithms:

- **A simple greedy algorithm without data replication:** This algorithm chooses the node with the less waiting jobs and with the most computation power. This algorithm does not make replications. (Figure 5)

- **A Minimum Completion Time algorithm:** The jobs are sent to the nodes which ensure the minimum completion time. It takes into account the time needed to replicate a data, the waiting jobs and the estimated execution time for a job. (Figure 6)
- **The SRA algorithm with synchronous replications:** When a job starts, if the database is not stored on the node, it downloads it from the nearest source in term of network capabilities. (Figure 7)
- **The SRA algorithm with asynchronous replications:** Transfers start as soon as the data distribution is computed (Figure 8).

Some parameters depend on the Grid network capacity (topology and performance), size of the databases, and the computation power of the nodes. Indeed, the move of a data from a node to another one can be much more costly than to maintain the distribution unchanged if the frequency variation is not too important. Thus, we have to fix a variation threshold beyond which a data should be redistributed. Another aspect, depending on the platform dynamicity is the number of requests to take into account for the frequency measurement. By simulating a realistic infrastructure, based on the Grid'5000 (Bolze, Cappello, Caron, Daydé, Desprez, Jeannot, Jégou, Lanteri, Leduc, Melab, Mornet, Namyst, Primet, Quetier, Richard, Talbi, & Touché 2006) experimental Grid, we evaluated these two parameters:

- A frequency variation threshold between 3.5% and 5% gives good results avoiding useless data transfers.
- A number of requests to take into account which is about 25 times the number of possible different requests kinds is sufficient to detect a significant frequency variation.

In Figure 5, we can see that, without replication, the nodes are quickly overloaded and the average execution time of the jobs increases linearly with the number of jobs submitted. The MCT algorithm (Figure 8) is better than a simple greedy scheduling without data replication. Moreover, when too many jobs are submitted, the platform saturates and the average waiting time increases with the number of jobs submitted. The use of SRA algorithm gives good performance when the number of jobs submitted is sufficient for the replications to have an effect on the throughput of the platform (Figures 7 and 8). With our simula-

Figure 7. SRA – synchronous replications

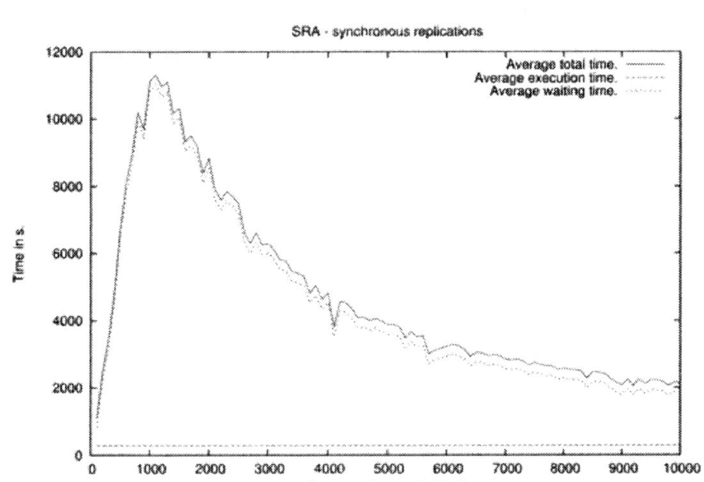

Figure 8. SRA – asynchronous replications

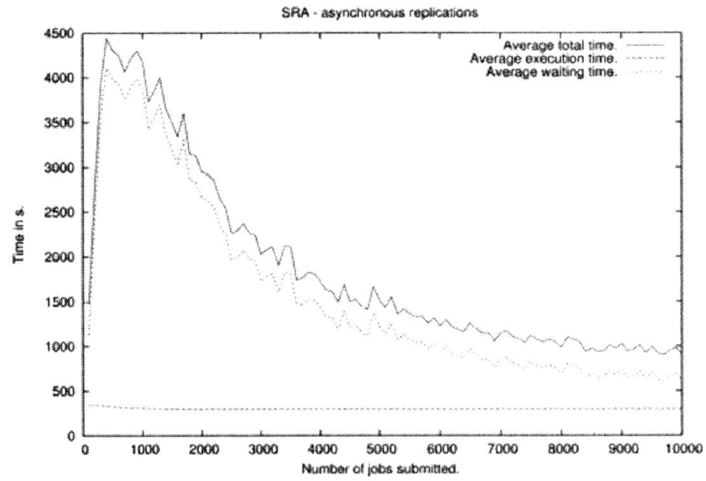

Figure 9. Frequencies variations without redistribution (SRA)

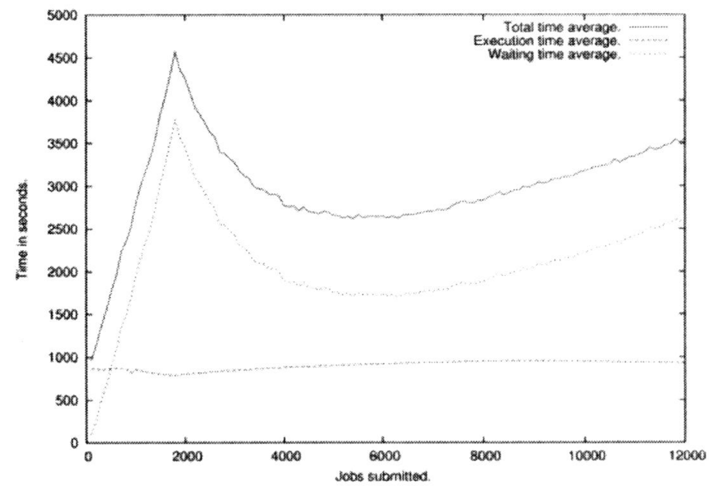

tion parameters, the average total time of each job increases with the number of jobs submitted for sets of jobs smaller than 1000 jobs (Figure 7). For larger sets, the data replications have an effect before the end of all the jobs and the last one can benefit of replication. Using asynchronous data replication, the jobs can take benefit from the data replication earlier and the average total time of the jobs decreases with sets of more than 500 jobs (Figure 8).

The first experiments showed that the SRA algorithm gives very good results when the frequencies of the jobs are constant. Figure 9 shows what happens when the frequencies vary along the process. In Figure 10 we present the performance of the dynamic data redistribution using the frequencies variations detection approach (Dynamic SRA).

Without correction of the data replication and job distributions over the nodes, the frequencies variations decrease the efficiency of the SRA algorithm (see Figure 9). The frequencies variations have little effect when the jobs submitted are not

Figure 10. Frequencies variations with redistribution (Dynamic SRA)

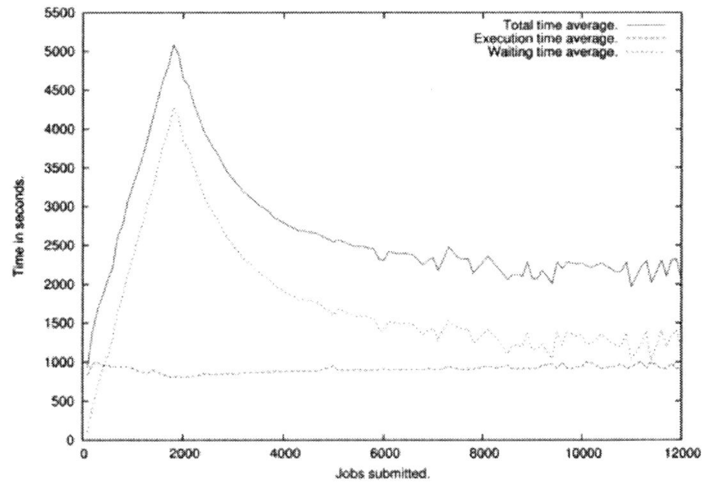

numerous. But, after some time, with a large set of jobs, the platform starts to saturate in spite of data replication. Then, the average waiting time of the jobs increases with the number of jobs submitted. By correcting the data and job distributions when detecting a significant change in the frequencies, our algorithm limits the platform saturation (see Figure 10).

DATA MANAGEMENT MIDDLEWARES AND THEIR USE FOR THE PARALLELIZATION OF BLAST-RELATED APPLICATIONS

In this last section, we will discuss implementation issues and survey existing middleware used to develop high performance bioinformatics grids.

Existing Middleware Frameworks

Data management is of course one of the most important part of bioinformatics grids as large databases need to be shared and replicated over the network (Venugopal, Buyya, & Ramamohanarao, 2006) by data management systems (Beck, Moore, & Plank 2002) (Kosar, & Livny, 2004). Among them, some of them directly implement replication like GDMP (GDMP website. http://project-gdmp.web.cern.ch/project-gdmp).

In (Salzemann, Jacq, & Breton, 2007), the databases update and replication are managed using web-services. In this system, databases are considered as resources and are briefly and efficiently described by a XML sheet storing the information needed to manage their update and retrieval. These services were designed to be used in a particular Grid architecture but are currently re-engineered to be used on the EGEE (Gagliardi, Jones, Grey, Bégin, & Heikkurinen, 2005) middleware.

Data Management in EGEE with gLite

EGEE uses the gLite middleware as an interface between resources and applications. To manage the data, gLite uses the Resource Management Services and the data Management Services that we will describe in the following.

In the gLite middleware, the user can define an "input sandbox" and an "output sandbox" to transmit data from/to the job. However, input/output sandboxes have to be managed by the

Resource Broker and they are limited to 10 MB. The user has thus to copy the data on its local file system and the Resource Broker has to copy them into the Grid. To avoid these problems, the EGEE jobs can use data stored on Storage Elements of the Grid. Using this solution has several advantages:

- The data are accessible from anywhere in the Grid by the users of the same Virtual Organisation.
- The Workload Management System can locate a data to schedule the job more efficiently.
- A data can be replicated to increase the performance.

This data management needs some advanced Grid services. These services are provided through the gLite Workload Management Service and Data Management Service.

GridRPC Approach with DIET

DIET Architecture

DIET (Caron, & Desprez, 2006) is built upon the client/agent/server paradigm. A Client is an application that uses DIET to solve problems. Different kinds of clients should be able to connect to DIET: from a web page or from a program written in C, C++, Java, or FORTRAN. Computations are done by servers running a Server Daemons (SeD). A SeD encapsulates a computational server. For instance it can be located on the entry point of a parallel computer. The information stored by a SeD is a list of the data available on its server, all information concerning its load (for example available memory and processor) and the list of problems that it can solve. The latter are declared to its parent agent. The hierarchy of scheduling agents is made of a Master Agent (MA) and Local Agents (LAs) (Figure 11).

When a Master Agent receives a computation request from a client, agents collect computation abilities from servers (through the hierarchy) and choose the best one according to some scheduling heuristics. The MA sends back a reference to the chosen server. A client can be connected to a MA by a specific distributed name server.

The information stored on an agent is the list of requests, the number of servers that can solve a given problem and information about the data distributed in its sub-tree. For performance reasons, the hierarchy of agents should be deployed depending on the underlying network topology.

Figure 11. DIET architecture

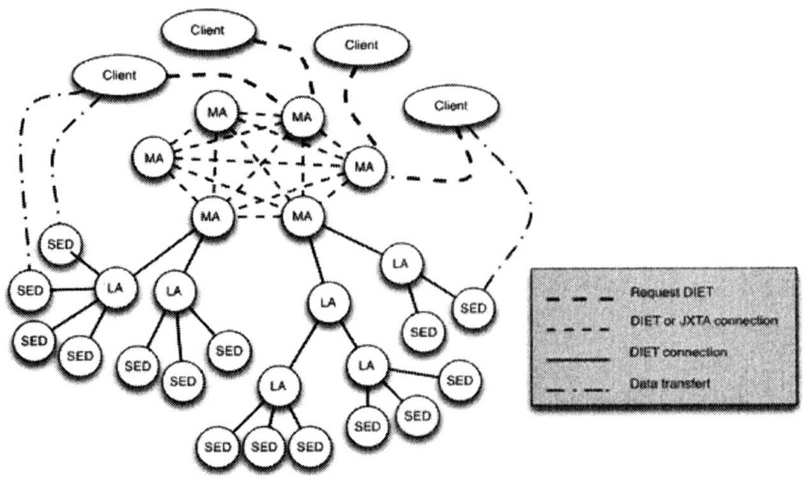

Moreover DIET provides a special feature for scheduling through its plug-in schedulers. As the applications that are to be deployed on the Grid vary greatly in terms of performance demands, the DIET user is provided with the possibility of defining requirements for the scheduling of tasks by configuring the appropriate scheduler. The performance estimation values to be used for scheduling are stored in a performance estimation vector by the SeDs as a response to a client call propagated from the master agent to local agents and finally to the server level. The values to be stored in this structure can be provided by CoRI (Collector of Resource Information).

CoRI is a platform performance subsystem that we have implemented to enable versions of the DIET framework to interface with third-party performance monitoring and prediction tools more easily. CoRI provides access to different collectors, which are software components that can provide information about the system. This modular design decouples the choice of measurement facilities and the utilization of such information in the scheduling process. Even if the manager should aggregate across performance data originating from different resource information services, the raw trace of data remains, and so its origin can be determined. Furthermore, the modular design of the CoRI Manager also permits a great deal of extensibility, in that additional collectors based on systems such as Ganglia or NWS can be rapidly implemented via relatively thin software interface layers. This capability enables DIET users to evaluate prototype measurement systems more easily even before they reach full production quality.

From CoRI, the standard values are to be identified based on standard estimation tags given in Table 3. Application developers may also define performance values to be included in a SeD response to a client request. For example, a DIET SeD that provides a service to query particular databases may need to include information about which databases are currently resident in its disk cache so that data transfer times can be minimized.

Application developers can define their own performance estimation routine or function when developing the application-specific portion of the SeD. At this point, any services added to the SeD will be associated with the performance estimation routine declared. In the performance estimation routine, the SeD developer should store in the provided estimation vector any performance data to be used in the server response aggregation methods. At the time a DIET service is defined,

Table 3. Standard estimation tags used in DIET

Information tag starts with EST_	multi-value	Explanation
TCOMP		the predicted time to solve a problem
TIMESINCELASTSOLVE		time since last solve started (seconds)
FREECPU		amount of free CPU (fraction between 0 and 1)
LOADAVG		CPU load average
FREEMEM		amount of free memory (Mb)
NBCPU		number of available processors
CPUSPEED	x	frequency of CPUs (MHz)
TOTALMEM		total memory size (Mb)
BOGOMIPS	x	the BogoMips
CACHECPU	x	cache size CPUs (Kb)
TOTALSIZEDISK		size of the partition (Mb)
FREESIZEDISK		amount of free space on partition (Mb)
DISKACCESREAD		average time to read from disk (Mb/sec)
DISKACCESWRITE		average time to write to disk (Mb/sec)
ALLINFOS	x	[empty] fill all possible fields

an aggregation method - the logical mechanism by which SeD responses are sorted - is associated with the service. If application-specific data are supplied (i.e., the estimation function has been redefined), an alternative method for aggregation is needed. Currently, a basic priority scheduler has been implemented, enabling an application developer to specify a series of performance values that are to be optimized in succession. From the point of view of an agent, the aggregation phase is essentially a sort of server responses from its children. A priority scheduler logically uses a series of user-specified tags to perform the pair-wise server comparisons needed to construct the sorted list of server responses.

DAGDA: A New Data Manager for DIET

DAGDA (Data Arrangement for Grid and Distributed Applications) is a new data manager for the DIET middleware, which allows data explicit or implicit replications and advanced data management on the Grid. It was designed to be backward compatible with previously developed applications for DIET, which benefit transparently of the data replications. Moreover, DAGDA limits the data size loaded in memory to a user-fixed value and avoids CORBA errors when transmitting too large data in relation to the ORB configuration.

DAGDA offers a new way to manage the data on DIET. The API allows the application developer to replicate, move, add, or delete a data to be reused later or by another application. Each component of DIET can interact with DAGDA. The data manipulation can be done from a client application, a server or an agent through a plug-in scheduler. That is very useful for BLAST application as shown with the DMR scheduler presented in the last section.

A DAGDA component is associated to each element in a DIET platform (client, Master Agent, Local Agent, SeD). These components are connected following the DIET deployment topology. Figure 12 shows how the DAGDA and DIET classical components are connected. Contrary to a DIET architecture, each DAGDA component has the same role. It can store, transfer, or move a data. The client's DAGDA component is isolated from the architecture and communicates only with the chosen SeDs DAGDA components when necessary. When searching for a data, DAGDA uses its hierarchical topology to contact the data managers. Among the data managers having one replicate of the data, DAGDA chooses the "best" source to transfer it. To make this choice DAGDA uses some statistics collected from previous data transfers between the nodes. By not using dynamic information, it is unsure that DAGDA really chose the "best" nodes for the transfers. In a future version, we plan to introduce some facilities to estimate the time needed to transfer a data and to improve the choice of a data stored on the Grid. To do the data transfers, DAGDA uses the pull model: it is the destination node that ask for the data transfer.

Figure 12 presents how DAGDA manages the data when a client submits a job. In this example, the client wants to use some data stored on the Grid and some personal data. He wants to obtain some results and to store some others on the Grid. Some of these output data are already stored on

Figure 12. DAGDA architecture in DIET

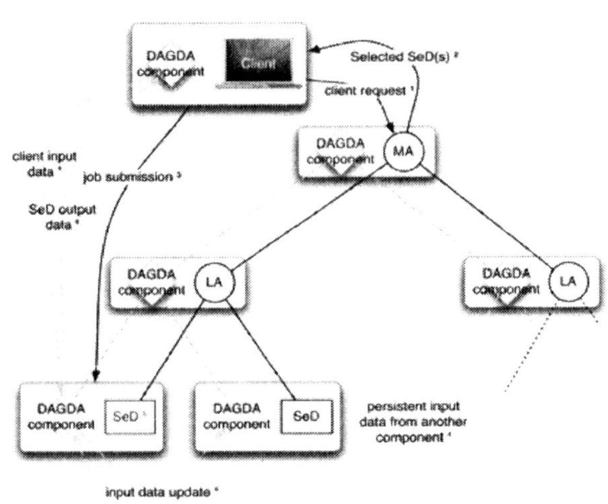

the platform and they should be updated after the job execution.

1. The client sends a request to the Master Agent.
2. The Master agent returns one or more SeD references.
3. The client sends its request to the chosen node. The parameters data are identified by a unique ID and the problem profile contains a reference to the client's data manager.
4. Receiving the request the SeD asks the client to transfer the data of the user and it asks to the DAGDA architecture to obtain the persistent data already stored on the platform.
5. The SeD executes the job. After the execution, the SeD stores the output data and it informs the client that the data are ready to be downloaded. It also asks to the architecture to update the modified output data.
6. The client uploads its results and the data are updated on the nodes.

Using DAGDA, each data stored on the platform can be replicated as many time as there is free space on the platform. To avoid DIET to take up all the storage resources (disk or memory) of the nodes, DAGDA allows to fix a limit to the data size that can be stored for each DAGDA component. When a job needs a data, it asks DAGDA for it using its unique ID and gets it from the "best source". DAGDA uses some statistic informations to determine which node should be used for the transfers, but could be easily extended to use some dynamic metrology parameters. DAGDA also offers a way to manage automatically the data by choosing a cache replacement algorithm for each node. The user can choose a persistence mode for each replica:

- **Persistent:** The data is kept as long as possible on the node but can be deleted when the node needs some storage resources.
- **Sticky:** The data will stay on the node until a user chooses to delete it explicitly.

Then, the three cache replacement algorithms implemented in DAGDA work as follows:

- **Least Recently Used (LRU):** The least recently used persistent data of sufficient size is deleted.
- **Least Frequently Used (LFU):** The least frequently used persistent data of sufficient size is deleted.
- **First In First Out (FIFO):** Among the persistent data of sufficient size, the oldest is deleted.

None of the above algorithms controls if another replica of the data exists on the platform before to delete it. The only way to ensure at least one replica of a data is still on the platform is to declare one of them as sticky data.

Extending the DIET API, DAGDA allows to explicitly manage the data in the platform by communicating directly with the DAGDA hierarchy. However, DAGDA does not deal with the replicas consistency and the data availability. The user is responsible of what he does with the data. Even if DAGDA offers a way to update a persistent data, if a job or a user modifies the data while the updating process is executing, some side effects can appear.

EXAMPLE OF INTEGRATED SOLUTION THROUGH DIET

Integrated scheduling and replication can also be implemented like in the DIET middleware (Caron, & Desprez, 2006), which implements the GridRPC API from the OGF (Seymour, Lee, Desprez, Nakada, & Tanaka, 2004) and an efficient data manager and replication system. We will describe how a bioinformatics Grid can be implemented using such a middleware and we

Figure 13. Different interaction layers between DIET core and application view

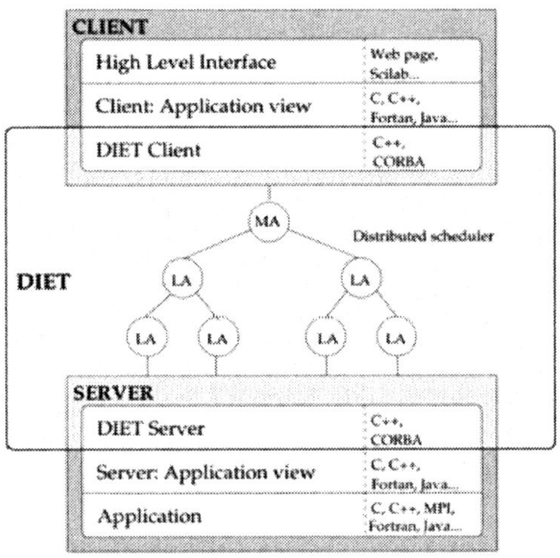

will provide performance results of experiments done within the Grid'5000 platform (Bolze & al. 2006).

How to Add a BLAST Application within DIET?

The main idea is to provide some integrated level for a Grid application. Figure 13 shows these different kinds of levels. The application server must be written to give to DIET the ability to use the application. A simple API is available to easily provide a connection between the DIET server and the application. The main goals of the DIET server are to answer to monitoring queries from its responsible Local Agent and launch the execution of a service, upon an application client request.

The application client is the link between high-level interface and the DIET client, and a simple API is provided to easily write one. The main goals of the DIET client are to submit requests to a scheduler (called Master Agent) and to receive the identity of the chosen server, and final step, to send the data to the server for the computing phase. Agents must find a convenient server or a list of convenient servers for a specific request (problem type, resource capacities, server status, etc.) and send this information to the DIET client.

The user who wrote the application client writes the corresponding server part called application server. A simple API is also available to easily provide a connection between the DIET server and the application.

DIETBlast Features

DIETBlast is a C++ program that allows to submit large sets of BLAST requests on a chosen biological database. It was designed to be as transparent as possible for the user, easy to use even for somebody that never heard anything about the Grid and, also, to permit to configure finely each step of the submission process for the programmers. DIETBlast can be as well a productivity tool for bioinformatics scientists and an experimental platform to test some scheduling strategies or divisible load heuristics for computer scientists. Each sensitive step of the submission process have been developed to be easily modified or replaced. We also modified DIET itself to allow the dynamic loading of a scheduling algorithm for each agent of the platform.

A request is a set of sequences to compare against one database. All the comparisons are independent and can be executed in any order on any site. We can divide the main request in as many sub-requests as we want. By dividing the request in *n* sub-requests, we can execute n BLAST operations in parallel. An atomic request is a sub-request that cannot be divided. We can consider that an atomic request and the result produced by the BLAST search have negligible sizes compared to the databanks sizes. Thus, we consider that the transmissions of the requests and the results are negligible in comparison with the databanks transfers on the network. Then, the BLAST submission process consists in the following steps:

1. Division of the main request into *n* subrequests.
2. Parallel submission of the subrequests.

3. Parallel execution of the subrequests.
4. Results aggregation.

We consider that for one atomic request each step of the process is negligible in comparison with the BLAST process itself (step 3.). However, if there is a large number of small subrequests and if the number of nodes available to execute the process is very low, the network overhead could exceed the execution time of the BLAST process if we divide the main request into too many parts. So, the division strategy is an important step of the process that should take into account how many BLAST servers are available.

On the client side, the user enter a request file, a database name, and if the database is not present on the platform, the formatted database files gathered in a compressed archive. The name of the database is used to store and locate a database on the platform. When a database is used for the first time, DIET uploads it inside the architecture, after the registration process, this database is accessible from everywhere using only its name.

Figure 14. General operation of DIETBlast

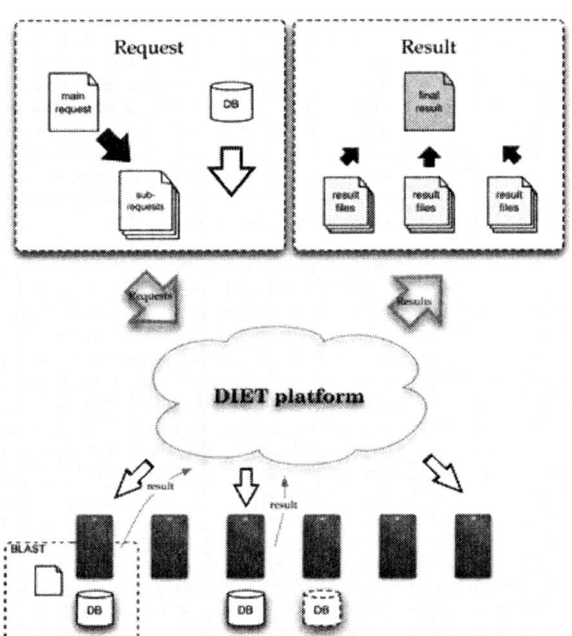

DIET then manages itself the data replications on the nodes that need them. Figure 14 presents how DIETBlast performs the BLAST requests over the Grid.

The first step of a submission consists in dividing the main request into some subrequests. Different strategies are possible depending on the way the requests will be scheduled and the number of available computing nodes:

- **Maximum division strategy:** The main request that contains *n* subrequests is divided into *n* "atomic" requests. If the main request contains too many small requests, the time to send data and to divide the requests into sub-requests added to the network overhead exceeds the BLAST treatment time itself. However, if the main request contains only big sequences, the maximum division strategy is the best way to parallelize the BLAST process over the Grid.
- **Fixed number of sub-request division:** To avoid the transmission of too small requests to the nodes, we can choose to divide the main request into *n* sub-requests where *n* is the number of available nodes in the platform.

These two different strategies have been evaluated on a real experiment performed on the Grid'5000 platform. Figures 15 and 16 show the performance of these two strategies with 300 nodes distributed over 9 sites.

Schedulers for DIETBlast

DIET provides a special scheduling feature through its plug-in schedulers (Caron, Chis, Desprez, & Su, 2008). As the applications that are to be deployed over the Grid vary greatly in terms of performance demands, the DIET user is provided with the possibility of defining requirements for scheduling of tasks by configuring the appropriate scheduler. The performance estimation values to be used for scheduling are stored in

Figure 15. Maximum request division

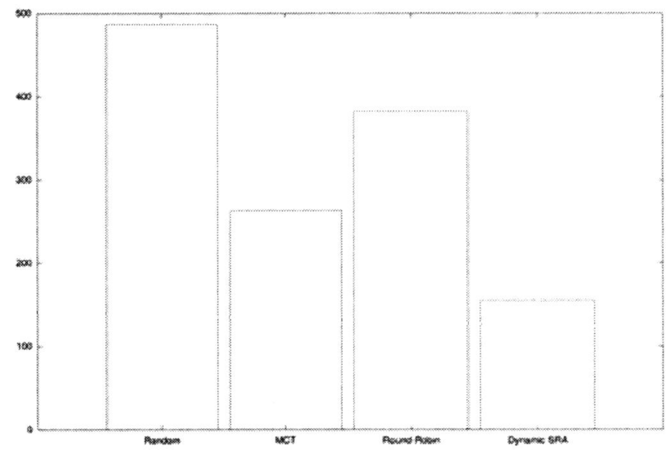

Figure 16. Division into n requests for a n nodes platform

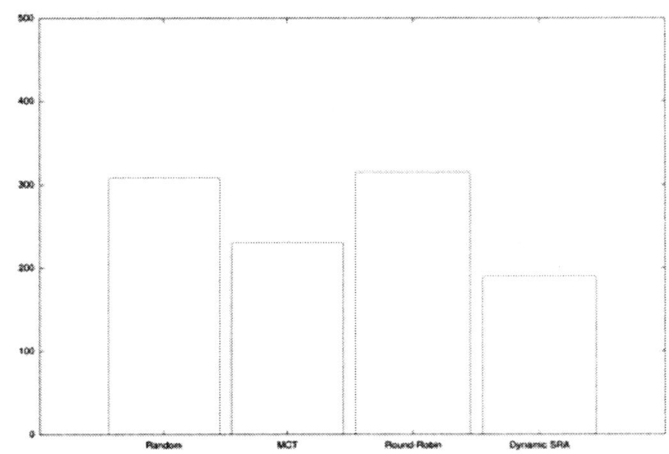

a performance estimation vector by the SeDs as a response to a client call propagated from master agent to local agents and finally to the server level. The values to be stored in this structure can be provided by CoRI.

Thus some schedulers can be easily implemented and compared:

- **Random Strategy (RS):** With a lot of computing nodes, this strategy gives better results that we could expect and is near to be equivalent to the Round-Robin strategy. If the platform is highly heterogeneous, this strategy can be better than the simple Round-Robin because it uniformly distributes the load on the nodes when Round-Robin will concentrate the load on a cluster at a time.
- **Round-Robin (RR):** The node is chosen randomly between the available nodes for the BLAST service.This strategy concentrates the requests on the same cluster at the beginning of the process and can suffer of its potentially bad performance.
- **Minimum Completion Time (MCT):** Because this strategy takes into account the load of the nodes, the power and the data availability, it gives good results in term of makespan. However, because this strategy concentrates the tasks on the best nodes and

gives advantage to the firsts requests, if the platform is very heterogeneous, there is a lot of inequity between nodes and between users.

- **Simultaneous Replication Algorithm (SRA):** this strategy tries to take into account the specificities of bioinformatics job submissions. It can give very good result as long as the jobs submission are sufficiently homogeneous.
- **Data manager with replication (DMR):** Data are kept as long as possible on (or near to) the computational servers on which they have been computed to minimize transfer times. This kind of optimization is mandatory to obtain performance on a wide-area network. DIET is also designed to take into account the data location during scheduling.

We proceeded to two experiments using the Random and MCT scheduling policies. The first experiment involved 173 nodes distributed over 3 clusters on the Grid'5000 platform. For this experiment, we used a 4MB request to align with a 400 MB database. We measured the total execution time needed to obtain the result for different sub-request sizes and compare the two jobs distribution heuristics. Figure 17 shows the total execution time in function of sub-request sizes.

As shown in the previous (Figures 15 and 16), for MCT, the best execution times are obtained when the number of sub-requests is near the number of available nodes. Globally, the MCT heuristic gives better results than the Random one. With only 173 nodes, to divide the request into more than 173 sub-requests increases the total execution time because of the network latency. In such a case, the jobs are enqueued on the nodes, waiting for the previous ones to end. As we will see, when we have enough computing resources, the network latency caused by a numerous jobs submission is negligible regarding to the improvement of the parallelization. Figure 18 shows the results obtained with 755 nodes distributed over 11 clusters on the Grid'5000 platform.

On Figure 18, we can see that with more nodes, the Random policy does not gives so different results than the MCT one. However, MCT is always better and the more jobs are distributed (small sub-requests size) the less time is needed to complete the BLAST search.

To take benefit of the DMR and SRA scheduling heuristics we have developed a new data manager for DIET which was presented in the previous section.

Figure 17. Execution time for two scheduling heuristics using 173 nodes

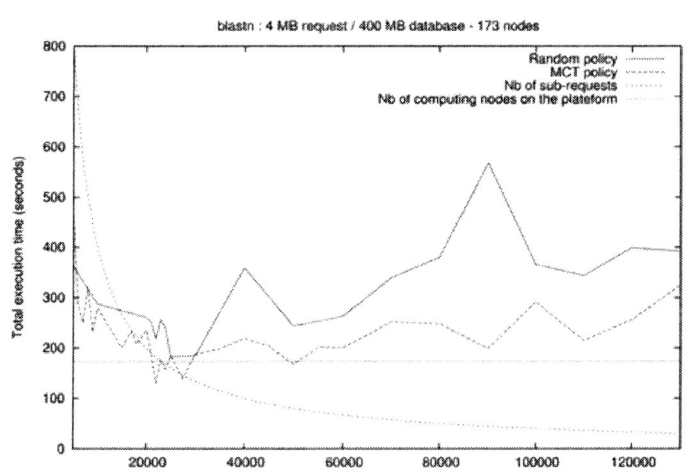

Figure 18. Execution time for two scheduling heuristics using 755 nodes

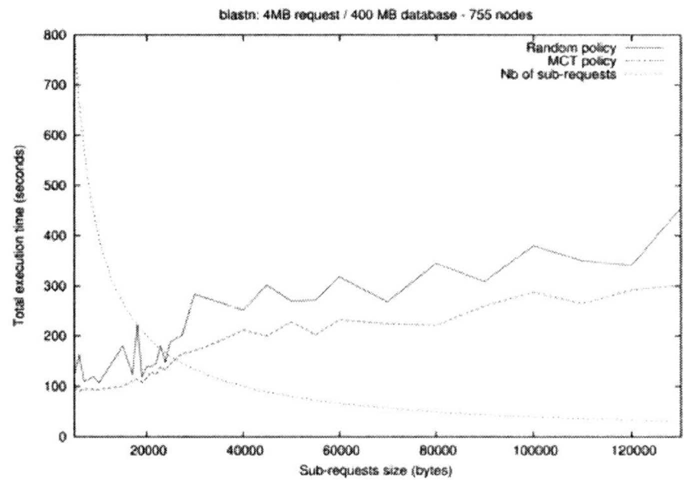

FUTURE TRENDS

Large scale application, including bioinformatics ones, have been successfully ported over clusters, clusters of clusters, grids, and desktop computing platforms. However, the use of such high performance distributed platforms remains tedious for non-expert application developpers. The future of bioinformatics over large scale Grid platforms will thus include the notion of services. Service Oriented Architectures (SOA) will allow the nasty technical details of grids to be hidden to end users with the promise of high performance, reliability, and ease of use.

CONCLUSION

The goal of this chapter was to review existing parallel approaches for large scale BLAST operations both on clusters and grids. We have shown that several levels exist that can be combined to get the best results depending on the performance of the platform and the problem size.

We also described algorithmic problems related to scheduling and data replication. This problem is particularly important for the gridification of such bioinformatics application. Again, several heuristics need to be tested for different conditions. Middleware for resource allocation and data management have been described.

Finally, we have described implementation issues of the gridification of the BLAST kernel using the DIET middleware. Experiments were done over the Grid'5000 platform.

ACKNOWLEDGMENT

We would like to thanks the smart students, researchers, postdoc, and engineers who have contributed to the development of DIET and its tools. We also want to thank Antoine Vernois for the design and development of heuristics for join scheduling and replication, which are presented in the fourth section.

DIET was developed with financial support from the French Ministry of Research (RNTL GASP and ACI ASP) and the ANR (*Agence Nationale de la Recherche*) through the LEGO project referenced ANR-05-CIGC-11 and the Gwendia project referenced ANR-06-MDCA-009. We would also like to thank the partners of the Décrypthon project (AFM, IBM, and CNRS).

REFERENCES

Carey, L., Darling, A. E., & Chun Feng, W. (2003). The Design, Implementation, and Evaluation of mpiBLAST. In *Proceedings of ClusterWorld 2003*. San Jose, CA: IEEE.

Aloisio G., Cafaro M., Fiore S., & Mirto M. (2005). ProGenGrid: A Workflow Service Infrastructure for Composing and Executing Bioinformatics Grid Services. In *18th IEEE Symposium on Computer-Based Medical Systems* (pp. 555-560). Dublin, Ireland: Trinity College..

Altschul S., Gish W., Miller E., Myers E., & Lipman D. (1990). A Basic Local Alignment Search Tool. *Journal of Molecular Biology, 215*, 403-410.

Angele, J., & Weiten, M. (2005). Semantic Technologies in the SIMDAT Grid Project. In C. Goble, C. Kesselman, and Y. Sure, (Ed.), *Semantic Grid: The Convergence of Technologies, number 05271 in Dagstuhl Seminar Proceedings*. Schloss Dagstuhl, Germany: Internationales Begegnungs und Forschungszentrum fuer Informatik (IBFI)..

Bala, P., Pytlinski, J., Nazaruk, M., Alessandrini, V., Girou, D., Erwin, D.W., et al. (2002). BioGRID: An European Grid for Molecular Biology. *11th IEEE International Symposium on High Performance Distributed Computing* (p. 412)., Edinburgh, UK: IEEE Computer Society.

Beck M., Moore T., & Plank J.S. (2002 July). An End-To-End Approach to Globally Scalable Network Storage. In *SIGCOMM '02: Proceedings of the 2002 conference on Applications, technologies, architectures, and protocols for computer communications*, (pp. 339-346).

Bell, W.H., Cameron, D.G., Capozza, L., Millar A.P., Stockinger, K., & Zini, F. (2002 November). Simulation of Dynamic Grid Replication Strategies in OptorSim. *In Proceedings of the 3rd Int'l. IEEE Workshop on Grid Computing (Grid'2002)*, Lecture Notes in Computer Science. Berlin, Germany: Springer Verlag.

Bolze, R., Cappello, F., Caron, E., Daydé, M., Desprez, F., Jeannot, E., et al. (2006). Grid'5000: A Large Scale and Highly Reconfigurable Experimental Grid Testbed. *International Journal of High Performance Computing Applications, 20*(4) 481-494.

Braun, R.C., Pedretti, K.T., Casavant, T.L., Scheetz T.E., Birkett, C.L., & Roberts, C.A. (2001). Parallelization of Local BLAST Service on Workstation Clusters. *Future Genereration Computers Systems, 17*(6), 745-754.

Buetow, K.H. (2003). Cyberinfrastructure: Empowering a "Third Way" in Biomedical Research. *Science*, 44, 417.

Cannataro, M., Comito, C., Lo Schiavo, F., & Veltri, P. (2004). *Proteus, A Grid Based Problem Solving for Bioinformatics: Architecture and Experiments.* IEEE Intelligent Informatics Bulletin, *3*(1), 7-18.

Caron, E., Chis, A., Desprez, F., & Su, A. (2008, January). Design of Plug-In Schedulers for a GridRPC Environment. *Future Generation Computer Systems, 24*(1), 46-57.

Caron, E., & Desprez, F. (2006). DIET: A Scalable Toolbox to Build Network Enabled Servers on the Grid. *International Journal of High Performance Computing Applications, 20*(3), 335-352.

De Carvalho, R.L., & Lifschitz, S. (2003). Database Allocation Strategies for Parallel BLAST Evaluation on Clusters. *Distributed and Parallel Databases, 13*, 99-127.

Chakrabarti, A., Dheepak, R.A., & Sengupta S. (July 2004). *Integration of Scheduling and Replication in Data Grids* (Tech. Rep. TR-0407-001). New York: Infosys Tech. Ltd.

Covitz, P.A., Hartel, F., Schaefer, C., De Coronado, S., Fragoso, G., Sahni H., Gustafson, S., & Buetow K.H. (2003). *caCORE: A Common Infrastructure for Cancer Informatics*.

DATAGRID. The European DataGrid Project. Retrieved from http://www.eu-datagrid.org

Desprez, F., & Vernois, A. (2006, March). Simultaneous Scheduling of Replication and Computation for Data-Intensive Applications on the Grid. *Journal Of Grid Computing, 4*(1), 19-31.

Desprez, F. & Vernois, A. (2007, September). Semi-Static Algorithms for Data Replication and Scheduling Over the Grid. In *IEEE 3rd International Conference on Intelligent Computer Communication and Processing, workshop on Grid Computing*, Cluj-Napoca, Romania.

Carvajal-Schiaffino, R., Cameron, D.G., Millar, A.P., Nicholson, C., Stockinger, K. & Zini F. (2003, November). Evaluating Scheduling and Replica Optimisation Strategies in OptorSim. In *4th International Workshop on Grid Computing (Grid2003)*. New York: IEEE Computer Society Press.

Embrace. Embrace website. Retrieved from http://www.embracegrid.info

Frey, J., Tannenbaum, T., Livny, M., Foster, I., & Tuecke, S. (2002). Condor-G: A Computation Management Agent for Multi-Institutional Grids. *Cluster Computing, 5*(3), 237-246.

Gagliardi, F., Jones, B., Grey, F., Bégin, M.E., & Heikkurinen M. (2005). Building an Infrastructure for Scientific Grid Computing: Status and Goals of the EGEE Project. *Philosophical Transactions: Mathematical, Physical and Engineering Sciences, 363*(1833), 1729-1742.

GDMP. Retrieved from http://project-gdmp.web.cern.ch/project-gdmp

Globus. Retrieved from http://www.globus.org

Goad, W.B., & Kanehisa, M.I. (1982). Pattern Recognition in Nucleic Acid Sequences: A General Method for Finding Local Homologies and Symmetries. *Nucleic Acids Research, 10*(1), 247-263.

Grethe, J.S., Baru, C., Gupta, A., James, M., Ludaescher, B., Martone, M.E., Papadopoulos, P.M., Peltier, S.P., Rajasekar, A., Santini, S., et al (2005). Biomedical Informatics Research Network: Building a National Collaboratory to Hasten the Derivation of New Understanding and Treatment of Disease. *Stud Health Technol Inform, 112*, 100-109.

Henikoff, S., & Henikoff, J.G. (1992). Amino Acid Substitution Matrices from Protein Blocks. *Proceedings of the National Academy of Sciences of the United States of America, 89*(22), 10915-10919.

Jacq, N., Blanchet, C., Combet, C., Cornillot, E., Duret, L., Kurata, K., Nakamura, H., Silvestre, T., & Breton, V. (2004). Grid as a Bioinformatic Tool. *Parallel Computing, 30*(9-10), 1093-1107.

Kim, J., Kim, H., & Han, D. (2006). Performance Evaluation of BLAST on SMP Machines. *Lecture Notes in Computer Sciences*, 4331, 668.

Korf. I., Yandell, M., & Bedell, J., (2003). BLAST. Sebastopol, CA: O'Reilly.

Kosar, T. & Livny M. (2004). Stork: Making Data Placement a First Class Citizen in the Grid. In *Proceedings of 24th IEEE International Conference on Distributed Computing Systems (ICDCS2004)*, Tokyo, Japan.

Krishnan, A. (2004). A Survey of Life Sciences Applications on the Grid. *New Generation Computing, 22*, 111-126.

Lamehamedi, H., Szymanski, B., Zujun Shentu, & Deelman E. (2002, October). Data Replication Strategies in Grid Environments. In *Proceedings of 5th International Conference on Algorithms and Architecture for Parallel Processing,*

ICA3PP'2002, (378-383). New York: IEEE Computer Science Press.

Lavenier, D., Leroy, H., Mac Wing, M., Andonov, R., Hurfin, M., Raipin-Parvedy, P., Mouchard, L., & Guinand, F. (2003, January). GénoGRID: An Experimental Grid for Genomic Applications. In *HealthGrid 2003* Lyon.

Liu, W., Schmidt, B., Voss, G., Schroder, A., & Muller-Wittig, W. (2006). Bio-Sequence Database Scanning on a GPU. *20th IEEE International Parallel & Distributed Processing Symposium (IPDPS 2006) (HICOMB Workshop)*, Rhode Island, Greece.

Milanesi, L., & Merelli, I. (2006). High Performance GRID Based Implementation for Genomics and Protein Analysis. *Studies in health technology and informatics, 120*, 374-380.

Mohamed, H.H., & Epema, D.H.J. (2004). An Evaluation of the Close-to-Files Processor and Data Co-Allocation Policy in Multiclusters. In *Cluster*, (pp. 287-298). New York: IEEE Computer Society Press.

NCBI. Ncbi website. Retrieved from http://www.ncbi.nih.gov/

Needleman, S.B., & Wunsch, C.D. (1970). A General Method Applicable to the Search for Similarities in the Amino Acid Sequences of Two Proteins. *Journal of Molecular Biology*, 443-453.

Pearson, W.R., & Lipman, D.J. (1988). Improved Tools for Biological Sequence Comparison. *Proceedings of the National Academy of Sciences of the United States of America, 85*(8), 2444-2448.

Pedretti, K.T., Casavant, T.L., Braun, R.C., Scheetz, T.E., Birkett, C.L., & Roberts, C.A. (1999). Three Complementary Approaches to Parallelization of Local BLAST Service on Workstation Clusters. In *PaCT '999: Proceedings of the 5th International Conference on Parallel Computing Technologies*, (pp. 271-282). London: Springer-Verlag.

Ranganathan, K., & Foster, I. (2003). Simulation Studies of Computation and Data Scheduling Algorithms for Data Grids. *Journal of Grid Computing, 1*(1), 53-62.

Rangwala, H., Lantz, E., Musselman, R., Pinnow, K., Smith, B., & Wallenfelt, B. (2005). Massively Parallel BLAST for the Blue Gene/L. *High Availability and Performance Workshop.*

Saltz, J., Oster, S., Hastings, S., Langella, S., Kurc, T., Sanchez, W., et al. (2006). *caGrid:* Design and Implementation of the Core Architecture of the Cancer Biomedical Informatics Grid. *Bioinformatics, 22*(15), 1910.

Salzemann, J., Jacq, N., & Breton, V. (2007, June). Replication and Update of Molecular Biology Databases. *IEEE Transactions on Nanobioscience, 6*(2).

Seymour, K., Lee, C., Desprez, F., Nakada, H., & Tanaka, Y. (2004, September). The End-User and Middleware APIs for GridRPC. In *Workshop on Grid Application Programming Interfaces, In conjunction with GGF12*, Brussels, Belgium.

Smith, T.F., & Waterman, M.S. (1981). Identification of Common Molecular Subsequences. *Journal of Molecular Biology, 147*, 195-197.

Stark, C., Breitkreutz, B.-J., Reguly, T., Boucher, L., Breitkreutz, A., & Tyers, M. (2006). BioGRID: A General Repository for Interaction Datasets. *Nucleic Acids Research, 34*, 535-539.

Stevens, R.D., Robinson, A.J., & Goble, C.A. (2003). *myGrid: Personalised Bioinformatics on the Information Grid.*

Venugopal, S., Buyya, R., & Ramamohanarao, K. (2006). A Taxonomy of Data Grids for Distributed Data Sharing, Management, And Processing. *ACM Computing Surveys, 38*(1), 3.

Wu, X., & Tseng, C.W. (2006). Searching Sequence Databases Using High-Performance BLASTs. In A.Y. Zomaya, (Ed.), *Parallel Computing for Bioinformatics and Computational Biology*, (pp. 211-232). Hoboken, NJ: John Wiley & Sons, Inc.

Yang, C.T., Luo, T.-L., & Lai, C.-L. (April 2005). Designing Computing Platform for BioGrid. *International Journal of Computer Applications in Technology, 22*, pp. 3-13.

Zomaya, A.Y. (Ed.) (2006). *Parallel Computing for Bioinformatics and Computational Biology.* Hoboken, NJ: John Wiley & Sons, Inc.

KEY TERMS AND DEFINITIONS

Biological Database: Library of annotated biological data collected from experiments or computational analyses, which are marshalled by type, by nature or by origin.

BLAST: Basic Local Alignment Search Tool is a bioinformatics application designed to find regions of local similarities between biological sequences like protein sequences or DNA sequences.

DAGDA: The new data manager for the DIET middleware which allows data explicit or implicit replications and advanced data management on the Grid. It was designed to be backward compatible with previously developed applications for DIET which benefit transparently of the data replications.

Data Replication: The way to copy and distribute the data among the storage resources.

DIET: DIET consists of a set of elements that can be used together to build applications using the Grid-RPC paradigm. This middleware is able to find an appropriate server according to the information given in the client's request, the performance of the target platform and the local availability of data stored during previous computations

Grid Middleware: Set of services used to federate a computing Grid resources, taking into account, the security, the data management, the tasks submissions and the results retrieval transparently from the users point of view.

Sequences Alignment: Sequences alignments are used to underline the similarities of two different sequences. In biology, the alignments between two proteins or DNA sequences can traduce functional or evolutionary relationships.

Tasks Scheduling: It refers to the way processes are assigned priorities in a priority queue. This assignment is carried out by software known as a scheduler. It consists of mainly, CPU utilization - to keep the CPU as busy as possible.

Chapter VIII
Functional Genomics Applications in GRID

Luciano Milanesi
Istituto di Tecnologie Biomediche – Consiglio Nazionale delle Ricerche, Italy

Ivan Merelli
Istituto di Tecnologie Biomediche – Consiglio Nazionale delle Ricerche, Italy

Gabriele Trombetti
Istituto di Tecnologie Biomediche – Consiglio Nazionale delle Ricerche, Italy

Paolo Cozzi
Istituto di Tecnologie Biomediche – Consiglio Nazionale delle Ricerche, Italy

Alessandro Orro
Istituto di Tecnologie Biomediche – Consiglio Nazionale delle Ricerche, Italy

ABSTRACT

A common ongoing task for Functional Genomics is to compare full organisms' genome with those of related species, to search in huge database for functional annotation of novel sequences and to identify specific patterns of them, such as ESTs, genes, and microRNA. The prediction of these patterns has a relevant computational cost, while public genome archives exceed one billion sequence traces from over 1,000 organisms and this number is increasing rapidly as costs decline, but powerful solution must be enabled in order to perform efficient searches. This means that Functional Genomics applications require significant computational infrastructures, where reusable tools and resources can be accessed. In particular, grid computing seems to fulfill both the computational and data management requirements, even if porting applications on this infrastructure can be difficult. The implementation of a suitable environment for the management of distributed computations can provide reliable advantage, reducing the gap between the requirements of the functional genomic domain and the potential of this technology.

INTRODUCTION

The Human Genome Project was just the first step in understanding humans at the molecular level. In the post genomic era, many questions still remain unanswered, including functions, regulations and interactions of the estimated 30,000 human genes. Moreover, breakthrough technologies such as the High Throughput ones allow, nowadays, the fast generation of enormous amounts of data related to different organisms. This is particularly true for the sequencing, where Roche and Illumina technologies enable the possibility of sequencing many bacterial genomes and even the production of multiple individual references, which allow reconsidering statistically the features of the genetic sequences.

The bottleneck of these studies is now related to the computational scalability of the downstream analysis. Researchers must work through the genome assemble, the prediction of genes and microRNA and the annotation of sequences against the available databases. Then, much work needs to be accomplished to understand the role of the expressed sequence tag (ESTs), which are short sub-sequences of a transcribed spliced nucleotide sequence, and of microRNA, which are single-stranded RNA molecules, which regulate gene expression.

The genome-wide analysis of gene and the related translated proteins, using this functional perspective, is usually referred to as Functional Genomics. In particular, this field of molecular biology attempts to make use of the vast wealth of data produced by genomic projects to describe gene and protein functions, regulations and interactions. In other words, Functional Genomics uses mostly high-throughput techniques to characterize the abundance gene products and clearly it deals with huge quantity of data from which is difficult to turn into real knowledge.

As a consequence, the merging between molecular biology and computer science fields is becoming a key point to mine information from the huge amount of data produced in laboratories. A computer science approach plays a crucial role to deal with the complexity of this scenario and in particular High Performance Computing is suitable to face these challenges. The exploitation of parallel computing and distributed solutions through the porting of bioinformatics applications and developing new systems on computer clusters and grid infrastructure improves performance when addressing genome wide analysis.

In this chapter we are going to discuss an infrastructure we developed to enable the management of biological large scale computations on the EGEE projects grid implementation, which relies on the *gLite* middleware. This infrastructure is useful to manage data on the grid, in terms of database replications and updating and for the submission and monitoring of the jobs. The idea is to provide an example of how a common infrastructure can be used on the top of a grid computing infrastructure to deal with different computation in the context of the Functional Genomics. The background section concerns a general introduction to grid computing technologies and some details related to the specific grid infrastructure employed in this work. In the next section a presentation of the bioinformatics software tested in our framework is provided. The core of the discussion is about the infrastructure developed to deal with large scale challenges, both in terms of computational management and data handling. The last couple of sections present respectively some challenges performed with this infrastructure and the performance figure obtained with a detailed description of the problems that researchers can encounter in using this platform.

BACKGROUND

Grid technology is a very important step forward from the Web, which simply allows the sharing of information over the internet. This paradigm of distributed computing aims to promote the development and advancement of technologies that provide seamless and scalable access to wide area distributed resources. Computational grids enable the sharing, selection, and aggregation of a wide variety of geographically distributed computational resources, such as supercomputers, compute clusters, storage systems, data sources, instruments and people. This idea is analogous to the electric power grid network, in which power generators are distributed but the users are able to access electric power without bothering about the source of energy and its location. Indeed, grid infrastructures present themselves as single, unified resources for solving large-scale compute and data intensive computing applications.

In particular, the grid paradigm represents a great opportunity to share data and solve intensive computing problems: for this reason it is nowadays largely considered a valid solution for Functional Genomics applications. Grid technologies have been formally defined as infrastructures for supporting the sharing and the coordination of diverse resources in dynamic, distributed virtual organizations (VOs) (Foster, 2002). Nowadays grids are implementations of a computational model which introduced key requirements for protocols and services (Foster, 2001), distinguishing among connectivity protocols, concerned with communication and authentication, resource protocols, concerned with negotiating access to individual resources, and collective protocols and services, concerned with the coordinated use of multiple resources. This integration can be technically challenging because of the need to achieve various qualities of service when running on top of different native platforms.

Nevertheless, grid computing is becoming more and more attractive for coordinating large-scale computations, even if the exploitation of grid solutions is not a trivial pursuit for many users. Nowadays, an important issue is to provide a simplified use of grid resources, enabling the use of tools that scientists usually employ and hiding in the meantime the complexity of the infrastructure. The gap between grids and application domains which require high computation power, such as Functional Genomics, must be progressively closed by researchers and developers who are working to improve the reliability and the usability of these infrastructures. Grid technology allows data intensive challenges to be accomplished, but a capable submission and monitoring environment should be set-up in order to deal with the volume of output data.

The Different Grid Paradigms

Most of the modern grid infrastructures rely on a standard implementation of middleware which is the *Globus Toolkit* (Foster, 2006). This middleware relies primarily on this open source software for building grid systems and applications. On the top of this middleware different grid implementations have been proposed, which usually follow two main directions. The first is mainly application oriented and uses *Grid Services* to provide utilities such as authenticated and session enabled *Web Services*, which provide high portability, but low scalability (Goble, 2003). The second is a set of blank computational resources on which both data and software must by copied to perform computations which present high scalability, but lower portability (Burke, 2008).

These implementations cover different aspects, which can be both useful in bioinformatics. The first is mainly related to the accessibility of distributed service, while the second concerns the possibility to perform wide range computations. While the service oriented implementation can be useful in terms of workflow composition which can be remotely executed on distributed resources, we believe that the added value of the grid relies in the second implementation which can be effectively used to mine the huge quantity of data that bioinformatics is currently producing. In the rest of the chapter we will discuss this second aspect, as this is highly innovative, and with a particular attention to the EGEE project grid implementation, which is becoming the de-facto standard for grid computing in Europe.

The EGEE Grid Platform

The EGEE project infrastructure is a wide area grid platform for scientific applications composed of thousands of CPUs. This platform is a network of several *Computing Elements*, which constitute the gateways for the computer clusters on which jobs are performed, and an equal number of *Storage Elements*, that implement a distributed file system on which databases are stored. The grid core is a set of *Resource Brokers* delegated for controlling submission and execution of the different jobs.

The computational resources are connected to a *Resource Broker* that routes each job on a specific *Computing Element* taking into account the directives of the submitting script, called JDL (Job Description Language), and implements a load balancing policy. Each *Computing Element* submits the incoming job to a batch system queue hiding a *Working Nodes* farm. A set of tools is used to manage data as in a distributed file system. These tools allow the data to be replicated to different *Storage Elements* easily, and help using this information in an efficient way. The *Resource Broker*, in fact, is able to redirect the execution of an application to a *Computing Element* located as near as possible to the data being used, hence minimizing the communication time.

Apart from uploading data from the *Storage Elements*, which typically is reserved for the database management, in the JDL script it is possible specify an *InputSandBox*: a collection of files which are sent directly on the computational resources. This aspect is very important because it allow the subdivision of the input sequences to be analyzed and the direct upload of them with the related specific job to be computed. At the same way output data can be downloaded from the *Working Nodes* without copying data on the *Storage Elements* by using the *OutputSandBox*.

The EGEE grid enables an efficient data management coupling the *Storage Element* services with the *LHC File Catalogue* where the physical handlers of files which rely on the storage facilities are recorder and associated to logical file names. This system allow to have more replicas for each logical file name, a feature that is crucial to have more replicas of reference database minimizing the transfer rate at running time. The system composed by the *Storage Elements* and the related file catalogue implement an effective virtual distributed file system, which allows redundancy.

All the elements of this implementation of the grid computing paradigm are deployed as *Grid Services*, which envelope the main provided functionality enabling secure communications among the grid components. In particular, the *gLite* middleware must be installed on a local server within the institution which wants to join the EGEE grid in order to establish secure communications between the institution itself and the distributed infrastructure. This local server, which is usually called *User Interface*, through the *gLite* API's enables to submit jobs, to monitor the state of advancement of the jobs, to retrieve the outputs when the computations have a normal termination and to resubmit the jobs in case of failure.

Due to the use of remote computational resources, the grid communication software must offer an efficient security system. Simply stated, security is guaranteed by Grid Security Infrastructure (GSI) which uses public key cryptography to recognize users. GSI certificates are encoded in the X.509 format and accompany each job to authenticate the user. In other words, the access to remote clusters is granted by a *Personal Certificate*, which accompanies each job to authenticate the user. Moreover, users must be authorized to job submission by a *Virtual Organization*, a grid community having similar tasks, which grants for him. This procedure is indispensable for maintaining a high security level.

PORTING GENOMICS APPLICATIONS ON THE GRID PLATFORM

The access to several tools and large databases in genomics forms the basis of research in computational biology. Genome software including gene finders, homology comparison, multiple alignment tools, and phylogenetic comparison often work on subsets of genome sequence, and collate results. Promising newer genome tools draw data from several sources: cross-species homologies, large scale functional and interaction data and primary genome sequences. A practice common in genome analyses is the ad hoc development of data processing scripts to split and collate genome data and results and this approach can be adopted and automated working on infrastructure of grid computing. Depending on the analysis, splitting and collation operations can be handled in a generic manner.

Then, Functional Genomics applications are typically data driven and have long running times because it is necessary to use many different tools and integrate information from many different biological databases. Users of the genomics applications need to find and discriminate among the most appropriate solution in terms of applications and databases, and may need to be notified when resources are changed or updated. From the computational point of view, most genomics applications are surprisingly parallel to a very high extent. That is, they consist of doing thousands of independent computations. The real problems for computer architectures oriented to genomics applications are related to handling large amount of data. Indeed, data management plays an important role in performing Functional Genomics applications on distributed architecture such as grid.

In the following paragraphs some Functional Genomics suite of tools, which have been ported on the grid infrastructure, are introduced, to provide some hints of how it is possible to deploy bioinformatics applications using the distributed computing paradigm.

Gene Prediction

Different tools are available for genes identification in datasets of genome wide sequences. Advanced gene finders for both prokaryotic and eukaryotic genomes typically use complex probabilistic models, such as Hidden Markov Models, in order to combine information from a variety of different signal and content measurements. This approach is particularly effective with prokaryotes, as implemented for example in GLIMMER (Delcher, 1999), which is a widely used and highly accurate gene finder. On the other hand, Eukaryotic ab initio gene prediction is more complex, and finders have achieved partial success. A notable example is the GENSCAN (Burge, 1997) program that, starting from a set of nucleotide sequences, looks for typical gene patterns, translating the different exons into the corresponding protein sequences. In particular, this tool searches in the nucleotide sequences for the key components of the gene expression like promoters, TATA box sequences and exons patterns. To accomplish this task GENSCAN comes with different databases of gene signatures each one related to a different organism.

The algorithm for searching gene signature is quite complex and takes long time in predicting exons and introns within multiple input sequences, because it has to work on both the strands while looking for genes, which implies to reverse all the sequences. Clearly, working at genome scale, it is computationally intensive to predict coding proteins, in particular when the input sequences are more the 1Mbp long. In this context we used the grid computing to scale computations, by distributing the load in different resources. To our knowledge, in the state of art there are no references to grid implementation of software for gene prediction at genome scale. However, the pure data parallel approach can be adopted in these cases because there is an effective independence among the computations, and the whole task can be viewed as the sum of repeated operation on single sequences. The distribution over the grid of GENSCAN is very efficient because the reference databases are quite small, in terms of KB, and we use to upload it directly on the *Working Nodes* through the *InputSandBox*, instead of storing them on the *Storage Element*, a procedure which prevents some failures from file unavailability.

Local Alignment Tools

The Basic Local Alignment Search Tool, or BLAST (Altschul, 1990), is algorithm for comparing primary biological sequence information. BLAST is one of the most widely used Functional Genomics applications, because it addresses the fundamental problem of comparing sequences against references databases. In particular, a BLAST search enables a researcher to compare a query sequence with a database of sequences, and identify library sequences that resemble the query sequence above a certain threshold. Consequently, the possibility to perform BLAST on the grid infrastructure is very important to parallelize the executions of time-consuming analysis exploiting the use of remote resources efficiently.

Two main implementation of the BLAST algorithm are available for researchers, the NCBI BLAST (National Centre for Biotechnology Information) and the WU-BLAST (Washington University). In both cases, the easiest and most effective way to use BLAST on the grid platform is to divide the input file into a series of smaller multifasta containing a balanced number of sequences. In this way the sequences are divided into a series of files that are submitted to the different grid nodes (Merelli, 2005), even if other solutions have been proposed (Moscicki, 2004). The main issue is related to the management of the reference databases, which must be uploaded together with part of the input sequence on each Computing Element. A strategy of data replica is also implemented in the MPI-BLAST to avoid the problem of the multiple accesses on distributed file systems, as in the NFS, typical of parallel computations.

Clearly, the main issue of porting BLAST on the grid infrastructure is the distribution and the replication of the databases over the *Storage Elements*, which involves a considerable effort, in terms of transfer time, bandwidth, and occupied disk space. The strategy used to replicate and update databases over the grid is a key factor to obtain good performance, because it is crucial to find a trade off between the number of replicas, the database usage, and the time needed to upload databases on the Working Nodes. Moreover, the use of BLAST in grid becomes convenient if the number of sequences to be elaborated, and therefore the computation time, justify the time employed for all the steps needed to parallelize the execution, in particularly those concerning the uploading of the databases on the computational resources.

Sequences Multiple Alignment

Multiple alignments are a systematic way of simultaneously comparing many sequences by disposing the sequences in a matrix and adding space (gap) in order to match amino acids with high degree of similarity. Many multiple alignment algorithms exist and can be classified in many ways depending of the target sequences (protein/DNA), the strategy (progressive/iterative), the underlying homology and gap model and the size of the resulting alignment (local/global).

Common applications for multiple alignments are ClustalW (Thompson, 1994), TCoffee (Notredame, 2000) and MAFFT (Kazutaka, 2002). ClustalW uses a progressive technique in which a sequence tree is built according to a distance measure and then the tree is used as an ordering for grouping sequences into the final alignment. TCoffee uses a similar approach but a more sophisticated scoring schema based on a library of pairwise alignments of the target sequences previously calculated. Finally, MA-FFT uses Fast Fourier transformation in order to encode physical properties of the amino acids along the sequence and uses the resulting signals to progressively align sequences. Parallelization paradigms for multiple alignment programs are usually not trivial, because different solutions must be studied for each specific algorithm: for this reason currently there are no grid implementations available of these applications.

MicroRNA

MicroRNAs (miRNAs) are post-transcriptional regulators of gene expression in animals and plants. Functional genomic computational methods have been developed to predict new miRNAs in worms, flies, and humans. In particular, a very useful suite of tools for this analysis is findMiRNA (Adai, 2005), which predicts potential miRNAs within candidate precursor sequences that have corresponding target sites within transcripts. The findMiRNA algorithm uses annotated transcripts to search the intergenic regions for candidate miRNAs directed against the transcripts. The search process is conducted by successive filtering of the genomic sequence. The procedure is organized in four steps: 1) fast estimation of secondary structure potential by computation of nucleotide scores; 2) search for hairpins and computation of their energies; 3) estimation of thermodynamic probability of the hairpin structure found; 4) search for miRNAs in the candidate hairpin.

The prediction power of this software is impressive, but all the steps by which the algorithm is composed are very time-consuming. Moreover, this kind of analysis is typically genome wide and require powerful computational infrastructure. The distribution paradigm for this application is similar to the one presented for other tools and relies in splitting the genetic sequence in order to distribute them in different computation resources. The main difference is that the dataset of intergenic sequence must be available to all the distributed processes that are involved in the computation and due to its specificity in the analysis it can not be uploaded a priori like a BLAST database, but should be provided at run time. In literature, there are no reference related to grid analysis on microRNA even if an interesting experience is reported about the identification of other genomic structures (nucleosome) using a pilot grid architecture (Radwan1, 2008).

Phylogenetic Applications

Molecular evolution is based on the principle that the level of similarity between genes displays the degree of evolutionary relationship between them. Today molecular evolution is a widely applied study, touching diverse areas like ecology and organisms development. It is commonly known that the unraveling of phylogenetic relationships between organisms' gene sequences is an important first step towards the reconstruction of their evolution. Huson and Bryant define a phylogenetic network as any network in which taxa are represented by nodes and their evolutionary relationships. The phylogenetic tree can be thought of as a structure that provides a meaningful way to order data by containing huge amounts of evolutionary data within its branches.

Building phylogenetic trees is a good starting point for infer the function of a particular gene. In order to improve the quality of this prediction it is important to annotate the tree with information about the evolutionary events that occur along the tree. In particular speciation events, that represent relationship between orthologous sequences, are deemed to be more conservative for the molecular function. On the other side, duplication events allow sequences to evolve in different functions respect to the original without loose the specificity of the original function. The FORESTER package (Zmasek, 2008) contains many tools for the analysis of phylogenetic tree and in particular it implements an algorithm for the annotations of the tree in terms of speciation/duplication events in all the nodes of the tree. It use a search algorithm that match the phylogenetic tree with the "tree of life" that is a general tree that represents the relationship between the species involved in the analysis.

PHYLIP (the PHYLogeny Inference Package) is a package of programs for inferring phylogenies (evolutionary trees) and consist of 35 programs (Swofford, 1996). Programs that are available in the package include parsimony, distance matrix, and likelihood methods, including bootstrapping and consensus trees. Data types that can be handled include molecular sequences, gene frequencies, restriction sites and fragments, distance matrices, and discrete characters. The porting to the grid infrastructure entails the running of the bootstrap algorithm on several nodes. In particular, the neighbor joining algorithm NEIGHBOR takes as input a matrix of pairwise distances between taxa and outputs a tree connecting the taxa.

ENABLING BIOINFORMATICS COMPUTATIONS OVER THE GRID

The experience we gained in performing Functional Genomics applications on grid show how the implementation of a flexible environment on the top of the *gLite* middleware greatly improves the performance in genome scale computations. According to our experience, the submission and the monitoring of all the jobs in which the computation is divided and the collection of the output results takes advantage of an efficient coordination system. Also advantageous is to provide an efficient post processing procedure of the output results directly on the *Worker Nodes* and a reliable system to manage databases on the *Storage Elements*.

The system has been implemented to work on the top of the *gLite* middleware, even if it can be customized to work with other infrastructure, and it is able to completely coordinate a challenge running on a single *User Interface*. Form the applications point of view, the system is flexible and can be applied to different Functional Genomics software with a similar distribution paradigm. Clearly, we refer to applications that are data parallel or with low requirements of inter-processes communications, for which the distribution approach is to split the computation into a set of independent subtasks.

Challenges Control System

Since bioinformatics analyses present huge computational load working at genome scale, the grid platform can be used to achieve time-consuming tasks subdividing the whole computation into a set of small jobs in order to execute them on different *Computing Elements*. As previously discussed, regarding job submission, monitoring the status and retrieving the results is usually a complex issue to address using the grid platform. The implementation of an efficient infrastructure, both from the computational and the data point of view, is crucial, in order to limit the time needed for the execution of the whole task and to optimally exploit the power of the grid.

From the computational point of view the greatest problem of the grid is the dynamic behavior of the available resources. Due to network and system errors or in relation to the global computational load, the available resources are continuously reshaped and the rate of failure in computations is quite high. The crucial point is then the creation of a fault tolerant infrastructure: by tracing the status of the job constantly it is possible to immediately resubmit every task that presents problems or inconsistent status.

The implementation our *Challenge Control System* can be viewed ad a double layered infrastructure (Fig. 1) that can be customized for different applications. The first layer is designed to control the execution of each single job and works directly on the top of the *gLite* middleware. For this reason it is installed on a grid *User Interface* where users that have a valid *Personal Certificate* can be authorized to submit the jobs on a valid *Virtual Organization*.

The second layer coordinates the distribution of the whole challenge by monitoring the Input/Output consistence of each job computed on the grid platform. The two layers can communicate through a relational database that stores the challenge's crucial information: the monitoring activity of jobs and files distribution, for the first layer, and the Input/Output status of the application and the parsed results, for the second layer. Using this approach the system is very flexible while a strict control of the data integrity can be performed by checking the consistency between the information of the two layers.

Post Processing

When dealing with biological challenges in which a huge quantity of data is involved, the analysis of the results can be computationally very expensive and in some cases as labour-intensive as the computation itself, if an adequate system to collect results has not been previously established. A lot of effort is usually needed to parse results and to collect them into a relational database after the completion of a grid challenge. We propose to perform both the parsing of the output results and the storage of data in the *output database* directly from the grid computational resources. By designing an analysis system that can be run immediately after the computation in order to parse the results, the grid can be fully exploited and the post processing problem is overcome.

According to this approach, we suggest a system that coordinates all the steps needed by the

Figure 1. The system proposed to perform functional genomics computation on grid platforms like the EGEE project infrastructure. The software is structured into two different layers: the topmost one is highly application oriented while the bottom one is closely connected with the implementation of the grid. The system relies on a database that allows the grid Input/Output coordination and the monitoring of the job status.

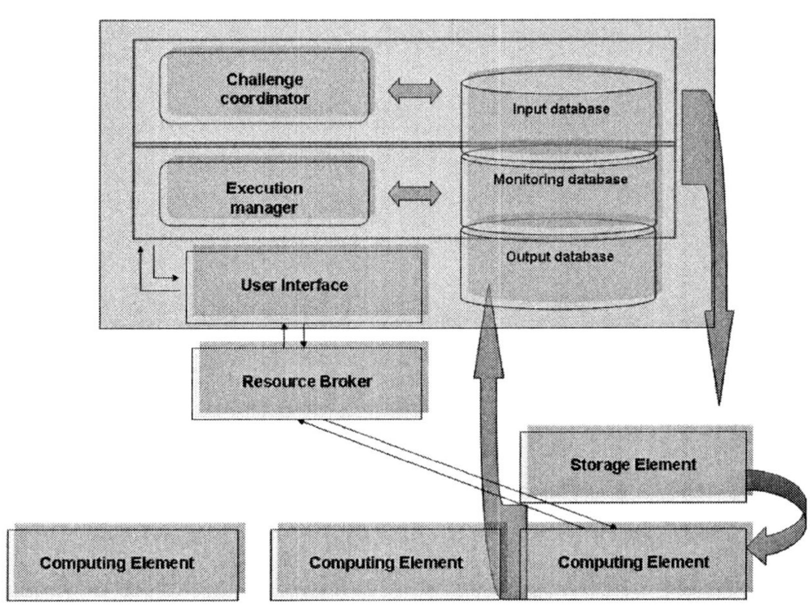

application to execute on the *Computing Elements* and, as a last step, establishes a connection to the *output database* in order to store the generated and post-processed results. Using this approach it is possible to store the key results of each computation locally and to evaluate them in the global context of the analysis. Using this approach, the implementation of an *output database* in which results can be collected for each Functional Genomics application plays a crucial role. However, it is still possible to retrieve the complete results to perform further analysis, whenever it is necessary, by downloading the output files from the *Storage Elements*.

Protein Database Handling

Functional Genomics applications are frequently based on sequence alignment or signature pattern matching against reference database in order to establish correspondences, which can be useful to infer information about functionality. To perform this computation efficiently on the grid platform, it is necessary to allocate these databases on different *Storage Elements*. In this way it is possible to use a high number of different *Computing Elements* on which the computation performs efficiently in a distributed way. In order to raise performance there is the necessity to control the number of replicas for each database in relation to its usage, minimizing the bandwidth occupation and the data load on the *Storage Element*.

In order to perform Functional Genomics database search an efficiently a system for the replication of the data has been implemented to minimize the time needed for uploading the databases onto each computational resource during the execution. This system allows each database to be replicated dynamically in relation with its usage in order to balance the number of replicas. The idea is to work on the statistical analysis of databases usage by the grid jobs, working on data acquired after each job execution, regarding grid queue times, database set up times and overall job computation, in order to decide the replication ratio in relation to the occupied disk space over the grid (Fig. 2). The system constantly monitors FTP and HTTP sites for newest versions of each database uploaded into the grid. When a new timestamp of a specific database is detected, the newest version will be automatically downloaded in order to replace the older version on the grid.

Figure 2. Bioinformatics databases management over the EGEE grid infrastructure. Databases are downloaded from public mirrors to a repository (Rep)and a number of replicas, proportional to the database usage, are stored in different Storage Elements.

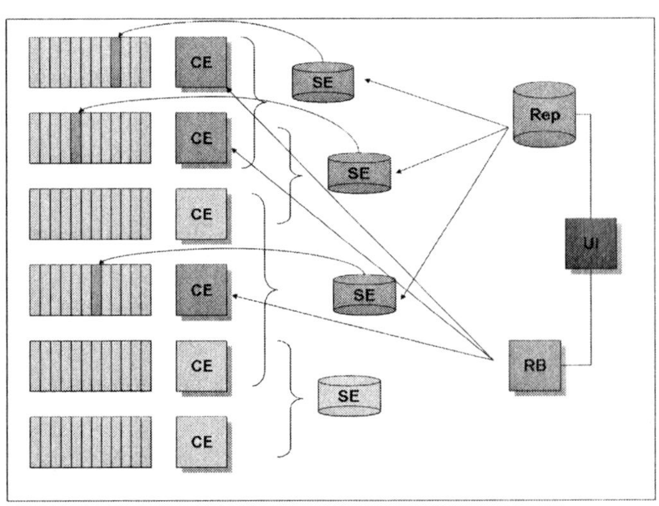

Moreover, the implemented system computes an xdelta patch to enable the possibility to regress the newer version of the downloaded database to an earlier version now being replaced, and uploads such xdelta patch on a predefined *Storage Element.* Such xdelta patches are many times smaller than any version of the databases they refer to, and this makes the storage costs of these patches reasonable. At the same way, tanks to the small dimension of the patches, it makes sense to have a single replica of them, because the transfer time to all the *Computing Elements* that run jobs that require the regression of the database is very low.

In the following tables we report for all the reference databases of nucleotide and protein sequences we manage with the implemented system, the average download time required to copy them on a *Computing Elements* which runs a BLAST jobs. As reported in table 1 and 2, there is great differences in the downloading a database from a local resources or from a remote *Storage Elements*, which is considered in the algorithm to balance the number of replicas required to obtain good scalability in relation to database usage and the occupied disk space.

EXPERIENCES IN FUNCTIONAL GENOMICS CHALLENGES

In this section we will present some results accomplished in performing typical Functional Genomics challenges related to genome scale analysis. All the computations have been performed on the EGEE grid platform and have been managed using the developed Challenges Control System, while all databases have been replicated and updated by the presented infrastructure.

EST Annotation Using BLAST

The first test is related to a typical sequence based analysis using the BLAST, version 2.2.11, for the annotation of human coding regions for proteins. In this challenge, we have performed the annotation of a dataset of human Expressed Sequence Tags (EST), which are pieces of expressed genes, against a database of protein sequences, UNIPROT, annotated using Gene Ontology (GO) terms. This analysis allows to identify the function of genomic sequences that will be translated into effective proteins completely annotated in UNIPROT. Moreover, considering the GO references available in the database, it is possible to classify the EST in different functional clusters using the ontology description.

Table 1. Protein database average time statistics used to trade off replication and average download time

Database name	Size (MB)	Average download time (close SE)	Average download time (remote SE)
nr	1231 1	m44s 1	6m10s
pdbaa	7.38 <	1sec 6	sec
uniprot_sprot	40.9 3	sec	32sec
uniprot_trembl	655	56sec	8m36s
ecoli_aa	0.97 <	1sec <	1sec
yeast_aa	1.95 <	1sec 2	sec

Table 2. Nucleotide database average time statistics used to trade off replication and average download time

Database name	Size (MB)	Average download time (close SE)	Average download time (remote SE)
nt	4483 6	m22s 5	8m53s
human_genomic	1381 1	m58s 1	8m08s
est_human	1116 1	m35s 1	4m39s
est_mouse	731	1m02s	9m36s
ecoli_nt	1.12 <	1sec <	1sec
yeast_nt	2.83 <	1sec 2	sec

The computation of this task is very time consuming because the UNIPROT annotated database, which relies on the SWISSPROT and TrEMBL protein collections, version May 2007, contains about 2.319.487 proteins sequences for a total of 1.06 GB and the human EST dataset, July 2007 version, was composed by 7.895.635 short sequences for a total of 5.12 GB. The computational time needed to perform this analysis is increased due to the translation that the BLAST algorithm has to perform while confronting genomics and proteins sequences.

The BLAST challenge for EST annotation has been defined subdividing the whole computation in 1.580 jobs each one consisting of 5.000 sequences to be screened against the UNIPROT database. In a period of about four weeks we collected all the results covering about 2 years of computer activity (Fig. 3). During the analysis 6.946.250 hits have been detected for a total of 577 MB of output results. The time needed for the whole challenge has to be valued taking into account the job submission rate, which was lowered in consideration to the high quantity of

Figure 3. Ratio of job accomplishment for the BLAST challenge. The step-trend of the graph can be in part explained with the submission policy aimed to reduce the Resource Broker overload.

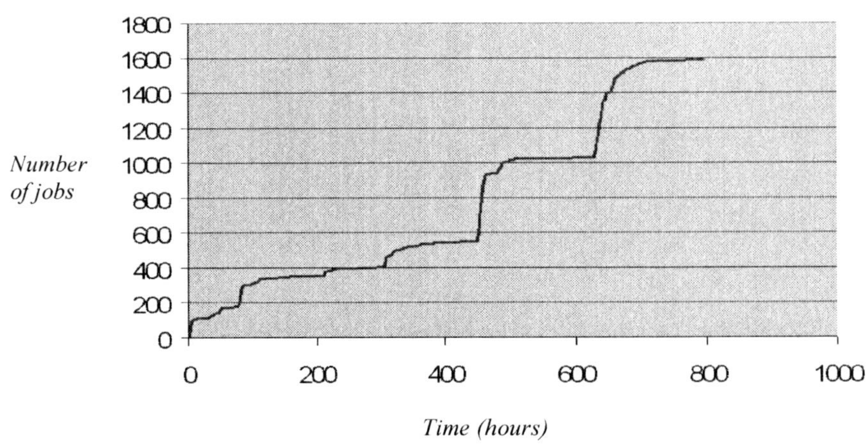

data to transfer between the User Interface, the *Resource Broker* and the *Computing Elements* for the input and output management and between the *Storage Element* and the *Computing Element* for uploading the database.

In order to perform the post-processing analysis directly onto the grid resources as mentioned, we have developed a simple parser to analyze the BLAST output in the tabular form in order to upload directly the results into the output database, which has been set-up on the same User Interface from which the challenge has been coordinated. Using this approach it is possible to store the key results of each computation locally and to evaluate them in the global context of the analysis.

MicroRNA Prediction

Another analysis performed on the grid platform using the developed infrastructure concerns the identification of the possible microRNA within the grapes genome, which has been recently sequenced in the frame of the International Grape Genome Program (IGGP). The prediction of these patterns relies on the use of findMiRNA, which requires as input the genetic and intergenic sequences. The genes prediction has been performed with GENSCAN using an ad hoc developed database of promoter sequences: the genome search and the following annotation have produced 33233 genetic sequences and 59140 intergenic sequences. This huge quantity of sequences implies the use of an HPC performance computational resource because the estimated running time of findMiRNA can be predicted in about 1 CPU year.

In order to accomplish this prediction of microRNA on the grapes genome 560 jobs of about 14 hours each have been submitted, splitting the genetic sequences and keeping the intergenic as reference. In a period of about two weeks we collected all the results covering about 1 years of computer activity (Fig. 4). During the analysis about 2.142.000 microRNA have been predicted for a total of 714 MB of output results. According to the computation statistics the efficiency of this challenge is better than the previous one, a result that can be explained considering the dimension of the intergenic sequences database, approximately of 500MB, which is about the half than in the previous challenge and implies less time to load the database on the computational resources.

Phylogenomic Analysis of Gene Function

Another important application that has been ported and tested in the EGEE grid environment is the functional annotation of gene within a genome

Figure 4. Ratio of job accomplishment for the microRNA challenge

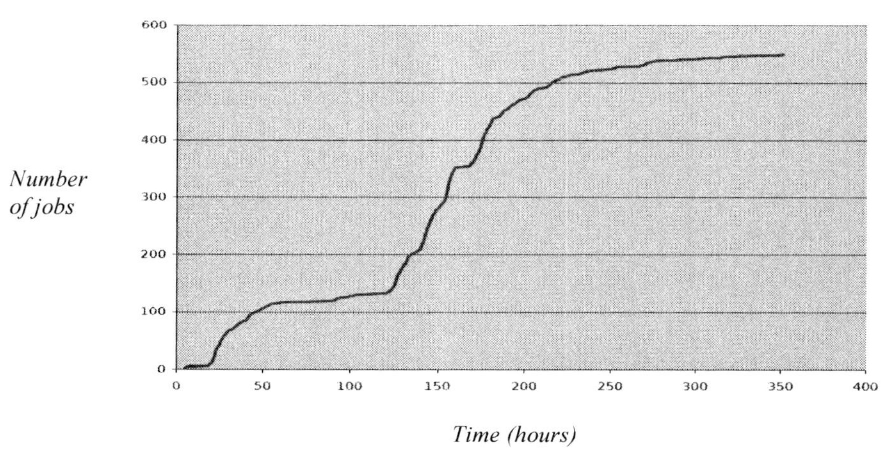

using a five steps pipeline. A target sequence is queried against a database by similarity search in order to find out potential related sequences. Then the cluster of sequences is globally aligned to highlight regions of high similarity. The alignment is used to produce a phylogenetic tree that is annotated with functional (e.g. Gene Ontology terms) and evolutionary (e.g. duplication/speciation events) information. Finally an inference rule is applied in order to transfer functions from the nodes in the tree to the target sequence.

This pipeline has been deployed in the *gLite* infrastructure in order to distribute computation and data and thus exploiting the intrinsic parallelism of the application. The programs involved in the pipeline are the BLAST program for the first step (search of homologous sequences), the ClustalW for the multiple alignment of the sequences, the PHYLIP program for building the phylogenetic tree and FORESTER for the final annotation of the tree.

The proposed pipeline has been tested in the annotation of the protein in the Swissprot database (human genome) that contains about 14000 annotated human sequences. The number of jobs to be submitted to the grid for each sequence is ideally one for each program in the pipeline. The real number increases due to the rate of failures in the grid system. The jobs has been distributed as single job (1 job for 1 sequence) and submitted in groups of 100 jobs. Table 3 (a) shows the number of jobs for each experiment (user jobs), the total number of jobs taking into account job resubmission (real jobs) and the overall execution and waiting time of jobs submitted to the grid. Execution times are compared with the running time of the jobs on a single local machine (in this case the waiting time is negligible). Table 3 (b) shows similar information grouped on the basis of the type of program used in the analysis.

PERFORMANCE DISCUSSION

In relation to the results presented some considerations can be drawn about the efficiency of this particular implementation of grid infrastructure in the context of Functional Genomics analysis. Clearly, these results are very important while taking into account the grid infrastructure as a possible solution to reduce the computational time of a computation. In particular we would discuss the job efficiency, the overhead imposed by the grid and the global scalability of the infrastructure.

Job Efficiency and Latency time

In the activity we carried out within Functional Genomics computations over the grid, we obtained about 63% of the jobs reported as successfully finished according to the status logged in the *Resource Broker*, however, the ratio went down

Table 3. (a) Parameters related to the execution of the phylogenomics pipeline on the swissprot database. The overall execution time if 158h (exec time + wait time) is compared to the local time of 1533h needed to execute the same jobs on a single local machine. (b) The same information grouped by the three main applications.

User jobs	41901
Real jobs	50281
Execution time (h)	98
Waiting time (h)	113
Local time (h)	1533

	BLAST	**ClustalW**	**Phylip**
Execution time (h)	18 4	8	26
Waiting time (h)	23 5	2	33

to 57% after checking the existence of the output. The main cause of job failure seems to be the data transfer between the *Storage Elements* and the *Computing Elements* and, in few cases, the communication between the *Computing Elements* and the local results database. This can be inferred by the differences in the challenge statistics: the challenges in which the data transfer rate is relatively important present a higher number of failures. Another frequent problem is related to errors during the job scheduling and a very high ratio of these failures came from the mis-configuration of *Computing Elements*. On the other hand, the number of jobs reported as correctly finished by the grid, but resubmitted because the results didn't match with the database information was pretty low.

A very important factor for the reliability of the grid is the average latency time, which can be defined as the sum of the time spent for establishing communication with the *Resource Broker* and the scheduling time on the Computing Element queue. Considering the 95% of the jobs that come to completion in both the proposed challenges, the average latency time is about 5 hours (Fig. 5). Clearly this latency time has low impact on long term jobs, but heavily influences short jobs, making the grid quite unsuitable for short computations.

Scalability

A key parameter for the evaluation of the distribution of the computational load is the estimated duration of a single grid job. Apart from job failures due to data transfer or internal errors, the time needed for completing the whole computation can also be elevated by the execution of jobs too short, which greatly increase the latency in the execution queue. On the other hand the submission of a small number of jobs limits the scalability of the system and in case of failure may impose to wait long time for the fulfillment of few resubmitted jobs. Hence, the length of the submitted jobs should be carefully planned. In these analyses, we choose to run smaller jobs for example in the surface computation challenge and this was done to increase the performance, but we built bigger jobs in the surface matching challenge, in order to avoid the overload of the grid. According to our opinion these choice well suited the different structure of the applications, improving the scalability as much as possible.

Figure. 5. Analysis of the job latency time on the EGEE grid platform. The latency time is intended as the time spent both for the communication for the job submission and on the clusters queues. To obtain a more readable graph the axes have been log scaled.

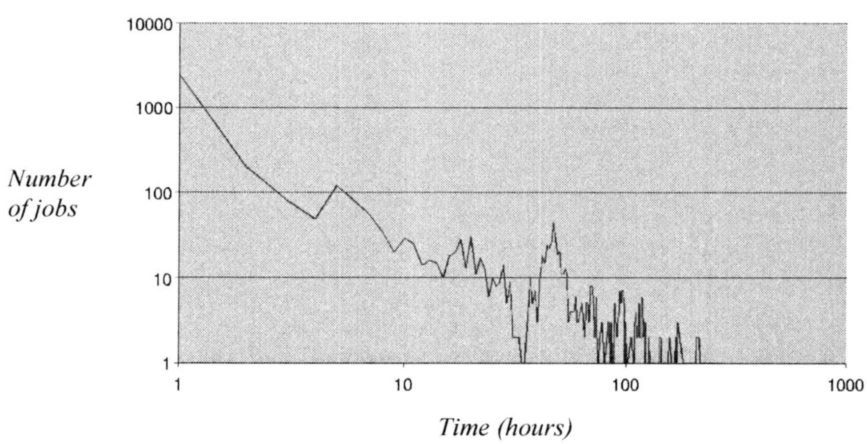

FUTURE TRENDS

The grid computing well suits bioinformatics requirements of computational power and scalability. Results are encouraging and we presume that an increasing number of applications will be ported to the grid platform in the near future. Although developing new applications for the grid computing appears difficult at first, since the distributed architecture has to be learned, some advices can be followed to reduce the gap from this technology.

The grid technology has not yet reached enough maturity that any life science application can be deployed on the existing grid infrastructures. Distributed computing is now widely available on grids like EGEE or myGRID but distributed data management is still in its infancy. Projects very close to the end users where biological concepts are managed in a very conceptual manner are not yet interfaced to the grid services so they can take advantage of the grid resources. The applications successfully deployed on the existing grids have a clear biological relevance but they were selected and designed after a careful evaluation of the services available through the EGEE middleware. This bottom-up approach is the key to avoiding disappointment and frustration.

The direct porting of an application on the grid computing should be valued in relation to the effective CPU demands of the computations and to the effort in deploying the system. Many clusters are now available for the bioinformatics community in Europe and their use is significantly simpler than the use of grid platforms. On the other hand, the grid itself can be considered as an improvement in the accessibility and in the load balancing of these infrastructures. Indeed, there is a clear interest to develop portals accessing the grid resources in a way which is transparent to the user. The effort to develop such portals should as much as possible be shared between different institutes to avoid duplications.

While the necessity of porting an application on the grid platform arises, choose the grid technology for which you have the easiest access and the best support. Once it is clear that grids are the right environments for the considered application, the next step is to choose the grid infrastructure and its middleware. It must be clear that both the amount of resources offered by the infrastructure and the services offered by the grid middleware must be taken into account. But this is not sufficient: a proper user support is mandatory at this early stage to help the new user in the learning phase.

In any case, the advice is to avoid a person implementation of core grid services, use the standard middleware distributions. This principle is very important to reduce the human cost to the deployment. The cost to develop a grid service is high but the cost to maintain it is even higher. Despite the efforts of standardization bodies like the Open Grid Forum, the interfaces to the grid services are not stabilized so any new service on top of a middleware, such as *gLite*, will have to adapt to the evolution of the interfaces. If a given application depends on such a service, it will not run on the next middleware release unless additional effort is made to adapt the home made service. To avoid this situation, it is better to only use standard middleware releases. It is even recommended not to use newly developed services which are not already tagged for integration in the next middleware release.

If there is a need to use the grid for a given application but the grid services do not allow to its complete management, the most effective approach is to use the grid for the subset of the data analysis that can be accomplished, if this is already significant. Another option could be to wait for the next middleware release in the hope for new functionalities. But the enrichment of the portfolio of functionalities for the grid user has been progressing very slowly especially in the field of data management. As a consequence, waiting for the next release is often disappointing. The grid user must be willing to adapt its application to the grid rather than adapt the grid to its application. This may sound unsatisfactory but it is the reality of today.

CONCLUSION

In the last years, the role of High Performance Computing has become increasingly important in scientific research in general and in biological and medical fields in particular. A number of bioinformatics applications need, in fact, large computational resources to analyze the increasing quantity of data produced by the emerging high throughput techniques in biotechnology. Functional Genomics analyses in particular are continuously demanding for more processing power specifically when dealing with structural analyses.

Grid technology allows computational intensive challenges to be accomplished, but a capable submission and monitoring environment should be set-up in order to appropriately manage the volume of data. According to our experience, in order to work with a grid infrastructure like the one from the EGEE project, which relies on the *gLite* middleware, a fault tolerant system able to accomplish the management of the jobs is needed to perform bioinformatics high throughput analysis.

Indeed, our intention was to show how the creation of a flexible environment can play an important role in the quality of the results obtained, and how a relational database can be used to manage the Input/Output with the grid in an effective way to avoid data loss during the communication with remote computational resources. Particular attention has been paid to the post process analysis of large amount of data: the exploitation of the distributed resources can be extremely useful to parse the important information from the results in order to screen output data hence minimizing the post process computational time.

The grid infrastructure can be successfully implied to face large scale computations using different Functional Genomics applications, giving results that are encouraging for future developments. In particular the use of the grid seems to fulfill both the computational and data management requirements even if this technology is still complex to use, in particular for non expert users, due to problems encountered during the computation phases of the challenges.

These problems are mostly related to the dynamic nature of the grid in which the status of the resources changes constantly and transient problems can happen, provoking the failure of jobs. Although the monitoring and logging systems of the grid already does a lot in the direction of limiting these problems, due to the sheer size of the infrastructure and its geographical dispersed nature the consistency between the effective status of the system and the provided information is a very difficult task to accomplish.

Moreover, virtual file systems which can be used over grid such as the one of the EGEE project, typically implemented associating a file catalog to the storage resources, have still a lot of inconsistencies, caused by temporary unavailability of resources which discontinue the accessibility of files. A system integrated in the middleware to constantly monitor the consistency between files registered in the catalog and replicas effectively available in grid would be extremely useful.

In conclusion, the grid is an effective system for coping with the increasing demanding of computational power of bioinformatics, and in particular in genomics field. The system scalability in case of independent computations is very high, even taking into account the time needed for scheduling jobs and transferring data. Such computing power is a concrete possibility to face challenges which have been thought to be impossible just few years ago.

ACKNOWLEDGMENT

This work has been supported by the European Projects, BioinfoGRID and EGEE III and by MIUR-FIRB Italian projects LITBIO and ITALBIONET.

REFERENCES

Adai, A., Johnson, C., Mlotshwa, S., Archer-Evans, S., Manocha,V., Vance, V. & Sundaresan, V. (2005). Computational prediction of miRNAs in Arabidopsis thaliana. *Genome Res., 15*, 78-91.

Altschul, S.F., Gish, W., Miller, W., Myers, E.W. & Lipman, D.J. (1990). Basic Local Alignment Search Tool. *J. Mol. Biol., 215*, 403-410.

Burge, C. & Karlin, S. (1997). Prediction of complete gene structures in human genomic DNA. *J. Mol. Biol., 268*, 78-94.

Burke, S. Campana, S., Mendez Lorenzo, P., Nater C., Santinelli, R., Sciabà, A. (2008). gLite 3.1 User Guide. Retrieved June 6, 2008, from https://edms.cern.ch/file/722398//gLite-3-UserGuide.pdf

Delcher, A.L., Harmon, D., Kasif, S., White, O. & Salzberg, S.L. (1999). Improved microbial gene identification with GLIMMER. *Nucl. Acids Res., 27*(23), 4636-4641.

Foster. I. (2006). Globus Toolkit Version 4: Software for Service-Oriented Systems. *LNCS*, 3779, 2-13.

Foster, I., Kesselman, C., Nick, J. & Tuecke, S. (2002). Grid Services for Distributed System Integration. *Computer, 35*(6), 37-46.

Foster, I., Kesselman, C. & Tuecke, S. (2001). The Anatomy of the Grid. Retrieved june 6, 2008, from http://www.globus.org/alliance/publications/papers/anatomy.pdf

Goble, C.A., Pettifer, S., Stevens R. & Greenhalgh, C. (2003). Knowledge Integration: In silico Experiments in Bioinformatics. In I. Foster, C. Kesselman (Eds.), *The Grid: Blueprint for a New Computing Infrastructure* (pp. 121-124). San Francisco, CA: Morgan Kaufmann Publishers Inc.

Kazutaka, K., Kazuharu, M., Keichi, K. & Takashi, M. (2002). MAFFT: A novel method for rapid multiple sequence alignment based on fast Fourier transform. *Nucl. Acids Res., 30*(14), 3059–3066.

Merelli, I. & Milanesi, L. (2005) *High performance implementation of BLAST using GRID technology.* Paper presented at the meeting of the Italian Bioinformatics Society, Milan, Italy.

Moscicki, J.T., Lee, H.C., Guatelli, S., Lin, S.C. & Pia, M.G. (2004) *A Biomedical Applications on the GRID: Efficient Management of Parallel Jobs.* Paper presented at the Nuclear Science Symposium, Geneve, Switzerland.

Notredame, C., Higgins, D.G. & Heringa, J. (2000). T-coffee: a novel method for fast and accurate multiple sequence alignment. *J. Mol. Biol., 302*, 205–217.

Radwanl, A., Younis, A., Luykx, P. & Khuri, S. (2008). Prediction and analysis of nucleosome exclusion regions in the human genome. *BMC Genomic, 9*, 186.

Swofford, D., Olsen, G., Waddell, P. & Hillis, D.M. (1996). Phylogenetic inference. In D.N. Hillis, C. Moritz, & B.K. Mable, (Eds.), *Molecular Systematics*, (pp. 407-514). Sunderland, MA: Sinauer Associated.

Thompson, J.D., Higgins, D.G. & Gibson, T.J. (1994). CLUSTAL W: improving the sensitivity of progressive multiple sequence alignment through sequence weighting, position-specific gap penalties and weight matrix choice. *Nucl. Acids Res., 22*(22), 4673-4680.

Zmasek, C.M. (n.d.). *Forester: software libraries for evolutionary biology research.* Retrieved June 6, 2008, from http://www.phylosoft.org/forester.

KEY TERMS AND DEFINITIONS

Functional Genomics: A field of molecular biology that attempts to make use of the vast wealth of data produced by genomic projects.

Sequence Alignment: A way of arranging the sequences of DNA, RNA, or protein to identify regions of similarity that may be a consequence of functional, structural, or evolutionary relationships.

Phylogenetics: The study of evolutionary relatedness among various groups of organisms.

microRNA: Are single-stranded RNA molecules encoded by genes, transcribed from DNA but not translated into protein, which regulate gene expression.

gLite: A set of components designed to enable communications and resource sharing over geographical dispersed computing facilities, usually referred as middleware.

EGEE: The acronym of Enabling Grids for E-sciencE, which is a European Commission founded project which aims to develop a grid service infrastructure working on the gLite middleware.

Virtual Organization: A consortium that uses informatics tools to enable, maintain and sustain the sharing of resources in distributed work environments.

Xdelta: Compute the differences between two versions of the same file, providing the possible to downgrade the latest version to a previous one.

Chapter IX
PheGee@Home: A Grid-Based Tool for Comparative Genomics

Bertil Schmidt
Nanyang Technological University, Singapore

Chen Chen
Nanyang Technological University, Singapore

Weiguo Liu
Nanyang Technological University, Singapore

Wayne P. Mitchell
Experimental Therapeutics Centre (ETC), Singapore

ABSTRACT

In this chapter we present PheGee@Home, a grid-based comparative genomics tool that nominates candidate genes responsible for a given phenotype. A phenotype is the physical manifestation of the interplay of genetic, epigenetic and environmental factors. Our tool is designed to facilitate the discovery and prioritization of candidate genes controlling or contributing to the genetically determined portion of a specified phenotype. However, in order to make reliable nominations of candidate genes from sequence data, several genome-size sequence datasets are required. This makes the approach impractical on traditional computer architectures leading to prohibitively long runtimes. Therefore, we use a computational architecture based on a desktop grid environment and commodity graphics hardware to significantly accelerate PheGee. We validate this approach by showing the deployment and evaluation on a grid testbed for the comparison of microbial genomes.

INTRODUCTION

High-throughput techniques for DNA sequencing have led to an enormous growth in the amount of publicly available genomic data. As of February 2008, 716 complete genome sequences are available and another 2,756 genome-sequencing projects are in progress (www.genomesonline.org). As the sequences of more and more genomes become available, we have reached a critical mass where, instead of focusing on a subset of sequences, we can use entire genome data sets to derive global inferences and metadata. Comparative genomics refers to the study of relationships between the genomes of different species or strains. It is currently being used for ortholog detection (Itoh, Goto, Akutsu & Kanehisa, 2005) bacterial pharmacogenomics (Fraser, et al., 2000), clustering of similar protein sequences (Itoh, Akutsu & Kanehisa, 2004), etc. Unfortunately, comparative genomics applications are highly computationally intensive tasks due to the large sequence data sets involved and typically take a few months to complete. These runtime requirements are likely to become even more severe due to the rapid growth in the size of genomic databases.

The objectives of this chapter are therefore two-fold:

1. The presentation of a new comparative genomics tool called PheGee (*Phenotype Genotype Explorer*). PheGee nominates candidate genes responsible for a certain phenotype π given genomic sequence datasets of phenotype positive (π+) and phenotype negative (π–) species.
2. The proposition of a hybrid computational grid platform to accelerate PheGee.

The proposed hybrid grid architecture efficiently combines desktop grid computing with GPGPUs (General-Purpose computation on Graphics Processing Units). The driving force and motivation behind this architecture is the price/performance ratio. Using desktop grids as in the volunteer computing approach is currently one of the most efficient and simple ways to gain supercomputer power for a reasonable price. Installing in addition massively parallel processor add-on boards such as modern computer graphics cards within each desktop can further improve the cost/performance ratio significantly. We show how this architecture can be used to accelerate PheGee efficiently. Moreover, the proposed grid approach is flexible and is therefore applicable to a variety of compute-intensive genomics applications.

BACKGROUND

Biological Motivation

PheGee exploits the simple observation: Given a set of species, *A*, which evince a particular phenotype, π, and a second set of species, *B*, which do not evince π, then in general genes responsible for π will be present in *A* and absent from *B*. Given incomplete knowledge, in the universe of sequenced organisms there will be a third set, *C*, generally the largest of the three, comprising those organisms undefined in terms of π. To cast this in symbolic terms, if a particular process is the outcome of three genes $x \rightarrow y \rightarrow z$, then in general homologs of *x*, *y*, and *z* will be found in *A* but not in *B*. Those genes found in all members of *A* and not found in *B*, are therefore the best candidates for genes responsible for the process resulting from *x*, *y*, *z*. However, a perfect concordance between phenotype and genotype in sets of organisms is not guaranteed. Apart from epigenetic and environmental factors that can shape phenotype, the following genetic phenomena might disrupt the relationship.

- *Gene Replacement* (Achtmann & Wagner, 2008): Gene *y* can be lost in some members of *A*, say *A**. In *A** the function of *y* may have been replaced by some other gene, *w*.

Thus, although the phenotype is preserved in all members of *A*, and although *y* is a gene that contributes to the *A* phenotype, *y* will nevertheless not map to every member of *A*. In particular, *y* is not found in *A**.

- *Horizontal Gene Transfer* (Achtmann & Wagner, 2008): Genes *x*, *y*, or *z* may be adopted by a new species through the process of "horizontal gene transfer". In this scenario some members of *B*, say *B**, will contain a gene correctly associated with phenotype π, even though no member of *B* evinces the phenotype.
- *Function driven by small protein motif* (Ponting & Russell, 2002): A third source of ambiguity emerges in the case of proteins where biochemical function is driven by a relatively small motif, with the bulk of the protein evolving freely. This situation can produce genes which are functionally linked to phenotype π, but which will only be identified as homologs by very sensitive sequence analysis methods.
- *Systemic convergent evolution* (Wood, Burke & Rieseberg, 2005): The phenotype chosen to parse the species may not be genetically "pure". This can happen when a macro process has evolved independently. Partitioning species according to the observed phenotype then will create a genetically mixed set of π+ organisms. In other words, *A* will actually represent the union of two or more genetically distinct species subsets which evolved independently along distinct lineages.

Because of the above processes a useful identification of gene candidates for a given genotype cannot demand that each π+ genome contains a homolog of the same π+ linked gene, or that every π- genome lack all π+ linked genes. Each absence of a homology in a π+ genome, or presence of a π+ homolog in a π- genome, can be termed an 'exception'. One way to prioritize groups of homolog candidates is by the number of exceptions generated for a all given genes for a given *π+/π- partition*. Despite the exceptions, we think that a phenotype to genotype strategy can successfully identify underlying genetic structure given that reasonably large numbers of sample genomes are available.

Background on Desktop Grid Computing with BOINC

The power and mass appeal of volunteer computing for implementing task parallel problems has been demonstrated in projects like SETI@home (Anderson, et al., 2002) and Predictor@home (Taufer, An, Kerstens & Brooks, 2006) that are based on *BOINC* (Berkeley Open Infrastructure for Network Computing (Anderson, 2004; Anderson & Fedak, 2006)). The basic difference between volunteer computing and desktop grid computing is that the former utilizes public resources, whereas the latter employs resources in an organization. Desktop grid computing is enabled using BOINC by a method known as cycle scavenging or cycle harvesting, wherein spare CPU cycles are used to perform a sub-task of a parallel application. BOINC was originally designed for a volunteer computing model with a loose framework of volunteered PCs connected to the internet, but can easily be customized to a desktop grid computing model by removing inbuilt features that are meant to address trust issues inherent in utilizing public resources.

BOINC is an open source middleware platform. The BOINC system architecture is illustrated in Figure 1. A BOINC project consists of a centralized server (or set of servers) connected to the internet, and is identifiable by a single master URL which can be accessed by client machines connected to the internet. A project can run one or more scientific applications. The BOINC server complex has as its core a relational database that stores information on platforms, *hosts* (client computers), users, applications, *workunits* (jobs), *results* (job instances), etc. The server functions are performed by a set of web services and daemons, and consist of:

1. Breaking down the problem into tasks that can be distributed to clients for computation.
2. Distributing the tasks to different clients depending upon the capabilities of the host system and task requirements.
3. Assimilating and processing the result files returned by the clients.
4. Managing data communication, i.e., uploading and downloading of files to and from the server.
5. Managing scheduling related communication between the clients and the server.
6. Managing the web interface.
7. Managing the common BOINC database.
8. Managing the project-specific database.

The functions (1), (3) and (8) are project-specific, and the rest are part of core BOINC functionality. Certain aspects of the core functions of task distribution and file handling are configurable.

The BOINC client, also known as the core client, can be downloaded from any BOINC project website, and can be attached to multiple projects. The BOINC client can be configured based on a set of user-specified preferences to restrict the resource usage and frequency of connections. BOINC follows a client-pull model to address disconnects and overcome firewall barriers, etc.

Figure 1. BOINC system architecture

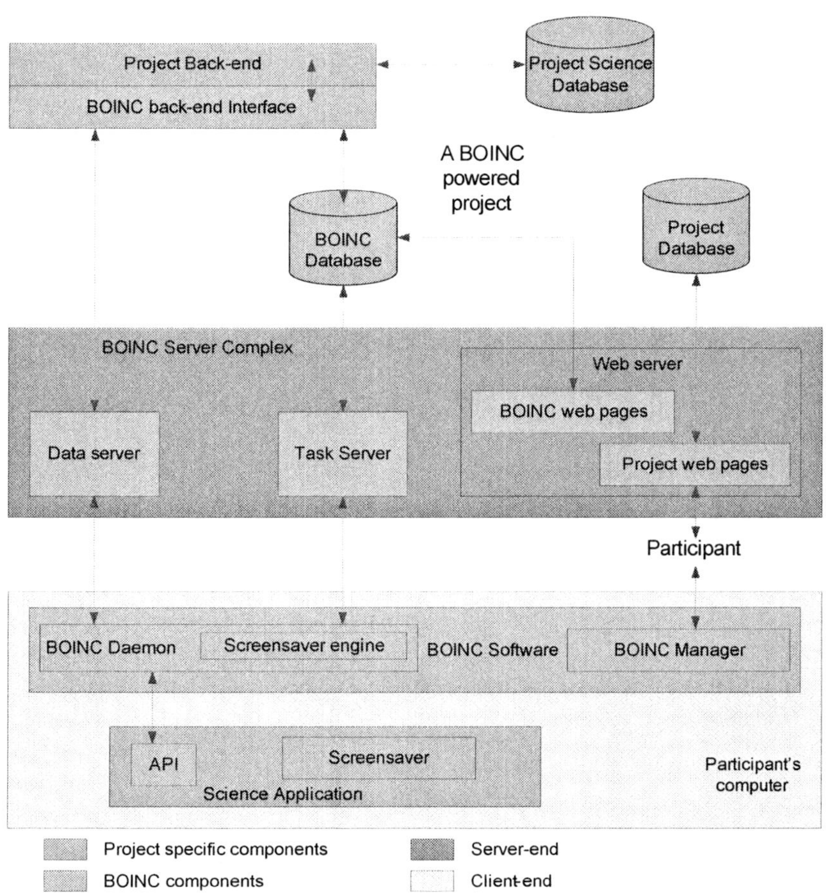

The BOINC client benchmarks the host computer using the Whetstone and Dhrystone benchmarks (Weicker, 1984). This information, along with other resource information like available hard disk space, memory, etc., is communicated to the server and facilitates intelligent task distribution based on host capabilities. The client also calculates how much CPU time was spent on each result. This information is sent to the server and is used in the granting of credit. The client is capable of displaying application-specific graphics which can be used as a screensaver.

Background on GPGPU

The highly competitive and dynamic nature of the computer game market has led to the commoditization of modern day graphics processing units (GPUs), leading to a relatively low price/unit and rapid developments of next generation processors. Due to their highly parallel architecture, the programmable pixel pipelines of modern GPUs are capable of a theoretical peak performance that is an order of magnitude higher than a CPU (Figure 2). Furthermore, GPU performance has been increasing by a factor of 2 to 2.5 per year, which is faster than the increase in CPU performance as predicted by Moore's law (see Figure 2). The high performance-to-price ratio, rapid increase in performance and widespread availability of GPUs, which can deliver several times the performance of a single CPU, have propelled them to the forefront of high performance computing. The combination of volunteer computing and GPU computing holds the promise of making high performance computing accessible on every desktop of the future. In particular, many organizations already have access to desktops with modern graphics cards. For these labs this direction provides a zero-cost solution. Even if a graphics card has to be bought, the installation of such a card is trivial (plug and play).

With their enhanced programmability, the utility of GPUs has expanded beyond traditional

Figure 2. Peak performance comparison of GPUs and CPUs in recent years (adapted from NVIDIA, 2007)

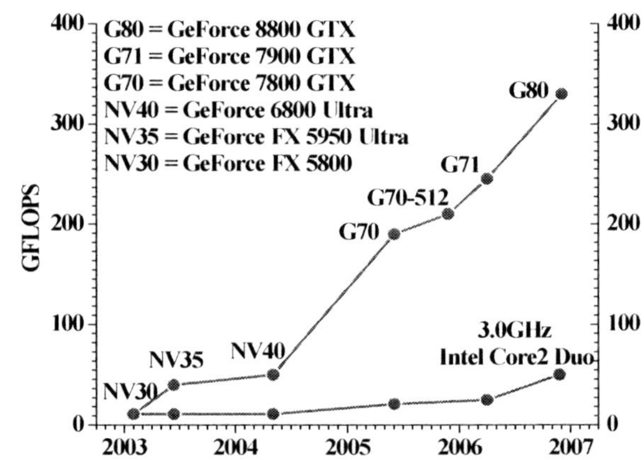

graphics rendering into a variety of general-purpose parallel applications, and GPGPU (General-Purpose computation on a GPU) is an active area of research in high performance computing, e.g. (Manocha, 2005; Owens et al. 2007). Thus, the evolution of GPUs which has hitherto been driven by the gaming market is now also influenced by the demands of GPGPU applications. The unavailability of high-level programming languages suited to high-performance general purpose programming on the GPU platform means that existing algorithms and data structures pertaining to any given problem need to be reformulated in terms of graphics primitives.

Computation on a GPU follows a three-stage process known as the graphics pipeline. These are vertex processing, rasterization and fragment processing. At the vertex processing stage, 3-D vertex world co-ordinates are transformed into 2-D vertex screen co-ordinates. The rasterizer converts the geometric vertex representation into an image fragment representation consisting of pixels. The fragment processor then computes the color for each pixel upon reading texels (texture pixels) from the texture memory. Modern GPUs allow programmability of the vertex and fragment processor and use two types of parallelism. The operations on different vertices and fragments

Figure 3. The GPU graphics pipeline

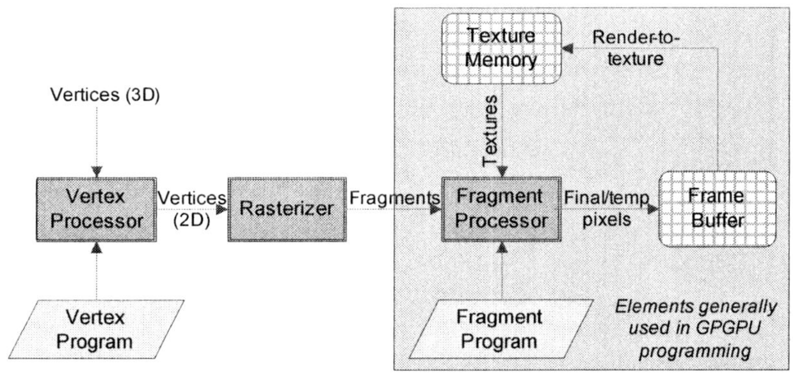

are done in parallel using multiple processors. For example the NVIDIA 7900 GTX has 8 vertex and 24 fragment processors. Another finer level of parallelism is due to native support of operations on 4-dimensional vectors (corresponding to RGBA channels). GPGPU applications typically use only fragment programs for their computation.

In this paper we are using a GPU desktop grid to accelerate biological sequence analysis with the Smith-Waterman algorithm. There have been a variety of techniques used to accelerate the Smith-Waterman (SW) algorithm on parallel computer architectures. They range from typical high performance computing (HPC) strategies such as clustering to novel processing architectures. We broadly classify these strategies into two categories: *coarse-grained* and *fine-grained*. The architectures in the coarse-grained category include general-purpose supercomputers and PC clusters. Computer architectures in the fine-grained category include reconfigurable architectures and GPUs. Reconfigurable solutions such as (Oliver, Schmidt & Maskell, 2005) are based on FPGAs and can achieve a high performance. The main advantage of GPUs compared to the FPGA approach is that they are true commodity components. Therefore, most users have already access to PCs with modern graphics cards making the GPU approach idea for desktop grid computing. Manavski also parallelized the SW algorithm on a GPU (Manavski & Valle, 2008). However, his solution is based on the CUDA programming language. Unfortunately, there is only a very limited number of CUDA-compatible GPUs available at the moment making this approach unsuitable for desktop grids. The approach presented in this paper is based on the OpenGL shading language and has been tested on a wide variety of GPUs.

THE PHEGEE ALGORITHM

Comparative genomics applications are typically based on all-against-all comparisons, i.e. each gene of each given genome is compared to all genes of all other genomes. In order to compare two gene sequences we have decided on using *E*-values computed from pairwise local sequence alignments scores as distances. The pairwsie alignment score is computed by the Smith-Waterman algorithm (Smith & Waterman, 1981). The choice of the Smith-Waterman algorithm is motivated by its superior sensitivity for low-scoring alignment (Brenner, Chothia & Hubbard, 1998) as compared to other, faster algorithms such as BLAST (Altschul et al., 1990).

Given a number of n genomes $G_1,\ldots,G_n$ with genome i containing $\phi(i)$ genes, i.e. $G_i = \{g_{i,1},\ldots,g_{i,\phi(i)}\}$, where $g_{i,k}$ denotes the kth gene of genome i. The result of the all-against-all comparison can then be described as a set of $n{\cdot}(n-1)/2$ distance matrices $D(i,j)$ of size $\phi(i)\times\phi(j)$ containing all pairwise distances $d(g_{i,k},g_{j,l})$, $1 \leq k \leq \phi(i)$, $1 \leq l \leq \phi(j)$, for each pair of genes $g_{i,k}$ and $g_{j,l}$ in genome i and genome j.

The PheGee algorithm initially computes the set of distance matrices $D(i,j)$ as described above. The *π+/π- partition* is subsequently computed using the following method:

- **INPUT:** A number of n genomes $G_1,\ldots,G_n$ and a set of $n\cdot(n-1)/2$ distance matrices $D(i,j)$ as described above. Additionally, the given genomes are divided into a phenotype positive set $\pi+ = \{G_1,\ldots,G_m\}$ and a phenotype negative set $\pi- = \{G_{m+1},\ldots,G_n\}$
- **OUTPUT:** All genes $g_{i,k}$ with the maximal partitioning score $P(g_{i,k})$ among all genes
- **METHOD:** The partitioning score $P(g_{i,k})$ is computed for each gene, $1 \leq i \leq n$, $1 \leq k \leq \phi(i)$), as follows.
- **CASE 1:** $G_i \in \pi+$; i.e. the given gene is part of a phenotype positive genome. Then the best match of the gene in the phenotype negative set $\varepsilon(g_{i,k})$ is identified, which is calculated by finding the smallest E-value in the distance matrix:

$$\varepsilon(g_{i,k}) = \min_{m+1\leq j\leq n, 1\leq l\leq \varphi(j)} \{d(g_{i,k}, g_{j,l})\}.$$

This value is then used as a threshold to determine partitioning score $P(g_{i,k})$, which is the number of genomes in $\pi+$ that contain a matching gene better than this score:

$$P(g_{i,k}) = \frac{\sum_{j=1}^{m} \delta(g_{i,k}, j, \varepsilon(g_{i,k}))}{m}$$

where

$$\delta(g_{i,k}, j, \varepsilon(g_{i,k})) = \begin{cases} 1 & \text{if } \min_{1\leq l\leq \varphi(j)} d(g_{i,k}, g_{j,l}) \leq \varepsilon(g_{i,k}) \\ 0 & \text{otherwsie} \end{cases}$$

- **CASE 2:** $G_i \in \pi$-: similar (but with positive and negative sets interchanged and n replaced by $n-m$)

MAPPING OF PHEGEE ONTO THE HYBRID GRID PLATFORM

The most time-consuming part of PheGee is the computation of the set of $n\cdot(n-1)/2$ distance matrices $D(i,j)$ containing all pairwise distances between sequences using the Smith-Waterman algorithm. In this section we describe how this all-against-all comparison can be mapped onto our hybrid grid architecture. Our approach consists of two parallelization steps.

- *Task parallelization.* The task parallelization splits the problem into smaller chunks of all-against-all comparison tasks at the server end.
- *Data parallelization.* The data parallelization processes these smaller tasks by calculating the individual Smith-Waterman scores at the client end, which can be implemented in a data parallel manner on the GPU. To allow for flexibility, a sequential version of the client application also needs to be implemented so that it can be run on the CPU.

These two parts are described in detail in the following subsections.

Task Parallelization Using BOINC

The various considerations that need to be taken into account at the server end for implementing the task parallelization are as follows:

- *The input data files for the client application should be optimized for the GPU code.* The GPU code runs optimally if all the sequences in the input files are of comparable lengths. To ensure this, we sort the original file based on sequence length before splitting it into smaller chunks.
- *All work units should roughly involve the same amount of computation.* This needs to be ensured for efficient task distribution.

For a CPU implementation (sequential algorithm), the amount of computation depends on the total number of residues in the input files. Thus, to ensure uniform computation on a CPU, total number of residues in each of the files should be approximately same. This number has been parameterized as the *lfactor*, and can be specified as an input to the work generation process. Uniform computation time across tasks in a GPU implementation is tougher to achieve, since the computation time depends upon the way the sequential algorithm has been parallelized. However, we can still take *lfactor* as a rough indicator.

- *The size of the input data files should be optimized such that the amount of computation should be sufficient to justify the data communication overhead.* This can be done by adjusting the *lfactor* parameter mentioned above. A higher *lfactor* reduces the number of tasks and increases the computation time per task, and therefore leads to greater efficiency.
- *The sequence size restriction for the GPU code needs to be addressed.* The GeForce 6 and 7 series graphics cards can only handle query sequences up to a length of 4096 as they are restricted by the size of the texture buffer. Thus, query sequences of length greater than 4096 must be run on the CPU rather than the GPU. The data files are ordered by sequence length, thus, for the workunits containing sequences greater than 4096, a parameter can be passed to the client application to force it to run on the CPU. Since there are very few sequences that are as long (less than 0.2% in Swissprot), this will not have a significant effect on the overall performance.

Keeping the above considerations in mind, the problem dataset, consisting of all the protein sequences of a set of genomes concatenated into a single file, is split into a number of smaller datasets, such that each smaller dataset has *lfactor* number of residues. Given *k* such small datasets,

$$k = \left\lceil \frac{\text{Total number of residues}}{\text{lfactor}} \right\rceil,$$

each dataset needs to be compared to itself and every other dataset, therefore, there are $k{\cdot}(k{+}1)/2$ number of total comparisons (see Figure 4).

Since each *workunit* involves one such comparison, the *work-generator* daemon creates $k{\cdot}(k{+}1)/2$ *workunits* in a steady stream. Once a host returns back a *result*, the SW (Smith-Waterman) scores are converted into *E*-values and stored in the project database. Figure 5 shows all steps involved in mapping all-against-all comparison onto the hybrid grid architecture.

Figure 4. Partitioning a dataset into k parts results in k(k+1)/2 number of comparison tasks

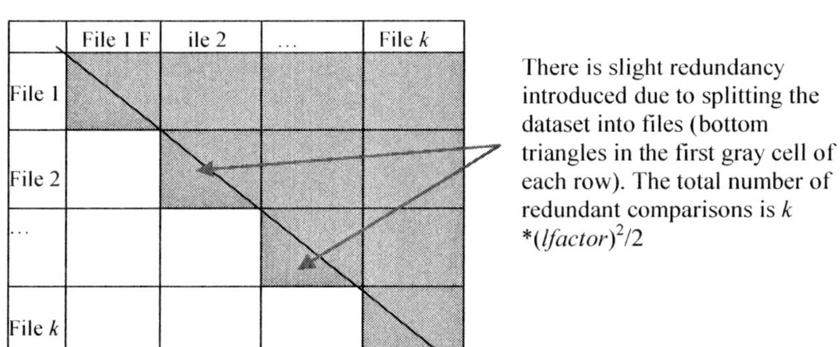

Figure 5. Mapping of all-against-all genome comparison onto the hybrid grid architecture.

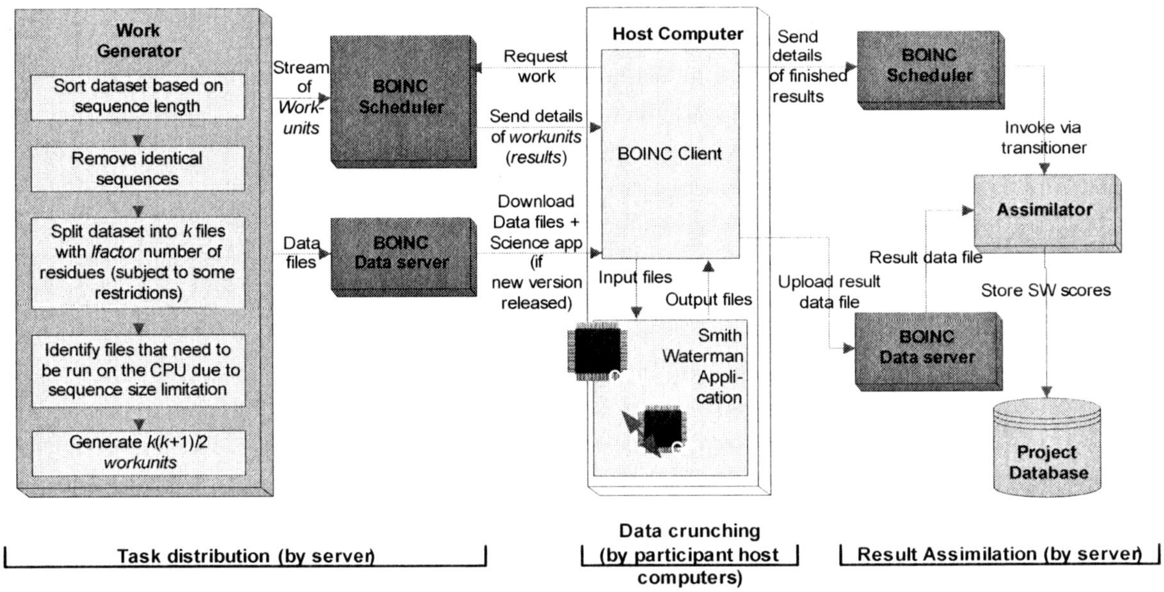

Data Parallelization on a GPU

Given two sequences S_1 and S_2 of length l_1 and l_2, the Smith-Waterman (SW) algorithm (Smith & Waterman, 1981) computes their similarity $H(i,j)$ ending at position i and j in S_1 and S_2 respectively using the following recurrence relations for the affine gap penalty:

$$H(i, j) = \max\{0, E(i, j), F(i, j), H(i-1, j-1) + sbt(S_1[i], S_2[j])\}$$
$$E(i, j) = \max\{H(i, j-1) - \alpha, E(i, j-1) - \beta\}$$
$$F(i, j) = \max\{H(i-1, j) - \alpha, F(i-1, j) - \beta\}$$

where sbt is the substitution matrix, α is the gap penalty for the first gap and β is the gap penalty for subsequent gaps.

In order to map the SW algorithm efficiently mapped onto the GPU architecture, we take advantage of the fact that all elements in the same anti-diagonal of the DP (dynamic programming) matrix can be computed independent of each other in parallel (see Figure 6). Therefore, parallelizing the algorithm involves computing the DP matrix in an anti-diagonal order. The anti-diagonals are stored as textures in the texture memory. Fragment programs are used to implement the arithmetic operations specified by the recurrence relation.

To align two sequences of lengths l_1 and l_2 using affine gap penalties on a GPU, both the sequences and the substitution matrix are loaded into the texture memory as a preprocessing step. The number of anti-diagonals in the DP matrix is l_1+l_2-1, and therefore, the matrix is computed in as many rendering passes. In rendering pass k, where $1 \leq i \leq l_1$ and $1 \leq j \leq l_2$ and $k = i+j-1$, the values for $H(i,j)$, $E(i,j)$ and $F(i,j)$ are computed by the fragment processors. As diagonal k depends upon the diagonals $k-1$ and $k-2$, each rendering pass needs to read the two previous anti-diagonals from the texture memory (see Fig. 6).

All the three diagonals are stored as separate buffers. A cyclic method is employed to change the buffer function. Another concern is how to map each diagonal in the DP matrix into a *quadrilateral* (*quad*). To invoke a GPU kernel, the range of the computation (or the size of the output stream) must be specified. The programmer does this by passing vertices to the GPU. A typical GPGPU invocation is a quadrilateral oriented parallel to the image plane, sized to cover a rectangular region of pixels matching the desired size of the

Figure 6. Data dependency relationship in the SW DP matrix: each cell (i,j) depends on its left neighbor (i,j–1), upper neighbor (i–1, j), and upper-left neighbor (i–1,j–1). Therefore, all the cells along anti-diagonal k can be computed in parallel from anti-diagonals k–1 and k–2.

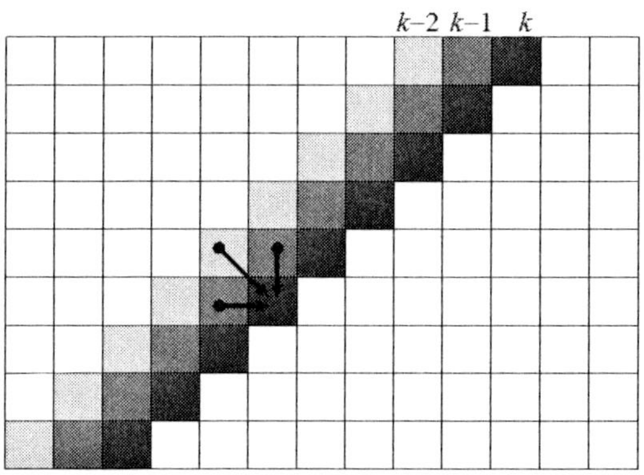

output array. As illustrated in Figure 7, drawing a diagonal quad that covers the diagonal in the buffer does not yield the desired result, since all the pixels touched by the quad are rendered instead of just the pixels on the diagonal. To resolve this, we transform the matrix such that the cells of each diagonal are brought into a single column. This is done by shifting each row of the matrix to the right by its row number *i* (Figure 8).

Using this method, a $1 \times L(k)$ quad can be rendered in each iteration, where $L(k)$ denotes the length of the diagonal *k*. Moreover, it is possible to perform *N* pair-wise comparisons at the same time by using two-dimensional buffers instead of one-dimensional buffers. Each two-dimensional buffer contains *N* diagonals of length $L(k)$. The computation is invoked by drawing an $N \times L(k)$ quad in each iteration. Therefore, in one batch, a given reference sequence can be compared against *N* database sequences, and *N* is known as the *batchsize*.

Optimizing GPU Performance Using Variable Batchsizes

For a given set of database sequences, the sequences need to be split into a number of batches since there is an upper limit on the batchsize *N* depending upon the memory size of the GPU. The total processing time for comparing a given reference sequence against a batch of database sequences roughly depends on the following:

- *Initialization overhead*: This is the time required to initialize the framebuffer, binding the textures, etc. A longer batchsize (higher throughput) minimizes the number of batches in a given set of database sequences, and therefore minimizes the initialization overhead.

Figure 7. Drawing a quad over the anti-diagonal results in rendering too many pixels

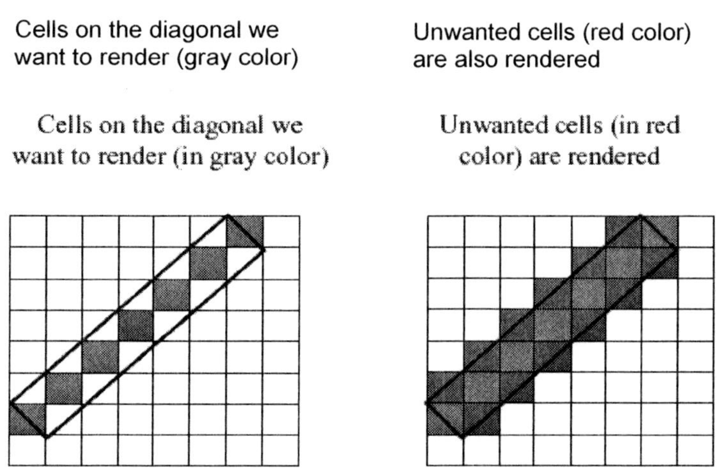

Figure 8. Shifting the rows by row number i results in aligning the diagonals along columns

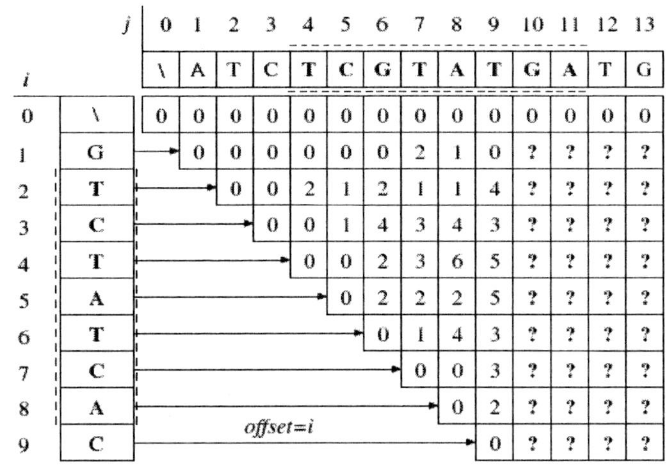

- *Rendering time*: This is the time required to render the texture data. The rendering time in turn depends mainly on two factors: the length of the diagonal $L(k)$, which is the smaller of the reference sequence and database sequence lengths; and the rendering efficiency, which can be defined as the ratio of the number of relevant cells to the total number of cells rendered. Efficiency loss happens when there is a difference in length between the first and last sequences in a batch. Since N diagonals are rendered at one time, not all the cells in the diagonals associated with the smaller sequences need to be rendered, but have to be rendered nevertheless. A shorter batchsize maximizes the efficiency.

Optimizing the batchsize involves a trade-off between throughput and efficiency. Based on these factors, given a sequence length profile of a set of database sequences, a shorter batchsize should be used for areas with high variation in sequence lengths, and a longer batchsize should be used for areas with lesser variation in sequence lengths. We have empirically determined the optimal batchsizes for different slope ranges (Figure 9), and implemented this in our algorithm.

PERFORMANCE EVALUATION

We have set up a BOINC project with BOINC server version 5.2.8 on an Intel P4, 3GHz server running Fedora Core 4 Linux with an Apache 2.0 web server. To benchmark the performance of our hybrid computing model, we have set up a campus-wide GPU-accelerated BOINC desktop grid with the following configuration of host computers:

PC configurations:

- Intel Pentium 4, 3.0 GHz, 1GB RAM, Windows XP, BOINC client version 5.4.11
  GPU configurations:
- NVIDIA GeForce 8800GTX, 575 MHz core clock rate, 1.3 GHz shader clock, 900 MHz memory clock rate, 128 stream processors, 768 MB
- NVIDIA GeForce 7900 GTX, 622 MHz core clock rate, 1.83 GHz memory clock rate, 8 vertex processors, 24 fragment processors, 512 MB
- NVIDIA GeForce 7800 GTX, 550 MHz core clock rate, 1.7 GHz memory clock rate, 8 vertex processors, 24 fragment processors, 512MB
- NVIDIA GeForce 7950 GT, 600 MHz core clock rate, 1.45 GHz memory clock rate, 8

Figure 9. Optimizing the batch size depending upon the sequence length profile for a given set of database sequences. A shorter batchsize is used for areas with greater variation, whereas a longer batchsize is used for areas with smaller variation in sequence lengths.

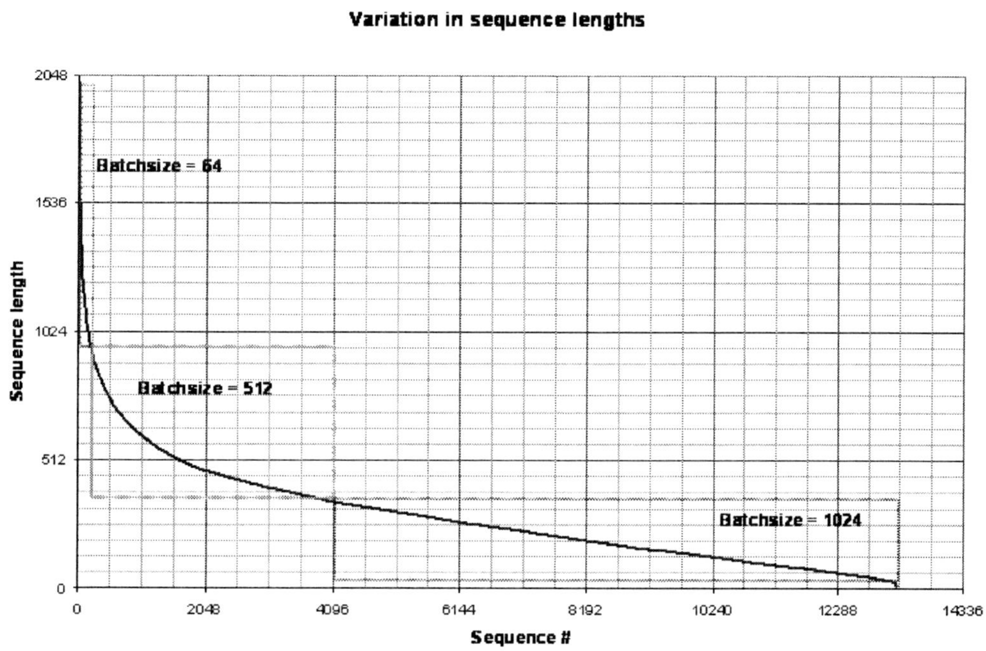

vertex processors, 24 fragment processors, 512MB

We then deployed the client code on a varying number of such hosts and ran an all-against-all genome comparison test dataset consisting of the following 6 microbial genomes: Escherichia coli K12-MG1655, Bacillus anthracis Sterne, Mycobacterium avium paratuberculosis, Vibrio cholerae el tor n16961, Yersinia pestis KIM, and Salmonella typhimurium LT2 SGSC1412. This dataset consists of *26,763* gene sequences. Since these are annotated genes, we used the corresponding protein sequence for the comparison. The corresponding maximum and average sequence lengths are 6384 and 308.

In order to measure the efficiency of the system, we have compared the runtimes for executing an all-against-all genome comparison using the above datasets on dedicated single systems to the runtime on our grid testbed. This comparison was performed using FASTA OSEARCH (Pearson, 2000) on the standalone CPU, and our GPU code using the high-level GPU programming language GLSL. The runtimes for the dedicated systems are shown in Table 1. A performance measurement commonly used in computational biology is *cell updates per second* (*CUPS*). A CUPS represents the time for a complete computation of one entry of the similarity matrix, including all comparisons, additions and maxima operations. The CUPS performance values are also shown in Table 1. As mentioned, the maximum texture buffer width on these graphics cards is 4096. Therefore, comparisons with sequences longer than 4096 are performed on the CPU. The CPU works together with GPU in a multithreading style.

From Table 1 we can conclude that for sequences shorter than 4096, the GPU achieves more than 10 times speedup over the CPU-only dedicated system. For sequences longer than 4096, the performance reduces depends on the

Table 1. Performance comparison between a single dedicated CPU and GPU

	Runtime (in mins)	Speedup	Mega CUPS
P4 3GHz	16235.52	1	44.02
GeForce 8800 GTX	1634.33	9.94	437.30
GeForce 7900 GTX	1806.63	8.99	395.59
GeForce 7800 GTX	1852.92	8.76	385.71
GeForce 7950 GT	1914.57	8.48	373.29

(Including sequences longer than 4096)

	Runtime (in mins)	Speedup	Mega CUPS
P4 3GHz	14905.77	1	44.02
GeForce 8800 GTX	731.39	20.38	897.14
GeForce 7900 GTX	1000.39	14.90	655.91
GeForce 7800 GTX	1176.46	12.67	557.73
GeForce 7950 GT	1183.00	12.60	554.64

(Excluding sequences longer than 4096)

length distribution in the dataset. In our dataset, the amount of cells involving sequences longer than 4096 is about 8% of the total cells to be compared.

Table 2 shows the runtime comparison on our grid testbed for a varying number of clients and varying client configurations. We have used a value of *lfactor* = 500,000 for this comparison. The client configuration (x_1, x_2, x_3, x_4) denotes x_1 hosts with GeForce 8800 GTX, x_2 hosts with GeForce 7900 GTX, x_3 hosts with GeForce 7800 GTX and x_4 hosts with GeForce 7950 GT. The maximal achievable Mega CUPS is calculated based on the dedicated system performances in Table 1. Efficiency is then calculated as "Achieved Mega CUPS" divided by "Maximal Achieved Mega CUPS)". It can be seen that the grid testbed achieves a speedup of almost two orders of magnitude compared to the CPU-only dedicated system.

In order to better understand efficiency on our grid testbed, we have identified the following factors that can influence the applications' performance.

- *Startup time delay.* This delay is introduced by the *client-pull* strategy utilized by BOINC to connect to clients at the beginning of a computation.
- *Time delay in-between workunits.* Once a client is processing, there is a time delay in-between two consecutive workunits on the same BOINC client.

In order to measure the effect of these factors, we have run another all-against-all comparison on our grid testbed using a larger sequence dataset with a homogeneous lengths distribution. This datasets consists of *59,364* sequences with maximum, minimum and average lengths of 424, 267, and 342, respectively. Table 3 reports the runtime, the average startup delay per client, the average (accumulated) delay in-between workunits per client and the efficiency for varying client configurations. The client configurations and the method to calculate efficiency are identical to the ones used in Table 2. From Table 3 we can make the following observations.

Table 2. Performance of GPU-accelerated BOINC testbed for varying client configurations

Client configuration	(1, 1, 1, 1)	(1, 2, 1, 2)	(1, 5, 1, 3)
Runtime (mins)	501	370	248
Speedup over dedicated CPU	32.41	44.00	65.47
Achieved Mega CUPS	1426.69	1936.82	2881.81
Maximal achievable Mega CUPS	1591.89	2360.77	3920.83
Efficiency	89.62%	82.04%	73.50%

(Including sequences longer than 4096)

Client configuration	(1, 1, 1, 1)	(1, 2, 1, 2)	(1, 5, 1, 3)
Runtime (mins)	287	206	147
Speedup over dedicated CPU	51.92	72.28	101.32
Achieved Mega CUPS	2285.33	3181.78	4460.28
Maximal achievable Mega CUPS	2665.42	3875.97	6398.34
Efficiency	85.74%	82.10%	69.71%

(Excluding sequences longer than 4096)

- Besides computing time, the startup delay and the delay in-between workunits are the main factors influencing runtime and efficiency (since subtracting the average startup delay per client and the average delay in-between workunits per client from the runtime would lead to efficiencies of close to 100%).
- The delay in-between workunits decreases with an increasing number of clients (since there are less workunits per client). This leads to an approximately constant percentage of the overall runtime spent on this type of delay for varying client configurations.
- The average startup delay per client remains approximately constant with an increasing number of clients. This in turn results in (i) a loss of efficiency for a larger number of clients (since a larger percentage of the runtime is spent on the startup). Because the startup time is also independent of the total amount of computations done, higher efficiencies can be achieved for larger datasets (i.e. efficiencies in Table 3 are higher than in Table 2).

COMPARATIVE GENOMICS STUDY USING POSSESSION OF TRNA SUBSTRATE

Bacterial tRNA Adenosine Deaminase activity is highly specific for its substrate, a single arginine decoding tRNA. The gene and primary transcript of this tRNA has an adenosine at the anti-codon "wobble" position which must be deaminated to inosine in order for the tRNA function productively in translation of mRNA to proteins. Inspection of microbial tRNA distributions as predicted from whole genome sequences reveals that some microbes have the appropriate adenosine deaminase substrate while others do not (Sprinzl & Vassilenko, 2005). We can regard possession of the tRNA substrate as a molecular phenotype, and therefore classify the substrate positive microbes as phenotype positive and the substrate negative microbes conversely, inferring that the phenotype positive set must genetically encode the functional capacity to convert the wobble adenosine to inosine.

Table 3. Performance analysis for varying client configurations on a larger dataset

Client configuration	(1, 1, 1, 1)	(1, 2, 1, 2)	(1, 5, 1, 3)
Runtime (mins)	1432	984	609
Average startup delay (mins) [percentage of runtime]	37 [2.58%]	28 [2.85%]	32 [5.25%]
Average delay in-between workunits (mins) [percentage of runtime]	51 [3.56%]	37 [3.76%]	22 [3.61%]
Efficiency	93.0%	93.3%	91.1%

Thus, the requirements for a genotype to phenotype analysis are met in this case, and we can proceed to assemble the data as an input to PheGee@Home. We first selected a corresponding set of ten phenotype positive microbial genomes (π+) and ten phenotype negative microbial genomes (π–), which is shown in Table 4. The compilation of tRNA sequences and sequences of tRNA genes from http://www.tRNA.uni-bayreuth.de (Sprinzl & Vassilenko, 2005) was used to classify each genome. The complete set of annotated protein sequences from each genome was obtained from TIGR (www.tigr.org). We then performed an all-against-all comparison on our grid testbed as described in the previous section. The results were input to PheGee, which then

Table 4. The selected phenotype positive/negative genomes

Microbial Genome Name	TIGR Genome Reference	Encodes tRNA substrate for tRNA Adenosine Deaminase (+/–)
Escherichia coli K12	Escherichia coli K12-MG1655	+
Bacillus subtilis	Bacillus subtilis 168	+
Haemophilus influenzae	Haemophilus influenzae 86 028NP	+
Vibrio cholerae	v_cholerae_el_tor_n16961	+
Mycobacterium tuberculosis	M. tuberculosis H37Rv (lab strain)	+
Staphylococcus aureus	Staphylococcus aureus N315	+
Buchnera aphidicola	Buchnera aphidicola Sg	+
Bartonella henselae	Bartonella henselae Houston-1	+
Caulobacter crescentus	C_crescentus_cb15	+
Chlamydia trachomatis	Chlamydia trachomatis serovar D	+
Mycoplasma pneumoniae	Mycoplasma pneumoniae M129	–
Borrelia burgdorferi	b_burgdorferi_b31	–
Campylobacter jejuni	c_jejuni_rm1221	–
Helicobacter pylori	h_pylori_26695	–
Mycoplasma genitalium	m_genitalium_g_37	–
Pyrococcus furiosus	Pyrococcus furiosus DSM 3638	–
Methanococcus jannaschii	m_jannaschii_dsm2661	–
Archaeoglobus fulgidus	a_fulgidus_dsm4304	–
Thermotoga maritima	t_maritima_msb8	–
Sulfolobus tokodaii	Sulfolobus tokodaii strain 7	–

reported 163 genes to have the highest partitioning score in π+. We further analyzed these genes for clusters of homologues. This was done by constructing a graph $G = (V,E)$ where V is the set of 163 genes and

$$E = \{(g_{i,k}, g_{j,l}) : \arg\min_{1 \le x \le \varphi(j)} d(g_{i,k}, g_{j,x}) = l\}.$$

G was then searched for cliques of maximal size. The four identified cliques with their gene members are shown in Table 5.

We further ranked the cliques by their average E-value distance to the top hits in the phenotype negative sets. Thus, a set of yfhC homologues genes is identified as the most of significant clique. The bacterial tRNA adenosine deaminase has been actually biochemically characterized in *E.coli* (Wolf, Gerber & Keller, 2002). The enzyme is encoded by the *tadA* gene (called *yfhC* prior to experimental functional attribution). PheGee is able to discover a cluster of homologous genes (Clique #1 in Table 5) in the phenotype positive set that includes *tadA* (*yfhC*) from *E.coli*. Thus, we are able infer the gene responsible for an enzymatic activity from the phylogenetic distribution of the enzyme substrate.

We remind readers that PheGee is conceived as a prioritization tool, useful in establishing an ordered list of gene candidates responsible for a particular phenotype. These candidates can, indeed *must*, be taken back to the bench for experimental validation. PheGee therefore exemplifies a philosophy of genomic science grounded in the fruitful interplay of theory and experiment. In the Adenosie Deaminase example presented, PheGee performs perfectly, and returns the authentic gene as the top candidate, but, for the genetic reasons discussed above, there is no guarantee that such will always be the case. Nevertheless, the partitions emergent in subsidiary cliques may point to unsuspected relationships to the subject phenotype. For instance – on a purely speculative level – it is interesting that the second best scoring clique contains a co-partioning ribosomal protein. A molecular interaction between this ribosomal protein and the tRNA substrate (or product) or the enzyme is not unimaginable.

FUTURE TRENDS

We have demonstrated the usefulness of PheGee@Home using as set of twenty microbial genomes. In this experiment we have chosen a sharp molecular feature (i.e. the possession of the tRNA substrate) to define the phenotype. Other such features can be imagined, for example, the possession/lack of characteristic metabolites (likely the products of enzymatic pathways), or, the existence/absence of landmark cellular structures such as flagella assemblies. We note further that conceptually the genotype-phenotype approach is not inherently limited to sets of discrete organisms; it could equally be applied to sets of "genomic communities" such are emerging from metagenomic sequencing studies. Although, currently, the data sets for community comparisons are too sparse, rapid advances in DNA sequencing efficiency may soon enable P-G analysis at meta-scale.

CONCLUSION

In this paper, we have introduced a new computational grid system based on a hybrid computing model. The architecture consists of two types of parallelism: (1) a desktop grid for coarse-grained parallelization; and (2) GPUs for fine-grained parallelization. We have shown how a comparative genomics application can be efficiently mapped onto this type of architecture and biologically validated the produced results using a new phenotype-genotype exploration method called *PheGee@Home*. Our implementation achieves significant speedups on our grid prototype, which consists of easily programmable mass-produced graphics hardware available for less than US$300

Table 5. The identified four cliques in $\pi+$ with corresponding genes (TIGR reference number) in each phenotype positive genome. The cliques are ranked by their average distance to the phenotype negative set.

	Clique #1 (yfhC)	Clique #2 (rspN)
Avg. Distance	2.24	1.21
E. coli K12	b2559,putative deaminase (yfhC)	b3307,30S ribosomal subunit protein S14 (rpsN)
Vibrio cholerae	VC0864,zinc-binding domain protein (yfhC)	VC2583,ribosomal protein S14 (rpsN)
Bacillus subtilis	Bsu0018,similar to hypothetical proteins (yaaJ)	Bsu0887,similar to ribosomal protein S14 (yhzA)
Haemo. influenzae	NTHI1073,conserved hypothetical protein	NTHI0954,30S ribosomal protein S14 (rpsN)
M.tuberculosis	Rv3752c,hypothetical protein Rv3752c	Rv2056c, Probable ribosomal protein S14
Buchner. aphidicola	BUsg246,hypothetical 20.0 kD protein in purL-dpj (yfhC)	BUsg492,30S ribosomal protein S14 (rpsN)
Bartonella henselae	BH02600,Nitrogen fixation protein	BH10380,30S ribosomal protein s14 (rpsN)
Chlam. trachomatis	CT844,cytosine deaminase (yfhC)	CT787,S14 Ribosomal Protein (rs14)
Caulob. crescentus	CC0231,cytidine and deoxycytidylate deaminase family protein	CC1261,ribosomal protein S14 (rpsN)
Staphyloc. aureus	SA0516, hypothetical protein, similar to Cu binding protein	SA1171,30S ribosomal protein S14 homolog (rpsN)

	Clique #3 (sucA)	Clique #4 (lipA)
Avg. Distance	0.578	0.539
E. coli K12	b0726, 2-oxoglutarate dehydrogenase (decarboxylase component) (sucA)	b0628, lipoate synthesis, sulfur insertion (lipA)
Vibrio cholerae	VC2087,2-oxoglutarate dehydrogenase, E1 component (sucA)	VC0943, lipoic acid synthetase (lipA)
Bacillus subtilis	Bsu1937,2-oxoglutarate dehydrogenase (E1 subunit) (odhA)	Bsu3230, similar to lipoic acid synthetase (yutB)
Haemo. influenzae	NTHI1964,2-oxoglutarate dehydrogenase E1 component (sucA)	NTHI0033, Lipoic acid synthetase (lipA)
M.tuberculosis	Rv1248c, 2-oxoglutarate dehydrogenase (sucA)	Rv2218, Probable lipoic acid synthetase (lipA),
Buchner. aphidicola	BUsg292,2-oxoglutarate dehydrogenase E1 component (sucA)	BUsg259, lipoic acid synthetase (lip-syn) (lipA)
Bartonella henselae	BH16540,Alpha-ketoglutarate dehydrogenase (sucA)	BH05790, Lipoic acid synthetase (lipA)
Chlam. trachomatis	CT054,Oxoglutarate Dehydrogenase (sucA)	CT558, Lipoate Synthetase (lipA)
Caulob. crescentus	CC0339,2-oxoglutarate dehydrogenase, E1 component (sucA)	CC1735, lipoic acid synthase (lipA)
Staphyloc. aureus	SA1245,2-oxoglutarate dehydrogenase E1 (odhA)	SA0785, lipoic acid synthetase (lipA

at any local computer outlet. This direction even provides a zero-cost solution for labs with access to desktops with modern graphics cards. The very rapid growth of genomic databases demands even more powerful high performance solutions in the near future. Hence, our results are especially encouraging since GPU performance grows faster than Moore's law as it applies to CPUs. Other bioinformatics algorithms that can benefit from this type of compute power include molecular dynamics simulations (Liu, Schmidt, Voss & Müller-Wittig, 2008) and motif finding in DNA sequences (Chen, Schmidt, Liu & Müller-Wittig, 2008). Two corresponding applications that could run on the presented GPU desktop grid could for instance be protein-ligand docking searches or the whole-genome discovery of transcription factor binding sites.

ACKNOWLEDGMENT

The work was supported by the A*Star BMRC research grant No. 04/1/22/19/375.

REFERENCES

Achtman, M., & Wagner, M. (2008). Microbial Diversity and genetic nature of microbial species. *Nat Rev Microbiology*, *6*, 431-440.

Altschul, S.F., Gish, W., Miller, W., Myers, E.W, & Lipman, D.J. (1990). Basic local alignment search tool. *Journal of Molecular Biology*, *215*, 403-410

Anderson, D.P., Cobb, J., Korpela, E., Lebofsky, M., & Werthimer, D. (2002). SETI@home: an experiment in public resource computing. *Communications of the ACM*, *45*(11), 56-61.

Anderson, D.P. (2004). BOINC: A System for Public-Resource Computing and Storage. In: *Proceedings of the 5th IEEE/ACM International Workshop on Grid Computing*, (pp. 4-10) New York: IEEE Computer Society.

Anderson, D., & Fedak, G. (2006). The Computational and Storage Potential of Volunteer Computing, In *Proceedings of the 6th IEEE/ACM International Symposium on Cluster Computing and the Grid*, (pp. 73-80. New York: IEEE Computer Society.

Brenner, S., Chothia, C., & Hubbard, T. (1998). Assessing sequence comparison methods with reliable structurally identified distant evolutionary relationships. *Biochemistry*, *95*(11) 6073-6078.

Chen, C., Schmidt, B., Liu, W., & Mueller-Wittig, K.W. (2008). GPU-MEME: Using Graphics Hardware to Accelerate Motif Finding in DNA Sequences. *Lecture Notes in Bioinformatics*, *5295*, 448-459.

Fraser, C.M., Eisen, J., Fleischmann, R.D., Ketchum, & K.A., Peterson, S. (2000). Comparative Genomics and Understanding of Microbial Biology. *Emerging Infectious Diseases*, *6*(5), 505-512.

Itoh, M., Goto, S., Akutsu, T., & Kanehisa, M. (2005). Fast and accurate database homology search using upper bounds of local alignment scores. *Bioinformatics*, *21*(7), 912-921.

Itoh, M., Akutsu, T., & Kanehisa, M. (2004) Clustering of database sequences for fast homology search using upper bounds on alignment score. *Genome Informatics*, *15*(1), 93-104.

Manavski, S.A., & Valle, G. (2008). CUDA compatible GPU cards as efficient hardware accelerators for Smith-Waterman sequence alignment. *BMC Bioinformatics*, *9*(S2), S10.

Manocha, D. (2005). General Purpose computations using graphics processors. *IEEE Computer*, *20*(7), 85-88.

NVIDIA Compute Unified Device Architecture-Programming Guide (2007). Retrieved from http://developer.download.nvidia.com/compute/

cuda/1_0/NVIDIA_CUDA_Programming_Guide_1.0.pdf.

Oliver, T., Schmidt, B., & Maskell, D. (2005). Reconfigurable Architectures for Bio-sequence Database Scanning on FPGAs. *IEEE Transactions on Circuits and Systems II, 52*(12), 851-855.

Owens, J., Luebke, D., Govindaraju, N., Harris, M., Krüger, J., Lefohn, A.E., & Purcell, T.J. (2007). A survey of general-purpose computation on graphics hardware. *Computer Graphics Forum, 26*(1), 80-113.

Ponting C.P., & Russell R.R. (2002). The natural history of protein domains. *Annu Rev Biophys Biomol Struct. 31*, 45-71.

Liu, W., Schmidt, B., Voss, and G., Mueller-Wittig, K.W. (2008). Accelerating Molecular Dynamics Simulations Using Graphics Processing Units with CUDA. *Computer Physics Communications, 179*(9), 634-641.

Pearson, W.R. (2000). Flexible sequence similarity searching with the FASTA3 program package. *Methods in Molecular Biology, 132*, 185-219.

Smith, T.F., & Waterman, M.S (1981). Identification of common molecular subsequences, *Journal of Molecular Biology, 147*, 195-197.

Sprinzl, M., & Vassilenko, K.S. (2005). Compilation of tRNA sequences and sequences of tRNA genes. *Nucleic Acids Research, 33*, D139–D140.

Taufer, M., An, C., Kerstens, A., & Brooks, C.L. (2006). Predictor@Home: Protein Structure Prediction Supercomputer Based on Global Computing. *IEEE Transactions on Parallel and Distributed Computing, 17*(8), 786-796.

Weicker, R. (1984). Dhrystone: A Synthetic Systems Programming Benchmark. *Communications of the ACM, 27*(10), 1013-1030.

Wolf, J., Gerber, A.P., & Keller, W. (2002). tadA, an essential tRNA-specific adenosine deaminase from Escherichia coli. *The EMBO Journal, 21*, 3841–3851.

Wood T.E., Burke J.M., & Rieseberg L.H. (2005). Parallel genotypic adaptation: when evolution repeats itself. *Genetica, 123*(1-2), 157-70.

KEY TERMS AND DEFINITIONS

Comparative Genomics: Comparative genomics refers to the study of relationships between the genomes of different species or strains.

Desktop Grid Computing: Desktop grid computing is a form of distributed computing in which an organization uses its existing computers to handle long-running computational tasks

General-purpose computing on graphics processing units (GPGPU): GPGPU is the technique of using a GPU, which typically handles computation only for computer graphics, to perform computation in applications traditionally handled by the CPU.

Genotype: A genotype is the genetic constitution of a cell, an organism, or an individual.

Phenotype: A phenotype is any observable characteristic of an organism, such as its morphology, development, biochemical or physiological properties, or behavior.

Sequence Alignment: In bioinformatics, a sequence alignment is a way of arranging the biological strings – such as DNA or protein – to identify regions of similarity that may be a consequence of functional, structural, or evolutionary relationships.

Smith-Waterman Algorithm: The Smith-Waterman algorithm is a well-known algorithm for performing local sequence alignment; that is, for determining similar regions between two nucleotide or protein sequences.

Chapter X
High-Throughput GRID Computing for Life Sciences

Giulia De Sario
Istituto di Tecnologie Biomediche, CNR, Italy

Angelica Tulipano
Istituto di Tecnologie Biomediche, CNR, Italy

Giacinto Donvito
INFN, Sezione di Bari, Italy

Giorgio Maggi
INFN Bari, Italy & Università e Politecnico di Bari, Italy

Andreas Gisel
Istituto di Tecnologie Biomediche, CNR, Italy

ABSTRACT

The number of fully sequenced genomes increases daily, producing an exponential explosion of the sequence, annotation and metadata databases. Data analysis on a genome-wide level or investigation within a specific data repository has become a data- and calculation-intensive process occupying single computers and even larger computer clusters for month or even years. In most cases such applications can be subdivided into many independent smaller tasks. The smaller tasks are particularly suited to distribution over a computational GRID infrastructure, which drastically reduces the time to reach the final result. In our analysis of gene ontology data and their associations to gene products of any kind of organism in a search to find gene products with similar functionalities, we developed a system to divide the full search into a large number of jobs and to submit these jobs to the GRID infrastructure as long as all jobs are processed successfully, guaranteeing an analysis of the data without missing any information.

INTRODUCTION

Data analysis in bioinformatics—due to the drastically high rate of increase in the sheer volume of data, not only in size but also in diversity—is becoming a very complex and data-intensive procedure occupying large numbers of computer units which often, for a typical user under normal condition, are not available. Sequencing projects all over the world, including high-throughput approaches such as the microarray technology and next-generation sequencing or the large scale mass spectrometry analysis, are responsible for this exponential growth of complex biological data sets. Data analysis within such projects and even more complex projects, such as comparisons and integration processes between such projects, often involves the examination of several big data sets with sizes on the order of hundreds of gigabytes. Fortunately, many of these analyses can be divided into many small tasks, producing the possibility of distributing the workload on an infrastructure such as the computational GRID. However, when the number of jobs necessary to carry on a particular analysis becomes huge, controlling the full production is not a simple enterprise. It is very important to carefully monitor each job, watching the success of each submitted job in order to be able to complete the full analysis without any missing data.

The biological task we are describing in this chapter is the comparison of gene products in a new, non-conventional way to find gene products with similar functionality. Usually gene products are compared by aligning sequences and looking for sequence similarity with the assumption that a high similarity corresponds to similar functionalities (Skolnick and Fetrow 2000). However, the relation sequence-function is not always true and often only small differences in the sequence may result in drastic changes in functionality. Those sequence differences are hardly detectable within a conventional sequence alignment.

Further, several gene products can have similar functionality but the active site or the conformation can be absolutely different, originating from an absolutely different primary sequence. It has also been shown that often the same functionality might have a similar tertiary structure but a similar tertiary structure doesn't need to have a similar primary sequence.

For this reason we propose a different approach to finding gene products which have similar functionalities. In most sequence databases there are keywords describing the functionality and localisation on a cellular level of the gene product they harbour. Those keywords and other functional annotations in other biological databases were converted to a standardised vocabulary describing the gene products on the level of the molecular functions and biological processes with which they are involved and the cellular components where they are localised, i.e., the gene ontology (GO (Ashburner, Ball et al. 2000)). The GO consortium (GOC) took the lead on this effort, so that we now have a repository of more than 3,6 million gene products described by more than 25000 GO terms creating more than 19,1 million associations (01_2008). All this information is available as a relational database (MySQL) called GO database (GODB (GOC 2001)), which occupies more than 2,7 gigabytes. The vocabulary within GO is well structured and well adapted to computational processing. This standardised vocabulary is perfectly suited to the comparison of the functionality of gene products and to the search for 'functional analogue' gene products, i.e., gene products which have the same functionalities but possibly a clearly different sequence, so that it is unlikely that they are evolutionarily related, such as homologous gene products.

This so called convergence seems to be ubiquitous throughout the tree of life. The 'evolution of similar features independently in different evolutionary lineages' has been found in many organismal phenotypes, from ecology, behavior, and morphology (Blackledge and Gillespie,

2004; Langerhans et al., 2006; Melville et al., 2006; Wiens et al., 2006; Wittkopp et al., 2003) to genes, proteins, enzymes or enzyme active sites (Charnock et al., 2002; Govindarajan and Goldstein, 1996; Lawn et al., 1997; Mattevi et al., 1996). Although studies that have found evidence of convergence are probably published more often than studies that have searched for convergence and not found it, this phenomenon has certainly been documented many times by many different groups. This search of functional analogues will definitively help to find more cases of convergence of gene products and help to understand better the mechanism of evolution.

To find the closest functional analogues of a gene product of interest one has to perform a comparison between the particular gene product and all others, of which there are more than 3,6 million with more than 19,1 million associations. Depending on the processor type, such a comparison takes between 45 and 60 minutes, which is too long for an 'on the fly' calculation. Therefore we decided to calculate all comparisons between all gene products and offer the data as an additional database dump with the data from the GO database.

However, this comparison approach asks for an enormously large amount of computation. Since this task can be divided into smaller components, it is suited for distribution over a GRID infrastructure. The GRID infrastructure is a network of geographically independent computers or clusters of computers which are administratively independent units but can be accessed via a well-described protocol. Normally the user enters the infrastructure via a user interface and submits the jobs to the GRID. The Workload Management Service (WMS) searches the available computing resources and assigns the job to one of the so called computing elements, (CE) the gateway to each GRID-cluster. The CE then distributes the jobs on the internally available computers, the worker nodes (WN). As soon as the job is terminated the user can retrieve the results from the WNs. Unique middleware makes this procedure possible and supports the successful management of each job. As simple as this procedure sounds, in reality the user is confronted with difficulties to be coped with.

In this chapter we would like to describe in detail how we analysed the data and data structure to extract the information for the functional analogue search to optimally use the GRID infrastructure, and to describe the steps required for the successful submission of a search so that no information is lost.

BACKGROUND

Functional Analogous Gene Products

An extensive comparison of different gene products is quite a complex procedure which is based mainly upon sequence analysis. The sequences of the gene products under examination are compared by aligning them and investigating their similarities. One approach is alignment on a global level where the algorithm searches for a similarity taking into account the whole stretch of sequence of the gene product (Needleman and Wunsch 1970). In this way one can find gene products with high similarities, in most cases homologous gene products, i.e., gene products which are evolutionarily related and have changed their sequences minimally while largely conserving functionality over time and during speciation. If sequences have very low global sequence similarity, then the adequate comparison is a sequence alignment on the local level, where the algorithm finds short sequence fragments within the total sequence with a higher similarity than the global similarity (Smith and Waterman 1981). Such sequence fragments are often structurally similar and represent a domain within the gene product with a probable specific functionality, which can be verified using specialised domain databases (Mulder, Kersey et al. 2008). These domain databases can also be used to search all gene products with the same domain which have a certain function associated with them. Local sequence similarity can also indicate a similar active site that again might point to similar functionality (Skolnick and Fetrow 2000). With these two approaches gene products are compared using sequence similarity, so that certain similar functionality can be proposed. However, the relation sequence-function is not always true. Small sequence differences within the functional domain can abolish or change functionality, even while sequence similarity remains high (Skolnick and Fetrow 2000). Among other important biological processes which can drastically impact functionality—especially on the localisation of gene products—are post-translational modifications whose sites, but not their final presence within the gene product, can be predicted (Seo and Lee 2004). On the other hand, nature has found several different solutions for some functions, and these gene products in most cases will have a totally different sequence and sometimes a different structural property. These gene products are called functional analogue gene products since they have achieved identical functionality through different evolutionary paths and are therefore evolutionarily totally independent. Such solutions would never have been found by a comparison at the sequence level (Skolnick and Fetrow 2000).

The only way to compare gene products at the functional level is to compare their descriptions. This, of course, works only if such descriptions are available and if these descriptions are not created by sequence comparison mechanisms as described above. Clearly, such an approach does not work with de novo sequenced gene products; in this case, only sequence analysis can produce first hints of what might be the functionality of these gene products. However, most sequence databases grant the sequence submitter the option of describing the functionality of the submitted gene product with keywords according to their experience investigating that gene product. This information is mostly experimentally achieved and is well suited to such comparisons on the level of gene products description. The only problem with this keyword system as applied to different databases is that keywords are not standardised; one function may correspond to several different keywords (synonymy) or different functions may be described by the same keyword (polysemy), which can lead to confusion and to incorrect functional associations. The gene ontology (GO) database partially solves this problem and was the information repository for our functional analogy search. GO is structured as a tree and organised as a direct acyclic graph, such that each node is a term of the vocabulary which has either zero or several children and one or more parents, based on two unconstrained relationships such as 'is-a' and 'part-of' (Figure 1).

Further, the ontology is split by 'term type' into three trees with root terms such as molecular function, biological process and cellular component. Each term can have zero, one or several gene products associated with it. The term to which a gene product is associated is called the directly associated term and all upstream terms towards the root term are called indirectly associated terms. The sequential list of all terms between the directly associated term and the root is called the path of the proper term. Since a term can have more than one parent a term can have more than one path associated with it. The nature of the association between a term and a gene product is indicated with an 'evidence code' (http://www.geneontology.org/GO.evidence.shtml) that indicates how this association was made (Table 1). Those associations are created by the corresponding database curators (Camon, Barrell et al. 2004; Camon, Magrane et al. 2004). A gene product can have more than one GO term associated with it, preferably at least one term from each branch of the GO tree.

Often there is more than one term from a GO branch associated to a gene product because

Figure 1. Part of the GO tree

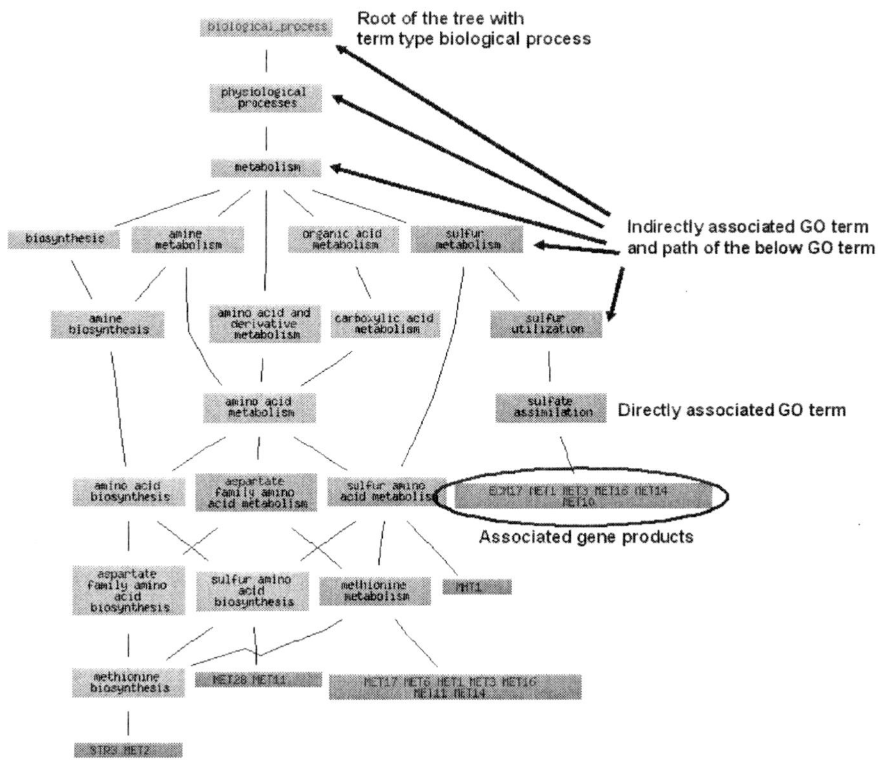

it might have several functions or might be involved in several different biological processes. Unfortunately, another reason is that terms of different levels of a path of an associated term are associated to the same gene product, producing a certain redundancy. Another impurity within the GO database is that often terms on one or even two branches are missing; an especially severe deficiency is that often the corresponding terms between molecular function and biological process are missing. A certain biological process can be split into several molecular functions and therefore there exists a relationship between the terms of the branch molecular function and the biological process. Actually, a bioinformatics group in Norway is creating some of those relationships using the information of pathway data repository, called the second layer of GO (Myhre, Tveit et al., 2006).

Returning to the description of a gene product, for our comparison we used as the description of a gene product the whole, non-redundant list of directly and indirectly associated GO terms for this gene product. Since biological information is changing and improving daily, the GO database—including the associations—is updated on a weekly basis, creating flat files of different formats according to the new entries in the databases and the parallel work of the internal curators who analyse the literature. With the same frequency the relational database, the GO database (GODB (Harris, Clark et al. 2004)) is updated, hosting all the GO terms, relationships, the associated gene products from any kind of organism and the associations between gene products and GO terms with the corresponding evidence codes. The first repository, circa late 2002 (go_12_02), contained 13166 GO terms describing 1,378 million gene products with 5,174 million asso-

Table 1. Evidences and their codes

Curator-assigned Evidence Codes	
Experimental Evidence Codes	
IDA	Inferred from Direct Assay
IPI	Inferred from Physical Interaction
IMP	Inferred from Mutant Phenotype
IGI	Inferred from Genetic Interaction
IEP	Inferred from Expression Pattern
Computational Analysis Evidence Codes	
ISS	Inferred from Sequence or Structural Similarity
IGC	Inferred from Genomic Context
RCA	inferred from Reviewed Computational Analysis
Author Statement Evidence Codes	
TAS	Traceable Author Statement
NAS	Non-traceable Author Statement
Curator Statement Evidence Codes	
IC	Inferred by Curator
ND	No biological Data available
Automatically-assigned Evidence Codes	
IEA	Inferred from Electronic Annotation
Obsolete Evidence Codes	
NR	Not Recorded

ciations and occupied about 1 GB storage. We recently received the latest version (go_01_08), with 25798 terms describing 3,593 million gene products with 19,093 million associations, occupying about 2,7 GB storage, representing an enormous quantity of outstanding work from the GO consortium. Whereas the analysis of gene products for human and model organisms such as mouse, rat, drosophila, etc, is very extensive, similar efforts for non-model organisms are largely non-existent, so often only sparse information is available. However, this sparse information of non-model organisms still represents significant knowledge, apart from the data from human and model organism annotations, and necessitates a search at the level of functionality. This is valuable information from outside the 'model world' which confirms or complements the data of the 'model world'.

The proposed approach described herein takes advantage of exactly this information from non-model organisms in comparison with the human and model organism data. If we have a gene product with a certain description and would like to know whether there is additional information from functionally similar, functional analogue gene products, we need to run a comparison between this gene product and all others within the GODB. Such a single search is already quite time- and data-consuming, since the algorithm has to access 19,1 million data points to fulfil this query. Running each gene product against each of the whole database turns into a several hundred CPU-year task. In addition to the fact that the calculation time for this search greatly exceeds a human life span, it is important to mention again that the GODB is updated monthly with new and more precise information. These updates are essential for our analysis, since they improve the accuracy of our functional analogy predictions. Therefore we should re-run the search with each upgrade of the GODB. With such long processing times, also in the CPU-month range, we would not be able to offer such a service. Reducing the execution time of the search to days or even hours would enable us to provide these important data to the scientific community in synchrony with the GODB upgrade.

Each comparison of one gene product against all others is the smallest independent element of the whole search; therefore the whole search can be split into any number of jobs and distributed over several computational units to reduce the actual execution time to a fraction of the original.

The GRID Infrastructure

The adequate infrastructure for such a workload distribution is the GRID, where several thousand computer units (CPUs) can be available under the same middleware to unify them into a single virtual calculating unit. The GRID infrastructure is a computer structure which emerged in the 1990's. Here in Europe—as CERN (European Organisation for Nuclear Research at Geneva) is a major promoter of the development of the GRID infrastructure, supported by the European Community, which finances the EGEE (Enabling Grids for E-sciencE) project—the GRID, now a large, distributed, reliable network of available computers, has become an important resource.

The structure of this GRID is organised such that the smallest unit is the worker node (WN), which is the actual place of the processing of the job. A group of WNs, preferably geographically co-localised, is represented by a computing element (CE) which is the point of reference for a job submission request. The jobs are submitted via the user interface (UI), which sends the submission request to the Workload Management Service (WMS), optimally several geographically distributed, which contact the different CEs to request final submission for processing. If data-intensive jobs are submitted, the data may be deposited at a central location, the storage element (SE), where the WNs then access the data for processing. The results of the successfully terminated job are located on the WN and need to be requested via a job identifier. To ensure efficient and optimal use of these resources, larger submissions require that an automated system be used, since it is impossible to manually submit, control, and analyse the jobs and retrieve the results. Further, often jobs are not properly submitted, or they may fail at any stage of the submission protocol; such jobs must be identified and subsequently resubmitted for a successful termination of the task. In general, there are two basic approaches to the submission of jobs. One is the 'push approach' where the UI submits fully formulated jobs and waits for the results to be retrieved. The other is the 'pull approach' where the UI submits a request for a job; when a WN gets assigned to this job, it asks (pulls) the full description of the job to be processed. The second approach, the pull model, has the great advantage that the job submitters have more control of there jobs. If jobs are sent (pushed) into the GRID infrastructure, they can easily find themselves queued for a very long time due to problems with priority and scheduling, and there is no control over when the job is done. Use of a mix of the push and the pull models sends only a request for job submission and not the actual job itself, so potential queuing is not a matter of concern; when the request is successful the WN pulls the actual job in and information about exactly which jobs are running is readily available. With a simple monitoring system, the submitters control which jobs are not yet assigned and which are running; they do not have jobs waiting. The pull model also guarantees the submitters the highest performance, since they have only running jobs. We used the pull model for our functional analogue search; in the following chapter we describe in more detail how we solved this problem.

GENE ANALOGUE FINDER

Gene Ontology (GO)

The gene ontology (GO), with its vast amount of knowledge of more than 3,6 million gene products of more than 150000 species, is the optimal dataset to investigate for functional analogues. Further analysis of the data revealed how to optimally use this data to search for functional analogues. One analysis we had to perform to understand the typology of the data was to calculate the usage frequency of the evidence codes. Most evidence codes indicate the experimental background of the association of the gene products. However, there are codes such as computational analysis evidence codes—including ISS, IGC and RCA or the automatically-assigned evidence codes IEA—which in most cases are based directly or indirectly upon sequence analysis and therefore add only low level information to our functional analogue study. From the 19,1 million associations, 97% are IEA associations and only 403399 associations are experimental in nature (IDA, IPI, IMP, IGI, IEP, TAS, NAS and IC). From the point of view of annotated gene products, the same statistics show that from the 3,6 million gene products, 95,3% contain at least one association as IEA, but only 1,8% (65786) contain an association which is experimental in nature; all the rest have only IEA associations. The reason for this large quantity of only IEA-associated gene products is the automated procedure to annotate new genomes using bioinformatics approaches to assign GO terminology. Gene products which also have experimental associations cover 705 species, but 64% of these gene products originate from the 7 most-studied organisms: human, C.elegans, mouse, zebrafish, baker's yeast, rat and fruit fly (in descending order of experimental annotation). Besides the predominant use of the evidence IEA, there is a significant number of gene products which have only few and often general GO terms associated with them. In a prior functional analogue search of a subset of the GODB data, we found that many gene products had only a few directly- and indirectly-associated GO terms, so we decided to set a minimum threshold for the number of associated terms so as to select gene products with adequate annotation for inclusion in the final search. The analysis of the list of selected gene products, a set of 2,189 million gene products (61% of the original list), shows that all excluded gene products only have associations with evidence code IEA. However, with the percentage of non-computer-analysed association evidences we will be able to provide reasonable quality results. It must be emphasised that although the associations with IEA evidence provide information which improves the search results, these results must be evaluated with more care, since the data are from sequence comparisons. Besides homologous gene products, which were purely electronically associated, we also found gene products which were, on the level of sequence similarity, absolutely different, yet the electronic assignment found functionality similar to the gene product in search (see http://www.geneontology.org/GO.evidence.shtml#iea).

A further important analysis was to understand how many gene products had exactly the same associated GO terms and therefore identical descriptions, or how many unique descriptions existed to describe the remaining 2,0 million gene products. Surprisingly, there were only 92723 different descriptions for all these gene products. This fact changed the strategy of our search, wherein we initially wanted to compare each gene product with all others. Since these 93000 descriptions represented the selected 2,0 million gene products, and the search used these descriptions for the comparison, the processing was reduced from a 2,0 million x 2,0 million matrix to a 93000 x 93000 matrix working directly with the descriptions; the number of comparisons

Table 2. Examples of p(term) values

Name	Level	Term-type	Count	p(term)
molecular_function	1	molecular_function	9902069	0.518606
biological_process	1	biological_process	5335494	0.279438
cellular_component	1	cellular_component	3856064	0.201956
hormone activity	4	molecular_function	7441	0.000389711
gliogenesis	4	biological_process	295	1.54502e-05
cell fate specification	5	biological_process	344	1.80165e-05
angiogenesis	7	biological_process	725	3.79708e-05

was thus reduced by a factor of about 400—a very significant reduction, produced by simply analysing and understanding the data.

Data Analysis

After the analysis of the data to optimise job submission, a further analysis must be undertaken to understand how to formulate the algorithm to obtain the correct information about the functional analogue. As mentioned above, the GO is organised as a tree structure with three main branches: (1) the molecular functions and (2) the biological processes in which the gene products are involved, and (3) the cellular components in which they are localised. The root term (e.g., molecular function) is the most global definition of a gene product associated to one of the terms of the tree. The more distant a term is from the root term the more specific is its description, and the more terms are included within a path of a term the larger is the knowledge about this specific term. For these reasons, we decided to include all directly- and indirectly-associated GO terms in the comparison, to take advantage of the whole knowledge around those gene products, but to weight each term according to its specificity. Some approaches, weighing the terms according to their specificity, propose to use the level of the term, assuming the root terms as level 1. This approach, however, is only partially valid, since level does not correctly represent an increase in specificity. The path length of two terms with apparently similar specificity can vary dramatically. For this reason, we introduced the usage of the semantic similarity measurement (Resnik 1999; Lord, Stevens et al. 2003) to compute the specificity of a GO term independently from its level within the GO tree. The semantic similarity measurement uses the frequency with which a term is associated to different gene products, considering the notion of 'information content'. The less a term of a vocabulary is used, the higher the information content and the more descriptive it is. We therefore calculated the probability p(term) for every term by counting the number of gene products associated with the term or any of its children, divided by the number of total associations between the GO terms and gene products. Information content (detail level of the description) is inversely proportional to p(term) (Table 2). The highest p(term) values are found for GO terms 'molecular function', 'biological process' and 'cellular component', which are the roots of the three DAGs; however, their p(term) value is not 1 (Table 2), since not every gene product has an association with all three term types, as we pointed out earlier. Further, Table 1 demonstrates the fact that terms of the same level in different paths can have clearly different information content and therefore different p(term) values. The p(term) value was then used as a weight in the statistical analysis we used for our functional analogue search; it was entered as 1–p(term) to obtain an additive weight system for high 'information content' GO terms.

Gene Analogue Finder

The actual comparison between two gene products was computed with a simple non-parametric χ^2-test. The non-parametric χ^2-test is a typical statistical test used to compare two different samples, in our case two gene products A and B, according to particular characteristics or aspects, in our case the GO terms directly- or indirectly-associated with the gene products. Since every gene product can be described by a different number of GO terms, the table of values was designed as a 2 x 2 array containing the sum of 1–p(term) of present or absent GO terms according to the two compared gene products (Table 3, Tulipano et al., 2007).

In this way we have the same degree of freedom for all gene-product-to-gene-product comparisons; therefore the χ^2-values are comparable. All three term types were taken into account for the main search to increase the precision of the description and to enable separation of, say, two gene products with the same function acting in different cellular components.

Results

To understand the topology of the potential functional analogue gene products, we analysed BCL2_HUMAN [UniProt: P10415], a well-studied apoptosis gene. The search resulted in 15 significant, functionally similar descriptions which are linked with 23 analogue gene products of BCL2_HUMAN (Table 4). Firstly, BCL2_HUMAN was found as the only best hit and therefore is a well-described gene product. Within the 23 best hits we found 12 (54,5%) other gene products which, according to the protein family database (Pfam, (Finn, Tate et al. 2008)) belong to the same family, the Bcl-2 family. The rest, 10 gene products (45,5%), belong to other protein families. This demonstrates how the Gene Analogue Finder can find members of the same or of a similar protein family, which is an important confirmation of our algorithm. It is important to mention that members of gene families in Pfam are assigned by sequence similarity and therefore can be found by sequence-based approaches. Focusing on the Bcl-2 family, Gene Analogue Finder found 3 out of 6 BCL2 orthologous gene products present in the SwissProt database (Gasteiger, Jung et al. 2001) with high sequence similarity. Most Bcl2 family members, however, have a significantly lower-than-maximum χ^2-value. In Table 4 we note the number of common and non-common terms, indicating that these family members do not have nearly as many associated terms as do the BCL2_HUMAN and the rat orthologues. An examination of the evidence code for these asso-

Table 3. The 2 x 2 array used for the χ2-calculation to compare gene product A with gene product B

	Term present in gene product A	Term not present in gene product A
Term present in gene product B	$O_{11} = \left(\sum_{\substack{all\ terms \\ common \\ to\ A\ and\ B}} (1 - p(term)) \right)$	$O_{12} = \left(\sum_{\substack{all\ terms \\ present\ in\ B \\ but\ not\ in\ A}} (1 - p(term)) \right)$
Term not present in gene product B	$O_{21} = \left(\sum_{\substack{all\ terms \\ present\ in\ A \\ but\ not\ in\ B}} (1 - p(term)) \right)$	$O_{22} = \left(\sum_{all\ terms} (1 - p(term)) - \sum_{\substack{all\ terms \\ in\ A\ and\ B}} (1 - p(term)) \right)$

Table 4. Example BCL2_HUMAN

Functional Analogue	Protein familiy	χ^2– value	Number common terms	Average p(term)	Number non-common terms	Organism	Common name
BCL2_HUMAN	bcl2	25789,79	193	0,020204	0	Homo sapiens	human
Q7WEY5_BORBR		22634,87	170	0,022843	0	Bordetella bronchiseptica	bacteria
Q7W3L0_BORPA		22634,87	170	0,022843	0	Bordetella parapertussis	bacteria
Bcl2	bcl2	20308,28	154	0,025084	1	Rattus norvegicus	Norway rat
BCL2_CHICK	bcl2	13206,1	101	0,037823	0	Gallus gallus	chicken
Q923R6_CRILO	bcl2	11717,28	90	0,041547	0	Cricetulus longicaudatus	hamster
Q8I008_FELCA	bcl2	11717,28	90	0,041547	0	Felis catus	domestic cat
BCLX_HUMAN	bcl2	10374,6	84	0,045444	4	Homo sapiens	human
Bcl2	bcl2	9814,676	76	0,048772	0	Mus musculus	house mouse
BAK_HUMAN	bcl2	8649,128	73	0,050633	6	Homo sapiens	human
BNIP3_HUMAN		8532,739	99	0,038871	49	Homo sapiens	human
Bnip3		8532,739	99	0,038871	49	Rattus norvegicus	Norway rat
Q9MZS6_SHEEP	bcl2	8053,33	65	0,056837	2	Ovis aries	sheep
INI2_HUMAN		7954,163	66	0,054293	4	Homo sapiens	human
debcl		7844,056	68	0,044104	8	Drosophila melanogaster	fuit fly
Q8SQ43_FELCA	bcl2	7765,406	99	0,038582	63	Felis catus	domestic cat
Q8HYU5_CANFA	bcl2	7765,406	99	0,038582	63	Canis lupus familiaris	dog
BAXB_HUMAN	bcl2	7765,406	99	0,038582	63	Homo sapiens	human
Nfs1		7615,217	115	0,033579	108	Rattus norvegicus	Norway rat
Bax	bcl2	7305,738	102	0,037456	80	Mus musculus	house mouse
LIRB4_HUMAN		7094,369	63	0,057724	8	Homo sapiens	human
EIN2		3235,432	46	0,037509	37	Arabidopsis thaliana	plant
NARJ_ECOLI		2902,651	90	0,039677	253	Escherichia coli K12	bacteria

ciations shows that the orthologous ones inherited the GO terms from well-described orthologues, often via common functional domains or high global sequence similarity, and are not yet as well-described by experimental evidence as are the human and rat Bcl2. Table 4 also shows that most of the gene products found have 50% or more terms in common with BCL2_HUMAN, but the χ^2-value is reduced because of the high number of non-common terms, pointing out other functions than those shared with BCL2_HUMAN. Some potential functional analogue gene products have quite a low number of common terms, but the average p(term) of the common terms is quite low, indicating that these common terms are of high specificity. Worthwhile to mention is that with our functional analogue search we found hits covering 12 different species, of which 7 are clearly from non-model organisms. Interestingly, within this list of found organisms we found gene products from 3 different bacteria and one plant. All these non-bcl2 family hits of our search would have been very difficult or impossible to find with a sequence-based approach.

JOB SUBMISSION PROBLEMS AND SOLUTIONS

In order to execute the functional analogue search, the entire problem was subdivided into atomic tasks. We chose to define an atomic task as the comparison of a single description against all the others (93000); the atomic task consists therefore of a total of 93000 comparisons. Since the atomic task requires on the order of a few minutes complete, 20 atomic tasks were assigned to a single job. Thus a job lasts about 2,5 hours, which is a good compromise between execution time and time spent by the worker node to recreate the running environment, i.e., for the input data retrieval from the SE. The full search therefore required the execution of 4650 jobs.

After subdividing the entire search into many small tasks, one needs to:

a. submit the resulting number of jobs to the GRID, and
b. make sure that all jobs correctly execute the atomic tasks assigned to them.

It is therefore necessary to keep track of all submitted jobs and understand if they have reached completion. Complications may arise, because when submitting jobs to a large and distributed environment like the EGEE GRID infrastructure, there is always the possibility that some jobs will fail, either during the submission phase or during the input data retrieval or job execution phases. The operator needs to promptly recognise failed jobs in order to resubmit them. This operation, when the number of jobs required to complete the full search is quite high, must be done by means of an automatic tool.

The Job Submission Tool (JST)

A job submission tool (JST, http://webcms.ba.infn.it/~pierro/JST) was developed to allow submission of large numbers of jobs and to keep track of all of them in an almost unattended way. At this point all the atomic tasks were inserted into a monitoring DB (MDB) server which controls the assignment of tasks to jobs and monitors the job to which that specific task has been assigned (Figure 2).

For proper monitoring of task assignment and job termination, several parameters are associated with each task:

- **Status.** Possible values: 'Free' (not assigned), 'Running' (assigned) or 'Done' (job terminated). If the task's status is 'Free', it can be assigned to a job. If the status persists as 'Running' for more then a fixed time interval, it is assumed that the job has probably failed, and the task will be reassigned to a new job. If the status is 'Done' or 'Running' for less then the fixed time interval, the task is ignored during the task assignment process.
- **Dependencies.** If the task is dependent upon the successful completion of one or more other tasks, it can be flagged as dependent upon those tasks; only tasks with no dependencies or those with dependencies whose status is 'Done' will be executed.
- **Priority.** An arbitrary priority value can be assigned to the task; tasks are executed in priority order.
- **Job Provenance.** A notation of which job has actually performed the task.
- **Description.** May be a string or a link to a specific script to execute. This provides better control over what is executed on the WN. Indeed, it is possible to interrupt and change a job's execution (e.g., if a bug is discovered, or there is a need for revised optimisation) after its submission.
- **Failure Count.** The system logs task failures to avoid repeated resubmission of consistently failing tasks. The maximum resubmission count is a settable parameter.
- **Date and Time of Execution.**

In the submission phase all jobs are identical; indeed, when a job is submitted it does not know which task(s) it has to execute. A background daemon was used to submit at a given rate, which could be tuned to consistently provide the same

JDL to the GRID. Indeed, several daemons were used in parallel, each one pointing to its own gLite WMS to prevent the failure of a single WMS from stopping the submission procedure. The daemons automatically cease job submission when no more unassigned tasks are found in the MDB.

A job submitted to the GRID is enclosed in a job wrapper. The job wrapper and the script that contains the executable can send to the MDB server all information that controls which jobs are running, as well as which task and which operation is executing on the WN. This DB server can be separated from the one hosting the task list, the MDB, if the scalability of the DB itself becomes an issue.

The schema of the MDB allows the user to accept any kind of information, including:

- The date and time on the WN that sends the monitoring information.
- The host that is sending the information.
- The name of the variable to be monitored.
- The value of the variable.
- The JOBID of the job that is sending the information.
- The date (and time), on the server, when the information is stored.

This general schema allows the user to monitor any job operation by adding new variables as needed, without changing the DB schema. This feature provides information on how many jobs are running at a given time if the job-wrapper is configured to regularly send monitoring information at fixed time intervals.

Only when a job starts running on a WN does it request of the central MDB the assignment of a task for execution. Information on the execution of each task is logged in the central MDB according to the parameters mentioned above. Only when all steps are correctly executed is the status of that particular task updated to 'Done'. Thus the MDB provides complete monitoring of task assignment and job execution; no manual intervention is required to follow each step or to manage the eventual resubmission of failed tasks. Usually we see two kinds of failed jobs: (1) jobs that were killed by the queue manager, and (2) jobs which failed due to internal problems (i.e., an operation failed). In the first case the wrapper cannot update the MDB, while in the second case the wrapper script updates the MDB and increments the 'FAILURE counter'.

To deal with the first kind of failure, tasks which are found in a 'RUNNING' state for more than a fixed amount of time are considered failed and are automatically reassigned to a new job.

Figure 2. Diagram of the JST

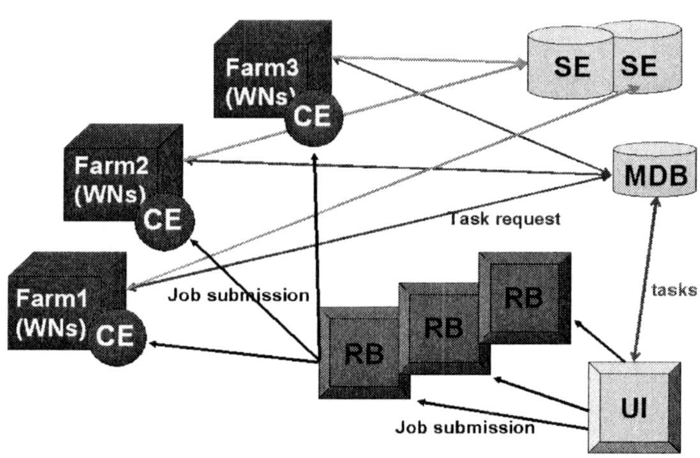

To optimise the input and output operations and to avoid bottlenecks and failures, two procedures were set up to randomly choose the SE source of the job input file as well as the SE on which the job output is stored. Thus if one SE fails temporarily, the other can continue running the task and store the output.

Data and Software Management

In a challenging run like the Gene Analogue Finder, the problem of data management and software installation is also an interesting aspect to consider in order to find optimal conditions to minimise execution time. For example, much effort was invested in the optimisation of input data retrieval. The original application used to be relational DB-driven, but to run the tasks more efficiently over the GRID, a flat-file dump of the DB query was used; the flat file dump was created once before initiating the search and stored on at least two SE's space in the GRID.

To ensure a high success rate, we chose to install the libraries required for each job in the local directory in which the jobs runs. Since it is not known a priori that the required library will be available on the WN, combining the libraries with the jobs guarantied that a missing library would not cause a failure. The library loaded onto the WN local disk does not remain there after the job's execution, since the job's local directory is cleaned automatically by the middleware when the job exits.

Both the job input data and the libraries need to be downloaded to the WN where one of the search jobs will be executed. The download procedure was tuned to minimise the time required in comparisons with the execution time of the job submitted. All input files were compressed to reduce transfer time, and the time needed to set up the environment was measured. Using to this measurement, the number of comparisons assigned to a given job was optimised to maintain good balance between the time spent in calculation and the time spent in setting up the environment. A balance of about 5% of the total time spent in 'non-CPU' bound operations was considered optimal.

As a further precaution to keep the success rate as high as possible, a technique was implemented for every operation in which an external server was required. For example, the input files are available on two different storage elements; the JST procedure wrapper uses the two storage elements in a round-robin manner—unless one of them fails, in which case only the functioning one is used.

During the challenge for the gene analogue search, the MDB was replicated in real time by the 'master-slave' method, to avoid a single point failure. This was done using the standard MySQL procedure, which has proven to be sufficiently stable and very fast. In order to test the replication procedure, a bulk update of 3Gbytes of SQL files was applied to the MDB; the replica was correctly updated in less than 5 minutes. With this replication approach we are guaranteed that if the primary MDB server fails, the jobs can always use the replica to continue the computation, thus avoiding any down time during the search.

Results

The clear advantage of the Job Submission Tool is that its automated processes make it easy to trace the execution of all tasks, while the operator is freed from manually controlling and resubmitting failed jobs. Job control and validation is done by the job itself; the operator need only be sure that the control procedure is well tuned to successfully recognise the state of a job in order to flag the corresponding task accordingly within the MDB server.

Tests were executed in a controlled environment to verify that one can easily deal with all requirements, such as monitoring a high number of distributed jobs on the GRID. In particular, the limits of the system were investigated by running

Table 5. Performance test of the monitoring database (MDB)

Number of tasks in the list	Number of concurrent Jobs	Query Execution Time
5000	72000	0.01
50000	23000	0.03
150000	2000	0.70

several jobs which made the same queries to the MDB as would be submitted by the job wrapper under normal conditions. This gives a measure of how many queries the system can serve in a given time window. We found that the performance of the system is strongly affected by the number of tasks in the to-do list of the MDB server. For this reason the results are reported by calculating the number of jobs that can run concurrently over the GRID as a function of the number of records in the MDB (Table 5). The relationship between number of tasks in the MDB and the number of concurrent jobs is determined by the time needed for the RDBMS to search for the right task among all available tasks; the larger the to-do list the longer the time to find the right task. In the table we indicate the task list size and the corresponding time to execute the query. From this time the number of jobs that can be executed concurrently was calculated, assuming the execution time of each task to be about one hour. Clearly better performance can be produced for longer tasks. Since the number of concurrent jobs possible on the GRID is primarily limited by the number of CPUs available on the GRID (VO biomed can access about 5000 CPUs in total), it appears that given the actual size of the EGEE infrastructure, the MDB will not be the limiting factor of the JST job submission procedure. The JST has been tested with up to 2000 concurrent jobs running over the GRID and no limitation due to MDB access was found, with typically 4500 MDB entries. If a higher number of concurrent jobs is necessary, it is always possible to split the full task list into several partial lists, each served by an MDB server; the job can be instructed to randomly choose one of the DB servers from a given list.

Using this JST it was possible to run the complete comparison without any manual intervention to check the status of the tasks or to monitor the status of all of the jobs. The challenge was executed using the VO 'biomed' on the EGEE infrastructure, without experiencing any kind of performance loss due to the monitoring activity of the JST.

The complete search, which would take about 1,25 CPU-years on a single CPU, was performed in less than one day on the GRID using about 950 different WN's. About 50 Million DB entries, corresponding to approximately 1,5 GB of data, were produced in the search.

The MDB consisted of 4700 different tasks, each of which comprised 20 elementary comparisons (one description against all the others). To submit the required jobs, two daemons were used to submit jobs to two different WMSs. Since the task list was not so big, a single instance of the MDB server was used; obviously this instance was replicated to avoid single point failure as described before.

FUTURE TRENDS

As mentioned above, the size of the GO—and especially the annotation of gene product using the GO terminology—is increasing daily. Every experiment in the laboratory and every bioinformatics analysis reported to the GO consortium helps to improve the total knowledge of gene products of any organism. Therefore, our search needs to be re-run after every update of the GO, and since the amount of data to be processed increases with every release, we are more and more dependent upon distributed processing on the GRID. The functional analogue search is important for the scientist; it is a way for the laboratory researcher to take direct advantage of the knowledge in the GO. However, it is important that not only the quantity, but also the quality, of the GO knowledge is increasing. Therefore future efforts must be directed toward the drastic increase of associations with experimental evidence codes. Only this information is capable of detailed individuation of all existing gene products. Of course, there are homologous gene products which have very similar functionality, but often slight changes of function or localisation occurred during evolution; in most cases this is not (yet) clearly detectable at the level of the sequence. Often, there are post-translational modifications which strongly influence the functionality or localisation of the gene product, but these modifications are not visible in the sequence and can only be guessed at. For this reason we will not only frequently provide the latest update of the functional analogue search, but will also actively search for methods for the successful integration of experimental data from gene products of any kind of organism, to take advantage of the vast amount of information in the laboratory which is not yet published or not publishable in a regular way.

A MySQL dump of the search results will be publicly available; it is planned that it be distributed together with the MySQL dumps of the GODB. Further, we are working on a graphical user interface to access these functional analogue data together with information from the GO and additional properties of the gene products under examination, such as sequence, sequence similarity, domains and other yet-to-be-defined information.

The job submission tool JST, developed within the bioinformatics application, turned out to be an important application to facilitate access to the GRID infrastructure and to successfully submit jobs to analyse complex and data-intensive biological problems. The tool was used in other situations, which verified its potential and philosophy in the efficient processing of the requested resources. Therefore it is important to continue the development of this tool, so that the end user may submit an analysis by providing an algorithm and a data set without requiring that he know about the GRID infrastructure and its properties. This seems to be a most important step within the GRID technology; it will make its resources available even to a user with little or no technological knowledge.

CONCLUSION

With Gene Analogue Finder we demonstrate that GO has the potential to be used not only for analysis of biological data but also directly for comparative research like the search for functional analogues we implemented in Gene Analogue Finder. When compared with other, similar tools—in particular with the GO-family tool (Martin et al., 2004)—we find that Gene Analogue Finder provides different and complementary information. However, it is important to mention that although there is a large space of knowledge provided by the GO, there are still limitations for an optimal comparative analysis. Too many gene products—mainly from non-model organisms—have low level and/or prevalently electronically-evaluated associations. For this reason, too many gene products have identical descriptions; only additional experimental knowledge will differentiate these gene products. There is also a certain redundancy, where the same biological abstraction described with different words leads to a separation of the same biological function and therefore also to the separation of corresponding gene products. This limitation of the GO has yet to be taken into account, both in our work and by all other tools using this knowledge. Since GO is updated daily, it is conceivable that most of these problems will be progressively reduced. It is worth stating again that additional effort is required to improve the

GO; adding more experimental information for non-model organisms will really push the GO over the threshold for use in optimal comparative analysis. Running Gene Analogue Finder over the GRID proved highly effective: the functional analogue search was divided into a large number of jobs and distributed over the EGEE infrastructure, which reduced its processing time from what would have been about 1,25 years on a single machine to about 1 day on up to 950 networked worker nodes, an acceleration factor of 500. The prototype framework used by Gene Analogue Finder for the remote installation of data and of the Perl Library, as well as the application submission and monitoring mechanism supplied by JST, proved highly reliable and general enough to be applied in other bioinformatics applications as well as in other research fields.

AKNOWLEDGMENT

We are grateful to Dr. Domenico Catalano for his careful reading, correcting, and making constructive comments on the manuscript. We also thank Stanley G. Clark for the English corrections. We would also like to thank the INFN production GRID Resource Centres administrators and the INFNGRID/GRID.IT management.

This development was partially supported by grants for ITB and INFN from MIUR: FIRB 2003 art. 8 D.D. 2187 del 12-12-2003 LIBI, and by the BIOINFOGRID project (INFSO-RI-026808) and for INFN by the INFN-GRID project and by the EGEE project (INFSORI-508833).

This work makes use of results produced by the Enabling Grids for E-sciencE (EGEE) project, a project co-funded by the European Commission (under contract number INFSO-RI-031688) through the Sixth Framework Programme. EGEE brings together 91 partners in 32 countries to provide a seamless GRID infrastructure available to the European research community 24 hours a day.

REFERENCES

Ashburner, M., Ball, C.A., et al. (2000). Gene Ontology: tool for the unification of biology. The Gene Ontology Consortium. *Nat Genet 25*(1): 25-9.

Blackledge, T.A., Gillespie, R.G. (2004). Convergent evolution of behavior in an adaptive radiation of Hawaiian web-building spiders. *Proc Natl Acad Sci USA, 101*(46):16228-33.

Camon, E., Barrell, D., et al. (2004). The Gene Ontology Annotation (GOA) Database--an integrated resource of GO annotations to the UniProt Knowledgebase. *In Silico Biol 4*(1): 5-6.

Camon, E., Magrane, M., et al. (2004). The Gene Ontology Annotation (GOA) Database: sharing knowledge in Uniprot with Gene Ontology. *Nucleic Acids Res, 32*, D262-6.

Charnock, S.J., Brown, I.E., Turkenburg, J.P., Black, G.W., Davies, G.J. (2002). Convergent evolution sheds light on the anti-beta-elimination mechanism common to family 1 and 10 polysaccharide lyases. *Proc Natl Acad Sci USA, 99*(19),12067-72.

Finn, R.D., Tate, J., et al. (2008). The Pfam protein families database. *Nucleic Acids Res, 36*, D281-8.

Gasteiger, E., Jung, E., et al. (2001). SWISS-PROT: connecting biomolecular knowledge via a protein database. *Curr Issues Mol Biol,* 3(3), 47-55.

Gene Ontology Consortium (2008). The Gene Ontology project in 2008. *Nucleic Acids Res, 36*, D440-4.

Gene Ontology Consortium (2001). Creating the Gene Ontology resource: design and implementation. *Genome Res, 11*(8), 1425-33.

Govindarajan, S., Goldstein, R.A. (1996). Why are some proteins structures so common? *Proc Natl Acad Sci USA, 93*(8), 3341-5.

Harris, M. A., Clark, J., et al. (2004). The Gene Ontology (GO) database and informatics resource. *Nucleic Acids Res, 32*, D258-61.

Langerhans, R.B., Knouft, J.H., Losos, J.B. (2006). Shared and unique features of diversification in Greater Antillean Anolis ecomorphs. *Evolution, 60*(2), 362-9.

Lawn, R.M., Schwartz, K., Patthy, L.(1997). Convergent evolution of apolipoprotein(a) in primates and hedgehog. *Proc Natl Acad Sci USA,* 94(22), 11992-7.

Lord, P. W., Stevens, R.D., et al. (2003). Investigating semantic similarity measures across the Gene Ontology: the relationship between sequence and annotation. *Bioinformatics,* 19(10), 1275-83.

Martin, D., Brun, C., Remy, E., Mouren, P., Thieffry, D., Jacq, B. (2004). GOTool-Box: functional analysis of gene datasets based on Gene Ontology. *Genome Biol, 5*, R101.

Mattevi, A., Vanoni, M.A., Tosone, F., Rizzi, M., Teplyakov, A., Coda, A., Bolognesi, M., Curti, B.(1996). Crystal structure of D-amino acid oxidase: a case of active site mirror-image convergent evolution with flavocytochrome b2. *Proc Natl Acad Sci USA, 93*(15), 7496-501.

Melville, J., Harmon, L.J., Losos, J.B. (2006). Intercontinental community convergence of ecology and morphology in desert lizards. *Proc Biol Sci., 273*(1586), 557-63.

Mulder, N. J., Kersey, P., et al. (2008). In Silico Characterization of Proteins: UniProt, InterPro and Integr8. *Mol Biotechnol, 38*(2), 165-77.

Myhre, S., Tveit, H., et al. (2006). Additional gene ontology structure for improved biological reasoning. *Bioinformatics, 22*(16), 2020-7.

Needleman, S., Wunsch, C. (1970). A general method applicable to the search for similarities in the amino acid sequence of two proteins. *J Mol Biol, 48*(3), 443-53.

Resnik, P. (1999). Semantic similarity in a taxonomy: an information based measure and its application to problems of ambiguity in natural language. *J Artif Intelligence Res, 11,* 95–130.

Seo, J. and Lee, K.J. (2004). Post-translational modifications and their biological functions: proteomic analysis and systematic approaches. *J Biochem Mol Biol, 37*(1), 35-44.

Skolnick, J. and Fetrow, J.S. (2000). From genes to protein structure and function: novel applications of computational approaches in the genomic era. *Trends Biotechnol, 18*(1), 34-9.

Smith, T.F., and Waterman, M.S. (1981). Identification of common molecular subsequences. *J Mol Biol, 147*(1), 195-7.

Tulipano, A., Donvito, G., Licciulli, F., Maggi, G., Gisel, A.(2007). Gene analogue finder: a GRID solution for finding functionally analogous gene products. *BMC Bioinformatics, 8*, 329.

Wiens, J.J., Brandley, M.C., Reeder, T.W.(2006). Why does a trait evolve multiple times within a clade? Repeated evolution of snakelike body form in squamate reptiles. *Evolution,* 60(1),123-41.

Wittkopp, P.J., Williams, B.L., Selegue, J.E., Carroll, S.B.(2003). Drosophila pigmentation evolution: divergent genotypes underlying convergent phenotypes. *Proc Natl Acad Sci USA,100*(4), 1808-13.

KEY TERMS AND DEFINITIONS

Analogy: A structure performs the same or similar function by a similar mechanism but evolved separately. Similar structures may have evolved through different pathways, a process known as convergent evolution.

gLite: Lightweight Middleware for GRID Computing – middleware stack developed inside the EGEE project.

GRID Computing: The creation of a 'virtual supercomputer' by using a network of geographically dispersed computers.

Homology: In evolutionary biology, homology has come to mean any similarity between characters that is due to their shared ancestry.

Job Description Language (JDL): The JDL is used in gLite to specify desired job characteristics and constraints, which are used by the match-making process to select the resources that the job can use.

Job: A scheduled and/or automated task for a WN in a GRID processing environment. It is defined by the executable with its parameters, and by the input and output files.

Ontology: A study of conceptions of reality and the nature of being. In computer science, in fact, ontologies are representations of a set of concepts within a domain and the relationships between these concepts.

Semantics: Relation between signs and the things they refer to, their denotata.

Sequence Alignment: A way of arranging the primary sequences of DNA, RNA, or protein to identify regions of similarity that may be a consequence of functional, structural, or evolutionary relationships between the sequences.

Task: The work assigned to a job. It can contain several elementary steps or can consist of a single operation.

Workload Management Service (WMS): The service that takes care of submitting and distributing jobs over the gLite infrastructure.

Chapter XI
Management and Analysis of Mass Spectrometry Proteomics Data on the Grid

Mario Cannataro
University Magna Graecia of Catanzaro, Italy

Pietro Hiram Guzzi
University Magna Graecia of Catanzaro, Italy

Giuseppe Tradigo
University Magna Graecia of Catanzaro, Italy

Pierangelo Veltri
University Magna Graecia of Catanzaro, Italy

ABSTRACT

Recent advances in high throughput technologies analysing biological samples enabled the researchers to collect a huge amount of data. In particular, mass spectrometry-based proteomics uses the mass spectrometry to investigate proteins expressed in an organism or a cell. The manual inspection of spectra is unfeasible, so the need to introduce a set of algorithms, tools and platforms to manage and analyze them arises. Computational Proteomics regards the computational methods for analyzing spectra data in qualitative (i.e. peptide/protein identification in tandem mass spectrometry), and quantitative proteomics (i.e. protein expression in samples), as well as in biomarker discovery (i.e. the identification of a molecular signature of a disease directly from spectra). This chapter presents main standards, tools, and technologies for building scalable, reusable, and portable applications in this field. The chapter surveys available solutions for computational proteomics and includes a deep description of MS-Analyzer, a Grid-based software platform for the integrated management and analysis of spectra data. MS-Analyzer provides efficient spectra management through a specialized spectra database, and supports the semantic composition of pre-processing and data mining services to analyze spectra on the Grid.

INTRODUCTION

Proteomics is about the study of the proteins expressed in an organism or a cell. Computational Proteomics regards the computational methods, algorithms, databases, and methodologies used to manage, analyze and interpret the data produced in proteomics experiments (Cannataro, 2008).

The broad application of proteomics and the increasing resolution offered by technological platforms, especially in Mass Spectrometry-based high-throughput proteomics, make the analysis of proteomics experiments difficult and error prone without efficient algorithms and easy-to-use tools.

Mass Spectrometry-based proteomics (Aebersold and Mann, 2003) requires computational methods for analyzing data in qualitative (e.g. peptide/protein identification in tandem mass spectrometry) and in quantitative proteomics (e.g. protein expression in samples). Other important issues regard the way how spectra are cleaned, pre-processed, organized, and stored in an efficient way. The data heterogeneity introduced by several available platforms, as well as the need to conduct repeatable biomedical experiments on large populations (e.g. for outcome research), demand for suitable standards for the representation, storage, transmission, and sharing of proteomics data among different research centers.

Main research fields of computational proteomics are: (i) *qualitative proteomics*, i.e. the identification of proteins expressed in a cell (ii) *quantitative proteomics*, the determination of the abundance of protein expressed, and (iii) *biomarker discovery*, e.g. finding most discriminant features among spectra coming from two different populations (e.g. healthy samples versus diseased ones).

In summary, Computational Proteomics requires a combination of (i) recognized standards for data representation and exchange, (ii) efficient algorithms and tools constituting the building blocks of proteomics studies, (iii) computer science technologies enabling an easy composition of tools, sharing of data, and repeatability of experiments, (iv) large storage repositories for storing and sharing the ever increasing large amount of proteomics data, (v) high-performance implementations of data pre-processing, analysis and visualization algorithms.

After presenting requirements of Computational Proteomics applications, with a special focus on Mass Spectrometry-based Proteomics, the chapter describes main standards, tools, and technologies for building scalable, reusable, and portable applications in this field. Relevant distributed and Grid-based Computational Proteomics platforms are surveyed, including a deep description of MS-Analyzer, a Grid-based software platform for the integrated management and analysis of spectra data. MS-Analyzer provides efficient spectra management through a specialized spectra database, and supports the semantic composition of spectra pre-processing and data mining services to analyze spectra on the Grid

BACKGROUND

Recent developments in high throughput technologies have enabled the study of the cell on a large scale. The resulting scenario is characterized by the birth of new disciplines leading to the start of a new era, the so-called *omics* era. The suffix omics identifies a discipline that analyzes the relation of various biological compounds in their environment. For instance, proteomics is the study of the proteins expressed in a cell (Aebersold and Mann, 2003).

Mass Spectrometry (MS) has become a core technology in proteomics. The application of techniques based on mass spectrometry to analyze samples has a big impact in understanding both qualitative and quantitative distribution of proteins in a sample and finally to individuate potential biomarkers (Listgarten et al., 2005; Petricoin et al., 2002; Morris et al., 2005) .

Mass Spectrometry is a technique that is known since the early twentieth century. Through the growing understanding of electromagnetic forces, J.J. Thomson was able to separate the two neon isotopes of mass 20 and 22 with his parabola mass spectrograph in 1913. A mass spectrometer takes advantage of a fundamental phenomenon that is the force balance established when a charged specie in a gas phase travels through an electric or magnetic field (Roboz, 1968). The trajectory of this specie will depend on its mass-to-charge ratio (m/z). This is still today the working principle of all mass filters or mass analyzers. A mass spectrometer can be structured in different components as depicted in Figure 1:

- **A sample preparation phase** that consists of two optional blocks: sample fractionation and sample labeling. The first one separates a sample into smaller fractions and then a process known as chromatography separates these fractions in their constitutive elements, i.e. different proteins, on the basis of different biochemical properties. Finally samples can be optionally labeled using specific reagents or tags, (e.g. ICAT or SILAC labeling) enabling the quantitative measuring of them.
- **The ion source:** It brings the sample into ion phase.
- **The mass analyzer:** In this stage, the formed ions are separated on the basis of their mass-to-charge ratios.
- **The detector:** It measures a physical property of an ion related to its mass and turns it into an electrical signal.

Generally, all the instruments are integrated with a data acquisition system that stores in a digital form all the electrical signals generated from the detector. Currently, the existing instruments use different strategies to ionize molecules (for example, proteins), such as Matrix Assisted Laser Desorption Ionization (MALDI) or Electrospray Ionization (ESI) coupled to different mass analyzers, for example, time of flight (TOF). Mass spectrometry analysis can be organized in two major groups: qualitative and quantitative proteomics considering the characteristics of information extracted from technological platforms.

Spectra data can be analyzed for different aims such as biomarker discovery, (e.g. the identification of a signature of a disease), peptide/protein identification (i.e. the identification of proteins contained in a biological sample), peptide/protein quantification (i.e. the identification of the relative abundance of proteins). Such applications are described in the rest of the section, after a brief introduction to the main technological platforms for Computational Proteomics.

MALDI-TOF

Matrix-Assisted Lased Desorption/Ionization (MALDI) is an ionization technique that, as the name implies, uses a photon-absorbing matrix for the desorption and ionization of the analyte. In preparation for a MALDI analysis, the analyte is mixed with matrix and co-crystallized on an electrically conducting target, for example, a stainless steel plate. Then steel plate is placed in a vacuum

Figure 1. Constitutive blocks of a mass spectrometer

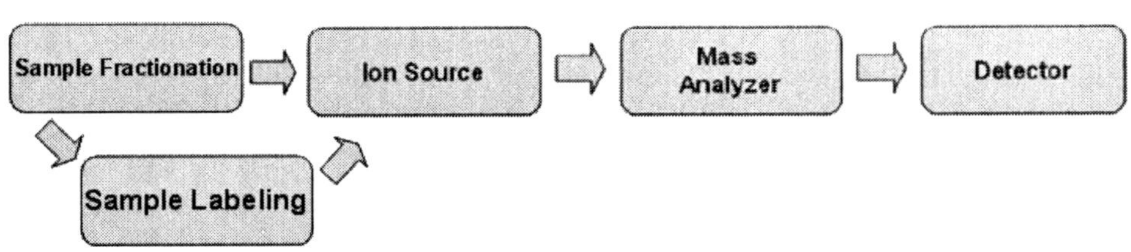

Figure 2. Mass spectrum of a biological sample

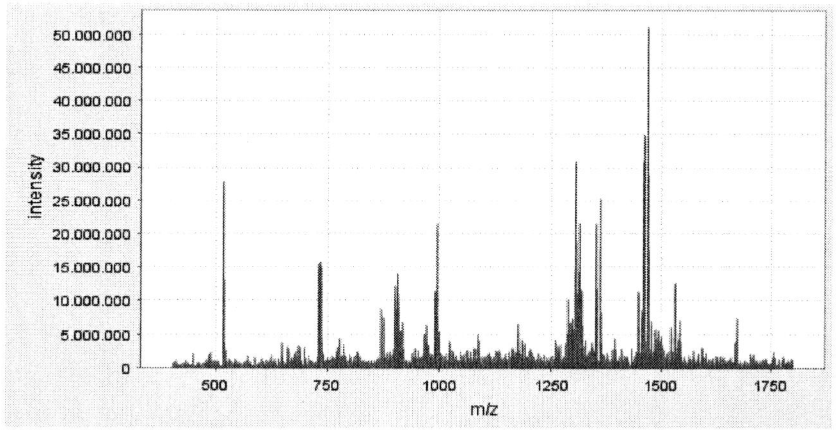

and energy, in the form of a short laser pulse, is applied to the sample and matrix crystal. The matrix will absorb the photon energy and a rapid heating will occur to create a micro-explosion. In this rapid process, the analyte is brought into gas phase and ionized.

The Time of Flight (TOF) mass analyzer measures the time it takes for the ions to fly from one end of the analyzer to the other one and strikes the detector. The flying speeds of ions are proportional to their mass-to-charge ratios. In an ideal case, all the ions with the same mass-to-charge ratio impact to the detector at the same time, and produce a single electrical pulse. Moreover, there is a substantial variability in the technology, so this technique can be utilized only for qualitative studies, (e.g. the individuation of a molecular signature), not for quantitative, (e.g. the determination of the proteins and their abundance in a sample).

MS instruments generate spectra that have peculiar characteristics with respect to the technological platform used. For instance, MALDI spectra consist of a huge number of couples (mass to charge ratio, intensity) as depicted in Figure 2. Each couple is related to the presence of a biological molecule that has a certain mass to charge ratio (m/Z) and abundance (intensity).

Separation and Labeling Techniques

Biological samples contain a mixture of different elements, for instance, a typical sample contains up to one thousand of different proteins. In order to better analyze these, a preliminary phase that separates proteins and introduces them one by one into mass analyzer is required. A set of well defined chromatographic techniques are commonly used to reduce the complexity of samples by fractionating their compounds on the basis of some characteristic. The term chromatography is used to indicate a set of laboratory techniques used to separates different substance in a mixture to enhance subsequent phases of analysis. Using chromatography researchers are able to separate a substance of interest, said analyte, from other substances.

A common separation technique is Liquid Chromatography in which molecules are inserted in to a liquid substance and then separated on the basis of their molecular weight. Liquid chromatography can be carried out either in a column or a plane. Present day liquid chromatography that generally utilizes very small packing particles and a relatively high pressure is referred to as high performance liquid chromatography (HPLC).

The Isotope Coded Affinity Tag (ICAT) LC-MS/MS (Gygi et al., 1999) is a labeling technique that allows analyzing the relative amounts of peptides/proteins in a sample with a respect to a control one by using a specific reagent, the ICAT reagent, produced by Applied Biosystems, and a successive tandem mass spectrometry. The resulting ICAT data provides the corresponding ratio for the level of expression of the parent protein in the control versus experimental sample.

Another common labeling technique is SILAC (Stable isotope labeling with amino acids in cell culture) (Ong et al., 2002) that is a mass spectrometry-based technique developed to improve the detection of differences in protein abundance between two (or more) samples (Ong et al., 2002). SILAC generally requires the use of two populations of cells that are tagged during their proliferation and growth in medium. One of those contains normal amminoacids, while the second one contains amminoacid labeled with stable non radioactive isotopes. Finally, when MS analyzes the proteins contained in cells coming from two different groups, pairs of chemically identical peptides of different stable-isotope composition can be differentiated in a mass spectrometer owing to their mass difference. The identification of correspondent peaks and of their ratio follows similar principles to those used in ICAT labeling.

LC-MS and LC-MS/MS

Liquid chromatography-mass spectrometry (LC-MS) (Politi et al., 2005) is an analytical chemistry technique that combines the physical separation capabilities of liquid chromatography with the mass analysis capabilities of mass spectrometry. The chromatographic column is used to separate proteins according to physical properties, and then proteins are inserted into the mass spectrometer at different times. With respect to MALDI-MS, it can provide a wealth of information on individual compounds even when present in complex mixtures, including identification by molecular mass, structure through fragmentation, and quantitation by using some specific protocols. The combination of two techniques improves analysis because LC both enables the analysis of single peptides and increases the efficiency of ionization phase resulting in an improvement of the analysis of single peptide.

Liquid Chromatography coupled to Tandem MS (LC-MS/MS) (Caas et al., 2006) uses two stages of mass analysis, the first one to pre-select an ion and the second one to analyze fragments induced, for instance, by collision with an inert gas like argon or helium. This dual analysis can be tandem in space or tandem in time. Tandem mass spectrometry is used to produce structural information about a compound by fragmenting specific sample ions inside the mass spectrometer and identifying the resulting fragment ions. This information can then be pieced together to generate structural information regarding the intact molecule. Tandem mass spectrometry also enables specific compounds to be detected in complex mixtures on account of their specific and characteristic fragmentation patterns. A tandem mass spectrometer is a mass spectrometer that has more than one analyzer, in practice usually two. On the other hand, LC/MS-MS uses two stages of mass analysis, thus produces a set of spectra for each sample, as depicted in Figure 3. MS/MS spectra may lead gigabyte of raw data that are often mined to extract qualitative or quantitative information, e.g. protein/peptide identification or determination of the abundance.

2D-Gel Electrophoresis

This method aims to separate and identify proteins contained in a sample by the displacement in a two dimensional Grid on the basis of two characteristics: (i) molecular weight (MW) and (ii) isoelectric point (pI). At the end of a 2D-gel experiments proteins are displaced on a grid on the basis of their properties and this displacement is recorded on a bi-dimensional image. Computational methods aim to enhance the image quality, to automatically extract relevant features from images and finally to compare different images.

Figure 3. An example of LC/MS-MS spectrum

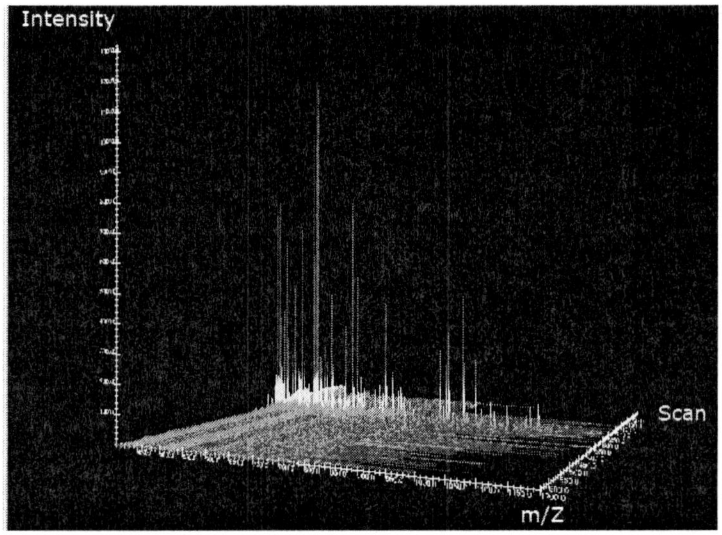

Compared to MALDI-MS, the analysis of protein using 2D-Gel Electrophoresis is a low-throughput technique. In proteomics, 2D gel electrophoresis is often used coupled to MS. In fact the former is used to isolate proteins, while the latter provides further characterization of proteins. In fact once the individual proteins are extracted from a 2D-gel sample, they can be more readily identified through the mass spectrometry data.

ANALYSIS OF MASS SPECTROMETRY DATA

Research focused on three main aspect for data analysis: (i) protein identification, i.e. the individuation of the proteins contained in a sample, (ii) protein quantification, i.e. the identification of the absolute or relative amount of different proteins, and (iii) biomarker discovery, i.e. the identification of a set of discriminant features in spectra useful to discriminate, for instance, healthy samples from diseased ones.

Protein Identification

A main objective of protein identification is the identification of the proteins contained in a sample. The early work of (Perkins, et. al 1999) showed the way to individuate the proteins contained in a sample starting from spectra. Actually, protein identification strategies using mass spectrometry are categorized in two main classes: (i) peptide mass fingerprinting (PMF) from MS and (ii) protein identification from peptide sequence information obtained by MS/MS, called peptide fragmentation fingerprinting (PFF). The methodologies belonging to the first class consist of a specific enzymatic cleavage of a protein followed by mass spectrometric measurement of the resulting peptides. So, the masses of generated peptides form a unique map in the resulting spectra called peptide mass fingerprinting. If these spectra are compared to a database of masses of theoretical peptides obtained by "*in silico*" digestion, then protein identification will be possible. The second strategy requires the acquisition of the information of the sequence of the peptide through Tandem-MS by selection, isolation and fragmentation of peptides (precursor ion). The fragmentation permits both the de novo sequencing, the identification of a protein not contained in a database of theoretical model, and the correlation of these data with sequence databases

Protein Quantification

Often, a key information in the explication of protein functionalities is not lied to their absence but in the modification of their relative abundance. For instance, a pathological case can be the result of a modification of the activation of a gene resulting in the synthesis of a greater or fewer quantity of the same protein. In such a way, the introduction of different techniques to infer the relative abundance, i.e. the abundance of a protein in a sample compared to a control one, can gain biological relevant information. Main algorithms in this area can be subdivided in two classes : (i) label-based methods, i.e. algorithms that mine data coming from labeled samples, and (ii) label-free algorithms that analyze spectra of non-labeled samples.

Biomarker Discovery

Early detection of cancer is currently a critical issue in medicine. In fact, it is fundamental for improving patient survival rates. Thus, the need for the discovery of novel biomarkers arises. Different works have proposed the use of MS to find possible novel biomarkers. These studies are based on various pre-processing algorithm to clean data and in the successive classification of data. Classification, also known as supervised learning, is a traditional data mining task that aims to find discriminant between data coming from different classes. Thus numerous classifiers based on MS data and multivariate statistics have been developed and applied in the analysis of mass spectra.

For instance, regarding prostate cancer, previous studies (Adam, et al. 2002; Yasui et al., 2003) have implemented various pre-processing algorithms and have applied different classification techniques for proposing biomarkers. The contribution of these studies lies in proposing biomarkers related to prostate cancer that discriminate between normal and cancerous cases and further between cases of high but different PSA values. For the ovarian cancer, analogously, many algorithms have been proposed.

STANDARDS FOR COMPUTATIONAL PROTEOMICS

Raw spectra generated both from MALDI-MS and LC-MS instruments are usually stored in flat files that are encoded in a proprietary format. The heterogeneity of available formats causes a set of issues such as data portability and experiment reproducibility that have been faced with the introduction of a standardized data format. Standard format can be used by different algorithms allowing an easy data exchange between geographically disseminated laboratories and can be stored in different databases for further use.

There are two main efforts toward the standardization of MS data format both based on XML. The first one, *mzXML*, is being developed at the Seattle Proteome Center of the Institute for Systems Biology (Pedrioli et al., 2004). The second one, *mzData* (Orchard et al., 2005), has been developed by the Proteomics Standard Initiative of the Human Proteome Organization. Currently there exists a new standard format, namely *mzML* (Martens et al., 2006) that merges together the two previous standards. Most mass spectrometers do not directly produce mzXML or mzData data, but there are several tools available that generate these files from native acquisition files. For instance, an open source project known as Sashimi offers a collection of converter programs for some common mass spectrometric file formats.

Common characteristics of these format are: (i) the usage of XML, (ii) the possibility to annotate spectra with information about instrument, laboratory and sample preparation, (iii) the compression of spectra (e.g. using Base64 format in mzData) with respect to the flat text format.

Another issue in spectra management is data dimension. File dimensions could span a broad range: from a few kilobytes to a few Gigabytes for each spectrum, depending both on the instrument type and parameters settings. Thus, the need to develop specialized techniques to manage huge data sets arises.

Finally, data produced by mass spectrometers are affected by errors and noise introduced during different experiment phases, e.g. sample preparation, and instrument distortion. Thus, another issue is the use and documentation of well-recognized, standard, pre-processing techniques (Yasui, 2003, Cannataro et al., 2007b).

DATABASES FOR COMPUTATIONAL PROTEOMICS

This section introduces the main repositories for spectra data produced through the experimental techniques introduced so far. Each repository is generally available through a web based interface and is freely downloadable. It is possible to induce a first categorization considering the kind of data stored distinguishing between databases storing data coming from Mass Spectrometry from those storing data from 2D-gel Electrophoresis.

Global Proteome Machine Database

The Global Proteome Machine Database[1] (gpmDB) (Craig et al., 2004) was constructed to utilize the information obtained by different servers included into Global Proteome Machine project (GPM), to validate peptide MS/MS spectra and protein coverage. GPM is a system for analyzing, storing, and validating proteomics information derived from tandem mass spectrometry. The system is based on a relational database, on different servers for data analysis, and on a user-friendly interface to retrieve and analyze data. This database has been integrated into GPM server pages.

The gpmDB data model is based on a modification of the HUPO-PSI Minimum Information About Proteomic Experiment (MIAPE) (Taylor et al., 2006) scheme. With respect to the proposed standard, the database is conceived to hold only the information needed in certain bioinformatics-related tasks such as sequence assignment validation. Data are mainly held in a set of XML files: the database serves as an index to those files. This combination of a relational database with XML is called by the authors XIAPE (Xml Information About a Proteomics Experiment). The system is available through a web interface and as a standalone application, allowing users to compare their experimental results with those that have been previously observed by other scientists.

PeptideAtlas

PeptideAtlas[2] (Desiere et al, 2004) is a database that aims to annotate the human genome with protein-level information. It contains data coming from identified peptides analyzed by liquid chromatography tandem mass spectrometry and thus mapped onto the genome. PeptideAtlas is not a simple repository for mass spectrometry experiments, but uses spectra as primary information source to annotate the genome, combining different information. Consequently, the population of this database involves two main phases: (i) a proteomic phase in which samples are analyzed through LC-MS/MS, and resulting spectra are mined to identify the contained peptides, (ii) an *in silico* phase in which peptides are processed by applying a bioinformatics pipeline and each peptide is used to annotate a genome. Resulting derived data, both genomics and proteomics, are stored in the PeptideAtlas database. Data submitted by the researchers are organized in a relational model.

PeptideAtlas is based on the Systems Biology Experiment Analysis Management System (SBEAMS) Project, which is a framework for collecting, storing, and accessing data produced by a variety of experiments. It combines a relational database management system (RDBMS) back end providing integrated access to remote data sources. The user can query data through a web interface or download whole datasets organized by the original publications.

PRIDE

The PRoteomics IDEntifications database[3] (PRIDE) (Martens et al., 2005) is a database of protein and peptide identifications that may be annotated with supporting mass spectra. PRIDE stores information about a complete proteomic experiment, starting from the title and a brief description of the experiment itself. Each experiment is annotated by a description of the sample under analysis and the instrumentation used to perform the analysis. The core elements of each entry are the protein identifications sorted by unique accession numbers and supported by a corresponding list of one or more peptide identifications. Optionally, an entry can contain a reference to any submitted mass spectra that form the evidence for the peptide identification, encoded in the versatile PSI mzData format. Users can directly submit protein and peptide identification data to be published in peer-reviewed publications using the PRIDE 2.1 XML schema.

SpecDB

SpecDB (Cannataro et al., 2007d) is a framework able to collect and manage spectra data in a Grid-based environment composed of several laboratories (nodes), where each laboratory run its local database. Each node contains an application programming interface and a data storage service, SpecDB, hosting spectra. System stores spectra in three different formats:

- **Raw files:** Original data produced by the MS instrument and encoded in its native format;
- **Tuples in a relational database:** Raw spectra are easily translated into a relational schema that enables the annotation with meta information about instrument, experiment, results and clinical/biological information;
- **mzData files:** that encodes both spectra and annotation in a single XML document, allowing a compression of spectra.

SpecDB can be used in a distributed environment where a set of SpecDB nodes, each one associated to a laboratory, form a comprehensive Grid environment built on top of the Globus Toolkit. Each node of the Grid runs: (i) a data service, called SpecDB, storing data produced in that laboratory; (ii) a query service that enables the searching among nodes; and (iii) a data processing service allowing data pre-processing or analysis using both local and remote data. Thus, each laboratory node stores and publishes spectra originated from its instruments and may access to both local and remote data. Access to the Grid by both nodes (laboratories) and clients is managed by using the security mechanism offered by the Globus infrastructure.

SWISS-2D PAGE

Databases produced by using gel electrophoresis generally store the metadata of the experiment, as well as gel images and protein identifications. Metadata are relative to the parameters of the experiment whereas core data store the proteins contained in the associated image. The SWISS-2DPAGE[4] (Hoogland et al 1998) database is maintained collaboratively by the Swiss Institute of Bioinformatics (SIB) and the Biomedical Proteomics Research Group (BPRG) of the Geneva University Hospital. Current content includes identified spots and corresponding protein entries in 36 reference maps from *Human, Mouse, Arabidopsis thaliana, Dictyostelium discoideum, Escherichia coli, Saccharomyces cerevisiae* and *Staphylococcus aureus origins.*

The protein entries in SWISS-2DPAGE are structured text files. Each entry is composed of defined lines that are used to record various kinds of data. For standardization purposes, the format of SWISS-2DPAGE entries is similar to the SWISS-PROT database, in addition to specific lines dedicated to the 2-DPAGE data: (i) the master line (MT) lists the reference maps where the entry has been identified, (ii) the images line (IM) lists the 2-D PAGE images available for the entry, (iii) the 2D lines group different topics including the mapping procedure, spot co-ordinates, protein amino acid composition, protein expression levels and modifications.

SWISS-2DPAGE is available through the ExPASy molecular biology server. The SWISS-2DPAGE top page provides text searches, and displays results with active links to other databases. It is also possible to get a local copy of SWISS-2DPAGE via ftp from the ExPASy FTP server. On the ExPASy Web server, the data image associated with a protein entry displays the experimental location of the protein on the chosen map, in addition to a theoretical region computed from the protein sequence.

TOOLS FOR COMPUTATIONAL PROTEOMICS

Tools presented in this section implements algorithms belonging to one of research area so far discussed. In such a way it is possible to discriminate between tools that search proteins in spectra, tools that try to identify the abundance of protein and finally complex tools that aims to identify the molecular signature of diseases.

Tools for Peptide/Protein Identification

MASCOT

The MASCOT[5] (Perkins et al., 1999) protein identification tool performs the protein identification through PMF and scores its results using a scoring based on the MOWSE algorithm. It compares the spectra generated from the user against the theoretical spectra. These spectra are obtained by an enzymatic digestion of each protein sequence in a reference database. In this way for each observed peak in a spectrum the system can calculate the probability that a match between a peak and a peptide is purely random.

Aldente

Aldente[6] (Toulup et al., 2003) is a PMF based protein identification system comprised in a suite of bioinformatics tools available at the ExPASy Proteomics Server. It allows the user to specify a set of constraints in order to restrict the dimensions of the search space. Moreover, differently from most other PMF packages, Aldente enables the user to define the parameters that contribute to the final score and their weights. After the identification, the tool calculates a threshold for identification by processing random sequences in the same parent mass range. The random sequence with the highest score becomes the threshold above which a protein is said to be identified.

X!Tandem

X!Tandem[7] (Craig et al., 2004) is an open source tool able to match tandem mass spectra with peptide sequences performing the PMF protein identification. X!Tandem takes in input the spectrum and an XML files describing the identification parameters and the search database, then it produces an XML files containing the identified proteins and the associated score.

swissPIT (Swiss Protein Identification Toolbox)

The swissPIT[8] (Swiss Protein Identification Toolbox) (Quandt et al., 2008) is a project that uses the Swiss BioGRID infrastructure to run workflows of analysis of mass spectrometry data. The rationale of swissPIT is to use different peptide identification tools as workflow's elements. In this way a user can automatically submit his/her experimental data to multiple analysis tools to improve the identification of proteins from spectra. A single input data can be submitted to the same tool and then analyzed with different parameter settings or can be submitted to different tools. Clearly, the identification of best setting parameters and tools may require the combinations of several executions of different tools employing different parameter settings. So as to be able to perform these calculations within a reasonable amount of time, a high-throughput computing infrastructure has to be used. The parallel execution of these tools requires the use of a high throughput computing infrastructure.

The swissPIT project uses the Swiss BioGRID (http://www.swissbiogrid.org) that brings together the resources of five partners: the Swiss Institute of Bioinformatics (SIB), the Swiss National Supercomputing Centre (CSCS), the Biozentrum at Basel University, the Friedrich Miescher Institute in Basel and the Functional Genomics Centre Zurich. The Grid uses the NorduGrid Advanced Resource Connector ARC middleware (http://www.nordugrid.org/) to manage the submission of jobs for swissPIT. Users can submit their jobs through a web-based front end.

Swiss Tandem

Swiss Tandem (Arlnond et al., 2007) is a Grid-based application service for the analysis of tandem mass spectra. It is based on the X!Tandem open-source algorithm and uses a PC Grid, so it exploits idle CPUs of desktop PCs to perform tasks. In particular, a submitted job is a protein identification from MS spectra using X!Tandem. After the completion of the job, results are collected in a web server and then are accessible to the users. Moreover, the server offers visualization capabilities and a set of tools allowing further evaluation and interpretation.

EIPEPTIDI

As introduced in a previous section, the ICAT method allows the identification and relative quantification of proteins present in two samples. Nevertheless, when the ICAT is coupled to LC-MS/MS the whole pipeline of analysis suffers of a main drawback: the insufficient sample-to-sample reproducibility on peptide identification. This problem results, in particular, in a low number of peptides commonly identified in different experiments. EIPeptiDi (Cannataro et al., 2007e) is a tool that may increase the number of peptides identified in a set of samples, thus improving the identification process. The tool implements a method for improving peptide identification using the cross validation of MS/MS results. The EIPeptiDi tool is available at http://bioingegneria.unicz.it/~veltri/projects/eipeptidi/ with a demo data set.

Tools for Peptide/Protein Quantitation

While the process of identifying the sequence of a protein in a compound is named protein identification, the process of measuring the (relative) quantities of such proteins is named protein quantitation.

The need to analyze the abundance of the proteins in a proteome has led the introduction of different technological platforms and subsequent computational methods. Currently, there exist two main approaches: (i) the labeling of samples using different markers, (e.g. ICAT, or SILAC), and (ii) label-free quantitation, **i.e. without using differential labels.**

The labeling strategies have a common workflow: specific reagents, referred as tags, are conjugated to the peptides such that properties of the tagged peptides, i.e. the masses, are different between two samples (e.g., in diseased patients versus healthy patients). Thus subsequent analysis can identify the corresponding peptides in two samples and reveal the relative abundance of a peptide. Main differences between labeling methods are the biochemical characteristics of the tag and the relative computational methods.

Conversely, label-free quantitation methods can be grouped in two main classes: (i) spectra comparison methods, and (ii) spectral counting methods. Methods belonging to the first class are based on the evaluation and comparison of the area under the different peaks within a MS scan. Differently, methods belonging to the second class perform a so called spectral counting, i.e. they tabulate the number of MS/MS scans that are attributed to the same precursor ion (i.e., peptide) and analyze the frequency of these MS/MS scans associating this value to the abundance of that peptide in the sample.

Label-Based Methods

The Automated Statistical Analysis on Protein Ratio (ASAPRatio) (Li et al., 2003) method calculates the relative abundances of proteins from ICAT-data. The software initially pre-processes spectra using a smoothing filter, then subtracts background noise from each spectrum, and finally calculates the abundance of each couple of corresponding peptides in spectra. At the end of this phase ASAPRatio groups all the peptides that belong to the same protein calculating for each one the relative abundance and finally the abundance of the whole protein. The software finally evaluates all the obtained results using statistical tests in order to show the confidence for each result, e.g. the confidence associated to each protein.

The XPRESS (Han et al., 2001) software calculates the relative abundance of proteins, from an ICAT-labeled pair of samples, by reconstructing the light and heavy elution profiles of the precursor ions and determining the elution areas of each peak.

ProICAT[9] is a commercial software provided by Applied Biosystems that performs identification and quantification of proteins analyzed through ICAT-based LC-MS/MS. ProICAT uses a 3-dimensional reconstruction algorithm to locate and accurately determine *experimental:control* (*heavy:light*) peak ratios in complex proteomic samples.

Label-Free Methods

As introduced, label-free methods implement two main strategies: spectra comparison and spectral counting. Example of methods in the first class are MSQuant (Schulze e Mann 2003), XCMS (Smith et al., 2006), and OpenMS (Kohlbacher et al., 2007) . MSQuant processes LC-MS spectra investigating deeply LC runs. Trough automated investigation it performs a spectra comparison and quantitation of proteins and peptides. A system for the identification of peptide based on spectral counting tool is presented in (Gao et al.,2003). Moreover, methods that count the number of peptides, such as PeptideProphet (Keller et al., 2002), can be adapted to perform spectral counting. In fact these methods perform the counting of peptides and these values can be used to determine the frequency of peptides and their abundance. The interested reader can find a more detailed discussion of these methods in (Wong et al., 2008).

Tools for Biomarker Discovery

Biomarker Discovery in spectra data aims to find a sub set of features, i.e. peaks or peptide/proteins, sufficient to discriminate a set of spectra in classes (usually two classes), e.g. to differentiate spectra from healthy and diseased subjects. A main issue of this field is the huge number of features, i.e. peaks or peptides, with respect to the small number of available samples. The rest of the section describes some important proteomic pipelines used for managing and analysing spectra data with a special focus on biomarker discovery.

BioDCV

BioDCV[10] (Barla et al., 2008) is a distributed computing framework composed of a suite of modules enabling the researchers to plan and execute an analysis of both proteomics and genomics data. The BioDCV system uses the high computational capabilities of Grid environment to select the discriminant features of a set of spectra building the correspondent predictive model. BioDCV faces the selection bias problem, i.e. the individuation of the correct set of discriminating features, using a learning process coupled to a validation method. The system was originally conceived for the complete validation of class prediction of gene expression data. It was based on the use of Support Vector Machines to avoid selection bias effects. Successively authors used main concepts of BioDCV to build a general analysis protocol for proteomic data. Currently, BioDCV can be used to setup a complete experiment of predictive biomarker discovery on proteomics data.

Proteios

PROTEIOS[11] (Gärdér et al., 2005) is a system based on the client/server architecture for the management of both MS and 2D-gel data. It has been designed to manage the whole workflow of the production of both mass spectra data and 2D-gel experiments. PROTEIOS, in fact manages not only raw spectra data and images but yet information about biomaterials and analysis. It allows the enhancement of analysis capabilities through plug-in in order to support classical task such as protein identifications. All the data stored in PROTEIOS are annotated with the information relevant to an experiment, e.g. experimenter, processing and identified proteins. User can upload data using both raw format and one of the existing standards. Once uploaded, data can be shared with a community of researcher or made private. PROTEIOS can be currently used through a client that runs on the researcher workstation.

Trans Proteomic Pipeline

The Trans Proteomic Pipeline[12] is a comprehensive suite of programs, maintained by the Seattle Institute of Systems Biology, related to the pipeline of proteomic data analysis. In this way, the pipeline contains programs for: (i) spectra handling, i.e. interfacing with the different binary formats and translating them into a standard format, (ii) probabilistic peptide identification, (iii) protein quantification, (iv) result validation, i.e. to avoid identification bias, and (v) biological interpretation of results on a systems scale.

myGrid

myGrid[13] (Stevens et al., 2004) is a Grid-based toolkit for composing and executing in silico experiments. myGrid initially focused on bioinformatics applications, but currently users can define their own web services without any restriction on the type of science. myGrid allows the user to assembly tools provided by the Taverna workbench. In such a way user can build and execute workflows that execute web services and legacy applications incorporated using wrapper tools. Taverna (Oinn et al., 2004) is an open-source workflow management system based on a workflow language, called Scufl (Simple Conceptual Unified Flow Language), and on a graphical interface to easily build workflows that bring together different services.

ProGenGrid

The Proteomic Genomic Grid[14] (ProGenGrid) (Aloisio et al., 2005) is a project that aims to deploy different bioinformatics tools over the Grid. It is accessed through a Grid portal that uses the well known GRB (Grid Resource Broker) middleware, a system used in the Globus architecture to hide the Grid middleware details and providing a standardized interface to Grid resources. ProGenGrid provides several Grid-enabled bioinformatics tools for sequence alignment, secondary and tertiary structure prediction. Moreover, such tools can be combined by using the provided workflow engine. In this way the researcher can build a Grid-based bioinformatics experiment without facing with the Grid complexity.

MS-ANALYZER: AN ONTOLOGY-BASED SYSTEM FOR COMPUTATIONAL PROTEOMICS

As introduced in previous sections, computational proteomics is a multidisciplinary area that involves: (i) the planning and the execution of several experiments, (ii) the organization and the storage of collected data in specialized databases, and (iii) the extraction of knowledge from these databases. The resulting pipeline of data analysis, from samples to knowledge, involves different types of expertises and different people.

Such a paradigm of computation which brings together different heterogeneous activities, with a coarse granularity, is known as workflow (Lawrence, 1997). A workflow is a partial or total automation of a process, in which a collection of activities must be executed according to certain procedural rules. Workflows have been introduced in the business environment to integrate enterprises but have found a very broad field of application in science, aiming to be used in problem-solving environments. Workflow Management Systems (WfMSs) allow the design of a workflow, supporting its enactment, by scheduling different activities on available entities (Yu and Buyya, 2005).

MS-Analyzer (Cannataro, Veltri 2007) is a tool for the management and analysis of proteomics data that takes into account such requirements. MS-Analyzer uses ontologies to model the proteomics domain and its data analysis methodologies, and a novel application designer based on workflows to design proteomics experiments, i.e. workflows of activities involving different bioinformatics tools (e.g. pre-processing, data mining, visalization).

In particular, MS-Analyzer provides an ontology-based workflow editor where the user, by browsing the ontology, can choose the pre-processing and analysis tools, compose them using simple *drag&drop* functions, and finally execute them on the Internet or over the Grid. Using MS-Analyzer, the user can design different analysis experiments where pre-processing is a new design variable, whose effectiveness can be evaluated not only in terms of memory occupations and performance (e.g. binned vs. not-binned spectra), but also in terms of data analysis quality (e.g. predictivity of the data mining classification with respect to pre-processing type).

The designed tool sits in the middle of proteomic facilities and advanced data analysis software tools, thus the main requirements of MS-Analyzer are: interfacing with the proteomics facility; storing and managing MS proteomics data; and interfacing with off-the-shelf data mining software platforms. In particular, MS-Analyzer provides the following logical layers and functions: (i) Data Acquisition and Management, (ii) Data Pre-processing and Preparation, (iii) Enhanced Composition of Bioinformatics Tools.

Data Acquisition and Management. This layer collects data coming from MS and stores them in a Spectra Repository database. There exist different spectrometers (e.g. MALDI-TOF, LC, ICAT), and different data formats, even for the same instrument. The Spectra Repository database has to manage these different formats and must to be able to export spectra in one of the different existing standards such as mzData, mz-XML, or the recent mzML. Nevertheless, the data storage system has to efficiently support the pre-processing and analysis phases. Each of these requires a different portion of data, for example a single spectrum, a set of spectra or a window of one or more spectra. Consequently, MS-Analyzer provides a set of API to retrieve these data efficiently. The typical application in this field comprises several components that process large data sets. They are often divided in smaller sets and are processed concurrently in different locations or merged in larger sets to be processed sequentially. MS-Analyzer takes care of moving these data reassembling them in the final stage, also ensuring semantic consistency. Such functions have been implemented in a specialized spectra database named SpecDB (Cannataro et al., 2007d) described in a previous section.

Data Pre-processing and Preparation. This phase is undoubtedly crucial for the whole pipeline of analysis. MS-Analyzer, starting from the data stored in SpecDB, executes different pre-processing techniques from those that are made available by the system and classified by the ontology. This phase aims to filter the noise present in the spectra. Nevertheless it has to prepare data to be given in input to different kind of data mining tools. For instance Weka, a popular data mining suite (Witten and Frank 2005), requires that MS spectra being organized in a unique file containing a metadata header.

Enhanced Composition of Bioinformatics Tools. The pre-processing and analysis activities can be structured autonomously as steps of a whole pipeline. This structure is usually defined as a workflow which involves the invocation of many tools according to a logical and temporal order. The easiest way to compose a workflow, also known as composition from scratch, requires that the user develops ad hoc scripts to handle data movement and has to manually check the logical soundness of the generated application. This solution tends to be impracticable when the size of the workflow increases. From this scenario, the need for a system that assists users in the creation, soundness check and management of workflow arises. In order to satisfy these requirements, MS-Analyzer uses ontologies to enhance and simplify the composition of workflows and to check constraints about tools composition (Cannataro et al., 2007b; Cannataro et al., 2006).

MS-Analyzer Architecture

The architecture of MS-Analyzer comprises two main modules: (i) the Proteomics Data Storage System, and (ii) the Ontology-based Workflow Editor. The data storage system of MS-Analyzer, named SpecDB, implements basic spectra management functions. Data stored in such a system are organized in three different repositories on the basis of the processing phase (raw, pre-processed, prepared) where they are involved, so it includes three main repositories: the **Raw Spectra Repository** (RSR), the **Pre-processed Spectra Repository** (PSR), the **Pre-processed and Prepared Spectra Repository** (PPSR), as well as a **Metadata repository.** The first one (RSR) stores the raw spectra coming from different experiments and instruments. The second one (PSR) stores the spectra that have been pre-processed. It stores both spectra and information about the applied algorithms. The third one (PPSR) stores spectra that have been pre-processed and are ready to be analyzed by data mining platforms, so that it stores spectra in a format readable by data mining tools such as Weka.

Ontology-Based Workflow Editor

The main idea underlying the workflow editor of MS-Analyzer is to take advantage of ontologies to support the specification and annotation of bioinformatics workflows. MS-Analyzer faces the problem of building scientific workflows, under the assumption that they are executed in a distributed (laboratory) environment. Each activity within such a workflow can be executed either by invocating a web service or another (sub) workflow. For these aims MS-Analyzer has an Ontology-based Workflow Editor that comprises a Dataset Manager, an Ontology Manager, and a Graphical Editor as depicted in Figure 4.

The Dataset Manager allows to combine in a logical datasets raw spectra provided bt the mass spectrometers, associating them information about the instruments and classes (e.g. diseases vs healthy). Using such tool the user can choose the input dataset where to perform pre-processing, preparation and data mining analysis.

The Ontology Manager allows for the browsing, searching and selection of bioinformatics tools and it is based on two ontologies: **ProtOntology,** an ontology for mass-spectrometry proteomics, and **WekaOntology,** an ontology for the data mining domain specialized to model the tools of the popular WEKA suite (Witten and Frank 2005).While ProtOntology contains a high level description of concept and tools useful to model a proteomic application, WekaOntology covers in detail all the concept related to data mining analysis of proteomic data.

When a user has to design a novel data analysis experiment, he/she starts by browsing the activities performed by the tasks modeled by ProtOntology, thus he/she finds suitable pre-processing algorithms. Then, he/she searches and finds in WekaOntology all the details about such algorithms, such as: (i) a reference to the Web Service Description Language (WSDL) file, (ii) a short description, (iii) preconditions, and (iv) post-conditions.

The first one is a file that contains all the information needed to invoke the implementations of algorithms as a web service. The second one holds a short description of the algorithm explaining its main characteristics (e.g. functionalities, field of application, or their rationale). The third one contains a pre-condition for the algorithms, expressed through the relation (receives in input). This precondition is semantically related to the kind of data that the service has to receive. As introduced before we make a distinction of the data in three main classes: (i) raw, (ii) pre-processed and (iii) prepared data. All the pre-processing algorithms take as input a raw or pre-processed datum and produce as output a pre-processed datum.

Usually data mining algorithms need a prepared data (i.e., a datum that is encoded in a format suitable for the data mining algorithm), that requires the application of further data conversion tools. These requirements have been translated in concepts of the ontology and then been linked using non taxonomic relations. Finally, a post-condition (produces as output) specifies what kind of output the algorithm produces. These last conditions are very important in the workflow composition phase because, during this step, for each service added to the workflow, MS-Analyzer automatically checks the soundness of the juxtaposition, avoiding any inconsistency.

For example, let us consider two cases: (i) the wrong juxtaposition of a pre-processing software (e.g. peak extraction) followed by a data mining software (e.g. J48) and (ii) the wrong juxtaposition of a pre-processing software after a data preparation software. In the first case the ontology manager of MS-Analyzer finds the mismatch between the properties *receives in input* of J48, linked

Figure 4. Architecture of MS-Analyzer

to *prepared data* and *produces as output* of the pre-processing one that is linked to *pre-processed data*. In the second case the ontology manager finds the mismatch between the properties *receives in input* of pre-processing software, linked to *raw data* and *produces as output* of the data preparing software that is linked to *prepared data*.

The Graphical Editor (see Figure 5), allows users to design bioinformatics workflows by using a notation based on the Unified Modeling Language (UML), providing basic control blocks of workflows (e.g., start/end, fork/join). The user can drag and drop the fundamental blocks, such as bioinformatics tools and data sources, and then can associate each block with a specific data source or a bioinformatics tool selected, respectively, through the Dataset Manager or the Ontology Manager. For example, the user first inserts a blank (empty) service block in the work space, then he/she retrieves in the ontology an available tool (e.g., binning), and then associates it to the empty block.

FUTURE TRENDS

Future trends are related to the realization of the "genomic medicine" concept, where genomics, proteomics and clinical data will be used together to enable translational medicine, aiming to apply research to clinical practice and improve health-care. The semantic integration and further analysis of genomic, proteomics, and clinical data is a first step towards the deployment of novel biomedical applications involving research-oriented competence centers, specialized proteomics facilities, and health centers where clinical guidelines are applied.

In order to make this possible, the following future trends can be envisioned for computational proteomics:

1. **Development of federated proteomics databases** enabling the automatic and easy sharing of proteomics data. The integration of such data will be based on a large adoption of a common accepted standard for data sharing

Figure 5. The MS-Analyzer graphical user interface

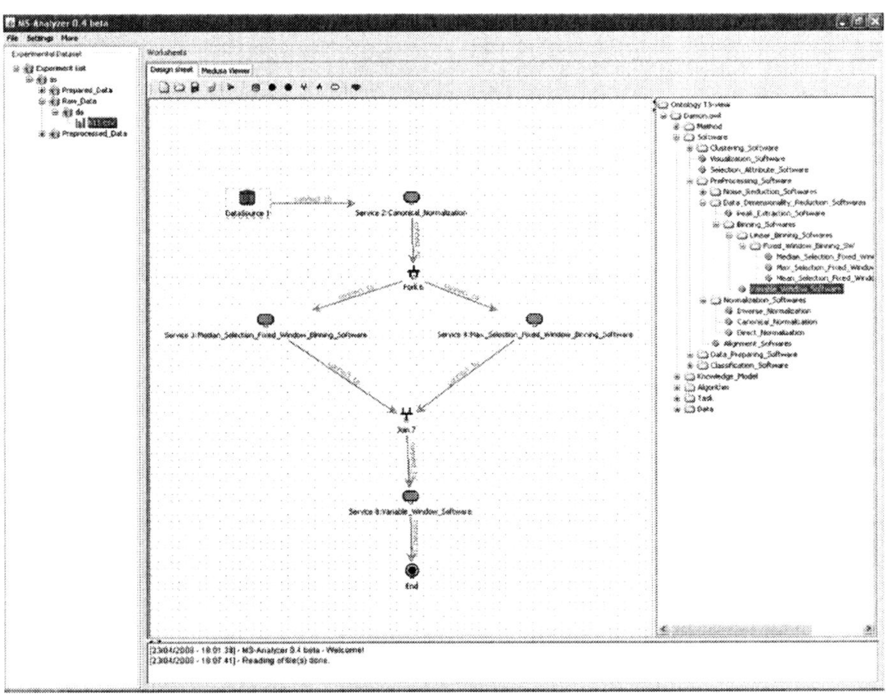

and annotation including controlled vocabularies. mzML will certainly play a central role in such task (Orchard and Hermjakob, 2008). Nevertheless the semantic annotation of data and analysis through ontologies, describing for instance the pre-processing and the various phases of the applied proteomics pipeline, could help researchers to plan and execute different analysis.

2. **Enhancement of protein identification and quantitation.** The availability of larger datasets of proteomics data (e.g. see the emerging PRIDE and PeptideAtlas initiatives) will enable the development of comprehensive validation tools for the improvement of protein identification and quantifications. In fact, all the validation methods reside on the comparison of results against a database of already verified ones. Existing tools, like EIPEPTIDI, could compare spectra data against larger spectra and peptides datasets, improving the identification and quantification performances. The sharing of a larger amount of data and results will be a key point in this area.
3. **Explicit management of quality in proteomics.** Computational proteomics encompasses many phases and the presence of low quality data may prevent the reuse of data for experiment validation. The explicit management of information quality of proteomics experiments and proteomics data repositories will be an important issue in the future (Stead et al, 2008) as well as the interpretation of proteomics data on a systems level. The integration and interpretation of proteomics and genomics data (e.g. microarray) with the methodologies provided by systems biology will be more and more important and is still in its beginning. For example, a promising research field will regard the study of the abundance of the proteins, their functions and the metabolic pathways where they act.

CONCLUSION

Mass spectrometry-based proteomics is the main technological platform for producing proteomics data, and is currently used for qualitative and quantitative proteomics, as well as for biomarker discovery. Computational Proteomics regards the data, algorithms, and databases used to manage and analyze the data produced in proteomics experiments, with a special focus on mass spectrometry-based proteomics.

Computational Proteomics studies how spectra data are cleaned, pre-processed, organized, and mined in an efficient way. To achieve its goals, computational proteomics requires a combination of (i) standards for data representation and exchange, (ii) algorithms for data cleaning and analysis, (iii) technologies enabling an easy composition of such tools offering yet high-performance.

The chapter introduced technologies of Mass Spectrometry-based Proteomics and described requirements for Computational Proteomics, including main standards, tools, and technologies for building scalable, reusable, and portable applications in this field. After describing main standards and databases, the chapter surveyed relevant distributed and Grid-based Computational Proteomics platforms, including a deep description of MS-Analyzer, a Grid-based software platform for the integrated management and analysis of spectra data. MS-Analyzer provides efficient spectra management through a specialized spectra database, and supports the semantic composition of spectra pre-processing services and data mining services to analyze spectra on the Grid.

REFERENCES

Aebersold, R., & Mann, M., (2003). Mass spectrometry-based proteomics. *Nature, 422*(1), 198–207.

G., Cafaro, M., Fiore, S., & Mirto, M. (2005). ProGenGrid: A Workflow Service Infrastructure for Composing and Executing Bioinformatics Grid Services. In A. Tsymbal & P. Cunningham (Eds.), *18th IEEE Symposium on Computer Based Medical Systems* (Vol. 1) (pp. 555-560). New York: IEEE Press

Arnold K., Schwede, T., & Podvinec, M. (2007). SWISS-TANDEM: A Web-based Workspace for MS/MS Protein Identification on PC Grids. In Dominique Zosso, (Ed.) *20th IEEE International Symposium on Computer-Based Medical Systems* (Vol. 1) (pp 406-411). New York: IEEE press.

Adam, B.L Qu, Y.. Davis, J.W. Ward, M.D Clements, M.A Cazares, et al. (2002). Serum Protein Fingerprinting Coupled with a Pattern-matching Algorithm Distinguishes Prostate Cancer from Benign Prostate Hyperplasia and Healthy Men. *Cancer Research, 62*(13), 3609–3614.

Barla, G. Jurman, S. Riccadonna, M. Chierici, S. Merler, & Furlanello,C. (2008). Machine learning methods for predictive proteomics. *Briefings in Bioinformatics*, *9*(2), 119-128.

Caas, D. Lpez-Ferrer, A. Ramos-Fernndez, E. Camafeita, & E. Calvo. (2006). Mass spectrometry technologies for Proteomics. *Briefing in Functional Genomics and Proteomics, 4*(4), 295–320.

M Cannataro (2008). Computational proteomics: management and analysis of proteomics data. *Briefings in Bioinformatics, 9*(2), 9-10.

Cannataro, M Barla, A. Flor, R. Jurman, G. Merler, S. Paoli, S. Tradigo, G. Veltri, P & Furlanello, C (2007). A grid environment for high-throughput proteomics. *IEEE Transaction on NanoBiosciences*, *6*(2),117-123.

Cannataro, M., Cuda, G., Gaspari, M., Greco, S., Tradigo, G., & Veltri, P. (2007). The EIPeptiDi Tool: Enhancing Peptides Discovering in ICAT-based LC MS/MS Experiments. *BMC Bioinformatics, 8*(255).

Cannataro, M., Guzzi, P.H., Mazza, T., Tradigo, G., & Veltri, P. (2007). Using Ontologies for Preprocessing and Mining Spectra Data on the Grid. *Future Generation Computer Systems, 23*(1), 55-60

Cannataro, M & Veltri, P. (2007). MS-Analyzer: Composing and Executing Preprocessing and Data Mining Services for Proteomics Applications on the Grid. *Concurrency and Computation: Practice and Experience, 19*(15), 2047-2066,

Cannataro, M Tradigo, G. & Veltri, P. (2007). Sharing Mass Spectrometry Data in a Grid-based Distributed Proteomics Laboratory. *Information Processing & Management, 43*(3), 577-591

Cannataro, M., Guzzi, P.H., Mazza, T., Tradigo, G., & Veltri, P. (2006). Managing Ontologies for Grid Computing, Multiagent and Grid Systems, *2*(1), pp. 29-44.

Craig, R., Cortens, J.P., & . Beavis, R.C, (2004). Open source system for analyzing, validating, and storing protein identification data. *Journal of Proteome Research* *3*(6) 1234–1242.

Desiere, F., Deutsch, E., Nesvizhskii, A., Mallick, P., King, N., Eng, J., et al. (2004). Integration with the human genome of peptide sequences obtained by high-throughput mass spectrometry. *Genome Biology*, *6*(1).

Hoogland, C., Sanchez, J.C., Tonella, L., Bairoch, A., Hochstrasser, D.F., & Appel, R.D.(1998). Current status of the SWISS-2DPAGE database, Nucleic. *Acids Research 26*(1), 332–333

Gao, J., Opiteck, G. J., Friedrichs, M. S. & Haefta, S.. (2003). Changes in the Protein Expression of Yeast as a Function of Carbon Source. *Journal of Proteome Research,* 643-649.

Gärdén , P., Alm, R., & Häkkinen, J. (2005). PROTEIOS: an open source proteomics initiative. *Bioinformatics, 21*(9), 2085-2087.

Gygi, S.P., Rist, B., Gerber, S.A., Turecek, F., Gelb, M.H., & Aebersold, R. (1999). Quantitative analysis of complex protein mixtures using isotope-coded affinity tags. *Nature Biotechnology, 17*(10), 994–999

Han, D.K., Eng, J., Zhou. H. & Aebersold, R. (2001). Quantitative profiling of differentiation-induced microsomal proteins using isotope-coded affinity tags and mass spectrometry. *Nature Biotechnology, 19*, 946-51.

Jeffries N.(2005). Algorithms for alignment of mass spectrometry proteomic data. *Bioinformatics, 21*(14), 3066–3073.

Keller, A., Nesvizhskii, A. I., Kolker, E., & Aebersold R., (2002). Empirical statistical model to estimate the accuracy of peptide identifications made by MS/MS and database search. *Analytical. Chemistry, 74*(20), 5383-92

Kohlbacher, O., Reinert, K., Gröpl, C., Lange, E., Pfeifer, N., Schulz-Trieglaff, O., & Sturm, M. (2007). TOPP - The OpenMS Proteomics Pipeline. *Bioinformatics,23*(1), 191-197.

Lawrence, P. (1997). *Workflow handbook*. New York: John Wiley & Sons.

Li, X.-J. Zhang, H. Ranish, J.R. & Aebersold, R. (2003). Automated statistical analysis of protein abundance ratios from data generated by stable isotope dilution and tandem mass spectrometry. *Analytical Chemistry, 75*(23), 6648-57

Listgarten, J. & Emili, A. (2005). Statistical and computational methods for comparative proteomic profiling using liquid chromatography-tandem mass spectrometry. *Molecular Cellular Proteomics, 4*, 419–434.

Martens, L., Hermjakob, H. Jones, P. Adamski,M. Taylor, C. States, D. Gevaert, K. Vandekerckhove, J. & Apweiler, R. (2005). Pride: the proteomics identifications database. *Proteomics, 5*(13), 3537-3545

Jeffrey, S., Morris, M., Kevin, R., Coombes, J., Koomen, K., Baggerly, K., & Kobayashi, R. (2005). Feature extraction and quantification for mass spectrometry in biomedical applications using the mean spectrum. *Bioinformatics, 21*(9) 1764–1775.

Oinn, T.M., Addis, M. Ferris, J., Marvin, D. Greenwood, R.M. Carver, T. Pocock, M.R. Wipat, A., & Li P.(2004) Taverna: A tool for the composition and enactment of bioinformatics workflows. *Bioinformatics, 20*(17), 3045–3054

Yu, J., Buyya, R., *Ong, SE., Blagoev, B., Kratchmarova, I., Kristensen, DB., et al. (2002).* Stable isotope labeling by amino acids in cell culture, SILAC, as a simple and accurate approach to expression proteomics. *Molecular & Cell Proteomics, 1, 376–86.*

Orchard, S., & Hermjakob, H., (2008) The HUPO proteomics standards initiative easing communication and minimizing data loss in a changing world. *Briefing in Bioinformatics, 9*(2), 166-173

Orchard, S., Hermjakob, H., Binz, P.A., Hoogland, C., Taylor, C.F., Zhu, W., Julian, R.K, & Apweiler R., (2005) Further steps towards data standardization: The Proteomic standard initiative. *Proteomics, 5*(2), 337-339

Pedrioli, P. G. A., Eng, J. K., Hubley, R., Vogelzang, M., Deutsch, E. W., Raught, B., et al. (2004). A common open representation of mass spectrometry data and its application to proteomics research. *Nature Biotechnology, 22*(11), 1459+.

Perkins, D. N., Pappin, D. J., Creasy, D. M., & Cottrell, J. S. (1999). Probability-based protein identification by searching sequence databases using mass spectrometry data. *Electrophoresis*, 20(18), 3551-3567.

Petricoin, E. F., Ardekani, A. M., Hitt, B. A., Levine, P. J., Fusaro, V. A., Steinberg, S. M., et al. (2002). Use of proteomic patterns in serum to identify ovarian cancer. *Lancet, 359*(9306), 572-577

Politi L, Groppi, A. & Polettini, A. (2005) Review: Applications of liquid chromatography- mass spectrometry in doping control. *Journal of Analytical Toxicology, 29,* (1), 1–14

Quandt, A., Hernandez, P., Kunzst, P., Pautasso, C., Tuloup, M., Hernandez, C., & Appel, R.. (2007) Grid-based Analysis of Tandem Mass Spectrometry Data in Clinical Proteomics. In (Eds.) N. Jacq, N. Müller, I. Blanquer, Y. Legré, V. Breton, D. Hausser, V. Hernández, T. Solomonides, M. Apitius, *From Genes to Personalized HealthCare: Grid Solutions for the Life Sciences - Proceedings of HealthGrid 2007,* (126), (pp.13-22).

J. Roboz, (1968). *Mass spectrometry instrumentation and techniques.* New York: Intersience Publisher.

Schulze, W & Mann, M (2003) A novel proteomic screen for peptide-protein interactions. *Journal of Biological Chemistry, 279*(11), 10756-64

Smith, C. A., Want, E. J., O'Maille, G., Abagyan, R., & Siuzdak, G. (2006). Xcms: processing mass spectrometry data for metabolite profiling using nonlinear peak alignment, matching, and identification. *Analytical Chemistry, 78*(3), 779-787.

Steadm, D.A., Paton, N.W., Missier, P., Embury, S.M. Hedeler, C. Jin, B. Brown, A.J., & Preece, A. (2008) Information quality in proteomic *Briefing in Bioinformatics,* 9(2), 174-188.

Stevens, RD, Robinson, AJ, & Goble CA (2004). myGrid: Personalised bioinformatics on the information Grid. *Bioinformatics,* 19(1), 302–302C.

Taylor, C. F., Hermjakob, H., Julian, R. K., Garavelli, J. S., Aebersold, R., & Apweiler, R. (2006). The work of the human proteome organisation's proteomics standards initiative (hupo psi). *OMICS, 10*(2), 145-151.

Tuloup, M., Hernandez, C., Coro, I., Hoogland, C., & Binz, P.,-A,. (2003) Aldente and BioGraph: An improved peptide mass fingerprinting protein identification environment. In Ducret A, Eberle AN, Faupel MD, Hofsteenge J, Jenö P, Langen H, et al. (Eds.), *Proceedings of the Swiss Proteomics Proceedings of the Swiss Proteomics Society, 2003 Congress, Understanding biological systems* (pp. 174–176). Lausanne, Switzerland: FontisMedia Publishers.

Witten, I. H., & Frank, E. (2005). *Data mining: Practical machine learning tools and techniques.* San Francisco: Morgan Kaufmann.

Wong, J. Cagney, G, & Cartwright, H.M (2005). Specalign—processing and alignment of mass spectra datasets. *Bioinformatics, 21* (9), 2088–2090.

Wroe, C., Stevens, R., Goble, C., Roberts, A., & Greenwood, M. (2003). A suite of DAML+OIL Ontologies to Describe Bioinformatics Web Services and Data. *International Journal of Cooperative Information Systems, 12*(2), 441-445.

Yasui Y., Pepe, M., Thompson, M.L., Adam, B.L., Wright, G.L., Qu, Y., et al. (2003). A data-analytic strategy for protein biomarker discovery: profiling of high-dimensional proteomic data for cancer detection. *Biostatistics, 4*(3), 449–463.

Yu, J. and Buyya, R. (2005). A taxonomy of workflow management systems for grid computing. *Journal of Grid Computing, 3*(3-4), 171-200.

KEY TERMS AND DEFINITIONS

Biomarker Discovery: The process to identify a possible molecular signature for a disease of a specific condition starting from proteomics data.

Computational Proteomics: Regards the use of computational science methods and tool to extract information from proteomics data.

LC-MS (Liquid Chromatography-Mass Spectrometry): The combined use of a chromatographic technique (LC) followed by a mass-spectometer. The former enables the separation of substances analyzed by MS improving the quality of overall process.

LC-MS/MS (Liquid Chromatography-Tandem Mass Spectrometry): The combined use of Liquid Chromatography coupled to two mass-analyzer. In the first stage a ion, namely precursor ion is selected and fragmented. In the second stage these fragments are analyzed. This techniques enable in such a way the identification of proteins in a sample starting from the constitutive peptides analyzed in the second stage.

MALDI-TOF (Matrix Assisted Laser/ Desorption Ionization – Time of Flight): Mass Spectrometry Is a kind of spectrometer. Information gained from using MALDI-TOF MS are mainly qualitative, e.g. the presence of a protein. Main advantage of such spectrometry relies in the relative.

Mass Spectrometry (MS): An analytical technique that aimis to identifies the chemical composition of a compound. In proteomics MS is used to identify the proteins contained in a sample, providing also some quantitative information, i.e. the relative abundance.

Peptide/Protein Identification: The process that aims to identify the proteins in a biological sample analyzed using MS. There exists many identification algorithms that can be grouped in: (i) database-based method, that match a real spectra with a database of theoretic ones; and (ii) de novo sequencing that aims to infer the presence of unknown proteins in a sample.

Peptide/Protein Quantitation: The process of the calculation of the relative abundance of peptides and proteins. It relies mainly on the kind of technology used to preprocess samples. There exist two possible approaches: (i) label based methods that pre-process sample with specific reagents, and (ii) label-free method that do not use labels.

Proteomics: The study of the proteome, i.e. the whole set of proteins expressed in a cell or in a tissue.

ENDNOTES

1 http://www.thegpm.org/GPMDB/index.html
2 http://www.peptideatlas.org/overview.php
3 http://www.ebi.ac.uk/pride
4 http://www.expasy.ch/ch2d/ch2d-top.html
5 http://www.matrixscience.org
6 http://expasy.org/tools/aldente/
7 http://www.thegpm.org/tandem/
8 http://www.swisspit.cscs.ch.
9 http://www.appliedbiosystems.com
10 http://biodcv.fbk.eu/
11 http://www.proteios.org
12 http://tools.proteomecenter.org/TPP.php
13 http://www.mygrid.org.uk/
14 https://sara.unile.it/cgi-bin/bioinfo/about

Chapter XII
High-Throughput Data Analysis of Proteomic Mass Spectra on the SwissBioGrid

Andreas Quandt
Swiss Institute of Bioinformatics, Switzerland

Sergio Maffioletti
ETH Zurich (Swiss National Supercomputing Centre), Switzerland

Cesare Pautasso
University of Lugano, Switzerland

Heinz Stockinger
Swiss Institute of Bioinformatics, Switzerland

Frederique Lisacek
Swiss Institute of Bioinformatics, Switzerland

Peter Kunszt
ETH Zurich (Swiss National Supercomputing Centre), Switzerland

ABSTRACT

Proteomics is currently one of the most promising fields in bioinformatics as it provides important insights into the protein function of organisms. Mass spectrometry is one of the techniques to study the proteome, and several software tools exist for this purpose. The authors provide an extendable software platform called swissPIT that combines different existing tools and exploits Grid infrastructures to speed up the data analysis process for the proteomics pipeline.

INTRODUCTION

Life sciences have become one of the major application areas in Grid computing due to several computing and/or data intensive challenges found in many bioinformatics domains. CPU intensive sequence search and analysis, structural analysis as well as drug discovery and protein docking are currently the most popular applications ported to Grid environments (Stockinger 2006). Bioinformatics applications, in the particular domain of proteomics, are not often described in the Grid and e-science literature. Here, we present one of the first initiatives in that direction.

Proteomics combines protein separation techniques such as two-dimensional gel electrophoresis (2-DE) or liquid chromatography (LC) with mass spectrometry (MS). Experimental data generated with this technology is processed and analysed to identify, quantify and characterise proteins. We have implemented a solution, the Swiss Protein Identification Toolbox (swissPIT), for the analysis of mass spectrometry data that leverages a Grid infrastructure dedicated to life sciences. The platform combines a range of proteomic tools which are usually executed in high throughput environments. The user interface is a high level Web portal which hides all details of the Grid specific implementation and therefore makes it an attractive tool for scientists dealing with large sets of mass spectrometry data.

There exist a large number of commercial and open-source software packages for protein analysis by mass spectrometry. Such an analysis has many research and biomedical applications: diagnosis of early-stage breast cancers, investigation of protein changes induced by drugs, identification of the proteome of organisms, identification of protein interactions, and more. Many different tasks are often involved in such studies, such as identifying and/or quantifying proteins and characterizing post-translational modification events (PTMs) that is, the addition of a chemical group that may affect the protein function. Some tools are specific for selecting high quality mass spectra for further analysis, others for assigning a statistical value to the obtained results. Often, a complete analysis requires the use of several analysis tools in sequence or in parallel. This leads to two main issues. First, the results of a tool usually have to be interpreted and reformatted in order to be used with the next tool. This task is often done manually by the user and is therefore time consuming and prone to errors. A second issue is due to the variety of existing tools that can perform a similar task (in our chase, there are N different tools dedicated to protein identification). When used in similar conditions (same input data and similar parameter sets), the results of these tools differ because of certain specificities of their algorithms (for example, their scoring functions). It consequently requires a large effort for the user to compare the obtained results and decide which one is best. With swissPIT, it is possible to analyze a given input set with various tools using a single user interface. The results are automatically reformatted in order to be easily compared or used with other tools. The platform also provides several analysis workflows, where tools are called in sequence and/or in parallel on the input data and where the whole analysis process is handled in a fully automated way. This way, user intervention is largely minimized while computing time is optimized by parallelization on the Grid computing infrastructure.

BACKGROUND

In the past, biology was not closely linked to computer science but the rising of molecular biology introduced a major change, and nowadays, biological experiments produce terabytes of data so that data analysis requires large infrastructures. Grids are a promising solution to this issue. Grids enable job distribution, task parallelization as well as data and results sharing between geographically scattered groups. Grid computing has been used in scientific domains for several years. Whereas physics (and in particular high energy physics) was one of the main early drivers, other scientific domains with Bioinformatics in particular, started to leverage Grid computing technologies for certain computing and/or data intensive applications. However, using this technology is far from trivial, Grids being heterogeneous, geographically distributed resources. Grids are more complicated to maintain, the organization and monitoring of the computation steps and the secure storage and distribution of data require additional knowledge from the user. These computing related issues may become a burden to non computer-specialists and, therefore, need to be hidden as much as possible from the end-users.

Grid Computing and Supercomputing

The idea of resource sharing is a central concept in the domain of parallel and distributed computing. The invention of the Web in the early 1990s provided the pathways for sharing data, information and knowledge over the Internet. The vision of Grid computing is built on the same ideas and technologies but extends them towards the idea of giving access to computing power and other resources at any given time from any given place.

In analogy to the electric power grid, a computing Grid aims to provide computing power without the need to know where it comes from. For instance, if a person plugs an electronic device into a socket, it is not mandatory to know if the power is produced by an atomic power plant, a water power plant or through alternative energy sources such as solar energy. Grid computing follows the same concept: a Grid user who is asking for computing power does not need to bother if the submitted job is executed on a supercomputer, a cluster or a set of workstations. However, this vision has not been fully achieved yet in practice, mostly because the usage policy models of supercomputers and those of clusters and workstations are still different.

The Grid computing concept has not changed much since its original. However, trends in Web services technologies have strongly influenced the way the current Grid is perceived and implemented. The most common building blocks of a typical Grid infrastructure are the following:

- Computing resources are an abstract layer accessible via standardized interfaces and service components which provide access to computing power. Often, the term "Computing Element" is used to describe this concept.
- Similar to the abstraction of computing resources, the service abstraction can be used for all other services provided in a Grid. One of the most common one is the Data Storage Element that provides a service abstraction for a (distributed) storage system offered in a Grid environment. A similar concept can be used for a biological database, a specific scientific instrument or similar.

In the literature, the bundle of software and hardware infrastructures that provides transparent access to computing resources is often described as "Computing Grid" or "Computational Grid". Computing Grids aim to provide homogeneous access to various resources by abstracting from service details and providing easy interfaces to Grid users, applications and other services. The actual software behind a Grid infrastructure is typically used "between" the application software and the operating system and is therefore called middleware. Several Grid middleware systems have been developed in the past years but in general they provide the following list of features:

- **Resource management, job submission and execution:** This comprises a set of client tools and services that are used to describe job requirements, the resource selection by a specialized service as well as the submission and the execution of user jobs. These "resource brokers" are often used to select adequate resources that are registered in so-called "information services" where all Grid resources are listed.

- **Data and storage management:** The data stored in a Grid often exceeds the volumes available on standard hard disks. Therefore, special storage hardware (often disk and tape based systems) and transfer protocols are necessary to handle such volumes. Another important aspect of the "Data Grid" is also the backup of data and the management of replicated data in different locations.
- **Development of Grid-based services:** Several Grid toolkits provide computational scientists with helpful high-level tools to design, create, and develop additional Grid-based services that can be integrated into existing infrastructures. This is particularly important if there is a need to develop additional functionality which is not yet available in the existing services structure.

However, there is no one-size-fits-all solution, therefore, each project and each application domain still needs to adapt the existing Grid middleware solutions to its needs, improving on them and extending them in the process. Therefore, also in the development of swissPIT there have been components and additions necessary to interface with and integrate into the existing middleware services.

Life Sciences and the Grid

Like many other sciences, life sciences make extensive use of information technologies such as database systems, information retrieval, and image processing. Due to the characteristics of certain application domains (often remote resources need to be accessed, additional computation power is required, etc.) life sciences have started to use Grid computing in various domains. Grid technologies are especially used for the analysis of large amounts of data in Genomics (Blanchet 2006) and Proteomics (Dowsey 2004, Cannataro 2005, Quandt 2008, Zosso 2007) but they have been also used for pathway studies, and are also considered for data analysis in Systems Biology (Burrage 2006). However, in contrast to other scientific domains, life sciences often have particular requirements for data security and privacy. For instance the strict privacy of patient data in hospitals is one of the challenges for existing Grid systems that did not provide such fine grained security mechanisms, and had to be implemented later (Frohner 2007).

Mass Spectrometry-Based Peptide Identification

Within the last decade the field of proteomics has significantly evolved. The completion of the human genome project has driven research towards functional genomics. Proteomics as part of this endeavour focuses on the large-scale determination of protein function directly at the protein level. A proteome has been defined as the whole set of proteins, coded by the genome of an organism. Proteomics technology is fast progressing, and many techniques have been developed to challenge the analysis of cellular proteomes showing a high degree of complexity. These techniques include imaging of X-ray crystallography, NMR spectroscopy or 2D-PAGE, protein chips, and mass spectrometry (MS) based technologies.

Today, mass spectrometry has increasingly become the method of choice to identify, characterize and quantify large-scale complex protein samples. A typical proteomic analysis using MS includes the following steps:

1. Given a sample containing a mixture of proteins to identify, the proteins in the mixture are cut into pieces (peptides) using a specific proteolytic enzyme and leading to an even more complex peptide mixture;

2. A first MS is performed on the peptide mixture. The aim here is to separate the peptides according to their mass-to-charge ratio (m/z); The next steps are repeated for each selected peptide (MS/MS):
3. The peptide is fragmented by collision with gas molecules and the obtained fragment masses are measured in a second MS stage.
4. The output is a mass spectrum (MS/MS spectrum) composed of fragment masses and associated intensities (for a particular peptide).
5. The MS/MS spectra are correlated (or matched) with theoretical peptide sequences (de-novo sequencing) or peptide sequences extracted from DNA and protein databases (classical database search). The matching procedure can be successful (one of the theoretical peptide confidently matches the spectrum) or unsuccessful.
6. Finally, a list of identified proteins is build from the set of confidently matched peptides.

Over the past 20 years, several tens of programs have been developed to improve the spectra identification process (steps 5 and 6). These programs can be categorized in four groups (Shadforth 2005, Nesvizhskii 2007):

- The classical database search (CDS) tools,
- De-novo sequencing tools,
- Open-modification search (OMS) tools,
- and spectra library search (SLS) tools.

Classical Database Search (CDS)

In general, a CDS performs the following steps for each protein sequence entry of a database:

1. In silico-digestion of the protein sequence by known cleavage rules.
2. Computation of the peptide masses.
3. Finding matches between experimental masses and computed masses.
4. Calculation of theoretical fragmentation spectra.
5. A scoring scheme is used to rate the match between theoretical and experimental spectra. Typically, the database entry with the highest score is taken as the correct identification.
6. The Peptide scores are then combined into protein scores.

CDS tools can be labeled as heuristic or probabilistic methods (Kapp 2003). (Kapp 2003) describe the difference between the two types as follow: Heuristic algorithm "...correlate the acquired experimental MS/MS spectrum with a theoretical spectrum and calculate a score based on the similarity between the two". Probabilistic algorithms "...model to some extent the peptide fragmentation process (e.g., ladders of sequence ions) and calculate the probability that a particular peptide sequence produced the observed spectrum by chance". In the following, two representatives of each type are described.

Sequest

Sequest (Eng 1994) is the most popular representative implementing the heuristic model. Basically, the algorithm contains four steps.

1. First, it determines the parent ion molecular mass of the MS/MS spectrum.
2. Then, it searches a sequence database to select the 200 closest peptide matches to this mass. With a second scoring function, these 200 spectra are re-scored and a theoretical spectrum is created for each of them by using known digestion rules.
3. This set of theoretical spectra is compared with the experimental spectrum and a value for the cross correlation, XCorr value, is assigned.

4. In the last step, these scores are ranked and used for the protein identification.

The XCorr value has been the basis for many modifications and extensions (Nesvizhskii 2003) of the algorithm and lead to the first pipeline in MS/MS-based protein identification (Keller 2005).

X! Tandem

X! Tandem is the second heuristic approach to describe (Craig 2004). Like Sequest, X! Tandem is also often used as reference for comparing identification results (Kapp 2003; Keller 2005). The approach is one of the few open-source tools which is widely used in the community. The algorithm as has been developed to reduce "the time necessary to match a large set of peptide tandem mass spectra with a list of protein sequences" (Craig 2003). Similar to Sequest it generates theoretical spectra and correlates them with the experimental data but in difference it uses intensity patterns which are associated with particular amino acid residues for this generation and also uses a dot product to create an expectation value instead of the XCorr value (Kapp 2003).

Mascot

Mascot is a probabilistic approach and today the most widely adopted program in MS/MS identification. It is based on the MOWSE algorithm (Pappin 1993) which has been initially developed for Peptide Mass Fingerprinting (PMF). Mascot extends this algorithm by a probability-based score to support MS/MS identification. The software is commercially distributed but its algorithm has, in contrast to Sequest, never been described in literature, nor has it been patented. In their original paper, Perkins et al. only describe the probability calculation as based on the observed match between the experimental data set and each database entry, and the best match reports the lowest probability score (Perkins 1999). The missing description of the algorithm makes it difficult to better understand in which situations the algorithm identifies spectra correctly and why in some other situations, false-positive hits are produced. Here, Perkins et al. only mentioned that the scoring depends on the size of the database used.

Phenyx

As Mascot, Phenyx is also commercially distributed. Phenyx is based on OLAV, a probabilistic identification algorithm (Colinge 2003) which has been also adapted for PMF-identification. In comparison to other approaches of this category, OLAV tries to outperform these algorithms by using additional information and to simultaneously apply an efficient scoring scheme to approximate a likelihood ratio, which is the most powerful statistic for distinguishing correct from wrong peptide matches (Colinge 2003).

Issues with CDS

The major problem of CDS tools is filtering peptides with molecular weight to derive candidate sequences from the database. Therefore, an incorrect molecular weight will only provide incorrect sequences. However, unless the program is not 'aware' of such modifications, the match will not be made. Consequently, the user needs to specify modifications 'a priori' in order to correctly identify spectra carrying these modifications with a CDS approach. This is possible for the a few modifications which are likely to occur (e.g. Oxidation, Deamidation, Carbamidomethylation) but not for all potential candidates.

Open-Modification Search (OMS)

One of the largest problems in MS/MS identification is the correct determination of spectra of modified peptides. In theory, de novo sequencing would be a solution for this problem as it does not rely on the prior specification of (potential) modifications but in practice these tools are too computer intensive to be applied on large scale data. This drawback was tentatively solved by combining the advantages of CDS and de-novo sequencing in some sort of hybrid application, the open-modification search. Tools implementing an OMS algorithm identify the best match between the database entries and the experimental spectra by considering any kind of unexpected mass shift within the theoretical spectra. There are only a few existing OMS tools. In the following we describe Popitam and InsPecT, the two most frequently used tools.

Popitam

Popitam is freely available after request and was one of the earliest attempts to tackle the OMS problem. This may be why the search algorithm is similar to the spectrum graph approach generally used in de novo sequencing. In computer science, a graph is a common abstract data representation consisting of vertices (connections) and edges (information). Popitam uses such graphs to model all potential combinations of AA-sequences, the spectrum graph, and to score these sequences due to their similarity to the unidentified experimental spectrum. The spectrum graph is created by using an Ant colony algorithm in order to emphasize relevant sequence tags for each theoretical sequence. This approach leads to a list of peptides ranked by a correlation score (Hernandez 2003).

(Hernandez 2003) describe the Popitam approach in four steps:

1. A given number of peaks in the experimental spectrum is selected according to their intensities.
2. The MS/MS spectrum is transformed into a graph to capture the relationship between the peaks.
3. The spectrum graph is compared to the theoretical spectra created from the entries of the sequence database which is used.
4. A ranked list of scored candidate peptides is created.

In contrast to other programs, Popitam's scoring scheme does not rely on the usual tag extraction followed by sequence alignment to calculate peptide scores. Instead, the program uses the sequence database to direct and emphasize specific sequence tags of the spectrum graph to calculate the peptide scores from this information.

InsPecT

InsPecT is an open-source OMS-tool which, in contrast to Popitam, does not create a spectrum graph to identify spectra on a reliable level. InsPecT belongs to the group of sequence tag approaches which has been introduced by (Pappin 1993) and (Eng 1994) and which is in general based on the incorporation of partial sequence data alongside the mass for database searching (Shadforth 2005). The algorithm consists of the following steps (Tanner 2005):

1. Local de novo sequencing is used to generate sequence tags.
2. These tags are used as PTM-aware text-based filters that reduce a large database to a few peptide candidates.
3. Application of a probabilistic score to rank the peptide candidates and to assess their quality.

The group argues that filtering is the key to modified peptide identification using a sequence database because the application of "... more sophisticated and computationally intensive scoring to the few remaining candidates" is less computer intensive. In their perspective, the correlation of masses to identify matches is more CPU-expensive than sequence-based searches since less high-score matches occur by chance.

Identification Workflows

Depending on the quality of the spectra and the complexity of the protein mixture only a limited part of a set of spectra is usually confidentially identified (Keller 2005, Hernandez 2006). Therefore, a major goal in MS/MS identification studies is the matching of as many spectra as possible while keeping the false positive rate as low as possible.

Beside the problem of validating the proposed identifications another problem is the correct identification of peptides carrying one or more post-translational modifications (PTM).

The most promising way to solve simultaneously both problems is the combination of several search strategies and identification tools in so-called identification workflows (Hernandez 2006, Quandt 2008). These workflows consist of multiple tools running in parallel and/or sequential order whereas information is passed from one workflow part to the other. This knowledge transfer helps to meaningfully align different search strategies. For instance a CDS can be used as pre-filter for an OMS which often cannot be applied without this filtering.

Grid Approaches for Mass Spectrometry Identification

Typical proteomics experiments generate tens of thousands of MS/MS spectra which are processed by comparing their similarities with hundreds of thousands of sequence entries in protein databases. All these need to be processed by the algorithms described above. Since none of the tasks is coupled to another one, the analysis can be done simultaneously on as many resources as are available. Dedicated resources are of course desirable, and having access to even more through resource sharing is also very attractive. Therefore, Grid computing is an attractive model to handle the computational needs of this scientific domain, as it gives access to a large pool of shared resources.

There are currently two specific projects using the Grid as computational backbone in Switzerland:

- The X! Tandem PC-Grid project of Biozentrum Basel and the Novartis research facility (Zosso 2007) and
- swissPIT, the Swiss Protein Identification Toolbox (Quandt 2008).

Issues Specific to PC-Grids

The idea of using free resources on already available workstations is interesting but is limited in its use for scientific computing because of the way data are processed.

Usually, the analysis involves the comparison (the matching) of the experimental mass spectral data with theoretical examples (e.g. protein sequences) stored in biological databases. In practice, the statistics behind the matching process require that all data are typically processed by the same application on a single processing unit, i.e., the data processing cannot be easily parallelized. There have been attempts to parallelize software for MS/MS data analysis in order to run it on a PC Grid (Zosso 2007) in screen-saver-mode such as SETI@home. This approach raises two main concerns:

- Using PC-Grids for mass spectrometry analysis raises the need to adapt the source code of the identification programs in order to introduce parallel code execution and therefore to benefit from the computational infrastructure. The parallelization of software code is not only difficult due to the potential impact on the statistics, i.e. for X! Tandem, but also because most programs used for the analysis of mass spectral data are commercial software. These proprietary programs cannot be parallelized due to the restricted access to the source code. However, parallelization of open-source programs is also complex as they partially undergo very short update cycles which requires each time to re-adapt the original code.

- Another limitation of PC-Grids is that they cannot be applied to process large data files as this requires more local hardware resources (e.g. for RAM, CPU power, and network capacity). Therefore, this type of approach tends more towards high-throughput analysis rather than high-performance computing. The software code does not need to be parallelized, but it can be run on multiple Grid resources to analyze many small data files in parallel.

Due to the previously described problems in using PC-Grids for high-throughput computing, the Swiss Protein Toolbox (swissPIT) is using resources of several research and supercomputing centers (Swiss National Supercomputing Centre, Vital-IT, University of Zurich).

In the following we are going to describe how we designed and deployed swissPIT to achieve a user-friendly platform that shields users from the complexity of a computing Grid.

THE SWISSPIT PLATFORM

The swissPIT is a software platform for the analysis of large-scale mass spectrometry data via a Web portal connected to a Grid infrastructure (Quandt 2008). The portal provides access to several protein identification and characterization tools that can be executed in parallel on the Grid. In the following sections we will first motivate a typical biological use case before we give details on the software platform as well as its design and implementation in a Grid environment.

The Swiss Protein Identification Toolbox

The swissPIT (Swiss Protein Identification Toolbox) provides an expandable multi-tool platform capable of executing workflows to analyze Tandem MS-based data (Quandt 2008). The identification and characterisation of peptides from tandem mass spectrometry (MS/MS) data represents a critical aspect of proteomics. Today, tandem mass spectrometry analysis is often performed by only using a single identification program achieving identification rates between 10 – 50 %. One of the major problems in proteomics is the absence of standardized workflows to analyze the produced data. This includes the pre-processing part as well as the final identification of peptides and proteins. Applying multiple search tools on a data set is recommended for producing more confident matches by cross-validating the matches of each search engine. The strategy of running parallel searches has been often highlighted in literature (Kapp 2005, Keller 2005, Hernandez 2006). However, the combination of identification tools with different strategies analyzing workflows is a novel approach with large scientific impact. The main idea of swissPIT is not only the usage of different identification tools in parallel but also the meaningful concatenation of different identification strategies at the same time.

Currently, four identification software packages can be used within swissPIT: Phenyx (GeneBio SA) (Colinge 2003), Popitam (Hernandez 2003), X! Tandem (Craig 2004), and InsPecT (Tanner 2005). The choice of these first four algorithms has been motivated by a number of factors, including their popularity, their known efficiency and their implementation of various search strategies (Hernandez 2005). Furthermore, two protein databases can be used: UniProtKB/SwissProt and UniProtKB/TrEMBL.

Architecture and Implementation Details

The core of swissPIT is implemented in programming languages using the Java virtual machine (Java, Groovy) and Java-related techniques such as Struts and Java Server Pages (JSP). This allows for both programming the interactive Web interface as well as all the core business logic of swissPIT. In order to deal with workflow creation and execution, the workflow engine JOpera (Pautasso 2006) is used. The web interface then also allows for status updates on the analysis jobs. The actual interface to the Grid is realized via predefined and standardized system calls to external scripts which then call Grid client tools to submit and monitor jobs. swissPIT therefore makes use of a high-level standardized Grid interface which allows to make use of several Grid infrastructures through various middleware implementations such as the Advanced Resource Connector ARC (http://www.nordugrid.org) or gLite (http://www.glite.org). ARC and gLite both implement the Grid middleware services as described in Section 'Grid and supercomputing'. User interfaces to these middleware implementations are standardized in the Open Grid Forum OGF, so in principle swissPIT can simply make use of the standardized client interfaces to make use of the various Grid infrastructures. In the current implementation, swissPIT is interfacing to ARC for executing its jobs on the Grid.

The selection of the underlying Grid middleware has consequences on the security model used for the swissPIT front-end. Most of the current Grid middleware tools (including ARC and gLite) make use of the Grid Security Infrastructure (GSI) that is based on Public Key Cryptography which is applied in the back-end of swissPIT. The users need to register and log-in to the Web interface but currently do not have to have their own X.509 user certificates. In contrast, the swissPIT back-end uses a certificate that is bound to the service itself and executes all Grid jobs on behalf of the user. As a result, scientists do not need to deal with credential delegation and renewal. While this is an acceptable solution in our Swiss Grid environment, it may not be acceptable on larger infrastructures where the resource providers expect individual authentication of 'real' people as a precondition to make use of their computational resources.

Deployment in the Grid

The interaction and the actual deployment on the Grid require several steps:

- **Integration at the interface level:** The implementation of the Grid interface needs to interact with the job submission and execution interface of the specific Grid middleware. In our first implementation, we have selected the ARC middleware and its basic interface (ngsub, ngstatus, etc.) for job submission and monitoring. However, since the interface of gLite is rather similar, it can be easily replaced as demonstrated in (Stockinger, 2006). Currently, swissPIT uses the command line interface of ARC rather than any programming language binding.
- **Usage of computational resources in the Grid:** ARC is the middleware which then needs to be deployed on several sites (mainly computing resources) in order to provide the basic Grid infrastructure. In our case, we have selected the SwissBioGrid (Podvinec 2006) infrastructure that spans several high performance computing centers in Switzerland and provides access to the life sciences community. In fact, swissPIT is one of the sample applications in the SwissBioGrid (SBG). More recently, since the unfunded SwissBioGrid project has come to an end, these resources are being provided through a funded effort, the Swiss Multi-Science Computing Grid, which is also embedded as a working group of the Swiss National

Grid Association SwiNG (http://www.swing-grid.ch).

- **Deployment of bioinformatics applications:** A central question in Grid computing is the deployment of end-user applications that are used on top of Grid middleware systems and executed on computing resources and clusters. In case the particular application is rather small and does not require many dependencies (statically linked executables in the best case), applications can be directly submitted with the user job. In case of more complex application and runtime environments, applications need to be pre-installed at certain sites and registered with the Grid information service to be found in the resource selection process. In our case, the protein identification applications that swissPIT provides have been pre-installed and configured in the infrastructure relying on the ARC Runtime Environment services (RTE), thus, relieving the high-level application services as well as the workflow manager to directly deal with application discovery and validation. The RTE information is then used in the matchmaking scheduling algorithm to select the best candidate resource to host the user request.
- **Deployment of biological databases:** Since the proteomics applications run in different sites and require input from biological databases, the access to biological data needs to be carefully prepared. Although Grid tools and services provide data management tools to locate and transfer data, they often do not provide a single, distributed file system view (with the exception of tools such as Sybase's Avaki Information Integration System that we have used in earlier case studies). Therefore, it is required that specific versions of the required database are installed at the sites. Additionally, the used protein identification tools need to be aware of the location. In order to allow identification applications to access the UniProtKB/SwissProt and UniProtKB/TrEMBL datasets, a specific RTE has been defined to identify those sites providing a local copy of the entire dataset. In this prototype solution, indeed, all sites providing resources for swissPIT must support the whole UniProtKB/SwissProt and UniProtKB/TrEMBL dataset providing local copies and granting direct access from all computing nodes in the computing cluster.

These four steps are taken care of by the swissPIT portal. The end-user will therefore not have to be concerned with these complexities.

Submission Wrapping

Instead of directly accessing the ARC middleware, each workflow component has a dedicated wrapper script. This allows more flexibility in changing the used infrastructure without the need to also adapt the interface to swissPIT. This has been very useful by for instance including the pre- and prost-processing cluster which has been done to optimize the execution of so-called short-time jobs which run in most cases only a few seconds or minutes but are not possible to execute directly on the front-end server because of the high demand on RAM.

Another advantage of using such wrapper scripts is the possibility to better monitor the job execution and to automatically retrieve all result files and system information produced during the execution. Since we are integrating several service calls, the return codes also have to be managed to provide a unified feedback to the user. By using the wrapper script, it is possible to monitor the program execution and for instance to redirect the content of error text files to the standard error stream or to filter log messages and to stop the program with a specific error code.

Last but not least, the wrapper scripts not only automate the job submission and result retrieving but they also combine the tasks in a single command which is to execute from swissPIT. That makes it easier to configure the job execution in swissPIT by choosing a local program execution, the processing on a computer cluster or a Grid infrastructure.

The 2-Step Workflow as Example Approach

Ordinary static workflows can be easily connected by using scripts. However, this simplistic solution is not sufficient to realize more complex workflows with decision trees, information extraction steps, and iterations. For such workflows a workflow management system is required, especially if the monitoring and visualization of long-running workflows is also an important requirement. In swissPIT, workflows are created, monitored and visualized by using JOpera (http://www.jopera.org).

Figure 1 shows the scheme of a 2-step workflow, one of several workflows available in swissPIT. The 2-step approach is an example of an integrated workflow that invokes multiple identification programs. This workflow combines the advantages of classical search tools and open-modification search tools to improve the identification rate. In this workflow, a 2-step identification strategy is applied where first some classical search tools are applied followed by a set of OMS-tools. When the workflow starts, all peak list files are converted into an MGF file and parent charge gets corrected. Then the parameter files for the programs are created. After this preparation step, the main identification workflow starts with the parallel execution of Phenyx and X! Tandem, two classical search tools. With this CDS, aim is to identify as many spectra as possible in a reasonable amount of time.

As already described, classical search tools are able to search large sequence databases quickly to identify experimental spectra, but they are unable to recognize spectra with unexpected modifications. Therefore, open-modification search tools are applied in a second step to identify further spectra from the data set. One drawback using OMS tools is the difficulty to query a large database. Therefore, we introduce two parameters to meaningfully reduce the search space from step 1 to step 2. First, the search space is reduced by applying a database filter based on the assumption that all unidentified peptides belong to a protein already identified in the first step. Therefore, we extract the protein accession numbers (ACs) from the result files of Phenyx and X! Tandem. Then

Figure 1. 2-step workflow realized with JOpera. Each of the nodes in the control-flow graph represents a software tool. The intermediate converting steps (Pidres2AC, Pidres2UnmatchedSpectra, AcList2Trie) between the two stages are used to transform the output of the first stage into a format suitable to be consumed by the tools used in the second. Each tool execution step is CPU intensive and can be executed on a Grid with the exception of Phenyx that runs on a separate cluster due to license restrictions.

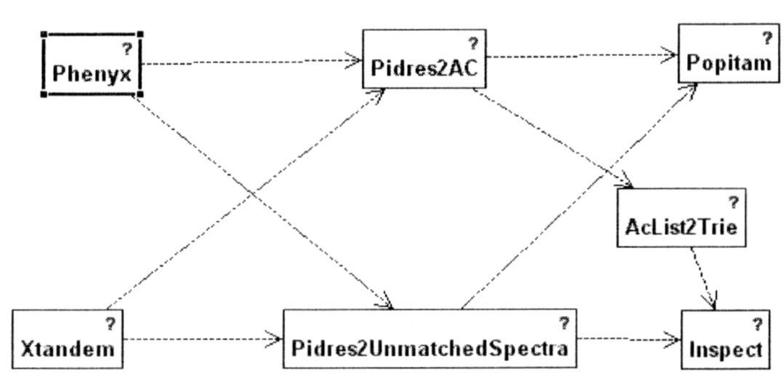

we create a list of their 'union' and use this list as input parameter for the programs of step 2. For Popitam it is possible to directly use the AC list as input parameter. InsPecT does not have such an input parameter. Here, we dynamically create a new database file from the AC list which is then used during the identification process of InsPecT. A second reduction of the search space is made by preventing the re-analysis of already identified spectra. Therefore, we extract all spectra from the result files of Phenyx and X! Tandem which have not been identified by these programs. Only this list of unmatched spectra is then used in the open-modification search in step 2.

Experimental Results

A prototype of the swissPIT toolbox (including the Web interface) is installed on a Linux-based machine at the Swiss National Supercomputing Centre in Manno (Switzerland). The Grid infrastructure was provided via the SwissBioGrid, now the Swiss Multi-Science Computing Grid and is based on ARC. In particular, three sites (CSCS in Manno, Vital-IT in Lausanne, and University of Zurich) provide access to the computing clusters. Note that the current prototype is not yet optimised for high performance. In particular, we are interested to use swissPIT for an improved peptide identification compared to the common single tool approach. However, we present a few preliminary results that show the functionality of the system (cf. Figure 2).

To test the functionality of the overall workflow in swissPIT we used a specific data set that has been created to evaluate the identification rate of MS/MS tools. In other words, we are interested in the number of correctly identified spectra. In total, the dataset contains 185 spectra where each of them contains a biological modification (http://bioinfo2.ucsd.edu/MSAlignmentExamples/Lens/index.html). These spectra are organized in several small files which represent the main input to the analysis and are uploaded via the Web interface. To detect any spectra in the classical database search we predefine carbamidomethylation and deamidation as expected modifications. Therefore, Phenyx and X!Tandem are only able to identify spectra with these modifications. For the test data set Phenyx identified 30 spectra and X! Tandem

Figure 2. Experiment with 185 spectra. In stage 1 ("classical search"), 35 spectra were identified and three new modifications were found; 150 spectra remained unidentified. In Stage 2 an additional 72 spectra was identified.

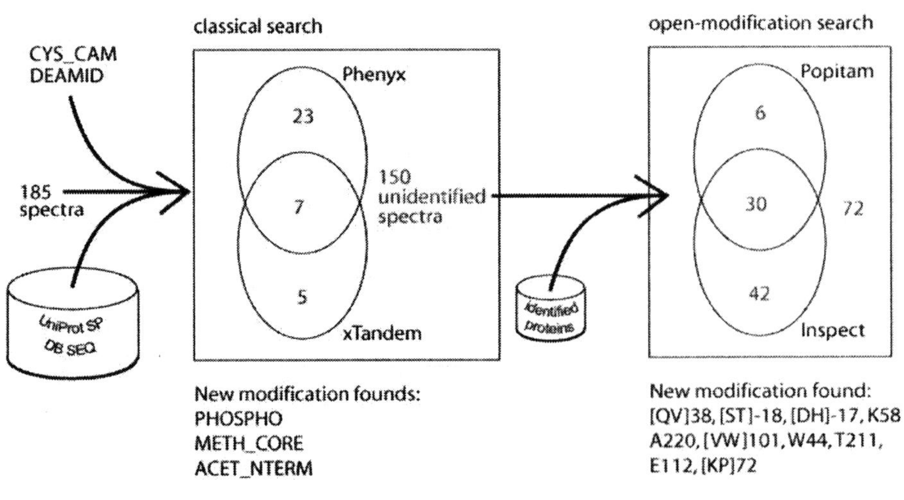

12 spectra (cf. Figure 2). Identification in that sense means the matching of specific spectra to a protein entry in the databank. The list of protein matches found by Phenyx and X!Tandem is used as preliminary information in stage 2 so that InsPecT and Popitam could only identify modified spectra also matching these proteins. For our test data set both programs identified 78 new spectra which increases the identification rate from less than 20% in stage 1 to over 60% after stage 2. This shows a clear identification improvement of swissPIT over only a single identification tool. Overall, for this rather small experiment the runtime does not increase fundamentally if the user runs one or several programs per workflow stage but the identification rate is improved significantly.

We also experimented with larger data sets containing about 10'000 spectra per file. The analysis time for such an analysis strongly depends on the parameter chosen for each program and on the time the queuing time of sub-jobs in the workflow. For processing a file with 20'000 spectra swissPIT currently needs about 5 hours using a 2-stage workflow (either with 2 or 4 identification programs).

FUTURE TRENDS

Grid infrastructures continue to evolve with the increased usage by many scientific domains, not only life sciences. While the services mature, additional complexities will be introduced through new and more advanced services. The most important technological developments will be in the area of data management since most sciences are now producing large amounts of experimental data through the latest-generation of digital instruments. Reducing the raw data to a well-analyzed set of usable information will need the development of an advanced set of distributed data management tools, from which also swissPIT should profit.

At the same time, the Grid communities are in the process to better organize themselves. The establishment of National Grid Initiatives will facilitate the availability of resources and will standardize the access to them. swissPIT intends to be one of the main driving applications in the Swiss National Grid Association SwiNG, assuring that the requirements of this domain of bioinformatics are properly addressed also in the future.

CONCLUSION

With swissPIT we show that a Grid-based system can successfully provide the computational infrastructure for analyzing data of high-throughput experiments in future. Especially for the queuing system, the easy usage of additional computational resources and the reduction of downtimes due to the availability of several resources are the obvious advantages of using a Grid of resources rather than single computer clusters to analyze the hundreds of thousands of spectra produced daily by high-throughput tandem-MS. This point is important since biology users often experience difficulties with the complexity of large computer systems and how to use them.

With swissPIT we also show that the Grid can be used by non-experienced users who do not know how it is accessed. To do so we developed a Web portal for users to maintain their experimental data and to prepare the analysis workflows before submitting data. A user can retrieve status information about job submissions directly from within the portal (the overall status of a submission plus status of all single jobs belonging to a workflow). In the background, swissPIT collects all user data and prepares an automated Grid submission via program-specific wrapper scripts. These scripts are also used to monitor the job status and to report it back to the Web portal. Therefore, swissPIT hides the Grid from the user and reduces the complexity of its usage.

In the future, we will investigate a series of problems that still need appropriate solutions. For instance, using the Grid for short-term jobs where the computational time is less than the time needed to submit them to the Grid. Solving further issues related to data management in general and in particular with the availability of reference databases will be a key to the success of the system. We also plan to integrate the SWITCH-AAI user authentication and authorization system to use individual user credentials to log into the web portal (currently, we rely on a single service certificate). SWITCH-AAI is interfacing to the regular authorization mechanisms that the users already have through their local institutions, so they can use their local username and password also to access swissPIT. In addition, it is important to better integrate the interaction with the Grid middleware to better monitor jobs, stop them, and also allow their resubmission and/or continuation of a workflow. This mandates the improvement of the wrapper scripts, probably extending them to proper modules calling the middleware interfaces directly instead of scripting.

REFERENCES

Blanchet, C., Lefort, V., Combet, C. & Deléage, G. (2006). GPS@ bioinformatics portal: from network to EGEE grid. *Stud Health Technol Inform, 120*, 187-193.

Burrage, K., Hood, L. & Ragan, M. A. (2006). Advanced computing for systems biology. *Brief Bioinform, 7*(4), 390-398.

Cannataro, M., Barla, A., Flor, R., Jurman, G., Merler, S., Paoli, S., Tradigo, G., Veltri, P. & Furlanello, C. (2007). A grid environment for high-throughput proteomics. *IEEE Trans Nanobioscience, 6*(2),117-123.

Colinge, J., Masselot A., Giron, M., Dessingy, T., & Magnin, J. (2003). OLAV: towards high-throughput tandem mass spectrometry data identification. *Proteomics, 3*(8), 1454-1463.

Craig, R. & Beavis, R. C. (2004). TANDEM: matching proteins with tandem mass spectra. *Bioinformatics, 20*(9), 1466-1467.

Dowsey, A. W., Dunn, M. J. & Yang, G. (2004). ProteomeGRID: towards a high-throughput proteomics pipeline through opportunistic cluster image computing for two-dimensional gel electrophoresis. *Proteomics, 4*(12), 3800-3812.

Eng, J. K., McCormack, A. L. & Yates III, J. R. (1994). An approach to correlate tandem mass spectral data of peptides with amino acid sequences in a protein database. *Journal of the American Society for Mass Spectrometry, 5*(11), 976-989.

Frohner, A., Jouvenot, D., Kunszt, P., Montagnat, J., Pera, C., Koblitz, B., Santos, N, Loomis, C. (2007). A Secure Grid Medical Data Manager Interfaced to the gLite Middleware. *Journal of Grid Computing, 6*(1), 45-59.

Hernandez, P., Gras, R., Frey, J., & Appel, R. D. (2003). Popitam: towards new heuristic strategies to improve protein identification from tandem mass spectrometry data. *Proteomics, 3*(6), 870-878.

Hernandez, P., Müller, M. & Appel, R. D. (2006). Automated protein identification by tandem mass spectrometry: issues and strategies. *Mass Spectrom Rev, 25*(2) 235-254.

Kapp, E. A., Schütz, F., Connolly, L. M., Chakel, J. A., Meza, J. E., Miller, C. A., et al.(2005). 'An evaluation, comparison, and accurate benchmarking of several publicly available MS/MS search algorithms: sensitivity and specificity analysis. *Proteomics, 5*(13), 3475-3490.

Keller, A., Eng. J., Zhang, N., Li. X., & Aebersold, R. (2005). A uniform proteomics MS/MS analysis platform utilizing open XML file formats. *Mol Syst Biol, 1*, 2005.0017

Lane, C. S. (2005). Mass spectrometry-based proteomics in the life sciences., *Cell Mol Life Sci, 62*(7-8), 848-869.

Nesvizhskii, A. I., Keller, A., Kolker, E. & Aebersold, R. (2003). A statistical model for identifying proteins by tandem mass spectrometry. *Analytical Chemistry, 75*(17): 4646-4658.

Nesvizhskii, A. I., Vitek, O. & Aebersold, R. (2007). Analysis and validation of proteomic data generated by tandem mass spectrometry. *Nat Methods, 4*(10). 787-797.

Pappin, D. J., Hojrup, P. & Bleasby, A. J. (1993). 'Rapid identification of proteins by peptide-mass fingerprinting. *Curr Biol 3*(6), 327-332.

Pautasso, C., Bausch, W., & Alonso, G. (2006), Autonomic Computing for Virtual Laboratories. In J. Kohlas, B. Meyer, A. Schiper (Eds.), *Dependable Systems: Software, Computing, Networks*, (LNCS 4028). Berlin, Germany: Springer Verlag.

Perkins, D. N., Pappin, D. J., Creasy, D. M. & Cottrell, J. S. (1999). Probability-based protein identification by searching sequence databases using mass spectrometry data. *Electrophoresis, 20*(18), 3551-3567

Podvinec, M., Maffioletti, S., Kunszt, P., Arnold, K., Cerutti, L., Nyffeler, B., et al. (2006). The SwissBioGrid Project: Objectives, Preliminary Results and Lessons Learned. *2nd IEEE International Conference on e-Science and Grid Computing (e-Science 2006) - Workshop on Production Grids.*

Quandt, A., Hernandez, P., Masselot, A., Hernandez, C., Maffioletti, S., Pautasso, C., et al. (2008). swissPIT: A novel approach for pipelined analysis of mass spectrometry data. *Bioinformatics,* 24(11), 1416-1417.

Shadforth, I., Crowther, D. & Bessant, C. (2005). Protein and peptide identification algorithms using MS for use in high-throughput, automated pipelines. *Proteomics, 5*(16), 4082-4095.

Stockinger, H., Pagni, M., Cerutti, L., & Falquet, L. (2006). Grid Approach to Embarrassingly Parallel CPU-Intensive Bioinformatics Problems. *2nd IEEE International Conference on e-Science and Grid Computing.*

Tanner, S., Shu, H., Frank, A., Wang, L., Zandi, E., Mumby, M., et al. (2005), InsPecT: identification of post-translationally modified peptides from tandem mass spectra. *Analytical Chemistry* 7(14), 4626-4639.

Zosso, D., Arnold, K., Schwede, T., Podvinec, M., (2007). SWISS-TANDEM: A Web-Based Workspace for MS/MS Protein Identification on PC Grids. *CMBS,* 406-411.

KEY TERMS AND DEFINITIONS

Bioinformatics: Comprises the management and the analysis of biological databases.

Grid Workflow: In general, a workflow can be considered as the automation of a specific process which can further be divided into smaller tasks. A Grid workflow consists of several tasks that need to be executed in a Grid environment but not necessarily on the same computing hardware.

Grid Job Submission and Execution: Workflows are typically expressed in certain languages and then have to be executed. Often, the entire workflow is called a "job" which needs to be submitted to the Grid and executed on Grid computing resources.

High Performance Computing (HPC): HPC is a particular field in computer science that deals with performance optimization of single applications, usually by running parallel instances on high performance computing clusters or supercomputers.

High Throughput Computing: In contrast to HPC, high throughput computing does not aim to optimize a single application but several users and applications. In this way, many applications share a computing infrastructure at the same time – in this way the overall throughput of several applications is supposed to be maximized.

Mass Spectrometry: In the field of proteomics, mass spectrometry is a technique to analyze, identify and characterize proteins. In particular, it measures the mass-to-charge ratio.

Proteomics: The large-scale study of proteins, their functions and their structures. It is supposed to complement physical genome research. It can also be defined as the qualitative and quantitative comparison of proteomes under different conditions to further unravel biological processes (http://www.expasy.ch/proteomics_def.html).

Chapter XIII
Data Mining in Proteomics Using Grid Computing

Fotis E. Psomopoulos
Aristotle University of Thessaloniki, Greece

Pericles A. Mitkas
Aristotle University of Thessaloniki, Greece

ABSTRACT

The scope of this chapter is the presentation of Data Mining techniques for knowledge extraction in proteomics, taking into account both the particular features of most proteomics issues (such as data retrieval and system complexity), and the opportunities and constraints found in a Grid environment. The chapter discusses the way new and potentially useful knowledge can be extracted from proteomics data, utilizing Grid resources in a transparent way. Protein classification is introduced as a current research issue in proteomics, which also demonstrates most of the domain – specific traits. An overview of common and custom-made Data Mining algorithms is provided, with emphasis on the specific needs of protein classification problems. A unified methodology is presented for complex Data Mining processes on the Grid, highlighting the different application types and the benefits and drawbacks in each case. Finally, the methodology is validated through real-world case studies, deployed over the EGEE grid environment.

INTRODUCTION

Although computational biology and bioinformatics are often confused as the same interdisciplinary field, they do have several distinguishing differences. Bioinformatics is mainly concerned with the analysis and processing of data, and therefore the advancement in both algorithmic and technical level of the techniques and theories to solve formal and practical data management problems. On the other hand, computational biology aims to solve specific biological problems, utilizing computers to test and evaluate hypotheses. The working definitions of these two fields, provided by National Institutes of Health (NIH, 2000), are the following:

"***Bioinformatics:*** *Research, development, or application of computational tools and approaches for expanding the use of biological, medical, behavioral or health data, including those to acquire, store, organize, archive, analyze, or visualize such data.*"

"***Computational Biology:*** *The development and application of data-analytical and theoretical methods, mathematical modeling and computational simulation techniques to the study of biological, behavioral, and social systems.*"

However, it is also emphasized that "*although bioinformatics and computational biology are distinct, there is also significant overlap and activity at their interface*". Proteomics is one of the key fields that exist in that overlapping area. In a nutshell, proteomics is the large-scale study of proteins, ranging from the structural and functional analysis to the construction of protein-protein interaction networks and phylogenetic trees. Proteins are large organic molecules composed of amino acids arranged in a linear chain and held together by peptide bonds. They are essential part of organisms, participating in all processes within cells; catalyzing biochemical reactions (enzymes), maintaining the cell shape serving as scaffolds, providing the means of signaling between cells, etc. The term proteome denotes the entire complement of proteins expressed by a genome at a given time and under defined conditions. The word itself is a portmanteau of "protein" and "genome".

There has been a recent shift in focus from genomics to proteomics, due to the fact that many consider proteomics to be the next step in the study of biological systems. The genome of an organism is fairly stable, showing little variation throughout its cells in comparison with the proteome, which is highly differentiated from cell to cell. One of the more significant insights that have emerged from proteomics is the nature of relationship between genes and proteins. The study of the mouse proteome (Gauss, 1999) has demonstrated that a protein can be considered as the expression of not one but many genes (Klose, 1999). Correspondingly, a single mutation in a gene can affect many proteins. Moreover, using the yeast proteome, the essential-essential protein interaction network has been proposed to form a generic scaffold around which organism-specific and taxon-specific proteins and interaction coalesce (Pereira-Leal, 2005).

BACKGROUND

In this section, some insight into the main data acquisition methods in proteomics will be provided, in order to present the common difficulties that may arise during data analysis. As far as the actual analysis is concerned, the main focus will be on the protein classification problem, due to the fact that it exhibits several of the issues common in other bioinformatics areas. Finally, after defining the concepts of Grid and Grid Computing, an overview of the current status concerning the symbiosis of bioinformatics and grid computing will be discussed.

Proteomics

There are several primary proteomics techniques for data acquisition; gel electrophoresis and mass spectrometry. Two-dimensional polyacrylamide gel electrophoresis (2D-PAGE) separates proteins in the first dimension according to charge and in the second dimension according to molecular mass (O'Farell, 1975). One of the many variations of this technique is two-dimensional difference in-gel electrophoresis (2D-DIGE), in which proteins from two samples (e.g., normal vs. diseased) are differentially labeled using fluorescent dyes and simultaneously electrophoresed (Unlu, 1997). Mass spectrometry (MS) on the other hand, is essential for protein identification (Glich, 2003; Honore, 2004). Utilizing ionization of the crystallized protein, and based on the applied voltage and ion velocity, a mass spectrometer can determine the m/z of each ion.

Analysis of proteomic data is much more demanding compared to genomic studies for many reasons; proteins are much more complicated, have numerous functions, and can be potentially modified by the environment (e.g. there are proteins that are not activated until they are phosphorylated). In previous years, genomics and proteomics could only focus on a single gene or protein at a time. However, the advancement in the technologies in biosciences is providing an exponentially rising amount of data. For this reason there is a shift in research from hypothesis-driven to data-driven studies.

This shift is presenting scientists with new challenges in data analysis. On one hand, most proteomics studies have a limited number n of instances (i.e. samples or records). Usually n ranges from tens to hundreds of cases, opposed to the thousands or millions of instances in a typical engineering or finance application. On the other hand, high-throughput techniques in life sciences yield several thousand variables per case, in sharp contrast with the traditional data mining scenarios. This problem is known as the *curse of dimensionality*, or *small-n-large-p problem* (Berrar, 2007). In proteomics data sets, the number of variables-attributes p can be in the order of 10^4, whereas the number of cases-instances n is usually in the order of 10^2.

Protein Classification

The curse of dimensionality is especially evident when dealing with protein classification studies. Large-scale sequencing projects throughout the world keep turning out new protein sequences at an ever increasing rate. Many of these sequences do not have fully documented functions or properties, creating the need for automated or semi-automated tools for data analysis and organization. Although there is a plethora of tools, techniques and databases for data analysis of MS-based proteomics data (Matthiesen, 2007), there are far fewer computational tools and methods available for proteomics in a higher level. Given a new protein sequence, the common approach to predicting its function and analyzing its properties relies on pairwise comparison with other well-documented protein sequences. However, there exist many proteins that do have the same or similar functionality, but have very low sequence similarity. In these cases, the pairwise similarity method fails to provide useful information regarding the properties of the protein in question.

For this reason, much effort has been directed into searching for significant patterns in protein sequences. The different methodologies have yielded various representations of motifs and domains, and consequently several different classification systems. On one hand there are methods, such as PROSITE (Hoffmann, 1999), PRINTS (Attwood, 2000), Blocks (Henikoff, 2000) and IDENTIFY (Nevill-Manning, 1998), that aim to discover short motifs and patterns in a protein sequence. Other methods focus on identifying whole domains and domain interactions, such as Pfam (Bateman, 2000), ProDom (Corpet, 2000), Domo (Gracy, 1998) and SMART (Ponting, 1999). Finally there are studies which aim to extract information on a higher level, including the analysis of complete protein sequences, such as COGs (Tatusov, 1997), PIR (Barker, 1996) and Systers (Krause, 1998), or combining measures of similarity on protein sequences, groups of homologous proteins (clusters) and groups of related clusters, such as ProtoMap (Yona, 2000).

Overall, the major methods for protein classification can be distinguished into three basic groups: pairwise sequence comparison algorithms (Altschul, 1990, Smith, 1981), generative models for protein families (Krogh, 1994; Park, 1998;

Psomopoulos, 2005) and discriminative classifiers (Jaakkola, 2000; Leslie et al, 2002; Weston, 2005). Despite the fact that the methods utilize either bottom level information (i.e. the primary aminoacid protein sequence) or higher level abstraction (i.e. the motifs present in a protein sequence), they all share a common factor; building an accurate protein classification system depends critically upon choosing a good representation of the input sequences, be they primary (aminoacid) or meta (motif). Moreover, there are relatively few fully labeled data available (~40,000 records) as opposed to the unlabeled sequences which number over a million, and for each instance the number of attributes range from several dozen (when using meta-info, such as motifs) to hundreds (when utilizing the primary aminoacid sequence).

Grid Computing

In order to cope with this situation, most machine learning methods focus on specific groups of proteins or reduce either the size of the original data set or the number of attributes involved. Grid computing could potentially provide an alternate approach to this issue, by combining multiple approaches in a seamless way. Although the term "Grid" has been conflated, at least in popular perception, to embrace anything from advanced networking to artificial intelligence, Grid computing has emerged as an important new field in computer science. Foster (1998) gives the following definition of the Grid:

"A computational grid is a hardware and software infrastructure that provides dependable, consistent, pervasive, and inexpensive access to high-end computational capabilities."

The key concept of a Grid is the ability to negotiate resource-sharing arrangements among a set of participating parties, and then to use the resulting resource pool for some purpose. It is noted that:

"The sharing that we are concerned with is not primarily file exchange but rather direct access to computers, software, data, and other resources, as is required by a range of collaborative problem-solving and resource-brokering strategies emerging in industry, science, and engineering. This sharing is, necessarily, highly controlled, with resource providers and consumers defining clearly and carefully just what is shared, who is allowed to share, and the conditions under which sharing occurs. A set of individuals and/or institutions defined by such sharing rules form what we call a virtual organization." (Foster, 2001)

Although in practice a Grid environment can be viewed as a virtual computing architecture that provides the ability to perform higher throughput computing by taking advantage of many computers geographically distributed and connected by a network, Grid computing is not Parallel Computing. Parallel computing is a form of computing in which many instructions are carried out simultaneously, and it exists in several different forms; bit-level parallelism, instruction level parallelism, data parallelism and task parallelism. In a Grid environment it exists in the form of computer clusters that are available as resources. On the other hand, due to the fact the resources on a Grid are connected via a comparatively low bandwidth network, Grid computing shows its true potential when dealing with embarrassingly parallel problems. Therefore it can be argued that Grid computing is a superset containing the capabilities of both Parallel computing (utilizing the clusters available) and the ability to use geographically distributed computer resources.

From the previous it is obvious that bioinformatics applications stand to gain in a Grid environment both in regards of computational resources available, but in reliability and efficiency as well. There is already a lot of research in progress towards applying parallel computing techniques on bioinformatics methods, such as multiple sequence alignment (Data, 2006), gene expression analysis (Martino, 2006) and phylogenetic studies (Stamatakis, 2006) among others. Moreover, there is a recent trend in utilizing the Grid as a platform for complex approaches to bioinformatics problems, such as reverse-engineering gene-regulatory networks (Swain, 2005) and the Grid Basic Local Alignment Search Tool (Grid-Blast) (Krishnan, 2005), and in building specific frameworks over the Grid that target genomics and proteomics issues, such as the Biotechnology Information and Knowledge Grid (BioGrid) (Bata, 2002) and the Bioinformatics and Genomics Grid for European Research (BIG-GER) (Cameron, 2003).

It is evident from the ongoing research in the field that there is a lot of potential in employing the Grid as a means to enhance most of the current approaches in bioinformatics and proteomics analysis studies. However, most gain can be found in combining the existing data mining techniques with the distributed environment of the Grid in order to boost the efficiency and overall performance of proteomics algorithms.

DATA MINING IN PROTEOMICS

Data Mining is defined as the process of discovering patterns in data. This process must be automatic or (more usually) semi-automatic. Data mining is an inductive, iterative process that extracts information or knowledge patterns from volumes of data (Witten, 2005). Although many regard data mining as form of "advanced" statistics, in truth there are many differences between these two fields. Usually statistics are concerned with evaluating hypotheses over a specific dataset, whereas data mining often focuses on searching a generalization model throughout a set of working hypotheses. However it would be unjust and oversimplifying to argue that there are no common elements in the two fields. On the contrary, very similar schemes have evolved in parallel in machine learning and statistics, with many standard statistical techniques being adapted in data mining methods and vice versa.

One way of interpreting the problem of data mining, simultaneously distinguishing it from statistical approaches, it to imagine a search through a space of possible concept descriptions, expressed as rules, for one that fits the data. The process of generalization can be regarded as a search through an enormous, but finite, search space. In principle, the problem can be solved by enumerating descriptions and ruling out those that do not fit the examples presented. The descriptions remaining at the end of the process can be used to create a working model of the initial data.

In practice it is usually inefficient, or downright impossible, to enumerate all possible descriptions that can be extracted from a specific dataset. If each data record is represented with a vector x of p attributes, then the number of possibilities for each rule is $\prod(|p_i|)$, where $|p_i|$ is the number of distinct values that p_i attribute can have. It is obvious that the size of the search space rises dramatically with the number of attributes involved. Moreover, there are many problems, especially in proteomics and protein classification issues, where the attributes do not have distinct values but rather take their values from a continuous space. In these cases direct enumeration of the rules-space is impossible without first discretizing the attribute involved.

The generalization process in data mining, as presented so far, shows clearly that the most important decisions in the system are the following:

1. The concept description language
2. The order in which the space is searched
3. The way that overfitting to the data is avoided

Focusing on the field of proteomics, with emphasis on the protein classification problem, the above decisions can be further analyzed. First of all, as mentioned previously, protein classification problems suffer from the curse of dimensionality. This translates into a vast rules search space with a comparatively very small instance set, against which the final model should be verified. The common approach in proteomic studies is to determine which features are important and work with those ignoring the rest. However, high-throughput experiments are often of an explanatory nature, and highly focused hypotheses may not always be desired or possible. For this reason, and based on the detail level of the generalization model, the concept language should range from the low level sequence information available for the protein data set, to the higher level motif representation.

On the other hand, including all information available carries the risk of overfitting the model to the data available. Especially when dealing with the aminoacid protein sequences the risk is much greater, due to the fact that low level data is inherently "noisy". In order to balance search space complexity and model efficiency, together with maintaining a meaningful (for the expert) representation of the initial data, data description through motif sequences is the most promising. Substituting motif sequences instead of aminoacid sequences in a protein dataset leads to a smaller search space (but still suffering from the dimensionality problem) together with a comparatively minimal loss of information.

Finally, one of the most important issues in the protein classification problem is the lack of a globally accepted classification scheme. In the context of classification, the class labels are discrete variables. Given n instances (protein sequences), the set of k class labels $C = \{c_1, c_2, c_3, \ldots, c_k\}$ is usually arbitrarily defined based on the specific problem at issue; for example protein families are defined in a different way in the case of enzyme classification and in the case of homology classification. As the total number of sequenced proteins increases and interest expands in proteome analysis, there is an ongoing effort to organize proteins into families and to describe their component domains and motifs, the Gene Ontology Project (GO Consortium, 2000) being one of those efforts. Moreover, in sharp contrast with traditional classification scenarios, the great majority of proteins do not belong into a single functionally family, due to the fact that they can be distinguished on several different functionality and processing levels.

Attempting to formally define the protein classification problem, each protein x_i is described by p attributes, which represent the feature space, i.e. $x_i = (x_{i1}, x_{i2}, x_{i3}, \ldots, x_{ip})$. With each instance, a number $m < k$ of class labels is associated, i.e. $(x_i, c_{i1}, c_{i2}, \ldots, c_{im})$. Classification involves a process of learning-from-examples, in which the objective is to classify a case into one or more of the k classes on the basis of an observed measurement, i.e., to predict $(c_{i1}, c_{i2}, \ldots, c_{im})$ from x_i. Summarizing the domain-specific issues that emerge in the protein classification problem, the following points can be made:

1. There is no generally accepted protein classification scheme. Protein families are defined every time within the specific domain context.
2. Each protein can belong in more than one family (*multi-labeling problem*)
3. The attributes involved are numerous (*small-n-large-p problem*), especially in the aminoacid sequence representation, less in the motif representation.

Another issue that arises from the dataset representation is that in every case the length of the attributes present in each instance-protein is variable. From the traditional data mining point of view, this would mean that the dataset would contain a lot of missing values, especially in the end attributes. Again, the common approach in this case is either to transform the nominal attributes to binary, or to introduce an "empty" value as symmetrical padding at the beginning and at the end of the shorter sequences. An alternate approach is to use custom data mining algorithms that can process sequence instances (i.e. order of appearance is important) as opposed to set instances (i.e. knowledge that a value exists is enough).

There are many classification algorithms in literature which address the protein classification problem. Each of them is characterized by a set of idiosyncratic requirements in terms of data pre-processing, parameter configuration, and result evaluation and interpretation. However, as stated by the *No Free Lunch* theorem (Wolpert, 1997), no classifier in inherently superior to any other. The first step in each data mining approach is the "*Question* → *Task* → *Method*" chain. It is the nature of the problem and the actual dataset in hand that determines which classifier is better suited for the specific approach.

However, one of the key concepts in data mining approaches, which is definitely applicable on the protein classification problem, is the postulate known as *Occam's razor*, or else *lex parsimoniae*, which suggests that it is usually advisable to prefer the simplest model that fits the data well. Support Vector Machines (SVMs) are considered by many as one of the most sophisticated methods, with empirical evidence showing remarkably good results for complex datasets involving two classes. Still, in theory there is no sufficient evidence to support that SVMs should have an edge over other algorithms when dealing with the dimensionality problem (Hastie, 2002).

Nevertheless, protein classification is a computationally demanding problem due to its complexity. The use of simple classifiers, such as 0-R, k-nearest neighbors or C4.5, usually yields poor results as opposed to more sophisticated methods, such as SVMs. On the other hand, execution times tend to rise with the more complex classifiers, leading researchers to locate the golden medium in the balance between accurate results and efficient timeframe. At this point it must be noted that in most traditional classifier systems, there is a significant pre-processing step where the protein data set has to be transformed, and essentially simplified, in order for the classifier to be applied. This may lead to some information loss, which could become a drawback especially in cases of sparse datasets. For this reason, there has been significant ongoing research in the development of domain specific algorithms targeting protein classification issues. There exist in literature several techniques that can be applied both at the primary (aminoacid) sequence level, such as algorithms evolving from kernel theory (Weston, 2005; Leslie Eskin, 2002), and at the motif sequence level, such as stochastic finite state automata algorithms (Psomopoulos, 2004).

However, all the aforementioned classifier systems have a common drawback; due to the complexity of the protein classification problem and the fact that the input protein data is located on various, geographically distributed locations, the overall learning process is very time-consuming. In this aspect, the Grid environment is especially suited for these systems. There are several recent attempts in integrating in a seamless way these two fields: data mining and grid computing (Stankovski, 2008). Nevertheless, a unifying methodology that couples them together, and at the same time taking into account the specific needs and constraints of the protein classification problem will potentially open the way for close collaboration between domain experts and computer scientists.

PROTEIN CLASSIFICATION ON THE GRID

Analyzing a typical protein classification process, five distinct phases emerge:

1. Data acquisition
2. Data pre-processing
3. Classifier construction
4. Model evaluation
5. Model interpretation

With the exception of the final step, which usually can be performed by the expert end user, in each phase specific questions are posed to the user, based on the characteristics discussed previously:

1. Data are required from a single or multiple sources?
2. Data are required to be reformatted? Are all attributes required for the study? If not, should feature selection be performed automatically or by the user (custom attribute selection)?
3. Which type of model is preferred (rule-based, tree-based, neural network etc)? What is the nature of the algorithm involved (single or multiple processes)? Which are the protein families to be discriminated?
4. Which method of evaluation is preferred (statistical hypothesis, testing, cross-validation)?

In the existing data mining approaches, the success or failure of a distributed data mining project depends highly on a particular person or a tool. Therefore, data mining need standard protocols in terms of methodology and technology used. The objective of these standards should be to help users translate a protein classification problem into a set of data mining tasks at a higher abstraction level with less effort.

When attempting to define an experiment design workflow, one must take into account the different processing paradigms that can exist in a Grid context:

a. **Single Process applications:** This category includes use cases where the application consists of a single computationally intensive process. It includes most of the traditional data mining algorithms. Programs in this category tend to show only slightly better efficiency when used in a Grid environment, as opposed to a local workstation. However they are the most commonly used, mainly due to the fact that they present easy Grid deployment for the end user.
b. **Multiple Process applications:** This paradigm includes all use cases where the application can be analyzed into a number of repetitive processes, with little to no dependency between each repetition and process. Good examples of this category are the parameter sweep and single-program-multiple-data applications. They typically tend to be embarrassingly parallel in nature. These applications usually present maximum flexibility to the end user, allowing for multiple experimental setups to be executed simultaneously instead of sequentially on a local workstation. However, they often require some knowledge on the part of the end user for the submission on the Grid.
c. **Parallel applications:** This class includes all use cases that have fine-grained parallelism. Typically they are MPI implementations of parallel algorithms and they require a cluster environment in order to be executed. Characteristic member of this category is the parallel CLUSTALW application. Despite the fact that these applications are the most efficient among the three categories in terms of processing time, they are also the most rarely used. The main reason is that they require special implementation, and therefore special knowledge on the part of the end user. Although there are several algorithms available, end users still have to know how to approach such methods.

It is obvious that the ideal scenario would be to have the flexibility of the Multiple Process class, with the efficiency of the Parallel category and the ease-of-use of the Single Process paradigm. Although the whole combination is fairly difficult to achieve, it is within reason to combine flexibility with convenience for the end user. In any case it can be argued that running several different configurations simultaneously on several remote units instead of sequentially in a local unit, is a huge gain.

The first step in defining an experiment design process, it to specify the data involved. At this point the significance of having publicly available protein databases which can be queried directly through application interfaces must be underlined. Although there are several protein databases accessible by means of provided web-services, such as InterProScan and eFamily, their number is not even close to the number of the highly detailed and fully documented public databases available. This fact forces researchers either to maintain a local mirror of the data or to use only the provided web interfaces, thus countering the potential of a Grid application. In any case, for the workflow process, the input data can be acquired from:

- A user provided file (either flat or formatted)
- A user-defined machine using traditional database access protocols
- A public database, by means of a web-services interface
- A file already existing in the Grid, using a unique grid identifier

In all these cases there is no limit to the number of sources included, provided that they have either the same format or that a format-converter is available. In practice, the latter case is not an issue, since most of the protein databases provide both XML and FASTA formatted output.

The next step is to define whether some data pre-processing should take place. Again, there are several cases to be distinguished:

- All protein attributes are requested to be included in the study
- Automated feature selection is to be performed
- Custom attribute selection is to be provided by the user

All these cases can be easily processed by requesting from the user to add a corresponding entry during the experiment design process. Specifically in the last case, the user should provide a list with the attributes that will be taken into account in the classifier training process.

After defining data and pre-processing preferences, the main phase of the workflow is to specify the classifier system. It must be noted that in most cases this step is closely coupled with the previous phases; several data mining algorithms require certain data types in attribute values (e.g. some algorithms are designed to work only with real values), etc. Moreover, despite the fact that the end user can provide the data mining library, several algorithms are ready available on the Grid as resources. This allows for more flexibility and provides the user with an arsenal of tools without the need for an in-depth knowledge of each method.

Besides the classifier system, the protein classes and the application type have to be defined. The protein classes are easily selected from among the attributes of the previous step. The application type can either be explicitly defined, or to be implicitly inferred from the classifier system selected. However, manually defining the application type could potentially allow, under certain conditions, for Single Process application to be executed as Multiple. Although this exchange is requested by the end user, the actual transformation is performed during execution, where the input dataset is split and processed simultaneously.

The final step in the workflow is to specify the evaluation method. Since most classifier systems contain also several evaluation processes, this phase is simply a method selection. However, it is possible to request the resulting classifier model to be tested against a custom (user provided) dataset.

After setting all the parameters, the application is submitted on the Grid, which in turn takes on the responsibility of executing and returning the requested output. It must be noted that in the case of selecting multiple protein families and the actual classifier does not have inherently the corresponding mechanism, the workflow is automatically transformed into a Multiple Process paradigm, in order to cope with the multi-label problem. In this case, each process is assigned to create a model describing one of the families defined in the application description, and the models are combined in the end either by means of ensemble methods, or by simple voting mechanisms.

The goals of the workflow presented are two-fold; to introduce a protocol accepted and validated against scientific approaches of bioinformatics and computer science, and to provide seamless integration and ease of use in a Grid context. Table 1 summarizes the workflow phases together with the most common options.

CASE STUDIES OVER THE EGEE GRID

In order to evaluate the methodology discussed in the previous sections, two use-cases have been developed and deployed over the EGEE Grid Environment (EGEE Consortium, 2008). The EGEE project brings together scientists and engineers from more than 240 institutions in 45 countries world-wide, providing a seamless Grid infrastructure available to scientists 24 hours-a-day. The EGEE Grid consists of 41,000 CPU in addition to about 5 PB of disk storage and can potentially maintain 100,000 concurrent jobs.

The following diagram (Figure 1) shows schematically the architecture of the EGEE grid. Entrance point to the Grid infrastructure is the User Interface (UI), where a user creates and submits the jobs (i.e. the tasks that will be performed on the provided resources). The Resource Broker (RB) is the most important system of the architecture: it receives the jobs submitted by the users and matches them with the most appropriate available resources. It must be noted that this matching process relies heavily on the job description provided from the user with each job. This means that, from the end user point of view, there is a lot of room for resource allocation optimization during the creation of each job. An accurate description of the application requirements using JDL (job description language) at design time, can lead to dramatic improvements in overall efficiency. Finally, every Computational Node (or Site) consists of three main systems: the Computing Element, the Storage Element and the Worker Nodes. The Storage Element (SE) is simply the storage area of the Grid. Data required for the

Table 1. Workflow phases and options

	Phase	Options
1.	Data Acquisition	a. File (*Local / Grid*) b. Database access (*User / Public*) c. Application access (*Web service*)
2.	Data Pre-processing	a. No Feature Selection b. Automatic Feature Selection (*PCA, etc.*) c. User – selected features
3.	Classifier System	Algorithm application (*Local / Grid*)
4.	Protein Class	Gene Ontology, PFam, Prosite, SCOP, etc.
5.	Application Type	a. Single b. Multiple c. Parallel (*MPI only*)
6.	Evaluation Method	User – selected method (*Cross-validation, user – defined test set, etc.*)

applications can be stored here until required. The Computing Element (CE) is the entrance point for each Site; it receives the application from the Resource Broker, and assigns a Worker Node to it. The Worker Node (WN) is in charge of the actual application execution, sending the results to the Computing Element and from then on to the Resource Broker and the User Interface.

There are several advantages of this architecture. It allows for multiple applications to run simultaneously, independent of the type of applications involved, it provides easy access over the standard network, and requires little knowledge of the actual architecture from the end user. Moreover, it reinforces collaboration among scientists across multiple disciples through easy resource sharing, resources being either data or libraries that can exist in the storage element. In fact, sharing is as easy as storing the data and publishing the corresponding unique grid identifier.

Next, two case studies in protein classification will be discussed. In order to avoid repetition, the tools and algorithms mentioned are physically located in Storage Elements on the EGEE and retrieved directly on demand, unless stated otherwise.

Case Study 1: G-Class

G-Class is a divide-and-conquer data mining methodology which utilizes existing data mining algorithms in a parallel-enabled environment in order to create a single protein classification model (Polychroniadou, 2006). Using motif-based representation of the protein sequences, the goal is to evaluate the effect of data parallelization on the overall performance, accuracy and efficiency of the final classifier. The input dataset is retrieved from PROSITE (Hoffmann, 1999), a supervised motif database.

The key concept here is the use of existing classifier systems, which are not optimized for protein classification. Moreover, standard classifiers cannot deal with variable feature lengths, so a custom converter has been developed that transforms motif sequence into a binary vector. Each position in the vector is set according to the existence of the corresponding motif in the initial sequence. It is evident that this transformation drastically changes the search space of the classifier; instead of having a sequence with average length approximately 15-20 motifs, a binary vector corresponding to all available motifs (approximately over a thousand) is used. For this reason,

Figure 1. EGEE architecture

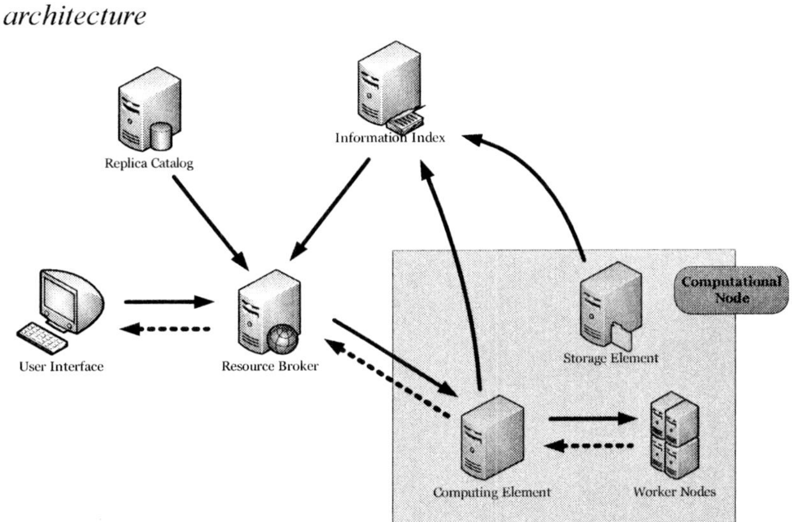

and when using all data available, most algorithms present low performance and the execution time is usually prohibitively long.

An overview of the workflow created in this case is shown in the figure above (Figure 2). Based on the methodology discussed previously, the first phase in deploying the whole process on the EGEE Grid is to define the input data source. In this case, the data are located on a local flat file, reformatted from the PROSITE data format to ARFF (Attribute Relation File Format) used by WEKA (Witten, 2005) data mining software package. One of the requirements of the procedure is to utilize all possible information, so in the next step no data preprocessing methods are specified. This in turn implies that the application

Figure 2. G-Class experiment design workflow

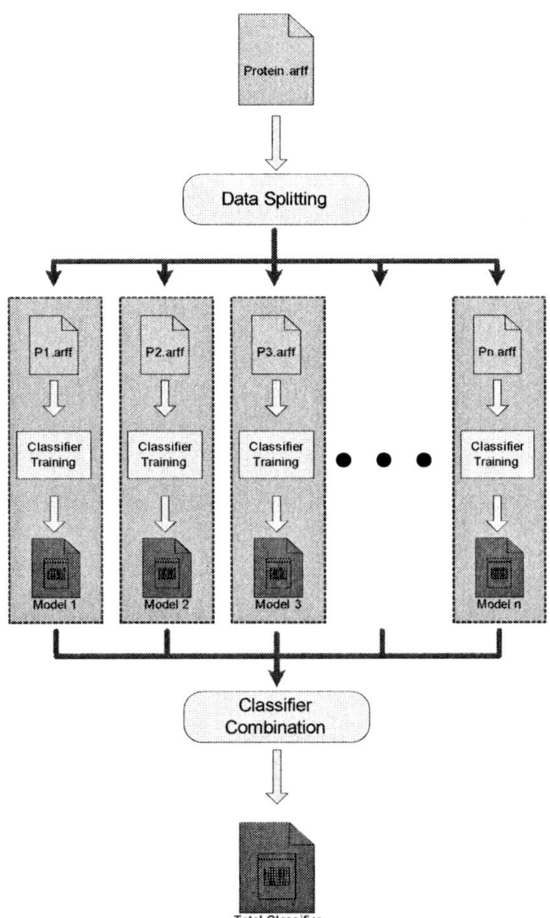

will be deployed not as a Single process, which is the default choice of standard classification algorithms, but as a Multiple process. The choice in application type is reinforced by the fact that PROSITE database contains numerous proteins families (approximately 1100), which leads to the creation and eventually a combination procedure of multiple classifiers. In order to ensure that all processes have access to the same amount of information, and therefore increasing the efficiency of the combination procedure, the original dataset is automatically split into k disjoint datasets, where k is the number of processes spawned.

Using two small datasets (one containing instances from 5 distinct classes and one containing instances from 70), the diagrams in Figure 3 clearly show that the actual distribution of instances across the classes remains the same before and after the splitting process. In each of the four diagrams, the top / top left figure (in red) shows the distribution of the instances in the original dataset, and the rest present the distribution in each of the "new" split datasets. It is obvious, especially in the 5-Classes case (first column of the Figure 3 diagrams), that the distribution shows little to no change in the new datasets, as opposed to the actual number of instances involved. For example, Class 1 contains approximately 1300 instances in the original dataset, where in each of the produced datasets it contains approximately 260 instances.

Based on the specifications discussed above, the file (Figure 4) describing the application and its parameters is created by a prototype local application and transformed into the corresponding jobs to be submitted through the UI on the EGEE Grid. This "internal" file is based on the JDL standard, with the necessary additions for the custom parameters. It is created automatically based on the user specifications, but it can also be constructed manually in order to allow for custom changes from the more experienced users.

The internal file is used in order to describe a number of jobs to be submitted, the number in

Figure 3. Instance distribution in classes

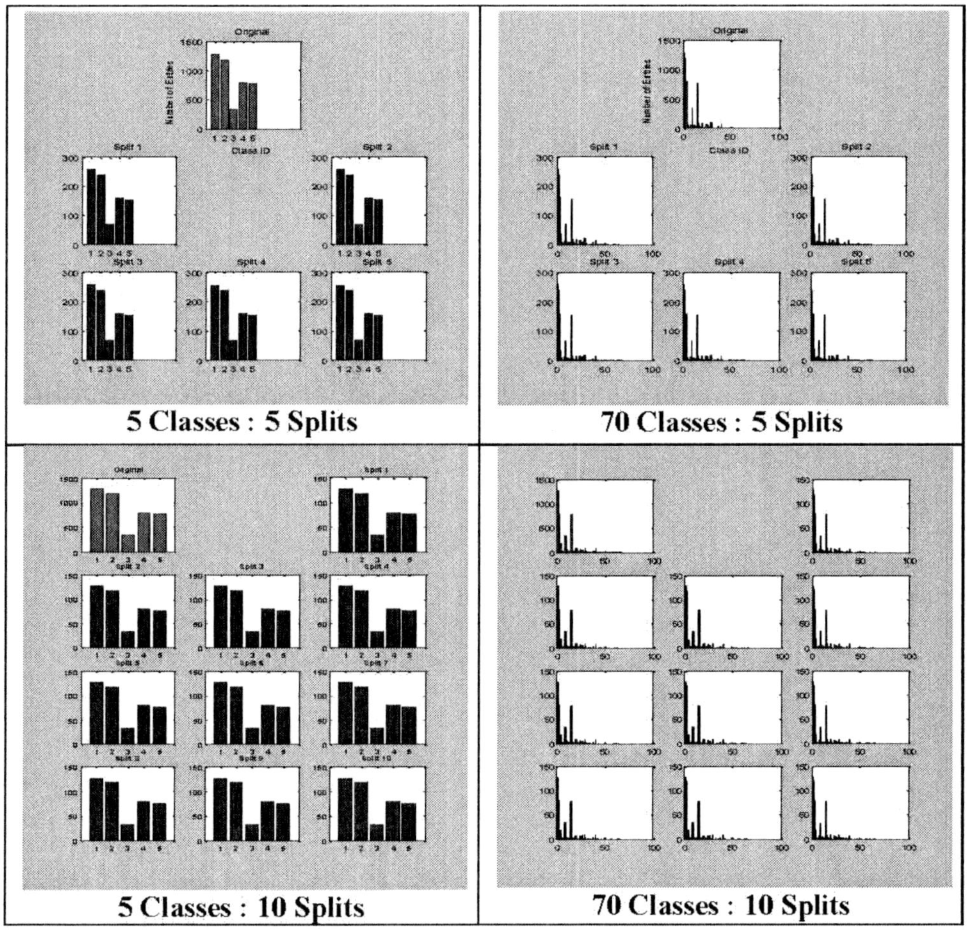

Figure 4. G-Class internal workflow representation

```
AppType = "Multiple";
NumCPU = 20;
PreProc = "NaN";
Class = {"Family"};
Algorithm = "weka.classifiers.trees.J48";
LocalData = {"Prosite.arff"};
RemoteData ={"lfn:/grid/see/fpsom/weka.jar"};
Output = {"gClass _ models.tar.gz"};
Eval = {{"cross-validation", 10}};
```

this case being 20. For each job, a pair of files is created: one shell script that is the higher level executable and the JDL file that specifically describes the corresponding process. An example of such a pair is shown in Figures 5 and 6.

When all process have finished, the produced models are returned to the end user, where they are combined and evaluated using 10-fold cross-validation. Generally, in *k*-fold cross-validation the original sample is partitioned into *k* subsamples. Of

Figure 5. G-Class sample JDL file

```
JobType = "normal";
VirtualOrganisation = "see";
Executable = "gClass5.sh";
StdOutput = "gClass5.out";
StdError = "gClass5.err";
InputSandBox = {"gClass5.sh", "Prosite_split5.arff"};
OutputSandBox = {"gClass5.out", "gClass5.err", "gClass5.model"};
InputData ={"lfn:/grid/see/fpsom/weka.jar"};
DataAccessProtocol = {"rfio"};
RetryCount = 5;
MyProxyServer = "myproxy.grid.auth.gr";
Requirements = (other.GlueCEPolicyMaxCPUTime > 1440) &&
                        (other.GlueHostProcessorClockSpeed >= 3000);
```

Figure 6. G-Class sample shell script file

```
#!/bin/sh
/bin/date
/bin/hostname -f
JAVA_BIN=`/usr/bin/which java`
$JAVA_BIN -Xmx1024M -cp .:`pwd`/weka.jar weka.classifiers.trees.J48
                -t `pwd`/Prosite_split5.arff -d `pwd`/gClass5.model
/bin/date
```

the *k* subsamples, a single subsample is retained as the validation data for testing the model, and the remaining $k-1$ subsamples are used as training data. The cross-validation process is then repeated *k* times (the folds), with each of the *k* subsamples used exactly once as the validation data. The *k* results from the folds then can be averaged (or otherwise combined) to produce a single estimation. Some results, using various subsets of the original data and different number of splits, are shown in the Figure 7. From the diagrams it is evident that there is a significant gain in execution time with the increase of the number of subsets, maintaining at the same time the overall classifier accuracy. Accuracy is defined as the proportion of the correctly classified instances in the total instances population.

Finally, Listing 1 shows a sample of one of the created C4.5 tree models. Using the motif-sequence representation of a protein on the model, a PROSITE family prediction can be made. For example, the motif sequence "PS00190 PS00192 PS50099 PS50016" is assigned to class "PDOC50016", and sequence "PS00027 PS00211 PS01187 PS01033" is assigned to "PDOC00027".

Case Study 2: Proteas

Proteas is a novel parallel methodology for protein function prediction (Gkekas, 2008). Data mining techniques are employed in order to construct a model for each Gene Ontology term, based on data generated from already annotated protein sequences. The models created can then be used to predict the annotation of new protein sequences. In more detail, the motif sequence of the protein under examination is extracted and run through all available Gene Ontology term models. This process generates similarity scores, which con-

Figure 7. G-Class experimental results

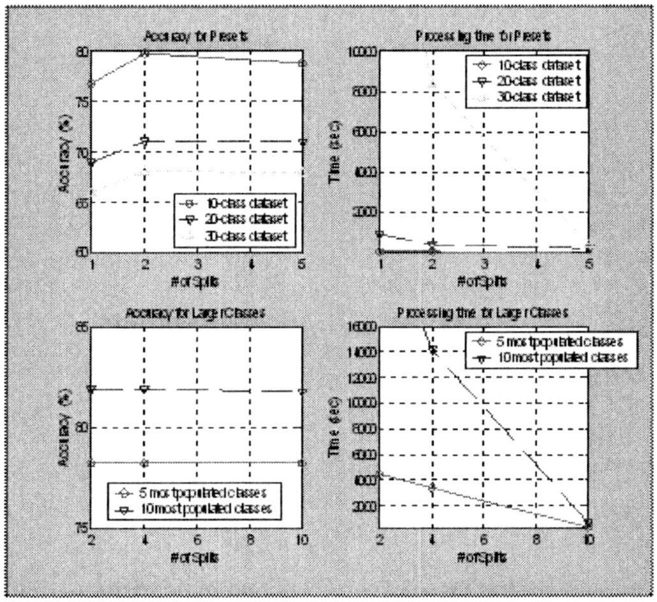

Listing 1. Part of a C4.5 tree model

```
PS00027 = YES: PDOC00027 (10.0/2.0)
PS00027 = NO
|    PS00086 = YES: PDOC00081 (5.0)
|    PS00086 = NO
|    |    PS00190 = YES
|    |    |    PS50016 = YES: PDOC50016 (2.0/1.0)
|    |    |    PS50016 = NO: PDOC00169 (5.0/1.0)
|    |    PS00190 = NO
|    |    |    PS00192 = YES: PDOC00171 (7.0/1.0)
|    |    |    PS00192 = NO
|    |    |    |    PS00211 = YES: PDOC00185 (5.0)
|    |    |    |    PS00211 = NO
|    |    |    |    |    PS01033 = YES: PDOC00793 (9.0/1.0)
|    |    |    |    |    PS01033 = NO
|    |    |    |    |    |    PS01187 = YES
|    |    |    |    |    |    |    PS50099 = YES: PDOC00010 (2.0/1.0)
|    |    |    |    |    |    |    PS50099 = NO: PDOC00913 (3.0/1.0)
|    |    |    |    |    |    PS01187 = NO
|    |    |    |    |    |    |    PS00018 = YES
.    .    .    .    .    .    .    ......
```

stitute an accurate prediction of the protein's annotation. The overall algorithm architecture is shown in the following figure (Figure 8).

Briefly the methodology consists of the following steps: first, the motifs present in already annotated protein sequences are obtained. This is achieved by the use of the UNIPROT code of each protein and the InterProScan tool, taking under consideration every available sequence database (e.g. PRODOM, PROFILE, PFAM, etc).

Figure 8. Proteas algorithm design

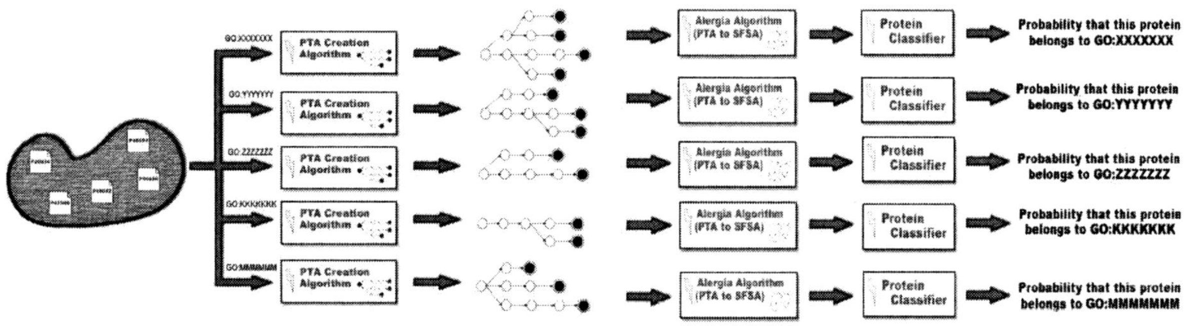

Through this process the initial protein data set is constructed, which contains for every protein its motif sequence and all the terms it has been annotated with. In the next step, the Training Set is further divided into several smaller protein data sets with each of them consisting solely of protein sequences that have been annotated with the same term. For each of the protein data sets created during the previous step, a Prefix Tree Acceptor (PTA) is constructed using the motif sequence of the proteins in the set. Utilizing the Alergia algorithm (Carrasco, 1994), this PTA is consequently transformed into a more generalized Stochastic Finite State Automaton (SFSA), which essentially models the corresponding term. Therefore, through this process, one model for each term is derived. The final step of the methodology is executed independently and in parallel. It involves the utilization of all previously constructed models in order to predict the annotation of unknown proteins. To this end, the motif sequence of each of the unknown proteins is extracted and run through every available model. This process produces similarity scores for every term, which altogether constitute an accurate prediction of the proteins' annotation.

The key point in this case-study is the custom MPI-enabled classification algorithm. It has been designed focusing on the protein classification problem, it's constrains and idiosyncrasy. The main advantage is flexibility in terms of usability and data access. However, it still shares some of the drawbacks of Parallel applications as mentioned previously, which means that its configuration, beyond basic parameters, requires some special knowledge from the part of the end user.

Data acquisition in this case is performed semi-automatically, utilizing the InterProScan

Figure 9. Proteas internal workflow representation

```
AppType = "Parallel";
NumCPU = 16;
PreProc = "NaN";
Class = {"GO"};
Algorithm =
        "lfn:/grid/see/fpsom/proteas";
LocalData = {"NaN"};
RemoteData =
   {"lfn:/grid/see/fpsom/input _ data.tar.jar"};
Output = {"Proteas _ models.tar.gz"};
Eval = {{"test _ ratio", 20}};
```

(Zdobnov, 2001) web-service tool. InterProScan allows data requests from multiple protein databases, using only the UniProt code of the proteins involved, while at the same time acquiring motif information from multiple sources. Motif representation of the protein sequences is selected, without any preprocessing involved. However, the number of motifs included is greater in comparison with the G-Class procedure (e.g. there are proteins containing over 30 motifs), and the sequences are approached not as motif set, but as true, ordered sequences. The involved amount of information can be efficiently managed due to the fact that the actual algorithm is optimized for protein classification, and is implemented for fine-grained parallel execution in a cluster environment.

The next phase in the workflow is to define the evaluation process, which will be a percentage of the original dataset. The internal file created locally is shown in the Figure 9. This time, both data and algorithm exist as Grid resources. The application is defined as belonging in the Parallel class, which means that it requires an MPI environment for execution. The evaluation is performed using a percentage (20%) of the original dataset, and the rest (80%) is used to construct the classifier models. This time, there is only a single JDL file created, due to the specific requirements of an MPI job submission on the EGEE Grid (Figure 10).

The following diagrams (Figures 11 and 12) show the execution time and the classifier accuracy over a dataset including approximately 20,000 protein sequences, using different number of CPUs. It is evident that there is an overall gain in time consumption, especially when using 8 or 16 CPUs, whereas the accuracy of the classifier remains constant.

Figure 11 demonstrates the time required to process all protein files of the Test Set. It is obvious that the larger the Test Set is the more protein files it contains and thus, more time is required for the application to produce results. It is illustrated that significant reduction of the processing time can be achieved if the application utilizes more than a single CPU. Specifically, the processing time of the experiment where 8 processors were used is approximately half in comparison to the processing time of the experiment were 4 processors where used. A similar relationship is demonstrated also between the experiments using 16 and 8 processors.

The mean accuracy for every Training Set / Test Set ratio is presented in Figure 12. Therein, small variance in accuracy is observed regarding the percentage of the dataset used for training,

Figure 10. Proteas sample JDL file

```
JobType = "mpich";
VirtualOrganisation = "see";
NodeNumber = 16;
Executable = "proteas";
StdOutput = "proteas.out";
StdError = "proteas.err";
InputSandBox = {"parameters.conf"};
OutputSandBox = {"proteas.out", "proteas.err",
                                        "Proteas _ models.tar.gz"};
InputData ={"lfn:/grid/see/fpsom/proteas",
                          "lfn:/grid/see/fpsom/input _ data.tar.jar"};
DataAccessProtocol = {"rfio"};
RetryCount = 5;
MyProxyServer = "myproxy.grid.auth.gr";
```

Figure 11. Proteas execution time

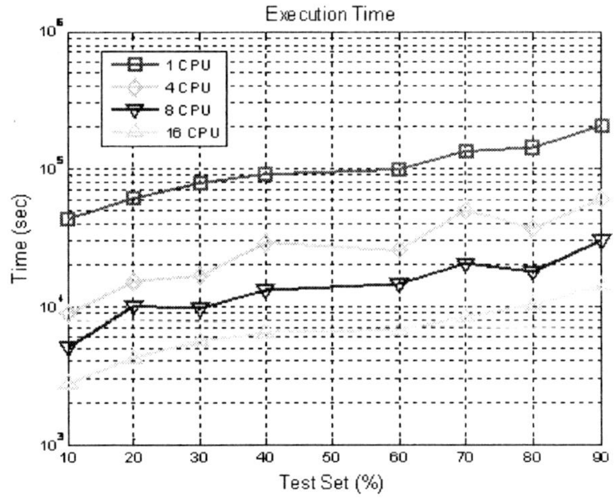

Figure 12. Proteas classifier accuracy

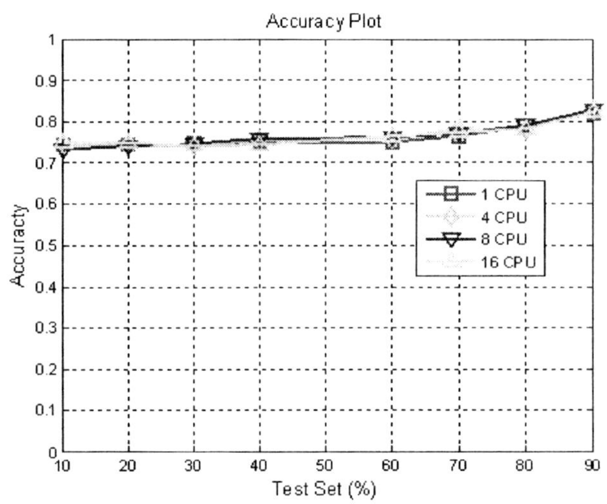

fluctuating between 0.7542 and 0.8364. The fact that the value of accuracy remains fairly constant across various Training Set / Test Set ratios shows that the methodology produces very accurate results even if the Training Set is a very small portion of the initial protein dataset and relatively small compared to the test set. Thus, the algorithm is not affected by the Training Set / Test Set ratio and is able to work flawlessly under conditions where information is rather limited.

Finally, Listing 2 presents some of the results produced using the Proteas algorithm. The protein family (class) selected is Gene Ontology (GO) and in the training set all databases available through InterProScan were utilized. Similarly with the G-Class methodology, the target "unknown" protein is represented with the corresponding motif sequence.

FUTURE TRENDS

The prototype workflow presented in this chapter is only the first step towards the unification of the

Listing 2. Sample Proteas output file

```
PROTEIN FILE:  ./test _ data/P01222 _ result.xml
Motif Sequence: SM00068 PR00438 PS00261 PS00689 PR00438 GO:0005179


GO term:  GO:0005576
Probability:  0.625
Support:  0.0322581
Confidence:  0
Motif combination:  PR00438 PF00207
Listing 2: Sample Proteas output file
```

end user's perceptions and needs, and the complexity of real world data mining applications over a grid environment. There is considerable ongoing research towards the development of more efficient and accurate data mining algorithms with emphasis on domain-specific solutions. Since the majority of these solutions have a somewhat limited scope, there still remains the goal of bridging the gap between the different viewpoints. Admittedly this gap tends to shrink, at least in recent years, however, the main issue remains unresolved; is it possible to provide state-of-the-art data mining applications while at the same time introducing a level of abstraction that would allow researchers to perform complex tasks without in-depth knowledge of the application itself?

This question in the context of bioinformatics points to two major directions. On one hand, the large scale integration and unified annotation of gene and protein data will allow for the construction of formal mathematical descriptions of the data. These descriptions can, in turn, be used to produce algorithms closely coupled with the domain specific traits, as they are portrayed in the mathematical concepts. The first step in this process is the integration of several different databases into a virtual repository accessible through a single interface. This non-trivial and highly parallel task could be greatly facilitated in a grid environment, where several of the technical issues that may arise have already been resolved.

On the other hand, the formalization of the scientific methodologies employed in various bioinformatics applications will provide the groundwork for the development of new algorithms that better mirror the actual human-performed procedures. Although there are several attempts toward this direction, scientific progress is hindered by the fact that each researcher operates in relative isolation, regarding datasets and methods. To this end, the Grid has the potential to act as a catalyst and foster stronger collaboration among researchers, through the establishment of universally accepted workflows and procedures in the form of common grid resources.

CONCLUSION

As the demand for automated analysis of large and distributed protein data grows, new challenges emerge both in the area of Grid Computing and in the collaboration among scientists of different fields. The typical end user is mainly driven by concepts, issues, tasks and requirements arising from his or her application domain. Besides efficiency and effectiveness, the main concerns of such a user include ease-of-use, responsiveness, interactivity, flexibility and conceptual simplicity of a grid-enabled data mining application. A prototype workflow has been presented which can be perceived as a first step towards this direction.

Despite the fact that the scientific community has realized the need for transparent and, especially, easy access to technologies and methodologies for high-throughput experimentation, the evolving process is hampered by the fact that scientists often work in isolation. Future developments in the area of seamless user interaction with the Grid will certainly overcome such difficulties and present a whole new way in how grids are used in Life Sciences.

REFERENCES

Altschul, S.F. et al. (1990). A basic local alignment search tool. *Journal of Molecular Biology, 215*, 403-410.

Attwood, T.K., Flower, D.R., Lewis, A. P., Mabey, J.E., Morgan, S.R., Scordis, P, Selley, J., & Wright, W. (2000). PRINTS-S: the database formerly known as PRINTS. *Nucleic Acids Research, 27*, 220-225.

Barker, W.C., Pfeiffer, F., & George, D.G. (1996). Superfamily classification in FIR-international protein sequence database. *Methods in Enzymology, 266*, 59-71.

Bata, P., Alessandrini, V., Girou, D., MacLaren, J., Brooke, J., Pytlinski, J., et al. (2002). BIOGRID-A European grid for molecular biology. In *Proceedings of the 11th IEEE International Symposium on High Performance Distributed Computing (HPDC-11)*, Edinburgh, UK.

Bateman, A., Birney, E., Durbin, R., Eddy, S.R., Finn, R.D., & Sonnhammer, E.L. (2000). The Pfam protein families database. *Nucleic Acids Research, 28*, 263-266.

Berrar, D., Granzow, M., & Dubitzky, W. (2007). Introduction to Genomic and Proteomic Data Analysis. In D. Berrar, M. Granzow, & W. Dubitzky (Eds.), *Fundamentals of Data Mining in Genomics and Proteomics,* (pp. 1-37). Berlin, German: Springer Science.

Cameron G., (2003). Bioinformatics and Genomics Grid for European Research (BIGGER). In *Proceedings of the First European Health Grid Conference (HealthGrid'03)*, Lyon, France.

Carrasco, R. C., & J. Oncina, J. (1994). Learning stochastic regular grammar by means of state merging method, In *Proceedings of the 2nd International Colloquium on Grammatical Inference (ICGI '94)*, Alicante, Spain. (Lecture Notes in Artificial Intelligence LNAI 862), (pp. 139 – 152). Berlin, Germany: Springer - Verlag.

Corpet, F., Gouzy, J., & Kahn, D. (2000). ProDom and ProDom-CG: tools for protein domain analysis and whole genome comparisons. *Nucleic Acids Research, 28*, 267-269.

Data, A. (2006). Multiple Sequence Alignment in Parallel on a Cluster of Workstations. In A. Zomaya (Ed.), *Parallel Computing for Bioinformatics and Computational Biology*, (pp. 193-210). Hoboken, NJ: Wiley.

Enabling Grids for E-Science (n.d.). Retrieved March 17, 2008, from http://www.eu-egee.org

Foster, I. (Ed.), Kesselman, C. (Ed.). (1998). *The Grid: Blueprint for a New Computing Infrastructure*. San Francisco, CA: Morgan Kaufmann Publishers.

Foster, I., Kesselman, C., & Tuecke, S. (2001). The anatomy of the grid: enabling scalable virtual organizations. *International Journal of High Performance Computing Applications, 15*(3), 200-222.

Gauss, C., Kalkum, M., Lowe, M., Lehrach, H., & Klose, J. (1999). Analysis of the mouse proteome. (I) Brain proteins: separation by two-dimensional electrophoresis and identification by mass spectrometry and genetic variation. *Electrophoresis, 20*, 575-600.

Gkekas, C., Psomopoulos, F., & Mitkas, P. (2008). *A Parallel Data Mining Application for Gene On-*

tology Term Prediction, 3rd EGEE User Forum, Clermont-Ferrand, France.

Glich, G.L., & Vachet, R.W. (2003). The basics of mass spectrometry in the twenty first century. *Nature Reviews Drug Discovery*, *2*(2), 140-150.

Gracy, J., & Argos, P. (1998). Automated protein sequence database classification. *Bioinformatics*, *14*, 164-187.

Hastie, T., Tibshirani, R., & Friedman, J. (2002). *The Elements of Statistical Learning*. New York: Springer.

Henikoff, J.G., Henikoff, S., & Pietrokovski, S. (2000). Increased coverage of protein families with the Blocks Database servers. *Nucleic Acids Research*, *28*, 228-230.

Hoffmann, K., Bucher, P., Falquet, L., & Bairoch, A. (1999). The PROSITE database, its status in 1999. *Nucleic Acids Research*, *27*, 215-219.

Honore, B., Ostergaard, M., & Vorum, H. (2004). Functional genomics studied by proteomics. *Bioessays*, *26*(8), 901-915.

Jaakkola, T., et al. (2000). A discriminative framework for detecting remote protein homologies. *Journal of Molecular Biology*, *7*, 95-114.

Klose, J. (1999). Genotypes and phenotypes. *Electrophoresis*, *20*, 643-652.

Krause, A., & Vingron, M. (1998). A set-theoretic approach to database searching and clustering. *Bioinformatics*, *14*, 430-438.

Krishnan, A. (2005). GridBLAST: a Globus-based high-throughput implementation of BLAST in a Grid Computing framework. *Concurrency and Computation: Practice and Experience*, *17*(13), 1607-1623.

Krogh, A., et al. (1994). Hidden Markov models in computational biology: applications to protein modeling. *Journal of Molecular Biology*, *235*, 1501-1531.

Leslie, C., Eskin, E., & Noble W.S. (2002). The Spectrum Kernel: A string kernel for SVM protein classification. *Pacific Symposium on Biocomputing*, *7*, 566-575.

Leslie, C., et al. (2002). Mismatch string kernels for SVM protein classification. *Advanced Neural Information Processing Systems*, *15*, 1441-1448.

Martino, R. L. (2006). Parallel Computing in the Analysis of Gene Expression Relationships. In Zomaya A. (Ed.), *Parallel Computing for Bioinformatics and Computational Biology*, (pp. 265-284). Berlin, Germany: Wiley.

Matthiesen, R. (2007). Methods, algorithms and tools in computational proteomics: A practical point of view. *Proteomics*, *7*, 2815-2832.

National Institutes of Health (2000). *Working definition of bioinformatics and computational biology*. Retrieved March 17, 2008, from http://www.bisti.nih.gov/CompuBioDef.pdf

Nevill-Manning, C.G., Wu, T.D., & Brutlag, D.L. (1998). Highly Specific Protein Sequence Motifs for Genome. *Proceedings of the National Academy of Sciences USA*, *95*, 5865-5871.

O'Farell, P.H. (1975). High-resolution two-dimensional gel electrophoresis of proteins. *Journal of Biolical Chemistry*, *250*(10), 4007-4021.

Park, J., et al. (1998). Sequence comparisons using multiple sequences detect twice as many remote homologues as pairwise methods. *Journal of Molecular Biology*, *284*, 1201-1210.

Pereira-Leal, J. B., Audit, B., Peregrin-Alvarez, J. M., & Ouzounis C. A. (2005). An Exponential Core in the Heart of the Yeast Protein Interaction Network. *Molecular Biology and Evolution*, *22*(3), 421-425.

Polychroniadou, H., Psomopoulos, F., & Miktas, P. (2006). G-Class: A divide and conquer application for protein classification. In *Proceedings of the 2nd ADBIS Workshop on Data Mining and*

Knowledge Discovery, Thessaloniki, Greece, (pp. 121-132).

Ponting, C.P., Schultz, J., Milpetz, F., & Bork, P. (1999). SMART: identification and annotation of domains from signalling and extracellular protein sequences. *Nucleic Acids Research, 27*, 229-232.

Psomopoulos, F., Diplaris, S., & Mitkas, P. (2004). A finite state automata based technique for protein classification rules induction. In *the Proceedings of the Second European Conference on Data Mining and Text Mining in Bioinformatcs*, ECML/PKDD, Piza, Italy, (pp. 54-60).

Psomopoulos, F., & Mitkas, P. (2005). A protein classification engine based on stochastic finite state automata. *Lecture Series on Computer and Computational Sciences VSP/Brill, 4B*, 1371-1374.

Smith, T., & Waterman, M. (1981). Identification of common molecular subsequences. *Journal of Molecular Biology, 147*, 195-197.

Stamatakis, A. (2006). Parallel and Distributed Computation of Large Phylogenetic Trees. In Zomaya A. (Ed.), *Parallel Computing for Bioinformatics and Computational Biology*, (pp. 327-346). Hoboken, NJ: Wiley.

Stankovski, V., Swain, M., Kravtsov, V., Niessen, T., Wegener, D., Kindermann, J., & Dubitzky, W. (2008). Grid-enabling data mining applications with DataMiningGrid: An architectural perpective. *Future Generation Computer Systems, 24*, 259-279.

Swain, M., Hunniford, T., Mandel, J., Palfreyman, N., & Dubitzky, W., (2005). Reverse-engineering gene-regulatory networks using evolutionary algorithms and grid computing, *Journal of Clinical Monitoring and Computing, 19*(4-5), 329-337.

Tatusov, R.L., Eugene, V.K., & David, J.L. (1997). A genomic perspective on protein families. *Science, 278*, 631-637.

The Gene Ontology Consortium (2000). Gene Ontology: tool for the unification of biology, *Nature Genetics, 25*, 25-29.

Unlu, M., Morgan, M.E., & Minden, J.S. (1997). Difference gel electrophoresis: A single gel method for detecting changes in protein extracts. *Electrophoresis, 18*(11), 2071-2077.

Weston, J., Leslie, C., Ie, E., Zhou, D., Elisseeff, A., & Noble William S. (2005). Semi-supervised protein classification using cluster kernels. *Bioinformatics, 21*(15): 3241-3247.

Witten, I., Frank, E. (2005). *Practical Machine Learning Tools and Techniques*, (2nd Ed.). San Francisco, CA: Morgan Kaufmann.

Wolpert, D., & Macready, W. (1997). No free lunch theorems for optimization. *IEEE Transactions on Evolutionary Computing, 1*(1), 67-82.

Yona, G., Linial, N., & Linial, M. (2000). ProtoMap: automatic classification of protein sequences and hierarchy of protein families. *Nucleic Acids Research, 28*(1), 49-55.

Zdobnov, E.M., & Apweiler, R. (2001). InterProScan: protein domains identifier. *Bioinformatics, 17*, 847-848.

KEY TERMS AND DEFINITIONS

Data Mining (DM): The process of knowledge discovery, consisting of applying data analysis and discovery algorithms that, under acceptable computational efficiency limitations, produce a particular enumeration of patterns over the data.

Grid Computing: The process of utilizing the existing infrastructure of geographically distributed and autonomous resources in order to perform parallel and/or distributed computing with non-trivial quality of service.

Proteome: The set of proteins expressed in a certain cell at a specified time given specific conditions.

Proteomics: The field of bioinformatics which focuses on the study of the proteome.

Protein Family: A set of proteins which show a specific structural and/or functional similarity.

Protein Classification: The process of assigning a single or multiple Protein Family labels to a protein sequence with known aminoacid composition but unknown functions and/or properties.

Scientific Workflow: A reliable pattern of activity enabled by a systematic organization of resources and information flows into a work process that can be documented and learned.

Section III
Grid-Based Bioinformatics Environments

Chapter XIV
ProGenGrid:
A Grid Problem Solving Environment for Bioinformatics

Maria Mirto
University of Salento, Lecce & SPACI Consortium, Italy

Italo Epicoco
University of Salento, Lecce & SPACI Consortium, Italy

Massimo Cafaro
University of Salento, Lecce & SPACI Consortium, Italy

Sandro Fiore
University of Salento, Lecce & SPACI Consortium, Italy

Marco Passante
University of Salento, Lecce & SPACI Consortium, Italy

Alessandro Negro
University of Salento, Lecce & SPACI Consortium, Italy

Giovanni Aloisio
University of Salento, Lecce & SPACI Consortium, Italy

ABSTRACT

In this chapter, the ProGenGrid (Proteomics and Genomics Grid) research project, which started in 2004, is described. It is a Grid Problem Solving Environment, specialized for the Bioinformatics domain, which aims at providing an integrated environment in order to compose, schedule and monitor biological applications in a Computational Grid. The main feature offered by this environment is the possibility to use a friendly web interface for composing workflow jobs, scheduled on different grid middleware.

INTRODUCTION

Bioinformatics applications require high computing power, owing to the large size of data sets and the complexity of basic computations, may access heterogeneous data, where heterogeneity is in data format, access policy, distribution etc., and require a secure infrastructure, because private data may also be accessed. A Problem Solving Environment (PSE) approach, exploiting many technologies, can fulfill such bioinformatics requirements. A PSE can be used for the definition and composition of complex applications, hiding programming and configuration details to the user that is free to concentrate only on her specific problem. Moreover, Grids can be used for building geographically distributed collaborative problem solving environments and Grid aware PSEs can search and use dispersed high performance computing, networking, and data resources.

In this chapter the architecture of ProGenGrid (Proteomics and Genomics Grid), a Grid PSE for supporting complex "in silico" experiments in bioinformatics, is presented. In this system, tools and data are combined into a graphical tool and a complex biological process is presented as a workflow represented by an arbirary graph in which vertices are bioinformatics processes and whose edges may give rise to loops. When a workflow is created and all of the parameters of each vertex are set, the workflow is executed through an engine that schedules the execution of each process on different Grid middleware such as gLite, Unicore and Globus.

Finally, several case studies about the protein tertiary structure prediction are presented.

BACKGROUND

PSEs have been investigated over the past years. Culler and Fried (Culler, 1963) initiated to investigate automatic software systems for solving mathematical problems with computers focusing primarily on applications issues instead of programming issues. At that time, the term applications indicated scientific and engineering applications that were generally solved using mathematical solvers or scientific algorithms managing vectors and matrices.

Despite the time passed from that early research work there in not a precise definition of what a PSE is. The following well-known definition was given by Gallopoulos, Houstis, and Rice: "*A PSE is a computer system that provides all the computational features necessary to solve a target class of problems. . . . PSEs use the language of the target class of problems*" (Houstis, 1997).

According to Walker et al., "*A PSE is a complete, integrated computing environment for composing, compiling, and running applications in a specific area*" (Walker, 2000).

In (Schuchardt, 2002), PSEs are defined as "*Problem-oriented computing environments that support all the scientific computational problem-solving activities ranging from problem formulation, to algorithm selection, to simulation execution, to solution visualization. PSEs link a heterogeneous mix of resources including people, computers, data, and information within a seamless environment to solve a problem*". In 2003 Cunha described a PSE as "*An integrated environment for solving a class of related problems in an application domain; easy to use by the end-user; based on state-of-the-art algorithms. It must provide support for problem specification, resource management, execution services*" (Cunha, 2003).

Such definitions agree on the basic features of PSEs, but provide different approaches about how PSEs can be composed and used. PSEs can benefit from advancements in hardware/software solutions achieved in parallel and distributed systems and tools.

One of the most interesting models in the area of parallel and distributed computing is the Grid Computing paradigm.

Grid computing (Berman, 2003) represents an opportunity for PSE designers and users. It can provide an high-performance infrastructure for running PSEs and, at the same time, a valuable source of resources that can be integrated in PSEs. Grids can be used for building geographically distributed collaborative problem solving environments, and Grid aware PSEs (G-PSE) can search and use dispersed high performance computing, networking and data resources.

So we define a Grid PSE as a grid-based PSE that integrates heterogeneous components into an environment providing transparent access to distributed resources, collaborative modeling and simulation, and grid portals.

From a Grid PSE designer's perspective,the Grid model is based on the Service Oriented Architecture (SOA) (Channabasavaiah, 2003; Krafzig, 2004), so that each component is exposed as a service using standard protocols such as SOAP. Indeed, the difficulty to integrate dispersed and de-coupled components and resources of Grid-based PSEs has lead to redefining each component as a Web Service entity, (WS), loosely coupled and independent of the architecture and programming language, and based on standard protocols for exchanging messages. So the Grid-based PSE architecture follows the SOA paradigm for organizing and utilizing distributed capabilities that may be under the control of different administrative domains.

The great consensus reached by SOA is even wider than the one about object orientation: SOA is endorsed by a number of cross company groups, gathering all of the major players of the IT world, that are strongly committed to guarantee to customers a uniform approach and minimize the dependencies on proprietary vendor technologies.

The Open Grid Services Architecture (OGSA, Foster 2001), developed under the auspices of the Global Grid Forum and primarily led by IBM and the Globus team, aims at extending the capabilities of WSs in order to support more sophisticated services. Indeed, WSs do have certain limitations: plain WSs (as currently specified by the World Wide Web Consortium, W3C) wouldn't be very helpful for building a grid application.

By 2004, the Web Service Resource Framework (WSRF Foster, 2005) specification, released by OASIS, improved several aspects of Web Services to make them more adequate for grid applications.

The main improvement in WSRF is the statefulness of grid resources.

Plain WS are usually stateless (even though, in theory, there is nothing in the WS Architecture that says they can't be stateful).

This means that a WS can't "remember" information, or keep state, from one invocation to another (a WS can use whatever format and related technology to save data; in this context, even adopting a standard such as SQL does not count as a standard, since what is required is that all of the WS agree to conform and to adopt to a particular standard, such as WSRF). There are plenty of applications which have no need for statefulness. For example, a Weather WS is an example of a WS which has no need to know what happened in previous invocations.

However, Grid applications do generally require state. Giving WS the ability to keep state information while still keeping them stateless seems like a complex problem. In WSRF, the WS and state information are completely separate. Instead of putting the state in the WS (thus making it stateful, which is generally regarded as a bad thing) it is put in a separate entity called a resource, which will store all of the state information. Each resource will have a unique key, so if a stateful interaction with a WS is required, the WS have to be instructed to use a particular resource.

According to the SOA architecture, in Figure 1 the architecture of a Grid PSE, specialized for Bioinformatics, is showed.

On the top layer a Grid portal provides a standard entry point and seamless access to the Grid PSE exposing services and resources provided by the Computational Grid.

The application services can run separately or can be inserted into a graph representing a workflow, by using an editor. This allows specifying complex (distributed) applications, integrating and composing individual simple services. The applications access the high level services namely Resource Management and Data Management. The Resource Management services implement the functionalities for job submission, job monitoring, file transfer and management of user's credentials. This layer is built on top of basic services offered by different grid middleware such as Globus Toolkit, a "de facto" standard for the deployment of a Grid (Foster, 1997), gLite (Laure, 2004) and Unicore (Almond, 1999).

Additional functionalities provided by the Resource Management services are scheduling and monitoring of the user's requests, the optimal allocation of resources in a distributed environment, etc.

The Data Management services provide a set of advanced data grid services to transparently, efficiently and securely manage databases and file on the Grid. This layer also includes secure mechanisms for data transfer on the Grid through advanced protocols to guarantee a high level of efficiency. The security layer concerns managing and accessing both services and resources within the Grid PSE environment. This layer is built on top of basic services offered by different grid middleware.

The Grid PSE Meets the Bioinformatics Requirements

Managing the deluge of biological data produced by large-scale experiments such as genome projects is one of the major global challenges of bioinformatics. Biological datasets present challenges from many different perspectives. The scale of data to be processed requires access to large collections of widely distributed resources such as those found in computational grids.

The analysis of these data needs to be farmed out to hugely provisioned computing resources with large and efficient storage devices. Indeed, bioinformaticians need for their daily analyses of genomes efficient access to these biological data and bioinformatics programs. Moreover, these datasets are not static: as discoveries are made, new entries are added to the database and existing ones are updated.

Complicating the problem of data management is the fact that most bioinformatics applications are not designed with grid computing in mind. Indeed, many valuable applications were designed, tested, and validated long before grid computing arose. As a result, such applications are designed to perform simple local I/O and have no facility for attaching to grid data systems. Nevertheless, these applications need both high-throughput computing and huge data storage (Breton, 2005; Jacq, 2004).

Should we simply rewrite such applications to take advantage of the grid? Although this might be possible for a small number of applications, rewriting would be an enormous amount of work to address all of the legacy bioinformatics codes in use today. Many are commercial codes for which source is not available. In some cases, the construction and validation of codes is tightly integrated into an audited scientific process; changing the code for grid deployment would invalidate the application, or at least require a new validation. To make grid computing easy, we must find a way to access data through familiar interfaces without changing the applications.

PROGENGRID PROJECT

The ProGenGrid (Proteomics & Genomics Grid) system is a distributed and ubiquitous grid environment ("virtual laboratory") based on Workflow, for supporting the design, execution and monitoring of "in silico" experiments in bioinformatics (Aloisio, 2005a).

Figure 1. Grid problem solving environment for bioinformatics

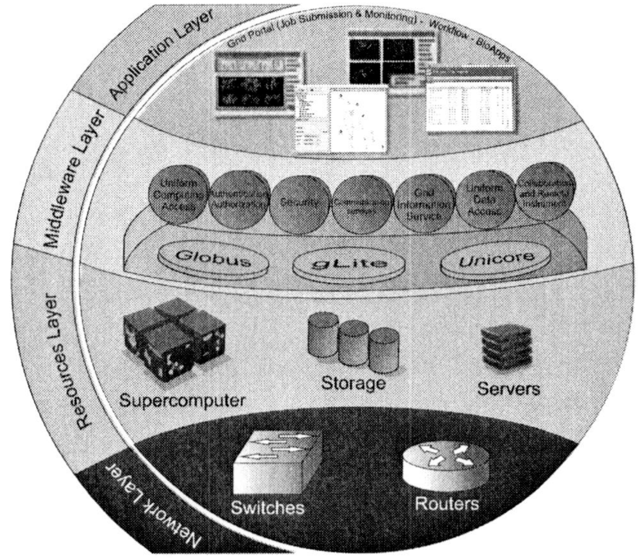

ProGenGrid is a Grid-based Problem Solving Environment that allows the composition of data sources and bioinformatics programs wrapped as Web Services. The use of WS provides ease of use and re-use.

The resulting workflow of Web Services is then scheduled on the Grid, leveraging some Grid-middleware services.

In particular, ProGenGrid offers a bag of services in order to obtain the biological simulation of two important bioinformatics problems: prediction of the secondary and tertiary structure of proteins, and sequence alignment of proteins. Both services are based on an enhanced data access service for biological data banks.

They overcome the problems of currently available bioinformatics tools, such as the use of legacy formats and the need for the user to manage explicitly the results of computations, since these have been wrapped as WS.

Moreover, since their bioinformatics functions are data and CPU intensive, they have been implemented in a distributed manner on the Grid. In fact, Computational Grids allow:

- The deployment, distribution and management of needed biological software components;
- Harmonized standard integration of various software layers and services;
- Computing power, flexible policy definition, control and negotiation mechanism for a collaborative environment.

Main features of ProGenGrid are:

- An integrated environment, where the e-scientist can discover available tools and compose them in a graph (Workflow editor). The system guides the user in the selection of these tools, interacting with her, if the graph is not correct;
- A system for running the experiments on a computational grid and managing the results (Workflow enactment service);
- Enhanced services for accessing the data, grid-based sequence alignment and grid-based secondary and tertiary structure prediction.

A fundamental feature of this system is that when the basic legacy software packages are updated by third parties, the new versions can simply replace the older ones, since these tools have been embedded in high level services and exposed through a WS interface.

Moreover, existing alignment services at the NCBI such as BLAST (Altschul, 1990), PSI-BLAST (Altschul, 1997), etc., although efficient, allow submitting only a sequence at a time whereas ProGenGrid allows carrying out more submissions in one shot, for instance specifying more sequences as input.

Finally, the implemented Grid-based BLAST and PSI-BLAST services exploit parallelism on clusters.

The Data Access Service and Workflow scheduler have been jointly implemented with the Grid Relational Catalog (GRelC) (Aloisio, 2005b) and the Grid Resource Broker (GRB) (Aloisio, 2007a) projects, developed at the HPC (High Performance Computing) of the University of Salento, Lecce.

Finally, the first version of ProGenGrid has been developed as a client-server application (Aloisio, 2005c). Then a Grid Portal version has been integrated based on GRB technology, to leverage its middleware services; recently a meta scheduler has been implemented in order to allow the interoperability among Globus, gLite and Unicore.

ARCHITECTURE

Following the requirements of a Grid PSE architecture, ProGenGrid comprises the following logical layers, each one offering specific services:

- Application Services,
- Core Grid Services.

Application Services comprise biological tools and the Workflow editor. Such services are based on Grid Middleware services, named "Core Grid Services" that offer:

- Data Management service for accessing distributed biological data banks;
- Resource Management service for scheduling, submitting and monitoring jobs;
- Discovery and use of existing analysis tools made available as Web Services, using an Information System named iGrid (Aloisio, 2005d);
- Access control list for secure access to specific grid resources, through the GSI plug-in (Aloisio, 2005e).

DATA MANAGEMENT SERVICE

The community of biologists is distributed among different laboratories with the particularity of sharing large quantities of rapidly evolving information. However, it is difficult to develop a global bioinformatics infrastructure to assist scientists in their research.

Indeed, each area of biology generates its own data repositories such as flat files, web pages and so on, and a wide range of specialized query and analysis tools are commonly used to access those resources.

Therefore, a data access service that hides the heterogeneity of data and offers a common interface to all of the steps related to an experiment is also useful.

One of the main requirements for a data access service is to provide a common interface to the available biological databases to different bioinformatics tools used by the biologists,. Since these databases are heterogeneous in format and access policy, this requirement results in data access transparency.

A second requirement for a data access service is the possibility to integrate data coming from different databases (e.g. the primary sequence information coming from UniProtKB (Bairoch,

2005) and the structure information coming from PDB (Berman, 2000) for a same protein).

The integration of biological data is possible either by linking different parts of data coming from different databases, or by discovering new knowledge,semi-automatically, from data semantically correlated. The goal of the integration is to develop the technology to grant a user access to multiple information systems uniformly, bridging the gap among information and knowledge. The first complexity in achieving this goal is that the information sources are often independent and autonomous, have completely different schema structures and use different data formats. Additionally, there is a significant degree of semantic heterogeneity among biological data. The integration system must provide a mechanism to cross this semantic difference.

Finally, a third requirement is related to the fact that information sources are distributed among the networks and are huge (they contain millions of record describing different aspects of biological data).

Remarkably, computational Grids could potentially satisfy the requirements described so far on large scale, because grids can face distribution of information sources and offer the computational power needed to access and process data. Moreover, grids also allow the distribution of the software components used by biologists thus fostering collaborative computing, and the sharing of available resources and results of analysis among many users.

Our goal has been investigating how to build a virtual database in a Grid environment hiding the heterogeneity of several biological data banks, and offering high performance access primitives for supporting large scale "in silico" experiments.

In achieving this goal, two aspects have been distinguished:

1. How to access geographically distributed data banks through a common mechanism while taking into account that data are not structured;
2. How to integrate several kinds of data that could be semantically correlated.

Regarding the first point, an enhanced data access service (DAS), named Split&Merge, for accessing biological data has been implemented. Such a system is based on the WS approach and offers several features such as searching of:

- A list or single protein sequence from Nr (NCBI BLAST databases. ftp://ftp.ncbi.nlm.nih.gov/blast/db/), PDBaa (Wang, 2003), PDB (Berman, 2000) and UniProt NREF (Bairoch, 2005) data banks, with different options for saving the result in several formats such as flat file, XML and FastA;
- Data from annotated data banks such as UniProtKB (Bairoch, 2005) and UTRdb (Pesole, 1999), using an opportune SQL-based filter.

Such a component has been thought for supporting high-throughput applications in order to extract chunks of data faster. Alignment tools represent an example of such applications. For each computation, these need to access large data banks (e.g. Nr contains about three millions of sequences). Splitting the computation on several machines, pooled in a Grid, allows reducing the response time (Mirto, 2008a).

The proposed Split&Merge system allows retrieving sequences and splitting them in blocks in order to distribute the workload on several machines in the sequence alignment problem.

It is worth noting here that an efficient mechanism for extracting the data is needed. This feature can be provided by a relational DBMS that provisioned with parallel hardware and software to support efficiently concurrent requests. However, often data banks are available as flat file, usually annotated.

We must consider that the relational approach is without doubt the more complete and efficient way for managing data, in fact the majority of administrators of biological databases use this approach for storing the data but also to provide efficient mechanisms for searching, whereas only flat files or XML data banks are released publicly. So, for improving the performances of Split&Merge, a relational model has been chosen for extracting the data by annotated data banks, whereas for data banks that contain just sequences, a library has been developed.

Hence an automatic system that translates unstructured biological data banks in a structured format, i.e. a relational scheme, has been developed. The prototype has been implemented for UniProtKB and UTRdb data banks and this choice is due to two reasons:

- UniProtKB is the most complete repository of proteins whereas UTRdb contains mRNA nucleotides and although the relation among nucleotides and proteins could be important for several studies, an explicit management of such relationship (cross-referenced link) is not yet available;
- A publicly available relational schema of the UniProtKB and UTRdb do not exist.

This activity has requested, respectively: (i) the creation of a metadata set for the common scheme, (ii) a component, named wrapper, that implements a library (an API) for accessing data of specific sources, (iii) a query formalism based on XML, and (iv) a mechanism for translating the data from XML to SQL.

An overview of the architecture of the Split&Merge is presented in the Figure 2 and in the following, a description of the system is described.

Split&Merge

The proposed Split&Merge component is a system that allows retrieving biological data and splitting them in order to distribute the computation on several grid nodes.

The Split&Merge architecture (Figure 2) is made of the following services:

- GRelC Data Access (GDA);
- GRelC Data Storage (GDS);
- Splitter/Merger;

Figure 2. Split&Merge architecture

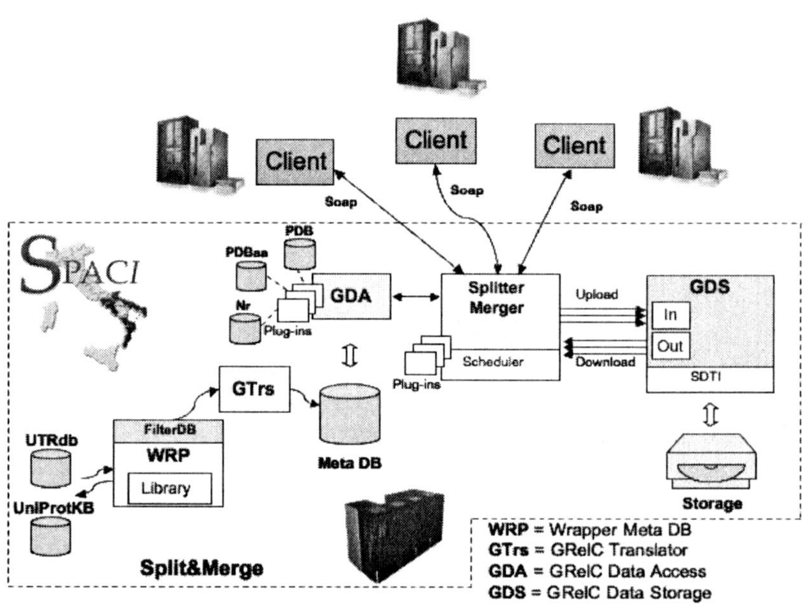

- Wrapper (WRP);
- GRelC Translator (GTrs).

The main feature of the system is a query protocol that manages the result, distributing the entire record set among several computational grid nodes and processing the subsets locally.

Before we delve into the various services, we briefly describe the query protocol that is based on the following phases:

- Query Submission;
- Fragment Download;
- Partial Result Upload;
- Result retrieval;

In the first phase, the client can submit a query thus obtaining a Query Identifier (QI). In the second one, a generic client can require a data fragment connected with a specific QI whereas in the third phase, the same client can upload the partial result obtained after performing a data computation on the input fragment. In the fourth phase, the client can retrieve the merged results from the Split&Merge system.

Detailing each service, the GDA acts as a standard front-end for each data bank. This service provides the basic primitives to transparently get access and interact with different data sources, concealing the back-end heterogeneity and other low level details (it is built on top of the Globus Toolkit (Foster, 1997), Libxml2 and DBMSs proprietary libraries).

The GDA is developed as a GSI enabled Web Service. To date, the GDA provides (by means of wrappers written using proprietary libraries) dynamic binding to several physical RDBMSs such as Oracle or PostgreSQL.

The security of the connection between DBMSs and GDA relies on Secure Socket Layer (SSL). Wrappers for data sources allow the GDA also accessing textual databases.

Currently, GDA provides ad hoc wrappers for the UniProtKB, UTRdb, PDBaa, Nr and PDB databases.

GDS provides a large set of services to manage group of files on a storage resource (i.e. a Disk Resource Manager DRM) and to share/make them available among/to a community of users.

The main services offered by this component are related to:

1. Data transfer (upload and download facilities),
2. User management (add/delete/update user information),
3. Project management (to define and manage projects, that is a way to group users with some common interests),
4. Group management (to define and manage groups, that is subset of users within a project with particular privileges w.r.t. the others) and
5. File system functions (make/delete folder, rename/delete file etc.).

Other minor services are also available within this component but they are not strongly connected/used with/by the overall architecture.

The GDS is developed as a GSI enabled WS. Currently, the GDS provides support for several data transport protocols such as DIME attachment and GridFTP. In the future support for sftp and scp will be also provided.

The interfaces related to data transfer belong to the Standard Data Transfer Interface (SDTI), which by means of a plug-in module architecture, based on dynamic libraries, allows easily extending the existing framework by adding new data transport protocols as they become available.

Splitter/Merger performs a two phase (split & merge) data management activity (interacting with the GDS). In the former phase the record set is split into several fragments (input data) and uploaded to the GDS, whereas in the latter this component downloads the partial results (output data) from the GDS and merges them composing the final user response.

Fragments and partial results are stored on the GDS in a couple of folders (input/output) different for each QI.

Wrapper and GTrs are used for translating flat files into relational DBMS.

The wrapper is a system that implements a library, named ProteinLib, for extracting the data by UniProtKB and UTRdb.

A filter, built on top of the Wrapper, allows associating the data to the chosen metadata set.

Moreover opportune wrappers based on the ProteinLib are added in the GDA as plug-ins, allowing also extracting chunks of sequences by PDBaa, Nr, UniProt NREF and retrieving proteins by PDB.

Translating Flat File to Relational DBMS

As described in (Aloisio, 2007b), our system has three main features:

- **Data download:** It downloads flat files from the repository of data banks using a transfer protocol (e.g. ftp);
- **Data format conversion:** It uses specific wrappers, built on top of data banks, to translate the data into a standard format such as XML;
- **Data ingestion:** It performs the ingestion of XML files into the relational database.

As shown in the Figure 2, the system is based on wrappers, built on top of specific data sources and several services of the GRelC toolkit (Aloisio, 2005b).

These wrappers implement a set of metadata and relations among these and the tags of the flat files. The wrappers allow the user/application extracting a metadata set, selected by data stored using the same tag. A wrapper (WRP) is made of two components: i) a library (ProteinLib, that we have implemented), containing all of the functions for opening/closing the data bank, pointing at a specific entry, selecting a field, etc., and ii) a filter (FilterDB) that offers commands for selecting the data specifying several criteria. In particular, the ProteinLib library has been developed according to a level-based stack, is built on top of core libraries (C and XML) and has a further level based on the applications. The library hides implementation details about flat files because a metadata set has been defined allowing extracting several information from data banks.

The library has been developed in C using Libxml2 for building XML files (e.g., translation of flat files in an XML file), with the gSOAP toolkit and GSI plug-in (Aloisio, 2005e) for exchanging data with the Globus Toolkit GSI (Grid Security Infrastructure) protocol.

Through ProteinLib, the user can make the most of a huge set of APIs, querying and filtering the data of flat files. Three dynamic libraries for accessing UTRdb and UniProtKB are available inside the library (ProteinLib) in order to simplify the access to flat files and to provide transparent access. By using the FilterDB component, a set of data is retrieved from the data sources and formatted into XML files, applying a custom formalism. This is a proprietary kind of query/format used for efficiently inserting/updating into a data source a massive amount of data.

The GRelC Translator (GTrs) takes in input the XML files and translates them into SQL instructions, carrying out the ingestion into a meta database (MetaDB). It is a relational data bank that maps the metadata set extracted by the flat files.

Finally, in order to exploit the functionalities of the ProGenGrid system, the MetaDB is accessed through the GRelC Data Access service (GDA), a Web Service that allows grid applications advanced access to Database Management Systems. GDA is a GSI/VOMS enabled web service addressing extreme performance, interoperability and security. It efficiently and transparently manages databases on the grid across VOs, with regard to emerging and consolidated grid standards (W3C and OASIS) and specifications (OGF) as well as production grid middleware (gLite & Globus).

The main advantages of this system are:

- Reducing the redundancy present into the flat file;
- Reducing the inconsistency of data that could have different values in the flat file;
- Better performance of searches when querying the relational MetaDB, by using the GDA service;
- Complex queries by using a standard language such as SQL.

RESOURCE MANAGEMENT SERVICE

An efficient resource management service represents one of the most important features in all Grid computing middleware. Resources may include processing facilities, storage elements, and network bandwidth that are required by a Grid-enabled application (also referred to as a job).

Currently available Grid middleware carries out the task of resource management such as job scheduling over multiple computational facilities, selecting and accessing datasets from suitable storage elements, out of sight from the user.

The complexity of submitting a job in a computational grid increases if two factors are considered: i) the job involves more applications and data i.e. it is made of several sub-jobs, whose flow is established according to a set of procedural rules. It is defined as workflow job; ii) the sub-jobs could involve resources managed by different grid middleware.

A system that defines, creates and manages the execution of workflows through the use of grid computing technologies, running on one or more workflow engines, which is able to interpret the process definition, interact with workflow participants and, where required, integrate distributed resources and legacy software modules, is defined as WorkFlow Management System (WFMS).

It is a useful tool that brings the power of computational grid resources to the desktop, and allows the users to conveniently assemble and run their own scientific workflows.

There are important and generic capabilities supported by well-recognized complete workflow systems (Good, 2005; Oinn, 2004; Shah, 2004; BioWBI, 2004; Lee, 2007), (e.g., workflow design, workflow repositories, information management, workflow execution, workflow scheduling, fault tolerance, and data management). However, many of these tools provide common functionalities and in general they are based on WS or built on specific grid middleware.

The motivation to create a WFMS in order to submit jobs on different grid middleware was born as requirement in the Italian LIBI (International Laboratory of Bioinformatics) project, in order to provide the interoperability among different grid resources managed by gLite, Globus and Unicore middleware.

The project, which started in 2005 and will terminate in 2010, aims at building a laboratory "without wall" focusing on the central activities of basic and applied research in modern Biology and Biotechnologies. One of the goals is designing and implementing the technological platform of the Laboratory, i.e. a Grid Problem Solving Environment, providing the interoperability among different grid resources, consisting in more than 1500 CPUs gLite based of the "libi" and "biomed" Virtual Organizations (VOs) with the support of the INFN sections in Bari, Bologna and Catania and more than 200 CPUs for the submission of parallel jobs on Globus and Unicore machines belonging to the SPACI and Cineca high perfomance computing centers.

Before the project, we had implemented a WFMS featuring an editor for composing and monitoring the workflow of jobs wrapped as Web Services and an engine for scheduling the jobs in a Computational Grid. Then, we extended the engine's capabilities to include also batch, parameter sweep and MPI jobs, by supporting arbitrary graphs containing cycles in our scheduler called GRB (developed in the homonymous project) (Aloisio, 2007a). The GRB scheduler submits jobs to Globus Toolkit 4 based grids.

As an activity of the LIBI project, for guaranting the interoperability with other grid middleware, we have developed a meta scheduler that supports the submission and monitoring of e-Science workflows distributed on gLite (Laure, 2004) and Unicore (Almond, 1999) based grids. In particular, we have developed several libraries, that are plugged into the meta scheduler, in order to provide core functions for job submission/monitoring and data transfer by using the GridFTP protocol (Allcock, 2002). The meta scheduler is interoperable with the Workload Management System (WMS) and Logging & Bookkeeping (L&B) services of gLite, with the Network Job Supervisor (NJS) service, through the Gateway, of Unicore and with the GRB scheduler for Globus Toolkit, by using a standard language, the Job Submission Description Language (JSDL) (Anjomshoaa, 2005), which is an extensible XML specification from the Open Grid Forum for the description of simple or complex tasks to non-interactive computer execution systems.

Moreover, the ProGenGrid WFMS also allows the composition of data tasks exploiting the GRelC Data Access Service (DAS) (Fiore, 2007).

In the following a description of the WFMS architecture is showed.

WFMS Architecture

The architecture of the ProGenGrid WFMS is shown in Figure 3.

It is composed of the following components:

- Editor;
- Application Manager (AM);
- Engine (WF engine).

The editor allows the composition of the applications and input data through an arbitrary graph that can contain cycles and saves the workflow in two XML files (abstract workflow), by using our own format. These files contain respectively the information about the graphical position of each component (view file) and the JSDL file with the

Figure 3. WorkFlow management system architecture

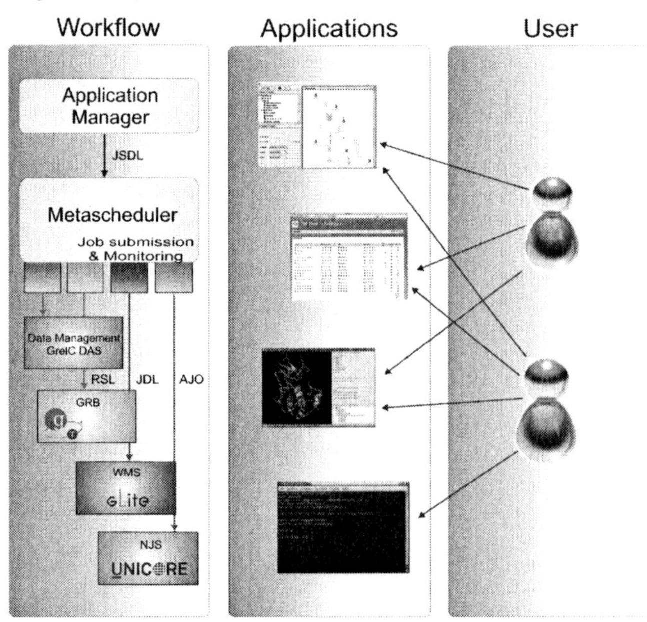

information related to both control and data flow of the application component and input data. The editor has been implemented as a signed Java applet so that the view file can be saved onto the local machine by the user whereas the JSDL file is saved on the web server from which the applet is downloaded, through the Application Manager service.

The Application Manager (AM) is an intermediate service between the editor and the engine, allowing resource discovery (hostname of the worker nodes, logical name of the available applications etc). Moreover, it handles user's account and application's profile, checks data consistency and saves JSDL files.

The engine (WF engine) is based on a meta scheduler that completes the JSDL files with the references to actual executables and data files as well as specific resources (concrete workflow), chooses the opportune resource broker on the basis of the selected application in the workflow and translates the JSDL in the proper language depending on the specific middleware (JDL/DAG for gLite and AJO for Unicore).

In particular, we have developed several libraries, that are plugged into the meta scheduler, in order to provide core functions for job submission/monitoring and data transfer by using the GridFTP protocol.

We developed a gLite library in C language that allows ProGenGrid to submit jobs to EGEE Grid by means of gLite Workload Management System (WMS) and to retrieve their status by means of Logging and Bookkeeping Service (L&B). In the gLite middleware, the WMS accepts requests for job submission and management and takes the appropriate actions to satisfy them. The interaction with the WMS is limited to the description of the characteristics and requirements of the request via a high-level, user-oriented specification language, the Job Description Language (JDL) and to the submission of it through the provided interfaces. The WMS then translate these abstract resource requirements into a set of actual resources, taken from the overall grid resource pool, to which the user has access permission. The L&B service provides job monitoring functionalities: it stores logging and bookkeeping information concerning events generated by jobs. Using this information, the LB service keeps a state machine view of each job.

The gLite library interacts with the WMS by means of its WMProxy interface. The WMProxy (Workload Manager Proxy) is a simple service providing access to the WMS functionality through a Web Services based interface. WMProxy accepts job submission requests described by the job description language (JDL) and other job management and control requests such as job cancellation, job file perusal, job output retrieval etc. WMProxy provides C++, Java and Python API, but we used the provided WSDL definition to develop a custom C library using the gSOAP Toolkit, relying on its SSL support for establishing a secure connection to the remote web service.

The L&B service offers a web service interface on a GSI channel, and we use this interface in our library to know the status of the submitted jobs.

The file transfer operations from and to WMS are accomplished by means of GridFTP using the inbox and outbox space created at job registration time.

The Unicore library includes several functions in java for the submission and monitoring of a job. These functions include the client of the Unicore Network Job Supervisor (NJS) service. A wrapper in C, through the Java Native Interface (JNI) functions, has been added and will be included in the meta scheduler (implemented in the C programming language).

The SSL and GSI protocols are used to provide security and the GridFTP protocol is used for data transfer.

For the Globus Toolkit, the GRB scheduler takes as input JSDL files and translates them in the RSL language. Finally, for applications with Web Service interface the WSDL is parsed by the editor to import both available methods and types of the input data, whereas the engine parses the WSDL to create a stub and a service's client. The JSDL description can also involve data retrieval and access. The user can specify data sources, filters and data related operation useful for the computation. This kind of task is processed by the WF engine through the GRelC DAS.

APPLICATIONS SERVICES

Our system offers the following services:

- Protein tertiary structure prediction, based on homology modelling;
- Secondary structure prediction based on a neural network;
- Sequence Alignment tools, such as BLAST and PSI-BLAST.

These tools are accessibile by means a Grid portal, based on GRB technology that also provides the following services:

- Grid Configuration;
- Resource Status;
- File Transfer.

The Grid Configuration section allows dealing with user's grid profiles, it retrieves the user's credentials (Figure 4a), inserts a new VO, configures the grid associating grid machines to a user and finally the Information Services monitors the grid status (Figure 4b).

It is possible to configure biological data banks, application tools and associate several databases to a software.

This option allows (through the Alignment web interface) having a list of possible data banks to use.

Finally job status can be visualized. Resource Status allows querying several information services to know the resource status, whereas File Transfer allows using GridFTP for transferring files specifying source and destination machines.

Alignment jobs allows submitting sequences for BLAST and PSI-BLAST (Figure 4c), using opportune interfaces (Mirto, 2007a).

This system uses the Split&Merge solution for retrieving the data and submitting the jobs. For the secondary structure prediction, it is possible to use an interface to submit the sequences whose structures need to be predicted. The predictor is based on a neural network that has been described in (Mirto, 2007b).

The tertiary structure prediction service is presented in the next section.

Finally, a workflow interface allows composing several applications in a graph, using drag-and-drop of the nodes on the left window (Fig. 4d). The parameters of each application can be specified using an opportune window. The system is available at the following url: https://sara.unisalento.it/cgi-bin/bioinfo/enter. In order to access the system, each user needs valid credentials (login and password).

Case Study: Tertiary Structure Prediction DIcarboxylate Carrier (DIC) of Saccharomyces Cerevisiae

Protein structure prediction is of paramount importance in medicine and biotechnology. From this point of view, an important aspect regards the function of metabolite mitochondrial carriers, a family of intrinsic proteins that mediate the flux of several metabolites from the matrix to the cytosol and vice versa, in relation to the role that such proteins cover in some mitochondrial pathologies. Along with the fast progress in the identification of novel mitochondrial carriers over the last few years, an increasing number of genes encoding mitochondrial carriers have been identified whose defects cause various inherited diseases. Sequence studies have shown that all carrier proteins have a highly conserved sequence motif, the carrier signature.

Figure 4. Grid Portal implementation: a) Users Credential; b) Grid status; c) PSI-BLAST application form; d) Workflow Editor

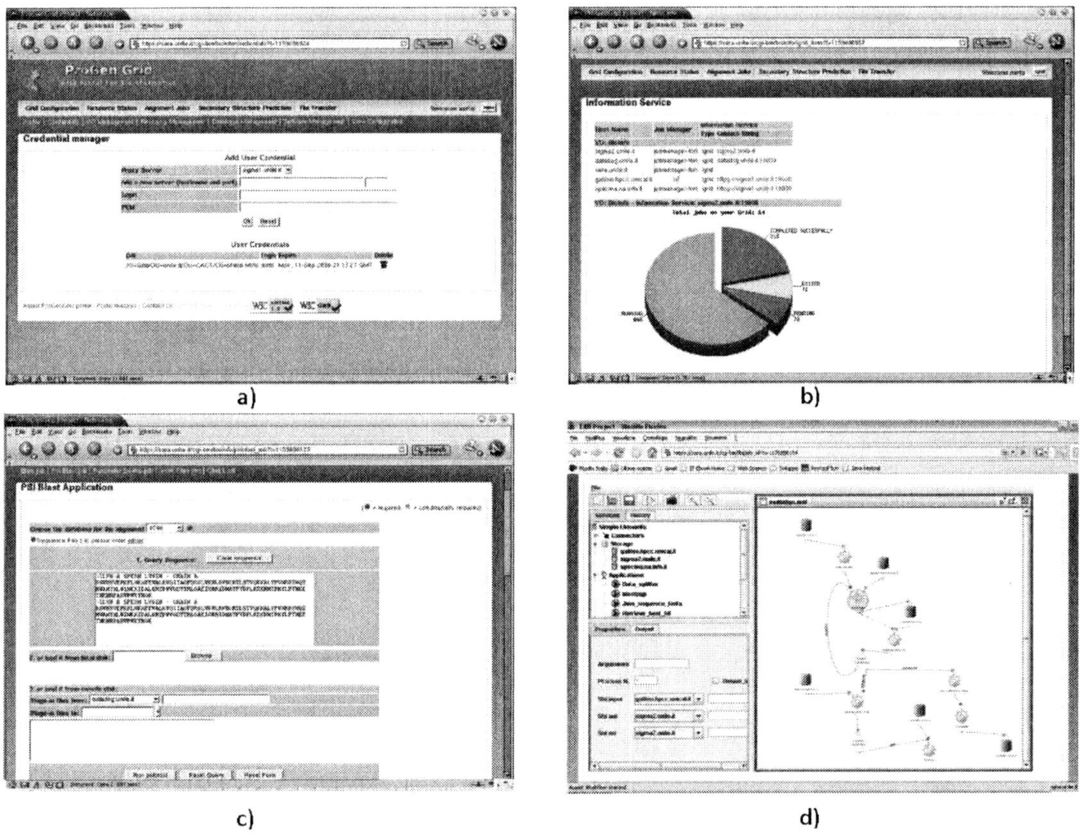

The protein tertiary structure prediction process involves several data and software tools that need to be integrated in a platform, taking into account that several steps are computationally intensive. The use of parallel computing associated to a Computational Grid is becoming crucial to maintain and improve the effectiveness of protein structure prediction process. Moreover, the possibility to use an integrated environment based on grid resources for submitting, monitoring and analyzing the results of a tertiary structure prediction process is very important for hiding implementation details to the user.

To determine the tertiary structure prediction of several proteins of interest, we integrated a routinely expert dependent strategy in a automatic tool that may facilitate the generation of carrier models at low resolution, good enough for serving as template models in several occasions, including site directed mutagenesis to prove and/or disprove the involvement of a computed topological structure in a specific functional process.

Therefore, we developed a new service, that uses applications deployed in a Grid environment, in order to automate the prediction procedure and integrate the data produced by the in vitro and in silico analysis.

This service has been implemented in the ProGenGrid project and it integrates several tools embedded in the in silico analysis that use information coming from in vitro analysis for carrying out the protein tertiary structure prediction.

As case study we have studied a protein of the carrier family. The inner mitochondrial membrane contains a family of intrinsic proteins that mediate the flux of several metabolites from the matrix to the cytosol and vice versa (Palmieri, 2004). Sequence studies have shown that these carrier proteins have a tripartite structure, made of related sequences of about 100 amino acids in length. Each repetitive element contains two membrane-spanning helices, connected by a loop at the matrix side. The members of the carrier family contain the signature sequence motif PxD/ExxK/R at the C-terminal end of the first helix of each of these modules, in proximity of the matrix exposed loop. Symbols represent the following amino acids: P, Proline; D, Aspartate; E, Glutammate; K, Lysine; R, Arginine; x, an unspecified amino acid. The recent resolution of the crystal structure of the AAC revealed the precise location of the carrier signature within the three-dimensional structure (Pebay, 2003); however, the function of the residues is still enigmatic.

To determine the possible role of the carrier signature in the function of a carrier protein, we used the dicarboxylate carrier (DIC) of Saccharomyces cerevisiae as a model protein (our target). The DIC is a protein of 298 residues, containing six putative membrane spanning different domains. It mediates the exchange of dicarboxylates such as malate and succinate against phosphate (Palmieri, 1996). A carrier signature is found in module I (helix 1) and module III (helix 5). In vitro experiments carried out in reconstituted systems revealed that DIC carrier signature has a functional role in the transport activity of the protein (Zara, 2007).

The in silico analysis is based on the homology modeling procedure for the tertiary structure prediction starting by the amino acid sequence. This technique constructs an all-atom model of the sequence with unknown structure (target) from one/more template/s structure. Templates are chosen from a database of structures, starting from the knowledge that the target sequence with template/s sequence/s have a sequence identity percentage above 30-40%.

As case study, we model the structure of DIC, using the only carrier so far solved with atomic resolution (0.22 nm) in the PDB data bank: the ADP/ATP carrier protein of Bos Taurus heart mitochondria (AACp; 1okc). The rationale for chosing as a template AACp is that DIC shows 40,4% identity and sequence homology with AAC and that they both belong to the same GO functional classification (www.geneontology.org). The DIC model was computed and described in (Mirto, 2008b).

Regarding the tertiary structure prediction process (Figure 5), our service integrates several software and data.

In particular starting by the target sequence in the FASTA format, the multi alignment between the target sequence and the PDBaa (that contains just the sequences of the PDB data bank) is carried

Figure 5. In silico analysis for the protein tertiary structure prediction

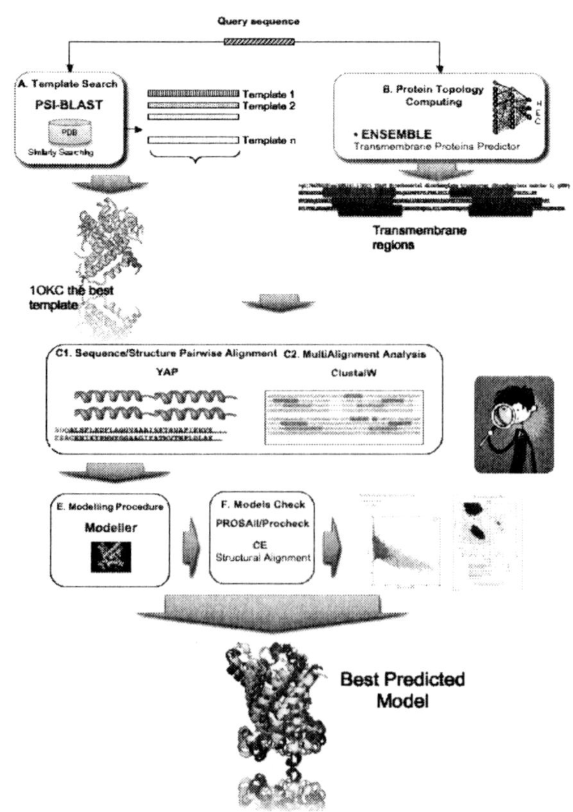

out and the template/s is/are found. In the DIC case, the template is known so we have ignored this step. In parallel, the DIC topological model is calculated by using ENSEMBLE (Martelli, 2003). As output, six putative transmembrane helices have been predicted by ENSEMBLE.

Following our test case the next step compares the alignment between the topological model of the target and that of the secondary structure of the ADP/ATP template as evaluated by the DSSP program (Kabsch, 1983). For obtaining the best alignment the three topological models have been overlapped with the secondary structure of the ADP/ATP. This was obtained with the YAP application (http://gpcr.biocomp.unibo.it/cgi/predictors/aligns/aligns.cgi) taking as input two sequences and the relative secondary structure. In the alignment, experimental results have been considered.

This is possible via a request to the user who selects the best conditions. The sequence motifs, PxD/ExxK/R, that represents a fundamental requirement of this protein class, is constrained on the odd alpha-helices of the target whereas glycines are constrained in the N-terminal portion of the even alpha helices. The alignment is then translated in a file in PIR format that is given as input to the Modeller program (v.9.4) (Eswar, 2000). The generated alignment can be checked by the user in order to verify the experimental consistency with the literature data. The models generated by Modeller (15 models in our experiment) have been checked by using Procheck (Laskowski, 1993) and ProsaII (Markus, 2007) for calculating geometry and energy constraints. Moreover, a structural alignment between these models and the ADP/ATP has been computed by using the CE tool (Shindyalov, 1983) and finally the best scoring model has been selected.

Mutations can be included in the original sequence and the computing process may start again.

In principle the effect of a mutation on the computed model structure may be also evaluated by encapsulating in the platform other predictors, such as I-Mutant (Capriotti, 2005) that predicts the effect of the mutations on protein stability after considering chemico-physical properties of the folded versus unfolded protein structure.

Adopting the DIC modeling procedure as a test case we also developed a new service, in order to automate the prediction procedure, predict the effect of single point mutations on protein stability and integrate the data produced by the "in vitro" and "in silico" analysis.

The service offers the following features:

- Prediction of three-dimensional structures;
- Optimization of the structures;
- Visualization of the structures, taking into account several formats such as PDB and X3D.

The service allows visualizing the molecules, that can be rotated, translated or extended, and visualizing the main and lateral chains of the molecules according to several modalities (wireframe, ribbons, sticks, backbone, etc.) and with different colours on the basis of temperature, group, etc.

RELATED WORK

The Grid is proposed as the next generation infrastructure necessary to support and enable the collaboration of people and resources through highly capable computation and data management systems (Foster and Kesselman, 1998). A number of BioGrid projects are underway, including myGrid (Stevens, 2004), GeneGrid (Kelly, 2005) and Proteus (Cannataro, 2004).

These primarily focus on the sharing of computational resources, large-scale data movement and replication for simulations, remote instrumentation steering or high throughput sequence analysis.

In the following these BioGrid projects are briefly described.

myGrid (Stevens, 2004) is a project targeted at developing open source high-level middleware to support personalised in silico experiments in biology on a Grid. myGrid is building services for integration such as resource discovery, workflow enactment and distributed query processing. Additional services are needed to support the scientific method and best practices found at the bench but often neglected at the workstation, notably provenance management, change notification and personalisation.

The target users of myGrid are tool and service providers who build applications for a community of biologists. Early prototypes of myGrid services were developed and tested with use cases based on the functional analysis of clusters of proteins identified in a microarray study of genes showing circadian rhythms in Drosophila melanogaster. Following this, a distributed system has been developed to meet the requirements of researchers studying the genetics of Graves' disease.

GeneGrid (Kelly, 2005) is a collaborative industrial grid computing R&D project initiated by the Belfast e-Science Centre (BeSC) under the UK e-Science programme and supported by the Department of Trade &Industry.

The project aims at providing a platform for scientists to access their collective resources, skills, experiences and results in a secure, reliable and scalable manner by creating a "Virtual Bioinformatics Laboratory".

GeneGrid provides seamless integration of a myriad of heterogeneous applications and datasets that span multiple administrative domains and locations across the globe, and presents these to scientist through a simple user friendly interface. A major challenge during GeneGrid development is to expose the full computational and collaborative potential of the Grid, while keeping the users' need for knowledge of the underlying technology to a minimum. GeneGrid meets this challenge through the GeneGrid Portal – the user interface for the project.

Proteus, a Grid based Problem Solving Environment for Bioinformatics (Cannataro, 2004) is a software architecture allowing to build and execute bioinformatics applications on Computational Grids by using an ontology-based methodology to describe bioinformatics applications as distributed workflows of software components.

It can be used to assist users in i) formulating problems, allowing to compare different available applications (and choosing among them) to solve a given problem, or to define a new application as composition of available software components; ii) running an application on the Grid, using the resources available in a given moment thus leveraging Grid scheduling and load balancing services; iii) viewing and analyzing results, by using high level graphic libraries, steering interfaces (that allows changing interactively the way a computation is conducted), and accessing the past history of executions, i.e. the past results, that form a knowledge base.

FUTURE TRENDS

Future work includes, as a first step, the testbed of the workflow management system in the LIBI project. We are going to add 1) the support for integrating the information coming from different grid information services in order to use efficient scheduling policies; 2) the development of a plug-in to support the CREAM-based Computational Elements of gLite; 3) the development of a GUI for job monitoring.

CONCLUSION

This chapter describes the ProGenGrid research project at the University of Salento in Lecce. It is a Grid Problem Solving Environment, specialized for the Bioinformatics domain, which aims at providing an integrated environment in order to compose, schedule and monitor biological applications in a Computational Grid. The main feature offered by this environment is the possibility to use a friendly web interface for composing workflow jobs, scheduled on different grid middleware such as gLite, Unicore and Globus. Other important features involve advanced data management solutions for accessing biological data.

ACKNOWLEDGMENT

This work was supported by the MIUR (Italian Ministry for Education, University and Research) in the LIBI "International Laboratory of BioInformatics" project, F.I.R.B. 2003, under grant RBLA039M7M.

REFERENCE

Allcock, W., Bester, J., Bresnahan, J., Chervenak, A., Liming, L., & Tuecke, S. (2002, April). *GridFTP: Protocol Extensions to FTP for the Grid.* Argonne, IL: Argonne National Laboratory.

Almond, J. & Snelling, D. (1999). UNICORE: uniform access to supercomputing as an element of electronic commerce. *Future Generation Computer Systems, 15*(5-6), 539–548.

Aloisio, G., Cafaro, M., Fiore, S. & Mirto, M. (2005a). ProGenGrid a Grid-enabled platform for Bioinformatics, Studies in Health Technology and Informatics, From grid to Healthgrid. In T. Solomonides et al. (Eds.), *Proceedings of HealthGrid 2005,* (Vol. 112) (pp. 113-126). Amsterdam, the Netherlands: IOS Press..

Aloisio, G., Cafaro, M., Fiore, S. & Mirto, M. (2005b). The Grid Relational Catalog Project. In L. Grandinetti (Ed.), *Advances in Parallel Computing, "Grid Computing: The New Frontiers of High Performance Computing"* (pp.129-155). Philadelphia: Elsevier.

Aloisio, G., Cafaro, M., Fiore, S., & Mirto, M. (2005c, June 23-24). ProGenGrid: a Workflow Service Infrastructure for Composing and Executing Bioinformatics Grid Services. *Proceedings of the 18th IEEE International Symposium on Computer-Based Medical Systems* (IEEE CBMS 2005), Dublin, Ireland.

Aloisio, G., Cafaro, M., Epicoco, I., Fiore, S. Lezzi, D., Mirto, M., & Mocavero, S. (2005d). iGrid, a Novel Grid Information Service. In P.M.A. Sloot et al. (Eds.), *Proceedings of the First European Grid Conference (EGC) 2005,* (LNCS 3470)(pp. 506-515). Berlin, Germany: Springer-Verlag,.

Aloisio, G., Cafaro, M., Lezzi, D. & Van Engelen, R. (2005e). The GSI plug-in for gSOAP: Enhanced Security, Performance, and Reliability. In *Proceedings of Information Technology Coding and Computing (ITCC 2005),* (Vol. I), (pp. 304–309). New York: IEEE Press.

Aloisio, G., Cafaro, M., Carteni, G., Epicoco, I., Fiore, S., Lezzi, D., Mirto, M., & Mocavero, S. (2007a). The Grid Resource Broker Portal. *Concurrency and Computation: Practice and Experience, 19*(12), 1663-1670.

Aloisio, G., Cafaro, M., Fiore, S., Mirto, M. (2007b). A Grid System for the Ingestion of Biological Data into a Relational DBMS. In *Proceedings of IEEE International Symposium on Bioinformatics and Life Science Computing (BLSC-07),* (pp. 713-718), Niagara Falls, Canada.

Altschul, S.F., Gish, W., Miller, W., Myers, E.W. & Lipman D.J. (1990). Basic local alignment search tool. *Nucleic Acids Res., 3*(215), 403–410. Retrieved from http://www.ncbi.nlm.nih.gov/BLAST/.

Altschul, S.F., Madden, T.L., Schffer, A.A., Zhang, J., Zhang, Z., Miller, W. & Lipman D.J. (1997). Gapped BLAST and PSI-BLAST: a new generation of protein database search programs. *Nucleic Acids Res., 17*(25), 3389-3402. Retrieved from http://www.ncbi.nlm.nih.gov/cgi-bin/BLAST/nph-psi.

Anjomshoaa A., Brisard F., Drescher M., Fellows D., Ly A., McGough S., Pulsipher D., Savva A. (2005). *GFD-R-P.056 Job Submission Description Language (JSDL) Specification v1.0*. Retrieved from http://forge.gridforum.org/projects/jsdl-wg.

Bairoch, A., Apweiler, R., Wu, C.H., Barker, W.C., Boeckmann, B., Ferro, S. & et al (2005). The Universal Protein Resource (UniProt). *Nucleic Acids Res.*, (33), 154–159. Retrieved from http://www.uniprot.org.

Berman, F., Hey, A. J. G., & Fox, G. (2003). *Grid Computing: Making The Global Infrastructure a Reality*. Hoboken, NJ: Wiley & Sons.

Berman, H.M., Westbrook, J., Feng, Z., Gilliland, G., Bhat, T.N. & et al. (2000). The Protein Data Bank. *Nucleic Acids Res.*, (28), 235–242. Retrieved from www.rcsb.org/pdb/.

BioWBI (2004). *Bioinformatic Workflow Builder Interface*. Retrieved from http://www.alphaworks.ibm.com/tech/biowbi

Breton, V., Blanchet, C., Legré, Y., Maigne, L. and Montagnat, J. (2005). Grid Technology for Biomedical Applications. In M. Daydé et al. (Eds.), *VECPAR 2004*, (LNCS 3402), (pp. 204–218).

Cannataro, M., Comito, C., Lo Schiavo, F., & Veltri, P. (2004, February). Proteus, a grid based problem solving environment for bioinformatics: Architecture and experiments. *IEEE Computational Intelligence Bulletin*, 3(1), 7–18.

Capriotti, E., Fariselli, P. & Casadio, R. (2005). I-Mutant2.0: predicting stability changes upon mutation from the protein sequence or structure. *Nucleic Acids Research,* (33), 303–305.

Channabasavaiah, K., Holley, K., & Tuggle, E. (2003). Migrating to a service-oriented architecture. *IBM DeveloperWorks*. Retrieved from http://www-128.ibm.com/developerworks/library/ws-migratesoa/

Culler, G.J. & Fried, B.D. (1963). An On-Line Computing Center for Scientific Problems. In *Proc. 1963 Pacific Computer Conf.*, (pp. 221–242). , Piscataway, NJ: IEEE.

Cunha, J. C. (2003). *Future Trends in Distributed Applications and PSEs*. Talk held at the Euresco Conference on Advanced Environments and Tools for High Performance Computing, Albufeira, Portugal.

Deelman, E., Singh, G., Su, M., Blythe, J., Gil, Y. Kesselman, C., Mehta, G., Vahi, K., Berriman, G.B., Foster, I., & Kesselman, C. (1997). Globus: A Metacomputing Infrastructure Toolkit. *Intl J. Supercomputer Applications*, 11(2), 115–128.

Eswar, N., Marti-Renom, M. A., Webb, B., Madhusudhan, M. S., Eramian, D., Shen, M., Pieper, U. & Sali, A. (2000). Comparative Protein Structure Modeling With MODELLER. *Current Protocols in Bioinformatics* (Supplement 15, 5.6.1-5.6.30). Hoboken, NJ: John Wiley & Sons, Inc..

Fiore, S., Negro, A., Vadacca, S., Cafaro, M., Mirto, M. & Aloisio, G. (2007). Advanced Grid DataBase Management with the GRelC Data Access Service. *Proceedings of ISPA 2007 - August 29-31, 2007 - Niagara Falls, Ontario, Canada,* (LNCS 4742) (pp. 683-694).

Foster, I., Kesselman, C., & Tuecke, S. (2001). The Anatomy of the Grid: Enabling Scalable Virtual Organizations. *International Journal of High Performance Computing Applications, 15*(3), 200–222.

Foster, I. & Kesselman, C. (2005). Globus Toolkit Version 4: Software for Service-Oriented Systems. In IFIP *International Conference on Network and Parallel Computing*, (LNCS 3779) (pp. 2–13). Berlin, Germany: Springer-Verlag.

Good, J., Laity, A., Jacob, J.C., & Katz, D.S. (2005). Pegasus: a framework for mapping complex scientific workflows onto distributed systems. *Scientific Programming, 13*(3), 219-237.

Houstis, E., Gallopoulos, E., Bramley, R., & Rice, J. (1997). Problem-Solving Environments for Computational Science. *IEEE Comput. Sci. Eng., 4*(3), 18–21.

Jacq, N., Blanchet, C., Combet, C., Cornillot, E., Duret, L., Kurata, K., Nakamura, H., Silvestre, T., Breton, V. (2004). Grid as a bioinformatics tool. *Parallel Computing, Special Issue: High-performance parallel biocomputing, 30.*

Kabsch, W. & Sander, C. (1983). Dictionary of protein secondary structure: pattern recognition of hydrogenbonded and geometrical features. *Biopolymers, 22*, 2577–2637.

Kelly, N., Jithesh, P.V., Wasnik, S., McLaughlin, R., Fragoso, F., Donachy, P., et al. (2005). The GeneGrid Portal: A User Interface for a Virtual Bioinformatics Laboratory. In *UK e-Science All Hands Meeting (AHM 2005).*

Laskowski, R.A., MacArthur, M.W., Moss, D.S. & Thornton, J.M. (1993). PROCHECK: a program to check the stereochemical quality of protein structures. *J. Appl. Cryst.* (26), 283–291.

Laure, E., Hemmer F. et al., (2004). Middleware for the Next Generation Grid Infrastructure. *Computing in High Energy and Nuclear Physics* (CHEP), Interlaken, Switzerland.

Lee, S. et al (2007). Bio-STEER: A SemanticWeb workflow tool for Grid computing in the life sciences. *Future Generation Computer Systems 23*, 497-509.

Martelli, P.L., Fariselli, P. & Casadio, R. (2003). An ENSEMBLE machine learning approach for the prediction of all-alpha membrane proteins. *Bioinformatics 19*(1), 205–211.

Markus, W. & Manfred, J. (2007). ProSA-web: interactive web service for the recognition of errors in three-dimensional structures of proteins. *Nucleic Acids Res.* (35), 407–410.

Mirto, M., Rossi, I., Epicoco, I., Fiore, S., Fariselli, P., Casadio, R. & Aloisio, G. (2007a). High Throughput Protein Similarity Searches in the LIBI Grid Problem Solving Environment. *Proceedings of the 5th International Symposium on Parallel and Distributed Processing and Applications (ISPA07)*, (pp. 414-423), Niagara Falls, Canada,.

Mirto, M., Cafaro, M., Fiore, S., Tartarini, D. & Aloisio, G. (2007b, June). A Grid-enabled Protein Secondary Structure Predictor. *IEEE Transactions on NanoBioscience, 6* (2),124-130.

Mirto, M., Fiore, S., Epicoco, I., Cafaro, M., Mocavero, S., Blasi, E. & Aloisio G. (2008a). A Bioinfomatics Grid Alignment Toolkit. *Future Generation Computer Systems, 24*(7), 752-762. Philadelphia: Elsevier

Mirto, M., Ferramosca, A., Tartarini, D., Romano S., Negro A., Tasco, G., Fiore, S., Zara, V., Casadio, R., & Aloisio, G. (2008b). A Protein Structure Prediction Service in the ProGenGrid System. *Global HealthGrid: e-Science Meets Biomedical Informatics Proceedings of HealthGrid 2008*, (pp 135-146), Chicago, June 2-4.

National Center for Biotechnology Information. Retrieved from http://www.ncbi.nlm.nih.gov/BLAST/.

Oinn, T., Addis, M., Ferris, J., Marvin, D., Senger, M., Greenwood, M., et al (2004). Taverna: A tool for the composition and enactment of bioinformatics workflows, *Bioinformatics, 20*(17), 3045-3054.

Palmieri, L., Palmieri, F., Runswick, M. & Walker, J. (1996). Identification by bacterial expression and functional reconstitution of the yeast genomic sequence encoding the mitochondrial dicarboxylate carrier protein. *FEBS Lett.* (399), 299–302.

Palmieri, F. (2004). The mitochondrial transporter family (SLC25): physiological and pathological implications. *Pflugers Arch., 447*, 689–709.

Pebay-Peyroula, E., Dahout-Gonzalez, C., Kahn, R., Trezeguet, V., Lauquin, G. & Brandolin, G. (2003). Structure of mitochondrial ADP/ATP carrier in complex with carboxyatractylosider. *Nature* (426), 39–44.

Pesole, G., Liuni, S. & et al (1999). UTRdb: a specialized database of 5' and 3' untranslated regions of eukaryotic mRNAs. *Nucleic Acids Res.*, (27):188–191. http://www.ba.itb.cnr.it/UTR/.

Schuchardt, K., Didier, B., & Black, G. (2002). Ecce - a problem-solving environment's evolution toward Grid services and a Web architecture. *Concurrency and Computation: Practice and Experience*, *14*,1221–1239.

Shah, S.P., He, D.Y.M., Sawkins, J.N., Druce, J.C., Quon, G., Lett, D., Zheng, G.X.Y., Xu, T., Quellette, B.F.F. (2004). Pegasys: Software for executing and integrating analyses of biological sequences. *BMC Bioinformatics*, *5*, 40.

Shindyalov, I.N. & Bourne, P.E. (1983). Protein structure alignment by incremental combinatorial extension (CE) of the optimal path. *Protein Engineering 11*(9), 739–747.

Stevens, R.D., Robinson, A. J. & Goble C. A. (2004). myGrid: Personalised Bioinformatics on the Information Grid. *Bioinformatics*, *19*(1), 302–302. Retrieved from http://www.mygrid.org.uk/.

Krafzig, D., Banke, K., & Slama, D. (2004). *Enterprise SOA: Service-Oriented Architecture Best Practices (The Coad Series)*. Upper Saddle River, NJ: Prentice Hall PTR.

Walker, D., Rana, O. F., Li, M., Shields, M. S., & Huang Y. (2000). The Software Architecture of a Distributed Problem-Solving Environment. *Concurrency: Practice and Experience*, *12*(15),1455–1480

Wang, G., & Dunbrack, R.L. (2003). PISCES: improvements to a protein sequence culling server. *Bioinformatics*, *19*,1589-1591. http://dunbrack.fccc.edu/Guoli/pisces_download.php

W3C. *XML Protocol Working Group*. Retrieved from http://www.w3.org/2000/xp/Group/

Zara, V., Ferramosca, A., Capobianco, L., Baltz, K.M., Randel, O., Rassow, J., Palmieri, F. & Papatheodorou, P. (2007). Biogenesis of yeast dicarboxylate carrier: The carrier signature facilitates translocation across the mitochondrial outer membrane. *J. Cell Science* (120), 4099–4106.

KEY TERMS AND DEFINTIONS

Grid Portal: A Grid Portal provides an efficient infrastructure to put Grid-empowered applications on corporate Intranet/Internet. Application interfaces can be tailored to the specific user's skills or access rights. Users can therefore access and control their computing and engineering resources via an intuitive, standards compliant Web interface, virtually from anywhere using a standard Web browser.

Grid File Transfer Protocol (GridFTP): It is a protocol for high-performance, secure, reliable data transfer.

Grid Security Infrastructure (GSI): It is a specification for secret, tamper-proof communication between software in a grid computing environment. Secure, authenticatable communication is enabled using both symmetric and asymmetric encryption. Delegation of credentials is also supported.

gLite: It is a middleware for grid computing. Born from the collaborative efforts of more than 80 people in 12 different academic and industrial research centers as part of the EGEE Project, gLite provides a framework for building grid applications tapping into the power of distributed computing and storage resources across the Internet.

Globus: It is a middleware used for building Grid systems and applications. It is being developed by the Globus Alliance and many others all over the world. A growing number of projects and companies are using the Globus Toolkit to unlock the potential of grids for their cause.

OGSA (Open Grid Service Architecture): It is the blueprint specification for standards-based grid computing. "Open" refers to both the standards-development process and the standards themselves. OGSA is "service-oriented" because it delivers functionality among loosely-coupled interacting services that are aligned with industry-accepted Web service standards. "Architecture" defines the components, their organizations and interactions, and the overall design philosophy.

Protein Tertiary Structure Prediction: It has the aim of determining the three-dimensional structure of proteins from their amino acid sequences. In more formal terms, this is the prediction of protein tertiary structure from primary structure.

UNICORE (Uniform Interface to Computing Resources): It offers a ready-to-run Grid system including client and server software. UNICORE makes distributed computing and data resources available in a seamless and secure way in intranets and the internet.

VOMS (Virtual Organization Membership Service): It serves as a central repository for user authorization information, providing support for sorting users into a general group hierarchy, keeping track of their roles, etc.

Web Service: A Web service (also Web Service) is defined by the W3C as "a software system designed to support interoperable machine-to-machine interaction over a network". Web services are frequently just Web APIs that can be accessed over a network, such as the Internet, and executed on a remote system hosting the requested services.

WorkFlow Management System: WorkFlow Management Systems (WFMSs) are software systems that support the management of workflow in an organization. According to the Workflow Management Coalition, a WFMS system consists of three functional components: a *definition tool* that is used to define business processes; a *worklist handler* through which interaction between users and their personal worklists is managed; and a *workflow engine*, that provides a runtime execution environment.

WSRF: The WS-Resource Framework (WSRF) is a set of six Web services specifications that define what is termed the *WS-Resource approach* to modeling and managing state in a Web services context. To date, drafts of three of these specifications have been released, along with an architecture document that motivates and describes the WS-Resource approach to modeling stateful resources with Web services.

Chapter XV
A Graphical Workflow Modeler for Docking Process in Drug Discovery

Qiang Wang
Harbin Institute of Technology, China

Yunming Ye
Harbin Institute of Technology, China

Kunqian Yu
Chinese Academy of Science (CAS), China

Joshua Zhexue Huang
University of Hong Kong, China

ABSTRACT

A drug discovery process is aimed to find from a large set of molecules the candidate leads that have strong interaction with the target proteins. The process of drug discovery is characterized by its complexity in data and computation. A useful tool to simplify the handling of intensive data and complex algorithms is necessary for domain scientists to build proper drug discovery procedures, carry through the data intensive computation tasks and produce fruitful results. This chapter presents a graphical workflow modeler for domain scientists to perform drug discovery tasks on high performance grid computing grid platforms. A client/server system is described as the platform for implementation of the graphical workflow modeler. A case study on drug discovery for avian influenza virus is given to demonstrate the use of this tool in drug discovery research.

INTRODUCTION

Scientific research today has changed its way. Computational science, which focuses on simulation of complex phenomena in the last few decades, is being replaced today by e-science or data-centric science where scientific discoveries often result from data intensive computation and analysis. In this area, Grid computing is playing a key role, owing to its characteristic that it allows geographically distributed scientists to work collaboratively in a networked environment and share resources and expertise to solve, in a large scale, difficult scientific and engineering problems. The typical scientific research areas cover a wide range including gravitational-wave science (Deelman, Kesselman, Mehta, Pearlman, Blackburn, Ehrens, Lazzarini, Williams, & Koranda, 2002), neuroscience (Lathers, Su, Kulungowski, Lin, Mehta, Peltier, Deelman, & Ellisman, 2006), astronomy (Katz, Anagnostou, Berriman, Deelman, Good, Jacob, Kesselman, Laity, Prince, Singh, Su, & Williams, 2006), high energy physics (Deelman, Blythe, Gil, & Kesselman, 2004), etc. All are distinctly characterized with complex research problems that involve huge amount of data and requires a lot of computing resources to analyze.

Drug discovery is another such area that grid computing is used to carry out various computational processes for identification of active compounds against a given target from a large number of chemical compounds in molecule databases (Richards, 2002; Ren, Zhang, Wan, Huang, Xie, & Yang, 2006). In modern drug discovery, molecule docking is an important process which aims to determine the candidate lead components that have the strongest interaction with the target component by a series of complex computations. Docking process involves multiple steps, including 3D molecule modeling and representation, search of molecule databases for candidate molecules that best match the receptor structure, evaluation of candidate molecules with scoring functions, conformation determination, hit identification, and lead optimization. In each step, multiple techniques and algorithms are developed, and different techniques are implemented in different systems for drug discovery research and applications. Nevertheless, these techniques and algorithms are used in setting parameters, integrating data sources and performing data analysis tasks. It is time-consuming and costly for domain scientists to learn and master multiple techniques and systems to solve drug discovery problems. Therefore, development of an easy-to-use collaborative workflow modeler is needed by drug discovery practitioners as part of their sophisticated problem solving environment.

In this chapter, we present a software platform and a graphical workflow modeler for drug discovery. The goal of our work is to develop an effective and efficient system that supports drug discovery practitioners to build docking processes for virtual high throughput screening (VHTS) (Yoon, 2005) of potential inhibitors through drag-and-drop and connecting operations in a graphical interface with a mouse. In this graphical working environment, a docking process is presented as a workflow consisting of a set of functional nodes connected in a directed acyclic graph (DAG). Each function node is an implementation of a particular algorithm to perform one docking function such as search or scoring. When a workflow is formed and all parameters of each function node are set, the workflow is executed through a workflow engine that schedules the execution of each function in the underlying computing infrastructure, e.g., a grid computing infrastructure (Foster, & Kesselman, 1998; Berman, Hey, & Fox, 2003). Such a system could provide an efficient and easy way for modeling and managing scientific processes of experimental investigation, evidence accumulation and result validation. Processes themselves can then be modified, reused and shared through collaboration between interdisciplinary scientists.

The rest of this chapter is organized as follows. Section 2 gives the background information on

China's Drug Discovery Grid (DDGrid[1]) project on which this work is rooted. Section 3 presents the client/server platform on which the graphical workflow modeler is implemented. The system architecture and functionality are described. The workflow patterns for supporting drug discovery processes are defined. Section 4 discusses implementation of the graphical workflow modeler from the viewpoint of users who would use this tool to carry out drug discovery research. A case study is presented in Section 5 to show how to use this tool to carry out a drug discovery process for avian influenza virus. Section 6 briefs the future trends in this work. And the work is concluded in Section 7.

BACKGROUND

Unlike the traditional method of drug discovery, which is characterized by the trial-and-error testing of the effects of chemical substances to treatments, modern drug discovery is involved in a complex computational process assisted by the knowledge of biomolecules, including three-dimensional information about bimolecular structure, biochemical properties, activity of a drug at its binding site, and molecule's druglikeness, etc. The process of finding a new drug against a chosen target for a particular disease usually involves high-throughput screening (HTS), wherein large libraries of chemicals are tested for their ability to modify the target.

To meet this challenging trend, China has initiated the Drug Discovery Grid (DDGrid) project that has been carrying out in China for more than 5 years. The purpose of this project is to set up the first nationwide grid-computing environment with aggregated computing and data resources to provide drug virtual screening and pharmaceutical chemistry information services. The objective of this work is to develop a graphical workflow modeler for DDGrid users to flexibly develop docking processes with different techniques implemented in DDGrid. At the time of writing, the graphical workflow modeler is being implemented on a client/server platform and will be integrated to DDGrid afterwards.

CLIENT/SERVER PLATFORM

In this section, we present the general features of the client/server platform that is used to implement the graphical workflow modeler for drug discovery. We briefly discuss the system architecture and functionality and define the workflow patterns for supporting drug discovery processes.

General Features

As a general platform for development of systems and software for different data analysis and mining purposes, the client/server system is designed and implemented with the following features:

- **Rich functionality:** Full-featured functionalities are provided to satisfy the requirements of scientific data analysis at all steps, including data access, preprocessing, process modeling, result evaluation and visualization. Functions of system management and monitoring are also provided to facilitate the use of this system. New functions can be added for specific applications such as drug discovery.
- **R Integration:** R is an open source system that is widely used in statistical analysis and biomedical research[2]. R contains a rich set of statistical functions which aim at both common and specific statistical analyses and provides a language environment for supporting customized statistical computing. In this platform, an integration interface is implemented to enable a full access to R statistical functions. With this integration, statistical functions of R can

be easily added to the analytical process for data analysis, e.g. drug discovery. This feature provides much flexibility and ease for user applications.

- **Extensibility:** Plug-in-based technology is used in the platform to support flexible and easy extensions with user's applications. Virtually, all elements in this platform such as user's customized function nodes, models and views are regarded as plug-ins which can be implemented under the instruction of the plug-in wizard.
- **Platform-independence:** The platform is developed in Java under the Eclipse open development platform[3] to ensure computing platform-independence on the underlying OS and hardware systems.
- **Easy-to-use:** Process-oriented work environment with user friendly graphical interfaces in which the user can use a mouse to create data analysis processes through drag-and-drop operations. In this environment, the user can concentrate on building the data analysis processes for solving his problems, rather than struggling with computer programs.

System Architecture

Figure 1 illustrates the system architecture of the client/server platform with three major components, the client, the server and the communication layer. We briefly describe the three components below.

Client

The client layer is composed of the User Interface (UI) layer and the underlying Workflow Modeling Operator (WoMO) layer. The UI layer provides users a graphical work environment for creating analytical processes with provided operations and conducting data analysis tasks, while WoMO layer contains a set of process modeling operators to support creation and execution of analytical processes in the UI layer. These operators include Workflow Definition Operator, Workflow Validation Operator, Workflow Storage Operator, Workflow Execution Operator, and Workflow Monitoring Operator.

Workflow Definition Operator is used to support operations of creating a workflow for an analytical process. The details of components and logical construction of a created workflow are described in Workflow Definition Operator.

Figure 1. System architecture of the client/server platform

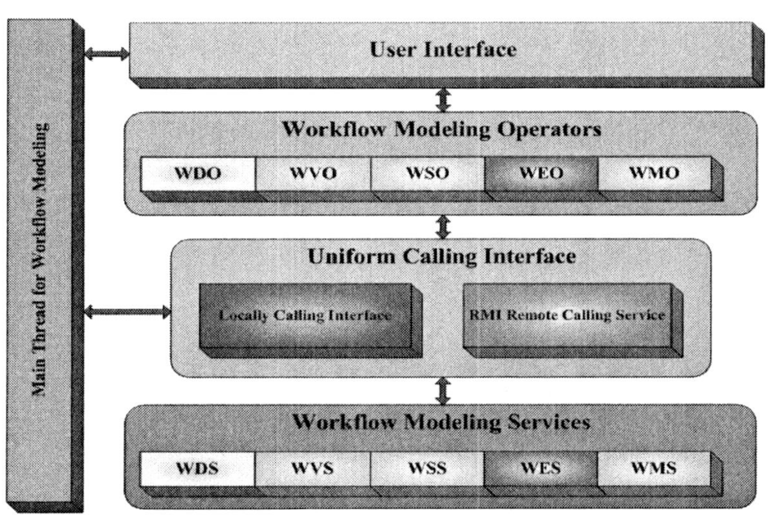

Workflow Validation Operator is responsible for operations to validate the logical validity of a workflow, especially the control and data interdependence between connected function nodes. Workflow Storage Operator is used to support the saving and accessing operations of a workflow. Workflow Execution Operator provides functions for controlling and managing the execution process of a workflow. And Workflow Monitoring Operator is responsible for support of operations to monitor the execution process and provides execution status information of a workflow.

Server

The server component consists of a workflow modeling service (WorMS) layer that provides various services for supporting the workflow modeling operators in the WoMO layer from the client end. These services include Workflow Definition Service, Workflow Validation Service, Workflow Storage Service, Workflow Execution Service, and Workflow Monitor Service. In general, these services are used to transform user operations supported by corresponding operators in the WoMO layer from the client end into internal system implementations which can be executed to generate the final results. Thus, the server component can be looked as a workflow execution environment where all workflows of analytical processes created at the client are submitted, managed and executed. The services are executed on a single machine or a grid computing environment.

Communication Layer

Between the client and the server is a middle layer called the Uniform Interface Calling (UnIC) layer. This layer is designed to facilitate the communication between the front-end client and the back-end server and increases the flexibility of the server implementation. This layer provides a uniform interface for calling services in the WorMS layer from the server end either locally or remotely. The UnIC layer masks the underlying complexity of service calling and result dissemination and enables a seamless integration between the user's graphical operations at the UI layer and the corresponding function services at the WorMS layer.

Workflow Patterns

A workflow pattern is a graphical representation of the business process schema which is composed of a set of function nodes that represent different business tasks and are connected by directed lines indicating the control flow and the data flow. Workflow patterns have been developed for more than two decades (White, 2004; Aalst, Hofstede, Kiepuszewski, & Barros, 2003) and current workflow patterns range from very simple to very complex ones, covering the business behaviors that can be captured within most business process models. In our system, we define five workflow patterns that occur in the majority of drug discovery processes. These workflow patterns are defined as follows.

- **Sequence Pattern:** This pattern is defined as a sequence of ordered activities that are connected one by one. Many analytical processes in data analysis including drug discovery fall in this pattern. In many cases, analysis starts from a single data set. A sequence of data operations are performed on the data set and some intermediate data set may be created by some operations to facilitate the later operations. At the end, the final result is generated for the user, which is usually illustrated by a visualization tool or saved as a result file. In the pattern, the output of the preceding operation is the input to the current operation.
- **Parallel Split Pattern or Fork Pattern:** This pattern is defined as being a mechanism that will allow activities to be performed

concurrently. In data analysis, this pattern occurs when an input data set is divided into two or more subsets or more than one operation is applied to one data set.

- **Join Pattern or Synchronizer:** This pattern brings together multiple workflow segments into a single activity to process. In data analysis, this pattern occurs when two or more data sets need to be merged into one data set or two or more output results from preceding operations are aggregated into one result. For example in data mining, two algorithms can be applied to the same training data to build two classification models. The model assessment operation joins the outputs of the two models to generate evaluation report on these two models.
- **Exclusive or Split (Choice):** This pattern represents a point in the workflow process where one of several branches is chosen based upon the results of a condition. In data analysis, this function can be implemented in the operations in the above Parallel Split pattern with additional conditions to control the workflow execution.
- **Exclusive or Join (Merge) Pattern:** This pattern allows alternate branches to come together without synchronization. In data analysis, the function is implemented in the operations in the above Join Pattern with conditions to selectively join the result in the operation.

The constructs of these five workflow patterns are illustrated in Figure 2.

IMPLEMENTATION

Our client/server platform is implemented in Java under the Eclipse open development platform. XML is used in the message exchange between client and server. A pluggable mechanism is implemented to allow easy integration of plug-ins of domain specific functions, for instance, a new search algorithm or a scoring function that is used in the docking process of drug discovery. The plug-in functions are executable programs that can be coded in any programming language such as C or C++. At execution, they are running as independent processes that are coordinated by the workflow engine of the platform. This design and implementation enables the platform to be easily customizable to different domain applications.

In this section, we focus on discussions of the user environment and implementation of the graphical workflow modeler.

User Environment

Similar to the Eclipse development environment, the user environment of the graphical workflow modeler for drug discovery is composed of a set of views (windows or panels), which provide different tools and features for specific applications, as illustrated in Figure 3. The user, at the client end, uses a mouse to select and work in these

Figure 2. Constructs of workflow patterns

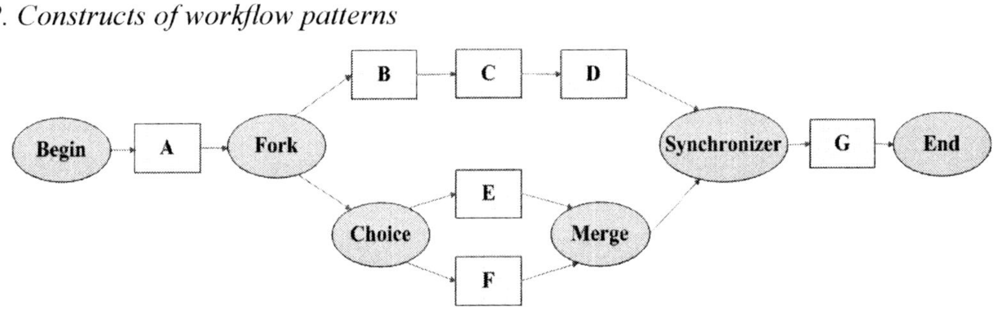

Figure 3. User environment with a workflow process for discovering drug candidates for AIV

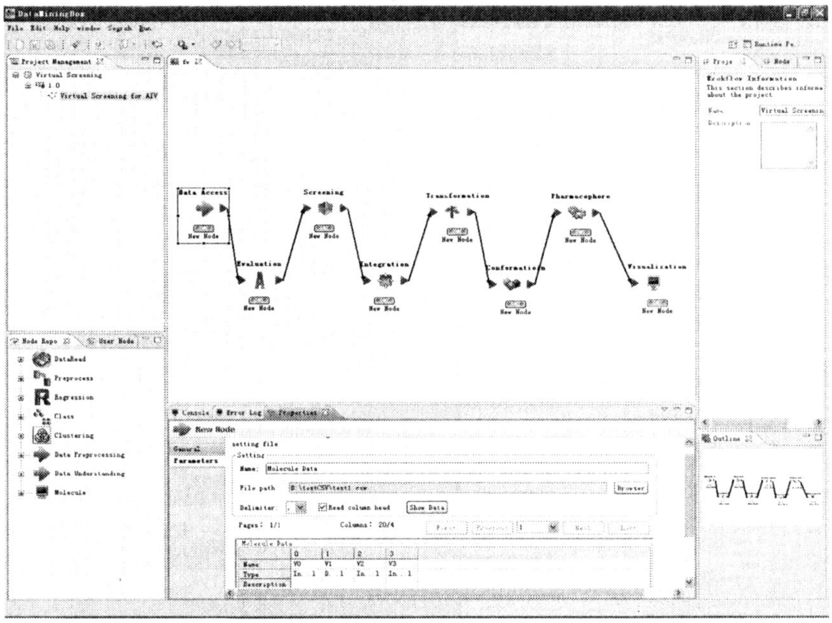

views to create, edit, submit and monitor a drug discovery process, and explore the results. The real execution of the process is performed in the backend server that can be a single machine or a grid cluster.

Node Repository View

Node repository view, illustrated at the lower left in Figure 3, is the place that hosts all the function nodes which are used to perform different operations on data in a drug discovery process. The node repository is organized into three sub repositories, the System Node sub-Repository (SNR), the User Node sub-Repository (UNR), and the Customized Node sub-Repository (CNR). The SNR contains all the nodes that the platform has implemented. From this aspect, it represents the overall functionalities of the system. The UNR contains the nodes which are often used by many different users in a specific application domain. From the UNR, users can quickly locate the function nodes that are needed in creating an analytical process. When necessary, a function node, which does not currently exist in UNR, can be selected from SNR and added in UNR to facilitate users' later use. The CNR contains the nodes that are created by the user to perform specific tasks. These nodes are the super-nodes that wrap two or more other function nodes to perform a more complex operation. On the one hand, a super-node represents a workflow segment that can be added in a workflow. On the other hand, a super-node is used just like an ordinary node from the user aspect.

Workflow View

Workflow view, which is the main area in the middle in Figure 3, is an operation panel in which the user designs, edits and saves the workflows of drug discovery processes. In this graphical environment, to create a workflow is very simple if the user knows the process and the operations that are needed. The user simply uses a mouse to select a function node from the node repository view, drag it into the workflow view and drop it by releasing the mouse button. In the same way, the user drags the second node to the workflow view. Then, the user connects the two nodes by drawing a directed line from the first node to the second one so that a workflow segment is created and the data and control flow from the first node to the second node is defined. The user can double click a node to open a window to set initial parameters or specify a data source. The user can continue this process until the workflow of the whole process is created.

When a workflow is created in the workflow view, it can be edited, saved for later use, and submitted to the server for execution. A knowledge repository can also be created to store the workflows of typical drug discovery processes for reuse and sharing with peer researchers, which is a valuable feature for domain knowledge accumulation and sharing.

Progress View

Progress view is the right panel in the user environment in Figure 3. The progress view provides for the user a flexible control of the workflow execution and status information of the workflow execution progress. In experimental research of drug discovery, scientists often try a process with different parameters and data sources, and revise the design of the workflow according to the current results when dealing with complex tasks. The workflow control operations such as start, stop, pause, cancel and resume give the user a flexibly control on each phase of the workflow execution. For instance, when finding something wrong with the current workflow, the user can cancel the execution of the workflow process. A lot of time can be saved by not executing the later function nodes. Furthermore, detail information on the workflow execution progress is helpful for the user to keep track of the workflow execution status and make corresponding decisions.

Other Views

To facilitate the use of the user environment and increase user's productivity, we also design several other useful views organized at the lower part in Figure 3. The **information view** provides two kinds of information on details of the workflow cases and nodes, such as case/node name, case/node task description, case/node version, case/node creator, time of case/node revision, etc. The **resource view** is a friendly explorer for the user to explore resources (files, data, results, etc.) by some predefined criteria, e.g., by project. The **search view** is a helpful tool to search for resources by the user defined criteria so the user can quickly locate the resources the user needs. **The thumbnail view** gives an outline of a workflow and highlights the current execution position with a shaded area. This is helpful for the user to master the overall sketch and rapidly locate where they want to go, when dealing with a very large workflow. The **console view** presents information on exception or errors.

XML Representation of Processes

In the user environment, a drug discovery task is managed as a project. A project consists of definitions of workflows and related resources in each workflow such as data sources. A workflow is represented as a direct acyclic graph (DAG) which consists of connected function nodes with directed links. A function node is a wrapper of implementation of an algorithm. All these above elements are described in a corresponding XML-like markup language so as to support the storage, management and communication.

It is common in the research community that scientific applications are shared by scientists of different disciplines. In this system, the workflows of analytical processes can be reused by other scientists and revised when dealing with their own research problems. We use a set of XML type markup languages in representation of the project data, i.e., the Data Mining Project Markup Language (DMPML), the Predictive Model Markup Language (PMML) and the Uniform Model Markup Language (UMML), to support import and export of workflow projects and models.

DMPML is an XML-based standard that is supported by several data mining systems. We use DMPML to support import and export of workflow projects, including data sources, property files, models and actions. With the help of DMPML, share and collaboration of workflow applications can be realized between different scientists.

PMML and UMML are two XML-based standards used for workflow model description and exchange in the system. PMML is used to define statistical and data mining models and to share models between PMML compliant applications. UMML has a stronger capability of describing more types of models and thus is used to describe the workflow process. A workflow process is represented as an .umml file which describes the workflow definition in details, including the input and output data types, function nodes and the sequence of these function nodes, etc. The workflow in an .umml file is then prepared for later submission to the workflow engine which provides a workflow execution environment for workflow execution. Furthermore, the other scientist who is doing a similar research can import this .umml file into his/her own system to reuse or make some revision to satisfy his/her own research needs.

A CASE STUDY SCENARIO

In this section, we present a case study scenario to show how to use the graphical workflow modeler to build a process to find the drug candidates for killing avian influenza virus. Avian influenza virus, especially H5N1 virus is a deadly virus that can infect humans. Search for drugs to kill avian influenza virus in human bodies is a diligent effort in medical research. The research work in this case study was conducted at the Shanghai Institute of Materia Medica (SIMM) of Chinese Academy of Science (CAS) in China. SIMM is a leading research institute in the area of drug discovery and drug design in China. Drug discovery for avian influenza virus is one of the major research areas at SIMM and a significant progress has been made in the past few years.

The case study was based on the receptor-based virtual high throughput screening method and the ligand-based drug discovery method, especially the pharmacophore method, and made use of the complementary advantages of each method (Shoichet, 2004). VHTS is characterized for the excellent matching rate between the conformation of small molecules and the protein active site, while the pharmacophore method has the distinctive advantage of high accuracy of the prediction for molecule activity. Thus, the combined consideration on both molecule structure and activity can help to improve the accuracy rate of finding appropriate leads for drug candidates. Furthermore, in order to reduce the false positive rate of screened molecules and improve the shoot straight of selected molecules, the variation of protein conformation must be considered in the process of virtual screening. The commercial software FlexX (Rarey, Kramer, & Lengauer, 1996) was used in docking protein molecules.

Preparation of Data and Discovery Process

The early research explored the structure and function of avian influenza virus (AIV), especially the active site of the nerve aminophenol (NA) of H5N1 AIV (Jong, & Hien, 2006; Sims, 2005; Taubenberger, 2005), and obtained eight different structures of mutation. These structures were used as the protein targets for molecule virtual screening.

Four small molecule databases, ACD, CNPD, MDDR and SPECS were chosen in our flexible docking and virtual screening experiment. These databases contain detail information of millions of small molecules about the molecule structure, molecule activity and molecule research literatures, etc. Duplicate compounds were filtered out based on Lipinski's rules (Lipinski, Lombardo, Dominy, & Feeney, 1997).

A drug discovery workflow process was designed to identify the lead components for NA drug candidates. The process was divided into two phases, each consisting of multiple steps.

The first phase was based on the principle of structure-based drug design (Shen, J., Xu, X., Cheng, F., 2003) for flexible molecule docking and scoring so as to select the molecules with the highest scores. This process can be accomplished in the following steps:

- *Multiple molecule databases integration*: In this step, the four databases were integrated to remove duplicates and provide a uniform database access interface for query and retrieval.
- *Molecule evaluation*: In this step, molecules in the databases are evaluated and those that fit the right structure and have the right properties of being a drug candidate are filtered out. We can use the drug like evaluation and ADME/T evaluation approaches to extensively evaluate molecule properties to improve the screening accuracy.
- *Virtual screening by multi mutation structures*: In this step, structure-based high throughput virtual screening (Alvarez, 2004) is performed to find the molecules that best match the protein structures of the 8 mutations of NA obtained in the early research.
- *Virtual screening results integration*: Hundreds of molecules may result from *Step* 3, containing similar molecules in different formats. In this step, the molecules from different databases are integrated and organized by category.

The second phase was based on the principle of the ligand-based drug design (Hawkins, & Skillman, 2006) to screen, in depth, molecules obtained from the previous phase by pharmacophore.

- *Molecule format transformation*: In this step, molecule formats are transformed from the mol2 format to the sd format to satisfy the subsequent processing requirements.
- *Molecule conformation generation*: In drug discovery research, molecule conformation is essential for pharmacophore modeling. In this step, a database is built to store different conformations of each selected molecule.
- *Screening by pharmacophore*: By extensively exploring the structure and function of existing NA inhibitors, a pharmacophore model can be created and from the model, ligand-based virtual screening is performed to find molecules with more bioactivity.
- *3-D visualization and analysis*: In this step, screening results are visualized in 3-D graph and analyzed.

Besides the above steps, a next step is to manually test the chemical, biological properties of the molecules selected previously in order to find the best leads for drug candidates.

Discovery Process

Following the above two phase discovery process, we can use the graphical workflow modeler to create and execute the workflow to find drug candidates for avian influenza virus. We can use the user environment to monitor the execution and explore the results.

Process Design

The workflow for drug discovery process is usually designed by domain scientists, such as biologists or bioinformaticians, who have necessary expertise and domain knowledge. They are familiar with particular task requirements and have the capability to capture the high-level requirements for identification of drug candidates and translate such requirements into a workflow schema.

In process design, a drug discovery task is specified at a sequence of operations on data by a domain expert. The sequence and operations are explained in a document for an analyst to use.

Workflow Creation

After the process of a drug discovery task is specified, the specification document can be used as the instruction to create a workflow with the graphical workflow modeler. The user needs to relate the operations in the specification document to the function nodes in the node repository view. The corresponding function nodes are selected to create a workflow in the workflow view. At each node, the parameters are set and the data sources are specified. Figure 3 shows an example of a workflow of discovering drug candidates for avian influenza virus. In practice, when the discovery process and operations are well understood by the user, it is very easy to create a workflow with the graphical workflow modeler. However, the user may try different processes in a research work. The graphical workflow modeler can release the user from tedious work on managing different kind of computer programs and enable the user to concentrate his work on testing different drug discovery processes.

When a workflow is created, it is saved in a UMML file that can be submitted to the server for execution.

Process Execution

The user submits the UMML file of a workflow to the server for execution. In the server, the workflow engine schedules the computing resources and the execution process. From the user environment, the user can interact with the workflow engine to start, pause, resume, stop, and cancel the execution of a workflow.

At beginning, the workflow XML file is processed by an XML parser that translates the XML file into a workflow definition file. Then, a workflow instance, representing the workflow during run time, is instantiated in accordance with the workflow definition file. Operations in the workflow are in turn identified and passed on to the appropriate service in the workflow modeling service (WMS) layer that contains the resources for execution in parallel or sequentially based on the interdependency of operations. Once the WMS service receives a task in the form of an XML file, it creates the actual commands to be executed using the parameters from the task XML and a local XML configuration definition specific to the resource. Input sequences are accessed via the uniform data service (UDS). The intermediate and final results of each task are stored in an embedded database for later use in the next task.

Process Monitoring

Once execution of a workflow starts, its execution status can be monitored from the user environment. In the workflow view, the execution status of a node in the workflow is characterized with three traffic lights of red, yellow and green colors. The red light indicates that the node is not ready for execution for some reasons, e.g., the input data file is not specified or a parameter is not set. The yellow light indicates that the node is ready for execution but not being executed now. The green light represents that the node has been executed successfully. Traffic lights of a function node will be triggered right after its previous node has been executed. Therefore, through the traffic lights being triggered one after another, the user can keep track of the execution status and progress of a workflow.

When the user wants to see the detail information of a function node or even the whole workflow, the user can use a monitoring tool to view two types of information, i.e., the static information and the dynamic information. The static information covers all necessary details of a function node or a workflow, e.g., node name, node version, node creation date, input and output file, and node description, etc. The dynamic information mainly includes the run-time details of a function node or a workflow, e.g., workflow execution status, error or exception warning, and results description, etc.

Result Exploration

Once the execution of a workflow is finished, the results are saved in the result files on the server. The results include the results generated by each node and the final results by the whole workflow. The results can be exported to other commonly used formats, such as .xls, .csv and .txt. These results can be explored by various interactive visualization tools such as Discovery Studio VisualizerPro for visualization and analysis. Figure 4 is an example of visualization of the docking conformation of a small molecule binding to the active site of the NA structure. Figure 5 shows the pharmacophore mapping of a small molecule that fits all the fingerprints of the pharmcophore model.

Figure 4. Visualization of the docking conformation of a small molecule binding to the active site of the NA structure

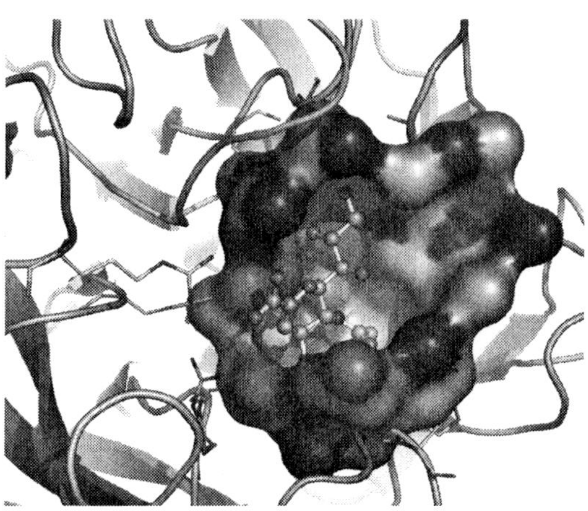

Figure 5. Visualization of the pharmacophore mapping of a small molecule that fits all the fingerprints of the pharmcophore model

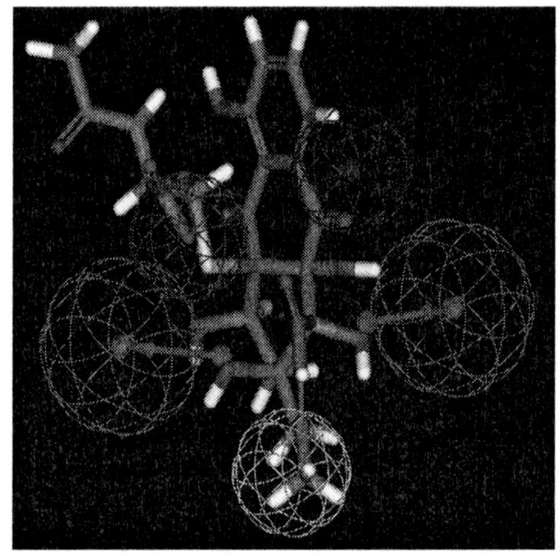

FUTURE TRENDS

Workflow modeling is a classical issue both in research and application, however, there are still some challenging issues coming up with new demands. In general, the future trends of the graphical workflow modeling include two aspects:

On the one hand, a full-featured task-oriented software suit should be implemented to support the processing and analyzing tasks of drug discovery. In the study of drug discovery, every possible algorithms, tools and operations may be tried as a pipeline across each phase of the processing procedure, including the simulation of molecular surface, the finding of best binding sites, the construction of the molecule conformation, and the evaluation of the affinity between the ligand and the receptor molecules, etc. Hence, a full-featured software suits may provide much flexibility and strength for the biologists and pharmaceutists.

On the other hand, a web-based workflow modeling environment enabled by the web service technology is increasingly emphasized in the workflow research community. This new trend results in a SOA architecture which provides more extensibility for the developers, as well as more flexibility and accessibility for end users.

CONCLUSION

Drug discovery is a hot research area with problems that require a lot of data and computing resources to solve. Grid computing that integrates both data and computing resources into a collaborative computing environment provides a new platform to solve drug discovery problems. Drug discovery involves many complex analytical processes and computation intensive operations on large complex data. As such, different algorithms and software tools are developed and used. In practice, drug discovery researchers often spend a lot of time on mastering different algorithms and tools in their research. A user friendly tool can facilitate the use of different drug discovery algorithms.

In this chapter, we have presented a graphical workflow modeler, a user tool for drug discovery. The case study scenario on finding drug candidates for avian influenza virus has demonstrated that with this tool, the creation of a workflow for a drug discovery process becomes a sequence of drag-and-drop mouse operations. The algorithms are hidden from the user so the user can concentrate on the discovery process instead of algorithms.

The graphical workflow modeler is being implemented in a client/server platform for data analysis. Our future work is to integrate this tool in DDGrid, a grid computing platform for drug discovery that is being developed in China.

ACKNOWLEDGMENT

The research work was supported by the National Nature Science Foundation of China (NSFC) under the grants No.60603066, No.2007AA01Z436 and No.2006AA01A124. The fourth author's research work was supported by the matching grant (10207635) of The University of Hong Kong.

REFERENCES

Alvarez, J. (2004). High-throughput docking as a source of novel drug leads. *Curr Opin Chem Biol*, *8*, 365-370.

Aalst, W., Hofstede, A., Kiepuszewski, B., & Barros, A. (2003). Workflow Patterns. *Distributed and Parallel Databases*, *14*(3), 5-51.

Berman, F., Hey, A.J.G., & Fox, G. (Ed.). (2003). *Grid Computing: Making the Global Infrastructure a Reality*. West Sussex, UK: John Wiley & Sons.

Deelman, E., Blythe, J., Gil, Y., & Kesselman, C. (2004). *Workflow management in GriPhyN. In Grid resource management: state of the art and future trends*. Norwell, MA: Kluwer Academic Publishers.

Deelman, E., Kesselman, C., Mehta, G., Meshkat, L., Pearlman, L., Blackburn, K., et al. (2002). Griphyn and ligo, building a virtual data grid for gravitational wave scientists. *In HPDC* (pp. 225-234). New York: IEEE Computer Society.

Foster, I., & Kesselman, C. (Ed.). (1998). *The Grid: Blueprint for a New Computing Infrastructure*. San Francisco, CA: Morgan-Kaufmann.

Hawkins, P., & Skillman, G. (2006). Ligand-Based Design Workflow.

Jong, M.D., & Hien, T.T. (2006). Avian influenza A (H5N1). *J Clin Virol, 35*(1), 2-13.

Katz, D.S., Anagnostou, N., Berriman, G.B., Deelman, E., Good, J., Jacob, J.C., Kesselman, C., et al. (2006). Astronomical Image Mosaicking on a Grid: Initial Experiences. In Martino, B.D., Dongarra, J., Hoisie, A., Yang, L., & Zima, H. (Ed.), *Engineering the Grid-Status and Perspective*. Valencia, CA: American Scientific Publishers.

Lathers, A., Su, M.H., Kulungowski, A., Lin, A.W., Mehta, G., Peltier, S.T., Deelman, E., & Ellisman, M.H. (2006). Enabling Parallel Scientific Applications with Workflow Tools. *Proceedings of Challenges of Large Applications in Distributed Environments* (pp. 55-60). New York: IEEE Computer Society.

Lipinski, C.A., Lombardo, F., Dominy, B.W., & Feeney, P.J. (1997). Experimental and computational approaches to estimate solubility and permeability in drug discovery and development settings. *Adv. Drug Delivery Rev, 23*, 3-25.

Rarey, M., Kramer, B., & Lengauer, T. (1996). A fast flexible docking method using an incremental construction algorithm. *J Mol Biol, 261*, 470-489.

Ren, Y., Zhang, Q., Wan, J., Huang, X., Xie, P., & Yang, G. (2006). Implementation of CCNUGrid-Based Drug Virtual Screening Applications Using Workflow Techniques. *Proceedings of the Fifth International Conference on Grid and Cooperative Computing Workshops* (pp. 371-374). New York: IEEE Computer Society.

Richards, W.G. (2002). Virtual Screening Using Grid Computing: the Screensaver Project. *Nat Rev Drug Discov, 1*, 551-555.

Shen, J., Xu, X., & Cheng, F. (2003). Virtual screening on natural products for discovering active compounds and target clues. *Curr Med Chem, 10*, 2327-2342.

Shoichet, B. (2004). Virtual screening of chemical libraries. *Nature, 432*, 862-865.

Sims, L.D., et al. (2005). Origin and evolution of highly pathogenic H5N1 avian influenza in Asia. *Vet Rec. 157*(6), 159-64.

Taubenberger, J.K., et al. (2005). Characterization of the 1918 influenza virus polymerase genes. *Nature, 437*(7060), 889-893.

White, S.A. (2004). *Process Modeling Notations and Workflow Patterns*. Retrieved from http://BPTrends.com

Yoon, S., Smellie, A., Hartsough, D., & Filikov, A. (2005). Surrogate Docking: Structure-based Virtual Screening at High Throughput Speed. *Jouranl of Computer-Aided Molecular Design, 19*, 483-497.

KEY TERMS AND DEFINITIONS

Drug Discovery: From computer science aspect, drug discovery is a complex computational process by which drugs are discovered and/or designed based on the knowledge that how disease and infection are controlled at the molecular and physiological level. The process of drug discovery involves the identification of candidates, synthesis, characterization, screening, and assays for therapeutic efficacy, most of which need to deal with large amount of data and computation. Once a compound has shown its value in these tests, it will be regarded as a drug candidate for drug development.

E-Science: E-Science (or eScience) is used to describe computation intensive or data intensive science that is carried out in highly distributed network environments, especially grid computing. This term was first presented by John Taylor, the Director General of the United Kindom's Office of Science and Technology in 1999, to describe "the global collaboration in key areas of science, and the next generation of infrastructure that will enable it."

Grid Computing: The term grid computing describes a distributed computing platform which integrates distributed computing resources such as CPUs and data to support computationally-intensive and/or data intensive scientific tasks. In general, grid computing is divided into two subtypes, i.e., data grid and computational grid. Data grids provide controlled sharing and management of large amounts of distributed data, while computational grid acts as a "virtual supercomputer" composed of a network of loosely-coupled computers, acting in concert to perform very large tasks. What distinguishes grid computing from typical cluster computing systems is that grids tend to be more loosely coupled, heterogeneous, and geographically dispersed. Meanwhile, it is often constructed with the aid of general purpose grid software libraries and middleware.

Molecular Docking: Molecular Docking is a method which is frequently used to predict the binding orientation of small molecule drug candidates to their protein targets in order to in turn predict the affinity and activity of the small molecule.

Workflow: A workflow serves as a model of virtual representation of real work which depicts a sequence of operations specific to certain processing. In real implementation, a workflow illustrates a graphical representation of the specific application process in which a node represents corresponding function and a directed link form one node to another represents the data flow or control flow between two nodes.

Workflow Engine: Workflow engine is a software application to manage and execute a workflow process. The workflow engine interprets tasks submitted to it, creates run time instances of these tasks, and act on these tasks according to the defined application processes. In general, workflow engine facilitates the flow of data, tasks, and control of a workflow application.

Workflow Modeler: Workflow modeler is a graphical tool that provides function nodes and connections for constructing the graphical model of a specific application process, i.e., the workflow.

ENDNOTES

1 http://www.ddgrid.ac.cn/p-grid/
2 http://www.r-project.org/
3 http://www.eclipse.org/

Chapter XVI
BioSimGrid Biomolecular Simulation Database

Kaihsu Tai
University of Oxford, UK

Mark S. P. Sansom
University of Oxford, UK

ABSTRACT

BioSimGrid is a distributed biomolecular simulation database. It is a general-purpose database for trajectories from molecular dynamics simulations. Though initially designed as a distributed data grid, BioSimGrid allows for installation as a stand-alone instance. This can later be integrated into a wider, networked system. This presentation of BioSimGrid follows a scenario in biological research to demonstrate how to install the system, and how to deposit, query, and analyze trajectories in this system, with real Python code examples for each step. What then follow are explanations of the underlying concepts in the implementation of BioSimGrid: relational database, distributed computing, and the input/output (deposit and analysis) modules. Finishing the presentation is a discussion of the emerging trends and concerns in the further development of BioSimGrid and similar biological databases. This discussion touches on quality-assurance issues and the use of BioSimGrid as a back-end for other speciality databases. The experience of developing BioSimGrid compels the conclusion: In the development and maintenance of biomolecular simulation databases, it is essential that sustainability be asserted as a key principle.

1. INTRODUCTION AND BACKGROUND: A REPOSITORY OF BIOMOLECULAR SIMULATIONS

Since the first application of the molecular dynamics on proteins in 1976 (Adcock and McCammon, 2006), this simulation methodology has added value to experimental structural biology by making biomolecules 'come alive' and by compensating in the nanosecond time-scale where experimental methods are only beginning to be able to access. Adding to this, we have the method of comparison, a precursor to the process of classification, which is fundamental to biology (Brooks and McLennan, 2006). Insights into the internal motions of proteins can come from comparing the results of molecular dynamics simulations, namely the trajectories (Pang et al., 2005; Tai et al., 2007). This process can be facilitated by having a database of trajectories. We have developed in the past few years (2003 to 2006) such a database called BioSimGrid (http://www.biosimgrid.org/; Tai et al., 2004).

Comparable endeavours to BioSimGrid include the Ascona B-DNA Consortium (http://humphry.chem.wesleyan.edu:8080/MDDNA/; Dixit et al., 2005), Dynameomics (http://www.dynameomics.org/; Scott et al., 2007), GEMS (http://gipse.cse.nd.edu/GEMS/; Wozniak et al., 2005), and SimDB (http://simdb.bch.msu.edu/; Feig et al., 1999). To differentiate, BioSimGrid is a general-purpose database for trajectories from molecular dynamics simulations. It is free software licensed under the terms of GNU General Public License (Stallman 2002). It can take advantage of distributed ('grid') computing, to enhance reliability and ensure longevity of the trajectory content (Berman et al., 2003a).

By 'general-purpose', we mean the following. Firstly, BioSimGrid can admit trajectories generated by different simulation packages, such as Amber (Pearlman et al., 1995), Gromacs (Lindahl et al., 2001), Charmm (Brooks et al., 1983), NWChem (Straatsma et al., 2000), and NAMD (Kalé et al., 1999). Secondly, BioSimGrid is not restricted to a special kind of system or granularity. It can host simulations for nucleic acids, proteins, small molecules, or even non-biological polymers; simulations at the all-atom level or coarse-grain molecular dynamics (Bond and Sansom, 2006; Marrink et al., 2004; Nielsen et al., 2004). It can also store non-sequential 'trajectories', or rather ensembles, generated by other methods such as Monte Carlo, homology modelling (Šali and Blundell, 1993), and CONCOORD (de Groot et al., 1997).

Here we briefly describe the implementation of BioSimGrid, so the reader can have a conceptual understanding of the system: aspects such as installation, interfaces, architecture, data schema, and the deposit and analysis modules. We present an end-to-end case scenario where a scientist can deposit a trajectory into BioSimGrid, query the database, and analyze a trajectory. Finally, we discuss the prospects of such a database: the quality-assurance and sustainability issues, and customization of BioSimGrid as 'back-ends' to specialist databases.

2. BIOSIMGRID: INSTALLATION AND INTERFACES

As we introduced above, BioSimGrid is a distributed database for biomolecular simulations. The BioSimGrid system was initially designed to be a distributed system, now deployed over a internet-based consortium including universities of Oxford, Southampton, Bristol, Nottingham, and York. A node has been added in the Pacific Northwest National Laboratory in Richland, Washington.

In the larger grid context, the National Grid Service (NGS, United Kingdom; http://www.grid-support.ac.uk/) provides the Storage Resource Broker (SRB) as a middleware service. Since Many large-scale grids elsewhere in the world also provide such a service for SRB. Since SRB is the principal back-end data-management middleware for BioSimGrid, we have had discussions with NGS to host BioSimGrid on such a grid, as one of the front-end data-management services for biomolecular scientists who use NGS; this is to be explored further. For more information about SRB, the reader is referred to Baru et al. (1998) and the SRB website (http://www.sdsc.edu/srb/).

2.1 Installation of a Stand-Alone Instance

In addition to the distributed model, a stand-alone adaptation has also been developed. This allows individual users to maintain their own instance of BioSimGrid which is disjunct from the original BioSimGrid consortium, yet still provides the same deposition and analysis functionality. For this, an entry-level or mid-range commodity personal computer should suffice. This stand-alone version requires that a variant of Linux be installed on the system (OpenSuSE 10.0 http://www.opensuse.org/ is recommended), and the resources meet those of the supporting software packages, primarily the relational database management system Oracle XE.

After the installation of OpenSuSE, the supporting software can then be obtained: The package for Oracle XE can be downloaded from Oracle's website (http://www.oracle.com/), and SRB from the San Diego Supercomputing Center website (http://www.sdsc.edu/srb/). (These packages are not distributed along with BioSimGrid due to their licensing conditions.) Once these supporting packages are placed in the installation directory along with the BioSimGrid package (available from the BioSimGrid website), we can proceed to install BioSimGrid along with the supporting software.

BioSimGrid supplies a default parameter file, which should be edited to reflect appropriate values for their node. Once this has been completed, the install script can be run, which verifies the prerequisites, then installs and configures a number of packages automatically, including Oracle XE, SRB, netCDF (network common data form), f2py (Fortran to Python interface generator), and MMTK (Molecular Modelling Toolkit). The database is automatically populated with the required basic data and an initial BioSimGrid user account is set-up with Oracle XE, based on the details supplied in the parameter file. Towards the end of the installation, some manual steps (walked through in the BioSimGrid stand-alone installation guide, also available from the website) are required to finalize SRB's configuration.

The resulting BioSimGrid setup is one that closely mirrors the distributed BioSimGrid system, but is effectively 'looped back' onto itself. It uses similar distributed technologies (SRB, Oracle XE) but available locally only: SRB and Oracle XE are configured without replication. This provides an effective, local platform for storing biomolecular simulation results and performing analyses thereon. This is particularly suitable in circumstances where is a requirement to keep data private: for example, drug development in a pharmaceutical-industry setting. When appropriate, it is possible to then integrate the stand-alone instance of BioSimGrid into a shared network of other instances.

2.2 Python and Web Interfaces

BioSimGrid provides two principal ways of access to the stored trajectories: One, the web interface, is a brief view for browsing the metadata (data describing the parameters and provenance of the stored trajectory). The other is a comprehensive interface via the Python language (van Rossum and Drake, 2003), providing full access to any aspect of the stored trajectory data, including full export of the trajectory in formats such as Protein Data Bank (PDB) files (Berman et al., 2003b) or Gromacs trajectory format (files usually suffixed '.xtc').

BioSimGrid assigns a global, unique identification string (GID) for each trajectory stored, in the form of BioSimGrid_*node_number*, where *node* is a code derived from the ISO 3166-2 standard (International Organization for Standardization 2007) identifying the originating BioSimGrid node, and *number* a unique number identifying the trajectory within the originating node. For example, BioSimGrid_GB-OXF_34 identifies a trajectory which was first deposited in Oxford, numbered 34 within the Oxford node. Taking advantage of this, a viable strategy for using the two interfaces in synergy can be: first using the light-weight, publicly-accessible web interface to search and identify the GIDs of the trajectories of interest, then use the Python interface to perform the analyses in detail.

In Figure 1, we show how a page describing the trajectory BioSimGrid_GB-OXF_34 from the web interface is displayed in a web browser. This happens to be a simulation for the outer-membrane protein OpcA (PDB ID: 1K24) from *Neisseria meningitidis* (Bond et al., 2007).

Having identified this trajectory, we could write an analysis script, using the BioSimGrid application programming interface, to do further detailed analysis. Writing this script can be made even simpler by using a 'meta-script' to generate the script (or a sample template script which can then be customized further).

In Scheme 1, we show an actual session to bootstrap an analysis using precisely this 'meta-script' approach. We run a Python script which asks us a series of questions about the script we want, including asking about the trajectory we are interested in. This desired script is generated. It can create a three-dimensional down-sampled 'video' of the trajectory, being a PDB file containing a series of snapshots (containing just the protein, but not the membrane and solvent environment)

Figure 1. The BioSimGrid web interface displaying the metadata (truncated at the bottom) which describes the trajectory BioSimGrid_GB-OXF_34. © 2009 Kaihsu Tai, et al. Used with permission.

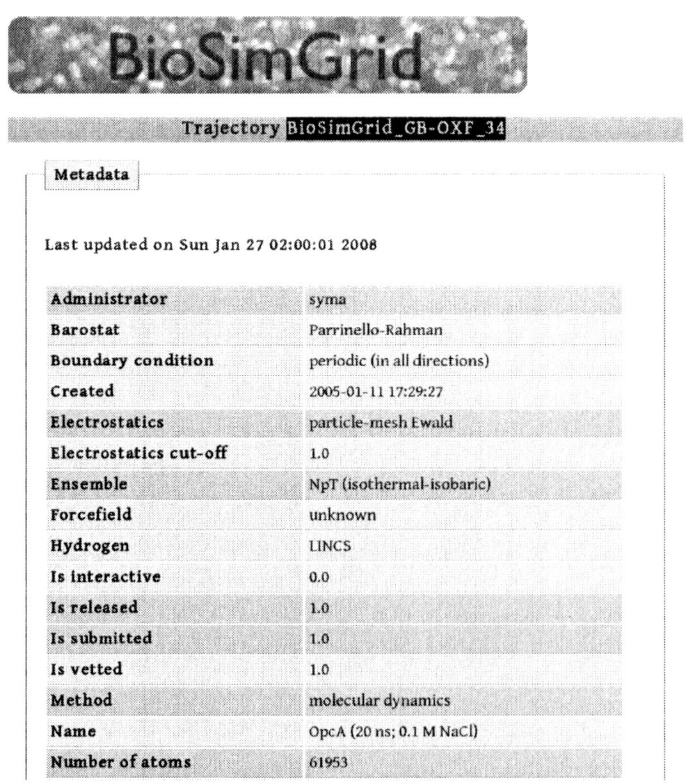
BioSimGrid

Trajectory BioSimGrid_GB-OXF_34

Metadata

Last updated on Sun Jan 27 02:00:01 2008

Administrator	syma
Barostat	Parrinello-Rahman
Boundary condition	periodic (in all directions)
Created	2005-01-11 17:29:27
Electrostatics	particle-mesh Ewald
Electrostatics cut-off	1.0
Ensemble	NpT (isothermal-isobaric)
Forcefield	unknown
Hydrogen	LINCS
Is interactive	0.0
Is released	1.0
Is submitted	1.0
Is vetted	1.0
Method	molecular dynamics
Name	OpcA (20 ns; 0.1 M NaCl)
Number of atoms	61953

Scheme 1. An actual session generating a script to create a PDB video

```
$ python /usr/lib/python/site-packages/BioSim/UserInterfaces/CommandLine/bsg-cmd.py
Please enter your user ID [guest] :kaihsu

0: Radial Distribution Function
1: Radius Of Gyration
2: Torsion Angles
3: COM
4: Cross Correlation
5: Bond Angles
6: Distance Matrix
7: Grace Analysis Comparison
8: Root Mean Square Fluctuation
9: Average Structure
10: Contact Order
11: Bond Lengths
12: Interatomic Distances
13: Domain Analysis
14: COMDifference
15: Center Of Geometry
16: PDB
17: Distance Matrix Error
18: Root Mean Square Displacement
19: PDBVideo
20: TrajectorySmoothing
21: Internal Angles
22: Surface And Volume

Enter a tool number [<Tool Number>? for description] :19

0: createFile

Enter a method number [<Method Number>? for description] :0

Enter the number of FCSettings you wish to configure for this tool [min=1, max=2]1
Press enter to configure FCSettings 1 of 1:

Please enter the trajectory [e.g. BioSimGrid_GB-STH_1]:BioSimGrid_GB-OXF_34
Please enter the frame[e.g. 12,13,14 or 12-14]:1-4000

Please enter the time step (default = 1):100
0 Atom Type
1 Residue Serial Number
2 Chain ID
3 Atom Serial Number
4 Residue Type
Select aditional settings? (<Setting Number>? for more info)[IGNORE]
2
Chain IDMethod: Chain ID

  This method sets restriction on frame collection such that atom(s) are retrieved
based on their atom serial number(s). It accepts a string of chain id in various
forms.

Example: ['U']

Please enter the valuses for this method: ['A']

(/home/kaihsu/tmp/test/Output/User-14/defaultScriptName.py)

Would you like to execute the generated script  : Yes/[No]?
```

taken during the simulation. This can then be animated using visualization software such as VMD (Humphrey et al., 1996).

The script generated is shown in Scheme 2.

Running this script (by issuing the command `python defaultScriptName.py`) gives a PDB file 'moleculeVideo.pdb' containing 40 models. Figure 2 shows VMD visualizing the animated video (frame 17 of 40 shown).

A more comprehensive web-based 'portal' can be set up to enable complicated analyses through the web. This is essentially a web front-end for the 'meta-script' approach mentioned above. As Python is an easy-to-use scripting language, with molecule-handling extensibility provided by packages such as the VMD Python module or PyMOL (http://pymol.sourceforge.net/), it is possible to extend the text and web interfaces described above to give a graphical user interface if necessary. This is to be developed by the user community as their needs evolve. Since running the generated scripts may require substantial

Scheme 2. Automatically-generated script `defaultScriptName.py` for creating a PDB video

```
"""

Author: team@biosimgrid.org
Please report any bugs to: bugs@biosimgrid.org

Descriptions:

This file is AUTO-generated, so make sure if you edit it that it will not be over written

"""

from BioSim.DataRetrieval import FrameCollection
from BioSim.DataRetrieval import FCSettings
from BioSim.Settings import UserSettings
from BioSim.Analysis import PDBVideo

#Create the usersettings
userSettings = UserSettings.UserSettings("kaihsu")

#Create the frame collections
framesForAnalysis0 = FCSettings.FCSettings(userSettings,[['BioSimGrid_GB-
OXF_34',range(1,4001,100)]])
frameCollection0 = FrameCollection.FrameCollection(framesForAnalysis0)


# PDBVideo:
#   This class takes either one or two Frame Collections and has methods to produce a
video format for the frames. If there is only one frame a PDB file for a single Frame is
produced. This is primarily for use with the visualisation package VMD. If the second
Frame Collection is used then an SVD minimisation may be done as the video is produced.

PDBVideoTool  = PDBVideo.PDBVideo(frameCollection0 )

# createFile:
#   This method creates the PDB formatted video file for use with a visualiser.

PDBVideoTool.createFile()

#Show where the output files will be stored
print "Output Dir :" +  str(userSettings.getOutputPath())
```

Figure 2. Visualizing an animated video of the molecular dynamics trajectory of protein OpcA. Frame 17 of 40 is shown. The protein is displayed in a cartoon representation to show the beta-barrel fold of OpcA. © 2009 Kaihsu Tai, et al. Used with permission.

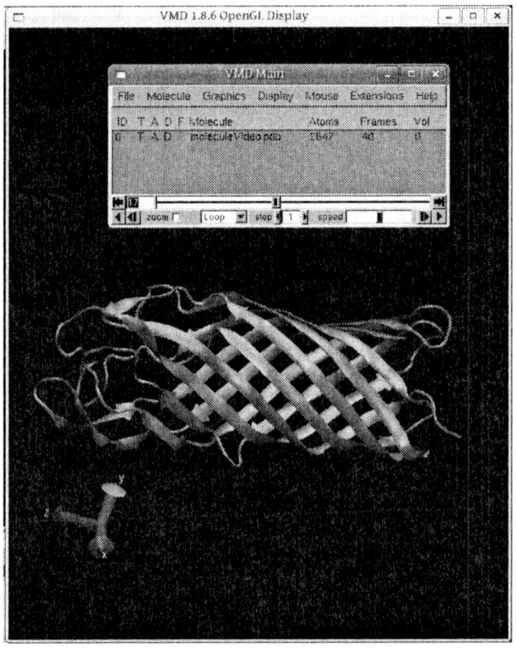

computational power on the node, it is not advisable that this portal be publicly accessible without limit or authentication. Instead, it is a good idea to enclose such web portal within an authenticated session (either by password or by digital certificate), so the computational requirement can be limited and tracked. In the next section, we see how these interfaces can be put to work.

3. USE CASE SCENARIO: COMPARATIVE ANALYSES OF SIMULATIONS

In this section, we show an end-to-end case scenario in which a scientist can deposit a Gromacs trajectory into BioSimGrid, query the database, and analyze a set of trajectories. This is all done using the web- and Python-based interfaces to the database. Each subsection will have Python code examples.

The scenario generally follows our earlier work on the active-site conformations of three hydrolases and a transferase (Tai et al., 2007). Here we deposit a trajectory for the transferase PagP (PDB ID: 1THQ), locate the other relevant trajectories already in the database, and measure the metrics of active-site conformational integrity.

3.1 Account Settings

After the installation of a BioSimGrid node, the administrator can give a Linux account to a user on that node. The user should be assigned an account for the BioSimGrid Python interface. Though the web interface is publicly accessible (no password is necessary) for only browsing purposes, everything that follows has to be done while logged into that node using the Linux account.

There are a few things that has to be done only once after the user acquires the account. The lines in Scheme 3 set up environment variables for the supporting software Oracle XE and SRB for the user. Ensuring that these lines are in the file `~/.bash_profile` has the advantage of not having to type them each time after logging in and before starting to use BioSimGrid.

The settings can then be activate this either by issuing the command '`source ~/.bash_profile`' or by logging out and back in again. Then, the SRB client environment files should be copied into the account's home directory: This can be done by issuing the commands '`scp -r /.srb ~/. && scp -r /srb ~/.`'.

3.2 Depositing a Trajectory

Now we proceed where the user deposits a molecular dynamics trajectory has obtained by running the Gromacs package. (Procedure for results from other molecular-dynamics software packages are similar, and documented in the user guide available on the website.) Three files associ-

Scheme 3. Setting up Oracle XE and the Storage Resource Broker

```
# Oracle XE setup
export ORACLE_BASE=/u01/app/oracle
export ORACLE_SID=BioSimDB
export ORACLE_HOME=$ORACLE_BASE/product/10.1.0/db_1
export PATH=$PATH:$ORACLE_HOME/bin
. /home/oracle/.bashrc

# SRB setup
PATH=$PATH:/srb/bin
export PATH SRB_ADMIN_HOME
alias Sman="man -M /srb/man"
```

ated with each trajectory, generated by Gromacs, are required for a trajectory to be deposit into BioSimGrid, namely:

- The PDB file containing the topology of the molecules: atom and residue names. This can be generated from the Gromacs native-format topology file (suffix '.gro') using the `editconf` command provided by Gromacs.
- The coordinate file. Gromacs native-format trajectory files are admissible (suffix '.xtc'). More than one file can be admitted at the same time: BioSimGrid will concatenated them.
- The molecular-dynamics parameters ('canonized'; that is, containing all information about all the parameters, including the default values) generated and returned by the preprocessor `grompp` of Gromacs (suffix '.mdp').

In addition, BioSimGrid requires a descriptive name for the trajectory, and a source PDB code ('`NONE`' should be entered if not applicable). Other optional files associated with the trajectory can be deposited along with the required trajectory files. Using the secure file transfer command `scp`, the user can upload these files to the node. It is advisable to keep everything in one directory. For example:

```
scp pagp.{pdb,mdp} pagp[0-2].xtc user@
node.biosimgrid.org:deposit
```

Then the user can create a new file in the same directory, called something like `myDeposit.py` (Scheme 4), to deposit this simulation of PagP, an outer-membrane enzyme which transfers a palmitate chain from a phospholipid to lipid A (Tai et al., 2007).

The user can then run this file by issuing the command `python myDeposit.py`. Depending on the size of the trajectory, this can take a few hours. BioSimGrid then reports that the depositing process completed with a success, returning a GID ('BioSimGrid_GB-OXF_31' in this case). The user can then ask a curator of the node to release the trajectory for analysis by running the script to publish (Scheme 5).

3.3 Database Query

Shortly after the trajectory has been published (within 24 hours), it will appear in the trajectory index on the web interface. We find an entry 'PagP (without charges) (BioSimGrid_GB-OXF_31)' among the other trajectories (Figure 3).

Following the link for this trajectory leads to a page (similar to that in Figure 1) with the detailed metadata for the trajectory. Other trajectories can be deposited and identified with a similar procedure.

Scheme 4. Deposit script `myDeposit.py`

```
# Importing the BioSimGrid application programming interface.
from BioSim.Settings import UserSettings
from BioSim.Deposit.GromacsDeposit import GromacsDeposit

# Setting up the names.
n = "pagp"
q = "PagP (without charges)"

# Specifying to BioSimGrid where to find the files and names.
filenames = {
  'topology': n + ".pdb",
  'coordinates': [n + "0.xtc", n + "1.xtc", n + "2.xtc"],
  'parameters': n + ".mdp",
  'otherFiles': ["additional_notes.txt"],
  'pdb_code': "1THQ",
  'name': q
}

# Setting up the user environment, specifying to commit to database.
s = UserSettings.UserSettings("kaihsu", databaseCommitSetting = True)

# Depositing the trajectory.
g = GromacsDeposit.GromacsDeposit(s, filenames)
```

Scheme 5. Curator's script to publish the trajectory

```
# Importing the BioSimGrid application programming interface.
from BioSim.DataPopulation.Curation import MetadataCuration

# Publishing the trajectory specified with authentication.
mc = MetadataCuration.MetadataCuration()
mc.publishTrajectory("kaihsu", "password.for.kaihsu", "BioSimGrid_GB-OXF_31")
```

3.4 Analyzing Trajectories

The protein PagP, the subject of this simulation BioSimGrid_GB-OXF_31, is an outer-membrane enzyme which transfers a palmitate chain from a phospholipid to lipid A. In a previous work (Tai et al., 2007), we investigated the intactness of active sites for this and more distantly-related enzymes, whose active sites are similar. In the case of PagP, it consists of three residues forming a catalytic triad: histidine 33, aspartic acid 76, and serine 77. Here we provide a step-by-step procedure to analyze the intactness of this catalytic triad, comparing an intactness metric under different simulation conditions.

To provide metrics for the intactness of PagP's catalytic triad, we measured the pairwise distance between three atoms: the gamma-carbon atom in the carboxyl group of Asp76, the gamma-carbon atom in His33 (placed between the other two catalytic-triad residues), and the gamma-oxygen atom in Ser77. The Python script for BioSimGrid, `triad_metric.py` (Scheme 6), retrieves these distance time series.

By looking at the trajectory index (Figure 3), we notice that there are more than one trajectories for PagP, already deposited. In addition to BioSimGrid_GB-OXF_31 which we have just deposited in Section 3.2, there is another, BioSimGrid_GB-OXF_88. The difference between

Figure 3. The trajectory 'PagP (without charges) (BioSimGrid_GB-OXF_31)' among others in the index.

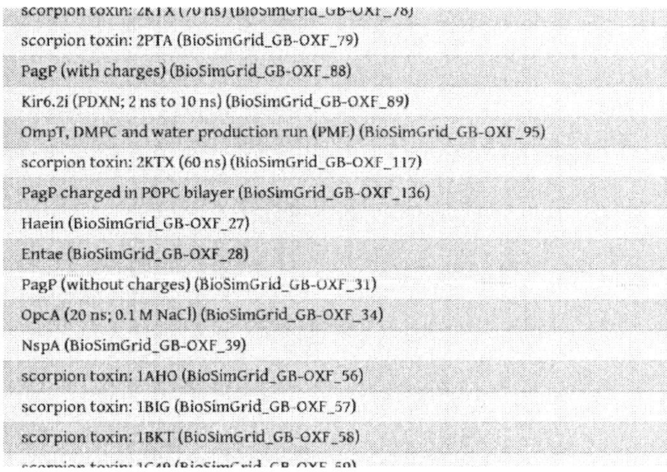
scorpion toxin: 2PTA (BioSimGrid_GB-OXF_79)
PagP (with charges) (BioSimGrid_GB-OXF_88)
Kir6.2i (PDXN; 2 ns to 10 ns) (BioSimGrid_GB-OXF_89)
OmpT, DMPC and water production run (PMF) (BioSimGrid_GB-OXF_95)
scorpion toxin: 2KTX (60 ns) (BioSimGrid_GB-OXF_117)
PagP charged in POPC bilayer (BioSimGrid_GB-OXF_136)
Haein (BioSimGrid_GB-OXF_27)
Entae (BioSimGrid_GB-OXF_28)
PagP (without charges) (BioSimGrid_GB-OXF_31)
OpcA (20 ns; 0.1 M NaCl) (BioSimGrid_GB-OXF_34)
NspA (BioSimGrid_GB-OXF_39)
scorpion toxin: 1AHO (BioSimGrid_GB-OXF_56)
scorpion toxin: 1BIG (BioSimGrid_GB-OXF_57)
scorpion toxin: 1BKT (BioSimGrid_GB-OXF_58)

these two is that in the former ('31), only the delta-nitrogen atom in His33 is protonated (hence the residue is neutral), but in the latter ('88), both the delta- and the epsilon-nitrogen atoms are protonated (hence positively charged). In both cases, the Asp76 is ionized.

The Python script in Scheme 6 is in several paragraphs. Paragraph 1 imports the necessary parts of the BioSimGrid interface, and also the Python 'sys' system library which we use in the exception handling. Paragraph 2 prepares the user setting and authentication; it also sets the log reporting level to only report critical events. Paragraph 3 determines the parameters of the measurements: the step size between sampled snapshots in the trajectory, the names under which each distance measurement is to be reported, the trajectory GIDs, and the indices for the atoms in question. Paragraph 4 generates the time series and associated plots with a subparagraph to handle possible exceptions and errors.

This script can be run by issuing the command `python triad_metric.py`. The time-series plots (in portable network graphics 'png' format; an example is shown in Figure 4) are generated along with the time series data (in human-readable 'agr' format, suitable for the plotting software Grace http://plasma-gate.weizmann.ac.il/Grace/ and others).

With the time series 'agr' files, we can compare the intactness of the catalytic triads by plotting the progression of each of the two metrics on a scatter plot (Figure 5). In discussing this plot, we express the coordinates as an ordered pair of two distances, (Asp76-His33, His33-Ser77). For the PagP simulation with charged His33 in the catalytic triad, the metrics are around (0.6 nm, 0.8 nm), but from time to time venture into (1.0 nm, 1.2 nm). For the neutral His33, (1.0 nm, 1.2 nm) is a typical value, and the sample data-points are more spread out than in the case of charged His33. This suggests that the electrostatic interactions present in the charged His33 may drive the association of the active-site loops and help stabilize the catalytic triad in PagP.

As can be seen in Scheme 1, other analysis tools are available in BioSimGrid in addition to interatomic distance measurements. These include root mean square displacement and fluctuation, secondary structure analysis, radial distribution function, radius of gyration, centre of geometry or mass, and principal component analysis. A user guide (more than 100 pages) explaining each of these in detail, with script examples for the use of each function, is available on the website.

Scheme 6. A Python program to extract distance-metrics in two trajectories of PagP from BioSimGrid, `triad_metric.py`.

```
# A Python program to extract distance-metrics in trajectories of PagP
# from BioSimGrid.

# Paragraph 1. Importing the necessary parts of the BioSimGrid interface.
from BioSim.DataRetrieval import FrameCollection
from BioSim.DataRetrieval import FCSettings
from BioSim.Settings import UserSettings
from BioSim.Analysis import InteratomicDistances
from BioSim.Logger import BioSimLog
import sys

# Paragraph 2. Setting up the environment.
uS = UserSettings.UserSettings("kaihsu")
log = BioSimLog.BioSimLog()
log.setLevel("CRITICAL")
thisLog = log.getLogHandler()

# Paragraph 3. Setting up the measurements.
stepSize = 1
desc = ['Asp', 'His', 'Ser']
pairs = [[0, 1], [1, 2]]
sets = [
  { 'name': 'neutral',
    'traj':[['BioSimGrid_GB-OXF_31', range(1, 4001, stepSize)]],
    'atoms': [868, 371, 877] },
  { 'name': 'charged',
    'traj':[['BioSimGrid_GB-OXF_88', range(1, 3001, stepSize)]],
    'atoms': [869, 372, 878] },
  # More trajectories can be added here.
]

# Paragraph 4. Obtaining the time series.
for set in sets:
  for pair in pairs:

    # Subparagraph 4.1. Actually generating the time series.
    try:
      nowPair = set['name'] + ": " + desc[pair[0]] + " and " + desc[pair[1]]
      thisLog.critical("doing: " + nowPair)
      toDo0 = FCSettings.FCSettings(uS, set['traj'])
      toDo0.setAtomSerialNo([set['atoms'][pair[0]]])
      FC0 = FrameCollection.FrameCollection(toDo0)
      toDo1 = FCSettings.FCSettings(uS, set['traj'])
      toDo1.setAtomSerialNo([set['atoms'][pair[1]]])
      FC1 = FrameCollection.FrameCollection(toDo1)
      myIAD = InteratomicDistances.InteratomicDistances(FC0, FC1)
      myIAD.graphFilename = set['name'] +"_"+ desc[pair[0]] +"_"+ desc[pair[1]]
      myIAD.graphyLabel = 'distance / nm'
      myIAD.graphxLabel = 'time / ps'
      myIAD.graphTitle = nowPair
      myIAD.createAsPNG()

    # Subparagraph 4.2. Handling exceptions.
    except:
      thisLog.critical(sys.exc_info()[0])

    thisLog.critical("done: " + nowPair)
```

Figure 4. A time-series plot `charged_His_Ser.png`*, showing the distance between the gamma-carbon atom in His33 and the gamma-oxygen atom in Ser77*

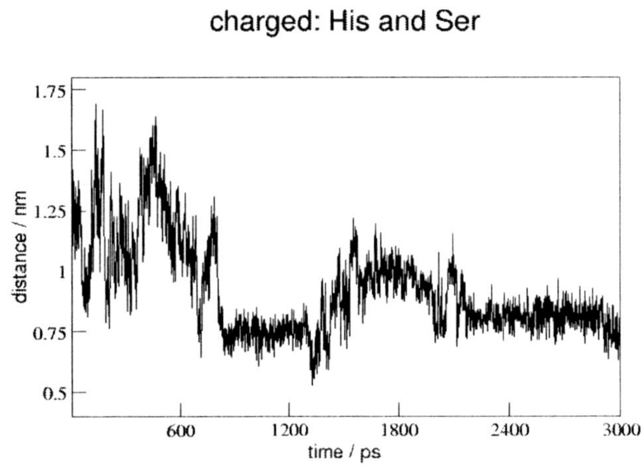

Figure 5. Comparing the intactness of the catalytic triads from the charged and neutral PagP simulations

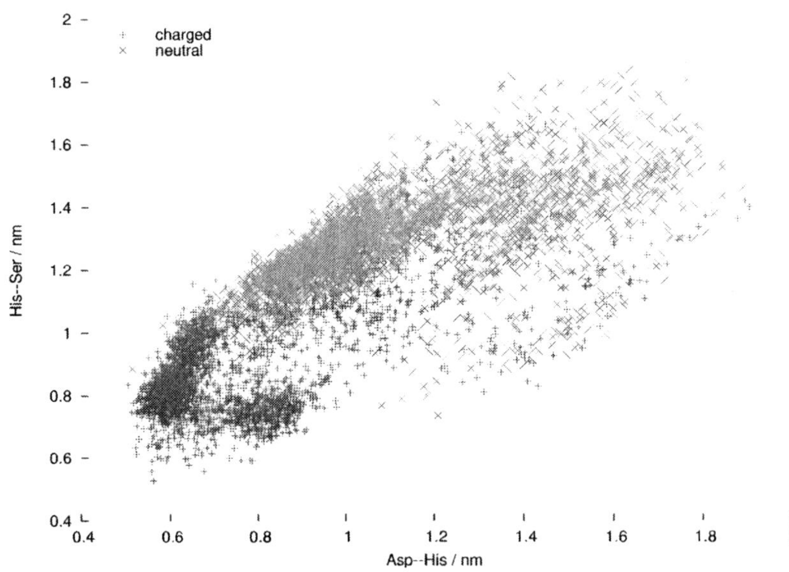

4. CONCEPTS IN THE IMPLEMENTATION OF BIOSIMGRID

Here we briefly describe the implementation of BioSimGrid, so the reader have a conceptual understanding of the architecture, data schema, and the deposit and analysis modules.

4.1 Relational Database: Data Schema

The schema of a relational database is a conceptual abstraction of a database. For our trajectory database, its schema (Scheme 7) frees us from having at all times to consider the physical storage aspect.

Scheme 7. The abridged database schema for BioSimGrid (Tai et al., 2004). Each of the boxes represent a table of relations. The arrows represent keys and referential links between tables. © 2009 Kaihsu Tai, et al. Used with permission.

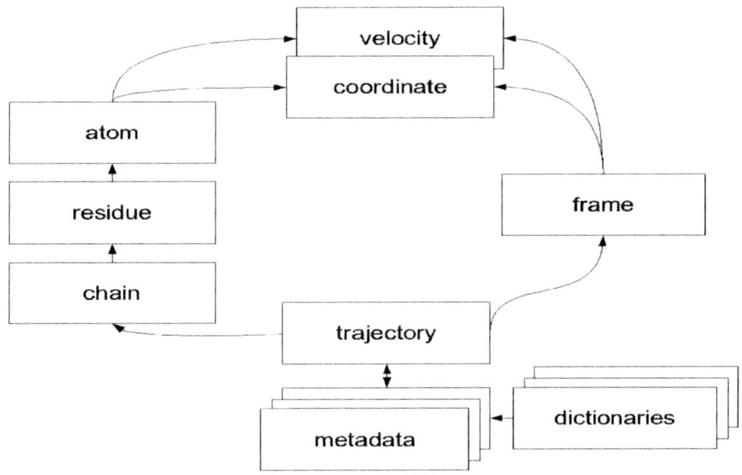

Most of the data (data 'in' the trajectories) are stored in the tables 'trajectory', 'frame', 'coordinate', and 'velocity'. The topology of the biomolecular system is contained in the tables 'chain', 'residue', and 'atom'; this mirrors the conventional PDB hierarchy. A 'chain', in this sense, is defined as a molecule which is internally connected by chemical bonds. A 'residue' may represent a monomer in a polymer, not necessarily a residue in a polypeptide/protein chain. An 'atom' in a coarse-grain simulation can be a particle representing more than one atom.

The rest of the metadata (data 'about' the trajectories) are in the other tables, with supporting dictionary tables. The table 'trajectory' brings together the data and the metadata. An entry in this table (a trajectory) may own a certain number of frames. Each frame may own a certain number of coordinates. The 'frame' table also stores properties with a time-series nature, such as volume, pressure, and temperature. Each coordinate entry stores the three-dimensional position of an atom in a frame.

4.2 Distributed Database: Resilience by Replica

BioSimGrid, as its name may suggest, is a 'data grid' in the taxonomy of grid computing, which itself is a subspecies of distributed computing. Grid computing offers computational power (for example, a large number of floating-point calculations or a large amount of data) as a commodified utility, in the same sense as the electrical grid.

Contrary to the abstract schema for BioSimGrid introduced previously, in practice we use heterogeneous the relational database and the structured flat-file storage. The tables with the most entries, namely 'coordinate' and 'velocity', are in an in-house flat-file format (in distributed duplicate for resilience). This design takes into consideration of the more-efficient access for the nature of trajectory data. The other tables are replicated at all nodes in relational database instances.

Soon after a new trajectory is deposited, BioSimGrid arranges for the flat files to be copied from the originating node to another node (with the help of the Storage Resource Broker), and for the relational database instances to be synchronized with the latest content (with the help of the relational database management system's distributed computing functions). With this arrangement, each piece of information is stored in at least two nodes. If one node happens to fail, at least one other node has the data required for any possible user query and for the restoration to integrity. We may call this architecture 'resilience by replica'.

4.3 Deposit Modules: Dealing with File Formats

Four deposit modules are available in BioSimGrid, each dedicated to a trajectory format for a particular molecular dynamics package. These are the four most popular packages used in the simulation community: Amber, Charmm, Gromacs, and NAMD. Additional deposit module for the PDB, NWChem, and other formats are in the prototype stage of development.

As shown in Section 3.2, the deposit module takes in the parameter, topology, and coordinate files (the mandatory files) which have been produced earlier by the simulation package. The mandatory raw data files are both stored in their original form and digested into the relational tables and in-house flat-files. Any other optional files, if nominated by the depositor, are stored but undigested. From then on, the process of 'homogenization' by BioSimGrid is complete, and any further action by BioSimGrid is agnostic to the originating file format.

4.4 Analysis Modules

As mentioned in Section 3.4, a large set of analysis tools are available in the BioSimGrid application programming interface. Most of these were implemented from scratch. Some others were wrappers for third-party software packages, so the packages can perform their specialized analysis on the BioSimGrid trajectory format. All of the tools can be applied to any trajectory stored, regardless of the originating simulation package. This capability was demonstrated in an earlier publication (Tai et al., 2007).

5. FUTURE TRENDS

Here we highlight three aspects of the future with BioSimGrid. In our capacity as curators of the stored trajectories, we have come to consider the quality assurance for these. Further, our colleagues have built specialist databases on top of BioSimGrid. We share some lessons learned from our experience in the implementation of BioSimGrid. Finally, we weigh up the sustainability prospects of biological databases, of which BioSimGrid is one.

5.1. Quality-Assurance Issues of Biomolecular Simulation Trajectories

Though the policy of BioSimGrid is to admit, deposit, and store any contributions which have been made in good faith, we have come to consider the quality-assurance issues of biomolecular simulation trajectories, and other existing databases based on BioSimGrid. In an earlier work (Murdock et al., 2006), we enumerated suitable measures for the quality assurance of molecular dynamics simulations of biomolecules. Therein we asserted that 'these measures are designed to be simple, fast, and general'; and that 'reporting of these measures in simulation papers should become an expected practice, analogous to the reporting of comparable quality measures in protein crystallography'.

We categorized these measures into four categories: (1) provenance metadata, whose reporting should be obligatory; (2) thermodynamic measures of quality, whose reporting should be obligatory; (3) convergence and fluctuation measures, whose reporting should be obligatory; and (4) structural measures, whose reporting is not obligatory but informative. We have implemented tools for the calculation of all these measures in BioSimGrid. The reported measures should then gives the researcher, upon encountering a stored trajectory in the database, a reasonable overview to decide whether to investigate further.

5.2 'Skins' of BioSimGrid: Specialty Databases for Biomolecular Simulations

BioSimGrid, unlike some of the comparable endeavours, is a general purpose database for trajectories. However, as our colleagues have found, it can serve as the 'back-end' of specialist databases. For example, there is a database for coarse-grain simulations of inserted membrane proteins (http://sbcb.bioch.ox.ac.uk/cgdb/; Sansom et al., 2008); an outer-membrane protein database (http://sbcb.bioch.ox.ac.uk/ompdb/) and a database of potassium ion channel molecular dynamics simulations (http://sbcb.bioch.ox.ac.uk/kdb/; Fowler et al., 2008). Moreover, it is conceivable that BioSimGrid can be extended to admit results from mixed methods such as quantum mechanics-molecular mechanics (QM/MM) simulations.

5.3 Lessons Learned

It is salutary to consider what we could have done differently or better, given the experience of our implementation of BioSimGrid. We offer two 'lessons learned' here, one about the flat-file format, the other on the relational database management system.

One, we could use the native file format of a particular molecular dynamics simulation package, say Gromacs or NAMD, as the internal flat-file trajectory representation, rather than our in-house format. Such a package would already come with a set of robust and easy-to-use analysis tools, which we could wrap into our analysis modules. We could then attach a light-weight relational database to manage the trajectory metadata. Further, we could gently extend the deposit modules to accommodate the digestion of trajectories from other simulation packages. We could then fit this framework into a distributed architecture, perhaps with the help of the Storage Resource Broker. We did not opt to do this in our early design stage, due to the project requirements of our funding supporter.

Two, we could have used an open-source relational database management system such as MySQL (http://www.mysql.com/) rather than a product such as Oracle XE as our relational database management system. This would help BioSimGrid approach a full status as free software, both *gratis* and *libre*. But this would require us to devise and test our own middleware for distributed relational database management, which is already mature in Oracle products. Therefore, we did not choose this option.

5.4 Sustainability of Databases

Recently, the scientific community has given some considerations about the sustainability of databases. This is perhaps an afterthought of the 'database boom' following the successes of sequencing projects in genomics in the 1990s, and the (unexpected?) longevity of the Protein Data Bank, a database of crystallographic models. Take an editorial in the journal *Nature,* which opined in early 2007: 'Funding agencies face conflicting challenges in supporting the databases essential to modern biology. [...] To obtain value for money, it will be vital for funding agencies to carefully select the databases they choose to support and then to support them for the long term' (Nature 2007).

Such considerations are sensible. At the very least, for scientists who are at the 'user' end of the database, it is fair for them to rely on that the data on which their research and publication bases will remain available after a few years, perhaps more. This also applies for those who contribute and deposit data. These expectations can only be wishful thinking if sustainability of the databases are not planned for and ensured.

With resilience as one of the early design goals, the BioSimGrid team also valued sustainability and worked for it with the resources available. Here we share a few interesting facts for BioSimGrid in the last 18 months since June 2006, when the funding for this project stopped. First, BioSimGrid continued to operate in full (with zero funding), where the curation is done by volunteers on a pro bono basis. We continue to curate the 100-some old trajectories, and 40-some new trajectories have been deposited, mainly for simulations of potassium channels, both atomistic and coarse-grain.

Likely future developments include a more robust web portal, a better interface for trajectory export, extending the database to include QM/MM simulations, new deposit modules for other simulation packages, and replacing Oracle XE with a free (*libre*, not just *gratis*) relational database management system. This lack of development is due to considerations of stability of system: We did not wish to compromise this for the software development cycle, which may give an unstable system at times.

6. CONCLUSION

In this chapter, we have introduced the BioSimGrid system, a database for biomolecular simulations. We explained the process to install a stand-alone instance of BioSimGrid, which can be integrated with a larger networked system when appropriate. We showed how to query the database with its web interface, and how to deposit trajectories and analyze them with the Python interface. In all of this, we followed a realistic scenario in scientific research, and supplied actual Python scripts for the analyses carried out.

Then we discussed the concepts in BioSimGrid's implementation: We touched upon the theoretical idea of a relational database, and explained that the actual BioSimGrid implementation has a mixed relational/flat-file model for efficiency. We offered the principle of 'resilience by replica', of which BioSimGrid takes advantage. The input of BioSimGrid are handled by the deposit modules; the output by the analysis modules.

We then discussed the future trends for BioSimGrid and similar databases for biology. We proposed a set of quality-assurance measures for biomolecular simulations. We showed that BioSimGrid now serves as a back-end for a handful of specialty databases, covering coarse-grain, outer-membrane protein, and potassium ion channels. We offered some lessons learned from our experience in implementing BioSimGrid. We asserted that sustainability is the most important issue in the initiation and maintenance of biological databases: Without considering it, the efforts of many scientists would be in vain.

7. ACKNOWLEDGMENT

We thank Kia Balali-Mood, Richard P. Boardman, Leo Caves, Simon Cox, Jonathan W. Essex, Hans Fangohr, Jesús Izaguirre, Paul Jeffreys, Steven J. Johnston, Syma Khalid, Canhai Lai, Charles Laughton, J. Andrew McCammon, David Moss, Adrian Mulholland, Stuart Murdock, Muan Hong Ng, Eric G. Stephan, Tjerk Straatsma, and Bing Wu for their help and efforts during the development, documentation, testing, and maintenance of BioSimGrid. We thank the Biotechnology and Biological Sciences Research Council (United Kingdom) for funding the BioSimGrid project.

8. REFERENCES

Adcock, S. A., & McCammon, J. A. (2006). Molecular dynamics: Survey of methods for simulating the activity of proteins. *Chemical Review, 106*, 1589-1615.

Baru, C., Moore, R., Rajasekar, A., & Wan, M. (1998) The SDSC Storage Resource Broker. In *Proceedings of the 1998 Conference of the Centre for Advanced Studies on Collaborative Research* (p. 5). Toronto, Canada: IBM Press.

Berman, F., Fox, G., & Hey, T. (Ed.). (2003a). *Grid computing: Making the global infrastructure a reality.* Chichester, UK: Wiley.

Berman, H. M., Henrick, K., & Nakamura, H. (2003b). Announcing the worldwide Protein Data Bank. *Nature Structural Biology, 10,* 980.

Bond, P. J., & Sansom, M. S. P. (2006). Insertion and assembly of membrane proteins via simulation. *Journal of the American Chemical Society, 128,* 2697-2704.

Bond, P. J., Derrick, J. P., & Sansom, M. S. P. (2007). Membrane simulations of OpcA: Gating in the loops? *Biophysical Journal, 92,* L23-L25.

Brooks, B. R., Bruccoleri, R. E., Olafson, B. D., States, D. J., Swaminathan, S., & Karplus, M. (1983). CHARMM: a program for macromolecular energy, minimization, and dynamics calculations. *Journal of Computational Chemistry, 4,* 187–217.

Brooks, D. R., & McLennan D. A. (2002). *The Nature of diversity: An evolutionary voyage of discovery.* Chicago, IL: University of Chicago Press.

Date, C. J. (2000). *An Introduction to database systems.* Reading, MA: Addison-Wesley.

de Groot, B. L., van Aalten, D. M. F., Scheek, R. M., Amadei, A., Vriend, G., & Berendsen, H. J. C. (1997). Prediction of protein conformational freedom from distance constraints. *Proteins, 29,* 240-251.

Dixit, S. B., Beveridge, D. L., Case, D. A., Cheatham III, T. E., Guidice, E., Lankas, F., Lavery, R., Maddocks, J. H., Osman, R., Sklenar, H., Thayer, K. M., & Varnai, P. (2005). Molecular dynamics simulations of the 136 unique tetranucleotide sequences of DNA oligonucleotides. II: Sequence context effects on the dynamical structures of the 10 unique dinucleotide steps, *Biophysical Journal, 89,* 3721-3740.

Feig, M., Abdullah, M., Johnsson, L., & Pettitt, B. M. (1999). Large scale distributed data repository: Design of a molecular dynamics trajectory database. *Future Generation Computer Systems, 16,* 101-110.

Fowler, P., Khalid, S., Mokrab, Y., Sansom, M. S. P., Stansfeld, P., & Tai, K. (2008). KDB: A potassium channel model and simulation database. *Biophysical Journal, 94,* 2038.

Humphrey, W., Dalke, A., & Schulten, K. (1996). VMD: Visual molecular dynamics. *Journal of Molecular Graphics, 14,* 33-38.

International Organization for Standardization (2007). *Codes for the representation of names of countries and their subdivisions – Part 2: Country subdivision code.* Geneva, Switzerland: International Organization for Standardization.

Kalé, L., Skeel, R., Bhandarkar, M., Brunner, R., Gursoy, A., Krawetz, N., Phillips, J., Shinozaki, A., Varadarajan, K., & Schulten, K. (1999). NAMD2: Greater scalability for parallel molecular dynamics. *Journal of Computational Physics, 151,* 283-312.

Lindahl, E., Hess, B., & van der Spoel, D. (2001). GROMACS 3.0: a package for molecular simulation and trajectory analysis. *Journal of Molecular Modeling, 7,* 306-317.

Marrink, S. J., de Vries, A. H., & Mark, A. E. (2004). Coarse grained model for semiquantitative lipid simulations. *Journal of Physical Chemistry B, 108,* 750-760.

Murdock, S. E., Tai, K., Ng, M. H., Johnston, S., Wu, B., Fangohr, H., Laughton, C. A., Essex, J. W., & Sansom, M. S. P. (2006). Quality assurance for biomolecular simulations. *Journal of Chemical Theory and Computation, 2,* 1477-1481.

Nature (2007). Editorial: The database revolution. *Nature, 445,* 229-230.

Ng, M. H., Johnston, S., Wu, B., Murdock, S. E., Tai, K., Fangohr, H., Cox, S. J., Essex, J. W., Sansom, M. S. P., & Jeffreys, P. (2006). BioSimGrid: Grid-enabled biomolecular simulation data storage and analysis. *Future Generation Computer Systems, 22,* 657-664.

Nielsen, S. O., Lopez, C. F., Srinivas, G., & Klein, M. L. (2004). Coarse grain models and the computer simulation of soft materials. *Journal of Physics: Condensed Matter, 16,* R481-R512.

Pang, A., Arinaminpathy, Y., Sansom, M. S. P., & Biggin, P. C. (2005). Comparative molecular dynamics – similar folds and similar motions? *Proteins: Structure, Function, and Bioinformatics, 61,* 809-822.

Pearlman, D. A., Case, D. A., Caldwell, J. W., Ross, W. S., Cheatham III, T. E., DeBolt, S., Ferguson, D., Seibel G., & Kollman, P. (1995). AMBER, a computer program for applying molecular mechanics, normal mode analysis, molecular dynamics and free energy calculations to elucidate the structures and energies of molecules. *Computer Physics Communications, 91,* 1-41.

Šali, A., & Blundell, T. L. (1993).Comparative protein modelling by satisfaction of spatial restraints. *Journal of Molecular Biology, 234,* 779-815.

Sansom, M. S. P., Scott, K. A., & Bond, P. J. (2008). Coarse-grained simulation: A high-throughput computational approach to membrane proteins. *Biochemistry Society Transactions, 36,* 27-32.

Scott, K. A., Alonso, D. O. V., Sato, S., Fersht, A. R., & Daggett, V. (2007). Conformational entropy of alanine versus glycine in protein denatured states. *Proceedings of the National Academy of Sciences USA, 104,* 2661-2666.

Stallman, R. M. (2002). *Free software, free society.* Boston, MA: GNU Press.

Straatsma, T. P., Philippopoulos, M., & McCammon, J. (2000). NWChem: Exploiting parallelism in molecular simulations. *Computer Physics Communications, 128,* 377-385.

Tai, K., Murdock, S., Wu, B., Ng, M. H., Johnston, S., Fangohr, H., Cox, S. J., Jeffreys, P., Essex, J. W., & Sansom, M. S. P. (2004). BioSimGrid: Towards a worldwide repository for biomolecular simulations. *Organic and Biomolecular Chemistry, 2,* 3219-3221.

Tai, K., Baaden, M., Murdock, S., Wu, B., Ng, M. H., Johnston, S., Boardman, R., Fangohr, H., Cox, K., Essex, J. W., & Sansom M. S. P. (2007). Three hydrolases and a transferase: Comparative analysis of active-site dynamics via the BioSimGrid database. *Journal of Molecular Graphics and Modelling, 25,* 896-902.

van Rossum, G., & Drake, Jr, F. L. (2003). *An Introduction to Python.* Bristol, UK: Network Theory.

Wozniak, J. M., Brenner, P., Thain, D., Striegel, A., & Izaguirre, J. A. (2005). Generosity and gluttony in GEMS: Grid Enabled Molecular Simulations. In *Proceedings of the 14th IEEE International Symposium on High Performance Distributed Computing.* Los Alamitos, CA: IEEE Computer Society Press.

9. KEY TERMS AND DEFINITIONS

Coarse-Grain Molecular Dynamics: A coarse-grain model is one where some of the fine details of the modelled system has been smoothed over, grouped together, or averaged out. In molecular dynamics, 'coarse-grain', as opposed to 'atomistic', is a model where all atoms are not represented, but some atoms are grouped together to form a (larger) particle. To justify, the atoms grouped together within the larger particle may have behaviours that are too detailed and of no interest to the present investigation. Such grouping therefore saves computational time, so that longer-timescale 'interesting' events can be more readily observed. An early example of coarse-grain MD is the 'united-atom' forcefields, where non-ionizable hydrogen atoms are 'united' with the heavy atom they bond to. Recent examples include lipid and biomolecular simulations, where the nanosecond time-scale afforded by atomistic MD has been surpassed by microsecond-scale coarse-grain MD simulations.

Forcefield: In molecular dynamics, a forcefield is the collection of potentials of interaction between all possible pairs of atom/particle types within the modelled system. This includes short-range interactions such as covalent and ionic bonds, and non-bonded interactions (such as van der Waals); and long-range interactions such as electrostatics. Two examples are the eponymous forcefields in the molecular dynamics packages Amber and Gromacs.

Metadata: 'A set of data that describes and gives information about other data' (Oxford English Dictionary). In the context of a database of trajectories, this may provide the name of the trajectory, the molecules included in the simulation system, related scientific publications, and the provenance (ontogeny) of the trajectory. For the last item, it may include the original molecular dynamics parameters with which the trajectory was generated.

Molecular Dynamics Simulation (MD): A method to simulate of molecules in which atoms (or particles in general) within molecules interact under laws of physics (traditionally the classical Newtonian laws) for some time, giving a view of the motion of the molecule. The underlying framework of physical (Newtonian) laws is called 'molecular mechanics' (MM). This simulation can be constructed using real 'balls' (spheres representing the atoms/particles) and appropriate 'springs' (possibly non-Hookean, to represent the interactions between the particles); but more conveniently, it is commonly done *in silico*, using a model of the molecules in the computer to progress through time according to the physical laws. Though usually only the classical interactions are considered, it is possible to mix in quantum mechanics considerations for certain areas where bond-breaking and bond-forming activities are of interest. This mixed method is called 'quantum mechanics-molecular mechanics' (QM/MM).

Relational Database: A relational database is a database based on predicate logic and set theory, or in other words, on tables of relations. An example table of relations is the simple publication database shown in the table below. Further relations can be expressed by adding links and keys between tables, and performing mathematical set operations on the tables. For an excellent introduction on the theory of relational databases, see the C. J. Date's book (Date, 2000). Software exist to model, create, and manage relational databases. Such a software package is called a relational database management system (RDBMS).

Trajectories. The results of the molecular dynamics simulation is called a trajectory. This is a time-series of molecular conformations, or in other words, a movie of three-dimensional snapshots for the molecules in the system. For a system of N particles and m frames, a trajectory can be seen conceptually as a $3N \times m$ matrix: On one side, it contains the $3N$ Cartesian coordinates, three scalars for each atom. On the other,

Title	Author	Publisher	City	Year
Free software, free society	Stallman, R. M.	GNU Press	Boston	2002
An Introduction to database systems	Date, C. J.	Addison-Wesley	Reading, Massachusetts	2000

it contains a list of frames each associated with a time-point. Often the time-step for Newtonian integration for an atomistic simulation is 1 fs or 2 fs, and the snapshots are recorded at the rate of 1 ps per frame. For a 1 ns trajectory, this gives 10^6 time-steps and 10^3 frames.

Chapter XVII
Molecular Structure Determination on the Grid

Russ Miller
Hauptman-Woodward Medical Research Institute, USA
SUNY-Buffalo, USA

Charles M. Weeks
Hauptman-Woodward Medical Research Institute, USA

ABSTRACT

Grids represent an emerging technology that allows geographically- and organizationally-distributed resources (e.g., compute systems, data repositories, sensors, imaging systems, and so forth) to be linked in a fashion that is transparent to the user. The New York State Grid (NYS Grid) is an integrated computational and data grid that provides access to a wide variety of resources to users from around the world. NYS Grid can be accessed via a Web portal, where the users have access to their data sets and applications, but do not need to be made aware of the details of the data storage or computational devices that are specifically employed in solving their problems. Grid-enabled versions of the SnB and BnP programs, which implement the Shake-and-Bake method of molecular structure (SnB) and substructure (BnP) determination, respectively, have been deployed on NYS Grid. Further, through the Grid Portal, SnB has been run simultaneously on all computational resources on NYS Grid as well as on more than 1100 of the over 3000 processors available through the Open Science Grid.

1. INTRODUCTION

The Grid is a rapidly emerging and expanding technology that allows geographically-distributed resources that extend across administrative boundaries to be linked together in a transparent fashion (www.gridcomputing.com/; www.globus.org/; Berman *et al.*, 2003; Foster & Kesselmann,

1999). These resources include compute systems, data storage devices, sensors, imaging systems, visualization devices, and a wide variety of Internet-ready instruments. The concept and terminology of the Grid is borrowed from the electrical grid, where utility companies have the ability to share and move resources (electricity) in a fashion that is transparent to the consumer. With rare exception, the view taken by a consumer is that they are able to plug a piece of equipment into a power outlet in order to obtain electricity and do not need to know, and in fact, do not want to know, the details pertaining to the manner in which electricity makes its way to the outlet. Similarly, the power of both computational grids (*i.e.*, seamlessly connecting compute systems and their local storage) and data grids (*i.e.*, seamlessly connecting large storage systems) lies not only in the aggregate computing power, data storage, and network bandwidth that can readily be brought to bear on a particular problem, but also on its ease of use.

Numerous government-sponsored reports state that grid computing is a key to 21st century discovery by providing seamless access to the high-end computational infrastructure that is required for revolutionary advances in contemporary science and engineering. In fact, National Science Foundation Director Arden Bement stated that "*leadership in cyberinfrastructure may determine America's continued ability to innovate – and thus our ability to compete successfully in the global arena.*"

Grids are now a viable solution to certain computationally- and data-intensive computing problems for reasons that include the following.

- Users can access many grids through a Web portal from virtually anywhere in the world. That is, a user only needs an account on a Grid administrative server in order to use a grid. This is similar to how someone uses a search engine or large e-business system. One only needs access to a gateway and not specifically to each individual server that the company has configured to be able to handle the requests/queries/business. For most grids, a user needs access to a Grid Portal, but does not need to be logged in to a site that hosts a particular Grid resource, does not need to be logged in to a computer that is on a Grid, and does not need to install any additional software on their Web-accessible system (workstation, cellular phone, laptop, etc.) in order to be able to use a Grid.
- The Internet is mature and able to serve as the fundamental infrastructure for network-based computing. In fact, network bandwidth, which has been doubling approximately every 12 months, has increased to the point of being able to provide efficient and reliable services for the vast majority of Grid applications.
- Storage capacity, which has been doubling approximately every 9 months, has now reached commodity levels, where one can purchase a terabyte of disk for roughly the same price as a high-end PC.
- Many instruments are Internet-aware.
- Clusters, supercomputers, storage and visualization devices are becoming more mainstream in terms of their ability to host scientific applications.

As grid computing initiatives move forward, issues of interoperability, security, performance, management, and privacy need to be carefully considered. In fact, security is concerned with various issues relating to authentication in order to insure application and data integrity. Grid initiatives are also generating best practice scheduling and resource management documents, protocols, and API specifications to enable interoperability. Several layers of security, data encryption, and certificate authorities already exist in grid-enabling toolkits such as Globus Toolkit (www.globus.org/toolkit/).

In this chapter, we present the *New York State Grid* (*NYS Grid*), which represents a major effort by our research group. The design, implementation, and support of this Grid has provided a wide variety of opportunities for institutions in New York State in terms of science, engineering, scholarship, education, outreach, and training. Indeed, the New York State Grid consists of a heterogeneous set of resources from institutions throughout New York State. These managed resources are available in a simple and seamless fashion to users worldwide. One of the major features of our grid is that it integrates a computational grid with a data grid so that a user may deploy computationally-intensive applications that read or write large data files in a very simple fashion. In particular, we designed this grid so that users do not need to know where data files are physically stored or where an application is physically deployed. In this chapter, we will discus the use of an active Web portal for deploying applications, dynamic resource allocation so that clusters and networks of workstations can be scheduled to provide resources on demand, and a scalable and dynamic scheduling system, to name a few.

In Section 2, we present background material as it pertains to both cyberinfrastructure, in general, and grid computing, in particular. Further, in section 2, we introduce and give an overview of the New York State Grid and a key application in molecular structure determination that we have ported to this Grid. In Section 3, we discuss details of the web interface and the underlying technology that allows for the *Shake-and-Bake* method of molecular structure determination to be ported to the New York State Grid. We focus on the Web Portal and its ease of use in manipulating key parameters to the *Shake-and-Bake* routine. In Section 4, we discuss future trends in molecular structure determination and how it dovetails with the maturing of cyberinfrastructure. Section 5 serves as the conclusion.

2. BACKGROUND

2.1 Cyberinfrastructure and Grid Computing

Cyberinfrastructure sits at the core of modern simulation and modelling, providing entirely new methods of investigation to scientists, engineers, and scholars in their attempt to address previously unsolvable problems or investigate problems in previously unexplored domains. In fact, new insights that are derived from advances in cyberinfrastructure might well lead to new "grand challenges" as well as providing solutions of previously stated "grand challenge problems." Specifically, the development of necessary software, algorithms, portals, and interfaces will enable research and scholarship by freeing end-users from dealing with the complexity of various computing environments. This is critical to extending the reach of high-end computing, storage, networking, and visualization to the general user community. In fact, globally dispersed and interconnected resources, coupled with advances in middleware and computational methodologies, may well enable new ways of thinking about research and enhance traditional means of discovery to advance knowledge. It might also provide a means for a non-professional scientist or scholar, the so-called "citizen-scientist" to drive scientific discovery. (CDI, 2007)

While cyberinfrastructure, or e-science as it is sometimes called, has existed for decades, the term "cyberinfrastructure" became commonplace when introduced in the "Atkins Report" (Atkins, et. al., 2003), which was produced by a National Science Foundation blue-ribbon panel. This report, which radically changed the vision and funding directions at NSF, recommended that existing barriers to the rapid evolution of high performance be removed so that such infrastructure would be easy to use by scientists, engineers, scholars, and non-professional citizen-scientists. The term cyberinfrastructure was introduced to describe a research environment that supports advanced data acquisition, data storage, data management, data integration, data mining, data visualization and other computing and information processing services through a network of geographically distributed resources, including people, compute systems, storage devices, visualization systems, sensors, and other Internet ready devices. More specifically, the goal of cyberinfrastructure is to advance the technology so that geographically-distributed high-end systems can be used in a seamless fashion and integrated in a ubiquitous fashion into scientific discovery and innovation.

While standard definitions for both "cyberinfrastructure" and "grid" do not exist, Yafchak & Trauner (2008) propose a simple definition that states that "a grid consists of shared heterogeneous computing and data resources networked across administrative boundaries." While we would like to see other Internet-ready instruments included in such a definition, it should be noted that a grid can refer to a set of workstations distributed throughout a company, where not all workstations are under control of the same IT manager, or it can refer to a set of supercomputers located around the world in different organizations, or some combinations thereof, including other Internet-ready devices such as data storage devices, visualization engines, and sensors.

As stated in the Introduction, grids are starting to move out of the research laboratory and into early-adopter production systems. Numerous grid projects have been initiated, many of which are discussed in Yafchak & **Trauner (2008)** [*e.g.*, GriPhyN (www.griphyn.org/), PPDG (www.ppdg.net/), EGEE (www.eu-egee.org/), EU DataGrid (eu-datagrid.web.cern.ch/eu-datagrid/), NASA's Information Power Grid (IPG) (www.gloriad.org/gloriad/projects/project000053.html), TeraGrid (www.teragrid.org/), Open Science Grid (www.opensciencegrid.org/), and caBIG (https://cabig.nci.nih.gov/), to name a few]. However, the construction of a real general-purpose grid is in its infancy since a true grid requires coordinated resource sharing and problem solving in a dynamic, multi-institutional scenario using standard, open, general-purpose protocols and interfaces that deliver a high quality of service. The immediate focus of grid deployment continues to be on the difficult issue of developing high quality middleware (www.nsf-middleware.org/).

2.2 New York State Grid

In order to support research, scholarship, education, and community outreach, Miller's Cyberinfrastructure Laboratory (MCIL; www.cse.buffalo.edu/faculty/miller/CI) at SUNY-Buffalo and the Hauptman-Woodward Institute focuses on the integration of research in disciplinary domains with research in enabling technologies and interfaces. The goal of MCIL is to provide systems that allow users to transparently collect, manage, organize, analyze, and visualize data without having to worry about details such as where the data is physically stored, processed, rendered, and so forth. This ease of use and high availability of data and information processing tools is expected to stimulate revolutions in science, engineering, and beyond.

MCIL was responsible for the design, development, and deployment of the *New York State Grid* (*NYS Grid*), which includes resources from institutions throughout New York State and is available in a simple and seamless fashion to users worldwide. NYS Grid contains a heterogeneous set of resources and utilizes general-purpose IP networks (Green & Miller, 2003, 2004a-c). A major feature of NYS Grid is that it integrates a computational grid with a data grid so that a user may deploy computationally-intensive applications that read or write large data files in a very simple fashion. In particular, NYS Grid was designed so that a user does not need to know where data files are physically stored or where an application is physically deployed, while providing the user with easy access to their files (uploading, downloading, editing, viewing, etc.). The core infrastructure for NYS Grid includes the installation of standard grid middleware, the use of an active Web portal for deploying applications, dynamic resource allocation so that clusters and networks of workstations can be scheduled to provide resources on demand, and a scalable and dynamic scheduling system, to name a few.

Several key packages were used in the implementation of NYS Grid and other packages have been identified in order to allow for the anticipated expansion of the system. The Globus Toolkit (www.globus.org/toolkit/) provides APIs and tools using the Java SDK to simplify the development of OGSI-compliant services and clients. It supplies database services and Monitoring & Discovery System index services implemented in Java (www.globus.org/toolkit/mds/), GRAM service implemented in C with a Java wrapper (www.globus.org/toolkit/docs/2.4/gram/), GridFTP services implemented in C (www.globus.org/grid_software/data/gridftp.php), and a full set of Globus Toolkit components. The recently proposed Web Service-Resource Framework provides the concepts and interfaces developed by the OGSI specification exploiting the Web services architecture (www.globus.org/wsrf/).

NYS Grid is the current incarnation of an MCIL-led grid that progressed from a Buffalo-based grid (ACDC-Grid) to a persistent, hardened, and heterogeneous Western New York Grid (WNY Grid), before being enhanced, expanded, and deployed throughout New York State. This series of Grids was largely funded by the National Science Foundation through ITR, MRI, and CRI grants. NYS Grid currently supports a variety of applications and users from NYS Grid institutions and the Open Science Grid. In addition, a grass-roots cyberinfrastructure initiative in New York State has been granted access to NYS Grid. NYS Grid also serves as a gateway to the Open Science Grid, TeraGrid, MCEER (mceer.buffalo.edu/), and NEES (it.nees.org/ and nees.buffalo.edu/), to name a few.

2.3 Molecular Structure Determination by Direct Methods

No microscope is powerful enough to let us observe biological molecules and their interactions with each other. Instead, structural biologists use tools provided by the science of X-ray crystallography. In a crystallographic experiment, a single crystal of a purified substance such as a protein is irradiated, and the incident radiation is scattered in many directions to produce a diffraction pattern. A complex mathematical analysis of the diffraction data (a process known as "solving" the structure) is then undertaken to determine the shape and atomic arrangement of the molecules comprising the crystal. Once the structure has been solved, molecular models can be constructed and examined for insight into how the protein molecules function, what might be happening when disease occurs, and what compounds might be designed as drugs to modify activity.

The pathway from diffraction data to atomic coordinates is not straightforward. Atomic coordinates can be computed readily from complex-valued quantities known as structure factors, but only the magnitudes of the structure factors can be measured in the diffraction experiment. The phases of the structure factors are lost, and this gives rise to the so-called "phase problem" of X-ray crystallography. However, in most cases, mathematical techniques known as direct methods provide a way to solve the phase problem. Direct methods use probabilistic relationships among the phases to derive the values of individual phases from the measured amplitudes. Conventional direct methods, implemented in computer programs prior to 1992 (*e.g.*, MULTAN (Main *et al.*, 1980), SHELXS (Sheldrick, 1990), SAYTAN (Debaerdemaeker *et al.*, 1985) and SIR (Burla *et al.*, 1989)), provided computationally efficient solutions for structures containing less than approximately 100 unique non-hydrogen atoms. However, few larger structures (*i.e.*, structures with more than 200 unique equal atoms) have ever been solved using these programs. The computer-intensive direct methods *Shake-and-Bake* approach to molecular structure determination, developed by the authors (Weeks *et al.*, 1993; Miller *et al.*, 1993) and described in detail in the next section, provided the breakthrough needed to achieve automated direct-methods solutions for much larger structures.

2.4 *Shake-and-Bake*

Shake-and-Bake is a powerful algorithmic formulation of direct methods that, given accurate diffraction data to 1.2Å or better resolution, has made possible the *ab initio* phasing of complete crystal structures containing as many as ~2000 independent non-H atoms (Frazão, *et al.*, 1999). It has also been used to determine the anomalously scattering substructures of selenomethionyl-substituted proteins containing as many as 160 selenium sites using 3-4Å data (Von Delft *et al.*, 2003). *Shake-and-Bake* belongs to the class of phasing methods known as 'multisolution' procedures (Germain & Woolfson, 1968) in which multiple sets of trial phases are generated in the hope that one or more of the resultant combinations will lead to a solution. Solutions, if they occur, are identified on the basis of the value of a suitable figure of merit, such as the minimal function (see below) or the crystallographic *R* value. Since each of the sets of trial phases can be processed independently, the *Shake-and-Bake* algorithm can be easily adapted to a coarse-grained parallel processing approach and implemented on a computational grid.

The distinctive feature of *Shake-and-Bake* is the repeated and unconditional cyclical alternation of reciprocal-space phase refinement with a complementary real-space process that seeks to improve phases by imposing constraints through a physically meaningful interpretation of the electron density (Miller, *et al.*, 1993; Weeks, *et al.*, 1994). First, a random number generator is used to assign initial coordinates to the atoms comprising the trial structures, and structure-factor calculations are performed to generate the corresponding sets of starting trial phases. Then, phases are refined either by the tangent formula (Karle & Hauptman, 1956) or by constrained minimization of the so-called minimal function (DeTitta, *et al.*, 1994) using the parameter-shift algorithm (Bhuiya & Stanley, 1963). Following Fourier transformation to real space and computation of an electron-density map, peak picking is used to impose the atomicity constraint. Peaks are located on the map, an appropriate number of the largest of these maxima are assumed to be atoms, and the cycle is completed by using inverse Fourier transformation (in the form of a structure-factor calculation) to generate phases for another round of refinement. The entire process is repeated for a predetermined (by the user) number of cycles. The success rate of this process (*i.e.*, the percentage of trial structures that converge to solution) depends on data quality and the size of the structure.

As mentioned, the *Shake-and-Bake* procedure has been implemented in a computer program, *SnB*, in a manner convenient for both protein substructures and complete structures (Miller *et al.*, 1994; Weeks & Miller, 1999; Rappleye *et al.*, 2002). (The *Shake-and-Bake* algorithm has also been implemented independently in the program *SHELXD* (Schneider & Sheldrick, 2002)). The *SnB* graphical user interface (GUI), written in Java, controls not only the main phasing program but also the *DREAR* program suite (Blessing & Smith, 1999) that computes the normalized structure-factor magnitudes ($|E|$) required for direct-methods calculations. In addition, the two-step process of substructure determination and protein phasing has been combined in the program *BnP* (Weeks *et al.*, 2002), which provides a common interface for *SnB* and components of the *PHASES* suite (Furey & Swaminathan, 1997). Thus, *BnP* provides an automated pathway from processed intensities to an unambiguous protein electron-density map. This pathway includes

SnB substructure determination, heavy-atom site validation, enantiomorph determination, substructure and protein phase refinement, and solvent flattening.

The repetitive shuttling of trial structures between real and reciprocal space gives the *Shake-and-Bake* algorithm its power, but the need to perform two Fourier transformations in each cycle yields a computationally-intensive procedure. In fact, the running time for *SnB* or *BnP* varies widely - from just a few seconds to many hours for large structures or structures with diffraction data of marginal quality that typically require a large number of trial structures to be examined before a solution is found. In such cases, the ability to increase throughput by processing many trial structures simultaneously on a cluster or a computational grid is invaluable.

3. STRUCTURE DETERMINATION ON THE GRID

NYS Grid may be accessed via a browser through a web portal, as shown in Fig. 1a, or through a standard command-line submission. Versions of the *SnB* and *BnP* programs capable of parallel processing have been equipped with web-compatible PHP interfaces and incorporated into the web portal for ease of use using Grid-enabling Application Templates (Green & Miller, 2004c). The traditional stand alone Java versions and the new grid-enabled PHP versions are compatible in the sense that files created in both versions are interchangeable and can be uploaded to and downloaded from the grid as needed. As mentioned previously, *SnB* has been run on NYS Grid and Open Science Grid through the Web Portal, as well as through command-line submission, to compute hundreds of thousands of trial structures using an aggregate of nearly 100,000 CPUs.

BnP combines the *Shake-and-Bake* substructure determination step, which can easily take advantage of a parallel computing environment, with protein phasing calculations that are applied to a successful *Shake-and-Bake* substructure trial. The approach that has been taken is to spread the processing of trial substructures among a large number of computational nodes. In most cases, it is possible for *BnP* to determine when a substructure solution has occurred and to move on to the next step. Jobs that complete the first step without identifying a solution can (optionally) be terminated. Since jobs on different machines are independent once they have been spawned, multiple solutions are possible - an advantage if any jobs produce false solutions or solutions of marginal quality. It was decided that this execution scenario maximized the probability that at least one solution would be found for a difficult structure in a modest amount of time.

Figure 1 illustrates the use of the grid-enabled version of *BnP* to phase the protein mannoheptose 6-epimerase (Deacon *et al.*, 2000) after first solving the 70-Se substructure. Fig. 1a shows the Grid portal as it initially appears to a user. After login, the user is presented with screens representing the various stages of the GAT, as indicated by the workflow defined at the top of the screen. In each case, the current stage (1b-1g) is indicated by the red rectangle. To move to the next stage, the user must click the "Continue" button at the bottom of the screen. Several "Help" buttons are available along the way.

First, the *BnP* software is selected (Fig. 1b). At the next stage (Fig. 1c), the user is reminded to upload input data files to the grid (not illustrated) and then instructed to choose how additional required information is to be supplied. If "Enter" (or "Continue") is selected, a blank version of screen 1d will appear, and the information requested about the structure (*e.g.*, space group) and its datasets must be entered manually. If "Import" is chosen, most or all of the information will be extracted from the headers of the input data files, and "Open" will restore information saved from a previous job.

Figure 1. Job submission, monitoring, and examination of results using the grid portal. (a) Login to the SUNY-Buffalo grid portal. (b) Select BnP as the software application to be used. (c) Select execution options and upload files from the home computer. (d) Input additional information about the structure.

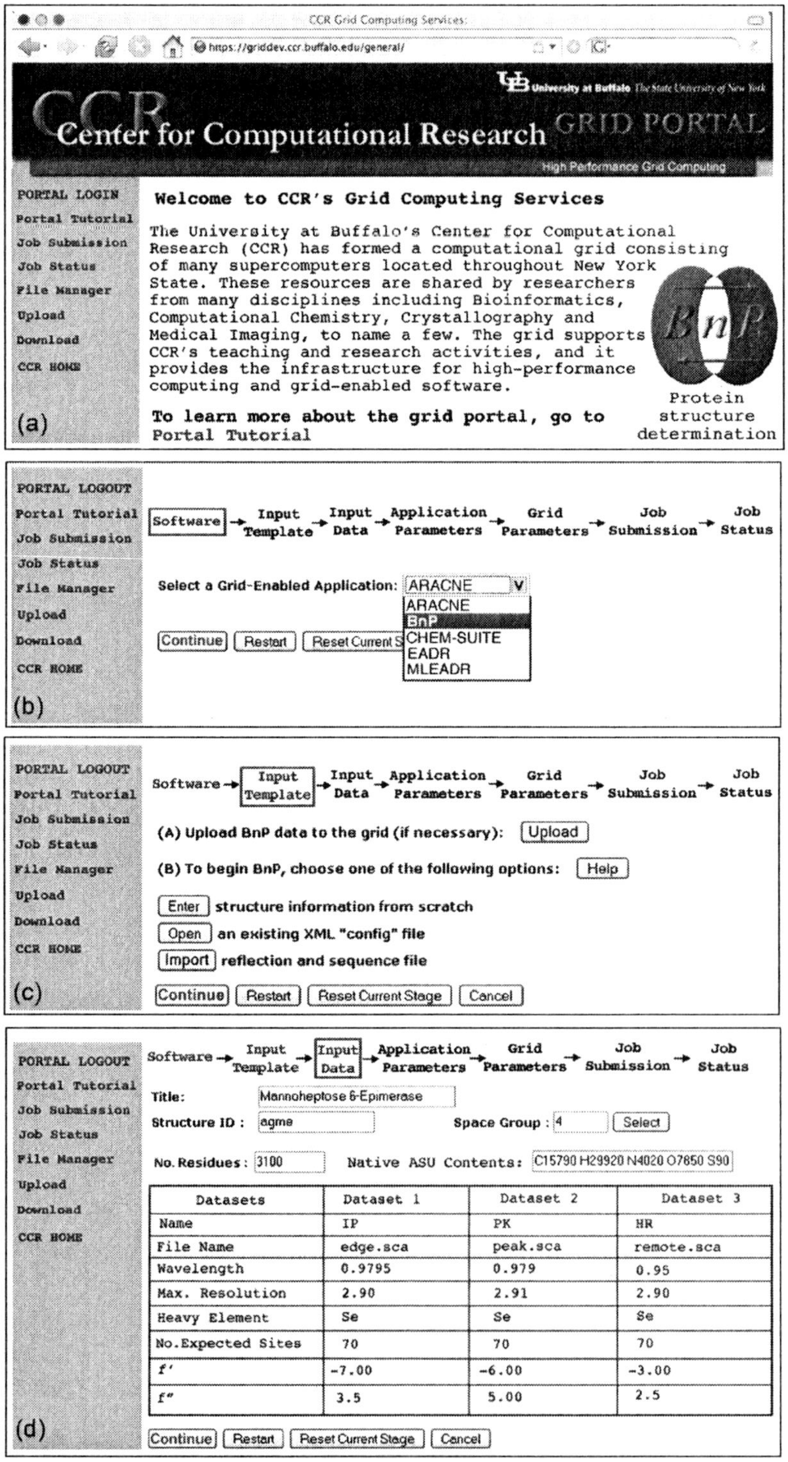

The next step is to supply the values for the parameters required to execute the *BnP* application (Fig. 1e). Experience has shown that it is sometimes important to vary the values of certain key parameters when performing a *Shake-and-Bake* calculation in order to ensure that a solution will be found. These parameters include the space group (specified by giving its number in *International Tables*), the maximum resolution of the diffraction data to be used for the substructure determination, the maximum number of expected heavy-atom or anomalously scattering sites, and the type of normalized difference data to be used for *Shake-and-Bake* (*e.g.*, peak wavelength anomalous differences or maximum dispersive differences). The interface computes the number of different parameter combinations that have been specified, and the user must then indicate the number of different substructure trials that are to be processed on the grid for each set of parameter values. The final choice the user has to make at this stage is to decide if the default values of other *BnP* parameters (*e.g.*, the numbers of phases, peaks, and refinement cycles) - that are to be fixed for all jobs - should be changed (option not illustrated).

Parameters such as the computational resource (*e.g.*, a particular computer cluster) on which the jobs are to be run and the maximum time allowed per job are specified at the "Grid Parameters" stage (Fig. 1f). The user can also specify whether or not jobs are to be terminated after the substructure determination (*Shake-and-Bake*) step if no solution has been detected automatically based on the figure-of-merit calculation. Otherwise, the job will continue using the trial substructure with the best figure of merit even though the program does not regard it as a solution. Making the decision to terminate frees resources for other users, but there is a small risk that an actual solution might be missed.

The next screen (Fig. 1g) permits the user to review all parameter choices before the job(s) are submitted. The progress of the jobs can then be monitored from the "Job Status" screen (Fig 1h). A drilldown button (shown in red) provides access to more information about an individual job or multiple job set, and Fig. 1i shows details about the status of the multiple job (number 44267) set up in step 1e and selected for viewing in step 1h. The values of the variable parameters for each job are given in the table, which indicates that the jobs were grouped in pairs (differing by the random number seed). In the case illustrated, all jobs that used the peak anomalous difference data gave recognizable solutions and proceeded to perform all *BnP* tasks up to and including protein phasing and solvent flattening (note indications that stage 5 was completed). The other jobs did not find solutions and were killed (terminated) after the *Shake-and-Bake* stage (#2) had processed all 500 trial structures assigned to them. If any jobs had failed because of *BnP* errors or for other reasons, that fact would have been indicated in the status column.

Drilling down one level further (indicated here by the red symbol on the line for job 1.1) provides access to the results for a single job (Fig. 1j). Those results include, for example, a *Shake-and-Bake* histogram (Fig. 1k) of minimal function values for the 89 trials that were processed in job 1.1 before the program automatically detected that a probable solution had been found. Finally, clicking the "Download results" button (on Fig. 1j) causes the screen shown in Fig. 1l to be displayed with options for downloading the files resulting from the *BnP* calculations to the home computer.

4. FUTURE TRENDS

The future of molecular structure determination lies in an automated workflow that will progress from crystal growth through the determination of a refined structure. In the near term, such an effort will rely on a workflow where the instrument used to measure diffraction data, data storage devices, compute systems, collaborative visualization

Figure 1 (continued). (e) Assign values to program parameters that will be varied in different jobs. Some additional parameters (constant for all jobs submitted simultaneously) are accessible from a pop-up window. (f) Supply information controlling job execution on the grid. (g) Review all parameters and start the jobs. (h) Check whether the jobs are still running.

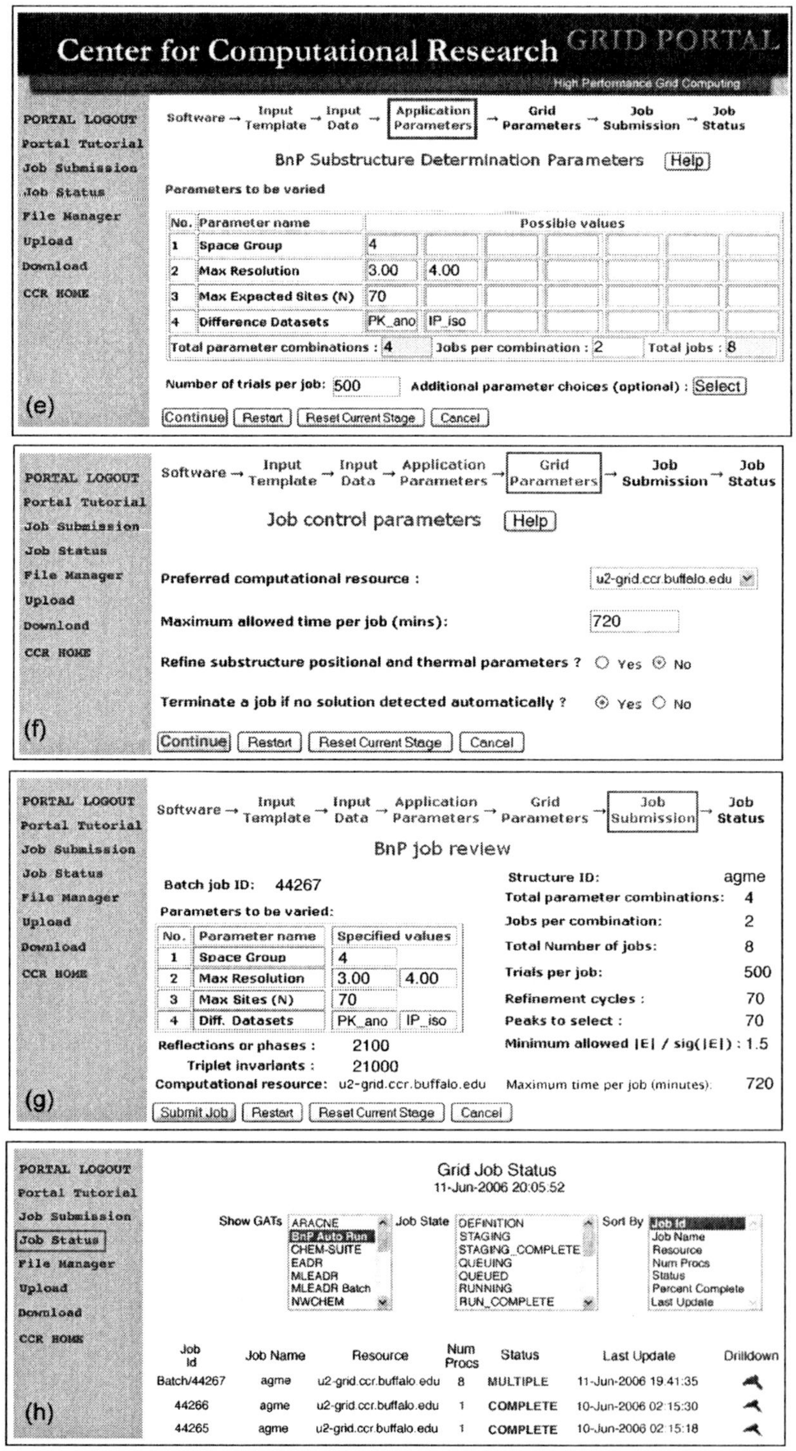

Figure 1 (continued). (i) Drilldown to check the execution status of individual components of a multiple job set. (j) Drilldown to inspect the results of an individual job. (k) Figure-of-merit histogram for the selected job. (l) Download the output files to the home computer.

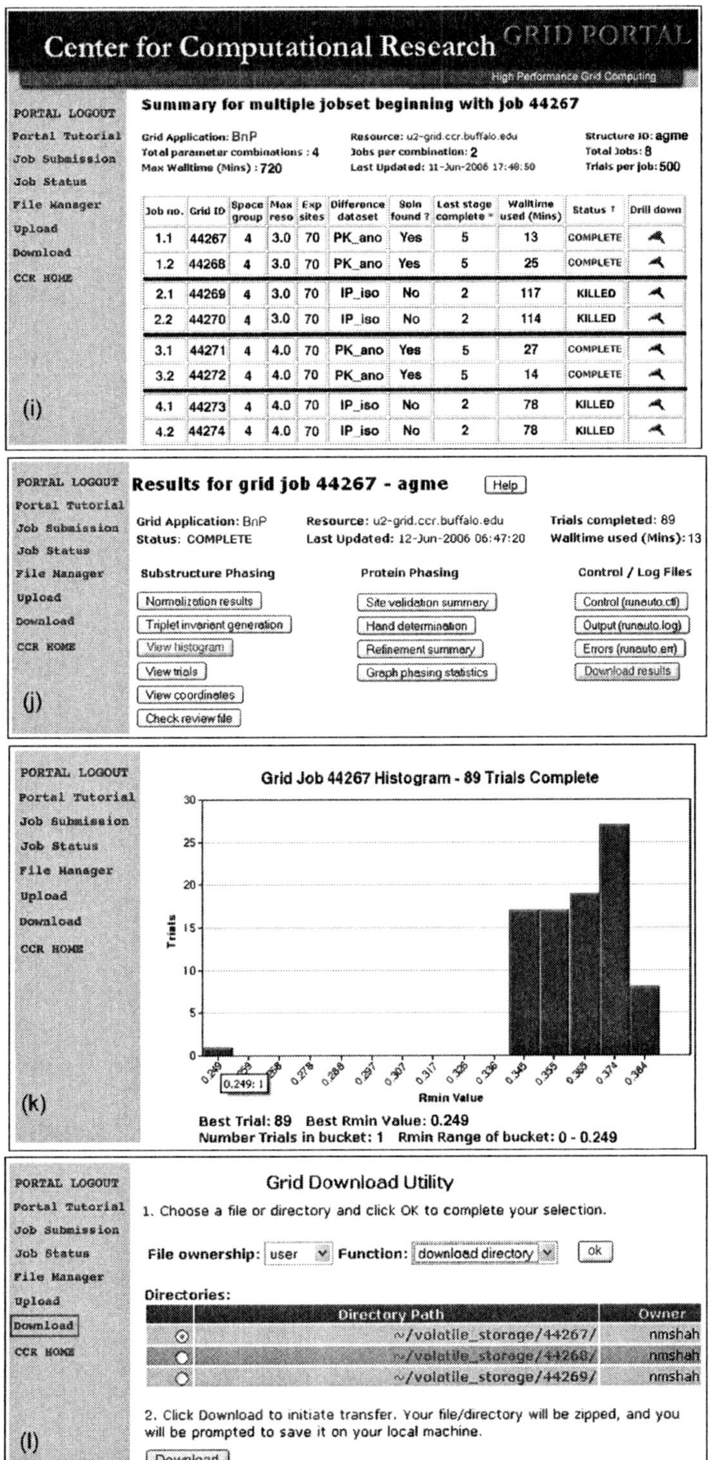

systems, and final data repositories are all on a grid and are able to be harnessed into a solution strategy. Currently, many of the individual pieces exist. However, the design, development, and deployment of an entire workflow will require a significant effort. In particular, many imaging systems are on the Internet. The Protein Data Bank (PDB; http://www.wwpdb.org/, Berman, Henrick, & Nakamura, 2003), which serves as a repository for 3-D structural data of proteins and nucleic acids, typically determined through X-ray crystallography or NMR spectroscopy methodologies, is freely available via the Internet and is also available through teragrid. The *SnB* and *BnP* programs have been ported to several grids and a wide variety of compute platforms. Some of the post-processing and refinement routines are Internet accessible. Collaborative visualization facilities for molecular structures have been developed in Buffalo. So, a project similar to NIH's cancer Biomedical Informatics Grid, caBIG (https://cabig.nci.nih.gov/), which is dedicated to accelerating research discoveries and improving patient outcome by linking researchers, patients, and physicians throughout the cancer community, should be sufficient to put such a black-box system for molecular structure determination in place. Note that the molecular structure determination effort would be a much smaller effort in terms of funding, equipment, and personnel than the caBIG project.

5. CONCLUSION

NYS Grid is a state-of-the-art, integrated computational and data grid. Using Grid-enabling Application Templates, the *SnB* and *BnP* programs have been adapted to this grid in a straightforward fashion. The web-based Grid portal provides convenient access to these programs for users of the grid. Users do not need to know details of where their files are maintained, nor where the computations are performed, although details are always available for those who wish to be informed. The grid-enabled versions allow for a much shorter wall time to solution for large and difficult structures.

ACKNOWLEDGMENT

This material is based upon work supported by the National Science Foundation under Grant Nos. CNS-0454114, CCF-0204918, EIA-0101244, ANI-0124929, ACI-9721373, and NIH Grant No. EB002057. The authors would like to thank members of Dr. Miller's Cyberinfrastructure Laboratory, including Steve Gallo, Jason Rappley, Cathy Ruby, Mark Green, Jon Bednasz, Tony Kew, Sam Guercio, Adam Koniak, Martins Innus, Dori Macchioni, Amin Ghadersohi, and Cynthia Cornelius, for their contributions to the efforts described in this paper. Much of this research was performed while Dr. Miller was Director of the Center for Computational Research at SUNY-Buffalo. The authors would also like to thank Herb Hauptman, George DeTitta, and Billy Furey for their contributions to this work.

REFERENCES

Atkins, D.E., Droegemeier, K.K., Feldman, S.I., Garcia-Molina, H. Klein, M.L., Messerschmitt, D.G., et al. (n.d.). *Revolutionizing Science and Engineering Through Cyberinfrastructure: Report to the National Science Foundation Blue-Ribbon Advisory Panel on Cyberinfrastructure.* Retrieved on January 15, 2008, from http://www.community.technology.org/nsf_ci_report/report.pdf.

Berman, F., Hey, A. J. G & Fox, G. C. (Eds.), (2003). *Grid Computing: Making the Global Infrastructure a Reality.* New York: John Wiley.

Berman, H.M., Henrick, K., & Nakamura, H. (2003). Announcing the worldwide Protein Data Bank. *Nature Structural Biology, 10* (12), 980.

Bhuiya, A. K. & Stanley, E. (1963). The refinement of atomic parameters by direct calculation of the minimum residual, *Acta Crystallogr, 16*, 981-984.

Blessing, R. H. & Smith, G. D. (1999). Difference Structure Factor Normalization for Determining Heavy-Atom or Anomalous Scattering Substructures, *J. Appl. Cryst. 32*, 664-670.

Burla, M.C., Camalli, M., Cascarano, G., Giacovazzo, C., Polidori, G., Spagna, R. & Viterbo, D. (1989). SIR88 - a direct-methods program for the automatic solution of crystal structures. *J. Appl. Cryst. 22*, 389-393.

CDI (2007). *Cyber-Enabled Discovery and Innovation (CDI), Computer and Information Science and Engineering Directorate, National Science Foundation*. Paper presented at the National Science Foundation.

Chang, C.-S., DeTitta, G., Hauptman, H., Jones, R., Miller, R., Thuman, P., & Weeks, C. (1993). Solving the phase problem of x-ray crystallography on parallel machines. In R.F. Sincovec, D.E. Keyes, M.R. Leuze, L.R. Petzold, and D.A. Reed, (Eds.) *Proceedings of The Sixth SIAM Conference on Parallel Processing for Scientific Computing*, (pp. 304-307).

Chang, C.-S., DeTitta, G.T., Miller, R. & Weeks, C.M. (1994). On the application of parallel genetic algorithms in x-ray crystallography. *Proceedings of the 1994 Scalable High Performance Computing Conference*, (pp. 796-802). New York: IEEE Computer Society Press.

Deacon, A. M., Ni, Y. S., Coleman, W. G. Jr. & Ealick, S. E. (2000). The crystal structure of ADP-L-glycero-D-mannoheptose 6-epimerase: catalysis with a twist. *Structure, 8*, 453-462.

Debaerdemaeker, T., Tate, C. & Woolfson, M.M. (1985). On the application of phase relationships to complex structures (XXIV: The Sayre tangent formula). *Acta Cryst.*A41), 286-290.

DeTitta, G., Hauptman, H., Miller, R., Pagels, M. Sabin, T., Thuman, P., & Weeks, C. (1991). Parallel solutions to the phase problem in x-ray crystallography: an update. In *Proceedings of The Sixth Distributed Memory Computing Conference*, (pp. 587-594). New York: IEEE Computer Society Press.

DeTitta, G. T., Weeks, C. M., Thuman, P., Miller, R. & Hauptman, H. A. (1994). Structure solution by minimal function phase refinement and Fourier filtering, (I. Theoretical basis). *Acta Crystallogr.* A50, 203-210.

Foster, I. & Kesselmann, C. (Eds.) (1999). *The Grid: Blueprint for a New Computing Intrastructure*. San Francisco, CA: Morgan Kaufmann Publishers.

Frazão, C., Sieker, L., Sheldrick, G. M., Lamzin, V., LeGall, J. & Carrondo, M. A. (1999). Ab initio structure solution of a dimeric cytochrome c3 from Desulfovibrio gigas containing disulfide bridges. *J. Biol. Inorg. Chem. 4*, 162-165.

Furey, W. & Swaminathan, S. (1997). PHASES-95: A Program Package for the Processing and Analysis of Diffraction Data from Macromolecules. *Meth. Enzymol. 277*, 590-620.

Germain, G. & Woolfson, M. M. (1968). On the application of phase relationships to complex structures, *Acta Crystallogr. B24*, 91-96.

Green, M. L. & Miller, R. (2003). Grid computing in Buffalo. *Annals of the European Academy of Sciences*, 191-218.

Green, M. L. & Miller, R. (2004a). Molecular structure determination on a computational & data grid. *Parallel Computing, 30*, 1001-1017.

Green, M. L. & Miller, R. (2004b). Evolutionary molecular structure determination using grid-enabled data mining. *Parallel Computing, 30*, 1057-1071.

Green, M. L. & Miller, R. (2004c). A client-server prototype for application grid-enabling template design. *Parallel Processing Letters*, *14*, 241-253.

Karle, J. & Hauptman, H. A. (1956). A theory of phase determination for the four types of non-centrosymmetric space groups 1*P*222, 2*P*22, $3P_12$, $3P_22$. *Acta Crystallogr.*, *9*, 635-651.

Main, P., Fiske, S.J., Hull, S.E., Lessinger, L., Germain, G., Declercq, J.P. & Woolfson, M.M. (1980). *MULTAN80: a system of computer programs for the automatic solution of crystal structures from x-ray diffraction data* (Universities of York and Louvain).

Miller, R., Bednasz, J.J., Chiu, K., Gallo, S.M., Govindaraju, M., Lewis, M., Ruby, C., & Weeks, C.M. (2008). Grid-based research, development, and deployment in New York State, In *Proceedings of the International Parallel and Distributed Processing Symposium,* in press.

Miller, R., DeTitta, G. T., Jones, R., Langs, D. A., Weeks, C. M. & Hauptman, H. A. (1993). On the application of the minimal principle to solve unknown structures, *Science,* 259, 1430–1433.

Miller, R., Gallo, S. M., Khalak, H. G. & Weeks, C. M. (1994). *SnB*: Crystal structure determination via *Shake-and-Bake. J. Appl. Cryst.. 27*, 613–621.

Rappleye, J., Innus, M., Weeks, C. M. & Miller, R. (2002). *SnB* v2.2: An Example of Crystallographic Multiprocessing. *J. Appl. Cryst., 35*, 374-376.

Schneider, T. R. & Sheldrick, G. M. (2002). Substructure solution with *SHELXD. Acta Crystallogr., D58*, 1772-1779.

Sheldrick, G.M. (1990). Phase annealing in SHELX-90: direct methods for larger structures. *Acta Cryst. A46*, 467–473.

Von Delft, F., Inoue, T., Saldanha, S. A., Ottenhof, H. H., Schmitzberger, F., Birch, L. M., et al. (2003). Structure of E. coli Ketopantoate Hydroxymethyl Transferase Complexed with Ketopantoate and Mg2+, Solved by Locating 160 Selenomethionine Sites. *Structure*, *11*, 985-996.

Weeks, C. M., Blessing, R. H., Miller, R., Mungee, R., Potter, S. A., Rappleye, J., Smith, G. D., Xu, H. & Furey, W. (2002). Towards automated protein structure determination: *BnP*, the *SnB*-PHASES interface. *Z. Kristallogr., 217*, 686-693.

Weeks, C. M., DeTitta, G. T., Hauptman, H. A., Thuman, P. & Miller, R. (1994). Structure solution by minimal function phase refinement and Fourier filtering (II. Implementation and applications). *Acta Crystallogr., A50*, 210-220.

Weeks, C.M., DeTitta, G.T., Miller, R. & Hauptman, H.A. (1993). Applications of the minimal principle to peptide structures. *Acta Cryst., D49*, 179-181.

Weeks, C. M. & Miller, R. (1999). The design and implementation of *SnB* v2.0. *J. Appl. Cryst., 32*, 120-124.

Yafchak, M.F. & Trauner, M., (Eds.) (2008). *The Grid Technology Cookbook.* Retrieved January 15, 2008, from http://www.sura.org/cookbook/gtcb/index.php?topic=12&mlevel=1&parent=0

KEY TERMS AND DEFINITIONS

Cyberinfrastructure: Provides for the transparent and ubiquitous application of technologies central to contemporary science and engineering, including high-end computing, networking, and visualization, data warehouses, science gateways, and virtual organizations, to name a few. It is a comprehensive phenomenon that involves creation, dissemination, preservation, and application of knowledge.

Diffraction: In very simple terms, diffraction is the bending or spreading of waves as they pass through an obstruction or gap. The gaps or distances between molecules in a crystal are such that X-rays (with a wavelength of $\sim 10^{6}$ mm) are diffracted by crystals. The X-rays scattered in a diffraction experiment produce a distinctive pattern that is related to the atomic arrangement within the crystal that was irradiated. Diffraction patterns can be recorded on photographic film or a suitable electronic recording device.

Direct Methods: A mathematical approach that makes it possible to glean phase information from the diffraction magnitudes. For example, the fact that molecules consist of atoms, and that atoms are small, discrete points relative to the spaces between them, creates certain constraints. Since there are many more reflections in a diffraction pattern than there are independent atoms in the corresponding crystal, the phase problem is overdetermined, and the existence of relationships among the measured magnitudes is implied. Certain linear combinations of three phases have been identified as relationships useful for determining unknown structures, and direct methods use probabilistic techniques to exploit these relationships. Herbert Hauptman and Jerome Karle won a Nobel prize in 1985 for their work in developing direct methods.

Grid Computing: Computational efforts involving computing, networking, storage, or visualization that involves geographically-distributed and independently-operated resources that are linked together in a transparent fashion.

GUI: A graphical user interface allows people to interact with a computer and computer-controlled devices. Instead of offering only text menus or requiring typed commands, graphical icons, visual indicators or special graphical elements are presented. Often the icons are used in conjunction with text, labels or text navigation to fully represent the information and actions available to a user. The actions are usually performed through direct manipulation of the graphical elements.

Molecular Structure Determination: The study of the three-dimensional architecture of molecules and the arrangement of their component atoms in space.

The Phase Problem: A diffraction pattern consists of a set of spots called *reflections* that result from capturing the image of an X-ray beam as it is scattered by the atoms in a crystal. Each reflection has a *magnitude*, which is experimentally accessible, and a *phase*, which is not. The inability to measure phase angles experimentally is known as the *phase problem*. Together, the magnitude and phase of a reflection constitute a quantity known as a *structure factor*. The set of structure factors provides the *reciprocal-space* representation of the structure, and the set of atomic coordinates provides the *real-space* representation. A Fourier transformation of the complex-valued structure factors leads directly to the real-valued electron density, which, after suitable interpretation, reveals the atomic positions and describes the molecular architecture of the crystalline material responsible for the scattering. Therefore, a solution to the phase problem requires an algorithm that will recover phase information that cannot be directly measured in the diffraction experiment.

Shake-and-Bake: A powerful algorithmic formulation of direct methods that, given complete and accurate diffraction data, has made possible the *ab initio* phasing of crystal structures containing as many as ~2000 independent non-hydrogen atoms. The distinctive feature of this algorithm is the cyclical alternation of phase refinement with the imposition of atomicity constraints.

SnB and BnP: *SnB* is a computer program developed in Buffalo (at the Hauptman-Woodward Medical Research Institute and SUNY-Buffalo) that provides an efficient implementation of the *Shake-and-Bake* method of molecular structure determination. This program has been distributed worldwide from the website http://www.hwi.buffalo.edu/SnB/ and is available for workstations, networks of workstations, clusters, and grids. *BnP* is a computer program that combines the direct-methods program *SnB* with components of Bill Furey's (University of Pittsburgh) PHASES suite. *BnP* targets the determination of large protein structures. First, the *SnB* component is used to find a substructure consisting of heavier atoms, and then the substructure is used as a starting point for phasing the complete protein.

X-Ray Crystallography: The scientific method most commonly used to determine molecular structures. Individual molecules cannot be seen under a light microscope because the wavelength of visible light is larger than the molecular size. However, crystals are made up of an array of many ($\sim 10^{11}$-10^{12}) identical, regularly-spaced molecules, and the regular spacing allows a technique called X-ray diffraction to be used to "see" the molecules that comprise the crystal.

APPENDIX

In this section, we provide a timeline and history with regard to Miller's Cyberinfrastructure Laboratory (MCIL), the ACDC-Grid, the WNY Grid, and NYS Grid. We also discuss the evolution and status of a grass-roots cyberinfrastructure initiative within New York State.

Timeline (MCIL and NYS Grid)[1]. In the early 1990s, in an effort to collaborate with scientists at the Hauptman-Woodward Institute in their effort to solve the phase problem of crystallography (Weeks, *et al.*, 1993), Miller's research group began to use commercial parallel computers, including shared- and distributed-memory machines from Thinking Machines Corporation, Encore, SGI, nCube, Intel, Alliant, and Sequent, to name a few (DeTitta, *et al.*, 1991; Chang, *et al.*, 1993; Chang, *et al.*, 1994). In addition, Miller's group used a laboratory of Sun workstations and ran their *Shake-and-Bake* method of structure determination using RPC (remote procedure call) to employ a master/worker solution of *Shake-and-Bake*. In 1998, with funding from the Department of Energy, Miller's group worked with Ian Foster's group at Argonne National Laboratory in order to use Globus to port *Shake-and-Bake* to early grids and clusters in an effort to solve larger molecular structures. In 1999, with funding from the National Science Foundation (NSF), Miller's group initiated a Buffalo-based grid research project that included the Hauptman-Woodward Institute, SUNY-Buffalo, and several Buffalo-area colleges, representing the genesis of a Western New York Grid.

In 2001, an NSF MRI grant funded a significant storage system that was incorporated into this existing grid, serving as a data repository for *Shake-and-Bake* results. This work led to an NSF ITR grant, funded in 2002, which focused on the deployment and efficient implementation of *Shake-and-Bake* on clusters and grids. In fact, this ITR grant also funded critical aspects of the design, development, deployment, and hardening of the aforementioned Buffalo-based grid (ACDC-Grid) and Western New York Grid (WNY Grid). In addition, these funds led to the expansion of WNY Grid to include institutions throughout New York State, which resulted in the establishment of the New York State Grid (NYS Grid) in 2004. The number of sites and variety of resources available on NYS Grid has grown substantially since 2004 and now includes a heterogeneous set of compute and storage systems throughout New York State. The institutions include academic and non-profit organizations, though NYS Grid is not restricted to such institutions. NYS Grid has been used extensively by the *Shake-and-Bake* team and numerous other users from Buffalo and the Open Science Grid.

A virtual organization, called GRASE (Grid Resource for Advanced Science & Engineering), was established by Miller's CI Laboratory in the early 2000's in order to support general science and engineering users on Open Science Grid and NYS Grid.

Given the success of the Buffalo-based ACDC-Grid, the WNY Grid, and the establishment of the NYS Grid, the NSF provided CRI funds in order to provide significant resources to the core sites in Western New York (SUNY-Buffalo, Niagara University, Hauptman-Woodward, and SUNY-Geneseo). More details, including publications, presentations, and the current status of these grids and their associated projects, as described earlier in this paper, are available at www.cse.buffalo.edu/ faculty/miller/CI/.

NYS Grid, Miller's CI Laboratory, and the Grass-Roots NYS CI Initiative. Beginning in 2000, SUNY-Buffalo incorporated requests from Miller in terms of high-end computing in its annual requests for funding to the State of New York. Miller's group also responded to requests from elected officials in the State of New York for a vision of high-end computing to support science and engineering in the state. Initial requests focused on the establishment of a NYS Supercomputing Center. These requests morphed into a request to establish NYS Technology Park that would include a New York State High-End

Computing Center, including Computing, Storage, Networking, and Visualization for NYS. Beginning in 2004, the requests focused on establishing an all encompassing New York State Cyberinstitute that would include funding to enhance and expand the New York State Grid that MCIL had established. In 2006, the Senior Vice Provost of SUNY-Buffalo supported and made announcements confirming the establishment of a Cyberinstitute of the State of New York (CSNY).

Subsequently, in July of 2006, a group gathered in Ithaca, New York, to discuss the possibility of initiating a state-wide effort in cyberinfrastructure. At this meeting, entitled a "New York State Workshop on Data-Driven Science and Cyberinfrastructure," Miller presented CSNY, which by then was being established within SUNY-Buffalo's Bioinformatics Center (BCOEBLS). Per the announcement of CSNY by SUNY-Buffalo's Senior Vice Provost, and with his approval, it was disclosed that CSNY would include i) the Center for Computational Research, ii) faculty working in computational science and engineering, iii) faculty working on fundamental problems in cyberinfrastructure, as well as iv) enabling staff, including programmers, GUI designers, and personnel focused on integrating middleware with applications. Miller also presented an overview of the Center for Computational Research and on-going efforts in MCIL.

At the end of the meeting, the attendees decided to schedule a meeting to continue to explore possibilities within NYS. In addition, the membership asked Miller's permission (granted) for *NYS Grid to serve as the underlying grid architecture for this potential state-wide CI initiative* so that this new initiative could focus on higher-level, non-grid, issues. Following the meeting, MCIL personnel worked with sites to educate system administrators on deploying and maintaining a node on a grid and obtaining a GRASE certificate.

In September of 2006, Miller gave a presentation at the second exploratory NYS Cyberinfrastructure meeting. In this talk, Miller discussed the status of NYS Grid and related information (www.cse.buffalo.edu/faculty/miller/talks.shtml). At the end of the September meeting, an inaugural board was voted on by the membership. It was agreed by the membership that the board would have a term of 1 year and would be required to a) propose a set of by-laws to be voted on by the membership, b) present a mission statement and vision for the initiative, and c) provide a status report of activities. Each initial board member was in charge of a working group: i) resource provider group, ii) user group, iii) technical working group, iv) communications group, v) education, outreach and training group, vi) funding group, and infrastructure group. Subsequent to this meeting, the group established a web site and chose the name NYSGrid for this organization, causing significant confusion nationally between this grass-roots effort and MCIL's well-established New York State Grid (NYS Grid).

In January of 2007, at the third Workshop, Miller and members of MCIL presented the status of NYS Grid to the membership, along with details and demonstrations. Critical results of this meeting included the recognition that NYS Grid was stable and serving numerous users from outside of New York State, that high-end users from New York State required assistance in order to move them onto a Grid, and that education and outreach to faculty, students, and staff throughout New York State was required.

Subsequent to this meeting, the board of this grass-roots initiative, which decided to call itself NYSGrid (www.nysgrid.org/), decided that this initiative was now an initiative of NYSERNet[2], that the board was now a steering committee within NYSERNet, to add new members to the steering committee, to create a position of "program manager" for one of the members of the steering committee, and to remove the executive director. They also received permission from OSG to create a VO called NYSGRID, informed all users of NYS Grid that they were being moved by NYSGrid from GRASE to NYSGRID, and stated that this NYSERNet initiative entitled NYSGrid was now in charge of NYS Grid. In addition, and still without input from the membership, the steering committee/board sent a proposal to the State of New York requesting funds for several member institutions of the steering committee/board. Needless to say, many of these actions created concern and confusion within New York State.

ENDNOTES

1 Information in this and the next section is available in more detail in (Miller, *et. al.*, 2008).

2 NYSERNet is a non-profit organization that focuses on networking concerns throughout New York State.

Section IV
Grids for Medical Informatics

Chapter XVIII
Aspects of Visualization and the Grid in a Biomedical Context

Ian Greenshields
University of Connecticut, USA

Gamal El-Sayed
University of Connecticut, USA

ABSTRACT

This chapter introduces some aspects of visualization and the grid. Visualization --the art and science of representing data visually-- is now recognized as an equal partner in the conduct of science via the simulation and modeling paradigm. Although not usually associated with Grid-scale problems, there are a number of Grid-dominant issues which subtend visualization. Evidently, certain data grooming issues (e.g., image preprocessing/analysis, certain computational geometric processes, certain computational topological processes) are amenable to deployment over compute Grids, but there has been equal focus on the collaborative aspect of Grid computing which is driving collaboration-based visualization systems. Here we survey some of the roles of visualization as they relate to the role of Grid computing within a biomedical context. We conclude by examining certain scheduling strategies we believe to have value in terms of the distribution of visualization tasks over Grid fabrics.

INTRODUCTION

From the earliest days of computational medicine, visualization in some form has always played a role in the biomedical sciences, with a particularly strong role in clinical medicine through the requirements of radiology. The advent of PACS (Picture Archiving and Communication Systems) (Dwyer III, 2000) introduced the bulk of radiologists to the digital delivery of image data, and concomitantly to the role of image processing in radiology. In fact, in as much as radiological applications of visualization are concerned, it can be difficult to separate the visualization component of the display process from the imaging component which grooms images prior to display. Even with the advent of the routine use of 3D (perhaps more properly, pseudo-3D) imaging modalities such as CT (computed tomography) and MRI (Magnetic Resonance Imaging) the visualization aspect of the imaging process was largely inseparable from the purely imaging aspects(Udupa and Hermann, 1991), although purely visualization concepts associated with the display of surfaces from clinical images begin to play a role quite separate from those traditionally associated with the imaging/image analysis tasks(Udupa, 1991; Vannier et al., 1991; Hohne and Bernstein, 1986). These early papers reflect the genesis of visualization as an adjunct to imaging, especially as regards the problems of visualizing surfaces represented by sampled data in the image lattice, and indeed even at that early stage notions of the role of topology began to appear(Peters et al., 1992). Most practitioners in the field would agree that a watershed was reached in 1987 with the publication of Lorensen and Cline's marching cubes algorithm(Lorensen and Cline, 1987). As the cost of graphics hardware plummeted and the range of visualization/imaging/computational geometric algorithms expanded, so too did the role of visualization expand into science areas external to the purely clinical field, so that today drug design, genomics, proteomics, large-scale simulations of cells (NRCAM) (to name but a few) would be unthinkable without the direct support of visualization.

As alluded to above, there is no strict line of demarcation between visualization and image processing when it comes to image data. For example, Figure 1 shows the surface of a skull deduced from the CT dataset of the Visible Human Female from the Visible Human Project (NLM), and there would be universal agreement that the image shown is created from purely visualization processes (in this case, an application of the marching cubes algorithm), even given that the raw image data was denoised using classical imaging techniques. Figure 2, on the other hand, shows polyhedral surfaces (the kidneys) embedded in (fused into) a standard volume visualization (data again from the Visible Human Project (op. cit.)). The situation here is somewhat more complex, in that the deduction of the polyhedral surfaces comes from a long line of image and cluster/classification algorithms (which constitute the bulk of the computation time taken to produce the image shown), while the "visualization component" (rendering the polyhedral surfaces and rendering the volume) is, while not trivial, certainly orders of magnitude less demanding than the imaging component.

Figure 1. Isosurface (bone)

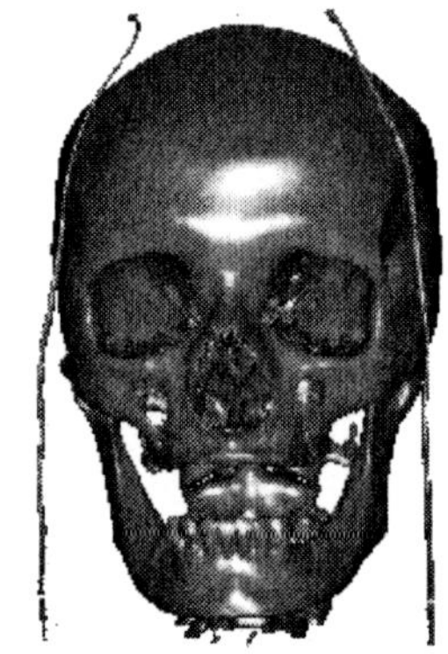

The reason we introduce this issue at this stage is because many of the final components of a visualization stream are often not truly Grid-scale in nature; certainly, handling ultra-large volumetric data is complex and demanding, but in truth few such image datasets (at the scale of the datasets seen in physics) are common (at least in clinical practice). The Visible Human Dataset alluded to above is possibly one of the more massive image datasets that can be encountered in biomedical computing (although there are others, such as data generated by (e.g.) spiral CT), and while it is large (the Visible Female dataset is some 40GB in extent), it is not Grid-massive in the same sense of the size of datasets generated by physicists, which

Figure 2. Polyhedral data fused into volumetric data

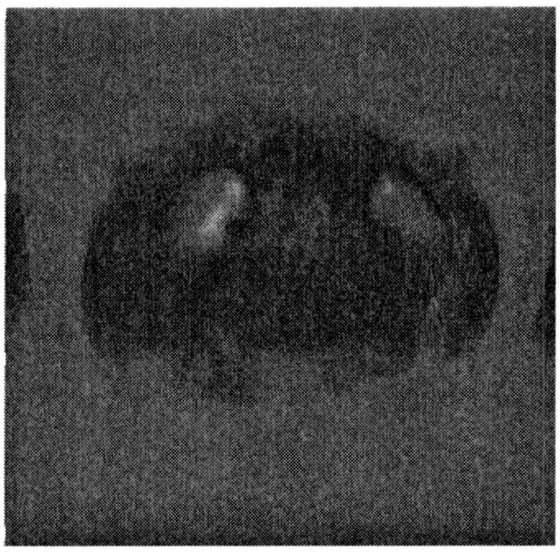

can be of petabyte scale(GriPhyN). Thus, while not dismissing nor minimizing the role of Grid computing (taken in its older sense of massively parallel computing) in visualization, the reality is that many of the final components in a visualization stream may simply be better handled locally (at the end-stage graphics server). However, if the line between the data grooming components (often classical imaging/pattern recognition processes, in addition to computed geometric and/or computed topologic processes) and the purely visualization components is blurred, then visualization (as an exercise in high-performance computing) becomes more clearly Grid-compliant. We return to this issue when we consider the dual role of the Grid (as a compute fabric or as a collaboration fabric) vis-a-vis visualization below. Although we address other aspects of biomedical visualization, our contention throughout the remainder of this chapter will be that those data grooming components of the visualization process (especially imaging components) should be considered as an integral part of the visualization process.

In order to set the scene for what follows, we consider a simple motivating example. The goal will be to derive a surface from a volumetric image dataset (see below for more explicit details). We imagine that the image dataset consists a set of *n* slices taken through some volume of interest (e.g., CT). *Data grooming* in this context may include any or all of the following pre-processing steps: (a) removal of noise from the images; (b) segmentation of the data in the images i.e., the assignment of an interpretation (such as tissue type) to the pixels in the images; (c) simple topological characterization of the results of the segmentation (e.g., derivation of connected components); (d) surface determination (e.g., by an isosurface process acting over the segmented images); (e) simplification of the resultant polyhedral surfaces (e.g., by smoothing and decimating the resultant triangles); (f) topological and/or geometric characterization of the deduced surface data (e.g., computation of inter- or intra-component metrics); and finally (g) fusion of the resultant surfaces back into the original volume. Actual *display* happens as the last stage.

BACKGROUND

Visualization Algorithms

Having somewhat conflated imaging with visualization above, we turn now to a quick survey of some of the standard data types and associated algorithms encountered in what might be referred to as "pure" visualization. This is the scope of standard texts in the area(Hansen and Johnson, 2005; Schroeder et al., 2006) and is covered here only briefly in order to provide a common terminology.

Perhaps the most common data sets encountered in visualization are those derived from field data (scalar, vector or tensor fields). In engineering and science, a scalar field is usually identified with some function $f: Q \subseteq R^n \rightarrow R$ and might correspond to the distribution in space-time of (e.g.) pressure, temperature or some other scalar property associated to each point of Q. Vector fields can be defined more or less the same way, so that the vector field is defined by the (vector-valued) function $v : Q \subseteq R^n \rightarrow R^m$, although the situation becomes somewhat more complicated in a differential geometric setting. Classical vector fields include velocity, and it can also be convenient to consider multivalent (multicomponent) images (e.g., RGB color images) to be vector fields. Tensor fields (treated generally) do not typically admit of the kind of simple notation given above; typically a more tedious formalism is followed (Dodson & Poston, 1991), but informally a tensor field associates a tensor t with each point $q \in Q \subseteq R^n$ Standard examples of tensors include the stress tensor or the elasticity tensor. More interestingly, diffusion tensor imaging -- DTI(Mori, 2007; Laidlaw et al., 1998) -- is a direct example of the role of tensor data in clinical imaging. In almost all practical applications, the field data to be visualized is sampled over Q, either on a regular sampling lattice, or (less frequently) over an irregular sampling lattice. Algorithms are typically tailored to the sampling lattice structure, and it is generally true that regular sampling lattices present as simpler data structures and have simpler traversal algorithms than irregular sampling lattices. Evidently, the fields may vary in time, and for the most part this complicates the visualization process only by the addition of discrete repetitions of the visualization stream over the sampled time interval. (This is not true if the evolution of some geometrical or topological property is to be visualized; in this case, the visualization (geometric/topologic) algorithms naturally are conditioned for this time evolution).

For the time being, we make the simplifying assumption that field data lies in R^2 or R^3 i.e., in the plane or in the volume. In this case, the visualization of scalar data is often quite straightforward, and can be as simple as color mapping perhaps combined with opacity mapping (see Figure 2). The data grooming component here can be trivial (the assignment by lookup table and/or interpolation of a color to the scalar, and the assignment by lookup table and/or interpolation of an opacity to the scalar), with the complications arising at the terminal end of the visualization stream in visualizing the volume when the scalar field lies in R^3. (Volume visualization, in fairness, can be quite tricky). Although this latter component can be hardware-assisted, it is more common to cast rays into the volume and deduce (from the field points intersected by the ray and the associated colors and opacities) what will actually be rendered on the screen. It is possible to construct the other major data structure associated with visualization (polyhedral data i.e., a dataset defining a polyhedral object) from scalar data by seeking level sets in the data. In the volume, such level sets are more commonly referred to as isosurfaces (surfaces constructed so that the surface passes through points of equal scalar value in the field). Since a great deal of visual control can be assigned to a polyhedral object (including such attributes as materials properties (color, opacity, transparency etc.) among other things) there is a tendency today to represent isosurfaces in terms of their approximating polyhedral data structures. Marching cubes (Lorensen and Cline, 1987) is the *locus classicus* example of such a conversion process, and the kidneys seen in Figure 2 are polyhedral objects deduced from a (kind of) isosurface. The process of developing isosurfaces from scalar field data is considerably more compute-intensive than the (usually) trivial process of assigning color/opacity data to the scalar field (and is often followed by further data grooming of the resultant polyhedral datasets, including polygon decimation and smoothing), but the process of rendering an isosurface is often considerably simpler than the ray-casting associated with rendering volume data. This raises an important point: compute-intensive processes associated with the terminal end of the visualization stream *may* not be good candidates for distribution in a Grid environment (up to all of the caveats attached to that statement, including communication latencies etc. (Neumann, 1994)); however, those associated with more distal layers of the visualization stream (such as isosurface determination) may be better suited to distribution in a Grid environment (again, up to usual process/communication metrics)(JaJa et al., 2008).

Vector fields are not quite so simple to visualize, particularly if they lie in the volume R^3. The classical picture one has of a 3-vector (a field $v: R^3 \to R^3$) is that of the directed arrow, but images which are made up of lots of arrows are typically worse than useless for dense vector fields (they can be impossible to comprehend). To an extent, this is where the art of visualization meets with the science, and there have been a number of clever strategies which address the visualization of vector fields(Crawfis et al., 1994). Often, it is convenient to formulate a visualization based on the specific nature of the vector being visualized.

Tensor data represents one of the most difficult challenges in visualization, and has attracted significant interest. One method that has shown promise is the use of so-called *glyphs* (Jankun-Kelly and Meta, 2006; Kindlmann, 2004; Ennis et al., 2005). These are usually implicit surfaces (such as ellipsoids or superquadrics) whose parameters are varied either directly by components of the tensor, or indirectly by properties of the tensor (such as principal components of the tensor). When the number of (independent) components of a tensor exceed the number of free parameters in the glyph, other characteristics of the glyph can be used to encode the unmapped tensor components (e.g., three components can be mapped to *r,g,b*-color, another to opacity etc.).

Polyhedral data, at least in biomedicine, is usually a constructed (derived) data type rather than a native data type. The whole gamut of classical computer graphics methods with application to polyhedral data is applicable to this type, and the fundamental algorithms are usually well-treated in texts on graphics (Foley et al., 1993) or computational geometry (Preparata and Shamos, 1990).

Mention should also be made of the role of implicit surfaces in Constructive Solid Geometry (CSG), in which entities are represented by Boolean combinations of a few simple primitives (such as spheres, cones, cylinders etc.). The CSG approach to visualization has applications in molecular modeling, especially in the modeling of large biochemical molecules (Figure 3). A wide variety of visualization tools have been written specifically for the purpose of visualizing protein data encapsulated in pdb (Protein Data Bank (Berman et al., 2000) format.

Finally, we mention the issue of visualizing data which is inherently very high dimensional. In this case, data is typically vector data in R^n(for large n) and arises frequently in data mining applications(Ammar et al., 2005), including genomics (and the more prosaic routine data mining problems associated with (for example) article searches within an online database). A typical problem encountered here is the visualization of clusters of data points in the high-dimensional space (or, almost equivalently, the visualization of clusters generated in a lower dimensional space obtained by projecting data from the original space into one of lower dimension (using, for example, Singular Value Decompositions or more computationally-efficient projections such as the Random Projection (Achlioptas, 2001)).

Figure 3. Protein (1QM2, a prion protein) from PDB file

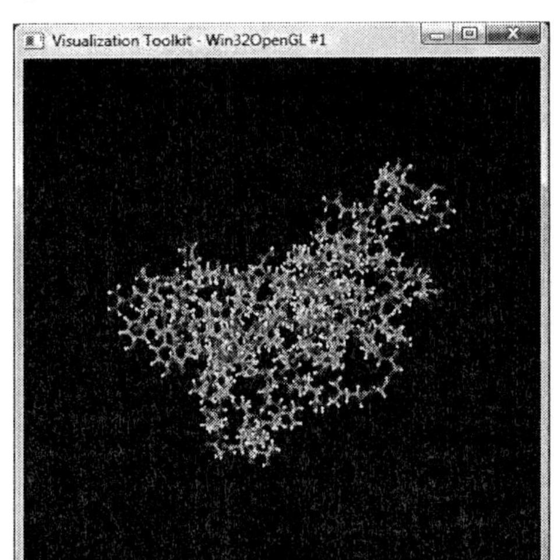

Care has to be taken here, both in the way the original projections are taken and in the way that the lower-dimensional spaces are eventually projected down onto the 2D viewing plane of the screen.

Collaborative Visualization and the Grid

There are two major areas of interest in the intersection of visualization and the Grid. We have touched on the first such issue above (i.e., the role of the Grid in high-performance computing), and expand on that issue below. Here we concentrate on one of the other major roles of the Grid i.e., as a fabric enabling virtual communities and collaboration. Foster(Foster et al.; Foster, 2002) and coworkers have made a persuasive case for the Grid being considerably more than an HPC (High-Performance Computing) engine, and it appears that collaborative visualization will fit well into the context of the Access Grid(AG; Childers et al., 2000). Collaborative visualization -- the process whereby many (remote) viewers share the same visual information remains an area of active research, with a number of groups identifying either Grid-specific protocols (e.g., GVK, see below) or more generic protocols(Manssour and Dal Sesso Freitas, 2000; Wood et al., 1997; Bajaj and Cutchin, 1999). However, one seminal paper by Johnson(Johnson, 1998) succinctly identifies the major architectural features of collaborative visualization systems and highlights basic structural issues raised by such systems.

Johnson (op. cit.) identifies four generic architectures into which any collaborative visualization system is likely to fall, depending on the level of distributed control permitted in the collaborative environment. In the most basic of these architectures, the Local Control Architecture (LCA), only the originating site has control over the visualization product. All remote sites may view the product, but exert no control over how the visualization is produced. In effect, the LCA structure devolves into a simple image mirroring facility, where the visualization product is mirrored to the remote sites more or less as a fully-formed image, lacking any further structure. In the Local Control with Shared Data Architecture (LCSDA), collaborating sites can share data from any step in the visualization pipeline, but control over the visualization process remains

Figure 4. Visualization Pipeline: Distribution of tasks. Task complexity usually decreases down the pipeline. The brackets indicate where such tasks may be allocated (in a single client, one large system computing a simulation, or distributed over a variety of servers).

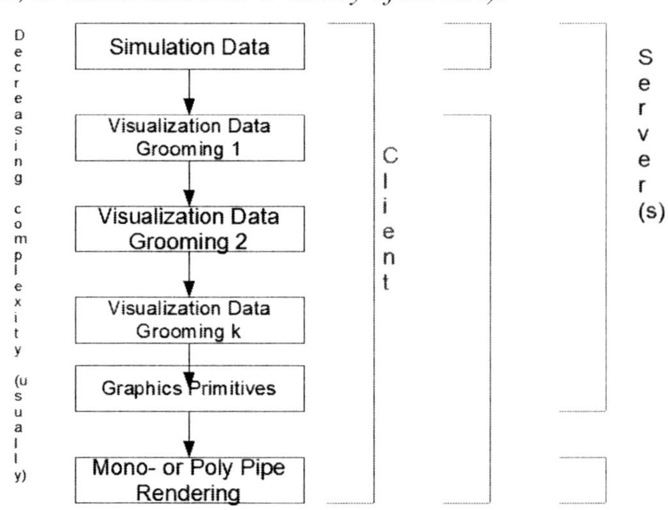

purely local. An extension to this architecture is the Limited Shared Control Architecture (LSCA), in which all participants in the collaborative visualization process can exert a small measure of control over the jointly viewed visualization product by (for example) annotating the product or exerting influence over the (joint) viewpoint. The culmination of these architectural steps is the Fully Shared Control Architecture (FSCA), in which every participant in the collaborative visualization process can affect any parameter of the visualization stream/pipeline. Overlaid on these architectural structures are three obvious requirements: *coherency, synchronicity* and *bandwidth/latency optimization*. Evidently, any distributed collaborative visualization system has to ensure that data coherency is maintained over all participating sites. The obvious approach here (as mentioned by Johnson) is to track the state of all shared data and to ensure that update order is enforced to prevent any new state change from overwriting a current state value until such a current value has at least propagated to all participating sites sharing the variable. Equivalently, synchronization ensures that contention for shared entities does not degenerate into worst-case scenarios. In a collaborative visualization environment, synchronization can often be assisted visually, by (for example) rendering synchronizable objects (e.g., camera viewpoint) differently depending on whether the viewer has or does not have control over the object. The issues of bandwidth and latency are, however, often out of the control of end users (and indeed the developers of collaborative visualization systems). Despite the promise of the QoS-based network, much work remains to be done in this area, and it is likely that many participants in a collaborative visualization environment will be using commodity-based networks rather than the more sophisticated academic networks such as Internet-2. Evidently, in a visualization pipeline which is parameter driven, it is relatively easy to broadcast a few KB of parameter data to all participants essentially simultaneously. It is also evidently impracticable to disseminate large volumes (GB or TB) of visualizable data across a collaborative group when anything like near-real-time responses are desired.

Visualization and the Computational Grid

We turn our attention to the role of the Grid as an HPC environment. Within the context of Grid-based visualization protocols, perhaps the best-known is GVK (Grid Visualization Kernel) (Kranzlmueller et al., 2004; Kranzlmueller, 2004). GVK is a protocol suite sitting on top of standard Grid middleware (specifically Globus (GLOBUS)) and permits the mapping of components of the visualization pipeline onto the Grid. The initial visualization system used by GVK was OpenDX (OpenDX), but it is clear that GVK could be extended to handle essentially arbitrary visualization pipelines. In particular, GVK appears amenable to handling the widely-used open-source Visualization Toolkit (VTK) (Schroeder et al., 2006; VTK) for which a parallel solution already exists (Ahrens et al., 2000). VTK is a conventional pipelined visualization suite, with modules interconnecting via output-to-input ports. It provides for a massive array of visualization primitives, including a significant number of data grooming primitives, is portable, relatively user-friendly, and can be accessed via conventional compiled languages (C++) or scripted languages (Tcl, Python) as well as web-friendly languages (Java). Parallelism is via MPI. In Ahrens (Ahrens et al., 2000), the authors identify three distinct types of parallel activity supported by (parallel) VTK: data parallelism, task parallelism and pipeline parallelism. Simplistically, data parallelism refers to the standard SPMD (Single Program, Multiple Data) parallel model, with merging as the final step. Task parallelism is the execution in parallel of a variety of subtasks, and is valuable whenever a visualization pipeline can be represented as a tree. Pipeline parallelism is, as its name suggests, the execution of stages of the pipeline in parallel.

Figure 4 shows a typical single visualization pipeline driven from a simulation environment. (More generically, the diagram would show a tree rooted at the mono- or poly-pipe rendering node). In the simulation context, data is generated by the simulation and is then groomed successively throughout the pipeline until the last stage produces the graphics primitives which are fed to the graphics hardware which drives the display device. Typically, the very lowest levels of the pipeline -- particularly those tied to the graphics primitives -- are best located locally, unless the local demands are so enormous (e.g., volume visualization of massive datasets) that the expense of communication between remote sites is negligible in comparison to the cost of a purely local computation. The issue of process distribution arises more naturally in the upper levels of the pipeline (pipeline tree), and it is within these levels that the role of the Grid as an HPC engine becomes more important. We illustrate this by means of an example below. Finally, we note that significant new work has appeared in parallelizing graphics environments, including Chromium (Humphreys et al., 2002) and a parallel sort-last procedure (Eilemann and Pajarola, 2007). Chromium permits the deployment of lower-level graphics primitives over clusters of workstations, and can be employed in cluster rendering. Eilemann and Pajorola's work exploits the parallel rendering system Equalizer(EQU) to exploit multi-pipe systems or graphics clusters for parallel rendering.

VIRTUAL ANATOMY FROM THE VISIBLE HUMAN PROJECT

The Visible Human Project from the U.S. National Library of Medicine(NLM) comprises sets of image data collected from a male and a female cadaver. For both cadavers, CT, MRI and cryosectioned imagery (full-color images made by sectioning the cadavers) were obtained. These image datasets find a role in a wide variety of projects, including virtual anatomy(NLM). Following its initial work with the Visible Male, the National Library of Medicine produced a corresponding Visible Female dataset, which corrected certain resolution issues arising in the Visible Male dataset. Both datasets include whole-body CT and MR images, but attention has been focused on the cryosectioned full-color datasets. In the Visible Male dataset, the cryosectioned full-color images are resolved to 0.33 x 0.33 x 1 mm^3 in the coronal (x-z), sagittal (y-z) and transverse (x-y) axes (respectively). The female dataset (the Visible Female, hereafter VF) corrected this resolution anomaly by offering a true cubic voxel representation of 0.33 x 0.33 x 0.33 mm^3 resolution in all three anatomic planes. The VF cryosectioned dataset comprises 5189 rgb images sectioned transversely from head to feet, with a (current) resolution of 2048 x 1216 pixels per image. A standard exemplar of the male cryosectioned dataset is given in Figure 5.

We became interested in mirroring aspects of the Stanford Lucy project (LUCY; Heinrichs, 1999) but with the NLM Visible Female (VF) dataset (NLM), concentrating on the renal system (kidneys, ureters, bladder and urethra). The goal was to represent and visualize these structures as polyhedral objects either embedded in the volumetric image data from which they arose (Figure 2) or fused into bony structures deduced from the CT dataset (Figure 6). The processes involved are primarily data parallel and pipeline parallel, and begin with the problem of segmenting the original data.

The fundamental classification tool we use is the Gibbs Classifier, from whose use we attempt to find the MAP (Maximum A Posteriori) estimate of the class structure of the image given some prior Π, which is usually expressed structurally. (The prior Π typically captures some prior model of the image). Gibbs classification is described in detail elsewhere(Geman and Geman, 1984; Winkler, 1995; Yang and Greenshields, 2001;

Figure 5. Visible human dataset: male thorax

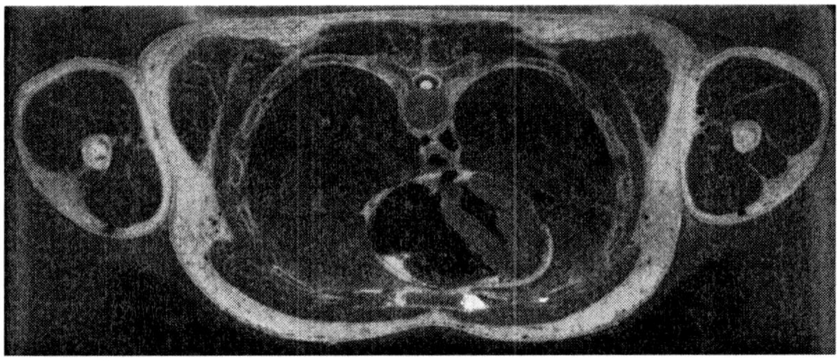

Figure 6. Kidneys and pelvic girdle, viewed inferiorly

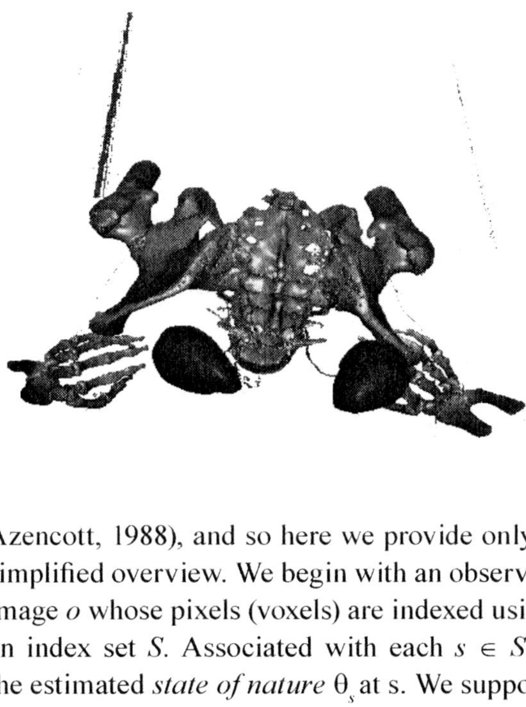

Azencott, 1988), and so here we provide only a simplified overview. We begin with an observed image o whose pixels (voxels) are indexed using an index set S. Associated with each $s \in S$ is the estimated *state of nature* θ_s at s. We suppose that θ_s can assume values in the (finite) set χ, and we call $\Omega = \chi^S$ the set of configurations $\omega = (\theta_s)_{s \in S}$. ω is therefore an assignment of classes (states of nature) to the observed image o. Let $H: \Omega \rightarrow R$ define some function (often referred to as an energy function) on the configurations; then the goal is to find a configuration $\omega_{min} \in \Omega$ such that $H(\omega_{min}) = \min_{\omega \in \Omega} H(\omega)$. Under certain circumstances, H can be expressed as a sum of local contributions $H = \sum_{c \in C} U_c(\omega)$ where C is a set of cliques of S. Cliques are subsets of S usually built from the neighborhood structure $N(S)$ of S and for all sites $s \in S$ one has that $s \notin N(s)$ and for any two sites $s, s' \in S$ that $s \in N(s') \Leftrightarrow s' \in N(s)$. Now consider the joint distributions $P(\Theta = \omega)$. The underpinning image process is said to be Markovian whenever these joints are always strictly positive (non-zero) and the conditional $P(\Theta_s = \theta_s | \Theta_t = \theta_t \quad t \neq s) = P(\Theta_s = \theta_s | \Theta_t = \theta_t \quad t \in N(s))$. Entropy arguments lead to a distribution on the configurations Ω having the Gibbs representation $G(\omega; T) = (1/Z_T)\exp[-H(\omega)/T]$ where $Z_T = \sum_{\omega \in \Omega} \exp[-H(\omega)/T]$ is called the partition function, and T is referred to as temperature (or "artificial temperature"). If we set $\Omega_{min} = \{\omega \in \Omega \mid H(\omega) = \min_{\omega' \in \Omega} H(\omega')\}$, then it can be shown that there is a sampling strategy which converges to the uniform distribution on Ω_{min} i.e., to the uniform distribution on the minimizers of H. The goal, then, is to generate a (random) configuration $\omega \in \Omega$ given the distribution $G(\omega; T)$ (for small T) such that ω is a minimizer (or more properly, a near minimizer) of H.

This cannot be done exhaustively by direct enumeration of the configurations. If we presume there are (say) only 5 states of nature (classes) in a given Visible Female image, then there are $5^{(1216x2048)}$ possible configurations for any single image. Instead, approaches based on stochastic relaxation are typically taken. Again, these approaches are detailed in the literature, and covered here only briefly. In single-flip Metropolis algorithms(Metropolis et al., 1953), a new configuration ω^{t+1} is produced from the current configuration ω^{t} by making a single change of state at one and only one site s and the energy change $\Delta H = H(\omega^{t+1}) - H(\omega^{t})$ is computed. Then the expression $\beta = \exp\left[{}^{-\Delta H}/_{T}\right]$ is computed, and the change to ω^{t+1} is permitted provided $\beta > 1$; when $\beta < 1$ the change is accepted with probability β. (This avoids getting stuck in local minima). In annealing(Kirkpatrick et al., 1982), a decreasing sequence of temperatures (an annealing schedule) is followed and the Metropolis algorithm is deployed throughout this schedule. Finally, the Gibbs sampler (Geman and Geman, 1984) is probably the most well-known case of Metropolis-like algorithms.

The algorithmic view is simpler; we use the simulated annealing method(Kirkpatrick et al., 1982) as an example. We begin by defining a prior and energy function, which naturally conditions the resultant posterior. The prior may be subjective; for example, we might define the prior so that it preferentially weights regions of constant class value (which will tend to smooth the image), or we might use prior classifications to deduce class transition probabilities relative to the cliques. Next we define an annealing (temperature) schedule. Starting at T_0 (the initial temperature) and given some starting configuration ω_0 we cycle through the image visiting each site in the image periodically (e.g., by simple repeated raster scan) making single flip changes, computing the resultant energy change, and accepting or rejecting the change as per the description above. This repeats for the given temperature until some form of stable state is reached in the configurations. The temperature is then lowered, and the process repeats (in theory, until a global energy minimum is reached). In practice, of course, only a finite amount of time can be allocated to the procedure -- the so-called finite-time annealing process. We accept that the resultant classification may not be a minimum energy classification, but is at least (hopefully) better than the originating configuration ω_0. Evidently, this is a compute-intensive process, and can run for hours on a modern processor. There are a variety of ways of deploying parallel processing in simulated annealing(Azencott, 1988; Azencott, 1992; Ram et al., 1996), but we consider a simple example here.

The idea is to replicate the annealing process over a variety of different starting configurations ω_0^i, and at the end, choose the final configuration having least energy. There is empirical evidence that starting the process with a configuration likely to be near a minimum energy configuration speeds the process up, and so we begin by generating n independent initial configurations using a variety of standard cluster/classification techniques. These become the initial configurations $\omega_0^0, \omega_0^1, \ldots, \omega_0^{n-1}$ for n independent runs of the Gibbs process. Again, there are a number of different strategies that can be deployed in parallelizing the process. The simplest is to let each annealing process run to its maximum number of N steps, and thereafter to choose the configuration having lowest energy. Alternatively(Ram et al., 1996), one can pass partial results between each of the annealing processes after some fixed number of iterations, so that all processes restart with the currently best solution. However, to motivate the scheduling algorithm of the next section, we focus on a strategy we first proposed in Yang(Yang and Greenshields, 2001), which relies on empirical evidence of the nature of the convergence of the image configurations.

As the process converges, it becomes clear that large homogeneous sections of the images settle in to a constant class and are thereafter rarely (if ever) changed. The simplest example is the background of images, which tends to converge quickly to the background class and subsequently does not change. The impact of this kind of homogeneous convergence is on the visitation schedule i.e., on the site-by-site scan of the images during the annealing process. In the VF images, initial scans may take into consideration all 1216 x 2048 sites in the image. If, after K iterations, M sites in the image have not changed in the last $L < K$ iterations, a decision can be taken to ignore these sites for future visitation. (Note that this breaks the guaranteed convergence to the/a global minimum; but since we are deploying finite-time annealing anyway, we start out with no guarantee of reaching such a minimum). We refer to sites whose class assignments Θ_t have not changed after some number of sweeps L as having *conditionally converged.* Now suppose that we assign a maximum of P sweeps (complete site visits) per invocation of the process; P (in the VF dataset case) will be some number k x 1216 x 2048 and we expect k to be quite large. Generally, assume that $|S|$ is the cardinality of the site index set. If, at the Q'th sweep ($Q<P$) the subset $I_j \subset S$ of sites in the j'th invocation has conditionally converged, and we determine not to visit the sites in I_j for the remaining sweeps associated with the j'th invocation, then the number of free sweeps (work no longer being done by the j'th invocation) is just

$$F_j = (P - Q)\left(1 - \frac{|S - I_j|}{|S|}\right)$$

These free sweeps can be distributed in a number of ways; clearly, they could be retained by the j'th invocation (effectively increasing P for that invocation), or they could be offered to the k'th invocation having $I_k < I_n$ (i.e., that invocation having the most remaining work) to assist in an innate parallel computation of its annealing schedule. If we adopt the latter approach, we need some way of redistributing processes around the grid fabric dynamically. We address this below.

DYNAMIC PROCESS MIGRATION IN THE GRID

In a purely locally-controlled environment, we would adopt any of the multiplicity of dynamic scheduling algorithms available in the literature. In the case of the Grid, however, we do not have the luxury of the guarantee of control over the components of the grid fabric. We must therefore design any dynamic process allocation protocol to be sensitive to the potential exercise of local authority within the Grid, and in particular, to the possibility that such long-time processes as the Gibbs classification described above might overstay their allotted time on a remote portion of the Grid fabric, and be requested by the local schedulers to cease computation on that portion of the fabric. Evidently, losing partial results of such long-haul processes is an unwelcome outcome, especially given that the processes may be executing for hours at a time. We elected therefore to explore dynamic migration protocols which carried within them checkpoint recovery mechanisms, and given the particular nature of the Gibbs processes described above, to design such a protocol with these kinds of tasks explicitly in mind.

The specifics of the tasks described above are that they are long-haul (i.e., compute for hours) and communicate with each other infrequently (for example, only to pass current configuration data after a number of iterations). Inter-task communication may therefore only occur between time intervals on the order of 10's of minutes. We characterize such tasks as having very low-frequency communication. Setting $T_p(t_i)$ as the interval between communication for the i'th task t_i, $T_m(t_i)$ as the actual inter-task messaging time for the i'th task and $T_{cc}(t_i)$ as the coordinator messaging time (see below) for the i'th task, we have (for all tasks t_i and t_j) $T_p(t_i) \approx T_p(t_j)$, $T_m(t_i) \approx T_m(t_j)$, $T_{cc}(t_i) \approx T_{cc}(t_j)$ and the inequality $T_p(t_i) >> T_m(t_i) >> T_{cc}(t_i)$. With these constraints in mind, we set about designing a robust non-blocking checkpointing algorithm having globally-consistent checkpoints, able to work in a non-FIFO network environment and having a simple coordinated protocol. The complete algorithm is described in (El Sayed and Greenshields, 2008). Here we summarize its salient features, especially as they apply to the application given above. The protocol, called Low-

FM (Low-Frequency Messaging), is a dynamic task migration protocol designed to operate over the Grid sitting one protocol layer below a more generic superscheduler, which is itself tasked with identifying dynamic changes in the Grid fabric such as the addition of fabric components, or (more likely) task pre-emption on a local fabric by local tasks having higher priority. While the goal of Low-FM is to be able to migrate tasks over the fabric, it is not a fault-tolerant protocol, in that (for example) the failure of a task on a node is not accounted for. (Such a failure could not, in general, be recovered by Low-FM). Low-FM keeps track of the state of its task set both by monitoring the state of the processes themselves (i.e., by saving classical process state) and by monitoring the state of the communication channels between processes. A global coordinator task serves to coordinate the taking of checkpoints (and obviously, Low-FM is not tolerant of failure of the global coordinator). Checkpoints are taken by the global coordinator at intervals corresponding roughly to $T_p(t_i)$ (in the specific case of the Gibbs classifiers described above).

The protocol requires a number of strong assumptions: task pre-emption is handled by the superscheduler coordinating with the checkpoint coordinator; messages are never lost; intertask messages may be out-of-order (non FIFO messaging fabric); communication is point-to-point; broadcast messages are counted as multiple single-issue messages; the coordinator and tasks are failsafe; all messages route via the checkpoint layer protocol; and the checkpoint layer protocol keeps track of all committed and tentative checkpoints.

Each task t_i is associated with a local checkpoint c_i^l which is the l'th checkpoint for t_i. Messages between tasks t_i and t_j are annotated with the index of the local checkpoint of the originating task, so that a message from i to j is denoted $m_{i,j}^l$. Tasks retain a count of the number of messages sent (S_i^l) and received (R_i^l) during the extent of the validity of checkpoint l, and the coordinator will keep track (during the checkpoint process) of the global sends and receives

$$S^l = \sum_i S_i^l \text{ and } R^l = \sum_i R_i^l.$$

The protocol requires the use two sets of adjunct counters to ensure that the checkpointing process commits robustly; these are the temporary counters tS_i^l and tR_i^l, used as temporary send and receive counters during tentative checkpoint taking, and the back-one counters tS_i^{l-1} and tR_i^{l-1}, used to handle lost commits from the coordinator. A message is called a *lost* message when it is counted by the sender (but not the receiver), and is called an *orphan* message if it is counted by the sender but not the receiver (see below for clarification).

The protocol functions as follows. Following the time-out of the checkpoint interval timer $\Delta T_c \approx T_p$, the checkpoint layer (coordinator) issues a checkpoint request q^l to all tasks t_i. At the same time, the coordinator starts a validation timer T_v^l. On receipt of the checkpoint request q^l, the task t_i takes a tentative checkpoint image c_i^l and forwards its counts R_i^l and S_i^l to the coordinator. During this *pre-commit* period, all subsequent messages from t_i to t_j are annotated with the current checkpoint index l. During this pre-commit period, if task t_j sees a message $m_{i,j}^l$ but has not yet seen q^l, t_j counts $m_{i,j}^l$ as received and commits to its tentative checkpoint c_j^l. At the same time (during the pre-commit), t_i (having sent $m_{i,j}^l$) discards c_i^l and instead generates an updated c_i^l which counts $m_{i,j}^l$ as having been sent. The temporary counters tS_i^l and tR_i^l are updated during this pre-commit period for all tasks sending and/or receiving messages (as are the general counters). In the event that t_j receives $m_{i,j}^l$ prior to receiving q^l from the coordinator, then the q^l from the coordinator is discarded by t_j (notification and checkpointing having been accomplished by receipt of $m_{i,j}^l$). The decision to commit to the checkpoint is made by the coordinator whenever it deems the global state to be consistent. On receipt of the commit, tasks commit to c^l and the protocol resumes awaiting c^{l+1}. If a commit is delayed, so that task t_j receives the commit c^{l-1} after receipt of a checkpoint request l, then t_j commits to c_j^{l-1} and recalculates its send/receive counters by subtracting the temporary counters from the general counters ($S_j^l = S_j^l - tS_j^{l-1}$) and resetting the temporary counters back to 0. Messages received within the duration of the checkpoint validation interval T_v^l are counted, and (as seen) may induce the receiver to take a checkpoint. However, if a message is not received within the validation interval, the checkpoint will fail (the global state is inconsistent) and the current committed checkpoint remains the previous committed checkpoint (c^{l-1}). Low-FM is built as a state machine, and is provably correct up to its (rather stringent, we concede) requirements.

FUTURE TRENDS

Perhaps the most promising future trend relating Grid computing and visualization is the promise of further integration of the Access Grid with the visualization stream (see Childers et al., 2000). Recall that the Access Grid is a collaborative environment built over the Grid fabric, and while it has long been possible for the Access Grid to support the dissemination of visualization products, it is currently not possible to support a completely immersive, collaborative environment. In the future, we might expect that remote immersive environments (such as set of remotely connected CAVEs) will be collaboratively linked via the Grid fabric in such a fashion that participants in each of the co-joined CAVEs will be able to interact with each other (and the shared virtual environment displayed in each of the remote CAVEs) in exact real-time.

CONCLUSION AND ACKNOWLEDGMENT

This short chapter has surveyed some issues relating visualization to the Grid in a biomedical context. Two major issues (collaborative visualization and computational visualization) have been addressed, with specific results given in the area of computational visualization. This work was partially funded under an NSF grant (NSF 0138760).

REFERENCES

Achlioptas, D. (2001). Database Friendly Random Projections. *Proc. Symposium on Principals of Database Systems (PODS01)*, (pp. 274-281).

AG. *The Access Grid.* Retrieved from http://www.accessgrid.org/

Ahrens, J., Law, C., Schroeder, W., Martin, K. & Papka, M. (2000). *A Parallel Approach for Visualizing Extremely Large, Time-Varying Datasets*(Tech. Rep. LAUR-00-1620). Los Alamos, NM: Los Alamos National Laboratory. Retrieved from http://www.vtk.org/pdf/pvtk.pdf

Ammar, R.A., Demurjian, S.A., Greenshields, I.R., Pattipatti, K. & Rajasekaran, S. (2005). Analysis of Heterogeneous Data in Ultrahigh Dimensions. In R. Popp and J. Yen (Eds.), *Emergent Information Technologies and Enabling Policies for Counter-Terrorism*, (pp. 105-124). Hoboken, NJ: Wiley Interscience.

Azencott, R. (1988). Simulated Annealing. *Seminaire Bourbaki 40eme annee, 1987-88, No. 697*, (pp. 223-237).

Azencott, R.(1992). *Simulated Annealing: Parallelization Techniques.* Hoboken, NJ: Wiley.

Bajaj, C. and S. Cutchin (1999). Web Based Collaborative Visualization of Distributed and Parallel Simulation. In *Proc. of the 1999 Sym-*

posium on Parallel Visualization and Graphics, San Francisco, (pp. 47-54).

Berman, H.M., Westbrook, J., Feng, Z., Gilliland, G., Bhat, T.N., Weissig, H., Shindayalov, I.N. & P.E. Bourne (2000). The Protein Data Bank. *Nucleic Acids Research, 28,* 235-242.

Childers, L., Disz, T., Olson, R., Papka, M.E., Stevens, R. and T. Udeshi (2000). Access Grid: Immersive Group-to-Group Collaborative Visualization. In *Proc. 4'th International Projection Technology Workshop.* Retrieved from http://www-unix.mcs.anl.gov/fl/publications/childers00.pdf

Crawfis, R., Max, N. and B. Becker (1994). Vector Field Visualization. *IEEE Computer Graphics and Applications, 14 (*5), 50-56.

Dodson, C.T.J. and T. Poston (1991). *Tensor Geometry: The Geometric Viewpoint and its Uses.* Berlin, Germany: Springer-Verlag.

Dwyer III, S.J. (2000). A personalized view of the history of PACS in the USA. In G. J. Blaine & E. L. Siegel, *Proceedings of the SPIE, "Medical Imaging 2000: PACS Design and Evaluation: Engineering and Clinical Issues"*, (Vol. 3980), (pp. 2-9).

Eilemann, S. & Pajorola, R. (2007). Direct Send Compositing for Parallel Sort-Last Rendering. *Eurographics Symposium on Parallel Graphics and Visualization*, (pp. 29-36).

El Sayed, G. and I. Greenshields (2008). Low FM: A Robust Non-Blocking Low-Frequency Messaging Checkpoint for Parallel/Grid Applications. In *Proc. 23rd International Conference on Computers and Their Applications, Cancun, Mexico, April 9-11.*

Ennis, D.B., Kindlmann, G., Rodriguez, I., Helm, P.A. & McVeigh, E.R. (2005).Visualization of Tensor Fields Using Superquadric Glyphs. *Magn. Resonance Med., 53*(1) 169-176.

EQU. (n.d.). *Equalizer.* Retrieved from http://www.equalizergraphics.com/

Foley,J.D., van Dam, A., Feiner, S.K., Hughes, J.F. & Philips, R.L. (1993). *Introduction to Computer Graphics.* Upper Saddle River, NJ: Addison-Wesley.

Foster, I., Kesselman, C. & Tuecke, S. (n.d.). *The Anatomy of the Grid: Enabling Scalable Virtual Organizations.* Retrieved from http://www.globus.org/alliance/publications/papers/anatomy.pdf

Foster, I. (2002). *The Grid: Enabling Resource Sharing With Virtual Organizations.* Keynote speech at WWW2002, Honolulu, HI. Retrieved from http://www2002.org/foster.pdf

Geman, S. and D. Geman (1984). Stochastic Relaxation, Gibbs Distributions, and the Bayesian Restoration of Images. *IEEE Trans. PAMI, 6*, 721-741.

GLOBUS (n.d.). *The Globus Alliance.* Retrieved from http://www.globus.org/

GriPhyN (n.d.). *The Grid Physics Network.* Retrieved from http://www.griphyn.org

Hansen, C.D. & C.R. Johnson (2005). *The Visualization Handbook.* Burlington, MA: Elsevier Butterworth-Heineman..

Heinrichs, W.L. (1999). The Critical Path from Tissue Slices to Surgical Simulation: What do Surgeons Want? In *Proc. 12'th IEEE Symposium on Computer-Based Medical Systems, June 18-20, Stamford, CT,* (pp. 118-123).

Hohne, K.H. & Bernstein, R. (1986). Shading 3D-Images from CT Using Gray-Level Gradients. *IEEE Transactions on Medical Imaging, 5*(1), 45-47.

Humphreys, G., Houston, M., Ng, R., Frank, R., Ahern, Kirchner, P.D. & J.T. Klosowski (2002). Chromium: A Stream-Processing Framework for Interactive Rendering on Clusters. *ACM Transactions on Graphics, 21*(3), 693-702.

JaJa, J., Varshney, A. & Shi, Q. (2008). Parallel Algorithms for Volumetric Surface Construction. In S. Rajasekaran & J. Reif (eds.), *Handbook of Parallel Computing*, (pp. 27-17). Boca Raton, FL: Chapman and Hall.

Jankun-Kelly, T.J. & K. Meta (2006). Superellipsoid-based, real Symmetric Traceless Tensor Glyphs Motivated by Nematic Liquid Crystal Alignment Visualization. *IEEE Transactions on Visualization and Computer Graphics*, (pp. 1197-1204).

Johnson, G. (1998). Collaborative Visualization 101. *ACM SIGGRAPH, 32*(2), 8-11.

Kirkpatrick, S., Gellatt Jr., C.D. & Vecchi, M.P. (1982). *Optimization by simulated annealing.* Yorktown Heights, NY: IBM T.J. Watson Research Center.

Kindlmann, G. (2004). Superquadric Tensor Glyphs, Joint Eurographics. In *IEEE TCVG Symposium on Visualization*, (pp. 147-154).

Kranzlmueller, D., Heinzlreiter, P. and J. Volkert (2004). *Grid-Enabled Visualization with GVK, in Grid Computing* (LNCS Vol. 2970) (pp. 139-146). Berlin, Germany: Springer-Verlag.

Kranzlmueller, D. (n.d.). *The GVK/GVid Approach: Visualization on the Grid.* Retrieved from http://www.ogf.org/OGF19/materials/562/kranzlmueller.pdf

Laidlaw, D.H., Ahrens, E.T., Kremers, D., Avalos, M.J., Jacobs, R.E. & Readhead, C. (1998). Visualizing Diffusion Tensor Images of the Mouse Spinal Cord. In *IEEE Visualization '98*, (pp. 127-134).

Lorensen, W.E. & Cline, H. (1987). Marching cubes: A High Resolution 3D Surface Construction Algorithm. *ACM SIGGRAPH, 21*(4), 163-169.

LUCY (n.d.). *Stanford Visible Female.* Retrieved from http://lucy.stanford.edu/home.html

Manssour, I.H. & Dal Sasso Freitas, C.M. (2000). Collaborative Visualization in Medicine. In *Proc. 8'th International Conference in Central Europe on Computer Graphics, Visualization and Interactive Digital Media (WSCG2000), Plzen*, (pp. 266-273).

Metropolis, N., Rosenblath, A.W., Rosenblath, M.N., Teller A.H., & Teller, E. (1953). Equations of State Calculations by Fast Computing Machines. *J. Chem. Physics, 21*, 1087-1091.

Mori, S. (2007). *Introduction to Diffusion Tensor Imaging.* Amsterdam, the Netherlands: Elsevier.

Neumann, U. (1994). Communication Costs for Parallel Volume-Rendering Algorithms. *IEEE Computer Graphics and Applications, 15*(4), 49-58.

NRCAM (n.d.). *The National Resource For Cell Analysis and Modeling.* Farmington, CT: University of Connecticut Health Center. Retrieved from http://www.nrcam.uchc.edu

NLM (n.d.). *Visible Human Project.* Bethesda, MD: National Library of Medicine., Retrieved from http://www.nlm.nih.gov/research/visible/visible_human.html

OpenDX. Retrieved from http://www.opendx.org/

Peters, T.J., Greenshields, I.R. & Dorney, S. (1992). Topological Fidelity in Surface Reconstruction, Curves and Surfaces in Computer Vision and Graphics III. In *Proc. SPIE* (Vol. 0277-786X), (pp. 221-225).

Preparata, F.P. & Shamos, M.I. (1990). *Computational Geometry.* Berlin, Germany: Springer-Verlag.

Ram, J. D., Sreenivas, T.H., & Subramaniam, K. G. (1996). Parallel Simulated Annealing Algorithms. *Jnl. Parallel and Distributed Computing, 37*, 207-212.

Schroeder, W., Martin, K. & B. Lorensen (2006). *Visualization Toolkit: An Object-Oriented Approach to 3D Graphics*. Clifton Park, NY: Kitware.

Udupa, J.K. & Hermann, G.T. (1991). 3D Imaging in Medicine. Boca Raton: CRC Press..

Udupa, J.K. (1991). Computer Aspects of 3D Imaging in Medicine: A Tutorial. In J.K. Udupa and G.T. Hermann (Eds.), *3D imaging in Medicine*, (pp. 1-70).

Vannier, M.W., Hildebolt, C.F., Gayou, D.E., & Marsh, J.L. (1991). Introduction to 3D Imaging. In J.K. Udupa and G.T. Hermann (Eds.), *3D imaging in Medicine*, (pp. 71-88).

VTK, (n.d.). *The Visualization Toolkit*. Clifton Park, NY: Kitware. Retrieved from http://www.vtk.org/

Winkler, G. (1995). *Random Fields and Dynamic Monte Carlo Methods*. Berlin, Germany: Springer-Verlag.

Wood, J., Wright, H. & K. Brodie (1997). Collaborative Visualization. *Proc. Visualization '97*, (pp. 253-259).

Yang, Z. & Greenshields, I. (2001). *Ensemble Classification of the VF Dataset with Limited Merging*. In *Proc. 14'th IEEE Symposium on Computer-Based Medical Systems*, Bethesda, MD, (pp. 117-122).

KEY TERMS AND DEFINITIONS

Classification: Process by which image data is analyzed into ground-truth categories, such as tissue type.

Grid: Collection of hardware and software resources enabling collaborative computing.

Process Migration: The act of migrating a process from one part of a compute fabric to another.

Scheduling: Assignment of tasks/processes onto a computational fabric, such as the Grid

Visualization: Art and science of representing data visually

Visible Human Project: Project initiated by the US National Library of Medicine enabling a variety of applications (such as virtual anatomy) to deploy high-resolution imagery of the human body.

Compilation of References

Aalst, W., Hofstede, A., Kiepuszewski, B., & Barros, A. (2003). Workflow Patterns. *Distributed and Parallel Databases, 14*(3), 5-51.

Abraham, A., & Nath, B. (2000). Hybrid heuristics for optimal design of artificial neural networks. In R. John & R. Birkenhead (Eds.), *Advances in Soft Computing Techniques and Applications* (pp. 15-22). Berlin, Germany: Springer-Verlag.

Abraham, A., Grosan, C., & Ramos, V. (Eds.). (2006). *Swarm Intelligence in Data Mining, Studies in Computational Intelligence*. Berlin, Germany: Springer-Verlag.

Abramoff, M., Magalhaes, P., & Ram, S. (2004). Image Processing with ImageJ. *Biophotonics International* (pp 36-42).

Abramson D., Lynch A., Takemiya H., Tanimura Y., Date S., Nakamura H., Jeong K., Lee H.-C., Arzberger P. A., Baldridge K. K., Li W. W., & Molina T. (2006). Deploying Scientific Applications to the PRAGMA Grid Testbed: Lessons Learned. In *IEEE/CCGRID International Grid Computing, Singapore*.

Abramson D., Roe P., Kotler L., & Mather D. (2001). *ActiveSheets: Super-Computing with Spreadsheets. 2001 High Performance Computing Symposium (HPC'01)*. Advanced Simulation Technologies Conference, April 22-26, Seattle, WA (pp. 110–115). Retrieved from http://www.csse.monash.edu.au/~davida/papers/ActiveSheets.pdf.

Abramson, D, Buyya, R. & Giddy, J. (2002). A Computational Economy for Grid Computing and its Implementation in the Nimrod-G Resource Broker. *Future Generation Computer Systems, 18*(8) 1061-1074.

Abramson, D, Lewis, A. & Peachy, T. (2001). *Case Studies in Automatic Design Optimisation using the P-BFGS Algorithm*. 2001 High Performance Computing Symposium (HPC'01), Advanced Simulation Technologies Conference, April 22-26, , Seattle, WA (pp.104 – 109). Retrieved from http://www.csse.monash.edu.au/~davida/papers/CaseStudies.pdf

Abramson, D., Sosic, R., Giddy, J., & Hall, B. (1995). Nimrod: a tool for performing parametrised simulations using distributed workstations. *Fourth IEEE International Symposium on High Performance Distributed Computing* (pp 112-121).

Achlioptas, D. (2001). Database Friendly Random Projections. *Proc. Symposium on Principals of Database Systems (PODS01)*, (pp. 274-281).

Achtman, M., & Wagner, M. (2008). Microbial Diversity and genetic nature of microbial species. *Nat Rev Microbiology, 6*, 431-440.

Adai, A., Johnson, C., Mlotshwa, S., Archer-Evans, S., Manocha,V., Vance, V. & Sundaresan, V. (2005). Computational prediction of miRNAs in Arabidopsis thaliana. *Genome Res., 15*, 78-91.

Adam, B.L Qu, Y.. Davis, J.W. Ward, M.D Clements, M.A Cazares, et al. (2002). Serum Protein Fingerprinting Coupled with a Pattern-matching Algorithm Distinguishes Prostate Cancer from Benign Prostate Hyperplasia and Healthy Men. *Cancer Research, 62*(13), 3609–3614.

Adcock, S. A., & McCammon, J. A. (2006). Molecular dynamics: Survey of methods for simulating the activity of proteins. *Chemical Review, 106*, 1589-1615.

Aebersold, R., & Mann, M., (2003). Mass spectrometry-based proteomics. *Nature, 422*(1), 198–207.

Aflori, C., & Craus, M. (2007). Grid implementation of the Apriori algorithm. *Advanced Engineering Software, 38*(5), 295-300.

AG. *The Access Grid.* Retrieved from http://www.accessgrid.org/

Agrawal, R., Mannila, H., Srikant, R., Toivonen, H., & Verkamo, A. I. (1996). Fast discovery of association rules. In , U. Fayyad et al. (Eds.), *Advances in Knowledge Discovery and Data Mining* (pp. 307-328). Menlo Park, CA: AAAI Press.

Ahrens, J., Law, C., Schroeder, W., Martin, K. & Papka, M. (2000). *A Parallel Approach for Visualizing Extremely Large, Time-Varying Datasets*(Tech. Rep. LAUR-00-1620). Los Alamos, NM: Los Alamos National Laboratory. Retrieved from http://www.vtk.org/pdf/pvtk.pdf

Aiftimiei, C., Andreozzi, S., Cuscela, G., Donvito, G., Fantinel, S., Fattibene, E., Misurelli, G., Pierro, A., Dudhalkar, V. & Maggi, G.P. (2007). Recent Evolutions of GridICE: a Monitoring Tool for Grid Systems. In C. Kesselman, J. Dongarra, D. Walker (Eds.), *International Symposium on High Performance Distributed computing (HPDC 2007)* (pp. 1-8). New York: ACM.

Ajayi, O., Sinnott, R. O., Stell, A. J. & Young, A. (2008). *Towards a Virtual Anonymisation Grid for Unified Access to Remote Clinical Data.* Proceedings of 6th International HealthGrid conference, Chicago.

Ajayi, O., Sinnott, R. O., Watt, J., Jiang, J. & Stell, A. J. (2006). *Single-Sign on and Authorization for Dynamic Virtual Organizations.* Proceedings of International Conference on Virtual Enterprises, (PRO-VE'06), Helsinki, Finland.

Aktas, M. S., Fox, G. C., & Pierce, M. (2007). Fault tolerant high performance Information Services for dynamic collections of Grid and Web services. *Future Generation Computer Systems*, 23(3), 317-337.

Alfieri, R., Cecchini, R., Ciaschini, V., dell' Agnello, L., Frohner, Á,. Gianoly, A., et al. (2003). VOMS, an Authorization System for Virtual Organizations. *1st European Across Grid Conference.*

Alfieri, R., Cecchini, R., Ciaschini, V., dell'Agnello, L., Frohner, A., Gianoli, A., et al. (2004). VOMS, an Authorization System for Virtual Organizations. In F. Fernández Rivera (Ed.), *Grid computing: first European Across Grids Conference, Santiago de Compostela, Spain, February 13-14, 2003 : revised papers. Lecture notes in computer science, 2970* (pp. xi, 328 p.). New York: Springer.

Alfieri, R., et al (2003). *VOMS: an authorization system for virtual organizations.* Proceedings of 1st European across Grids conference, Santiago de Compostela, Spain.

Ali, A. S., & Taylor, I. J. (2005). Web services composition for distributed data mining. In *The 2005 IEEE International Conference on Parallel Processing Workshops* (pp. 11-18).

Allaert, F. A. (2002). *Security standards for healthcare information systems: a perspective from the EU ISIS MEDSEC Project.* Amsterdam, the Netherlands: IOS Press.

Allcock W., Bresnahan J., Kettimuthu R., & Link M. (2005). *The Globus Striped GridFTP Framework and Server.* ACM/IEEE SC 2005 Conference (p. 54).

Allcock, B., Bester, J., Bresnahan, J., Chervenack, A. L., Foster, I., Kesselman, C., et al. (2002). Data management and transfer in high-performance computational grid environments. *Parallel Computing, 28*(5), 749-771.

Allcock, B., Bester, J., Bresnahan, J., Chervenak, A.L., Foster, I., Kesselman, C., et al. (2002). Data Management, Transfer in High Performance Computational Grid Environments. *Parallel Computing Journal, 28* (5), 749-771.

Allcock, W., Bester, J., Bresnahan, J., Chervenak, A., Liming, L., & Tuecke, S. (2002, April). *GridFTP: Protocol Extensions to FTP for the Grid.* Argonne, IL: Argonne National Laboratory.

Allcock, W., Bester, J., Bresnahan, J., Chervenak, A., Liming, L., & Tuecke, S. (2001, March). In GridFTP: Protocol Extension to FTP for the Grid. Unpublished Internet-Draft, Grid Forum.

Allcock, W., Bresnahan, J., Kettimuthu, R., & Link, M. (2005). The Globus Striped GridFTP Framework and Server. (p.54) *Proceedings of the 2005 ACM/IEEE conference on Supercomputing.* Washington, DC: IEEE Computer Society.

Almeida, J., et al (2003). EURISWEB – Web-based epidemiological surveillance of antibiotic-resistant pneumococci in Day Care Centers. *BMC Medical Informatics and Decision Making, 3*(9).

Almond, J. & Snelling, D. (1999). UNICORE: uniform access to supercomputing as an element of electronic commerce. *Future Generation Computer Systems, 15*(5-6), 539–548.

Aloisio G., Cafaro M., Fiore S., & Mirto M. (2005). ProGenGrid: A Workflow Service Infrastructure for Composing and Executing Bioinformatics Grid Services. In *18th IEEE Symposium on Computer-Based Medical Systems* (pp. 555-560). Dublin, Ireland: Trinity College..

Aloisio, G., Breton, V., Mirto, M., Murli, A. & Solomonides, T. (2008). Special section: Life science Grids for biomedicine and bioinformatics. *Future Generation Computer Systems, 23*(3), 367-370

Aloisio, G., Cafaro, M., Carteni, G., Epicoco, I., Fiore, S., Lezzi, D., Mirto, M., & Mocavero, S. (2007). The Grid Resource Broker Portal. *Concurrency and Computation: Practice and Experience, 19*(12), 1663-1670.

Aloisio, G., Cafaro, M., Carteni, G., Epicoco, I., Fiore, S., Lezzi, D., Mirto, M., & Mocavero, S. (2007). The Grid Resource Broker Portal. *Concurrency and Computation: Practice and Experience, 19*(12), 1663-1670.

Aloisio, G., Cafaro, M., Epicoco, I., Fiore, S. Lezzi, D., Mirto, M., & Mocavero, S. (2005). iGrid, a Novel Grid Information Service. In P.M.A. Sloot et al. (Eds.), *Proceedings of the First European Grid Conference (EGC) 2005*, (LNCS 3470)(pp. 506-515). Berlin, Germany: Springer-Verlag,.

Aloisio, G., Cafaro, M., Fiore, S. & Mirto, M. (2005a). ProGenGrid a Grid-enabled platform for Bioinformatics, Studies in Health Technology and Informatics, From grid to Healthgrid. In T. Solomonides et al. (Eds.), *Proceedings of HealthGrid 2005*, (Vol. 112) (pp. 113-126). Amsterdam, the Netherlands: IOS Press..

Aloisio, G., Cafaro, M., Fiore, S. & Mirto, M. (2005a). The Grid Relational Catalog Project. In L. Grandinetti (Ed.), *Advances in Parallel Computing, "Grid Computing: The New Frontiers of High Performance Computing"* (pp.129-155). Philadelphia: Elsevier.

Aloisio, G., Cafaro, M., Fiore, S. & Mirto, M. (2005). ProGenGrid a Grid-enabled platform for Bioinformatics In T. Solomonides et al. (Eds.), *Studies in Health Technology and Informatics, From grid to Healthgrid: Vol. 112*, (pp. 113-126). Amsterdam: IOS Press.

Aloisio, G., Cafaro, M., Fiore, S. & Mirto, M. (2005). The Grid Relational Catalog Project. In L. Grandinetti (Ed.), *Advances in Parallel Computing, "Grid Computing: The New Frontiers of High Performance Computing"* (pp.129-155). Philadelphia: Elsevier.

Aloisio, G., Cafaro, M., Fiore, S., & Mirto, M. (2005). ProGenGrid: a Workflow Service Infrastructure for Composing and Executing Bioinformatics Grid Services. *Proceedings of the 18th IEEE International Symposium on Computer-Based Medical Systems* (IEEE CBMS 2005), Dublin, June 23-24 2005, Ireland. Washington, DC: IEEE Computer Society.

Aloisio, G., Cafaro, M., Fiore, S., & Mirto, M. (2005, June 23-24). ProGenGrid: a Workflow Service Infrastructure for Composing and Executing Bioinformatics Grid Services. *Proceedings of the 18th IEEE International Symposium on Computer-Based Medical Systems* (IEEE CBMS 2005), Dublin, Ireland.

Aloisio, G., Cafaro, M., Fiore, S., Mirto, M. (2007). A Grid System for the Ingestion of Biological Data into a Relational DBMS. In *Proceedings of IEEE International Symposium on Bioinformatics and Life Science Computing (BLSC-07)*, (pp. 713-718), Niagara Falls, Canada.

Aloisio, G., Cafaro, M., Lezzi, D. & Van Engelen, R. (2003). Secure Web Services with Globus GSI and gSOAP. In *Euro-Par*: (LNCS 2790, pp.421-426). Berlin/Heidelberg: Springer.

Aloisio, G., Cafaro, M., Lezzi, D. & Van Engelen, R. (2005). The GSI plug-in for gSOAP: Enhanced Security, Performance, and Reliability. In *Proceedings of Information Technology Coding and Computing (ITCC 2005)*, (Vol. I), (pp. 304–309). New York: IEEE Press.

Aloiso, G., Benkner, S., Bilofsky, H., Blanquer Espert, I., Breton, V., Cannataro, M., et al. (2005). *HealthGrid White Paper*. Retrieved from http://whitepaper.healthgrid.org.

Alsairafi, S., Emmanouil, F. S., Ghanem, M., Giannadakis, N., Guo, Y., Kalaitzopoulos, D., et al. (2003). The design of discoverynet: Towards open grid services for knowledge discovery. *International Journal of High Performance Computing and Applications, 17*(3), 297-315.

Altekar, G., Dwarkadas, S., Huelsenbeck, J.P. & Ronquist, F. (2004). Parallel Metropolis coupled Markov chain Monte Carlo for Bayesian phylogenetic inference. *Bioinformatics, 20* (3), 407–415.

Altintas, I., Barney, O., Cheng, Z., Critchlow, T., Ludaescher, B., Parker, S., Shoshani, A., & Vouk, M. (2006). Accelerating the scientific exploration process with scientific workflows. *J. Phys.: Conf. Ser., 46*(1), 468+.

Altintas, I., Berkley, C., Jaeger, E., Jones, M., Ludascher, B., & Mock, S. (2004). Kepler: an extensible system for design and execution of scientific workflows. Proceedings of the 16th International Conference on Scientific and Statistical Database Management (SSDBM), Santorini Island, Greece. Kepler website at http://kepler-project.org

Altschul S. F., Madden T. L., Schaffer A. A., Zhang J., Zhang Z., Miller W., & Lipman D. J. (1997). Gapped BLAST and PSI-BLAST: a new generation of protein database search programs. *Nucleic Acids Research, 25*(17), 3389-3402.

Altschul, S.F., Gish, W., Miller, W., Myers, E.W. & Lipman, D.J. (1990). Basic Local Alignment Search Tool. *J. Mol. Biol., 215*, 403-410.

Altschul, S.F., Madden, T.L., Schaffer, A.A., et al. (1997). Gapped BLAST and PSI-BLAST: a new generation of protein database search programs. *Nucleic Acids Research*, 25(17):3389-3402.

Altschul, S.F., Madden, T.L., Schaffer, A.A., Zhang, J., Zhang, Z., Miller, W., Lip-man, D.J. (1997). Gapped BLAST and PSI-BLAST: a new generation of protein database search programs. *Nucleic Acids Res., 25*(17), 3389-402.

Altschul, S.F., Madden, T.L., Schffer, A.A., Zhang, J., Zhang, Z., Miller, W. & Lipman D.J. (1997). Gapped BLAST and PSI-BLAST: a new generation of protein database search programs. *Nucleic Acids Res., 17*(25), 3389-3402. Retrieved from http://www.ncbi.nlm.nih.gov/cgi-bin/BLAST/nph-psi.

Alur, N., et al. (2005). *Patterns: Information Aggregation and Data Integration with DB2 Information Integrator.* Armonk, NY: IBM Redbook.

Alvarez, J. (2004). High-throughput docking as a source of novel drug leads. *Curr Opin Chem Biol, 8*, 365-370.

Alves, D., Hass, V. & Caliri, A. (2003). The predictive power of R0 in an epidemic probabilistic model. *Journal of Biological Physics, 29*(1), 63.

Amazon Elastic Compute Cloud (2007). Retrieved from www.amazon.com/ec2

Ammar, R.A., Demurjian, S.A., Greenshields, I.R., Pattipatti, K. & Rajasekaran, S. (2005). Analysis of Heterogeneous Data in Ultrahigh Dimensions. In R. Popp and J. Yen (Eds.), *Emergent Information Technologies and Enabling Policies for Counter-Terrorism*, (pp. 105-124). Hoboken, NJ: Wiley Interscience.

Andersen, A. H., & Kak, C. A. (1984). Simultaneous algebraic reconstruction technique (SART): a superior implementation of the art algorithm. *Ultrasonic imaging, 6*(1), 81-94.

Anderson, D. (2006). Boinc: A System for public-resource computing and storage. *Proceedings of the 2006 ACM/IEEE conference on Supercomputing* (a. 126). New York: ACM.

Anderson, D., & Fedak, G. (2006). The Computational and Storage Potential of Volunteer Computing, In *Proceedings of the 6th IEEE/ACM International Symposium on Cluster Computing and the Grid*, (pp. 73-80. New York: IEEE Computer Society.

Anderson, D.P. (2004). BOINC: A System for Public-Resource Computing and Storage. In: *Proceedings of the 5th IEEE/ACM International Workshop on Grid Computing*, (pp. 4-10) New York: IEEE Computer Society.

Anderson, D.P., Cobb, J., Korpela, E., Lebofsky, M., & Werthimer, D. (2002). SETI@home: an experiment in public resource computing. *Communications of the ACM, 45*(11), 56-61.

Andronico, G., Ardizzone, V., Barbera, R., Catania, R., Carrieri, A., Falzone, A., et al. (2005). GILDA: the Grid INFN virtual Laboratory for Dissemination Activities. *Proceedings of First International Conference on Test-*

beds and Research Infrastructures for the Development of NeTworks and COMmunities (pp. 304-305).

Angele, J., & Weiten, M. (2005). Semantic Technologies in the SIMDAT Grid Project. In C. Goble, C. Kesselman, and Y. Sure, (Ed.), *Semantic Grid: The Convergence of Technologies, number 05271 in Dagstuhl Seminar Proceedings*. Schloss Dagstuhl, Germany: Internationales Begegnungs und Forschungszentrum fuer Informatik (IBFI).

Anjomshoaa A., Brisard F., Drescher M., Fellows D., Ly A., McGough S., Pulsipher D., Savva A. (2005). *GFD-R-P.056 Job Submission Description Language (JSDL) Specification v1.0*. Retrieved from http://forge.gridforum.org/projects/jsdl-wg.

Antonioletti, M., Atkinson, M., Baxter, R., Borley, A., Hong, N. P. C., Collins, B., et al. (2005). The design and implementation of Grid database services in OGSA-DAI. *Concurrency and Computation-Practice & Experience, 17*(2-4), 357-376.

Antonioletti, M., Atkinson, M., Baxter, R., Borley, A., Hong, N.P.C., Collins, B., et al. (2005). The design and implementation of grid database services in OGSA-DAI. *Concurrency Computation Practice and Experience*, 17(2-4), 357-376.

Arabia, M., & Akutsu, T. (1984). A New Test Circulatory System for Research in Cardiovascular Engineering. *Annals of Biomedical Engineering, 12*, 29-48.

Arbona, T., Benkner, S., Fingberg, J., Engelbrecht, G., Hof-mann, M., Kumpf, K., Lonsdale, G., & Woehrer, A. (2007). A Service-oriented Grid Infrastructure for Biomedical Data and Compute Services. *IEEE Transaction on NanoBioscience*, *6*(2), 136-141.

Arnold K., Schwede, T., & Podvinec, M. (2007). SWISS-TANDEM: A Web-based Workspace for MS/MS Protein Identification on PC Grids. In Dominique Zosso, (Ed.) *20th IEEE International Symposium on Computer-Based Medical Systems* (Vol. 1) (pp 406-411). New York: IEEE press.

Arsanjani, A. (2004). *Service-oriented modeling and architecture*, (IBM technical article). Retrieved October 20, 2008, from http://www.ibm.com/developerworks/library/ws-soa-design1/.

Artacho E., Anglada E., Di´eguez O., Gale J. D, Garc´ıa A., Junquera J., Martin R. M., Ordej´on P., Pruneda J. M., S´anchez-Portal D., & Soler J. M. (2008). The SIESTA method; developments and applicability. *J. Phys.: Condens. Matter, 20*, 064208.

Arzberger, P. W., & Papadopoulos, P. (Feb 2006). PRAGMA: Example of Grass-Roots Grid Promoting Collaborative e-Science Teams. *CTWatch Quarterly, 2* (1).

Arzberger, P., Farazdel, A., Konagaya, A., Ang, L., Shimojo, S. & Stevens, R. (2004). Life sciences and cyberinfrastructure: dual and interacting revolutions that will drive future science. *New Generation Computing, 22* (2), 97-110.

Ashburner, M., Ball, C.A., et al. (2000). Gene Ontology: tool for the unification of biology. The Gene Ontology Consortium. *Nat Genet 25*(1): 25-9.

Assel, M., & Kalyoncu, O. (2008). Dynamic Access Control Management for Distributed Biomedical Data Resources. In Cunningham, P. & Cunningham, M. (Eds.), *Collaboration and the Knowledge Economy: Issues, Applications, Case Studies* (pp. 1593-1599). IOS Press.

Astakhov, V., Gupta, A., Grethe, J., Ross, E., Little, D., Yilmaz, A., Qian, X., Santini, S., Martone, M., Ellisman, M. (2006). Semantically Based Data Integration Environment for Biomedical Research. In *Proceedings of the 19th IEEE Symposium on Computer-Based Medical Systems (CBMS'06)*. New York: IEEE Computer Society Press.

Atkins, D. (2003). *Report of Blue-Ribbon Advisory Panel on Cyberinfrastructure*. Retrieved from http://www.nsf.gov/publications

Atkins, D., Kroegemeier, K., Feldman, S., Garcia-Molina, H., Klein, M., Messerschmitt, D., et al. (2003). Revolutionizing Science and Engineering Through Cyberinfrastructure. In *Report of the National Science Foundation Blue-Ribbon Advisory Panel on Cyberinfrasturcuture*. National Science Foundation: Arlington, VA.

Atkins, D.E., Droegemeier, K.K., Feldman, S.I., Garcia-Molina, H. Klein, M.L., Messerschmitt, D.G., et al. (n.d.). *Revolutionizing Science and Engineering Through Cyberinfrastructure: Report to the National Science Foundation Blue-Ribbon Advisory Panel on*

Cyberinfrastructure. Retrieved on January 15, 2008, from http://www.communitytechnology.org/nsf_ci_report/report.pdf.

Attwood, T.K., Flower, D.R., Lewis, A. P., Mabey, J.E., Morgan, S.R., Scordis, P, Selley, J., & Wright, W. (2000). PRINTS-S: the database formerly known as PRINTS. *Nucleic Acids Research, 27,* 220-225.

Austin, J., Jackson, T., Fletcher, M., Jessop, M., Cowley, P., & Lobner, P. (2004). Predictive maintenance: distributed aircraft engine diagnostics. In Elsevier (Ed.), *The grid 2* (2nd Ed., pp. 69-79). San Francisco, CA: Morgan Kaufmann.

Axelson, O. (1999). The character and the public health implications of ecological analyses. In A. Lawson, A. Biggeri, D. Böhning, E. Lesaffre, J.-F. Viel & R. Bertollini (Ed.), *Disease Mapping and Risk Assessment for Public Health,* (pp. 301-309). New York: Wiley and Sons.

Azencott, R. (1988). Simulated Annealing. *Seminaire Bourbaki 40eme annee, 1987-88, No. 697,* (pp. 223-237).

Azencott, R.(1992). *Simulated Annealing: Parallelization Techniques.* Hoboken, NJ: Wiley.

Badia, R.M., Labarta, J.S., Sirvent, R.L., Perez, J.M., Cela, J.M., & Grima, R. (2003). Programming Grid Applications with GRID Superscalar. Journal of Grid Computing, vol. 1, pp. 151-170.

Bagnasco, A., Poggi, A., Parodi, G., & Scapolla, A.M. (2006). Exposing Measurement Instruments As Grid Services. In *Distributed Cooperative Laboratories: Networking, Instrumentation, and Measurements* (pp. 321-329). New York: Springer.

Bailey, T. C., Gatrell, A. C. (1995). *Interactive Spatial Data Analysis.* Essex, UK: Longman Scientific and Technical.

Bairoch, A., Apweiler, R., Wu, C.H., Barker, W.C., Boeckmann, B., Ferro, S. & et al (2005). The Universal Protein Resource (UniProt). *Nucleic Acids Res.,* (33), 154–159. Retrieved from http://www.uniprot.org.

Bajaj, C. and S. Cutchin (1999). Web Based Collaborative Visualization of Distributed and Parallel Simulation. In *Proc. of the 1999 Symposium on Parallel Visualization and Graphics, San Francisco,* (pp. 47-54).

Baker N. A., Sept D., Joseph S., Holst M. J., & McCammon J. A. (2001). *Electrostatics of nanosystems: application to microtubules and the ribosome.* Proc. Natl. Acad. Sci. USA 98, (pp. 10037-10041).

Baker, M., & Lakhoo, R. (2006). Application Reuse Through Portal Frameworks. *Proceedings of the UK e-Science All Hands Conference* (pp. 510-518).

Baker, N. A., Sept, D., Joseph, S., Holst, M. J., & McCammon, J. A. (2001). Electrostatics of nanosystems: application to microtubules and the ribosome. *Proceedings of the National Academy of Sciences of the United States of America 2001,* (Vol. 98), (pp. 10037-41).

Baker, S. (2007). Google and the Wisdom of Clouds. Business Week, Dec. 13. Retrieved from www.businessweek.com/magazine/content/07_52/b4064048925836.htm

Bala, P., Pytlinski, J., Nazaruk, M., Alessandrini, V., Girou, D., Erwin, D.W., et al. (2002). BioGRID: An European Grid for Molecular Biology. *11th IEEE International Symposium on High Performance Distributed Computing* (p. 412)., Edinburgh, UK: IEEE Computer Society.

Baldridge K. K., Greenberg J. P., Elbert S. T., Mock, S., & Papadopoulos, P. (2002). *QMView and GAMESS: Integration into the World Wide Computational Grid.* Supercomputing, ACM/IEEE 2002 Conference, 64.

Baldridge, K. K., & Greenberg, J. P. (1995). QMView: a computational chemistry three-dimensional visualization tool at the interface between molecules and mankind. *Journal of Molecular Graphics, 13,* 63-66).

Baldridge, K., Bhatia, K., Greenberg, J. P., Stearn, B., Mock, S., Sudholt, W., Krishnan, S., Bowen, A., Amoreira, C., & Potier, Y. (2005). In GEMSTONE: Grid Enabled Molecular Science Through Online Networked Environments. *Life Sciences Grid Workshop, Singapore.* Singapore: World Scientific Press.

Balis, B., Bubak, M., & Wach, J. (2007) User-Oriented Querying over Repositories of Data and Provenance. In G. Fox, K. Chiu, & R. Buyya, (Eds.), *Third IEEE International Conference on e-Science and Grid Computing, e-Science 2007, Bangalore, India, 10-13 December 2007,* (pp.77-84). New York: IEEE Computer Society.

Balis, B., Bubak, M., & Wach, J. (2008). Provenance Tracking in the ViroLab Virtual Laboratory. In *Proceedings of PPAM 2007, Seventh International Conference on Parallel Processing and Applied Mathematics, Lecture Notes in Computer Science*, (Vol. 4967), (pp. 381-390). New York: Springer.

Balis, B., Bubak, M., & Wach, K. (2007). User-Oriented Querying over Repositories of Data and Provenance. In *Proceedings of the 3rd IEEE International Conference on e-Science and Grid Computing, e-Science 2007*, Bangalore, India, Dec. 2007, (pp. 187-194). New York: IEEE Computer Society.

Balis, B., Bubak, M., & Wach, K.(2008). Provenance Tracking in the ViroLab Virtual Laboratory. In R. Wyrzykowski, (Ed.), *Proceedings of the Seventh International Conference on Parallel Processing and Applied Mathematics*, LNCS, Gdansk, Poland, Sept. 2007. Berlin, Germany: Springer.

Baloa, L., Boston, J., & Antaki, J. (2001). Elastance-Based Control of a Mock Circulatory System. *Annals of Biomedical Engineering, 29*, 244-251.

Banks, D. (2001). *Game Theory and. Risk Analysis for. Counterterrorism*. Retrieved November 2008 from http://www.galaxy.gmu.edu/interface/I03/I2003Proceedings/BanksDavid/Banks David.presentation.pdf

Barker, W.C., Pfeiffer, F., & George, D.G. (1996). Superfamily classification in FIR-international protein sequence database. *Methods in Enzymology, 266*, 59-71.

Barla, G. Jurman, S. Riccadonna, M. Chierici, S. Merler, & Furlanello,C. (2008). Machine learning methods for predictive proteomics. *Briefings in Bioinformatics, 9*(2), 119-128.

Barrett C. L., Eubank S. G., and Smith J. P. (2005). If Smallpox Strikes Portland. *Scientific American Magazine*. Retrieved July 2008 from http://www.sciam.com/article.cfm?id=if-smallpox-strikes-portl

Baru, C., Moore, R., Rajasekar, A., & Wan, M. (1998) The SDSC Storage Resource Broker. In *Proceedings of the 1998 Conference of the Centre for Advanced Studies on Collaborative Research* (p. 5). Toronto, Canada: IBM Press.

Baruah, S., Cohen, N., Plaxton, C. & Varvel, D. (1996). Proportionate Progress: A Notion of Fairness in Resource Allocation. *Algorithmica, 15*(6), 600-625.

Basney J., Humphrey M., & Welch V. (2005). The MyProxy online credential repository. *Software: Practice and Experience, 35*(9) 801-816.

Basney, J., Livny, M., & Tannenbaum T. (1997, June). High Throughput Computing with Condor. *HPCU news, 1*(2).

Basu, S., Talwar, V., Agarwalla, B., & Kuma, R. (2003). *Interactive Grid Architecture for Application Service Providers* (HP Technical Report HPL-2003-84R1).

Bata, P., Alessandrini, V., Girou, D., MacLaren, J., Brooke, J., Pytlinski, J., et al. (2002). BIOGRID-A European grid for molecular biology. In *Proceedings of the 11th IEEE International Symposium on High Performance Distributed Computing (HPDC-11)*, Edinburgh, UK.

Bateman, A. & Valencia, A. (2006). Structural genomics meets computational biology. *Bioinformatics, 22*, 2319.

Bateman, A., Birney, E., Durbin, R., Eddy, S.R., Finn, R.D., & Sonnhammer, E.L. (2000). The Pfam protein families database. *Nucleic Acids Research, 28*, 263-266.

Bates, D. G. A. (2003). Improving safety with information technology. *New England Journal of Medicine, 348*, 2526-2534.

Bavier M., Bowman M., Chun B., Culler D., Karlin S., Muir S., Peterson L., Roscoe T., Spalink T., & Wawrzoniak M. (2004). *Operating System Support for Planetary-Scale Network Services*. NSDI '04 Proceedings. Retrieved from http://www.cs.princeton.edu/nsg/papers/plos_nsdi_04/nsdi.pdf

Bechhofer, S., Carr, L., Goble, C., Kampa, S., & Miles-Board, T. (2002). The semantics of semantic annotation. In Meersman, R. & Tari, Z., (Ed.), *On the Move to Meaningful Internet Systems 2002: CoopIS, DOA, and ODBASE*, , London, (pp. 1152-1167). Berlin, Germany: Springer-Verlag.

Beck M., Moore T., & Plank J.S. (2002 July). An End-To-End Approach to Globally Scalable Network Storage. In *SIGCOMM '02: Proceedings of the 2002 conference on*

Applications, technologies, architectures, and protocols for computer communications, (pp. 339-346).

BEinGRID (2008). Business Experiments in Grids. Retrieved from www.beingrid.com.

Belhajjame, K., Embury, S.M., Paton, N. (2008). Automatic annotations of web services based on workflow definitions. *ACM Transactions on the Web, 2* (2), 1-34.

Bell, W.H., Cameron, D.G., Capozza, L., Millar A.P., Stockinger, K., & Zini, F. (2002 November). Simulation of Dynamic Grid Replication Strategies in OptorSim. *In Proceedings of the 3rd Int'l. IEEE Workshop on Grid Computing (Grid'2002)*, Lecture Notes in Computer Science. Berlin, Germany: Springer Verlag.

Bellotti, R., Cerello, P., Tangaro, S., Bevilacqua, V., Castellano, M., Mastronardi, G., et al. (2007). Distributed medical images analysis on a Grid infrastructure. *Future Gener. Comput. Syst., 23* (3), 475-484.

Beltrame, F., Maggi, P., Melato, M., Molinari, E., Sisto, R., & Torterolo, L. (2006). SRB Data Grid and Compute Grid Integration via the EnginFrame Grid Portal. Proceedings of the 1st SRB Workshop, 2-3 February 2006, San Diego, USA. PDF retrieved from www.sdsc.edu/srb/Workshop/SRB-handout-v2.pdf

Benson, D.A., Karsch-Mizrachi, I., Lipman, D.J., Ostell, J. & Wheeler, D.L. (2007). GenBank. *Nucleic Acids Res., 35*(Database issue), D21-5.

Bergamaschi, S., Castano, S., Vincini, M., & Beneventano, D. (2001). Semantic integration of heterogeneous information sources. *Data and Knowledge Engineering, 36*(3), 215-249.

Bergmann, J., Bott, O. J., Hoffmann, I., & Pretschner, D. P. (2005). An eConsent-based System Architecture Supporting Cooperation in Integrated Healthcare Networks. *Stud Health Technol Inform, 116*, 961-966.

Berman F. Fox G., & Hey A. (2003). *Grid Computing: Making the Global*. Hoboken, NJ: John Wiley and Sons.

Berman H. M., Westbrook J., Feng Z., Gilliland G., Bhat T. N., Weissig H., Shindyalov I. N., & Bourne P. E. (2000). The Protein Data Bank. *Nucleic Acids Research, 28*, 235-242.

Berman, F. (2001). Viewpoint: From teragrid to knowledge grid. *Commun. ACM, 44*(11), 27-28.

Berman, F. Wolski, R. & Casanova, H. (2003). Adaptive Computing on the Grid Using AppLeS. *IEEE Transactions on Parallel and Distribributed Systems, 14*(4), 369–382.

Berman, F., Fox, G. &, Hey, A.J.G. (2003). *Grid Computing: Making the Global Infrastructure a Reality.* Hoboken, NJ: John Wiley.

Berman, F., Fox, G., & Hey, T. (Ed.). (2003). *Grid computing: Making the global infrastructure a reality.* Chichester, UK: Wiley.

Berman, F., Hey, A. J. G & Fox, G. C. (Eds.), (2003). *Grid Computing: Making the Global Infrastructure a Reality.* New York: John Wiley.

Berman, H.M., Henrick, K., & Nakamura, H. (2003). Announcing the worldwide Protein Data Bank. *Nature Structural Biology, 10*(12), 980.

Berman, H.M., Westbrook, J., Feng, Z., Gilliland, G., Bhat, T.N. & et al. (2000). The Protein Data Bank. *Nucleic Acids Res.*, (28), 235–242. Retrieved from www.rcsb.org/pdb/.

Berman, H.M., Westbrook, J., Feng, Z., Gilliland, G., Bhat, T.N., Weissig, H., Shindayalov, I.N. & P.E. Bourne (2000). The Protein Data Bank. *Nucleic Acids Research, 28*, 235-242.

Berrar, D., Granzow, M., & Dubitzky, W. (2007). Introduction to Genomic and Proteomic Data Analysis. In D. Berrar, M. Granzow, & W. Dubitzky (Eds.), *Fundamentals of Data Mining in Genomics and Proteomics,* (pp. 1-37). Berlin, German: Springer Science.

Berry, D., Germain-Renaud, C., Hill, D., Pieper, F. & Saltz, J. (2004). Report on the Workshop IMAGE'03: Images, medical analysis and grid environments (TR UKeS-2004-02). UK National e-Science Centre, Glasgow, UK.

Bhandarkar M., Budescu G., Humphrey W. F., Izaguirre J. A., Izrailev S., Kalé L. V., Kosztin D., Molnar F., Phillips J. C., & Schulten K. (1999). BioCoRE: A collaboratory for structural biology. In A. G. Bruzzone, A. Uchrmacher, & E. H. Page (Eds.), *Proceedings of the*

SCS International Conference on Web-Based Modeling and Simulation (pp. 242-251).

Bhatia, K., Chandra, S., & Mueller, K. (2006). *In GAMA: Grid Account Management Architecture.* 1st IEEE International Conference on e- Science and Grid Computing, Melbourne, Australia.

Bhuiya, A. K. & Stanley, E. (1963). The refinement of atomic parameters by direct calculation of the minimum residual, *Acta Crystallogr, 16*, 981-984.

Bidgood, W.D., Horii, S.C., Prior, F.W., & Van Syckle, D.E. (1997). Understanding and Using DICOM, the Data Interchange Standard for Biomedical Imaging. *Journal of the American Medical Informatics Association, 4*, 199-212.

Bilbao-Castro, J. R., Carazo, J. M., García, I., & Fernandez, J. J. (2006). Parallelization of Reconstruction Algorithms in Three-Dimensional Electron Microscopy. *Applied Mathematical Modelling, 30*(8), 688-701.

Bilbao-Castro, J. R., García, I., & Fernández, J. J. (2007). EGEETomo: a user-friendly, fault-tolerant and grid-enabled application for 3D reconstruction in electron tomography. *Bioinformatics, 23*(24), 3391-3393.

Bilbao-Castro, J. R., Merino, A. J., García, I., Carazo, J. M., & Fernández, J. J. (2007b). Parameter optimization in 3D reconstruction on a large scale grid. *Parallel Computing, 33*(4-5), 250-263.

Bilgin, A., Zweig, G., & Marcellin, M.W. (2000). Three-dimensional image compression with integer wavelet transforms. *Applied Optics, 39*, 1799-1814.

Bilykh, I., Bychkov, Y., Dahlem, D., Jahnke, J.H., McCallum, G., Obry, C., et al. (2003). Can grid services provide answers to the challenges of national health information sharing? *In Proceedings of the 2003 conference of the Centre for Advanced Studies on Collaborative research (CASCON '03)* (pp. 39-53). Armonk, NY: IBM Press.

BioWBI (2004). *Bioinformatic Workflow Builder Interface.* Retrieved from http://www.alphaworks.ibm.com/tech/biowbi

BIRN (2008). Biomedical Informatics Research Network. Retrieved from www.nbirn.net/index.shtm

BIRN. (2000). *Biomedical Informatics Research Network.* Retrieved November 2008 from http://www.nbirn.net/

Biton, O., Cohen-Boulakia, S. & Davidson, S. (2007). Zoom* userviews: Querying relevant provenance in workflow systems. In *VLDB* (pp. 1366–1369).

Biton, O., Cohen-Boulakia, S., Davidson, S. & Hara, C. (2008) Querying and managing provenance through user-views in scientific workflows. In *ICDE*. Los Alamitos, CA: IEEE.

Black, D., Morris S. N., Smith, C., Townsend, P. (1982). The pattern of present health inqualities. In P. Townsend & N. Davidson (orgs.), *Inequalities in health: the black report* (pp. 51-64). New York: Penguin Books.

Blackledge, T.A., Gillespie, R.G. (2004). Convergent evolution of behavior in an adaptive radiation of Hawaiian web-building spiders. *Proc Natl Acad Sci USA, 101*(46):16228-33.

Blanchet, C., Lefort, V., Combet, C. & Deléage, G. (2006). GPS@ bioinformatics portal: from network to EGEE grid. *Stud Health Technol Inform, 120*, 187-193.

Blanquer, I., Hernandez, V., & Segrelles, D. (2006). TRENCADIS – A Grid Architecture for Creating Virtual Repositories of DICOM Objects in an OGSA-Based Ontological Framework. In *Biological and Medical Data Analysis* (pp. 183-194).

Blessing, R. H. & Smith, G. D. (1999). Difference Structure Factor Normalization for Determining Heavy-Atom or Anomalous Scattering Substructures, *J. Appl. Cryst. 32*, 664-670.

Bobashev, G. V., Goedecke, D. M., Yu, F., Epstein, J. M. (2007). A Hybrid Epidemic Model: Combining the Advances of Agent-Based and Equation-Based Approaches. In *Proceedings of the 2007 Winter Simulation Conference*. New York: IEEE Press.

Boccia, V., Guarracino, M.R., D'Amore, L. & Laccetti, G. (2005). A grid enabled PSE for medical imaging: experiences on MedIGrid. In *Proceedings of the 18th IEEE Symposium on computer-based medical systems* (pp. 529-536). Washington, DC: IEEE Computer Society.

Bodenreider, O. (2004) The Unified Medical Language System (UMLS): integrating biomedical terminology. *Nucleic Acids Research, 32*, D267-D270.

Boguski, M.S., Lowe T. M. & Tolstoshev, C.M. (1993). dbEST- database for "expressed sequence tags." *Nat. Genet., 4*(4), 332-333.

Bohr, J. et al. (1993). Protein structures from distance inequalities. *J. Mol. Biol., 231*, 861-869.

Boinc (2008). Retreived March 2008 from http://boinc.berkeley.edu/

BoincGamess (2008). Retreived March 2008 from http://cvs.rocksclusters.org/viewcvs/viewcvs.cgi/gamess/boincgamess-doc/

Bolze, R. Cappello, F., Caron, E., Daydé, R., Desprez, F., & Jeannot, E., et al. (2006) Grid'5000: a large scale and highly reconfigurable experimental Grid testbed. *International Journal of High Performance Computing Applications, 20*(4), 481-494.

Bolze, R., Cappello, F., Caron, E., Daydé, M., Desprez, F., Jeannot, E., et al. (2006). Grid'5000: A Large Scale and Highly Reconfigurable Experimental Grid Testbed. *International Journal of High Performance Computing Applications, 20*(4) 481-494.

Bonabeau, E., Dorigo, M., & Theraulaz, G. (Eds.). (1999). *Swarm Intelligence: From Natural to Artificial Systems*, New York, USA: Oxford University Press.

Bond, P. J., & Sansom, M. S. P. (2006). Insertion and assembly of membrane proteins via simulation. *Journal of the American Chemical Society, 128*, 2697-2704.

Bond, P. J., Derrick, J. P., & Sansom, M. S. P. (2007). Membrane simulations of OpcA: Gating in the loops? *Biophysical Journal, 92*, L23-L25.

Borking, J. J. (2005, October 3, 2005). *Privacy Standards for Trust.* Paper presented at the 27th International Conference on Privacy and Personal Data Protection, Montreux, Switzerland.

Boser, B. E., Guyon, I. M., & Vapnik, V. N. (1992). A training algorithm for optimal margin classifiers. *In Proceedings of the fifth annual workshop on Computational learning theory* (pp. 144-152). New York: ACM.

Bowers, S., McPhillips, T. M., Ludascher, B., Cohen, S., & Davidson, S. B. (2006). A Model for User-Oriented Data Provenance in Pipelined Scientific Workflows. In L. Moreau & I. T. Foster, (Eds.), *Provenance and Annotation of Data, International Provenance and Annotation Workshop, IPAW 2006, Lecture Notes in Computer Science*, (Vol. 4145) (pp. 133-147). Chicago, IL: Springer.

Brady, J. M., Gavaghan, D. J., Simpson, A. C., Parada, M. M. & Highnam, R. P. (2003). eDiaMoND: A grid-enabled federated database of annotated mammograms. In Berman, F., Fox, G. C. & Hey, A. J. C. (Eds.). *Grid computing: making the global infrastructure a reality*, (3rd Ed.) (pp. 923-943). Hoboken, NJ: Wiley.

Brandon D., Vandivort K., Brunner R. & Bach M. (2005). *2005 BioCoRE Survey Report.* Retrieved from http://www.ks.uiuc.edu/Research/biocore/eval/surveyData/BioCoRE_2005_Survey_Report%20.pdf

Braun, R.C., Pedretti, K.T., Casavant, T.L., Scheetz T.E., Birkett, C.L., & Roberts, C.A. (2001). Parallelization of Local BLAST Service on Workstation Clusters. *Future Genereration Computers Systems, 17*(6), 745-754.

Brenner, S., Chothia, C., & Hubbard, T. (1998). Assessing sequence comparison methods with reliable structurally identified distant evolutionary relationships. *Biochemistry, 95*(11) 6073-6078.

Breton V., Kasam V., & Jacq N. (2006). *High Throughput Grid Enabled Virtual Screening.* Proceedings of the NETTAB2006: Santa Margherita, 14–18.

Breton, V. (2007). A perspective on biomedical applications on Grids. In *eScience Conference 2007.*

Breton, V., Blanchet, C., Legré, Y., Maigne, L. and Montagnat, J. (2005). Grid Technology for Biomedical Applications. In M. Daydé et al. (Eds.), *VECPAR 2004*, (LNCS 3402), (pp. 204–218).

Breton, V., Dean, K. & Solomonides, T. (2005). The HealthGrid White Paper. In *Proceedings of Third HealthGrid Conference: Studies in Health Technology and Informatics*, Vol. 112. Amsterdam, The Netherlands: IOS Press (pp 249–321). Retrieved from http://www.healthgrid.org

Breton, V., Dean, K. & Solomonides, T. (2005). The HealthGrid white paper. *Studies in Health Technology and Informatics, 112*, 249-321.

Breu, H. & Kirkpatrick, D.G. (1998). Unit disk graph recognition is NP-hard. *Computational Geometry, 9*, 3-24.

Brevik, J.,Nurmi, D. & Wolski, R. (2006, March*). Predicting bounds on queuing delay for batch-scheduled parallel machines.* In *11th ACM SIGPLAN symposium on principles and practice of parallel programming* (pp 110–118). New York, NY: ACM Press.

Brezany, P., Hofer, J., Tjoa, A., & Wohrer, A. (2003). Gridminer: An infrastructure for data mining on computational grids. *In Data Mining on Computational Grids APAC'03.*

Brin, S., Motwani, R., Ullman, J. D., & Tsur, S. (1997). Dynamic itemset counting and implication rules for market basket data. In J. Peckham (Ed.), *International Conference on Management of Data* (pp. 255-264). New York: ACM Press.

Brooks, B. R., Bruccoleri, R. E., Olafson, B. D., States, D. J., Swaminathan, S., & Karplus, M. (1983). CHARMM: a program for macromolecular energy, minimization, and dynamics calculations. *Journal of Computational Chemistry, 4*, 187–217.

Brooks, D. R., & McLennan D. A. (2002). *The Nature of diversity: An evolutionary voyage of discovery.* Chicago, IL: University of Chicago Press.

Brownstein, J.S., Freifeld, C.C. (2007). HealthMap: the development of automated real-time internet surveillance for epidemic intelligence. *Euro Surveill 12: E071129 071125.* Retrieved February 28, 2008 from http://www.eurosurveillance.org/ew/2007/071129.asp#5.

Bruno, G., Katz, M., Sacerdoti, F. D., & Papadopoulos, P. M. (2004) *In Rolls: Modifying a Standard System Installer to Support User-Customizable Cluster Frontend Appliances.* IEEE International Conference on Cluster Computing, San Diego, CA (pp 421-430).

Bubak, M., Gubala, T., Kapalka, M., Malawski, M., and Rycerz, K. (2005). Workflow composer and service registry for grid applications. *Future Generation Computer Systems, 21*(1), 79-86.

Bubak, M., Gubala, T., Kasztelnik, M., Malawski, M., Nowakowski, P. & Sloot, P. M. A. (2007). Collaborative virtual laboratory for e-health. In *Expanding the Knowledge Economy: Issues, Applications, Case Studies, eChallenges e-2007 Conference Proceedings* (pp. 537-544). Amsterdam, the Netherlands: IOS Press.

Budura, A., Cudré-Mauroux, P., & Aberer, K. (2007). From bioinformatic web portals to semantically integrated Data Grid networks. *Future Gener. Comput. Syst., 23* (3), 485-496.

Buetow, K.H. (2003). Cyberinfrastructure: Empowering a "Third Way" in Biomedical Research. *Science*, 44, 417.

Bulger, E. & Nathens, A. (2002). Management of severe head injury: institutional variations in care and effect on outcome. *Critical Care Med, 30*, 1870-1876.

Bundesministerium für Bildung und Forschung (BMBF). (2008). *3. D-Grid call: BMBF-Förderbekanntmachung „Grid-Dienste für Wirtschaft und Wissenschaft".* Retrieved May 26, 2008, from http://www.bmbf.de/foerderungen/12378.php

Buneman, P., Chapman, A. & Cheney, J. (2006) Provenance management in curated databases. In Chaudhuri, et al.

Burge, C. & Karlin, S. (1997). Prediction of complete gene structures in human genomic DNA. *J. Mol. Biol., 268*, 78-94.

Burke, S. Campana, S., Mendez Lorenzo, P., Nater C., Santinelli, R., Sciabà, A. (2008). gLite 3.1 User Guide. Retrieved June 6, 2008, from https://edms.cern.ch/file/722398//gLite-3-UserGuide.pdf

Burke, S., Campana, S., Delgado Peris, A., Donno, F., Mendez Lorenzo, P., Santinelli, R., & Sciaba, A. gLite 3 User Guide. *gLite 3.1 User Guide Manual Series.*

Burla, M.C., Camalli, M., Cascarano, G., Giacovazzo, C., Polidori, G., Spagna, R. & Viterbo, D. (1989). SIR88 - a direct-methods program for the automatic solution of crystal structures. *J. Appl. Cryst. 22*, 389-393.

Burnum, J.F. (1989). The misinformation era; the fall of the medical record. *Annals of Internal Medicine, 110*(6), 482-484.

Burrage, K., Hood, L. & Ragan, M. A. (2006). Advanced computing for systems biology. *Brief Bioinform, 7*(4), 390-398.

Butler, D. (2006) Disease surveillance needs a revolution. *Nature, 440*(March 2), 6-7.

Butte, A. J., & Kohane, I. S. (2006). Creation and implications of a phenome-genome network. *Nat Biotechnol, 24*(1), 55-62.

Buyya, R., Abramson, D., & Giddy, J. (2000). Nimrod/G: an architecture for a resource management and scheduling system in a global computational grid", Proceedings of the 4th International Conference on High Performance Computing in the Asia-Pacific Region. www.cssc.monash.edu.au/~davida/nimrod/nimrodg.htm

Caas, D. Lpez-Ferrer, A. Ramos-Fernndez, E. Camafeita, & E. Calvo. (2006). Mass spectrometry technologies for Proteomics. *Briefing in Functional Genomics and Proteomics, 4*(4), 295–320.

Cafaro, M., Epicoco, I., Mirto, M., Lezzi, D. & Aloisio, G. (2008). The Grid Resource Broker Workflow Engine. *Concurrency and Computation: Practice and Experience, Special Issue: 2nd International Workshop on Workflow Management and Applications in Grid Environments (WaGe2007), 20*(15), 1725-1739.

Calanducci, A., Cherubino, C., Ciuffo, L.N., & Scardaci, D. (2007). A Digital Library Management System for the Grid. *Proceedings of the 16th IEEE International Workshops on Enabling Technologies: Infrastructure for Collaborative Enterprises* (pp. 269-272). Washington, DC: IEEE Computer Society.

Callahan, S.P., Freire, J., Santos, E., Eduardo Scheidegger, C., Silva, C.T. & Vo, H.T. (2006) Vistrails: visualization meets data management. In Chaudhuri, et al. (pp. 745–747).

Cameron G., (2003). Bioinformatics and Genomics Grid for European Research (BIGGER). In *Proceedings of the First European HealthGrid Conference (HealthGrid'03)*, Lyon, France.

Camon, E., Barrell, D., et al. (2004). The Gene Ontology Annotation (GOA) Database--an integrated resource of GO annotations to the UniProt Knowledgebase. *In Silico Biol 4*(1): 5-6.

Camon, E., Magrane, M., et al. (2004). The Gene Ontology Annotation (GOA) Database: sharing knowledge in Uniprot with Gene Ontology. *Nucleic Acids Res, 32*, D262-6.

Cannata, N., Merelli, E. & Altman, R., (2005). Time to organize the bioinformatics resourceome. *PLoS Computational Biology, 1*(7), e76.

Cannataro, M. (2008). Computational proteomics: management and analysis of proteomics data. *Briefings in Bioinformatics, 9*(2), 9-10.

Cannataro, M Barla, A. Flor, R. Jurman, G. Merler, S. Paoli, S. Tradigo, G. Veltri, P & Furlanello, C (2007). A grid environment for high-throughput proteomics. *IEEE Transaction on NanoBiosciences, 6*(2),117-123.

Cannataro, M & Veltri, P. (2007). MS-Analyzer: Composing and Executing Preprocessing and Data Mining Services for Proteomics Applications on the Grid. *Concurrency and Computation: Practice and Experience, 19*(15), 2047-2066,

Cannataro, M Tradigo, G. & Veltri, P. (2007). Sharing Mass Spectrometry Data in a Grid-based Distributed Proteomics Laboratory. *Information Processing & Management, 43*(3), 577-591

Cannataro, M. & Talia, D. (2003). The knowledge grid: An architecture for distributed knowledge discovery. *Commun. ACM, 46*(1), 89-93.

Cannataro, M. (2006). Next-generation Grids: requirements and knowledge-based services: Research Articles, *Concurrency and Computation: Practice & Experience, 18*(8), 887-898.

Cannataro, M., Barla, A., Flor, R., Jurman, G., Merler, S., Paoli, S., Tradigo, G., Veltri, P. & Furlanello, C. (2007). A grid environment for high-throughput proteomics. *IEEE Trans Nanobioscience, 6*(2),117-123.

Cannataro, M., Comito, C., Lo Schiavo, F., & Veltri, P. (2004). *Proteus, A Grid Based Problem Solving for Bioinformatics: Architecture and Experiments.* IEEE Intelligent Informatics Bulletin, *3*(1), 7-18.

Cannataro, M., Comito, C., Lo Schiavo, F., & Veltri, P. (2004, February). Proteus, a grid based problem solving environment for bioinformatics: Architecture and experiments. *IEEE Computational Intelligence Bulletin*, 3(1), 7–18.

Cannataro, M., Cuda, G., Gaspari, M., Greco, S., Tradigo, G., & Veltri, P. (2007). The EIPeptiDi Tool: Enhancing

Peptides Discovering in ICAT-based LC MS/MS Experiments. *BMC Bioinformatics, 8*(255).

Cannataro, M., Guzzi, P. H., Mazza, T., Tradigo, G., and Veltri, P. (2007). Using ontologies for preprocessing and mining spectra data on the grid. *Future Generation Computer Systems, 23*(1), 55-60.

Cannataro, M., Guzzi, P.H., Mazza, T., Tradigo, G., & Veltri, P. (2006). Managing Ontologies for Grid Computing, Multiagent and Grid Systems, *2*(1), pp. 29-44.

Cannataro, M., Guzzi, P.H., Mazza, T., Tradigo, G., & Veltri, P. (2007). Using Ontologies for Preprocessing and Mining Spectra Data on the Grid. *Future Generation Computer Systems, 23*(1), 55-60

Capriotti, E., Fariselli, P. & Casadio, R. (2005). I-Mutant2.0: predicting stability changes upon mutation from the protein sequence or structure. *Nucleic Acids Research,* (33), 303–305.

Cardoso, J. (2007). The semantic web vision: Where are we? *IEEE Intelligent Systems, 22*(5), 84-88.

Carey, L., Darling, A. E., & Chun Feng, W. (2003). The Design, Implementation, and Evaluation of mpiBLAST. In *Proceedings of ClusterWorld 2003*. San Jose, CA: IEEE.

Caron, E., & Desprez, F. (2006) DIET: A Scalable Toolbox to Build Network Enabled Servers on the Grid. *International Journal of High Performance Computing Applications, 20*(3), 335-352.

Caron, E., Chis, A., Desprez, F., & Su, A. (2008, January). Design of Plug-In Schedulers for a GridRPC Environment. *Future Generation Computer Systems, 24*(1), 46-57.

Carota, L., Bartoli, L., Fariselli, P., Martelli, P.L., Montanucci, L., Maggi G. & Casadio, R. (2008, May). *High Throughput Comparison of Prokaryotic Genomes* (LNCS 4967, pp 1200-1209). Berlin/Heidelberg: Springer.

Carrasco, R. C., & J. Oncina, J. (1994). Learning stochastic regular grammar by means of state merging method, In *Proceedings of the 2nd International Colloquium on Grammatical Inference (ICGI '94)*, Alicante, Spain. (Lecture Notes in Artificial Intelligence LNAI 862), (pp. 139 – 152). Berlin, Germany: Springer - Verlag.

Carvajal-Schiaffino, R., Cameron, D.G., Millar, A.P., Nicholson, C., Stockinger, K. & Zini F. (2003, November). Evaluating Scheduling and Replica Optimisation Strategies in OptorSim. In *4th International Workshop on Grid Computing (Grid2003)*. New York: IEEE Computer Society Press.

Castrignanò, T., Canali, A., Grillo, G., Liuni, S., Mignone, F. & Pesole, G (2004). CSTminer: a web tool for the identification of coding and noncoding conserved sequence tags through cross-species genome comparison. *Nucleic Acids Res., 32*(Web Server issue),W624-7.

Catlett C. (2002). The philosophy of TeraGrid: Building an open, extensible, distributed TeraScale facility. In *Cluster Computing and the Grid 2nd IEEE/ACM International Symposium CCGRID2002*.

Catlett, C. et al (2007). TeraGrid: Analysis of Organization, System Architecture, and Middleware Enabling New Types of Applications. In L. Grandinetti (Ed.), *Advances in Parallel Computing*. Amsterdam, the Netherlands: IOS Press.

CDC (2007). *Center for Disease Control of Atlanta*. Retrieved November 2008 from http://www.cdc.gov/

CDI (2007). *Cyber-Enabled Discovery and Innovation (CDI), Computer and Information Science and Engineering Directorate, National Science Foundation*. Paper presented at the National Science Foundation.

CDO2 (2008). CDOSheet for Pricing and Risk Analysis. Retrieved from www.cdo2.com

Chadwick, D. W., & Otenko, A. (2002). The PERMIS X.509 role based privilege management infrastructure. *Proceedings of the seventh ACM symposium on Access control models and technologies*. Monterey, CA: ACM.

Chadwick, D. W., Otenko, O. (2002). The PERMIS X.509 Role Based Privilege Management Infrastructure. *Future Generation Computer Systems, 936*, 1–13.

Chadwick, D., & Otenko, O. (2003). *A Comparison of the Akenti and PERMIS Authorization Infrastructures in Ensuring Security in IT Infrastructures*. Paper presented at the Proceedings of the ITI First International Conference on Information and Communications Technology (ICICT 2003) Cairo University, Egypt.

Chakrabarti, A. (2007). *Grid computing security*. New York: Springer.

Chakrabarti, A., Dheepak, R.A., & Sengupta S. (July 2004). *Integration of Scheduling and Replication in Data Grids* (Tech. Rep. TR-0407-001). New York: Infosys Tech. Ltd.

Chang, C.-S., DeTitta, G., Hauptman, H., Jones, R., Miller, R., Thuman, P., & Weeks, C. (1993). Solving the phase problem of x-ray crystallography on parallel machines. In R.F. Sincovec, D.E. Keyes, M.R. Leuze, L.R. Petzold, and D.A. Reed, (Eds.) *Proceedings of The Sixth SIAM Conference on Parallel Processing for Scientific Computing*, (pp. 304-307).

Chang, C.-S., DeTitta, G.T., Miller, R. & Weeks, C.M. (1994). On the application of parallel genetic algorithms in x-ray crystallography. *Proceedings of the 1994 Scalable High Performance Computing Conference*, (pp. 796-802). New York: IEEE Computer Society Press.

Channabasavaiah, K., Holley, K., & Tuggle, E. (2003). Migrating to a service-oriented architecture. *IBM DeveloperWorks*. Retrieved from http://www-128.ibm.com/developerworks/library/ws-migratesoa/

Charleston, M.A., & Perkins L. (2006). Traversing the tangle: Algorithms and applications for cophylogenetic studies. *Journal of Biomedical Informatics*, 39 (2006):62-71.

Charnock, S.J., Brown, I.E., Turkenburg, J.P., Black, G.W., Davies, G.J. (2002). Convergent evolution sheds light on the anti-beta-elimination mechanism common to family 1 and 10 polysaccharide lyases. *Proc Natl Acad Sci USA, 99*(19),12067-72.

Chaubal, Ch. (2003). Sun Grid Engine, Enterprise Edition—Software Configuration Guidelines and Use Cases. Sun Blueprints. www.sun.com/blueprints/0703/817-3179.pdf

Chaudhuri, S., Hristidis, V. & Polyzotis, N. (2006) *Proceedings of the ACM SIGMOD International Conference on Management of Data, Chicago, Illinois, USA, June 27-29, 2006*. New York: ACM.

Cheeseman, P., & Stutz, J. (1996). Bayesian classification (autoclass): Theory and results. *In Advances in Knowledge Discovery and Data Mining* (pp. 153-180).

Chen, C., Schmidt, B., Liu, W., & Mueller-Wittig, K.W. (2008). GPU-MEME: Using Graphics Hardware to Accelerate Motif Finding in DNA Sequences. *Lecture Notes in Bioinformatics, 5295*, 448-459.

Chen, L., Reddy, K., & Agrawal, G. (2004). Gates: A grid-based middleware for processing distributed data streams. *In the 13th International Symposium on High Performance Distributed Computing* (pp. 192-201). New York: IEEE Computer Society.

Cheng, Y., Zak, O., Aisen, P., Harrison, S., & Walz, T. (2004). Structure of the human transferrin receptor-transferrin complex. *Cell, 116*(1), 565-576.

Chervenak, A., Foster, I., Kesselman, C., Salisbury, C., & Tuecke, S. (2001). The Data Grid: Towards an Architecture for the Distributed Management and Analysis of Large Scientific Datasets. *Journal of Network and Computer Applications, 23*, 187-200.

Cheung, D. & Xiao, Y. (1998). Effect of data skewness in parallel mining of association rules. *In the Second Pacific-Asia Conference on Research and Development in Knowledge Discovery and Data Mining* (pp. 48–60). Berlin: Springer-Verlag.

Childers, L., Disz, T., Olson, R., Papka, M.E., Stevens, R. and T. Udeshi (2000). Access Grid: Immersive Group-to-Group Collaborative Visualization. In *Proc. 4'th International Projection Technology Workshop*. Retrieved from http://www-unix.mcs.anl.gov/fl/publications/childers00.pdf

Cho, S., Kim, D., & Pearlman, W.A. (2004). Lossless compression of volumetric medical images with improved three-dimensional SPIHT algorithm. *Journal of Digital Imaging, 17*, 57-63.

Cho, V., & Wüthrich, B. (2002). Distributed mining of classification rules. *Knowledge Information Systems, 4*(1), 1-30.

Chor, B., & Tuller, T. (2005). Maximum likelihood of evolutionary trees: hardness and approximation. *Bioinformatics*, 21(1):97-106.

CHRONOS Portal (2004). Retrieved from http://portal.chronos.org/gridsphere/gridsphere

Chun B., Culler D., Roscoe T., Bavier A., Peterson L., Wawrzoniak M., & Bowman M. (2003). PlanetLab: An

Overlay Testbed for Broad-Coverage Services. *ACM Computer Communications Review, 33*(3).

Churches, D., Gombas, G., Harrison, A., Maassen, J., Robinson, C., Shields, M., Taylor, I., & Wang, I. (2005). Programming scientific and distributed workflow with triana services. *Concurrency and Computation: Practice and Experience, 18*(10), 1021-1037.

Ciepiela, E., Kocot, J., Gubala, T., Malawski, M., Kasztelnik, M., & Bubak, M. (2007). Virtual laboratory engine – gridspace engine. In Bubak, M., Turala, M., and Wiatr, K., (Eds.), *Proceedings of Cracow Grid Workshop - CGW'07,* Krakow, Poland, (pp. 53-58). ACC CYFRONET AGH.

Cirne, W., Paranhos, D., Costa, L., Santos-Neto, E., Brasileiro, F., Sauve, J., Silva, F.A.B., Barros, C. O. & Silveira, C. (2003) Running Bag-of-Tasks Applications on Computational Grids: The MyGrid Approach. In *Proc. International Conference on Parallel Processing (ICPP'03)*, pp.407-416.

Clark, T. (2003). Editorial: Identity and interoperability in bioinformatics. *Briefings in Bioinformatics, 4*(1),4–6.

Clementi, L., Ding, Z., Krishnan, S., Wei, X., Arzberger, P. W., & Li, W. W. (2007). *In Providing dynamic virtualized access to grid resources via the web 2.0 paradigm.* Grid Computing Environments Workshop (Supercomputing 2007), Reno, NV.

CloudCamp (2008). Retrived from http://www.cloudcamp.com/

CNRS – IBCP (2006). *Grid Genomic Web Portal.* Retrieved November 2008 from gpsa.ibcp.fr.

CODESSA PRO (2008). Retrieved March 2008 from http://www.codessa-pro.com/

Cohen, W. W. (1995). Fast effective rule induction. In A. Prieditis and S. Russell (Eds.), *12th International Conference on Machine Learning,* (pp 115-123). San Francisco, CA: Morgan Kaufmann.

Coiera E. (2003). *Guide to health informatics.* London: Hodder Arnold.

Colacino, F.M., Arabia, M., Danieli, G.A., Moscato, F., Nicosia, S., Piedimonte, F., et al. (2005). Hybrid test bench for evaluation of any device related to mechanical cardiac assistance. *The International Journal of Artificial Organs, 28*(8), 817-826.

Colacino, F.M., Moscato, F., Piedimonte, F., Danieli, G., Nicosia, S., & Arabia, M. (2008). A modified elastance model to control mock ventricles in real-time: numerical and experimental validation. *Asaio J, 54*(6), 563-573.

Colinge, J., Masselot A., Giron, M., Dessingy, T., & Magnin, J. (2003). OLAV: towards high-throughput tandem mass spectrometry data identification. *Proteomics, 3*(8), 1454-1463.

Collobert, R. , Bengio, S., & Bengio, Y. (2002). A parallel mixture of svms for very large scale problems. In *Advances in Neural Information Processing Systems.* Cambridge, MA: MIT Press.

Congiusta, A., Talia, D., & Trunfio, P. (2007). Distributed data mining services leveraging WSRF. *Future Generation Computing Systems, 23*(1), 34-41.

Corcho, O., Alper, P., Kotsiopoulos, I., Missier, P., Bechhofer, S., & Goble, C. (2006). An overview of s-ogsa: A reference semantic grid architecture. *Web Semantics: Science, Services and Agents on the World Wide Web, 4*(2), 102-115.

Corpet, F., Gouzy, J., & Kahn, D. (2000). ProDom and ProDom-CG: tools for protein domain analysis and whole genome comparisons. *Nucleic Acids Research, 28*, 267-269.

Council of Europe (1997). *Recommendation No. R (97) 5 on the Protection of Medical Data of Feb. 13, 1997.*

Covitz, P.A., Hartel, F., Schaefer, C., De Coronado, S., Fragoso, G., Sahni H., Gustafson, S., & Buetow K.H. (2003). *caCORE: A Common Infrastructure for Cancer Informatics.*

Craig, R. & Beavis, R. C. (2004). TANDEM: matching proteins with tandem mass spectra. *Bioinformatics, 20*(9), 1466-1467.

Craig, R., Cortens, J.P., & . Beavis, R.C, (2004). Open source system for analyzing, validating, and storing protein identification data. *Journal of Proteome Research 3*(6) 1234–1242.

Crawfis, R., Max, N. and B. Becker (1994). Vector Field Visualization. *IEEE Computer Graphics and Applications, 14 (*5), 50-56.

Crespo, A., & Garcia-Molina, H. (2002). *Routing indices for peer-to-peer systems*. Paper presented at the 22nd International Conference on Distributed Computing Systems (ICDCS 2002), Wien, Austria.

Crippen, G. M. (1981). *Distance Geometry and Conformational Calculations*. New York: Research Studies.

Crompton, S. Y., Matthews, B. M., Gray, W. A., Jones, A. C., White, R. J. & Pahwa, J. S. (2006). OGSA-DAI and bioinformatics grids: Challenges, experience and strategies. In *Proceedings of the 6th IEEE International Symposium on Cluster Computing and the Grid (CCGRID 2006)*, (pp. 193-200). Washington, DC: IEEE Computer Society.

CrossGrid.org. (2008). *CrossGrid project homepage*. Retrieved March 8th, 2008, from http://www.crossgrid.org/main.html

Cui, Y. & Widom, J. (2003). Lineage tracing for general data warehouse transformations. *VLDB J*, *12*(1),41–58.

Culler, G.J. & Fried, B.D. (1963). An On-Line Computing Center for Scientific Problems. In *Proc. 1963 Pacific Computer Conf.*, (pp. 221–242). , Piscataway, NJ: IEEE.

Cunha, J. C. (2003). *Future Trends in Distributed Applications and PSEs*. Talk held at the Euresco Conference on Advanced Environments and Tools for High Performance Computing, Albufeira, Portugal.

CVIMO (2007). Retrieved July 2008 from http://www.grycap.upv.es/cvimo

Dasgupta, P. (2004). *Intelligent Agent Enabled P2P Search Using Ant Algorithms*, Paper presented at the 8th International Conference on Artificial Intelligence, Las Vegas, USA.

Data, A. (2006). Multiple Sequence Alignment in Parallel on a Cluster of Workstations. In A. Zomaya (Ed.), *Parallel Computing for Bioinformatics and Computational Biology*, (pp. 193-210). Hoboken, NJ: Wiley.

DATAGRID. The European DataGrid Project. Retrieved from http://www.eu-datagrid.org

DataTag (2008). *DataTag project description*. Retrieved March 8th 2008, from http://datatag.web.cern.ch/datatag/project.html

Date, C. J. (2000). *An Introduction to database systems*. Reading, MA: Addison-Wesley.

DDQB: IBM tool designed to help medical providers and researchers perform complex search queries without programming knowledge or IT backgrounds. Retrieved 2006, from http://www-03.ibm.com/systems/deepcomputing/solutions/ddqb.html.

De Beenhouwer, J., Staelens, S., Vandenberghe, S., & Lemahieu, I. (2008). Acceleration of GATE SPECT simulations. *Med Phys, 35*(4), 1476-1485.

De Carvalho, R.L., & Lifschitz, S. (2003). Database Allocation Strategies for Parallel BLAST Evaluation on Clusters. *Distributed and Parallel Databases, 13*, 99-127.

de Groot, B. L., van Aalten, D. M. F., Scheek, R. M., Amadei, A., Vriend, G., & Berendsen, H. J. C. (1997). Prediction of protein conformational freedom from distance constraints. *Proteins, 29*, 240-251.

de la Calle, G., Maojo, V., & Benito, M. (2006). The INFOBIOMED Network of Excellence: facilitating training and mobility in biomedical informatics in Europe. *Stud Health Technol Inform, 124*, 893-898.

De Roure, D., & Frey, J. (2007). Three Perspectives on Collaborative Knowledge Acquisition in e-Science. In *Proceedings of Workshop on Semantic Web for Collaborative Knowledge Acquisition* (SWeCKa) 2007, Hyderabad, India.

De Roure, D., Jennings, N.R., & Shadbolt, N. (2003): The Semantic Grid: A future e-Science infrastructure. In *Grid Computing - Making the Global Infrastructure a Reality*, (pp. 437-470). Hoboken, NJ: John Wiley and Sons Ltd.

De Roure, D., Jennings, N.R., & Shadbolt, N.R. (2005). The Semantic Grid: Past, Present, and Future. In *Proceedings of the IEEE*, *93*(3), 669-681.

de Vienne, D.M., Giraud, T., & Martin, O.C. (2007). A congruence index for testing topological similarity between trees. *Bioinformatics*, 23(23):3119-3124.

Deacon, A. M., Ni, Y. S., Coleman, W. G. Jr. & Ealick, S. E. (2000). The crystal structure of ADP-L-glycero-D-mannoheptose 6-epimerase: catalysis with a twist. *Structure*, *8*, 453-462.

Dean, J., & Ghemawat, S. (2004). *In MapReduce: Simplified Data Processing on Large Clusters*. Sixth Symposium on Operating System Design and Implementation (OSDI'04), San Francisco, CA.

Debaerdemaeker, T., Tate, C. & Woolfson, M.M. (1985). On the application of phase relationships to complex structures (XXIV: The Sayre tangent formula). *Acta Cryst.*A41), 286-290.

DECI (2008). DEISA Extreme Computing Initiative. Retrieved from www.deisa.eu/science/deci

Deelman, E., Blythe, J., Gil, Y., & Kesselman, C. (2004). *Workflow management in GriPhyN. In Grid resource management: state of the art and future trends*. Norwell, MA: Kluwer Academic Publishers.

Deelman, E., Blythe, J., Gil, Y., Kesselman, C., Mehta, G., Patil, S., Su, M. H., Vahi, K. & Livny, M. (2004). Pegasus: Mapping scientific workflows onto the Grid. In Grid *Computing: Second European AcrossGrids Conference, AxGrids*, (LNCS 3165, pp 11-20). Berlin, Germany: Springer.

Deelman, E., Kesselman, C., Mehta, G., Meshkat, L., Pearlman, L., Blackburn, K., et al. (2002). Griphyn and ligo, building a virtual data grid for gravitational wave scientists. *In HPDC* (pp. 225-234). New York: IEEE Computer Society.

Deelman, E., Singh, G., Su, M., Blythe, J., Gil, Y. Kesselman, C., Mehta, G., Vahi, K., Berriman, G.B., Foster, I., & Kesselman, C. (1997). Globus: A Metacomputing Infrastructure Toolkit. *Intl J. Supercomputer Applications*, 11(2), 115–128.

Degtyarenko K. et al. (2008). ChEBI: a database and ontology for chemical entities of biological interest. *Nucleic Acids Research, 36*. Retrieved from http://www.ebi.ac.uk/chebi/

DEISA. (2008). Distributed European Infrastructure for Supercomputing Applications. Retrieved from www.deisa.eu

Delcher, A.L., Harmon, D., Kasif, S., White, O. & Salzberg, S.L. (1999). Improved microbial gene identification with GLIMMER. *Nucl. Acids Res., 27*(23), 4636-4641.

Delhay, B., Clarysse, P., Pera, C. & Magnin, I. E. (2006). A spatio-temporal deformation model for dense motion estimation in periodic cardiac image sequence. In *From Statistical Atlases to Personalized Models: Understanding Complex Diseases in Populations and Individuals, Satellite Workshop MICCAI 2006*, Copenhagen, Denmark.

Delhay, B., Clarysse, P., Pera, C. & Magnin, I. E. (2007, June). Locally adapted spatio-temporal deformation model for dense motion estimation in periodic cardiac image sequences. In *Functional Imaging and Modeling of the Heart*. Salt Lake City, UT, (LNCS 4466, pp. 393-402). Berlin, Germany: Springer-Verlag.

Deneubourg, J. L., Goss, S., Franks, N., Sendova-Franks, A., Detrain, C., & Chretien, L. (1990) *The dynamics of collective sorting robot-like ants and ant-like robots.* Paper presented at the 1st International Conference on Simulation of Adaptive Behavior on From Animals to Animats, Cambridge, USA.

DESHL. (2008). DEISA Services for Heterogeneous management Layer. http://forge.nesc.ac.uk/projects/deisa-jra7/

Desiere, F., Deutsch, E., Nesvizhskii, A., Mallick, P., King, N., Eng, J., et al. (2004). Integration with the human genome of peptide sequences obtained by high-throughput mass spectrometry. *Genome Biology, 6*(1).

Desprez, F. & Vernois, A. (2007, September). Semi-Static Algorithms for Data Replication and Scheduling Over the Grid. In *IEEE 3rd International Conference on Intelligent Computer Communication and Processing, workshop on Grid Computing*, Cluj-Napoca, Romania.

Desprez, F., & Vernois, A. (2006, March). Simultaneous Scheduling of Replication and Computation for Data-Intensive Applications on the Grid. *Journal Of Grid Computing, 4*(1), 19-31.

DeTitta, G. T., Weeks, C. M., Thuman, P., Miller, R. & Hauptman, H. A. (1994). Structure solution by minimal function phase refinement and Fourier filtering, (I. Theoretical basis). *Acta Crystallogr.* A50, 203-210.

DeTitta, G., Hauptman, H., Miller, R., Pagels, M. Sabin, T., Thuman, P., & Weeks, C. (1991). Parallel solutions to the phase problem in x-ray crystallography: an update. In

Proceedings of The Sixth Distributed Memory Computing Conference, (pp. 587-594). New York: IEEE Computer Society Press.

D-Grid (2008). Retrieved from www.d-grid.de/index.php?id=1&L=1

D-Grid. (2008). *D-Grid*. Retrieved July 21, 2008, from www.d-grid.de

Dhillon, I. S., & Modha, D. S. (2000). A data-clustering algorithm on distributed memory multiprocessors. In *Large-Scale Parallel Data Mining* (pp. 245-260). Lecture Notes in Artificial Intelligence.

Dietterich, T. G. (2000). An experimental comparison of three methods for constructing ensembles of decision trees: Bagging, boosting, and randomization. *Machine Learning, 40*, 139-157.

DIMACS (2002) *Special Focus on Computational and Mathematical Epidemiology*. Retrieved July from http://dimacs.rutgers.edu/SpecialYears/2002_Epid/

Dinda, P. (2006). Design, Implementation, and Performance of an Extensible Toolkit for Resource Prediction In Distributed Systems. *IEEE Transactions on Parallel and Distributed Systems, 17*(2), 160-173.

Ding, Z., Luo, Y., Wei, X., Misleh, C., Li, W. W., Arzberger, P. W., & Tatebe, O. (2006). *In My WorkSphere: Integrative Work Environment for Grid-unaware Biomedical Researchers and Applications*. Supercomputing Conference 2006, SC06, 2nd Grid Computing Environment Workshop, Tampa, FL.

Ding, Z., Wei, X., Tatebe, O., Papadopoulos, P. M., Arzberger, P. W., & Li, W. W. (2008). Cyberinfrastructure for Biomedical Applications: Metascheduling as Essential Component for Pervasive Computing. In, J. Cao, (Ed.), *Cyberinfrastructure Technologies and Applications*,. Hauppauge, NY: Nova Science.

Dixit, S. B., Beveridge, D. L., Case, D. A., Cheatham III, T. E., Guidice, E., Lankas, F., Lavery, R., Maddocks, J. H., Osman, R., Sklenar, H., Thayer, K. M., & Varnai, P. (2005). Molecular dynamics simulations of the 136 unique tetranucleotide sequences of DNA oligonucleotides. II: Sequence context effects on the dynamical structures of the 10 unique dinucleotide steps, *Biophysical Journal, 89*, 3721-3740.

Docking@Home. Retrieved March 5, 2008 from http://docking.cis.udel.edu/

Dodson, C.T.J. and T. Poston (1991). *Tensor Geometry: The Geometric Viewpoint and its Uses*. Berlin, Germany: Springer-Verlag.

Dolin, R. H., Alschuler, L., Beebe, C., Biron, P. V., Boyer, S. L., Essin, D., et al. (2001). The HL7 Clinical Document Architecture. *Journal of the American Medical Informatics Association, 8*(6), 552-569.

Dolin, R. H., Alschuler, L., Boyer, S., Beebe, C., Behlen, F. M., Biron, P. V., et al. (2006). HL7 Clinical Document Architecture, Release 2. *J Am Med Inform Assoc, 13*(1), 30-39.

Dolinsky, T. J., Czodrowski, P., Li, H., Nielsen, J. E., Jensen, J. H., Klebe, G., & Baker, N. A. (2007). PDB-2PQR: expanding and upgrading automated preparation of biomolecular structures for molecular simulations. *Nucleic Acids Res, 35*, 522-525.

Dongarra, J., Foster, I., Fox, G., Gropp, W., Kennedy, K., Torczon, L., & White, A. (2003). Sourcebook of Parallel Computing. Morgan Kaufmann Publishers.

Donnely, C. (2002). FMD 2001: Using Statistics and Mathematics for Outbreak Control and Eradication. *International Conference on Computational and Mathematical Epidemiology*. Retrieved July 2008 from http://dimacs.rutgers.edu

Donovan, F.J. (1975). Design of a hydraulic analog of the circulatory system fo evaluating artificial Hearts. *Biomaterials, Medical Devices, and Artificial Organs, 3*(4), 439-449.

Dowsey, A. W., Dunn, M. J. & Yang, G. (2004). ProteomeGRID: towards a high-throughput proteomics pipeline through opportunistic cluster image computing for two-dimensional gel electrophoresis. *Proteomics, 4*(12), 3800-3812.

Droegemeier, K. K., Gannon, D., Reed, D. A., Plale, B., Alameda, J. Baltzer, T., et al. (2005). Service-oriented Environments in Research and Education for Dynamically Interacting with Mesoscale Weather. *Computing in Science and Engineering 7*(6)12-29.

Droegemeier, K. K., Gannon, D., Reed, D., Plale, B., Alameda, J., Baltzer, T., et al. (2005). Service-oriented environments for dynamically interacting with mesoscale weather. *Computing in Science and Engineering, 7*(6), 12-29.

Dubitzky, D., McCourt, D., Galushka, M., Romberg, M., & Schuller, B. (2004). Grid-enabled Data Warehousing for Molecular Engineering. *Parallel Computing, 30*, 1019-1035.

Dublin Core Metadata Initiative (2008). Retrieved July 20, 2008 from http://www.dublincore.org

Dubochet, J., Adrian, M., Chang, J. J., Home, J. C., Lepault, J., McDowall, A. W., et al. (1988). Cryo-electron microscopy of vitrified specimens. *Quarterly Review of Biophysics, 21*(1), 129-228.

Dwyer III, S.J. (2000). A personalized view of the history of PACS in the USA. In G. J. Blaine & E. L. Siegel, *Proceedings of the SPIE, "Medical Imaging 2000: PACS Design and Evaluation: Engineering and Clinical Issues"*, (Vol. 3980), (pp. 2-9).

Dzenowagis, J. & Kernen, G. (2005). *Connecting for Health: global vision, local insight* (Report for the World Summit on Information Society). Tunis, Tunisia: World Health Organization Press

ECOTOX (ECOTOXicology) database. Retrieved March 4, 2008 from http://www.epa.gov/ecotox/

Edgar, R. C. (2004), MUSCLE: multiple sequence alignment with high accuracy and high throughput. *Nucleic Acids Research, 32*(5), 1792-97.

EGEE Grid infrastructure. Retrieved from http://www.eu-egee.org/.

EGEE SDJ WG (2006). *Report of the Short Deadline Jobs Working Group*. Retrieved from http://egee-na4.ct.infn.it/wiki/index.php/ShortJobs.

EGEE, Enabling the Grid for E-sciencE 2007 Conference, Live Demo Competition (2007). Retrieved October 20, 2008, from http://indico.cern.ch/contributionDisplay.py?contribId=174&sessionId=23&confId=18714.

EGEE-SEE ROC (2007). *Web Authentication and Authorization using MyProxy APIs and Grid Portals*. Retrieved July 2008 from http://wiki.egee-see.org/index.php/Web_Authentication_and_Authorization_using_MyProxy_APIs_and_Grid_Portals

EGI Consortium (2007). *European Grid Initiative*. Retrieved July 2008 from http://www.eu-egi.org

Eilemann, S. & Pajorola, R. (2007). Direct Send Compositing for Parallel Sort-Last Rendering. *Eurographics Symposium on Parallel Graphics and Visualization*, (pp. 29-36).

Ekman A., Klint A., Dickman P.W. , Adami H.O., Litton J. (2007). Optimizing the design of web-based questionnaires - experience from a population-based study among 50.000 women. *European Journal of Epidemiology 22*, 293-300.

Ekman A., Litton J. (2007). New times, new needs; e-epidemiology. *European Journal of Epidemiology, 22*, 285-292.

El Sayed, G. and I. Greenshields (2008). Low FM: A Robust Non-Blocking Low-Frequency Messaging Checkpoint for Parallel/Grid Applications. In *Proc. 23rd International Conference on Computers and Their Applications, Cancun, Mexico, April 9-11*.

Ellisman, M. & Peltier, S. (2004). Medical data federation: the biomedical informatics research network. In Foster, I. & Kesselman, C. (Eds.), *The Grid: Blueprint for a New Computing Infrastructure* (2nd Ed.). San Francisco, CA: Morgan Kaufmann.

Embrace. Embrace website. Retrieved from http://www.embracegrid.info

Emmen, A. (2008). National Digital Mammography Archive announces installation at University of Pennsylvania Health System. *EnterTheGrid - PrimeurMonthly* Retrieved March 8, 2008 from *http://www.hoise.com/primeur/05/articles/monthly/AE-PR-02-05-1.html* .

Enabling Grids for E-Science (n.d.). Retrieved March 17, 2008, from http://www.eu-egee.org

Eng, J. K., McCormack, A. L. & Yates III, J. R. (1994). An approach to correlate tandem mass spectral data of peptides with amino acid sequences in a protein database. *Journal of the American Society for Mass Spectrometry, 5*(11), 976-989.

EnginFrame (2008). Grid and Cloud Portal. Retrieved from www.nice-italy.com

Englander, R. (1997). *Developing Java Beans*. Sebastopol, CA: O'Reilly & Associates, Inc. Bellia, M., Ochiutto, M.E. (2005). Higher order Programming in Java: Introspection, Subsumption and Extraction. *Concurrency Specification and Programming* (pp. 29-44). Amsterdam, The Netherlands: IOS Press.

Ennis, D.B., Kindlmann, G., Rodriguez, I., Helm, P.A. & McVeigh, E.R. (2005).Visualization of Tensor Fields Using Superquadric Glyphs. *Magn. Resonance Med., 53*(1) 169-176.

Epstein, J. M., Cummings, D. A., Chakravarty, S., Singa, R. M., Burke, D. S. (2002, December). *Toward a Containment Strategy for Smallpox Bioterror: An Individual-Based Computational Approach. Brookings Institution, Center on Social and Economic Dynamics* (Working Paper No. 31).

EQU. (n.d.). *Equalizer*. Retrieved from http://www.equalizergraphics.com/

Erberich, S., Silverstein, J., Chervenak, A., Schuler, R., Nelson, M. & Kesselman, K. (2007). Globus MEDICUS – Federation of DICOM Medical Imaging Devices into Healthcare Grids. *Studies in Health Technology and Informatics, 126*, 269-278.

Erberich, S., Silverstein, J., Kratz M., Chervenak, A., Schuler, R., Nelson, M. & Kesselman, K. (2007, April). *Globus MEDICUS – Standards Based Enterprise for Medical Image Publication, Discovery, and Archiving in HealthGrids.* Paper presented at the Open Source Solutions for Multi-Center Information Management. Saint Louis, MO.

Eres, M.H., Pound, G.E., Jiao, Z., Wason, J.L., Xu, F., Keane, A.J. & Cox, S.J. (2005) Implementation and utilisation of a Grid-enabled problem solving environment in Matlab. *Future Generation Computer Systems, 21* (6) 920-929.

Erickson, B. (2002). Irreversible Compression of Medical Images. *Journal of Digital imaging, 15*(1), 5-14.

Erwin D. W, & Snelling D. F. (2001). UNICORE: A Grid Computing Environment. In. J. Darlington et al. (Eds.) *Euro-Par 2001 Parallel Processing*. Berlin, Germany: Springer.

Estrella, F., Hauer, T., McClatchey, R., Odeh, M., Rogulin, D. & Solomonides, T. (2007). Experiences of engineering Grid-based medical software. *International Journal of Medical Informatics, 76*(8), 621-32.

Eswar, N., Marti-Renom, M. A., Webb, B., Madhusudhan, M. S., Eramian, D., Shen, M., Pieper, U. & Sali, A. (2000). Comparative Protein Structure Modeling With MODELLER. *Current Protocols in Bioinformatics* (Supplement 15, 5.6.1-5.6.30). Hoboken, NJ: John Wiley & Sons, Inc..

Etzold, T., Harris H. & Beulah, S. (2003). SRS: An integration platform for databanks and analysis tools in bioinformatics. *Bioinformatics Managing Scientific Data*, 35–74.

Eubank, S. (2002). *Scalable, Efficient, Epidemiological Simulation*. New York: ACM. Retrieved March 1, 2008 from http://www.ccs.lanl.gov/ccs5/documents/pdf/2001/LA-UR-01-5513.pdf.

Eubank, S., Guclu, H.,Kumar, V. S. A., Marathe, M. V., Srinivasan, A., Toroczkai, Z.& Wang, N. (2004). Modeling disease outbreaks in realistic urban social networks. *Nature*, 429(May 13), 180.

Eugster, P., Guerraoui, R., Kermarres, A.M., & Massoulieacute, L. (2004). *Epidemic Information Dissemination in Distributed System*, IEEE Computer, 37(5), 60-67

EUROCARE Consortium. (2008). *EUROCARE*. Retrieved July 2008 from http://www.eurocare.it/

European Parliament. (1995). *Official Journal of the European Communities 281*, 31.

European Parliament. (2002). *Official Journal 201*, 0037 - 0047.

European Union (2002). *Resource Book of IST Projects Relating to Health (1998-2002)*. Retrieved from http://cordis.europa.eu/ist/ka1/health/projectbooklet/index.htm.

EUROREC (2002). Retrieved July 2008 from http://www.eurorec.org/

EU-SHARE. (2006). *SHARE - Roadmap for a European Health Grid*. Retrieved 28.06.2006, from http://www.eu-share.org/

Falkner, J., & Sax, U. (2008). *Service Certificates and Portal Security - Longterm requirements towards security mechanisms (German).* Paper presented at the D-Grid Security Workshop 2008, Göttingen.

Fan, W., Stolfo, S. J., & Zhang, J. (1999). The application of adaboost for distributed, scalable and on-line learning. In *The fifth ACM SIGKDD international conference on Knowledge discovery and data mining* (pp. 362-366). New York: ACM.

FDA. (2004, April 1). *FDA 21 CFR Part 11.* Retrieved September 21, 2005, from http://www.accessdata.fda.gov/scripts/cdrh/cfdocs/cfcfr/CFRsearch.cfm?CFRPart=11

Federrath, H. (2005, 22-26 August). *Privacy Enhanced Technologies: Methods – Markets – Misuse* Paper presented at the Trust, Privacy and Security in Digital Business Copenhagen, Denmark.

Feig, M., Abdullah, M., Johnsson, L., & Pettitt, B. M. (1999). Large scale distributed data repository: Design of a molecular dynamics trajectory database. *Future Generation Computer Systems, 16,* 101-110.

Fensel, D. (2003). *Ontologies: Silver Bullet for Knowledge Management and Electronic Commerce (*2nd Ed.). Berin, Germany: Springer.

Ferguson N. M., Cummings, D. A.T., Cauchemez, S., Fraser, C., Riley, S., Meeyai, A., Iamsirithaworn, S., Burke, D. S. (2005). Strategies for containing an emerging influenza pandemic in Southeast Asia. *Nature 437,* 209 - 214.

Fernández, J. J., Bilbao-Castro, J. R., Marabini, R., Carazo, J. M., & García, I. (2005). On the Suitability of Biological Structure Determination by Electron Microscopy to Grid Computing. *New Generation Comput, 23*(1), 101-112.

Fernández, J. J., García, I., Carazo, J. M., & Marabini, R. (2007). Electron tomography of complex biological specimens on the Grid. *Future Generation Computing Systems, 23*(3), 435-446.

Fernández, J. J., Lawrence, A. F., Roca, J., García, I., Ellisman, M. H., & Carazo, J. M. (2002). High-performance electron tomography of complex biological specimens. *Journal of Structural Biology, 138*(1-2), 6-20.

Ferrari, G., Kozarski, M., De Lazzari, C., Gorczynska, K., Tosti, G., & Darowski, M. (2005). Development of a hybrid (numerical-hydraulic) circulatory model: Prototype testing and its response to IABP assistance. *International Journal of Artificial Organs, 28*(7), 750-759.

Fielding, R.T. (2005). JSR170 Overview. *Stardandizing the Content Repository Interface.*

Finn, R.D., Tate, J., et al. (2008). The Pfam protein families database. *Nucleic Acids Res, 36,* D281-8.

Fiolet, V., Olejnik, R., Lefait, G., & Toursel, B. (2006). Optimal grid exploitation algorithms for data mining. *In the Fifth International Symposium on Parallel and Distributed Computing* (pp. 246-252). New York: IEEE Computer Society.

Fiore, S., Negro, A., Vadacca, S., Cafaro, M., Mirto, M. & Aloisio, G. (2007). Advanced Grid DataBase Management with the GRelC Data Access Service. *Proceedings of ISPA 2007 - August 29-31, 2007 - Niagara Falls, Ontario, Canada,* (LNCS 4742) (pp. 683-694).

Foley, J.D., van Dam, A., Feiner, S.K., Hughes, J.F. & Philips, R.L. (1993). *Introduction to Computer Graphics.* Upper Saddle River, NJ: Addison-Wesley.

Forestiero, A., Mastroianni, C., & Spezzano, G. (2005). *A Multi Agent Approach for the Construction of a Peer-to-Peer Information System in Grids,* Paper presented at the International Conference on Self-Organization and Adaptation of Multi-agent and Grid Systems (SOAS 2005), Glasgow, United Kingdom.

Forestiero, A., Mastroianni, C., & Spezzano, G. (2007). *A Decentralized Ant-Inspired Approach for Resource Management and Discovery in Grids,* Multiagent and Grid Systems - An International Journal, 3(1), 43-63.

Forge, S., & Blackman, C., (2006). *Commercial Exploitation of Grid Technologies and Services. Drivers and Barriers, Business Models and Impacts of Using Free and Open Source Licensing Schemes,* (Final Report by SCF Associates for European Commission DG Information Society and Media). Retrieved March 9, 2008 from IST website http://cordis.europa.eu/ist/grids/pub-report.htm

Foster I. & Kesselman C. (1999). *The Grid: Blueprint for a Future Computing Infrastructure.* San Francisco, CA: Morgan Kaufmann Publishers.

Foster I., Kesselman C., & Tuecke S. (2001). The Anatomy of the Grid: Enabling Scalable Virtual Organizations. *International J. Supercomputer Applications, 15*(3).

Foster I., Kesselman C., Nick J., & Tuecke S. (2002). *The Physiology of the Grid: An Open Grid Services Architecture for Distributed Systems Integration. Open Grid Service Infrastructure WG, Global Grid Forum.* Retrieved from http://www.globus.org/alliance/publications/papers/ogsa.pdf

Foster, I. & Kesselman, C. (2005). Globus Toolkit Version 4: Software for Service-Oriented Systems. In IFIP *International Conference on Network and Parallel Computing*, (LNCS 3779) (pp. 2–13). Berlin, Germany: Springer-Verlag.

Foster, I. & Kesselman, C. (2006). Scaling system-level science: Scientific exploration and it implications. *Computer, 39*(11), 31-39.

Foster, I. & Kesselman, K. (Eds.), (2003). *The Grid 2: Blueprint for a New Computing Infrastructure.* San Francisco, CA: Morgan Kaufmann Publishers.

Foster, I. & Kesselmann, C. (Eds.) (1999). *The Grid: Blueprint for a New Computing Intrastructure.* San Francisco, CA: Morgan Kaufmann Publishers.

Foster, I. (2000). Internet Computing and the Emerging Grid. Nature. Retrieved from www.nature.com/nature/webmatters/grid/grid.html

Foster, I. (2002) What is the Grid? A Three Point Checklist. *GRID Today, 1*(6). Retrieved February 13, 2008 from http://www.gridtoday.com/02/0722/100136.html.

Foster, I. (2002). *The Grid: Enabling Resource Sharing With Virtual Organizations.* Keynote speech at WWW2002, Honolulu, HI. Retrieved from http://www2002.org/foster.pdf

Foster, I. (2005). Globus Toolkit Version 4: Software for Service-Oriented Systems. *IFIP International Conference on Network and Parallel Computing*, (LNCS 3779, pp. 2-13). Berlin, Germany: Springer-Verlag.

Foster, I. (2005). Service-oriented science. *Science, 308*, 814-817.

Foster, I. (2006). *Globus Toolkit Version 4: Software for Service-Oriented Systems.* Paper presented at the IFIP International Conference on Network and Parallel Computing.

Foster, I. (2006). Globus Toolkit Version 4: Software for Service-Oriented Systems. In *IFIP International Conference on Network and Parallel Computing* (LNCS 3779, pp. 2-13). Berlin, Germany: Springer-Verlag.

Foster, I. (2006, October). Globus Toolkit Version 4: Software for Service-Oriented Systems. *IFIP International Conference on Network and Parallel Computing*, Tokyo, Japan (LNCS 3779, pp 2-13). Berlin, Germany: Springer-Verlag.

Foster, I. (Ed.), Kesselman, C. (Ed.). (1998). *The Grid: Blueprint for a New Computing Infrastructure.* San Francisco, CA: Morgan Kaufmann Publishers.

Foster, I., & Kesselman C. (1997). Globus: A metacomputing infrastructure toolkit. *International Journal of Scientific Applications, 11*(2) 115-128.

Foster, I., & Kesselman, C. (2004). *The Grid 2: Blueprint for a New Computing Infrastructure*, (2 Ed). San Francisco, CA: Morgan Kaufmann Publishers, Inc..

Foster, I., & Kesselman, C. (Ed.) (1999). *The Grid: Blueprint for a New Computing Infrastructure.* San Francisco, CA: Morgan Kaufmann Publishers, Inc..

Foster, I., & Kesselman, C. (Eds.). (2004). The Grid 2: Blueprint for a New Computing Infrastructure. Morgan Kaufmann Publishers.

Foster, I., Czajkowski, K., Ferguson, D. E., Frey, J., Graham, S., & Maguire, T., et al. (2005). Modeling and managing state in distributed systems: The role of OGSI and WSRF. *Proceedings of the IEEE, 93*(3), 604-612.

Foster, I., Kesselman, & C., Tuecke, S. (2002), The Anatomy of the Grid: Enabling Scalable Virtual Organizations. Retrieved from www.globus.org/alliance/publications/papers/anatomy.pdf

Foster, I., Kesselman, C. & Tuecke, S. (1999). *The Grid.* San Francisco: Morgan Kaufmann Publishers.

Foster, I., Kesselman, C. & Tuecke, S. (2001). The Anatomy of the Grid. Retrieved june 6, 2008, from http://www.globus.org/alliance/publications/papers/anatomy.pdf

Foster, I., Kesselman, C. & Tuecke, S. (2004). *The Grid 2.* San Francisco: Morgan Kaufmann Publishers.

Foster, I., Kesselman, C. & Tuecke, S. (n.d.). *The Anatomy of the Grid: Enabling Scalable Virtual Organizations.* Retrieved from http://www.globus.org/alliance/publications/papers/anatomy.pdf

Foster, I., Kesselman, C. (2006). Scaling system-level science: Scientific exploration and its implications. *IEEE Computer, 39* (11), 31-39.

Foster, I., Kesselman, C., & Tuecke, S. (2001). The anatomy of the grid: enabling scalable virtual organizations. *International Journal of High Performance Computing Applications, 15*(3), 200-222.

Foster, I., Kesselman, C., & Tuecke, S. (2001). The Anatomy of the Grid: Enabling Scalable Virtual Organizations. *International Journal of High Performance Computing Applications, 15*(3), 200–222.

Foster, I., Kesselman, C., Nick, J. & Tuecke, S. (2002). Grid Services for Distributed System Integration. *Computer, 35*(6), 37-46.

Foster, I., Kesselman, C., Nick, J., Tuecke, S. (2002). The Physiology of the Grid: An Open Grid Services Architecture for Distributed Systems Integration. Retrieved February 14, 2008 from http://www.globus.org/alliance/publications/papers/ogsa.pdf.

Foster, I., Kesselman, C., Nick, J.M., & Tuecke, S. (2002). Grid Services for Distributed Systems Integration. *Computer 35*(6).

Foster, I., Kesselman, C., Tsudik, G., & Tuecke, S. (1998). *In A Security Architecture for Computational Grids.* ACM Conference on Computer and Communications Security, San Francisco, CA.

Foster, I., Kesselman, K. & Tuecke, S. (2001). The Anatomy of the Grid: Enabling Scalable Virtual Organizations. *International Journal Supercomputer Applications, 15*(3), 200-222.

Foster, I., Kishimoto, H., Savva, A., Berry, D., Djaoui, A., Grimshaw, A., et al. (2005).The Open Grid Services Architecture, Version 1.0. *Informational Document, Global Grid Forum (GGF).*

Foster. I. (2006). Globus Toolkit Version 4: Software for Service-Oriented Systems. *LNCS*, 3779, 2-13.

Foti, D., Lipari, D., Pizzuti, C., & Talia, D. (2000). Scalable parallel clustering for data mining on multicomputers. *In the 15th Workshops on Parallel and Distributed Processing* (pp. 390-398). Berlin, Germany: Springer-Verlag.

Fowler, P., Khalid, S., Mokrab, Y., Sansom, M. S. P., Stansfeld, P., & Tai, K. (2008). KDB: A potassium channel model and simulation database. *Biophysical Journal, 94,* 2038.

Fox, G. C., Guha, R., Mcmullen, D. F., Mustacoglu, A. F., Pierce, M. E., Topcu, A. E., & Wild, D. J. (in print, 2007). Web 2.0 for grids and e-science. In *Proceedings of INGRID 2007 - Instrumenting the Grid 2nd International Workshop on Distributed Cooperative Laboratories*, S. Margherita Ligure Portofino, Italy.

Fox, G., Williams, R., & Messina, P. (1994). Parallel Computing Works! Morgan Kaufmann Publishers.

Fox, P., McGuinness, D., Raskin, R., and Sinha, K. (2007). A volcano erupts: semantically mediated integration of heterogeneous volcanic and atmospheric data. In Mitra, P., Giles, L. C., & Carr, L., (Eds), *CIMS '07: Proceedings of the ACM first workshop on CyberInfrastructure: information management in eScience*, (pp. 1-6). New York: ACM.

Frank, J. (1992). *Electron Tomography: Three-Dimensional Imaging whith the Transmission Electron Microscope.* New York: Plenum Press.

Frank, J. (2006). *Three-Dimensional Electron Microscopy of Macromolecular Assemblies: Visualization of Biological Molecules in Their Native State.* Oxford, UK: Oxford University Press.

Fraser, C.M., Eisen, J., Fleischmann, R.D., Ketchum, & K.A., Peterson, S. (2000). Comparative Genomics and Understanding of Microbial Biology. *Emerging Infectious Diseases, 6*(5), 505-512.

Frazão, C., Sieker, L., Sheldrick, G. M., Lamzin, V., LeGall, J. & Carrondo, M. A. (1999). Ab initio structure solution of a dimeric cytochrome c3 from Desulfovibrio gigas containing disulfide bridges. *J. Biol. Inorg. Chem. 4*, 162-165.

Freitas, A. A. (1998). A survey of parallel data mining. In *2nd International Conference on the Practical Applications of Knowledge Discovery and Data Mining* (pp. 287-300). London: The Practical Application Company.

Freitas, A. A. (Ed.). (2002). *Data Mining & Knowledge Discovery with Evolutionary Algorithms*. Berlin, Germany: Springer-Verlag.

Freund, J., Comaniciu, D., Ioannis, Y., Liu, P., McClatchey, R., Morley-Fletcher, E., et al. (2006). Health-e-child: an integrated biomedical platform for grid-based paediatric applications. *Stud Health Technol Inform, 120*, 259-270.

Freund, Y. & Schapire, R. E. (1997). A decision-theoretic generalization of on-line learning and an application to boosting. *Journal of Computing System Science, 55*(1), 119-139.

Frey, J., Mori, T., Nick, J., Smith, C., Snelling, D., Srinivasan, L., & Unger, J. (2005). The Open Grid Services Architecture, Version 1.0. www.ggf.org/ggf_areas_architecture.htm

Frey, J., Tannembaum, T., Foster, I., Livny, M., & Tuecke, S. (2001). Condor-G: A Computation Management Agent for Multi-Institutional Grids. Proceedings of the *Tenth IEEE Symposium on High Performance Distributed Computing* (pp. 237-246). Hingham, MA: Kluwer Academic Publishers.

Frey, J., Tannenbaum, T., Livny, M., Foster, I., & Tuecke, S. (2002). Condor-G: A Computation Management Agent for Multi-Institutional Grids. *Cluster Computing, 5*(3), 237-246.

Frist, W. (2005). Health care in the 21st century. *New England Journal of Medicine, 352*, 267-272.

Friston, K. (1996). Statistical Parametric Mapping and Other Analysis of Functional Imaging Data. In *Brain Mapping The Methods*, (pp. 363-385). New York: Academic Press.

Frizziero, E., Gulmini, M., Lelli, F., Maron, G., Oh, A., Orlando, S., et al. (2006). *Instrument Element: a new Grid component that enables the control of remote instrumentation*. Paper presented at the Proceedings of the Sixth IEEE International Symposium on Cluster Computing and the Grid Workshops (CCGRIDW'06).

Frohner, A., Jouvenot, D., Kunszt, P., Montagnat, J., Pera, C., Koblitz, B., Santos, N, Loomis, C. (2007). A Secure Grid Medical Data Manager Interfaced to the gLite Middleware. *Journal of Grid Computing, 6*(1), 45-59.

Funika, W. & Pegiel, P. (2007). GScript Editor as Part of the ViroLab Presentation Layer. In *Proceedings of Cracow Grid Workshop 2006 (CGW06)*, (pp. 320-327). Krakow, Poland: ACC CYFRONET AGH.

Furey, W. & Swaminathan, S. (1997). PHASES-95: A Program Package for the Processing and Analysis of Diffraction Data from Macromolecules. *Meth. Enzymol. 277*, 590-620.

G., Cafaro, M., Fiore, S., & Mirto, M. (2005). ProGenGrid: A Workflow Service Infrastructure for Composing and Executing Bioinformatics Grid Services. In A. Tsymbal & P. Cunningham (Eds.), *18th IEEE Symposium on Computer Based Medical Systems* (Vol. 1) (pp. 555-560). New York: IEEE Press

Gagliardi, F., Jones, B., Grey, F., Bégin, M. E., Heikkurinen, M. (2005). Building an infrastructure for scientific grid computing: status and goals of the EGEE project. *Phylosophical Transactions: Mathematica, Physical and Engineering Sciences, 363*(1833), 1729-1742.

Gagliardi, H. F., Silva, F. A. B., Alves, D. (2006). Automata Network Simulator Applied to the Epidemiology of Urban Dengue Fever. *ICCS 2006, Part III*, (LNCS 3993, pp. 297 – 304).

Galaktionov, S.G. & Marshall, G.R. (1994). Properties of intraglobular contacts in proteins: an approach to prediction of tertiary structure. *System Sciences, Vol. V:, Proceedings of the Twenty-Seventh Hawaii International Conference on Biotechnology Computing*: Vol. 5, (pp. 326 – 335). Washington, DC: IEEE Computer Society Press.

Galiardi F. (2005). EGEE – a first results and future plans. In M. Bubak, M. Turała, K. Wiatr (Eds.) Proceedings of Cracow'04 Grid Workshop.

Galizia, A., Viti, F., Orro, A., D'Agostino,D. Merelli, I., Milanes, L., & Clematis, A. (2008). TMAinspect, an EGEE Framework for Tissue MicroArray Image Handling. *Proceedings of 8th Internal Symposium on Cluster Computing and the Grid* (pp 741-746), IMATI-CNR, Genoa, Italy.

Gallo, E., Gagliardi, H.F., Neto, V. C., Alves, D., Silva, F.A.B. (2007) GISE: A Data Access and Integration Service for a Grid-Based Epidemic Surveillance System. *Proceedings of the 40th Annual Simulation Symposium*. New York: IEEE Computer Society Press.

Gao, H., Sengupta, J., Valle, M., Korostelev, A., Eswar, N., Stagg, S. M., et al. (2003). Study of the structural dynamics of the E. coli 70S ribosome using real-space refinement. *Cell, 113*(1), 789-801.

Gao, J., Opiteck, G. J., Friedrichs, M. S. & Haefta, S.. (2003). Changes in the Protein Expression of Yeast as a Function of Carbon Source. *Journal of Proteome Research*, 643-649.

Garcia, J. L., Dikaiakos, M. D., Kyprianou, T., Bilas, A., & Marazakis, M. (2008). Data Privacy considerations in Intesive Care Grids. *HealthGrid Conference 2008*. Chicago, IL.

Gardarin, G.; Sha, F.; Ngoc, T. (1999) XML-based Components for Federating Multiple Heterogeneous Data Sources. In *Proc. of International Conference on Conceptual Modeling*, Paris, France.

Gärdén, P., Alm, R., & Häkkinen, J. (2005). PROTEIOS: an open source proteomics initiative. *Bioinformatics, 21*(9), 2085-2087.

Gardner, S. P. (2005). Ontologies and semantic data integration. *Drug Discovery Today, 10*(14), 1001-1007.

Garduño, E., Gabor, G. T., & Herman, G. T. (2004). Optimization of Basis Functions for Both Reconstruction and Visualization. *Discrete Applied Mathematics, 139* (1-3), 95-111.

Gasteiger, E., Jung, E., et al. (2001). SWISS-PROT: connecting biomolecular knowledge via a protein database. *Curr Issues Mol Biol,* 3(3), 47-55.

GAT. (2005). Grid Application Toolkit. www.gridlab.org/WorkPackages/wp-1/

Gaudeau, Y., & Moureaux, J.M. (2007, November). *Lossy compression of volumetric medical images with 3D dead zone lattice vector quantization.* 26th Picture Coding Symposium, Lisboa, Portugal.

Gauss, C., Kalkum, M., Lowe, M., Lehrach, H., & Klose, J. (1999). Analysis of the mouse proteome. (I) Brain proteins: separation by two-dimensional electrophoresis and identification by mass spectrometry and genetic variation. *Electrophoresis, 20*, 575-600.

Gaussian 03 (2004). Frisch, M. J.; Trucks, G. W.; Schlegel, H. B.; Scuseria, G. E.; Robb, M. A.; Cheeseman, J. R. et al. Wallingford, CT: Gaussian, Inc.

GDMP. Retrieved from http://project-gdmp.web.cern.ch/project-gdmp

g-Eclipse project (2008). Retrieved October 20, 2008, from http://www.geclipse.eu/

Geddes, J., Lloyd, S., Simpson, A., Rossor, M., Fox, N., Hill, D., Et al. (2005). NeuroGrid: using grid technology to advance neuroscience. In *Proceedings of the 18th IEEE symposium on computer-based medical systems*, (pp. 570-572). Washington, DC: IEEE Computer Society.

Geiser, D.M., Jiménez-Gasco, M., Kang, S., Makalowska, I., Veerarahavan, N., Ward, T.J., Zhang, N., Kuldau, G.A. & O'Donnell, K. (2004). FUSARIUM-ID v.1.0: A DNA sequence database for identifying Fusarium. *European Journal of Plant Pathology, 110*, 473-479.

Geman, S. and D. Geman (1984). Stochastic Relaxation, Gibbs Distributions, and the Bayesian Restoration of Images. *IEEE Trans. PAMI, 6*, 721-741.

Gemstone (2008). Retrieved March 3, 2008 from http://gemstone.mozdev.org/

Gene Ontology Consortium (2001). Creating the Gene Ontology resource: design and implementation. *Genome Res, 11*(8), 1425-33.

Gene Ontology Consortium (2008). The Gene Ontology project in 2008. *Nucleic Acids Res, 36*, D440-4.

Gentzsch (2008). Top 10 Rules for Building a Sustainable Grid. In: Grid Thought Leadership Series. Open Grid Forum website at www.ogf.org/TLS/?id=1

Gentzsch, W. (2002). Response to Ian Foster's "What is the Grid?" GRIDtoday. August 5. www.gridtoday.com/02/0805/100191.html

Gentzsch, W. (2004). Enterprise Resource Management : Applications in Research and Industry. In I. Foster & C. Kesselman (Eds.). The Grid 2: Blueprint for a New Computing Infrastructure (pp. 157 – 166). Morgan Kaufmann Publishers.

Gentzsch, W. (2004). Grid Computing Adoption in Research and Industry. In A. Abbas (Ed.), Grid Computing: A Practical Guide to Technology and Applications (pp. 309 – 340). Charles River Media Publishers.

Gentzsch, W. (2007a). Grid Initiatives: Lessons Learned and Recommendations, RENCI Report retrieved from www.renci.org/publications/reports.php

Gentzsch, W. (Ed.) (2007b). A Sustainable Grid Infrastructure for Europe, Executive Summary of the e-IRG Open Workshop on e-Infrastructures, Heidelberg, Germany. Retrieved from the e-IRG Website www.e-irg.org/meetings/2007-DE/workshop.html

GEONgrid (2008). Retrieved from www.geongrid.org

GeoServer website (2008). Retrieved July 20, 2008 from http://geoserver.org

Germain, G. & Woolfson, M. M. (1968). On the application of phase relationships to complex structures, *Acta Crystallogr. B24*, 91-96.

Germain-Renaud, C., Loomis, C., Moscicki, JT., & Texier. R. (2008). Scheduling for responsive grids. *Journal of Grid Computing*, *6*(1), 15-27.

Germain-Renaud, C., Perez, J., Kégl, B. & Loomis, C. (2008, May). Grid Differentiated Services: a Reinforcement Learning Approach. *8th IEEE Symposium on Cluster Computing and the Grid*. Lyon, France: IEEE Computer Society.

Gervasi O., Dittamo C., & Laganà A. (2005). A Grid Molecular Simulator for E-Science. In *Advances in Grid Computing - EGC 2005* (LNCS 3470, pp. 16-22). Berlin, Germany: Springer.

Geyer, C.J. (1991). Markov chain Monte Carlo maximum likelihood. In Keramidas (ed.), *Computing Science and Statistics: Proceedings of the 23rd Symposium on the Interface* (pp. 156–163). Fairfax Station, VA: Interface Foundation.

Giannadakis, N., Rowe, A., Ghanem, M., & Guo, Y. (2003). Infogrid: providing information integration for knowledge discovery. *Information Sciences 155*, 199-226.

Gil, Y., Deelman, E., Ellisman, M., Fahringer, T., Fox, G. Gannon, D., Goble, C., Livny, M., Moreau, L., & Myers J.(2007). Examining the Challenges of Scientific Workflows. *IEEE Computer, 40*(12), 24-32

Gil, Y., Deelman, E., Ellisman, M., Fahringer, T., Fox, G., Gannon, D., Goble, C., Livny, M., Moreau, L., & Myers, J. (2007). Examining the challenges of scientific workflows. *IEEE Computer*, *40*(12), 24-32.

Gkekas, C., Psomopoulos, F., & Mitkas, P. (2008). *A Parallel Data Mining Application for Gene Ontology Term Prediction, 3rd EGEE User Forum*, Clermont-Ferrand, France.

Glatard, T., Montagnat, J., Emsellem, D. & Lingrand, D. (in press). A Service-Oriented Architecture enabling dynamic services grouping for optimizing distributed workflows execution. In *Future Generation Computer Systems*.

Glatard, T., Montagnat, J., Erberich S., Carmeli B. & Kaufman, J. (2007, November). *Cross Enterprise Document and Image Exchange (XDS/-I) in HealthGrids*. Paper presented at the Radiology Society of North America (RSNA'07), Chicago, IL.

Glatard, T., Montagnat, J., Lingrand, D. & Pennec, X. (in press). Flexible and efficient workflow deployement of data-intensive applications on grids with MOTEUR. *International Journal of High Performance Computing and Applications*.

Glatard, T., Pennec, X. & Montagnat, J. (2006). Performance evaluation of grid-enabled registration algorithms using bronze-standards. In *Proc. of MICCAI'06, Part II*, (LNCS 4191, pp. 152-160).

Glich, G.L., & Vachet, R.W. (2003). The basics of mass spectrometry in the twenty first century. *Nature Reviews Drug Discovery*, *2*(2), 140-150.

gLite middleware. Retrieved from http://glite.web.cern.ch/glite/.

glite, A lightweight middleware for grid computing (2008). Retrieved October 20, 2008, from http://glite.web.cern.ch/glite/.

GLOBUS (n.d.). *The Globus Alliance*. Retrieved from http://www.globus.org/

Globus Alliance (2008). Retrieved July 20, 2008 from http://www.globus.org

Globus. Retrieved from http://www.globus.org

Goad, W.B., & Kanehisa, M.I. (1982). Pattern Recognition in Nucleic Acid Sequences: A General Method for Finding Local Homologies and Symmetries. *Nucleic Acids Research, 10*(1), 247-263.

Goble, C. A., & De Roure, D. (2004). The Semantic Grid: Myth Busting and Bridge Building. In *Proceedings of the 16th Eureopean Conference on Artificial Intelligence, ECAI'2004, including Prestigious Applicants of Intelligent Systems, PAIS 2004, Valencia, Spain, August 22-27, 2004* (pp. 1129-1135). Amsterdam, the Netherlands: IOS Press.

Goble, C., Corcho, O., Alper, P. & De Roure, D. (2006). E-science and the semantic web: A symbiotic relationship. In *Discovery Science 2006, Barcelona, Spain, LNAI* (Vol. 4265)(pp. 1-12). New York: Springer.

Goble, C.A., Pettifer, S., Stevens R. & Greenhalgh, C. (2003). Knowledge Integration: In silico Experiments in Bioinformatics. In I. Foster, C. Kesselman (Eds.), *The Grid: Blueprint for a New Computing Infrastructure* (pp. 121-124). San Francisco, CA: Morgan Kaufmann Publishers Inc.

Goloboff, P. (1999). Analyzing Large Data Sets in Reasonable Times: Solutions for Composite Optima. *Cladistics,* 15(4): 415-428.

Gonzales, J., Pomel, S., Breton, V., Clot, B., Gutknecht, J.L., Irrthum, B. & Legré, Y. (2005) Empowering humanitarian medical development using grid technology. *Methods of Information in Medicine, 44*, 186-189.

Good, J., Laity, A., Jacob, J.C., & Katz, D.S. (2005). Pegasus: a framework for mapping complex scientific workflows onto distributed systems. *Scientific Programming, 13*(3), 219-237.

Goodale, T., Jha, S., Kaiser, H., Kielmann, T., Kleijer, P., Merzky, A., Shalf, J., Smith, Ch. (2008). A Simple API for Grid Applications (SAGA). Grid Forum Document GFD.90. Open Grid Forum. Retrieved from www.ogf.org/documents/GFD.90.pdf

Goodhew, P. J. (1984). *Specimen Preparation for Transmission Electron Microscopy of Materials.* Oxford, UK: Oxford University Press.

Google (2008). Google App Engine. Retrieved from http://code.google.com/appengine/

Google Groups (2008). Cloud Computing. Retrieved from http://groups.google.ca/group/cloud-computing

Gordeeva E., Katritzky A. R.. Shcherbukhin, V. V., & Zefirov, N. S. (1993). Rapid Conversion of Molecular Graphs to Three-dimensional Representation Using the MOLGEO Program. *J. Chem. Inf. Comput. Sci., 33,* 102-111.

Gordon, P. M., Trinh, Q., & Sensen, C. W. (2007). Semantic web service provision: a realistic framework for bioinformatics programmers. *Bioinformatics, 23*(9), 1178-1180.

Gordon, R., Bender, R., & Herman, G. T. (1970). Algebraic reconstruction techniques (ART) for three-dimensional electron microscopy and x-ray photography. *Journal of Theoretical Biology, 29*(3), 471-481.

Gottschling, M., Stamatakis, A., Nindl, I., et al. (2007). Multiple Evolutionary Mechanisms Drive Papillomavirus Diversification. *Molecular Biology and Evolution,* 24(5):1242-1258.

Govindarajan, S., Goldstein, R.A. (1996). Why are some proteins structures so common? *Proc Natl Acad Sci USA, 93*(8), 3341-5.

Gracy, J., & Argos, P. (1998). Automated protein sequence database classification. *Bioinformatics, 14,* 164-187.

Grassè, P. (1959). *La reconstruction du nid et les coordinations inter-individuelles chez belicositermes natalensis et cubitermes sp. la theorie de la stigmergie: Essai d'interprtation du comportement des termites constructeurs*, Insectes Sociaux 6, 41–84.

Green, M. L. & Miller, R. (2003). Grid computing in Buffalo. *Annals of the European Academy of Sciences,* 191-218.

Green, M. L. & Miller, R. (2004). Molecular structure determination on a computational & data grid. *Parallel Computing, 30,* 1001-1017.

Green, M. L. & Miller, R. (2004). Evolutionary molecular structure determination using grid-enabled data mining. *Parallel Computing, 30,* 1057-1071.

Green, M. L. & Miller, R. (2004). A client-server prototype for application grid-enabling template design. *Parallel Processing Letters, 14*, 241-253.

Grethe, J. S., Baru, C., Gupta, A., James, M., Ludaescher, B., Martone, M. E., Papadopoulos, P. M., et al. (2005). Biomedical informatics research network: Building a national collaboratory to hasten the derivation of new understanding and treatment of disease. *Stud Health Technol Inform, 112*, 100-109.

Grethe, J.S., Baru, C., Gupta, A., James, M., Ludaescher, B., Martone, M.E., Papadopoulos, P.M., Peltier, S.P., Rajasekar, A., Santini, S., et al (2005). Biomedical Informatics Research Network: Building a National Collaboratory to Hasten the Derivation of New Understanding and Treatment of Disease. *Stud Health Technol Inform, 112*, 100-109.

Grid Engine (2001). Open Source Project. Retrieved from http://gridengine.sunsource.net/

GridSphere (2008). Retrieved from www.gridsphere.org/gridsphere/gridsphere

GridWay (2008). Metascheduling Technologies for the Grid. Retrieved from www.gridway.org/

Grillo, G., Licciulli, F., Liuni S., Sbisà, E. & Pesole G. (2003). PatSearch: a program for the detection of patterns and structural motifs in nucleotide sequences. *Nucleic Acids Res., 31*, 3608–3612.

Grimshaw A. S., Natrajan A., Humphrey M. A., Lewis M. J., Nghuyen-Thong A., Karpovich J. F., Morgan M. M., & Ferrari A. (2003). J. From Legion to Avaki. In Berman F. Fox G., Hey A. (Eds.) *Grid Computing: Making the Global*. Hoboken, NJ: John Wiley and Sons.

GriPhyN (n.d.). *The Grid Physics Network*. Retrieved from http://www.griphyn.org

Groth, P., Jiang, S. Miles, S. Munroe, S., Tan, V., Tsasakou, S., & Moreau, L. (2006). *An Architecture for Provenance Systems* (Tech. Rep.). Southampton, UK: University of Southampton..

GRyCAP (2007). *Grid Biocomputation Portal*. Retrived July 2008 from http://www.grycap.upv.es/bio

Guan, J. W., & Bell, D. A. (1991). *Evidence Theory and Its Application*. Amsterdam, the Netherlands: Elsevier Science.

Guardian (2007). *Family doctors shun national database of patients' records, 20 Nov 2007*. Retrieved from www.guardian.co.uk/society/2007/nov/20/nhs.health

Guarracino, M. R., Cifarelli, C., Seref, O., & Pardalos, P. M. (2006). A Parallel Classification Method for Genomic and Proteomic Problems. *In 20th international Conference on Advanced information Networking and Applications* (Vol.02) (pp. 588-592). AINA. Washington, DC: IEEE Computer Society,.

Gubala, T. & Hoheisel, A. (2006). Highly dynamic workflow orchestration for scientific applications. In *CoreGRID Integation Workshop 2006 (CIW06)*, (pp. 309-320). Krakow, Poland: ACC CYFRONET AGH.

Gubala, T., Balis, B., Malawski, M., Kasztelnik, M, Nowakowski, P., Assel, M., et al. (2008). ViroLab Virtual Laboratory. In *Proceedings of Cracow Grid Workshop '07, ACC CYFRONET AGH, Krakow*.

Gubala, T., Balis, B., Malawski, M., Kasztelnik, M., Nowakowski, P., Assel, M., et al. (2007). ViroLab Virtual Laboratory. In Bubak, M., Turala, M., and Wiatr, K., (Eds.), *Proceedings of Cracow Grid Workshop - CGW'07*, (pp. 35-40). Krakow, Poland: ACC CYFRONET AGH

Gubala, T., Kasztelnik, M., Malawski, M., & Bubak, M. (2007). Development and execution of collaborative application on the virolab virtual laboratory. In Bubak, M., Turala, M., and Wiatr, K., editors, *Proceedings of Cracow Grid Workshop - CGW'07*, (pp. 41-46). Krakow, Poland: ACC CYFRONET AGH.

Gustafson, J. (1987). Reevaluating Amdahl's Law, CACM, vol. 31, pp. 532-533.

Guyton, A.C., Jones, C.E., & Coleman, T.G. (1973). *Circulatory Physiology: Cardiac Output and its Regulation* (2nd Ed.). Philadelphia: W.B. Saunders.

Gwak. K.-W., Paden, B.E., Noh, M.D., & Antaki, J.F. (2006). Fluidic Operational Amplifier for Mock Circulatory Systems. *IEEE Transactions of Control Systems Technology, 14*(4), 602-612.

Gygi, S.P., Rist, B., Gerber, S.A., Turecek, F., Gelb, M.H., & Aebersold, R. (1999). Quantitative analysis of complex protein mixtures using isotope-coded affinity tags. *Nature Biotechnology, 17*(10), 994–999

H. F. Gagliardi, F. A.B. da Silva, & D. Alves. (2006) Automata Network Simulator Applied to the Epidemiology of Urban Dengue Fever. In *Proceedings of the ICCS'06, Lecture Notes in Computer Science, 3993* (pp. 297-304). Berlin, Germany: Springer

Hall, L. O., Chawla, N., & Bowyer, K. W. (1998). Combining decision trees learned in parallel. In *Working Notes of the KDD-97 Workshop on Distributed Data Mining* (pp. 10-15).

Hamari, Z., Juhász, A. & Kevei, F. (2002). Role of mobile introns in mitochondrial genome diversity of fungi (a mini review). *Acta Microbiol Immunol Hung. 49*(2-3), 331-5.

Han, D.K., Eng, J., Zhou. H. & Aebersold, R. (2001). Quantitative profiling of differentiation-induced microsomal proteins using isotope- coded affinity tags and mass spectrometry. *Nature Biotechnology, 19*, 946-51.

Handschuh, S., Staab, S., & Ciravegna, F. (2002). S-CREAM - semi-automatic creation of metadata. In Gómez-Pérez, A. and Benjamins, R. V., (Eds.), *EKAW '02: Proceedings of the 13th International Conference on Knowledge Engineering and Knowledge Management. Ontologies and the Semantic Web*, London, (pp. 358-372). Berlin, Germany: Springer-Verlag.

Hansen, C.D. & C.R. Johnson (2005). *The Visualization Handbook*. Burlington, MA: Elsevier Butterworth-Heineman..

Hansen, H., Bachmann, L., & Bakke, T.A. (2003). Mitochondrial DNA variation of *Gyrodactylus* spp. *Monogenea, Gyrodactylidae* populations infecting Atlantic salmon, grayling, and rainbow trout in Norway and Sweden. *International Journal of Parasitology*, 33(13): 1471-1478.

Harris, M. A., Clark, J., et al. (2004). The Gene Ontology (GO) database and informatics resource. *Nucleic Acids Res, 32*, D258-61.

Hastie, T., Tibshirani, R., & Friedman, J. (2002). *The Elements of Statistical Learning*. New York: Springer.

Havel, T.F. (1998). *Distance Geometry: Theory, Algorithms, and Chemical Applications in the Encyclopedia of Computational Chemistry*. Hoboken, NJ: J. Wiley & Sons.

Hawkins, P., & Skillman, G. (2006). Ligand-Based Design Workflow.

Health Level 7 (1997). Retrieved July 2008 from http://www.hl7.org/

Health Level Seven Inc. (2008). *Health Level Seven homepage*. Retrieved March 8, 2008 from http://www.hl7.org/

Health-e-Child, The EU FP6 Information Societies Technology Project (2008). Retrieved October 20, 2008, from http://www.health-e-child.org/.

Heath, T. & Motta, E. (in press). Ease of interaction plus ease of integration: Combining web2.0 and the semantic web in a reviewing site. *Journal of Web Semantics*, (Special Issue on Web2.0 and the Semantic Web).

Hebert, P.D., Ratnasingham, S., deWaard J.R. (2003). Barcoding animal life: cytochrome c oxidase subunit 1 divergences among closely related species. *Proc Biol Sci., 270*, (1), S96-9.

Heinicke, M.P., Duellman, W.E., & Hedges, S.B. (2007). From the Cover: Major Caribbean and Central American frog faunas originated by ancient oceanic dispersal. *Proceedings of the National Academy of Sci*

Heinrichs, W.L. (1999). The Critical Path from Tissue Slices to Surgical Simulation: What do Surgeons Want? In *Proc. 12'th IEEE Symposium on Computer-Based Medical Systems, June 18-20, Stamford, CT*, (pp. 118-123).

Heinzlreiter, P. & Kranzlmueller, D. (2003). Visualization Services on the Grid: The Grid Visualization Kernel. *Parallel Processing Letters, 13*(2), 135–148.

Henderson, R. (1995). The potential and limitations of neutrons, electrons and x-rays for atomic resolution microscopy of unstained biological molecules. *Quarterly Review of Biophysics, 28*(1), 171-193.

Henderson, R. (2004). Realising the potential of electron cryomicroscopy. *Quarterly Review of Biophysics, 37*(1), 3-13.

Henderson, R.L. (1995). Job Scheduling Under the Portable Batch System. *Proceedings of the workshop on Job Scheduling Strategies for Parallel Scheduling* (pp. 279.294). London: Springer-Verlag.

Hendrix, E. M. T., Ortigosa, P. M., & García, I. (2001). On success rates for controlled random search. *Journal of Global Optimization, 21*(3), 239-263.

Henikoff, J.G., Henikoff, S., & Pietrokovski, S. (2000). Increased coverage of protein families with the Blocks Database servers. *Nucleic Acids Research, 28*, 228-230.

Henikoff, S., & Henikoff, J.G. (1992). Amino Acid Substitution Matrices from Protein Blocks. *Proceedings of the National Academy of Sciences of the United States of America, 89*(22), 10915-10919.

Hennessy, J. L., & Patterson, D. A., (2006). *Computer Architecture (4th Ed.): A Quantitative Approach*. San Francisco, CA: Morgan Kaufmann.

Hernandez, P., Gras, R., Frey, J., & Appel, R. D. (2003). Popitam: towards new heuristic strategies to improve protein identification from tandem mass spectrometry data. *Proteomics, 3*(6), 870-878.

Hernandez, P., Müller, M. & Appel, R. D. (2006). Automated protein identification by tandem mass spectrometry: issues and strategies. *Mass Spectrom Rev, 25*(2) 235-254.

Herveg, J. A. M., & Poullet, Y. (2003,January 16-172003). *Directive 95/46 and the use of GRID technologies in the heathcare sector: selected legal issues.* Paper presented at the Healthgrid 2003, Lyon, France.

Hes, R., & Borking, J. J. (1998). *Privacy-enhancing technologies: the path to anonymity* (Rev. Rd.). The Hague, the Netherlands: Registratiekamer.

Hey, A. J. G., & Trefethen, A. E. (2002). The UK e-Science Core Programme and the Grid. *Future Generation Computer Systems 18*(8), 1017-1031.

Hey, T., & Trefethen, A. E. (2005). Cyberinfrastructure for e-Science. *Science, 308*, 817-821.

Hibbett, D. S., Nilsson, R. H., Snyder, M., Fonseca, M., Costanzo, J. & Honfeld M. (2005). Automated phylogenetic taxonomy: an example in the homobasidiomycetes (mushroom-forming fungi). *Syst. biol., 54*(4), 660–668.

HIPPA (2006). Retrieved July 2008 from http://www.hhs.gov/ocr/hipaa/

HL7. (2005). HL7 Receives ANSI Approval of Three Version 3 Specifications Including CDA, Release 2.

Hluchy, L., Habala, O., Babik, M., Laclavik, M., Balogh, Z., & Gatial, E. (2007). Knowledge-based platform for environmental risk management. In Volkert, J., Kranzlmüller, D., & Schreiner, W., (Eds.), *ISPDC '07: Proceedings of the Sixth International Symposium on Parallel and Distributed Computing*, Washington, DC, (p.1). New York: IEEE Computer Society.

HmtDB: Human Mitochondrial Genomic Resource Based on Variability Studies. Retrieved from http://www.hmtdb.uniba.it/.

Hofer, J. (2004). *Distributed decision tree induction within the grid data mining framework gridminer-core* (Tech. Rep. No. 2004-04). Vienna, Austria: University of Vienna, Institute for Software Science.

Hoffmann, K., Bucher, P., Falquet, L., & Bairoch, A. (1999). The PROSITE database, its status in 1999. *Nucleic Acids Research, 27*, 215-219.

Hohne, K.H. & Bernstein, R. (1986). Shading 3D-Images from CT Using Gray-Level Gradients. *IEEE Transactions on Medical Imaging, 5*(1), 45-47.

Honore, B., Ostergaard, M., & Vorum, H. (2004). Functional genomics studied by proteomics. *Bioessays, 26*(8), 901-915.

Hoogland, C., Sanchez, J.C., Tonella, L., Bairoch, A., Hochstrasser, D.F., & Appel, R.D.(1998). Current status of the SWISS-2DPAGE database, Nucleic. *Acids Research 26*(1), 332–333

Housley, R. & Polk, T. (2001). *Planning for PKI: Best Practices Guide for Deploying Public Key Infrastructures*. Hoboken, NJ: Wiley Computer Publishing.

Houstis, E., Gallopoulos, E., Bramley, R. & Rice J. (1997). Problem-Solving Environments for Computational Science. *IEEE Comput. Sci. Eng., 4*(3), 18–21.

Houstis, E., Gallopoulos, E., Bramley, R., & Rice, J. (1997). Problem-Solving Environments for Computational Science. *IEEE Comput. Sci. Eng., 4*(3), 18–21.

Huang, X., Huang, L. & Li, M. (2006). Grid-enabled medical image processing application system based on OGSA-DAI techniques: In Proceedings of the International workshops on advanced Web, network technologies and Applications, pp. 460-464. In Shen, H. T., Li, J., Li, M., Ni, J. & Wang, W. (Eds.), *Lecture notes in Computer Science,* (Vol. 3842). Aachen, Germany: Springer-Verlag.

Huber V. (2001). Supporting Car-Parrinello Molecular Dynamics with Unicore. In *Computational Science ICCS - San Francisco, USA, Proceedings, Pt. 1,* (Vol. 2073), (pp. 560-567). Berlin, Germany: Springer Verlag.

Huedo, E., S. Montero, R., & M. Llorente, I. (2004). A framework for adaptive execution in grids. *Software: Practice and Experience, 34*(7), 631-651).

Huhns, M. N. & Singh, M. P. (2005). Service-oriented computing: Key concepts and principles. *IEEE Internet Computing, 9*(1), 75-81

Humphrey, W., Dalke, A., & Schulten, K. (1996). VMD: Visual molecular dynamics. *Journal of Molecular Graphics, 14,* 33-38.

Humphreys, G., Houston, M., Ng, R., Frank, R., Ahern, Kirchner, P.D. & J.T. Klosowski (2002). Chromium: A Stream-Processing Framework for Interactive Rendering on Clusters. *ACM Transactions on Graphics, 21*(3), 693-702.

Hwang, S., & Kesselman, C. (2004). A Flexible Framework for Fault Tolerance on the Grid. *Journal of Grid Computing, 1*(3), 251-272.

I. Andoulsi, J. Herveg, V. Stroetmann, K. Stroetmann, A. Dobrev, C. Van Doosselaere, P. Wilson (2007). *Bottlenecks & Challenges and RTD Responses for Legal, Ethical, Social, and Economic Aspects of Healthgrids – Roadmap II.* Retrieved July 2008 from http://eu-share.org/about-share/deliverables-and-documents.html

I. Blanquer, V. Hernandez, D. Segrelles (2007). TRENCADIS – Secure Architecture to Share and Manage DICOM Objects in a Ontological Framework Based on OGSA. *From Genes to Personalized HealthCare: Grid Solutions for the Life Sciences, Studies in Health Technology and Informatics, 126,* 115-124.

Iakovidis, I. (2005). *HealthGrid – 3 sided concept: ICT for Health, ISTAG WG.* Retrieved August 30, 2007, from http://www.who.int/classifications/terminology/iakovidis.pdf

Iamnitchi, A., & Foster, I. (2005). *Interest-aware information dissemination in small-world communities,* Paper presented at the 14th IEEE International Symposium on High Performance Distributed Computing (HPDC 2005), Research Triangle Park, USA.

Iavindrasana, J., Loiacono, L., Müller, H., Periz, I., Summers, P., & Wright, J. (2008). Access to Clinical Information Systems for Research in the Life Sciences - Security and Privacy Considerations, *HealthGrid Conference 2008.* Chicago, IL.

Ichikawa, K., Date, S., Krishnan, S., Li, W. W., Nakata, K., Yonezawa, Y., Nakamura, H., Shimojo, S. (2007). *In OPAL OP: An Extensible Grid-Enabling Wrapping Tool For Legacy Applications.* 3rd International Workshop on Grid Computing & Applications (GCA2007), Singapore.

INFN Production Grid. Retrieved from http://grid-it.cnaf.infn.it/.

INSPIRE Project (2008). Retrieved July 12, 2008 from http://www.inspire-geoportal.eu

Institut Pasteurand the HKU-Pasteur Research Centre (2005). *Information on the international outbreak of the illness known as severe acute respiratory syndrome (SARS).* Retrieved July 2008 from http://www.pasteur.fr/recherche/unites/REG/SARS.html

Integrating the Healthcare Enterprise (2008). Retrieved July from 2008 from http://www.ihe.net/

International Organization for Standardization (2007). *Codes for the representation of names of countries and their subdivisions – Part 2: Country subdivision code.* Geneva, Switzerland: International Organization for Standardization.

Internet2 Shibboleth, (2006). *Shibboleth Architecture Protocols and Profiles.* Retrieved from http://shibboleth.internet2.edu/docs/draft-mace-shibboleth-arch-protocols-latest.pdf

Internet2 Shibboleth, (2006). *Shibboleth Architecture Technical Overview.* Retrieved from http://shibboleth.internet2.edu/docs/draft-mace-shibboleth-tech-overview-latest.pdf

Itoh, M., Akutsu, T., & Kanehisa, M. (2004) Clustering of database sequences for fast homology search using upper bounds on alignment score. *Genome Informatics, 15*(1), 93-104.

Itoh, M., Goto, S., Akutsu, T., & Kanehisa, M. (2005). Fast and accurate database homology search using upper bounds of local alignment scores. *Bioinformatics, 21*(7), 912-921.

ITU-T Recommendation X.509 (2001) | ISO/IEC 9594-8: 2001, Information technology – Open Systems Interconnection – Public-Key and Attribute Certificate Frameworks.

ITU-T Recommendation X.812 (1995). | ISO/IEC 10181-3:1995, Security Frameworks for open systems: Access control framework.

J. Roboz, (1968). *Mass spectrometry instrumentation and techniques*. New York: Intersience Publisher.

Jaakkola, T., et al. (2000). A discriminative framework for detecting remote protein homologies. *Journal of Molecular Biology, 7*, 95-114.

Jacob, B., Ferreira, L., Bieberstein, N., Gilzean, C., Girard, J.-Y., Strachowski, R., & Yu, S. (2003). Enabling applications for Grid computing with Globus. IBM Redbook, retrieved from www.redbooks.ibm.com/abstracts/sg246936.html?Open

Jacq N., Breton B., Chen H., Ho L., Hofmann M., Lee H., et al. (2006). *Large Scale In Silico Screening on Grid Infrastructures*. Proceedings of the LSGRID2006, Yokohama, Japan (pp.123–136).

Jacq N., Salzemann J., Jacq F., Legré Y., Medernach E., Montagnat J., Maaß A., et al. (2008). Grid-enabled Virtual Screening Against Malaria. *J Grid Computing 6*,29–43.

Jacq, N., Blanchet, C., Combet, C., Cornillot, E., Duret, L., Kurata, K., Nakamura, H., Silvestre, T., & Breton, V. (2004). Grid as a Bioinformatic Tool. *Parallel Computing, 30*(9-10), 1093-1107.

Jacq, N., Blanchet, C., Combet, C., Cornillot, E., Duret, L., Kurata, K., Nakamura, H., Silvestre, T., Breton, V. (2004). Grid as a bioinformatics tool. *Parallel Computing, Special Issue: High-performance parallel biocomputing, 30*.

Jacq, N., Salzemann, J., Legre, Y., Reichstadt, M., Jacq, F., Zimmermann, M., et al. (2006). Demonstration of in silico docking at a large scale on Grid infrastructure. *Stud Health Technol Inform, 120*, 155-157.

JaJa, J., Varshney, A. & Shi, Q. (2008). Parallel Algorithms for Volumetric Surface Construction. In S. Rajasekaran & J. Reif (eds.), *Handbook of Parallel Computing*, (pp. 27-17). Boca Raton, FL: Chapman and Hall.

Jankun-Kelly, T.J. & K. Meta (2006). Superellipsoid-based, real Symmetric Traceless Tensor Glyphs Motivated by Nematic Liquid Crystal Alignment Visualization. *IEEE Transactions on Visualization and Computer Graphics*, (pp. 1197-1204).

Januzaj, E., Kriegel, H. P., & Pfeifle, M. (2004). Scalable density-based distributed clustering. *In the 8th European Conference on Principles and Practice of Knowledge Discovery in Databases* (pp. 231-244). Berlin, Germany: Springer-Verlag.

Jeannot, E. (2005, April), Improving middleware performance with AdOC: an Adaptive Online Compression library for data transfer. *IEEE International Parallel and Distributed Processing Symposium* (p. 70). Denver, CO. New York: IEEE Computer Society.

Jeffrey, S., Morris, M., Kevin, R., Coombes, J., Koomen, K., Baggerly, K., & Kobayashi, R. (2005). Feature extraction and quantification for mass spectrometry in biomedical applications using the mean spectrum. *Bioinformatics, 21*(9) 1764–1775.

Jeffries, N. (2005). Algorithms for alignment of mass spectrometry proteomic data. *Bioinformatics, 21*(14), 3066–3073.

Jensen, ED., Locke, CD. & Tokuda, H. (1985, December). *A time driven scheduling model for real-time operating systems* (pp. 112–122). IEEE Real-Time Systems Symposium, San Diego, CA. New York: IEEE Computer Society.

Jha, S., Kaiser, H., El Khamra, Y., & Weidner, O. (2007). Design and Implementation of Network Performance Aware Applications Using SAGA and Cactus. 3rd IEEE Conference on eScience and Grid Computing, Bangalore, India, 10-13 Dec, pp. 143 – 150.

Jithesh, P. V., Kelly, N., Wasnik, S., Donachy, P., Harmer, T., Perrott, R., et al. (2005). Bioinformatics application integration in genegrid. *In the UK e-Science All Hands Meeting.*

Johnson, G. (1998). Collaborative Visualization 101. *ACM SIGGRAPH, 32*(2), 8-11.

Jones, D. M., Fenner, J. W., Berti, G., Kruggel, F., Mehrem, R. A., Backfrieder, W., Moore, R. and Geltmeier, A. (2004). The GEMSS Grid: An evolving HPC environment for medical applications. In *Proceedings of HealthGrid 2004*. Retrieved March 8, 2008, from http://www.ccrl-nece.de/gemss/Reports/Jones-healthgrid2004.pdf

Jones, K. E., Pattel, N. G., Levy, M. A., Storeygard, A., Balk, D., Gittleman, J. L., Daszak, P. (2008). Global Trends in Emerging Infectious Diseases. *Nature, 451*(February 21), 990-994.

Jong, M.D., & Hien, T.T. (2006). Avian influenza A (H5N1). *J Clin Virol, 35*(1), 2-13.

Joshi, A., Karypis, G., & Kumar, V. (1998). ScalparC: A new scalable and efficient parallel classification algorithm for mining large datasets. In *11th Inter national Parallel Processing Symposium*. New York: IEEE Computer Society Press.

Jourdan L., Dhaenens C., Talbi E.-G. (2004). A Parallel adaptive GA for linkage disequilibrium in genomics. NIDISC'2004 Workshop. *IEEE International Parallel & Distributed processing symposium (IPDPS 2004),* Santa Fe, NM.

JSDL 1.0. Job Submission Language Specification (2006). Retrieved March 6, 2008 from http://www.gridforum.org/documents/GFD.56.pdf

Kabsch, W. & Sander, C. (1983). Dictionary of protein secondary structure: pattern recognition of hydrogen-bonded and geometrical features. *Biopolymers, 22,* 2577–2637.

Kakazu, K., Cheung, L., & Lynne, W. (2004). The Cancer Biomedical Informatics Grid (caBIG): pioneering an expansive network of information and tools for collaborative cancer research. *Hawaii Med J, 63*(9), 273-275.

Kal L., Skeel R., Bhandarkar M., Brunner R., Gursoy A., Krawetz N., et al. (1999). NAMD2: Greater Scalability for Parallel Molecular Dynamics. *J. Comp. Phys., 151,* 283–312.

Kalé, L., Skeel, R., Bhandarkar, M., Brunner, R., Gursoy, A., Krawetz, N., Phillips, J., Shinozaki, A., Varadarajan, K., & Schulten, K. (1999). NAMD2: Greater scalability for parallel molecular dynamics. *Journal of Computational Physics, 151,* 283-312.

Kalfoglou, Y. & Schorlemmer, M. (2003). Ontology mapping: the state of the art. *The Knowledge Engineering Review, 18*(1), 1-31.

Kandaswamy, G., Fang, L., Huang, Y., Shirasuna, S., Marru, S., & Gannon, D. (2006). Building web services for scientific grid applications. *IBM Journal of Research and Development, 50*(2/6), 249-260.

Kapp, E. A., Schütz, F., Connolly, L. M., Chakel, J. A., Meza, J. E., Miller, C. A., et al. (2005). 'An evaluation, comparison, and accurate benchmarking of several publicly available MS/MS search algorithms: sensitivity and specificity analysis. *Proteomics, 5*(13), 3475-3490.

Karasavvas, K., Antonioletti, M., Atkinson, M., Hong, N. C., Sugden, T., Hume, A., et al. (2005). Introduction to OGSA-DAI Services. In *Scientific Applications of Grid Computing, First International Workshop* (Vol. 3458)(pp. 1-12). Beijing: Springer Berlin / Heidelberg.

Kargupta, H., & Chan, P. (Eds.) (2000). *Advances in Distributed and Parallel Knowledge Discovery.* Cambridge, MA: MIT Press.

Karle, J. & Hauptman, H. A. (1956). A theory of phase determination for the four types of non-centrosymmetric space groups $1P222$, $2P22$, $3P_12$, $3P_22$. *Acta Crystallogr., 9,* 635-651.

Karonis, N.T., Toonen, B., & Foster, I. (2002). MPICH-G2: A Grid-Enabled Implementation of the Message Passing Interface. Journal of Parallel and Distributed Computing (JPDC), vol. 63, pp. 551-563.

Kasztelnik, M., Bubak, M., Gorka, C., Malawski, M., & Gubala, T. (2007). Fault tolerant grid registry. In Kacsuk, P., Fahringer, T., & Nemeth, Z. (Eds.), *Distributed and Parallel Systems*, Distributed and Parallel Systems, (pp. 135-144). Berlin, Germany: Springer.

Katz, D.S., Anagnostou, N., Berriman, G.B., Deelman, E., Good, J., Jacob, J.C., Kesselman, C., et al. (2006). Astronomical Image Mosaicking on a Grid: Initial Experiences. In Martino, B.D., Dongarra, J., Hoisie, A., Yang, L., & Zima, H. (Ed.), *Engineering the Grid-Status and Perspective*. Valencia, CA: American Scientific Publishers.

Kazutaka, K., Kazuharu, M., Keichi, K. & Takashi, M. (2002). MAFFT: A novel method for rapid multiple sequence alignment based on fast Fourier transform. *Nucl. Acids Res.*, *30*(14), 3059–3066.

Keahey, K., Foster, I., Freeman, T. & Zhang, X. (2006). Virtual Workspaces: Achieving Quality of Service and Quality of Life in the Grid. *Scientific Programming, 13*(4), 265-275.

Keller, A., Eng. J., Zhang, N., Li. X., & Aebersold, R. (2005). A uniform proteomics MS/MS analysis platform utilizing open XML file formats. *Mol Syst Biol, 1*, 2005.0017

Keller, A., Nesvizhskii, A. I., Kolker, E., & Aebersold R., (2002). Empirical statistical model to estimate the accuracy of peptide identifications made by MS/MS and database search. *Analytical. Chemistry, 74*(20), 5383-92

Kelly, N., Jithesh, P.V., Wasnik, S., McLaughlin, R., Fragoso, F., Donachy, P., et al. (2005). The GeneGrid Portal: A User Interface for a Virtual Bioinformatics Laboratory. In *UK e-Science All Hands Meeting (AHM 2005).*

Ken Lin, Booz Allen Hamilton, Gary Daemer. (2006) *caBIG Security Technology Evaluation.*(White Paper). Retrieved July 2008 from https://cabig.nci.nih.gov/workspaces/Architecture/Security_Tech_Eval_White_Paper/.

Kesselman, C., Foster, I., & Prudhomme, T. (2004). Distributed Telepresenze:The NESSgrid Earthquake Engineering Collaboratory. In Elsevier (Ed.), *The grid 2* (2nd Ed, pp. 81-93). San Francisco, CA: Morgan Kaufmann.

Khoussainov, R., Zuo, X., & Kushmerick, N. (2004). Grid-enabled weka: A toolkit for machine learning on the grid. *ERCIM News, 59.*

Kikinis, R. Warfield, S. & Westin, C. (1998). *High Performance Computing (HPC) in Medical Image Analysis (MIA) at the Surgical Planning Laboratory (SPL).* Paper presented at the 3rd High Performance Computing Asia Conference & Exhibition, Singapore, Malaysia.

Kim, J., Deelman, E., Gil, Y., Mehta, G., & Ratnakar, V. (2007). Provenance trails in the Wings/Pegasus system. In *Concurrency and Computation: Practice and Experience*, San Francisco, CA: Wiley InterScience, *20*(5), 587-597.

Kim, J., Kim, H., & Han, D. (2006). Performance Evaluation of BLAST on SMP Machines. *Lecture Notes in Computer Sciences,* 4331, 668.

Kindlmann, G. (2004). Superquadric Tensor Glyphs, Joint Eurographics. In *IEEE TCVG Symposium on Visualization*, (pp. 147-154).

Kirkpatrick, S., Gellatt Jr., C.D. & Vecchi, M.P. (1982). *Optimization by simulated annealing.* Yorktown Heights, NY: IBM T.J. Watson Research Center.

Kiryakov, A., Popov, B., Terziev, I., Manov, D., & Ognyanoff, D. (2004). Semantic annotation, indexing, and retrieval. *Web Semantics: Science, Services and Agents on the World Wide Web, 2*(1), 49-79.

Kleinberg, J. (2000). Navigation in a Small World , Nature 406(2000), 845.

Klose, J. (1999). Genotypes and phenotypes. *Electrophoresis, 20*, 643-652.

Kluszczyński R., & Bała P. (2008). Supporting NAMD Application on the Grid using GPE. To appear in LNCS.

Kluszczyński R., Borcz M., & Bała P. (2007). BLAST Application on the GPE/UnicoreGS grid. In *Euro-Par 2006 Workshops* (pp. 245-253). Berlin, Germany: Springer-Verlag.

Knierbein, B., Reul, H., Eilers, R., Lange, M., Kaufmann, R., & Rau, G. (1992). Compact mock loops of the systemic and pulmonary circulation for blood pump testing. *The International Journal of Artificial Organs, 15*(1), 40-48.

Kohlbacher, O., Reinert, K., Gröpl, C., Lange, E., Pfeifer, N., Schulz-Trieglaff, O., & Sturm, M. (2007). TOPP - The OpenMS Proteomics Pipeline. *Bioinformatics, 23*(1), 191-197.

Kolff ,W.J. (1959). Mock circulation to test pumps designed for permanent replacement of damaged hearts. *Cleve Clin Q, 26*, 223-226.

Kollman P. (2001). *AMBER (Assisted Model Building with Energy Refinement)*. San Francisco, CA: University of California..

Konagaya, A. (2006). Trends in life science grid: from computing grid to knowledge grid. *BMC Bioinformatics, 7 Suppl 5*, S10.

Kondylakis, H., Analyti, A., & Plexousakis, D. (2007). *Quete: Ontology-based query system for distributed sources*. (*LNCS* 4690, pp. 359-375).

Konishi, F., Yagi, T., & Konagaya, A. (2006). MolWorks+G: Integrated Platform for the Acceleration of Molecular Design by Grid Computing. In *Grid Computing in Life Science* (pp. 134-141). Singapore: World Scientific.

Korf. I., Yandell, M., & Bedell, J., (2003). BLAST. Sebastopol, CA: O'Reilly.

Korhonen, J. (2006). *Finpro, Post workshop report, Fifth EGEE industry day concentrated on Challenges in Bioscience*, November 2006, Helsinki, Finland.

Kosar, T. & Livny M. (2004). Stork: Making Data Placement a First Class Citizen in the Grid. In *Proceedings of 24th IEEE International Conference on Distributed Computing Systems (ICDCS2004)*, Tokyo, Japan.

Krafzig, D., Banke, K., & Slama, D. (2004). *Enterprise SOA: Service-Oriented Architecture Best Practices (The Coad Series)*. Upper Saddle River, NJ: Prentice Hall PTR.

Kranzlmueller, D. (n.d.). *The GVK/GVid Approach: Visualization on the Grid*. Retrieved from http://www.ogf.org/OGF19/materials/562/kranzlmueller.pdf

Kranzlmueller, D., Heinzlreiter, P. and J. Volkert (2004). *Grid-Enabled Visualization with GVK, in Grid Computing* (LNCS Vol. 2970) (pp. 139-146). Berlin, Germany: Springer-Verlag.

Krause, A., & Vingron, M. (1998). A set-theoretic approach to database searching and clustering. *Bioinformatics, 14*, 430-438.

Krefting, D., Bart, J., Beronov, K., Falkner, J., Hartung, M., Hoheisel, A., et al. (2008). MediGRID: A Userfriendly Secured Grid Infrastructure. *Future Generation Computer Systems Journal - International Journal of Grid Computing.*

Krishnan, A. (2004). A Survey of life sciences applications on the grid. *New Generation Computing, 22*(2), 111-126.

Krishnan, A. (2005). GridBLAST: a Globus-based high-throughput implementation of BLAST in a Grid Computing framework. *Concurrency and Computation: Practice and Experience, 17*(13), 1607-1623.

Krishnan, S., & Bhatia, K. (2007). SOAs for Scientific Applications: Experiences and Challenges. In *the 3rd International Conference on e-Science and Grid computing.*

Krishnan, S., Stearn, B., Bhatia, K., Baldridge, K. K., Li, W. W., and Arzberger, P. (2006). Opal: Simple Web Services Wrappers for Scientific Applications. In *IEEE International Conference on Web Services*, (pp 823-832).

Krogh, A., et al. (1994). Hidden Markov models in computational biology: applications to protein modeling. *Journal of Molecular Biology, 235*, 1501-1531.

Kryza, B., Dutka, L., Slota, R., & Kitowski, J. (2008). Supporting management of dynamic virtual organizations in the grid through contracts. In Bubak, M., Turala, M., and Wiatr, K., (Eds.), *Proceedings of Cracow'07 Grid Workshop*, (pp. 140-147). Krakow, Poland: ACC CYFRONET AGH.

Kuperman, G. J. (1999). Improving response to critical laboratory values with automation: results of a randomised controlled trial. *J Am Med Inform Assoc, 6*, 512-522.

L. Moreau, et al. (2007). The first provenance challenge. *Concurrency and Computation: Practice and Experience, 20*(5), 409-418. San Francisco, CA: Wiley InterScience.

Laidlaw, D.H., Ahrens, E.T., Kremers, D., Avalos, M.J., Jacobs, R.E. & Readhead, C. (1998). Visualizing Diffusion Tensor Images of the Mouse Spinal Cord. In *IEEE Visualization '98*, (pp. 127-134).

Lamehamedi, H., Szymanski, B., Shentu, Z., & Deelman, E. (2002). Data replication strategies in grid environments. In *Proceedings Of the Fifth International Conference on Algorithms and Architectures for Parallel Processing*, Beijing, China, (pp. 378-383). New York: IEEE Computer Society Press.

Lamehamedi, H., Szymanski, B., Zujun Shentu, & Deelman E. (2002, October). Data Replication Strategies in Grid Environments. In *Proceedings of 5th International Conference on Algorithms and Architecture for Parallel Processing, ICA3PP'2002*, (378-383). New York: IEEE Computer Science Press.

Lanave, C., Preparata, G., Saccone, C. & Serio, G. (1984). A new method for calculating evolutionary substitution rates. *Journal of Molecular Evolution, 20*, 86-93.

Lane, C. S. (2005). Mass spectrometry-based proteomics in the life sciences. *Cell Mol Life Sci, 62*(7-8), 848-869.

Langerhans, R.B., Knouft, J.H., Losos, J.B. (2006). Shared and unique features of diversification in Greater Antillean Anolis ecomorphs. *Evolution, 60*(2), 362-9.

Laskowski, R.A., MacArthur, M.W., Moss, D.S. & Thornton, J.M. (1993). PROCHECK: a program to check the stereochemical quality of protein structures. *J. Appl. Cryst.* (26), 283–291.

Lathers, A., Su, M.H., Kulungowski, A., Lin, A.W., Mehta, G., Peltier, S.T., Deelman, E., & Ellisman, M.H. (2006). Enabling Parallel Scientific Applications with Workflow Tools. *Proceedings of Challenges of Large Applications in Distributed Environments* (pp. 55-60). New York: IEEE Computer Society.

Laure, E. et al. (2004, September). *Middleware for the next generation Grid infrastructure*. Paper presented at the 2004 Conference for Computing in High Energy and Nuclear Physics, Interlaken, Switzerland.

Laure, E., Fisher, S. M., Frohner, A., Grandi, C., Kunszt, P., Krenek, A., et al. (2006). Programming the Grid with gLite. *Computational Methods in Science and Technology, 12*(1), 33-45.

Laure, E., Hemmer F. et al., (2004). Middleware for the Next Generation Grid Infrastructure. *Computing in High Energy and Nuclear Physics* (CHEP), Interlaken, Switzerland.

Lavenier, D., Leroy, H., Mac Wing, M., Andonov, R., Hurfin, M., Raipin-Parvedy, P., Mouchard, L., & Guinand, F. (2003, January). GénoGRID: An Experimental Grid for Genomic Applications. In *HealthGrid 2003* Lyon.

Lavorgna, G., Dahary, D., Lehner, B., Sorek R., Sanderson, C.M. & Casari, G. (2004). In search of antisense. *Trends Biochem. Sci., 29*, 88–94.

Lavorgna, G., Triunfo, R., Santoni, F., Orfanelli, U., Noci, S., Bulfone, A., Zanetti, G. & Casari, G. (2005). AntiHunter 2.0: increased speed and sensitivity in searching BLAST output for EST antisense transcripts. *Nucleic Acids Res., 1*(33)W665-8.

Lawford-davies, J., Jenkins, J. (2005). Legal Issues to Address when Managing Clinical Information across Europe: The ECIT Case Study. *Healthgrid 2005, Studies in Health Technology and Informatics, IOS Press; 120*: 210-221.

Lawn, R.M., Schwartz, K., Patthy, L.(1997). Convergent evolution of apolipoprotein(a) in primates and hedgehog. *Proc Natl Acad Sci USA*, 94(22), 11992-7.

Lawrence, P. (1997). *Workflow handbook*. New York: John Wiley & Sons.

Lazarevic, A., & Obradovic, Z. (2002). Boosting algorithms for parallel and distributed learning. *Distributed and Parallel Databases, 11*(2), 203-229.

Lazarevic, A., Pokrajac, D., & Obradovic, Z. (2000). Distributed clustering and local regression for knowledge discovery in multiple spatial databases. *In 8th European Symposium on Artificial Neural Networks* (pp. 129-134).

Lee H., Salzemann J., Jacq N., Ho L., Chen H., Breton,V., Merelli L., Milanesi L., Lin S., & Wu Y. (2006). *Grid-enabled High Throughput in-silico Screening Against Influenza A Neuraminidase*. Proceedings of the NETTAB2006, Santa Margherita, 19–25

Lee, C. (2003). Grid Programming Models: Current Tools, Issues and Directions. *Grid Computing*. In G. F. Fran Berman, Tony Hey, Ed., pp. 555–578, Wiley Press, USA.

Lee, D., Redfern, O. & Orengo, C. (2007). Predicting protein function from sequence and structure. *Nat Rev Mol Cell Biol., 8*, 995-1005.

Lee, S., Wang, T. D., Hashmi, N., & Cummings, M. P. (2007). Bio-STEER: A semantic web workflow tool for Grid computing in the life sciences. *Future Gener. Comput. Syst., 23*(3), 497-509.

Leenaars, M., Heikkurinen, M., Louridas, M., & Karayannis, F. (2005). e-infrastructures roadmap. Espoo, Finland: e-IRGSG. Retrieved from http://www.e-irg.org/roadmap/eIRG-roadmap.pdf

Legendre, P., Anderson, M.J. (1998). DistPCOA program description, source code, executables, and documentation: http://www.bio.umontreal.ca/Casgrain/en/labo/distpcoa.html

Legendre, P., Desdevises, Y., & Bazin, E. (2002). A Statistical Test for Host-Parasite Coevolution. *Systematic Biology*, 51(2):217-234.

Legré, Y., Y., M., Viezens, F., Rienhoff, O., & Sax, U. (2008). HealthGRID/ SHARE: Grids for Health - International Interoperability. In *eHealth Conference 2007 - Experts' Special Interest Sessions* (Vol. 58) (pp. 81-99). Bonn, Germany: nanos Verlag oHG.

Lelli, F., Frizziero, E., Gulmini, M., Maron, G., Orlando, S., Petrucci, A., et al. (2007). The many Faces of the Integration of Instruments and the Grid. *International Journal of Web and Grid Services, 3*(3), 239-266.

Leo, P., Scioscia, G., Pappadà, G. & Quinto, V. (2008). *EMBL Wrapper for WebSphere Information Integrator.* Retrieved from http://www.alphaworks.ibm.com/tech/emblwrapper.

Lesk, A. (2006). *Introduction to Bioinformatics.* New York: Oxford University Press.

Leslie, C., Eskin, E., & Noble W.S. (2002). The Spectrum Kernel: A string kernel for SVM protein classification. *Pacific Symposium on Biocomputing, 7*, 566-575.

Leslie, C., et al. (2002). Mismatch string kernels for SVM protein classification. *Advanced Neural Information Processing Systems, 15*, 1441-1448.

Letondal, C. (2001). A Web interface generator for molecular biology programs in Unix. *Bioinformatics, 17*, 73-82).

Lewitt, R. M. (1992). Alternatives to voxels for image representation in iterative reconstruction algorithms. *Physics in Medicine and Biology, 3*(12), 705-716.

Li, A.W. & Murphy, P.R. (2000). Expression of alternatively spliced FGF-2 antisense RNA transcripts in the central nervous system: regulation of FGF-2 mRNA translation. *Mol. Cell. Endocrinol., 170*, 233–242.

Li, W. W., Baker, N. A., Baldridge, K., McCammon, J. A., Ellisman, M. H., Gupta, A., et al. (2006). National Biomedical Computation Resource (NBCR): Devloping End-to-End Cyberinfrastructure for Multiscale Modeling in Biomedical Research. *CTWatch Quarterly, 2*, 6-17.

Li, W. W., Krishnan, S., Mueller, K., Ichikawa, K., Date, S., Dallakyan, S., et al. (2006). *In Building Cyberinfrastructure for Bioinformatics Using Service Oriented Architecture: Cluster Computing and the Grid Workshops.* Sixth IEEE International Symposium on Cluster Computing & the Grid, Singapore.

Li, W., Byrnes, R., Hayes, J., Birnbaum, A., Reyes, V., Shahab, A., et al. (2004). The encyclopaedia of life project: grid software and deployment. *New Generation Computing, 22*, 127-136.

Li, X.-J. Zhang, H. Ranish, J.R. & Aebersold, R. (2003). Automated statistical analysis of protein abundance ratios from data generated by stable isotope dilution and tandem mass spectrometry. *Analytical Chemistry, 75*(23), 6648-57

Liberman, M.A., & Lichtenberg, A.J. (1992). *Regular and Chaotic Dynamics.* Berlin, Germany: Springer-Verlag.

Libin P., Deforche K., Van Laethem K., Camacho R. & Vandamme A-M. (2007). RegaDB: An Open Source, Community-Driven HIV Data and Analysis Management Environment. In *Reviews in Antiretroviral Therapy* (Vol. 2007-2). Fifth European HIV Drug Resistance Workshop, Cascais, Portugal, March 2007.

Lickerman, E., Bry, L., Herring, N., Pandelidis, Y., Afsheen, N., & Sharif, A. (2006). *Grid-Enabled Bio-Sample Management Application for Data-Intensive Biomarker Analysis.* Paper presented at the Proceedings of the 19th IEEE Symposium on Computer-Based Medical Systems.

Lin, B. & Dinda, PA. (2005, November). *VSched: Mixing Batch And Interactive Virtual Machines Using Periodic Real-time Scheduling*. ACM/IEEE conference on Supercomputing. New York: IEEE Computer Society

Lindahl, E., Hess, B., & van der Spoel, D. (2001). GROMACS 3.0: a package for molecular simulation and trajectory analysis. *Journal of Molecular Modeling, 7,* 306-317.

Lipinski, C.A., Lombardo, F., Dominy, B.W., & Feeney, P.J. (1997). Experimental and computational approaches to estimate solubility and permeability in drug discovery and development settings. *Adv. Drug Delivery Rev, 23,* 3-25.

Lippmann, R. P. (1987). An introduction to computing with neural nets. *IEEE ASSP Magazine*, *4*(22), 7-25.

Listgarten, J. & Emili, A. (2005). Statistical and computational methods for comparative proteomic profiling using liquid chromatography-tandem mass spectrometry. *Molecular Cellular Proteomics, 4,* 419–434.

Liu, W., Schmidt, B., Voss, and G., Mueller-Wittig, K.W. (2008). Accelerating Molecular Dynamics Simulations Using Graphics Processing Units with CUDA. *Computer Physics Communications, 179*(9), 634-641.

Liu, W., Schmidt, B., Voss, G., Schroder, A., & Muller-Wittig, W. (2006). Bio-Sequence Database Scanning on a GPU. *20th IEEE International Parallel & Distributed Processing Symposium (IPDPS 2006) (HICOMB Workshop)*, Rhode Island, Greece.

Lord, P. W., Stevens, R.D., et al. (2003). Investigating semantic similarity measures across the Gene Ontology: the relationship between sequence and annotation. *Bioinformatics,* 19(10), 1275-83.

Lorensen, W.E. & Cline, H. (1987). Marching cubes: A High Resolution 3D Surface Construction Algorithm. *ACM SIGGRAPH, 21*(4), 163-169.

Lučić, V., Förster, F., & Baumeister, W. (2005). Structural studies by electron tomography: From cells to molecules. *Annual Review of Biochemistry, 74*(1), 833-865.

LUCY (n.d.). *Stanford Visible Female.* Retrieved from http://lucy.stanford.edu/home.html

Ludäscher, B., Altintas, I., Berkley, C., Higgins, D., Jaeger, E., Jones, M., et al. (2006). Scientific workflow management and the kepler system. *Concurrency and Computation: Practice and Experience, 18*(10),1039–1065.

Luengo, F., Cofino, A. S., & Gutierrez, J. M. (2004). Grid oriented implementation of self-organizing maps for data mining in meteorology. *Lecture Notes in Computer Science, 2970,* 163-171.

Luo, C., Pereira, A. L., & Chung, S. M. (2006). Distributed mining of maximal frequent itemsets on a data grid system. *Journal of Supercomputing, 37*(1), 71-90.

Luther, A., Buyya, R., Ranjan, R., & Venugopal, S. (2005). Peer-to-Peer Grid Computing and a .NET-based Alchemi Framework. High Performance Computing: Paradigm and Infrastructure. In Guo, M. (Ed.). Wiley Press, USA. www.alchemi.net

Luz, P.M., Codeço, C.T., Massad, E., Struchiner C.J. (2003, October). Uncertainties Regarding Dengue Modeling. *Mem Inst Oswaldo Cruz, 98*(7), 871-8.

Lv, C., Cao, P. Cohen, E. Li, L., & Shenker, S. (2002). Search and *replication in unstructured peer-to-peer networks*, ACM, Sigmetrics.

Macfarlane, A.; Mccann, Liddell, H. (1996). A Common Data Model For Meta-Data Interoperable Environments. In *Proceedings of II International Baltic Workshop on Database and Information Systems.*

Madoff, L., Woodall, J. (2005). The Internet and the Global Monitoring of emerging Diseases: Lessons from the First 10 Years of ProMED-mail. *Archives of Medical Research, 36,* 724-730.

Main, P., Fiske, S.J., Hull, S.E., Lessinger, L., Germain, G., Declercq, J.P. & Woolfson, M.M. (1980). *MULTAN80: a system of computer programs for the automatic solution of crystal structures from x-ray diffraction data* (Universities of York and Louvain).

Malawski, M., Bartynski, T., & Bubak, M. (2008). A tool for building collaborative applications by invocation of Grid operations. In Bubak, M., van Albada, D. G., Dongarra, J., and Sloot, P. M. A., (Eds.), *Computational science - ICCS 2008, 8th International Conference, Kraków, Poland, June 23-25, 2008, Proceedings, part III*, (LNCS 5103 pp. 243-252). Berlin, Germany: Springer.

Malawski, M., Bubak, M., Placek, M., Kurzyniec, D. & Sunderam, V. (2006). Experiments with distributed component computing across Grid boundaries. In *Proceedings of the HPC-GECO/CompFrame workshop in conjunction with HPDC 2006*, (pp. 109-116), Paris, France.

Malawski, M., Gubala, T., Kasztelnik, M., Bartynski, T., Bubak, M., Baude, F., & Henrio, L. (2008). High-level scripting approach for building component-based applications on the grid. In *CoreGRID Workshop on Grid Programming Model Grid and P2P Systems Architecture Grid Systems, Tools and Environments*, Heraklion, Crete. Berlin, Germany: Springer.

MammoGrid (2008). *European federated mammogram database implemented on a grid structure.* Retrieved March 8, 2008 from http://mammogrid.vitamib.com

Manavski, S.A., & Valle, G. (2008). CUDA compatible GPU cards as efficient hardware accelerators for Smith-Waterman sequence alignment. *BMC Bioinformatics*, *9*(S2), S10.

Manocha, D. (2005). General Purpose computations using graphics processors. *IEEE Computer*, *20*(7), 85-88.

Manssour, I.H. & Dal Sasso Freitas, C.M. (2000). Collaborative Visualization in Medicine. In *Proc. 8'th International Conference in Central Europe on Computer Graphics, Visualization and Interactive Digital Media (WSCG2000), Plzen*, (pp. 266-273).

Maojo V, de la Calle G, Martín-Sánchez F, Díaz C, Sanz F. (2005). INFOBIOMED: European Network of Excellence on Biomedical Informatics to support individualised healthcare. *AMIA Annual Symposium Proceedings:* 1041.

Marabini, R., Herman, G. T., & Carazo, J. M. (1998). 3D reconstruction in electron microscopy using ART with smooth spherically-symmetric volume elements (blobs). *Ultramicroscopy, 72*(1), 53-65.

Maran U., Sild S., Kahn I., & Takkis K. (2007). *Future Gen. Comput. Syst. 23*, 76-83

Maran U., Sild S., Mazzatorta P., Casalegno M., Benfenati E., & Romberg M. (2007). In Dubitzky W., Schuster A., Sloot P. M. A., Schhroeder M., Romberg M. (Eds.), *Grid Computing in Computational Biology*, (LNCS 4360, pp. 60–74). Berlin Heidelberg: Springer-Verlag.

Maran, U., Sild, S., Mazzatorta, P., Casalegno, M., Benfenati, E., & Romberg, M. (2007). Grid Computing for the Estimation of Toxicity: Acute Toxicity on Fathead Minnow (Pimephales promelas). In *Distributed, High-Performance and Grid Computing in Computational Biology* (pp. 60-74).

Marinescu, D. C., & Ji, Y. (2003). A computational framework for the 3D structure determination of viruses with unknown symmetry. *Parallel and Sitributed Computing, 63*(1), 738-758.

Markus, W. & Manfred, J. (2007). ProSA-web: interactive web service for the recognition of errors in three-dimensional structures of proteins. *Nucleic Acids Res.* (35), 407–410.

Marrink, S. J., de Vries, A. H., & Mark, A. E. (2004). Coarse grained model for semiquantitative lipid simulations. *Journal of Physical Chemistry B, 108,* 750-760.

Marsden, R.L., Lewis, T.A. & Orengo, C. (2007). Towards a comprehensive structural coverage of completed genomes: a structural genomics viewpoint. *BMC Bioinformatics. 8*, 86-103.

Martelli, P.L., Fariselli, P. & Casadio, R. (2003). An ENSEMBLE machine learning approach for the prediction of all-alpha membrane proteins. *Bioinformatics 19*(1), 205–211.

Martens, L., Hermjakob, H. Jones, P. Adamski, M. Taylor, C. States, D. Gevaert, K. Vandekerckhove, J. & Apweiler, R. (2005). Pride: the proteomics identifications database. *Proteomics, 5*(13), 3537-3545

Martin, D., Brun, C., Remy, E., Mouren, P., Thieffry, D., Jacq, B. (2004). GOTool-Box: functional analysis of gene datasets based on Gene Ontology. *Genome Biol, 5*, R101.

Martino, R. L. (2006). Parallel Computing in the Analysis of Gene Expression Relationships. In Zomaya A. (Ed.), *Parallel Computing for Bioinformatics and Computational Biology*, (pp. 265-284). Berlin, Germany: Wiley.

Martonne, M. (2005). The BIRNLex: Principles and practices of community ontology development. *Re-use or*

Re-invention - a Roadmap for Data Integration (Schizophrenia as a Test Case). NeSC Events. Retrieved July 2008 from http://www.nesc.ac.uk/action/esi/download.cfm?index=3359

Mastroianni, C., Talia, D. & Trunfio, P. (2004). Metadata for managing grid resources in data mining applications. *Journal of Grid Computing, 2*(1), 85-102.

Mattevi, A., Vanoni, M.A., Tosone, F., Rizzi, M., Teplyakov, A., Coda, A., Bolognesi, M., Curti, B. (1996). Crystal structure of D-amino acid oxidase: a case of active site mirror-image convergent evolution with flavocytochrome b2. *Proc Natl Acad Sci USA, 93*(15), 7496-501.

Matthiesen, R. (2007). Methods, algorithms and tools in computational proteomics: A practical point of view. *Proteomics, 7*, 2815-2832.

Mawudeku A., Blench M. (2006). Global Public Health Intelligence Network (GPHIN). In *Proceedings of the 7th Conference of the Association for Machine Translation in the Americas.* Retrieved February 11, 2008 from www.mt-archive.info/MTS-2005-Mawudeku.pdf

McClatchey, R., Manset, D., & Solomonides, T. (2006). Lessons learned from MammoGrid for integrated biomedical solutions. In *19th IEEE Int. Symposium on Computer-Based Medical Systems, CBMS 2006.*

McGinnis, L., Wallom, D., & Gentzsch, W. (Eds.) (2007). 2nd International Workshop on Campus and Community Grids. Report retrieved from http://forge.gridforum.org/sf/go/doc14617?nav=1

McIlraith, S.A., Son, T.C. & Zeng, H. (2001). Semantic web services. *IEEE Intelligent Systems, 16*(2), (46–53.

McMullen, D.F., Devadithya, T., & Chiu, K. (2005). *Integrating Instruments and Sensors into the Grid with CIMA Web Services.* Paper presented at the Proceedings of the Third APAC Conference on Advanced Computing, Grid Applications and e-Research.

Medical Dictionary for Regulatory Activities (2008). Retrieved March 8, 2008, from http://www.meddramsso.com/MSSOWeb/index.htm

MediGRID. (2005). *www.medigrid.de.* Retrieved March 14, 2007.

Meier-Kolthoff, J.P, Auch, A.F., Huson, D.H., & Göker, M.(2007). COPYCAT: Co-phylogenetic Analysis tool. *Bioinformatics*, 23(7):898-900.

Meinilä, M., Kuusela, J., Zietara, M.S., & Lumme, J. (2004). Initial steps of speciation by geographic isolation and host switch in salmonid pathogen *Gyrodactylus salaris (Monogenea: Gyrodactylidae). International Journal of Parasitology,* 34(4):515-526

Melville, J., Harmon, L.J., Losos, J.B. (2006). Intercontinental community convergence of ecology and morphology in desert lizards. *Proc Biol Sci., 273*(1586), 557-63.

Merelli, I. & Milanesi, L. (2005) *High performance implementation of BLAST using GRID technology.* Paper presented at the meeting of the Italian Bioinformatics Society, Milan, Italy.

Merkle, D., & Middendorf. M. (2005). Reconstruction of the cophylogenetic history of related phylogenetic trees with divergence timing information. *Theory in Biosciences,* 123(4):277-299.

Metropolis, N., Rosenblath, A.W., Rosenblath, M.N., Teller A.H., & Teller, E. (1953). Equations of State Calculations by Fast Computing Machines. *J. Chem. Physics, 21*, 1087-1091.

Mignone, F., Grillo, G., Licciulli, F., Iacono, M., Liuni, S., Kersey, P.J., Duarte, J., Saccone C. & Pesole G. (2005). UTRdb and UTRsite: a collection of sequences and regulatory motifs of the untranslated regions of eukaryotic mRNAs. *Nucleic Acids Res., 33*, D141-D146.

Mikler, S. Venkatachalam, A.& (2005). Towards computational epidemiology: Using stochastic cellular automata in modeling spread of diseases. *Proceedings. of the 4th Annual International Conference on Statistics.* Retrieved July 2008 from http://www.cerl.unt.edu/publications/2005/slides/ppt/Hawaii.ppt

Milanesi, L., & Merelli, I. (2006). High Performance GRID Based Implementation for Genomics and Protein Analysis. *Studies in health technology and informatics, 120*, 374-380.

Miles, S., Wong, S.C., Fang, W., Groth, P.T., Zauner, K.P. & Moreau, L. (2007) Provenance-based validation of e-science experiments. *J. Web Sem., 5*(1), 28–38.

Miller, R., Bednasz, J.J., Chiu, K., Gallo, S.M., Govindaraju, M., Lewis, M., Ruby, C., & Weeks, C.M. (2008). Grid-based research, development, and deployment in New York State, In *Proceedings of the International Parallel and Distributed Processing Symposium*, in press.

Miller, R., DeTitta, G. T., Jones, R., Langs, D. A., Weeks, C. M. & Hauptman, H. A. (1993). On the application of the minimal principle to solve unknown structures, *Science*, 259, 1430–1433.

Miller, R., Gallo, S. M., Khalak, H. G. & Weeks, C. M. (1994). *SnB*: Crystal structure determination via *Shake-and-Bake*. *J. Appl. Cryst.*. *27*, 613–621.

Minar, N., Burkhart, R., Langton, C., & Askenazi, M. (1996). *The swarm simulation system, a toolkit for building multi-agent simulations*, from http://www.santafe.edu.

Mirto, M., Cafaro, M., Fiore, S., Tartarini, D. & Aloisio, G. (2007, June). A Grid-enabled Protein Secondary Structure Predictor. *IEEE Transactions on NanoBioscience*, *6* (2),124-130.

Mirto, M., Ferramosca, A., Tartarini, D., Romano S., Negro A., Tasco, G., Fiore, S., Zara, V., Casadio, R., & Aloisio, G. (2008). A Protein Structure Prediction Service in the ProGenGrid System. *Global HealthGrid: e-Science Meets Biomedical Informatics Proceedings of HealthGrid 2008*, (pp 135-146), Chicago, June 2-4.

Mirto, M., Fiore, S., Epicoco, I., Cafaro, M., Mocavero, S., Blasi, E. & Aloisio G. (2008). A Bioinfomatics Grid Alignment Toolkit. *Future Generation Computer Systems*, *24*(7), 752-762. Philadelphia: Elsevier

Mirto, M., Rossi, I., Epicoco, I., Fiore, S., Fariselli, P., Casadio, R. & Aloisio, G. (2007). High Throughput Protein Similarity Searches in the LIBI Grid Problem Solving Environment. *Proceedings of the 5th International Symposium on Parallel and Distributed Processing and Applications (ISPA07)*, (pp. 414-423), Niagara Falls, Canada,.

Missier, P., Belhajjame, K, Zhao, J. & Goble, C.A. (2008). Data lineage model for Taverna workflows with lightweight annotation requirements. *IPAW 2008*. New York: Springer.

Missier, P., Embury, S.M. & Stapenhurst, R. (2008). Exploiting provenance to make sense of automated data acceptance decisions in scientific workflows. *IPAW 2008*. Springer.

MitoRes: bio-sequences resource for mitochondria-related genes, transcripts and proteins. Retrieved from http://www2.ba.itb.cnr.it/MitoRes/.

Mohamed, H.H., & Epema, D.H.J. (2004). An Evaluation of the Close-to-Files Processor and Data Co-Allocation Policy in Multiclusters. In *Cluster*, (pp. 287-298). New York: IEEE Computer Society Press.

Mohammed, Y. (2006). Erweiterte Sicherheit und Datenschutz Techniken. In U. Sax, Y. Mohammed, F. Viezens & O. Rienhoff (Eds.), *Grid-Computing in der Biomedizinischen Forschung – Datenschutz und Datensicherheit* (Vol. 90, pp. 70-77). München, Germany: Urban &Vogel.

Mohammed, Y., Sax, U., Viezens, F., & Rienhoff, O. (2007). Shortcomings of Current Grid Middlewares Regarding Privacy in HealthGrids. *Stud Health Technol Inform*, *126*, 322-329.

Montagnat J., Frohner A., Jouvenot D., Pera C., Kunszt P., Koblitz B., et al. (2008). A Secure Grid Medical Data Manager Interfaced to the gLite Middleware. *Journal of Grid Computing (JGC)*, *6* (1): 45-59.

Montagnat, J., Bellet, F., Benoit-Cattin, H., Breton, V., Brunie, L., Duque, H., et al. (2004). Medical images simulation, storage, and processing on the European datagrid testbed. *Journal of Grid Computing*, *2* (4), 387-400.

Montagnat, J., Breton, V., & Magnin, I. (2003). Using Grid technologies to face medical image analysis challenges. In *Proceedings of IEEE/ACM CCGrid'03*, (pp. 588-593). New York: IEEE Computer Society.

Montagnat, J., Frohner, Á., Jouvenot, D., Pera, C., Kunszt, P., Koblitz, B., et al. (2008). A Secure Grid Medical Data Manager Interfaced to the gLite Middleware. *Journal of Grid Computing*, *6*(1), 45-59.

Moore R. (2001). Data Management Systems for Scientific Applications. In Boisvert R., Tang P. (Eds.) *The Architecture of Scientific Software* (pp. 273-284). Norwell, MA: Kluwer Academic Publishers.

Moore, GA. (1991). *Crossing the Chasm: Marketing and Selling High-Tech Products to Mainstream Customers*. New York: Harper Business.

Morgan, M., Chue-Hong, N., & Drescher, M., (2006). *ByteIO Specification 1.0*. Retrieved on July 10, 2008 from https://forge.gridforum.org/sf/docman/do/download/Document/ projects.byteiowg/docman.root.current_documents/doc13719/1

Mori, S. (2007). *Introduction to Diffusion Tensor Imaging*. Amsterdam, the Netherlands: Elsevier.

Morris G. M., Goodsell D. S., Halliday R.S., Huey R., Hart W. E., Belew R. K., & Olson A. J. (1998). Automated Docking Using a Lamarckian Genetic Algorithm and and Empirical Binding Free Energy Function. *J. Computational Chemistry, 19*, 1639-1662.

Moscicki, J.T., Lee, H.C., Guatelli, S., Lin, S.C. & Pia, M.G. (2004). *A Biomedical Applications on the GRID: Efficient Management of Parallel Jobs*. Paper presented at the Nuclear Science Symposium, Geneve, Switzerland.

Mozilla (2008). Retrieved on July 10, 2008, from http://www.mozilla.org/newlayout/doc/

Mulder, N. J., Kersey, P., et al. (2008). In Silico Characterization of Proteins: UniProt, InterPro and Integr8. *Mol Biotechnol, 38*(2), 165-77.

Murdock, S. E., Tai, K., Ng, M. H., Johnston, S., Wu, B., Fangohr, H., Laughton, C. A., Essex, J. W., & Sansom, M. S. P. (2006). Quality assurance for biomolecular simulations. *Journal of Chemical Theory and Computation, 2*, 1477-1481.

MyGrid (2008). Retrieved from www.mygrid.org.uk

Myhre, S., Tveit, H., et al. (2006). Additional gene ontology structure for improved biological reasoning. *Bioinformatics, 22*(16), 2020-7.

Nagesh, H., Goil, S., & Choudhary, A. (1999). *Mafia: Efficient and scalable subspace clustering for very large data sets* (Tech. Rep. No. 9906-010), Center for Parallel and Distributed Computing.

Narayan, R. M. (2002). Clinical Trials in Head Injury. *Journal of Neurotrauma, 19*, 503-557.

Naseer, A. & Stergioulas, L. K. (2006). Resource discovery in grids and other distributed environments: States of the art. *Multiagent and Grid Systems - An International Journal, 2*(2), 163-182.

Naseer, A. & Stergioulas, L. K. (2006). "A Taxonomy of HealthGrids and HealthGrid Resources". *11th International Symposium on Health Information Management Research* (iSHIMR2006), Halifax, Canada (pp. 93-99).

Naseer, A. and Stergioulas, L. K. (2006). Discovering HealthGrid services. In *Proceedings of the IEEE International Conference on Services Computing (SCC'06), Chicago, IL*, (pp.301-306).

Naseer, A. and Stergioulas, L. K. (2007). Combining web services and grid services: practical approaches and implications to resource discovery. In Periorellis, P. (Ed.), *Securing Web Services: Practical Usage of Standards and Specifications*, (pp. 298-317). Hershey, PA: IGI Global..

National Center for Biotechnology Information. Retrieved from http://www.ncbi.nlm.nih.gov/BLAST/.

National Institutes of Health (2000). *Working definition of bioinformatics and computational biology*. Retrieved March 17, 2008, from http://www.bisti.nih.gov/CompuBioDef.pdf

National Toxicology Program server at http://ntp-server.niehs.nih.gov/

Nature (2007). Editorial: The database revolution. *Nature, 445*, 229-230.

NCBI Blast (2007). Retrieved July 2008 from http://www.ncbi.nlm.nih.gov/blast/Blast.cgi.

NCBI. Ncbi website. Retrieved from http://www.ncbi.nih.gov/

NCSA (2007). *MyProxy Credentials Management Service*. Retrieved July 2008 from http://grid.ncsa.uiuc.edu/myproxy/

Needleman, S., Wunsch, C. (1970). A general method applicable to the search for similarities in the amino acid sequence of two proteins. *J Mol Biol, 48*(3), 443-53.

NEESgrid (2008). Retrieved from www.nees.org/

NEMA – ACR (2008). *DICOM Standard*. Retrieved July 2008 from http://medical.nema.org/

Nesvizhskii, A. I., Keller, A., Kolker, E. & Aebersold, R. (2003). A statistical model for identifying proteins by tandem mass spectrometry. *Analytical Chemistry, 75*(17): 4646-4658.

Nesvizhskii, A. I., Vitek, O. & Aebersold, R. (2007). Analysis and validation of proteomic data generated by tandem mass spectrometry. *Nat Methods, 4*(10). 787-797.

Neto V. C., Gagliardi H. F., Rezende A., Gonçalves E. S. , Gallo E. , Barbosa F. A., Torres I., Alves D. (2006). Data Access Service in a Computational Grid Platform Applied to the Monitoring and Control of Epidemics on Georeferenced Dynamic Maps. *Proceedings of the e-Science conference,* (p. 142).

Neto, V.C., Gagliardi, H. F., Rezende, A., Gonçalves, E.S., Gallo, E., Silva, F.A.B., Pisa, I.T., Alves, D. (2006). Data Access Service in a Computational Grid Platform Applied to the Monitoring and Control of Epidemics on Georeferenced Dynamic Maps. In *Proceedings of the 2nd IEEE Conference on e-Science and Grid Computing*. New York: IEEE Computer Society Press.

NeuGrid Project (2008). Retrieved October 20, 2008, from http://www.neugrid.eu/.

Neumann, U. (1994). Communication Costs for Parallel Volume-Rendering Algorithms. *IEEE Computer Graphics and Applications*, *15*(4), 49-58.

Neuroth H., Kerzel, M., & Gentzsch, W. (Eds) (2007). *Die D-Grid Initiative*. Goettingen, Germany: Goettingen University Press. Retrieved from http://webdoc.sub.gwdg.de/univerlag/2007/D-Grid_de.pdf

Neuroth, H., Kerzel, M., & Gentzsch, W. (Eds.). (2007). German Grid Initiative D-Grid. Universitätsverlag Göttingen Publishers. Retrieved from www.d-grid.de/index.php?id=4&L=1

Nevill-Manning, C.G., Wu, T.D., & Brutlag, D.L. (1998). Highly Specific Protein Sequence Motifs for Genome. *Proceedings of the National Academy of Sciences USA*, *95*, 5865-5871.

Ng, M. H., Johnston, S., Wu, B., Murdock, S. E., Tai, K., Fangohr, H., Cox, S. J., Essex, J. W., Sansom, M. S. P., & Jeffreys, P. (2006). BioSimGrid: Grid-enabled biomolecular simulation data storage and analysis. *Future Generation Computer Systems, 22,* 657-664.

Niederberger R. & Alessandrini V. (2004). The DEISA Project: Motivations, Strategies, Technologies. In *International Supercomputer Conference*, Heidelberg, Germany.

Nielsen, S. O., Lopez, C. F., Srinivas, G., & Klein, M. L. (2004). Coarse grain models and the computer simulation of soft materials. *Journal of Physics: Condensed Matter, 16,* R481-R512.

NIH. (2008). *Models of Infectious Disease Agent Study*. Retrieved July 2008 from http://www.nigms.nih.gov/Initiatives/MIDAS

NLM (n.d.). *Visible Human Project.* Bethesda, MD: National Library of Medicine., Retrieved from http://www.nlm.nih.gov/research/visible/visible_human.html

Notredame, C., Higgins, D.G. & Heringa, J. (2000). T-coffee: a novel method for fast and accurate multiple sequence alignment. *J. Mol. Biol., 302*, 205–217.

Novotny J., Tuecke S., & Welch V. (2001). *An Online Credential Repository for the Grid: MyProxy*. HPDC, 10th IEEE International Symposium on High Performance Distributed Computing (HPDC-10 '01) (p. 0104).

Novotny, J. (2003). The grid portal development kit. In F. Berman, G. Fox, & A. Hey, *Grid Computing: Making the Global Infrastructure a Reality* (pp. 657-674). Hoboken, NJ: John Wiley and Sons.

Novotny, J., Russell, M., & Wehrens, O. (2004). GridSphere: a portal framework for building collaborations. *Concurrency and Computation: Practice and Experience, 16,* 503-513.

Novotny, J., Russell, M., & Wherens, O. (2004). Gridsphere: A Portal framework for building collaborations. *Concurrency and Computation: Practice and Experience* (pp. 503-513). Hoboken, NJ: Wiley & Sons.

Noy, N. F. (2004). Semantic integration: A survey of ontology-based approaches. *SIGMOD Record*, *33*(4), 65-70.

NRCAM (n.d.). *The National Resource For Cell Analysis and Modeling.* Farmington, CT: University of Connecticut Health Center. Retrieved from http://www.nrcam.uchc.edu

NVIDIA Compute Unified Device Architecture-Programming Guide (2007). Retrieved from http://developer.download.nvidia.com/compute/cuda/1_0/NVIDIA_CUDA_Programming_Guide_1.0.pdf.

O'Farell, P.H. (1975). High-resolution two-dimensional gel electrophoresis of proteins. *Journal of Biolical Chemistry, 250*(10), 4007-4021.

O'Reilly, T. (2005). *What is Web 2.0: Design patterns and business models for the next generation of software.* Retrieved on April 10, 2008 from http://www.oreillynet.com/pub/a/oreilly/tim/news/2005/09/30/what-is-web-20.html.

OASIS (2003). *Assertions and Protocol for the OASIS Security Assertion Markup Language (SAML) v1.1.* Retrieved from http://www.oasis-open.org/committees/security/

OASIS Consortium (2008). Retrieved July 20, 2008, from http://www.oasis-open.org.

OGF - Open Grid Forum (2008). Retrieved July 20, 2008, from http://www.ogf.org.

OGF (2008). Open Grid Forum. Retrieved from www.ogf.org

OGSA-DAI (2008) Open Grid Service Architecture–OGSA-DAI Documentation. Retrieved February 20, 2008, from http://www.ogsadai.org.uk/documentation/.

OGSA-DAI project (2007). *OGSA-DAI.* 2007. Retrieved July 2008 from http://www.ogsadai.org.uk/

Oinn, T., Addis, M., Ferris, J., Marvin, D., Senger, M., Greenwood, M., et al. (2004). Taverna: A tool for the composition and enactment of bioinformatics workflows. *Bioinformatics, 20*(17), 3045-3054.

Oinn, T., Greenwood, M., Addis, M., Nedim Alpdemir, M., Ferris, J., et al. (2006). Taverna: lessons in creating a workflow environment for the life sciences. *Concurrency and Computation: Practice and Experience, 18*(10) 1067-1100.

Oinn, T.M, Greenwood, R.M., Addis, M., Alpdemir, N., Ferris, J., Glover, K., et al. (2006). Taverna: lessons in creating a workflow environment for the life sciences. *Concurrency and Computation: Practice and Experience, 18*(10) 1067–1100.

Oinn, T.M., Addis, M. Ferris, J., Marvin, D. Greenwood, R.M. Carver, T. Pocock, M.R. Wipat, A., & Li P.(2004) Taverna: A tool for the composition and enactment of bioinformatics workflows. *Bioinformatics, 20*(17), 3045–3054

Olesen, J. L. M. (2003). The burden of brain disease in Europe. *European Journal of Neurology, 10*, 471-477.

Oliveira, M. C., Cirne, W., & de Azevedo Marques, P. M. (2007). Towards applying content-based image retrieval in the clinical routine. *Future Gener. Comput. Syst., 23* (3), 466-474.

Oliver, T., Schmidt, B., & Maskell, D. (2005). Reconfigurable Architectures for Bio-sequence Database Scanning on FPGAs. *IEEE Transactions on Circuits and Systems II, 52*(12), 851-855.

Open Grid Services Architecture - Data Access and Integration Services. Retrieved March 13, Accessed 2006, from www.ogsadai.org.uk/

OpenDX. Retrieved from http://www.opendx.org/

OpenEHR (2007). Retrieved July 2008 from http://www.openehr.org/home.html

Operation and maintenance manual for the mock circulatory system. (1965). Rockville, MD: Hydrospace Research Corporation.

Orchard, S., & Hermjakob, H., (2008) The HUPO proteomics standards initiative easing communication and minimizing data loss in a changing world. *Briefing in Bioinformatics, 9*(2), 166-173

Orchard, S., Hermjakob, H., Binz, P.A., Hoogland, C., Taylor, C.F., Zhu, W., Julian, R.K., & Apweiler R., (2005) Further steps towards data standardization: The Proteomic standard initiative. *Proteomics, 5*(2), 337-339

Oren, E., Haller, A., Hauswirth, M., Heitmann, B., Decker, S., & Mesnage, C. (2007). A flexible integration framework for semantic web 2.0 applications. *IEEE Software, 24*(5), 64-71.

Osorio A., Valette PJ., Mihalcea A., Atif J. & Ripoche X. (2002) *A new PC based software to perform semi-automatic hepatic segmentation using CT and MR images.* Presented at InfoRAD-RSNA, Chicago, IL.

Oster, S., Langella, S., Hastings, S., Ervin, D., Madduri, R., Phillips, J., et al. (2007). caGrid 1.0: An Enterprise Grid Infrastructure for Biomedical Research. *Journal of the American Medical Informatics Association, 15,* 138-149.

Ott. E. (1993). *Chaos in Dynamical Systems.* New York: Cambridge University Press.

Ourselin, S., Roche, A., Prima, S. & Ayache, N. (2000). Block Matching: A General Framework to Improve Robustness of Rigid Registration of Medical Images. In proc. of *MICCAI 2000*, , Pittsburgh, PA, (LNCS 1935, pp. 557-566). Berlin, Germany: Springer.

Owens, J., Luebke, D., Govindaraju, N., Harris, M., Krüger, J., Lefohn, A.E., & Purcell, T.J. (2007). A survey of general-purpose computation on graphics hardware. *Computer Graphics Forum, 26*(1), 80-113.

Palmieri, F. (2004). The mitochondrial transporter family (SLC25): physiological and pathological implications. *Pflugers Arch., 447,* 689–709.

Palmieri, L., Palmieri, F., Runswick, M. & Walker, J. (1996). Identification by bacterial expression and functional reconstitution of the yeast genomic sequence encoding the mitochondrial dicarboxylate carrier protein. *FEBS Lett.* (399), 299–302.

Pang, A., Arinaminpathy, Y., Sansom, M. S. P., & Biggin, P. C. (2005). Comparative molecular dynamics – similar folds and similar motions? *Proteins: Structure, Function, and Bioinformatics, 61,* 809-822.

Pantalos, G., Koenig, S., Gillars, K., Giridharan, G., & Ewert, D. (2004). Characterization of an Adult Mock Circulation for Testing Cardiac Support Devices. *ASAIO Journal, 50*(1), 37-46.

Papazoglou, M.P. (2003). Service-oriented computing: concepts, characteristics and directions. In *Proceedings of the 4th International Conference on Web Information Systems Engineering, WISE 2003,* Roma, Italy.

Pappin, D. J., Hojrup, P. & Bleasby, A. J. (1993). 'Rapid identification of proteins by peptide-mass fingerprinting. *Curr Biol 3*(6), 327-332.

Park, J., et al. (1998). Sequence comparisons using multiple sequences detect twice as many remote homologues as pairwise methods. *Journal of Molecular Biology, 284,* 1201-1210.

Parker, J. (2007). A Flexible, Large-Scale, Distributed Agent Based Epidemic Model. *Proceedings of the 2007 Winter Simulation Conference.* New York: IEEE Press.

Paterson, S., Viney M. E. (2000). The interface between epidemiology and population genetics. *Parasitology Today, 16*(12), 528.

Patil, A.A., Oundhakar, S.A., Sheth, A.P. & Verma; K. (2004) Meteor-s web service annotation framework. In S. I. Feldman, M. Uretsky, M. Najork, & C. E. Wills (Eds.), *WWW,* (pp. 553–562). New York; ACM.

Pautasso, C., Bausch, W., & Alonso, G. (2006). Autonomic Computing for Virtual Laboratories. In J. Kohlas, B. Meyer, A. Schiper (Eds.), *Dependable Systems: Software, Computing, Networks,* (LNCS 4028). Berlin, Germany: Springer Verlag.

Peachey T., Abramson D., & Lewis A. (2006). *Model Optimization and Parameter Estimation with Nimrod/O.* The International Conference on Computational Science, May 28-31, University of Reading, UK. Retrieved from: http://www.csse.monash.edu.au/~davida/papers/iccs06.pdf

Pearlman, D. A., Case, D. A., Caldwell, J. W., Ross, W. S., Cheatham III, T. E., DeBolt, S., Ferguson, D., Seibel G., & Kollman, P. (1995). AMBER, a computer program for applying molecular mechanics, normal mode analysis, molecular dynamics and free energy calculations to elucidate the structures and energies of molecules. *Computer Physics Communications, 91,* 1-41.

Pearlman, L. W., V. Foster, I. Kesselman, C. Tuecke, S. (2002). A Community Authorization Service for Group Collaboration. In IEEE Computer Society, TC on Distributed Processing, Naval Postgraduate School (U.S.) & United States, Office of Naval Research. (Eds.), *Third International Workshop on Policies for Distrib-*

uted Systems and Networks, June 5-7, 2002, Monterey, California, USA: proceedings (pp. 50-59). Los Alamitos, CA.: IEEE Computer Society.

Pearson, W.R. (2000). Flexible sequence similarity searching with the FASTA3 program package. *Methods in Molecular Biology, 132*, 185-219.

Pearson, W.R., & Lipman, D.J. (1988). Improved Tools for Biological Sequence Comparison. *Proceedings of the National Academy of Sciences of the United States of America, 85*(8), 2444-2448.

Pebay-Peyroula, E., Dahout-Gonzalez, C., Kahn, R., Trezeguet, V., Lauquin, G. & Brandolin, G. (2003). Structure of mitochondrial ADP/ATP carrier in complex with carboxyatractylosider. *Nature* (426), 39–44.

Pedretti, K.T., Casavant, T.L., Braun, R.C., Scheetz, T.E., Birkett, C.L., & Roberts, C.A. (1999). Three Complementary Approaches to Parallelization of Local BLAST Service on Workstation Clusters. In *PaCT '999: Proceedings of the 5th International Conference on Parallel Computing Technologies*, (pp. 271-282). London: Springer-Verlag.

Pedrioli, P. G. A., Eng, J. K., Hubley, R., Vogelzang, M., Deutsch, E. W., Raught, B., et al. (2004). A common open representation of mass spectrometry data and its application to proteomics research. *Nature Biotechnology, 22*(11), 1459+.

Peltier, ST. & Ellisman MH. (2003). The Biomedical Informatics Research Network (BIRN). In: I Foster and C Kesselman (Eds.), *The Grid: Blueprint for a New Computing Infrastructure* (pp 109-120). San Francisco, CA: Morgan Kaufmann Publishers.

Penedo, M., Souto, M., Tahoces, P.G., Carreira, J.M., Villalón, J., Porto, G., et al. (2005). Free-response receiver operating characteristic evaluation of lossy JPEG2000 and object-based set partitioning in hierarchical trees compression of digitized mammograms. *Radiology, 237*, 450-457.

Pennec, X & Thirion, J-P. (1997). A Framework for Uncertainty and Validation of 3D Registration Methods based on Points and Frames. *Int. Journal of Computer Vision, 25*(3) 203-229.

Pennec, X., Ayache, N. & Thirion, J.P. (2000). Landmark-based registration using features identified through differential geometry. In I. Bankman, (Ed.), *Handbook of Medical Imaging*, (pp. 499-513). Academic Press.

Pereira-Leal, J. B., Audit, B., Peregrin-Alvarez, J. M., & Ouzounis C. A. (2005). An Exponential Core in the Heart of the Yeast Protein Interaction Network. *Molecular Biology and Evolution, 22*(3), 421-425.

Perelmutov, T., Petravick, D., Gu, J., Barring, O., Baud, J.-P., De Witt, S., et al. (2002, May). *SRM Interface Specification v2.2* (FNAL technical report). Retrieved from http://sdm.lbl.gov/srm-wg/doc/SRM.v2.2.html

Pérez, M. S., Sànchez, A., Herrero, P., Robles, V., & Pena, J. M. (2005). Adapting the weka data mining toolkit to a grid based environment. *In 3rd Atlantic Web Intelligence Conference* (pp. 492-497).

Perkins, D. N., Pappin, D. J., Creasy, D. M., & Cottrell, J. S. (1999). Probability-based protein identification by searching sequence databases using mass spectrometry data. *Electrophoresis*, 20(18), 3551-3567.

Perlman, L., Welch, V., Foster, I., Kesselman, C., & Tuecke, S. (2002). *In A Community Authorization Service for Group Collaboration.* IEEE 3rd International Workshop on Policies for Distributed Systems and Networks.

Pesole, G., Liuni, S. & et al (1999). UTRdb: a specialized database of 5' and 3' untranslated regions of eukaryotic mRNAs. *Nucleic Acids Res.*, (27):188–191. http://www.ba.itb.cnr.it/UTR/.

Peters, T.J., Greenshields, I.R. & Dorney, S. (1992). Topological Fidelity in Surface Reconstruction, Curves and Surfaces in Computer Vision and Graphics III. In *Proc. SPIE* (Vol. 0277-786X), (pp. 221-225).

Petersen, K., Spreitzer, M., Terry, D., Theimer, M., & Demers, A. (1997). *Flexible Update Propagation for Weakly Consistent Replication*, Paper presented at the 16th ACM Symposium on Operating System Principles, Saint-Malo, France.

Petricoin, E. F., Ardekani, A. M., Hitt, B. A., Levine, P. J., Fusaro, V. A., Steinberg, S. M., et al. (2002). Use of proteomic patterns in serum to identify ovarian cancer. *Lancet, 359*(9306), 572-577

P-GRADE (2003). Parallel Grid Run-time and Application Development Environment. Retrieved from www.lpds.sztaki.hu/pgrade/

Phillips, J. C., Braun, R., Wang, W., Gumbart, J., Tajkhorshid, E., Villa, E., Chipot, C., Skeel R. D., Kale, L. & Schulten, K (2005). Scalable molecular dynamics with NAMD. *Journal of Computational Chemistry, 26*, 1781-1802.

Piggott, D., Teljeur, C. & Kelly, A. (2004). Exploring the potential for using the grid to support health impact assessment modelling. *Parallel Computing, 30*(9-10), 1073-1091.

Piper, I., Citerio, G., & the BrainIT Group (2003). The BrainIT Group: Concept and Core Dataset Definition. *Acta Neurochirurgica, 145*, 615-629

Podvinec, M., Maffioletti, S., Kunszt, P., Arnold, K., Cerutti, L., Nyffeler, B., et al. (2006). The SwissBioGrid Project: Objectives, Preliminary Results and Lessons Learned. *2nd IEEE International Conference on e-Science and Grid Computing (e-Science 2006) - Workshop on Production Grids.*

Podvinec, M., Schwede, T., Cerutti, L., Kunszt, P., Nyffeler, B., Stockinger, H., et al (2007). The SwissBioGrid Project: Objectives, Preliminary Results and Lessons Learned. Proceedings of the *2nd IEEE International Conference on e-Science and Grid Computing*, Amsterdam, Netherlands

Politi L, Groppi, A. & Polettini, A. (2005). Review: Applications of liquid chromatography- mass spectrometry in doping control. *Journal of Analytical Toxicology, 29*, (1), 1–14

Pollastri, G., Vullo, A., Fiasconi, P. & Baldi, P. (2006). Modular DAG-RNN Architectures for Assembling Coarse Protein Structures. *J. Comp. Biol., 13*(3), 631-650.

Polychroniadou, H., Psomopoulos, F., & Miktas, P. (2006). G-Class: A divide and conquer application for protein classification. In *Proceedings of the 2nd ADBIS Workshop on Data Mining and Knowledge Discovery*, Thessaloniki, Greece, (pp. 121-132).

Pommerening, K. (2006). Grundlagen des sicheren Grid-Computing. In *Grid-Computing in der biomedizinischen Forschung* (Vol. 90, pp. 63-69). München, Germany: Urban & Vogel.

Ponting C.P., & Russell R.R. (2002). The natural history of protein domains. *Annu Rev Biophys Biomol Struct. 31*, 45-71.

Ponting, C.P., Schultz, J., Milpetz, F., & Bork, P. (1999). SMART: identification and annotation of domains from signalling and extracellular protein sequences. *Nucleic Acids Research, 27*, 229-232.

PostGIS website (2008). Retrieved July 20, 2008 from http://postgis.refractions.net.

Potter, C. S., Zhu, Y., & Carragher, B. (2004). Automated particle selection for cryo-electron microscopy. *Journal of Structural Biology, 145*(1), 1-180.

Power, D. J., Politou, E. A., Slaymaker, M. A. & Simpson, A. C. (2006). Securing web services for deployment in HealthGrids. *Future Generation Computer Systems, 22*(5), 547-570.

PRACE (2008). Partnership for Advanced Computing in Europe. Retrieved from www.prace-project.eu/

Pragma 2008, http://www.pragma-grid.net/

Preparata, F.P. & Shamos, M.I. (1990). *Computational Geometry*. Berlin, Germany: Springer-Verlag.

Prieto, F., Cardenas, M., & Rubio, M. (2008). gPhase: A GRID-based technology for the analysis of phase space structure in nonlinear dynamics. Applications to rocking systems under harmonic loading. *Proceedings of the 2nd Iberian Grid Infrastructure Conference.*

Proactive. (2005). "Proactive Manual REVISED 2.2", Proactive, INRIA. Retrieved from http://www-sop.inria.fr/oasis/Proactive/

Provost, F. P., & Hennessy, D. N. (1996). Scaling up: Distributed machine learning with cooperation. *In the Thirteenth National Conference on Artificial Intelligence* (74-79). Menlo Park, CA: AAAI Press.

Psomopoulos, F., & Mitkas, P. (2005). A protein classification engine based on stochastic finite state automata. *Lecture Series on Computer and Computational Sciences VSP/Brill, 4B*, 1371-1374.

Psomopoulos, F., Diplaris, S., & Mitkas, P. (2004). A finite state automata based technique for protein classification rules induction. In *the Proceedings of the Second European Conference on Data Mining and Text Mining in Bioinformatcs*, ECML/PKDD, Piza, Italy, (pp. 54-60).

PsyGrid Consortium (2008). *PsyGrid*. Retrieved July 2008 from http://www.psygrid.org/

Pujol, S, Kikinis, R & Gollub, R. (2008). Lowering the barriers inherent in translating advances in neuroimage analysis to clinical research applications. *Acad Radiology, 15*(1), 114-118.

Pytliński J., Skorwider Ł., Benedyczak K., Wroński M., Bała P., & Huber V. (2003). Uniform Access to the Distributed Resources for the Computational Chemistry Using UNICORE. In: Sloot M. A et al. (Eds.) Computational Science - ICCS 2003 (pp.307-315). Berlin, Germany: Springer-Verlag.

Pytliński J., Skorwider Ł., Huber V., Benedyczak K., & Bała P. (2002). UNICORE - An Uniform Platform for Chemistry on the Grid. *Journal of Computational Methods in Science and Engineering 2* (3s-4s), 369-376.

Quandt, A., Hernandez, P., Kunzst, P., Pautasso, C., Tuloup, M., Hernandez, C., & Appel, R. (2007). Grid-based Analysis of Tandem Mass Spectrometry Data in Clinical Proteomics. In (Eds.) N. Jacq, N. Müller, I. Blanquer, Y. Legré, V. Breton, D. Hausser, V. Hernández, T. Solomonides, M. Apitius, *From Genes to Personalized HealthCare: Grid Solutions for the Life Sciences - Proceedings of HealthGrid 2007*, (126), (pp.13-22).

Quandt, A., Hernandez, P., Masselot, A., Hernandez, C., Maffioletti, S., Pautasso, C., et al. (2008). swissPIT: A novel approach for pipelined analysis of mass spectrometry data. *Bioinformatics*, 24(11), 1416-1417.

Quinlan, J. R. (Ed.). (1993). *C4.5: Programs for Machine Learning*. San Francisco, CA: Morgan Kaufmann.

Radwan1, A., Younis, A., Luykx, P. & Khuri, S. (2008). Prediction and analysis of nucleosome exclusion regions in the human genome. *BMC Genomic, 9*, 186.

Rajasekar, A., Wan, M., Moore, R., Schroeder, W., Kremenek, G., Jagatheesan, A., et al. (2003.). Storage Resource Broker - Managing Distributed Data in a Grid. *Computer Society of India Journal, 33*(4), 42-54

Ram, J. D., Sreenivas, T.H., & Subramaniam, K. G. (1996). Parallel Simulated Annealing Algorithms. *Jnl. Parallel and Distributed Computing, 37*, 207-212.

Ranganathan, K., & Foster, I. (2003). Simulation Studies of Computation and Data Scheduling Algorithms for Data Grids. *Journal of Grid Computing, 1*(1), 53-62.

Rangwala, H., Lantz, E., Musselman, R., Pinnow, K., Smith, B., & Wallenfelt, B. (2005). Massively Parallel BLAST for the Blue Gene/L. *High Availability and Performance Workshop*.

Rappleye, J., Innus, M., Weeks, C. M. & Miller, R. (2002). *SnB* v2.2: An Example of Crystallographic Multiprocessing. *J. Appl. Cryst., 35*, 374-376.

Rarey, M., Kramer, B., & Lengauer, T. (1996). A fast flexible docking method using an incremental construction algorithm. *J Mol Biol, 261*, 470-489.

Ratnasingham, S. & Hebert, P. (2007). BOLD: The Barcode of Life Data System. *Molecular Ecology Notes, 7*(3), 355-364. Retrieved from www.barcodinglife.org.

Rauhut G. (2002). Web Site: Your Web Browser as a Computational Chemistry Frontend. *Angewandte Chemie International Edition, 41*(12), 2205.

Reed D. A. (2003). Grids, the TeraGrid, and Beyond. *Computer, 36*, 62-68. http://portal.acm.org/citation.cfm?id=642202.

Reed, D. A., Mendes, C. L., Lu, C., Foster, I., & Kesselman, C. (Eds.). (2003). *The Grid 2: Blueprint for a New Computing Infrastructure - Application Tuning and Adaptation*. San Francisco, CA: Morgan Kaufman.

Reference Model for Service Oriented Architecture (2006). Committee Draft 1.0. Retrieved from http://www.oasis-open.org/committees/download.php/16587/wd-soa-rm-cd1ED.pdf

Ren, Y., Zhang, Q., Wan, J., Huang, X., Xie, P., & Yang, G. (2006). Implementation of CCNUGrid-Based Drug Virtual Screening Applications Using Workflow Techniques. *Proceedings of the Fifth International Conference on Grid and Cooperative Computing Workshops* (pp. 371-374). New York: IEEE Computer Society.

Resnik, P. (1999). Semantic similarity in a taxonomy: an information based measure and its application to problems of ambiguity in natural language. *J Artif Intelligence Res, 11,* 95–130.

Richards, W.G. (2002). Virtual Screening Using Grid Computing: the Screensaver Project. *Nat Rev Drug Discov, 1,* 551-555.

Ricklefs, R.E., Fallon, S.M., & Birmingham, E. (2004). Evolutionary relationships, cospeciation, and host switching in avian malaria parasites. *Systematic Biology,* 53(1):111-119.

Roche A., Pennec X., Malandain G. & Ayache N. (2001). Rigid Registration of 3D Ultrasound with MR Images: a New Approach Combining Intensity and Gradient Information. *IEEE Transactions on Medical Imaging, 20*(10), 1038-1049.

Romberg, M. (Ed.) (2005). *OpenMolGRID - Open Computing Grid for Molecular Science and Engineering,* Final Report, NIC Directors, Germany, Jülich, (Vol. 29).

Romei, A., Ruggieri, S., & Turini, F. (2006). Kddml: a middleware language and system for knowledge discovery in databases. *Data Knowledge Engineering, 57*(2), 179-220.

Romei, A., Sciolla, M., Turini, F., & Valentini, M. (2007). Kddml-G: a grid-enabled knowledge discovery system. *Concurr. Comput. : Pract. Exper., 19*(13), 1785-1809.

Ronquist, F. & Huelsenbeck, J. P. (2003). MRBAYES 3: Bayesian phylogenetic inference under mixed models. *Bioinformatics, 19,* 1572-1574.

Rosenberg, G., Phillips, W., Landis, D., & Pierce, W. (1981). Design and Evaluation of the Pennsylvania State University Mock Circulation System. *ASAIO Journal, 4*(1), 41-49.

Rosenfield, B. A. & Dorman, T. (2000). Intensive care unit telemedicine; alternative paradigm for providing continuous intensive care. *Crit. Care Med, 28,* 3925-3931.

Rosenthal, A., Seligman, L., & Renner, S. (2004). From semantic integration to semantics management: Case studies and a way forward. *SIGMOD Record, 33*(4), 44-50.

Rushing, J., Ramachandran, R., Nair, U., Graves, S., Welch, R., & Lin, H. (2005). ADaM: a data mining toolkit for scientists and engineers. *Computers and Geosciences, 31,* 607-618.

Rutten, M., de Hart, J., de Mol, B., van de Molengraft, R., & de Vosse, F. (2007). Analysis of LVADs in a controlled mock systemic circulation. *ASAIO Journal, 53*(2), 26A.

Rycerz, K., Bubak, M., Sloot, P., Getov, V. (2008). Problem solving environment for distributed interactive simulations. In Gorlatch, S., Bubak, M., Priol, T., (Eds.), *Achievements in European Reseach on Grid Systems. CoreGRID Integration Workshop 2006 (Selected Papers),* (pp. 55—66). Berlin, Germany: Springer.

Saara Väärtö, S. (Ed.). (2008). Advancing Science in Europe. DEISA – Distributed European Infrastructure for Supercomputing Applications. EU FP6 Project. Retrieved from www.deisa.eu/press/DEISA-AdvancingScienceInEurope.pdf

SAGA (2006). SAGA implementation home page. http://fortytwo.cct.lsu.edu:8000/SAGA

Šali, A., & Blundell, T. L. (1993).Comparative protein modelling by satisfaction of spatial restraints. *Journal of Molecular Biology, 234,* 779-815.

Saltz, J., Oster, S., Hastings, S., Langella, S., Kurc, T., Sanchez, W., et al. (2006). *caGrid:* Design and Implementation of the Core Architecture of the Cancer Biomedical Informatics Grid. *Bioinformatics, 22*(15), 1910.

Salzberg, S.L., Kingsford, C., Cattoli, G., et al. (2007). Genome analysis linking recent European and African influenza (H5N1) viruses. *Emerg Infect Dis,.* 13(5):713-8.

Salzemann, J., Jacq, N., & Breton, V. (2007, June). Replication and Update of Molecular Biology Databases. *IEEE Transactions on Nanobioscience, 6*(2).

Samatova, N. F., Ostrouchov, G., Geist, A., & Melechko, A. (2002). RACHET: An Efficient Cover-Based Merging of Clustering Hierarchies from Distributed Datasets. *Distributed and Parallel Databases, 11*(2), 157-180.

Sandhu, R., Coyne, E. J., Feinstein, H. L. & Youman, C.E. (1996). Role-Based Access Control Models. *IEEE Computer 29* (2), 38-47.

Sanner, M. F., Duncan, B. S., Carrillo, C. J., & Olson, A. J. (1999). Integrating computation and visualization for biomolecular analysis: an example using python and AVS. *Pacific Symposium in Biocomputing, 99,* 401-412.

Sansom, M. S. P., Scott, K. A., & Bond, P. J. (2008). Coarse-grained simulation: A high-throughput computational approach to membrane proteins. *Biochemistry Society Transactions, 36,* 27-32.

Santos, N. & Koblitz, B. (2006, June). *Distributed Metadata with the AMGA Metadata Catalog.* Paper presented at the workshop on Next-Generation Distributed Data Management, Paris, France. Retrieved from http://project-arda-dev.web.cern.ch/project-arda-dev/metadata

Santos, N., & Koblitz, B. (2006). Distributed Metadata with the AMGA Metadata Catalog. *In workshop on Next Generation Distributed Data Management*

Sax, U., Mohammed, Y., Viezens, F., & Rienhoff, O. (2006). Grid-Computing in der biomedizinischen Forschung. In K. Überla, O. Rienhoff & N. Victor (Eds.), *Medizinische Informatik, Biometrie und Epidemiologie* (Vol. 90). München, Germany: Urban & Vogel.

Scardaci, D. & Scuderi, G. (2007). A Secure Storage Service for the gLite Middleware. In the *proceedings of third International Symposium on Information Assurance and Security* (pp. 261-266), Manchester, UK. Washington, DC: IEEE Computer Society..

Scheres, S. H. W., Merino, A. J., Sorzano, C. O. S., & Carazo, J. M. (2005). Grid Computing in 3D-EM Image Processing Using Xmipp. In *Proceedings of the 18th IEEE Symposium on Computer-Based Medical Systems,* Washington, D.C., (pp. 561-563).New York: IEEE Computer Society.

Scheres, S. H. W., Valle, M., Nuñez, R., Sorzano, C. O. S., Marabini, R., Herman, G. T., et al. (2005). Maximum-likelihood Multi-reference Refinement for Electron Microscopy Images. *Journal of Molecular Biology, 348*(1), 139-149.

Schiffers, M. (2008). VO Management, *IVOM Workshop.* Regionales Rechenzentrum für Niedersachen, Hannover, Germany.

Schmidt, M. W., Baldridge, K. K., Boatz, J. A., Elbert, S. T., Gordon, M. S., Jensen, J. H., et al. (1993). General atomic and molecular electronic structure system. *Journal of Computational Chemistry, 14,* 1347-1363.

Schneider, T. R. & Sheldrick, G. M. (2002). Substructure solution with *SHELXD. Acta Crystallogr., D58,* 1772-1779.

Schroeder, W., Martin, K. & B. Lorensen (2006). *Visualization Toolkit: An Object-Oriented Approach to 3D Graphics.* Clifton Park, NY: Kitware.

Schuchardt, K., Didier, B., & Black, G. (2002). Ecce - a problem-solving environment's evolution toward Grid services and a Web architecture. *Concurrency and Computation: Practice and Experience, 14,*1221–1239.

Schuller B., Romberg M., & Kirtchakova L. (2005). Application Driven Grid Developments in the OpenMolGRID project. In P.M.A. Sloot, A.G. Hoekstra, T. Priol, A. Reinefeld, M. Bubak (Eds.), *Advances in Grid Computing* (Vol. 3470) (pp. 23–29). Berlin, Germany: Springer-Verlag.

Schulze, W & Mann,M (2003) A novel proteomic screen for peptide-protein interactions. *Journal of Biological Chemistry, 279*(11), 10756-64

Schuster, A., & Wolff, R. (2001). Communication-efficient distributed mining of association rules. *In ACM SIGMOD international conference on Management of data* (pp. 473-484). New York: ACM.

Schuster, A.,Wolff, R., & Trock, D. (2003). A High-Performance Distributed Algorithm for Mining Association Rules. In *Third IEEE International Conference on Data Mining.*

Scifo, S. (2007). GSAF - Grid Storage Access Framework. In the *proceeding of 16th IEEE International Workshops on Enabling Technologies: Infrastructure for Collaborative Enterprises* (pp.296-297), Evry, France, June 2007. Washington, DC: IEEE Computer Society.

Scott, K. A., Alonso, D. O. V., Sato, S., Fersht, A. R., & Daggett, V. (2007). Conformational entropy of alanine versus glycine in protein denatured states. *Proceedings of the National Academy of Sciences USA, 104,* 2661-2666.

Sederberg, T. W., & Parry, S. R. (1986). Free-form deformation of solid geometric models. *ACM Computer Graphic, 20,* 151-160.

Seifert, K.A., Samson, R.A., Dewaard, J.R, Houbraken, J., Lévesque, C.A., Moncalvo, J.M., Louis-Seize G. & Hebert, P.D. (2007). Prospects for fungus identification using COI DNA barcodes, with Penicillium as a test case. *Proc Natl Acad Sci U S A*, *104*(10), 3901-6.

Seitz, L., Montagnat, J., Pierson, J. M., Oriol, D., & Lingrand, D. (2005). Authentication and Authorisation Prototype on the µgrid for Medical Data Management. In T. Solomonides, R. McClatchey, V. Breton, Y. Legré & S. Nørager (Eds.), *From Grid to Healthgrid - Proceedings of Healthgrid 2005* (pp. 222 - 233). Washington, DC: IOS Press.

Seo, J. and Lee, K.J. (2004). Post-translational modifications and their biological functions: proteomic analysis and systematic approaches. *J Biochem Mol Biol. 37*(1), 35-44.

Seymour, K., Lee, C., Desprez, F., Nakada, H., & Tanaka, Y. (2004, September). The End-User and Middleware APIs for GridRPC. In *Workshop on Grid Application Programming Interfaces, In conjunction with GGF12*, Brussels, Belgium.

Seymour, K., Nakada, H., Matsuoka, S., Dongarra, J., Lee, C., & Casanova, H. (2002). Overview of GridRPC: A Remote Procedure Call API for Grid Computing. Proceedings of the Third International Workshop on Grid Computing, Lecture Notes in Computer Science, Springer, ed 2536, pp. 274–278, Baltimore, USA.

Shadforth, I., Crowther, D. & Bessant, C. (2005). Protein and peptide identification algorithms using MS for use in high-throughput, automated pipelines. *Proteomics*, *5*(16), 4082-4095.

Shafer, J. C., Agrawal, R., & Mehta, M. (1996). SPRINT: A scalable parallel classifier for data mining. In T. M. Vijayaraman, A. P. Buchmann, C. Mohan, & N. L. Sarda (Eds.), *22nd Int. Conf. Very Large Databases* (pp. 544-555). San Francisco, CA: Morgan Kaufmann.

Shah, S.P., He, D.Y.M., Sawkins, J.N., Druce, J.C., Quon, G., Lett, D., Zheng, G.X.Y., Xu, T., Quellette, B.F.F. (2004). Pegasys: Software for executing and integrating analyses of biological sequences. *BMC Bioinformatics*, *5*, 40.

Shamir, A. (1979). How to share a secret. *Communications of the ACM*, *22*, 612-613.

SHARE Consortium (2006). *Structuring Healthgrid Activities and Research in Europe: Developing a Roadmap*. Retrieved July 2008 from http://www.eu-share.org

Shaw, M. & Piper, I. (2008). The Brain Monitoring with Information Technology (BrainIT) collaborative network: Data Validation Results. *Acta Neurochirurgica, ICP Symposium Journal*.

Sheldrick, G.M. (1990). Phase annealing in SHELX-90: direct methods for larger structures. *Acta Cryst. A46*, 467–473.

Shen, J., Xu, X., & Cheng, F. (2003). Virtual screening on natural products for discovering active compounds and target clues. *Curr Med Chem*, *10*, 2327-2342.

Shibboleth Project (2006). Retrieved July 2008 from http://shibboleth.internet2.edu/

Shindyalov, I.N. & Bourne, P.E. (1983). Protein structure alignment by incremental combinatorial extension (CE) of the optimal path. *Protein Engineering 11*(9), 739–747.

Shoichet, B. (2004). Virtual screening of chemical libraries. *Nature*, *432*, 862-865.

Siebenlist, F., Magaratnam, N., Welch, V., & Neuman, C. (2004). Security for Virtual Organizations: Federating Trust and Policy Domains. In *The Grid: Blueprint for a New Computing Infrastructure* (2nd Ed.). Amsterdam: Morgan Kaufmann.

Signorini, M., Paulesu, E., Friston, K., Perani, D., Colleluori, A., Lucignani, G., et al (1999). Rapid assessment of regional cerebral metabolic abnormalities in single subjects with quantitative and nonquantitative [18F] FDG PET: A clinical validation of statistical parametric mapping. *Neuroimage*, *9*(1), 63-80.

Sild S., Maran U., Lomaka A., & Karelson M. (2006). Open Computing Grid for Molecular Science and Engineering. *J. Chem. Inf. Model. 46*, 953–959

Sild S., Maran U., Romberg M., Schuller B., & Benfenati E. (2005). OpenMolGRID: Using Automated Workflows in GRID Computing Environment. In P.M.A. Sloot, A.G. Hoekstra, T. Priol, A. Reinefeld, M. Bubak (Eds.),

Advances in Grid Computing (Vol. 3470) (pp. 464–473). Berlin, Germany: Springer-Verlag.

Silva, F.A.B., Gagliardi, H. F., Gallo, E., Madope, A.M. Neto, V.C. , Pisa, I.T., Alves, D.(2007) IntegraEPI: a Grid-based Epidemic Surveillance System. In N. Jacq, Y. Legré, H. Muller, I. Blanquer, V. Breton, D. Hausser, V. Hernández, T. Solomonides and M. Hofman-Apitius (Eds.), *From Genes to Personalized HealthCare: Grid Solutions for the Life Sciences - Proceedings of HealthGrid 2007.* Amsterdam, the Netherlands: IOS Press.

Sim, A. & Shoshani, A. (2007) *The Storage Resource Manager Interface Specification Version 2.2.* Retrieved from https://forge.gridforum.org/ sf/projects/gsm-wg

SIMDAT (2008). Grids for Industrial Product Development. Retrieved from www.scai.fraunhofer.de/about_simdat.html

Simmhan, Y. L., & Plale, B., & Gannon, D. (2007). Query capabilities of the Karma provenance framework. In *Concurrency and Computation: Practice and Experience, 20*(5) 441-451. San Francisco, CA:Wiley InterScience

Simmhan, Y. L., Plale, B., & Gannon, D. (2005). A survey of data provenance in e-science. *SIGMOD Rec. 34* (3), 31-36.

Simmhan, Y., Plale, B. & Gannon, D. (2005) A survey of data provenance in e-science. *SIGMOD Record, 34*(3) 31–36.

Simmhan, Y., Plale, B., Gannon, D. & Marru, S. (2006) Performance evaluation of the karma provenance framework for scientific workflows. In L. Moreau and I. T. Foster (Eds.), *IPAW* (pp. 222–236). New York: Springer.

Sims, L.D., et al. (2005). Origin and evolution of highly pathogenic H5N1 avian influenza in Asia. *Vet Rec. 157*(6), 159-64.

Sinnott, R. O., Jiang, J., Stewart, G., Stell, A. J., Martin, D., Doherty, T., Su, L. & Watt, J. (2008). *Advanced Security for Virtual Organizations: Exploring the Pros and Cons of Centralized vs Decentralized Security Models.* Proceedings of 8th IEEE International Symposium on Cluster Computing and the Grid (CCGrid 2008), Lyon, France.

Sinnott, R. O., Stell, A. J., Chadwick, D. W., & Otenko, O. (2005). Experiences of Applying Advanced Grid Authorisation Infrastructures. In P.M.A. Sloot, et al, (eds.), *Proceedings of European Grid Conference (EGC)*, Amsterdam, the Netherlands (pp. 265-275).

Sinnott, R. O., Stell, A. J., Chadwick, D. W., Otenko, O., Watt, J., Koetsier, J. & Nguyen, T. A. (2006). *Supporting Decentralized, Security focused Dynamic Virtual Organizations across the Grid.* Proceedings of 2nd IEEE International Conference on e-Science and Grid Computing, Amsterdam, Holland.

Sinnott, R. O., Watt, J., & Jiang, J. (2007). *The GLASS Project: Supporting Secure Shibboleth-based Single Sign-On to Campus Resources.* Proceedings of UK e-Science All Hands Meeting, Nottingham, UK.

Sinnott, R. O., Watt, J., Ajayi, O. & Jiang, J. (2006). *Shibboleth-based Access to and Usage of Grid Resources.* Proceedings of IEEE International Conference on Grid Computing, Barcelona, Spain.

Sinnott, R., Atkinson, M., Bayer, M., Berry, D., Dominiczak, A., Ferrier, M., Gilbert, D., Hanlon, N., Houghton, D., Hunt, E., & White, D. (2004). Grid services supporting the usage of secure federated, distributed biomedical data. *In the UK e-Science All Hands Conference. Engineering and Physical Sciences Research Council.*

Sinnott, R.O. Ajayi,O. Stell,A.J. (2006). Development of Grid Frameworks for Clinical Trials and Epidemiological Studies. *HealthGrid 2006 Conference, Studies in Health Technology and Informatics, 120,* 117-130.

Skillicorn, D. B. (2002). The case for datacentric grids. *In the 16th International Parallel and Distributed Processing Symposium* (p. 313). New York: IEEE Computer Society.

Skolnick, J. and Fetrow, J.S. (2000). From genes to protein structure and function: novel applications of computational approaches in the genomic era. *Trends Biotechnol, 18*(1), 34-9.

Slaughter, L. A., Soergel, D. & Rindflesch, T. C. (2006). Semantic representation of consumer questions and physician answers. *International Journal of Medical Informatics, 75*(7), 513-529.

Sloot, P. M. A., Ramos, A. T., Altintas, I., Bubak, M., & Boucher, C. (2006). From molecule to man: Decision support in individualized e-health. *Computer, 39*(11), 40-46.

Smarr, L., & Catlett, C. E. (June 1992). Metacomputing. *Communications of the ACM, 35*(6), 44-52).

Smith, C. A., Want, E. J., O'Maille, G., Abagyan, R., & Siuzdak, G. (2006). Xcms: processing mass spectrometry data for metabolite profiling using nonlinear peak alignment, matching, and identification. *Analytical Chemistry, 78*(3), 779-787.

Smith, M., Barton, M., Bass, M., Branchofsky, M., McClellan, G., Stuve, D., Tansley, R., & Harford J. (2003). An Open Source Dynamic Digital Repository. *D-Lib Magazine, 9.*

Smith, T., & Waterman, M. (1981). Identification of common molecular subsequences. *Journal of Molecular Biology, 147,* 195-197.

Smith, T.F., & Waterman, M.S. (1981). Identification of Common Molecular Subsequences. *Journal of Molecular Biology, 147,* 195-197.

Snell, Q., Clement, MJ., Jackson DB. & Gregory, C. (2000, May). *The Performance Impact of Advance Reservation Meta-scheduling.* 6th workshop on Job Scheduling Strategies for Parallel Processing, Cancun, Mexico (LNCS 1911, pp 137–153). Berlin, Germany: Springer-Verlag.

Soh, H., Shazia Haque, S., Liao, W., & Buyya, R. (2006). Grid Programming Models and Environments. In Yuan-Shun Dai, et. al. Advanced Parallel and Distributed Computing (pp. 141–173). Nova Science Publishers.

Sorzano, C. O. S., Marabini, R., Herman, G. T., & Carazo, J. M. (2005b). Multi-objective algorithm parameter optimization using multi-variate statistics in three-dimensional electron microscopy reconstruction. *Pattern Recognition, 38*(12), 2587-2601.

Sprinzl, M., & Vassilenko, K.S. (2005). Compilation of tRNA sequences and sequences of tRNA genes. *Nucleic Acids Research, 33,* D139–D140.

SRB – The SDSC Storage Resource Broker. (2008). Retrieved October 20, 2008, from http://www.sdsc.edu/srb/index.php/Main_Page.

Srivastava, A., Han, E., Kumar, V., & Singh, V., (1999). Parallel formulations of decision-tree classification algorithms. *Data Mining and Knowledge Discovery, 3,* 237-261.

Stallman, R. M. (2002). *Free software, free society.* Boston, MA: GNU Press.

Stamatakis, A. (2006). Parallel and Distributed Computation of Large Phylogenetic Trees. In Zomaya A. (Ed.), *Parallel Computing for Bioinformatics and Computational Biology,* (pp. 327-346). Hoboken, NJ: Wiley.

Stamatakis, A. (2006). RAxML-VI-HPC: maximum likelihood-based phylogenetic analyses with thousands of taxa and mixed models. *Bioinformatics,* 22(21): 2688-2690.

Stamatakis, A., Auch, A.F., Meier-Kolthoff, J., & Göker, M. (2007). *AxPcoords & parallel AxParafit: statistical co-phylogenetic analyses on thousands of taxa. BMC Bioinformatics,* 8, 405.

Stamatakis, A., Hoover, P., & Rougemont, J. (2008). A Rapid Bootstrapping Algorithm for the RAxML Web Servers. *Systematic Biology,* in press.

Stankovski, V., Swain, M., Kravtsov, V., Niessen, T., Wegener, D., Kindermann, J., & Dubitzky, J. (2007). Grid-enabling data mining applications with datamining-grid: An architectural perspective. *Future Generation Computing Systems, 24*(4), 259-279.

Stankovski, V., Swain, M., Kravtsov, V., Niessen, T., Wegener, D., Kindermann, J., & Dubitzky, W. (2008). Grid-enabling data mining applications with DataMiningGrid: An architectural perpective. *Future Generation Computer Systems, 24,* 259-279.

Stark, C., Breitkreutz, B.-J., Reguly, T., Boucher, L., Breitkreutz, A., & Tyers, M. (2006). BioGRID: A General Repository for Interaction Datasets. *Nucleic Acids Research, 34,* 535-539.

Starling, E. (1918). *The linacre lecture on the law of the heart.* London: Green and Co.

Steadm, D.A., Paton, N.W., Missier, P., Embury, S.M. Hedeler, C. Jin, B. Brown, A.J., & Preece, A. (2008) Information quality in proteomic *Briefing in Bioinformatics,* 9(2), 174-188.

Stein, L. D. (2008). Towards a cyber-infrastructure for the biological sciences: progress, visions and challenges. *Nature Rev. Genet., 9*, 678-688.

Stell A. J., Sinnott R. O., Watt J. P. (2005). Comparison of advanced authorisation infrastructures for grid computing. *International Symposium on High Performance Computing Systems and Applications 2005 (HPCS2005)*, (pp. 195-201).

Stell, A. J., Sinnott, R. O. & Ajayi, O. (2008). *Supporting Nationwide e-Clinical Trials and Studies*. Proceedings of 15th Mardi Gras Conference, Baton Rouge, LA.

Stell, A. J., Sinnott, R. O. & Ajayi, O. (2006). *Secure, Reliable and Dynamic Access to Distributed Clinical Data*. Proceedings of Life Science Grid Conference, Yokohama, Japan.

Stell, A. J., Sinnott, R. O., & Watt, J. (2005). *Comparison of Advanced Authorisation Infrastructures for Grid Computing*. Proceedings of International Conference on High Performance Computing Systems and Applications, Guelph, Canada.

STEP (2002). *Seeding the EuroPhysiome: A Roadmap to the Virtual Physiological Human*. Retrieved from http://www.biomedtown.org/biomed_town/STEP/Reception/step_presentations/RoadMap/vph_roadmap_printed_3.pdf

Sterzel M., & Szepieniec T. (2007). Enabling commercial chemical software on EGEE Grid – Gaussian VO. In M. Bubak, M. Turała, K. Wiatr (Eds.) *Cracow'06 Grid Workshop. Proceedings* (pp. 499–504).

Stevens, R, Robinson, A., & Goble, C. (2003). myGrid: personalised bioinformatics on the information Grid. *Bioinformatics, 19*(Suppl. 1), i302-i304.

Stevens, R. C. (2006). Trends in Cyberinfrastructure for Bioinformatics and Computational Biology. *CTWatch Quarterly, 2*, 6-17.

Stevens, R., Goble, C., Paton, N.W., Bechhofer, S., Ng, G., Baker, P., and Brass A. (2003. May). Complex Query Formulation Over Diverse Information Sources in TAMBIS. In Z. Lacroix & T. Critchlow, (Eds.), *Bioinformatics: Managing Scientific Data*. San Fransisco, CA : Morgan Kaufmann.

Stevens, R., McEntire, R., Goble, C., Greenwood, M., Zhao, J., Wipat, A. & Li, P. (2004). myGrid and the drug discovery process. *Drug Discovery Today: BIOSILICO, 2*(4), 140-148.

Stevens, R.D., Robinson, A. J. & Goble C. A. (2004). myGrid: Personalised Bioinformatics on the Information Grid. *Bioinformatics, 19*(1), 302–302. Retrieved from http://www.mygrid.org.uk/.

Stevens, R.D., Robinson, A.J., & Goble, C.A. (2003). *myGrid: Personalised Bioinformatics on the Information Grid*.

Stevens, R D, Robinson, A J, & Goble C A (2004). myGrid: Personalised bioinformatics on the information Grid. *Bioinformatics*, 19(1), 302–302C.

Stewart J. J. P. (1990). MOPAC: a Semiempirical Molecular Orbital Program. *J. Comput.-Aided Mol. Des., 4*, 1-45.

Stewart, C. A. (2004). Introduction to the edition. *Communications of the ACM, 47*(11), 30-33.

Stewart, G., Cameron, D., Cowan, G., & McCance, G. (2007). Storage and Data management in EGEE. *Proceedings of Fifth Australasian Symposium on Grid Computing and e-Research* Australian Computer Society, Inc., Darlinghurst, Australia, (pp. 69-77).

Stockinger, H., Pagni, M., Cerutti, L., & Falquet, L. (2006). Grid Approach to Embarrassingly Parallel CPU-Intensive Bioinformatics Problems. *2nd IEEE International Conference on e-Science and Grid Computing (e-Science 2006)*, IEEE Computer Society Press, Amsterdam, The Netherlands.

Straatsma, T. P., Philippopoulos, M., & McCammon, J. (2000). NWChem: Exploiting parallelism in molecular simulations. *Computer Physics Communications, 128*, 377-385.

Sun Network.com (2008). retrieved from www.network.com/

SunGrid (2005). Sun Utility Computing. Retrieved from www.sun.com/service/sungrid/

SURA -Southeastern Universities Research Association. (2007). The Grid Technology Cookbook. Programming

Concepts and Challenges. Retrieved from www.sura.org/cookbook/gtcb/

Swain, M., Hunniford, T., Mandel, J., Palfreyman, N., & Dubitzky, W., (2005). Reverse-engineering gene-regulatory networks using evolutionary algorithms and grid computing, *Journal of Clinical Monitoring and Computing, 19*(4-5), 329-337.

Switula, D. (2000). Principles of good clinical practice (GCP) in clinical research. *Sci Eng Ethics, 6*(1), 71-77.

Swofford, D., Olsen, G., Waddell, P. & Hillis, D.M. (1996). Phylogenetic inference. In D.N. Hillis, C. Moritz, & B.K. Mable, (Eds.), *Molecular Systematics,* (pp. 407-514). Sunderland, MA: Sinauer Associated.

Sycara, K. (1998). *Multiagent systems*, Artificial Intelligence Magazine, Association for the Advancement of Artificial Intelligence, 10(2), 79–93.

Tai, K., Baaden, M., Murdock, S., Wu, B., Ng, M. H., Johnston, S., Boardman, R., Fangohr, H., Cox, K., Essex, J. W., & Sansom M. S. P. (2007). Three hydrolases and a transferase: Comparative analysis of active-site dynamics via the BioSimGrid database. *Journal of Molecular Graphics and Modelling, 25,* 896-902.

Tai, K., Murdock, S., Wu, B., Ng, M. H., Johnston, S., Fangohr, H., Cox, S. J., Jeffreys, P., Essex, J. W., & Sansom, M. S. P. (2004). BioSimGrid: Towards a worldwide repository for biomolecular simulations. *Organic and Biomolecular Chemistry, 2,* 3219-3221.

Talia, D., Trunfio, P., & Verta, O. (2005). Weka4WS: A WSRF-Enabled Weka Toolkit for Distributed Data Mining on Grids. *In the 9th European Conference on Principles and Practice of Knowledge Discovery in Databases*, *3721*, 309-320.

Tan, H., Chen, X. & Gu, J. (2007). Semanteme-based processing mechanism under data grid environment. *Journal of Huazhong University of Science and Technology (Natural Science Edition), 35*(4), 22-25.

Tan, W. C. (2007). Provenance in Databases: Past, Current, and Future. In *IEEE Data Eng. Bull. 30*(4),3-12.

Tanner, S., Shu, H., Frank, A., Wang, L., Zandi, E., Mumby, M., et al. (2005), InsPecT: identification of post-translationally modified peptides from tandem mass spectra. *Analytical Chemistry* 7(14), 4626-4639.

Tatusov, R.L., Eugene, V.K., & David, J.L. (1997). A genomic perspective on protein families. *Science, 278,* 631-637.

Tatusova T. A., & Madden T. L. (1999). BLAST 2 Sequences, a new tool for comparing protein and nucleotide sequences. *FEMS Microbiology Letters, 174* (2), 247–250.

Taubenberger, J.K., et al. (2005). Characterization of the 1918 influenza virus polymerase genes. *Nature, 437*(7060), 889-893.

Taufer, M., An, C., Kerstens, A., & Brooks, C.L. (2006). Predictor@Home: Protein Structure Prediction Supercomputer Based on Global Computing. *IEEE Transactions on Parallel and Distributed Computing, 17*(8), 786-796.

TAVERNA (2008). The Taverna Workbench 1.7. Retrieved from http://taverna.sourceforge.net/

Taylor, C. F., Hermjakob, H., Julian, R. K., Garavelli, J. S., Aebersold, R., & Apweiler, R. (2006). The work of the human proteome organisation's proteomics standards initiative (hupo psi). *OMICS, 10*(2), 145-151.

Taylor, I. (2006) Triana generations. In *e-Science*, (pp. 143). Los Alamitos, CA: IEEE Computer Society.

Taylor, I., Shields, M., Wang, I., & Harrison, A. (2005). Visual Grid workflow in Triana. *Journal of Grid Computing, 3*(3-4), 153-169.

Tesauro, GJ. & Kephart, J. O. (2004, May). *Utility functions in autonomic systems.* 1st International Conference on Autonomic Computing (pp. 70–77). New York: IEEE Computer Society.

Texier, R (2007) *Systèmes de globalisation des ressources données-calculs. Application à l'imagerie médicale.* Unpublished doctoral dissertation, Université Paris-Sud, France.

The Gene Ontology Consortium (2000). Gene Ontology: tool for the unification of biology, *Nature Genetics, 25,* 25-29.

The Globus Security Team. (2005). *Globus Toolkit Version 4 Grid Security Infrastructure: A Standards Perspective.*

Thomas M., & Boisseau J. (2003). Building Grid Computing Portals: The NPACI Grid Portal Toolkit. In F. Berman, G. Fox & T. Hey, (Eds). *Grid Computing: Making the Global Infrastructure a Reality*. Hoboken, NJ: John Wiley & Sons.

Thomas M., Boisseau J., Mock S., Dahan M., Mueller K., & Sutto D. (2003). *The GridPort Toolkit Architecture for Building Grid Portals*. Proceedings of the 10th IEEE Intl. Symp. on High Perf. Dist. Comp.

Thomas, P.L., & Menzies, J.G. (1997). Cereal smuts in Manitoba and Saskatchewan, 1989-95. *Canadian Journal of Plant Pathology*, 19(2):161-165.

Thompson, J.D., Higgins, D.G. & Gibson, T.J. (1994). CLUSTAL W: improving the sensitivity of progressive multiple sequence alignment through sequence weighting, position-specific gap penalties and weight matrix choice. *Nucl. Acids Res.*, *22*(22), 4673-4680.

Tian C., Ding W., Cao R. and Jiang S. (2007). Extensive Epidemic Spreading Model Based on Multi-agent System Framework (LNCS 4490, pp. 129-133).

Ting, K. M., & Witten, I. H. (1997). Stacked generalizations: When does it work? *In International Joint Conference on Artificial Intelligence* (pp. 866-873).

Tishkoff S.A., Varkonyi R., Cahinhinan N., Abbes S., Argyropoulos G., Destro-Bisol G., et al. (2001). Haplotype diversity and linkage disequilibrium at human G6PD: recent origin of alleles that confer malarial resistance. *Science, 293.*

TMF. (2006). *AG Datenschutz*. Retrieved March 16, 2006, from www.tmf-ev.de/site/DE/int/AG/DS/container_ag_ds.php

Tohsato, Y., Kosaka, T., Date, S., Shimojo, S. & Matsuda, H. (2005). Heterogeneous database federation using grid technology for drug discovery process,. In *Lecture notes in Computer Science*, (Vol. 3370) (pp. 43-52). Aachen,Germany: Springer-Verlag.

Toivonen, H. (1996). Sampling large databases for association rules. In T. M. Vijayaraman, A. P. Buchmann, C. Mohan, and N. L. Sarda (Eds.), *International Conference on Very Large Data Bases*. San Francisco, CA: Morgan Kaufman.

Torres E., De Alfonso C., Blanquer I., Hernandez V. (2006). Privacy Protection in HealthGrid: Distributing Encryption Management over the VO. *Studies in Health Technology and Informatics, 120*, 131-41.

Traxer, O., Osorio A., Pasqui, F., Sebe, P., Merran, S., Ripoche, X., et al. (2005, March). *New PC based software using an Augmented Reality system to simulate PCNL*. 20th European Association of Urology (EAU) Congress, Istanbul, Turkey.

TRIANA (2003). The Triana Project. Retrieved from www.trianacode.org/

Tsaregorodtsev, A., Garonne, V. & Stokes-Rees, I. (2004). DIRAC: A Scalable Lightweight Architecture for High Throughput Computing. In *5th IEEE/ACM International Workshop on Grid Computing*, (pp 19-25). New York: IEEE Computer Society.

Tsiknakis, M., Kafetzopoulos, D., Potamias, G., Analyti, A., Marias, K., & Manganas, A. (2006). Building a European biomedical grid on cancer: the ACGT Integrated Project. *Stud Health Technol Inform, 120*, 247-258.

Tsoumakas, G. & Vlahavas, I. (2002). Distributed data mining of large classifier ensembles. *In the Second Hellenic Conference on Artificial Intelligence* (pp. 249-256).

Tsoumakos, D., & Roussopoulos, N. (2003). *A Comparison of Peer-to-Peer Search Methods*, Paper presented at the 6th International Workshop on the Web and Databases (WebDB), San Diego, USA.

Tsymbal, A., Huber, M., Zillner, S., Hauer, T., & Zhou, K. (2007a). Visualizing patient similarity in clinical decision support. In *Lernen - Wissensentdeckung - Adaptivität, LWA 2007, Joint Workshops of the German Society for Informatics (GI), Workshop on Knowledge & Experience Management, FGWM'07.*

Tsymbal, A., Zillner, S., & Huber, M. (2007b). Ontology - supported machine learning and decision support in biomedicine, In *Data Integration in the Life Sciences, 4th International Workshop, DILS 2007, Philadelphia, USA*, (LNCS 4544, pp. 156-171). Berlin-Heidelberg: Springer.

Tudor DICOM Viewer. *Main page*. Retrieved February 12, 2008 from http://www.santec.lu/projects/optimage/dicomtools

Tuecke, S., Czajkowski, K., Foster, I., Frey, J., Graham, S., & Kesselman, C., et al. (2003). Open grid services infrastructure (OGSI) (No. Ver. 1.0)., *Global Grid Forum (GGF)*. Retrieved March 8, 2008 from http://www.globus.org/alliance/publications/papers.php#GSSpec

Tulipano, A., Donvito, G., Licciulli, F., Maggi, G., Gisel, A.(2007). Gene analogue finder: a GRID solution for finding functionally analogous gene products. *BMC Bioinformatics, 8*, 329.

Tuloup, M., Hernandez, C., Coro, I., Hoogland, C., & Binz, P.,-A,. (2003) Aldente and BioGraph: An improved peptide mass fingerprinting protein identification environment. In Ducret A, Eberle AN, Faupel MD, Hofsteenge J, Jenö P, Langen H, et al. (Eds.), *Proceedings of the Swiss Proteomics Proceedings of the Swiss Proteomics Society, 2003 Congress, Understanding biological systems* (pp. 174–176). Lausanne, Switzerland: FontisMedia Publishers.

Tveit, A., & Engum, H. (2003). Parallelization of the Incremental Proximal Support Vector Machine Classifier using a Heap-based Tree Topology. In *Parallel & Distributed computing for Machine Learning.*

Udupa, J.K. & Hermann, G.T. (1991). 3D Imaging in Medicine. Boca Raton: CRC Press..

Udupa, J.K. (1991). Computer Aspects of 3D Imaging in Medicine: A Tutorial. In J.K. Udupa and G.T. Hermann (Eds.), *3D imaging in Medicine*, (pp. 1-70).

Ullman, C. & Dykes, L. (2007). *Beginning Ajax*. Hoboken, NJ: John Wiley & Sons. Retrieved from www.Wrox.com.

Umetsu, R., Ohki, S., Fukuzaki, A., Konagaya, A., Shinbara, D., Saito, M., Watanabe, K., Kitagawa, T., & Hoshino, T. (2006). An Architectural Design of Open Genome Services. In *Grid Computing in Life Science* (pp. 87-98). Singapore: World Scientific.

UNICORE (2008). UNiform Interface to COmputing Resources. Retrieved from www.unicore.eu/

UNICORE (Uniform Interface to Computing Resources) (2008). Retrieved October 20, 2008, from http://www.unicore.eu/.

United States. Dept. of Defense. (1999). *Common criteria & methodology for information technology security evaluation*

Unlu, M., Morgan, M.E., & Minden, J.S. (1997). Difference gel electrophoresis: A single gel method for detecting changes in protein extracts. *Electrophoresis, 18*(11), 2071-2077.

Uren, V., Cimiano, P., Iria, J., Handschuh, S., Vargas-Vera, M., Motta, E., and Ciravegna, F. (2006). Semantic annotation for knowledge management: Requirements and a survey of the state of the art. *Web Semantics: Science, Services and Agents on the World Wide Web, 4*(1), 14-28.

UTREF: untranslated regions database from RefSeq. Retrieved from http://www.ba.itb.cnr.it/BIG/UTRHome/.

UTRsite: collection of functional sequence patterns located in 5' or 3' UTR sequences. Retrieved from http://www2.ba.itb.cnr.it/UTRsite/.

Valle, M., Zavialov, A., Li, W., Stagg, S. M., Sengupta, J., Nielsen, R. C., et al. (2003). Incorporation of aminoacyl-tRNA into the ribosome as seen by cryoelectron microscopy. *Nature Structural Biology, 10*(1), 899-906.

Van Dyke Parunak, H., Brueckner, S. A., Matthews, R., & Sauter, J. (2005). *Pheromone Learning for Self-Organizing Agents*, IEEE Transactions on Systems, Man, and Cybernetics, Part A: Systems and Humans, 35(3).

Van Engelen R. et al. (2003). Developing Web Services for C and C++. *IEEE Internet Computing Journal*, 53–61.

van Heel, M. (1987). Similarity measures between images. *Ultramicroscopy, 21*, 95-100.

van Heel, M., & Schatz, M. (2005). Fourier Shell Correlation threshold criteria. *Journal of Structural Biology, 151*(3), 250-262.

van Rossum, G., & Drake, Jr, F. L. (2003). *An Introduction to Python.* Bristol, UK: Network Theory.

Vandenberghe, S., Segers, P., Meyns, B., & Verdonck, P. (2001). Hydrodynamic Characterization of Ventricular Assist Devices. *International Journal of Artificial Organs, 24*(7), 470-477.

Vannier, M.W., Hildebolt, C.F., Gayou, D.E., & Marsh, J.L. (1991). Introduction to 3D Imaging. In J.K. Udupa and G.T. Hermann (Eds.), *3D imaging in Medicine*, (pp. 71-88).

Vassura, M., Margara, L., Di Lena, P., Medri, F., Fariselli, P. & Casadio, R. (2007). Reconstruction of 3D Structures From Protein Contact Maps. In D. I. Mandoiu, A. Zelikovsky (Eds), *Third International Symposium (ISBRA 2007)*: Vol. 4463. *Bioinformatics Research and Applications*, (pp. 578-589)., Berlin, Germany: Springer.

Vassura, M., Margara, L., Di Lena, P., Medri, F., Fariselli, P. & Casadio, R. (2007). Fault Tolerance for Large Scale Protein 3D Reconstruction from Contact Maps. In R. Giancarlo and S. Hannenhalli (Eds.) *Seventh International Workshop on Algorithms in Bioinformatics (WABI 2007)*, (Lecture Notes in Bioinformatics 4645, pp. 25-37). Berlin, Germany: Springer-Verlag.

Vassura, M., Margara, L., Di Lena, P., Medri, F., Fariselli, P. & Casadio, R. (2008). Reconstruction of 3D Structures From Protein Contact Maps. *Computational Biology and Bioinformatics*, *5*(3), 357-367.

Vendruscolo, M., Kussell, E. & Domany, E. (1997). Recovery of protein structure from contact maps. *Folding and Design*, *2*(5), 295-306.

Ventrone, V.; Heiler, S., (1991, December). Semantic Heterogeneity as a Result of Domain Evolution. *ACM SIGMOD Record*, *20*(4), 16-20.

Venugopal, S., Buyya, R., & Ramamohanarao, K. (2006). A Taxonomy of Data Grids for Distributed Data Sharing, Management, And Processing. *ACM Computing Surveys*, *38*(1), 3.

Venugopal, S., Buyya, R., & Winton, L. (2004). A grid service broker for scheduling distributed data-oriented applications on global grids. Proceedings of the 2nd workshop on Middleware for grid computing, pp. 75–80, Toronto, Canada. www.Gridbus.org/broker

Verschelde, J.L., Santos, M.C.D., Deray, T., Smith, B., & Ceusters, W. (2004) Ontology-Assisted Database Integration to Support Natural Language Processing and Biomedical Data-mining. *Journal of Integrative Bio-informatics*.

Von Delft, F., Inoue, T., Saldanha, S. A., Ottenhof, H. H., Schmitzberger, F., Birch, L. M., et al. (2003). Structure of E. coli Ketopantoate Hydroxymethyl Transferase Complexed with Ketopantoate and Mg2+, Solved by Locating 160 Selenomethionine Sites. *Structure*, *11*, 985-996.

von Laszewski, G., Foster, I., Gawor, J., & Lane, P. (2001). A Java Commodity Grid Kit. *Concurrency and Computation: Practice and Experience*, *13*, 643-662.

Von Laszewski, G., Foster, I., Gawor, J., Lane, P., Rehn, N. & Russell, M. (2001). Designing Grid-based Problem Solving Environments and Portals. In *Proceedings of International Conference on System Sciences* (HICSS-34).

VOTES project. (2007). *Virtual Organisations for Trials and Epidemiological Studies*. Retrieved July 2008 from http://labserv.nesc.gla.ac.uk/projects/votes/index.html

VTK, (n.d.). *The Visualization Toolkit*. Clifton Park, NY: Kitware. Retrieved from http://www.vtk.org/

W3C (2001). *Web Services Description Language (WSDL)*. Retrieved from http://www.w3.org/TR/wsdl. Warren, R. et al (2007a). A Prototype Distributed Mammographic Database for Europe 2007. *Clinical Radiology 62*(11) pp. 1044-51. Amsterdam, The Netherlands: Elsevier.

W3C (2004). *OWL Web Ontology Language*. Retrieved July 2008 from http://www.w3.org/TR/owl-features/

W3C (2007). *RDF Resource Description Framework*. Retrieved July 2008 from http://www.w3.org/RDF/

W3C. *XML Protocol Working Group*. Retrieved from http://www.w3.org/2000/xp/Group/

Wache, H., Vogele, T., Visser, U., Stuckenschmidt, H., Schuster, G., Neumann, H., and Hubner, S. (2001). Ontology-based information integration: A survey of existing approaches. In Gomez-Perez, A., Gruninger, M., Uschold, M., & Stuckenschmidt, H., (Eds.), *International Joint Conference on Artificial Intelligence; Workshop: Ontologies and Information Sharing*, (pp. 108-117). New York: ACM.

Wagner, E.G., Altuvia, S. & Romby, P. (2002). Antisense RNAs in bacteria and their genetic elements. *Adv. Genet.*, *46*, 361-398.

Wagstaff, A. (2000). Socioeconomic inequalities in child mortality: comparisons across nine developing countries. *Bull World Health Organ 2000, 78,*19-29.

Walker, D., Rana, O. F., Li, M., Shields, M. S., & Huang Y. (2000). The Software Architecture of a Distributed Problem-Solving Environment. *Concurrency: Practice and Experience*, *12*(15),1455–1480

Wang, G., & Dunbrack, R.L. (2003). PISCES: improvements to a protein sequence culling server. *Bioinformatics*, *19*,1589-1591. http://dunbrack.fccc.edu/Guoli/pisces_download.php

Warfield S.K., Zou, K.H, & Wells, W.M. (2004). Simultaneous Truth and Performance Level Estimation (STAPLE): An Algorithm for the Validation of Image Segmentation. *IEEE Transactions on Medical Imaging*, *23*(7) 903-921

Warren R., Solomonides A.E., del Frate C., Warsi I., Ding J., Odeh M., et al. (2007). MammoGrid — a prototype distributed mammographic database for Europe. *Clinical Radiology, 62* (11), 1044-1051.

Warren, R. et al (2007b). A Comparison of Some Anthropometric Parameters Between an Italian and a UK Population: "proof of principle" of a European project using MammoGrid. *Clinical Radiology 62*(11), pp. 1052-60. Amsterdam, The Netherlands: Elsevier

Watry, P., & Larson, R. (2005). Implementing support for digital repositories in a data grid environment. *Local to Global data interoperability – Challenges and technologies* (pp. 60-64) LGDI.

Watt, J., Sinnott, R. O., Ajayi, O., Jiang, J. & Stell, A. J. (2006). *User Oriented Access to Secure Biomedical Resources through the Grid.* Proceedings of Life Science Grid Conference, Yokohama, Japan.

Watt, J., Sinnott, R. O., Jiang, J., Stewart, G., Stell, A. J., Martin, D. & Doherty, T. (2007). Federated Authentication and Authorisation for e-Science. In *Proceedings of APAC conference*, Perth, Australia.

WebMo (2008). Retrieved from http://www.webmo.net

Weeks, C. M. & Miller, R. (1999). The design and implementation of *SnB* v2.0. *J. Appl. Cryst., 32*, 120-124.

Weeks, C. M., Blessing, R. H., Miller, R., Mungee, R., Potter, S. A., Rappleye, J., Smith, G. D., Xu, H. & Furey, W. (2002). Towards automated protein structure determination: *BnP*, the *SnB*-PHASES interface. *Z. Kristallogr., 217*, 686-693.

Weeks, C. M., DeTitta, G. T., Hauptman, H. A., Thuman, P. & Miller, R. (1994). Structure solution by minimal function phase refinement and Fourier filtering (II. Implementation and applications). *Acta Crystallogr., A50*, 210-220.

Weeks, C.M., DeTitta, G.T., Miller, R. & Hauptman, H.A. (1993). Applications of the minimal principle to peptide structures. *Acta Cryst., D49*, 179-181.

Weicker, R. (1984). Dhrystone: A Synthetic Systems Programming Benchmark. *Communications of the ACM*, *27*(10), 1013-1030.

Welch, V. (2005). *Globus Toolkit Version 4 Grid Security Infrastructure: A Standards Perspective*. Retrieved from http://www.globus.org/toolkit/docs/4.0/security/GT4-GSI-Overview.pdf.

Welch, V., Siebenlist, F., Foster, I., Bresnahan, J., Czajkowski, K., Gawor, J., et al. (2003). Security for Grid services. *Proceedings of the 12th IEEE International Symposium on High Performance Distributed Computing, 2003*, (pp. 48-57).

Westerhof, N., Elzinga, G., & Sipkema, P. (1971). An Artificial Arterial System for Pumping Hearts. *Journal of Applied Physiology, 31*(5), 776-781.

Weston, J., Leslie, C., Ie, E., Zhou, D., Elisseeff, A., & Noble William S. (2005). Semi-supervised protein classification using cluster kernels. *Bioinformatics*, *21*(15): 3241-3247.

Whiddett, R., Hunter, I., Engelbrecht, J. & Handy, J. (2006). Patients' attitudes towards sharing their health information. *International Journal of Medical Informatics, 75*(7), 530-541.

White, S.A. (2004). *Process Modeling Notations and Workflow Patterns*. Retrieved from http://BPTrends.com

WHO (2007). *The world health report 2007 - A safer future*. Geneva, Switzerland: WHO publications. Retrieved July from http://www.who.int/whr/2007/en/index.html

WHO (2008). *International Classification of Diseases*. Retrieved July 2008 from. http://www.who.int/classifications/icd/en/.

Wiens, J.J., Brandley, M.C., Reeder, T.W.(2006). Why does a trait evolve multiple times within a clade? Repeated evolution of snakelike body form in squamate reptiles. *Evolution*, 60(1),123-41.

Wikipedia (2007). *Ontology*. Retrieved July 2008 from http://en.wikipedia.org/wiki/Ontology_%28computer_science%29

Wikipedia (2008). *Epidemiology*. Retrived July 2008 from http://en.wikipedia.org/wiki/Epidemiology

Williams, D. B., & Carter, C. B. (2004). *Transmission Electron Microscopy: A Textbook for Materials Science*. Berlin, Germany: Springer.

Winkler, G. (1995). *Random Fields and Dynamic Monte Carlo Methods*. Berlin, Germany: Springer-Verlag.

WISDOM (2006). *Wide In-Silico Docking On Malaria*. Retrieved from http://wisdom.eu-egee.fr/

Witten, I. H., & Frank, E. (2005). *Data mining: Practical machine learning tools and techniques*. San Francisco: Morgan Kaufmann.

Witten, I. H., & Frank, E. (Eds.). (2005). *Data mining: Practical machine learning tools and techniques*. San Francisco, CA: Morgan Kaufmann.

Witten, I., Frank, E. (2005). *Practical Machine Learning Tools and Techniques*, (2nd Ed.). San Francisco, CA: Morgan Kaufmann.

Wittkopp, P.J., Williams, B.L., Selegue, J.E., Carroll, S.B.(2003). Drosophila pigmentation evolution: divergent genotypes underlying convergent phenotypes. *Proc Natl Acad Sci USA*, *100*(4), 1808-13.

Witzig, C. (2008). Interoperability Shibboleth - gLite, *IVOM Workshop*. Regionales Rechenzentrum für Niedersachen, Hannover, Germany.

Wolf, J., Gerber, A.P., & Keller, W. (2002). tadA, an essential tRNA-specific adenosine deaminase from Escherichia coli. *The EMBO Journal*, *21*, 3841–3851.

Wolniewicz, P., Meyer, N., Stroiński, M., Stuempert, M., Kornmayer, H., Polak, M., et al. (2007). Accessing Grid Computing Resources with g-Eclipse Platform. *Computational Methods in Science and Technology*, *13*(2), 131-141.

Wolpert, D. H. (1992). Stacked generalization. *Neural Networks*, *5*, 241-259.

Wolpert, D., & Macready, W. (1997). No free lunch theorems for optimization. *IEEE Transactions on Evolutionary Computing*, *1*(1), 67-82.

Wong, J. Cagney, G, & Cartwright, H.M (2005). Specalign—processing and alignment of mass spectra datasets. *Bioinformatics*, *21* (9), 2088–2090.

Wong, S.C., Miles, S., Fang, W., Groth, P.T. & Moreau, L. (2005) Provenance-based validation of e-science experiments. In Y. Gil, E. Motta, V. R. Benjamins, & M. A. Musen (Eds.), *International Semantic Web Conference: Lecture Notes in Computer Science*, (Vol. 3729), (pp. 801–815). New York: Springer.

Wood T.E., Burke J.M., & Rieseberg L.H. (2005). Parallel genotypic adaptation: when evolution repeats itself. *Genetica*, *123*(1-2), 157-70.

Wood, J., Wright, H. & K. Brodie (1997). Collaborative Visualization. *Proc. Visualization '97*, (pp. 253-259).

Wozniak, J. M., Brenner, P., Thain, D., Striegel, A., & Izaguirre, J. A. (2005). Generosity and gluttony in GEMS: Grid Enabled Molecular Simulations. In *Proceedings of the 14th IEEE International Symposium on High Performance Distributed Computing*. Los Alamitos, CA: IEEE Computer Society Press.

Wroe, C., Stevens, R., Goble, C., Roberts, A., & Greenwood, M. (2003). A suite of DAML+OIL Ontologies to Describe Bioinformatics Web Services and Data. *International Journal of Cooperative Information Systems*, *12*(2), 441-445.

WSRF Web Services Resource Framework. Retrieved from http://docs.oasis-open.org/wsrf/wsrf-primer-1.2-primer-cd-02.pdf

Wu, X., & Tseng, C.W. (2006). Searching Sequence Databases Using High-Performance BLASTs. In A.Y. Zomaya, (Ed.), *Parallel Computing for Bioinformatics and Computational Biology*, (pp. 211-232). Hoboken, NJ: John Wiley & Sons, Inc.

Wüthrich, B. (1997). Discovering probabilistic decision rules. *International Journal of Intelligent Systems in Accounting, Finance and Management, 6*(4), 269-277.

Wypychowski J., Pytlinski J., Skorwider Ł, Nazaruk M., Benedyczak K., Wroński M., Bała P. (2004). Life Sciences Grid in EUROGRID and GRIP projects. *New Generation Computing, 22* (2), 147-156.

Xiong, Z., Wu, X., Cheng, S., & Hau, J. (2003). Lossy-to-Lossless compression of medical volumetric data using three-dimensional integer wavelet transform. *IEEE Transactions on Medical Imaging, 22*, 459-470.

Yafchak, M.F. & Trauner, M., (Eds.) (2008). *The Grid Technology Cookbook.* Retrieved January 15, 2008, from http://www.sura.org/cookbook/gtcb/index.php?topic=12&mlevel=1&parent=0

Yang, C. T., Tsai, S. T., & Li, K. C. (2005). Decision tree construction for data mining on grid computing environments. *In the 19th International Conference on Advanced Information Networking and Applications* (pp. 421-42). New York: IEEE Computer Society.

Yang, C.T., Luo, T.-L., & Lai, C.-L. (April 2005). Designing Computing Platform for BioGrid. *International Journal of Computer Applications in Technology, 22*, pp. 3-13.

Yang, Z. & Greenshields, I. (2001). *Ensemble Classification of the VF Dataset with Limited Merging.* In *Proc. 14'th IEEE Symposium on Computer-Based Medical Systems*, Bethesda, MD, (pp. 117-122).

Yang, Z. (1994). Maximum likelihood phylogenetic estimation from DNA sequences with variable rates over sites: Approximate methods. *Journal of Molecular Evolution, 39*, 306-314.

Yasui Y., Pepe, M., Thompson, M.L., Adam, B.L., Wright, G.L., Qu, Y., et al. (2003). A data-analytic strategy for protein biomarker discovery: profiling of high-dimensional proteomic data for cancer detection. *Biostatistics, 4*(3), 449–463.

Yona, G., Linial, N., & Linial, M. (2000). ProtoMap: automatic classification of protein sequences and hierarchy of protein families. *Nucleic Acids Research, 28*(1), 49-55.

Yoon, S., Smellie, A., Hartsough, D., & Filikov, A. (2005). Surrogate Docking: Structure-based Virtual Screening at High Throughput Speed. *Jouranl of Computer-Aided Molecular Design, 19*, 483-497.

Yu, J. and Buyya, R. (2005). A taxonomy of workflow management systems for grid computing. *Journal of Grid Computing, 3*(3-4), 171-200.

Yu, J., Buyya, R., *Ong, SE., Blagoev, B., Kratchmarova, I., Kristensen, DB., et al. (2002).* Stable isotope labeling by amino acids in cell culture, SILAC, as a simple and accurate approach to expression proteomics. *Molecular & Cell Proteomics, 1, 376–86.*

Zaki, M. J. (1999). Parallel and Distributed Association Mining: A Survey. *IEEE Concurrency.*

Zaki, M. J., & Ho, C. T. (Eds.). (2000). *Large-Scale Parallel Data Mining.* Lecture Notes in Computer Science. Berlin, Germany: Springer.

Zara, V., Ferramosca, A., Capobianco, L., Baltz, K.M., Randel, O., Rassow, J., Palmieri, F. & Papatheodorou, P. (2007). Biogenesis of yeast dicarboxylate carrier: The carrier signature facilitates translocation across the mitochondrial outer membrane. *J. Cell Science* (120), 4099–4106.

Zdobnov, E.M., & Apweiler, R. (2001). InterProScan: protein domains identifier. *Bioinformatics, 17*, 847-848.

Zeilinger, G. (2006). dcm4che, a DICOM implementation in Java and dcm4chee J2EE based PACS. *Main page*. Retrieved February 14, 2008 from http://www.dcm4che.org

Zhang, J. (2008). *Visualization for information retrieval.* Berlin-Heidelberg: Springer.

Zhao, J., Goble, C., Stevens, R., & Turi, D. (2007). Mining Taverna's Semantic Web of Provenance. In *Concurrency and Computation: Practice and Experience, 20*(5), 463-472.

Zhao, J., Wroe, C., Goble, C. A., Stevens, R., Quan, D., Greenwood, R. M. (2004). Using Semantic Web Technologies for Representing E-science Provenance. In *International Semantic Web Conference, LNCS* (Vol. 3298), (pp. 92–106). New York: Springer.

Zhao, Y., Wilde, M., & Foster, I. T. (2006). Applying the virtual data provenance model. In L. Moreau and I. T. Foster (Eds.) *Provenance and Annotation of Data, International Provenance and Annotation Workshop, IPAW 2006, Chicago, IL, USA, May 3-5, 2006, Revised Selected Papers, Lecture Notes in Computer Science,* (Vol. 4145) (pp. 148–161). New York: Springer.

Zmasek, C.M. (n.d.). *Forester: software libraries for evolutionary biology research.* Retrieved June 6, 2008, from http://www.phylosoft.org/forester.

Zomaya, A.Y. (Ed.) (2006). *Parallel Computing for Bioinformatics and Computational Biology.* Hoboken, NJ: John Wiley & Sons, Inc.

Zosso, D., Arnold, K., Schwede, T., Podvinec, M., (2007). SWISS-TANDEM: A Web-Based Workspace for MS/MS Protein Identification on PC Grids. *CMBS,* 406-411.

Zwickl, D. (2006). Genetic algorithm approaches for the phylogenetic analysis of large biological sequence datasets under the maximum likelihood criterion. *PhD Thesis,* The University of Texas at Austin.

About the Contributors

Mario Cannataro is an associate professor of Computer Engineering at University Magna Græcia, Catanzaro, Italy, since 2002. He received his Laurea Degree Magna cum Laude in Computer Engineering from the University of Calabria, Rende, Italy, in 1993. He was a researcher at the CRAI Research Consortium and a senior researcher at the Italian National Research Council (ICAR-CNR), Rende, Italy, where he worked on parallel computing, massively parallel architectures, and parallel implementation of logic programs and cellular automata. He is currently the responsible of the Bioinformatics Laboratory at University Magna Græcia of Catanzaro. His current research interests are on grid computing, bioinformatics, computational genomics and proteomics, and ontologies. He authored a book on parallel implementation of logic programs and over 150 scientific papers published on international journals and conference proceedings. He is a member of ACM, IEEE Computer Society, BITS (Italian Bioinformatics Society), and HealthGrid. He is a co-founder and a member of Exeura (www.exeura.com). Mario Cannataro can be reached at cannataro@unicz.it.

* * *

Giovanni Aloisio is full professor of Information Processing Systems at the Engineering Faculty of the University of Salento, Lecce. His research interests are in the area of High Performance Computing, Distributed and Grid Computing. He has founded SPACI Consortium (www.spaci.it). He is also the Vice-director of the NanoSciences and Grid Computing Section of the "Scuola Superiore ISUFI" and CEO of the SPACI Consortium. He is also director of the Division "ICT and Operations" of the Euromediterranean Center for Climate Chances (CMCC). He is the author of more than 100 papers in refereed journals on parallel & grid computing.

Domingos Alves obtained his doctoral degree on Biologic Physics in São Paulo University, USP in 1999. He is currently a professor of the Social Medicine Department at São Paulo University, Brazil. He has been working in the area of Theoretical Biology, Modeling and Simulation. His current research focuses on the transmission dynamics and within-host population biology of infectious disease. It combines viral quasispecies modeling with epidemic modeling to address questions in these areas.

Isabelle Andoulsi is a lawyer at the Brussels Bar, working within the law firm Exelia, as well as a researcher and at doctoral candidate at the Institute of European Studies of the Brussels University, Belgium (www.iee-ulb.eu). She currently works on the assumption of a use or a transposition of the assets as regards data protection within the framework of the first pillar in the third pillar of the European Union Treaty. She also undertook research concerning the possibility of an innovating use of information technologies in order to allow a broad access of patients to effective, profitable and of high-quality health care services. She is thus particularly interested in personal data processing and in personal data transfers, but also in the questions of interoperability of personal databases, of shared responsibilities within the framework of telematics and telemedicine networks, and in intellectual property rights and competition law.

Ashiq Anjum has recently completed a PhD on data-intensive and network-aware Grid scheduling and previously held a MSc in Computer Science.He has been influential in the Grids community and has responsibility for UWE's involvement in the Grid aspects of the Health-e-Child project. in addition Anjum has led research into the provision of generic, reusable services in the context of a service-oriented architecture for the FP7 neuGRID project. Dr Anjum has a growing publications list with over 30 papers in international journals or conferences. His current research is in Grid-based medical systems and the role of SOAs and Cloud computing in applied computer science.

Anna Anselmo obtained her Master's degree in Functional Genomics and Bioinformatics in 2005 carrying out a thesis on alternative splicing prediction at the University of Milan, Italy. She is now concluding her PhD in "Molecular and Cellular Biology" at the University of Milan. The main topic of her PhD project has been the in silico characterization of cancer specific isoforms. Her research interests include also the study of coding and non-coding conserved sequence tags in human and mouse genomes.

Maurizio Arabia received the Laurea Degree in Electrical Engineering in 1957 and specialized in Nuclear Engineering in 1959 from the University of Rome (Italy). He worked in the bioengineering field and particularly in adaptive control applied to cardiovascular system, cardiac mechanics, total artificial heart, ventricular assist devices, and mock circulatory systems. Together with Prof. Westerhof (Institute for Cardiovascular Research, Free University of Amsterdam, Netherland) he launched the European project HEART. He collaborated with the Univ. of Rome Italy (Department of Applied Mechanics Dept.), ENEA Agency, University of Mississippi Medical Center (Jackson, Mississippi, USA), Texas Heart Institute (Houston, USA), National Cardiovascular Center (Osaka, Japan), and University of Calabria (Department of Mechanical Engineering, Rende, Italy). He is now working on volumetric continuous flow-high efficiency pumping devices for industrial applications.

Matthias Assel received his diploma degree in computer science from the Technische Universität München in 2005 and has joined the High Performance Computing Center of University Stuttgart (HLRS) in 2006 in order to write his PhD thesis in the field of (bio)medical IT solutions. He is currently working on Grid-related European projects on research and development level, examining and making use of emerging technologies. His actual research focus rests on secure environments for distributed data access, distributed data and text mining in biomedicine, but also intelligent infrastructures including e.g. the automatic reconfiguration of relevant services.

Piotr Bala is professor at ICM Uniwersity od Warsaw and at Faculty of Mathematics and Computer Science N. Copernicus University, director of Department of Parallel and Distributed Processing. He obtained PhD in Physics in 1993 in Institute of Physics, N. Copernicus University and in 2000 habilitation in physics for development and application of new methods to study enzymatic reactions in biological systems. He has published over 60 papers. His main research interest is in development of parallel and grid tools for scientific applications. He has been involved in number of European grid projects, currently acts as coordinator of Chemomentum project.

Kim Baldridge, professor of Theoretical Chemistry at the University of Zürich. She graduated with a double BA in chemistry and mathematics, received MA in mathematics and a PhD in theoretical chemistry in 1988 in the USA. She is a theoretical and computational chemist with expertise in the design, development, implementation, and application of computational chemistry methodology. She is the coauthor, developer, and supporter of the computational quantum chemistry software, GAMESS, in addition to several key associated tools for Grid computing, analysis and visualization. She has over 100 publications that cover topics in these areas and directs a programme in computational Grid applications and computing.

Bartosz Balis obtained his MSc degree in Computer Science from the AGH University of Science and Technology, Krakow, Poland. He is employed as a research and teaching assistant at the Institute of Computer Science, AGH. Co-author of about 50 international publications. His research interests include monitoring and performance analysis of applications, Grid computing, scientific workflows, adapting legacy codes to modern environments, and provenance tracking. Participant of several international research projects including EU-IST CrossGrid, CoreGRID, K-Wf Grid, ViroLab, Gredia. Member of Program Committee for ICCS conference and a reviewer at journals: Future Generation Computer Systems, Scientific Programming, and conferences: ICCS, EuroPVM/MPI, and e-Science 2006.

Roberto Barbera graduated in Physics "cum laude" at the University of Catania in 1986 and since 1990 he holds a PhD in Physics from the same University. He is currently Associated professor at the Department of Physics and Astronomy. His main research activity has been done in the realm of Experimental Nuclear and Particle Physics. Since 1997 he is involved in the NA57 Experiment at CERN SPS and in the ALICE Experiment at CERN LHC. Since late 1999 he is interested in Grid Computing. He is a member of the Technical Board of the Italian INFN Grid Project.

Tomasz Bartynski is a computer scientist at the Academic Computer Center CYFRONET AGH in Krakow. He received his MSc degree in computer science from the AGH University of Science and Technology in Krakow. Since 2006 he has been working for ACC Cyfronet AGH in Krakow. Currently he is involved in EU-funded projects: ViroLab and GREDIA as a designer and developer. His scientific interests involve distributed systems, parallel computing, Web technologies and programming paradigms.

Lisa Bartoli has a degree in Physics. Actually she is a PhD student in Physics at the Biocomputing Group, University of Bologna. Her main research interests concern Computational Biology, in particular protein structure prediction, prediction of protein-protein interactions, study of protein contact maps with machine learning methods.

Khalid Belhajjame is a researcher at the school of computer science of the University of Manchester. He obtained his PhD from the University of Grenoble in 2004. His research interests include semantic annotations of web services, scientific workflows, dataspaces, and data mapping and integration.

Emilio Benfenati has been researcher at Stanford University, CA, in 1983-84. In that period he also had a collaborative research at the Berkeley University, CA. He has gained a considerable experience in the management of international programs, coordinating 11 European projects and participating to 16 others. Many of them are on toxicity and environmental modelling. He has been Member of the National Commission for the evaluation of pesticides, Italian Ministry of Health, 1997-1999. His research activities include: computer chemistry, QSAR, toxicity prediction, risk assessment and many others. He is author or co-author of about 200 papers in international journals and books.

Margherita Berardi is a computer science researcher in the field of data mining. She received the PhD Degree (Curriculum of Artificial Intelligence) from the University of Bari by discussing a work entitled "Towards Semantic Indexing of documents: a Data Mining perspective" under the supervision of Prof. D. Malerba. She worked as a research consultant for Exhicon s.r.l. and the University of Bari within projects for IBM Innovation Centre focusing on unstructured data management for Bioinformatics.

José Román Bilbao Castro received the MSc and PhD degrees in computer science from the University of Almería, Spain, in 2001 and 2006, respectively. As a PhD student he worked at the BioComputing Unit of the National Center for Biotechnology, Spain. Since 2007 he holds a postdoctoral position at the University of Almería under the "Juan de la Cierva" contract. He is a member of the Supercomputing-Algorithms group at the Department of Computer Architecture and Electronics and his research interests include parallel computing in 3D electron microscopy, grid computing and image processing algorithms.

Ignacio Blanquer (PhD) is assistant professor of the Computer System Department (DSIC) of the Polytechnic University of Valencia (UPV) since 1999. He has been involved in Parallel Computation and Medical Image processing since 12 years ago participating in 17 national and European Research Projects. He is also a research fellow of the Institute for the Applications of Advanced Information and Comunication Technologies (ITACA) and Network Centre for Biomedical Engineering (CRIB) and member of the board of directors of HealthGrid association.

Vincent Breton received his engineer degree from Ecole Centrale de Paris in 1985 and his PhD in Nuclear Physics from the University of Paris XI- Orsay in 1990. Since 1990, he has been a research associate at the French National Centre for Scientific Research (CNRS). In 2001, he founded a research group (http://clrpcsv.in2p3.fr) on the application to biomedical sciences of the IT technologies and tools used in high energy physics. He is also investigating the relevance of grid technology for biomedical applications. He co-founded of the Healthgrid initiative, chaired the first European conferences on grids for health in January 2003 and January 2004. He is leading application support and identification in the EGEE european project (FP6). He is involved in various European initiatives in the biomedical area.

Marian Bubak, PhD, is an adjunct at the Institute of Computer Science AGH, a staff member at the ACC CYFRONET AGH, and the professor at the Informatics Institute of the Universiteit van Amsterdam. His research interests include distributed and grid systems for scientific simulations. He co-authored about 230 papers. He lead the architecture team of the EU IST CrossGrid Project, he was the Scientific Coordinator of K-WfGrid Project and the member of the Integration Monitoring Committee of CoreGRID. He served as a program committee member, chairman and organizer of several international conferences (HPCN, Physics Computing, EuroPVM/MPI, SupEur, HiPer, ICCS, HPCC, e-Science'2006); he is co-editor of 17 proceedings of international conferences.

Massimo Cafaro is an assistant professor at the Department of Innovation Engineering of the University of Salento. His research covers High Performance, Distributed and Grid Computing. He is the principal investigator and developer of the Grid Resource Broker, a web portal to computational grids and of the GSI plug-in for gSOAP. He received a PhD in Computer Science from the University of Bari, is a member of ACM, IEEE and IEEE Computer Society, and holds a visitor position at CACR Caltech. He authored more than 70 refereed papers on parallel, distributed and Grid Computing. He co-authored and holds a patent on distributed database technologies.

Antonio Calanducci graduated in Computer Science "cum laude" at the University of Catania in 2004 and his final dissertation was carried out at NYU in the field of BioInformatics. Since his graduation, he has been involved in the realm of grid computing: starting as a collaborator at INAF, since 2005 he has been working at the INFN Catania within EGEE projects. His main activities are related to the dissemination and training of Grid computing, focusing on gLite. He also has carried out research activities on Grid-based Digital Libraries, and has a particular expertise on data and metadata management in grids.

Barbara Canesi received the master degree cum laude in Biomedical Engineering from the University of Genoa, Italy, in 2004. In 2004-2005 she collaborated to many biomedical projects, aimed to support diagnosis of neurodegenative diseases. Since 2006, she has been a PhD student at DIST, University of Genoa, Italy. In 2007-2008 she was visiting student at University of Maastricht, The Netherlands. Her research activity is focused on neuroimaging and integration of different kinds of biomedical signals and data.

Eddy Caron is an assistant professor at Ecole Normale Supérieure de Lyon and holds a position with the LIP laboratory (ENS Lyon, France). He is a member of GRAAL project and technical manager for the DIET software package. He received his PhD in C.S. from University de Picardie Jules Verne in 2000. His research interests include parallel libraries for scientific computing on parallel distributed memory machines, problem solving environments, and grid computing. He is co-chair of the GridRPC working group in OGF. He is the coordinator of LEGO a french ANR project (ANR-CICG05-11). See http://graal.ens-lyon.fr/~ecaron for further information.

Luciana Carota is working in grid tecnology from 2004 at the National Institute of Nuclear Physics (INFN), Bologna, Italy. She received a PhD in Physics in January 2004 from L'Aquila University, Italy, working in collaboration with INFN, Rome, on neural dynamics and learning in VLSI recurrent neural networks. She received a master degree in Physics in 2000 and a degree in Electronic Engineering in 2006.

Rita Casadio is Full professor of Biochemistry/Biophysics at the University of Bologna and Group Leader of the Biocomputing Group (www.biocomp.unibo.it). She has a degree in Physics and has been working mainly in the fields of membrane and protein Biophysics both experimentally and theoretically. Presently she is interested in computer modelling of relevant biological processes and her researches are devoted to different aspects of protein modelling, including prediction of secondary and tertiary structures with neural networks, hidden Markov models, genetic algorithms, molecular docking, protein reaction mechanisms and de novo design of helical peptides. RC is the author of about 200 scientific papers and presented her work at several (over 300) national and international meetings.

Mosè Casalegno is senior research fellow at Mario Negri Institute for Pharmacological Research. In 2002 he obtained a PhD degree in theoretical and computational chemistry from the University of Milan. As a post-doctoral fellow he has been working on quantum Monte Carlo methods. In 2003 he joined the department of Health and Environment at Mario Negri. His main research interest is the development of algorithms and models in the field of chemical information and QSAR. He has been involved in different European funded projects in the area of computational chemistry, toxicity prediction and Grid computing.

Tiziana Castrignanò is a bioinformatics senior technologist at CASPUR (www.caspur.it). She received her PhD in Biophysics in 1999 from the University of Rome "La Sapienza". She has developed bioinformatics web tools and databases largely used by the scientific community. Her primary activity is the development of high performance bioinformatics services and databases and leads the technological support activity to many national and international bioinformatics research projects.

Chen Chen received his Bachelor from the Fudan University, China in 2003, and the PhD degree from the Nanyang Technological University (NTU), Singapore, in 2008. His research interests include 3D modeling and reconstruction, computer animation, physical simulation and general purpose GPU computing.

Eryk Ciepiela received his MSc title in Computer Science from the AGH University of Science and Technology in Kraków, Poland for the thesis "Monitoring of component-based applications". In 2005 - 2006 he was a memeber of Technologies Group in Reseach and Development departament in ComArch, Kraków. Since 2006 he has worked for CYFRONET supercomputing centre in Kraków where he has been involved in EU-funded project ViroLab and Gredia working on GridSpace Engine. He's also a founder of IT enterprise it-xpander.pl. His area of interest ranges from enterprise systems, web technologies to distributed systems, monitoring, programming paradigms, models and methodologies.

Patrick Clarysse received a MSc degree in 1987 and a PhD in 1991 from the Scientific and Technological University of Lille, France. Since 1992, Patrick Clarysse is with the French National Center for Scientific Research (CNRS) at CREATIS-LRMN, CNRS UMR #5220, Inserm U630, Lyon, France. His primary research interests are in the bioengineering and medical image analysis fields, and include multimodal image segmentation and registration, motion estimation and deformable models with applications to the analysis of the heart functions and the motion of thoracic structures.

Luca Clementi received his MSc in Software Engineering in 2005 from the University of Bologna, Italy. He worked form 2005 till 2007 as a researcher at CINECA where he was involved in several EU founded projects on Grid computing (DEISA and A-WARE). Since 2007, he is working at the University of California, San Diego in the National Biomedical Computation Resource project where he is in charge of the development of the SOA infrastructure which integrates biomedical applications with high end computational resources.

Francesco Maria Colacino received the Laurea Degree (summa cum laude) in Mechanical Engineering in 2000 from the University of Calabria (Rende, Italy), and the PhD in 2004 from the same University with a thesis entitled "An Elastance-Based Artificial Ventricle for Testing any Cardiac Assist Device". From 2005 to 2008 he was Research Fellow at the University of Calabria as well as at the Department of Surgery, University of Kentucky Chandler Medical Center (Lexington, KY - USA). He is currently Research Fellow at the Institute of Sound and Vibration Research (ISVR), University of Southampton (UK). His present research interests deal with the applications of bio-feedback to skeletal muscle models.

Antonio Congiusta received his PhD in systems and computer engineering from University of Calabria, Italy, in 2006. From 2002 to 2003 he was research fellow at ICAR-CNR, Institute of the Italian National Research Council, working on Grid-based data mining systems. From 2004 to 2006 he was research fellow at University of Calabria, working on the Knowledge Grid project. Currently he is research fellow at University of Salerno, Italy. His research interests are focused on Grid programming environments, workflow-based Grid systems, and virtual organizations.

Luca Corradi received his master degree in Biomedical Engineering at the University of Genoa, Italy, in 2006 and gained admission into the engineering profession, in 2007. Since 2007, he has been a PhD student at DIST, University of Genoa, Italy. His research activity is focused on data and application integration applied to bioinformatics and neuroinformatics fields.

Paolo Cozzi received the Laurea degree in Bioinformatics from the university of "Milano – Bicocca", Italy, in 2007 with a thesis on the development of databases for the correlation between protein functions and molecular surfaces. His research activity concerns the development of software for functional analysis in genomics and proteomics. Moreover, he has implemented HPC solutions for grid computing, with a particular interest in searching for surface similarities and molecular visualization. At the moment he works at Institute of Biomedical technology supported by a fellowship of the PhD School of Molecular Medicine at the University of Milan.

Guido Cuscela has a software engineer degree from the "Politecnico di Bari", Italy. He is a member of EGEE I and EGEE II project since March 2006 as SA1 and JRA1. He has worked for GridICE project (INFN Grid monitoring tool) deploying specific sensors for jobs and batch systems monitoring. He is now a member of LIBI project .

Bertrand Delhay received an engineering degree and a PhD degree in Image Processing from the National Institute of Applied Sciences (INSA) of Lyon, France in 2002 and 2006, respectively. He is currently a postdoctoral research fellow at the CREATIS-LRMN lab. His research interests are medical image processing, image registration, spatio-temporal motion estimation, object tracking and intensive computing.

Domenica D'Elia received a summa cum laude degree in Biological Sciences in 1991 and a PhD in Biochemistry and Molecular Biology in 1996 from the University of Bari (IT). She presently works as a researcher at the Institute of Biomedical Technologies of the National Research Council (CNR-IT). Her research interests include bioinformatics (database and analysis tools development) applied to the study of mitochondrial biogenesis and to genome expression studies. She is node manager of the European Molecular Biology Network node (http://www.embnet.org/) in Italy, member of the editorial board of EMBner.news journal and of EMBnet Publicity and Public Relations Project Committee.

Giulia De Sario is a researcher of the Italian CNR (National Research Council). She graduated in Physics from the University of Bari in 2003. She works on GRID infrastructure applied on bioinformatics, in particular her research field is the implementation of bioinformatic tools and databases on grid.

Frédéric Desprez is a director of research at INRIA and holds a position at LIP laboratory (ENS Lyon, France). He received is PhD in C.S. from the Institut National Polytechnique de Grenoble in 1994 and his MS in C.S. from the ENS Lyon in 1990. His research interests include scheduling for large scale distributed platforms, SOA architectures, data management, and grid computing. See http://graal.ens-lyon.fr/~desprez for further information.

Luigi Di Pace is the director of the IBM GBS Innovation Center. He began his career at the IBM Scientific Center based in Rome in 1986. Since 1991 he was a researcher involved in complex AI projects for large customers. In 1995 he was the funder of the Innovation Center located in Bari starting a team of 16 researchers working on Object-Oriented and Internet technologies. He currently heads, directly and/or receive his functional direction, an organization of more than 150 researchers and specialists providing innovative services projects to all relevant IBM customers for all main industrial sectors and business areas.

Frank Dickmann is research assistant in the department of Medical Informatics, Georg-August-University Göttingen, Germany. In 2005 he finished his diploma degree in Information Systems at the Georg-August-University Göttingen. Since 2006 he is a PhD-student at the Georg-August-University Göttingen. His main scope of study is strategic IT management of universities. From 2006 to 2007 he worked as academic assistant in the department of IT Strategy at the University Medical Center Göttingen. His main field of work is the development of economic sustainability for biomedical Grid Computing.

Zhaohui Ding was born in 1979. He received the BS and MS degrees in Computer Science from Jilin University, in 2001 and 2006, respectively. Currently, he is a PhD student of College of Computer Science and Technology of Jilin University, P.R.C. His research fields include grid computing, meta-scheduling and distributed system.

Alexander Dobrev holds a Bachelor of Arts Honours Degree in Philosophy, Politics and Economics from the University of Oxford (2005). He has an extensive record of experience in international work and education environments, including international projects and conferences across Europe. He joined the *empirica* eHealth team as a full time consultant in 2005. Alexander is focusing on the socio-economic and commercial aspects of healthcare and eHealth. Recent EU studies he has been involved in include Scenarios4Health, a foresight study on scenarios for ICT-enabled new models of healthcare; eHealth ERA, a study analysing priorities and strategies for eHealth in 32 European countries; eHealth Indicators, a representative survey of European GPs. Being responsible for Financing eHealth – a study on investment decision making in eHealth, Alexander investigated decision making processes and success factors for eHealth investments. Alexander has gained extensive experience and expertise with good practice case identification, selection, description and presentation through his substantial contribution to a number of studies, including eHealth IMPACT, a Study on the Economic Impact of eHealth and Good eHealth, Study of best practice across Europe in providing innovative health-related services, both for European Commission, DG INFSO. Through his contributions to SHARE, he has also explored the socio-economic and business issues concerning HealthGrids. Currently, he is leading the EHR IMPACT study on the economic impact of interoperable electronic health records and ePrescribing systems and is responsible for the socio-economic impact evaluations in several large EU research projects.

Giacinto Donvito is currently working as technical coordinator for LIBI (International Laboratory of Bioinformatics) a National project for supporting bioinformatics application over gLite grid. As system administrator of the Bari Grid farm since 2004, has gained experience in deploying and using several storage software solutions and database grid access software tools.

Celine Van Doosselaere has a background in liberal arts and in development management. Since 2000, she has worked in Brussels on European policies impacting health - including the fascinating field of eHealth. At the time of her collaboration with the SHARE team, Celine worked for the European Health Management Association (EHMA), a non-profit membership association bringing together academics & researchers, policy-makers, trainers and managers to improve health management with a view to raising standards in care.

Gamal El-Sayed is an assistant professor in the Department of Electrical Engineering at Assiut University in Egypt. He obtained his PhD in computer science at the University of Connecticut, where he also completed his post-doctoral training. His research interests include general grid computing with specific interests in dynamic task migration over the Grid.

Andrew Emerson received his PhD in computational chemistry in 1991 from the University of Southampton (UK). In 2000 he took up a post at the CINECA supercomputer centre (Bologna) where apart from performing research, he also provides support and gives courses to researchers using the centre's facilities. His current interests include the molecular dynamics of biomolecular systems, systems biology and grid computing.

Italo Epicoco is an assistant professor at the University of Salento, Italy. He received PhD in "Materials and Innovative Technologies" in June 2003 at University of Salento. He is staff member of the Center for Advanced Computational Technology (CACT) of the National Nanotechnology Laboratory (NNL CNR-INFM). He is responsible of the High Performance Computing group within the Scientific Computing and Operation division of the Euro-Mediterranean Center for Climate Changes (CMCC). His research interests include high performance computing, distributed and grid computing. He is author of about 40 papers in conference proceedings, journals and book chapters.

Francesco Falciano obtained his master degree in Biology with a specialization in Bioinformatics from the university of Milano Bicocca. After taking up a position at the CINECA supercomputing center he obtained a PhD in Biochemistry and Molecular Biology from the University of Ferrara. At CINECA he works in the high performance computing group with interests in computational biology and bioinformatics.

Juergen Falkner works in the business unit Software Technology at the Fraunhofer Institute for Industrial Engineering (IAO). He has been a key contributer to the development and setup of the Fraunhofer Resource Grid. He has been working within several Grid-related projects on European as well as the German national level (D-Grid) with focus on Grid user interfaces such as Grid portals. Within MediGrid he gathered experience on a variety of biomedical applications. Juergen Falkner has also been working in the field of IT security and incident response with a focus on VO and user management in Grids, authentication mechanisms and tools and the procedural aspects of IT security.

Piero Fariselli has a PhD in Biophysics and a Laurea in Physics. Author of over 80 publications he is a permanent researcher at the Biocomputing Group, University of Bologna. His main research interests concern Computational Biology and Machine Learning. PF main achievements are in the field of bioinformatics, where he developed machine learning codes, such as neural networks and hidden Markov models, and other algorithms for the prediction of structural and functional features of biomolecules.

Marco Fato is associate professor in Biomedical Engineering at School of Engineering of the University of Genoa since 2006. He has a career covering 20 years as a Research Collaborator at DIST and other national and international laboratories. He held a research position at DIST from 2001 to 2005, during which time he authored more than research 50 papers.

José-Jesús Fernández received the MSc and PhD degrees in computer science from the University of Granada, Spain, in 1992 and 1997, respectively. He was a PhD student at the BioComputing Unit of the National Center for Biotechnology (CNB), Spanish Research Council (CSIC), Madrid, Spain. He became an assistant professor in October 1997 and since 2000 he is an Associate professor of Computer Architecture at the University of Almería, Almería, Spain. He is a member of the Supercomputing-Algorithms Research Group at the University of Almería. His current research interests include high performance computing, image processing, and tomographic reconstruction.

Alessandra Ferramosca is researcher of Biochemistry at the University of Salento (Lecce). Her research includes the following topics: study of the import pathway of some proteins, in particular the metabolite carriers, into isolated mitochondria; purification of carrier proteins of the inner mitochondrial membrane and characterization of their structural and functional properties; effect of different nutritional states on the hepatic fatty acid biosynthesis; energetic metabolism of spermatozoa. She is the author of 43 scientific articles, most of them published in specialized international journals. She has participated as a speaker at various national conferences.

Sandro Luigi Fiore was born in Galatina, Italy (1976). He received a summa cum laude Laurea degree in Computer Engineering from the University of Lecce (Italy), (2001). He received a PhD degree in Informatic Engineering from the ISUFI-University of Lecce, Italy, (2004). Since 2001 he is the Principal Investigator of the GRelC project. He was involved in EGEE and EGEE-II projects and currently in EGEE-III. Since 2006, leads the Data Grid Group at the Euro-Mediterranean Centre for Climate Change (CMCC). He is author/co-author of about 45 papers on parallel & grid computing and he got 1 patent on data management.

Raffaele Fronza has a Degree in Molecular Biology. He worked for 5 years on cancer research. Now is Phd student at the University of Bologna and is involved in a project of applied genomics aimed to identify the basic biological process affecting the etiology and pathogenesis of complex congenital disorders.

Gionata Fragomeni graduated in Engineering at the University of Calabria in 1997. He received the PhD in Bioengineering from the University of Bologna in 2003. Gionata Fragomeni was a temporary chair in Cardiovascular Mechanics and Mechanics for Biomedical Applications at Calabria University and Magna Graecia University. He is an assistant professor at the Faculty of Medicine at Magna Graecia University of Catanzaro. He was the coordinator of 2 national research projects financed by the Italian Ministry of Research, and has authored over 30 publications on journals and international conference proceedings. His research interests include Biomechanics, Virtual Reality Applied to Orthopaedic, Bio-Fluid Dynamics, and Cardiovascular Mechanics.

Wlodzimierz Funika holds MSc degree in Mathematics and PhD in Computer Science. From 1978 to 1993 he was with Moldavian Academy of Sciences. At present he is with the Institute of Computer Science AGH. His research interests include parallel and distributed computing, development tool construction, performance monitoring, semantic-oriented systems, visualisation techniques. He has authored and co-authored ca. 100 papers. He participated in EU funded and national projects, e.g. CrossGrid, K-Wf Grid, CoreGRID. Currently he is involved in the EU IST ViroLab project where he is responsible for the presentation layer.

Henrique Fabricio Gagliardi has a master degree in Computer Science from Universidade Católica de Santos, Brazil. His recent work is related to grid computing and health care applications applied to the modeling of biological epidemic systems, design and development of grid based simulation services for epidemic forecasting tools and development of health based applications for PalmOS. His current work-in-progresses consists on the development of a peer-to-peer network-based communication service for Tidia-AE, a Brazilian e-learning project and Grid to Peer-to-Peer integration for use in health care applications.

Eduardo Gallo has a master degree in Computer Science from Universidade Católica de Santos (2006). He has also a specialist degree in Information Technology applied to Disabled People (Universidade Federal do Mato Grosso do Sul, 2000) and has a bachelor degree in Computer Science (UNIBAN, 2000). Currently he is System Manager at APRAESPI. He is interested in the following areas: grid computing, distributed systems, web development, database administration, database integration and epidemiological surveillance systems.

Inmaculada García Fernández received the BSc degree in physics in 1977 from the Complutense University of Madrid, Spain, and the PhD degree in 1986 from the University of Santiago de Compostela, Spain. From 1977 to 1987 she was an assistant professor at the University of Granada, from 1987 to 1996 an Associated professor at the University of Almería, and since 1997 she is a Full professor at the University of Almería, Spain. She is head of the Department of Computer Architecture and Electronics and the Supercomputing-Algorithms Research Group. Her research interest lies in the field of parallel algorithms for irregular problems related to image processing, global optimization, and matrix computation.

Yann Gaudeau received the engineering degree from Université Technologique de Troyes,France, in 2003, and, in 2006, the PhD degree in Automatic Control and Signal Processing from Université Henri Poincaré, Nancy, France. He is currently a postdoctoral fellow with the Centre de Recherche en Automatique de Nancy (CRAN, Nancy Université, UMR CNRS 7039). His research interests are image coding, lattice vector quantization and medical image compression.

Cécile Germain-Renaud received the agrégation de mathématiques in 1979, her PhD in Computer Science in 1989 and her Habilitation à Diriger les Recherches in 2005. She is a professor at Paris-Sud University (Laboratoire de Recherche en Informatique). Her research interests are in models and optimisation of high performance computing systems, parallel architectures and grids.

Andreas Gisel is a researcher of the Italian CNR (National Research Council). He graduated in molecular biology from the Swiss Federal Institute of Technology Zuerich in 1990 and obtained a PhD in 1995 from the same institution. His research field is bioinformatics with emphasize on biological ontologies, bio-sequence analysis, bio-databases and grid infrastructure applied to bioinformatics. He is involved in various European and National projects on those topics.

Tristan Glatard obtained a master degree from Ecole Centrale de Lyon (France) in 2004 and a PhD from the University of Nice - Sophia Antipolis in 2007. He is a researcher at CNRS (French National Scientific Research Center). Tristan Glatard is working on data-intensive workflows. He is the main developer of the MOTEUR grid-enabled workflow manager.

Carole Goble is a full professor in the the School of Computer Science in the University of Manchester, UK, where she has co-led the Information Management Group since 1997. She has an international reputation in the Semantic Web, e-Science and Grid communities and has led the application of Semantic Web technologies to both the Grid and e-Science, a fusion dubbed the Semantic Grid.

Ian Greenshields is an associate professor of Computer Science and Engineering at the University of Connecticut. He obtained his PhD in Electrical Engineering from Brunel University in the United Kingdom. His current research interests include general computational medicine, medical image processing, scheduling on the Grid and medical visualization.

Giorgio Grillo is a researcher at the Institute Biomedical Technologies CNR of Bari, Italy. He obtained a degree in Computer Science at University of Bari and since 1994 he has been involved in Bioinformatics and particularly in the development of software for genomics and proteomics. He takes part in many national and international research projects for his expertise in the analysis of biosequences and their functional characterization, in the linguistic and computational analysis in nucleotide sequences. Efforts include design, development and implementation of bioinformatic databases largely used by international researchers and in the management of query systems.

Tomasz Gubala received his MSc title in Computer Science from the AGH University of Science and Technology in Kraków, Poland. Since then he worked for CYFRONET supercomputing centre in Kraków as a researcher and for the Section of Computational Science (SCS) at the University of Amsterdam, the Netherlands as a scientific programmer and computer science research assistant. Currently he is involved in a major EU-funded project ViroLab as a chief designer of the virtual laboratory for infectious diseases. He is an external PhD student at the SCS (University of Amsterdam). His main scientific interests are Grid and distributed (cloud) computing, problem-solving environments and services for eScience.

Vicente Hernández García: Full professor in Computer Science and Artificial Intelligence and leader of the Grid and High Performance Computing Group (GRyCAP). He has large experience in Parallel and Grid Computing and Numerical Methods. He has managed and participated in more than 25 European projects, from the III to the VI Framework Programme, and national projects. He has been the Vice-chancellor of Research, Development and Innovation of the Polytechnic University of Valencia (UPV) during the period 2000-2005. He is currently the Scientific Coordinator of the Spanish e-Science Network (NGI).

Pietro Hiram Guzzi received his PhD in Biomedical Engineering in 2008, from Magna Græcia University of Catanzaro. He received his Laurea degree in Computer Engineering in 2004 from the University of Calabria, Rende, Italy. Currently, he is an assistant professor of the University of Catanzaro. His research interests comprise bioinformatics, the analysis of proteomics data, and the analysis of protein interaction networks. Pietro H. Guzzi can be reached at hguzzi@unicz.it.

Moez Ben Hadj Hamida received the Master degree in computer science from the University of Tunis El Manar, Tunisia, in 2005. He is currently pursuing the PhD in large computer sciences at the same university. He is currently lecturer assistant at the Department of Computer Science of Faculty of Sciences of Tunis, Tunisia. His current research interests include parallel, distributed and Grid-based data mining applications.

Daniel Harezlak is a computer scientist at the Academic Computing Center CYFRONET in Krakow. He has graduated from the AGH University of Science and Technology ang got MSc degree in computer science. He has been working on EU-funded projects such as K-WfGrid, ViroLab and GREDIA in the areas of workflow management and web computing, which resulted in many publications. He is also interested in building enterprise applications and agile programming.

Jean Herveg is a member of the Bar of Brussels. Since 2000, he has been with the Research Center on IT & Law (www.crid.be) at Facultés Universitaires de Notre-Dame de la Paix, Namur, Belgium (FUNDP). He has been a lecturer at FUNDP since September 2005, teaching a course on "Medical Law and ICT". He is also teaching a course on "Data Protection" in the post-graduate program "INFOSAFE". His research activity deals with Data Protection and ehealth products and services. His full biography is at: http://www.fundp.ac.be/universite/personnes/page_view/01004558/cv.html).

David Horner received a PhD in Biochemistry from the University of London in 1995. He worked for 7 years at the Natural History Museum in London on molecular sequence evolution and is currently a researcher at the Department of Molecular Biosciences and Biotechnology at the University of Milan. His research interests include comparative genomics and bioinformatics of non-coding RNAs.

Joshua Zhexue Huang is an assistant director at the E-Business Technology Institute (ETI), Honorary professor at the Department of Mathematics of the University of Hong Kong (HKU). Dr. Huang received his PhD degree from the Royal Institute of Technology in Sweden. Dr. Huang is acquired many years of experience in databases and data mining research and has contributed to the development of a series of k-means type algorithms in data mining, including k-modes, fuzzy k-modes and k-prototypes which are being widely used in research and real world applications. Dr. Huang has written more than 90 research papers published in international conferences and journals.

Martin Huber studied at the University of Karlsruhe, Germany and received a PhD degree in computer science in 1999. Since 1996 he has been with Siemens Corporate Technology and Siemens Medical Solutions including an assignment as scientific collaboration manager for computed tomography in the U.S.A. Currently, he is heading an interdisciplinary team in Erlangen, Germany consisting of specialists in imaging and machine learning focusing on information-driven software in healthcare. Further, he is technical coordinator of the EU funded project Health-e-Child as well as coordinator of THESEUS-MEDICO funded by the German ministy of economy. His research interests are in medical imaging and semantic data integration.

Nicolas Jacq is a consultant in industry in grid computing and data center automation. He received his Engineer degree from CUST, Clermont-Ferrand, France, in 2000. He achieved his PhD in Bioinformatics in 2006 in the University of Clermont-Ferrand, France, in collaboration with an industrial partner Communication & System. As research engineer, he deployed a genomic platform on a grid infrastructure within the framework of the DataGrid project (2000-2003). He was involved in the RUGBI project (2004-2006), suported by the French ministry of research, which aimed to create a set of grid services for the bioinformatics. In 2005 and 2006, he was in charge of the first and the second biomedical datachallenges, called WISDOM, in the EGEE and EGEE-II project about the drug discovery against malaria and avian flu. These experiences were the first large scale productions of biological results on several grid infrastructures. He was the technical coordinator of the European SHARE project for the HealthGrid association in 2007.

Emmanuel Jeannot is a researcher at INRIA (Institut National de Recherche en Informatique et en Automatique), LORIA laboratory. He got his PhD and master degree of computer science (resp. in 1996 and 1999) both from Ecole Normale supérieure de Lyon. His main research interests are scheduling for heterogeneous and large- scale environments, data redistribution, experimental tools, adaptive online compression and programming models.

Marek Kasztelnik graduated from the AGH University of Science and Technology in Kraków, Poland and received his MSc in Computer Science. He joined Technologies Group in Research and development department in ComArch in 2005 where he is responsible for creating IT solutions for telecommunication management. He has been also involved in EU-funded projects ViroLab, Gredia and CoreGRID in ACC CYFRONET AGH supercomputing centre in Kraków since 2005. He is applying his scientific knowledge there by developing the concept of resource discovery mechanisms. His scientific interests include enterprise systems, resource discovery, distributed computing, solutions for e-science and telecommunication applications.

Joanna Kocot received her MSc degree in computer science from the AGH University of Science and Technology in Krakow. Since 2006 she has been working for ACC Cyfronet AGH in Krakow. Currently she is involved in EU-funded projects: ViroLab and GREDIA as a designer and developer. Her scientific interests involve distributed computing, Web technologies, programming paradigms and e-science/e-health applications.

Dagmar Krefting is researcher and lecturer at the Institute of Medical Informatics of the Charité - Universitätsmedizin Berlin, Germany since 2004. She obtained her Ph.D in physics at the University of Göttingen in 2003 and worked as postdoctoral fellow at the Fritz-Haber-Institute, Germany. Her current research focus is on image processing of medical ultrasound data and she has taken special research interest in integration of grid based medical image processing applications. She heads of the medical image processing module of the german MediGrid project.

Sriram Krishnan is a Senior Distributed Systems Researcher at the San Diego Supercomputer Center. He received his Master's and PhD degrees in Computer Science from Indiana University in 2001 and 2004 respectively. He is currently the Project Manager for the NEES Cyberinfrastructure Center (NEESit) at SDSC, which is a service-focused organization created to deliver information technology tools to enable earthquake engineers to remotely participate in experiments, perform simulations, organize and share data, and collaborate with colleagues. He is also the lead architect for the Services Oriented Architecture for the National Biomedical Computation Resource (NBCR), which provides transparent access to emerging cyberinfrastructures by "grid-enabling" biomedical codes.

Dariusz Krol is a student at the Institute of Computer Science, AGH University of Science and Technology in Krakow. Since 2007 he has been working for ACC Cyfronet AGH in Krakow where he has been involved in EU-funded projects: ViroLab and Gredia as a scientific developer. He is particularly interested in Web technologies, distributed computing and semantic based monitoring.

Peter Kunszt has a PhD in theoretical physics (lattice QCD) from the University of Bern. He holds a builder status in the Sloan Digital Sky Survey SDSS project, where he was one of the developers of its science archive database at the Johns Hopkins University in Baltimore. After 2001, he has contributed to the EU DataGrid and EGEE projects at CERN, Geneva, where he was leading the development efforts for the data management middleware in both projects. Currently, he is responsible for several projects at the Swiss National Supercomputing Centre CSCS in Manno, an institution of the ETH Zurich.

Giuseppe La Rocca is doctor degree in Software Engineer from the University of Catania, Italy. Since 2004 he joined the Italian INFN of Catania as a scientific collaborator. He has been from 2004 to 2006 scientific collaborator for NA4 work package in the context of the Enabling Grid for E-sciencE (EGEE) Project. Afterward, he has been involved from 2006 to 2007 as a scientific collaborator in the WP6 of the BIOINFOGRID Project. Up to April 2008 he was involved in LIBI Project. Starting from May 2208 he's involved in EGEE-III project. He is author of 7 proceedings of international conferences.

Giovanni Lavorgna is research scientis at DIBIT- Ospedale San Raffaele and professor of Bioinformatics at the Vita-Salute University in Milan. After graduating, he has spent several years at the National Institutes of Health (NIH) in Bethesda, MD, working on transcriptional regulation. After moving back to

Italy, Giovanni has developed several innovative algorithms for the analysis of nuclei acid and proteins. He is the author of several bioinformatics papers and is reviewer for several journals in the field.

Yannick Legré is one of the co-founders and current president of the International HealthGrid Association. He works in the domain of Grid technologies (network grids for healthcare applications and contexts, application integration, service-oriented architectures and software engineering research and development projects involving health service providers and the software industry) since August 2000, in particular within the European Project named DataGrid and now within EGEE, as well as several others. Since early June 2008, he's employed as Consultant and International Relations Manager by Maat-G, a major Grid player (www.maat-g.com)

Pietro Di Lena is research assistant of Computer Science at the University of Bologna, Italy. He received the Laurea degree in Informatica (2003) and the Dottorato di Ricerca in Informatica (PhD in Computer Science, 2007) from the University of Bologna. His research supervisors are Prof. Luciano Margara and Prof. Rita Casadio. His research interests are in Combinatorial Optimization, Cellular Automata, and recently in Bioinformatics.

Pietro Leo is an executive IT architect at IBM GBS Innovation Center in Bari, Italy. He is a permanent member of the IBM Italy TEC, affiliated with the IBM Academy of Technology. His areas of expertise include data, information and application integration, unstructured information management, mining and semantic/conceptual indexing and search, bioinformatics, healthcare, and wireless solutions. He was the technical responsible in a number of multi-millions complex projects for IBM customers and for Industrial Research projects. He is the author or co-author of more than 45 scientific or technical publications, and is the co-author of two books edited by IBM Redbooks.

Daniele Lezzi gained a PhD in Information Technology Engineering in the Center for Advanced Computing Technologies division of the National Nanotechnology Laboratory of the University of Salento. His research covers High Performance, Distributed and Grid Computing and he is also interested in computer security. He is team member of the Grid Resource Broker, of the iGrid Information Service and of the GSI plugin for gSOAP projects. He has been a researcher in the Euro-Mediterranean Centre for Climate Changes involved in the design of the computational infrastructure and it is currently researcher in the Computer Science department of the Barcelona Supercomputing Center.

Wilfred W. Li received his PhD in Biochemistry and Molecular Biology from the University of Southern California (USC), and his postdoctoral training in signal transduction at the University of California, San Diego (UCSD). He is currently the Executive director of the National Biomedical Computation Resource (NBCR at http://nbcr.net), senior fellow in Bioinformatics at the San Diego Supercomputer Center, UCSD, and adjunct professor at the College of Computer Science and Technology, Jilin University, PRC. His current research interests include cluster and grid computing, cyberinfrastructure, high throughput proteome annotation, data mining, virtual screening, structure based drug discovery, multiscale modeling and visualization.

Vito Flavio Licciulli obtained a degree in Computer Science at University of Bari in 1992. Since 1994 he is involved in Bioinformatics with the Bari BioInformatics Group and he is a researcher at the

C.N.R.- Istitute for Biomedical Technologies (ITB). His main research activity are design, development, implementation and maintenance of biological databases, development of bioinformatics software and WEB interfaces, and he has published over 20 scientific publications in these fields. He is responsible for the database and operating systems management at ITB-CNR-Bari institute.

Thomas Lingner received the Diploma degree in computer science in 2005 from the University of Bielefeld, Germany. In 2005 he became a member of the MediGRID project and a group member of the Bioinformatics Department at the University of Göttingen, Germany. He is currently working towards his PhD degree with focus on the application of machine learning techniques for biological sequence analysis. His research interests also include Grid computing and high performance computing.

Frédérique Lisacek received a PhD in Computer Science (Artificial Intelligence) from the University Pierre & Marie Curie, Paris, France in 1984. From then on she has held research positions in bioinformatics in France, Japan and Australia working on knowledge representation and sequence analysis in various fields of biology. She has been involved in proteomics projects since 1999 through work with companies Proteome Systems (Australia) and GeneBio where she acted as CSO until 2005. She currently manages the Proteome Informatics Group of the SIB while being on leave from a tenured position in bioinformatics at the University of Versailles-Saint-Quentin, France.

Weiguo Liu received his bachelor and master degree from the Xi'an JiaoTong University, China in 1998 and 2002, and the PhD degree from the Nanyang Technological University (NTU), Singapore, in 2006. He is currently a research fellow with the Center for Advanced Media Technology at NTU. His research interests include computational biology, parallel algorithms and architectures, high-performance computing, and data mining.

Sabino Liuni has a degree in Biological Sciences at the University of Bari and is researcher at Institute for Biomedical Technologies in Bari (Italy). His interest has been focused mainly on the design and implementation of specialized databases of nucleotide sequences and on development of algorithms for the analysis of biosequence information. He is the coordinator of some research lines and member of research projects in the field of Bioinformatics. In particular his is the responsible of research unit into regional project for development of new bioinformatics and biotechnological tools to analyze gene profile expression of cancer using microarray data.

Maria Antónia Alberto Madope is presently a member of Ford Foundation Alumni Association - Mozambique, and a Computer Science Advisor for Transportes Públicos de Maputo Company. She has a master degree in Computer Science (2007) from Universidade Católica de Santos UNISANTOS, Brazil with a thesis about the IntegraEPI Project titled IntegrAnalysis - an Epidemic Analysis Service on the Grid. She has a bachelor degree in Computer Science (2000) from Universidade Eduardo Mondlane, Mozambique. Her areas of interest are health informatics, distributed systems and informatics systems analysis.

Sergio Maffiolettti has been working as scientist at the Swiss National Supercomputing Centre (CSCS) since June 2004. Previously, he worked in the Pervasive and Artificial Intelligence group at the University of Fribourg where he obtained his PhD in 2004. Currently, he is responsible for the infrastructure management and operation in national and international Grid related projects. He is also co-responsible for the design, planning and operation of the computing infrastructure for the WLCG project as Tier 2 center in Switzerland. Sergio is a member of the Scientific Council and active member of the Swiss National Grid Association (SwiNG).

Giorgio Maggi is professor at the Politecnico di Bari. He graduated in physics at the University of Bari (Italy) in 1972. His research field is in particle physics in close connection with the INFN research program. He has been involved in several high energy physic experiments both at the Fermilab and CERN laboratories, in particular in the Aleph experiment and, now days, in CMS. He is also collaborating to the EGEE european projects to the establishment of a Grid infrastructure and its exploitation both in high energy physics and bioinformatics.

Gaël Le Mahec is a PhD student in computer science at University of Clermont-Ferrand, France. He is a member of the PCSV team (IN2P3 Clermont-Ferrand) since 2005 and his research interests include data management and scheduling in parallel and distributed environments. He received his M.S. from Amiens University, France, in 2004.

Maciej Malawski MSc in Computer Science (AGH, Krakow) and in Physics (Jagiellonian University, Krakow). Since 2001 he is the research and teaching assistant at the Institute of Computer Science AGH. He was involved in EU-funded projects such as CrossGrid, K-WfGrid and CoreGRID at ACC CYFRONET AGH. In 2004-2005 he was a research fellow at Distributed Computing Laboratory at Emory University, Atlanta. Currently he is involved in EU IST ViroLab project, where he is responsible for component-based middleware. His scientific interests include parallel computing, Grid systems, distributed service- and component-based systems, and e-science applications.

David Manset, born in 1978, Mph in Computer Sciences. David Manset is CEO of MAAT France and director of Biomedical Applications for MAAT Gknowledge industrial group. During the last 7 years, he has contributed to European and International projects focussing on e-health with a special emphasis on the application of Grid technologies. His work and team have been awarded several times in major events (such as Salon International des Inventions de Genève 2007, EGEE 2007 Conference, EGEE 2008 User Forum, ICT 2008 etc). David Manset is specialized in the application of formal Model Driven Engineering techniques to Grid system architecture design and development.

Uko Maran is senior research fellow of the Molecular Technology of University of Tartu. He holds a PhD (since 1997) in theoretical and computational chemistry from University of Tartu. His research interests include molecular design and modelling of bio- and macromolecular systems, quantum chemistry, statistical analysis, development of molecular descriptors, distributed computing, etc. He has published more than 50 papers in international pre-reviewed journals related to these areas. He has gained experience as participant and/or site principal investigator in number of EU projects such as IMAGETOX, OpenMolGRID, NANOQUANT, Cardioworkbench and Chemomentum.

Luciano Margara is full professor of Computer Science at the University of Bologna, Italy. He received the Laurea degree in Scienze dell'Informazione (1991) and the Dottorato di Ricerca in Informatica (PhD in Computer Science, 1995) from the University of Pisa. He Joined the University of Bologna in

1995 (Research Associate from 1995 to 2000 and Associate professor from 2000 to 2005). He has been a visiting scientist at the International Computer Science Institute (Berkeley) and a visiting professor at the Cornell University (Department of Computer Science). His research interests are in Discrete Time Dynamical Systems, Optical Networks, Computational Complexity, and recently in Bioinformatics.

Osvaldo Marra received a Laurea Degree in Computer Science from the University of Bari, Italy, in 1994. From December 1994 to January 2002, he was the IT Systems Engineer of the Networks Lab at the University of Lecce. From February 2002 to date, he is the senior IT Systems Engineer of the High Performance Computing Lab at the University of Salento. Recently he has been involved in the SPACI Project (Southern Partnership for Advanced Computational Infrastructures), and he is currently leading the IT technical staff of the Lecce site of the SPACI Consortium. He has been also involved, as a site-manager and a IA64 supporter, in the European Project EGEE (Enabling Grids for E-Science in Europe), and now he is continuing his collaboration with the third phase of the project (EGEE-II).

Pier Luigi Martelli is researcher in Biochemistry at the University of Bologna. He has a degree in Physics and a PhD degree in Physics. His research is mainly focused on the design and implementation of systems based on neural networks, hidden Markov models and support vector machines for the prediction of different features in proteins. PLM is author of more than 30 paper on international peer-reviewed journals and presented the results of his research to more than 50 national and international meetings. He was awarded for the best paper at the ISMB (Intelligent Systems for Molecular Biology) 2002 meeting.

Juan Manuel Gonzalez Martin studied Computing at the University of Castilla La Mancha, now he finishes his studies at UNED. He worked as a programmer in Maat G Knowledge since 2003, until 2006 he was working at diverse proyects of this company, some related to grid tecnology . On March 2006 he joined CETA-CIEMAT, he employed at UAA(Unit of Applications Architecture). He is currently working in projects related to Grid Digital Repositories.

Mirco Mazzucato: Current Activities: Present position: INFN director of Research. CNAF (INFN Advanced Computing Centre) director, since 2004. INFN Grid Project Project Manager since 2000. INFN delegate and member of the Management Board in the (W)LCG, World Computing Grid for LHC Computing Grid Project at CERN since 2002 (~ 6000 physicists in Europe, USA and Asia). Member of the Project Management Board and co-ordinator of the Italian Federation in the FP6 European project EGEE (and its successor EGEE II, EGEE-III) MIUR Italian delegate in the e-Infrastructure Reflection Group.

Richard McClatchey is director of the Centre for Complex Co-operative Systems at the University of the West of England, Bristol and a CERN Associate. He is a Fellow of the Institute of Engineering and Technology (FIET) and of the British Computer Society (FBCS) as well as a Chartered Information Technology Practitioner and a Chartered Engineer. McClatchey established his Research Centre in 1997 and was given a Chair in Applied Computer Science in 2000. His specialities are in the management of distributed data and processes, Grids Computing and the design of data models for the capture of workflow information. He has published over 120 articles in journals or in international conferences and has been on programme committees of numerous conference series during his 18 years at UWE.

McClatchey holds a PhD in High Energy Physics (from University of Sheffield in 1982) and was awarded a DPhil by UWE in 2000 acknowledging his contribution to the field of Computer Science. He has successfully supervised ten PhD students and 12 Masters students and currently supervises six Doctoral students. He has considerable experience in large European projects (e.g. ARTEMIS, CICERO, CRISTAL, MAMMOGRID, STAFF, Health-e-Child and neuGRID) and has several years consultancy experience including ORACLE. McClatchey has held research awards from the Royal Society, the Royal Academy of Engineering and CERN and has been a Visiting professor at the University of the Savoie, France and the National University of Science & Technology at Rawalpindi, Pakistan.

Ivan Merelli received the Laurea degree in Biomedical Engineering from the Polytechnic University, Milan, Italy, in 2003 with a thesis about molecular surface modelling and analysis. His research activity concerns the development of software for sequence based genomics and for structural proteomics researches, with particular interest in protein-protein interactions. He works actively on the high performance implementation of bioinformatics components using parallel programming and distributed platforms. Currently, he is researcher at the Institute of Biomedical Technologies of National Research Council and works at the FIRB - MIUR projects "Italian Bioinformatics Network - ITALBIONET" and "Laboratory of Interdisciplinary Bioinformatics Technologies - LITBIO".

Flavio Mignone is assistant professor of Computer Science at the University of Milan, Italy. He received a PhD in "Molecular and Cell Biology" at the University of Milan in 2004 working on the in-silico characterization of untranslated sequences. He received a master degree in Biological Sciences in 1998 at the University of Milan and a degree in Information Engineering in 2008 at the Politecnico of Turin, Italy.

Luciano Milanesi is currently researcher of the Italian National Research Council (CNR). Since 1988 he is head of the Bioinformatics Division of CNR-ITB. He focused on developing tools for the genome sequence analysis and prediction of gene structure in different organisms. More recently his work has shifted towards studying the functions of genes, promoter prediction, gene expression analysis and the development of databases integrated with sequence analysis tools. He has been principle investigator for the European Projects: TRADAT "TRAnscription Database and Analysis Tools", ORIEL "an Online Research Information Environment for the Life Sciences" and Bioinfogrid "Grid for Bioinformatics applications". He has published contributions in books and scientific publications in bioinformatics.

Russ Miller is UB Distinguished professor of Computer Science and Engineering and senior scientist at the Hauptman-Woodward Medical Research Institute. He has co-authored seminal papers in the fields of parallel algorithms for the mesh, pyramid, hypercube, and reconfigurable architectures in domains that include computational geometry, image analysis, and fundamental data movement operations. He has published nearly 200 papers, chapters, and abstracts of presentations at national and international meetings and has co-authored two textbooks on parallel and sequential algorithms. His work in theoretical and applied parallel algorithms led to his work in molecular structure determination, which was recognized by the IEEE poster "Top Algorithms of the 20th Century." His current research interests are in cyberinfrastructure and grid computing.

Lukasz Miroslaw holds a post-doc position at the University of Zurich. He received a PhD (since 2006) in automation and robotics from Wroclaw University of Technology. His research interests include grid computing, image processing and statistical pattern recognition with expertise in the software design and development. As a grant holder of Max Planck Society, Herbert Quandt Stiftung and Ministry of Saxony he performed several research projects at Max Planck Institute of Cell Biology and Genetics in Dresden, Germany (with Dr. Frank Buchholz) and at Dresden University of Technology (with Prof. Siegfried Fuchs). Author and co-author of several papers in peer reviewed journals including Nature Cell Biology.

Maria Mirto received her Laurea Degree in Computer Engineering from the University of Lecce, Italy in 2002. She received the PhD degree in Computer Engineering from the ISUFI-University of Lecce, in 2006. Since 2006, she has a postdoc position in the LIBI project for the University of Salento. Since 2003 she is the Principal Investigator of the ProGenGrid project. Her research interests include High Performance, Distributed and Grid Computing, Bioinformatics and Workflow. She received the best paper award at the ITCC 2003 conference. She is co-author of more than 45 papers on parallel, grid computing and Bioinformatics.

Paolo Missier Since 2004 Paolo Missier has been a Research Associate at the School of Computer Science, University of Manchester, UK, where he is now a Resarch Fellow. Prior to joining academia, he was a Research Scientist at Telcordia Technologies (formerly Bellcore), NJ, USA from 1994 through 2001, where he gained research and industrial experience in the area of information management and software architectures. He has also been a lecturer in databases at Universita' of Milano Bicocca, in Italy; an independent IT consultant for the Italian public sector; a course instructor in Software Engineering for a number of private companies, and has contributed to research projects in Europe in the area of Information Quality management and information extraction from Web sources. He holds a PhD in Computer Science from the University of Manchester, UK, a MSc in Computer Science from University of Houston, Tx., USA and a BSc and MSc in Computer Science from Universita' di Udine, Italy.

Wayne Mitchell is group leader, Informatics at the ASTAR Experimental Therapeutics Centre, Biopolis, Singapore and Adjunct Associate professor, Division of Information Sciences, School of Computer Engineering, Nanyang Technological University, Singapore. He holds an BA from Harvard University, a BS from the University of Massachusetts, Boston and a PhD from the University of California, Berkeley. He was scientific founder and Chief Genomics Officer at Tao Biosciences, and has published in genomics, bioinformatics, and knowledge engineering.

Pericles A. Mitkas is a professor of Electrical and Computer Engineering at the Aristotle University of Thessaloniki (AUTH) in Greece. Between 1990 and 2000 he was a faculty member with the Department of Electrical and Computer Engineering at Colorado State University in USA. Dr. Mitkas' research interests include databases and knowledge bases, data mining, software agents, enviromatics and bioinformatics. His work has been published in over 150 papers, book chapters, and conference publications. Dr Mitkas received his Diploma of Electrical Engineering from AUTH in 1985 and an MSc and PhD in Computer Engineering from Syracuse University, USA, in 1987 and 1990, respectively.

Yassene Mohammed is research assistant in the department of medical informatics at the Georg-August-University Göttingen, Germany. Studied 1997-2002 biomedical engineering at the University of Damaskus, Licencié in September 2002. Studied 2002-2005 Photonics Engineering at the Faculty of Sciences and Technology University for Applied Science and Art, Göttingen, MSc, January 2005. 2002-2005 research assistant in the medical image processing group of the Department of Medical Informatics, University Göttingen. Since 2005 PhD-thesis in Medical Informatics, Georg-August University, Göttingen, Germany and research assistant in the project MediGRID.

Johan Montagnat is a researcher at CNRS (French National Scientific Research Center). He graduated from the Ecole Normale Supérieure de Cachan in 1996. He obtained a PhD in 1999 and an habilitation thesis in 2006, both from the University of Nice - Sophia Antipolis. His main research interests are medical imaging and grid infrastructures applied to biomedicine. He is involved in various European and national projects on these topics.

Ludovica Montanucci has a PhD in Physics and a Laurea in Physics. Actually she is post-doc reasearcher at the Biocomputing Group, University of Bologna. Her main research interests concern Computational Biology, in particular cross genome comparison by means of machine learning methods and multivariate data analysis techniques.

Jean-Marie Moureaux received his PhD degree in Electrical Engineering from the University of Nice-Sophia Antipolis, France in 1994. Since 1995, he is with the Centre de Recherche en Automatique de Nancy (CRAN, Nancy Université and CNRS). His research interests are image coding, watermarking, lattice vector quantization, multimedia transport over the Internet and medical image compression.

Salvatore My is researcher at Facoltà di Ingegneria of the Technical University of Bari since July 1st 2002 (scientific sector: Experimental Physics). He is member of the Bari's Physics Department. The main activity of Dr. S. My have been in the building of micro-strip silicon modules for the CMS tracker and he has especially been involved in the test of these modules. He is collaborating to the LIBI project, in particular to the porting of bioinformatics applications to grid.

Aisha Naseer is a research fellow working on an EPSRC-funded project RIGHT (Research into Global Healthcare Tools) at the School of Information Systems, Computing and Mathematics, Brunel University, UK. She received her PhD (2007) in Computer Science from Brunel University, UK, specialising in Grid-Based Semantic Integration of Heterogeneous Data Resources on HealthGrids. She has taught various Computer Science and Information Systems courses at both the undergraduate and postgraduate level. She has published several papers in journals and international conferences and also has chapters published in several books. She has chaired various conference sessions and reviewed many papers and book chapters. Her research interests include HealthGrids, semantic interoperability, intelligent data management, HealthGrid applications, health informatics, Grid computing, artificial intelligence and simulation & modelling of healthcare systems.

Alessandro Negro was born in S. Pietro Vernotico (BR) in 1981. He received a summa cum laude bachelor degree in Computer Engineering from the University of Salento, Lecce, Italy in February 2004 and a summa cum laude master degree in Computer Engineering from the same University in April

2006. His research interests include GUI Development, Data Management, Distributed, Grid Computing and Web Services. He is also interested in web and pattern-oriented design. Since 2003 he is a Team Member of the GRelC Project. Since September 2006 he holds a contract position in the LIBI project for the University of Salento.

Virgilio Cavicchioli Neto has a bachelor degree in computer engineering, and is currently pursuing a master degree in information technology applied to public health since 2006 at Federal University of Sao Paulo – UNIFESP, in the Health Informatics department. His thesis will be about georeferencing epidemic cases of dengue fever, tuberculoses and leprosies using open source tools and developing a service which make georreferenced data available to users.

Piotr Nowakowski holds an MSc degree from the AGH University of Science and Technology and is currently employed as a research associate at the Academic Computing Center CYFRONET AGH in Krakow. He has been involved in a number of Grid and distributed computing projects, including CrossGrid, K-WfGrid, ViroLab and GREDIA. He has authored or co-authored approximately 20 publications in the field of Grid and distributed computing. His interests revolve around problems of distributed data access and resource sharing.

Mark Olive is a doctoral candidate in Information Systems at the Bristol Institute of Technology. He graduated from the University of the West of England, Bristol, with a BA in Systems Analysis and was engaged by the SHARE project after a period of work in industry and independent technical consultancy. His subject is knowledge management in healthgrids, with particular emphasis on the interpretation of Integrated Care Pathways as a vehicle for knowledge representation and evolutionary learning.

Alessandro Orro received his PhD in Electronics and Computer Science Engineering in February 2005, after a three-year course at the University of Cagliari, Italy. He currently works at the Institute of Biomedical Technologies of the National Research Council in Segrate, Italy. His research interests are in the field of artificial intelligence (mainly soft computing and multi-agent systems) applied to problems in computational biology and medicine, in particular protein multiple alignment, protein structure prediction and protein function annotation, genetic and clinical data analysis. Presently, he collaborates in the BioinfoGRID project.

Angel Osorio is physician and physicist. He obtained his PhD in 1971. After working as an Associated professor in Madrid (Spain), San Francisco (USA) and London (UK) he joined the CNRS (French National Research Center) in Paris (France) in September 1973 and still working there as a Research director. Since 1995, his research activity is focusing on 3D medical imaging treatment. He is the head of many software packages about automatic 3D treatment of DICOM images. He directed 27 PhD and is currently directing 3 ones. He is the author of 151 published articles and scientific communications. He obtained 8 international awards for his scientific work.

Vitaliy Ostropytskyy is a research associate at the School of Biomedical Sciences at the University of Ulster. He holds a Degree in Computer Sciences, and a PhD (since 2000) in Progressive Information Technologies from Ukraine. He has published over 40 papers. His main research interests are database management, data warehousing and Internet technologies. He has been involved in different international projects.

Graziano Pappadà is a bioinformatic scientist in Italy responsible of the Exhicon s.r.l Bioinformatics unit. He graduated in biology from the Faculty of Science of the University of Bari in 2002 and since then has worked on several bioinformatics projects concerning biological database development, data integration and workflow design in collaboration with the CNR, University of Bari and IBM Innovation Centre.

Marco Passante is PhD student in Information Technology Engineering at the University of Salento, Lecce, Italy. He received a summa cum laude master degree in Computer Engineering from the same University, in October 2008. Since January 2009, he has a contract position in the LIBI International laboratory of BioInformatics, project FIRB 2003, at the University of Salento. His interests concern the area of Biological Data Management, Grid Problem Solving and Grid Workflow Management System.

Cesare Pautasso is assistant professor of Informatics at the University of Lugano, Switzerland. He was a researcher at the IBM Zurich Research Lab and a senior researcher at ETH Zurich. His research focuses on building experimental systems to explore the intersection of model-driven software composition techniques, business process modeling languages, and autonomic/Grid computing. He has developed an interest in Web 2.0 Mashups and Architectural Decision Modeling. He is the lead architect of JOpera, a Grid workflow management tool for Eclipse. His teaching and training activities both in academia and in industry cover Web Development, SOA and emerging Web services technologies.

Giulio Pavesi is assistant professor of Computer Science at the Department of Biomolecular Science and Biotechnology of the University of Milano, where he had obtained both his M.S. and PhD degrees in Computer Science and currently works in the Bioinformatics and Comparative Genomics lab.

Michal Pelczar holds an MSc degree from the AGH University of Science and Technology. He is currently involved in EU-funded ViroLab project. He has co-authored several publications in the field of collecting and mining information about virtual experiments. He is mostly interested in representation and processing of knowledge in Semantic Web.

Xavier Pennec is a research director at the ASCLEPIOS team at INRIA Sophia-Antipolis (France). He holds an Engineering degree from the French Ecole Polytechnique in 1992 and a PhD degree in Computer Science from the same institution in 1996. He was a post-doctoral associate at MIT AI Lab (Mass. USA) in 1997, before joining INRIA in 1998. His main research interests are about statistics on geometric data, in particular for medical image analysis, biomedical image registration and computational anatomy. He is associate editor of the Medical Image Analysis Journal (Elsevier) and chairs the International Workshop on Mathematical Foundations of Computational Anatomy since its inception in 2006.

Graziano Pesole is full professor of Molecular Biology at the University of Bari and leads a research team in Bioinformatics and Comparative Genomics at the Institute of Biomedical Technology (National Research Council). His research interests include bioinformatics, development of tools for genome annotation, comparative genomics and molecular evolution. He is a member of the Editorial Board of international journals (BMC Bioinformatics, BMC Genomics, Computational Biology and Chemistry,

Briefings in Bioinformatics; Comparative and Functional Genomics), author of over 130 papers published in international journals and co-author of books on Bioinformatics and Genomics published by italian (Zanichelli, Gnocchi) and international (Wiley) editors.

Ernesto Picardi is assistant professor at the University of Bari, Italy. He received a PhD in Plant biology in January 2006 at University of Calabria. His research interests are molecular evolution, comparative genomics and bioinformatics. He is the author of 8 papers in international peer-reviewd journals.

Viviana Piccolo is a PhD student of Molecular and Cellular Biology at the University of Milan. She obtained her Master's degree in Biology at the University ROMA TRE (Rome) and her post lauream Master in Bioinformatics at the University of BICOCCA (Milan). She is working in the functional and comparative genomics of non coding RNAs with particular interest in micro RNAs, small interfering RNAs and targets prediction in plants and animals. Her research interests include also the development of tools focused on Support Vector Machine classification for the ab initio discrimination of micro RNA precursors from other genomic hairpin structures.

Ian Piper was awarded a BSc in Biological Sciences (University of Edinburgh) with Honours in Physiology in 1979, an Electronics Diploma (Devry Institute of Technology: Toronto, Canada) in 1984 and a PhD (University of Edinburgh) Dept. Clinical Neurosciences, Faculty of Medicine. Ian has worked within the field of traumatic brain injury research since 1979 first at the National Hospital for Nervous Diseases, Queen Square, London; in Spinal Cord injury research at the University of Toronto, and for the last 20 years first with professor Douglas Miller in Edinburgh and latterly with professor Sir Graham Teasdale in Glasgow, Scotland. During this time he has been responsible for key developments in the field of computerised detection and quantification of adverse events and the measurement of intracranial compliance in patients with traumatic brain injury. He is the coordinator of the Brain Monitoring with information Technology (BrainIT) research group.

Ivan Torres Pisa is currently an associate professor at the Health Informatics Department, Federal University of São Paulo (UNIFESP) and vice-president of the department's graduate program. His areas of interest are Health Informatics, Telemedicine, Medical Decision Support Systems, PACS, and Distributed Systems. He has a PhD in Health Informatics, University of São Paulo (USP), 2003, with a thesis about the MIDster Project. Ivan Pisa also has a MSc degree in Health Informatics, USP, 1999, and a bachelor degree in Mathematics from São Paulo State University (UNESP), Rio Claro Campus, 1997.

Raul Ramos Pollan graduated as Computer Engineer from University of A Coruña, Spain in 1995 and obtained his MSc in Artificial Intelligence in 1997. He is since 2007 director of CETA-CIEMAT in Trujillo, Spain, a grid based computer centre of CIEMAT. He worked during 5 years at CERN (European Laboratory for Particle Physics) IT Division as a software engineer and lecturer. Between 2000 and 2004 he worked for Sun Microsystems as member of their Java Centre providing consultancy in software engineering, as responsible of software development processes and their methodology team. In 2004 he created an SME specialized in the field of aeronautics and aerospace and was in charge of their systems department. In 2006 he joined CETA-CIEMAT as Assistance director until 2007 when he took over his current possition.

Frederic Pourraz After five years of scientific studies, Frédéric Pourraz has joined the LISTIC Laboratory at the University of Savoie and has participated to the ArchWare European Project (IST-2001-32360). During this project, Frédéric has made a PHD in the domain of SOAs. He has then joined the maatG. Knowledge company and work as an SOA expert in the context of the Health-e-Child European Project (IST-2004-027749). Since 2008, Frédéric is an assistant professor of the University of Savoie affiliated to the Software an Systems group of the LISTIC Laboratory.

Francisco Prieto graduated at Universidad Autonoma de Madrid (UAM) in Theoretical Physics. He was granted by the European Commission for the Safety Assessment for Earthquake Risk Reductions (SAFERR) Project. Within SAFERR, he joined outstanding scientific institutions such as Joint Research Centre (JRC) at Ispra (Italy). He obtained a PhD in Physics at UAM in the field of complex systems and chaos theory. He has relevant contributions in prestigious scientific journals in the fields of physics and engineering. Recently leads the Science and Technology Unit at the Extremadura Research Centre for Advanced Technologies.

Fotis E. Psomopoulos is a PhD candidate with the Electrical and Computer Engineering Department of the Aristotle University of Thessaloniki, Greece, where he received his Diploma in 2004. He is a research associate of the Informatics and Telematics Institute of CERTH, where he is a member of the Intelligent Systems and Software Engineering Lab. His research interests include bioinformatics, data mining and grid computing. He has worked on various National and European projects, including "eTHMMY" and "ASSIST" among others. Also, he is a member of the IEEE Computer Society and the Technical Chamber of Greece.

Andreas Quandt received a PhD in bioinformatics from the University of Geneva. He studied Biotechnology in Berlin and was involved in several research projects focusing on tissue engineering (TU Hamburg-Harburg) and stem cells proliferation (Charite Berlin) before he moved more towards the data analysis aspect in life sciences. He continued studying Bioinformatics in Berlin (MSc) and worked in different research centers such as the Max Delbrück Centrum Berlin-Buch, the Max Planck Institute for Molecular Genetics, and the Karolinska Institute Stockholm. His current research is focusing on analysis workflows for mass-spectrometry data in combination with large-scale computing resources.

Vincenzo Quinto is an IT specialist with extensive experience in Java technologies and bioinformatic applications. He worked as an IT consultant for Exhicon s.r.l. within projects for IBM Innovation Centre focusing on concerns related to biological data access and integration. He has developed particular experience in application development, client/server architectures and web services.

Hanene Boussi Rahmouni is a doctoral candidate in Computer Science at the Bristol Institute of Technology. She gained a first class honours degree in Computer Science from the University of the West of England, Bristol, and joined the SHARE team on graduation. She has worked on issues of privacy, confidentiality and security, specialising in the automatic interpretation of policies into permissions and obligations at the operational level in healthgrids.

Alfredo Tirado-Ramos is a researcher at the University of Amsterdam's Informatics Institute, working on software architectures for distributed biomedical applications. He chairs committees in international conferences in topics ranging from Grids for healthcare to teaching computational science and electronic commerce. He is currently a member of scientific committees on health Grids, high performance medicine and biology, learning technologies, and computational science. Alfredo worked previously at Philips Research North America on medical informatics interoperability, and as modeling and methodology facilitator for the DICOM/HL7 joint committee. He has co-authored 3 patents pending and more than 30 peer-reviewed publications in refereed journals and conferences.

Katharina Rasch is a research fellow at the Center for Information Services and High Performance Computing at the Technische Universität Dresden. She holds a degree in Computer Science with a specialization in data management and communication systems. Before joining the Chemomentum project she has been researching music database systems and automatic transcription of music.

Matteo Rè, PhD in Molecular Biology, is a post doctoral fellow at the department of Information Science of University of Milano. He currently works in the Machine learning and Bioinformatics lab.

Jérôme Revillard, born in 1978, Doctor in Computer Sciences from the University of Savoie, in France. Before joining the company MAAT France, he has been working for MAAT Gknowledge for which he was responsible of the developments achieved in various bio-informatics European projects such as the EU FP6 Health-e-Child. Currently, R&D director of MAAT France, he is responsible for all the technical decisions and developments orientations of the company.

Alfonso Rios, born in 1972, electronic engineer and PhD in Physics at CERN (2000), he is the CTO of MAAT Gknowledge. He has been working in several R&D groups in Databases, Data Mining, supercomputers, semantic Web and their applications to health, anthropology, High Energy Physics etc. He is the main author of MAAT's core technology "G".

Elda Rossi has a degree in Chemistry from the University of Modena . Siince 1984 she has been working at CINECA where she is responsible for computational applications in fields related to Chemistry and Life Science. Her expertise is in support to the sceintific research using advanced computing tools, in addition to writing and integrating scientific programs in the ICT field. She has been the principal investigator of several projects funded by the European Community and she was the coordinator of three of them. The last projects (ImmunoGrid) started in February 2006 and is about the simulation of the human immune system.

Ivan Rossi holds a PhD in Computational Chemistry. He has been Adjunct professor of Computational Biology at the University of Bologna and authored around 40 peer-reviewed publications. His main interest in the field of bioinformatics is the application of bioinformatics methods to investigate problems in biology and biochemistry.

Cecilia Saccone has been full professor of Molecular Biology, University of Bari, Italy since 1973. She was President of the CNR Research Area of Bari (1987-2000). She has been and/or is a member of many scientific committees among which: EMBO (European Molecular Biology Organization), European Commission for Nucleotide Sequence Databases, Task Force Biodiversity Committee of the OCSE Forum of Megasciences, CBOL Scientific Advisory Board (Consortium for the Barcoding of Life). Her main research areas of interest are: Functional and Comparative Genomics, Molecular Evolution and Biodiversity, Bioinformatics and Biocomputing.

Mark S. P. Sansom is in the Department of Biochemistry at the University of Oxford, where he is head of the Structural Bioinformatics and Computational Biochemistry unit. He received his degree from Oxford in 1983, and worked for seven years in the University of Nottingham, before returning to Oxford. He has interests in many aspects of computational studies of membrane proteins with special interest in ion channels, bacterial outer-membrane proteins, and the development of e-science and high-end computing for simulation and modelling of complex membrane systems.

Monica Santamaria received a summa cum laude degree in Biological Sciences in 2000 and a PhD in Genetics and Molecular Evolution in 2004 from the University of Bari (IT). At present she works as a researcher at the Institute of Biomedical Technologies of the National Research Council (CNR-IT), where she is involved in Molecular Biodiversity studies and in particular in Fungi DNA barcode surveys. Furthermore she collaborates to human molecular variability studies focused on the pyrosequencing applications on mitochondrial genome.

Ulrich Sax is assistant professor for Medical Informatics at the Georg-August-University in Göttingen, Germany. He finished his PhD-thesis in Medical Informatics 2002 at the Georg-August University, Göttingen, Germany and was Postdoctoral Research Fellow in Children's HST Informatics Program, Harvard-MIT Division of Health Sciences & Technology, Harvard Medical School, Boston 2003-2005. He is head of the CIOffice Medical Research Networks, Göttingen, Germany, vice head of the BMBF-funded project MediGRID within D-Grid. His research is focused on Privacy and security aspects in Medical informatics, Electronic Patient Records, Personal Health Records and how to deal with genomic data in research and health care.

Diego Scardaci was born in Catania, Italy, the 11th September, 1975. He got the degree in Computer Science Engineering on 21th July, 2000. Since 2006 he is working as Researcher for the INFN (Italian National Institute of Nuclear Physics) involved in Grid Computing activities. He worked on Grid interoperability, Grid Security and Grid API development. Moreover, he has been involved in the Dissemination of Advanced Grid knowledge activities, organizing many induction courses. Since 2002-2005 he has worked as Researcher for Telecom Italia Lab and he has designed and developed innovative platforms and services for the mobile telephony operator TIM.

Joël Schaerer received an Engineering degree in Telecommunications from INSA Lyon, France in 2004, and a MSc in Signal and Image Processing also in 2004 from the same university. He is currently a PhD student at CREATIS-LRMN, Lyon, France. His research interests include medical image segmentation, deformable elastic models, and cardiac tractography.

Bertil Schmidt is an Associate professor at the School of Computer Engineering at NTU, Singapore. He received his Master degree in Computer Science from the Kiel University, Germany, in 1995 and his PhD degree from the Loughborough University, UK, in 1999. He has worked with the company ISATEC in the area of embedded parallel systems. His research interests include parallel algorithms and architectures, high-performance computing, and bioinformatics.

Robert Schöne is a research fellow at the Center for Information Services and High Performance Computing at the Technische Universität Dresden. He holds a degree in Computer Science from the Technische Universität Dresden. Before joining the Chemomentum project he was involved in analyzing the performance of parallel programs and in developing performance analysis tools.

Bernd Schuller is core UNICORE developer team and co-head of the open source UNICORE project at SourceForge, as well as a member of the Technical Board of the UNICORE Forum (www.unicore.org). He has extensive experience with development of Grid middleware, deployment and application integration. He has been involved in number of EU projects such as OpenMolGRID and Chemomentum.

Salvatore Scifo is graduated in Computer Science at the University of Catania in 2001. Since 2006 he was involved as researcher in the Grid R&D and Application Support fields with both National Institute of Nuclear Physics of Catania (2006-2007) and Cometa Consortium of Catania (since 2007). From 2000 to 2005 he was employed with several Italian SMEs: as a "Software Designing Engineer " (2000-2002) and as a "Senior Software Engineer" (2002-2005) on both TLC and ITC areas. He has also carried out research activities on Grid-based Digital Libraries and has a particular expertise on data and metadata management in grids.

Gaetano Scioscia is an IT architect at IBM GBS Innovation Centre in Bari, Italy. He graduated in 1995 in physics and earned a PhD in theoretical physics in 1999 in Bari. In 1998, Dr. Scioscia joined IBM, where he worked as an IT specialist and then an IT architect in such fields as data and application integration, and information and knowledge management. During recent years, his work has been focused on bioinformatics and currently he is the IBM Innovation Lab bioinformatic team technical leader.

Giordano Scuderi is graduating in Computer Science at the University of Catania, Italy. Since 1998 he has collaborated with several private companies in projects related to software development, security and IT Infrastructure design. Since 2006 he has been involved in the design and the development of the Secure Storage Service, in the context of Trigrid VL Project.

Giovanna Selvaggi is professor at the University of Bari. She graduated in physics at the University of Bari (Italy) in 1972. She has been involved in several high energy physic experiments both at the Fermilab and CERN laboratories, in particular in the Aleph experiment and, now days, in CMS. She is now collaborating to the LIBI project in particular to the porting of bioinformatics applications to grid.

Sulev Sild is a senior research fellow at the University of Tartu. He holds a PhD (since 2001) in theoretical and computational chemistry from University of Tartu. He has been participating in several international research projects (with prof. Katritzky at University of Florida, with prof. Diercksen at Max Planck Institute in Munich, Germany, and in the OpenMolGRID project funded by European Union). His research interests include molecular design and modelling, application of artificial neural networks, computational chemistry, and Grid computing.

Fabricio Alves Barbosa da Silva obtained his doctoral degree in Computer Science from Université Paris VI (Pierre et Marie Curie) in 2000. He is currently an assistant professor of the Informatics Department at Universidade de Lisboa, Portugal. His research interests are scientific computing, modeling and simulation, distributed computing and grid computing. Recently he has been working on the application of these technologies to the epidemiology and public health areas.

Richard O. Sinnott was awarded a BSc (Honours) in Theoretical Physics from the University of East Anglia, Norwich, England in 1988; an MSc in Software Engineering from the University of Stirling, Scotland in 1983 and a PhD also at Stirling in 1997 in the area of formal modeling of distributed systems. He has worked across a wide range of computing science research areas both in academia and in industry. He has over 100 publications across a range of computing science and application oriented fields, most recently in the area of Grid security and usability especially in the life sciences. He has edited several ITU-T/ISO international standards.

Konstantin Skaburskas, born in 1972, MSc in Biomedical Engineering and Physics from University of Tartu, Estonia. Before joining IT Department at CERN in February 2006, he has been working in the area of computational biomedical engineering, cluster and distributed computing and grids for about seven years. Currently, holding a Fellowship at CERN IT Grid Deployment group, he is involved in two projects: EGEE/WLCG by providing Grid middleware distributed monitoring solutions and in Health-e-Child by providing Grid middleware expertise to the project.

Peter Sloot, PhD, is the group leader of the Section Computational Science and director of the Computer Science Institute of the University of Amsterdam. He is an expert in computational science, modeling and simulation and Grid computing. He has over 200 publications in these fields, has laid the foundations for the conference series of computational science (http://www.iccs-meeting.org/), organized the first European Grid Conference, in February 2005 in Amsterdam (http://genias.biz/egc2005/) and the IEEE eScience Conference in 2006. Moreover, he is editor in chief of a new established international Elsevier-Science journal on grid computing (http://www.elsevier.com/locate/future).

Manuel Rubio del Solar graduated in Computing at the University of Extremadura in 2003. He worked as programmer in e-learning projects and also in the computing service at the University of Extremadura. On 2003 he began his Phd, currently he works on his thesis. In 2005 he began working as project manager in the computing service on the Agriculture department of the Extremadura regional government. On March 2006 he joined CETA-CIEMAT, where leads the Unit of Application Architecture, managing projects related to Grid Digital Repositories and Boinc middleware development. He is participating on the EELA2 and EDGeS projects.

Tony Solomonides is a Reader in Computer Science and Medical Informatics at the Bristol Institute of Technology, University of the West of England, Bristol. He switched from Mathematics to Computer Science in the 1980's and worked in clinical medical informatics with particular reference to diabetes and inflammatory bowel disease for a decade before beginning work on healthgrids. He was a coinvestigator (and briefly at the end of the project Scientific Coordinator) of MammoGrid, the principal investigator for the integrated road map in SHARE, and has continued to contribute to other healthgrid projects, notably Health-e-Child and neuGRID. He is also leading a non-framework European project on pre-implantation genetic diagnosis (PGD) in assisted reproduction. He recently completed a UK-funded project on knowledge sharing in the National Health Service.

Lampros K. Stergioulas is a Reader and director of Undergraduate Studies in the Department of Information Systems and Computing at Brunel University, UK. He is a qualified Chartered Engineer. He has published over 150 papers in journals and international conferences. He has supervised and examined numerous PhD dissertations in the areas of information systems, information processing, and computer science in general. He has held many National and European Grants in educational computing, human-centred communications and computing, information processing, modelling and simulation, and intelligent systems. He has chaired several International Conferences and has been the coordinator of interdisciplinary, international projects. His research interests include human-centred information systems, educational computing, intelligent data processing, information management, modelling and simulation of business processes, and healthcare information systems.

Karl A. Stroetmann, MBA PhD FRSM, is senior associate with *empirica* Communication & Technology Research in Bonn, Germany. Dr. Stroetmann received an MBA from the Free University in Berlin, Germany, and a PhD in Business Administration and Finance from the University of British Columbia in Vancouver, Canada. He has been a consultant to the European Commission, OECD and global industrial players. He was / is project coordinator or principal investigator in European projects on eHealth policy and strategic issues, on market validation and business development (various eTEN studies), economic impact assessment (eHealth IMPACT and EHR IMPACT socio-economic studies), and in health RTD studies (patient safety; interoperability; semantics; business case development for patient data mining applications to improve clinical use of antibiotics as well as of Virtual Physiological Human [VPH] simulation applications for osteoporosis diagnosis and treatment). Recent work also concerned a foresight study on scenarios for ICT-enabled new models of healthcare, research on financing models, and a representative survey of European GPs. In Nov. 2002, he and his wife, Dr. Veli Stroetmann, received an award for: „Telemedicine - Human Applications of IT in Medicine" (Integrata Foundation). They regularly publish in peer-reviewed journals and present at international conferences, including the European Union High Level Ministerial eHealth conferences, national eHealth symposia, and European and global scientific and professional meetings.

Velitchka N. Stroetmann, MD PhD, studied at the Academy for Medical Sciences, Sofia/Bulgaria, and obtained an M.D. degree (1984). After working as a family doctor she continued studies in Informatics and received a PhD in Medical Informatics (1992). After four years as a doctor and researcher in Medical and Health Information Science in Sofia, she came to Germany and joined *empirica* in 1992. There she has been involved in various telemedicine pilot and validation projects, research on older or disabled people, on eHealth applications and information policy issues. She was the chief editor of the highly acclaimed expert report on "Telematic Applications in the Health Sector", and principal investigator for the SATS "Study on the use of advanced telecommunications services by health care establishments in Europe". She was actively involved in the eHealth IMPACT and eHealth ERA projects. Presently, she coordinates a project for the European Commission to improve semantic interoperability among Member State health systems (Semantic Interoperability Deployment and Research Roadmap) and is involved in longer-term eHealth RTD studies (economic issues and business case development of patient data mining applications to improve clinical use of antibiotics [DebugIT] as well as of Virtual Physiological Human [VPH] simulation applications for osteoporosis diagnosis and treatment [VPHOP]). In Nov. 2002, she received an award for: „Telemedicine - Human Applications of IT in Medicine" (Integrata Foundation).

Heinz Stockinger has been working in Grid projects in Europe (SIB, CERN, etc.) and in the USA (Stanford Linear Accelerator Center) in various technical, scientific and management functions. Heinz is affiliated with the Swiss Institute of Bioinformatics where he works on diverse Grid subjects. He has been appointed "Privatdozent" at the University of Vienna - leading the Research Lab for Computational Technologies and Applications in 2005. Currently, he is also a lecturer at the Swiss Federal Institute of Technology in Lausanne (EPFL). Heinz holds a PhD degree in Computer Science and Business Administration from the University of Vienna, Austria.

Kaihsu Tai is a research associate in the Department of Biochemistry, University of Oxford. He led the team to develop BioSimGrid. He received a B.Sc. (Hon.) from California Institute of Technology (1998) and PhD in chemistry from the University of California, San Diego (2002) with Prof. J. Andrew McCammon. His research interests include simulations of biomolecular systems. He is a Chatered Chemist, a Member of the Royal Society of Chemistry, and an advisory member of the Green Economics Institute.

Daniele Tartarini is PhD student in "Interdisciplinary sciences and technologies" for the Nanoscience and Grid computing section of the "Scuola Superiore ISUFI", University of Salento, Lecce, Italy. He received a summa cum laude master degree in Computer Engineering from the University of Lecce, Italy, in July 2006. Since September 2006, he has a contract position in the LIBI International laboratory of BioInformatics, project FIRB 2003, at the University of Salento. His research interests concern the area of High Performance Computing, Distributed and Grid Computing, Neural networks, BioInformatics and NanoBiosciences. He is coauthor of some papers in refereed journals on Grid Computing & NanoBioscience.

Dorin Tcaci graduated in Mechanics at Moldova Technical University (Chişinău, Moldova) in 1998. He worked in Moldova's companies as software engineer on many projects related to geoinformational systems, e-commerce and e-learning services. From 2006 he started to work at "MAAT-G knowledge" and has been involved in projects like Health-e-Child and MammoGrid+. On 2006 he began his Phd at University Jaume I, Castellon. On January 2007, as a result of an egreement between "MAAT-G knowledge" and "CETA-CIEMAT", he started to work at CETA on Grid Digital Repositories especially on the DRI & MammoGrid+ projects.

Romain Texier completed his PhD in 2007 at University of Paris Sud, which focused on quality of service adapted to responsiveness for EGEE computing resources. He has been the main developer of the gPTM3D software. He joined the I3S laboratory as CNRS engineer since 2007 and currently contributes to the Medical Data Manager. Antonio Turi is working in bioinformatic from 2003 at the Institute for Biomedical Technologies, Bari, Italy. From 2007 he is interesting in grid technology at the University of Salento, Lecce & SPACI Consortium, Italy. He received a master degree in Computer Science in 2002 and, at present, he is last year PhD student in knowledge discovery in databases (KDD) at University of Bari, Italy.

Livia Torterolo received her master degree cum laude in Biomedical Engineering from the University of Genoa, Italy, in 2004. After her admission into the engineering profession in 2004, she started collaborating with several private companies related to grid computing technologies. Since 2006, she has been a PhD student at DIST, University of Genoa, Italy. In 2007-2008 she was visiting student at Trinity College in Dublin, Ireland. Her research activity is focused on grid portal technology and grid-enabled biomedical applications.

Giuseppe Tradigo received his Laurea degree in Computer Engineering from the University of Calabria, Rende, Italy, in 2003. Currently he is working with the bioinformatics group of University "Magna Græcia" of Catanzaro. His research interests are in geographical databases and GIS, grid computing, software engineering, protein structure prediction, and grid environments for bioinformatics applications. Giuseppe Tradigo can be reached at gtradigo@si.deis.unical.it.

Gabriele Trombetti received the PhD degree in Computer Science Engineering in 2008 at the University of Bologna, Italy with a thesis on Enabling Bioinformatics Applications on the Grid Platform. Previously he had received the Laurea (MS) degree with full marks (100/100 cum Laude) in Computer Science Engineering in 2004 at the same university with a thesis on Modelling and Validation Tools for QoS Enabled Distributed Systems and Applications. The thesis was developed at the DOC group at Vanderbilt University, Nashville, TN where he also worked as a Research assistant in 2004 for researches on the same topics. In 1996 he was graduated at the High school Liceo Scientifico Augusto Righi in Bologna, Italy with full marks (60/60). Since 2008 he is a researcher at the Institute for Biomedical Technologies, National Research Council, Italy.

Alexey Tsymbal is a research scientist at Siemens AG in Erlangen, Germany. He holds a university diploma with distinction in software engineering gained from National Tech University, Ukraine in 1997. He received his PhD in computer science in 2002 from the University of Jyväskylä, Finland. In 2003-2006 he also took part in research projects at Trinity College Dublin, Ireland. His research interests include machine learning and data mining, knowledge-based systems, and applications of AI techniques to biomedical domains. He has more than 50 peer-reviewed publications in these areas".

Angelica Tulipano graduated in Physics at the University of Bari in 2003 and obtained a PhD in 2008 from the same institution working on microarray data analysis using clustering methods developed in mechanical statistics. She was involved in an european project on grid applications to Bioinformatics.

Andrea Vanni gained a degree in Computer Science from Bologna University. He joined CINECA in 2002 where he works in Grid computing. He is involved in various European projects such: DEISA2, eDeisa, HPC-Europa, Unigrids and PRACE. He has a wide experience in grid solutions developed for Unicore, Globus and Gridsphere. Since 2006 he has been the co-chair of the DEISA2 WP4 and eDEISA Service Activity 3.

Marco Vassura is research assistant of Computer Science at the University of Bologna, Italy. He received the Laurea degree in Informatica (2001) and the Dottorato di Ricerca in Informatica (PhD in Computer Science, 2005) from the University of Bologna. His research supervisors are Prof. Luciano Margara and Prof. Rita Casadio. His research interests are in Combinatorial Optimization, Optical Networks, and recently in Bioinformatics.

Pierangelo Veltri is an assistant professor of Computer Engineering at University "Magna Græcia" of Catanzaro, Italy, since 2002. He received his PhD in computer science in 2002, from University of Orsay at INRIA Rocquencourt, France. From 1998 to 2000 he was member of the Verso group and he participated to the Chorochronous project (1998-1999) and at the Xyleme project (1999-2001). From 2002 he partecipate to several projects, whose main topics are pure computer science (such as Grid computing, geographical and temporal databases, XML views) or more recently bioinformatics topics (spectra data analysis for cancer prevention, proteomics, prediction of protein structures). His main research interests are: XML and semi-structured database, data modelling, spatial and multidimensional databases, protein matching structure, proteins structure prediction, data management on grid computing. Pierangelo Veltri can be reached at veltri@unicz.it

Saverio Vicario is a researcher the Institute of Biomedical Technologies of the National Research Council (CNR-IT), where he works on Molecular Biodiversity within the LIBI project. In 1999 he graduated in Biology at University of Rome "TorVergata" and got a PhD in 2006 from the Department of Ecology and Evolutionary Biology at Yale University. His research interests are molecular evolution and molecular systematics both in basic research, as evolution of codon usage, and applied research, as the barcode initiative.

Fred Viezens is research assistant in the department of medical informatics at the Georg-August-University Göttingen, Germany. Studied 1988-1993 Informatics at the Technical University "Otto-von-Guericke", Magdeburg. 1994-1997 Lecture at the DEKRA Academy GmbH. 1998-2001 Freelance Activity in an Engineer's Office. 2001-2002 research assistant at the Otto-von-Guericke University Magdeburg, Institute for Biometry and Medical Informatics. 2003 assistant in the MBR Computing Centre GmbH, Magdeburg. 2004-2005 research assistant at the Otto-von-Guericke University Magdeburg, Faculty for Mechanical Engineering, Institute for Material Engineering and Material Testing, Institute for Logistical Systems. Since 2006 research assistant at the University Medical Center Göttingen.

Jakub Wach holds and MSc degree from the AGH University of Science and Technology. He is currently employed as a Technical Consultant at the Volantis Systems, leader in mobile solutions middleware. He has also experience from other companies, including ACK Cyfronet AGH and GridwiseTech. His main fields of interest are large distributed systems and leading-edge novel solutions for e-Science. He has been invloved in EU-IST K-Wf Grid and ViroLab projects, primarily as an architect and leading engineer of provenance systems. He is a co-author of eight scientific papers.

Qiang Wang is now a PhD candidate student at the School of Computer Science of Harbin Institute of Technology (HIT). Mr. Wang's research interests include data mining and bioinformatics, especially the study and development of novel clustering algorithms for microarray data analysis. He is now taking active part in the research and development of China Drug Discovery Grid (DDG) project.

Charles Weeks is a senior research scientist in the Department of Structural Biology at the Hauptman-Woodward Institute. His research interests lie in computational crystallography, in particular the development of algorithms and software for automating the process of protein structure determination. He is a coauthor of the widely used programs *SnB* and *BnP*.

Nadya Williams is a software architect at the Institute of Organic Chemistry at the University of Zurich. She has a Master of Science degree in Oceanography from Far-East State University, Vladivostok, Russia, and a Master of Science in Computer Science from University of California, San Diego. She has worked at the University of California, San Diego for various projects involving oceanographic research, and dealing with all aspects of software and hardware installation and maintenance, and grid computing.

Petra Wilson is a director in the Healthcare team of Cisco's Internet Business Solutions Group (IBSG), focusing on helping healthcare providers and deciders develop their strategies and implementation plans for "Connected Health" information management and technology. Previously, she was deputy director of the European Health Management Association (EHMA), a membership organisation committed to improving healthcare in Europe by raising standards of managerial performance in the health sector. At EHMA she had responsibility for the association's Brussels' office and, in particular, EHMA's relationship with the European Commission. In that role, Petra advised a wide number of European healthcare organisations on EU-level health policy and assisted them in engaging with policy-makers in the Commission. Prior to EHMA, Petra worked for the European Commission in its eHealth directorate. She worked specifically on the European eHealth Action Plan, which commits all member states of the European Union to adopt road maps for the implementation of Connected Health solutions. Petra also advised the Commission on legal and ethical issues surrounding use of information technologies in healthcare. Alongside her role at Cisco Petra currently holds the chairmanship of the European Policy Workgroup of the Continua Health Alliance, as well as the chair of the Governing Council of Health Information and Management Systems Society Europe. She is also acts as an advisor to the Health on the Net Foundation. Originally trained in law, Petra spent several years as a lecturer and Senior Research Fellow at Nottingham University, UK, specializing in healthcare law and European law. She has a PhD in public health law from Oxford University, and has published widely on the legal aspects of using information technologies in healthcare. Originally from the UK, she is fluent in English, German, and French.

Yunming Ye is a vice director at the E-Business & Intelligent Enterprise Computing Research Center (ICES), Associate professor and PhD Supervisor at the School of Computer Science of Harbin Institute of Technology (HIT). Dr. Ye received his PhD degree form the Shanghai Jiao Tong University, visiting scholar at E-Business Technology Institute (ETI) of Hong Kong University. Dr. Ye has many years of experience in business intelligence, data mining and web searching research and has written more than 40 research papers published in international conferences and journals. Dr. Ye also is now responsible for 3 national projects.

Kunqian Yu is an associate professor of Shanghai Institute of Materia Medica (SIMM), CAS. He graduated from the State Key Lab of Institute of Theoretical Chemistry of Jilin University July 2002, and then worked as a post-doctor in Shanghai Institutes for Life Sciences (SIBS), CAS. Dr. Yu did much research work on Molecular Dynamics (MD) simulations for polymer, studied the recognition and interaction of ligands with proteins by computer simulation, and also preformed virtual drug screening and computer aided drug design for tens of important protein targets. He now focuses on the drug development for avian influenza virus, and also is responsible for the construction of China drug discovery grid (DDG).

Federico Zambelli is a PhD student of Molecular Biology at the Department of Biomolecular Science and Biotechnology of the University of Milano, where he currently works in the Bioinformatics and Comparative Genomics lab.

Vincenzo Zara is full professor of Biochemistry at the University of Salento. His research has included the following topics: study of the import pathway of metabolite carriers into isolated mitochondria; study of the assembly of the hydrophobic proteins in the inner mitochondrial membrane; purification and structural and functional characterization of mitochondrial carriers; effect of different hormonal and nutritional states on the hepatic fatty acid biosynthesis; study of the biogenesis of the yeast cytochrome bc_1 complex. He is the author of 157 scientific articles, most of them published in specialized international journals. He is also a reviewer for several international journals.

Index

C

D

H

I

J

K

L

M

N

O

P

Q

R

S

T

Handbook of Research on Computational Grid Technologies for Life Sciences, Biomedicine, and Healthcare

Mario Cannataro
University Magna Graecia of Catanzaro, Italy

Volume II

Medical Information Science
REFERENCE

MEDICAL INFORMATION SCIENCE REFERENCE
Hershey · New York

Director of Editorial Content: Kristin Klinger
Senior Managing Editor: Jamie Snavely
Managing Editor: Jeff Ash
Assistant Managing Editor: Carole Coulson
Typesetter: Jeff Ash
Cover Design: Lisa Tosheff
Printed at: Yurchak Printing Inc.

Published in the United States of America by
Information Science Reference (an imprint of IGI Global)
701 E. Chocolate Avenue
Hershey PA 17033
Tel: 717-533-8845
Fax: 717-533-8661
E-mail: cust@igi-global.com
Web site: http://www.igi-global.com/reference

and in the United Kingdom by
Information Science Reference (an imprint of IGI Global)
3 Henrietta Street
Covent Garden
London WC2E 8LU
Tel: 44 20 7240 0856
Fax: 44 20 7379 0609
Web site: http://www.eurospanbookstore.com

Library of Congress Cataloging-in-Publication Data

Handbook of research on computational grid technologies for life sciences, biomedicine, and healthcare / Mario Cannataro, editor.
p. cm.

Summary: "This book provides methodologies and developments of grid technologies applied in different fields of life sciences"--Provided by publisher.

Includes bibliographical references and index.
ISBN 978-1-60566-374-6 (hardcover) -- ISBN 978-1-60566-375-3 (ebook) 1. Bioinformatics. 2. Computational biology. 3. Problem solving--Data processing. I. Cannataro, Mario, 1964-

QH324.2.H3576 2009
570.285--dc22

2009001527

British Cataloguing in Publication Data
A Cataloguing in Publication record for this book is available from the British Library.

All work contributed to this book is new, previously-unpublished material. The views expressed in this book are those of the authors, but not necessarily of the publisher.

To Francesco and Matteo

Editorial Advisory Board

List of Contributors

Table of Contents

Detailed Table of Contents

Section I
Infrastructures and Services for HealthGrids and BioGrids

This section introduces the main concepts and defines technical, methodological and organizational challenges of HealthGrids and BioGrids. After discussing the roadmap toward future HealthGrids, the Section defines basic resources and their discovery in HealthGrids. Then, taking into account the large use of scientific workflows in BioGrids and HealthGrids, the Section introduces the data provenance concept and discusses ways to query and exploit provenance data. Data protection and security, an important aspect that must be faced when considering biomedical data, is also discussed. Finally, a review of distributed data mining and knowledge discovery systems useful for implementing the analysis layer of HealthGrids is also presented.

Section II
Grids for Genomics and Proteomics

This section discusses the use of the Grid for the management and analysis of genomics and proteomics data, the basic data at the biological level. After describing main issues for the parallel and distributed implementation of BLAST, a cornerstone of all genomics analysis, the Section discusses some significant Grid-based implementations of genomics applications. Then, the Section introduces proteomics with a special focus on mass spectrometry-based proteomics and different Grid-based proteomics applications, ranging from biomarker discovery to protein identification and protein classification, are discussed.

As grids become more and more attractive for solving complex problems with high computational and storage requirements, bioinformatics algorithms start to be ported on large scale platforms. The BLAST kernel was one the first application ported on such platform. However, although simple parallelization was enough for the first proof of concept, its use in production platform needed more optimized algorithms. The chapter reviews existing parallelization and "gridification" of the BLAST algorithm as well as related issues such as data management and replication. A case study using the DIET middleware over the Grid'5000 experimental platform is also presented.

A common ongoing task for Functional Genomics is to compare full organisms' genome with those of related species, to search in huge database for functional annotation of novel sequences and to identify specific patterns of them, such as ESTs, genes, and microRNA. The prediction of these patterns has a relevant computational cost, while public genome archives exceed one billion sequence traces from over 1,000 organisms and this number is increasing rapidly. Thus Functional Genomics applications require significant computational infrastructures, where reusable tools and resources can be accessed.

Section III
Grid-Based Bioinformatics Environments

This section discusses Grid-based software environments for bioinformatics. Grid-based Problem Solving Environments specifically devoted for bioinformatics, as well as Grid-based implementations of relevant bioinformatics applications, such as docking, bio-molecular simulation and molecule structure determinations, are presented.

This chapter describes ProGenGrid (Proteomics and Genomics Grid), a Grid Problem Solving Environment specialized for the Bioinformatics domain. The presented system provides an integrated environment for composing, scheduling and monitoring biological applications in a Grid. The main feature offered by this environment is the possibility to use an easy-to-use web interface for composing workflow jobs, scheduled on different grid middlewares.

A drug discovery process is aimed to find from a large set of molecules the candidate leads that have strong interaction with the target proteins. The process of drug discovery is characterized by its complexity in data and computation. Thus, the combination of complex algorithms and data management functions are necessary for domain scientists to build proper drug discovery procedures. This chapter presents a graphical workflow modeler for domain scientists to perform drug discovery tasks on high performance grid computing platforms. A client/server system is described as the platform for implementation of the graphical workflow modeler. A case study on drug discovery for avian influenza virus is presented to demonstrate the use of this tool in drug discovery research.

The chapter presents BioSimGrid, a distributed biomolecular simulation database. It is a general-purpose database for trajectories from molecular dynamics simulations. The presentation of BioSimGrid explains how to install the system, and how to deposit, query, and analyze trajectories with real Python code examples for each step. The chapter also presents the underlying concepts in the implementation of BioSimGrid: relational database, distributed computing, and the input/output (deposit and analysis) modules. The chapter concludes by discussing emerging trends in biomolecular simulations and concerns in the further development of BioSimGrid and similar biological databases.

The New York State Grid (NYS Grid) is an integrated computational and data grid that provides access to users from around the world to a wide variety of resources. This Grid can be accessed via a Web portal, where the users have access to their data sets and applications, but do not need to be made aware of the details of the data storage or computational devices that are specifically employed in solving their problems. The chapter presents a Grid-enabled version of the SnB and BnP programs which respectively implement the Shake-and-Bake method of molecular structure (SnB) and substructure (BnP) determination. The programs were run on the NYS Grid and their performance evaluation is presented. In particular, SnB has been run simultaneously through the Grid Portal on all the computational resources of the NYS Grid as well as on more than 1100 of the over 3000 processors available through the Open Science Grid.

Section IV
Grids for Medical Informatics

This Section focuses on Grid-based medical informatics applications. Since a core aspect of medical applications regards the acquisition and analysis of biomedical images, the Section first introduces different aspects of biomedical images visualization on the Grid. Then, as significant applications in the field, Grid analysis of radiological data and 3D electron microscopy reconstruction are discussed. Finally, as an example of biomedical instrumentation made available on the Grid, a hybrid mock circulatory system to test cardiovascular prostheses is presented.

Visualization, the art and science of representing data visually, is now recognized as an equal partner in the conduct of science via the simulation and modeling paradigm. Different aspects of visualization, such as image preprocessing/analysis, and some computational geometric and computational topological processes are amenable to deployment over compute Grids, but there has been equal focus on the collaborative aspect of Grid computing which is driving collaboration-based visualization systems. The chapter introduces some aspects of visualization and the grid. The chapter first surveys some of the roles of visualization as they relate to the role of Grid computing within a biomedical context. Then it examines certain scheduling strategies that are believed to have value in terms of the distribution of visualization tasks over Grid fabrics.

Volume II

Grid technologies and infrastructures can contribute to harnessing the full power of computer-aided image analysis into clinical research and practice. A main challenge in medical images analysis is to fill the gap between the grid middleware and the requirements of clinical applications, that can be summarized in large volumes of data, sensitivity of medical information, and complexity of medical datasets. This chapter reports on the goals, achievements and lessons learned from the AGIR (Grid Analysis of Radiological Data) project. AGIR faces these challenges by providing some core grid medical services (data management, responsiveness, compression, and workflows) supporting medical data processing, and by grid-enabling a panel of applications ranging from algorithmic research to clinical use cases.

Three-dimensional electron microscopy allows scientists to study biological specimens and to understand how they behave and interact with each other depending on their structural conformation. Electron microscopy projections of the specimens are taken from different angles and are processed to obtain

Grid systems provide mechanisms for single sign-on, and uniform APIs for job submission and data transfer, in order to allow the coupling of distributed resources in a seamless manner. However, new users face a daunting barrier of entry due to the high cost of deployment and maintenance. They are often required to learn complex concepts relative to Grid infrastructures (credential management, scheduling systems, data staging, etc). To most scientific users, running their applications with minimal changes and yet getting results faster is highly desirable, without having to know much about how the resources are used. Hence, a higher level of abstraction must be provided for the underlying infrastructure to be used effectively. The chapter describes the Opal toolkit, a framework for exposing applications on Grid resources as simple Web services. Opal provides a basic set of Application Programming Interfaces (APIs) that allows users to execute their deployed applications, query job status, and retrieve results. Opal also provides a mechanism to define command-line arguments and automatically generates user interfaces for the Web services dynamically. In addition, Opal services can be hooked up to a Metascheduler such as CSF4 to leverage a distributed set of resources, and accessed via a multitude of interfaces such as Web browsers, rich desktop environments, workflow tools, and command-line clients.

The LIBI project (International Laboratory of BioInformatics) aims to develop an advanced bioinformatics and computational biology laboratory, focusing on basic and applied research in modern biology and biotechnologies. The paper presents the core part of the system, a Grid Problem Solving Environment, built on top of EGEE, DEISA and SPACI infrastructures, allowing the submission and monitoring of jobs mapped to complex experiments in bioinformatics. Several case studies on different bioinformatics applications and related results which have been obtained using the LIBI platform are also reported.

Section VII
Building and Deploying HealthGrids

This Section describes main infrastructures, middleware, and tools for building and deploying HealthGrids. After introducing what are the main requirements on Grid middleware posed by biomedical applications and how these are satisfied by a well know Grid middleware, the Section describes a Grid-based portal that aims to promote collaboration and cooperation among scientists and healthcare research groups on the Grid. The management and deployment of Grid repositories for biomedical digital contents such as mammograms is also discussed.

Repositories are digital stores that manage data and metadata providing their access to users, offering an easy-to-use service and a powerful system to handle digital assets. This chapter presents the gLibrary/DRI (Digital Repositories Infrastructure) platform, a Grid-based digital repository that takes advantage of the Grid features such as VO authentication, file catalogues, and metadata services. The main goal of the platform is to reduce the cost in terms of time and effort that a repository provider spends to get its repository deployed. This is achieved by providing a common infrastructure and a set of mechanisms that repository providers use to define the data model, the access to the content and the storage model. Two use cases are also presented: a mammograms repository example that provides clinicians with a tool that eases diagnostics process and an algorithmic repository based on the Poincare Surface Section.

Section VIII
Selected Readings

This Section is a short collection of suggested readings of different authors, aiming to enrich this book with others knowledge, experience, thought and insight. After introducing Cloud Computing and describing the porting of applications to Grids and Clouds, the Section describes a bio-inspired approach for the construction of a self-organizing Grid information system, and concludes discussing Grid-based implementation of phylogenetic analysis.

Cloud Computing is an emerging style of computing in which dynamically scalable resources are provided as a service. The chapter describes main stages of implementing applications on Grid and Cloud infrastructures. As a case study, the chapter presents the Distributed European Infrastructure for Supercomputing Applications (DEISA) and describes the DEISA Extreme Computing Initiative (DECI) for porting and running scientific grand challenge applications on the DEISA Grid. The chapter concludes proposing the top ten rules of building a sustainable Grid.

Foreword

"Empirical science may have to be considered, not merely as a scientific theory – that is, a system of asserted statements arranged according to certain rules – but rather as a complex consisting partly of such statements and partly of human activities." Alfred Tarski, Introduction to Logic xi-xii.

"We don't really have a health care delivery system in this country. We have an expensive plethora of uncoordinated, unlinked, economically segregated, operationally limited micro systems, each performing in ways that too often create suboptimal performance both for the overall health care infrastructure and for individual patients." George Halverson, Health Care Reform Now!: A Prescription for Change, 2007.

The development of the grid since the early 1990s has led from a system of methods for sharing computation and information resources, to a system of active virtual communities: from computation through sharing, to sharing through computation. Our ability to extend and exploit both perspectives is of extraordinary relevance to biology and medicine. It gives us the opportunity to organize increasingly sophisticated, flexible and powerful systems of logic, data, rules, computation and human activities within social networks in which time and space constraints are reduced to a minimum. In so doing, we may hope to create the infrastructure, methods, and (virtual) organizations needed to deliver the benefits of evidence-based and personalized medicine to the seven billion inhabitants of planet Earth.

What are we sharing and distributing when we apply the grid to healthcare and biomedical research? Computation, software, and information, certainly – but also to a great extent we are enabled to share computed results, which given an appropriate infrastructure can be passed from process to process (workflows), stored and retrieved, layered in mutual support, interpreted, counterposed and resolved. The Grid now enables us to construct a connected system of computational "collaboratories" within virtual organizations (VOs) that embrace different specialties and disciplines in biomedicine and healthcare, as well as engineering and software, and that must ultimately also extend to healthcare providers, payers, and of course patiens. In this way, we create *an ecosystem of actions, tools and information*. The Grid as VO is a system of radical enabling technologies for this scientific knowledge ecosystem.

Virtual cooperation and complex layering of technologies and results are now available on a global scale, allowing interpretations and conclusions to be extracted from masses of data obtained in ultra high-throughput experiments. The results of one group's experiment, or data from huge numbers of clinical observations, can now be data mined, consumed by workflow processes and used as input to further interpretations and conclusions, with great rapidity and agility. These results can then drive further experiments, inform the design of, and participation in, clinical trials, and ultimately ensure that better care is delivered to more patients, faster than before. At the core of this knowledge ecosystem are grid-enabled virtual organizations, made possible through organized and uniform service models around computational, communications and storage resources, with shared commitments to common semantics and models of component behavior.

In perhaps no other area of endeavor does the emergence of virtual organizations present such complex issues – and striking opportunities – as in biomedical research and healthcare. In the U.S. alone, health care costs surpassed $2 trillion in 2006[1] with approximately 6% of these expenditures going to biomedical research[2]. It is widely accepted that this "nonsystem," while outstanding in its technical capabilities, is far from effective in delivering the right care to the right people at the right time. Grids and VOs promise to contribute significantly to productive organization of service delivery, improvement of quality of care, and more rapid translation of basic medical research into advances in clinical care. Yet realizing this vision involves a host of challenges, technical and social.

Technical challenges include achieving the uniformity of infrastructures and services, and network, computational and communication resources, required to make sharing feasible; balancing local versus global policies and requirements (access, security, trust, load, service provisioning); and developing computable semantic representations of data, metadata, policies and services. Social challenges include adequate funding of the research required to build out the grids and VOs that will support a healthcare revolution; balancing research with product development, and the quest for the perfect with the deployment of the good; the agile capture, mapping and translation of user requirements into working systems features with adequate fidelity; overcoming the substantial disincentives that currently hinder sharing across the healthcare nonsystem; and building methods and systems that can scale not only to many patients and providers, but also to many countries, societies, and economies.

This book provides a comprehensive and focused view of the state of research in the application of grids and virtual organizations to these and other challenges. Reporting in particular on the results of the European Union's substantial HealthGrid program, the book includes descriptions of both cutting-edge research and substantial deployments. We believe it will make a significant contribution to meeting the challenges that currently face healthcare, and will be of great importance in broadening the application of an extremely powerful system of technology and social organization in biomedicine.

ENDNOTES

1 U.S. Centers for Medicare and Medicaid Services, Office of the Actuary, National Health Statistics Group, 2006 National Health Care Expenditures Data, January 2008. Retrieved from http://www.cms.hhs.gov/nationalhealthcareexpenddata/01_overview.asp?

2 Moses, H. 3rd,, Dorsey, E.R., Matheson, D.H., Their, S.O.,(2005, September 21). Financial anatomy of biomedical research. *JAMA 294*(11), 1333-42.

Tim Clark & Ian Foster
March 23, 2009

Tim Clark *is director of Informatics at the MassGeneral Institute for Neurodegenerative Disease and an Instructor in Neurology at Harvard Medical School. Tim was formerly Vice President of Informatics at Millennium Pharmaceuticals, where his team built one of the first integrated bio- and chemi-informatics software platforms in the pharmaceutical industry. Tim is also a Founding Editorial Board member of the journal Briefings in Bioinformatics. His research group's program is focused on applications of semantic web and social computing to biomedicine and scientific publishing.*

***Ian Foster** is director of the Computation Institute, a joint institute of the University of Chicago and Argonne National Laboratory, where he is also the Arthur Holly Compton Distinguished Service professor of Computer Science and an Argonne Distinguished Fellow. He received a BSc (Hons I) degree from the University of Canterbury, New Zealand, and a PhD from Imperial College, United Kingdom, both in computer science. His research deals with distributed, parallel, and data-intensive computing technologies, and innovative applications of those technologies to scientific problems. Methods and software he has developed underpin many large national and international cyberinfrastructures. Dr. Foster is a fellow of the American Association for the Advancement of Science and the British Computer Society. His awards include the British Computer Society's award for technical innovation, the Global Information Infrastructure (GII) Next Generation award, the British Computer Society's Lovelace Medal, R&D Magazine's Innovator of the Year, and an honorary doctorate from the University of Canterbury.*

Preface

Grid technologies emerged from specific needs in basic sciences demanding high-computing power and large store repositories like physics, and from the availability of high-speed high-bandwidth networks and high-performance computers. In the last few years the term Grid has evolved toward a concept of ubiquitous and transparent computing that extends the network paradigm offered by the Internet and offers a new way to think and use computing resources and databases. As this evolution continued new terms like Data Grid, Knowledge Grid, and Semantic Grid emerged, including new functionalities and capabilities regarding, respectively, the efficient management of data stores, the discovery and sharing of knowledge, and the use of semantic-rich services and workflows.

Presently Grids start to be available as a commodity and have reached a maturity that are now available for different application domains, as it happened with the Internet in recent years. In particular, Grids are currently used in many sectors of Life Sciences, from basic sciences such as genomics, proteomics, and bioinformatics, to biomedical research and applications such as drug discovery, biomedical imaging, and biomedicine, and to clinical research and healthcare. So, the application of Grid technologies to Life Sciences, yielded new terms like Bioinformatics Grid, Biomedical Grid, and Health Grid.

The term "HealthGrid" is used to describe the application of Grid technology in the field of biomedical and healthcare informatics. The scientific and industrial community recognized the importance of governing and coordinating technical, standardization and organization activities of HealthGrids by establishing a set of international initiatives.

The international HealthGrid Association (http://www.healthgrid.org/) is an organization based in Europe that promotes the application of Grid technologies in the biomedical sector. With the HealthGrid White Paper (http://whitepaper.healthgrid.org), it established an agenda and defined the concepts, the opportunities and the likely benefits of such an approach. During the last few years it has been working to develop a strong relationship with clinicians and researchers to gain feedback from the use of Grid in the biomedical and healthcare sectors. The HealthGrid.US Alliance (http://www.healthgrid.us), an affiliate of the international HealthGrid Association, is a partnership of scientific, medical and technology professionals whose goal is to promote the application of advanced information technology to solve cutting-edge problems in biomedical science and Healthcare.

Starting from the HealthGrid vision, the European SHARE project (http://www.eu-share.org) identified the key developments – technical advances, social actions, economic investments and ethical or legal initiatives – needed to achieve wide adoption and deployment of HealthGrids throughout Europe. The full road map (http://roadmap.healthgrid.org) analyses several case studies and discusses technical, ethical, legal, social and economic issues which may impede early deployment of HealthGrids.

Relevant projects are addressing these challenges worldwide and include, among others, sharing datasets for cancer cure (caBIG – https://cabig.nci.nih.gov/, ACGT – http://www.eu-acgt.org/), science portals for better visualization of brain morphology (BIRN – http://www.nbirn.net/), collaborative sharing of mammograms (MammoGrid), 3D spatio-temporal visualization of the mammalian cells (Virtual Cell – http://www.visiblecell.com/), accurate multi-scale computational models of the heart and of cancer tumours (Integrative Biology – http://www.integrativebiology.ox.ac.uk/), "virtual" research repository of clinical, laboratory and genetic data sets spread within hospitals and research organisations (MMIM – http://www.mmim.org.au/). These and other projects have begun to demonstrate the power and potential of the Grid approach in biomedicine.

Among the others, HealthGrid is currently attempting to face the following questions:

- How to address the needs of the multiple laboratories collecting genomics and post-genomics data and willing to analyse them in an up-to-date and competitive environment?
- How to solve the need for large computation power needed in medical informatics?
- How to address the need for easy accessible data and computing power for both scientist and medical staff?
- How to make information on all levels, from molecule to population, accessible and understandable to the large variety of people which could benefit from such a knowledge?

As will be explained in the following, the book aims to give a contribution to those and other questions regarding the methods, techniques, benefits and problems of using Grid technology in the biological, biomedical and healthcare domains.

The book surveys state of the art research, methodologies and applications of Grid technology in different fields of Life Sciences, from basic science to clinical and healthcare applications. According to an agreed taxonomy, the fields where life science research takes place can be classified as molecule, cell, tissue, individual, and population levels.

Considering data and computational models, bioinformatics is related to the molecule and cell; medical informatics is related to tissue/organ and individual, while public health informatics (in short healthcare informatics) is related to population.

According to that taxonomy, the book discusses the application of Grid technology to each of the information levels where life science research takes place, by focusing on three main areas: bioinformatics for the study of molecules and cells, medical informatics for the study and simulation of tissues and organs, and healthcare informatics related to population.

On the other hand, the Grid is a valuable tool for collaboration and cooperation, enabling the remote use of medical instruments in the so called virtual laboratory. Thus, the use of the Grid to improve collaboration among scientists and health centers and to enhance the healthcare provision to the final users is another important topic that has been addressed by the book.

Finally, the application of Grid technology to Life Sciences poses news challenges about security and privacy of patient's data, provenance of data, protection and reliability of information, and collaborative analysis of biomedical data. The book introduces these requirements and discusses new methodologies and techniques, such as the Virtual Laboratory concept, used to realize HealthGrids and BioGrids. So, one of the main questions the book attempts to answer is "How current Grids have to be extended to support Life Sciences and become HealthGrids?".

Guided by these considerations, the book's contents are organized in a systematic way by considering different orthogonal dimensions: the complexity level of information being considered (molecule, cell, tissue/organ, individual, population), the support to collaboration and knowledge sharing (e.g. virtual laboratory), as well as the technology, infrastructures and middleware currently available for planning and building HealthGrids.

The chapters of the book, organized accordingly to those dimensions of analysis, are divided into the following eight sections:

- Section I Infrastructures and Services for HealthGrids and Biogrids
- Section II Grids for Genomics and Proteomics
- Section III Grid-Based Bioinformatics Environments
- Section IV Grids for Medical Informatics
- Section V Collaborative Grids for Healthcare and Clinical Applications
- Section VI Grid-Based Virtual Laboratories for Bioinformatics and e-Science
- Section VII Building and Deploying HealthGrids
- Section VIII Selected Readings

Section I introduces the main concepts and defines technical, methodological and organizational challenges of HealthGrids and BioGrids. After discussing the roadmap toward future HealthGrids (Chapter I), the Section defines basic resources and their discovery in HealthGrids (Chapter II). Then, taking into account the large use of scientific workflows in BioGrids and HealthGrids, the Section introduces the data provenance concept (Chapter III) and discusses ways to query and exploit provenance data (Chapter IV). Data protection and security, an important aspect that must be faced when considering biomedical data, is also discussed (Chapter V). Finally, a review of distributed data mining and knowledge discovery systems useful for implementing the analysis layer of HealthGrids is also presented (Chapter VI).

At the molecular level, common trends are the increasing size and complexity of biological data and the increasing number and heterogeneity of collected databases. In fact, biological data is being more and more produced by different high-throughput technologies, like micro array and mass spectrometry, and such novel data must be correlated to well known sequences and structures databases, like SwissProt and PDB. Such requirements can be faced by Grids that provide the storage and the computing power needed to store, analyse and correlate such data.

In particular, two current trends in developing BioGrids are reported in the book (Sections II and III): (a) well established bioinformatics tools like BLAST are being ported, eventually through parallelization, on the Grid, allowing to face the analysis issues of the overwhelming biological data; and (b) complete software environments allowing to combine basic bioinformatics tools through workflow of Web/Grid services are deployed on the Grid, supporting the complete life cycle of complex "in silico" experiments.

Section II discusses the use of the Grid for the management and analysis of genomics and proteomics data, the basic data at the biological level. After describing main issues for the parallel and distributed implementation of BLAST (Chapter VII), a cornerstone of all genomics analysis, the Section discusses some significant Grid-based implementations of genomics applications (Chapters VIII to X). Then, the Section introduces proteomics with a special focus on mass spectrometry-based proteomics (Chapter XI), and different Grid-based proteomics applications, ranging from biomarker discovery to protein identification (Chapter XII) and protein classification (Chapter XIII), are discussed.

Section III discusses Grid-based software environments for bioinformatics. Grid-based Problem Solving Environments specifically devoted for bioinformatics (Chapter XIV), as well as Grid-based implementations of relevant bioinformatics applications, such as docking (Chapter XV), bio-molecular simulation (Chapter XVI) and molecule structure determinations (Chapter XVII), are presented.

Section IV focuses on Grid-based medical informatics applications. Since a core aspect of medical applications regards the acquisition and analysis of biomedical images, the Section first introduces different aspects of biomedical images visualization on the Grid (Chapter XVIII). Then, as significant applications in the field, Grid analysis of radiological data (Chapter XIX) and 3D electron microscopy reconstruction (Chapter XX) are discussed. Finally, as an example of biomedical instrumentation made available on the Grid, a hybrid mock circulatory system to test cardiovascular prostheses is presented (Chapter XXI).

Considering the research at individual and population level, the Grid is used both as a technology to enable compute intensive applications such as simulation of an organ or organism, and also as a mean to enhance collaboration among individuals and to enable interoperability among applications and databases. As an example, personalized medicine will be based on the integration of genotype, phenotype and environment information (e.g. clinical, genomic, proteomic, and drug toxicity data).

Healthcare applications at the population level as well as applications used in the clinical context are the focus of Section V. These applications are the most difficult to port on the Grid, due to the high number of patients involved, the partitioning of data among different health centers, the privacy and security problems and the need of collaboration among various operators. The use of the Grid as a support for collaboration in epidemiology (Chapter XXII), as well as the integration and analysis of epidemiology data (Chapter XXIII) are fully described. Two important HealthGrids for paediatrics (Chapter XXIV) and brain injury (XXV) are presented showing how the Grid is a mature technology also for the clinical setting.

Section VI discusses concepts and properties of Virtual Laboratories, an abstraction for cooperative data analysis and distributed collaboration among scientists, and their applications in Life Sciences. After describing the foundations of modern virtual laboratories (Chapter XXVI), such as formalisms for representing domain knowledge, data integration, semantic annotations and shared vocabularies, the Section describes some emergent virtual laboratories that focus on distributed collaboration and use of provenance data (Chapter XXVII), transparent use of the Grid (Chapter XXVIII) and support for specific domains like bioinformatics (Chapter XXIX).

Section VII describes main infrastructures, middleware, and tools for building and deploying HealthGrids. After introducing the main requirements posed on Grid middleware by biomedical applications, and how these are satisfied by a well known Grid middleware (Chapter XXX), the Section describes a Grid-based portal that aims to promote collaboration and cooperation among scientists and healthcare research groups on the Grid (Chapter XXXI). The management and deployment of Grid repositories for biomedical digital contents such as mammograms is also discussed (Chapter XXXII).

Finally, Section VIII is a short collection of suggested readings of different authors, aiming to enrich this book with others knowledge, experience, thought and insight. Chapter XXXIII introduces Cloud Computing, an emerging style of computing in which dynamically scalable resources are provided as a service, and describes the porting of applications to Grids and Clouds. Chapter XXXIV describes a bio-inspired approach for the construction of a self-organizing Grid information system. Chapter XXXV discusses Grid-based implementation of phylogenetic analysis.

In summary, the book records the main experiences and best practices on using Grids in bioinformatics, genomics, proteomics, medical informatics, and healthcare informatics, allowing a better understanding of benefits and added values in concrete biomedical and clinical applications.

Readers of the book can explore the power and flexibility of Grid solutions with respect to either the complexity level of life exploration or the level of cooperation among scientists.

In particular, the book reports the main methodologies currently used for exploiting Grid's features in Life Sciences, and it may be an important aid for engineers, computer scientists, and clinicians working on biomedicine and healthcare.

The book contains also a comprehensive description of tools, infrastructures and organizational issues currently available for planning and building HealthGrids and BioGrids. Thus, healthcare practitioners may find several interesting cases and best practices useful to plan and build HealthGrids in real situations, as well as arguments to evaluate the impact on organization.

Many open research issues remain in these fields and new challenges emerge continuously. I hope that this book will serve to introduce readers to the potential developments and benefits of Grid technology for Life Sciences and to the challenges posed by biomedical and healthcare applications to current Grid.

Mario Cannataro
University Magna Graecia of Catanzaro, Italy
January 2009

Acknowledgment

I would like to acknowledge the help of all involved in the preparation, collation and review process of the handbook, without whose support the project could not have been satisfactorily completed.

Deep appreciation and gratitude is due to the members of the Editorial Advisory Board, who shared with the editor the main decisions about chapter selection and handbook organization and who also contributed in the review process. Special thanks go to Tim Clark and Ian Foster who kindly wrote the Foreword to this handbook.

Most of the authors of chapters included in this handbook also served as referees for chapters written by other authors. Thanks go to all of them who provided constructive and comprehensive reviews. Moreover, special thanks go to the following additional reviewers: Eugenio Cesario, Marcelo Lobosco, Gregorio Mercurio and Renato Pajarola.

Thanks also to University of Catanzaro and ICAR-CNR, Italy, who provided the WWW resources for the handbook web site. Deep appreciation is due to Pierangelo Veltri and all the members of the Bioinformatics Laboratory, University of Catanzaro, Italy, for their work on bioinformatics and biomedical applications. Special thanks go to Valeria Farrelli, who helped me in the recruitment of prospective authors by organizing the invitation emails, to Pietro Hiram Guzzi who developed the handbook web site pages, and to Antonio Oliverio who solved some connectivity problems.

Special thanks also go to the publishing team at IGI Global, whose contributions throughout the whole process from inception of the initial idea to final publication have been invaluable. In particular to Julia Mosemann and Beth Ardner, who continuously prodded via e-mail for keeping the project on schedule, to Lindsay Johnston, who first invited me to this exciting project, and to Jessica Thompson, Kristin M. Klinger, and Jan Travers who helped me in starting this editorial activity.

Special thanks go to Domenico Talia, University of Calabria, Italy, and to Giovanni Cuda, University of Catanzaro, Italy, who introduced me to the exciting fields treated in this handbook, respectively Grids and Life Sciences.

In closing, I wish to thank all of the authors for their insights and excellent contributions to this handbook.

Mario Cannataro
University Magna Graecia of Catanzaro, Italy
January 2009

Chapter XIX
Grid Analysis of Radiological Data

Cécile Germain-Renaud
Université Paris-Sud, CNRS, France

Vincent Breton
IN2P3, CNRS, Clermont-Ferrand, France

Patrick Clarysse
CNRS; Inserm; INSA-Lyon; Université Lyon 1, France

Bertrand Delhay
CNRS; Inserm; INSA-Lyon; Université Lyon 1, France

Yann Gaudeau
Université Strasbourg; CNRS, France

Tristan Glatard
Université de Lyon; CREATIS-LRMN, France

Emmanuel Jeannot
INRIA; CNRS; Université Henri Poincaré, France

Yannick Legré
CNRS; Université Blaise Pascal, France

Johan Montagnat
CNRS; Université Nice Sophia-Antipolis, France

Jean Marie Moureaux
CNRS; Université Henri Poincaré, France

Angel Osorio
CNRS, France

Xavier Pennec
INRIA Sophia-Antipolis, France

Joël Schaerer
CNRS; Inserm; INSA-Lyon; Université Lyon 1, France

Romain Texier
CNRS; Université Nice Sophia-Antipolis, France

ABSTRACT

Grid technologies and infrastructures can contribute to harnessing the full power of computer-aided image analysis into clinical research and practice. Given the volume of data, the sensitivity of medical information, and the joint complexity of medical datasets and computations expected in clinical practice, the challenge is to fill the gap between the grid middleware and the requirements of clinical applications.

This chapter reports on the goals, achievements and lessons learned from the AGIR (Grid Analysis of Radiological Data) project. AGIR addresses this challenge through a combined approach. On one hand, leveraging the grid middleware through core grid medical services (data management, responsiveness, compression, and workflows) targets the requirements of medical data processing applications. On the other hand, grid-enabling a panel of applications ranging from algorithmic research to clinical use cases both exploits and drives the development of the services.

INTRODUCTION

Harnessing the full power of computer-aided image analysis into clinical research and practice remains an open issue. Given the amount of data produced by X-ray Computed Tomography (CT), Magnetic Resonance Imaging (MRI), or PET-scan, and the difficulty to interpret medical images, algorithms for medical image analysis, processing, and diagnostic assistance have been developed these last 15 years or so. Some of these algorithms have reached a high level of usability and proved to have a real impact in the clinical domain. However, their widespread adoption by clinicians is not realized yet. Two stringent examples, amongst many others, are radiotherapy, which could greatly benefit of exploiting advances in segmentation and registration algorithms, and intra-operative situations as well as intervention planning, which could exploit modern high-performance computing systems for augmented reality (Kikinis, 1998). In the 90's, G.A Moore has coined the term "crossing the chasm" (Moore, 1991) for the issue of selling high-tech products to mainstream customers. Crossing the long-lasting chasm between, on one hand the advances in computer science and engineering in the field of medical images analysis, and on the other hand clinical research and practice, is a challenge of the same nature.

This chapter reports on the goals, achievements and lessons learned from the AGIR (Grid Analysis of Radiological Data) project. AGIR is a multi-disciplinary collaboration funded by the French ministry of research under the ACI/ANR scheme, for the 2004-2007 period. AGIR gathers collaborators from eight laboratories, including computer scientists, middleware experts and physicians. The central tenet of AGIR is that grids can help crossing the chasm, despite their recent apparition, because they introduce a change in paradigm in the access to high-end resources. Grids provide seamless scaling of the development, validation, and exploitation cycles of analysis methods: the same infrastructure allows multiple user communities to access and manipulate medical data, and to explore individual images or create augmented reality situations through compute-intensive visualization methods; the same infrastructure provides the computing power needed to validate algorithms on large datasets and to process complete databases for the most demanding applications such as epidemiology.

Filling the gap between the clinical applications and the grid middleware raises many specific issues, ranging from computer science basic research to legal concerns, as will be exemplified in section 2. Addressing all these issues is an active research and technology area, and a new scientific community is emerging (Berry, 2003). In this framework, the specific objectives of AGIR are to define and validate:

- New grid services that address some of the requirements of complex medical image processing and data manipulation applications; these services are described individually in section 3.
- New medical image processing algorithms making advantage of the underlying grid

infrastructure for compute and data intensive needs; sections 4 to 6 report on these developments.

The method is to confront the expertise of medical and computer science teams, specialized in clinical applications, medical images analysis algorithms, grids and distributed systems, around a few paradigmatic medical applications, in order to get a cross-section of the middleware, algorithmic and medical issues. A specific application (section 7) was a "bottom-up" approach targeting the immediate development of a telemedicine platform, whose requirements and results were to be confronted with the more long-term activities in AGIR.

GRID-ENABLING MEDICAL IMAGE ANALYSIS

Challenges and Issues

Grid computing has been considered to tackle some of the requirements of the medical image analysis community (Montagnat, 2004; Aloiso, 2005). The scenarios investigated in AGIR (see section 4-6) led to the identification of common area-specific grid services:

- Medical databases management in order to handle large amounts of sensitive data. Federation of distributed data sources is needed e.g. for large scale testing involved in algorithms validation.
- Optimized and reliable data transfers even on low-end networks, e.g. to support humanitarian medicine.
- Efficient execution of analysis pipelines on a distributed grid infrastructure in order to match the computing requirements of the most demanding applications such as cardiac sequences analysis.
- Responsiveness needed by human guided procedures such as assisted segmentation and surgery planning.
- Production quality support to ensure sustainable grid usability.

Overview of AGIR

The deployment of medical applications on grids can be structured following four layers: 1) general purpose grid services; 2) the core services dedicated to medical applications that are not available in general purpose middleware; 3) the medical image processing algorithms taking advantage of the underlying grid services to process large amounts of data or to run expensive computations; 4) clinical applications developed to address medical challenges.

AGIR mainly addresses core medical services and image processing algorithms and, to some extent, clinical applications. Figure 1 summarizes the various AGIR activities.

e- Infrastructures

Grid concepts and research are around since the 90's at least (Foster, 2003) (Foster, 2001). With the maturation of technologies, and the deployment of production grids, however, the focus has now shifted towards strategic issues, amongst which forecasting and shaping roadmaps towards future e-infrastructures (Atkins, 2003). Following the analysis of the e-IRG (e-infrastructures reflection group, a body of EU member states representatives advising the European Commission), sustaining the needs of science requires resource integration at the European scale not as a fully integrated ("monolithic") grid, but at the level of an interoperable grid of grids (Leenars, 2005).

At the beginning of AGIR, these issues were not fully formulated. The popularity of the Globus middleware (Foster, 2006) pushed towards autonomous developments. However, we already

Figure 1. AGIR structure

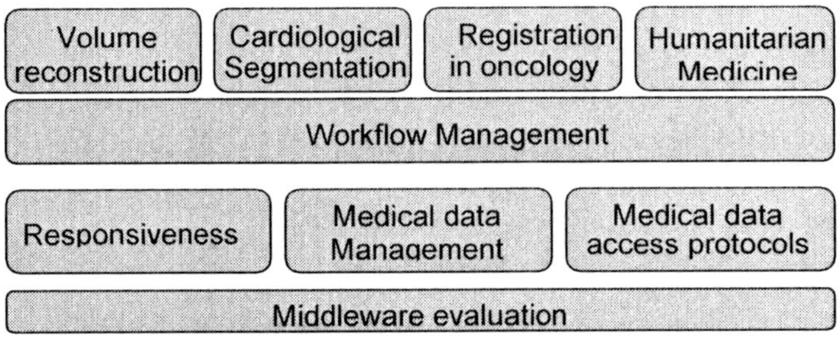

considered that it was critical for AGIR to refrain from developing middleware functionalities, which were already available, even if not perfectly suited for its needs. Throughout the project, we tried to keep a balance between neutrality with respect to middleware implementations on one hand, and strategic concerns on the other hand.

On the generic side, we conducted experiments and analysis in order to assess the compatibility of various middleware platforms, from research ones to production oriented ones. Amongst the numerous research suites, our experiments focused on DIET (The Distributed Interactive Engineering Toolbox) (Caron, 2006), which exemplifies remote procedure call (RPC)-based architectures. Another important component of our experiments is the French national grid, Grid'5000 (Bolze, 2006), which allows repeatable experiments in controlled settings.

On the strategic concerns side, AGIR has developed a close collaboration with the FP6 and FP7 infrastructure projects EGEE (Gagliardi, 2005), which is expected to move towards a permanent European Grid Infrastructure federating National Grid Infrastructures (NGI). EGEE is strongly involved in the various prospective bodies such as e-IRG, interoperability actions with other international grids (e.g. NorduGrid or Open Science Grid), and with the standardization efforts of the Open Grid Forum. EGEE is thus one of the key actors in the European e-infrastructures theater.

EGEE is a controlled but widely distributed environment over the Internet which usage for manipulating sensitive medical data imply the implementation of strong protection measures. However, the strong interaction with EGEE has a triple advantage. The first and most immediate one is the sheer availability of resources, showing that IT resources are readily available, which is a cornerstone for real-world usage scenarios. The second advantage is to interact with middleware developments, specifically gLite (Laure, 2004). On one hand, this process is considerably more difficult than it would be in an experimental grid. On the other hand, the impact of these requirements, when satisfied, benefits from the general impact of EGEE, and is sustainable.

An alternative might be to separate grids for health from other e-infrastructures. In the AGIR time frame, other large scale infrastructures targeting life sciences have been deployed world-wide, adopting this strategy. In particular, the BIRN (Biomedical Informatics Research Network) project (Peltier, 2003) has deployed a homogeneous hardware and software infrastructure all over the USA. Similarly on the EU side, Health-e-child[1] has deployed an EGEE-compatible but independent infrastructure. This has the advantage to minimize the legal and ethical issues as the nodes are owned and controlled by the hospitals themselves so that they comply to each site requirements. Moreover, the different sites are much more easily administrated by the consortium. We believe that this is a good way to expose the clinical community to grid technologies in order to facilitate in the future the acceptance and the use of more standard e-infrastructures. Conversely, the CaBIG[2] project has adopted a more open approach with the publication of different level compatibility recommendations for the different components, existing or to appear, of its grid infrastructure.

In both cases the logic of a production grid is the same: achieving a proper balance between, one hand quality and sustainability (24x7 availability, and compliance to standards), and on the other hand coping with scaling requirements, both in heterogeneity (workloads and use cases), and in geographical and institutional distribution.

CORE MEDICAL SERVICES

Grid Interface for Medical Data Management

Medical data management in clinics is a complex problem given the multiplicity of data sources, the sensitivity of the data, and the tremendously large data sets involved. As a consequence, medical information systems are often fragmented between different clinical sites where homogenization efforts are rarely invested. Even inside a single health enterprise, it is not rare to encounter several information systems partially overlapping and covering the different aspects of clinical information with more or less success.

Considering medical images, we have developed a complete solution, the *Medical Data Manager* (MDM). The MDM, interfaced to the EGEE middleware, bridges the DICOM storage used internally to clinical centres and the grid storage units (Montagnat, 2008). The MDM both ensures a transparent access to medical images by any grid service and does not interfere with the usual clinical workflow of images. Benefiting from the distributed storage capability provided by the grid middleware, it enables data sources federation: large image sets originating from different sites can be assembled for studies involving processing population of images such as the algorithm validation through statistical procedures as the Bronze standard (section 5).

Contexts

In health enterprises, the Digital Image and COmmunication in Medicine (DICOM) standard has become by far the widely available interface to access medical storage. Although a largely adopted standard, DICOM poorly matches the need for cross-health enterprise image exchanges. It was seminally developed for local network exchanges. It is known to provide both a low level of access control and a low efficiency in data transfers. Grids technologies can help on this issue: they not only provide federated data services but also reliable, secure and efficient protocols to transport data over the open Internet. However, an issue specific to medical data management is the fact that privacy preserving at the grid level must not hamper possible patient benefit. The challenge is thus to harness the DICOM access with the grid capacities regarding security, performance and storage resources.

DICOM is exploited by image device acquisition and post-processing vendors to ensure image communications in the clinical Picture Archiving and Communication Systems (PACS). DICOM defines both an image format and a transfer protocol which are completely integrated in standard clinical image workflows. DICOM stores structure medical images hierarchically in studies, series and individual image slices. On the other hand, grid technologies provide distributed file catalogs and remote file storage access protocols suited to build a file-based distributed data management system.

The Medical Data Manager

The MDM was designed to interface the EGEE grid infrastructure with state of the art DICOM compliant storage resources inside hospitals. It mostly targets the exploitation of medical image files as manipulated by image processing algorithms. It exploits a limited amount of associated metadata though to ease image retrieval in clinical use cases.

All medical data is considered as sensitive to preserve patient privacy. Nominative medical records are of course the most critical data and therefore, no binding between nominative data and images should be possible for non accredited users. On a grid, data distribution aggravates the security problems. To ensure patient privacy, the header of all DICOM images sent out of a DICOM server is removed, at least partially, to ensure patients' anonymity. All images stored out of the source center is encrypted to ensure that non accredited users cannot read the image content.

To fulfill both the needs of clinical practice and grid computing, the MDM:

- Provides access to medical data sources for computing without interfering with the clinical image workflow.
- Ensures transparency so that accessing medical data does not require any specific user intervention.
- Ensures a high data protection level to respect patients' privacy and data confidentiality constraints.

The MDM is thus compliant with DICOM clinical servers internally and interfaced to the EGEE grid data management system externally.

gLite Data Management Layer

The gLite middleware deployed on EGEE provides data management services for handling distributed files. Storage resources, known as *Storage Elements*, are distributed over the various sites participating in the infrastructure. All storage elements expose a common interface for interacting with the other middleware services: the *Storage Resource Manager* interface (SRM) that is standardized in the context of the Open Grid Forum (Sim, 2007). The SRM is handling local data at the file level. It does not implement data transfer by itself: additional services such as GridFTP (Allcock, 2002) or gLiteIO (EGEE JRA1 2005b) are coexisting within storage elements to provide transfer and data access capabilities. The gLite data management system uses a *File Catalog* providing a unique entry point for files distributed on all grid storage elements. Each file is uniquely identified through a *Global Unique IDentifier* (GUID). The file catalog contains an association between GUIDs and site-dependent Storage URLs (SURL) enabling distributed files location. The File Catalog and the Storage Element may implement different file access policies and it is one of the MDM challenge to ensure a coherent grid-wide access control through coordinated updates of the fine-grain Access Control Lists (ACLs) handled by the different services.

MDM Services

The MDM is a data management layer based on multiple grid services. It uses:

- A core MDM component exposing on SRM interface and querying the DICOM server.
- An ACL-based access right controller.
- A file catalog to index images available in the DICOM server.
- A secured file transfer service for sending images to requesting grid services.
- An access-controlled key store associating encryption keys to images.

In addition, the AMGA secured relational database interface is used for managing metadata associated to images (such as image metadata, patient metadata, radiology metadata, etc). AMGA (Santos, 2006) is a secured front-end to traditional databases back-ends (MySQL, Postgres, Oracle). It provides grid credential ACL-based access control and a secured communication layer. It relates medical information with grid file identifier (GUID).

MDM Security

The latest version of the MDM is based on the lightweight *Disk Pool Manager* (DPM) disk server which implements SRM version 2 (Perelmutov 2002). Access is controlled through the combined use of the *LFC* File Catalog and the backend SRM server. The gLite secured transport protocols (typically GridFTP) are used for data transfer. Encryption keys are stored in a *Hydra* distributed keystore exploiting the Shamir secret sharing scheme (Shamir,1979). Shamir's scheme consists in splitting keys into n fragments stored in different places. Only $m < n$ fragments are needed to reconstruct a complete key. Thus, the system security is improved as it is both resistant to attacks and provides redundancy. The standard symmetric key algorithm using AES cipher available in the OpenSSL distribution is used for encryption/decryption. All images are encrypted prior to communication thus ensuring on-transmission security and on-disk security once the data file is stored on a remote site. Only users with access right to the file key can access the data content. In addition, DICOM headers which contain nominative information are anonymized prior to transfer.

The MDM provides a much higher level of data security than the standard gLite Data Management System. It required the addition of specific components such as the coordinated access control and the Hydra key store for encryption.

MDM Service Set Up and Exploitation

To fulfill its role, the MDM service needs to be notified when files are produced by the imagers and stored into the DICOM server. The DICOM data triggering the operation is first stored into the hospital DICOM server as usual. The DICOM header is then analyzed to extract image identifying information. This DICOM ID is used to build a file identifier that is registered into the grid File Catalog. The other metadata extracted from the DICOM header is stored into the AMGA metadata server. Finally, an encryption key that is associated to the file and that will be used for data retrieval is stored into the Hydra key store.

Once DICOM data sets have been registered into the MDM, the server is able to deliver requested data to the grid. A client library is used for this purpose. To cover all application use cases, the MDM client library provides APIs for requesting files based on their grid file identifier or on the metadata attached to the file. In many cases, complete image volumes are needed rather than individual image slices. A grid file identifier is associated to each volume registered and the image volume can be reconstructed on the fly by the SRM-DICOM core on request. All image files are anonymized and encrypted before being sent out to ensure that no sensitive information is ever transferred or stored on the grid in a readable format.

Discussion

The MDM service is a gLite integrated DICOM-compliant storage resources manager. It provides transparent access to the clinical storage from other gLite services and it ensures a high security level. A similar service has been recently developed on the GLOBUS middleware stack. The MEDICUS incubator project (Erberich 2007a) is proposing a DICOM interface to the GLOBUS data manager. MEDICUS is based an OGSA-DAI compliant Meta Catalog Service (MCS) and a DICOM Grid Interface Service (DGIS). As compared to the MDM, MEDICUS does not directly index DICOM storage images into the file catalog. Instead, it explicitly exports images selected for grid access into the DGIS server deployed in demilitarized zone. The data is therefore replicated on a specific grid storage resource before being accessible. MEDICUS does not provide file encryption facility either. It relies on the GridFTP transport layer to ensure file security upon transfer. It does not provide on-disk file encryption though. MEDICUS provides an elaborated file access control mechanism through the use of Shibboleth (Erberich 2007b).

Responsiveness

Responsiveness is the support for on-demand access and interactive usage (Basu, 2003).The choice of targeting production grids makes responsiveness most challenging. Interaction and on-demand features must be built on top of the traditional grid scheduling tools, which are batch-oriented and dominated by fair-share policies at institutional time-scales. The associated constraints might be summarized as non-intrusiveness: delays incurred by non-interactive jobs must be bounded, resource utilization should not be degraded (e.g. by idling processors), and finally, the local policies governing resource sharing (Virtual Organizations, advance reservation, etc.) should not be impacted. Another related issue is the middleware latency: even if all the necessary resources are available, submitting, scheduling and mapping of jobs on EGEE remains in the order of minutes. Thus, two orthogonal issues must be addressed.

- Differentiated Quality of Service (QoS) at the workload management level, where batch jobs are serviced as best effort and responsive applications are guaranteed to be serviced without queuing.
- Coping with the middleware penalty, which requires bypassing the middleware stack.

Differentiated QoS

Intra-operative situations require advance reservation. For more routine clinical or research grid usage, advance reservation suffers from two drawbacks: first, planning is not consistent with the goal of seamless integration with everyday computing practice, and specifically medical usage; second, reservation is inherently not work-conserving, meaning that processors might idle while eligible jobs are queued (Snell, 2000), thus jeopardizing utilization. Differentiated QoS is the generic answer, and has been extensively explored in both the network and operating systems (OS) frameworks. Differentiated QoS usually relies on some implementation of Generalized Processor Sharing (GPS)(Baruah, 1996), where resources may be allocated by quanta. In grids, such quanta do not exist: except for the rare case of checkpointable jobs, a job that has started running cannot be suspended and restarted later; moreover, the execution times exhibit high variance (Germain, 2006).

We have defined and implemented the concept of *Virtual Reservation* (VRes) (Germain, 2006), which addresses both issues of advance reservation and scheduling quanta by allowing controlled time-sharing. VRes permits the definition of time quanta and their exposure at the grid level. At the site level, each of the p physical processors is virtualized into k virtual processors, providing pk slots to the site scheduler. When a virtual slot is unused, the computing bandwidth is transparently returned to the other classes sharing the same physical processor, by the virtue of time-sharing. Thus, a fraction of these slots can then be permanently reserved for the interactive applications without jeopardizing utilization, One of the advantages of this scheme, is that classes (batch and interactive) are isolated: resource allocation becomes internal to the classes. True QoS has the unavoidable counterpart of access control: the guaranteed bandwidth being limited by the number of available slots, a job requiring QoS may be rejected if the available resource is exhausted.

An implementation of VRes has been developed for the MAUI scheduler and the gLite middleware (EGEE SDJ WG, 2006). The Job Description language (JDL) has been modified to include a Boolean attribute SDJ (Short Deadline Job), denoting that the job must be executed immediately, or rejected. These sites also configure their scheduler with parameters controlling the computational bandwidth dedicated to SDJ. VRes has been put in production since May 2006. and is routinely used by several biomedical applications.

User-Level Scheduling

Submitting, scheduling and mapping of jobs on a grid take at least one order of magnitude more time than the execution time for SDJ even in absence of competition for resources. User-level scheduling is the most promising way to address this issue. User-level scheduling is a virtualization layer: instead of being executed directly, the application is executed via an overlay scheduling layer, which runs as a set of regular user jobs and therefore bypasses the most costly middleware services. User-level is here analogous to the user-level thread schedulers versus kernel-level ones in processor scheduling. Just as a user-level thread scheduler avoids the penalties of the kernel calls, a Grid application-level scheduler avoids the traversal of the Grid information systems for executing each and every small parallel task.

AGIR has embedded a user-level scheduler inside the gPTM3D application, corresponding to the interactive volume reconstruction use case. User-level scheduling can instead be factorized across applications through generic frameworks, also called overlay grids, such as AppLeS (Berman, 2003), DIRAC (Tsaregorodtsev, 2004), DIANE, or the more complete solution DIET presented above. While DIRAC is not integrated with gLite, which precluded its usage in AGIR, DIANE is. It would have thus be useful to capitalize over the DIANE generic framework. However, synthetic experiments (Texier, 2007) have shown that the latency of DIANE remains too high for the volume reconstruction use case. Finally, we have been able to experiment the DIET scheduler. A detailed experiment (Texier, 2007) run on the Grid 5000 resource has proved that DIET provides low latency for fine-grained tasks.

Discussion

An alternative approaches to grid scheduling for QoS is virtualization; the Virtuoso project has more specifically explored virtualization for differentiated services (Lin, 2005). Virtual machine can indeed be deployed in stable grid environments, e.g. the Virtual Workspaces project (Keahey, 2006) with Globus. Virtual machines provide complete freedom of scheduling and even migrating an entire OS and associated computations which considerably eases time-sharing between deadline-bound short jobs and long running batch jobs. On the other hand, the virtual machines strategy requires that all of, or a significant fraction of, the computations must be run inside virtual machines to provide scheduling opportunities, something for which traditional batch users have little incentive. As the role of interactive users is still modest in the grid community, the smoother path proposed by VRes represents a more acceptable tradeoff.

Compression

Transmission as well as storage of the tremendous amount of data involved in medical applications can easily saturate existing systems. Along the interaction of AGIR with clinicians, it appeared that a prominent requirement for any networked access to images is low response time. In the long-term vision of grids as the infrastructure for perennial conservation and pervasive access to medical data, compression is thus an unavoidable component of medical data access protocols.

In the following we give details about the two components of enhanced medical data access protocols we have explored: lossy compression algorithms, which are required for efficiency, but yet questioned by the medical community, and adaptive coupling of compression and transfer.

Compression Algorithms

The tolerance to lossy compression (defined as the maximal threshold of compression for which the decompressed image is still acceptable for interpretation) is related to the type of data. Digitized chest radiographs are very tolerant to compression (Erickson, 2002), at least 40:1 for 2D SPIHT (Cho, 2004), digitized bone films are moderately tolerant (between 20:1 and 40:1), and CT or MR images exhibit fairly low tolerance to compression (less than 20:1) (Erickson, 2002). In the following, we consider 3D images, or more exactly stacks of 2D images, like MRIs or CTs.

The basic idea of 3D medical image coding algorithms is to take advantage of the correlation among the data samples in the three dimensional space to improve compression performance. Most of the approaches combine a three-dimensional space decorrelating transform with the extension of a coding algorithm that has proven to be effective on 2D images. The reference methods are based on 3D Wavelet Transform (Bilgin, 2000).

Two strategies are available to encode wavelet subbands supplied by the 3D WT: the interband and the intraband approaches respectively. While the interband schemes, exemplified by 3D SPIHT (Cho, 2004), are generally more flexible and offer a lower computational cost than intraband schemes, they are sensitive to errors occurring during data transmission.

On the contrary, the intraband schemes encode the whole subbands separately and need a bit allocation procedure to share the binary resources among the subbands in order to minimize the overall distortion. Recently, we have proposed a new multidimensional intraband algorithm based on lattice vector quantization associated with a vector dead zone: 3D DZLVQ (Gaudeau, 2007). The corresponding codebook structure takes into account both the sparsity and the clustering properties which characterize the wavelet coefficients. Thanks to the proposed dead zone, this scheme allows to threshold non significant source vectors according to a criterion based on their norm, while putting more bits on significant ones. It can efficiently apply to medical images in the 3D wavelet domain which often contain huge non-significant areas.

Table 1 shows compression performance in terms of Peak Signal-to-Noise Ratio (PSNR) of DZLVQ compared to 3D SPIHT, as well as 3D ESCOT (Xiong, 2003), one of the most efficient intraband algorithms of the literature. The selected 256x256x192 8-bit CT image "Skull" of this experiment belongs to an abundantly used volumetric medical image data set given in (Bilgin, 2000). Furthermore, a four level 3D WT has been performed, using the well-known 9.7 floating-point filter. As we can see in the table, 3D DZLVQ slightly outperforms 3D SPIHT and 3D ESCOT. Note that the same tendency has been observed for the other images of the data set and for DICOM 12-bit ENT-CTs (Gaudeau, 2007).

PSNR, as a global criterion, is not sufficient to estimate the quality of reconstructed images, especially in medical applications. Figure 2 shows that 3D DZLVQ allows to better preserve fine structures and produces a better overall visual quality of the reconstructed image (Gaudeau, 2007).

Table 1. PSNR (dB) comparison on slice 128 of Skull image compressed by 3D SPIHT, 3D DZLVQ and 3D ESCOT at two bit rates: 0.5 bit/voxel and 0.1 bit/voxel

	3D SPIHT	3D DZLVQ	3D ESCOT
0.5 bpv	44.11 dB	**44.14 dB**	43,82 dB
0.1bpv	33.98 dB	**34.69 dB**	34,68 dB

Figure 2. Zoom of decoded slice 60 (Skull) when the target decoding bit rate is 0.1 bpv for the whole 128-slice unit. Clock-wise from the top to the bottom left: Original slice, 3D SPIHT, 3D DZLVQ.

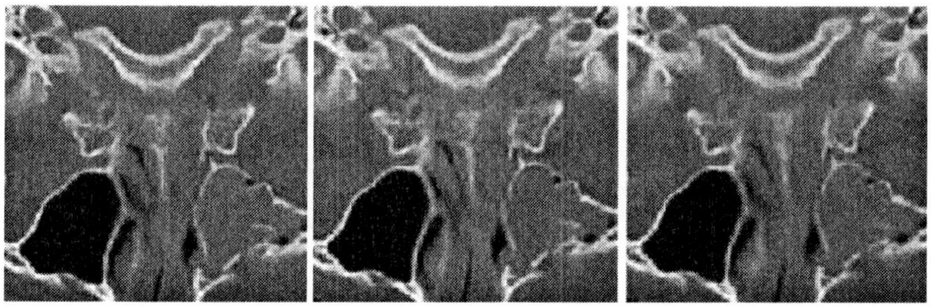

Adaptive On-Line Compression of Medical Images

While local area networks (e.g. hospital intranets) offer the bandwidth required for quasi instantaneous transfer of 3D images, widening the scope of access may involve poor (e.g. roaming user) or congested (e.g. wide area networks) infrastructures, resulting in low bandwidth. In a previous work, we developed AdOC (Adaptive Online Compression) (Jeannot, 2005), which enables on-the-fly compression, thus reducing the required bandwidth.

In AGIR, we have combined the principles of AdOC and wavelet transform based lossy compression. In this framework, each wavelet coefficient of a generated subband is approximated by quantization and then encoded. The adaptive techniques of AdOC are exploited to change the coding effort depending of the environment. If the network is very slow, more aggressive coding algorithms are used to compress the coefficients. Therefore, we are able to detect the best tradeoffs between spending some time for compressing the data and sending them to the network as fast as possible. Furthermore, in order to speed-up the whole process of compression and decompression we have pipelined each step (wavelet transform, quantization and coding); on a multi-core architecture, which is now the standard processor, these different steps are overlapped and data start to be sent to the network only after a minimum processing of the image.

We have enhanced the proposed framework to the case of DICOM 3D images. Each image of the 3D stack is processed separately in a pipelined manner. We have found that doing this way is much faster than performing 3D wavelet transform and enables real-time transfer.

We have tested this implementation on several images and several kinds of networks. Preliminary results show that the proposed approach is very promising for wide-area networks such as Internet both in terms of transfer speed (it can be divided by 3 on transatlantic connection compared to FTP transfer) and quality the images (they do not suffer visible degradation). Moreover, the progressive transmission of images allows the clinician to discard those of less interest without a complete upload.

Image Pipelines and Parallel Processing

A pipeline, better known as workflow in the grid community, enables the definition of processing chains composed of different image processing units (e.g., image filtering, intensity normalization, registration, segmentation, parameters quantification, interpretation...) and dependencies between these units. The dependencies constrain the order of execution of the pipeline's units. In the context of grid computing, these pipelines are also a convenient and user friendly way to decompose a complex application into more or less independent computing units which execution can be distributed over different computing resources.

In the context of AGIR we adopted a data-flow approach to describe workflow constraints that matches the traditional notion of image analysis and visualization pipelines well known from the medical image analysis community. In a data-flow, constraints between processing units are expressed through data links describing how the output of on processing unit are transferred to the input of a subsequent processing unit. Those data transfers enforce a partial order of execution for all units. Another critical requirement for the AGIR workflow manager was the ability to embed image processing units from various origins into the same application. We adopted a flexible service-oriented architecture to match this requirement. Business codes are embedded into a Web-Service (WS) wrapper and the AGIR workflow manager invokes the WS composing the pipeline in turn.

Workflow Representation

To describe AGIR workflows, we adopted Scufl (Oinn 2004), a data-flow language that combines a rich semantics a compact and user friendly language. Scufl features data composition operators,

Figure 3. Action of the one-to-one (left) and all-to-all (right) operators on the input data sets

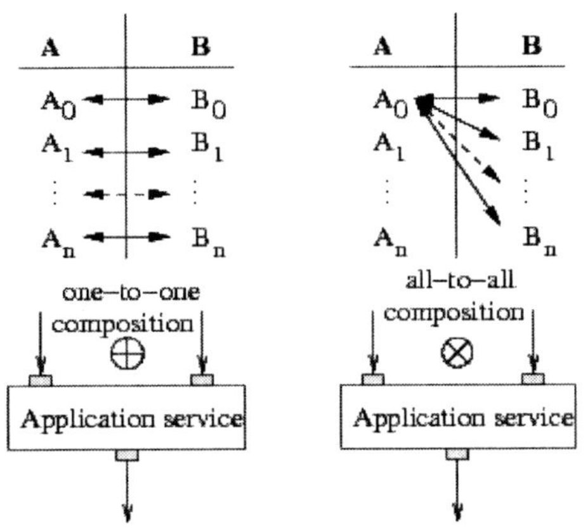

also called iteration strategies. Two main composition patterns are *one-to-one* composition pattern (left of Figure 3), where two input data sets are processed pairwise in their order of arrival and the *all-to-all* composition pattern (right of Figure 3), where all inputs in one data set are processed with all inputs in the other one. A common example is the case where all images in the first input set are to be processed with all parameter configurations defined in the second input set.

Enacting Workflows

The Scufl workflow enactor is in charge of invoking the various services composing a workflow and ordering the transfers between consecutive services. One limitation of Scufl, bound to the nature of WS, is that each invocation targets a specific service endpoint hosted on a particular host. If the computation is data intensive, e.g. if a single service has to be invoked several times for processing different data segments, the basic WS invocation procedure overloads the host endpoint. This limitation made the execution of the data intensive medical image analysis applications far too inefficient. We developed a solution based on two components:

- GASW: a grid compliant generic web service wrapper (Glatard, 2008a).
- The MOTEUR parallel enactor for Scufl workflows (Glatard, 2008b).

GASW, a generic application code wrapper compliant with the Web-Services specification enables the execution of a legacy executable through a standard service interface. This service is generic in the sense that it is unique and does not depend on the executable code. It exposes a standard interface that can be used by any Web-Service compliant client to invoke the execution. It completely hides the grid infrastructure from the end user as it takes care of the interaction with the grid middleware. To accommodate to any executable, the generic service is taking two different inputs: a descriptor of the legacy executable command line format, and the input parameters and data of this executable. The production of the legacy code descriptor is the only extra work required from the application developer. It is a simple XML file which describes the legacy executable location, command line parameters, input and output data (including access protocols).

MOTEUR is an efficient grid execution of service workflows exploiting the parallelism levels that are explicitly or implicitly expressed in a Scufl workflow. To enable parallelism during the workflow execution, multiple application services have to be invoked concurrently. The calls made from the workflow enactor to these services need to be non-blocking in order to exploit the potential parallelism. Web Services theoretically enable asynchronous calls. However, the vast majority of existing web service implementations does not cover the whole standard and none of the major implementations provides any asynchronous service calls for now. Alternatively, MOTEUR implements asynchronous calls to Web Services at the workflow enactor level, by spawning independent system threads for each service being executed. With asynchronous calls, three levels of parallelism can be exploited:

- **Workflow parallelism** is the intrinsic parallelism expressed through the workflow graph topology. If two services lay in two independent branches of the workflow graph (i.e., there are no data links imposing an execution order constraint between these services nor their ancestors), then they can be enacted concurrently.
- **Data parallelism.** The processing of several input data segments is needed in many applications. Services can be instantiated as several computing tasks running on different hardware resources and processing different input data segments in parallel. Data parallelism is the major source of parallel execution gain in medical image analysis applications.
- **Service parallelism** corresponds to the concurrent execution of two independent data segments by two different services that are sequentially linked. This pipelining model, very successfully exploited inside CPUs, can be adapted to sequential parts of service workflows.

Fully parallel workflow execution, handling the data composition operators described above is not straightforward as the order of data items may be completely disturbed by the concurrent execution of different jobs and a dedicated algorithm has to be designed in order to keep track of their provenance during the execution.

MOTEUR (Glatard, 2007) is interfaced to the EGEE grid and can exploit the production infrastructure resources for running applications. The Scufl language and the MOTEUR workflow manager have been used to describe and efficiently enact various medical image analysis applications on EGEE, in particular, the rigid registration algorithms evaluation application described in section 5 (Figure 6 shows the workflow).

INTERACTIVE VOLUME RECONSTRUCTION

This section exemplifies the case of grid-enabling the compute-intensive tasks involved in clinically oriented augmented reality applications.

Motivation

The original software, PTM3D, is a fully featured DICOM images analyzer developed at LIMSI (Figure 4).

The system currently runs on standard PC computers and it is used on line in radiology centres (Osorio, 2002). The most compute-intensive task in PTM3D is the computer-aided generation of

Figure 4. PTM3D snapshots

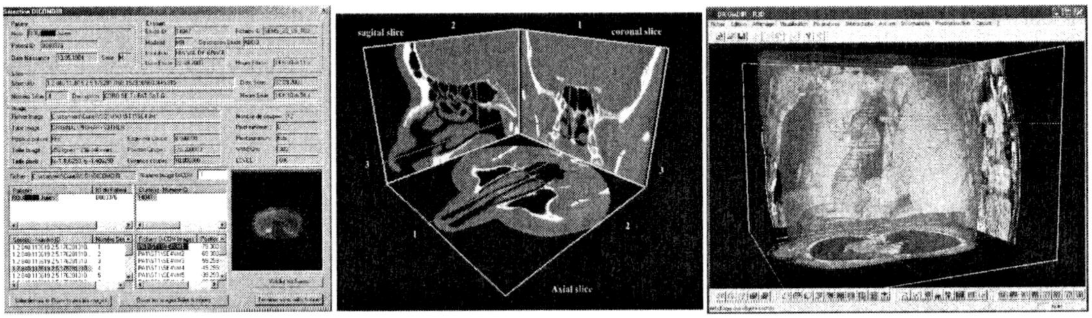

(a) DICOM interface *(b) Navigation* *(c) 3D reconstruction*

three-dimensional representations from CT, MRI, PET-scan, or echography 3D data.

3D-reconstructions targeting surgery planning and guided surgery through augmented reality can exceed the computing power of a workstation. An example is the augmented reality system developed with PTM3D for percutaneous nephrolithotomy (Figure 5) : renal stones, kidneys and external skin of the body are reconstructed from CT images; the real-time projection of organs and lesions on the patient in the surgical position leads the surgeon towards the target region and choosing the best route for percutaneous surgery (Traxer, 2005).

This kind of application does not require a grid as absolutely as e.g. analysis of distributed databases of medical images, because data are not initially distributed, and the computation is a classical parallel one. The motivation for seeking for grid resources is thus not different from many other medical or non-medical fields: to propose a high-end services that can be accessed freely by any authorized organization, and which provides the appropriate responsiveness without the burden of maintaining the resource at the individual organization level.

Requirements

The volume reconstruction algorithm includes a semi-automatic segmentation component based on an active contours method where the user initiates the segmentation, and can correct it anytime. It also includes a tessellation component, which is the compute-intensive part of the algorithm, and requires fine-grained parallelism.: the execution time of the majority of the tasks is in the order of a few hundreds of milliseconds but with high variability: when the geometry of the volume becomes complex, the reconstruction of the critical slices can last for 20 seconds or more.

Figure 5. Augmented reality in surgical planning for percutaneous nephrolithotomy

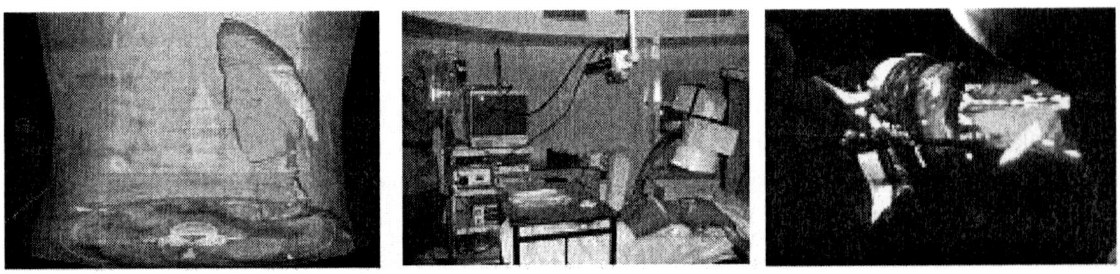

The requirements are quite new with respect to the grid most frequent use cases. The challenge is to smoothly integrate local and remote computing. The radiologist wants a unique tool, to browse the images, to measure a tumor, to reconstruct a complex organ or to create a virtual reality situation. The grid must be only an accelerator of the tool.

Method

A data-driven application level scheduler has allowed fully interactive remote execution of the volume reconstruction module on the EGEE grid. This development shows that complete grid accountability and compliance with the gLite middleware is compatible with keeping sophisticated and medically-oriented interaction schemes unmodified. As we explained in the section 3.2, the generic user-level scheduler DIET offers satisfactory performance. However, the most difficult challenge was to seamlessly integrate two heavily protected worlds: hospitals on one hand, and the grid, with minimal intrusiveness on both sides. This motivated a specialized architecture.

The architecture has two components: scheduler/worker agents at the user-level and the Interaction Bridge (IB) as an external service. The IB acts as a proxy between the PTM3D workstation, which is not gLite-enabled and the EGEE world. When opening an interactive session, the PTM3D workstation connects to the IB. In turn, the IB launches a scheduler and a set of workers on an EGEE site, through fully standard requests to an EGEE User Interface. A stream is established between the scheduler and the PTM3D front-end through the IB. When the actual volume reconstruction is required, the scheduler receives contours. The scheduler/worker agents follow a pull model with each worker computing one slice of the reconstructed volume at a time, and sending it back to the scheduler, which forwards them to IB from where they finally reach the front-end.

Results

The main result is that decoupling the local interaction and the grid computation gives an overall response time fully compatible with user requirements (less than 2 minutes), while the sequential time on a 3GHz PC with 2GB of memory can reach 20 minutes and more than 30 minutes on less powerful front-ends. The IB allows tunneling a TCP flow; as a consequence, the local interaction scheme, which includes stopping, restarting and improving locally the segmentation, is kept strictly unmodified: when a contour is manually reshaped on the front-end, the new contours will be processed on the grid.

EVALUATING THE IMPACT OF COMPRESSION ON RIGID REGISTRATION ALGORITHMS

Lossy compression of medical images is a disputed point as the impact of compression algorithms on the use of images has to be carefully evaluated. In this section, we are interested in evaluating the impact of lossy compression on the performance of rigid registration algorithms (RAs). To realize this evaluation, we first determined the reference registrations from the uncompressed images using a statistical procedure called the *Bronze standard*. Then, we evaluated to what extent the compression makes the registration results deviate from this reference.

The Bronze Standard Method

For the sake of analysis, one may consider registration algorithms as black boxes that take images as input and output a transformation. Considering the input data and the output transformation as random variables, the variability of the transformation (standard deviation or covariance matrix) can be propagated to some target points. Depending on the distance to the "true" transformation, one usually distinguishes gross errors (convergence to wrong local minima) and small errors around the exact transformation. The *robustness* quantifies the size of the basin of attraction of the right solution or the probability of false positives. The *accuracy* measures the error with respect to the truth (which may be unknown) for small errors. The *repeatability* or *precision* only measures the deviation from the average value, i.e. it does not take into account *systematic biases*, which are often hidden.

One of the simplest evaluation schemes is to simulate noisy data and to measure how far is the registration result from the true one (the ground truth is obviously known). The main drawback of synthetic data is that it is very difficult to identify and model faithfully all the sources of variability, and especially unexpected events (pathologies, artifacts, etc). Forgetting one single source of error (e.g. camera calibration errors in 2D-3D registration) automatically leads to underestimate the final transformation variability. Moreover, the faithful simulation of images is computationally extremely demanding. The second evaluation level is to use real data in a controlled environment, for instance imaging a physical phantom. However, one need to measure precisely the motion or deformation of the phantom with an external apparatus to obtain a gold standard registration. Moreover, it is difficult to test all the clinical conditions (e.g. different types or localizations of pathologies). Last but not least, it is often argued that these phantoms are not representative of real in vivo biological system.

Thus, we are left with the last level of evaluation methods which relies on a database of in-vivo real images representative of the clinical application. Such a database should be large enough to span all sources of variability. In such a database, images are very realistic but establishing a gold standard registration is barely feasible. To still evaluate the performances of the registration algorithms without any gold standard, a first idea is to compose transformations that form a closed circuit and to measure the difference of the composition from the identity. This criterion does not require any ground truth, but it only measures the repeatability as any bias will get once again unnoticed.

In order to robustify the results, the idea of the Bronze standard method is to built a statistical estimation of the ground truth by exploiting a large number of longitudinal image sequences and many different registration algorithms. The feasibility proof of the bronze standard technique was first established in (Roche, 2001), but a proper deployment on a large database was only realized in (Glatard, 2006) using the grid technologies presented in this paper. The principle is similar the STAPLE algorithm proposed for the validation of segmentation (Warfield 04): we consider the exact registration as a hidden variable that we have to estimate concurrently with the variability of the registration results (e.g. the variance or mean square distance) with respect to these references. This is realized using an Expected Minimization like algorithm (Roche, 2001).

Application to the Evaluation of the Impact of the Compression

The clinical problem context is the follow-up of brain radiotherapy. Monitoring tumor evolution requires a rigid registration of longitudinal images of the patient. The accuracy of this registration is directly related to the estimation of the tumor evolution. Experiments used a dedicated database of 65 images corresponding to 25 patients for which gadolinium injected MRIs have been acquired at several time points to monitor the tumors growth. A total of 126 registrations are required to relate all pairs images of the same patient. We used four different algorithms: *Baladin* (Ourselin 2000) and *Yasmina* (Roche, 2001) are intensity-based while *CrestLines* and *PFRegister* (Pennec, 2000) are feature-based (extracting crest lines). The application workflow is displayed in Figure 6.

Figure 6. The Bronze standard workflow

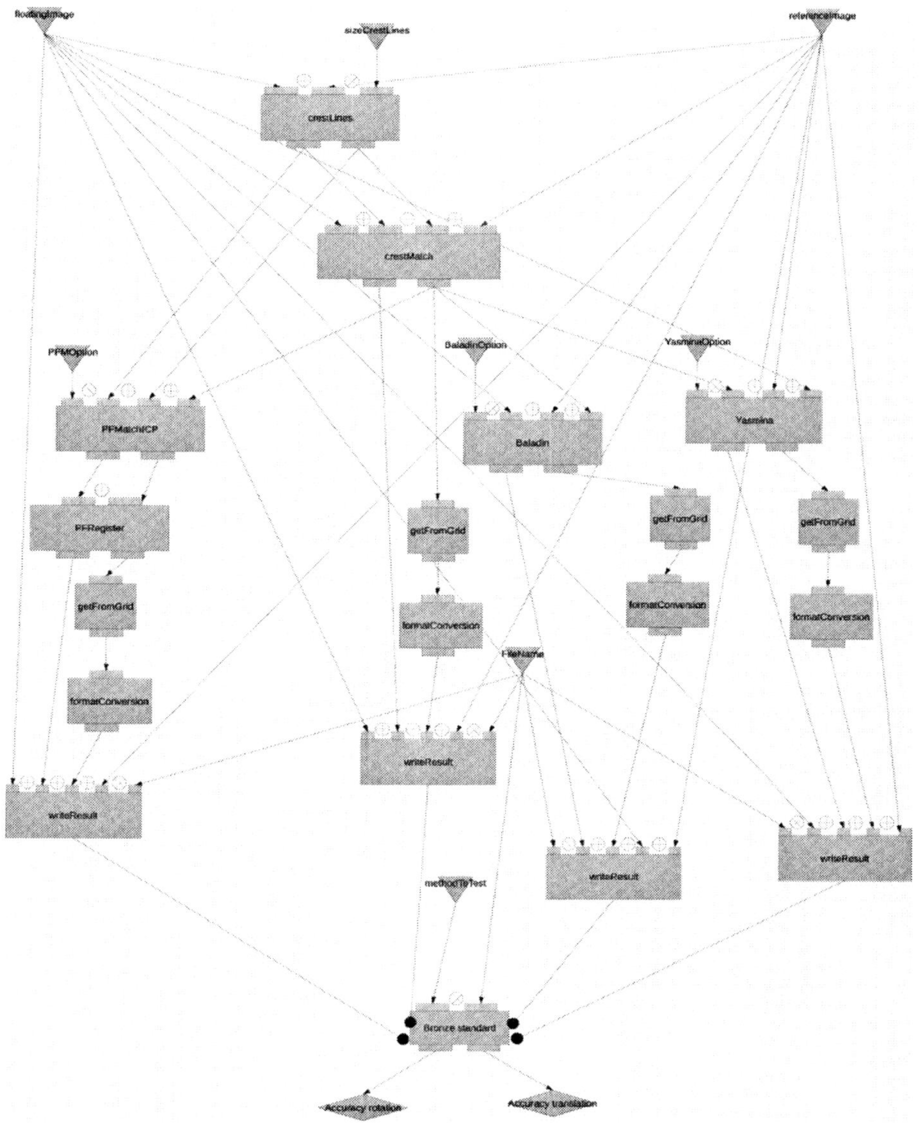

This database has been compressed at Compression Ratio (CR) = 6,12,24,48 and 64, with the 3D-SPIHT algorithm. A total of 3024 transformations (126 × 4 algorithms × 6 CRs) have to be computed. Given the four registration algorithms, the total sequential execution time of this experiment is about 7.5 days. Through MOTEUR, the total duration of the experiment reduces to 18 hours on the shared EGEE production grid.

The number of outlier transformations gives an estimation of the robustness of the algorithms with respect to the compression. For each CR, outliers have to be identified by comparison to the transformations obtained from uncompressed images. Among the rejected transformations, a visual inspection is performed to determine whether they correspond to wrong local minima (when it is obvious that a manual registration can lead to a better result) or not. When a transformation is found to be in a wrong local minimum, the whole patient is removed. Otherwise, the absence of a specific algorithm for a given patient could bias the quantification of the accuracy of the remaining ones.

For each CR, the repeatability is measured by the variances of the transformations obtained from the compressed images only, after having removed the patients leading to a wrong local minimum in one of the CR.

The transformations obtained from the uncompressed images are considered as the reference for the evaluation. The accuracy of each algorithm is computed by measuring the mean distance of compressed transformations to the uncompressed reference. To avoid biases, the evaluated RA is excluded from the algorithms used to build the uncompressed reference.

Among the 25 patients of the database, 6 were removed as outliers. For these, the rigidity assumption was hardly valid, for instance because of high deformations in tumor areas. The reference was built from the uncompressed images of the remaining 19 patients. The ratio of outlier in those 19 patients is plotted on left of figure 7 for each algorithm. *Baladin* and *Yasmina* are very robust. It may be a consequence of their multi-scale strategy (both use a pyramidal decomposition). For those two algorithms, it seems that the robustness even improves with the CR. For *CrestLines* and *PFRegister*, the robustness degrades significantly for a CR higher of 64. This was expected since the image derivatives used to extract features are highly impacted by the compression.

Figure 7 right displays the uncertainty of the algorithms measured by the standard deviation of the rotation and the translation components of the transformation. As it measures the distance to the uncompressed case, the smaller the better. The uncertainty of feature-based methods is highly augmented at CR=64. *Yasmina* is quite insensitive to the compression: its mean error only increases by 10% for the rotation and by 5.5% for the translation. More surprisingly, after a brief rise until CR 24, the accuracy of *Baladin* is improving: at CR=64, it is 34% better than without compression for the rotation and 18.5% better for the translation. This behavior may be due to the multi-scale handling compensating the effects of potential noise introduced in the images.

In summary, results tend to show that the impact of 3D-SPIHT compression on robustness, repeatability and accuracy is quite negligible until a significant CR (48). Beyond this threshold, feature-based methods based on crest-lines are highly penalized, while intensity based algorithm do not seem to be deeply impacted, probably

Figure 7. Left: Ratio of outlier patients with respect to the CR. One can see that the robustness does not significantly degrades until a CR of 64. Right: uncertainty of the algorithms measured by the standard deviation of the rotation (rot in degrees) and the translation (tr in millimeters) of each algorithm with respect to the uncompressed bronze standard.

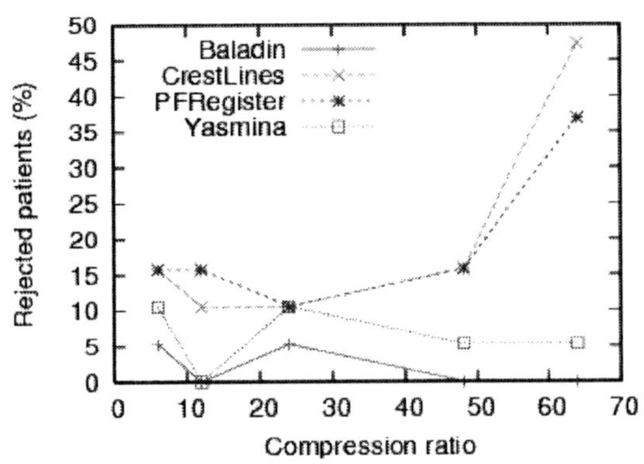

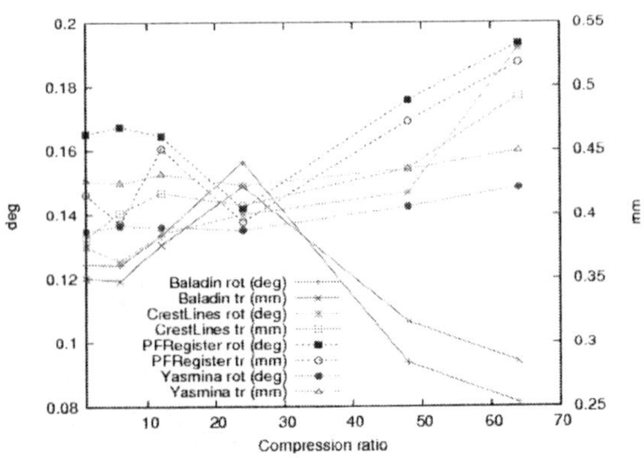

thanks to a good multi-scale handling. Results on *Baladin* are interesting as they tend to show that performances can even improve.

Drawing general conclusions about the impact of lossy compression methods on registration algorithm would still require confirming these trends with other registration problems and compression methods. Lossy compression in medical applications should be evaluated by experienced radiologists over a large data set. The fact that this kind of study is costly to set up, and time-consuming for the experts, explains the scarcity of the literature on the actual 3D lossy coders that uses a medical opinion. With the application of the Bronze standard method to evaluation of lossy compression, we propose an intermediate step, which automatically builds an approximation of the ground truth, and could be an incentive towards targeted studies involving clinicians.

CARDIAC IMAGE ANALYSIS

Motivation

Cardiac imaging has gained a central place in the diagnosis and treatment of cardiac pathologies, which are still the leading cause of death in industrial countries. Examinations in MRI or X-Ray CT typically result in a huge amount of 3D/4D image data delivering kind of snapshots of the heart status at successive time points of the cardiac cycle. In MRI, a set of 3 to 10 parallel slices are usually acquired over the heart. At each slice level, a dynamic series of frames is acquired during the cardiac cycle. This acquisition represents about 400 MBytes of 3D interpolated image data. A complete examination on a patient usually requires tens of them. The extraction of quantitative functional parameters, needed for cardio-vascular disease diagnosis, relies on sophisticated analysis programs that may require intensive computational resources in order to produce results in time compatible with the clinical practice. In particular, a method has been designed to retrieve the heart's motion from 3D dynamic MR imaging data. Indeed, motion appears to be one of the altered functions in severe cardiac ischemic diseases.

The motion estimation problem is a typical example of inverse problems which necessitates high level image processing methods with the introduction of *a priori* models and information. Whe have defined a method where the estimation is performed within a state space formulation and a recursive Kalman filtering approach (Delhay, 2006; Delhay, 2007). It takes into account the whole image sequence achieving a spatiotemporal (ST) modeling of the motion flow over the cardiac cycle. Moreover, it can be rather straightforwardly parallelized. With the aim of routinely processing cardiac examinations in a clinical setting, one needs high performance computing facilities such as those offered by computational grids.

The ST Model Principle

The principle of our motion estimation approach is closely related to the optical flow methods and their extension to multiple frames. The spatiotemporal motion is modeled by a parametric free form deformation (FFD) based model in both 3D space and time (Sederberg, 1986). A regular grid of control points (CP) encloses the initial space at rest. L is the number of grid nodes. The displacement of control points implies a dense spatial deformation the effect of which depends on the FFD function basis.

We proposed to extend the initial spatial approach to a spatio-temporal approach by attributing to each CP a continuous trajectory. The collection of all the CP trajectories forms the ST model parameters; each trajectory is individually modeled by a Fourier series of order N. This formulation insures two main properties of our model: the continuity and periodicity of the motion over the cardiac cycle. Therefore, the number of model parameters is $3xLx(2N+1)$ in 3D space.

At the finest transformation level the grid size is currently of size 19x19x19 and the ST model is completely defined by 185193 parameters with $N=4$, which makes the problem quite difficult to solve. Thanks to the periodization of the image sequence, the estimation is performed within a state space formulation and a recursive Kalman filtering approach (Delhay, 2006; Delhay, 2007). The goodness of fit is estimated through the sum of squared difference similarity criterion

Parallelization Strategies

Due to the high complexity of the problem, especially in the 3D+time space, several strategies have been proposed and implemented to reduce the computing time. First, a double multi-level approach has been adopted: a multi-resolution representation of the images (considering images with increasing details) and a multi-level representation for the transformations. Also, strategies have been set up to focus on moving regions and automatically tune the temporal model order through the computation of attention maps (Delhay, 2007). The process allows for a natural splitting for parallel computations without impacting the result accuracy. As shown in figure 8, the computational load can be distributed on multiple processors that all receive part of the image domain. Processes are synchronized to reconstruct the whole result. The partitioning is performed so as to balance the computational load of each processor in order to reduce latency as much as possible. Multi- threaded and message-passing (MPI) versions have been developed and deployed on a shared memory multiprocessors (SMP) machines and on PC clusters, resulting in an almost linear relation between computing time and CPU number (Figure 8).

In the view of going one step further in the context of multi-center clinical studies on large patient populations, the method has also been ported to the EGEE grid The grid version exploits trivial parallelism (across patients) as well as coupled parallelism (MPI on PC clusters for each patient dataset).

Results and Discussion

The experiments conducted both on synthetic and true MR image sequences demonstrated the good accuracy of the proposed method. On a 2D+time synthetic realistic sequence for instance, the accuracy (over the whole sequence)

Figure 8. Splitting computation load among processors obtained by partitioning the image data and an example of the computing time as a function of the processor number (here on a SMP architecture).

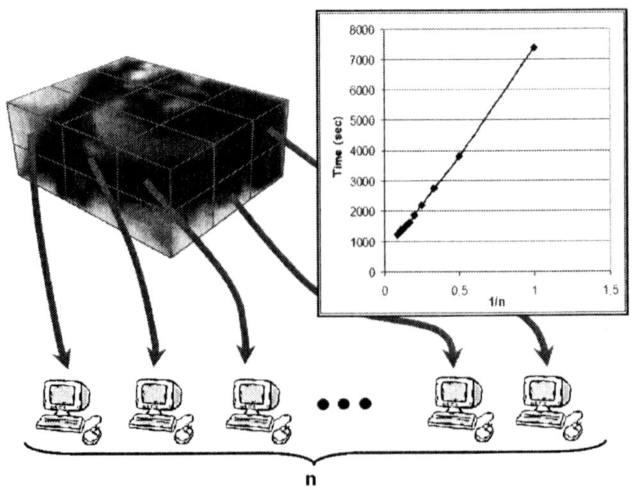

is 1,8±1,2 mm with $L = 21$ and $N = 5$. However, model identification is clearly a time-consuming process. A 4D experiment required 20 CPU hours for 160^3 image size and 21 frames, on a single processor machine. Experiments on a SMP-SGI system confirmed that the computation time reduces almost linearly as a function of the number of processors (approximately 2 hours in the previous case with 10 processors). Currently, the MPI version is evaluated on the EGEE grid. Our intention is to make the method available to the community through a web portal and evaluate its performance on a clinical database before deploying it for patient specific analysis and large multi-center studies on cardiac pathologies. Also, the complete sequence of computations, including the various image processing operators involved, has been reformulated within workflow representation languages and partly implemented on two workflow engines, including MOTEUR, in the context of the French ANR project GWENDIA (http://gwendia.polytech.unice.fr).

HUMANITARIAN MEDICINE

One of the greatest opportunities offered by grids is to build federation of databases where the medical data are kept locally in the different participating hospitals and can be queried remotely on request by clinicians for second diagnosis. This section describes a platform developed for this purpose on top of the gLite data management services using Web Service technology.

Introduction to Grid-Enabled Telemedicine

The concept of telemedicine covers all the potential means to exchange medical data or images whatever the distance between two physicians,. As stressed in a very recent report from the European Commission and the World Health Organization, "Information and communication technologies are changing health care delivery and are at the core of effective, responsive health systems. These technologies are key to connecting people, information and research to improve health in countries. They are also vital in enabling rapid response to global threats to health." (Dzenowagis, 2005)

One of the main added values of grids for telemedicine is to offer to healthcare professionals the capacity to query distant databases in a transparent and secure way (Gonzales, 2005).

- Patient data are kept in databases inside hospitals where patients are treated.
- Access for second diagnosis can be strictly controlled through the authorization and authentication mechanisms of the grid. Only a subset of the patient record can be exposed through the grid.
- In case additional information is needed, the team of clinicians requesting the second diagnosis can provide access to more patient information or perform new examinations and store the results on their local database.
- The data being distributed over the participating hospitals, there is no single point of failure

Platform Architecture and Design

The platform goal is to enable the exchange and usage of medical information over the grid in a secure way, which means exposing the information to the grid users which are allowed to access it and exploiting the medical information with analysis and job management facilities.

The platform was designed to fulfill additional requirements:

- The applications running on the platform must be easy to use as healthcare professionals are too busy to learn complex tools. In particular, the grid complexity must be completely hidden.

- The platform should be adaptable to any medical field. It must be flexible so it can be customized to adapt to the "modus operandi" of its users.
- The platform must be able to interact with existing hospital data and image archiving systems (PACS).
- The platform should be fault tolerant in the perspective of clinical routine
- The platform should not require global connectivity. Any network failure outside the hospital should not affect data management in the hospital, nor the exchange and treatment of the medical information outside the area affected by the failure.
- The platform should fulfill all legal requirements in relation to security

The platform architecture is built on three pillars:

- The medical data are stored in an AMGA server inside the hospital. AMGA is the metadata catalogue developed within the EGEE project (see section)
- The healthcare professional accesses all the platform services through a web portal developed with the GridSphere portlet container
- The grid infrastructure provides the storage and computing resources needed to store the medical images and to treat the data according to the user need.

Figure 9 displays the overall architecture of the platform (left vignette) and shows how information is shared between hospitals in different locations (right vignette).

In the design of the telemedicine, a separation was made between the presentation layer, developed using Gridsphere portal container, and the business logic part which was developed using web services to manage all the low level interaction with different grid services. This also allows the design of a personalized platform interface for each different customer without changing anything on data management or grid interaction side.

The Medical Data Manager described in section 3 was not deployed on the EGEE Infrastructure when the development of the platform was started. As a consequence, only one its component, namely the AMGA metadata catalogue, was used. However, the platform design allows to easily interface it to the MDM services.

Figure 9. (a) Information and image management inside a hospital; (b) information and image sharing between hospitals in different locations

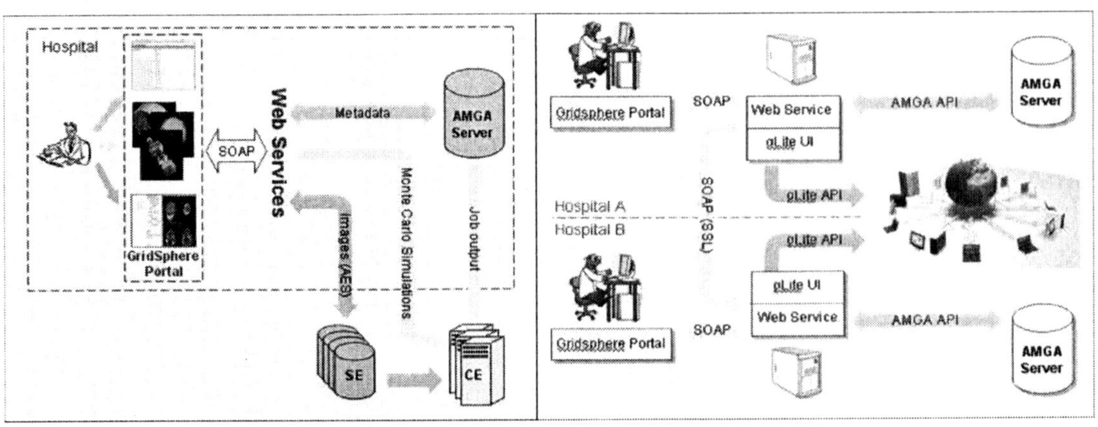

(a) (b)

The choice of web service technology came from the need to use standard and well supported communication protocols (SOAP in our case) to guarantee interoperability and security between different instances of the platform installed in remote hospitals. The web service layer manages all routing mechanisms used to exchange SOAP messages between remote platform instances allowing physicians belonging to different and remote hospitals to share patient files, clinical exams and medical images.

Medical information and DICOM metadata extracted from medical images are managed using AMGA metadata catalogue. This grid service provides access to different database back-ends in a uniform way and independently of the particular database system used. This makes AMGA a really interesting service to develop software solutions in a distributed and heterogeneous environment.

To adapt to the existing hospital information systems, the architecture should allow different hospitals to specify a different (and usually personalized) database structure for the same medical field. To make this possible we decided to use the XML mark-up language to represent all the information managed and exchanged in the platform. Together with SOAP protocol and the Web Service technology, the use of XML documents to store and exchange information makes the platform completely independent and decoupled from any particular database structure deployed inside each hospital.

Platform Deployment

The platform beta version is currently deployed in four sites; three of them are research laboratories while one is a hospital in Clermont-Ferrand. The platform has been customized for an application in the field of neurosurgery. Indeed, since March 2000, humanitarian missions have been conducted in neurosurgical unit of the Regional Hospital of District of Chuxiong (Yunnan, PRC) by French clinicians from healthcare centers in Clermont-Ferrand in order to operate neurosurgical pathologies including hydrocephalus in children which were not treated previously in this hospital. As these missions are limited in time, key issues for their success are the identification of patients needing surgery before the mission and their post-surgery follow-up after the mission. Another issue is related to the transfer of skills and techniques which should not be limited to the mission durations.

The HOPE platform was customized to the needs of the clinicians and designed in collaboration with them. A standard medical file was designed and translated in Chinese language. This file, subdivided in six sheets, gathers the relevant information at the different steps of the medical procedure (parental agreement for surgery, pre-operative status, operative report, follow-up, complications and re-operation, anesthetic examination). However, due to network problems, its deployment has not yet been possible in Chuxiong.

FUTURE TRENDS

Parallel computer architectures and algorithms have co-evolved with hard science applications, on an ascending ladder of trials and improvements. At the opposite, researchers in medical imaging know all too well that imperfect but promising, or simply new and not fully validated methods and algorithms have little hope to go from scholarly publication into clinical research. Considering the opportunities offered by modern computing infrastructures, clinical research and practice is not fully aware that high performance computing may be more than the limited parallelism offered by a high-end workstation.

Amongst the many causes of this sobering fact, the first one is sociological: given the academic structures, the opportunities to gather truly multi-disciplinary teams in medical image analysis like e.g. the BIRN project are much less frequent than in the hard science field. At the French level, AGIR has gathered a truly multi-disciplinary team, involving computer scientists with various specialties, and physicians. As a consequence, one of the basic options of AGIR is that the medical requirements regarding computing practice (dependability, interactivity, security and privacy protection, compliance to the DICOM standard) must be considered as first-order requirements, not as user interface or practical issues. In the limits of a 3-years project, AGIR has followed the path pioneered at the Surgical Planning Laboratory (Kikinis, 1998): "utilization of HPC technology (...) in an application driven clinical and research environment". Dependability requirements oriented AGIR to focus the core medical services towards production grids, and namely the EGEE grid, which design was largely driven by a very different community. The responsiveness and data management have successfully implemented the basic services complying with these requirements on top of the EGEE grid.

A second cause is that the medical constraints have very limited flexibility: a physicist or a chemist can appreciate the potential of prototype tools, and experiment on it, providing feedback to the computer scientist; a physician requires highly dependable, fully secure, user-friendly software, which has also to comply with complex standards. Software dependability involves both production quality at the usage level, and guarantees about the medical relevance of the embedded analysis tools, which is mainly about validation. Production grids change the conditions of validation of methods and algorithms. Validating new processing methods requires large computing power, large datasets, or both. With production grids, the infrastructure for sharing images and algorithms is here, together with the availability of computing power increased by orders of magnitude with respect to the usual workstation. At the same time, the development and generalization of service-oriented frameworks offers the perspective of integrating the grid resource as a backend of the usual interfaces and workflows of the physician. Ideally, the computer scientist will focus on its core contribution, methods and algorithms, and the physician could evaluate them into existing and appreciated environments. The application of the Bronze standard method to compression evaluation is an excellent example of integration of these multiple expertise, whose enactment is made possible by the contributions of the core medical services, and finally the grid infrastructure.

CONCLUSION

Overall, the main results of the AGIR project are twofold: real-world production oriented grids, and namely the EGEE grid, allow breakthrough in medical image processing research; second, some of the services required for clinical research and practice can be deployed on production grids.

The interaction with the medical partners has been pervasive all along the project, by providing real (and thus difficult) datasets, and has continuously steered the project towards the clinical constraints. To some extent, the value of their contribution is demonstrated by the fact that the experience and tools developed in AGIR has founded the participation of AGIR partners in two much larger-scale projects in the area of grid-enabling medical image processing: the FP6 IP Health-e-child, dedicated to clinical research in pediatric healthcare, and the French software R&D project NeuroLOG, dedicated to open-source grid-enabled tools for knowledge integration, with applications to neuro-imaging and oncology.

ACKNOWLEDGMENT

The AGIR project has been funded by the French Ministry of Research under the *Masses de Données* program. Developments has been partially supported by the EGEE-II project funded by the European Union INFSO-RI-031688, by other grants from the European Commission (BioinfoGRID, Embrace), and for the work in section 7 by regional authorities (Conseil Régional d'Auvergne, Conseil Général du Puy-de-Dôme, Conseil Général de l'Allier). The authors are indebted to M. Diarena, B. Grenier, A. Fessy, S. Nowak, N. Verhaeghe and N. Spalinger for their invaluable contribution to the development of the HOPE platform and its customization for telemedicine.

REFERENCES

Allcock, B., Bester, J., Bresnahan, J., Chervenak, A.L., Foster, I., Kesselman, C., et al. (2002). Data Management, Transfer in High Performance Computational Grid Environments. *Parallel Computing Journal*, *28* (5), 749-771.

Aloiso, G., Benkner, S., Bilofsky, H., Blanquer Espert, I., Breton, V., Cannataro, M., et al. (2005). *HealthGrid White Paper*. Retrieved from http://whitepaper.healthgrid.org.

Atkins, D. (2003). *Report of Blue-Ribbon Advisory Panel on Cyberinfrastructure*. Retrieved from http://www.nsf.gov/publications

Baruah, S., Cohen, N., Plaxton, C. & Varvel, D. (1996). Proportionate Progress: A Notion of Fairness in Resource Allocation. *Algorithmica, 15*(6), 600-625.

Basu, S., Talwar, V., Agarwalla, B., & Kuma, R. (2003). *Interactive Grid Architecture for Application Service Providers* (HP Technical Report HPL-2003-84R1).

Berman, F. Wolski, R. & Casanova, H. (2003). Adaptive Computing on the Grid Using AppLeS. *IEEE Transactions on Parallel and Distribributed Systems, 14*(4), 369–382.

Berry, D., Germain-Renaud, C., Hill, D., Pieper, F. & Saltz, J. (2004). Report on the Workshop IMAGE'03: Images, medical analysis and grid environments (TR UKeS-2004-02). UK National e-Science Centre, Glasgow, UK.

Bilgin, A., Zweig, G., & Marcellin, M.W. (2000). Three-dimensional image compression with integer wavelet transforms. *Applied Optics, 39*, 1799-1814.

Bolze, R. Cappello, F., Caron, E., Daydé, R., Desprez, F., & Jeannot, E., et al. (2006) Grid'5000: a large scale and highly reconfigurable experimental Grid testbed. *International Journal of High Performance Computing Applications, 20*(4), 481-494.

Brevik, J.,Nurmi, D. & Wolski, R. (2006, March). *Predicting bounds on queuing delay for batch-scheduled parallel machines.* In *11th ACM SIGPLAN symposium on principles and practice of parallel programming* (pp 110–118). New York, NY: ACM Press.

Caron, E., & Desprez, F. (2006) DIET: A Scalable Toolbox to Build Network Enabled Servers on the Grid. *International Journal of High Performance Computing Applications, 20*(3), 335-352.

Cho, S., Kim, D., & Pearlman, W.A. (2004). Lossless compression of volumetric medical images with improved three-dimensional SPIHT algorithm. *Journal of Digital Imaging, 17*, 57-63.

Delhay, B., Clarysse, P., Pera, C. & Magnin, I. E. (2006). A spatio-temporal deformation model for dense motion estimation in periodic cardiac image sequence. In *From Statistical Atlases to Personalized Models: Understanding Complex Diseases in Populations and Individuals, Satellite Workshop MICCAI 2006,* Copenhagen, Denmark.

Delhay, B., Clarysse, P., Pera, C. & Magnin, I. E. (2007, June). Locally adapted spatio-temporal deformation model for dense motion estimation in periodic cardiac image sequences. In *Functional*

Imaging and Modeling of the Heart. Salt Lake City, UT, (LNCS 4466, pp. 393-402). Berlin, Germany: Springer-Verlag.

Dinda, P. (2006). Design, Implementation, and Performance of an Extensible Toolkit for Resource Prediction In Distributed Systems. *IEEE Transactions on Parallel and Distributed Systems, 17*(2), 160-173.

Dzenowagis, J. & Kernen, G. (2005). *Connecting for Health: global vision, local insight* (Report for the World Summit on Information Society). Tunis, Tunisia: World Health Organization Press

EGEE SDJ WG .(2006). *Report of the Short Deadline Jobs Working Group*. Retrieved from http://egee-na4.ct.infn.it/wiki/index.php/ShortJobs.

Erberich, S., Silverstein, J., Chervenak, A., Schuler, R., Nelson, M. & Kesselman, K. (2007a). Globus MEDICUS – Federation of DICOM Medical Imaging Devices into Healthcare Grids. *Studies in Health Technology and Informatics, 126*, 269-278.

Erberich, S., Silverstein, J., Kratz M., Chervenak, A., Schuler, R., Nelson, M. & Kesselman, K. (2007b, April). *Globus MEDICUS – Standards Based Enterprise for Medical Image Publication, Discovery, and Archiving in HealthGrids*. Paper presented at the Open Source Solutions for Multi-Center Information Management. Saint Louis, MO.

Erickson, B. (2002). Irreversible Compression of Medical Images. *Journal of Digital imaging, 15*(1), 5-14.

Foster, I. & Kesselman, K. (Eds.), (2003). *The Grid 2: Blueprint for a New Computing Infrastructure*. San Francisco, CA: Morgan Kaufmann Publishers.

Foster, I., Kesselman, K. & Tuecke, S. (2001). The Anatomy of the Grid: Enabling Scalable Virtual Organizations. *International Journal Supercomputer Applications, 15*(3), 200-222.

Foster, I. (2006, October). Globus Toolkit Version 4: Software for Service-Oriented Systems. *IFIP International Conference on Network and Parallel Computing*, Tokyo, Japan (LNCS 3779, pp 2-13). Berlin, Germany: Springer-Verlag.

Gagliardi, F., Jones, B., Grey, F., Bégin, ME. & Heikkurinen, M. (2005). Building an Infrastructure for scientific Grid computing: status and goals of the EGEE project. *Philosophical Transactions of the Royal Society A, 363*(1833), 1729-1742.

Gaudeau, Y., & Moureaux, J.M. (2007, November). *Lossy compression of volumetric medical images with 3D dead zone lattice vector quantization*. 26th Picture Coding Symposium, Lisboa, Portugal.

Germain-Renaud, C., Perez, J., Kégl, B. & Loomis, C. (2008, May). Grid Differentiated Services: a Reinforcement Learning Approach. *8th IEEE Symposium on Cluster Computing and the Grid*. Lyon, France: IEEE Computer Society.

Germain-Renaud, C., Loomis, C., Moscicki, JT., & Texier. R. (2008). Scheduling for responsive grids. *Journal of Grid Computing, 6*(1), 15-27.

Glatard, T., Pennec, X. & Montagnat, J. (2006). Performance evaluation of grid-enabled registration algorithms using bronze-standards. In *Proc. of MICCAI'06, Part II*, (LNCS 4191, pp. 152-160).

Glatard, T., Montagnat, J., Erberich S., Carmeli B. & Kaufman, J. (2007, November). *Cross Enterprise Document and Image Exchange (XDS/-I) in HealthGrids*. Paper presented at the Radiology Society of North America (RSNA'07), Chicago, IL.

Glatard, T., Montagnat, J., Emsellem, D. & Lingrand, D. (in press). A Service-Oriented Architecture enabling dynamic services grouping for optimizing distributed workflows execution. In *Future Generation Computer Systems*.

Glatard, T., Montagnat, J., Lingrand, D. & Pennec, X. (in press). Flexible and efficient workflow deployement of data-intensive applications on grids with MOTEUR. *International Journal of High Performance Computing and Applications.*

Gonzales, J., Pomel, S., Breton, V., Clot, B., Gutknecht, J.L., Irrthum, B. & Legré, Y. (2005) Empowering humanitarian medical development using grid technology. *Methods of Information in Medicine, 44,* 186-189.

Heinzlreiter, P. & Kranzlmueller, D. (2003). Visualization Services on the Grid: The Grid Visualization Kernel. *Parallel Processing Letters, 13*(2), 135–148.

Jeannot, E. (2005, April), Improving middleware performance with AdOC: an Adaptive Online Compression library for data transfer. *IEEE International Parallel and Distributed Processing Symposium* (p. 70). Denver, CO. New York: IEEE Computer Society.

Jensen, ED., Locke, CD. & Tokuda, H. (1985, December). *A time driven scheduling model for real-time operating systems* (pp. 112–122). IEEE Real-Time Systems Symposium, San Diego, CA. New York: IEEE Computer Society.

Keahey, K., Foster, I., Freeman, T. & Zhang, X. (2006). Virtual Workspaces: Achieving Quality of Service and Quality of Life in the Grid. *Scientific Programming, 13*(4), 265-275.

Kikinis, R. Warfield, S. & Westin, C. (1998). *High Performance Computing (HPC) in Medical Image Analysis (MIA) at the Surgical Planning Laboratory (SPL).* Paper presented at the 3rd High Performance Computing Asia Conference & Exhibition, Singapore, Malaysia.

Laure, E. et al. (2004, September). *Middleware for the next generation Grid infrastructure.* Paper presented at the 2004 Conference for Computing in High Energy and Nuclear Physics, Interlaken, Switzerland.

Leenaars, M., Heikkurinen, M., Louridas, M., & Karayannis, F. (2005). e-infrastructures roadmap. Espoo, Finland: e-IRGSG. Retrieved from http://www.e-irg.org/roadmap/eIRG-roadmap.pdf

Lin, B. & Dinda, PA. (2005, November). *VSched: Mixing Batch And Interactive Virtual Machines Using Periodic Real-time Scheduling.* ACM/IEEE conference on Supercomputing. New York: IEEE Computer Society

Montagnat, J., Bellet, F., Benoit-Cattin, H., Breton, V., Brunie, L., Duque, H., et al. (2004). Medical images simulation, storage, and processing on the European datagrid testbed. *Journal of Grid Computing, 2* (4), 387-400.

Montagnat, J., Frohner, Á., Jouvenot, D., Pera, C., Kunszt, P., Koblitz, B., et al. (2008). A Secure Grid Medical Data Manager Interfaced to the gLite Middleware. *Journal of Grid Computing, 6* (1), 45-59.

Moore, GA. (1991). *Crossing the Chasm: Marketing and Selling High-Tech Products to Mainstream Customers.* New York: Harper Business.

Oinn, T., Addis, M., Ferris, J., Marvin, D., Senger, M., Greenwood, M., et al. (2004). Taverna: A tool for the composition and enactment of bioinformatics workflows. *Bioinformatics journal, 17* (20), 3045-3054.

Osorio A., Valette PJ., Mihalcea A., Atif J. & Ripoche X. (2002) *A new PC based software to perform semi-automatic hepatic segmentation using CT and MR images.* Presented at InfoRAD-RSNA, Chicago, IL.

Ourselin, S., Roche, A., Prima, S. & Ayache, N. (2000). Block Matching: A General Framework to Improve Robustness of Rigid Registration of Medical Images. In proc. of *MICCAI 2000,* , Pittsburgh, PA, (LNCS 1935, pp. 557-566). Berlin, Germany: Springer.

Peltier, ST. & Ellisman MH. (2003) The Biomedical Informatics Research Network (BIRN). In: I Foster and C Kesselman (Eds.), *The Grid: Blueprint for a New Computing Infrastructure* (pp 109-120). San Francisco, CA: Morgan Kaufmann Publishers.

Penedo, M., Souto, M., Tahoces, P.G., Carreira, J.M., Villalón, J., Porto, G., et al. (2005). Free-response receiver operating characteristic evaluation of lossy JPEG2000 and object-based set partitioning in hierarchical trees compression of digitized mammograms. *Radiology, 237*, 450-457.

Pennec, X & Thirion, J-P. (1997). A Framework for Uncertainty and Validation of 3D Registration Methods based on Points and Frames. *Int. Journal of Computer Vision*, *25*(3) 203-229.

Pennec, X., Ayache, N. & Thirion, J.P. (2000). Landmark-based registration using features identified through differential geometry. In I. Bankman, (Ed.), *Handbook of Medical Imaging*, (pp. 499-513). Academic Press.

Perelmutov, T., Petravick, D., Gu, J., Barring, O., Baud, J.-P., De Witt, S., et al. (2002, May). *SRM Interface Specification v2.2* (FNAL technical report). Retrieved from http://sdm.lbl.gov/srm-wg/doc/SRM.v2.2.html

Pujol, S, Kikinis, R & Gollub, R. (2008). Lowering the barriers inherent in translating advances in neuroimage analysis to clinical research applications. *Acad Radiology, 15*(1), 114-118.

Roche A., Pennec X., Malandain G. & Ayache N. (2001). Rigid Registration of 3D Ultrasound with MR Images: a New Approach Combining Intensity and Gradient Information. *IEEE Transactions on Medical Imaging, 20*(10), 1038-1049.

Santos, N. & Koblitz, B. (2006, June). *Distributed Metadata with the AMGA Metadata Catalog*. Paper presented at the workshop on Next-Generation Distributed Data Management, Paris, France. Retrieved from http://project-arda-dev.web.cern.ch/project-arda-dev/metadata

Sederberg, T. W., & Parry, S. R. (1986). Free-form deformation of solid geometric models. *ACM Computer Graphic, 20*, 151-160.

Shamir, A. (1979). How to share a secret. *Communications of the ACM, 22*, 612-613.

Sim, A. & Shoshani, A. (2007) *The Storage Resource Manager Interface Specification Version 2.2.* Retrieved from https://forge.gridforum.org/sf/projects/gsm-wg

Snell, Q., Clement, MJ., Jackson DB. & Gregory, C. (2000, May). *The Performance Impact of Advance Reservation Meta- scheduling.* 6th workshop on Job Scheduling Strategies for Parallel Processing, Cancun, Mexico (LNCS 1911, pp 137–153). Berlin, Germany: Springer-Verlag.

Tesauro, GJ. & Kephart, J. O. (2004, May). *Utility functions in autonomic systems.* 1st International Conference on Autonomic Computing (pp. 70–77). New York: IEEE Computer Society.

Texier, R (2007) *Systèmes de globalisation des ressources données-calculs. Application à l'imagerie médicale*. Unpublished doctoral dissertation, Université Paris-Sud, France.

Traxer, O., Osorio A., Pasqui, F., Sebe, P., Merran, S., Ripoche, X., et al. (2005, March). *New PC based software using an Augmented Reality system to simulate PCNL.* 20th European Association of Urology (EAU) Congress, Istanbul, Turkey.

Tsaregorodtsev, A., Garonne, V. & Stokes-Rees, I. (2004). DIRAC: A Scalable Lightweight Architecture for High Throughput Computing. In *5th IEEE/ACM International Workshop on Grid Computing*, (pp 19-25). New York: IEEE Computer Society.

Warfield S.K., Zou, K.H, & Wells, W.M. (2004). Simultaneous Truth and Performance Level Estimation (STAPLE): An Algorithm for the Validation of Image Segmentation. *IEEE Transactions on Medical Imaging, 23*(7) 903-921

Xiong, Z., Wu, X., Cheng, S., & Hau, J. (2003). Lossy-to-Lossless compression of medical volumetric data using three-dimensional integer wavelet transform. *IEEE Transactions on Medical Imaging*, *22*, 459-470.

KEY TERMS AND DEFINITIONS

Data Compression: Encoding information using fewer bits than the unencoded representation of data would. Lossy compression methods do not allow to fully retrieve the original data, while lossless compression methods do.

DICOM (Digital Imaging and Communications in Medicine): DICOM is standards to ensure the interoperability in the processing of medical image. In particular, DICOM defines standard for image and data formats, image transmission, query or store images, print and display.

Image Registration: Evaluating the transformation compensating for the difference in location, pose (rigid registration) and possibly scale or shape (non-rigid registration) between the images. Different images can thus be spatially aligned in one common coordinate system before further analysis. For instance,. rigid registration of images of the head is used to correct for the different positioning of the subject in the image acquisition device at two distinct times points.

Medical Imaging: The techniques and processes used to create digital images of the human body (or parts of it) for clinical purposes (diagnosis and therapy) or medical science (including the study of normal anatomy and function).

Performance Evaluation: Assessment of the technical characteristics of algorithms used in medical image analysis (e.g. registration accuracy). In some cases, a well established reference or gold standard exists and can be used for evaluation. In other cases, there is no direct way of comparing a result to the known reference. Statistical procedures can be used to define an approximate reference, or Bronze Standard.

Quality of Service: The ability to provide different priority to different applications, users, or data flows, or to guarantee a certain level of performance to a data flow.

Responsiveness: In the context of human-computer-interaction, responsiveness of an interactive system describes how quickly it responds to user input.

Telemedicine: Clinical practice where medical information is transferred via telephone, the Internet or other networks for the purpose of consulting, and sometimes remote medical procedures or examinations.

Workflow: A description of a process involving the ordered execution of different tasks. A workflow graph describes the tasks involved and their dependencies. The dependencies might be temporal or due to data exchanges needed between tasks (data flow).

Chapter XX
Grid Computing in 3D Electron Microscopy Reconstruction

J.R. Bilbao-Castro
University of Almería, Spain

I. García
University of Almería, Spain

J.J. Fernández
University of Almería, Spain

ABSTRACT

Three-dimensional electron microscopy allows scientists to study biological specimens and to understand how they behave and interact with each other depending on their structural conformation. Electron microscopy projections of the specimens are taken from different angles and are processed to obtain a virtual three-dimensional reconstruction for further studies. Nevertheless, the whole reconstruction process, which is composed of many different subtasks from the microscope to the reconstructed volume, is not straightforward nor cheap in terms of computational costs. Different computing paradigms have been applied in order to overcome such high costs. While classic parallel computing using mainframes and clusters of workstations is usually enough for average requirements, there are some tasks which would fit better into a different computing paradigm – such as grid computing. Such tasks can be split up into a myriad of subtasks, which can then be run independently using as many computational resources as are available. This chapter explores two of these tasks present in a typical three-dimensional electron microscopy reconstruction process. In addition, important aspects like fault-tolerance are widely covered; given that the distributed nature of a grid infrastructure makes it inherently unstable and difficult to predict.

INTRODUCTION

Studying the structure of biological specimens on different levels is of vital importance in understanding their functionality and interactions. Different techniques like X-ray crystallography, nuclear magnetic resonance and electron microscopy (EM) are used for such purposes.

EM relies on the transmission electron microscope to obtain projection images of the specimen under study (Williams & Carter, 2004). Such projections are somewhat similar to the well-known medical radiographs – but with X-rays being substituted by highly energetic electron beams. The projections of the specimen are processed, using complex computational algorithms, to create a three-dimensional (3D) representation of the specimen that can then be thoroughly studied. 3D EM is therefore the result of applying certain computing algorithms to the 2D projections obtained from the specimen under study through the electron microscope (Frank, 2006).

Depending on the nature of the specimen and the problem under consideration, different reconstruction techniques are used. For example, when studying specimens in the cellular range, the so-called electron tomography (ET) is used. On the other hand, specimens in the macromolecular domain are usually studied through electron crystallography and single particle techniques, depending on their structural features (like, for example, symmetry).

The objectives of the work presented here is to develop and test computational tools which cope with high computational demands existing at different steps of a 3D reconstruction process. In particular, grid computing will be the computational platform chosen to tackle two different tasks; parameters optimization in algebraic reconstruction algorithms and the application of such algebraic algorithms in the 3D ET field.

BACKGROUND

Three-Dimensional Electron Microscopy and Computational Demands

A complete reconstruction process, from the rough projections to the reconstructed volume, involves a set of complex subtasks oriented towards reaching as high a resolution as possible. Most such subtasks pose computational challenges. For example, ET studies usually involve a small number of projections - with those normally being of considerable dimensions. Consequently, the reconstructed volume will also be large, resulting in high memory and processing power demands. On the other hand, single particle studies typically comprise a large number of smaller projections and while the required processing power is still huge, because of the large number of projections, memory requirements are, in turn, smaller.

High computational demands have usually been dealt with by using parallel computing techniques like message passing and shared memory. Nevertheless, many research groups do not have unlimited access to parallel computing facilities or available resources are simply insufficient for some problems. Here is where grid computing techniques could play an important role.

Progress of 3D EM techniques goes hand in hand with the development of new, more powerful computational software and hardware. Current studies use up to tens of thousands of projections to reach resolutions between 6 and 10 Angstroms (Valle et al., 2003; Gao et al., 2003; Cheng, Zak, Aisen, Harrison, & Walz, 2004). New, automated techniques for image processing will soon allow the use of hundreds of thousands of projections in a single study (Potter, Zhu, & Carragher, 2004). The final objective of the technique is to reach atomic resolutions (less than 3 Angstroms) (Henderson, 2004). It is estimated that, for this purpose, millions of projections could be needed

(Henderson, 1995). With such an increase in the amount of data associated to a single experiment, computational demand requirements are expected to soar. But the increase in the data volume is not the only computational issue. Additionally, new, more complex but also more powerful algorithms have pushed the resolution to limits that were unthinkable some years ago (Scheres et al., 2005b). However, powerful algorithms usually mean they are more complex: which translates into even more computational requirements. What's more, as is usual, such requirements are growing faster than the mean available computing power. Therefore, many groups rely on parallel computing as a way to overcome such limitations. Parallel computing has been broadly applied in the field of 3D EM at different steps (Bilbao-Castro, Carazo, García, & Fernández, 2006; Fernández, Lawrence, Roca, García, Ellisman, & Carazo, 2002; Marinescu & Ji, 2003).

Nonetheless, the demand on parallel computers broadly exceeds the supply, making it impossible for all groups to have access to such platforms. Even when the platform is available, sometimes the machine does not fit the problem requirements. Here, the continuous availability of a computational grid makes it an interesting alternative to parallel computers for many groups and/or problems (Fernández, Carazo, García, & Marabini, 2007; Bilbao-Castro, García, & Fernández, 2007; Fernández, Bilbao-Castro, Marabini, Carazo, & García, 2005; Scheres, Merino, Sorzano, & Carazo, 2005). Although computational grids are still relatively new developments, they have proven useful for many purposes: tackling previously unexplored problems (Bilbao-Castro, Merino, Carazo, García, & Fernández, 2007b) and providing good speed-ups (Fernández et al., 2007).

Grid Computing on 3D EM

Despite the steady growth and development of grid computing infrastructures, they are still not well-suited to all kinds of problems. Grid computing is generally well-suited to high throughput applications, where the user wants to run a large number of independent tasks. Also, high network and grid service latencies preclude the user from using the grid like a big, typical parallel supercomputer. Therefore, the grid is an ideal platform for jobs where a large number of independent, long-lasting and low network-consuming tasks can be run.

Taking the last paragraph into account, various steps exist in the process of 3D reconstruction from projections where grid computing can be an attractive alternative to classical parallel computing. All such applications present a low requirement for transferred data compared to the required processing time. In some ET studies, as shown later, the volume to be reconstructed can be subdivided into smaller, independent volumes which can also be processed concurrently on the grid.

The grid has therefore proven that it is mature enough for production environments. But, with potential users still reluctant to adopt the grid as their usual researching framework, special efforts should be made to make it more accessible for the end users. Accessibility will be improved by making grid usage simpler and more user-friendly (Wolniewicz et al., 2007), by adopting new fault tolerance techniques (Hwang & Kesselman, 2004) and by hiding the grid complexity to the user; through, for example, easy-to-use graphical user interfaces like web portals (Novotny, 2003).

GRID APPLICATIONS DEVELOPED

Starting with the enhancement of electron microscope projections through image processing techniques and finishing with the segmentation of a reconstructed volume, 3D EM studies involve a long list of tasks to be performed. Some of these tasks require high computational requirements and therefore running times are relatively long. The first logical approach to solving this is to distribute computing load across multiple processors on a multiprocessor or multicomputer machine (Hennessy & Patterson, 2006). Nevertheless, an attractive and more affordable alternative is in using grid computing. This section covers the development and testing of two interesting tasks in the workflow of 3D EM studies. Firstly, the optimization of parameters in 3D algebraic reconstruction algorithms (Bilbao-Castro et al., 2007b) and secondly, the application of the reconstruction algorithm in electron tomography (ET) studies (Fernández et al., 2007; Bilbao-Castro et al., 2007). At the end of this section, a whole subsection is devoted to exploring (and proposing solutions to) the most common problems found during the development of such applications.

Optimization of Parameters in 3D Algebraic Reconstruction Algorithms

Filtered Back-Projection algorithm (FBP) is usually the chosen algorithm when performing 3D reconstructions from projections. It is the most commonly-used algorithm in medical computed tomography (CT), as well as in 3D EM, because of its computational simplicity. However, FBP does not behave so well when the set of projections suffers from a low signal-to-noise ratio and/or when such a set of projections does not sufficiently cover the space around the specimen. As both limitations usually occur in 3D EM studies, other families of reconstruction algorithms are used, which deal better with such conditions. This is the case with the algorithms based on algebraic reconstruction techniques, such as ART (Gordon, Bender, & Herman, 1970) and SART (Andersen & Kak, 1984). The two major drawbacks of such algorithms are their computational complexity (they are iterative algorithms, see Figure 1) and the need to correctly tune the so-called relaxation parameter to ensure a good convergence with the final result.

As shown in the following equation, the relaxation parameter (lambda), acts by modulating the update to apply to each volume element at each step of the algorithm evolution.

$$x_j^{k+1} = x_j^k + \lambda_k \sum_{s=1}^{S} \frac{y_i - \sum_{v=1}^{J} l_{i,v} x_v^k}{\sum_{v=1}^{J} w_v^b (l_{i,v})^2} l_{i,j}$$

Figure 1. Conceptual schema of an iterative reconstruction algorithm. The volume is updated based on error measurements between experimental and theoretical data

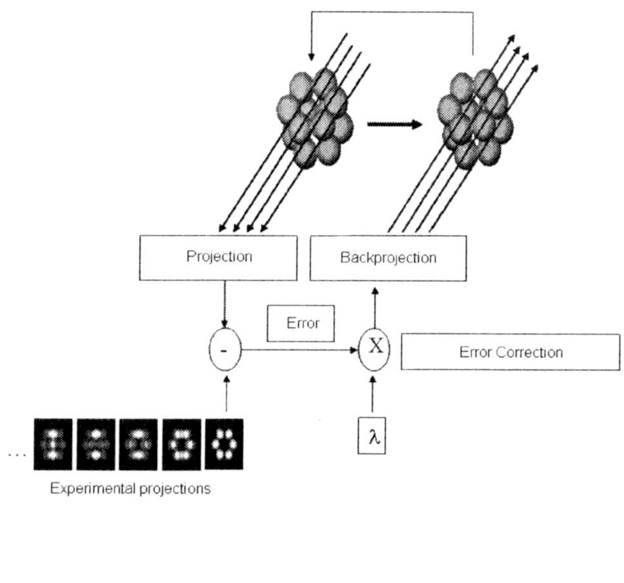

Where λ_k denotes the relaxation parameter, $b = (k \bmod B)$, represents the block index, $i=bS+s$, is the i-th equation of the system and w_v^b represents a weighting factor

Some authors suggest a statistical approach in proposing optimal relaxation parameters based on the different characteristics of the original projections, like their number and their signal-to-noise ratio (Sorzano, Marabino, Herman, & Carazo, 2005).

Apart from the relaxation parameter, other parameters exist whose optimization may have an influence on the final reconstruction quality. For example, different basis elements can be used to represent the reconstructed volume. As a pixel is the common basis of a two-dimensional picture, a voxel is the common basis for representing a volume. Nevertheless, when representing biological, smooth shapes, other more appropriate basis elements can be found (Marabini, Herman, & Carazo, 1998) (Lewitt, 1992). The so-called blob (see Figure 2) is one such well-adapted basis for the representation of biological specimens. In spite of its advantages, the blob is defined by a series of free parameters, which should be optimized to adapt it to different scenarios.

The following formula represents a blob.

$$b(m,\alpha,a,r) = \frac{I_m\left(\alpha\sqrt{1-\left(\frac{r}{a}\right)^2}\right)}{I_m(\alpha)}\left(\sqrt{1-\left(\frac{r}{a}\right)^2}\right)^m$$

when $0 \le r \le a$ and 0 elsewhere

Where r represents the radial distance from the center of the blob, I_m is the modified Bessel function of order m, a is the maximum radius and α is a parameter which controls the shape of the blob.

The values chosen for each of these parameters will strongly influence the visual quality of the resulting volume (Garduño, Gabor, & Herman, 2004).

As a result, the best reconstruction quality will be that obtained when using an optimal set of parameters, which includes not only the correct relaxation parameter but also the correct set of parameters for the blob. Such optimal sets will depend on many aspects of the problem under consideration, like the number and spatial distribution of the projections, their signal-to-noise ratio, etc... Therefore, in a field where a small improvement in achieved resolution is welcome, it would be desirable to use optimal parameters for each experiment. The work presented here aimed to provide a computational framework able to find the optimal set of parameters for each experiment.

Figure 2. Blob structure and projection mechanism. Blobs overlap each other as their value vary with the radius (bottom left)

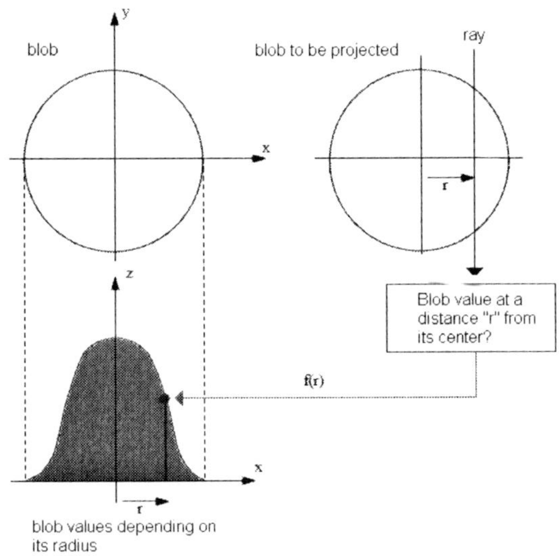

Different alternatives exist that can approximate the optimal set of parameters with reasonable accuracy. First, a brute force approach would consist of performing a large number of reconstructions using different sets of parameters and choosing that set which yields the highest resolution. Nonetheless, this is not a logical approach because (a), the search domain is comprised of a large multidimensional, continuous space and (b), computing requirements are far beyond what is reasonable. A more elegant and logical approach consists of using some kind of optimization algorithm to decide which sets of parameters to test (and which not) through the search domain. Despite this improvement, computing requirements are still too high and preclude the user from performing the optimization of parameters as resolution improvements probably will not overcome the costs involved. Here is where the grid emerges as a powerful tool. Performing reconstructions for different sets of parameters can be done in parallel using different nodes of a computing grid. Also, data (projections of the specimen) are uploaded to the grid only once and reused when testing different sets of parameters, which means low network overheads.

To tackle the optimization of parameters problem, an application was developed which joined the EGEE grid (Gagliardi, Jones, Grey, Bégin, & Heikkurinen, 2005) computing capabilities and a global optimization algorithm called Delayed Controlled Random Search (DCRS), based on CRS algorithms (Hendrix, Ortigosa, & García, 2001).

EGEE was the chosen grid infrastructure because of its worldwide deployment and great institution support. Originally conceived as an European infrastructure, it has finally expanded worldwide, reaching 250 institutions in around 45 countries. It consists of more than 41000 CPUs as well as a storage capacity in excess of five petabytes, reaching 100000 jobs running concurrently. On the other hand, DCRS was chosen basically because its computational simplicity and its inherent parallel nature. Here, parallelism denotes that the optimization algorithm allows the simultaneous evaluation of multiple sets of parameters. This is of great utility in a grid environment where thousands of machines are available for use at a time.

For the problem under consideration, DCRS tries to find the sets of parameters which maximize the reconstruction quality. Therefore, the reconstruction quality is the objective function and, in this case, it represents a resolution estimation of the reconstructed volumes. The resolution estimation can be done by means of the so-called Fourier Shell Correlation (FSC) (van Heel, 1987; van Heel & Schatz, 2005). This method involves the set of projections being divided into two equally-distributed subsets of randomly-chosen projections. For each subset, and using the same parameters for both, a reconstruction is performed. The two resulting volumes are compared through the FSC.

The application included an intuitive and easy-to-use windowed graphical interface. The inputs of the application were:

- **Projections of the specimen under study:** The set of projections of the specimen will be processed to obtain reconstructed volumes. This data is reused for each set of parameters. Therefore, uploading of the files to the grid is only needed once.
- **Executable to perform the reconstructions:** This is a statically-compiled executable. Using static executables avoids potential problems when using dynamically-linked libraries. This is because we can not guarantee that dynamic libraries are available at each computing node and that their versions are those required by our executables. The user can provide whatever program and/or algorithm he wants. Later on, the user can define different parameters as needed. This program will use the projections of the specimen to build a new volume using a set of parameters.
- **Executable to perform the resolution measurements:** A statically-compiled executable performing a FSC estimation. It will work on the two reconstructed volumes obtained with each half of the projections.
- **Parameters to optimize:** Each reconstruction algorithm and/or implementation will have its own set of characteristic parameters. The user must input a definition for each one and its values. For example, if a parameter's value is to be constant through

different experiments, it could be defined as "-p foo_constant", and even "-p" if it is non-valued. On the other hand, if we want to optimize a discrete parameter, we could define it as "-p [1,2,3]" or "-p [a,b,c]", with 1, 2 and 3 or a, b and c being the distinct possibilities. Another option is to optimize a parameter whose values are contained within a continuous range, for example: "-p [0.001,0.5]". The application allows the defining of such kinds of parameters as well as other auxiliary types, like filenames.
- **Stop conditions:** We do not want an optimization process to run forever. To prevent this, the user must provide some kind of stop condition. This application allows the user to decide between a series of them, such as the number of sets tested and the precision reached (resolution does not vary further than a certain threshold between new tested sets of parameters).

To run the application, the user needs access to a properly-configured user interface (UI) machine (the machine configured to access the grid). Also, valid credentials (user certificates) are needed to gain access to the EGEE grid.

The application automatically manages data (projections) upload to the grid; parameters selection via a global optimization algorithm (DCRS was used but any other allowing parallel testing of different points in the search domain could be used) and automatic grid job management. Also, some degree of fault-tolerance, regarding failed jobs resubmission and data replication capabilities, is provided.

As the output, the user obtains a list containing the best-performing sets of parameters and the resolution reached for each of them.

The parameters optimization application was tested using different biological models (virtual representations of well-known specimens) as a part of a PhD thesis. Projections of the models were obtained through computer simulation of an electron microscope, adding noise to simulate real-world conditions. Using known models is a vital step to validate the new procedure, given that there is something (the models) to compare results to.

Once the projections were ready, various optimization experiments were carried out. For each model, single (relaxation parameter) and multiple (relaxation parameter plus blob parameters) parameter optimization were performed. The main reason for performing single vs. multiple parameter optimization was to elucidate if the lambda parameter was, as previous works suggested, the most influencing one in terms of reconstruction quality.

The experiments, the results of which are presented in Bilbao-Castro et al. (2007b), allowed us to conclude that, for the studied data, the default parameters proposed in the literature were correct. Also, lambda ended up being, as expected, the most important parameter to optimize - with small or zero resolution improvements achieved through optimizing the blob parameters. Despite good reconstructions being provided by using standard parameters, some minor improvements were still achievable through lambda optimization.

In terms of computational speed-ups, when comparing optimization time on the grid with sequential estimations, the grid showed a discrete good behaviour, with values ranging from 10 to 19 - confirming this kind of optimization application was fully-transferable to the grid. Jobs resubmission rates ranged from 20% to 40%, which is quite high, meaning that a lot of useful time is wasted. Nevertheless, this negative aspect is well-managed thanks to the automatic job management shown later. More detailed information about application performance and experiments carried-out can be obtained from Bilbao-Castro et al. (2007b).

3D Electron Tomography

Certain biological specimens verify that different specimens of the same biological entity present structural homogeneity. This means that a virtually unlimited number of projections of the specimens can be taken in order to reach as far as atomic resolutions. On the other hand, when such structural homogeneity does not exist, the number of projections used for a study is limited to those obtained from a single specimen before it is damaged by the microscope's electronic radiation. In the latter case, a single specimen is rotated inside the electron microscope to obtain a limited number of projections. This process is performed using specially prepared specimens –the most usual technique, known as cryo-electron microscopy (Dubochet et al., 1988; Goodhew, 1984), involves the freezing of the specimens- and applying a reduced dose of radiation to avoid excessive damage to them. Nevertheless, preparation and low dose radiation mean that obtained projections will be extremely noisy. Also, technical limitations of the microscope limit the projection space around the specimen that can be explored. Both limitations, reduced number of low signal-to-noise ratio projections and incomplete covering of the projection space, have an impact and limit the quality and resolution reached in the final reconstruction. This process, in which a single specimen is used for a study, is called electron tomography (ET) (Frank, 1992; Lučić, Foster, & Baumeister, 2005). ET is frequently used for relatively big specimens like cellular organelles (like a mitochondrion, a nucleus, etc..) or even whole cells. Therefore, projections used on ET are usually large-scale, easily reaching sizes of up to 2048x2048 pixels - which means resulting volumes will be composed of up to 2^{33} volume elements (voxels). If we take into account that performing the reconstruction using ART (or other algebraic algorithm) is nothing but iteratively solving a system of equations with voxels as the unknowns, one can easily understand the profound computational implications - in terms of memory and CPU requirements - that the ET problem represents.

When rotating the specimen around a single axis, the volume to reconstruct can be divided into independent slabs (sets of neighbour slices) which can be reconstructed in parallel and put together later to compose the whole 3D reconstruction. This problem seemed to be well-suited to grid computing as data transfers for each task were small in comparison with the required processing time. Also, with the EGEE grid, we had virtually thousands of computers available round-the-clock, whereas access to a traditional supercomputer was not always possible nor was it cheap. Our group developed a test of concept application to study how the problem really adapted to grid computing.

The usage of the EGEE grid platform to perform tomography reconstructions on big volumes showed itself to be a good and feasible alternative (Fernández et al., 2007). However, due to the lifespan of a single tomography reconstruction, great effort for the automation and fault-tolerance was still needed before such an application was ready for production environments. As a result of these previous experiments, it was decided to go a step further and develop a production-enabled grid application for performing 3D ET reconstructions. The application, which is described in the next paragraphs, is called EGEETomo.

EGEETomo was designed to be a final-user application. Usually, grid computing applications suffer from poor usability from the point of view of a non-trained user. Ideally, grid applications should be as easy to use as desktop applications with the implication that grid complexity should be hidden to the user as much as possible. When EGEETomo was developed, the aim was to make it so easy to use that the final user would not need any special training on grid usage. With such a purpose in mind, two main aspects were taken especially into consideration:

- **A graphical user interface (GUI):** Two different approaches were studied; firstly, a web application that could be accessed from any point and secondly, a typical windowed application running on the user's UI machine. The first option was discarded given that, despite its important advantage of world-wide availability, it greatly increased development and maintenance complexity. We then decided to develop a classical, graphical user interface using a GUI development toolkit. The program was developed using C/C++ languages, Qt as the GUI toolkit and LCG/gLite commands as grid interaction tools. Grid commands were invoked through system calls, which would provide some degree of extra flexibility to adapt to future updates of the grid middleware and a higher degree of fault tolerance. This decision was made because some gLite API functions designed for data replication management (lcg-utils) proved not to be fault-tolerant, causing all the application to crash.
- **Automation and fault-tolerance:** From data provided by the user, job descriptions (needed by the grid to manage jobs) are automatically generated. In this case, for each slab, a job description is created. Also, for each job, a shell script is automatically generated which will run on a remote grid host. The shell scripts perform various tasks that can be outlined as
    - Retrieve data to process (projections affecting current slab) from storage elements on the grid.
    - Retrieve the executable, which will perform the reconstruction of the slab.
    - Perform the reconstruction by invoking the reconstruction executable and passing slab's data as parameters.
    - Publish results to be available for the user's application retrieval.

A special section, later on this chapter, explains in detail the advantages of using shell scripts to wrapping the execution of real applications on the grid.

Data used to perform reconstructions, as well as the executable that will perform them, must be uploaded to some storage element (SE, the machine specially configured to store user's data) on the grid before they can be used. This is also done automatically by the EGEETomo application. The user does not need to worry about where or how data is stored.

Regarding fault tolerance, all the automatic processes explained above need some mechanisms to recover from different sources of failure on the grid. Due to the importance of such an aspect in an inherently unstable environment like the grid, the next section of this chapter is devoted to covering fault-tolerance techniques used in our developments.

Figure 3 shows the schematic functionality of EGEETomo. Firstly, the user has a group of projections specially prepared as sinograms. A sinogram is a single structure (in this case a single file) containing those parts of all projections which affect a single slice. Using sinograms allows for more efficient data management on the grid. When processing a single slab, only the sinograms affecting it, and not all the projections, should be transferred. Also, the group of sinograms affecting a single slab can be packed and compressed together in order to increase storage and transfer efficiency (a single sinogram consisting on 200 projections of 2048x2048 pixels needs as much as 3 MB of storage capacity). Once the sinograms are grouped (depending on the desired slab size) and compressed, they are uploaded to the grid for subsequent use. To ensure that the files are available even in the case of possible SE failures, multiple copies of each group of sinograms are created on different SEs. This is possible thanks to replica catalogs, as seen in the next section of the chapter, which keep a record of logical versus physical files stored on the grid. Once the data

Figure 3. Conceptual schema of the reconstruction process on the EGEE grid infrastructure. The big gray box represents the grid domains (SEs and CEs) and outside of it is the UI (where EGEETomo runs).

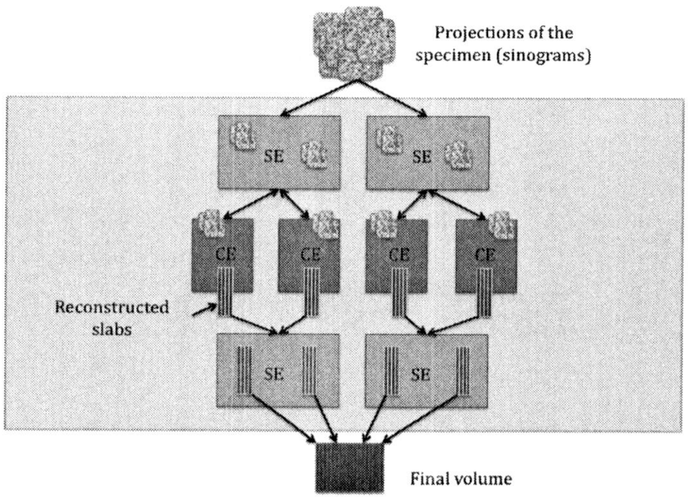

is replicated, the reconstructions of the different slabs can start. At this point, EGEETomo generates one job description file per slab and submits it to the grid. Each job descriptor contains information about the job to run - like the executable name, the kind of job (parallel, normal, etc...) and the hardware requirements. In the case of EGEETomo, the executable is not the reconstruction program itself but a shell script wrapper, which invokes the reconstruction program and does some necessary extra tasks. Such tasks include downloading needed files containing the sinograms affecting the slab to be processed as well as replicating the resulting slabs to different SEs.

Once the jobs have been submitted, EGEETomo monitors their status. When jobs are finished, the user can retrieve the whole set of resulting slabs and join them together, forming the whole reconstructed volume. It is important to note that none of the replication nor the job execution are failure-safe. At any moment, the replication process or a job execution can fail due to a variety of different reasons. Therefore, it is imperative for large grid jobs, like tomography reconstructions, to have different strategies for fault tolerance implemented. EGEETomo implements a series of fault tolerance techniques adapted to the grid that are thoroughly explored in the next section of this chapter.

For experiments performed on Fernandez (2007), speed-ups factors of 70 were reached, with numbers decreasing when slabs sizes decreased. This is due to the fact that big slabs will have a low transfer-to-processing times rate.

IMPLEMENTATION DETAILS

Implementation of the grid applications shown above was plagued with technical issues and difficulties. This section is aimed at helping prospective grid programmers identify and deal with potential sources of problems. Additionally, some helpful tips and tricks used on our applications are shown in order to provide some practical techniques.

Ideally, a grid infrastructure is intended to be secure, easy-to-use and reliable. Nevertheless, grid technologies and their different implementations are still not quite mature enough to provide such a stable framework. Some computational grids, like EGEE, are international and even intercontinental infrastructures, where tens of thousands of computers are shared between multiple organizations through different interconnected networks. In such a large and heterogeneous infrastructure, where management is highly-distributed, failures are relatively frequent and emerge randomly. Misconfiguration of sites, scheduled and non-scheduled shutdowns, electrical shortages and network failures, among other things, are common sources of problems.

The reader should bear in mind that although all this work is based on EGEE's grid, our experience shows that many different grid platforms rely on the same middleware or follow a similar paradigm: making the technical considerations shown here easily adaptable to other platforms.

Automatic Grid Job Management

A major focus of interest when working with grids is job management. Usually, a user writes a plain text file describing each job and its requirements. This is then submitted to the grid. Submitted jobs are scheduled to run on a specially configured grid site like, in the case of the EGEE grid, a computing element (CE). A CE usually consists of many worker nodes (WN), where jobs are effectively run. This is more or less like a typical parallel computer batch system, in which a front-end node manages different computing nodes (the WNs). Without going into too fine a detail, we can consider that, regardless of the grid infrastructure we are using, when a grid job entity is created, it follows a series of steps until it either fails or finishes correctly. At each and any of those steps, the job can fail for different technical reasons (memory limitations, machine failures, etc..) or simply be discarded by grid services (full queue, failed authorization, etc.). As fully preventing such failures is not possible, mechanisms to avoid constant human intervention are desirable so as to make grid applications useful in production environments. The *gLite* middleware (Laure et al., 2006), for example, can perform some automatic job submission retries if something fails although this mechanism is certainly simple and rather limited.

When the parameter optimization problem (shown above) was being ported to the grid, a first, simple implementation was limited to creating jobs, submitting them to the grid and getting results back. This was done through simple shell scripts. Of course, this first approach soon showed it was not practical because jobs usually failed and required manual intervention to solve the problem. In this case, the problem is especially important, given that for each set of parameters being tested, we have to create and manage three different grid jobs; one for each half of the projections and another to assess the achieved reconstruction resolution. Depending on the optimization problem under consideration, the number of parameters to test can rapidly burgeon into their thousands, which makes human intervention a complete and unacceptable waste of time and resources. To solve these important issues, a fully automated job manager was developed. It consists of a group of *C++* classes that automatically manage grid job submission and monitoring. The integration of this jobs manager within a newly-developed application is almost straightforward as it is completely encapsulated. The jobs manager communicates with the hosting application through notifications about the progress of jobs so that the programmer can decide what to do under different circumstances.

The implemented jobs manager works as follows:

- Firstly, the application creates a jobs group. A jobs group is the basic structure managed by our jobs manager. A job will be managed only in the context of a jobs group. The minimum number of jobs within a jobs group is one. By default, jobs groups are managed following a first-in first-out schema. Nevertheless, priorities to override such schema are also available.

- Secondly, the application populates the jobs group with different job descriptions. A job description is an object containing all the information the grid "needs" to know about a job. For example, we can assign a type to a job, which will be used by our application's logic to decide what to do with its results.
- When the jobs group is populated, it is set to "runnable" so the jobs manager can start submitting the jobs it contains to the grid.
- Once a job is submitted, it is monitored until it finishes correctly or fails. As already stated, the normal life cycle of a job is comprised of a series of steps; at each of which a job can fail. This jobs manager can detect, identify and solve most of these failures. It checks on the status of jobs with a certain periodicity and extracts important information regarding their progress. When a job is found to have failed, it is resubmitted and failure statistics are updated. Such statistics can be very useful in building adaptable grid job schedulers that avoid failure-prone resources and tend to use those generating fewer problems.
- This jobs manager also provides time-out mechanisms. Thus, it prevents jobs from running for too long a period (longer than that considered normal for the particular kind of job) and from stalling forever in the same status. The jobs manager allows the user to define three different time-outs: Submission time-out (time between submission and execution resource assignment), queuing time-out (how long a job waits in a remote resource before it is set to run) and running time-out (effective running time).

As commented on above, the implemented jobs manager also collects detailed execution statistics. Such statistics cover aspects like CE reliability (rate of failures), submission times (this represents grid service overheads), queuing times, effective running times, etc.. All this information could be used as input to adaptive scheduling mechanisms.

Automatic Files Replication

Another important mainstay in grid computing is data management (Allcock et al., 2002). Generally, the grids provide data storage capabilities in dedicated machines. In the EGEE grid, such functionality relies on the SEs. Such SEs are intended to store any user data to be used on the grid as well as data resulting from grid executions. Once more, a SE can fail for different reasons; either because the hardware fails, or because the middleware does.

Keeping multiple physical copies of the same logical file in different SEs is vital to ensure data accessibility at any moment. The process of maintaining multiple copies of the same file is known as data replication and is assisted by replica catalogs -a kind of database relating logical versus physical data files- (Lamehamedi, Szymanski, Shentu, & Deelman 2002). It is worth noting that data replication is not only a fault-tolerance technique but that it also has the potential to reduce massive data access bottle-necks. EGEE, like other grid infrastructures, provides replication mechanisms oriented towards managing multiple copies of the same file in different SEs of the grid. However, applications to upload files to the grid, as well as to copy files between different SEs, and to complete data replication (to multiple SEs), still needs user intervention and management.

As in the case of jobs management, we have developed a group of C++ classes to manage automatic data replication. This replication manager works as follows:

- The user's application makes use of the manager providing a local file path, pointing to the data file to replicate, and an integer representing the number of replicas for it.
- In a first replication phase (see Figure 4), the replication manager performs an upload of

the file from the UI to an SE. The upload of the file to the grid represents a critical step in the replication process as this first copy will be used as the data origin for the next replication phase.

- Once the first copy of the file is successfully uploaded, the manager starts a second phase where data is replicated between different SEs, reducing potential bottlenecks in the UI network. At this iterative step, each replicated SE will start a copy to a new SE until the number of success replicas is, at least, the number specified by the user. The complete replication process can be viewed as a tree-like structure.

The replication process is not free of problems. Things can fail and therefore, the replication manager performs continuous checks on the process. If, for example, the first phase fails, a new SE must be chosen to upload the file. On the other hand, if the first upload goes smoothly but subsequent attempts to replicate such copy to other SEs fail continually, the first copy is considered to be incorrect and the replication must be started again from the beginning. Once the file has been correctly replicated to two or more SEs, the chances of whole replication success improve greatly.

In spite of everything, it is possible that apparently correct replications might have failed in some way, meaning that data may be accessible but corrupted. To minimize the chances of this happening, when data is replicated for use on the grid, its MD5 checksum is computed and stored locally to the UI. When a job uses such data, it will compare the stored checksum against a new calculated one. If data is corrupted, the job will download the file from another SE until a copy is found to be correct. The reader might say: "Well, but what if the first replication has failed? - then all the subsequent copies of the file will be corrupted". This is true and it is very difficult to cover all possible failure scenarios. In this last case, extra checking could be done by downloading the file back from the first SE to the UI and checking the checksum. If it is correct, the probability of error is so low that, in practice, it can be safely considered zero.

Using Scripts to Enhance Grid Job Flexibility

A typical job description is usually comprised of a binary executable and a series of parameters to pass to it when running on a grid node. Nevertheless, this is quite a rigid mechanism as, often

Figure 4. Schematic replication processes

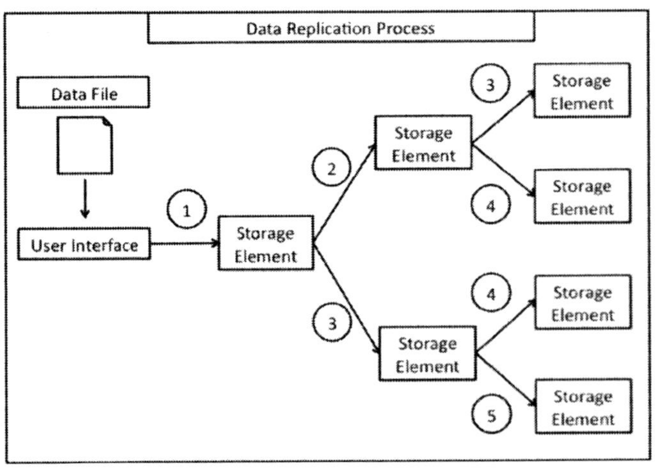

times, we also need some accessory tasks to be performed for our job to be useful. By not directly invoking the executable but instead calling a script - which performs auxiliary tasks and also invokes the executable - can easily achieve this. This strategy is very powerful and flexible, so it was applied both in parameter optimization and 3D ET grid applications.

The following is a typical workflow used for scripts developed for EGEETomo:

- **Data retrieval:** Our jobs will usually process some data. Such data could be stored on different SEs around the grid. Our script can contain the necessary instructions to successfully download required data from the SEs to the WN where it is needed. In the case of EGEETomo, both the executable performing the reconstruction and data files (sinograms) are downloaded to the WN.
- **Data manipulation:** Retrieved data is not always ready to be used by the executables. In fact it is quite common to replicate compressed data so it is more efficiently stored and transferred. Our scripts can decompress/compress data as required. EGEETomo data files are compressed (for optimal transfer times and storage) before they are uploaded to the grid. Therefore, sinograms are needed to be uncompressed before they can be used. Also, the executable permissions are set to "executable" before it is invoked.
- **Data checking:** In the previous section, it was shown that data stored on SEs can be corrupted, which is especially dangerous in some applications as their meaning can be altered. Using the script, we can compute a file checksum to confirm data integrity. If the checksum is not correct, the file can be retrieved from a different SE. During EGEETomo development, this step was added because some sinograms appeared to be corrupted without a reason. It resulted in a problem with some SEs being misconfigured. To avoid such problems (which can be hard to detect and debug), a checksum is always a good alternative.
- **Applications execution:** Once data is ready, not only a single executable can be subsequently invoked but many of them. Therefore, a powerful and adaptable workflow can be defined to work on the data. Here is where the reconstruction program was invoked. If a certain slab is comprised of multiple slices (as usual), a reconstruction program is invoked for each slice.
- **Results preparation and uploading:** Once results are ready, we will probably want to upload them to the grid for later use. Throughput the script, we can compress data and replicate it to different SEs. EGEETomo packed together and compressed the different generated slices comprising a slab. Then this file was replicated to at least three different SEs. Therefore, in the case of an SE failure we will still be able to retrieve the results.

The previous list shows only a small part of the great potential this technique represents. In fact, is can also be used for grid application debugging. The information about failures that can be obtained from classical job submission is limited to the exit code provided by the executable. Using this approach, we can get a much more detailed and personalized description for each failure and even analyze what caused it.

FUTURE TRENDS

After years of development, a lot of promises and, let us face it - a skeptical attitude about everything concerned with the grid - things seem to have started to clear up. New real-world grid applications are rapidly emerging and many of them are starting to be broadly used on production environments. For example, in the last EGEE user forum (celebrated in Clermont-Ferrand, France in 2008), lots of interesting demos were shown working under real conditions. Among the various uses: medical applications for image processing, collaborative tools for medical information-sharing and virtual laboratories showing their capabilities. All these tools were presented to the user through pleasant, user-friendly graphical interfaces and the complexity inherent in the grid was hidden. It seems this is the most probable scenario over the coming years - the creation and deployment of final-user applications, which will mean the grid has finally achieved its main objective - becoming an efficient and useful tool with which to do science. Of course, things will continue to improve: better response times, lower failure rates and improved usability (including infrastructure management) will be the key to extending grid applicability.

In the field of 3D EM, it is intended that the whole process, from the projections to the final reconstructed volume can be integrated onto the grid. Thus, projections for an experiment would be safely stored on the grid. Then, through any kind of user interface, such as a web portal, all successive tasks could be performed over the stored projections from any connected point. Looking further into the future, it would be an impressive utility to have all the related process integrated onto the grid. This would include connecting the previous application with specialized databases, meaning that a resulting new 3D structure could be visited and examined by other interested scientists.

CONCLUSION

The grid is proving to be of great potential in tackling scientific problems where other tools, like cluster computing, are falling short. Also, new applications are appearing as a result of the unique technology developed for the grid. The main advantages of grid computing are its affordability for small and medium-sized groups in comparison with costs derived from using other kind of computing systems, large data management capabilities and great CPU availability. Also, the grid is a secured environment, which is crucial for certain experiments where sensitive data is involved. However, along with the pros, there are also some cons: the grid is not yet well suited for all kinds of applications. Real-time applications can be difficult to implement due to high network and grid service latencies and some special implementation techniques must be followed. For the same reason, tasks involving low CPU time versus data transfer time will not achieve good performance from the grid. Despite this last disadvantage, the grid is still usable, in spite of low performance, when a supercomputer is not available or affordable.

Regarding 3D EM, it is one of the fields that can be greatly improved if grid technologies are adopted because many of the tasks involved present a good profile to be ported to the grid. For example, the presented application for 3D ET provides an easy-to-use tool that completely hides the complexity of the grid from the final user (a biologist) who will thus be able to analyze large data sets at any time without the need for either a dedicated supercomputer or the need to manually control the process.

As a general conclusion, it can be said that the adoption of grid technologies still requires great effort from the developer's point of view. They have the responsibility of making grid experience painless for end-users, who are reluctant to adopt new working tools and who do not want to have to worry about grid complexities. In this aspect, it is important to mention that grid technologies (meaning the middleware) still need to improve their reliability, performance and ease-of-use. The work presented in this chapter required special effort simply to make things work smoothly. Failures arose frequently and unexpectedly affecting data management and grid job management.

ACKNOWLEDGMENT

Work partially funded by grants from the Spanish MEC (TIN2005-00447, BIO2007-67150-C03-01/03), J. Andalucía (P06-TIC-01426), Fondo de Investigación Sanitaria (FIS 04/0683) and EU (LSHG-CT-2004-502828). JRBC is a fellow of the Spanish "Juan de la Cierva" post-doctoral contracts program, co-financed by the European social fund.

REFERENCES

Allcock, B., Bester, J., Bresnahan, J., Chervenack, A. L., Foster, I., Kesselman, C., et al. (2002). Data management and transfer in high-performance computational grid environments. *Parallel Computing, 28*(5), 749-771.

Andersen, A. H., & Kak, C. A. (1984). Simultaneous algebraic reconstruction technique (SART): a superior implementation of the art algorithm. *Ultrasonic imaging, 6*(1), 81-94.

Bilbao-Castro, J. R., Carazo, J. M., García, I., & Fernandez, J. J. (2006). Parallelization of Reconstruction Algorithms in Three-Dimensional Electron Microscopy. *Applied Mathematical Modelling, 30*(8), 688-701.

Bilbao-Castro, J. R., García, I., & Fernández, J. J. (2007). EGEETomo: a user-friendly, fault-tolerant and grid-enabled application for 3D reconstruction in electron tomography. *Bioinformatics, 23*(24), 3391-3393.

Bilbao-Castro, J. R., Merino, A. J., García, I., Carazo, J. M., & Fernández, J. J. (2007b). Parameter optimization in 3D reconstruction on a large scale grid. *Parallel Computing, 33*(4-5), 250-263.

Cheng, Y., Zak, O., Aisen, P., Harrison, S., & Walz, T. (2004). Structure of the human transferrin receptor-transferrin complex. *Cell, 116*(1), 565-576.

Dubochet, J., Adrian, M., Chang, J. J., Home, J. C., Lepault, J., McDowall, A. W., et al. (1988). Cryo-electron microscopy of vitrified specimens. *Quarterly Review of Biophysics, 21*(1), 129-228.

Fernández, J. J., Lawrence, A. F., Roca, J., García, I., Ellisman, M. H., & Carazo, J. M. (2002). High-performance electron tomography of complex biological specimens. *Journal of Structural Biology, 138*(1-2), 6-20.

Fernández, J. J., Bilbao-Castro, J. R., Marabini, R., Carazo, J. M., & García, I. (2005). On the Suitability of Biological Structure Determination by Electron Microscopy to Grid Computing. *New Generation Comput, 23*(1), 101-112.

Fernández, J. J., García, I., Carazo, J. M., & Marabini, R. (2007). Electron tomography of complex biological specimens on the Grid. *Future Generation Computing Systems, 23*(3), 435-446.

Frank, J. (2006). *Three-Dimensional Electron Microscopy of Macromolecular Assemblies: Visualization of Biological Molecules in Their Native State*. Oxford, UK: Oxford University Press.

Frank, J. (1992). *Electron Tomography: Three-Dimensional Imaging whith the Transmission Electron Microscope*. New York: Plenum Press.

Gagliardi, F., Jones, B., Grey, F., Bégin, M. E., Heikkurinen, M. (2005). Building an infrastructure for scientific grid computing: status and goals of the EGEE project. *Phylosophical Transactions: Mathematica, Physical and Engineering Sciences, 363*(1833), 1729-1742.

Gao, H., Sengupta, J., Valle, M., Korostelev, A., Eswar, N., Stagg, S. M., et al. (2003). Study of the structural dynamics of the E. coli 70S ribosome using real-space refinement. *Cell, 113*(1), 789-801.

Garduño, E., Gabor, G. T., & Herman, G. T. (2004). Optimization of Basis Functions for Both Reconstruction and Visualization. *Discrete Applied Mathematics, 139* (1-3), 95-111.

Goodhew, P. J. (1984). *Specimen Preparation for Transmission Electron Microscopy of Materials.* Oxford, UK: Oxford University Press.

Gordon, R., Bender, R., & Herman, G. T. (1970). Algebraic reconstruction techniques (ART) for three-dimensional electron microscopy and x-ray photography. *Journal of Theoretical Biology, 29*(3), 471-481.

Henderson, R. (2004). Realising the potential of electron cryomicroscopy. *Quarterly Review of Biophysics, 37*(1), 3-13.

Henderson, R. (1995). The potential and limitations of neutrons, electrons and x-rays for atomic resolution microscopy of unstained biological molecules. *Quarterly Review of Biophysics, 28*(1), 171-193.

Hendrix, E. M. T., Ortigosa, P. M., & García, I. (2001). On success rates for controlled random search. *Journal of Global Optimization, 21*(3), 239-263.

Hennessy, J. L., & Patterson, D. A., (2006). *Computer Architecture (4th Ed.): A Quantitative Approach.* San Francisco, CA: Morgan Kaufmann.

Hwang, S., & Kesselman, C. (2004). A Flexible Framework for Fault Tolerance on the Grid. *Journal of Grid Computing, 1*(3), 251-272.

Lamehamedi, H., Szymanski, B., Shentu, Z., & Deelman, E. (2002). Data replication strategies in grid environments. In *Proceedings Of the Fifth International Conference on Algorithms and Architectures for Parallel Processing,* Beijing, China, (pp. 378-383). New York: IEEE Computer Society Press.

Laure, E., Fisher, S. M., Frohner, A., Grandi, C., Kunszt, P., Krenek, A., et al. (2006). Programming the Grid with gLite. *Computational Methods in Science and Technology, 12*(1), 33-45.

Lewitt, R. M. (1992). Alternatives to voxels for image representation in iterative reconstruction algorithms. *Physics in Medicine and Biology, 3*(12), 705-716.

Lučić, V., Förster, F., & Baumeister, W. (2005). Structural studies by electron tomography: From cells to molecules. *Annual Review of Biochemistry, 74*(1), 833-865.

Marabini, R., Herman, G. T., & Carazo, J. M. (1998). 3D reconstruction in electron microscopy using ART with smooth spherically-symmetric volume elements (blobs). *Ultramicroscopy, 72*(1), 53-65.

Marinescu, D. C., & Ji, Y. (2003). A computational framework for the 3D structure determination of viruses with unknown symmetry. *Parallel and Sitributed Computing, 63*(1), 738-758.

Novotny, J. (2003). The grid portal development kit. In F. Berman, G. Fox, & A. Hey, *Grid Computing: Making the Global Infrastructure a Reality* (pp. 657-674). Hoboken, NJ: John Wiley and Sons.

Potter, C. S., Zhu, Y., & Carragher, B. (2004). Automated particle selection for cryo-electron microscopy. *Journal of Structural Biology, 145*(1), 1-180.

Scheres, S. H. W., Merino, A. J., Sorzano, C. O. S., & Carazo, J. M. (2005). Grid Computing in 3D-EM Image Processing Using Xmipp. In *Proceedings of the 18th IEEE Symposium on Computer-Based Medical Systems,* Washington, D.C., (pp. 561-563). New York: IEEE Computer Society.

Scheres, S. H. W., Valle, M., Nuñez, R., Sorzano, C. O. S., Marabini, R., Herman, G. T., et al. (2005). Maximum-likelihood Multi-reference Refinement for Electron Microscopy Images. *Journal of Molecular Biology, 348*(1), 139-149.

Sorzano, C. O. S., Marabini, R., Herman, G. T., & Carazo, J. M. (2005b). Multi-objective algorithm parameter optimization using multivariate statistics in three-dimensional electron microscopy reconstruction. *Pattern Recognition, 38*(12), 2587-2601.

Valle, M., Zavialov, A., Li, W., Stagg, S. M., Sengupta, J., Nielsen, R. C., et al. (2003). Incorporation of aminoacyl-tRNA into the ribosome as seen by cryoelectron microscopy. *Nature Structural Biology, 10*(1), 899-906.

van Heel, M. (1987). Similarity measures between images. *Ultramicroscopy, 21*, 95-100.

van Heel, M., & Schatz, M. (2005). Fourier Shell Correlation threshold criteria. *Journal of Structural Biology, 151*(3), 250-262.

Williams, D. B., & Carter, C. B. (2004). *Transmission Electron Microscopy: A Textbook for Materials Science*. Berlin, Germany: Springer.

Wolniewicz, P., Meyer, N., Stroiński, M., Stuempert, M., Kornmayer, H., Polak, M., et al. (2007). Accessing Grid Computing Resources with g-Eclipse Platform. *Computational Methods in Science and Technology, 13*(2), 131-141.

KEY TERMS AND DEFINITIONS

Data Replication: Data replication consists of keeping multiple physical copies of the same logical file. It is intended for fault-tolerance purposes and performance improvement on data access. Data replication is commonly used on grid computing.

EGEE: Enabling Grids for E-sciencE is a project funded by the European Commission which aims to promote grid technologies usage and implementation.

Electron Tomography: Technique to derive the three-dimensional structure of an unique specimen that is imaged in the electron microscope by tilting it around a single axis; this technique is similar to computerized axial tomography in medicine.

Fault Tolerance: In computing, is the ability of a code to manage and recover from a situation which, untreated, would make to application to fail and/or produce unexpected resultss.

gLite: The current middleware used for the EGEE grid. It joins, extends and enhances previously existing grid middleware initiatives like EDG and LCG.

Storage Element: On grid infrastructures, a storage element is where users data is stored. It usually hides the complexity of different types of storage technologies (pools of disks, backup units, etc.) making user interaction easier.

Three-Dimensional Electron Microscopy: Technique to elucidate the three-dimensional structure of biological specimens from projection images taken by electron microscopy.

Three-Dimensional Reconstruction: The computational process to derive the three-dimensional structure of an object from its two-dimensional projections (i.e. as radiographs).

Chapter XXI
Hybrid Mock Circulatory System to Test Cardiovascular Prostheses on the Grid

Francesco Maria Colacino
University "Magna Græcia" of Catanzaro, Italy & University of Calabria, Italy

Maurizio Arabia
University of Calabria, Italy

Gionata Fragomeni
University "Magna Græcia" of Catanzaro, Italy

ABSTRACT

In the last decades cardiovascular diseases greatly increased worldwide, and bioengineering provided new technologies and cardiovascular prostheses to medical doctors and surgeons. The design of active and passive devices aroused notable interests becoming more and more challenging as well as crucial. In this framework, it is important to faithfully reproduce the interaction between the prostheses and the cardiovascular system when in-vitro experiments are performed. For this reason, a new and improved kind of test benches becomes necessary. Purely hydraulic mock circulatory systems showed low flexibility to allow tests of different cardiovascular devices and low precision when a reference mathematical model must be reproduced. In this chapter a new bench is described. It combines the computer model of the cardiovascular system and its real-time interaction with the device to be tested. The solution adopted can be exploited in a Grid environment to allow remote experimentation.

INTRODUCTION

Mock circulatory systems (MCSs) are a very important tool to design and test cardiovascular prostheses (Arabia & Akutsu, 1984; Donovan, 1975; Knierbein et al., 1992; Kolff, 1959; Pantalos, Koenig, Gillars, Giridharan, & Ewert, 2004; Rosenberg, Phillips, Landis, & Pierce, 1981; Vandenberghe, Segers, Meyns, & Verdonck, 2001; Westerhof, Elzinga, & Sipkema, 1971). In-vitro tests provide a better control of working conditions with respect to animal tests while performing experiments. On the other hand, it is hard to correctly reproduce input and output hydraulic loads for the prostheses under test. This is mainly due to difficulties arising when hydraulic components must mimic mathematical models describing the cardiovascular system, since every hydraulic component is affected by spurious effects (Westerhof et al., 1971). Moreover, if different devices have to be tested by means of the same MCS or significant parameters have to be changed in real-time, costs and time to set up the MCS will rise significantly. In order to enhance and exploit the advantages of numerical simulations, even in presence of hydraulic components and prostheses, new test benches have been conceived which aimed at obtaining a more realistic behavior (Baloa, Boston, & Antaki, 2001; F M Colacino et al., 2005; F. M. Colacino et al., 2008; Ferrari et al., 2005; Gwak, Paden, Noh, & Antaki, 2006; Rutten, de Hart, de Mol, van de Molengraft, & de Vosse, 2007). These realizations fall into the framework of hybrid MCSs which, exploiting the interaction between a software environment and a hydraulic prosthesis, represent the new trend in terms of in-vitro experiments.

The development of cardiovascular prostheses is technological challenge, since current technologies allow to understand underlying mechanisms of devices' performance more deeply. All these factors raise both economic and computational costs. For this reason, it would be more useful to develop platforms for MCS sharing.

Grid computing has emerged as a way to take advantage of computing resources spread in various geographical areas. Current Grid technologies offer unlimited computational power and storage capacity for scientific research in heterogeneous fields (Frizziero et al., 2006).

In the past the Grid infrastructure allowed users to aggregate sophisticated resources, such as distributed computational power, data collections and analysis codes, in a flexible environment.

At first, most of the projects taking advantage of the Grid technology focused on the distributed management and analysis of datasets after acquisition, without worrying about the instruments.

Now an increasing attention is directed towards instruments. In the last years widely distributed systems have been developed which are able to remotely control and monitor complex instrumentation in many domains: Aerospace, Medicine, Transport, Earthquake Engineering, Astronomy and Physics (Austin et al., 2004; Kesselman, Foster, & Prudhomme, 2004; Lelli et al., 2007).

Scientific research does not start from data, but from the process of data collection. For this reason, recent projects are now investigating the possibility of exposing instruments directly as Grid services (Bagnasco, Poggi, Parodi, & Scapolla, 2006). Examples are the projects GRIDCC (Lelli et al., 2007), CIMA (McMullen, Devadithya, & Chiu, 2005) and CRIMSON (Lickerman et al., 2006). They aim at integrating the interactive and real-time management of remote instrumentation, storage and computational resources with real-time data analysis and visualization services.

The future trends in terms of Grid platform will be outlined thereinafter to show the feasibility of this technology for implementing a more computationally and logistically efficient MCS.

BACKGROUND

In this section the main MCSs available at the present time for use in functional experiments are briefly described.

Types of Existing Mock Circulatory Systems: Purely Hydraulic

Figure 1 shows the electric analog of Guyton's simplified model of the circulatory system. As it can be noticed, the systemic circulation (comprised between the left ventricle and the right atrium) and the pulmonary circulation (comprised between the right ventricle and the left atrium) are topologically represented by the same components connected likewise. The physiological role of each component has been highlighted by means of arrows indicating their regulation properties (Guyton, Jones, & Coleman, 1973). In particular, the elastic properties (compliance) of vessels are represented by means of capacitors, whose hydraulic realizations are made up of reservoirs, while the vessels' energy loss properties are represented by resistors, whose hydraulic realizations are generally made up of valves or clamps.

All the present purely hydraulic MCSs (briefly described below) are mainly based on Guyton's model and their architecture can be schematically described as shown in Figure 2.

By referring to Figure 2, the prosthesis to be tested is hydraulically integrated into a network where a fluid actuator, responsible for circulating the blood-analogous fluid, is attached to a hydraulic circuit made up of properly designed reservoirs and connections whose role is to mimic the vascular bed of the human body. In addition, pressure and flow sensors are used to monitor and record measurements in the proximity of the prosthesis. These measurements, acquired by a PC provided with a DAQ system, are usually visualized in a graphical format and/or stored for post-processing purposes. It is worth noting the "passive" role to which the computer is banished to in this configuration, being it reduced to a mere signal recorder.

Figure 1. Electric analog of Guyton's simplified model of the circulatory system. Symbols according to Guyton (Guyton et al., 1973). C_{ap}: arterial pulmonary compliance; C_{as}: arterial systemic compliance; C_{vp}: venous pulmonary compliance; C_{vs}: venous systemic compliance; P_{ap}: arterial pulmonary pressure; P_{as}: arterial systemic pressure; P_t: thoracic pressure; P_{vp}: venous pulmonary pressure; P_{vs}: venous systemic pressure; R_{ap}: arterial pulmonary resistance; R_{as}: arterial systemic resistance; R_{vp}: venous pulmonary resistance; R_{vs}: venous systemic resistance.

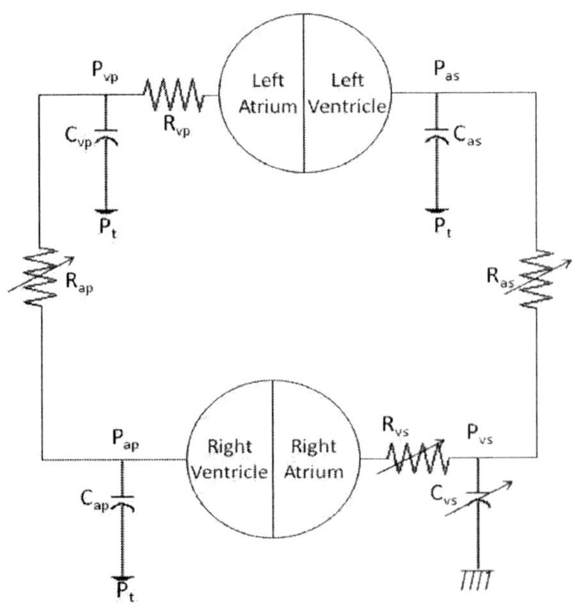

Figure 2. Scheme of present hydraulic MCSs. The fluid actuator is often a Total Artificial Heart (TAH), either electro-mechanic or pneumatic, or a Ventricular Assist Device (VAD). Pressure and flow sensors monitor the prosthesis inlet and outlet, while a DAQ system records the signals.

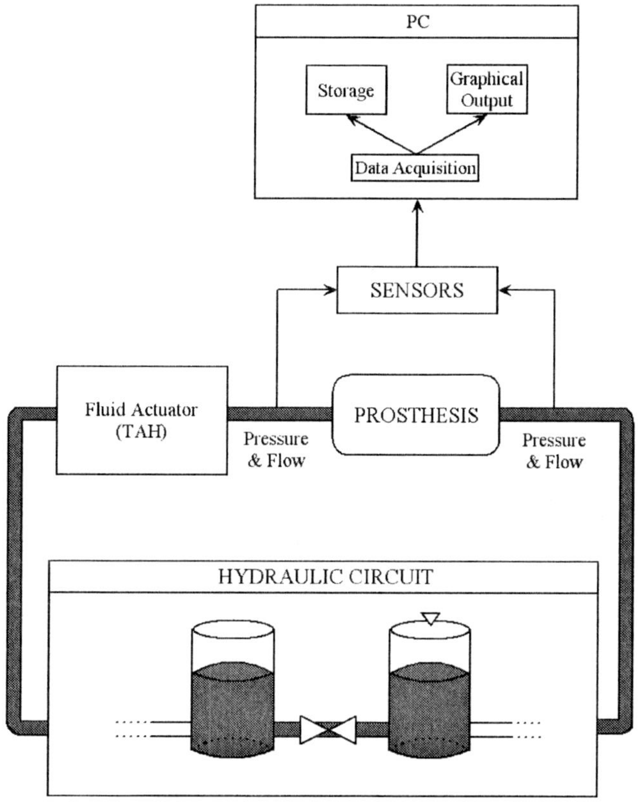

This type of purely hydraulic MCSs presents several limitations, indeed reproducibility of inlet and outlet pressures and/or flow conditions for prostheses to be tested by means of hydraulic components is a complex and challenging issue. Hydraulic feasibility of mathematical models describing the phenomenon under study is characterized by low flexibility when different devices must be tested or significant parameters changed in real time, and the total cost of hydraulic components with respect to those numerically simulated are the main limits of hydraulic MCSs.

The fluid actuator itself, which is characterized by limited responsiveness to changing working conditions, represents another limitation. In general, artificial circulation of blood-analogous fluids into mock circulatory systems is obtained by means of total artificial hearts (TAHs) or ventricular assist devices (VADs), which are reciprocating pumps (also known as 'ventricle pumps'). A reciprocating pump consists of a variable volume chamber completed by two valves allowing respectively, inflow and outflow of the blood. The pump chamber consists of two parts, namely a semi-rigid housing containing the inlet and outlet and the corresponding valves, while the other part is flexible and consists of a thin sheet of pliable material in the shape of either a sac or membrane (or diaphragm). In the majority of cases, pneumatic TAHs or VADs are used. The primary energy is thus provided by compressed air, produced by an external driving unit connected by means of flexible tubes, to the driving chamber of the mock ventricle. A system of electric valves connects, in turn, the driving chamber with a source of pressurized air (pressure tank) and a vacuum source (vacuum tank), to create the systole (ejection) and the diastole (filling). Sometimes the source of decompressed gas is constituted by the atmosphere. Most of the electric valves used so far have been of the on-off type, i.e. with uncontrolled opening and closing phases. However, with the on-off type electric valves is difficult to obtain a ventricular pressure pattern similar to that in physiological conditions. Moreover, with manual control of the air pressure level in the tanks cardiac output cannot be adapted to body requirements from one cycle to the next.

The passive hydraulic elements of the MCS representing the vascular network of the human body are usually realized by properly designed reservoirs and connections.

The Kolff's Simulator (Kolff, 1959) was the first MCS to be realized (1959). It is extremely simple, the diastolic pressure at the base of the aorta being obtained by means of an 82 cm water column (60 mmHg), while a non rigid tube is used in order to reproduce the compliance of the aorta. After passing through a flow meter, the liquid ends up in an open air reservoir that constitutes the venous system reserve. These three elements represent the systemic circulation. At the exit from the right ventricle a structure similar to the one just described reproduces the lung circuit (or pulmonary circulation). This type of bench was realized to test total artificial hearts (TAHs). If the TAH under test works correctly (balance of output from the two ventricles), one will not observe any level variations in the two venous reservoirs and the liquid volumes present in the two circuits will therefore remain stable. Kolff perfected his MCS in later years, introducing air compliances for the arterial branches.

The simulator of the Hydrospace Research Corporation (*Operation and maintenance manual for the mock circulatory system*, 1965) can be distinguished from the previous one by the fact that its design paid particular attention to reproducing the flow dynamics of the systemic and pulmonary circuits, using the characteristic hemodynamic data of a middleaged man as a reference. Numerous controls and adjustments of the circulatory parameters are possible in this MCS. The compliances are still of the air type, while the control system acts on one parameter at a time (i.e. arterial pressure, venous pressure, or cardiac output). Having chosen the parameter to be controlled, the others become dependent variables.

The Westerhof's simulator (Westerhof et al., 1971) dominated the subsequent MCS designs. It was built to faithfully reproduce the arterial input impedances, both systemic and pulmonary. The innovative concepts introduced by these authors are: the introduction of the aortic characteristic resistance, which takes also into account high frequency properties of the proximal aorta; the accurate design (with no spurious influences) of the three basic hydraulic components constituting the vascular network, be they resistance, capacitance or inertia, at the frequencies of interest for the human physiology.

The Donovan's simulator (Donovan, 1975) is the one that comes closest to the Guyton's simplified model of Figure 1. The compliances are of the air type, and regulation of the peripheral resistances is obtained by obstructing the tube opening by means of a diaphragm. It includes an automatism to vary the peripheral resistances in such a manner as to obtain physiological pressure/flow rates curves and time constants.

The Pennsylvania State University MCS (Rosenberg et al., 1981) is characterized by the entire arterial tree being reproduced by means of a windkessel (one resistance and one compliance). The compliance is of the spring type, while the peripheral resistance can be varied by flattening a bundle of tubes. This simulator is not equipped with any automatisms and permits only tests of the steady type.

The Arabia's simulator (Arabia & Akutsu, 1984) makes possible to verify the interaction between TAH and MCS, although only in steady conditions. The simulator is controlled by means of five variables (the arterial and pulmonary resistance, the venous systemic resistance, the unstressed volume of the systemic venous compliance, and the venous systemic compliance). These variables are used to control the cardiac output in accordance to Guyton's model. The arterial branches have been realized with four hydraulic cells in an attempt to obtain a correct hydraulic load that will ensure physiological impedances, but also a correct total compliance. Both these characteristics, together with correct values of the venous compliance, allow to obtain a physiological pressure distribution, as well as physiological average volumes and pressures in the various circuits of the system. The venous system reserve is realized by means of a single compliance with an adjustable rest volume. The peripheral resistances are of the laminar type for streamline motion and can be regulated by means of a slide that modifies the number of open capillaries. The compliances in this MCS are of the spring type.

The Knierbein's simulator (Knierbein et al., 1992) was constructed for the purpose of testing a TAH in steady conditions and obtaining information about the cardiac function curves. Particular attention was therefore paid to realizing impedances on the input and output side of the heart. The greater part of the hydraulic elements can be regulated in such a way as to make it possible to reproduce various load conditions for the TAH under test, but effects dependent on the propagation of pressure waves cannot be reproduced with this simulator.

A NOVEL HYBRID MOCK CIRCULATORY SYSTEM

As one can see from the descriptions referred to purely hydraulics MCSs, each adopted solution, with its relative configuration, was conceived to pay particular attention "solely" to specific aspects of the cardiovascular system, the complexity of the hydraulic setup increasing exponentially with the aspects one desires to mimic.

For these reasons and by considering how flexible, precise, and accurate mathematical models are when compared with analog hydraulic components, a new type of MCS shall be conceived to conjugate numerical simulations with the hydraulic environment the cardiovascular prostheses are going to operate in.

A scheme of this newly conceived hybrid MCS is shown in Figure 3.

This hybrid test bench comprises: a computer which solves the differential equations of the mathematical model describing the cardiovascular system; electro-hydraulic interfaces (EHIs), which, driven in real-time by the computer, convert electric signals (voltage and/or current) into hydraulic quantities (pressure and/or flow), in order to reproduce the desired working conditions for the prosthesis under test; sensors, placed at the inlet and outlet of the prosthesis, which feed back inlet and outlet measurements to the computer, that in turn performs the next step of computation.

Figure 3. Scheme of the Hybrid MCS (see discussion in text)

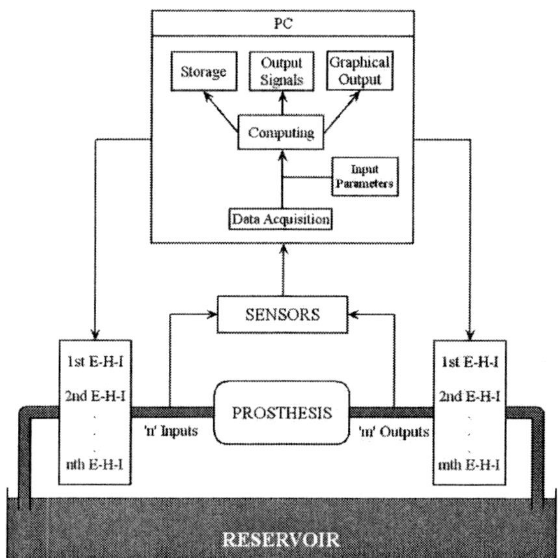

Typically, every cardiovascular prosthesis is a single input-single output system. As a consequence, at least two EHIs are necessary: one placed at the inlet and the other at the outlet of the prosthesis to be tested. Moreover, either passive or active cardiac assist device can be described by a relationship which links pressure and flow, that is P(t)=f[Q(t)], or its inverse Q(t)=g[P(t)]. Once the input variable to the system is chosen, which relationship must be used is determined as a consequence. For example, if pressure is chosen as the input variable to the hydraulic prosthesis, the software module, which in turn has instantaneous measures of flow Q(t) as an input, will provide the instantaneous values of pressure P(t) as the result of the numerical computation. Hence, an EHI will be inserted between the software module and the inlet of the device to be tested. This EHI will instantaneously generate the values of pressure P(t) resulting from the computation. In this case, a flow meter will be placed between the EHI and the prosthesis to feed back this measurement to the computer, and the EHI will work as a pressure controlled pump.

The hydraulic circuit is then completed by a very large reservoir (infinite compliance) filled with the working fluid.

When compared to the classical hydraulic one (Figure 2), Figure 3 schematically shows the step forward of the hybrid architecture. More tasks are now carried out by the computer, which is not used solely as a mere signal recorder: the main important improvement is the translation of the cardiovascular system from being hydraulic to being numeric. Indeed, the mathematical model of the cardiovascular system is now implemented on the computer (block Computing in Figure 3) instead of being realized hydraulically. A mathematical library (represented by the Input Parameters block in Figure 3) allows mimicking all the possible testing scenarios and the EHIs is used to link the mathematical environment and the prosthesis to be tested, which now represents the only hydraulic component of the entire loop (beside, of course, the EHIs). If, for example, control algorithms for VADs have to be developed, appropriate physiological and pathological conditions can be thus reproduced, being the VAD under test the only hydraulic component of the cardiovascular loop.

The Hybrid Mock Circulatory System Prototype

In between the two above mentioned solutions, that is purely hydraulic or hybrid mock circulatory systems, new test benches exploiting the hybrid solution have been recently developed (Baloa et al., 2001; F M Colacino et al., 2005; F. M. Colacino et al., 2008; Ferrari et al., 2005; Gwak et al., 2006; Rutten et al., 2007).

Figure 4 schematically shows the architecture of this newly developed MCSs. In this case, the prosthesis to be tested is connected to a hydraulic circuit mimicking the vascular system (as it is usually done in hydraulic MCSs - Figure 2). On the other hand, the fluid actuation is realized by an EHI mimicking the left ventricle. This fluid actuator is controlled in real-time by the computer according to the elastance-based mechanism of the native ventricle, thus reproducing the correct ventricular pressure-volume relationship and its correct interaction with the hydraulic circuit connected to it. Therefore, the component of the whole cardiovascular system being simulated by the computer is "only" the native ventricle, while the vasculature is still realized hydraulically.

Baloa et al. (Baloa et al., 2001) conceived a pressure controlled bellow pump driven in real time by the time-varying elastance function (real time control of the ventricular pressure-volume relationship of the ventricle). They tested the mock ventricle responsiveness to changes in preload, afterload, and contractility by means of the PV loop analysis. It must be noted that the presence of an elastic diaphragm inside the ventricular chamber filters the pressure disturbances, although makes difficult to exactly measure the instantaneous pump volume.

Figure 4. Scheme of present solutions of Hybrid MCSs (see discussion in text)

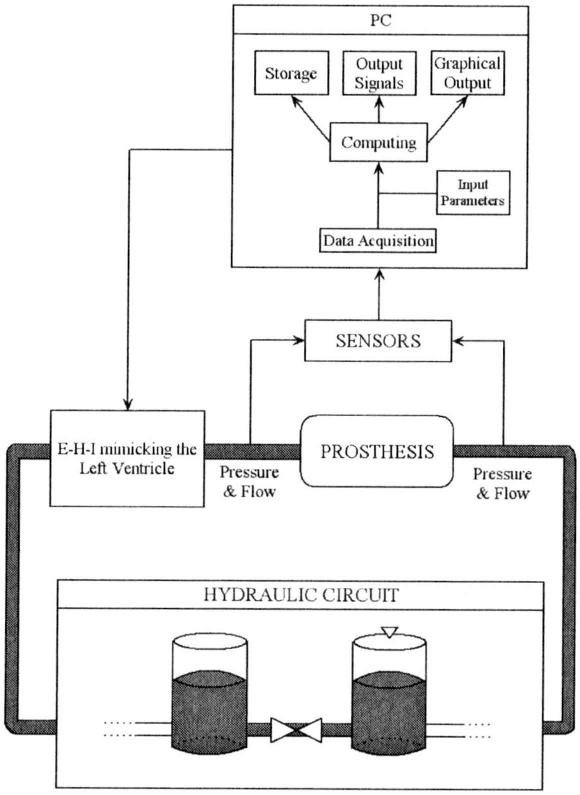

Ferrari et al.(Ferrari et al., 2005) showed the feasibility of the interaction between hydraulic and numerical environment. This interaction was obtained by electro-hydraulic interfaces driven by numerical simulations and mimicking appropriate flow and pressure signals at the connections with the hydraulic environment. Nevertheless, in their work an explicit real-time control of the elastance function is absent: the LV pressure-volume relationship is controlled by adaption to a fixed volume reference signal which does not change in real time with the working conditions.

Gwak et al.(Gwak et al., 2006) showed experimental preliminary studies on an elastance-based MCS equipped with gear pumps as fluid actuators. Their setup is at moment the realization which more closely reproduce the hybrid solution as depicted in Figure 3. The gear pumps were controlled to mimic the principal active and passive elements of the circulatory system, thus reproducing the impedances of the cardiovascular system. Their solution was successfully tested on a numerical basis, but the mock loop experiments (limited to only one working condition) showed the strong limitation of pronounced pressure ripples due to the type of pump used.

Rutten et al.(Rutten et al., 2007) used a pressure controlled positive displacement pump driven in real time by a Hill-type heart contraction model (instead of an elastance-based model). A baroreflex heart rate control mechanism was included as well. In this work the mock circulatory system interaction with different types of LVADs (Left ventricle VADs), both of the pulsatile and continuous flow type, was analyzed with appreciable results.

In the works from our group (F M Colacino et al., 2005; F. M. Colacino et al., 2008) a valved screw-driven piston pump was used to mimic the left ventricle. The pressure measured into the pump chamber was fed back into the mathematical model of the left ventricle and the calculated reference left ventricular volume was used to drive the piston. Moreover, a modified version of the classical elastance model of the native heart was used to drive in real-time the mock left ventricle and results show its suitability even in presence of sudden pressure disturbances (which represent a key issue and are not negligible). This real-time controlled mock ventricle is able to reproduce the elastance mechanism of a natural ventricle by mimicking its preload (mean atrial pressure) and afterload (mean aortic pressure) sensitivity, that is the Starling mechanism (Starling, 1918).

All the above described solutions, characterized by the real-time control of mock ventricles, fall within the framework of hybrid systems conceived to reproduce the correct ventricle-vascular system interaction aimed at improving design and test of cardiovascular prostheses. They represent a first step toward the fully hybrid setup as the one shown in Figure 3.

THE HYBRID MOCK CIRCULATORY SYSTEM ON THE GRID

A diverse community of scientists, medical doctors, surgeons and engineers studies the impact of cardiovascular prostheses on living organism by using MCSs, with the goal of making them efficient and safe. Advances in understanding and improving cardiac assist devices are directly linked to access to MCS testing facilities. Sharing both facilities and resulting experimental data across bioengineering community can enhance the usefulness of experimental studies and produce improved engineering results.

For the above mentioned reasons, significant breakthrough in prostheses' design and development could be achieved by introducing advanced networking, distributed computing and collaboration technologies into MCS testing procedures.

A distributed software platform that is suitable to host such a collaborative MCS is the Grid. The goal of a MCSgrid (MCS on the Grid) project is to establish a virtual organization able to share data repositories and equipment sites, to increase computational power and to broaden bioengineering community. MCSgrid builds on a standard Grid infrastructure, by providing simulation services for remote users.

A properly designed MCSgrid will be able to address three high-level requirements:

- Increase access to hybrid MCS equipment by integrating this platform with simulation and storage resources.
- Facilitate sharing of simulation codes and mathematical models needed to perform simulations, through Grid resources.
- Increase the opportunity for scientists to collaborate in research investigations.

By means of a cooperative environment combining simulations, experiments and stored data, MCSgrid will allow easier access to computers and storage systems, provide collaborative in-

Figure 5. MCS remote control for performing experiments. By means of a Graphical User Interface, the remote user can: set up the experiment and select the relative parameters; visualize results in a numeric format. The dashed box encloses blocks relative to local equipment and instrumentation.

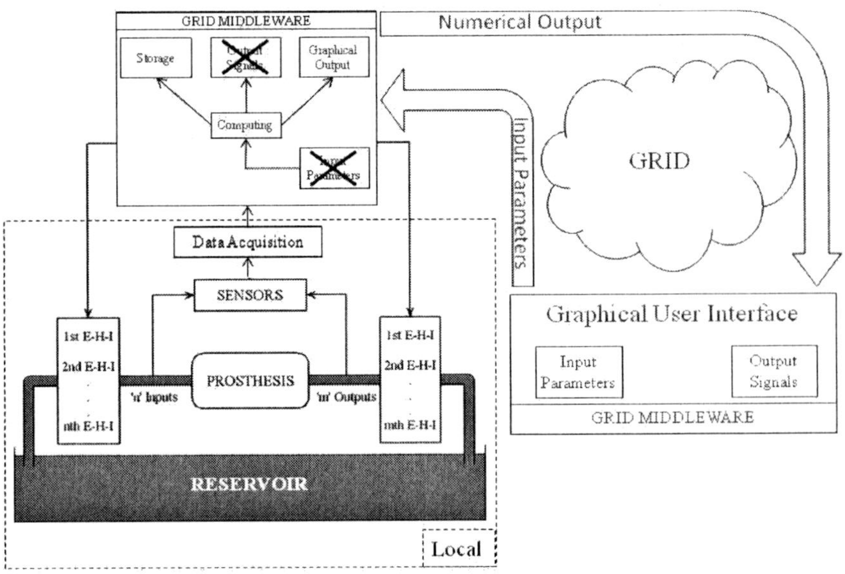

Table 1. Example of numerical results received by the remote user after a test. Both the input parameters and the resulting quantities are listed.

Parameters			***Description***
Symbol	***Units***	***PH***	
HR	*bpm*	80	Heart Rate (bpm)
Δ	*%*	0	Intraventricular phase lag percentage
Results			
η		0,76	Efficiency
dp/dt_{max}	*mmHg/s*	1609,01	Maximum value of dp/dt
$P_{Ao\text{-}ave}$	*mmHg*	99,51	Average aortic pressure (mmHg)
$P_{At\text{-}ave}$	*mmHg*	6,50	Average left atrial pressure (mmHg)
$Q_{A\text{-}ave}$	*l/min*	5,51	Average flow rate in Aorta (l/min)
r_{max}	*cm*	3.03	End diastolic radius
SV	cm^3	71,59	Stroke Volume
V_{AED}	cm^3	59,22	Maximum left atrial volume (cm^3)
V_{AES}	cm^3	27,62	Minimum left atrial volume (cm^3)
V_{mean}	cm^3	79,41	LV's mean volume
V_{VED}	cm^3	115,20	Left end-diastolic Volume (cm^3)
V_{VES}	cm^3	43,61	Left end-systolic Volume (cm^3)
w_{min}	*cm*	1.25	End diastolic wall thickness
σ	*mmHg*	94,39	Systolic ventricular wall stress
P_0	*mmHg*	0,00	y-coordinate of the ventricular elastance curve rotation point
V_0	cm^3	10,00	x-coordinate of the ventricular elastance curve rotation point
t_{IC}	*s*	0.11	Duration of the isometric contraction

Figure 6. MCS remote control with additional graphical output. In this case the remote user visualizes experiments' results also in a graphic format. The dashed box encloses blocks relative to local equipment and instrumentation.

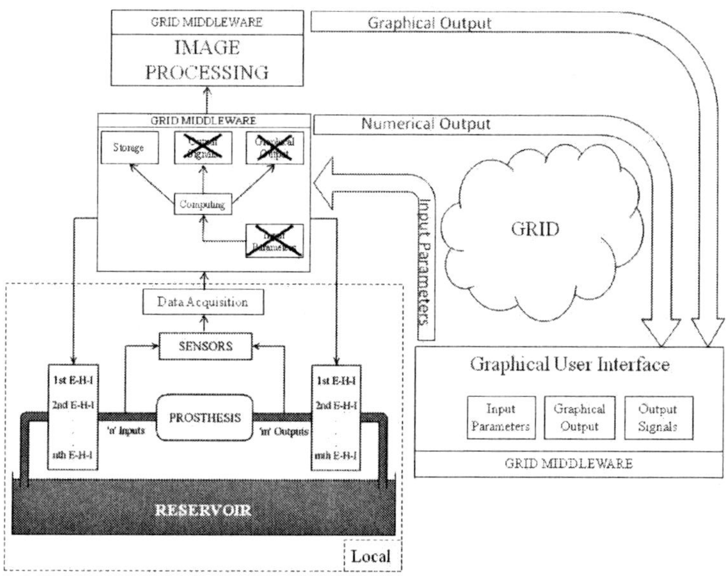

Figure 7. Example of graphical output received by the remote user after a test. In this case a real-time variation of the working conditions of the mock ventricle, mimicking the elastance mechanism of the native ventricle, is plotted in real-time (Left: pressure and volume waveforms; Right: pressure-volume loops).

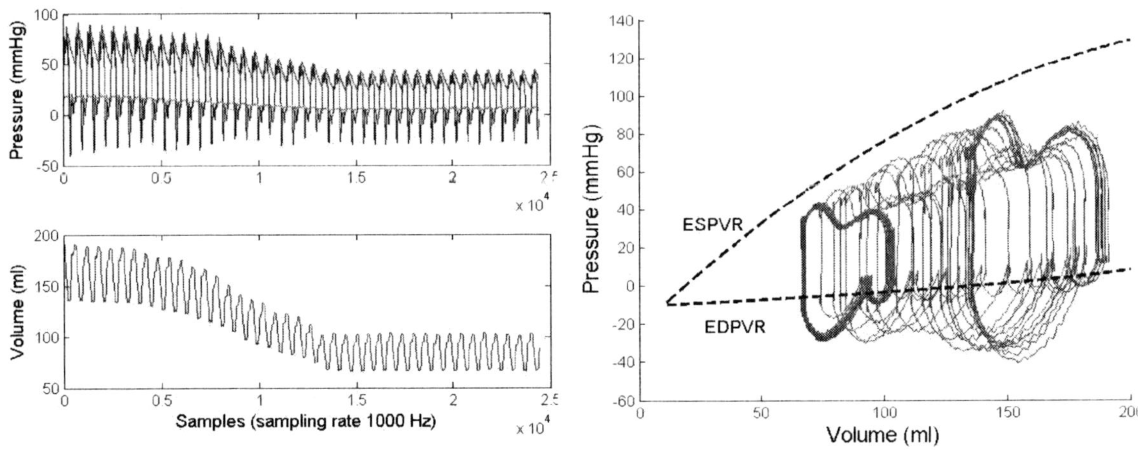

frastructure by which existing co-operations can interact, and increase the capabilities of existing Web-based tools. MCSgrid can thus improve the research in the cardiovascular field by stimulating the development of new methods, as illustrated by the two following scenarios.

MCS Remote Control and Test Performing

Performing tests on cardiovascular prostheses requires several resources. MCSgrid collaboration and teleoperation facilities, when applied to a specific MCS, will allow to share all the instrumentation and to utilize the mathematical models implemented accordingly, thus allowing the remote control of the MCS setup.

In this new approach, different experiments on a same device placed on a specific MCS, can be performed from different sites. Besides this, each remote user can improve a specific test by using a custom mathematical model in order to change the device working conditions.

Such a scenario can be used by a research institution or a company spread on a broad geographical area. Indeed the use of a limited number of MCS allows to reduce experiments' costs while enhancing their standardization.

Before starting an experimental session, the local MCS administrator selects and installs the prosthesis under test according to the indications of a remote user. After the setting of the hardware procedure has been completed, the remote user can access the simulation engine on a Grid node and, by introducing the parameters related to the mathematical model that mimics the desired working scenario, he can run the simulation which drives the EHIs to perform the test (Figure 5). Whenever the experiment is completed, he receives the corresponding results (one example is shown in Table 1) and may decide further actions (e.g. to carry out other experiments).

MCS Remote Control and Graphical Output Management

Another improvement one may seek about the use of MCSs concerns the elaboration of the graphical output for a specific test. Often, one single personal computer cannot elaborate and visualize the graphical output of a specific experiment, as the resources used while the experiment is in run are plenty and used for higher priority tasks (i.e. simulation and control of the test environment). For this reason, the software components that process in real-time the data to obtain the graphical output could be implemented on the Personal Computer of every single remote user, instead of residing only on the local machine.

In practice, in addition to the tasks shown in Figure 5, by means of the Graphical User Interface the remote user can also receive the graphical output of the experiment in order to better evaluate the test (Figure 6). One example of graphical output is shown in Figure 7.

CONCLUSION

The solution proposed is a basic step to radically improve in-vitro tests for cardiovascular prostheses. The scientific community can gain several benefits by coupling MCSs and Grid Computers technologies. Indeed, the creation of specific units specialized in testing cardiovascular devices permit the reduction of costs and time: different tests can be performed with a minimal or, in some cases, no setup time, while gaining in standardization.

Finally, the greatest advantage of this approach is the sharing of data repository and the improvement of collaboration within the scientific community.

REFERENCES

Arabia, M., & Akutsu, T. (1984). A New Test Circulatory System for Research in Cardiovascular Engineering. *Annals of Biomedical Engineering, 12*, 29-48.

Austin, J., Jackson, T., Fletcher, M., Jessop, M., Cowley, P., & Lobner, P. (2004). Predictive maintenance: distributed aircraft engine diagnostics. In Elsevier (Ed.), *The grid 2* (2nd Ed., pp. 69-79). San Francisco, CA: Morgan Kaufmann.

Bagnasco, A., Poggi, A., Parodi, G., & Scapolla, A.M. (2006). Exposing Measurement Instruments As Grid Services. In *Distributed Cooperative Laboratories: Networking, Instrumentation, and Measurements* (pp. 321-329). New York: Springer.

Baloa, L., Boston, J., & Antaki, J. (2001). Elastance-Based Control of a Mock Circulatory System. *Annals of Biomedical Engineering, 29*, 244-251.

Colacino, F.M., Arabia, M., Danieli, G.A., Moscato, F., Nicosia, S., Piedimonte, F., et al. (2005). Hybrid test bench for evaluation of any device related to mechanical cardiac assistance. *The International Journal of Artificial Organs, 28*(8), 817-826.

Colacino, F.M., Moscato, F., Piedimonte, F., Danieli, G., Nicosia, S., & Arabia, M. (2008). A modified elastance model to control mock ventricles in real-time: numerical and experimental validation. *Asaio J, 54*(6), 563-573.

Donovan, F.J. (1975). Design of a hydraulic analog of the circulatory system fo evaluating artificial Hearts. *Biomaterials, Medical Devices, and Artificial Organs, 3*(4), 439-449.

Ferrari, G., Kozarski, M., De Lazzari, C., Gorczynska, K., Tosti, G., & Darowski, M. (2005). Development of a hybrid (numerical-hydraulic) circulatory model: Prototype testing and its response to IABP assistance. *International Journal of Artificial Organs, 28*(7), 750-759.

Frizziero, E., Gulmini, M., Lelli, F., Maron, G., Oh, A., Orlando, S., et al. (2006). *Instrument Element: a new Grid component that enables the control of remote instrumentation.* Paper presented at the Proceedings of the Sixth IEEE International Symposium on Cluster Computing and the Grid Workshops (CCGRIDW'06).

Guyton, A.C., Jones, C.E., & Coleman, T.G. (1973). *Circulatory Physiology: Cardiac Output and its Regulation* (2nd Ed.). Philadelphia: W.B. Saunders.

Gwak. K.-W., Paden, B.E., Noh, M.D., & Antaki, J.F. (2006). Fluidic Operational Amplifier for Mock Circulatory Systems. *IEEE Transactions of Control Systems Technology, 14*(4), 602-612.

Kesselman, C., Foster, I., & Prudhomme, T. (2004). Distributed Telepresenze:The NESSgrid Earthquake Engineering Collaboratory. In Elsevier (Ed.), *The grid 2* (2nd Ed, pp. 81-93). San Francisco, CA: Morgan Kaufmann.

Knierbein, B., Reul, H., Eilers, R., Lange, M., Kaufmann, R., & Rau, G. (1992). Compact mock loops of the systemic and pulmonary circulation for blood pump testing. *The International Journal of Artificial Organs, 15*(1), 40-48.

Kolff,W.J. (1959). Mock circulation to test pumps designed for permanent replacement of damaged hearts. *Cleve Clin Q, 26*, 223-226.

Lelli, F., Frizziero, E., Gulmini, M., Maron, G., Orlando, S., Petrucci, A., et al. (2007). The many Faces of the Integration of Instruments and the Grid. *International Journal of Web and Grid Services, 3*(3), 239-266.

Lickerman, E., Bry, L., Herring, N., Pandelidis, Y., Afsheen, N., & Sharif, A. (2006). *Grid-Enabled Bio-Sample Management Application for Data-Intensive Biomarker Analysis.* Paper presented at the Proceedings of the 19th IEEE Symposium on Computer-Based Medical Systems.

McMullen, D.F., Devadithya, T., & Chiu, K. (2005). *Integrating Instruments and Sensors into the Grid with CIMA Web Services.* Paper presented at the Proceedings of the Third APAC Conference on Advanced Computing, Grid Applications and e-Research.

Operation and maintenance manual for the mock circulatory system. (1965). Rockville, MD: Hydrospace Research Corporation.

Pantalos, G., Koenig, S., Gillars, K., Giridharan, G., & Ewert, D. (2004). Characterization of an Adult Mock Circulation for Testing Cardiac Support Devices. *ASAIO Journal, 50*(1), 37-46.

Rosenberg, G., Phillips, W., Landis, D., & Pierce, W. (1981). Design and Evaluation of the Pennsylvania State University Mock Circulation System. *ASAIO Journal, 4*(1), 41-49.

Rutten, M., de Hart, J., de Mol, B., van de Molengraft, R., & de Vosse, F. (2007). Analysis of LVADs in a controlled mock systemic circulation. *ASAIO Journal, 53*(2), 26A.

Starling, E. (1918). *The linacre lecture on the law of the heart.* London: Green and Co.

Vandenberghe, S., Segers, P., Meyns, B., & Verdonck, P. (2001). Hydrodynamic Characterization of Ventricular Assist Devices. *International Journal of Artificial Organs, 24*(7), 470-477.

Westerhof, N., Elzinga, G., & Sipkema, P. (1971). An Artificial Arterial System for Pumping Hearts. *Journal of Applied Physiology, 31*(5), 776-781.

KEY TERMS AND DEFINITIONS

Afterload: In cardiovascular mechanics, is the myocardial wall stress during systolic ejection. If the muscle is allowed to shorten only after developing a certain force, this additional force is called afterload. Arterial input impedance is the ventricular afterload system. It influences, but is independent of, cardiac ejection of blood into the arterial vasculature and the resultant instantaneous pressure and volume of the ventricle. Instantaneous ventricular pressure and flow are the ventriculo-arterial coupling variables.

The arterial input impedance is defined as the ratio of arterial pressure to arterial flow, both of which are measured at the root of the ascending aorta and expressed as a complex variable in the frequency domain. Mean arterial pressures can be considered as appropriate measures of ventricular afterload.

Compliance: It is the tendency of the vessels to stretch in response to pressure variations. It has a large effect on perfusion and blood pressure. Compliance is computed as: $=\frac{\Delta V}{\Delta P}$, where ΔV and ΔP are the volume and pressure variations, respectively. It is the reciprocal of elastance.

Elastance: (Pressure-Volume Relationship, and Time-Varying Elastance): A powerful tool for studying cardiac mechanics and the interactions between ventricle and circulatory system is the pressure-volume (PV) relationship.

Cardiac contractions can be viewed "through the window of the pressure-volume diagram". Thus, it is possible to link the instantaneous pressure $P_{lv}(t)$ and the instantaneous blood volume $V_{lv}(t)$ inside the ventricle by means of the following pressure-volume relationship:

$$P_{lv} = P_0 + E(t)\cdot[V_{lv}(t) - V_0]$$

where P_0 is the external pressure, V_0 is the ventricular volume at zero transmural pressure $[P(t)-P_0]$, and E(t) is the time-varying ventricular elastance.

By neglecting the term P_0, the ratio:

$$E(t) = \frac{P_{lv}(t)}{[V_{lv}(t) - V_0]}$$

states that in one cardiac cycle the relationship between the instantaneous pressure $P_{lv}(t)$ and the instantaneous volume $V_{lv}(t)$ inside the ventricle is a time-dependent parameter, which is the time-varying elastance E(t). By connecting isochronous sets of instantaneous pressure-volume data points, a family of curves whose slope rise and fall can be plotted. This results in an increase and decrease of the elastance.

All the elastance curves E(t), if normalized with respect to the peak value E_{max} and the time to peak T_{max} would reduce a single curve which has unique shape for any heart, loading conditions, contractile state, or heart rate. As a consequence, E_{max} and T_{max} can serve as a reliable index of ventricular contractility and are independent of preload (end diastolic ventricular volume/atrial pressure) and afterload (aortic pressure/arterial impedance).

A more accurate formulation of the pressure-volume relationship is the non-linear time-varying elastance model, where a non-linear time varying function is used instead of the linear one, E(t). Nevertheless, the classical linear formulation of the elastance is still meaningful and much easier to use in most of the cases.

Mock Circulatory System (MCS): is a hydraulic test bench used to test cardiovascular prostheses. By mimicking the cardiovascular system, MCSs allow analyzing the functionality of each prostheses, their design, and their interactions with the mimicked environment.

Preload: In cardiovascular mechanics, is the myocardial wall stress at the end of diastole. As a measure of ventricular preload, end-diastolic (end-filling) volume is preferable to end-diastolic pressure for its closer relationship to muscle length, and thus to sarcomere length, but end-diastolic pressure can be used as preload as well. Mean atrial pressure can be considered as an appropriate measure of ventricular preload.

Pulmonary Circulation: It is the portion of the cardiovascular system comprised between right ventricle and left atrium. Its task is to transport blood poor of oxygen throughout the lungs and to return oxygenated blood to the heart.

Systemic Circulation: It is the portion of the cardiovascular system comprised between left ventricle and right atrium. Its task is to transport oxygenated blood throughout the peripheral circulation (except lungs) and to return deoxygenated blood to the heart.

Section V
Collaborative Grids for Healthcare and Clinical Applications

Chapter XXII
Grid Technologies in Epidemiology

Ignacio Blanquer
Universidad Politécnica de Valencia, Spain

Vicente Hernández
Universidad Politécnica de Valencia, Spain

ABSTRACT

Epidemiology constitutes one relevant use case for the adoption of grids for health. It combines challenges that have been traditionally addressed by grid technologies, such as managing large amounts of distributed and heterogeneous data, large scale computing and the need for integration and collaboration tools, but introduces new challenges traditionally addressed from the e-health area. The application of grid technologies to epidemiology has been concentrated in the federation of distributed repositories of data, the evaluation of computationally intensive statistical epidemiological models and the management of authorisation mechanism in virtual organisations. However, epidemiology presents important additional constraints that are not solved and harness the take-off of grid technologies. The most important problems are on the semantic integration of data, the effective management of security and privacy, the lack of exploitation models for the use of infrastructures, the instability of Quality of Service and the seamless integration of the technology on the epidemiology environment. This chapter presents an analysis of how these issues are being considered in state-of-the-art research.

INTRODUCTION

Epidemiology is defined as "the study of factors affecting the health and illness of populations, which serves as the foundation and logic of interventions made in the interest of public health and preventive medicine" (wikipedia, 2008). The epidemiology study is the main support for public health research which involves those issues that are relevant for a population as a whole, rather than the specific treatment of individuals. Epidemiology is closely related with public health, which uses epidemiology and geographic information as main tools to describe the outbreak, evolution and eradication of diseases.

The management of individuals' health is normally structured around the concept of Electronic Health Record (EHR) (EUROREC, 2002). Notwithstanding the commonalities between epidemiology and EHR, there are several important differences which should be considered:

- Specificity and Sensibility in epidemiology studies have different relevance in individual clinical practice and global population analysis, since one erroneous or missing register could not affect a global result, but can be critical for an individual.
- Epidemiology studies are not bounded by interactivity constraints since they are not used at the delivery of health.
- Confidentiality is considered at a different level, since data is anonymised or pseudo-anonymised, requiring different (additional in some cases) techniques.
- The level of detail of the information managed is coarser, considering that the aim is to find and study general patterns.
- Along with the clinical data, information from models, clinical trials, geographic data and social information is of high importance.

However, there are several issues that are shared with EHR:

- A large part of the data comes from clinical practice, and thus it goes through the needs and problems of data coding and integration.
- Integration of data around the identity of the patient. Although data is considered at the level of global population, information coming from individuals must be integrated in the form of complete entries.
- Need for normalisation, standard coding and standard interfaces, which are even more important in the case of epidemiology, as the data will be automatically processed.

The main needs of epidemiology are the integration and processing of clinical, social and geographical data through statistical methods and models. These issues require tackling the following technological challenges.

- Integrate the information from diverse sources of medical, social and geographical data for the construction of data warehouses, knowledge bases or similar data repositories to drive the epidemiology research.
- Use epidemiology models, data mining tools, biocomputing and other tools to extract correlations, predict effects and guide public health policies.
- Ensure a correct and legal use of the information, without compromising the privacy of the individuals.

Grid technologies constitute a successful alternative for solving those issues. This chapter presents the current status of grid technologies applied on epidemiology and how the above issues are solved in several state-of-the-art international projects.

The objective of the chapter is to make a survey of how state-of-the-art projects are solving the problems of data management, data processing and data security in epidemiology. Since there are many different use cases and approaches, different projects are considered and analysed. It does not aim at providing an exhaustive list of projects and papers, but rather to a list of relevant alternatives that can be considered as good success stories for the use of epidemiology and grids.

The chapter is structured as follows. Section 3 presents the current status of IT in epidemiology, discussing the different problems, approaches and needs. Section 4 presents a set of significant use cases that assist on defining the steps and technological needs. Section 5 deals with the needs on data integration and quality, describing the approaches followed in several projects on data federation, data collection, data knowledge representation and data extraction. Section 6 deals with the issue of processing, describing some cases on biocomputation, statistical modelling and data mining. Section 7 focus on the problem of security and privacy, which is a central problem in medical data management, analysing the approaches followed for authentication, authorisation and privacy protection. Section 8 describes the future trends in the use of epidemiology on Grids, using as reference roadmaps and other forecasting studies and section 9 ends with the conclusions of the chapter. Sections 10 and 11 include the references and key terms.

Finally, it is important to outline that epidemiology does not imply only clinical data. Social and geographical data are very important to explain the effect of the phenotype on patients' health. Geographical Information Systems are used for the forensic analysis of epidemiological burst and the localisation of their sources. Geographic analysis can point out water sources, air-condition machines or personal contacts (as the SARS outbreak in 2003 (Institut Pasteur, 2005)), as the main sources for an epidemic burst.

BACKGROUND

Many studies in public health use epidemiology as the main tool. The common denominator of all these applications is the objective of studying the population as a whole, although their main aims are quite different. The most frequent types of study are:

- **Surveillance networks.** They are organisations that collect on real time information about diseases of compulsory declaration (such as tuberculosis, smallpox, cholera, hepatitis, etc.) and react immediately providing the means for aborting the start-up of disease burst. Such networks require quick access to data, prediction capability and the ability to simulate scenarios ("what if"), such as the recent cases of SARS (Tian C., 2007) or avian flu (Neil M. Ferguson, 2005).
- **Cancer information systems.** They collect periodically information about the diagnosis and evolution of cancer patients in a wide population. The main aim is the exhaustively collection of all the cases. These systems need access to wide heterogeneous and distributed sources of data, intensive pre-processing capabilities for data quality and less intensive post-processing. Examples can be found on the EUROCARE study (EUROCARE, 2008).
- **Pharmacoepidemiology and pharmacoeconomy.** Pharmacoepidemiology consist on collecting the prescription of drugs and the analysis of the real efficiency of the treatments. Efficiency is the real effect of a treatment on a real, heterogeneous and exhaustive population. Pharmacoepidemiology is more focused on the quality on the therapy and Pharmaeconomy on the global efficiency on the use of resources. A specific area is the Vaccination monitoring. They need additional information about prescription and therapy follow-on, as well as on economical issues. The analysis of the data requires very intensive computing, both for preparation and processing. They usually make use of models for those variables that cannot be exhaustively obtained.
- **Molecular epidemiology.** The advent of the genomic and proteomic era has opened many new ways to study the diseases. The relation of the genomic profile of pathogens

and patients on the effect of treatments enables public health decision makers to adapt the therapy, to forecast resistance and to improve disease propagation models. The tools used are shared with the bioinformatics community and are mainly bounded by their computational cost and data access.

- **Environmental and biological risks.** The effect of natural and human disasters, contamination or even the global heating have a strong influence on the populations health (e.g. one of the current pathogens causing malaria appeared as a mutation in the Mediterranean area as a consequence of a global change on the temperature (Tishkoff S.A., 2001)). Thus, the modelling of the spreading of contaminant materials, floods, draughts and other situations, as well as the use of advanced geographic systems are necessary. The use of engineering and computer simulation is significantly recent in this epidemiology discipline, and clearly is requiring important computational capacities.

The use of ICT in medical research and healthcare is obvious. From the efforts on the EHR, throughout the development of standards (HL7 (Health Level 7, 1997), openEHR (OpenEHR, 2007), ENV 13606-4, DICOM (NEMA - ACR, 2008), IHE (Integrating the Healthcare Enterprise (IHE), 2008)), the collection and integration of clinical data for the purpose of patient health management is becoming a reality at different aggregation levels. Although there is no worldwide EHR format agreement, integration at regional and national level is habitual in developing countries.

However, many issues are still unsolved from the point of view of epidemiology. The use of ICT in epidemiology is traditionally considered from two complementary views. From one side, the collection of new data through electronic means (Ekman A. L. J., 2007) implies a big advance in the scalability, duration and cost of the process. The use of Internet reduces costs, increases the number of potential targets and enables a quick collection process (Ekman A., 2007). However, although some errors are minimised by the guidance of the electronic forms, the digital divide could imply an important bias in many studies. Moreover, there are authors that claim that digital collection of data is error prone by nature (Burnum, 1989). Using data from clinical practice has some caveats. Clinical data were obtained with the objective of gathering data around a patient for the interpretation of human experts. The exploitation of such information at the global level is reduced and not systematic. Clinical data is large, has personal implications and their structure is complex.

On the other side, ICT is used for developing and running mathematical methods to characterise and solve epidemiological models. The Centre for Discrete Mathematics and Theoretical Computer Science, created as a consortium of several North-American universities and research centres, started on 2002 a Special Focus on Computational and Mathematical Epidemiology (DIMACS, 2002). Statistical, dynamic, probabilistic, discrete math and theoretical computation methods are used to model populations, evolutions, interactions and risks.

The integration of data and the support for the computational requirements that advanced models imply are key issues that could be tackled from a grid computing paradigm. Throughout the next sections, those needs will be described and current solutions commented.

EPIDEMIOLOGY USE CASES

Epidemiology studies have in common the analysis of a large and distributed data through complex simulation, statistical procedures or knowledge discovery methods. These processes involve different use cases, such as the ones showed in the previous section.

The main users (from the highest-concept level to the lowest one) are public health authorities, epidemiologists and pharmaceutical companies. The data are normally owned by clinical care (both public and private).

The steps that the use case must go through (from the point of view of the user) are:

- **Automatic data gathering.** Data from different, geographically distributed sources (primary care, prescription, demographic information, hospital information, microbiological data, etc.) must be put together.
- **Data quality improvement.** Data are poorly coded and must be corrected, completed and improved. Aberrations, incoherent, incomplete or inaccurate fields must be revised and corrected.
- **Data processing.** From simple aggregation analysis to complex modelling, simulation and data-mining techniques, epidemiological data are used for prospective and retrospective analysis.
- **Presentation of results.** The analysis must provide, by the end, well-known indicators. Cancer survival rates, epidemic secure intervals and other measures are typically obtained by well-known and widely used computer programs that are fed upwards with the results of the analysis.

These steps are summarised in Figure 1.

THE CAPTURE AND INTEGRATION OF DATA

The problem of the monitoring of public health depends primarily on the availability of data (WHO, 2007). Data come from different sources and media. Clinical data, clinical trials or research studies provide data in electronic and printed format. Moreover, data comes from different distributed sources (Primary Care, Laboratory Information, Radiology Information, Social Data, etc.), with different formats and structures.

The increase in globalisation is requiring the consideration of the populations' health as a whole, and across the barriers of the different countries. Thus, epidemiology has to deal with an increasing amount of distributed, incomplete and inaccurate information. This process is demanding new data integration and intensive computing tools to deal with the identification and filtering of relevant data, the advanced processing and the visualisation (Mikler, 2005).

Data Capture

Automatic data gathering implies that data sources can be automatically fetched, the data properly retrieved and the integration of the different data fragments performed.

Regarding automatic fetching, this implies normally setting up agents and middleware on the local repositories. Setting up software will require

Figure 1. General view of the gathering and processing of data in epidemiology

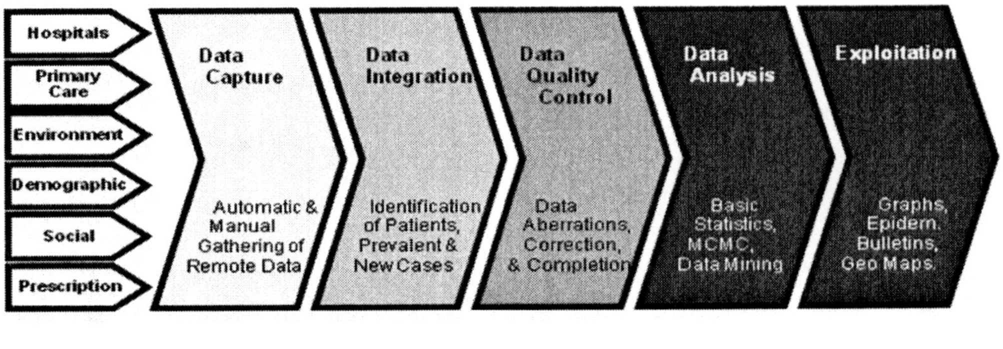

direct access to databases, which is not always feasible, especially in the context of proprietary systems. This could require the manual execution of data queries, sometimes exporting the data to plain formats that could be read by the mentioned agents. Automatic fetching could require remote-control applications that could, under a secure environment, log in and fetch the data safely. Examples of interaction components are available at TRENCADIS middleware (I. Blanquer, 2007) in the CVIMO (CVIMO, 2007) infrastructure, globus Medicus (Erberich S.G., 2007) or the PsyGrid (PsyGrid Consortium, 2008).

Along with the interaction components, it is necessary to set up a federated schema for accessing data. OGSA-DAI (OGSA-DAI project, 2007) and ad-hoc developments have been used in prototypes (Neto V. C., 2006). The adaptation to different formats, the delegation of authorization and the management of privacy are issues under research.

Another important source of information are clinical trials. Clinical trials focus on the analysis of a treatment on a selected subset of the population and under controlled conditions. This subset normally involves both ill and sound (controls) people. The recruitment and follow-on of the both controls and patients requires selecting the adequate patients through the analysis of their clinical data. The follow-on is normally a complex but important process, since it accuracy is key for the success of the trial. It is also important to outline that the financing of hospitals through clinical trials is quite important. Most successful hospitals are then preferred by pharmaceutical companies.

Thus, there are many efforts on improving these tools to increase the number of studies completed and with the maximum of information, without compromising patients' health. The application of Grids in this field is also relevant. The VOTES (VOTES project, 2007) project uses a grid infrastructure based on OGSA-DAI to access data from different hospital sources. It provides a framework for the follow-on of the patients and to inform pharmaceutical companies about their evolution. One of the key aspects of VOTES system is the authentication model, described in section 7.

Other approaches on data collection in what some authors call "e-epidemiology" (Ekman A. L. J., 2007), focus on using web interfaces for retrieving the data from distributed populations. This approach has proven to be successful when addressing large populations with enough skills on Internet communication. Some authors (Ekman A. L. J., 2007) conclude that the electronic collection of data does not decrease the average quality of the data collected.

Finally, both approaches are also considered altogether. PsyGrid is a project that aims at developing a platform for the collection and tracking of psychiatric patients structured in different research projects. PsyGrid provides access to the National Health Network and tools for defining and sharing electronic standardised forms in which the physicians collect and store patients' information. The grid concept in PsyGrid is mainly related to the collaboration environment, rather than on the data which is stored centrally.

Data Integration

Once the data are available in the distributed environment, data integration introduces additional problems. Data will rarely present unique keys, and private information will be required, raising legal and ethical issues. Moreover, unique keys could easily reveal a patient's identity. Pseudo-anonymisation techniques are used to create identifiers on trustful contexts, matching data from different sources. However, technological solutions might provide the answer: there are, for example, pseudo-anonymisation techniques that could be used to create identifiers on trustful contexts, matching data from different sources. For example, two databases could merge the results from complementary records matching a global identifier using asymmetric cryptography. This would ensure that no private information is released, that only trustful entities are connected, and that records are correctly matched.

The second important problem is data structuring. When dealing with large amount of information, the organisation of the data is a crucial issue that directly affects performance and usability. Clinical use of data requires identification of the different data fragments of a patient through their unique identifier. Indexing by practitioners and treatment devices and drugs is also an important issue for the sake of quality control.

However, from the perspective of epidemiology, the relevance of the patient identifier ends when the integration process is completed. From this point on, other factors (such as sex, age, diseases stages, geographical location, etc.) constitute the indexing and searching criteria. Thus, data integration should consider also these indexing criteria when creating the data warehouse or knowledge bases.

An interesting approach is the use of ontologies (I. Blanquer, 2007), (Martonne). In both computer science and information science, an ontology (Wikipedia, 2007) is a representation of a set of concepts within a domain and the relationships between those concepts. The relationships could be on the form of "contains", "is a part of", "is equivalent", etc. It is used to reason about the properties of that domain, and may be used to define the domain. An ontology can classify the data available with respect to complex criteria, such as the diagnosis. Diagnosis is coded using the International Classification of Diseases (ICD-10, (WHO, 2008)) in the form of description trees with different levels of detail. Along with the relations, each term has a unique identifier and a human-readable definition. An ontology can reflect the branches of the diagnosis coding that are common for a specific study. TRENCADIS (I. Blanquer, 2007) data integration model defines ontologies based on the content of structure documents and images. User groups are related to ontologies, which gave them access to a part of the complete set of data available. This reduces the data available to the subset of data for a community. Defining also in a second level the fields from the registers of the database which are relevant for searching, indexes for these dimensions can be created in a virtual multi-dimensional data warehouse.

On the other hand, even ensuring that coding schemas are used, there are many different terminologies and codifications for different areas and diseases. Moreover, terminologies could involve many different levels of detail (such in ICD-10), being different terms related by semantic relations. The BIRN project (BIRN, 2000) solves this issue by the use of wrappers that standardise the data to local and accepted standards and the integration through a mediator that contains the acceptable relations among the different terms. In this sense, compatible terms among standards and specialisation terms are defined through a knowledge base. Considering the presentation templates defined in a view integrator, consolidated virtual views are constructed dynamically.

Finally, along with the clinical data integration, integration with other sources is important. The use of ICT for GIS data integration is provided in many commercial and free software (ESRI ArcgGIS, IDRISI, MapInfo, EpiINFO, etc.). Graph topology and neighbouring weights are used to define the degree of connectivity (based on the movement of populations) to forecast and analyse the most probable distribution of a disease.

Along with the problem of integration of data, it will be important to develop gateways to standard formats of medical data exchange, such HL7, DICOM or CEN TC251 norms. This will ease the integration of the tools with infrastructures.

The most relevant interfaces needed are:

- **Hospital data.** Although medical databases have different storage formats, there exist de-facto standards and other standards under development that regulate the exchange of medical data. Hospital information, for example, is needed for epidemiological research. The availability of HL7 (de facto standard) and ENV 13606-4 (norm under development by the CEN TC251) gateways will ease the integration of the medical resources on the grid. Support to other vital signs exchange formats, such as ENV13734 will also be important.

- **Medical imaging.** Although not directly related with the epidemiology use case proposed, medical imaging is also widely used in epidemiology. Screening for early treatment of cancer in breast, colon, lung or prostatic cancer is habitual in many areas. There are several attempts to develop DICOM-conformant interfaces to grid-storage systems, such as DICOM-SRM (Montagnat J., 2008), MEDICUS (Erberich S.G., 2007) or TRENCADIS (I. Blanquer, 2007). However, DICOM is a large and complex specification, and current approaches only cover parts of the standard. Moreover, DICOM components need to be certified for their use in production. Finally, the interoperability of medical imaging is not only at the level of the data format, but at the level of the protocol acquisition and the device tuning. Images from the same patients performed at different locations and by different operators can be difficult to compare automatically. Transformation to standard formats (which also deal with the data content, as the SMF in MAMMOGRID (Warren R., Nov 2007)), are revealed to be very successful.
- **Statistical tools.** Along with the standardization of the interfaces to data, it is important to provide interfaces to the statistical tools most widely used in the epidemiological community. Along these programs, the tools of the Centres for Disease Control of Atlanta (CDC, 2007) and other related tools (EPIInfo, EPIMap, SIGEpi, EPIDAT) are widely spread. The support for software using "R" and "S" statistical languages will also improve the interoperability of the infrastructures.

The availability of those interfaces will ease the process of integrating grids for health infrastructures in the health environments without affecting severely the current processes, thus quickly providing enhanced performance.

Data Quality Problem

Provided that the data are properly gathered and the individuals and cases properly identified, and along with the technological difficulties integration of the information, there are additional problems related to the data themselves. The different pieces of information could be inconsistent, so data curation, i.e. keeping the most reliable and complete piece of data, must be considered. This requires solving the following issues:

- **Reliability.** Data provenance mechanisms are necessary. Data sources should be ranked according to factors such as nature of the source, level of certainty of the diagnosis, date, methods used, etc. This will require an advance on the reliability of the source and data in distributed database integration. Dynamic adjustment of the reliability will be important in production environments.
- **Coherence.** A detailed medical ontology of terms is needed. This ontology should be understandable by computers and should define the possible and impossible relations. It should be important to have weak and strong relations that could define terms that are normally compatible and rare associations. For examples, there should not be any relation between "cervix" and "male", there should be a strong relation between "brain" and "temporal lobe", and weak relation (only few cases) between "male" and "breast cancer".

- **Coding.** Since data are captured normally in the clinical environments, data is not always coded, but it could be correctly and accurately described in natural text. Specialized text mining tools should be able to propose codes, which should not replace, but complement natural text. Standardised nomenclatures do exist, although they are maintained and updated to cover advances in treatment and diseases. The coding is essential for large-scale computer-based automatic analysis (Maojo V, 2005).

The advances on the semantic web and semantic grid could partially solve the problems of characterizing relations and ontologies. Specifications such as OWL (Ontology Web Language) (W3C, 2004) RDF (Resource Description Framework) (W3C, 2007) are used to characterise the semantics of entities and relations. Although more research effort is needed on dynamic characterization and provenance, the main issue on this field lies on the deployment of these ontologies, rules and text mining tools.

COMPUTATIONAL PROBLEMS IN STATISTICAL MODELLING

The usage of real clinical databases has leaded to the development of new statistical tools and models to deal with large populations and correlation studies which require intensive computing. Many data-mining tools used in economics, resource-provider industry or other areas are being adapted. With the introduction of genomics, genetics and proteomics, new methods are being applied. The performance of applications related to health (and particularly to epidemiology) would require solving two main issues: the integration of the maximum amount of resources and the Quality of Service.

One difficulty on modelling epidemic outbreaks is the realistic characterisation of the social behaviour of actual populations with respect to the most probable transmission paths (Barrett C. L., 2005). Moreover, mobility of population in developed countries is much more complex to model, and the threat of bio-terrorism makes this activity of high interest for research. In those cases, the use of intensive computation is fundamental.

Understanding infectious systems requires being able to reason about highly complex biological systems, with hundreds of demographic and epidemiological variables. Intuition alone is insufficient to fully understand the dynamics of such systems (DIMACS, 2002).

Traditional statistics are used since long ago in epidemiology, especially on characterising the error and to evaluate the correlations among variables. New approaches make use of differential equations (modelling of dynamic systems), probabilistic methods (Markov Chains, Stochastic Methods, Monte Carlo), discrete math and theoretical computational Science (linked with Geographic Information Systems) and Data Mining.

Basic statistics (aggregation, averaging, specific indexes, regression, etc.) require a reasonable computational cost compared to the needs for data storage. However, the new methods that are being developed require large computational resources, which are needed both due to the methods themselves and to deal with the size of the large populations covered.

On one hand, model-based statistics (H. F. Gagliardi, 2006), such as Monte Carlo Markov Chain (MCMC) methods and Decision Trees define models of the diseases which are trained and filled-in with the data recorded in healthcare delivery. Those models define probabilities and transitions between states that can be used for prospective analysis. Although they have moderate computing requirements, these requirements increase with the complexity of the models and the data sizes.

On the other hand, Data Mining, such as Bayesian inference and Clustering Methods are well used in many other knowledge discovery processes. The computational requirements of those methods are far higher than basic statistics.

The Research network for Models of Infectious Disease Agent Study (MIDAS, (NIH, 2008)) organised by the National Institute of General Medical Sciences in the United States, developed the EpiSims project, aimed at tackling this problems. In the frame of EpiSims, an individual-based epidemic model on a scale of millions of people by using high-performance supercomputing clusters has been developed. This kind of models is mainly based on agents, contact networks (graphs) (Donnely, 2002) and even game theory computational models (Banks, 2001).

The development of mathematical models enables researchers on many tasks, such as increasing the understanding of the disease, comparing different alternatives, assist on making decisions, estimate the risk or predict future trends.

The size and overwhelming complexity of modern epidemiological problems calls for new approaches, that could deal with the dynamics of multiple interacting strains of viruses through construction and simulation of dynamic models; the spatial spread of disease through pattern analysis and simulation and the early detection of emerging diseases or bioterrorist acts through rapidly-responding surveillance systems.

There is a clear need for the creation of collaborative and multidisciplinary research groups and the involvement of computational experts (DIMACS, 2002).

Although the efficiency of fine-grain parallelism in grids has not been absolutely demonstrated, these currently unsolved technical issues are not critical in the epidemiology scenario. MPI (Message Passing Interface) on the grid versions faces many problems regarding the configuration of sites, network latencies and scheduling. Applications in model simulation or statistical inference normally rely on a high-throughput model, rather than on a fine-grain coupled parallel mode. Provided that the infrastructures solve the problems of Quality of Service and efficiency, current infrastructures could provide enough resources for solving the problems faced in epidemiology.

On the contrary, most implementations of currently popular software for epidemiology are not "grid-enabled". A few of them use parallel computing, although their computing model is fairly uncoupled. The migration of those applications into more decoupled high-throughput models will enable to adapt them and to run them easily on the grid.

In the simulation of epidemiological models, it will be necessary to access data and to perform complex calculations. Since data will normally be stored at the local repositories and made available to the grid under demand, network, storage and computing resources must be guaranteed for any processing to start. Epidemiology is not bound by interactive processing needs, although best-effort processing will be needed. The patterns of access to the resources will generally be predictable and regular (weekly or daily updates of the information and consolidation, periodic analysis), although there could be at least two additional cases: large, schedulable and computing intensive analysis could be requested by researchers on specific studies; and shorter, unpredictable, computing intensive and time-limited analysis could be requested by epidemiologists when an epidemic burst appears in order to evaluate counter means or predict the expansion of a disease.

In the case of predictable or regular executions, there is a need to provide resource reservation and Quality of Service. Grid middleware and infrastructures should provide the means to reserve, block or checkpoint resources to attend those demands. Network reservations should be necessary to fetch and load temporarily data on the grid. In the case of unpredictable, quick need for resources when facing an emergency, the system should allow for pre-emptive scheduling and priority management. This will enable higher-priority actions to be performed in an effective time.

Main problems on the implementation of such policies lie on local resources, which are owned and administered by the resource providers. This should be implemented at the level of the resource managers and pushed up to the middleware level.

This model of high-throughput computing requires effective and fault-tolerant scheduling. An important fraction of time is still consumed by resource brokers and workload managers to discover the resources and schedule the jobs. Moreover, fault-tolerance, although implemented, is rather inefficient, and normally the developers end up creating their own solutions. Computing in epidemiology, as described in the document, does not require generally interactive nor short deadline jobs (although it could benefit in specific cases). Current scheduling models can be considered sufficiently effective. Thus, efforts must be concentrated on reliability and integrity as in transactional systems. It must be ensured that a job is completely finished, or, if it has intrinsic errors, no effect has been produced over the system. A user must be sure that, when a job fails, the error is due to the job itself, and not to the infrastructure.

VO MANAGEMENT AND SECURITY

The use of clinical data for epidemiology implies important ethical and legal problems. Moreover, the distributed nature of grids, using different administrative domains increases the difficulties and risks. In principle, clinical data can only be used for the provision of healthcare to the individual. It can be used for research if an informed consent is subscribed. However, European and International regulations (I. Andoulsi, 2007) require important constraints on the use of the data. The reason of the experiment should be clearly depicted and it should not be used for any additional mean. The research must be of clear public interest and there are obligations with the distribution of the additional knowledge that could be derived (depending on the right of the patient of knowing or not-knowing). The effect of many of these issues can be minimised through the use of anonymisation and pseudoanonymisation (HIPPA, 2006). This fact is extremely severe when dealing with genetic data and medical imaging, in which anonymisation could not be sufficient. Finally, the legal obligations of hospitals in keeping the information within their boundaries could refrain from using International Grid infrastructures.

Along with the authentication, there are three main aspects to be considered in the epidemiology framework: Authorisation, delegation of credentials and privacy:

- **Authentication.** Although user authentication is a problem well solved in public key infrastructure (PKI) environments, in which most grid infrastructures sit, it will be important to analyze how these procedures are being implemented in health infrastructures. Normally, health users do have (or will very soon have) a means of digital identification used for accessing clinical records in their daily practice. Trusting different certifying authorities is feasible and should not present additional problems.
- **Authorisation.** Up to now, the management of the authorization (Shibboleth Project, 2006) has been managed with the use of attribute extensions in a central authorization system such as VOMS or through authorisation delegation schemas, combining Shibboleth and PERMIS systems, which reduce the problems in deploying large-scale VO membership and delegating on trusted authorization mechanisms.
- The delegation of credentials is a well-known problem that has been reasonably solved in many situations. Processes should be able to act on behalf of third parties who started them, with different levels of capabilities. Delegation could be full or limited and last

for a defined period of time. However, the delegation of authorization, of keys for accessing back-ends, and of roles has not been completely solved yet, and these issues have an important impact when accessing third-party applications whose security levels include additional features, such as login and password. Moreover, the risks on stealing credentials, which proxy-like mechanisms have, are much more critical in health environments. Single sign-on is another important issue regarding delegation. Users must be able to provide their credentials only once and let processes act on their behalf. The use of proxies, proxy repositories and X509 standard attribute extensions are sufficient to deal with these requirements, but imply new privacy threats, as most of grid environments store temporally the user's credentials on disk, opening a big threat for privacy. Users with administrative privileges at a specific resource could supplant any user's identity by copying and modifying the name and ownership of temporal proxy files.

- Privacy management is the hardest problem regarding security. Legal regulations, as the Directive 95/46/EC of the European Parliament and of the Council of 24 October 1995 on the protection of individuals with regard to the processing of personal data and on the free movement of such data (European Parliament, 1995) or the Directive on privacy and electronic communications (European Parliament, 2002) impose, from a technological perspective, data dissociation, pseudo-anonymisation and encryption. Since any medical data is potentially personal–since further research could discover particularities unique to a patient–and considering that the processing starts with the storing of the data, encryption, both on the storages and on the network, is used. Many schemas have been proposed (E. Torres, 2006) (J. Montagnat, 2007) aiming at data encryption and decryption on the fly, multiple key shares, reliable services, etc. Finally, the management of genetic data introduces more problems and difficulties due to the potential re-identification of data. Large-scale deployment of these techniques should be performed. It must be ensured that the issues outlined in Council of Europe Recommendation R (97) 5 (Council of Europe, 1997) on the Protection of Medical Data (Feb. 13, 1997), are taken into account.

However, there are doubts on the scalability and granularity of those well-established technologies (Stell A. J., 2005). The VOTES project proposes the use of Shibboleth (Shibboleth Project, 2006), which defines an authorisation mechanism based on the federation of security attributes. Authentication is performed at the local site and remote access is granted on the base of the security attributes presented by the requesting entity. This reduces the central issuing of credentials, but constitutes a more rigid mechanism, in which the agreements on security attributes are fixed and permanent on the time. In the opinion of the authors, this mechanism is efficient for its application in the health environment (Sinnott, 2006). Other security models dealing with identity federation and trust management have been developed in the frame of CaBiG (Ken Lin, 2006).

Finally, auditing, logging and monitoring are key issues to ensure compliance to regulations and proper response to adverse events. Current systems are mainly focused on the availability of the resources, rather than on the access policies. Intrusions, non-conformal uses or security leakages are detected through manual inspection. Since there are effective mechanisms in other technological areas, the adaptation of such techniques to grids for health should be considered and studied.

Last, but not least, there are two important ethical problems derived from the use of clinical data for research, which have important technological implications. Patients providing access to their clinical data, may have the right to know (and the right to "not to know") about results derived from the research that could affect their health. When using pseudoanonymised data, and if a study reveals a risk of suffering a disease, a hidden disease or any emergent information from the application of the research techniques, there must be a way to re-identify patients and inform (or not, considering their decision) about this issue (J. Lawford-davies, 2005).

Security Models Based on VOs

TRENCADIS (I. Blanquer, 2007) software architecture aims at federating distributed repositories of medical images and reports in DICOM format in a single view. The security model of TRENCADIS aims at sharing the data on different groups from Virtual Organisations.

Before the introduction of the concept of groups in the VOs, the membership of users was normally bound to a single VO. With the use of VOMS and similar systems, the membership to multiple VOs and groups enabled to design different views of permissions for the same user.

The classic approach of Access Control Lists (ACLs) requires that permissions are explicitly given to individuals or groups. If a piece of information should be made available to different VOs or VO groups (but not all of them), the data owner should explicitly indicate this when sharing the data. This could be tedious and complex if many data are created regularly. However, the metadata associated to a piece of information can have enough information to decide which groups should have the access granted. TRENCADIS aims at sharing, and thus it does not directly consider the "owner" access right, but only the group access rights.

Studies can be classified according to ontologies. Users bind their credentials to one of the available ontologies of the VO group they belong to. Thus, the images available by a specific user are those which match the criteria of the ontology and are stored on repositories that accept such VO group. Ontologies are created at a central level, and under the approval of the management of the system, so no privacy leakages appear. When a new image is inserted, it becomes available to all the groups which support ontologies matching the data of the image and the report. This approach is much more scalable than assigning access rights to groups, since it automatically manages the permissions for all the groups in the system. Finally, individual permissions can always be removed.

The Portal-Based Authentication

Certificates and credentials are issued for the identification of real subjects. A certificate contains the necessary information to identify a subject, such as name, surname, entity, organisation or country. Certificates are personal and unique, and require a several-step procedure that take days and requires the human intervention to guarantee their legitimacy.

On the other side, grid portals are becoming very popular. Grid portals give access to users without the burden of installing and configuring applications and data sources. They provide a simple and unified interface that can be learnt in a short time.

Epidemiology users and specially those dealing with biocomputing are very used to portals. There are many portals for Public Health and biocomputing and some of them involve grids (NCBI, 2007) (CNRS - IBCP, 2006) (GRyCAP, 2007). In these last cases, the executions made on the grid must be done under legitimate credentials issued for a real individual. Requesting all users of grid portals to ask for certificates and subscribe a compatible VO is unrealistic.

Current solutions (EGEE-SEE ROC, 2007) deal with the use of an alternative based on "portal administrator credentials". A user with valid grid credentials and membership to a specific VO stores medium-term living credentials on a credential repository (such as MyProxy (NCSA, 2007)) which can be fetched by a service that acts as a gateway between the grid and the portal. Users which do not own valid credentials can still execute specific jobs under the credentials of the portal administrator and only for the jobs authorised by it.

This schema transmits the responsibility of the anonym users to the responsibility of the portal administrator, who should ensure that the jobs executed under their name are for a specific authorised task and for a reasonable use of the resources. The exposure of medium-term credentials through a repository does not increase risk severely, and it constitutes a much more secure alternative to the storage of private keys on the computer accessed by the portal users. It also avoids the exchange of private keys and it only requires to periodically update the credentials hold in the repository.

FUTURE TRENDS

In the frame of the SHARE project (SHARE Consortium, 2006), a deep study of the current status, medium and long-term evolution of the application of grid technologies in health has been developed and published in the form of a roadmap. One of the areas of study has been epidemiology.

Research on the application of grid technologies on epidemiology will surely aim at the main unsolved problems in this field. Grid technologies have reached a degree of maturity which is close to sufficient in other scientific areas, such as High Energy Physics, Astrophysics or Bioinformatics, among others. However, technical problems on data availability and gathering; adaptation, implementation and deployment of models and tools and the fusion of e-Health and grid standards could be the most relevant issues from the technological point of view.

Along with the technical issues, there are other considerations that will surely be investigated in the medium-term. Issues regarding the management of consent and privacy on the grid, especially considering the transnational nature of grid infrastructures, require a careful interpretation of the law and the availability of tools. Finally, there are other important issues considering the real deployment of those tools and systems.

The consideration of the quality of service, which is insufficient in current e-infrastructures will be an important issue. It is forecasted that the grid will be a federation of many grids (EGI Consortium, 2007), and thus the grid infrastructures for health can grow similarly. This will be absolutely compatible with the current technological, managerial and human structure of health systems. The adoption of standards and the integration of grid services on the processing chains will leverage its use. Finally, certification of grid services will open the door to the clinical use of grid processing infrastructures to the health provision, bridging the gap between research and clinical practice.

CONCLUSION

Epidemiology constitutes a very promising application area for grids. From the problems of data integration to the computational requirements and the security constraints, epidemiology shares many challenges with other scientific disciplines that are being tackled by grid technologies. Although it presents many commonalities with other e-health disciplines, epidemiology implies new issues regarding the large, comprehensive and population-level views that are necessary for their automatic processing.

The distributed nature of data, whether coming from clinical management or by distributed questionnaires requires integration problems that are solved through the use of distributed database techniques, DICOM and HL7 gateways and specific recruitment tools. Data quality and data coding are extremely important in epidemiological studies, and constitute an issue in current clinical data. The integration with geographical data also increases the complexity and the distributed nature of the sources of data.

On the other side, the development of advanced tools and models for the simulation of epidemiological outbreaks, disease diffusion models and resistance to treatment have extremely benefited from the use of advanced computing and large-scale simulations. This development will surely increase in the future with the availability of larger amounts of more detailed medical data.

Finally, the ethical and legal issues that the management of medical data imposes have opened the doors to the development of many protocols, infrastructures and protocols focusing on managing privacy of data, integrity of the information, tracing and authorisation.

The access to population-level data in epidemiology, which was a main aim in the last years, is becoming a reality at a large scale in developed countries. The future trends in managing these data will surely go through the use of grid infrastructures for solving the challenges that this processing implies.

REFERENCES

Banks, D. (2001). *Game Theory and. Risk Analysis for. Counterterrorism*. Retrieved November 2008 from http://www.galaxy.gmu.edu/interface/I03/I2003Proceedings/BanksDavid/Banks David.presentation.pdf

Barrett C. L., Eubank S. G., and Smith J. P. (2005). If Smallpox Strikes Portland. *Scientific American Magazine*. Retrieved July 2008 from http://www.sciam.com/article.cfm?id=if-smallpox-strikes-portl

BIRN. (2000). *Biomedical Informatics Research Network*. Retrieved November 2008 from http://www.nbirn.net/

Burnum, J.F. (1989). The misinformation era; the fall of the medical record. *Annals of Internal Medicine, 110*(6), 482-484.

CDC (2007). *Center for Disease Control of Atlanta*. Retrieved November 2008 from http://www.cdc.gov/

CNRS – IBCP (2006). *Grid Genomic Web Portal*. Retrieved November 2008 from gpsa.ibcp.fr.

Council of Europe (1997). *Recommendation No. R (97) 5 on the Protection of Medical Data of Feb. 13, 1997.*

CVIMO (2007). Retrieved July 2008 from http://www.grycap.upv.es/cvimo

DIMACS (2002) *Special Focus on Computational and Mathematical Epidemiology*. Retrieved July from http://dimacs.rutgers.edu/SpecialYears/2002_Epid/

Donnely, C. (2002). FMD 2001: Using Statistics and Mathematics for Outbreak Control and Eradication. *International Conference on Computational and Mathematical Epidemiology*. Retrieved July 2008 from http://dimacs.rutgers.edu

Torres E., De Alfonso C., Blanquer I., Hernandez V. (2006). Privacy Protection in HealthGrid: Distributing Encryption Management over the VO. *Studies in Health Technology and Informatics, 120,* 131-41.

EGEE-SEE ROC (2007). *Web Authentication and Authorization using MyProxy APIs and Grid Portals*. Retrieved July 2008 from http://wiki.egee-see.org/index.php/Web_Authentication_and _Authorization_using_MyProxy_APIs_and_Grid_Portals

EGI Consortium (2007). *European Grid Initiative*. Retrieved July 2008 from http://www.eu-egi.org

Ekman A., Klint A., Dickman P.W. , Adami H.O., Litton J. (2007). Optimizing the design of web-based questionnaires - experience from a population-based study among 50.000 women. *European Journal of Epidemiology 22,* 293-300.

Ekman A., Litton J. (2007). New times, new needs; e-epidemiology. *European Journal of Epidemiology, 22,* 285-292.

Erberich S.G., Silverstein J.C., Chervenak A., Schuler R. , Nelson M. D. , Kesselman C. (2007). Globus MEDICUS - Federation of DICOM Medical Imaging Devices into Healthcare Grids. *Studies in Health Technology and Informatics, 126,* 269-278.

EUROCARE Consortium. (2008). *EUROCARE.* Retrieved July 2008 from http://www.eurocare.it/

European Parliament. (2002). *Official Journal 201,* 0037 - 0047.

European Parliament. (1995). *Official Journal of the European Communities 281,* 31.

EUROREC (2002). Retrieved July 2008 from http://www.eurorec.org/

GRyCAP (2007). *Grid Biocomputation Portal.* Retrived July 2008 from http://www.grycap.upv.es/bio

H. F. Gagliardi, F. A.B. da Silva, & D. Alves. (2006) Automata Network Simulator Applied to the Epidemiology of Urban Dengue Fever. In *Proceedings of the ICCS'06, Lecture Notes in Computer Science, 3993* (pp. 297-304). Berlin, Germany: Springer

Health Level 7 (1997). Retrieved July 2008 from http://www.hl7.org/

HIPPA (2006). Retrieved July 2008 from http://www.hhs.gov/ocr/hipaa/

I. Andoulsi, J. Herveg, V. Stroetmann, K. Stroetmann, A. Dobrev, C. Van Doosselaere, P. Wilson (2007). *Bottlenecks & Challenges and RTD Responses for Legal, Ethical, Social, and Economic Aspects of Healthgrids – Roadmap II.* Retrieved July 2008 from http://eu-share.org/about-share/deliverables-and-documents.html

I. Blanquer, V. Hernandez, D. Segrelles (2007). TRENCADIS – Secure Architecture to Share and Manage DICOM Objects in a Ontological Framework Based on OGSA. *From Genes to Personalized HealthCare: Grid Solutions for the Life Sciences, Studies in Health Technology and Informatics, 126,* 115-124.

Institut Pasteurand the HKU-Pasteur Research Centre (2005). *Information on the international outbreak of the illness known as severe acute respiratory syndrome (SARS).* Retrieved July 2008 from http://www.pasteur.fr/recherche/unites/REG/SARS.html

Integrating the Healthcare Enterprise (2008). Retrieved July from 2008 from http://www.ihe.net/

Lawford-davies, J., Jenkins, J. (2005). Legal Issues to Address when Managing Clinical Information across Europe: The ECIT Case Study. *Healthgrid 2005, Studies in Health Technology and Informatics, IOS Press; 120*: 210-221.

Montagnat, J., Frohner, Á., Jouvenot, D., Pera, C., Kunszt, P., Koblitz, B., et al. (2007). A Secure Grid Medical Data Manager Interfaced to the gLite Middleware. *Journal of Grid Computing, 6* (1), 45–59.

Ken Lin, Booz Allen Hamilton, Gary Daemer. (2006) *caBIG Security Technology Evaluation.* (White Paper). Retrieved July 2008 from https://cabig.nci.nih.gov/workspaces/Architecture/Security_Tech_Eval_White_Paper/.

Maojo V, de la Calle G, Martín-Sánchez F, Díaz C, Sanz F. (2005). INFOBIOMED: European Network of Excellence on Biomedical Informatics to support individualised healthcare. *AMIA Annual Symposium Proceedings:* 1041.

Martonne, M. (2005) The BIRNLex: Principles and practices of community ontology development. *Re-use or Re-invention - a Roadmap for Data Integration (Schizophrenia as a Test Case). NeSC Events*. Retrieved July 2008 from http://www.nesc.ac.uk/action/esi/download.cfm?index=3359

Mikler, S. Venkatachalam, A.& (2005). Towards computational epidemiology: Using stochastic cellular automata in modeling spread of diseases. *Proceedings. of the 4th Annual International Conference on Statistics*. Retrieved July 2008 from http://www.cerl.unt.edu/publications/2005/slides/ppt/Hawaii.ppt

Montagnat J., Frohner A., Jouvenot D., Pera C., Kunszt P., Koblitz B., et al. (2008). A Secure Grid Medical Data Manager Interfaced to the gLite Middleware. *Journal of Grid Computing (JGC), 6* (1): 45-59.

NCBI Blast (2007). Retrieved July 2008 from http://www.ncbi.nlm.nih.gov/blast/Blast.cgi.

NCSA (2007). *MyProxy Credentials Management Service*. Retrieved July 2008 from http://grid.ncsa.uiuc.edu/myproxy/

Ferguson N. M., Cummings, D. A.T., Cauchemez, S., Fraser, C., Riley, S., Meeyai, A., Iamsirithaworn, S., Burke, D. S. (2005). Strategies for containing an emerging influenza pandemic in Southeast Asia. *Nature 437,* 209 - 214.

NEMA – ACR (2008). *DICOM Standard*. Retrieved July 2008 from http://medical.nema.org/

Neto V. C., Gagliardi H. F., Rezende A., Gonçalves E. S., Gallo E., Barbosa F. A., Torres I., Alves D. (2006). Data Access Service in a Computational Grid Platform Applied to the Monitoring and Control of Epidemics on Georeferenced Dynamic Maps. *Proceedings of the e-Science conference,* (p. 142).

NIH. (2008) *Models of Infectious Disease Agent Study*. Retrieved July 2008 from http://www.nigms.nih.gov/Initiatives/MIDAS

OGSA-DAI project (2007). *OGSA-DAI*. 2007. Retrieved July 2008 from http://www.ogsadai.org.uk/

OpenEHR (2007). Retrieved July 2008 from http://www.openehr.org/home.html

PsyGrid Consortium (2008). *PsyGrid*. Retrieved July 2008 from http://www.psygrid.org/

SHARE Consortium (2006). *Structuring Healthgrid Activities and Research in Europe: Developing a Roadmap*. Retrieved July 2008 from http://www.eu-share.org

Shibboleth Project (2006). Retrieved July 2008 from http://shibboleth.internet2.edu/

Sinnott, R.O. Ajayi,O. Stell,A.J. (2006). Development of Grid Frameworks for Clinical Trials and Epidemiological Studies. *HealthGrid 2006 Conference, Studies in Health Technology and Informatics, 120,* 117-130.

Stell A. J., Sinnott R. O., Watt J. P. (2005). Comparison of advanced authorisation infrastructures for grid computing. *International Symposium on High Performance Computing Systems and Applications 2005 (HPCS 2005),* (pp. 195-201).

Tian C., Ding W., Cao R. and Jiang S. (2007). Extensive Epidemic Spreading Model Based on Multi-agent System Framework (LNCS 4490, pp. 129-133).

Tishkoff S.A., Varkonyi R., Cahinhinan N., Abbes S., Argyropoulos G., Destro-Bisol G., et al. (2001). Haplotype diversity and linkage disequilibrium at human G6PD: recent origin of alleles that confer malarial resistance. *Science, 293*.

VOTES project. (2007) *Virtual Organisations for Trials and Epidemiological Studies*. Retrieved July 2008 from http://labserv.nesc.gla.ac.uk/projects/votes/index.html

W3C (2004). *OWL Web Ontology Language*. Retrieved July 2008 from http://www.w3.org/TR/owl-features/

W3C (2007). *RDF Resource Description Framework*. Retrieved July 2008 from http://www.w3.org/RDF/

Warren R., Solomonides A.E., del Frate C., Warsi I., Ding J., Odeh M., et al. (2007). MammoGrid — a prototype distributed mammographic database for Europe. *Clinical Radiology, 62* (11), 1044-1051.

WHO (2008). *International Classification of Diseases*. Retrieved July 2008 from. http://www.who.int/classifications/icd/en/.

WHO (2007). *The world health report 2007 - A safer future*. Geneva, Switzerland: WHO publications. Retrieved July from http://www.who.int/whr/2007/en/index.html

Wikipedia (2008). *Epidemiology*. Retrived July 2008 from http://en.wikipedia.org/wiki/Epidemiology

Wikipedia (2007). *Ontology*. Retrieved July 2008 from http://en.wikipedia.org/wiki/Ontology_%28computer_science%29

KEY TERMS AND DEFINITIONS

Anonymisation: procedure by which the owner of a piece of information cannot be guessed by reasonable techniques using the information publicly available directly or available by third means.

Biostatistics: The application of statistics to a wide range of topics in biology. The science of biostatistics encompasses the design of biological experiments, especially in medicine and agriculture; the collection, summarization, and analysis of data from those experiments; and the interpretation of, and inference from, the results.

Data Integration: The process of combining data residing at different sources and providing the user with a unified view of these data.

Epidemiology: The scientific study of factors affecting the health and illness of populations, and serves as the foundation and logic of interventions made in the interest of public health and preventive medicine.

Mathematical Modelling: A representation of the essential aspects of an existing system (or a system to be constructed) which presents knowledge of that system in usable form and expressed using a Mathematical language. Mathematical models can take many forms, including but not limited to dynamical systems, statistical models, differential equations, or game theoretic models.

Ontology: A representation of a set of concepts within a domain and the relationships between those concepts. The relationships could be on the form of "contains", "is a part of", "is equivalent", etc. It is used to reason about the properties of that domain, and may be used to define the domain.

Public Health: The management and investigation of the health of a population at the global level, aiming at the early detection, promotion of health and study of the efficiency of treatments.

Chapter XXIII
IntegraEPI:
Epidemiologic Surveillance on the Grid

Fabrício Alves Barbosa da Silva
Universidade de Lisboa, Portugal

Henrique Fabrício Gagliardi
Instituto de Ensino Superior COC, Brasil

Eduardo Gallo
APRAESPI, Brasil

Maria Antónia Madope
Ford Foundation Alumni Association, Moçambique

Virgílio Cavicchioli Neto
Universidade Federal de São Paulo, Brasil

Ivan Torres Pisa
Universidade Federal de São Paulo, Brasil

Domingos Alves
Universidade de São Paulo, Brasil

ABSTRACT

The authors present in this work a large-scale system for space-time visualization, monitoring, modeling and analysis of epidemic data using a Grid platform. This system, dubbed IntegraEPI, is capable to integrate data from heterogeneous databases related to epidemic analysis and to make available analytical and computational methods to increase the predicting capability of the public heath system, in order to optimize its activities and resources when dealing with epidemic outbreak and prevention. This system, differently of what has been proposed before, is integrated and consequently it enables the construction of detailed predictive models of the dynamics of disease spreading. With the help of IntegraEPI, Health authorities will be able to decide about a set of possible actions that will be previously tested in a virtual population interacting in an urban infrastructure, considering its environmental factors, and finally compare the simulated data to consolidated data of real epidemic dynamics.

INTRODUCTION

Conventional epidemiology of infectious disease requires extensive collections of population, health and disease pattern data, as well as data related to environmental factors and social conditions. An epidemiologic study may focus on a particular region or a particular outbreak, or it may take as its theme the epidemiology of a condition across a wide area. The range and amount of data required will, therefore, vary depending on the type of study. Lack of data quality control, lack of definition about the content to be registered, storage heterogeneity and availability are some problems that must be solved to allow more precise and thorough studies in epidemiologic vigilance. In addition, public health data must be usually analyzed together with data from other fields, like data related to the environment, basic sanitation and the social services infra-structure. This set of databases has a variable degree of heterogeneity. As a consequence, it is cumbersome to work with these data in an integrated manner. Furthermore, analytical studies to identify risk factors related to epidemic development are eventually used by health agencies (Wagstaff, 2000, Axelson, 1999). These studies need several types of data, such as geo-referenced disease cases, space-temporal environmental data relevant to epidemic prevention and population data based on demographic and geographic information with territory expressiveness.

Considering this scenario, we present in this paper advances in the development of a large-scale system for space-time visualization, monitoring, modeling and analysis of epidemic data using a Grid platform (Foster, 2002). This system, dubbed IntegraEPI (Silva et al., 2007), is capable to provide the integration of heterogeneous databases related to epidemic analysis and to make available analytical and computational methods to increase the predicting capability of the public heath system, in order to optimize its activities and resources when dealing with epidemic outbreak and prevention. The main objectives of the IntegraEPI project are:

1. To develop a distributed computational laboratory to simulate, test and foresee the epidemic outbreak of selected diseases in a city using a computational grid;
2. To develop analytical methods and algorithms to visualize and analyze the available data in order to build detailed predictive models of the dynamics of epidemic outbreak; and
3. To integrate through the grid large space-temporal databases containing public health system notification data, environmental data and urban infrastructure data.

This system, differently of what has been proposed, is integrated and consequently it enables the construction of detailed predictive models of the dynamics of disease spreading. It is worth noting that simulators implementing these models will be able to use actual, updated data (Gallo et al., 2007, Neto et al., 2006). These simulators will be executed into distributed virtual laboratories where the health agents would test the efficacy of different strategies before the real outbreak occurrence. The health authority will be able to decide about a set of possible actions that would be previously tested in a virtual population interacting in an urban infrastructure, considering its environmental factors, and finally compare the simulated data with the consolidated data of real epidemic dynamics (Silva et al., 2007).

The development of the system described in this introduction with all its integration levels represents a great challenge from the computational standpoint. Moreover, the integration of the set of related databases and space-temporal analysis methods together with a computational model based on contact patterns among individuals and real urban structures demands a high-performance computational environment. Therefore, a logical choice for a computational platform for executing this system is the Grid. We have then chosen to implement the system on a grid platform (Foster,2002) based on standards defined by the Open Grid Forum (OGF, 2008), the Globus Alliance (Globus, 2008) and the OASIS consortium (OASIS, 2008).

BACKGROUND

Grid computing aims at the provision of a global ICT infrastructure that enables a coordinated, flexible, and secure sharing of diverse resources, including computers, applications, data, storage, and networks across dynamic and geographically dispersed organizations and communities (sometimes known as Virtual Organizations). Grid technologies promise to change the way organizations tackle complex problems by offering unprecedented opportunities for resource sharing and collaboration. Just as the World Wide Web transformed the way we exchange information, the Grid concept takes parallel and distributed computing to the next level, providing a unified, resilient, and transparent infrastructure, available on demand, in order to solve increasingly complex problems. In all these areas, Grid technology can either significantly reduce the cost or time to produce results, or even to provide resources that are able to deliver services that cannot be economically delivered using conventional networked information systems.

The implementation of a system on computational grid platform for epidemiologic studies open new perspectives for gathering data on large populations and - as a consequence - would allow stratification of large scale Metropolitan epidemiology studies.

On the other hand, as stated in the introduction, conventional epidemiology of infectious disease requires extensive collections of data. The range of data required will vary depending on the type of study, but certain elements are required: a degree of confidence in the data is essential, so its *provenance* has to be assured and the standards of clinical practice under which it was obtained have to be above a certain threshold. Despite the fact that the data may be gathered under different clinical regimes, it must be still possible to establish their semantic equivalence, to ensure that aggregation or comparison of datasets is legitimate.

Regarding a population health, recent developments of Information Technology (IT) in Brazil make possible the access to diverse and disaggregate information databases about born lives (SINASC), mortality (SIM), notification diseases (SINAN) and hospital internment (SHI-SUS), among others. These anonymized databases contain important data that is routinely used in scientific research in the Public Health field as, for instance, in studies about the public health quality of service and its usability (Wagstaff, 2000). In addition, among the consolidated health indicators, the mortality data, hospital internment and compulsory disease notification are the most important and relevant for population health (Wagstaff, 2000, Black, 1982, Axelson, 1999). However, these databases (excepting SINAN) carry in its conception a centralized view of health system, making its usability for metropolitan area studies very difficult. Moreover, health indicator gathering in general does not follow standards among the different services responsible to data capture and storage, compromising the quality of information, its usability and the comparison of data originated at different metropolitan areas. In addition, public health data must be usually analyzed together with data from other fields, like data related to the environment, basic sanitation and the social services infrastructure. This set of databases has a variable degree of heterogeneity. As a consequence, it is cumbersome to work with these data in an integrated manner. Finally, analytical studies to identify risk factors related to epidemic development are eventually used by the municipal district health services. These studies need several types of data, such as geo-referenced disease cases, space-temporal environmental data relevant to the epidemic prevention and population data based on demographic and geographic information with territory expressiveness.

Another central task for a grid project aimed at regional epidemiology studies, beyond data integration, is to enable and to promote interoperability of the predictive modeling tools and of the statistical analysis tools. Similar to initiatives in the field of systems biology, a repository service of statistical models based on a common understanding and classification scheme of statistical approaches is needed. A point to start with would be a "tool box" of statistical models which will permit the gathering of relevant meta-information on algorithms, modeling strategies and constrains related to application scenarios and evaluation of possible equivalence of statistical models. By using the concept of Grid services this toolbox would allow easy combination of methods and will improve interoperability of statistical models and modeling capabilities.

INTEGRAEPI SYSTEM ARCHITECTURE

The IntegraEPI system architecture is based on the Open Grid Services Architecture (OGSA) specification (Foster et al., 2002) and is organized as shown in Figure 1. Three main services were developed: the data integration service Integra-GISE (IntegraEPI-data Grid Integration SErvice), the simulation service Integra-Model and the analysis service Integra-Analysis.

Each one of the IntegraEPI services cover an important aspect of an epidemic surveillance system, such as: the integration of heterogeneous, geographically distributed population and disease notification databases (Integra-GISE), a simulation service based on epidemiological models for prediction of the forthcoming trends related the transmission of a particular disease over a city population (Integra-Model) and, at last, statistical tools to analyze and visualize the status of such diseases, which make possible the inference of risk indicators and to establish epidemiological thresholds to trigger alerts if some risk situation is detected (Integra-Analysis).

Figure 1. The IntegraEPI system architecture

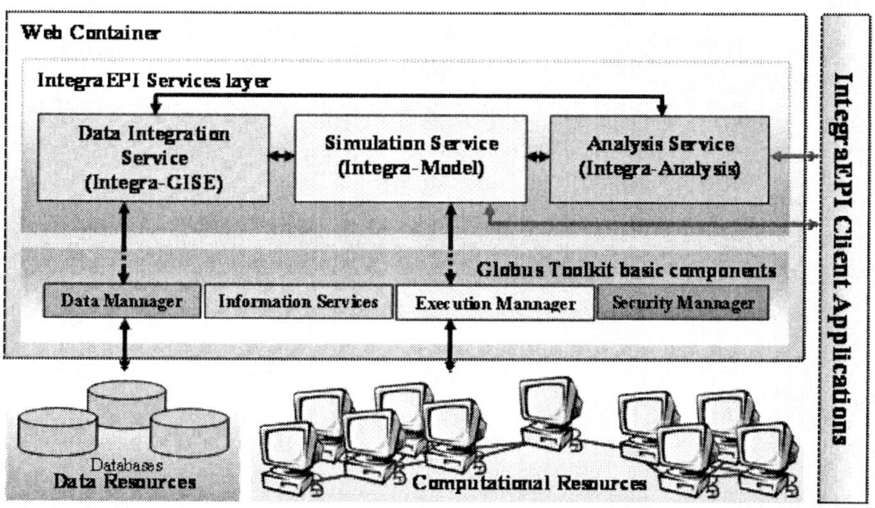

Due to the interoperability provided by the use of grid services and the definition of a Canonical Data Model, the IntegraEPI services are able to interact with each other. The integrated data provided by the Integra-GISE may be used by both the simulation service Integra-Model and the analysis service Integra-Analysis for estimator inference and detection of risk situations. At the same time, the simulation module can use any of the analysis tools provided by Integra-Analysis to study simulated patterns. Such interoperability allows a large degree of flexibility for service composition and implementation.

In the Figure 1 we can see (in the middle) the Globus Toolkit components which provide additional services used by IntegraEPI modules such as reliable file transfer, replica location service, security and authorization services, execution manager components, information services and common runtime components.

Note that there is one instance of the simulation service for each disease covered by this project. Only one Integra-GISE instance is needed for all instances of analysis and simulation services. However, the Integra-GISE service itself can be replicated in order to improve performance.

After this brief system description, we detail, in the following sections, the IntegraEPI main services. We start with a description of the data metamodeling process of IntegraEPI.

EPIDEMIC DATA METAMODELING AND INTEGRATION

Data source integration is a complex process and one of its phases is to convert the heterogeneous data source attributes to a Canonical Data Model (CDM). The CDM is a common model in which all involved data sources are transformed, hiding the syntax differences of local data sources (Macfarlane, 1996). The CDM allows users to make queries and data manipulation in a transparent way for multiple data sources.

There are several texts in the literature that describe CDM requirements (Macfarlane, 1996). Some of these requirements are described below:

- Express different abstraction levels to represent diverse semantic models;
- Possibility to include metadata not stored in local data sources;
- Show differences between schemas; and
- Provide semantic information to all users of the distributed infrastructure.

An initiative that is comparable to the metamodel proposed in this section is the Dublin core metadata initiative (Dublin, 2008). However, the Dublin Core was originally developed with an eye to describing document-like objects. Its suitability for use with particular non-document resources will depend to some extent on how closely their metadata resembles typical document metadata and also what purpose the metadata is intended to serve.

The approach used in the IntegraEPI project is the definition of a data metamodel suited for epidemic data stored in relational and XML sources, that is instantiated in order to generate a particular CDM for each disease considered in the project. In the following section we describe the main components of IntegraEPI data metamodel.

Epidemic Data Metamodel

The data integration process used in the Integra-GISE (or simply GISE) architecture is based on the development of canonical data models. A Canonical Data Model (CDM) for a particular disease maps all the data sources involved on the integration process into a single common model, hiding syntactic and structural differences generated during the database modeling process (Ventrone, 1991).

Furthermore, the creation of a CDM is a critical factor on the data integration. A semantically rich

CDM makes the discovery of similarity among the system data sources easier. Thus, the users can perform queries automatically on multiple data sources.

Disease Canonical Data Models are instantiations of a more general Metamodel, which is part of the GISE architecture. The GISE Metamodel is based on the eXtensible Markup Language (XML) to supply uniform and flexible representation of integrated and local heterogeneous databases (Gardarin,1999) and it is divided in two main parts: GISE-LOCAL-SCHEMA and GISE-INTEGRATED-SCHEMA. The first part is devoted to represent a local relational database. The second part is used to represent a virtual data source through the integration of two or more GISE-LOCAL-SCHEMAs. We shall discuss these two elements in more detail in the following paragraphs.

GISE-LOCAL-SCHEMA

The GISE-LOCAL-SCHEMA is used to represent a single local data source as a resource of a computational grid. In order to achieve standardization and database integration, GISE-LOCAL-SCHEMA is built through pre-established rules. The standardization of the description of local schemas is available with the utilization of XML schemas (Gallo, 2007).

The local schema stores database attributes involved in the integration process to resolve semantic or syntactic conflicts (format, type or unit conflicts) related to the integration of multiple data sources. Figure 2 illustrates a graphic representation for GISE-LOCAL-SCHEMA.

The root element of this base schema is the DataSource. The DataSource is composed of the following attributes: DataSourceAlias, DataSourceID, DataSourceDescription, DataSourceType, DataSourceOwner, ResourceName, Metadata and DataSet elements. The main attributes of the DataSet element are: DataSetId, DataSetDescription, Metadata and Field elements. The main attributes of a Field element are: FieldId, FieldDescription, FieldType, Format, Size, Unit and Metadata.

GISE-INTEGRATED-SCHEMA

GISE-INTEGRATED-SCHEMA is a XML metamodel used to represent a virtual data source, and it is derived from two or more GISE-LOCAL-SCHEMAs. Through such virtual data sources, generated from multiple databases, it is possible to obtain more precise and inclusive data to be used in the epidemic monitoring and simulation process on the grid.

In the case of GISE-INTEGRATE-SCHEMA, the root element is IntDatabase, which is composed of the following attributes: IntDatabaseId, IntDatabaseAlias, IntDatabaseDescription, IntDatabaseType, IntDataBase-Owner, Metadata and IntField elements. The main attributes of the IntField element are: IntFieldId, IntFieldDescription, IntFieldFormat, IntSize, IntUnit, IntType, Metadata and IntFieldComposition elements. The IntFieldComposition element is used to inform the metadata service about the local fields (represented in the GISE-LOCAL-SCHEMA) which composes the integrated field. The IntFieldComposition element main attributes are: IntFieldComposition-FieldId and IntFieldComposition-Function. The IntFieldCompositionFieldId element stores the identification of local fields and the IntFieldCompositionFunction element stores a mathematical function of local fields' conversion to a predefined standard unit.

The GISE-INTEGRATED-SCHEMA is the key to resolve several types of data conflicts, such as denomination, type and format conflicts. With the set of information stored in IntFieldComposition Element (see Figure 3) it is possible to identify and map integrated fields. In this way, the GISE sends a custom query operation for each local data source, respecting attribute names and resolving the denomination conflicts. Additionally, to resolve type conflicts, GISE also uses the information stored in GISE-INTEGRATED-SCHEMA.

Figure 2. Graphic representation of GISE local schema

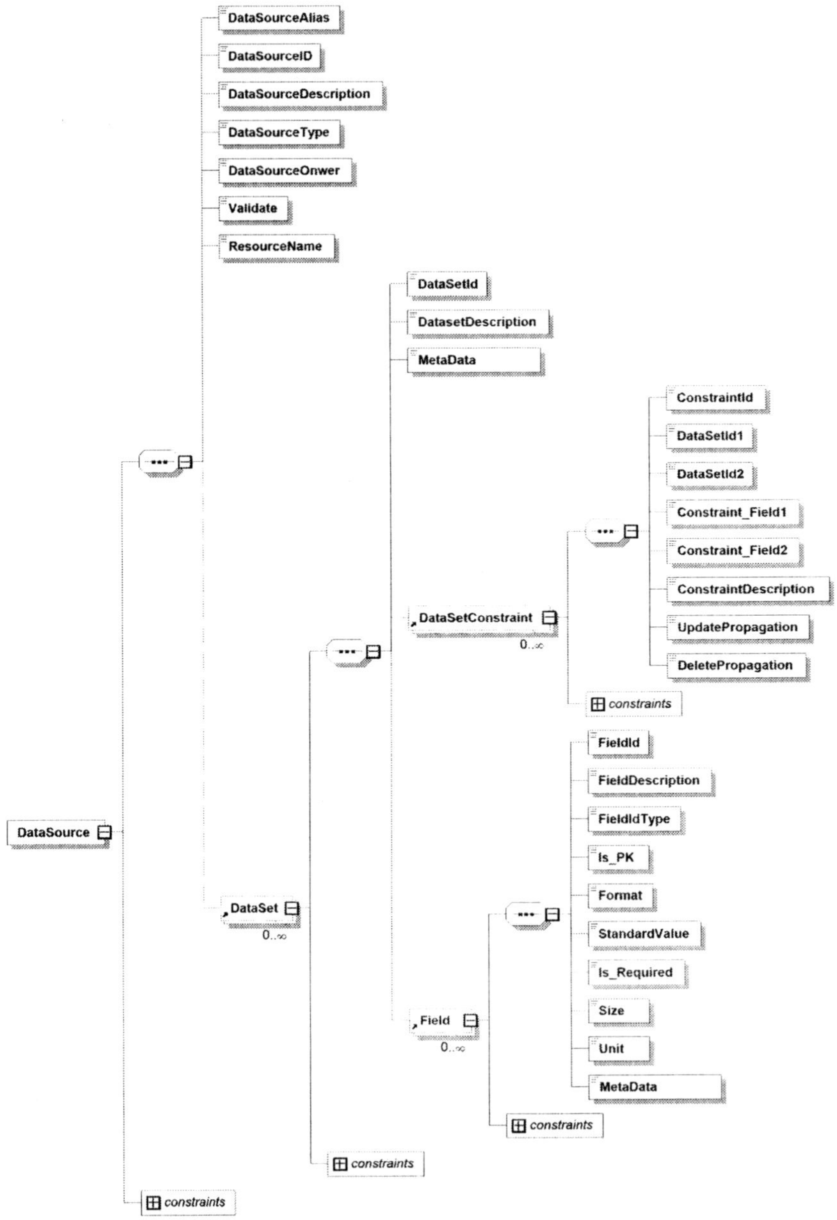

The information stored in InFieldIdType Element is used to convert the information stored in a local data source to a common type. This conversion is executed when the local query is built, using SQL type converter functions. The GISE executes this task using a component named Type Converter.

Finally, format conflicts are resolved using the information stored in IntFormat Element. With this information is possible the conversion of the format stored in a local data source to a standard format of an integrated data source (stored in IntFormat Element). GISE executes this task using the Format Converter component.

Figure 3. Graphic representation of GISE-INTEGRATED-SCHEMA

Epidemic Data Integration

The GISE architecture was developed to obtain data from multiple data sources through a single point of access. With the utilization of a grid infrastructure it is possible to access the several services implemented for this architecture through one single entry point.

The GISE architecture is divided into layers which are composed of components which reside on each layer. Figure 4 illustrates the GISE architecture, its layers and components. We describe the layers in more detail in the following paragraphs.

Figure 4. The layers and components of GISE architecture

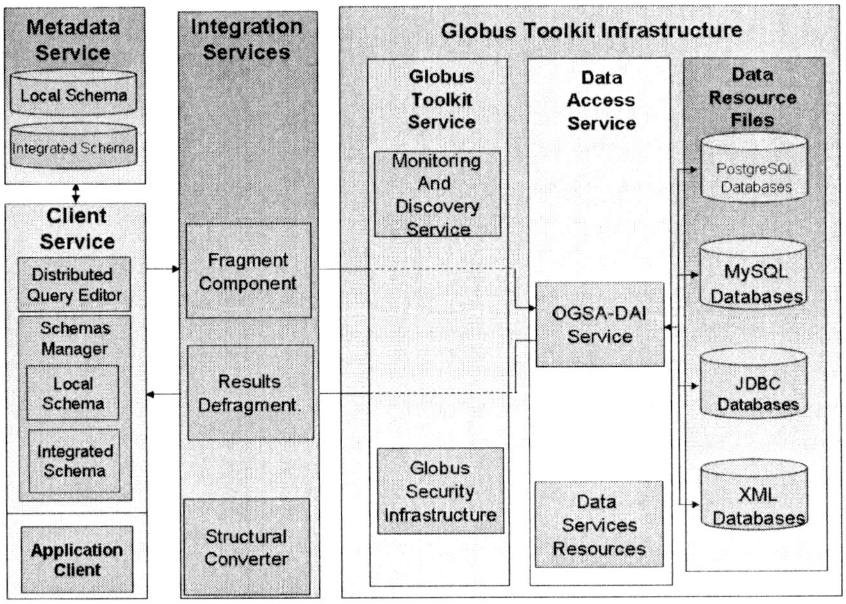

The Client Layer

The client layer supplies the necessary components and interfaces used by final users. The layer is composed of the local schema managers, integrated schema generator and the execution tool for composing distributed data queries. Furthermore, in this layer it is also implemented the Integration Helper. The Integration Helper is a tool used to search for syntactic similarities of integrated schemas attributes in the local schemas, helping the construction of integrated schemas. It is worth noting that, in the GISE, integrated schemas are generated semi-automatically whenever possible. A complete automatic procedure is subject of current research.

The client layer builds the query operation through the Query Editor and Publisher, forwarding it to the Integration Services Layer (Mediator), which is described later in this section.

Metadata Services Layer

The GISE Metadata Service stores database schemas and other metadata relevant to the data source integration process. The Metadata Service supplies information about the existent data sources for the Client Layer, the Mediator and other services of the epidemic monitoring and simulation system. The GISE Metadata Service is a grid service dubbed GISE-MCA (Grid Integration Service – Metadata CAtalog). The structure of Metadata storage is designed for maintaining the consistency between local and integrated schemas. Therefore, it is not possible that an element from the integrated schema establishes a reference with inexistent local elements.

The Integration Services Layer (Mediator)

GISE has a service layer with specialized components to perform data integration. This layer is named Integration Services Layer. The Integration Services Layer is the Mediator of GISE. The Mediator is responsible for data integration and conflict resolution among integrated data sources. The GISE mediator is composed of several components: a fragmentation component, a defragmentation component and several converters.

The *Fragmentation Component* receives an integrated query and divides it in a set of sub-queries according to the schemas stored by the Metadata Service. It receives an operation which will be executed based on a particular integrated schema and performs its decomposition. Therefore, a set of local query operations are generated.

The local queries are stored onto OGSA-DAI Activity Documents (OGSA-DAI, 2008). Such Activity Documents store an operation which will be performed on some local data source in an SQL (Structured Query Language) format. The Activity Documents are processed by the Data Access Layer Service.

The *Structural Converter* is the component of the Integration Service Layer (Mediator) which resolves type and format conflicts, which may occur during the data source integration process. It supports the Results Defragmentator component in returning the results formatted according to the CDM. Its main subcomponent, the *Format and Type Converter,* is responsible for the conversion of the data obtained from the Results Defragmentator in a unique standard, as defined in the Canonical Data Model.

Such conversion is performed by consulting the information about the local and integrated schemas on the Metadata Service and, in the sequence, the local database attribute values are adjusted according to the definition established on the integrated schema. In this way, the format and type conflicts are resolved.

The *Results Defragmentator* complements the Fragmentation Component by merging the fragmented query results into a single and unique result set. The Result Defragmentator integrates in a single view all data from local data sources according to the information stored in the integrated schema. Therefore, the query results are available in a single XML document.

Globus Toolkit Service Layer

The Globus Toolkit (Foster, 2005) supplies the services that enables the utilization of the grid by GISE. This layer of services is managed by the Globus Toolkit middleware (GT4). The main services used by GISE are security-related services and information services.

The Globus security stack provides authorization mechanisms that can be used for accessing data sources on the grid. GISE makes use of the GT4 grid security infra-structure, which provides security mechanisms on the transport and message levels. GT4 implements security on the grid with credential based components and standard protocols for message protection, authentication, delegation and authorization. On the GT4 standard distribution, every Globus user has a X.509 public credential and the protocols are implemented to validate it along the grid. Those credentials are used to establish a secure channel for message exchange. Note that the local data access policies of a given site can also be enforced with services provided also by the GT4, such as CAS (Community Authorization Service). Therefore it is possible to refine security constraints and policies for a particular site, if these additional measures are required to access clinical data from that site.

To grant local data source access, a user must have his X.509 certification included on the Globus access control list of the sharing resource. Therefore, only authorized users are granted access to the system resources.

The Information Service (MDS – Monitoring and Discovery Service) is responsible for supplying information to the other layers of GISE architecture about available Grid Services. The MDS4 is a GT4 service that provides information such as: available resources, the localization of such resources, propriety, costs, load, and others. MDS gather information directly on computational grid resources or services, offering access interfaces to the information about computational resources as well. GISE uses the MDS-Index as an aggregator service to gather the information provided by information sources and the WebMDS system to offer a query interface to the information gathered. It should be noted, however, that MDS is used by all services that compose the Integra-EPI architecture to obtain information about resources and other services.

The Data Access Services Layer

The Data Access Services Layer provides access to data sources on the grid. In this layer, the components are implemented based on the WSRF (Web Services Resource Framework) standard. This layer is implemented on GISE as OGSA-DAI services.

OGSA-DAI is a middleware which permits the data sources implemented with different technologies to be accessed on a grid. Furthermore, OGSA-DAI have a specialized service in its architecture that informs which data sources are available on the system, dubbed Grid Data Service Resource. This grid component is also responsible by the management of the life cycle of the instances of the Grid Data Services (wrappers) necessary to access the local databases.

Finally, OGSA-DAI deliver the results of the local operations performed by the Grid Data Services in XML files which are sent to the GISE Results Defragmentator component.

A Grid Data Service Resource is a used by GISE to inform which data sources can receive local requests. It is also used to inform about the respective Grid Data Services (wrappers) that access a specific data source. The Grid Data Service Resources are persistent services and their distribution are native on GT4.

A Grid Data Service is a transient service used by GISE and its life cycle is managed by OGSA-DAI. These services are called wrappers in GISE architecture and its objective is to provide access to a specific data source (Data Resource File) and to perform local operations. After obtaining the local query response, the Grid Data Service sends the result to OGSA-DAI and can be automatically destroyed, freeing the computational resources allocated for its execution.

The Data Source Layer (Data Resource Files)

The Data Source Layer (Data Resource Files) is the last layer defined by GISE wherein the available local databases for the integration are included. This layer just includes the data source repositories which will be accessed by the Grid Data Services (Wrappers).

In this layer, the heterogeneity among the several technologies used on databases development becomes clear.

Data Curation and Auditing

One of the most critical processes for database integration is the curation and auditing of available data. Often the contents of a database are not verified for redundant entries or have information that does not follow a standard nomenclature or filling process. This fact entails the loss of information and even inconsistency among databases, which reduces the reliability of the information provided by the system.

One important feature of our system is the capability to standardize the notification data through the use of "smart" interfaces which supports data filling and verification processes. Computers are distributed to key locations where new notification data arrives first (typically offices belonging to the public health authority) with the smart interface program installed. New data is entered through this interface, and is then sent to a local database managed by GISE. If any new data is missing or it is clearly incorrect, the interface is able to detect the problem and ask for intervention before the data is actually stored. Note that no personal information about patients is requested.

However, as there are still legacy data that have been registered over the years without standardization and with several inconsistencies with regard to the names of streets and typos in fulfillment of relevant information, semi-automatic procedures were created to scan that data and correct them. When the semi automatic procedures cannot correct the data, the system may call the attention of an data curator. This curated data can also be permanently stored by GISE as needed.

Geo-Referenced Data

Access to Geo-referenced data is supported by GISE through the instantiation of the metamodels presented in this section. Access to maps and other geo-referenced information is provided by a specialized GIS service named Integra-GIS, which interacts directly with GISE through an instance of the GISE-LOCAL-SCHEMA. The service IntegraEPI-GIS provides access to a set of geo-referenced data and maps for spatiotemporal-related analysis. For instance, it is possible through Integra-GIS to obtain - as a XML file - the set of geo-referenced hotspots related to a particular disease using some predefined coordinate system, such as, for instance, UTM (Universal Transverse Mercator). The IntegraEPI-GIS service also provides conversion functions among different coordinate systems. This service is build using open source software, such as GeoServer (GeoServer, 2008) and PostGIS (PostGIS, 2008).

One interesting future possibility is the integration with INSPIRE services (INSPIRE, 2008). INSPIRE aims at making available, in the European Union, quality geographic information to support formulation, implementation, monitoring and evaluation of policies and activities which have an impact on the environment. The integration with INSPIRE services and portals will be particularly important for future deployments in the European Union.

EPIDEMIC DATA ANALYSIS

The analysis of aggregated data requires the construction and use of sophisticated statistical tools. Therefore, the development of the INTEGRA-EPI project includes the design and the implementation of analysis tools to be used by general users and public health officers.

Particularly, the use of geo-processing techniques allows one to gather socioeconomic, health, and environmental data on a spatial basis. Indeed, interpretation of associations among epidemiological and environmental variables requires the utilization of geo-processing techniques. The choices of scale and objectives for a study precede the configuration of the system, conditioning the possible statistical results. The choice of scale must be compatible with the phenomenon which one intends to focus, in order to guarantee the adequacy of spatial units. The interdependency of spatial processes, reflected in the spatial configuration of social, environmental, and epidemiological data distribution, affects interpretation of the possible causes for simultaneous processes. Geo-processing techniques help the generation of knowledge related to the context or situational surroundings in which the damage to public health takes place.

Therefore, in this platform we incorporate several space-temporal analysis methods and algorithms from a Geographic Information System (GIS) perspective. These analytical services are developed to support methods of space-temporal cluster detection per area of data, analysis methods of temporal series (self-correlation and spectral analysis), methods of spatial interpolation for environmental and per-area data and Bayesian methods. It is worth noting that these analysis methods can be used in an interactive way, through a graphic interface provided by a client, for model calibration and to monitor risk indicators thresholds, epidemic dissemination and social vulnerability of a population.

Service Architecture

In order to provide mechanisms capable to discover infectious disease case patterns which would become future epidemics, the IntegraEPI system provides a set of analytical tools to be applied over a given set of data related to that disease. This analytical system, dubbed Integra-Analysis, is formed by a set of services implemented as OGSA grid services capable to interoperate with other grid services which compose the IntegraEPI system.

These tools are in its majority spatiotemporal and statistic methods deployed as services. These services are able to collect epidemic data and establish a correlation to observable patterns (e.g. cluster formation) of the disease cases in some area of the map. Therefore it is possible to generate risk indicators for such diseases or even detect historical patterns from the data sources. Integra-Analysis interaction with other IntegraEPI components is illustrated in Figure 5.

As shown in Figure 5, Integra-Analysis services are able to fetch specific disease information using the disease CDM from GISE. The IBGE, SINAN and curated, updated databases shown are, respectively, a Brazilian socio-geographic database, SINAN database itself, and updated data stored locally by GISE.

One example of spatiotemporal analysis method provided by Integra-Analysis, is the "Quadrat Analysis", that is a method of analysis of dot (notification cases) patterns that transforms data into counting of points for defined map areas, in order to define the occurrence of a specific pattern of points distribution (Bailey, 1995).

The number of points inside each cell that make up the area under study is used for the measurement of variance. The variance compares the number of points (notification cases) in each cell (or square) with the average number of points in all cells of the surface.

The Variance-mean-ratio (VMR) standardizes the degree of variability of the frequency of the points on the squares in relation to the average frequency throughout the area under study (in all the squares of the area).

The variance of the distribution is often compared to the characteristics of random distribution. A random distribution indicates that the variance and the mean have about the same value and therefore the rate VMR is close to 1. If VMR is different from 1 the observed pattern of distribution is not random. It then can be, for instance, uniform or clustered. In these cases, a more detailed analysis is made in order to ascertain whether the occurrence of the disease is related to some specific environmental condition, pol-

Figure 5. Interaction among Integra-Analysis and other services in the IntegraEPI system

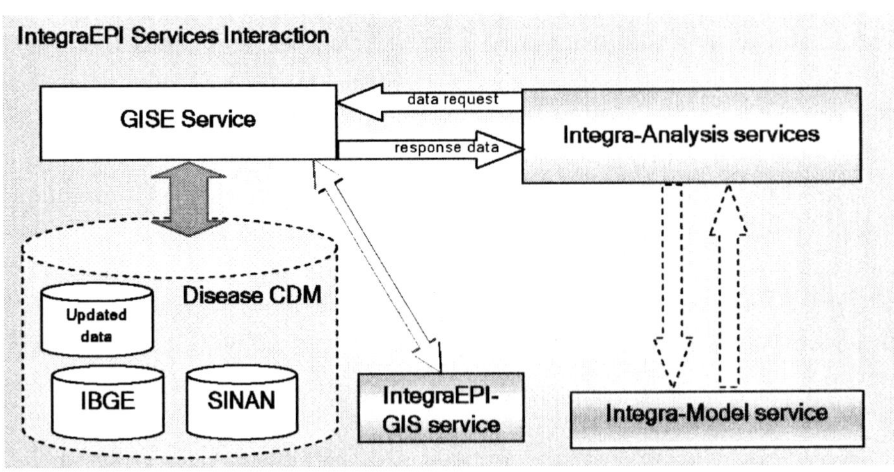

lution focus, urban characteristic, among other possible causes.

EPIDEMIC SIMULATION

The development of epidemic models is currently gaining momentum, since it deals with a subject of great social relevance. However, there are several ways of modeling the dynamic of an epidemic. The best solution for each case should be determined by trying to find the reasons behind the epidemiologic phenomenon, through the integration of spatial and temporal scales with the mathematical and computational model defined to describe the epidemic and to foresee its evolution (Paterson, Viney, 2000). Moreover, in order to these models be used in a predictive way to assist and to evaluate the decision making process made by public health agencies, they must be carefully parameterized and validated with actual epidemiologic data (Luz et al., 2003).

Indeed, the modeling of epidemics has often dealt with the predictions and predictability of outbreaks in real populations with complicated social and spatial structures and with heterogeneous patterns in the contact network (Alves et al., 2003, Eubank, 2002). All these factors have led to sophisticate modeling approaches including disease realism, meta-population grouping, stochasticity and more recently to agent-based numerical simulations that recreate entire populations and their dynamics down to the scale of the single individual (Eubank et al., 2004). In many instances, however, the introduction of the inherent complex features and emerging properties of the network in which epidemics occur implies the breakdown of standard homogeneous approaches and calls for a systematic investigation about the impact of the detailed system's characteristics in the evolution of the epidemic outbreak (Eubank, 2002, Epstein et al., 2002).

These considerations are particularly relevant in the study of the geographical spread of epidemics where various long-range heterogeneous connections typical of modern transportation networks naturally give rise to a very complicated evolution characterized by heterogeneous and seemingly erratic outbreaks (Luz et al., 2003, Alves et al., 2003). Normally, the building of such models does not take into account the synergism between the social processes and the ecosystems related to the spreading of specific contagious diseases. However, the networks that support daily activities are the same ones that infectious diseases explore to spread themselves among human hosts. By modeling them with a certain level of detail, we can understand its structures and discover how to modify them in order to hold the spread of a disease, at the same time that we inflict the smaller possible damage to the society.

Therefore, in order to control disease spreading, considering the limited resources and staff of public health services, it is necessary the development of a distributed virtual laboratory where scenarios of the type "what it would happen if" can be simulated in the most realistic way. In this section we describe the Integra-Model service, which enables the simulation of multiple models and scenarios on the grid.

Integra-Model Service Architecture

The Integra-Model system is composed of two classes of elements: the model simulators, which implement specific infectious disease behaviors, and the Integra-Model service, a simulation service capable to receive simulation requests from Integra-Model client applications.

Each particular infectious disease simulator can be implemented in any programming language, although a command-line interface must be provided, as well as a Integra-Model definition document describing the main parameters of the implemented model which will be managed by the Integra-Model simulation service. The simulation service was built to work as a wrapper for any simulators which are deployed in IntegraEPI. Therefore, model implementation aspects are black boxes from the simulation service perspective. In particular, this modular simulator solution enables Integra-Model system to deal with the plurality of techniques and approaches of epidemic modeling, as well as the possibility to compare and/or complement the analysis of forecast data generated by simulations.

It is worth noting that the set of simulations to be executed by Integra-Model, in most cases, composes a Bag-of-Tasks (BoT) application. A bag-of-tasks application is a set of independent tasks that can be executed in any order. For an example related to dengue fever see (Gagliardi, 2006). It is well know that BoT applications is specially well suited for the Grid (Cirne, 2003). Particularly, in the case of stochastic models, the number of simulations needed can easily be in the order of thousands. Beyond that, since most organizations do not have computing clusters large enough to timely execute a large number of simulations, the grid is an affective way to harvest idle computing power in multiple, geographically distributed organizations. A grid platform is then a natural choice to execute simulations through the Integra-Model service.

This service defines a new type of document named Integra-Model definition document. These documents are extended versions of GISE CDM documents, which contain also information related to specific parameters and configuration data of a given simulator. Like GISE CDM documents, the Integra-Model definition documents are also handled by GISE. The basic Integra-Model definition document schema is shown in Figure 6.

The simulation service is able to interact with the epidemic simulators through the respective Integra-Model definition document, which contain disease CDM information plus the basic information needed for each simulator be deployed as a grid resource: the simulator name, version, associated operating system, its logical filename (used to query the RLS Server to resolve the executable physical file path) and its particular parameters (with its constraints). In particular, the Logical Filename is one of the key elements of this document since it will be used to configure the path where the executable is to be transferred from in the process of job submission through Integra-Model.

On the other hand, Integra-Model is a standard OGSA grid service, so it is able to invoke other OGSA grid services to discover and manipulate the necessary data to prepare its job submission requests. Therefore, the Integra-Model service invokes GISE to retrieve Integra-Model definition documents, it transfer files via Globus GridFTP, queries MDS to obtain information about avail-

Figure 6. Graphic representation of the Integra-Model definition document schema

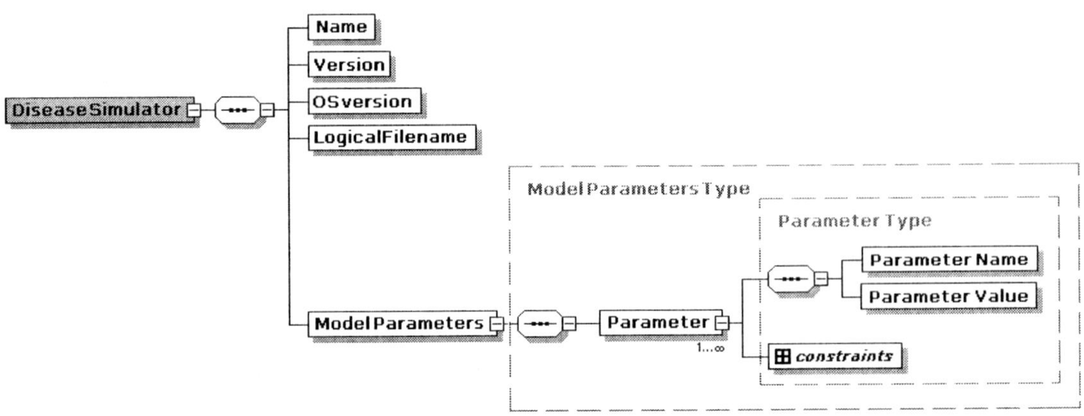

Figure 7. Graphic representation of Integra-Model service execution

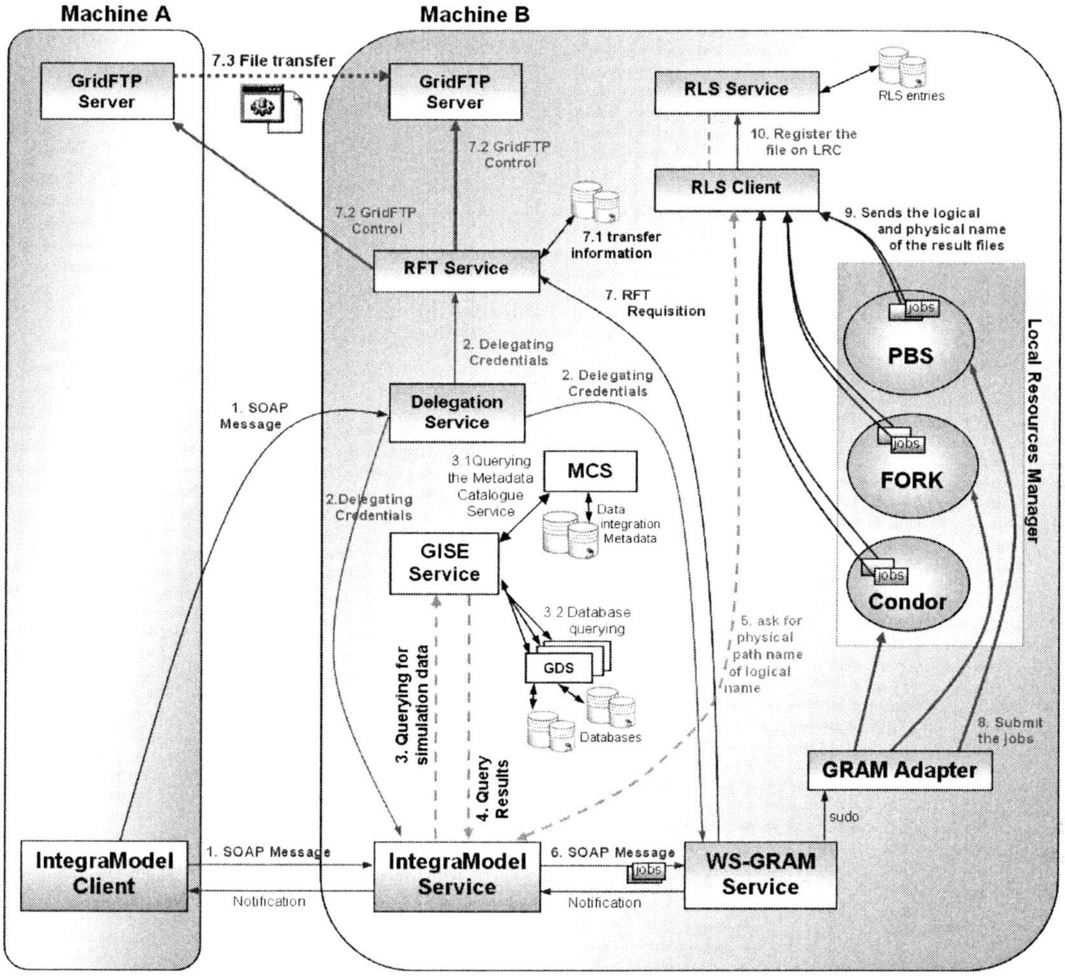

able resources and submits its jobs on Globus WS-GRAM service through a GRAM Adapter class. This process is detailed in Figure 7.

The diagram in Figure 7 basically depicts a scenario where a Machine A wants to start a simulation job by invoking the Integra-Model service instantiated on a Machine B. The whole process is executed typically in 10 steps:

1. A client application of Integra-Model service makes a simulation request for a specific infectious disease. In order to verify his authenticity, the client also sends information for the grid Delegation Service;
2. The Delegation Service delegates users credentials as an user capable to use Integra-Model service, WS-GRAM service and other related grid tools and services;
3. If the disease related data is not cached, the service queries GISE service for simulation data. GISE will gather the specific Disease CDM and the Integra-Model definition documents. This is done on step 3.1. After this, GISE is able to parse the CDM retrieved and create distributed queries to be processed on the integrated databases (step 3.2);
4. The data needed to configure the scenario wherein the simulations will occur is deliv-

ered to Integra-Model service with the Integra-Model definition document. Henceforth, the simulation service starts to manipulate the necessary data for the configuration of simulation jobs. At this moment, any GIS information needed to generate the model networks or other city social/environmental settings is also processed;

5. Meanwhile, the simulator parses the Integra-Model definition document and consults the grid Replica Location Service (RLS) in order to discover the physical path of the chosen simulation to be staged in.

In this point of execution the user/application can visualize the gathered information and can also configures the retrieved Integra-Model CDM parameters through a web portal or client application interface. Nevertheless, a user is able to execute the steps 1-5 again in order to change the simulator and respective parameters. The next steps are:

6. The Integra-Model service sends a job definition to the WS-GRAM service. Note that nodes are chosen for simulation execution after a querying the MDS service;
7. The stage-in process starts and asks for simulator executable file transfers through the RFT service;
8. A GRAM Adapter class submits the jobs to the available job schedulers;
9. When a job finishes its execution, its result files information is sent to a RLS client;
10. The client sends to the RLS service the logical and physical names of simulation results to be stored as RLS service entries.

After the completion of this process, the client is able to gather the simulation results via file transfer after resolving, through the RLS service, their physical locations in a GsiFTP URI format (e.g. gsiftp://integraepi.com.br:2811/tmp/Santos_02-02-2007_d844f4c0-d34a-11db-95fa-b82f185b40ce.xml).

We decided to use the RLS service as an index service for simulation results to increase the availability, scalability and reliability of the system, since result data would be replicated to other RLS services available on grid environment. It should be highlighted, however, that the RLS service is used only to store intermediate simulation results. Those simulation results are then aggregated and presented to the user as a set of statistics. The RLS service is not used to store epidemic data or final results.

As occurs with other IntegraEPI services, any simulation results can be used by other services, like Integra-Analysis services, in order to verify the prospected scenarios for an emerging epidemic situation or to be visualized by visualization tools. This possibility seems to be of great help for health officers, since they can have analytical and visual outputs which would help them to observe and detect the most critical regions or spreading patterns over the population and take the required countermeasures.

A result file format would vary from simulator to simulator since it depends on its implementation, but it is recommended to save metadata information about the simulation in the result file, as well as the number of elements in each modeled clinical state (as in SIR model, the susceptible, infected and recovered states). In individual-based models it is also interesting to save some information about geographic coordinate positioning.

INTEGRA-EPI CLIENT

This section describes the general architecture of the IntegraEPI client. Although individual services can have as clients specialized systems or even other services, the client described in this section accesses all three main services in an interactive and integrated way.

The client provides basically three main features: simulation, analysis and visualization. A configuration panel example is presented at Figure

Figure 8. The configuration panel presenting the configuration data for the urban Dengue fever simulator. © 2009 Fabrício Alves Barbosa da Silva et al. Used with permission.

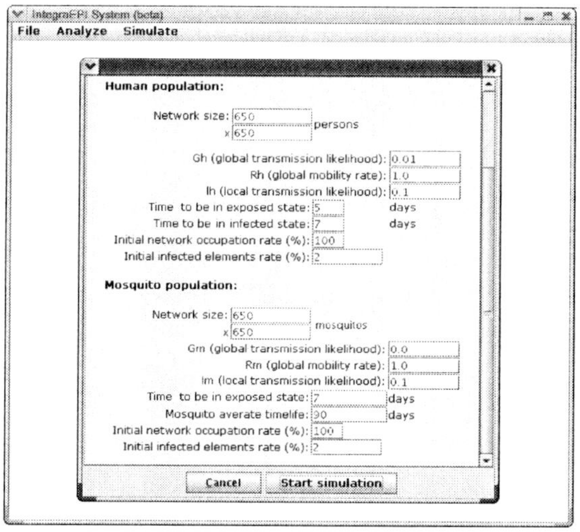

8 and some snapshots of the client interacting with the IntegraEPI services are shown in Figure 9 and Figure 10.

Just after launching the client, the user can choose between simulate or analyze disease data. When choosing the simulate option, the user can start to specify a simulation task through a configuration box. The data presented in Figure 8 comes from the GISE CDM files and Integra-Model definition files.

It is worth noting that this screen let users configure most of the simulator parameters. However, these parameters can be automatically configured by a configuration agent, which can apply statistical concepts for model parameterization.

After configuration, simulation tasks are ready to be submitted to Integra-Model service. A couple of simulation screenshots are presented in Figure 9.

At any time the user can pause, stop or continue the visualization of the simulated disease spreading over the city. At any moment the user also can apply an analysis tool over the displayed data for a better comprehension of what is happening over the city. Let's illustrate this feature by invoking the quadrat analysis command in the client. This is done by selecting in the top menu the option analysis, quadrat analysis.

While performing the quadrat analysis, the user is able to specify the number of quadrants for x and y axis. Henceforth, the Integra-Analysis quadrat analysis service is invoked and the returned data is used to build another visualization layer by the application visualization feature which can be hidden or displayed by the user. At

Figure 9. Simulated Dengue fever results over Santos (a Brazilian costal city) city map. © 2009 Fabrício Alves Barbosa da Silva et al. Used with permission.

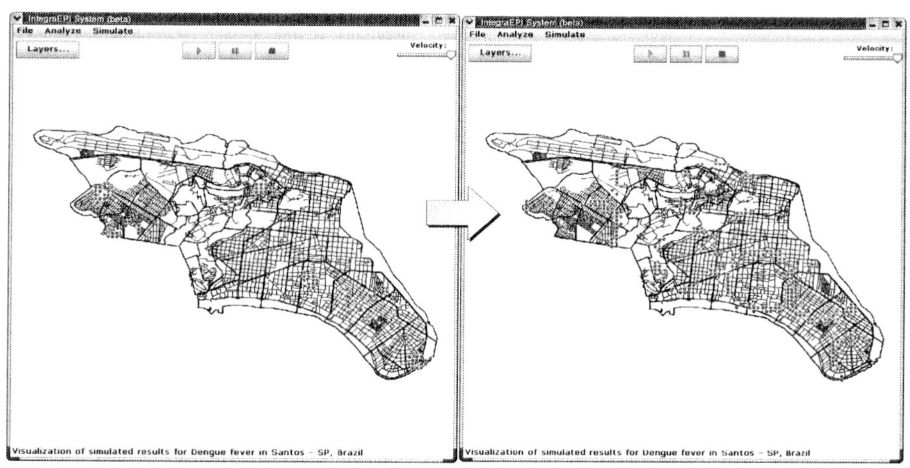

Figure 10. The quadrat analysis service from Integra-Analysis processing for simulated dengue fever data over Guarujá map, another Brazilian costal city. © 2009 Fabrício Alves Barbosa da Silva et al. Used with permission.

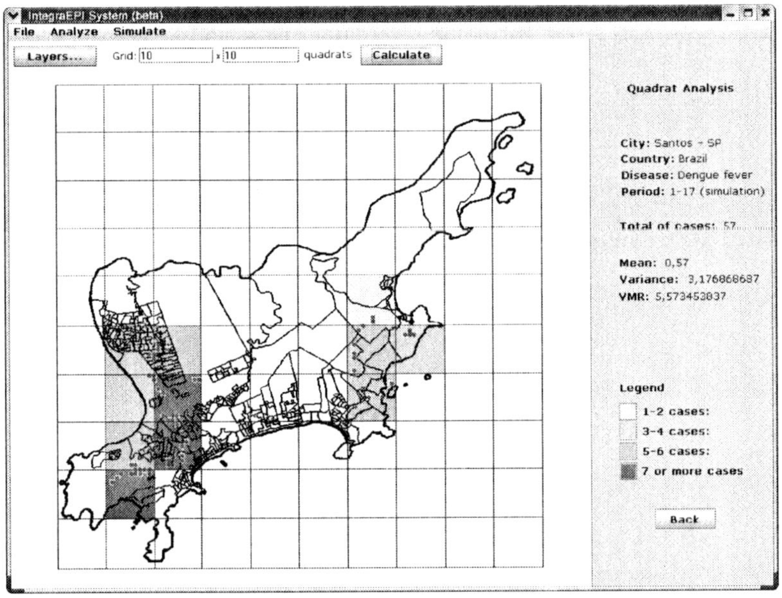

the same time, additional information provided by the Integra-Analysis service is displayed in the frame on the right of the window. Another interesting feature is the ability to change the color legend of dot counting to highlight its visualization over a map. At any time the user can switch again to the simulation results view and advance the animation of such results.

On the other hand, this client can invoke any of the available Integra-Analysis services from its main screen to analyze the disease data stored and accessed by the GISE service. The visualization module in this case can display, for the quadrat analysis, the resultant analysis for a specified period in time.

We detail below an example of interaction between the Integra-EPI client and the Integra-Model simulation service. A general client execution flow is presented in Figure 11.

As depicted in Figure 11, the presented client execution flow has 9 steps:

1. After a simulation is configured, the Integra-Model client submits jobs to the Integra-Model service;
2. When simulation tasks execution are complete, the Integra-Model service returns a Logical File Name (LFN) which represents the set of simulation result files stored and managed as RLS replicas;
3. The client queries RLS to discover the physical location of result files on the grid;
4. The physical path of each file is returned in a list to the client;
5. The client requests the transfer of the result files. Result files are transferred to the client machine through GridFTP;
6. GridFTP transfers the simulation result files from where they are stored to Integra-Model client machine;
7. After the transfer, the client receives the local file system location where those files were saved. Then the files are parsed and prepared by client to be visualized as an information data layer over the city maps;
8. The client retrieves city maps data and information. This is done by querying the IntegraEPI-GIS service through GISE;
9. After receiving map data, the client prepares a set of suitable map layers to be shown on

Figure 11. Graphic representation of the client interacting with the Integra-Model service

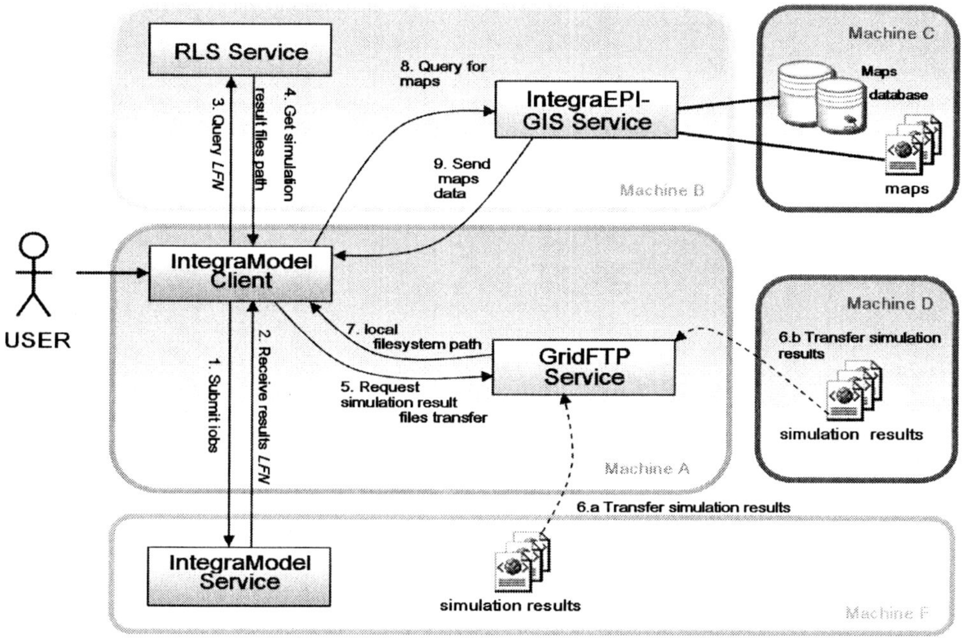

the background of the simulated data layer. The visualization then starts and an animation of the simulations is presented.

FUTURE TRENDS

One promising area of research is semantically based integration of georeferenced epidemiological data. Astakhov et al. (Astakhov et al., 2006) describe a data integration mediation system developed as part of the Biomedical Informatics Research Network. This system uses UMLS (Bodenreider,2004) to perform semantically-based integration and to provide the ability to query through ontological concepts. However, no support for geo-referenced data is presented. We believe that such support is fundamental for epidemic data integration and analysis.

Agent-based simulation is another area that we expect significant advances in the following years. Considering individual-based/agent-simulation, the trend in the following years is the availability of systems capable to simulate hundreds of millions of agents in COTS hardware. Computational requirements and characteristics of one of those systems are described by Parker (Parker,2007). It is worth noting that the system described by Parker was implemented in Java using Java RMI. Although Java RMI is not suitable to Grids, it is possible to implement equivalent systems using web services technology.

Another promising area of research is the development of Hybrid epidemic models, one that combines agent-based and equation-based approaches. The main advantages of hybrid models are computational efficiency and the allowance of mathematical analysis of emerging structures generated by agent-based models. Bobashev, Goedecke, Yu and Epstein (Bobashev et al., 2007) propose a conceptual model and initial examples of a threshold hybrid model that switches between agent-based and equation-based descriptions.

Several epidemic monitoring systems have been deployed in the last decade. Those systems inspired the proposal and development of IntegraEPI. One prominent characteristic of most systems is the capability of gathering epidemic information from several different sources in order to provide early detection (and, by consequence, early response) of a epidemic outbreak. Some representative examples are:

- **ProMED-Mail** (Madoff, Woodall, 2005) is a mediated e-mail list that acts as an early warning system for outbreaks of emerging diseases involving humans, plants and animals around the world. It operates since 1994 and connects more than 30.000 people over 180 countries.
- **GPHIN** (Global Public Health Intelligence Network) (Mawudeku, Blench, 2006) is an Internet-based early warning system that gathers preliminary reports of public health significance. It monitors global media sources such as news wires, web sites, local and national newspapers.
- **HealthMap** (Brownstein, Freifeld, 2007) is a web based systems designed to collect and visualize outbreak data according to time, geography and infectious disease agent, operating since September 2006. Healthmap objective is to bring structure to an information flow that otherwise would be overwhelming to the user. It is capable to integrate data available from news wires, RSS feeds, ProMED-mail messages and validated official alerts (WHO). Once collected, the data are aggregated by source, disease and geographic location, and overlaid on an interactive, user-friendly map. The map itself is available at http://www.healthmap.org.
- **Gripenet.pt/degrotegriepmeting.nl** is an Internet-based system relying on weekly questionnaires completed online. This system provides real-time, geo-referenced, incidence data about influenza cases in Portugal (www.gripenet.pt), Netherlands (http://www.degrotegriepmeting.nl/), and other European Countries.

Finally, recent papers (Butler, 2006, Jones et al., 2008) have been emphasizing the need for allocating resources for smart surveillance of emerging disease hotspots, particularly in lower latitudes, such as Latin America, tropical Africa and Asia. We believe that the Integra-EPI project is a step in this direction.

CONCLUSION

The essential proposal of this research project is to contribute for the modernization of the epidemiologic surveillance and monitoring system by comparing results of detailed simulations with the observed experimental data related to the spreading of a disease, considering both temporal and geographic aspects. More specifically, we are interested in the development of interactive algorithms and software that implement space-temporal predictive models of epidemic behavior in detailed geo-referenced maps of a metropolitan area. The resulting analysis is to be used by local public health agencies, assisting public health authorities in the decision making process related to a particular epidemic.

It is worth noting that the implementation of the IntegraEPI project over a computational grid platform open completely new perspectives for gathering data on large populations and - as a consequence - allow stratification of large scale Metropolitan epidemiology studies. Other advantages of this technology are the relatively small deployment cost and high processing and storage capacities.

It is important to emphasize that some of the expected results to be obtained during the development of this project do not apply solely to the public health sector and epidemiology is one of its applications. There is a whole class of problems of spatial and temporal nature, over which simulating, detecting, monitoring and visualizing patterns is part of the response to the problem. Moreover, with the advancement of this project, we believe that will be possible to create and connect simulations of other systems, including environmental and atmospheric pollution, transportation, water supplies and municipal district management networks, to make available virtual laboratories where one could explore solutions for a large variety of problems of the real world. The possibility to construct a family of applications related to metropolitan area management, containing more intelligence in its analysis, would be advantageous for a community of programmers, researchers and entrepreneurs. These communities want to have better support for the planning and operation of its services and products, which depends on segmentation, based on its territorial distribution.

ACKNOWLEDGMENT

Authors are grateful to FAPESP and CNPq for supporting the project.

REFERENCES

Alves, D., Hass, V. & Caliri, A. (2003). The predictive power of R0 in an epidemic probabilistic model. *Journal of Biological Physics, 29*(1), 63.

Astakhov, V., Gupta, A., Grethe, J., Ross, E., Little, D., Yilmaz, A., Qian, X., Santini, S., Martone, M., Ellisman, M. (2006). Semantically Based Data Integration Environment for Biomedical Research. In *Proceedings of the 19th IEEE Symposium on Computer-Based Medical Systems (CBMS'06).* New York: IEEE Computer Society Press.

Axelson, O. (1999). The character and the public health implications of ecological analyses. In A. Lawson, A. Biggeri, D. Böhning, E. Lesaffre, J.-F. Viel & R. Bertollini (Ed.), *Disease Mapping and Risk Assessment for Public Health,* (pp. 301-309). New York: Wiley and Sons.

Bailey, T. C., Gatrell, A. C. (1995). *Interactive Spatial Data Analysis.* Essex, UK: Longman Scientific and Technical.

Black, D., Morris S. N., Smith, C., Townsend, P. (1982). The pattern of present health inqualities. In P. Townsend & N. Davidson (orgs.), *Inequalities in health: the black report* (pp. 51-64). New York: Penguin Books.

Bobashev, G. V., Goedecke, D. M., Yu, F., Epstein, J. M. (2007). A Hybrid Epidemic Model: Combining the Advances of Agent-Based and Equation-Based Approaches. In *Proceedings of the 2007 Winter Simulation Conference.* New York: IEEE Press.

Bodenreider, O. (2004) The Unified Medical Language System (UMLS): integrating biomedical terminology. *Nucleic Acids Research, 32*, D267-D270.

Brownstein, J.S., Freifeld, C.C. (2007) HealthMap: the development of automated real-time internet surveillance for epidemic intelligence. *Euro Surveill 12: E071129 071125.* Retrieved February 28, 2008 from http://www.eurosurveillance.org/ew/2007/071129.asp#5.

Butler, D. (2006) Disease surveillance needs a revolution. *Nature, 440*(March 2), 6-7.

Cirne, W., Paranhos, D., Costa, L., Santos-Neto, E., Brasileiro, F., Sauve, J., Silva, F.A.B., Barros, C. O. & Silveira, C. (2003) Running Bag-of-Tasks Applications on Computational Grids: The MyGrid Approach. In *Proc. International Conference on Parallel Processing (ICPP'03)*, pp.407-416.

Dublin Core Metadata Initiative (2008). Retrieved July 20, 2008 from http://www.dublincore.org

Epstein, J. M., Cummings, D. A., Chakravarty, S., Singa, R. M., Burke, D. S. (2002, December). *Toward a Containment Strategy for Smallpox Bioterror: An Individual-Based Computational Approach. Brookings Institution, Center on So-*

cial and Economic Dynamics (Working Paper No. 31).

Eubank, S. (2002). *Scalable, Efficient, Epidemiological Simulation*. New York: ACM. Retrieved March 1, 2008 from http://www.ccs.lanl.gov/ccs5/documents/pdf/2001/LA-UR-01-5513.pdf.

Eubank, S., Guclu, H.,Kumar, V. S. A., Marathe, M. V., Srinivasan, A., Toroczkai, Z.& Wang, N. (2004). Modeling disease outbreaks in realistic urban social networks. *Nature*, 429(May 13), 180.

Foster, I. (2002) What is the Grid? A Three Point Checklist. *GRID Today, 1*(6). Retrieved February 13, 2008 from http://www.gridtoday.com/02/0722/100136.html.

Foster, I., Kesselman, C., Nick, J., Tuecke, S. (2002). The Physiology of the Grid: An Open Grid Services Architecture for Distributed Systems Integration. Retrieved February 14, 2008 from http://www.globus.org/alliance/publications/papers/ogsa.pdf.

Foster, I. (2005). Globus Toolkit Version 4: Software for Service-Oriented Systems. *IFIP International Conference on Network and Parallel Computing*, (LNCS 3779, pp. 2-13). Berlin, Germany: Springer-Verlag.

Gagliardi, H. F., Silva, F. A. B., Alves, D. (2006). Automata Network Simulator Applied to the Epidemiology of Urban Dengue Fever. *ICCS 2006, Part III*, (LNCS 3993, pp. 297 – 304).

Gallo, E., Gagliardi, H.F., Neto, V. C., Alves, D., Silva, F.A.B. (2007) GISE: A Data Access and Integration Service for a Grid-Based Epidemic Surveillance System. *Proceedings of the 40th Annual Simulation Symposium*. New York: IEEE Computer Society Press.

Gardarin, G.; Sha, F.; Ngoc, T. (1999) XML-based Components for Federating Multiple Heterogeneous Data Sources. In *Proc. of International Conference on Conceptual Modeling*, Paris, France.

GeoServer website (2008). Retrieved July 20, 2008 from http://geoserver.org

Globus Alliance (2008). Retrieved July 20, 2008 from http://www.globus.org

INSPIRE Project (2008). Retrieved July 12, 2008 from http://www.inspire-geoportal.eu

Jones, K. E., Pattel, N. G., Levy, M. A., Storeygard, A., Balk, D., Gittleman, J. L., Daszak, P. (2008). Global Trends in Emerging Infectious Diseases. *Nature, 451*(February 21), 990-994.

Luz, P.M., Codeço, C.T., Massad, E., Struchiner C.J. (2003, October). Uncertainties Regarding Dengue Modeling. *Mem Inst Oswaldo Cruz, 98*(7), 871-8.

Madoff, L., Woodall, J. (2005). The Internet and the Global Monitoring of emerging Diseases: Lessons from the First 10 Years of ProMED-mail. *Archives of Medical Research, 36*, 724-730.

Macfarlane, A.; Mccann, Liddell, H. (1996). A Common Data Model For Meta-Data Interoperable Environments. In *Proceedings of II International Baltic Workshop on Database and Information Systems*.

Mawudeku A., Blench M. (2006). Global Public Health Intelligence Network (GPHIN). In *Proceedings of the 7th Conference of the Association for Machine Translation in the Americas*. Retrieved February 11, 2008 from www.mt-archive.info/MTS-2005-Mawudeku.pdf.

Neto, V.C., Gagliardi, H. F., Rezende, A., Gonçalves, E.S., Gallo, E., Silva, F.A.B., Pisa, I.T., Alves, D. (2006). Data Access Service in a Computational Grid Platform Applied to the Monitoring and Control of Epidemics on Georeferenced Dynamic Maps. In *Proceedings of the 2nd IEEE Conference on e-Science and Grid Computing*. New York: IEEE Computer Society Press.

OASIS Consortium (2008). Retrieved July 20, 2008, from http://www.oasis-open.org.

OGF - Open Grid Forum (2008). Retrieved July 20, 2008, from http://www.ogf.org.

OGSA-DAI (2008) Open Grid Service Architecture – OGSA-DAI Documentation. Retrieved February 20, 2008, from http://www.ogsadai.org.uk/documentation/.

Parker, J. (2007) A Flexible, Large-Scale, Distributed Agent Based Epidemic Model. *Proceedings of the 2007 Winter Simulation Conference*. New York: IEEE Press.

Paterson, S., Viney M. E. (2000) The interface between epidemiology and population genetics. *Parasitology Today, 16*(12), 528.

PostGIS website (2008). Retrieved July 20, 2008 from http://postgis.refractions.net.

Ventrone, V.; Heiler, S., (1991, December). Semantic Heterogeneity as a Result of Domain Evolution. *ACM SIGMOD Record, 20*(4), 16-20.

Wagstaff, A. (2000). Socioeconomic inequalities in child mortality: comparisons across nine developing countries. *Bull World Health Organ 2000, 78*,19-29.

Welch, V. (2005). *Globus Toolkit Version 4 Grid Security Infrastructure: A Standards Perspective*. Retrieved from http://www.globus.org/toolkit/docs/4.0/security/GT4-GSI-Overview.pdf.

Silva, F.A.B., Gagliardi, H. F., Gallo, E., Madope, A.M. Neto, V.C. , Pisa, I.T., Alves, D.(2007) IntegraEPI: a Grid-based Epidemic Surveillance System. In N. Jacq, Y. Legré, H. Muller, I. Blanquer, V. Breton, D. Hausser, V. Hernández, T. Solomonides and M. Hofman-Apitius (Eds.), *From Genes to Personalized HealthCare: Grid Solutions for the Life Sciences - Proceedings of HealthGrid 2007*. Amsterdam, the Netherlands: IOS Press.

KEY TERMS AND DEFINITIONS

Canonical Data Model (CDM): The Canonical Data Model is a technique utilized to standardize the data source models. The Canonical Data Model hides syntactical differences of multiple heterogeneous data sources and provides a common model for data querying.

eXtensible Markup Language (XML): XML, describes a class of data objects called XML documents and partially describes the behavior of computer programs which process them. XML is an application profile or restricted form of SGML, the Standard Generalized Markup Language. By construction, XML documents are conforming SGML documents.

Globus Alliance: The Globus Alliance is a community of organizations and individuals responsible for the development of the Globus Toolkit.

Globus Toolkit: Open source software toolkit used for building grids.

OASIS: Not-for-profit consortium that drives the development, convergence and adoption of open standards for the global information society.

OGSA-DAI: OGSA-DAI is a middleware, integrated in the Globus Toolkit version 4 distribution, which permits the data sources implemented with different technologies to be accessed on a grid.

Open Grid Forum (OGF): International community dedicated to accelerating grid adoption by providing an open forum for grid innovation and developing open standards for grid software interoperability.

Web Services: A software system designed to support interoperable machine to machine interaction over a network through the use of common standards.

Web Service Resource Framework (WSRF): WSRF is a set of specifications for expressing the relationship between stateful resources and Web services. The specifications define specific message exchange formats and related XML definitions.

World Health Organization (WHO): Co-ordinating authority for health within the United Nations system.

Chapter XXIV
Gridifying Biomedical Applications in the Health-e-Child Project

David Manset
maat Gknowledge, France

Frederic Pourraz
maat Gknowledge, France

Alexey Tsymbal
Siemens AG, Germany

Jerome Revillard
maat Gknowledge, France

Konstantin Skaburskas
CERN, Switzerland

Richard McClatchey
University of the West of England, UK

Ashiq Anjum
University of the West of England, UK

Alfonso Rios
maat Gknowledge, Spain

Martin Huber
Siemens AG, Germany

ABSTRACT

The Health-e-Child project started in January 2006 with the aim of developing a Grid-based healthcare platform for European paediatrics and providing seamless integration of traditional and emerging sources of biomedical information. The objective of this chapter is to share experiences, and present major issues faced, solutions found and a roadmap for future work in developing the Grid infrastructure for interactive biomedical applications in the project, as Health-e-Child approaches its final phases. This chapter starts with a brief introduction of the project itself, followed by a description of its architecture on the Grid. It then illustrates the approach with the description of a concrete example of one integrated key application, the Health-e-Child CaseReasoner, which is intended for biomedical decision support over the Grid, and is based on similarity search and advanced data visualization techniques.

1. INTRODUCTION

In recent time demand has risen for more holistic views of patients' health so that healthcare can be delivered at the appropriate time, by the appropriate clinician, with the appropriate means at the level of individual patients. The Health-e-Child (HeC, pronounced "healthy child") project ("Health-e-Child," 2008) aims to provide data integration across heterogeneous biomedical information in order to facilitate improved clinical practice, scientific research and ultimately such personalised healthcare. As one of the largest integrated projects of the 6th Framework Programme of the European Commission, HeC brings together three major paediatric medical centres with several European companies, university groups and research institutions specialised in Grid-based biomedical information integration and related technologies.

The main objectives of the HeC project are:

- To gain a comprehensive view of a child's health by vertically integrating biomedical data, information and knowledge that spans the entire spectrum from the genetic through clinical to the epidemiological;
- To develop a biomedical information platform, supported by sophisticated search, optimisation and matching techniques for heterogeneous information, empowered by Grid technology;
- To build enabling tools and services on top of the HeC platform, that will lead to innovative and better healthcare solutions in Europe, based on:
    - Integrated disease models exploiting all available information levels.
    - Database-guided biomedical decision support systems provisioning novel clinical practices and personalized healthcare for children.
    - Large-scale, cross-modality, and longitudinal information fusion and data mining for biomedical knowledge discovery.

The realization of these project goals requires an infrastructure that is highly dependable and reliable. Indeed, physicians demand guarantees that the system will be always available and that the processes which integrate and manipulate patient data will be reliable, even in the case of failures. The infrastructure will have to allow for transparent access to distributed data, to provide a high degree of scalability, and to efficiently schedule access to computationally intensive services by applying sophisticated load-balancing strategies. Consider a scenario where a similarity search across the entire HeC patient population is needed to make a better decision over a critical case. In order to support such a search possibility on demand intensive query processing, feature extraction and distributed similarity calculations have to take place. All these steps require significant computing power, storage capacity and an acceptable quality of service (QoS) over the infrastructure resources.

Consequently, the HeC project has as one of its primary objectives, the delivery of a complete suite of Grid-based and cost-efficient tools for individualised disease prevention, screening, early diagnosis and therapy and associated follow-up for paediatric diseases across three different domains; cardiology (e.g. Right Ventricle Overload caused by Atrial Septal Defect or the Tetralogy of Fallot), rheumatology (Juvenile Idiopathic Arthritis), and neuro-oncology (e.g. Pilocytic Astrocytoma). To facilitate this, it has started building a gLite-enabled European network linking leading clinical centres to enable them to share and annotate biomedical data, to validate systems clinically and to disseminate clinical excellence across Europe by establishing new technologies, clinical workflows and standards in the domain.

The project brings together three heterogeneous communities, in a well-balanced configuration, which can be described as three equally important cornerstones:

- The collaboration of clinicians and health-care workers from the cardiology, rheumatology and neuro-oncology domains, bringing together the expertise that is crucial to identifying relevant clinical research directions;
- The cooperation between medical imaging and health IT experts, who are able to bridge the clinical and IT worlds; and
- The marriage of the Grid and distributed computing technologies, where experts harness the power of the Grid to solve requests coming from the other two communities.

This cross-discipline blend of expertise underpins the HeC project's developments and provides the match that is needed to ensure user requirements can become manifest in a delivered healthcare platform.

The main objective of this chapter is to report on the early experiences of the HeC partners in developing their component tools constituting the distributed biomedical platform (see the conceptual overview in Figure 1). In particular, it focuses on the underlying challenges of adopting Grid technologies in healthcare IT, i.e. the so-called gridification of healthcare applications. The next section therefore introduces the main issues related to porting applications to the Grid, whereas Section 3 discusses the approach and implementation of the HeC Gateway middleware, a stack of software services sitting on top of the Grid, which aims at simplifying resources integration within the infrastructure. Section 4 then illustrates the relevance of the adopted approach through the concrete gridification of the CaseReasoner, the cornerstone of HeC decision support. Section 5 identifies several lessons learned in gridifying such applications before Section 6 concludes on the added value of the Grid with a review of the recently achieved results and identifies future work and challenges to be addressed in the second phase of the project.

2. BACKGROUND

2.1. The Complexity of the Grid

Medical Images and clinical information in general represent huge quantities of data, which can amount to several Petabytes per year (Montagnat *et al.*, 2003). Studies in rare diseases, in aneurism research, in HIV drug resistance treatments and in biomedical modelling have demonstrated the need for large scale data storage and computation across

Figure 1. HeC conceptual overview ("Health-e-Child," 2008)

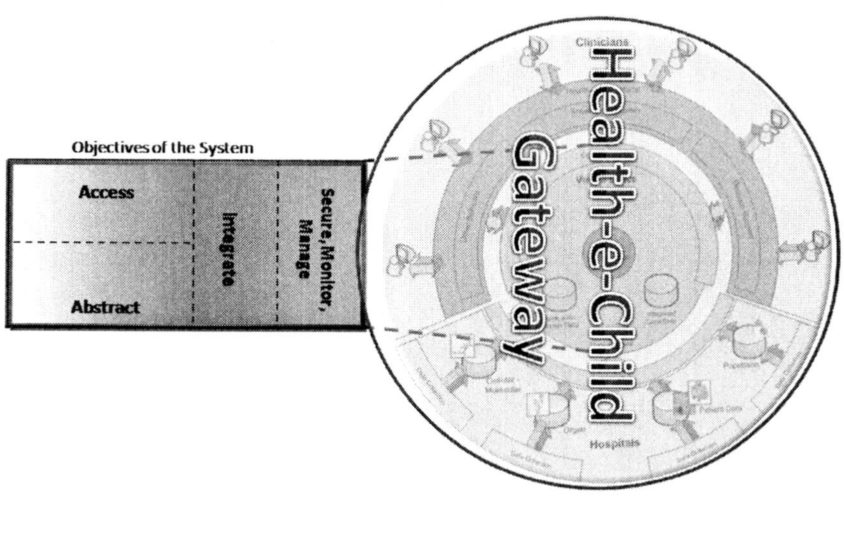

distributed sets of clinicians and researchers. Many of these investigations have also highlighted the need for personalised healthcare with treatment customised and targeted at the individual, requiring the capture and assimilation of complex biomedical data on a patient-by-patient basis. Furthermore the increasing mobility of patients from one healthcare centre to another, the need for clinicians to track their medical data and the increasing distributed nature of healthcare provision leads directly to requirements for collaboration and cooperation across healthcare networks. According to a recent journal report (Frist, 2005), "During the next decade, the practice of medicine will change dramatically, through personalized, targeted treatments that will enable a move beyond prevention to pre-emptive strategies."

Grid technology has been developed for applications addressing these kinds of requirements, i.e. with large storage and computation needs, though it was originally developed for High Energy Physics (HEP). Grids offer a promising tool to deal with the current needs in diverse medical domains involving complex anatomical and physiological modelling of organs from images or large image databases assembling and analysis. However, deploying Grid technologies for healthcare still presents technological challenges. Currently, the bottleneck is in organising data and assembling datasets and providing uniform access to these sources of information, hence the recent emphasis in the community on ontologies and distributed databases technologies. As it is, most of the increasing flood of information – the so-called "data deluge" - has not been adequately tackled, as there are often no systematic data accumulation practices nor metadata annotation which would describe how, where and when the data were collected (Korhonen, 2006). The effort of porting legacy applications remains equally demanding and rather complex and cumbersome. In spite of recent developments in supporting tools such as g-Eclipse ("g-Eclipse," 2008), or even Taverna (Oinn *et al.*, 2004), which can make life easier when exposing existing business logic as proper Web services or Grid jobs, the complexity related to optimizing such applications scheduling in a distributed environment remains unaddressed. Most of the time, biomedical projects rely on pre-existing applications and algorithms that deal with data integration, feature extraction and image processing, which have been developed for years by experts in the area, but use specific toolkits and libraries that the Grid is not forcefully aware of. Therefore, simply porting such applications to the Grid is not sufficient; such freshly ported applications would gain a lot from being reengineered to take advantage of the Grid.

In the HeC project various kinds of applications have been discovered from the user requirements analyses that potentially require different levels of integration. Some applications have short runtimes but are iteratively triggered on a large dataset, whereas others are executed less frequently but can take several hours to complete. Such applications are domain specific and originate from Grid-agnostic developer communities, which might not be ready (skills- and cost-wise) to optimise their solution. Thus, the Grid still remains rather immature, constantly evolving, and multi-disciplinary in nature. As a consequence, such applications' design and development is increasingly complex and the use of most classical engineering practices can be unsuccessful. Not only is the development of such applications a time-consuming, error prone and expensive task, but also the resulting applications are often hard-coded for specific Grid configurations, platforms and infrastructures. Porting applications to and/or developing in the Grid appear to be a barrier that few Grid-agnostic developers dare cross. The complexity goes even higher when interactivity in such applications is required. Current challenges being faced in the ubiquitous adoption of Grids for healthcare include (Breton, 2007; Korhonen, 2006):

- International Grid infrastructures are available for scientific research but these have not been adequately deployed into hospitals. Due to its confidential nature, healthcare IT needs to be marshalled by stronger security and privacy constraints than in all other fields of Science.

- Grid toolkits offering Grid services in a secure, interoperable and flexible manner have not been tested on a large scale with biomedical applications. Healthcare professionals use IT technologies differently, e.g. most of the time they access the infrastructure for short periods of time, but need immediate and reliable answers from the system.
- Some CPU intensive biomedical applications have been deployed worldwide but very few that involve the manipulation of distributed biomedical data, have been successfully demonstrated thus far.
- Recently created eScience environments such as MyGrid (Stevens *et al.*, 2003) where bioscientists can manipulate their own concepts, are not yet available in current Grid infrastructures.
- As of today, no middleware satisfies all the requirements of Life Sciences. Those that have demonstrated their scalability (e.g. gLite ("gLite," 2008), Unicore ("UNICORE," 2008)) need additional functionality in the area of data management, security, privacy, resource and workflow management. Some which offer powerful and demonstrated data management functionalities (e.g. SRB ("SRB," 2008)) have limited job management services as opposed to recent Grid middleware based on web services, which on the other hand have not demonstrated their robustness nor scalability.
- The definition and adoption of international standards and interoperability mechanisms is required for storing biomedical information in the Grid.
    - Standards for the exchange of medical images in the Grid (like DICOM and others) are still at the prototype level.
    - Standards for the exchange of Electronic Healthcare Records in the Grid are missing.
    - Standards for data anonymization are also missing though there are plenty of concurrent approaches demonstrated already.
- Beyond standards, agreed ontologies are also needed to align data and business logic.

The application of Grid technology in the healthcare domain has however spawned a new and exciting research area, that of so-called healthgrids. In the HealthGrid White Paper (Breton *et al.*, 2005), a healthgrid has been defined as 'an environment in which data of medical interest can be stored and made easily available to different actors in healthcare systems such as physicians, healthcare centres, patients and citizens'. Healthgrids focus equally on the sharing of data (and the associated issues of privacy and ethics) and on distributed health analysis across the biomedical spectrum from public health to patient care and from tissue/organ data to cellular and genomic information. For individualised healthcare, healthgrids are envisaged to facilitate access to biomedical information and ultimately knowledge, no matter where the requestor of that information may reside or where the relevant data is stored: biomedical information on demand. Much research activity continues in the field of healthgrids.

2.2. The Convergence of Web and Grid Services in SOA

With the recent convergence of Web and Grid services standards, as well as the wide recognition of the Service-Oriented Architecture (SOA) paradigm, the Grid vision has started materializing into concrete technologies finally bringing it closer to the needs of healthgrids. The benefits of this vision have already been demonstrated in several prototype systems across the globe. Indeed, in the last few years, there has been significant effort, resources and funding invested into national and international initiatives to investigate the development of healthgrid infrastructures, services and applications. A number of pioneering systems have been architected and tested which highlight the Grid add-on value within the area. For instance, as stated in (McClatchey *et al.*, 2006), it allows the federation of databases of mammograms from across Europe along with software tools to improve breast cancer screening and diagnosis, whereas in (Jacq *et al.*, 2006) it helps in discovering new drugs against emergent diseases like malaria. Other early initiatives like neuGRID ("NeuGrid Project", 2008) look quite promising in terms of the gridification of pipelines of algorithms respectively for diagnosing vascular stroke, multiple sclerosis and Alzheimer's disease. Finally, it is worth mentioning the cancer Biomedical Informatics Grid (caBIG) (Kakazu *et al.*, 2004), a network of networks for cancer researchers in the US. This style of architectures promotes and facilitates software reuse at the macro(service) level rather than at the micro(objects) level, making it quite successful despite its relative maturity.

Indeed, while the term SOA is being heavily used in the community to promote concurrent implementations of healthgrid platforms, very few of them fully respect its concepts. The SOA paradigm is not new; historically, the concept appeared with the client/server architecture; SOA is an evolution of distributed computing and modular programming. It provides a modularity of logic that can be presented as a service for a client (that is a client as in client-server architecture) and at the same time functions as a client for other services. Relative to earlier attempts to promote software reuse via the modularity of functions or classes, SOA's atomic level objects are hundreds of times larger, and are associated by an application designer or engineer using orchestration. While SOAs are made of services, the fact of exposing business logic as services does not transform a system in to a proper SOA. Indeed, there are rules that one has to enforce in order to fully benefit from the concept, as described below.

2.3. SOA Paradigm and Associated Rules

The main characteristics of an SOA are the loose coupling between services, the abstraction from technological aspects and its extensibility. The loose coupling property implies that services do not invoke each other directly but rather interactions are managed by an orchestration entity. SOA provides an easy way to reuse software artefacts through the concept of services that are not bound together. Technological abstraction is obtained from using service contracts that are platform-independent. Extensibility is finally reached through service discovery and composition at execution time. Several definitions of the concept can be found in the literature, however for the remainder of this document, only the three following were retained, as they are most relevant for this work:

- "A service-oriented architecture is a style of multi-tier computing that helps organizations share logic and data among multiple applications and usage modes" as stated in 1996 by the Gartner group.
- "Service Oriented Architecture is a paradigm for organizing and utilizing distributed capabilities that may be under the control of different ownership domains. It provides a uniform means to offer, discover, interact with and use capabilities to produce desired effects consistent with measurable preconditions and expectations" established in the OASIS reference model.
- "SOA enables flexible integration of applications and resources by: (1) representing every application or resource as a service with standardized interface, (2) enabling the service to exchange structured information (messages, documents, "business objects"),

and (3) coordinating and mediating between the services to ensure they can be invoked, used and changed effectively".

In spite of the absence of an officially and consensually single agreed definition for SOA, three elements are usually identified: service producers, service brokers and service consumers. The service producer's role is to deploy a service on a server and to generate the description of this service (the service contract), which defines available operations as well as invocation mode(s). This description is published in a directory of services inside a service broker. Thus, services consumers are able to discover available services and to obtain their description by interacting with the service directory. The obtained descriptions can then be used to establish a connection with the producer and to invoke the desired service operations.

Just as for the SOA concept, loose coupling does not benefit from a unique definition. The commonly adopted approach though is to introduce a minimum of dependencies between services in order to support better the reusability of existing ones (i.e. services which are already deployed). Moreover, these services should be combined in order to quickly and cost-efficiently respond to new demands. To achieve this goal, some engineering rules which are not always specific to SOA, have been identified (Papazoglou, 2003). Note that this chapter will not focus on these, but rather include appropriate references for the interest of readers.

Encapsulation and abstraction principles originally came from the world of object-orientation. The idea was to hide self-contained information of a service to end-users and to propose only one stable interface stressing the details considered to be necessary for handling it. A service is therefore seen as a black box from the outside, which makes it possible to separate its interface (its external description) from its actual implementation. One can thus modify a service implementation without changing its interface, which turns it into a sustainable model. The following rules are more specific to SOAs:

- A simple and ubiquitous interface must be provided by any service and must be universally accessible by all suppliers and all customers of services. Thanks to a generic interface, it is then possible to interconnect any services and to forward any messages between the various interfaces. The master word here is "decoupling" and it can take various roles: (1) to reduce the coupling between modules, for improved reuse, (2) to reduce the coupling with respect to the infrastructure and to the implementation platform, for improved interoperability and (3) to reduce the coupling between a service customer and a specific implementation of this service, for improved evolution. In Web service architecture, the consensus to achieve this rule goes throw the use of WSDL (i.e. Web Service Description Language).
- Messages delivered by a service should not contain business logic. On the contrary, they must be restricted to the transport of, and only of, data structures from one service to another. That makes it possible to modify or to add services without impacting the other services of the architecture. These data structures can nevertheless be very complex in order to deal with security management (i.e. authentication, encryption, authorization, etc.) or even file transfer. These aspects are addressed thanks to different specifications that strengthen the "standard" Web service architecture (e.g. WS-Security, SOAP-attachments, etc.).
- A well-formed service must be stateless. This rule, which can seem very constraining, must be moderated. It is recommended that the state conservation (i.e. the management of the context) as well as the action coordination (i.e. the management of the

transactions) are localised in a specific function of the SOA, such as the orchestration. The application of such a rule facilitates the reuse, the scalability and the robustness of services. Moreover, this rule allows to keep into account the main SOA focus: the lose coupling.

- Cohesion is a delicate rule to define. It translates the degree of operations and functional proximity inside a service. In other words, it aims at facilitating the comprehension and reusability of a service by gathering homogeneous operations corresponding to similar business logic.
- A service should be idempotent. That makes it possible to be unaware of multiple receptions of the same request. The idea is that the use of such a service makes it possible to slacken the assumptions of reliability on the communication layer. In Web service architecture, the WS-Addressing specification allows among others to take into account part of this rule.

If some of these rules can be moderated according to the system requirements, i.e. the stateless and the idempotent ones, all these recommendations remain vital to create an open, sustainable and standard healthgrid system. Indeed, these characteristics are mandatory in order to cope with heterogeneous resources ranging from data, to knowledge, to applications, and beyond software, to people. The SOA approach makes it possible for a wide range of people having different skills to collaborate in the development of a system covering different application areas, which is of absolute importance in the case of healthgrids.

Aiming at addressing these challenges, the HeC partners have therefore started complementing the Grid middleware services offered with domain specific logic following this SOA approach and respecting its cornerstones. They have engaged in the development of an upperware stack, the so-called Health-e-Child Grid Gateway, which is a thin layer of software services sitting on top of the Grid middleware that wraps up and abstracts from underlying technologies to deliver adapted functionality to end-users, while respecting the SOA model. The following section discusses the design and implementation of this Gateway and to what extent it conforms to the previously introduced rules for delivering a reusable healthgrid platform.

2.4. Related Work

Various projects around the globe are utilizing grid infrastructures to support biomedical applications. In the US, most notably the caBIG™ initiative (http://cabig.cancer.gov/) founded by the NCI (National Cancer Institute) in 2004 to speed discoveries in cancer research by linking research institutes and health care providers by enabling "the collection, analysis, and sharing of data and knowledge along the entire research pathway from laboratory bench to patient bedside". Similar to Health-e-Child, caBIG™ has a very strong focus data modeling and semantics to be able to create a shareable and interoperable infrastructure.

In Europe, Information and Communication Technologies (ICT) for Health is one key component of the Sixth and Seventh Framework Programmes by the European Commission. Several projects besides Health-e-Child combine biomedical applications and grid infrastructures, for example:

- @neurIST (http://www.aneurist.org) focuses on integrative decision support systems based on multi-scale computational suites for personalised brain aneurysm rupture risk assessment and treatment,
- ACGT (http://www.eu-acgt.org/) develops semantic and grid-based technologies in support of post genomic clinical trials in cancer research, targeting mainly breast cancer and paediatric nephroblastoma,

- ViroLab (http://www.virolab.org/) develops a virtual laboratory for infectious diseases that facilitates medical knowledge discovery and decision support by allowing, e.g. to relate genotypes to drug-susceptibility phenotypes, and
- neuGRID (http://www.neugrid.eu/pagine/home.php) aiming at the provision of sets of distributed medical and grid services for neuroscientists, with one focus being the gridification of algorithm pipelines for brain image analysis.

Further related research includes applying grid technologies to content based image retrieval in radiology (Oliveira *et al.*, 2007) and distributed medical image analysis (Bellotti *et al.*, 2007), as well as efforts aiming at semantic data integration allowing services to be composed to meaningful workflows (Budura *et al.*, 2007; Lee *et al.*, 2007).

3. HEALTH-E-CHILD GATEWAY

3.1. Gateway Design

The major goal that guided the Gateway design process throughout was to establish a common coarse-grained view of the system architecture, useful to identify inner constituents and corresponding interfaces, as well as to help in better splitting the work and responsibilities among the developers. Therefore, the methodology that the project has and still is following is based on standard requirements engineering practices and more specifically attempts to follow the Service Oriented Modelling and Architecture (SOMA) (Arsanjani, 2004) process. In short, SOMA is an IBM offering that defines the three familiar service modelling steps, namely: identification, specification, and realization. These steps consist of several sub-steps prescribing the various artefacts to be delivered and recommending appropriate techniques. SOMA is followed since it gives good coverage of the requirements engineering process, to provide a project-wide standard approach to requirements engineering and because of prior familiarity with that methodology.

For clarity, only a relevant subset of this methodology is presented here, to enable the reader to understand the gross system architecture, its layers as well as its key elements. Successful service modelling is not as simple as it might first appear; it is more than a simple drilling-down from business-level to IT implementation, and many SOA-specific architectural decisions have to be respected. Almost all cases require non-traditional modelling approaches as opposed to a simple top-down process. Indeed, while a top-down approach starting from a set of use cases, was adopted by the developers facing the end-users, and a bottom-up approach was undergone for understanding and analyzing existing IT assets such as the Grid middleware and databases technologies, the project partners in charge of the healthgrid platform developments aimed at satisfying both ends and therefore had to cope with requirements originating from the two (different) perspectives, as shown in Figure 2 (note: Figure 2 recalls the objectives formerly introduced in Figure 1).

The natural resulting approach is one of inside-out where the top-down and bottom-up approaches meet and a set of services and functionalities are identified, validated and grouped into logical categories – or layers, reconciling the needs between the features identified in the top-down analysis, and what is already provided from existing IT systems.

- The top-down approach concerned decomposing use cases and refining them for the elicitation of system functionality. During this approach, services and basic operations were identified, which helped in establishing the Gateway high level functionality.

Figure 2. Gateway design objectives and approach

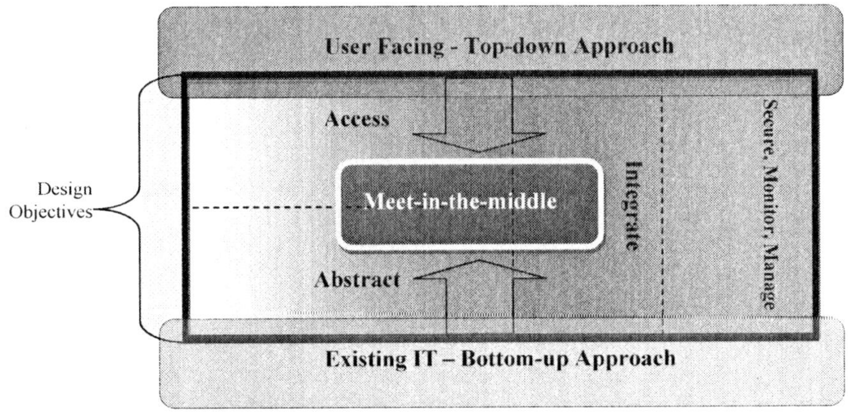

- The bottom-up approach consisted of analyzing existing IT assets and finding functionality that could be exposed, to be reused by many. Legacy applications (i.e. those that are already deployed and available) are the most valuable assets and therefore were most influential. In this process, existing IT capabilities were carefully selected, gradually exposed and mapped to higher level logic, to remain aligned with the project objectives.
- The meet-in-the-middle approach was about reconciling the needs between users and IT. It therefore required collaboration between domain specialists, software architects, and specialists of the legacy applications.

Based on this, a set of layers, services and, where possible, their operations were specified and grouped according to their nature, resulting in a gross system architecture description, sharable among the partners, as illustrated in Figure 3. This process was structured in three main tasks, which were not strictly separated, but rather overlapping; their goals can be summarized as follows:

- The extraction and understanding of the project objectives.
- The elaboration of a set of design objectives from the user requirements, and the evaluation of the corresponding design constraints.
- The specification of a coarse-grained system architecture.

3.2. Gateway Implementation

As stated earlier and graphically illustrated in Figures 1 and 2, the Gateway is a subset of the software solution being developed in the project, which aims at securing, enabling access to, abstracting from and integrating all forms of applications and data useful to end-users. The Gateway is designed as a SOA that respects the project guidelines and design objectives. More precisely it is a layered architecture of standard secure (medical) services running on top of a gLite-based ("glite," 2008) Grid infrastructure, which is physically installed at the clinical centres. This thin layer of software plays the role of an adapter/mediator between the end-users and the lower level IT. As such, it maximizes Grid functionality usage according to HeC's requirements, see Figure 3 above for an outline. This high level view of the system categorises the platform functionality and highlights the philosophy being chased, which consists of isolating the business logic from underlying substrates and therefore corresponding technologies. In this representation, the functionality is grouped in six major areas, respectively:

Figure 3. Gateway layered architecture

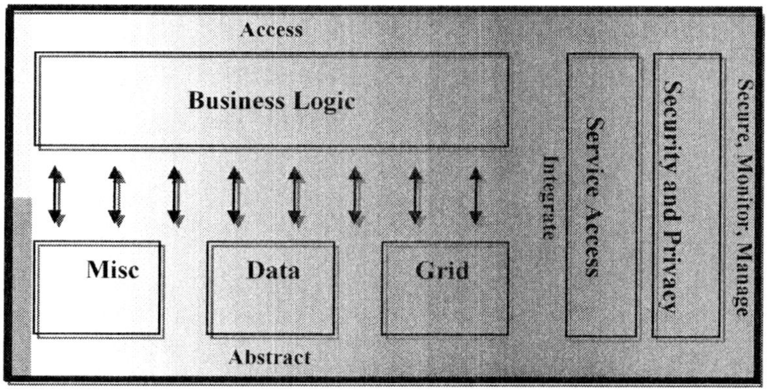

- **Security and Privacy Layer (SPL):** As its name implies this concerns all security and privacy features of the Gateway. This area covers aspects ranging from authentication, to authorization and to monitoring of users/services in the system as well as de-identification and sharing of sensitive data. All services delivering such functionality are grouped in this layer, which addresses cross-system needs, from the highest level of the system (close to end-users), to its very bottom (computing resources).
- **Service Access Layer (SAL):** Represents the services implementing the SOA concepts for allowing the publication, discovery and composition of functionality. This layer makes use of the Gateway information system to address requirements of the users.
- **Grid Access Layer (GAL):** The abstraction from the grid middleware (e.g., gLite). This area is concerned with all facilities related to the grid middleware exposure. One good example is the Grid interface which wraps the gLite functionality to make it programmatically available to services/users of the system. This layer is important since it guarantees the decoupling between the project business logic and technological constraints underneath.
- **Data Access Layer (DAL):** The abstraction from the databases manipulated by the Gateway. This area covers external (i.e. potential technological bridges) as well as internal data access (i.e. Gateway Information System Database, ISD, and Integrated Case Database, ICD). As with the Grid abstraction layer, this one also contributes to business logic decoupling.
- **Miscellaneous (or Interoperability Layer - IL):** Any other abstraction and interconnection with external systems, which can neither be attached to DAL nor to GAL layers. This area is concerned with technological bridges to other Grid platforms. One example of such a bridge could be a connection and query service for interacting with other infrastructures such as CaBIG.
- **Business Logic (or Service Resources Layer - SRL):** The whole domain specific functionality of the project. This area covers data integration as well as medical query processing, data mining and knowledge discovery.

Figure 4 illustrates the main services populating the architecture. While the lower part of the diagram in light grey, i.e. Miscellaneous, Data and Grid layers, is made of platform dependent services, the Business Logic layer in grey at the top remains platform independent. Indeed, the services provided in the lowest layer considerably simplify access to the underlying technologies and

Figure 4. Gateway system architecture

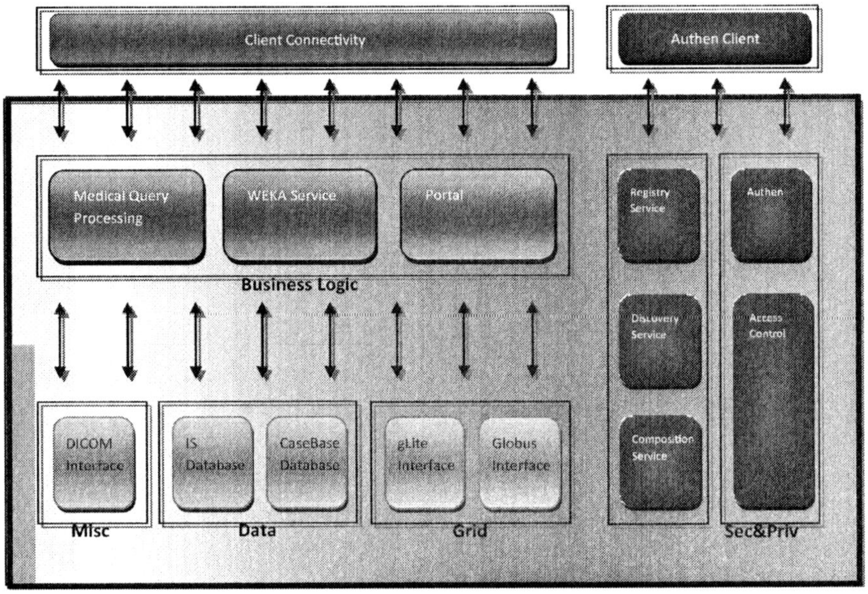

expose these facilities in a simpler and harmonized way to the rest of the system. The project business logic therefore remains isolated from the evolving Grid and can develop independently. On the other hand, the Access, Security and Privacy layers in dark grey, on the right, act as a "glue" to combine and secure all the resources together. Thus the entire system functionality is published and seen as a set of services that can be consulted from the Gateway ISD. Thanks to these layers, one can discover all the system capabilities ranging from Web and Grid services to simpler APIs and binary executables, these being documented with rich WSDL descriptions and stored in the ISD. Together these abstraction and glue layers are the pillars of the system, and can support the business logic of other projects. Among the Gateway services and as illustrated in the above figure, the following key entities can be found per layer:

Security and Privacy Layer (SPL)

- The Authentication service, which enforces the platform security model and allows users to login through a strong authentication (Two Factor Authentication) and single sign-on process (SSO). Users are provided with a USB key containing their credentials and authentication client, which can be executed under most popular operating systems. The authentication process is made secure through encrypted communications and results in the generation of a VOMS-enabled Grid proxy Gateway-side that services can use to further manipulate Grid resources on behalf of corresponding users.
- The Access Control service is restricted to the Gateway administrators. It allows system administrators to set access restrictions on all services and data resources hosted in the infrastructure. These restrictions can be set for a specific user or assigned to groups, using the VOMS information.

Service Access Layer (SAL)

- The Service Registry has a dual role. It allows uploading and registering new applications in the Gateway. New applications such as APIs, remote services or even fresh new ones can be added to the system portfolio. Once invoked, the registration service deploys the provided application as a new Web service in the Gateway. In parallel, the description of the service is added to the global WSDL repository (ISD).

- The Service Discovery is in fact an XPath query service facilitating the retrieval of any functionality exposed in the platform using the previously introduced ISD database. A query can be performed according to several criteria. Thus, one can look for particular operations by navigating through layers, services, functions, functions' inputs or even output. Requests can be specified as a string description (i.e. "a la Google"). Behind this query facility, complex registry mechanisms have been implemented to enforce access controls (as shown in Figure 5). After having retrieved the query result set under the form of an array of WSDL descriptions, an XSLT transformation is performed by the system, according to the user access rights to filter the resulting list of identified functionality. The Gateway uses an ACL-enabled database exposed under AMGA to achieve these access controls.
- The Composition Service provides facilities to enrich the Gateway functionality. It allows defining new BPEL processes - the current web services orchestration language standard - that execute workflows of Gateway services and/or any others deployed in the Internet. Moreover, this composition service supports short running as well as long running processes. In the latter case, the end user is periodically notified of the current process status while everything is being performed in an asynchronous way. At the backend of this service, the open-source ActiveBPEL engine is used. For the sake of the HeC Gateway, ActiveBPEL has been extended to cope with the latest services containers and to run processes either as new services or as local processes (i.e. on-the-fly execution). Similarly to all other resources of the system, newly created BPEL processes are ruled by access controls.

Figure 5. Gateway – service access layer

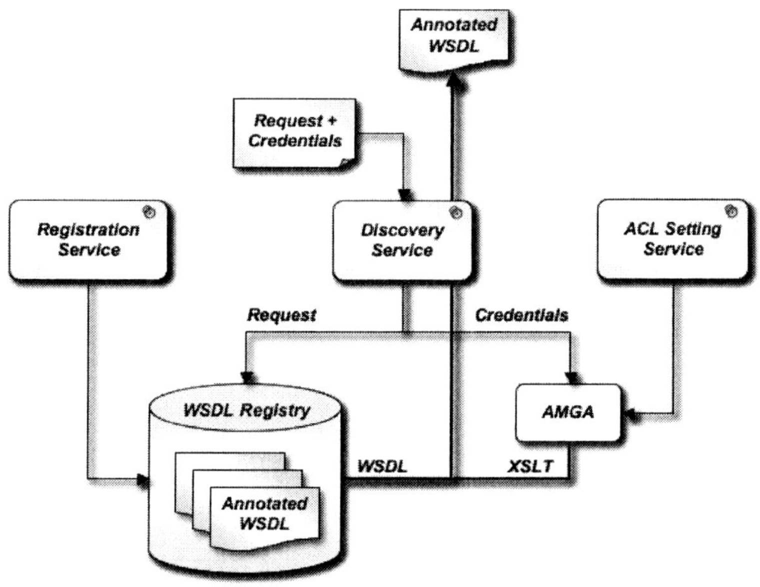

The system architecture and associated components are formalised using the UML deployment view, which illustrates the communications cross-centres. As formerly introduced, the SPL and SAL layers take care of maintaining security and providing appropriate resources from the SRL pool. While abstraction services from the GAL and DAL layers are duplicated since necessary on each site, those from the SRL can differ from one Gateway to another. Moreover, these services can be composed into cross-centre workflows. The proof of concept of the Gateway has been exercised and, at the time of writing, the Gateway is being tested for performance and for scalability with the intention of full deployment across all the partner hospitals in HeC by the end of 2009. The post 2009 period is envisaged as a period of clinical tests and evaluation on live data.

3.3. Gateway Conformance to SOA

As stated earlier, the Gateway architecture is developed following (and, when possible, enforcing) the SOA concepts. Thus, all services are exposed via simple and ubiquitous interfaces, which are published in the ISD for further discovery and composition. Messages between running services never contain any logic. In most cases, services interact using a well-defined and shared set of data structures, which reflect the Gateway inner data models, i.e. ISD and ICD. Unlike the Authentication service which requires such functionality all other resources in the system are stateless services, which expose the required functionality in an harmonized way. Indeed, since the very beginning of this development work, and based on the presented system architecture, naming conventions have been agreed among the developers making it easier to understand services roles, content and their placement in the project software packages. Last but not least, all operations triggered in the Gateway are treated uniquely and independently, guaranteeing a coherent overall behaviour.

As a direct result, the Gateway offers acceptable scalability; it promotes and facilitates software reuse which is of absolute importance in HeC to reduce development costs and to make the solution sustainable. The Gateway ultimately provides a robust and secure framework for integrating applications and data, by implementing a strong security model allowing fine-grained access controls on all resources, which is a vital requirement in order to maintain the integrity of the virtual organisation and to deal with sensitive clinical data. The downside however appears in user-perceived performances for some of the basic functionalities. Indeed, abstraction through services introduces an overhead in communications with an increase of nearly 700%. Workarounds have already shown that there are ways to circumvent this. For instance, improvements can be applied in services composition, with optimisation techniques to solve particular requirements, however for clarity, this paper will not expand on the detail of these aspects.

As a concluding statement, the Gateway, while still being a prototype system, already shows interesting and stable results based on which the HeC partners can develop advanced applications. The concepts and technologies used to achieve this makes it easier to integrate new applications, though one still has to carefully assess one's needs prior to applying a given integration model. This is what the next section discusses.

3.4. Medical Applications Integration: Gridification versus Servitization

3.4.1. Methodology

In HeC, a requirements analysis for applications integration was initiated at the outset. A technical questionnaire was circulated to the project's medical software developers, and the results were used to categorize applications based on their placement in the overall framework. In addition, a simple methodology was introduced to measure the level of integration of applications - both in terms of interactions with the various layers of the platform, and in terms of the physical location of its inner components. In this way, the work was prioritized, and developers were able to precisely understand the extent to which their applications could benefit from functionality offered by the system.

Indeed, given the type of application and final functionality required to be achieved on the Grid, gridification as such can either turn out to be a very simple or a rather complex problem. Thus, a number of factors have to be considered before actually choosing a gridification model. As an example, two extreme cases are considered (although they may not fully represent the broad spectrum of all possible cases in biomedicine). Consider the case of a genetics profiling (i.e. number-crunching) application that has been pre-compiled in advance for a known and available target platform and requiring as an input a data set that is not too large and is locally available at submission time to the Grid. For this the gridification process is simply a matter of specifying a job description along with its required resources and then submitting it to the Grid. However, in an alternative case, such as the reconstruction of a temporal Magnetic Resonance Images (MRI) series from various DICOM images, that:

- requires interaction with a user for steering its execution process,
- uses other (Grid-)services to obtain required functionalities,
- reads/writes files stored and replicated on remote Grid storage resources and
- alters metadata information stored on gridified databases,

the gridification process could require substantial re-engineering of the application. This could cause considerable development and integration work as well as deployment and operational effort. In such a case, the major question that could therefore arise is: what approach – whether gridification or servitization (or a certain mixture of both) – is the more relevant? A good understanding of the application's architecture, its hardware and networking requirements, its foreseen working environment and its use-cases are therefore vital to know before integrating.

3.4.2. Servitization versus Gridification

Gridification is concerned with porting project business logic to software jobs or executables that can further be scheduled in the Grid, whereas servitization is about wrapping the logic into proper web services, when such gridification is irrelevant or too invasive. These concepts are complimentary especially in projects like HeC where the distribution requirements are extremely diverse. However, they are quite different in implementation and application areas.

Selecting one integration model or the other should be straightforward for most applications. For example, if hundreds of services are being invoked at the same time by dozen of users with different priorities or the same service is being called by hundreds of users with different parameters; these invocations can be transformed into proper Grid jobs. If the query execution time is too high, the query might also be divided into subtasks which can then be scheduled according to some criteria to boost performance. This allocation could be latency aware (MPI) or High throughput computing (supported by Condor, SGE etc). On the other hand, if execution time is very low and the invocation frequency of a service/component/ database object is also low then simply sending the queries to that service or component might suffice. In the first case, the problem is too resource demanding to be solved in a single location, data could be distributed, users are distributed and they need intensive resource and knowledge sharing. In the second case, the problems are not so challenging (in terms of resources) and could be solved by a single site. Therefore the division is basically between resource sharing and no-sharing at all. Grids are essentially about resource sharing but services could equally be used for Grid and non-Grid applications.

From these facts, one can understand that integration has to be carefully evaluated by contrasting the needs of the concerned application with the capabilities of technologies underneath. Indeed, while the Grid offers a complete secure and distributed environment for scheduling jobs based on resource requirements, it mainly was designed and remains appropriate for workflows of non-interactive atomic and long runtime applications. In contrast, many applications in life sciences differ in nature, in the sense that they consist of (interactive) workflows of composite short runtime processes. For this reason, the notion of servitization has recently emerged in the community as a possible alternative, which at the same time opens the pathway to several new possibilities with hybrid gridification models. The next sections discuss some of these models through concrete examples.

3.4.3. Gridification of DICOM Files Registration and Thumbnailing

Among the functionality provided by the Gateway to client applications is the ability to compute DICOM files. Various operations are therefore available ranging from anonymization, to storing and to registration of images and corresponding metadata in the Grid storage elements and file catalogs. As a concrete example of the benefits of gridification, the HeC partners decided to simplify successive operations on their images – i.e. the browsing on the client side – by thumbnailing them. The resulting images can then be stored/registered in the Grid along with the initial ones for fast retrieval and preview. However, thumbnailing large sets of substantial files can put a high load on the Gateway and considerably lower its responsiveness to other requests. Indeed, while extracting a thumbnail from a single image is not a challenging issue, doing the same on large MRI scans is another story. Thus, when dealing with sets of images larger than a certain threshold (defined according to the access point hardware specifications), the Gateway offloads this processing to the Grid. To do so, ImageMagic was chosen as an application to do the actual manipulation with DICOM images. The application was installed on a number of Computing Elements available in the HeC Grid infrastructure (closed Grid infrastructure consisting of computing and storage resources deployed at hospitals participating in the HeC project) and was published to the Grid information system. This made the application available and discoverable in the Grid.

On the Gateway side the DICOM files registration process was augmented with the ability to send Grid jobs to the Grid infrastructure to perform the thumbnailing part, when the size of the set of images was larger than a certain predetermined value. For that, a script was written to:

- Retrieve DICOMs to be thumbnailed from the Grid (this assumes that the DICOM images were registered before the thumbnailing process started (this ensured by the registration workflow)),
- Call ImageMagic to extract thumbnails,
- Store thumbnails in the Grid Storage Elements and register them in File and Metadata Catalogs.

To be able to submit the script for execution on the Grid the job and required resources description file (JDL) was written which contained a special requirement for ImageMagic-<*version*> to be deployed on a targeted Computing Element. To actually submit the job to the grid and then get notified about its completion, appropriate calls to the GAL layer of the Gateway (see Figure 2) were added to the DICOM registration code.

The gridification process was simple enough to be accomplished within a couple of days. As a result we are able to offload the excessive computational load from the Gateway to the Grid and this has proven to be useful. However, there are applications for which pure gridification is neither obvious nor relevant. As an illustrative example, the next section describes the integration of the CaseReasoner application, an advanced case-based reasoning logic for measuring patient similarities within the HeC population.

4. DECISION SUPPORT WITH GRID-BASED SIMILARITY SEARCH AND ADVANCED DATA VISUALIZATION TECHNIQUES: HEC CASEREASONER

4.1. CaseReasoner: The Concept

There is a growing interest in the use of clinical decision support systems (DSS) to reduce medical errors and to increase healthcare quality and efficiency. One important DSS subclass well suited for leveraging the promise of the Grid is that of Case-Based Reasoning (CBR) systems – systems which have reasoning by similarity as the central element of decision support. CBR is a recognized and well established method for building reasoning systems. It is, however, commonly acknowledged that CBR has not yet become as successful in medicine as in other application areas. A commonly reported reason for the relatively slow progress of the field is the lack of transparency and explanation in medical case-based reasoning. We believe that one way to approach this problem is to visualize better the patient's clinical history and the underlying inter-patient similarities, which is the core concept of any clinical CBR system.

To do so, a number of visualization techniques which hold promise in being adopted in the clinical workflow have been evaluated. Three of these were found suitable for visualizing inter-patient proximity; i.e. treemaps, relative neighbourhood graphs and combined correlation plots/heatmaps (Tsymbal *et al.*, 2007a). Besides this, a novel graph-based patient data visualization technique that effectively depicts clinical history and related treatment workflow for each patient was developed. The visualization techniques were implemented in a prototype DSS the so-called CaseReasoner, which bases its similarity search on the Grid, since the process of extracting data and measuring similarity distances is quite demanding in most cases. This section therefore presents a brief comparative analysis of the implemented techniques, as well as their added value when jointly used with the Gateway and the Grid. The considered innovative visualization techniques help to not only visualize existing clinical data from the patient history and inter-patient similarity, but also to integrate the data with the knowledge base in the form of ontologies stored in the Gateway, which may lead to an improved understanding of the underlying problem domain, and to provide better opportunities for hypothesis verification and knowledge discovery.

The workflow within the CaseReasoner follows the information retrieval paradigm presented in Figure 6. According to it, there are two basic approaches to information retrieval: browsing and querying/similarity search. Querying/similarity search is the task of finding a set of cases similar to a provided reference case or corresponding to a particular query. Browsing refers to viewing, looking around, glancing over, and scanning information in an information environment (Zhang, 2008). Each of the two approaches has its own strengths and weaknesses but they remain complementary. As a result of the information retrieval process, users are provided with information at two different levels; the micro-level and the macro-level. Information at the micro-level refers to individual cases, while information at the macro-level refers to the aggregate information of cases in a data collection. Information at the micro-level is direct and obvious while information at the macro-level is indirect and sophisticated. The

Figure 6. The information retrieval process; adapted from (Zhang, 2008)

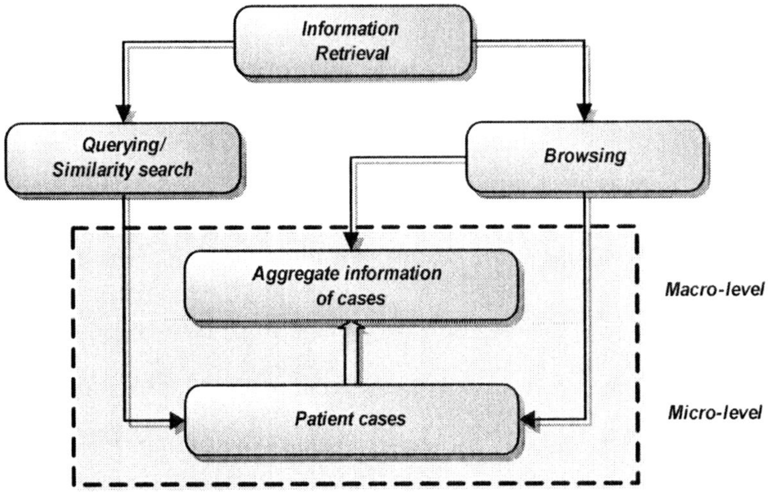

aggregate information is derived, or generated from individual cases in a data collection; it is an important asset of the data set and it is also vital and valuable for users.

4.2. CaseReasoner Implementation

The basic philosophy behind the design of the CaseReasoner is to provide clinicians with a flexible and interactive tool to enable operations such as data filtering and similarity search over a Grid of clinical centres (following the formerly introduced information retrieval paradigm), and also to facilitate the exploration of the resulting data sets. The aim is to let clinicians explore and compare the patients' records regardless of his/their geographical location, and to visualize their place in the distribution of both the whole population of patients, as well as in the distribution of its semantic subsets.

The selected visualization techniques are implemented to display and navigate through the results of similarity searches in the CaseReasoner. The distance function for similarity search is defined based on a similarity context, which is a subset of features of interest defined by the clinician. The features for each problem domain are organised into a so-called feature ontology, which represents the relationships between features (Tsymbal *et al.*, 2007b). Similar cases are found both in the whole ICD Grid database, and in some subsets of interest (e.g. high-grade tumours, males, a certain node in the Grid, etc), defined in the form of a simple filter. For each patient in the ICD, it is possible to visualize and compare related images from the patient history, thanks to the Gateway's abstracted accesses to backends, storage elements and file catalogs. In combination with the basic feature statistics, class distribution histograms and the scatter plots for the ICD under study, this will be a universal tool for Grid-based decision support in the diseases covered by HeC. Indeed, having a number of clinical centres connected and sharing their data gives the CaseReasoner significant added value. Not only can the CaseReasoner benefit from larger samples but also part of its reasoning logic can be made reusable and delegated to the Grid, with the complexity that it implies.

In short and as illustrated in Figure 7 below, after selecting a search context (1), the clinician can view basic statistics of the retrieved cases (2) as well as visualize them utilizing treemaps (3a), combined correlation plots/heatmaps (3b) and neighbourhood graphs (3b).

The process of finding on clinicians demand inter-patient similarities according to a selected set of features in the ICD database can be characterised as follows; it is a non CPU intensive operation triggered on a potentially unlimited population of patients managed by different medical centres. It therefore requires accessing intensively data records from the ICD databases and/or 3D meshes of patient's organs (e.g. right ventricle), as similarity searches can be defined on clinical records, genetics data, medical images or a mixture of these. The process is interactive in the sense that clinicians define the similarity criteria to be investigated and has to be asynchronous allowing various searches to be launched at the same time and giving feedback to users as they progress.

Thus the integration of this process was carefully planned and executed by partly offloading business logic to a dedicated service (servitization), in order to furnish an appropriate connectivity allowing interactivity and addressing the rather limited computing resources needs when calculating similarity distances; whereas all the access to data and files was left to the underlying Grid (gridification). By applying such a hybrid integration model, developers obtained very quickly a multi-centre and multi-level similarity search process making optimized use of the various technologies available. Moreover the process of porting this logic to the Grid revealed not to be too invasive, turning the resulting application into a reusable piece of software that one could eventually extract for other purposes.

5. EVALUATION OF GRIDIFYING APPLICATIONS IN HEC

The CaseReasoner that has been implemented in HeC is one example of an advanced gridified biomedical application that uses the Gateway fully and in developing it a number of observations can be made. Firstly, the design of the Gateway has simplified the process of connecting the CaseReasoner to the Grid infrastructure (i.e. initiating a session and interacting with the Grid) and making it able to query the distributed databases. In fact customizing the CaseReasoner for the gLite implementation of the Grid as used in HeC took developers less than one working day. Indeed, as described in former sections, the Gateway comes with a client framework for facilitating its connection and use, and which abstracts from the complexity of the underlying Grid. Thus, non-functional aspects related to interactivity

Figure 7. The HeC CaseReasoner application

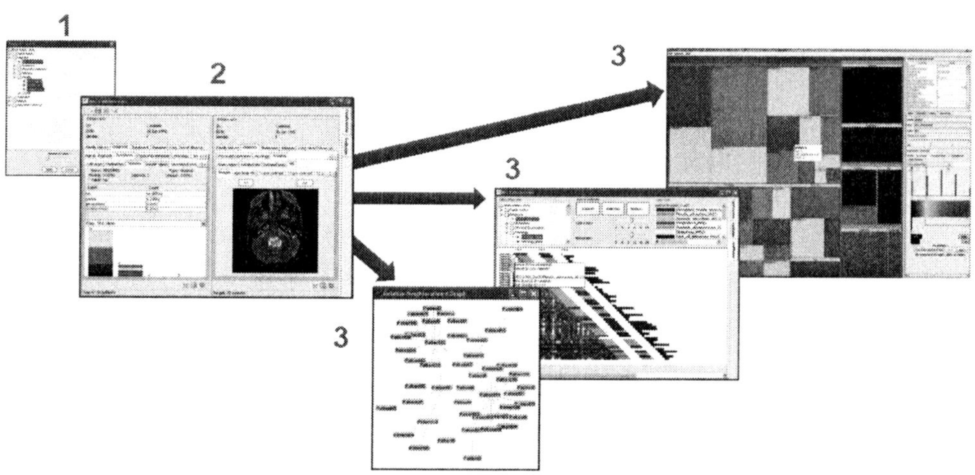

such as asynchronism and notification are natively handled through a software library which encompasses a well-defined application programming interface. With relatively little effort, the CaseReasoner with advanced connectivity was developed very quickly, giving it access to the rest of the Gateway functionality.

Secondly, the Gateway has acted as a suitable interface for the integration of the differing levels of data integration that were required in delivering the CaseReasoner. The next steps will be to explore the full distribution and optimization of the reasoning processes from simple interactions to advanced computations and that will require further integration studies. The lessons that have been and will be learned in providing that level of integration will be valuable in the latter stages of the HeC when the CaseReasoner will be used with ontology techniques to provide the integration required across very heterogeneous paediatric data to unleash innovative clinical applications in cardiology, rheumatology and neuro-oncology.

Thirdly, the investigations have shown that the Gateway can be used as the technological glue to interconnect several clinical partners, in particular for dealing with their heterogeneous information systems. Through the Gateway on-site access points users transparently utilize a number of computing resources ranging from local databases, to the distributed Grid infrastructure regardless of their location and available connectivity. Within the HeC project, the Grid forms the backbone for collaboration between clinical sites. It has been selected as a cost-effective solution which can appropriately support the end-user's needs in terms of computing power, storage capacity and security.

Finally, Grid technologies that have been implemented in the HeC studies thus far offer a virtual organisation which can use applications, store files or any kind of information, all of it using an unbounded cluster of geographically dispersed computational resources (e.g. hospitals). This provides the ability to solve more rapidly computation and intensive data sharing problems across a large scale distributed environment.

Moreover, we have found that using Web services technologies provides a simple way to deploy and to register new applications, new functionalities and new data. Each member of the virtual organisation can dynamically discover all these resources. This discovery also covers resource description as the mechanism to use these heterogeneous resources. All these functionalities are offered using Grid and SOA standards and in this sense can easily be integrated with any standards based applications and platforms.

6. FUTURE TRENDS

Currently HeC is well into the second half of its project lifetime. The main focus of the project remains in the integration of information across multiple heterogeneous data sources and in the use of that information in generating new knowledge that clinicians can reason about in order to resolve biomedical queries. As stated in the paper the project concentrates on three forms of diseases in children - cardiological diseases, rheumatological disorders and brain tumours – and these present very complex problems of integrating data that can be highly structured (test results, measurements and patient information), through semi-structured (annotation, graphical analyses) to very unstructured (clinician's notes, image data etc). The project is presently studying the use of ontologies and the matching of existing ontology fragments in order to facilitate data fusion, complex information processing and knowledge representation. Study is also being carried out on the generation and satisfaction of clinical queries based on an HeC ontology and on the optimisation of query resolution. We believe that the vertical integration of heterogeneous data in the life sciences will long remain an area of growing interest for researchers and developers, advancing from the latest developments in related fields including bioinformatics, medical imaging, data modeling, ontologies, complex query processing, and data mining. Vertical data integration is still unfortunately largely an unsolved problem for many practical domains, at least definitely not exploiting the potential of existing heterogeneous data sources to the full, although particular solutions and frameworks already appear. The extensive field of vertical data integration lies on the intersection and benefits from contributions from various areas, each of which is dramatically changing and quickly progressing today.

Disease modelling, data mining and knowledge discovery from data are the fruitful areas of HeC research, work on which is planned to be continued, in the context of the three diseases of focus. Our particular focus with this will be on data visualization. The importance of data visualization and visual data mining for knowledge discovery and decision support has long been underestimated; we believe this field has a great promise of being significantly advanced in the near future and of finding a variety of successful applications in different subject domains. Last but not least, it is expected that the project will have much to contribute to the study of healthgrids in the coming years and that it will continue to inform the process of Grid-based biomedical application provision.

7. CONCLUSION

The gridification of biomedical applications is fraught with difficulties not least of which are the confidential nature of the data, the nature of the user community, the heterogeneity of the information sources involved and the need to have an integrated, 'holistic' view of those data for the purposes of ultimately providing personalised healthcare. Biomedical data spans a wide range of data types from the genomic/cellular through the tissue and organ levels to the individual patient, the population of those patients and the epidemiological studies required by clinicians. The homogenisation of these diverse data types is one primary objective of the Health-e-Child project so that clinicians may have a holistic view across the Grid of biomedical data to enable them to provide individualised and customised healthcare for patients.

This paper has reported on a number of developments in the HeC project from which other Grid-based biomedical projects may benefit. In order to emphasise the problems that can be faced when gridifying biomedical applications, the requirements for, and the design philosophy that underpins, the development of the HeC Grid Gateway has been described with particular stress being placed on the challenges that must be addressed in healthgrid developments. It was shown that adopting a stack of software services sitting on top of the Grid, which aims at simplifying resources integration within the infrastructure, greatly eases the gridification of biomedical applications. This coupled with a service-oriented architecture have been the main reasons for the rapid development of applications in HeC.

One example for application development that was detailed was the concrete gridification of a CaseReasoner which has become the cornerstone of the HeC decision support system. Through the example of the CaseReasoner this paper has demonstrated the add-on value of the Gateway to standard Grid middleware and has established the working practice that will be followed in the latter stages of the HeC project. The Gateway software has been successfully demonstrated to the Grid community a number of times, and has recently and successively reached the top five live demonstration competition of the International Enabling the Grid for E-sciencEs Conference, EGEE 2007, in Budapest, Hungary ("EGEE," 2007); and been awarded as the best live demonstration at the 3rd International EGEE User Forum in Clermont-Ferrand, in France this January.

ACKNOWLEDGMENT

The authors would like to acknowledge the efforts of the complete HeC Consortium, without naming all 14 institutes or the 50 or so computer scientists, clinicians and engineers working on the project. Particular thanks go to those colleagues who have been working on the Grid aspects of the project and on those involved with the CaseReasoner as well as those involved with data integration. The HeC project is being co-funded by the European Commission under contract no. IST-2004-027749 and thanks are extended for this support. The project is being coordinated by Siemens AG and we acknowledge the support of Dr. Joerg Freund, the HeC Project Coordinator.

REFERENCES

Arsanjani, A. (2004). *Service-oriented modeling and architecture*, (IBM technical article). Retrieved October 20, 2008, from http://www.ibm.com/developerworks/library/ws-soa-design1/.

Bellotti, R., Cerello, P., Tangaro, S., Bevilacqua, V., Castellano, M., Mastronardi, G., et al. (2007). Distributed medical images analysis on a Grid infrastructure. *Future Gener. Comput. Syst., 23* (3), 475-484.

Breton, V. (2007). A perspective on biomedical applications on Grids. In *eScience Conference 2007.*

Breton, V., et al. (2005). The Healthgrid white paper. *Stud Health Technol Inform, 112*, 249-321.

Budura, A., Cudré-Mauroux, P., & Aberer, K. (2007). From bioinformatic web portals to semantically integrated Data Grid networks. *Future Gener. Comput. Syst., 23* (3), 485-496.

EGEE, Enabling the Grid for E-sciencE 2007 Conference, Live Demo Competition (2007). Retrieved October 20, 2008, from http://indico.cern.ch/contributionDisplay.py?contribId=174&sessionId=23&confId=18714.

Frist, W. (2005). Health care in the 21st century. *New England Journal of Medicine, 352*, 267-272.

g-Eclipse project (2008). Retrieved October 20, 2008, from http://www.geclipse.eu/

glite, A lightweight middleware for grid computing (2008). Retrieved October 20, 2008, from http://glite.web.cern.ch/glite/.

Health-e-Child, The EU FP6 Information Societies Technology Project (2008). Retrieved October 20, 2008, from http://www.health-e-child.org/.

Jacq, N., Salzemann, J., Legre, Y., Reichstadt, M., Jacq, F., Zimmermann, M., et al. (2006). Demonstration of in silico docking at a large scale on Grid infrastructure. *Stud Health Technol Inform, 120*, 155-157.

Kakazu, K., Cheung, L., & Lynne, W. (2004). The Cancer Biomedical Informatics Grid (caBIG): pioneering an expansive network of information and tools for collaborative cancer research. *Hawaii Med J, 63*(9), 273-275.

Korhonen, J. (2006). *Finpro, Post workshop report, Fifth EGEE industry day concentrated on Challenges in Bioscience*, November 2006, Helsinki, Finland.

Lee, S., Wang, T. D., Hashmi, N., & Cummings, M. P. (2007). Bio-STEER: A semantic web workflow tool for Grid computing in the life sciences. *Future Gener. Comput. Syst., 23* (3), 497-509.

McClatchey, R., Manset, D., & Solomonides, T. (2006). Lessons learned from MammoGrid for integrated biomedical solutions. In *19th IEEE Int. Symposium on Computer-Based Medical Systems, CBMS 2006.*

Montagnat, J., Breton, V., & Magnin, I. (2003). Using Grid technologies to face medical image analysis challenges. In *Proceedings of IEEE/*

ACM CCGrid'03, (pp. 588-593). New York: IEEE Computer Society.

NeuGrid Project (2008). Retrieved October 20, 2008, from http://www.neugrid.eu/.

Oinn, T., Addis, M., Ferris, J., Marvin, D., Senger, M., Greenwood, M., et al. (2004). Taverna: A tool for the composition and enactment of bioinformatics workflows. *Bioinformatics*, *20*(17), 3045-3054.

Oliveira, M. C., Cirne, W., & de Azevedo Marques, P. M. (2007). Towards applying content-based image retrieval in the clinical routine. *Future Gener. Comput. Syst.*, *23* (3), 466-474.

Papazoglou, M.P. (2003). Service-oriented computing: concepts, characteristics and directions. In *Proceedings of the 4th International Conference on Web Information Systems Engineering, WISE 2003,* Roma, Italy.

SRB – The SDSC Storage Resource Broker. (2008). Retrieved October 20, 2008, from http://www.sdsc.edu/srb/index.php/Main_Page.

Stevens, R, Robinson, A., & Goble, C. (2003). myGrid: personalised bioinformatics on the information Grid. *Bioinformatics*, *19*(Suppl. 1), i302-i304.

Tsymbal, A., Huber, M., Zillner, S., Hauer, T., & Zhou, K. (2007a). Visualizing patient similarity in clinical decision support. In *Lernen - Wissensentdeckung - Adaptivität, LWA 2007, Joint Workshops of the German Society for Informatics (GI), Workshop on Knowledge & Experience Management, FGWM'07.*

Tsymbal, A., Zillner, S., & Huber, M. (2007b). Ontology - supported machine learning and decision support in biomedicine, In *Data Integration in the Life Sciences, 4th International Workshop, DILS 2007, Philadelphia, USA,* (LNCS 4544, pp. 156-171). Berlin-Heidelberg: Springer.

UNICORE (Uniform Interface to Computing Resources) (2008). Retrieved October 20, 2008, from http://www.unicore.eu/.

Zhang, J. (2008). *Visualization for information retrieval.* Berlin-Heidelberg: Springer.

KEY TERMS AND DEFINITIONS

Case-Based Reasoning Systems: Case-Based Reasoning (CBR) systems is an important subclass of decision support systems using reasoning based on *similarity* as the central element of decision support. The basic components of a CBR system include a Case Base (serving as the Knowledge Base and including previously observed cases with known decisions), a Similarity Assessment element used for case retrieval, and a Solution Adaptation block for the ultimate decision making based on the decisions of similar cases stored in the case base. For example, HeC CaseReasoner can be regarded as a typical CBR decision support system where decisions are made based on inter-patient similarity. The main benefit of CBR systems is the inherent transparency of the case retrieval and decision support process that raises the trust of the user in suggested decisions, which is especially important in clinical decision support.

Clinical Decision Support Systems: Clinical decision support systems (CDSSs) are interactive computer programs, which are intended to assist physicians or other healthcare professionals with decision making. CDSS link clinical or health observations with accumulated facts and/or knowledge influence decision choices by clinicians for improved personalised healthcare. CDSS usually contains a (medical) knowledge base and a reasoning mechanism (often these are a set of rules derived from the experts and the evidence-based medicine principle), used to derive conclusions from future observations.

Typical examples of CDSS include rule-based expert systems, like the CDSS pioneer, MYCIN, and Computer-Assisted Detection (CAD) systems assisting radiologists in analyzing and evaluating medical images comprehensively in a short time, for example detecting a likely tumour. The objective and the main challenge of the HeC project is to develop a CDSS making use of (distributed) data of different modalities (clinical, imaging and genetic) and of different vertical levels (molecule, tissue, organ, individual, and population), as a part of the European paediatric platform being implemented.

Computing Element (CE): service representing a computing resource. Its main functionality is job management (job submission, job control, etc.).

Condor: specialized workload management system for compute-intensive jobs [http://www.cs.wisc.edu/condor/].

Data Visualization and Visual Data Mining: Data visualization is a field of study that concentrates on the use of computer-supported tools to explore and represent large amount of data. Data visualization focuses on the creation of approaches for conveying abstract information in intuitive ways.Visual representations take advantage of the human eye's broad bandwidth pathway into the mind. Practical application of data visualization in decision support involves selecting, transforming and representing data in a form that facilitates data exploration and understanding, often called *knowledge discovery* or *visual data mining*. Important aspects of data visualization are the interactivity and dynamics of visual representation. For example, CaseReasoner implements treemaps, relative neighbourhood graphs (RNGs) and heatmaps as techniques for the visualization of inter-patient similarity in clinical decision support, assisting also in clinical knowledge discovery. The importance of data visualization and visual data mining for knowledge discovery and decision support has long been underestimated; this field has a great promise of being significantly advanced in the near future and of finding a variety of successful applications in different subject domains.

File Catalog (FC): offers a hierarchical view of and UNIX-like interface to files stored on grid. FC catalog provides Logical File Name (LFN) to Storage URL (SURL) mappings and authorization for file access.

Healthgrid: A healthgrid has been defined as 'an environment in which data of medical interest can be stored and made easily available to different actors in healthcare systems such as physicians, healthcare centres, patients and citizens'. Healthgrids focus equally on the sharing of data (and the associated issues of privacy and ethics) and on distributed health analysis across the biomedical spectrum from public health to patient care and from tissue/organ data to cellular and genomic information. For individualised healthcare, healthgrids are envisaged to facilitate access to biomedical information and ultimately knowledge, no matter where the requestor of that information may reside or where the relevant data is stored: biomedical information on demand. Much research activity continues in the field of healthgrids.

High Throughput Computing: in computer science a term to describe the use of many computing resources over long periods of time to accomplish a computational task.

Metadata Catalog: A catalog to store any sort of "data about data".

Storage Element (SE): Service which provides storage resources and manages requests for storage space and files. The storage space managed can be disk space, tape space or a combination of the two.

Sun Grid Engine (SGE): Policy-based workload management and dynamic provisioning of application workloads [http://www.sun.com/software/gridwar].

Virtual Organisation Membership Service, VOMS: holds information about VO members and their membership with respect to groups they are assigned to.

Chapter XXV
e-Infrastructures Fostering Multi-Centre Collaborative Research into the Intensive Care Management of Patients with Brain Injury

Richard O. Sinnott
University of Glasgow, UK

Ian Piper
Southern General Hospital, Glasgow, UK

ABSTRACT

Clinical research is becoming ever more collaborative with multi-centre trials now a common practice. With this in mind, never has it been more important to have secure access to data and, in so doing, tackle the challenges of inter-organisational data access and usage. This is especially the case for research conducted within the brain injury domain due to the complicated multi-trauma nature of the disease with its associated complex collation of time-series data of varying resolution and quality. It is now widely accepted that advances in treatment within this group of patients will only be delivered if the technical infrastructures underpinning the collection and validation of multi-centre research data for clinical trials is improved. In recognition of this need, IT-based multi-centre e-Infrastructures such as the Brain Monitoring with Information Technology group (BrainIT - www.brainit.org) and Cooperative Study on Brain Injury Depolarisations (COSBID - www.cosbid.de) have been formed. A serious impediment to the effective implementation of these networks is access to the know-how and experience needed to install, deploy and manage security-oriented middleware systems that provide secure access to distributed hospital based datasets and especially the linkage of these data sets across sites. The recently funded EU framework VII ICT project Advanced Arterial Hypotension Adverse Event prediction through a Novel Bayesian Neural Network (AVERT-IT) is focused upon tackling these challenges. This chapter describes the problems inherent to data collection within the brain injury medical domain, the current IT-based solutions designed to address these problems and how they perform in practice. The authors outline how the authors have collaborated towards developing Grid solutions to address the major technical issues. Towards this end we describe a prototype solution which ultimately formed the basis for the AVERT-IT project. They describe the design of the underlying Grid infrastructure for AVERT-IT and how it will be used to produce novel approaches to data collection, data validation and clinical trial design.

1. INTRODUCTION

Traumatic brain injury (TBI), also known as head injury, is a significant clinical problem. The incidence of severe TBI is approximately 200 patients/100,000 population with the most common causes including road traffic accidents, falls and assaults. Males are more than twice as likely to receive a severe injury than woman and currently the reported mortality rate following severe TBI ranges from less than 10% up to 50% with the most common rate quoted between 20-30%. Although the incidence of TBI is significantly less than those of other major medical diseases such as cardiovascular disease, cancer and stroke, as TBI occurs mostly in the young and the resultant morbidity is severe and long-lasting, the burden of TBI to the individual, their carers and the society that supports them is as great if not greater than the other disease domains. On a European level, fifty percent of the years individuals spend with disability are caused by brain disease of which traumatic brain injury now carries an equal burden to patients as do those of cerebrovascular and depressive illness disorders (Olesen 2003).

After ten years of pharmaceutical industry sponsored drug development and despite promising pre-clinical data, most of the clinical trials of these agents have failed to show any significant improvement in patient outcome (Narayan 2002). Many researchers feel a significant cause underlying this lack of success is the poor resolution of paper based methods for detection of adverse events and poor methods for monitoring of and controlling for protocol violations and medication errors. These limitations combine to make it difficult to detect small but clinically important treatment effects in the general noise of the brain injured patient management environment. The poor success rate of TBI clinical trials combined with the high cost to the pharmaceutical industry to conduct phase III trials in brain injury has, in recent years, caused a reluctance of the pharmaceutical industry to bring forward promising compounds to clinical trial in the field of brain injury. The high cost of conducting clinical trials is due in large part for the need to hire specially trained staff to collect and validate data. If technical solutions could be developed to reduce, even partially, the need for human resources in the data collection/validation process, potentially enormous savings could be made by these organisations. These efficiencies would ensure the organisations' longer term sustainability, and the lower running costs would reduce the cost of service delivery. Above all, this would improve the overall patient care.

This chapter focuses upon how Grid based infrastructures can help to address these issues. We focus in particular on the aspects of usability and security of Grid based e-Infrastructures and illustrate with examples from a range of projects at the National e-Science Centre at the University of Glasgow, how the vision of the Grid in providing seamless access to a range of heterogeneous resources (such as a variety of neurological data resources) can be undertaken in a secure, ethical framework where information governance and associated policy is paramount.

2. BACKGROUND TO BRAIN TRAUMA RESEARCH AND BRAINIT

There is increasing evidence that targeted use of IT can improve patient health care. A review article by Bates and Gawande (Bates 2003) outline several trials of IT technology that have decreased medication errors, errors of omission from poor handoffs between clinical staff and providing earlier detection of adverse events. A good example of this can be found in the work by Kupermann (Kupermann 1999) in a randomised control trial of IT for the earlier detection of adverse events which showed an 11% reduction in the time to treatment and a 29% reduction in the overall duration of dangerous adverse conditions to patients. Rosenfield and colleagues (Rosenfield 2000) reported on a study of IT based remote monitoring of a multi-bed intensive care unit and found a reduction in mortality with a reduced length of stay of nearly 30% when compared to historical controls.

Critics of this type of research will point out that better resolution of events is of no value unless their direct management influences patient outcome. Providing this type of evidence from single-centre studies with small patient numbers is not an efficient approach to answer these questions, and multi-centre studies conducting trials of new IT driven management is not readily funded nor easily justified as a research priority. Paradoxically, the patient populations which might benefit most from better IT-driven event detection and management standardisation would be the TBI population but this is also the most challenging in which to conduct such studies due to the continuing inter-centre management variation fostered to a large part by a lack of evidence for any type of effective therapy (Bulger 2002).

With this background in mind, certain individuals working within the field of TBI research met to discuss the foundation and development of a network for creating an IT based infrastructure aimed at improving the standards for multi-centre studies of monitoring and managing patients with traumatic brain injury during their acute stay in intensive care. It was agreed a different approach was needed, one which focused on using IT-based methods towards not only increasing the resolution of data capture but also the quality and validation of data captured. The pervasive nature of the internet and extensive use by clinicians of email systems fostered the creation of such a network as an internet based e-Infrastructure: the BrainIT group (www.brainit.org).

The BrainIT group works collaboratively on developing standards for collection and analyses of data from brain injured patients towards providing a more efficient infrastructure for assessing new health technology. Over a period of 12 months and four international meetings, the group have defined a core dataset designed to be collected using PC based tools and providing a common minimal dataset for all studies, regardless of the underlying research question. This data definition period was funded as part of an EC study (QLGT-2000-00454). The meetings brought together clinical and scientific experts from the domain of TBI basic research and also in the conduct of multi-centre clinical trials such as the European Brain Injury Consortium (EBIC – www.ebic.nl) as well as representatives from the medical device and pharmaceutical industries. A series of meetings and workshops spread over one year enabled the group to define a minimum set of data that could be collected from all patients with TBI, which would be useful in most research projects conducted in this population of patients. To facilitate discussion, the core dataset was sub-divided into four logical groups: a) demographic and clinical Information, b) minute by minute monitoring information, c) intensive care management information, and d) secondary insult treatment information.

From these series of meetings a consensus dataset was formed which includes nine categories:

1. **Demographic** and one-off clinical data, e.g. pre-neurosurgical hospital data, first and worst CT scan data etc;
2. **Daily management** data, e.g. daily summary measures of the use of sedatives, analgesics, vasopressors, fluid input/output balance etc;
3. **Laboratory** data, e.g. blood gas, haematology, biochemistry data etc;
4. **Event** data, e.g. nursing manoeuvers, physiotherapy, medical procedures (line insertion), calibrations etc;
5. **Surgical** procedures;
6. **Monitoring** data summary, e.g. the type and placement location of Intra-Cranial Pressures (ICP) sensors, Blood Pressure (BP) lines, etc;
7. **Neuro-event** summary, e.g. Glasgow Coma Scores (GCS), pupil size and reactivity;
8. **Targeted therapies**, e.g. mannitol given for raised ICP, pressor given for arterial hypotension etc;
9. **Vital monitoring** data, e.g. minute by minute BP, ICP, SaO_2 etc collected from the bedside monitoring.

The full details of the core dataset definition and the collaboration structure of the group can be found in the BrainIT publication: BrainIT Group – core concept and data definition (Piper 2003). Unique to other dataset definitions, the BrainIT core dataset defined a special approach to quantify secondary insult management. This is medical management therapy given to patients specifically to treat secondary insults which occur to patients despite their baseline intensive care medical management. To distinguish therapy given to patients to treat secondary insults from those of baseline intensive care, we have devised a coding system which allows specific categories of therapy to be assigned a "therapy target". For example, arterial pressors may be given to treat systemic hypotension or to treat reduced cerebral perfusion pressure (CPP) secondary to raised ICP. Choosing an appropriate target for each secondary insult therapy will enhance the usefulness of the database on medical therapy. Each therapy must be assigned a "Target" chosen from a drop down list. If drugs are given then one can indicate continuous infusion if drugs are delivered by a continuous infusion pump or one can indicate boluses if it is delivered non-continuously. Figure 1 summarises the minimum choice of therapy categories and associated targets for the BrainIT core dataset. This therapy tracking model has been designed to be easily implemented in software.

It is one thing to define on paper a dataset and another to actually collect it. Although a paper based feasibility exercise established some baseline information, the acid test was still to develop a series of IT-based tools to collect the core dataset and to prospectively trial the collection of core data from a number of neuro-intensive care centres.

A three year follow up EC funded study (QLGC-2002-00160) enabled the group to develop IT methods to collect the core dataset and to assess the feasibility and accuracy for collection of this core-dataset from 22 neuro-intensive care centres. The main data collection instrument for the "episodic" non-monitoring data was a PDA based data collection tool.

In this system, clinical data is entered by bedside nursing staff on hand held PDA's which supported the BrainIT core dataset definition through a Java Struts-based tool. This allowed the core dataset to be entered by roaming research nurses using a set of PDA documents accessed via a series of buttons and tabs. With this system, indicators were present showing data documents which were fully complete, partially complete or totally incomplete. When convenient, the PDA was connected to a docking station and a client program allowed viewing and saving of patient data collected. An anonymisation routine removed patient identification elements from the collected data and labelled the patient data file with a unique BrainIT study code generated from the BrainIT

Figure 1. Tracking therapies and targets

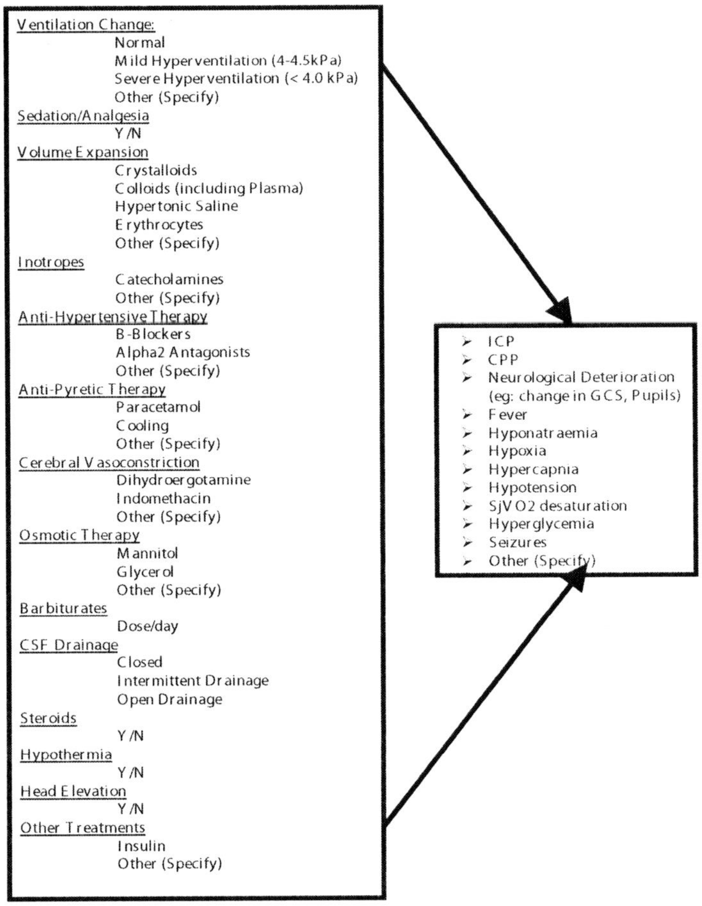

web-site. A local database held in each centre linked the anonymised data to local centre patient id information which was needed during the data checking stage of the study. The multi-centre ethics approval precluded connection of the PC client system holding the data to any computer connected to the Internet. We relied on local research nurses to download the anonymised data from the PDA system client PC onto a memory stick or CD and transfer the data to an Internet connected PC for upload of the data to the BrainIT database via the website data upload page.

As this was a multi-centre study collecting data from different countries with a range of languages, a multi-language implementation was needed to foster ease of use by local nursing staff.

A training course was held for the data validation nursing staff in Glasgow on the use of this data collection instrument which also included using their medical term and language expertise to translate all PDA labels and text output into six European languages (English, French, German, Spanish, Flemish and Italian). Data could be entered in the local language, exported in an XML file format where a table lookup driven by an XSL transformation converted the data into a standard English language version.

Data validation research nurse staff were hired on a country by country basis to check samples of the collected data against gold standard clinical record sources in order to quantify the accuracy for collection of the BrainIT core-dataset using the group IT based data collection methods. Anonymised data was uploaded via the BrainIT web upload services were a server side data converter tool converted data from centre-based formats into BrainIT data format generating data category files which were imported into the BrainIT database (SQL Server 2000). A validation request tool sampled 20% of the data sent for each data category and generated a validation request file listing the timestamps and data items to be checked by local data validators. Emails were generated to the data validation staff which contained the validation data requested documents listing the data items to be checked. Data validators entered into a data validation tool the requested data items for checking from source documentation held in each local centre. Validation data was then uploaded to the BrainIT data coordinating centre via the website and using data validation checking software tools, the validated data was checked against the data items originally sent from which percentage accuracy data was calculated. Figure 2 shows the flow of data between a remote centre and the BrainIT database with the validation procedure selecting a random sample of 20% of data items uploaded per data type which are sent to data validation (DV) nurses. The DV nurses enter requested data into a PC based validation data tool, upload the data to the data manager who can then check the accuracy of data for each data category and estimate an overall error rate.

As part of this validation process, in addition to the categorical and numeric clinical data being checked for accuracy, the BrainIT system also assessed the minute by minute monitoring data too. Random samples of monitoring data channels uploaded, e.g. ICP, SaO_2, were selected and validation staff asked to manually enter the hourly recorded values from the nurses chart (or local gold standard data source) for the first and last 24 hour periods of bedside monitoring for a given patient for a given channel. These "validation" values could then be compared with a range of summary measures, e.g. mean, median, from the computer based monitoring data acquired from the patient.

To date (mid-2008), 384 TBI patient's core data have been collected from 22 European neuro-intensive centres. The first 200 patient's data has been cleaned and validation analyses conducted. In total, 19,461 comparisons were made between collected data elements and source documentation data (Shaw 2008). The number of comparisons made per data category was in proportion to the

Figure 2. BrainIT data validation flow of information

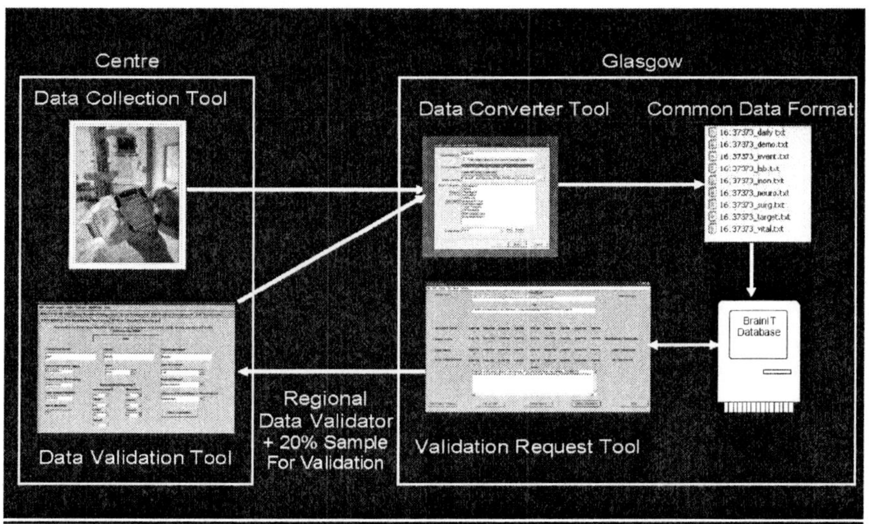

size of the data received for that category with the largest number checked in laboratory data (5,667) and the least in the surgery data (567). Table 1 summarises error rates by data class. Error rates were generally less than or equal to 6%, the exception being the surgery data class where an unacceptably high error rate of 34% was found.

With regard to the proportion of surgery errors, this was primarily due to the classification system used to simplify and thereby reduce the burden of data entry. As an example of this, through discussions with local nursing and data validation staff it was found that there was particular confusion over when to record ICP sensor placement and the presence of skull fractures as the primary surgical procedure. Typically, these procedures occur during the same operative procedure. As such, confusion over coding these two procedures by both the local research nurse and the data validation nurse accounted for the majority of errors in this data category.

To our knowledge, this study conducted by the BrainIT group is one of only a few projects to attempt to prospectively assess the data capture error rate within an academic environment. We have shown that it is feasible to collect the BrainIT dataset from multiple centres in an international setting with human-intensive (research nurse) IT based methods and the accuracy of the data collected is greater than or equal to 94%, with the exception of the surgery data type which must be revised. We have also shown that computer collected minute by minute vital signs data, summarised as end hour averages, correlate well with nursing chart end hour recordings. This allows the end hour averaged computer records to be used in database analyses assessing nurses chart recorded detection of events with computer based sampling. These validation results calculated on a subset of patients provides an estimate of the data quality for future analyses on the full patient cohort of 350 patients collected as part of the EC funded study which was conducted over the same time period by the same staff using the same data methods. Clearly though, future data collection projects will generate datasets under differing data collection conditions and will require a separate validation stage if we wish to maintain our confidence in the level of data accuracy. However, the costs of maintaining such a data validation network are prohibitively high. To maintain a full time data validation nurse within each participating country costs in excess of 1 Million Euro's per year. Such large running costs for an academic network are not sustainable in the long term and a more cost-effective solution for data validation must be found.

Table 1. percentage error rate by data type class with common error types

Data Class	Error Rate (%)	Common Errors
Laboratory	2	*pCO2, FiO2 value*
Demographic	4	*Monitoring on arrival at neurosurgery, intubation on arrival at neurosurgery*
Neuro Observations	5	*Pupil Size, GCSv (code 1 Vs Unknown)*
Monitoring Summary	5	*ICP type, ICP Location*
Daily Management Summary	5	*Infusion type (bolus vs infusion or both), drug number (1, > 1)*
Targeted Therapy	6	*Non-standard target, no Target specified*
Surgeries	34	*ICP placement, Skull #, mass lesion*

3. GRID-BASED SOLUTIONS TO BRAIN TRAUMA RESEARCH

Grid technology provides middleware that can allow distributed federation of data to occur. Different domains have their own requirements on how this federation can be achieved. The clinical domain in particular demands that strict adherence to ethics and information governance is achieved which in turn demands fine grained security.

It is the case that many IT and data storage solutions already exist crossing primary care, secondary care and a range of specialised resources such as disease registries in the clinical domain. Many of these solutions have been developed largely in isolation and as a result have widely differing data descriptions and associated security policies – or in many cases, no security policies other than protection at the firewall level provided through bodies such as the NHS in the UK. This situation is greatly magnified when crossing national boundaries. Dealing with such heterogeneity from the data perspective at least has been one of the drivers behind Grid technologies.

Due to the sensitivity of data, the establishment of security policies and their rigorous enforcement is of paramount importance in the clinical domain. It is clear that a single static system accessing a closed/fixed set of data will not meet the needs of healthcare providers nor researchers using the clinical data sets. Systems and data sets evolve. Different researchers may be allowed access to a given clinical data set for a given time after applying specifically for ethical approval to do so. Instead, software and tools are required to build infrastructures where a variety of data can be made accessible to different individuals for different times for different reasons. A cornerstone of these solutions is in ensuring fine grained security. However it is an unavoidable fact that the specific privileges required in a particular trial or study will not be known when the system is first created. Similarly a doctor in one hospital may have privileges to access various systems in that hospital, but these do not transfer directly when this doctor attends a different hospital for example. Therefore systems capable of adding and removing resources or privileges *"on the fly"* are necessary, where the corresponding allocation of privileges can be added or removed depending on the needs of different trials or healthcare systems. In Grid parlance, the framework by which such rules and regulations on the resources and the users that may access them and under what conditions is given by the concept of a *Virtual Organisation* (VO).

The basic models put forward for the majority of Grid-based security systems can be broadly broken down into the "AAA" categories:

- *Authentication* – establishing the identity of the person requesting access to a resource.
- *Authorization* – having established identity, establishing and enforcing what that person is allowed to do on a given resource.
- *Accountability* – being able to establish the activities, and time of activities, of a particular person on that resource (or resources) so that they can not subsequently deny potential misuse later on (non-repudiation).

Of course there are other important aspects when considering the wider challenges of building secure systems. For example in the clinical domain, confidentiality and data integrity are essential, but in this chapter we restrict ourselves to authentication and authorisation as these are arguably the most important things to get right in the first instance. Put another way if authentication and authorisation are not adequately addressed, then other aspects of security are largely redundant

Methods of authentication with grid technology tend to favour two methods, either username/password combinations or public key infrastructures (PKIs) (Housley 2001). The latter are used to set up and use safer, encrypted communication channels through trusting a third party root of trust Certification Authority (CA) such as the UK e-Science CA (www.grid-support.ac.uk/ca). Authentication is focused upon identity management and a process exists through which a user establishes their identity to obtain an X.509 digital certificate (ITU-T X509, 2001). Typically this is through showing some form of physical identification to a local registration authority at their institution. However there are issues with this process, not least of which is the complexity of converting certificates to formats suitable for usage on the Grid and with the lack of local identity management. Furthermore, a user might be expelled from their institution but still have access to a valid Grid certificate. These issues are described in more detail in (Sinnott, 2006; Stell, 2006; Watt, 2006; Ajayi, 2006).

One approach to overcoming the issues with PKI based identity management is through exploitation of local identity management systems. The Internet2 Shibboleth technology has been developed to fulfil this need (Internet2 ShibArch, 2006, Internet2 ShibProt, 2006). In essence, Shibboleth provides a method of securely transferring attributes between institutions subscribing to an over-arching federation. The basic model of Shibboleth is that institutions enter into federations, or more precisely federated access management federations. Users at those institutions attempting to access remote resources across the federation (typically service providers) are redirected to their home institution (typically through a Where Are You From service) where they log in locally. A digitally signed Security Assertion Markup Language (SAML) (OASIS SAML, 2003) assertion showing that the user has authenticated is then delivered to the target resource which may then decide whether access is granted or not. Often further information such as attributes for authorisation need to be returned. This whole process however is transparent to the end users who only log in to their local system with the normal usernames and passwords.

The key benefit of Shibboleth is that end users have simple ways to access resources. Furthermore depending upon local policies and trust relationships across the federation, users are able to access a range of distributed resources thereby supporting *single sign-on*. Once a secure session between a user and a resource is established, the user can access further resources (in the same federation) without the need to further authenticate – based upon browser-based information. Additionally since their authentication is tied to their local institution, they will have their federation privileges revoked, if they are revoked locally (which is the most likely scenario).

Authentication is only a starting point with regard to security however. Knowing that someone has authenticated at the University of Glasgow is not likely (indeed very unlikely) to allow access to a remote clinical data set. Instead authorisation is required. Unlike authentication which has a reasonably mature body of software and approaches in supporting the process of its application, authorisation is a much more fluid area. Numerous competing software solutions and standards exist with their own advantages and disadvantages. A comparison of these and overview of some of their advantages and disadvantages is given in (Sinnott, 2005; Stell, 2005). It is essential in the clinical domain that any solution put forward for security is simple both for the end users, but equally importantly for the clinical data provider IT-administrators. These will typically be unlikely to have any experience with Grid-based systems. As such, solutions are required which provide them with capabilities to easily manage secure access to their local data according to agreed information governance policies.

One of the more mature authorisation-based approaches is based upon role based access control (RBAC) (Sandhu, 1996). In this model, roles are created which offer a given privilege and subsequently assigned to trusted individuals. Several approaches for creation of these roles and their assignment are possible including federated models, centralised models or hybrid approaches of the two. The pros and cons of such role based creation and assignment is described in (Sinnott, 2008).

Once these roles are created and assigned to individuals they can be used to enforce finer grained authorisation at given data provider sites. To actually enforce these authorisation decisions, several software solutions can be adopted. The PERMIS RBAC (Chadwick, 2002) offers one such solution for role definition, assignment and its use in enforcing local access control. Anther popular solution is Virtual Organisation Membership Service (VOMS) (Alfieri, 2003) which provides a centralised repository of VO-roles and has been shown to interoperate with PERMIS in (Chadwick 2002). An essential part of authorisation is in understanding both when and how an authorisation decision needs to be made. The X812 standard (ITUT X812, 1995) defines concepts such as policy enforcement points (PEP) and policy decision points (PDP) which define a generic approach for deciding where (PEP) and how (PDP) authorisation decisions should be made. Trust plays a crucial role in any security system however, and the allocation of privileges and their use for authorisation needs to be augmented with user guidelines and agreements on best practices on accessing and using clinical data according to ethics and information governance.

To understand how these security concepts can be applied in the brain trauma domain, the National e-Science Centre (NeSC) at the University of Glasgow has undertaken a range of case studies exploring authorisation and Shibboleth technologies applied in the clinical domain. This work was undertaken as part of the Joint Information Systems Committee (JISC) funded GLASS (www.nesc.ac.uk/hub/projects/glass) and the Medical Research Council funded VOTES (www.nesc.ac.uk/hub/projects/votes) projects.

The VOTES project focused on realising a software solution meeting the challenges inherent in federating distributed clinical data, specifically in supporting the various stages of clinical trials and epidemiological studies such as patient recruitment; data collection and study management. The GLASS project was primarily focused upon the roll out of Shibboleth technology across the University of Glasgow and exploring its suitability in a range of e-Science and non-e-Science applications, e.g. to provide access to student records for authorised individuals in Glasgow.

Usability is at the heart of VOTES efforts and portal based solutions have been prototyped with this in mind. As identified previously, this usability should be for end users, administrators, investigators and the other stakeholders involved in clinical trials and studies. To support this, a

Figure 3. VOTES node infrastructure and clinical VO overlays

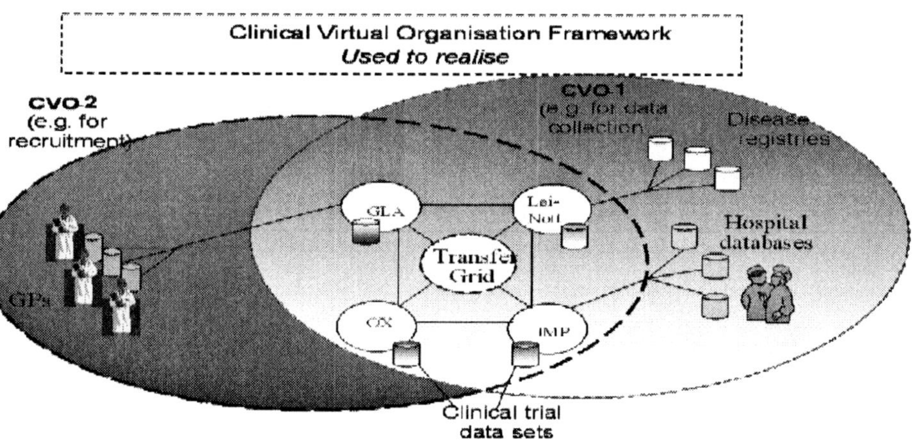

node infrastructure supporting the addition or removal of institutions and the resources that they wish to contribute is needed. This should support fine grained user-oriented VO-specific security. Figure 3 shows an overall schematic representation of such a VO.

In this infrastructure, each node provides access to its resources according to local policies. A key part of this infrastructure is that it supports both heterogeneous data resources but equally importantly heterogeneous security infrastructures. Thus sites may recognise certain roles associated with a given clinical trial or not as the case may be. Furthermore, these roles and the privileges that are associated with them will have different interpretations at each site. Thus a clinical nurse role at one node might be allowed to query a range of clinical data including identifying data, whilst the same nurse role at a different site may be restricted to only accessing statistical information, e.g. the number of patients with a particular Glasgow Coma Score without knowing any further information on those patients.

The VOTES infrastructure supports a variety of Grid components to support secure access to federated clinical data. These include Grid services developed with the Globus toolkit version 4 (www.globus.org) and data federation technology such as the Open Grid Service Architecture Data Access and Integration (OGSA-DAI) technology (www.ogsadai.org.uk). The interface to the systems themselves is through portal based technologies based upon GridSphere (www.gridsphere.org).

To understand the interplay of role based access to federated resources we consider the scenario of a particular BrainIT trial undertaken with Glasgow Southern General Hospital. In this model we identified different roles that could access different demographic, physiological and monitoring data. For simplicity we simply assigned these roles directly to known and trusted individuals at Glasgow. This assignment was based upon adding the appropriate signed attribute certificates to the appropriate individuals in the Glasgow Identity Provider (IdP) given as an LDAP server although other authentication systems are possible. Extensions and refinements to this scenario allowing for example the delegation of these privileges is also supported through the JISC funded DyVOSE project (www.nesc.ac.uk/hub/projects/dyvose). In this model, the privileges (roles) can be pushed to remote trusted administrators who can subsequently assign them to local staff or students involved in that particular VO for example. Similarly, adding the attributes to separate attribute authorities (AAs) is supported rather than to a *single* institutional IdP. In this case, a user would log in to their IdP and the appropriate attributes linked and pushed (or pulled) from separate AAs set up for specific VOs. This linkage is possible through having unique identities in place across institutions and is described in more detail in (Sinnott, 2007; Watt, 2007).

When a user attempts to access the BrainIT portal, they are redirected to their home institutions when they are asked to authenticate. After authenticating, the attributes needed for access to the BrainIT portal for authorised access to the federated clinical data are returned in a signed SAML assertion. We note that the tools for scoping of attributes, e.g. which sites to trust, and for defining and enforcing attribute release policies, e.g. to determine which attributes should be released to which sites have been implemented in the SPAM-GP project (www.nesc.ac.uk/hub/projects/spam-gp). The portal will then use these attributes to configure the contents of the portal, i.e. the users are restricted to see and do what their roles dictate. This is shown in Figure 4 where two different users have logged in to the system - one with the *brainit-investigator* role (right side) and one with the *brainit-nurse* role (left side). The former role has superior privileges in this particular trial and hence is allowed access to a wider range of information within the context of that trial, e.g. the GCS, the ICP, Glucose levels as well as identifying patient information. The nurse role on the other hand is restricted to a subset of non-patient identifying information.

For clinical data providers, it is unlikely that they would simply allow access to their resources based upon a role which has been used to configure a portal where the portal itself may well exist outside of their own domain of control. The VOTES e-Infrastructure has been designed with security throughout however. This includes security of the local Grid and data services as well as security at potentially remote data providers so that they are able to make their own autonomous authorisation decisions. Thus the roles are used to configure the portal, but then importantly the distributed data providers accessible from this portal have their own local authorisation policies which must be satisfied by the information provided to the portal. It is quite possible that further information than the supplied role needs to be obtained before a given access decision is made. This might be a notification from an ethical oversight committee for example. The infrastructure allows for a range of authorisation possibilities to be supported, including scenarios where additional attributes from one or more attribute authorities are required to authorise specific requests. Examples of these different kinds of model are described in (Stell, 2008).

We note that data security offers challenges that existing Grid security infrastructures are not well aligned to and hence cannot be easily supported. In large scale federated data infrastructures, it is often the case that access to individual subsets of data within different federated repositories is required. Many mainstream Grid based solutions today work primarily at securing services or methods that services support. Whilst it is possible to develop targeted services to specific data resources for a nurse or an investigator, a more scalable solution is to have one or more generic services that differ in the security models that they offer. Thus, to support a range of studies it should not be necessary to support individual services at each service provider for each study. Rather, the ideal scenario is to have a single data access service, which can distinguish between the different trial/study specific security identifiers. To support such scenarios, the VOTES project has

Figure 4. Role-based access to portal and restrictions to data access

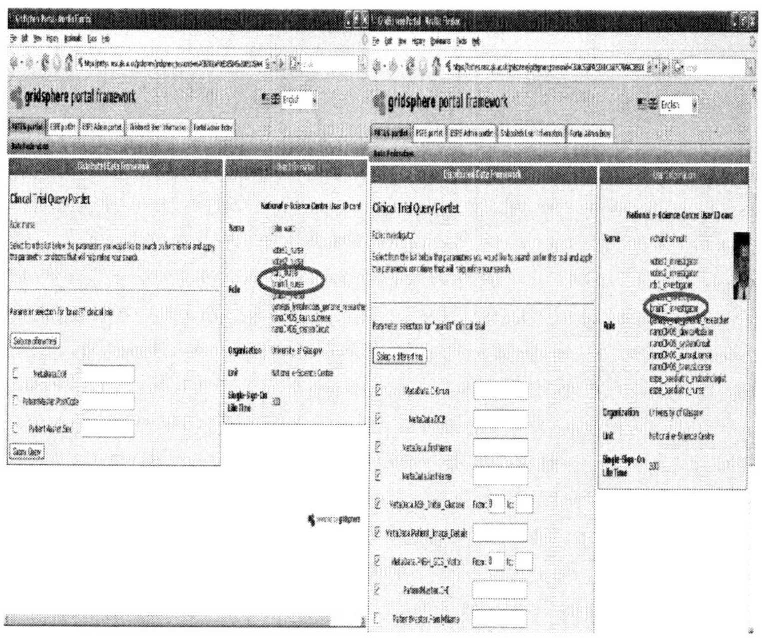

implemented RBAC models based upon access control matrices. With these models, users and roles are defined with specific relationships over the data models themselves at each data provider. Thus, a given data provider may decide that a subset of their tables should be made available with specific columns and rows accessible for different roles within a given study. Through tools and services implemented in VOTES it is possible to associate those users and their roles with specific subsets of the databases. Details on how this is supported, is described in (Sinnott, 2008) along with the benefits and drawbacks compared to other RBAC systems. This model is based on the premise that the access matrix will be available at every node in the VO and will be regularly updated. Consequently, every user that has access to the VO in some form will go through this matrix model to access any other resource, be it local or remote, within the VO with each data provider defining its own local access policies for each role.

Having defined the roles and local policies on how these are interpreted when enforcing policy decisions it is necessary to consider the ramifications of the data sets that are made available through the combination of the policies. Thus a given provider might be completely satisfied that the data sets that they release are in accordance with local policy and information governance frameworks. However, when linked or joined with other sets, the consequences need to be understood within the context of the VO. For example, a provider may be happy to release anonymised data but if this is linked with other data sets containing identifying data for given patients for example, then this needs to be clearly understood by all stakeholders across the VO. We note that it is also possible to link data and remove (anonymised) the fields that have been used for the data linkage. The algorithms for supporting these kinds of scenarios are described in detail in (Ajayi, 2008).

The typical interactions of the components involved in supporting the access to and usage of federated resources is as follows. Firstly the user must log into the portal either through Shibboleth or directly. In the Shibboleth case the various user roles and privileges are provided to personalise/authorise their access to a variety of services as described above. To achieve this, the portal server checks the local resource files to discover the available grid servers, data servers and associated databases accessible to this particular user with the associated privileges. The appropriate Grid service consults the local access matrix and returns the parameters for the resources that the user can query for that particular trial. These are presented to the user as a list of check-boxes, with the option to specify conditions if desired (as shown in Figure 4). Following this, the user makes their selection of parameters and submits them. These are constructed into a query which is distributed, i.e. federated, across the various resources associated with that particular trial. To support this query distribution, the query is sent from the Grid server to the data server, where it is wrapped as an OGSA-DAI service request before being passed to a local driving database. This driving database subsequently co-ordinates the execution of the distributed queries over the resources associated with that trial. Where possible, for example where web services are offered and no-direct connections to a remote database are offered or possible, the federating of the sub-queries can result in combinations of direct database queries through JDBC for example, as well as SOAP messaging to remote services. Once all sub-queries are returned the results are joined together and returned for display in the portal after conversion to HTML. We note that it is quite possible for the results to be stored into databases for future access by researchers. The actual joining of results itself is made possible through the unique identifier assigned to patients in Scotland – the Community Health Index number. This greatly simplifies the development of federated queries since a unique joining key is available. In other situations, this is not always available hence other mechanisms and identifiers have to be identified through which data linkage can be made. This might be based upon study specific identifiers generated for particular patients. With this approach, joining of data within a given neurological centre can be achieved however joining across centres is not possible. Since it is unlikely that the same individual will be found in different neurological data centres, this is not overly constraining. Rather in this case queries are federated to the different centres, the data joined on local identifiers, and then returned to the portal where the union for example of the data sets can be made. The result of a particular query is shown in Figure 5.

In the results shown in Figure 5 we see the DICOM images and various physiological data associated with a given patient that includes any changes in their condition or their overall Glasgow Coma Score in response to the specific treatments they may be receiving. A variety of other information can be returned in the existing systems including the current treatment, ICP measurements amongst other factors.

The above systems were developed in a controlled test-bed at the NeSC in Glasgow. This comprised a variety of SQL Server, MySQL and Oracle databases hosting the various neurological data sets as well as a consent database.

4. FUTURE TRENDS

The above systems have shown the proof of concept upon which clinical data can be accessed and used within a secure framework. There are numerous enhancements to this system that are being explored within the AVERT-IT project. We outline upon some of these here since we believe they capture future trends in Grid-enabled brain trauma research. Firstly, it is clear that the models and systems described here need to deal with the actual concerns regarding access to and usage of live systems supporting multi-centre trials. Proof of concept VOs and federated authorization systems as described here need to reflect the particular needs and requirements of particular neurological data centres. In particular it is essential to understand what data needs to be made accessible and

Figure 5. Results of executing a particular brain trauma query

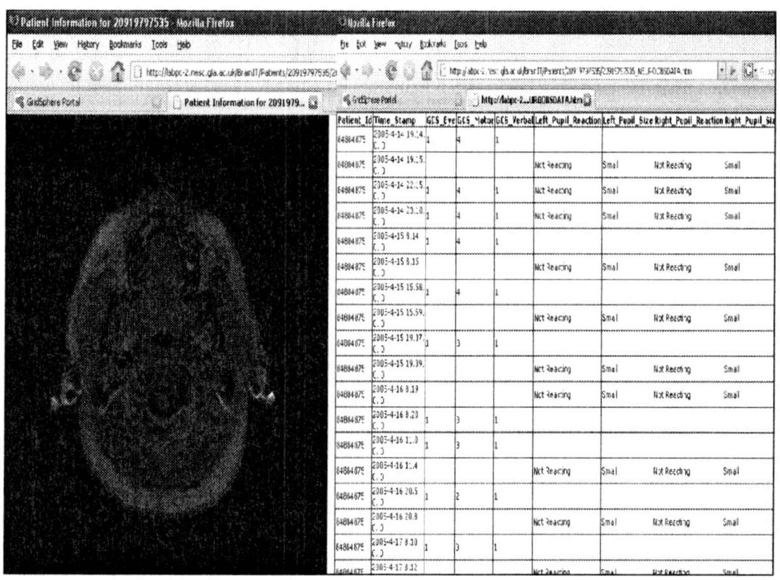

linked across the various centres and importantly for what reason. Within the AVERT-IT project we propose to capture near real time information from the various neurological data centres to explore the potential prediction of hypotensive events. To support this, we are exploring how determination of hypotensive event might be predicted via empirical analysis, univariate analysis and multivariable linear regression techniques of the weighted association between multiple patient parameters (drawn from demographic, periodic and episodic datasets) and subsequent arterial hypotension. This association will then form the basis for the initial definition of a novel Bayesian Neural Network, to be trained against the BrainIT dataset prior to undertaking a novel clinical trial demonstrating the effectiveness of the AVERT-IT project concept.

From a technological perspective, the AVERT-IT system will focus on the development of a novel IT-based decision support system ("HypoPredict"), appropriate for deployment within intensive and high dependency care units which is capable of:

- Automatically and continually monitoring at least four in-vivo patient parameters (ECG, arterial blood pressure, Oxygen saturation and core temperature), together with open interfaces providing input of key demographic data (age, gender etc.) and periodic data (clinical pathology results etc.) related to the patient.
- Outputting a continuous Hypotension Prediction Index (HPi) in the range 0 – 100 (0 = no risk; 100 = patient is currently hypotensive);
- Providing primary (P1) and secondary (P2) weighted (0-100; 0 = not considered relevant; 100 = critical importance) causal data (current values of input parameters) in parallel with HPi to facilitate appropriate intervention selection by clinician (for example, elevated core temperature could be indicative of sepsis, a common precursor to hypotension);
- Providing updated HPi, P1 and P2 values immediately changes are detected in the patient parameter input set.

Such predictive approaches offer one of the greatest opportunities from the application of Grid middleware to link different clinical data centres. Can we determine the best way to treat individuals suffering major brain trauma thereby extending the state of the art knowledge in brain trauma treatment from initial assessment, diagnosis and treatment to continued software support systems which augment the whole process of improved healthcare and treatment. We emphasis here that these systems are <u>not</u> and are never intended as a replacement for informed clinical decisions in brain trauma patient management, but as a guide and tool that helps intensivists and brain trauma specialists better understand the whole course of brain trauma treatment with improved monitoring and diagnostics.

However from past and on-going experience in projects such as VOTES, we recognise the direct transfer of Grid technology within a clinical setting often requires a degree of pragmatism and especially consideration of the clinical IT staff and their existing systems in place. As noted it is unrealistic to expect a hospital IT administrator to be knowledgeable in Grid and/or advanced authorisation technologies. Furthermore, one of the major issues with most Grid middleware today is their lack of stability. This in turn directly impacts upon the validation of the software. Before widespread take up of Grid middleware in the clinical domain can be achieved, software validation is needed. This has to ensure both that the middleware satisfies the basic tests that it is fit for purpose, but also that it is resilient enough to withstand attempts at breaking the software from third parties, e.g. hackers. This has not been the focus until now, but is clearly needed for the wider community to have faith. To address such aspects within the AVERT-IT systems we are initially proposing models based upon pushed data transfer only, i.e. the clinical data provider firewalls will reject all incoming data requests, but will push data out of their firewalls into demilitarised zones set up as a buffer between the Grid research domain and the domain of live healthcare provision. It is our intention, however to eventually develop systems that allow to integrate the various neurological data sets within a given clinical data provider domain and push them out of the firewall. In supporting this, we are acutely aware of the pragmatic approach that is needed and are not yet proposing any overly complicated middleware solutions, but instead focusing upon lighter-weight data integration clients that can query various clinical data resources directly, e.g. over JDBC or ODBC connections before anonymising, encrypting and ultimately sending them to the centralised AVERT-IT repository accessible to the partner sites. We are currently in the progress of evaluating the different data models and software used within the individual partner sites and how these map onto the BrainIT core data set identified previously so that the queries to extract the needed data sets can be defined and implemented in the lightweight clients. This is non-trivial given the international dimension of the AVERT-IT project which involves collaborators from the UK, Sweden, Italy, Germany and Lithuania with each site having their own software and data infrastructures.

One challenge that remains to be solved in the AVERT-IT system and within the wider e-Research community more generally is with regard to real time or near real-time data. Monitoring data that needs to be streamed to support the Bayesian adverse event prediction algorithms offers new challenges that have hitherto not been addressed within the VOTES project for example. In this case, it is likely that data streaming models with authorisation capabilities are supported, or enhancements to systems that allow for frequent periodic queries to be undertaken with periods of the order of 5 minutes or less being likely. This is made further challenging since these near real-time data sets need to be joined with other non-real time data sets each time. Obviously to predict the onset of an adverse hypotensive event, the closer to real time the information is the better and more accurate the identification will be and hence less false positives identified.

These challenges are not specific to the brain trauma domain but apply more generally to wider healthcare support where real time information to patient information is essential. However real time access to patient data even by clinical healthcare providers is often a fraught process as exemplified by the recent furore in the UK to establish a national consent database (Guardian, 2007). It is interesting to note here that many patients were largely in favour of such national data resources for improved healthcare, whereas general practitioners and trusts where more reluctant to make their patient data available to national-level resources.

5. CONCLUSION

The Grid paradigm provides a compelling model for secure access to clinical data and the proof of concept systems have shown that this vision of seamless access to federated data can be made a reality. We recognise that a proof of concept system and a live system that has been validated requires a step change in the way in which Grid based systems are currently developed. The Grid middleware evolution or revolution in many cases needs to solidify to hardened software stacks that clinical and healthcare providers understand and are comfortable in supporting. Indeed the Grid community as a whole need to agree upon best practice of establishing the numerous different flavours of Grids that exist. Why should software developed for distributing petabytes of elementary particle data from the Large Hadron Collider in CERN be applicable to accessing a hospital database in Scotland? Surprisingly, such applications of Grid middleware are still being explored. From direct experience of working with healthcare providers and their IT support staff, the deployment of complex open source Grid middleware stacks from a variety of sources will never be accepted. It is essential that in future health Grid efforts this is recognised and a common core set of functionality agreed upon, documented and the detailed pro's and con's of such software defined for others to use. It is the case that the weakest link in any system is the one that will be exploited. Ensuring that all nodes within a particular VO understand the consequences of being in that VO for all parties is crucial. Similarly, whilst compromises of Grid facilities can be accepted to a degree in some domains, the healthcare domain is not one of them. The reputation of the Grid is perhaps as important to convince healthcare providers to provide access to their resources via the Grid as are the Grid technologies themselves. The authentication-only based models of most large scale Grids such as the National Grid Service (www.ngs.ac.uk) in the UK will not convince these providers – thus it will never be the case that a researcher is allowed access to a clinical provider database to "*do stuff*". Rather, clearly defined rules and regulations on what can be accessed and used and in what context are needed, and Grid technologies need to *demonstrably* support this process.

6. REFERENCES

Ajayi, O., Sinnott, R. O., Stell, A. J. & Young, A. (2008). *Towards a Virtual Anonymisation Grid for Unified Access to Remote Clinical Data.* Proceedings of 6th International HealthGrid conference, Chicago.

Ajayi, O., Sinnott, R. O., Watt, J., Jiang, J. & Stell, A. J. (2006). *Single-Sign on and Authorization for Dynamic Virtual Organizations.* Proceedings of International Conference on Virtual Enterprises, (PRO-VE'06), Helsinki, Finland.

Alfieri, R., et al (2003). *VOMS: an authorization system for virtual organizations.* Proceedings of 1st European across Grids conference, Santiago de Compostela, Spain.

Bates, D. G. A. (2003). Improving safety with information technology. *New England Journal of Medicine, 348*, 2526-2534.

Bulger, E. & Nathens, A. (2002). Management of severe head injury: institutional variations in care and effect on outcome. *Critical Care Med, 30*, 1870-1876.

Chadwick, D. W., Otenko, O. (2002). The PERMIS X.509 Role Based Privilege Management Infrastructure. *Future Generation Computer Systems, 936*, 1–13.

Guardian (2007). *Family doctors shun national database of patients' records, 20 Nov 2007.* Retrieved from www.guardian.co.uk/society/2007/nov/20/nhs.health

Housley, R. & Polk, T. (2001). *Planning for PKI: Best Practices Guide for Deploying Public Key Infrastructures.* Hoboken, NJ: Wiley Computer Publishing.

Internet2 Shibboleth, (2006). *Shibboleth Architecture Protocols and Profiles.* Retrieved from http://shibboleth.internet2.edu/docs/draft-mace-shibboleth-arch-protocols-latest.pdf

Internet2 Shibboleth, (2006). *Shibboleth Architecture Technical Overview.* Retrieved from http://shibboleth.internet2.edu/docs/draft-mace-shibboleth-tech-overview-latest.pdf

ITU-T Recommendation X.509 (2001) | ISO/IEC 9594-8: 2001, Information technology – Open Systems Interconnection – Public-Key and Attribute Certificate Frameworks.

ITU-T Recommendation X.812 (1995). | ISO/IEC 10181-3:1995, Security Frameworks for open systems: Access control framework.

Kuperman, G. J. (1999). Improving response to critical laboratory values with automation: results of a randomised controlled trial. *J Am Med Inform Assoc, 6*, 512-522.

Narayan, R. M. (2002). Clinical Trials in Head Injury. *Journal of Neurotrauma, 19*, 503-557.

OASIS (2003). *Assertions and Protocol for the OASIS Security Assertion Markup Language (SAML) v1.1.* Retrieved from http://www.oasis-open.org/committees/security/

Olesen, J. L. M. (2003). The burden of brain disease in Europe. *European Journal of Neurology, 10*, 471-477.

Piper, I., Citerio, G., & the BrainIT Group (2003). The BrainIT Group: Concept and Core Dataset Definition. *Acta Neurochirurgica, 145*, 615-629

Rosenfield, B. A. & Dorman, T. (2000). Intensive care unit telemedicine; alternative paradigm for providing continuous intensive care. *Crit. Care Med, 28*, 3925-3931.

Sandhu, R., Coyne, E. J., Feinstein, H. L. & Youman, C.E. (1996). Role-Based Access Control Models. *IEEE Computer 29* (2), 38-47.

Shaw, M. & Piper, I. (2008). The Brain Monitoring with Information Technology (BrainIT) collaborative network: Data Validation Results. *Acta Neurochirurgica, ICP Symposium Journal.*

Sinnott, R. O., Jiang, J., Stewart, G., Stell, A. J., Martin, D., Doherty, T., Su, L. & Watt, J. (2008). *Advanced Security for Virtual Organizations: Exploring the Pros and Cons of Centralized vs Decentralized Security Models.* Proceedings of 8th IEEE International Symposium on Cluster Computing and the Grid (CCGrid 2008), Lyon, France.

Sinnott, R. O., Stell, A. J., Chadwick, D. W., & Otenko, O. (2005). Experiences of Applying Advanced Grid Authorisation Infrastructures. In P.M.A. Sloot, et al, (eds.), *Proceedings of European Grid Conference (EGC)*, Amsterdam, the Netherlands (pp. 265-275).

Sinnott, R. O., Stell, A. J., Chadwick, D. W., Otenko, O., Watt, J., Koetsier, J. & Nguyen, T. A. (2006). *Supporting Decentralized, Security focused Dynamic Virtual Organizations across the Grid.* Proceedings of 2nd IEEE International Conference on e-Science and Grid Computing, Amsterdam, Holland.

Sinnott, R. O., Watt, J., & Jiang, J. (2007). *The GLASS Project: Supporting Secure Shibboleth-based Single Sign-On to Campus Resources.* Proceedings of UK e-Science All Hands Meeting, Nottingham, UK.

Sinnott, R. O., Watt, J., Ajayi, O. & Jiang, J. (2006). *Shibboleth-based Access to and Usage of Grid Resources.* Proceedings of IEEE International Conference on Grid Computing, Barcelona, Spain.

Stell, A. J., Sinnott, R. O. & Ajayi, O. (2008). *Supporting Nationwide e-Clinical Trials and Studies.* Proceedings of 15th Mardi Gras Conference, Baton Rouge, LA.

Stell, A. J., Sinnott, R. O. & Ajayi, O. (2006). *Secure, Reliable and Dynamic Access to Distributed Clinical Data.* Proceedings of Life Science Grid Conference, Yokohama, Japan.

Stell, A. J., Sinnott, R. O., & Watt, J. (2005). *Comparison of Advanced Authorisation Infrastructures for Grid Computing.* Proceedings of International Conference on High Performance Computing Systems and Applications, Guelph, Canada.

Watt, J., Sinnott, R. O., Ajayi, O., Jiang, J. & Stell, A. J. (2006). *User Oriented Access to Secure Biomedical Resources through the Grid.* Proceedings of Life Science Grid Conference, Yokohama, Japan.

Watt, J., Sinnott, R. O., Jiang, J., Stewart, G., Stell, A. J., Martin, D. & Doherty, T. (2007).

Federated Authentication and Authorisation for e-Science. In *Proceedings of APAC conference*, Perth, Australia.

KEY TERMS AND DEFINITIONS

Adverse Hypotensive Event: Hypotension lasting 5 minutes or longer falling below the commonly accepted threshold of 90 mmHg systolic OR 70 mmHg mean pressure.

Authentication: The act of establishing or confirming something or someone as authentic. In the security domain this might for example involve electronically confirming the identity of a person wishing to access a given software or hardware resource. Authentication can be achieved in many ways, e.g. usernames/passwords, certificate based systems etc.

Authorisation: The process of restricting access to resources only to those permitted to use them. In the security domain this will typically entail the definition of security policies associated with resources, the assignment of privileges to individuals that should be able to access those resources and the subsequent enactment of those policies when requests for access are received by individuals. Standards have been identified to support both the definition of policies, where access decisions need to be enforced, and how such decisions are made. Authorisation typically augments authentication and allows finer grained access control to be supported.

Certificate Authority (CA): An entity which issues digital certificates for use by other parties (including individuals and computers). Through trusting a CA and the process by which it issues and revokes certificates, the certificates can be used for accessing multiple resources seamlessly to support one of the basic tenets of Grid: single sign-on.

Glasgow Coma Score (GCS): Widely used clinical score assessing brain stem function following a suspected brain injury.

Hypotension: Abnormally low blood pressure which can be especially dangerous for brain injury patients.

Pressor: Pharmacological agent which acts to increase the work of the heart and often includes actions on increasing systemic vascular resistance towards raising blood pressure when hypotensive.

Registration Authority (RA): Typically an individual at an institution that supports the processes required by a CA to verify the identity of individuals applying for digital certificates. Typically this is through the certificate requestor presenting in person a visual identity such as a passport or student matriculation card to the RA.

Single Sign-On: The ability to securely access and use a variety of distributed resources without the need for multiple usernames/passwords or authentication challenge/responses. In the Grid world this is typically achived through trust of the CA that issued the certificate and local policy on whether that individual with that certificate is allowed access. For many Grids, this can be through a mapping of the Distinguished Name associated with the certificate to a local system account.

Section VI
Grid-Based Virtual Laboratories for Bioinformatics and E-Science

Chapter XXVI
Semantic Integration for Research Environments

Tomasz Gubala
University of Amsterdam, The Netherlands & ACC CYFRONET AGH, Poland

Marian Bubak
Institute of Computer Science AGH, Poland & University of Amsterdam, The Netherlands

Peter M.A. Sloot
University of Amsterdam, The Netherlands

ABSTRACT

Research environments for modern, cross-disciplinary scientific endeavors have to unite multiple users, with varying levels of expertise and roles, along with multitudes of data sources and processing units. The high level of required integration contrasts with the loosely-coupled nature of environments which are appropriate for research. The problem is to support integration of dynamic service-based infrastructures with data sources, tools and users in a way that conserves ubiquity, extensibility and usability. This chapter presents a close examination of related achievements in the field and the description of proposed approach. It shows that integration of loosely-coupled system components with formally-defined vocabularies may fulfill the listed requirements. The authors demonstrate that combining formal representations of domain knowledge with techniques like data integration, semantic annotations and shared vocabularies, enables the development of systems for modern e-Science. For demonstration they present how several semantically-augmented experiments are modeled in the ViroLab virtual laboratory for virology.

INTRODUCTION

Modern research in life sciences that aims at understanding system-level phenomena imposes serious requirements on computer systems designed to support researchers. The growing complexity of scientific issues, combined with the collaborative aspects of virtual research environments, demands that developed systems be more aware of the modeled knowledge domain than at any point in the past. As a response, the virtualization and enhancement of human activities with new capabilities offered by cyberinfrastructure, becomes visible in life sciences, particularly in the construction of modern virtual laboratories, problem-solving environments and other types of systems for e-Science. Just like traditional scientific research facilities, such solutions aim at bringing together researchers in certain domains to achieve breakthrough discoveries (Foster, 2006). They also involve equipment and enable the users of this equipment to collaborate using a common language: definitions, terms and procedures. However, in contrast to traditional research facilities, modern virtual laboratories are geographically dispersed and usually integrate people at different institutions, with varying levels of technical and scientific expertise, working simultaneously on collaborative applications.

There is an obvious need for a mechanism that could maintain the integration of human knowledge, tools, data, tasks and results on a level that is understandable for both humans and computer systems. However, the problem is that current approaches to integration are not flexible enough to support loosely-coupled, dynamically-changing environments with multi-type users (i.e. numerous users performing different tasks on the same infrastructure) in a way that preserves a crucial set of features (Gil, 2007). The required features of such an integration mechanism are:

1. Ubiquity, since all resources form a single laboratory in the minds of its users;
2. Extensibility, as both users and their tools evolve together with new scientific challenges;
3. Friendliness for users who would rather communicate with systems using terms of their own domain of science.

In this chapter, basing on a thorough revision of related research endeavors and on our own experiences in the life sciences domain, we show that a system design methodology based heavily on domain semantics may provide a sound tool for building and operating modern e-Science systems. Moreover, we argue that such a system would support easier content creation (domain-specific services, in-silico experiments, data sources).

The core idea is to apply domain knowledge as an omnipresent integration glue that not only helps users interact and understand the system, but also (and this is where the novelty lies) interweaves interaction of distinct system modules. We consider a semantic layer, integrated with a system on the middleware level. As (Gardner, 2005, p.1003-1004) claims, a well devised set of ontologies may form a glue-like middle layer that integrates all possible data sources of a system (or of an organization, for that matter). The work also postulates (p.1007) that this kind of semantic-based middleware could be the future of knowledge-driven organizations (such as pharmaceutical companies) which try to overcome their intensive R&D data growth. We argue that a very similar approach may be successfully extended not only to data-related assets of an organization, but also to services, tools, messages, user interfaces and any elements of IT systems which deal with semantically-enriched data.

The innovation we propose comes as an evolution rather than a revolution of the general trend in semantic technologies. A thorough survey, executed by (Cardoso, 2007), indicates arising interest in semantic (web) technologies from early adopters, both in academia and industry. The entire field of ontological modeling and semantic systems integration is already shifting its focus from research to real market applications. As will be shown in this chapter, life sciences are no exception to this trend and are, in fact, poised to assume a leading role in the uptake of semantic technologies.

The following section provides an overview of the background of this research, including a discussion on related works and general trends in e-Science advancements. This is followed by Section 3, which describes (both on a general, conceptual level and in a more detailed manner) the approach we propose and implement for the stated objective. Further on, we provide a brief discussion on possible future trends in this research field (Section 4). The chapter is concluded with a summary wrap-up section (Section 5).

BACKGROUND

The Challenging Problem of System-Level Science

The challenges of scale that modern system-level science poses for information systems include the heterogeneity of data sources, analysis tools, supporting services and people. All these entities need to collaborate in a single, integrated (though highly distributed) installation (Foster, 2006, p.32). The emerging virtual laboratories that try to fulfill these demands tend to be complex and difficult to design. Moreover, the collaborative aspect of these solutions revolves around the notion of joining many users of a single class (e.g. many programmers who use the same component catalog or many medical doctors who share the same medical publication database). The collaborative research systems of tomorrow need to join multiple types of users, with various levels of expertise – for instance data administrators, tool programmers and simulation results interpretation teams assessing the likely scenarios of flood propagation (Hluchy, 2007) or potential ramifications of dynamically-changing weather patterns (Droegemeier, 2005).

Recently, we have observed the application of broadly-understood semantics to different tools and subsystems. The sources of these semantics were various: metadata provided by users, predefined ontological models, semiautomatically-extracted knowledge etc. While the exact origin of the knowledge modeled in the system is important, we prefer to focus on semantic exploitation. Strictly speaking, this section tries to overview modules, subsystems, tools and methods that have been "semantically augmented" with success.

Semantic Data and Services Integration

When considering modern semantics-driven systems, one has to devote special attention to mechanisms and procedures of enriching system assets with domain-related knowledge. In other words, there has to be a method by which the objects' (data, tools, modules, documents etc.) meaning is inserted into the system – not only to make it visible to human users but also (and most of all) to make it usable by artificial algorithms (such as semantic discovery, sorting, presentation, layout and hundreds of other potential applications). For our purposes, we call this procedure semantics injection.

Recent research on semantic integration has directed substantial attention to the issue of data integration. One of the most visible obstacles for further advancement in sciences (with life sciences being no exception here) is not only the sheer amount of data required for meaningful analysis – this is usually handled with modern instruments and data filtering/storage facilities – but most of all the variability and heterogeneity of different data sources that must be combined to achieve results. Semantic data integration is a sound and proven approach to this issue (see for instance (Cannataro, 2007), which described integration of medical instrument readings). An example of a successfully-designed and deployed data integration mechanism for life sciences is also reported by (Kondylakis, 2007). The authors combine two crucial methods – semantic data source annotation and semantics-based dynamic query augmentation – to deliver a system that combines heterogeneous data sources into an integrated database-like store.

An earlier work, (Bergamaschi, 2001, pp.218-220), proposes another method for integration of heterogeneous data sources: object patterns. These patterns are derived from semi-structured (meaning self-described – to a certain extent) data, by grouping available (usually repeating) labels of specific types of objects. An object pattern defines what attributes a particular concept carries and which ones are optional. Thus, it provides a basis for further translations and common schema development.

Another notion, somewhat parallel to the data integration idea, is semantic description of software elements that act on the data, usually to analyze and transform it for scientific purposes. An example of an activity that tries to provide semantic integration and interoperability among services is Semantic MOBY (Gordon, 2007). Its aim is to develop an easy-to-adopt mechanism for bioinformatics to integrate various tools and resources. It also provides an extension to prepare workflows based on semantic search capabilities. The main idea here is to have the semantic information attached to a service so that users are able to find and use the tool easier. An assistance tool may be designed to help the user properly and effectively combine available, semantically-described services into a workflow of data analysis and transformation (Bubak, 2005).

This area of semantic binding and annotation with respect to services is further explored in (Corcho, 2006) where authors propose the Semantic Open Grid Service Architecture that defines the semantic binding between a resource and its metadata as a separate entity which is the subject of storage, browsing and management. This introduces a new mediation point between augmented systems (tools, services) and the meta-information with which the systems are enriched – no injection technique is required here, since all the added information is stored elsewhere. This may also prove to be an important proposition in semantics-based system-to-system interaction.

Semantic Metadata and Annotations

From the above observations it is clear that significant advancements have occurred both in ontological modeling of entire laboratories and semantic description of external resources of an organization (Kryza, 2008). However, the question remains: what happens with these assets when and after they are used by complex middleware, employed by the organization? Does the information flow inside the system still carry over the additional information? Are the results produced by the tools and procedures inside the environment still described in a way that human users (and, possibly, other external entities) can understand? Below, we try to summarize the current state of research on this problem, i.e. on semantics injection inside the system.

One established method of injection of human knowledge into a system is through annotations. Let us define a semantic annotation as a piece of (structured or not) human-readable information that (1) constitutes a metadata description of the annotation subject and (2) relates to the knowledge domain of the reality modeled by the system. It is a simple and perhaps most widely used definition of the term (Kiryakov, 2004, p.54) (Bechhofer, 2002).

(Uren, 2006) provides a well-rendered and quite complete review of annotation techniques and approaches to knowledge management. The work is structured in a very pragmatic fashion: a list of core requirements of an "ideal semantic annotation infrastructure" is contrasted and reevaluated by a thorough analysis of available annotation creation and management solutions. While this point of view (in fact, very well represented in literature) captures what is probably the most technologically demanding side of the problem, much less attention is given to another issue: what tools exist to effectively make use of annotations and how to plan future semantics-driven information systems so as to maximize the value of such an (arguably very expensive) asset as annotations (particularly human-authored ones). In this work we try to focus on this part of the problem.

It is only when full attention is given to annotation (or, more generally, metadata) producers and consumers that a true *annotation economy* for metadata may evolve (Kiryakov, 2004, p.50). Proper, effective consumption of semantic annotations (provided either by humans or artificial entities) is not a trivial notion by any means: one needs to deal with important qualities such as disambiguation of meaning, detection of contradictions, proper context-wise interpretation, etc. In (Uren, 2006, p.16) the authors also acknowledge the need for careful annotation and continuous maintenance as the two main sources of expenses. Their prohibitive cost may render this approach to semantics injection too cumbersome and expensive for the industry to embrace.

It has to be noted that there are also other research initiatives, which deal semiautomation of the expensive process of annotation provision. An example of such a semiautomatic approach given by (Handschuh, 2002) is based on a discourse between an annotator and an artificial classification program. The latter one uses learning techniques: it may be trained for fully-automated information extraction. (Heath, 2008, pp.5-6) presents an example of a system where tagging-based classification, originating from human perception of a subject, is augmented with a past tagging record insight. The system uses both syntactic similarity detection and relation-based synonym notation. Heath's mechanism leads to better term reuse, stronger correlation of knowledge derived from users and, ultimately, dynamic emergence of taxonomy-like classifications. This approach overcomes notable shortcomings of the predefined taxonomy approach: inflexibility and immutability. Here, the classification bends to the described domain and not the other way around.

From Single Tool Augmentation to Complex Systems Integration

As one may clearly see, modern computer science methods are advanced enough to provide means for knowledge to be inserted, merged and reused inside a distributed system. There are ways of transcribing such knowledge into proper models and relating first-class organizational assets to this model. Solutions have been proposed to dynamically inject (in the form of ad-hoc metadata) further knowledge and expertise into a system where predefined semantic information is not present (for instance due to the fact that the annotated content is created inside the system).

What we observe (and what is the main thesis of this chapter, backed by our own research presented in the following section), is that the design and development of complex, distributed systems for life sciences (and, in fact, for e-Science in general) becomes more and more driven by a semantic-based integration methodology. By a semantic-based integration methodology we understand a usage of formally-modeled domain knowledge to augment both the static structure and the dynamic behavior of the designed system. Future "collaboratories" designed to fulfill yet unmet requirements of system-level science applications will use semantics as a cornerstone of their architecture, both for internal mechanisms and for user or external system interactions alike.

SEMANTICS AS AN INTEGRATION MECHANISM FOR COLLABORATIVE ENVIRONMENTS

Challenges

Before going into detail on the specifics of our solution, let us consider the application domain. The case we apply the method to is supporting science and medical care for HIV research and treatment centers (Sloot, 2006). The main objective of the project is to set up human collaboration assisted with a modern virtual laboratory environment to propel future joint research and discoveries in HIV virus resistance and treatment strategies for infected patients (Gubala, Balis, 2007). This objective poses several interesting challenges that we sum up as follows:

1. Joining many users with different backgrounds and roles inside one, integrated and cooperative environment (task-oriented design);
2. Combining heterogeneous data sources with a set of remote computation tools to deliver flexibility and dynamism into the virtual experimentation process;
3. Tracking complex, distributed execution of processes to gain (and retain) information on provenance of scientific results.

To correctly distinguish roles and tasks of distinct members of the virtual laboratory we have identified four classes of human users. Experiment developers plan in-silico experiments and decide what resources these will use and what their results are. Scientists use the experiment plans prepared by developers and execute them to produce real scientific results. The results are subject to storing, sharing, discussing and publishing. The third group consists of final "consumers" of the knowledge acquired by scientists through experimentation – in the case of ViroLab, these people are clinical virologists who use the decision support system to help them apply appropriate drugs (the support system, in turn, uses input provided by scientists). The fourth group includes administrators who monitor the entire installation, add new resources and react to failures. All these groups cooperate in the framework of a so-called experiment pipeline – a multidirectional process of planning, developing and using virtual experiments to acquire and apply scientific results (de Roure, 2007). It is, however, important to note that although all these groups differ with respect to levels expertise and tasks, their efforts need to be combined to achieve the main objective of the project. First of all, let us consider the top-level (conceptual) design of the system and distinction between what we call horizontal and vertical integration (regarding both internal subsystems and modules).

Conceptual View

The solution is built around several interconnected layers, which together deliver both horizontal and vertical integration (see Figure 1). The lowest level of logical integration involves the identification mechanism. Every entity of the environment is uniquely identifiable, for instance by an ID number assigned by middleware upon the birth of the entity (for volatiles, such as experiment result data) or is set by an administrator (for servers and services). Since the lowest-level resources are inactive by definition (they are instead acted upon by users and their tools), horizontal integration on this level is purely logical. It is important to understand that there is no strong uniformity constraint that may impair the extensibility of the solution – the only requirement is that the modules or tools that act upon a certain type of entity are able to comprehend its identification schema. The design of this level is dictated mainly by the infrastructure technology used to create the identified resources.

The middle layer of integration is delivered by a metadata mechanism. Since we have unique

Figure 1. Conceptual view of different integration layers: passive objects (resources), middleware services and user tools

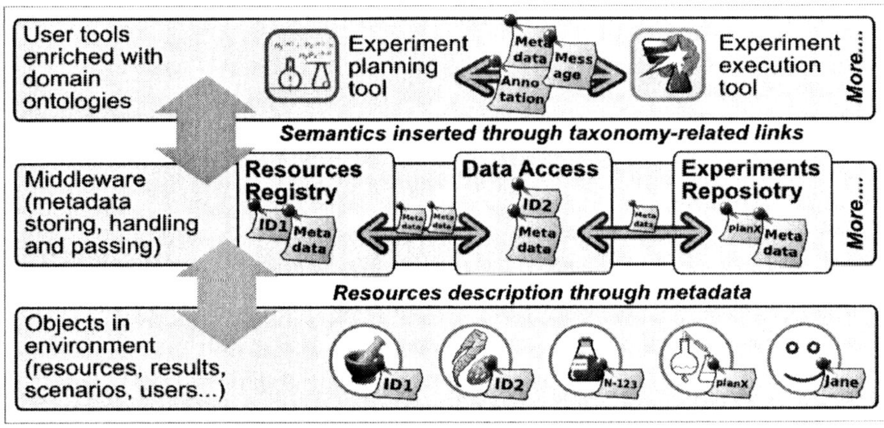

identification in place, all that is identifiable may be described. Descriptions, either provided by human users (i.e. annotations), or automatically created by a middleware module, are located in a dedicated metadata storage mechanism. Different stores can be used for different kinds of metadata: for instance a service registry for computational resources (Kasztelnik, 2007), an experiment repository for in-silico experiments or a result metadata store for the findings of scientists who use the virtual laboratory. Once a set of describing modules is in place, their interoperability is enabled with a module interaction mechanism – for instance using some message passing technology. As identification remains unique throughout the entire environment and all middleware modules are able to participate in the exchange of metadata, the environment is horizontally integrated in a physical sense. Again, there is no need for every element to understand all kinds of metadata – only those that are required for the task performed by a given element (and elements are in turn reused for various user tasks). The design of this layer is strongly dependent on the specific roles of modules of the virtual laboratory and may be technology-agnostic: information carried by metadata descriptions and annotations can be transferred via any kind of communication mechanism. With this layer in place, the system is ready for proper insertion of semantics.

The upper layer of integration that constitutes the presented solution is the semantic vocabulary layer. This level is strongly connected to the end users of the system and the tools they use. The main goal here is to overlay the lower levels and allow users to communicate with the system using the notions they are accustomed to. The proposed mechanism involves ontological taxonomies that form formally-transcribed vocabularies and naming conventions, and are complex enough to model human knowledge in a computer-legible format (Fensel, 2003). Ontological taxonomies also deliver the ability to evolve – they can be refined, republished and extended at runtime.

Ontological Model

Our approach exploits something that (Wache, 2001) calls the hybrid model for ontology-based integration (p.3). This involves a single topmost vocabulary of the most basic yet at the same time most pervasive terms, characterizing the domain of the system (see Figure 2). The crucial role of this most basic layer is twofold. First of all, it gathers all the root concepts and relations that constitute the main basis of inter-system integration (or "agreement"). On the other hand (from the technical point of view), it serves as an abstraction base that developers of separate components or protocols may extend for new purposes without sacrificing generality. The fact that an evolved, hybrid ontological integration model may be successfully applied to life sciences – see (Fox, McGuinness, 2007, p.5) for a similar conclusion about earth sciences – signifies the potential similarity between inter-system and inter-component solutions.

This hybrid methodology, based mainly on fine decomposition of the vocabulary body followed by incremental purpose-based extensions, is also more adaptive. (Rosenthal, 2004, p.47) uses examples from the military ontology field that a very small schema of well-devised terms and relations (attributes) is much more acceptable for external systems then any holistic approach. This is especially important for future integrations with external systems (or, similarly, with new components within the same system) – the easier the adoption of an information exchange schema, the easier the system-to-system integration. This characteristic is of one the main factors shaping open system design.

We assume the existence of a main ontological model of the application domain (namely virology) without getting into details of the process of formulation of this model. Nevertheless, it is worth mentioning that in the general case it is clearly a difficult task to initiate and maintain the process of ontological conceptualization of a knowledge area. As there are no well-established methodologies of taxonomy and terminology definition for life sciences, ad-hoc or informal methods are frequently used (e.g. see (Fox, McGuinness, 2007, p.2)). In the case of the ViroLab project, we have established a small multi-disciplinary team that gathered preexisting data entity relationship diagrams as a source of domain data taxonomy. The diagrams were extracted into OWL classes and relations. At the same time, the domain-agnostic parts of the hybrid model (such as the experiment model or the upper ontology model shown in Figure 2) were designed from scratch by developers of the virtual laboratory.

Vertical and Horizontal Integration for Task-Oriented Processing

Going back to the layered conceptual model (Figure 1), we may see that horizontal integration on the level of semantics is based on the fact that tools use integrated domain ontologies: while the exact models and taxonomies may differ (since the purpose of each given tool is specific), all are submerged in a more general (upper) domain ontology. For instance, a service browser tool may use semantic data type taxonomy to enable semantic service search, while an experiment execution history tracking tool uses the taxonomy activity types. Both taxonomies are, however, connected with semantic links that bind activities and data together. As the nature of these semantic links is defined by the upper, basic vocabulary, this ensures that both the user who browses the

Figure 2. Purpose-based separation of taxonomy models: basic models are extended to cover both application domains and task-related taxonomies

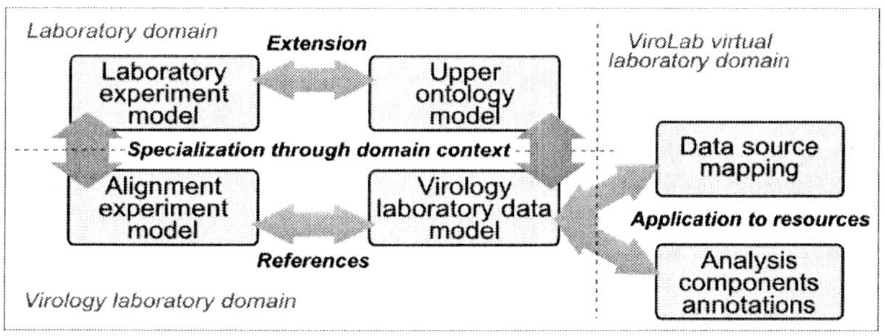

available services and another user who learns about services used in past experiments share a common understanding of the basic domain terms (and their associated tools) and reuse semantically equivalent data (Ciepiela, 2007).

Vertical integration, starting with basic resources and ending with human users, is accomplished by the laboratory middleware. The resources are identified (named) and described by description modules. The tools that use middleware modules to perform specific tasks on behalf of users are able to comprehend the descriptions and are also able to translate (or relate) the information in these descriptions to the domain ontologies they use. In this way, information propagates to the end users in a form that is understandable to them and to all their collaborators. However, if a user wishes to query for more technical, detailed information on some subject, the tools are also able to provide it. Additional facts may be acquired by users who express them through further semantic annotations – this leads to information and workspace evolution.

To explain how this conceptual view was realized and how the mechanisms we devised help build a semantics-aware collaborative environment, we will consider some processes (tasks) taking place in the virtual laboratory.

One of the leading challenges of the ViroLab project (see Section 3.1) is the integration of various data sources with different analysis software components to provide a flexible framework for formulation of *in-silico* experiments. The heterogeneity is twofold: not only do data sources rely on different storage structures, but the data processing units themselves are developed using different technologies. The solution (Gubala, Kasztelnik, 2007) involves middleware components that ensure interoperation of different data sources with various remote computation platforms. Figure 3 shows how semantic modeling helps in this process. The Experiment Planning Tool (the main workbench for experiment developers) fetches a

Figure 3. Semantic query formulation: the domain data model combined with concept-to-data mapping allows build queries using generic concepts for specific data sources

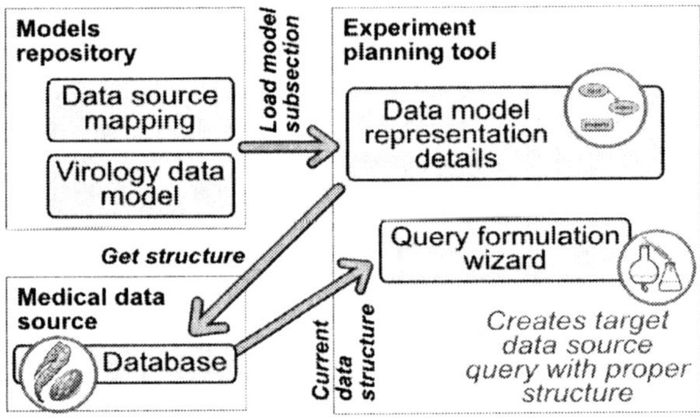

proper ontological submodel that defines domain data vocabularies (terms such as patient, viral isolate, sequence mutation etc.) and also contains mappings of these concepts onto real data sources structures (such as tables in a relational database or markup elements in an XML document store). When the user chooses the concept to query for, the tool retrieves (on an ad-hoc basis) the current structure of the data source that actually provides data related to this concept (for instance records of patients, nucleotide sequences of virus mutations). The mapping information is used for exact identification of data sources, hosts, tables etc. as well as for finding which parts of the source (tables, columns) represent the specified ontological concepts. By applying a source-specific mapping for each database, the user only works with the conceptual data model and the technology uses the mappings to perform queries. The mechanism operates on the description level (it only fetches data schema information, not the actual records themselves) as the tool is used at the experiment *design* time (not at runtime). Therefore, there are no scalability or efficiency issues to consider. This is an example of vertical integration in action: from ontological taxonomy, through mapping annotations, down to resource identifiers.

Use example. A virology experiment developer wants to perform some studies over HIV-infected patients in Italy, in particular on the genotype of their viruses (due to the high mutation factor, the HIV genotype is slightly different in each patient). While editing the experiment body, he opens the query formulation tool. The tool loads the ViroLab data taxonomy model (located in an RDF server), from which the developer chooses the term *Patient*. The mapping files are loaded (also formulated in RDF) and the tool follows the *mappedTo* relation to find out that there are two mappings to hospitals in Italy (partners from Rome and from Brescia are involved in the project), both of which store patient data in a single table named *Patient*. However, in the case of the database in Rome, the *NucleotideSequence* is mapped to the *nq_seq* column and in the other case, the column is called *sequences*. The tool allows the developer to choose which database (out of the two) he wants to query and then it creates a query stub with the correct database, table and column names (hence the developer does not have to remember this information by heart).

Another example of a semantically-enriched process inside the virtual laboratory is shown in Figure 4.

This process is, to some extent, analogous to the previous one. Although it deals with services, it may be seen as parallel to data management. The data model of the virology domain is fetched by the experiment planning tool where the developer may choose what type of data should be processed by an experiment. The following lookup procedure for the semantically-enriched component registry allows the system to propose a set of suitable candidates to take part in the newly created experiment. (By "component" we understand any remotely published software module that the middleware is able to contact and that may be asked to perform some well-defined processing on behalf of an executed experiment – currently web services, stateful remote services, WSRF, EGEE jobs and distributed components (Malawski, 2008) are supported). Now, the developer may decide which remote computation will best suit the user needs and includes its invocation into the body of the experiment plan – this model implements the concept of the Grid, understood as computation on distributed resources that crosses different administrative domains. Again, we go through all the layers of integration: from domain terms in the ontological data model, through semantic annotations of components up to remote endpoint identifiers inside the experiment plan.

Figure 4. Remote computation invocation: an application of the conceptual layer (represented in taxonomy) for easier component browsing

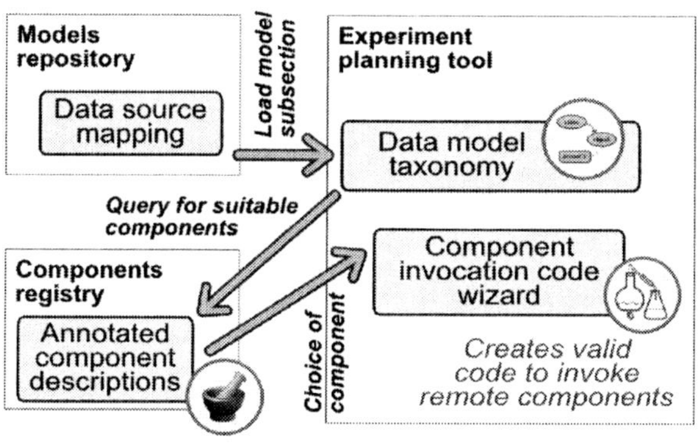

Use example. The experiment being designed is now able to extract HIV nucleotide sequences of Italian patients. One of the most frequent types of analysis in this case is the detection of mutations of the particular HIV viruses. The developer uses the same tool to load the ViroLab data model but this time he chooses the *NucleotideSequence* concept and performs a remote computation search. Specifically, he asks for all the processing components that are able to consume virus genotype sequences as their input. From among a set of findings, he chooses the alignment service from Belgium (by a partner in Leuven who develops virology tools). The tool inserts proper invocation details into the experiment body to automate the process.

Finally, we present a scenario for the experiment provenance tracking facility (as stated in

Section 3.1, this scenario realizes one of the main challenges of ViroLab). This time we consider the point of view of a virologist who uses defined experiment plan to perform some experimentation – see Figure 5. Using the experiment execution tool (accessed through a web interface, omitted in the figure for clarity) the person runs the experiment process. This, apart from usual in-situ processing, may include distributed computation and remote data access (both mechanisms were briefly described before) and, as stated, both types of resources carry semantic annotations that bind them to domain ontologies. Logging-like messages are generated by the runtime component and relayed (through a distributed monitoring middleware) to the Provenance tracking subsystem (Balis, 2008). The experiment execution tool takes care of proper annotation of these messages – full automation is possible because each message concerns specific activities performed on resources that are already annotated. Furthermore, the activities are unambiguously defined in the ontological experiment model – thus a "semantic agreement" between parties is maintained. An example of an event could be a superposition of the *DataRetrievalFinished* activity concept from the general experiment ontology model and the *ViralIsolate* data concept from the domain-specific (virology) data model. The provenance module uses the same experiment model with

Figure 5. Joint case of experiment execution and provenance information querying: at runtime concept-related events are reported to the aggregation store

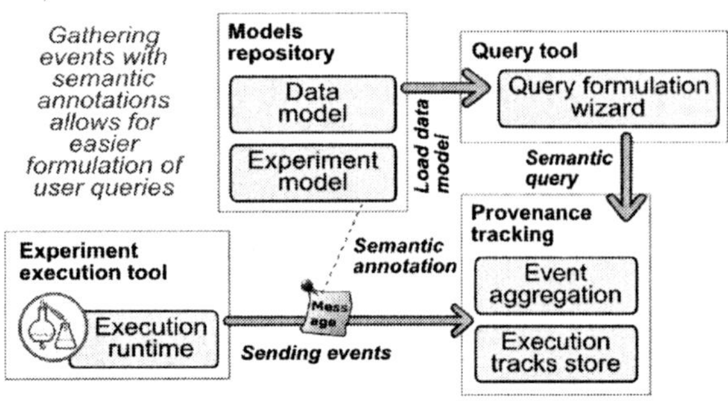

internal aggregation rules to effectively store and infer knowledge on the history of a certain experiment run.

Use example. As can be recalled, the developer designed an experiment which fetches HIV genotype information from a database in Italy and detects its mutations with an alignment tool. Now, the user executes the experiment. During execution, the enactment mechanism takes care of collecting runtime-related events – in this case, one such event is the invocation of the remote alignment tool to detect mutations. The event is of the *invocation* type and associates the identifier of the tool with the *NucleotideMutation* concept. Once the experiment finishes, another person may use a dedicated provenance browser to query for (or to browse through) all the recent experiments which involve *NucleotideMutation* as one of their results. This way, the user finds out that the mutations returned in this particular experiment are detected by the tool from Belgium, so he knows whom to cite in case he wishes to publish an article regarding the detected mutations.

The knowledge aggregated in this way may be subsequently used by other collaborators for various purposes. One goal could be to learn how a specific scientific result was obtained: to check what data sources were used and what computational models were applied to achieve a certain result. To this end (see again Figure 5), the interested person uses a query tool (Balis, 2007) to find out facts on experiment execution history. In order to sustain a conversation with the user that is understandable for both peers, the tool loads the data model taxonomy from the Models store – the human user constructs the query in the familiar domain terminology. The query is then dispatched to the provenance tracking module where the mining process is performed on gathered (and aggregated) information. The findings are returned to the query tool for presentation. In this more complex scenario one may see how a properly-used semantic model may serve as:

- A terminology model of a human-to-system interface (conversation);
- An established agreement for an internal protocol used between different modules of the system;
- An integration point between different use scenarios: different users perform distinct tasks in the semantically-integrated environment.

The above examples were chosen for two reasons: they present our point of view on the discussed matter and they are already implemented in the virtual laboratory of the ViroLab project (their current status may be checked on the virolab.cyfronet.pl website). Having analyzed the related work (Section 2) and following presentation of our own research in the field, we try to draw some conclusions on what the future may bring in terms of applying semantics to complex experimentation environments for life sciences.

FUTURE TRENDS

It is clear that semantically-enriched solutions are already established in the area. The emergence of solid and stable standards for transcription of knowledge, using formally-defined vocabularies and taxonomies, allows interoperability in a broader sense than previously possible. Such abstract models of conceptualization are now beginning to serve as the basis for agreements between people and artifacts in a common environment. The trend for formalized transcription of human knowledge, while already present, may be further strengthened by introduction of better, more sophisticated methodologies of collaborative ontological modeling, management, versioning and publication (Rosenthal, 2004, p.49). There is a broad spectrum of works on inter-system integration based on semantics. This area presents its own unique challenges, related not only to inter-system communication, but also to mapping of concepts and categories between different expressions of a similar domain of discourse. Both (Noy, 2004) and (Kalfoglou, 2003) present a good overview of methods proposed for overcoming obstacles in semantic inter-system integration (the second work stresses the so-called ontology mapping, or alignment, methods). As stated before, in this work we focus on the concept of internal integration of different layers, components and roles within a single modern, distributed system.

As discussed in Section 3, establishing a semantic background for the system may help design user interfaces in a more human-friendly way. However, further extensions towards semantics-driven web interfaces and tools are not only possible but also strongly pursued. Frameworks and methodologies for setting up user interfaces with semantic, and possibly dispersed, knowledge sources are already available in their early forms (Oren, 2007). We will inevitably witness the emergence of further technologies to integrate ontologies, vocabularies and web development programming tools – this is also one of main targets of Semantic Web research. Moreover, this trend will probably be only strengthened by the application of modern, content-rich Web 2.0 techniques for better over-the-web interactivity that is so important for researchers (Fox, Guha, 2007).

Another trend in building modern collaborative environments for system-level science that is arising is the gradual application of Grid computing solutions. Grid research currently focuses on semantic enrichment of middleware components (De Roure, 2003). This allows us to envision a shorter path to true interoperability of Grid computation frameworks with PSE systems designed for life sciences. Here, again, we see the semantic layer as a possible medium of agreement between these two parts that will eventually meld into virtual laboratories of tomorrow. Regardless of how computer scientists will name this kind of processing paradigm, the so-called Grid or cyber-infrastructure is already there and employing it as one of cornerstones of in-silico experimentation systems becomes a more common practice (in slight contrast to industrial applications, where demand for and assimilation of Grid solutions seem to be more grudging (Forge, 2006)).

Finally, trying to draw a slightly larger horizon, one may consider the possibility of inclusion of formally-modeled domain knowledge into the very heart of the distributed system design process. Similarly to the way developers use system modeling languages, like UML, one may envision ontologically-transcribed vocabularies and taxonomies as the main building block for system design. This may prove especially true for analyzing and developing protocols, interfaces and modes of conversation between subsystems of future frameworks. This approach may also blur the distinction between internal system integration mechanisms and external, system-to-system interoperability channels – the distinction that remains quite sharp in current solutions. Parallel research on service-oriented architectures (SOA) and computing (Huhns, 2005) may also lead system designers in the direction of fully open systems.

There are certainly considerable limitations to any technology based on the semantic (and, in particular, ontological) approach. From among them, the most important ones are:

- **Management issues:** As the semantic description logic stays on the metadata (or meta-resource) level, there is always the problem of aligning the resources and their descriptions (i.e., reflecting changes in the environment with proper updates of the description),
- **Creation issues:** The correct development of an ontological model is not easy to perform: even though the vocabulary and taxonomy design is relatively easy, the establishment and maintenance of a wide consensus over such a conceptualization remains an organizational challenge.

Such problems need to be solved (at least to some acceptable extent) so that semantic (web) technology becomes more applicable.

CONCLUSION

In recent years we have observed a convergence of two interesting trends that emerged in distributed systems for life sciences. First of all, since the modeled domains are complex by nature and involve combinations of various data sources with processing methods to achieve breakthroughs, the related PSE (Problem Solving Environment) or virtual laboratory systems, tend to be equally complex. On the other hand, we are witnessing a focal shift in approaches to supporting user tasks. Previous solutions were usually contained in a single tool, aimed for one type of activity. In spite of their limited usability, these mostly proof-of-concept realizations paved the way for future solutions that support multiple roles of multiple classes of users in a single, integrated environment. This challenge, in turn, brings high interoperability demands and requires tight logical integration in otherwise (by nature) loosely coupled systems. Both these converging trends, together with the observations resulting from ongoing research, indicate an arising need (and an opportunity as well) to apply semantic integration techniques on a broader scale than has been the case thus far.

In this chapter we supported this claim with both an overview of current research initiatives in the field and an example of our own results. The solution we propose is based on ubiquitous use of domain knowledge in a formalized form that is understandable for both humans and artificial entities. While many authors regard the process of building ontological models for selected science domains very tedious and difficult, this stands in stark contradiction with the fact that those expensive assets are applied to a single mode of system use (e.g. service search, query augmentation, security policies definition etc.) Instead of applying ontological models to a single task or tool in a system, we propose treating these high-value assets as one of main building blocks of future "collaboratories" for e-Science. In our case, we have observed successful reuse of the same domain knowledge model in various tasks performed by different classes of users in the same collaborative environment for virology. This enabled more effective use of applications such as the Drug Ranking System, which uses dispersed medical databases and rule-based decision support computation service to help clinicians treat HIV-positive patients more effectively. Examples of other experiments that were semantically enabled with the methodology we describe include patient data mining, HIV virus genotype analysis and molecular dynamics-based HIV protein behavior simulations (currently in progress). All these cases show that the approach based on semantic system integration may be both effective and rewarding. Therefore, we claim that devoting more attention to this direction of research may yield more fruitful achievements in future computer systems dedicated to life sciences.

ACKNOWLEDGMENT

This work was partially funded by the European Commission project ViroLab IST-027446 and the related Polish grant SPUB-M and ACC Cyfronet AGH grant 500-08. The authors are grateful to Piotr Nowakowski for his valuable comments and suggestions.

REFERENCES

Balis, B., Bubak, M., & Wach, K. (2007). User-Oriented Querying over Repositories of Data and Provenance. In *Proceedings of the 3rd IEEE International Conference on e-Science and Grid Computing, e-Science 2007*, Bangalore, India, Dec. 2007, (pp. 187-194). New York: IEEE Computer Society.

Balis, B., Bubak, M., & Wach, K.(2008). Provenance Tracking in the ViroLab Virtual Laboratory. In R. Wyrzykowski, (Ed.), *Proceedings of the Seventh International Conference on Parallel Processing and Applied Mathematics*, LNCS, Gdansk, Poland, Sept. 2007. Berlin, Germany: Springer.

Bechhofer, S., Carr, L., Goble, C., Kampa, S., & Miles-Board, T. (2002). The semantics of semantic annotation. In Meersman, R. & Tari, Z., (Ed.), *On the Move to Meaningful Internet Systems 2002: CoopIS, DOA, and ODBASE*, , London, (pp. 1152-1167). Berlin, Germany: Springer-Verlag.

Bergamaschi, S., Castano, S., Vincini, M., & Beneventano, D. (2001). Semantic integration of heterogeneous information sources. *Data and Knowledge Engineering*, *36*(3), 215-249.

Bubak, M., Gubala, T., Kapalka, M., Malawski, M., and Rycerz, K. (2005). Workflow composer and service registry for grid applications. *Future Generation Computer Systems*, *21*(1), 79-86.

Cannataro, M., Guzzi, P. H., Mazza, T., Tradigo, G., and Veltri, P. (2007). Using ontologies for preprocessing and mining spectra data on the grid. *Future Generation Computer Systems*, *23*(1), 55-60.

Cardoso, J. (2007). The semantic web vision: Where are we? *IEEE Intelligent Systems*, *22*(5), 84-88.

Ciepiela, E., Kocot, J., Gubala, T., Malawski, M., Kasztelnik, M., & Bubak, M. (2007). Virtual laboratory engine – gridspace engine. In Bubak, M., Turala, M., and Wiatr, K., (Eds.), *Proceedings of Cracow Grid Workshop - CGW'07*, Krakow, Poland, (pp. 53-58). ACC CYFRONET AGH.

Corcho, O., Alper, P., Kotsiopoulos, I., Missier, P., Bechhofer, S., & Goble, C. (2006). An overview of s-ogsa: A reference semantic grid architecture.

Web Semantics: Science, Services and Agents on the World Wide Web, 4(2), 102-115.

Droegemeier, K. K., Gannon, D., Reed, D., Plale, B., Alameda, J., Baltzer, T., et al. (2005). Service-oriented environments for dynamically interacting with mesoscale weather. *Computing in Science and Engineering, 7*(6), 12-29.

Fensel, D. (2003). *Ontologies: Silver Bullet for Knowledge Management and Electronic Commerce (*2nd Ed.). Berin, Germany: Springer.

Forge, S., & Blackman, C., (2006). *Commercial Exploitation of Grid Technologies and Services. Drivers and Barriers, Business Models and Impacts of Using Free and Open Source Licensing Schemes,* (Final Report by SCF Associates for European Commission DG Information Society and Media). Retrieved March 9, 2008 from IST website http://cordis.europa.eu/ist/grids/pub-report.htm

Foster, I. & Kesselman, C. (2006). Scaling system-level science: Scientific exploration and it implications. *Computer, 39*(11), 31-39.

Fox, G. C., Guha, R., Mcmullen, D. F., Mustacoglu, A. F., Pierce, M. E., Topcu, A. E., & Wild, D. J. (in print, 2007). Web 2.0 for grids and e-science. In *Proceedings of INGRID 2007 - Instrumenting the Grid 2nd International Workshop on Distributed Cooperative Laboratories*, S. Margherita Ligure Portofino, Italy.

Fox, P., McGuinness, D., Raskin, R., and Sinha, K. (2007). A volcano erupts: semantically mediated integration of heterogeneous volcanic and atmospheric data. In Mitra, P., Giles, L. C., & Carr, L., (Eds), *CIMS '07: Proceedings of the ACM first workshop on CyberInfrastructure: information management in eScience*, (pp. 1-6). New York: ACM.

Gardner, S. P. (2005). Ontologies and semantic data integration. *Drug Discovery Today, 10*(14), 1001-1007.

Gil, Y., Deelman, E., Ellisman, M., Fahringer, T., Fox, G., Gannon, D., Goble, C., Livny, M., Moreau, L., & Myers, J. (2007). Examining the challenges of scientific workflows. *IEEE Computer, 40*(12), 24-32.

Gordon, P. M., Trinh, Q., & Sensen, C. W. (2007). Semantic web service provision: a realistic framework for bioinformatics programmers. *Bioinformatics, 23*(9), 1178-1180.

Gubala, T., Balis, B., Malawski, M., Kasztelnik, M., Nowakowski, P., Assel, M., et al. (2007). ViroLab Virtual Laboratory. In Bubak, M., Turala, M., and Wiatr, K., (Eds.), *Proceedings of Cracow Grid Workshop - CGW'07,* (pp. 35-40). Krakow, Poland: ACC CYFRONET AGH

Gubala, T., Kasztelnik, M., Malawski, M., & Bubak, M. (2007). Development and execution of collaborative application on the virolab virtual laboratory. In Bubak, M., Turala, M., and Wiatr, K., editors, *Proceedings of Cracow Grid Workshop - CGW'07,* (pp. 41-46). Krakow, Poland: ACC CYFRONET AGH.

Handschuh, S., Staab, S., & Ciravegna, F. (2002). S-CREAM - semi-automatic creation of metadata. In Gómez-Pérez, A. and Benjamins, R. V., (Eds.), *EKAW '02: Proceedings of the 13th International Conference on Knowledge Engineering and Knowledge Management. Ontologies and the Semantic Web*, London, (pp. 358-372). Berlin, Germany: Springer-Verlag.

Heath, T. & Motta, E. (in press). Ease of interaction plus ease of integration: Combining web2.0 and the semantic web in a reviewing site. *Journal of Web Semantics*, (Special Issue on Web2.0 and the Semantic Web).

Hluchy, L., Habala, O., Babik, M., Laclavik, M., Balogh, Z., & Gatial, E. (2007). Knowledge-based platform for environmental risk management. In Volkert, J., Kranzlmüller, D., & Schreiner, W., (Eds.), *ISPDC '07: Proceedings of the Sixth International Symposium on Parallel and Distributed*

Computing, Washington, DC, (p.1). New York: IEEE Computer Society.

Huhns, M. N. & Singh, M. P. (2005). Service-oriented computing: Key concepts and principles. *IEEE Internet Computing, 9*(1), 75-81

Kalfoglou, Y. & Schorlemmer, M. (2003). Ontology mapping: the state of the art. *The Knowledge Engineering Review, 18*(1), 1-31.

Kasztelnik, M., Bubak, M., Gorka, C., Malawski, M., & Gubala, T. (2007). Fault tolerant grid registry. In Kacsuk, P., Fahringer, T., & Nemeth, Z. (Eds.), *Distributed and Parallel Systems*, Distributed and Parallel Systems, (pp. 135-144). Berlin, Germany: Springer.

Kiryakov, A., Popov, B., Terziev, I., Manov, D., & Ognyanoff, D. (2004). Semantic annotation, indexing, and retrieval. *Web Semantics: Science, Services and Agents on the World Wide Web, 2*(1), 49-79.

Kondylakis, H., Analyti, A., & Plexousakis, D. (2007). *Quete: Ontology-based query system for distributed sources.* (*LNCS* 4690, pp. 359-375).

Kryza, B., Dutka, L., Slota, R., & Kitowski, J. (2008). Supporting management of dynamic virtual organizations in the grid through contracts. In Bubak, M., Turala, M., and Wiatr, K., (Eds.), *Proceedings of Cracow'07 Grid Workshop*, (pp. 140-147). Krakow, Poland: ACC CYFRONET AGH.

Malawski, M., Gubala, T., Kasztelnik, M., Bartynski, T., Bubak, M., Baude, F., & Henrio, L. (2008). High-level scripting approach for building component-based applications on the grid. In *CoreGRID Workshop on Grid Programming Model Grid and P2P Systems Architecture Grid Systems, Tools and Environments*, Heraklion, Crete. Berlin, Germany: Springer.

Noy, N. F. (2004). Semantic integration: A survey of ontology-based approaches. *SIGMOD Record, 33*(4), 65-70.

Oren, E., Haller, A., Hauswirth, M., Heitmann, B., Decker, S., & Mesnage, C. (2007). A flexible integration framework for semantic web 2.0 applications. *IEEE Software, 24*(5), 64-71.

Rosenthal, A., Seligman, L., & Renner, S. (2004). From semantic integration to semantics management: Case studies and a way forward. *SIGMOD Record, 33*(4), 44-50.

De Roure, D., Jennings, N.R., & Shadbolt, N. (2003): The Semantic Grid: A future e-Science infrastructure. In *Grid Computing - Making the Global Infrastructure a Reality,* (pp. 437-470). Hoboken, NJ: John Wiley and Sons Ltd.

De Roure, D., & Frey, J. (2007). Three Perspectives on Collaborative Knowledge Acquisition in e-Science. In *Proceedings of Workshop on Semantic Web for Collaborative Knowledge Acquisition* (SWeCKa) 2007, Hyderabad, India.

Sloot, P. M. A., Ramos, A. T., Altintas, I., Bubak, M., and Boucher, C. (2006). From molecule to man: Decision support in individualized e-health. *Computer, 39*(11), 40-46.

Uren, V., Cimiano, P., Iria, J., Handschuh, S., Vargas-Vera, M., Motta, E., and Ciravegna, F. (2006). Semantic annotation for knowledge management: Requirements and a survey of the state of the art. *Web Semantics: Science, Services and Agents on the World Wide Web, 4*(1), 14-28.

Wache, H., Vogele, T., Visser, U., Stuckenschmidt, H., Schuster, G., Neumann, H., and Hubner, S. (2001). Ontology-based information integration: A survey of existing approaches. In Gomez-Perez, A., Gruninger, M., Uschold, M., & Stuckenschmidt, H., (Eds.), *International Joint Conference on Artificial Intelligence; Workshop: Ontologies and Information Sharing,* (pp. 108-117). New York: ACM.

KEY TERMS AND DEFINITIONS

Annotation Economy is freely used to denote the category which gathers together both producers (be it human or artificial) and consumers (also of both types) of broadly-understood annotations. The term is introduced to describe the future state of knowledge enriched computer systems where annotations, as entities that carry crucial information, will play a first-class role in building and operating systems. (Kiryakov, 2004, p.50)

Collaboratory is a short, handy name for "collaborative laboratory" and means a virtual location or a system where scientific experimental research is performed in a collaborative manner i.e. involving multiple individuals in one experiment.

Ontology is a formalized mean to express some part of reality in a well-organized manner. One of the most important qualities of ontologies is the fact they are legible for artificial entities like computer programs. Since this formalization brings inevitable simplification and classification of a part of the modeled domain, we say ontologies are a form of abstraction through conceptualization.

Semantic Annotation is a piece of (structured or not) human-readable information (which can also be computer-readable) that (1) constitutes a metadata description of the annotation subject and (2) relates to the knowledge domain of the reality modeled by the system (Bechhofer, 2002).

Semantic Data Integration is a mechanism which associates different sources of data on the basis of the meaning of data content. This is usually applied to merging (logically or physically) the content of different (distributed) data sources so that an end user may use all the sources through some unified mechanism. This mechanism sometimes also tackles the problem of data source heterogeneity.

Semantics Injection is any process that aims at attaching some knowledge to artificial entities modeled by some computer system. The knowledge is usually a piece of metadata that relates a virtual element inside the system to its real counterpart in the modeled domain of human knowledge.

Semantic (Resource) Discovery is a process where some entities (documents, web content, services) are browsed or searched with specific consideration of their meaning in a certain domain of human knowledge. For instance, one may semantically search for *"annual management report file"* (in contrast to searching for *".doc file"* which is a pure syntax-based search).

Semantic System Integration is used to descriptively express the methodology of building computer systems where the meaning of modeled entities (data, components, documents, actions etc.) has a significant impact on the way the realized system behaves and operates (also internally).

Semantic Web is a wide-ranging term describing one of the possible future steps of evolution of the World Wide Web. Though definitions of Semantic Web vary, they usually involve enriching documents and tools that form the web with domain-related information so their semantics is not only obvious to human users but is also comprehensible for other computer programs.

Task-Oriented Design is a descriptive term for a methodology of building computer systems which stresses the various tasks that users would like to perform inside the systems and then follows this separation during the development phase. While task-oriented design is not formally defined, the term is typically used to describe the approach itself, rather that any specific kind of design methodology.

Chapter XXVII
Virtual Laboratory for Collaborative Applications

Marian Bubak
Institute of Computer Science AGH, Poland & University of Amsterdam, The Netherlands

Maciej Malawski
Institute of Computer Science AGH, Poland

Tomasz Gubala
University of Amsterdam, The Netherlands
ACC CYFRONET AGH, Poland

Marek Kasztelnik
ACC CYFRONET AGH, Poland

Piotr Nowakowski
ACC CYFRONET AGH, Poland

Daniel Harezlak
ACC CYFRONET AGH, Poland

Tomasz Bartynski
ACC CYFRONET AGH, Poland

Joanna Kocot
ACC CYFRONET AGH, Poland

Eryk Ciepiela
ACC CYFRONET AGH, Poland

Wlodzimierz Funika
Institute of Computer Science AGH, Poland

Dariusz Krol
ACC CYFRONET AGH, Poland

Bartosz Balis
Institute of Computer Science AGH, Poland

Matthias Assel
University of Stuttgart, Germany

Alfredo Tirado-Ramos
University of Amsterdam, The Netherlands

ABSTRACT

Advanced research in life sciences calls for new information technology solutions to support complex, collaborative computer simulations and result analysis. This chapter presents the ViroLab virtual laboratory, which is an integrated system of dedicated tools and services, providing a common space for planning, building, improving and performing in-silico experiments by different groups of users. Within the virtual laboratory collaborative applications are built as experiment plans, using a notation based on the Ruby scripting language. During experiment execution, provenance data is created and stored. The virtual laboratory enables access to distributed, heterogeneous data resources, computational resources in Grid systems, clusters and standalone computers. The process of application development as well as the architecture and functionality of the virtual laboratory are demonstrated using a real-life example from the HIV treatment domain.

INTRODUCTION

Modern practices of science in such area as investigation of HIV virus drug resistance require collaborative sharing, processing and analysis of virological, immunological, clinical and experimental data, as well as advanced tools for (bio)statistical analysis, visualization, modeling and simulation (Sloot, 2006). The process of integrating these computational tools and data, which leads to results relevant for the application domain, is called an in-silico experiment. Within the context of a virtual laboratory, experiments should combine data and activities which are available in the distributed Web- and Grid-based infrastructure, and orchestrate them in potentially complex scenarios.

The objective of this chapter is to present the ViroLab Virtual Laboratory (VLvl), which is a platform enabling the development and execution of collaborative applications by a group of scientists with various levels of expertise, located in geographically-distributed institutions. The main goal of our work on VLvl was to provide an environment which would combine the efforts of computer scientists, virology and epidemiology experts, as well as experienced physicians, to support advanced HIV-related research (Bubak, 2007).

As a first scientific objective, we had to decide on the notation for specifying the experiment planning. This notation enables us to describe the experiments in the virtual laboratory, to define all data access and computational steps and to facilitate execution by the runtime system. It was necessary to analyze existing virtual laboratories and scientific workflow systems and to investigate the advantages and disadvantages of the experiment planning notations employed by them. Based on this analysis and on the requirements stemming from the types of experiments to be performed in the virtual laboratory, we decided to propose a new approach to in-silico experiment planning, utilizing a high-level scripting notation.

Once the notation was defined, the second objective was to propose and develop a virtual laboratory system, which would enable collaborative planning and execution of experiments. The architecture should support experiment execution along with distributed data access, and be capable of interfacing with various Web and Grid middleware technologies. To facilitate collaborative work, the system should also be extended with mechanisms for knowledge-based searching and browsing of available services, experiment sharing and reuse facilities, and provenance tracking and querying tools. To support various types of users – ranging from scientific programmers to domain specialists – the virtual laboratory should offer dedicated user interfaces, including Web-based access and a rich development environment. While dealing with confidential data obtained from thousands of treated patients, the system infrastructure should also consider security as a very significant issue and an essential requirement before any clinical studies can be performed on the integrated data resources respectively single patient data sets.

In this chapter, we discuss the challenges when creating a virtual laboratory for in-silico experiments and describe our approach for solving them based on our experience gained from the ViroLab project. The chapter is organized as follows: we begin with an introduction covering the background and related work in the area of virtual laboratories and scientific workflow systems. Next, we present the ViroLab virtual laboratory, emphasizing our approach to experiment planning. This is followed by a detailed description of issues related to various aspects of virtual laboratory and the modules responsible for handling them. The collaborative aspect of the virtual laboratory is also emphasized and followed by application examples. A discussion of future trends and conclusions wraps up the chapter.

BACKGROUND

In this section we begin with definitions of the most important terms we are going to use, which explain how we understand the virtual laboratory and related concepts. Subsequently, we analyze the state of the art and discuss other solutions, which influence the approach we are following in our research.

The **Virtual laboratory** is a set of integrated components, which used together, form a distributed and collaborative space for science. Multiple, geographically-dispersed laboratories and institutes use the virtual laboratory to plan, and perform experiments as well as share their results. The term experiment in this context means a so-called in-silico experiment – that is, a process that combines data and computations in order to yield new knowledge on the subject of an experiment.

Experiment is a process that combines data with a set of activities that act on that data in order to create results. The substrate data required for an experiment may be obtained from multiple resources in various possible forms. The activities may be manual, semi-manual or fully automatic, depending on their nature. No definite restrictions are imposed on the level of experiment complexity: it might be as simple as listing data inside some remote database, or much more complex algorithm, such as a set of simulators combined together to obtain some insight into complex phenomena. Moreover, experiments are not required to involve just a single, local computer – in fact, the power of the virtual laboratory comes from combining multiple distributed resources, even computing centers, dispersed over various geographical locations.

The purpose of the laboratory we present is to support collaborative work of all the people who are effectively involved in any stage of the experimentation process. For our purpose, we refer to this process as the experiment pipeline. The section below explains our view on this subject.

First of all, one has to decide what an experiment is about and how it should proceed. The part of the pipeline concerning design of the future experiment process is called experiment planning. During this phase the user has to decide upon the main subject of the planned experiment, its intended results and the means by which these results should be obtained. Such a detailed description of the experiment results in experiment execution (the middle part of the process). During this phase the user performs the experiment according to the plan developed at the planning stage, using the resources provided to that user within the virtual laboratory. The usual outcome of such execution is the result of the experiment. Since the result itself is of the highest importance for the user, special attention is given to it in the last phase of the pipeline, called result management. Here, the outcome of the experiment may be evaluated, described and stored. Given the strong collaboration aspect of a virtual laboratory, results can also be shared among the users of the laboratory. This stage concludes a single experiment pipeline.

As the virtual laboratory is meant to bring together people with various levels of expertise, we define two main classes of users who take part in the experiment pipeline.

Experiment developer is a person who designs experiments in a specific domain (e.g. virology). First of all, this means that the person is skilled enough to design and describe the way an experiment should precede. Apart from technical skills, the developer also possesses a certain level of domain-related knowledge to understand the nature of the processes the experiment should model – otherwise, the designed experiments will never be valid. However, it is typically not required for a developer to fully comprehend all the data and results of a specific experiment, this is provided through appropriate expert (scientist) assistance in order to evaluate the developer's work. This assistance gives initial requirements and descriptions of the desired experiment, and also feedback on the developed experiment.

Experiment user is any person who runs a previously prepared experiment in order to obtain results. The user may or may not be involved in the process of experiment preparation – in the latter case it is probable that (future) experiment users would support developers with their expert knowledge about the modeled phenomena. The main objective of the experiment user is to obtain valuable results that answer important questions. In the case of ViroLab, we can divide the experiment user class into specific subclasses of scientists and clinicians: the former are actively involved in the process of experiment planning and execution, while the latter are mainly interested in the specifically prepared experiments which they can use for treatment of their patients.

Experiment plan is a recipe that describes the process of experiment execution in the environment of the virtual laboratory. Physically, it is a set of files that, combined together, contain enough information for experiment users and their tools to proceed with experiment execution. The division of the plan into distinct parts is meant mainly to keep it well structured and easier to maintain by experiment developers. An important aspect of the virtualization part inside the laboratory is the fact that experiment plans are subject to storing, versioning, exchange, collaborative design and use etc. Experiment plans are the central focus of the virtual experimental space of the laboratory. The notation in which experiment plans are specified has a crucial influence on the whole operation of the virtual laboratory, both from the end-user and from the internal architecture point of view.

State of the Art

Problem-solving environments and virtual laboratories have been the subject of research and development for many years (Rycerz, 2008). A good overview of dedicated environments supporting development and execution of complex applications in biomedicine is presented in (Aloisio et al., 2008).

A common approach to experiment orchestration is to use one of many scientific workflow systems available for the Grid. This is why during the first phase of the ViroLab project we took a closer look at the existing solutions of virtual laboratories. They can be split into two groups: drag-and-drop workflow composition applications and systems that provide a high-level language which facilitates experiment creation. We also analyzed the methods which support the collaborative aspect of experiment planning and execution.

The Kepler system (Altintas, 2006) belongs to the first of the presented groups. It supports scientific workflow composition by providing a drag-and-drop application which allows non-programmers to create complex experiments. The workflows are represented in the MoML language (an XML-based modeling markup language) which enables building models as clustered graphs of entries. By using this notation the user is able to insert into a complex workflow such technologies as Web and Grid Services, Globus Grid Jobs or GridFTP operations. Additionally, the user can create custom connectors for other technologies. The execution engine, connected with the application, enables execution of workflows directly on the user's machine. Users can collaborate by sharing created workflows or connectors for new technologies, but this has to be done in a traditional fashion, for example by sending created workflows or a piece of software by email.

Another similar solution, the Taverna (Oinn, 2006) virtual laboratory, is a drag-and-drop system created to support exploratory, data-intensive, in-silico experiments in molecular biology. It has a graphical interface which enables users with a limited IT background to create complex workflows using public or private resources and computing power (e.g. Web Services). Similarly to the previously presented virtual laboratory, it uses the XML notation to save information about workflows, but instead of MoML, Simple Conceptual Unified Flow Language (Scufl) notation is used, which constitutes a high-level workflow description facility. This virtual laboratory provides many tools that enable user collaboration, e.g. browsing UDDI and GRIMOIRES registries for computing resources that can be used during workflow creation, URL submission (users can add new services by pointing to a URL containing WSDL files or scavenging local disks for WSDL files). What is more, the myExperiment (http://www.myexperiment.org/) project provides collaboration space for the Taverna workflow designers, allowing exchange of knowledge, experience and workflows. By using this collaboration space users can download existing workflows on their local machines and execute them using the Taverna application.

Triana (Taylor, 2005) is an open-source problem solving environment that combines an intuitive visual interface with powerful data analysis tools. It combines a workflow construction engine with a user-friendly GUI enabling the creation of experiments with minimum effort and no programming skills. Users can use the drag and drop method to insert building blocks into the workspace and connect them. This allows creating complex workflows that can access different remote resources, such as Web Services, Grid jobs, P2PS services, as well as local operations. Triana can be easily integrated with other systems, e.g. for the GriPhyN project a dedicated writer was created that allows generating Pegasus/Condor input files from workflows.

Geodise (Eres, 2005) is a Grid-enabled optimization and design search tool for engineering, which enables easy access to Grid resources such as jobs submission, GridFTP, Condor resources, relational or XML databases, etc. It belongs to the second group of the presented virtual laboratories, delivering a scripting language for experiment creation and, as a result, avoids problems with loops and conditional statements, typically encountered in workflow systems. It delivers many tools that help users explore domain knowledge about existing services (e.g. by searching for services using their semantic descriptions) and, finally, simplifies the process of experiment creation.

Apart from the presented solutions, there are many other systems that assist users during experiment creation and execution – for instance LEAD (Droegemeier, 2005), Pegasus (Deelman, 2004), K-WFGrid (Gubala, 2006) or DataMiningGrid (Stankovski, 2008) etc. Most problem solving environments and virtual laboratories are built on top of scientific workflow systems. Extensions of the expressiveness of their programming models, interoperability, and access to different computing resources is still a subject of research (Gil et al. 2007). In this chapter we present an alternative approach to development of collaborative applications and virtual laboratories, basing on the experience with workflow systems.

VIROLAB VIRTUAL LABORATORY

A New Approach to In-Silico Experiments Planning

As discussed in the introduction, selecting the notation for specifying experiment plans is one of the crucial decisions when designing a virtual laboratory. The notation has to be simple and clear, to enable non-expert programmers to design the plans, and on the other hand, powerful enough to express the potentially complex experiments, including e.g. iterations and conditional execution. Moreover, since the experiments integrate various data access and computational services, there is need for a "glue-type mechanism", e.g. providing local data conversion and computations which are too simple to be delegated to external (Grid) services. The notation should also be suitable for "rapid" development lifecycle of experiment plans, possibly allowing interactive experiment execution.

Having analyzed the existing virtual laboratories and scientific workflow systems, we have concluded that workflow languages, despite such advantages as the possibility of using a graphical notation which is easy to understand for non-expert users, also exhibit some important disadvantages. For more complex experiments, including nested loops and conditional statements, the graphical notation becomes too complex and unclear. Moreover, the syntax may not be flexible enough to express complex workflow patterns and some simple operations, e.g. data conversion, have to be implemented as specialized services.

To overcome the limitations of workflow systems, we have decided to define an experiment plan notation based on a high-level scripting language, namely Ruby. An experiment plan is a Ruby script which features a concise and clear syntax, combined with a full set of control structures, allowing to express experiments of an arbitrary complexity level. We have chosen Ruby because of its object-oriented and flexible syntax and growing popularity in Web development systems, although we considered other interpreted languages as well, e.g. Perl or Python.

Architecture of the Virtual Laboratory

Once the Ruby-based notation for specifying experiment plans was decided upon, the next challenge was to design the whole development and execution system. The resulting virtual laboratory should fulfill the following requirements:

- Provide a system for execution of experiments specified as Ruby scripts;
- Hide the details of underlying Grid infrastructure and allow programming experiments on a high level of abstraction;
- Access distributed and heterogeneous data sources;
- Interface with various Grid middleware technologies for accessing computational resources;
- Provide knowledge-based and semantic-rich mechanisms for browsing and searching the resources which are available for experiment developers;
- Provide user-friendly tools and interfaces for experiment developers and users;
- Facilitate collaboration by providing tools for experiment sharing, reuse and provenance tracking.

When designing the ViroLab virtual laboratory, we have built on the experience gained from such Grid systems as CrossGrid and K-WfGrid (Gubala, 2006). The resulting architecture of the virtual laboratory is presented in Figure 1. Following, we give a brief overview of the modules, while the detailed description of their underlying concepts, functionality and scientific challenges related to their development is left for the subsequent section.

The central module of the virtual laboratory is the GridSpace Engine, which includes an experiment script interpreter based on JRuby, a pure-Java implementation of Ruby. It includes the Computation Access subsystem responsible for interfacing underlying Grid middleware. The Virtual Laboratory is equipped with the Data Access Services (DAS) which acquire data from remote databases located in research institutions and hospitals.

The Grid Resources Registry holds information on which computing elements are available and where. The Experiment Repository stores experiment plans prepared by developers and

Figure 1. Architecture of ViroLab virtual laboratory

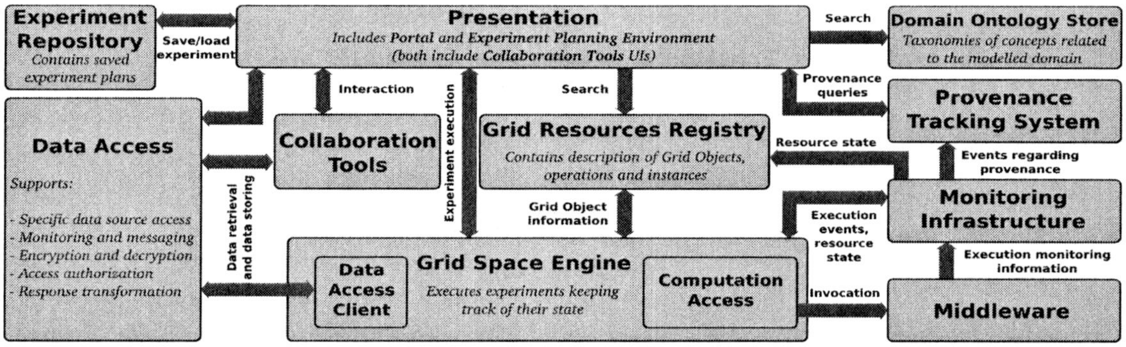

published for scientists. A provenance tracking, recording and publishing system is also provided so that users are able to find past experiments, browse archived data and track origins of certain experiment results.

There are also dedicated tools to help users perform their tasks. The Experiment Planning Environment is an Eclipse-based development tool supporting rapid experiment plan development: it uses the Ruby scripting language as the main development platform for future experiments. Scientists can use their Web browsers to access the Web-based Experiment Management Interface (EMI) where experiments may be loaded and executed, yielding results which can be collected for offline analysis. A set of Web-based collaboration tools that help scientists exchange such results, observations and opinions is also being prepared to enrich the functionality of the environment.

GridSpace Engine and the Access to Computational Resources

As we chose the scripting notation as a way of specifying experiment plans, there was a need to provide mechanisms that would retain a high level of abstraction for experiment developers, without the need to have them interface with specific technologies, middleware and the infrastructure details. The experiment scripts should be able to access computational resources as Ruby objects by simple method invocation. Some low-level decisions, e.g. finding the location of particular computational services or choosing the optimal resource, should be left for the system. However, for more advanced users it may be necessary to provide full control over experiment execution, if needed.

When analyzing the Grid middleware, we have identified four major types of technologies which can be used as the means to provide access to computation. The first one is the Web Services technology, based on SOAP and WSDL, providing stateless remote invocation and messaging semantics. Then there is WSRF, based on extensions to Web Services, providing standardized access to stateful entities (resources). The third supported technology is the component paradigm, which emerged as the research interest of the European CoreGRID project and is implemented by such toolkits as MOCCA (Malawski, 2006) and ProActive. Last but not least, access to computational resources in infrastructures such as EGEE and DEISA requires interfacing with their middleware, which is based on the common batch job processing model.

As a solution to the problem of interfacing diverse technologies, we have introduced a *Grid object* abstraction level hierarchy. Each Grid object class is an abstract entity which defines the operations that can be invoked from within the script. Each class may have multiple implementations, representing the same functionality, but built using different technologies. Moreover, each implementation may have multiple instances, possibly running on different resources and thus differing in performance. Grid objects may have different properties, such as stateless or stateful interaction mode, and they may also be either private or shareable between experiment runs and users.

In order to enable such high-level programming, the virtual laboratory engine, called GridSpace Engine, includes as its core a Grid Operation Invoker (GOI) (Malawski, 2008). Its goal is to instantiate Grid object representatives (proxies) and handle remote operation invocations using the appropriate technologies. GOI provides a set of pluggable adapters for invoking remote operations on Web services and distributed components, whereas for middleware technologies which are based on a job-processing model (e.g. gLite) we offer a programmer-friendly API which can be used to wrap such job-based applications into high-level Grid-objects. A specialized, built-in optimization module (GridSpace Application Optimizer) takes care of optimal load balancing on computational servers.

As an additional facility, a user data input library is provided to enable experiments to interactively request input from the user, which is especially desired in the case of decision support experiments. Furthermore, in order to deal with the outcome of experiments demands some means of streaming output of experiments and a library for sending results back to the user, with rendering modules on the client side.

The GridSpace Engine, including the JRuby interpreter, is accessed via an API and can be called from any Java application. In particular, considering the Virtual Laboratory, it is used by specialized tools such as the Experiment Planning Environment (EPE) dedicated for experiments developers, and the Experiment Management Interface (EMI), which allows experiment launching via the ViroLab Portal. Apart from these robust and powerful tools, the GridSpace Engine comes along with a simple client in a form of a command line tool.

To keep the GridSpace Engine operable in the environment of the Virtual Laboratory, and also to keep it generic, there are a number of ways of providing the experiment plan to the engine. First of all, it is possible to programmatically pass the experiment plan via an API. However, to facilitate retrieval of experiment plans from the Experiment Repository, a dedicated module called GridSpace Engine Application Repository Client is introduced. It enables plugging in clients for different implementations of Experiment Repositories.

A significant part of the functionality of the GridSpace Engine is to manage the session of the experiment execution. In this aspect, the engine preserves the scope of the user context (including security credentials) enabling Single Sign-On (SSO) access to Grid objects, as well as experiment execution context indispensable e.g. for the monitoring and provenance events correlation.

The GridSpace Engine is also intended to carry out on-line monitoring of experiment courses including invocation of Grid operations, access to data, current status of experiment plan execution, logging messages etc. This data is to be provided both to client tools requesting and tracing execution, and to the monitoring infrastructure. Thereafter, historical information can be used to further optimize execution by choosing the most efficient Grid objects by the GridSpace Application Optimizer (GrAppO).

Resource Registration

One of the objectives of the ViroLab virtual laboratory is to enable users to create complex experiments using different types of middleware without technical knowledge concerning the means with which these types of middleware can be invoked. To realize this requirement, the whole complexity of the underlying middleware has to be hidden from the user. To store information regarding all the available and potentially accessible computational resources and present them to the experiment developer in a user-friendly way, the Grid Resources Registry (GRR) has been created. It provides two layers of resource description:

- **Technology-independent:** Information about functionality exposed by the resource. It describes all methods, required input parameters and produced results. Additionally, this layer is strictly connected with ontology description of resources, thus allowing users to search for information on a higher level of abstraction.
- **Technology-specific:** Information required by other ViroLab components (such as GrAppO optimizer and Grid Operation Invoker) allows for choosing the best instance to invoke and to expose a proxy that delivers a uniform resource interface to the user.

An additional requirement comes from the fact that the Grid environment is dynamic: some computational resources can appear while others disappear. During experiment execution only working resources should be taken into account. To fulfill this requirement, the registry is connected to a monitoring system which sends out notifications not only about resource availability, but also about the environment where they operate (CPU usage, available RAM, etc.) This information is consumed by the GrAppO optimizer which, thanks to this data, is able to speed up experiment execution.

Another characteristic of the ViroLab virtual laboratory is its ability to store information about result provenance. To fulfill this requirement, information about computational resources used during script execution is needed. GRR is therefore connected with the PROToS provenance system (Balis, 2007) and logs information about newly registered resources.

For the experiment developer a dedicated user interface has been developed, allowing the developer to browse and manage registry content. It is an Eclipse plugin that can be easily installed into the Experiment Planning Environment (EPE). It is integrated with a script editor, which helps developers during the experiment creation phase and an ontology browser, which assists in searching for selected resources on the ontology level and vice versa.

Data Access in Virtual Laboratory

Data Access within the Virtual Laboratory consists of a developer interface, which is provided by the so-called Data Access Client and a backend, the Data Access Services that is used to aggregate medical data gathered from hospitals participating in the ViroLab project and exposes them as a unified data repository, with the use of the RegaDB schema, as depicted in Figure 2. RegaDB (Libin, 2007) provides a kind of data storage system including some software tools, which may be installed and managed locally, to store clinical data related to HIV treatment. It aims to support clinicians and researchers in their daily work by delivering a free and open source software solution. For researchers the objective is to offer several tools for HIV sequence analysis and to enable and disburden collaborations between researchers of different hospitals and institutions. Clinicians benefit from this system through the visualization of genetic data, relevant drugs, and algorithms in a well arranged form and the automatic generation of drug resistance interpretation reports.

Every data provider can either host a collaborative RegaDB installation within their trusted region, the so-called Demilitarized Zone (DMZ), outside their institutional firewall(s) or upload data onto one of the "centrally managed" collaborative RegaDB proxies installed at some trusted third parties. Direct external access into the hospitals' security regions is not required anymore and the existing database system can still be used independently from sharing any data within ViroLab's virtual laboratory. Additionally, a hospital can also set up a private RegaDB for its daily purposes and internal data management. To actually contribute data to the virtual workspace, the established database schemes need to be converted into the RegaDB schema before data can be stored within any of the RegaDB installations. The migration is done by exporting data from the original database and converting that data extract through a custom script into the latest schema. This procedure can be conducted repeatedly over time at the discretion of the corresponding database administrator(s) and occurs within each hospital. Data anonymization respectively pseudonymization can also occur during that data conversion or alternatively before transferring data from a private RegaDB onto a collaborative one.

Once the integration of heterogeneous data resources has been realized, the queried data sets need to be somehow aggregated and securely published for making them accessible within different applications and/or systems.

As mentioned earlier, a particular set of services (Assel, 2008) is required for dealing with multiple resources simultaneously and for ensuring transparent and secure access to corresponding data storages. This set of services, the so-called Data Access Services (DAS), has been designed and implemented according to the specific needs and requirements for biomedical data and confidential information shared among the laboratory users. Apart from the basic data access capabilities, these services are also responsible to control access to certain underlying data resources and to ensure that only authorized users are able to browse, query, and store particular data sets. This is achieved by making use of corresponding access control policies being stored and evaluated by a trusted third party component, a Policy Decision Point. These policies typically contain several access rules defined by the corresponding database administrator that specify the minimum set of user attributes obtained from the Shibboleth infrastructure and which are required for becoming authorized for a particular resource (Assel, 2008).

Complementing the Data Access Services is the Data Access Client; a library which provides experiment developers with access to remote data sources. The DAC is a part of the GridSpace engine devoted to communicating with the underlying data sources, which comprise the ViroLab architecture. Specifically, the Data Access Client currently enables access to the following:

- **Standalone databases:** Any SQL-compliant database can be directly referenced from within the experiment script. It is only necessary to supply the URI at which the DBMS is located and the name of the database schema in question. Communication is enabled via JDBC, through JRuby.
- **Data Access Services:** As explained above, the Data Access Services overlay a number of data sources which represent the hospitals

Figure 2. The ViroLab data access service concept

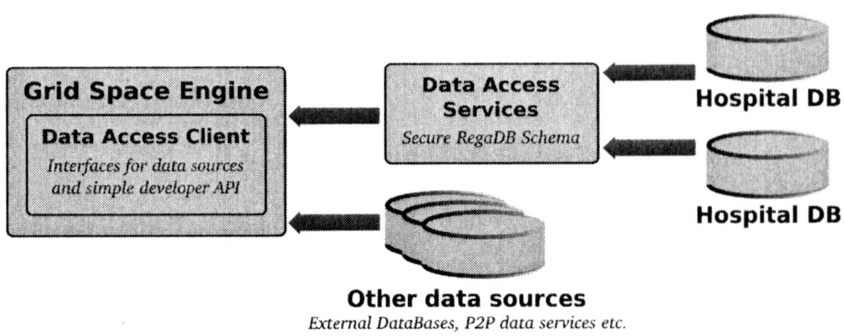

participating in the ViroLab project. They enable secure and uniform access to hospital laboratory data, using the REGA schema. As the DAS is exposed as a WSRF service, the DAC takes care of communicating with this service and formulating the proper queries. The experiment developer is provided with a simple API which enables submission of queries using a JRuby data access object. Security is supported, following integration of the Shibboleth authentication mechanism.

- **Proprietary data access infrastructures:** Work is underway to interface other types of data sources; specifically P2P architectures such as GridTorrent. This will further extend the range of technologies with which the Virtual Laboratory is compatible.

Presentation Layer

The presentation functionality in ViroLab is provided by two different layers. By using one layer accessible from within a Web browser, the user can run an experiment and view the results of its execution. The layer also comprises inspection of the experiment and support for tracing errors. The interface is implemented as a lightweight Web application built using the GridSphere portal technology. The concept of a thin client enables users to manage their experiments even through mobile devices. The environment used is a portal which is an extensible platform, where the basic functionality is provided by a portlet, representing a dedicated web application.

The main goal of the other layer is to provide the user with an extendible and helpful script code editor. This layer also involves an integrated preprocessor, semantic code inspection, code autocompletion and an ontology browser. Within this layer the user can access the ViroLab Grid Environment, execute script code (created in the script code editor) and trace the execution. The underlying technology is the Eclipse RCP platform (Funika, 2007).

The Experiment Management Interface supports the user in the following tasks. The first task is to locate a desired experiment by browsing various experiment repositories and then by selecting an experiment version. The repositories are populated by developers in the process of creating experiments with EPE. Different versions deliver different functionalities with respect to the same experiment plan (e.g. enhancements such as mail notification or a service implementing a different algorithm). Another task is to manage experiment execution and monitor its status. This part also includes interacting with the experiment by providing data when needed. Data requests are planned by the experiment developer during the implementation phase. The third task is to gather, store and share results. This involves building a collaboration space for the users executing the experiments, which enables sharing the data returned by an experiment and the knowledge which comes with the results. Finally, it is desirable to collect feedback on the quality of an experiment for its developers. Last but not least the user-friendliness of the interface, which refers to both accessibility and use, is taken into account.

In ViroLab, an experiment is planned with the Ruby script language. This technological choice determines the presentation layer functionality – its main purpose is to provide a simple but powerful and extensible script editor. As mentioned above, EPE is a dedicated tool for managing the script-based experiment development process that involves creating new experiments, developing experiments, running experiments with GSEngine, sharing experiments via the Experiment Repository etc. The idea behind EPE is to gather the above activities into a single integrated environment. EPE consists of so-called features; groups of plugins used to solve a certain problem. This approach simplifies organizing the internal structure of EPE and extending its functionality. The current version of EPE contains the following elements:

- *Workbench,* which is the central point of the application that manages the layout of each part of EPE. From the workbench one can choose a view or an editor to open.
- *Script editor,* which supports the experiment development process. It aims to provide such functions as syntax highlighting and code assistance.
- *Experiment Repository client* that supports the experiment sharing operations. Dedicated wizards enable exporting and importing experiments from the Experiment Repository.
- *Running experiments* with GSEngine. After setting up the experiment interpreter, the user can run an experiment with one click. Different versions of the interpreter can be applied along with different experiments.

A sample screenshot of EPE is shown in Figure 3, where the script editor (top part) is accompanied by a Registry Browser plugin (bottom). The registry browser allows locating all registered Grid Objects and their operations, and, by double clicking on a selected object, the experiment developer can insert a required code snippet into the experiment plan (as marked in the figure).

Another view on the experiment management is shown in Figure 4 where EMI is shown inside the ViroLab portal. A user can browse the repository for available experiments, select the required version and execute it on a remote engine by using three corresponding tabs. Such items as user input, status and results are dynamically updated while experiment execution is in progress. Additionally, the user can see the implementation code of the experiment, browse the experiment's license or provide feedback to experiment developers.

While the described solution delivers two separate graphical user interfaces, specific for each user group, there are some common fields of cooperation between these two groups. Those include reporting problems regarding the correctness of experiments, identifying new requirements, enhancements, etc. This issue imposes the need to establish a collaborative space between EMI and EPE users. This role is fulfilled by an experiment repository, used by EPE to store and manage versions of the developed experiments and by EMI to list and present details of the available

Figure 3. Experiment planning environment with a registry plugin

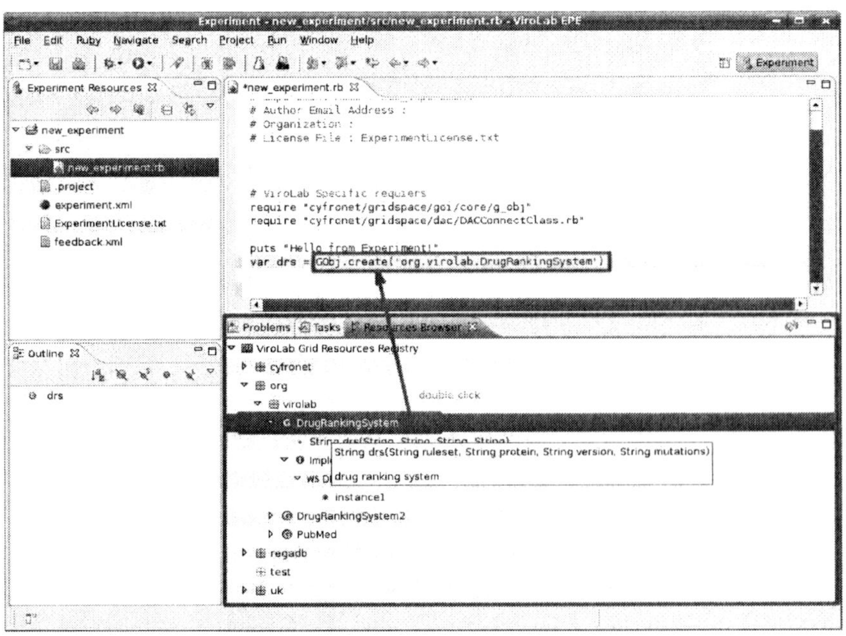

Figure 4. Experiment management interface (EMI) in action

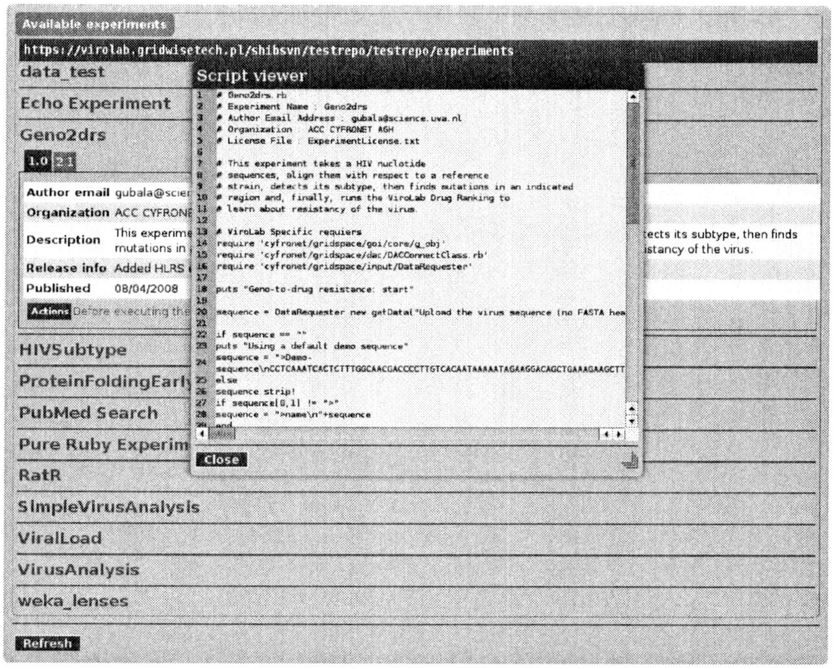

experiments. More than a single repository can be used; one per specific field of interest.

Provenance

Experiment results would not be reliable without provenance, i.e. information on how, where, by whom, etc., the results were obtained. In ViroLab, we follow a process-oriented approach of provenance wherein the complete records of the execution of experiments are captured and permanently stored in a provenance store. Our provenance model is based on ontologies which model (1) *in silico experiments* in general, as well as the domain-specific semantics of (2) *experiment data* used in the course of an experiment, and (3) *operations* executed as stages of an experiment.

The lifecycle of provenance in ViroLab consists of the following stages:

- *Experiment monitoring,* where the low-level events related to the execution of an experiment are captured;
- *Event aggregation,* when the low-level events collected in the previous phase are correlated, and translated into a high-level aggregate ontology representation;
- *Storage* of ontology-based provenance records;
- Provenance *querying.*

The basic experiment execution events are generated by GSEngine and passed via a monitoring infrastructure to a Semantic Aggregator component which is responsible for the aggregation and translation of the provenance records into the ontological representation, according to predefined aggregation rules. The experiment scripts are annotated in order to generate domain events. Thus, the provenance ontology representation is enhanced with domain-specific semantics of data and operations of the experiment. The ontology records of provenance, based on an OWL/RDF representation, are stored in a Provenance Tracking System (PROToS) which features a distributed storage and exposes interfaces to publish and query provenance records (Balis, 2007).

It is not only tracking but also *querying provenance* which is a non-trivial challenge. The reason for this is that provenance consumers are often people who are non-IT experts and do not know the technical details of database models, representations and query languages. To address this problem, we have created Query Translation Tools (QUaTRO) which enable end-users to construct powerful queries to provenance in an easy-to-use wizard-based manner. Our ontology-based provenance model enables us to exploit ontologies as a query language. As a result, end users compose queries to provenance using the concepts of the domain of their expertise. For example, in order to find drug rankings performed by a fellow researcher from another institute which resulted in a specific drug recommendation, one can compose a query such as "*Return all DrugRankings **made by** John Doe whose **result** contained recommendation for DrugX*". Furthermore, we have annotated the data ontology with mappings to underlying data sources and models (database address, table name). QUaTRO tools make use of those mappings and allow for browsing the structure of experiment data entities which is modeled in a database, not the ontology. Consequently, QUaTRO tools offer a very unique capability to construct queries composed of heterogeneous subqueries which refer to different repositories with different data models and query languages (for example XQuery and SQL) (Balis, 2007).

COLLABORATIVE ASPECT OF APPLICATION DEVELOPMENT AND EXECUTION

The process of experiment planning (development) and execution in the ViroLab virtual laboratory is collaborative in the sense that the virtual laboratory supports cooperation of multiple experiment developers and users.

Experiment Planning

When a developer prepares an experiment script, it can be published in the experiment repository and thus becomes available to others. Then, scientists who do not intend to get into the details of scripting, can access the virtual laboratory through a portal. Subsequently, they can execute the published experiments using a Web browser, providing only input data when asked for. The main idea is that the experiment in the repository can be shared and reused, what is a very efficient way of promoting collaboration between scientists. Provenance data related to the experiment is also recorded and available for queries, thus making the results more reliable, reproducible and scientifically relevant.

The first stage of experiment lifecycle is performed by the experiment developer, whose task (with the assistance of the domain researcher) is to develop an experiment plan using a scripting notation. To hide sophisticated details of the underlying Grid infrastructure, a high level object-oriented API has been introduced. It allows for defining "which" computational functionality is required, without a need to specify "how" to access it with available middleware. Uniform access to computational resources in a Grid environment is possible due to the Grid Object abstraction hierarchy. While creating experiments, only the highest level of resources' descriptions may be used – on the other hand, however, if the developer needs to retain a full control over the experiment plan, it is possible to specify all the technical details on one of the lower levels of abstraction. When a resource is registered in the virtual laboratory, it is available for the whole community and other developers can reuse it in new applications. Another feature, provided by the ViroLab virtual laboratory, which is useful during development, is the high-level DAC API that allows connecting and querying various databases.

When preparing the experiment plan, the experiment developer uses the Experiment Planning Environment (EPE) which offers a user-friendly, integrated with a set of tools, editor for writing scripts. The developer can use the semantic-web based Domain Ontology Store graphical browser to discover available data and computational services, coupled with Grid Resources Registry which provides the available operations that can be invoked directly from a script. To facilitate collaboration, EPE is integrated with the Experiment Repository based on the Subversion (SVN) version control system – as a result many developers can work on single experiment or share experiment codes. When the experiment is ready (it fulfills all requirements of an end user) the developer uses a dedicated EPE wizard to release it. When it is released, it instantly becomes available through the web interface for the scientific community. Moreover, clear releasing and versioning policy is very important for provenance data that has to be connected with a specific version of the experiment.

Experiment Execution

A script can be executed during development phase directly from the EPE which is integrated with the GSEngine (Figure 5). Once an experiment plan is developed, tested and released, a scientist can use a dedicated web interface (EMI) for executing the experiment. The main advantage of the Web-based interface is that the scientist does not have to have any additional software installed (only a Web browser with JavaScript support is required). Scientists may browse released applications, view their documentation and execute them. EMI is also connected with the GridSpace Engine, which is able to connect to the SVN repository, download the experiment and run it. While the application executes, data for the PROToS provenance system is recorded and stored in a dedicated provenance storage. This information can be queried by scientists through QUaTRO – a Web interface dedicated to browsing provenance data.

Figure 5. Application execution flow through Grid Space Engine

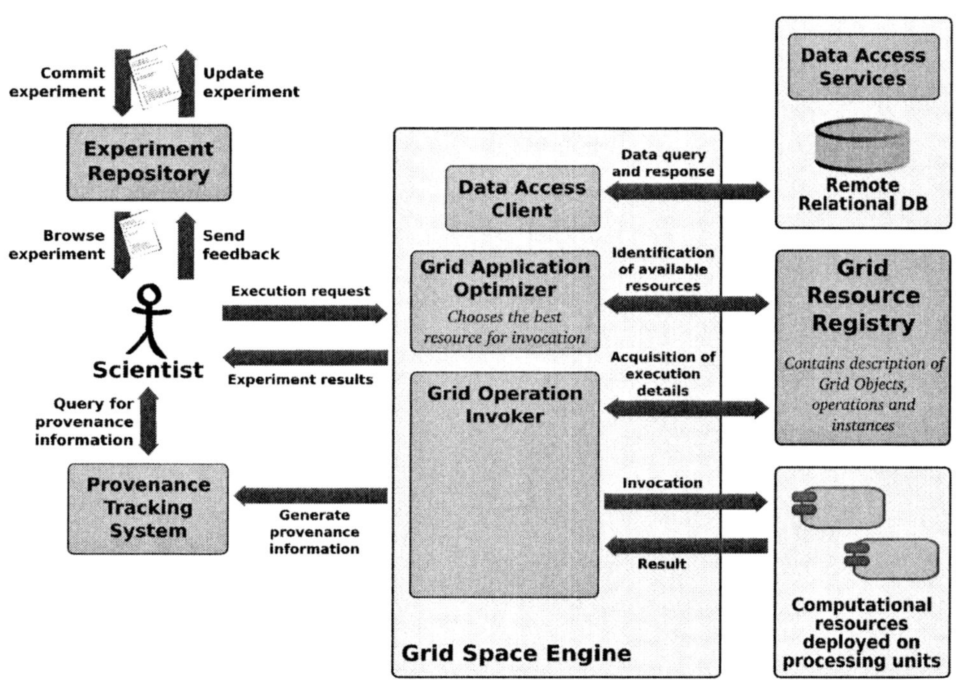

Feedback from the End User to the Experiment Developer

Experiment development is a long-lasting and iterative process: bugs may occur in the code, not every user's requirement is reflected correctly, additional requirements can emerge, etc. These problems, in most cases, are discovered by the end users of the application during its execution. The ViroLab virtual laboratory supports solving these kinds of problems by simplifying the process of communication between experiment users and developers. EMI allows the end user to submit feedback that can be useful for the developer who is able to browse it using EPE, respond to it and take it into account when creating a new version of the experiment. Scientists can track the progress of the new version of the experiment development using a simple SVN client (as a standalone or Web-based application) and help developers during this phase (with e.g. further comments).

ViroLab shares the vision of a collaborative working environment that advances the state of the art in biomedical Grid-based e-science. Previous large-scale efforts in collaborative laboratories for e-science have largely focused on learning environments, whereas in ViroLab the aim is to extend the possibilities of Grid technology to real-time working environments such as hospitals and virology departments that closely collaborate with each other.

The ViroLab Virtual Laboratory provides collaboration tools and services for users, interconnecting the presentation and data access layers of the architecture. A set of collaborative Grid services are contained within the Patient Treatment Support (PTS) tool, which provides a portal-based friendly interface for medical and other users to contact with clinical staff and acquire patient virological data from ViroLab hospital data repositories via the Data Access Services (DAS), using on-the-fly service instances. Once data, such as required sequences, mutation lists or FASTA files has been uploaded, an interactive Grid service provides connectivity to the meta Decision Support Ranking Service (DRS) which connects to a number of HIV drug-resistance ranking algorithms, such as Retrogram, REGA, Stanford HIV DB and ANRS, within specific regions of the genome such as Reverse Transcriptase or Protease. Each secure connection to the ranking Grid service is established via the ViroLab GridSpace Engine, using the Grid Object Interface multiprotocol runtime client to establish the connection and relay procedure calls.

APPLICATIONS WHICH ARE RUN IN THE VIROLAB VIRTUAL LABORATORY

The ViroLab virtual laboratory is deployed and operational for the project partners, which include users from such disciplines as virology, bioinformatics, as well as clinicians. The main experiments, which are available in the laboratory can operate on clinical databases and can combine the following services:

- HIV sequence alignment and mutations detection using RegaDB tools (Libin, 2007),
- HIV subtype identification,
- Drug resistance prediction using various rulesets (Retrogram, Stanford, RegaDB, ANRS),
- Data mining using various techniques (classification, clustering, association rules) like the Weka data mining toolkit,

We are also adding experimental support for computationally-intensive applications operating on the EGEE infrastructure, such as NAMD, Gaussian and TURBOMOLE.

Below we present two sample experiments demonstrating how the experiment plan is created and then evaluated.

Virological Analysis of HIV Virus Genotype

The experiment described below follows the short procedure from the samples of HIV virus nucleotide sequences (obtained from patient viral isolates) through some analysis of those sequences up to the list of rules regarding the resistance levels of those viruses to certain drugs.

The input to the experiment, a set of valid genetic virus mutations, is stored in a database.

In short, the process runs as follows:

- the **nucleotide** sequences of the HIV virus are obtained from a data source through the integrated, uniform protocol of DAC using the remote DAS,
- these sequences are subject to both **alignment** and virus **subtype detection**,
- a very important outcome of the alignment step is the **list of mutations,** a particular virus mutant has in comparison to some well-known reference sequences,
- these mutations are then uploaded to the **drug resistance system,** which returns the virus-to-drug susceptibility values as the final result.

The plan of the experiment specified as Ruby script is shown in Listing 1.

The output of the experiment can be viewed in the Experiment Management Interface (portal) or in Experiment Planning Environment, as seen in Listing 2.

Sample Data Mining Experiment Using Weka

The experiment presented below demonstrates how to create more complex workflows, including e.g. loops. We show it on a simplified example of a data mining experiment, which uses Weka associators wrapped as MOCCA components. The plan of the experiment is shown in Listing 3.

The `associator` object in the script is a representative of a `WekaAssociator` component, an instance of which is dynamically created on a remote computing resource. The component is a stateful entity, so it can be configured from the script: first we assign a specific algorithm (here we use three of the Weka algorithms), then we feed it with an input data set. The data set can be specified either as a local file, or as a remote one accessible via an URL. The latter case is useful if we do not want to pass large data sets through GSEngine, which executes the script – instead, the data is fetched by the computing component directly from the provided location.

Listing 1. Drug resistance experiment plan

```
patientID = GS _ DATA _ REQUESTER.getData("Provide patient\'s ID")
region = GS _ DATA _ REQUESTER.getData("Region (\"rt\" or \"pro\")")
nucleoDB = DACConnector.new("das", "virolab.hlrs.de:8080/wsrf/services/DataAccessService")
sequences = nucleoDB.executeDistributedQuery( "select nucleotides from nt _ sequence where
 patient _ ii=#{patientID.to _ s};")
subtypesTool = GObj.create("regadb.RegaHivSubtype")
subtypes = subtypesTool.subtype(sequences)
puts "Subtypes: #{subtypes}"
mutationsTool = GObj.create("regadb.RegaAlignment")
mutationsTool.align(sequences, region)
mutations = regaDBMutationsTool.getResult
drs = GObj.create("org.virolab.DrugRankingSystem2")
puts = drs.drs("ANRS", region, mutations.split(' '))
```

Listing 2. Drug resistance resulting from using ANRS rules, obtained as experiment output

```
-----------------------------------
Alignment information:
Status: Success, score: 2895, alignment range: 1:244
Mutations obtained for region rt:
P1M I2L S3T P4Q E6G T7C V8T P9L V10N K11F V35T T39K S68G E122K D123N S162C K173I Q174K
D177E I178M Q207A R211S L214F K238R
-----------------------------------
Virus subtype: HIV-1 Subtype A (01 _ AE)
-----------------------------------
DRS result: Using default ruleset: ANRS
-----------------------------------
3TC, 1
ABC, 1
D4T, 1
DDI, 1
EFV, 1
ETV, 1
FTC, 1
NVP, 1
TDF, 1
ZDV, 1
-----------------------------------
Geno-to-drug resistance: finish
```

Listing 3. Experiment plan demonstrating the use of multiple association rule mining algorithms

```
associators = ['weka.associations.Apriori',
 'weka.associations.PredictiveApriori',
 'weka.associations.Tertius']
associator = Obj.create('cyfronet.gridspace.gem.weka.WekaAssociator')
dataURL = 'contact-lenses'
i = 0
3.times do associator.assignAssociator(associators[i])
 rules = associator.createAssociationsURL(dataURL, '')
 puts 'Rules created are: '
 puts rules
 i = i + 1
end
```

FUTURE TRENDS

The concepts and methods, which are used in the existing virtual laboratories and scientific workflow systems, are of interest for many research groups around the world. Such issues as experiment planning, execution, provenance, and collaboration are the subject of numerous current and future investigations. In the case of the ViroLab virtual laboratory and the underlying GridSpace system, these issues and challenges take a new dimension which is implied by the specific scripting notation used for specifying experiment plans. Such topics as workflow execution, scheduling and instrumentation for monitoring and provenance tracing, are widely investigated in scientific workflow systems. For a scripting notation, however, specialized techniques should be used and these lead to interesting and challenging scientific questions and problems. Therefore, the approach we chose, as an alternative to the popular graphical workflow systems, opens a rich field of research and we believe that it can influence the developments of the workflow systems and virtual laboratories in general.

As specific subjects of future work, we will concentrate on providing additional methodologies and tools that allow creating applications based on the data, resources and experience of the community. Currently, work on management of results produced by experiments is in progress. Subsequently, data will be connected with ontological descriptions of the environment. Finding interesting information will thus be made easy and connecting it with provenance data will allow tracking the origin of such information. This functionality is very important for virologists who are the main end users of the ViroLab virtual laboratory. We are also working on supporting additional middleware technologies and underlying infrastructures, as well as introducing further refinements and enhancements in the API, to make experiment plan development even more user-friendly.

Recently, we observe a new approach to scientific investigations, which, besides analyses of individual phenomena, integrates different, interdisciplinary sources of knowledge about complex systems in order to acquire an understanding of such systems as a whole. This new way of conducting investigations is called system-level science (Foster, Kesselman, 2006). The ViroLab virtual laboratory may be considered as a good starting point for the development of a system supporting the system-level science paradigm. Its collaborative tools will be enhanced using recent achievements in Web 2.0 technologies.

CONCLUSION

The unique feature of the virtual laboratory developed for ViroLab is that by using a set of dedicated tools, both advanced experiment developers and domain scientists, can collaborate and conduct their research in a modern, highly-distributed environment. Thanks to the use of a scripting language, it is possible to define complex experiments easily, retaining a high-level of abstraction and concealing the details of the underlying Grid middleware. Initial prototypes of the integrated virtual laboratory have been released, installed and made accessible to experiment developers and the scientists who are beginning to set up communities which share the data, resources, experiments scripts and knowledge. In the future we plan to extend and improve the semantic descriptions of data, computations, applications, and system components.

ACKNOWLEDGMENT

This work was partially funded by the European Commission, Project ViroLab IST-027446, the related Polish grant SPUB-M, ACC Cyfronet AGH grant 500-08 and the Foundation for Polish Science. We wish to thank Tomasz Jadczyk for integrating Weka with the virtual laboratory.

REFERENCES

Aloisio, G., Breton, V., Mirto, M., Murli, A. & Solomonides, T. (2008). Special section: Life science Grids for biomedicine and bioinformatics. *Future Generation Computer Systems, 23*(3), 367-370

Altintas, I., Barney, O., Cheng, Z., Critchlow, T., Ludaescher, B., Parker, S., Shoshani, A., & Vouk, M. (2006). Accelerating the scientific exploration process with scientific workflows. *J. Phys.: Conf. Ser., 46*(1), 468+.

Assel, M., & Kalyoncu, O. (2008). Dynamic Access Control Management for Distributed Biomedical Data Resources. In Cunningham, P. & Cunningham, M. (Eds.), *Collaboration and the Knowledge Economy: Issues, Applications, Case Studies* (pp. 1593-1599). IOS Press.

Balis, B., Bubak, M., & Wach, J. (2007) User-Oriented Querying over Repositories of Data and

Provenance. In G. Fox, K. Chiu, & R. Buyya, (Eds.), *Third IEEE International Conference on e-Science and Grid Computing, e-Science 2007, Bangalore, India, 10-13 December 2007*, (pp.77-84). New York: IEEE Computer Society.

Bubak, M., Gubala, T., Kasztelnik, M., Malawski, M., Nowakowski, P. & Sloot, P. M. A. (2007). Collaborative virtual laboratory for e-health. In *Expanding the Knowledge Economy: Issues, Applications, Case Studies, eChallenges e-2007 Conference Proceedings* (pp. 537-544). Amsterdam, the Netherlands: IOS Press.

Deelman, E., Blythe, J., Gil, Y., Kesselman, C., Mehta, G., Patil, S., Su, M. H., Vahi, K. & Livny, M. (2004). Pegasus: Mapping scientific workflows onto the Grid. In Grid *Computing: Second European AcrossGrids Conference, AxGrids*, (LNCS 3165, pp 11-20). Berlin, Germany: Springer.

Droegemeier, K. K., Gannon, D., Reed, D. A., Plale, B., Alameda, J. Baltzer, T., et al. (2005). Service-oriented Environments in Research and Education for Dynamically Interacting with Mesoscale Weather. *Computing in Science and Engineering* 7(6)12-29.

Eres, M.H., Pound, G.E., Jiao, Z., Wason, J.L., Xu, F., Keane, A.J. & Cox, S.J. (2005) Implementation and utilisation of a Grid-enabled problem solving environment in Matlab. *Future Generation Computer Systems, 21* (6) 920-929.

Foster, I., Kesselman, C. (2006). Scaling system-level science: Scientific exploration and its implications. *IEEE Computer, 39* (11), 31-39.

Funika, W. & Pegiel, P. (2007) GScript Editor as Part of the ViroLab Presentation Layer. In *Proceedings of Cracow Grid Workshop 2006 (CGW06)*, (pp. 320-327). Krakow, Poland: ACC CYFRONET AGH.

Gil, Y., Deelman, E., Ellisman, M., Fahringer, T., Fox, G. Gannon, D., Goble, C., Livny, M., Moreau, L., & Myers J.(2007). Examining the Challenges of Scientific Workflows. *IEEE Computer, 40*(12), 24-32

Gubala, T. & Hoheisel, A. (2006). Highly dynamic workflow orchestration for scientific applications. In *CoreGRID Integation Workshop 2006 (CIW06)*, (pp. 309-320). Krakow, Poland: ACC CYFRONET AGH.

Libin P., Deforche K., Van Laethem K., Camacho R. & Vandamme A-M. (2007). RegaDB: An Open Source, Community-Driven HIV Data and Analysis Management Environment. In *Reviews in Antiretroviral Therapy* (Vol. 2007-2). Fifth European HIV Drug Resistance Workshop, Cascais, Portugal, March 2007.

Malawski, M., Bubak, M., Placek, M., Kurzyniec, D. & Sunderam, V. (2006). Experiments with distributed component computing across Grid boundaries. In *Proceedings of the HPC-GECO/CompFrame workshop in conjunction with HPDC 2006*, (pp. 109-116), Paris, France.

Malawski, M., Bartynski, T., & Bubak, M. (2008). A tool for building collaborative applications by invocation of Grid operations. In Bubak, M., van Albada, D. G., Dongarra, J., and Sloot, P. M. A., (Eds.), *Computational science - ICCS 2008, 8th International Conference, Kraków, Poland, June 23-25, 2008, Proceedings, part III*, (LNCS 5103 pp. 243-252). Berlin, Germany: Springer.

Oinn, T., Greenwood, M., Addis, M., Nedim Alpdemir, M., Ferris, J., et al. (2006). Taverna: lessons in creating a workflow environment for the life sciences. *Concurrency and Computation: Practice and Experience, 18*(10) 1067-1100.

Rycerz, K., Bubak, M., Sloot, P., Getov, V. (2008). Problem solving environment for distributed interactive simulations. In Gorlatch, S., Bubak, M., Priol, T., (Eds.), *Achievements in European Reseach on Grid Systems. CoreGRID Integration Workshop 2006 (Selected Papers)*, (pp. 55—66). Berlin, Germany: Springer.

Sloot, P. M. A., Ramos, A. T., Altintas, I., Bubak, M., & Boucher, C. (2006). From molecule to man: Decision support in individualized e-health. *Computer, 39*(11), 40-46.

Stankovski, V., Swain, M., Kravtsov, V., Niessen, T., Wegener, D., Kindermann, J., & Dubitzky, W. (2008). Grid-enabling data mining applications with DataMiningGrid: An architectural perspective. *Future Generation Comp. Syst., 24*(4), 259-279.

Taylor, I., Shields, M., Wang, I., & Harrison, A. (2005). Visual Grid workflow in Triana. *Journal of Grid Computing, 3*(3-4), 153-169.

KEY TERMS AND DEFINITIONS

Experiment is a process that combines data with a set of activities that act on that data in order to yield experiment results.

Experiment Plan is a recipe that describes the process of experiment execution in the environment of the virtual laboratory.

Experiment Management Interface (EMI) is a Web-based user interface used for browsing and executing experiments as well as managing their results.

Experiment Planning Environment (EPE) is an Eclipse-based interface used by developers to create experiment plans.

Grid Object is the abstraction of a Grid computational resource, accessible in the Grid Space environment as a simple object of the host language.

Grid Object Implementation is an abstract description of a Grid Object, taking into account the specific middleware technology in which this Grid Object is implemented (e.g. Web Service or MOCCA component).

Grid Object Instance is the specific instance of a service, component etc., which implements the functionality of a given Grid Object.

MOCCA is a CCA-compliant component framework based on H2O resource sharing platform, which enables building and running scientific applications on the Grid.

Ruby is a modern, interpreted, dynamically typed, object-oriented, scripting programming language.

Virtual Laboratory is a set of integrated components that, used together, form a distributed and collaborative space for e-science (see.http://virolab.cyfronet.pl).

Chapter XXVIII
Leveraging the Power of the Grid with Opal

Sriram Krishnan
University of California San Diego, USA

Luca Clementi
University of California San Diego, USA

Zhaohui Ding
Jilin University, China

Wilfred Li
University of California San Diego, USA

ABSTRACT

Grid systems provide mechanisms for single sign-on, and uniform APIs for job submission and data transfer, in order to allow the coupling of distributed resources in a seamless manner. However, new users face a daunting barrier of entry due to the high cost of deployment and maintenance. They are often required to learn complex concepts relative to Grid infrastructures (credential management, scheduling systems, data staging, etc). To most scientific users, running their applications with minimal changes and yet getting results faster is highly desirable, without having to know much about how the resources are used. Hence, a higher level of abstraction must be provided for the underlying infrastructure to be used effectively. For this purpose, we have developed the Opal toolkit for exposing applications on Grid resources as simple Web services. Opal provides a basic set of Application Programming Interfaces (APIs) that allows users to execute their deployed applications, query job status, and retrieve results. Opal also provides a mechanism to define command-line arguments and automatically generates user interfaces for the Web services dynamically. In addition, Opal services can be hooked up to a Metascheduler such as CSF4 to leverage a distributed set of resources, and accessed via a multitude of interfaces such as Web browsers, rich desktop environments, workflow tools, and command-line clients.

INTRODUCTION

In 1992, the term "Metacomputing" was coined by Smarr and Catlett to describe a connected network computing environment (Smarr, Catlett, 1992), which would eventually enable everyone to obtain information on demand "all from their own desktop workstations." Metacomputing has now been popularized as Grid computing with the emergence of the Globus toolkit (Foster, Kesselman, 1997) and other related software, with the goal of providing dynamic accessibility to computing resources like an electric power grid. With the release of the Atkins report in 2003 (Atkins et al, 2003), the term "cyberinfrastructure" is generally used to refer to the national and international network of computers, storage, software and human resources dedicated to support the advancement of science, engineering and medicine. Various countries have made major investments in the development of the Grid middleware, and software stacks that bridge the network and computing resources to the domain specific applications and users (Foster, Kesselman, 1999).

Since 1997, US funding agencies such as the National Science Foundation (NSF) and the Department of Energy (DOE) have funded the development of cyberinfrastructures through a series of initiatives including, but not limited to:

- The National Partnership for Advanced Computational Infrastructure (NPACI: http://www.npaci.edu/), National Middleware Initiatives (NMI), Software Development for Cyberinfrastructure (SDCI), and the Science Discovery through Advanced Computing (SciDAC) for software;
- The OptIputer (A Powerful Distributed Cyberinfrastructure to Support Data-Intensive Scientific Research and Collaboration: http://www.optiputer.net) and the Global Ring Network for Advanced Application Development (GLORIAD: http://www.gloriad.org) for networking;
- The TeraGrid (http://www.teragrid.org), the Open Science Grid (OSG: http://www.opensciencegrid.org), and the Petascale Computing Environment for high throughput and high performance computing.

A number of international Grid activities have also sprung up, such as UK e-Science Programme (Hey, Trefethen, 2005), The Enabling Grids for E-science (EGEE: http://public.eu-egee.org) from the European Union, and the Pacific Rim Grid Applications and Middleware Assembly (PRAGMA) (Arzberger, Papadopoulos, 2006) supported by NSF and member institutes in the Asian Pacific region. In the public health and biomedicine sector, the National Center of Research Resources (NCRR) at the National Institutes of Health (NIH) has funded the development of the Biomedical Informatics Research Network (BIRN) (Grethe et al, 2005) for data sharing, and the National Cancer Institute (NCI) has supported the development of the Cancer Biomedical Informatics Grid (caBIG) for Grid services for biomedical data models and integration (Oster, 2007).

The impact of cyberinfrastructures on e-science is most felt through the collaborative research it enables on a scale never imagined before. One of the key components is the ability to form "virtual organizations," where researchers and professionals alike are able to share data, applications, and visualization environment transparently and seamlessly with the necessary authentication and authorization (Siebenlist, Magaratnam, Welch, Neuman, 2004). For example, the BIRN project enables the biomedical researchers to share neuroscience imaging datasets, analysis tools, as well as access to TeraGrid resources through its portal environment. The OptIportals allow researchers to visualize large datasets over high speed optical networks. Similarly, the OSG supports high energy physics research, as well as computational biology virtual organizations (Stevens, 2006).

Users of e-science cyberinfrastructures or Grid systems fall under two main groups:

- Scientific application developers, who would like to provide their users access to their scientific applications running on Grid resources. However, they would much rather focus on improving the scientific functionality of their codes, rather than learn about the intricacies of the back-end infrastructure.
- Scientific end-users, who would like to access scientific applications and resources in a seamless way without having to learn concepts related to Grid systems. They would prefer to run the scientific codes via interfaces such as Web browsers or desktop clients, completely oblivious to the setup at the back-end.

Developers of Grid middleware tools should strive to accommodate both types of users. The key challenge in doing so is to provide novel functionality in the most easy to use manner, while shielding the scientific users from the complexity of the Grid resources that the applications are executing on.

BACKGROUND

From a computer science perspective, the architectural concept of service orientation provides several interesting features:

- Services reduce complexity through encapsulation of service implementation. A service publishes its interface and ensures that it provides the appropriate quality of service, presumably agreed to before the client accesses the service. Service implementations can leverage legacy applications, disparate resources, and even other services to implement their functionality; however, the key is that the implementation complexity is hidden from the client applications behind the service interface.
- Service interfaces enable interoperability across systems and architectures through the use of open standards. The service interface is specified by a Web Service Definition Language (WSDL) document that includes all information needed by client applications to interact with the service. The WSDL specification itself leverages Extensible Markup Language (XML) standards such as Simple Object Access Protocol (SOAP) for message exchange and XML Schemas for complex data structures. The broad adoption of these standards has greatly facilitated interoperability across systems and vendor products unlike in the past with other distributed systems standards. This lends itself very well to the creation of workflows, or composition of services for the creation of complex application pipelines, where the services themselves may be hosted at disparate locations.
- Services support a loosely coupled model where clients can bind to service endpoints at run-time. Such late binding provides greater fault tolerance and flexibility. Services can be discovered at run-time using metadata-based discovery mechanisms, and Quality-of-Service (QoS) can be negotiated dynamically.

The Open Grid Services Architecture (OGSA Foster et al, 2002) represents an evolution towards a Grid system architecture based on service orientation and Web services concepts and technologies. OGSA envisions the Grid to be an integration of "Grid services" across distributed, heterogeneous, dynamic "virtual organizations." It defines a Grid service to be a potentially transient stateful service instance that can support reliable and secure invocation, lifetime management, notification, policy management, credential management, and virtualization. At the core of OGSA is the Web Service Resource Framework (WSRF: http://www.oasis-open.org/committees/

tc_home.php?wg_abbrev=wsrf), which defines a set of standards for the definition and interaction with Grid services.

There are several tools that provide access to Grid systems (Foster, Kesselman, 2004), and several of them are being rewritten to be compliant with OGSA and WSRF. One of the most important and popular pieces of software in the world of Grid computing is the Globus toolkit (Foster, Kesselman, 1997). The Globus toolkit provides software services and libraries for resource monitoring, discovery, and management, as well as for security and file management. The Globus toolkit is packaged as a bag of services – and users are free to choose the services that they wish to use. This is one of the Globus toolkit's greatest advantages – and drawbacks. This makes it very convenient for tool developers to access the various components programmatically, but it also makes it difficult for scientific end-users to master due to the complexity and jargon of the various pieces. In practice, the Globus toolkit is often used as a building block by a number of Grid middleware toolkits. However, scientific end-users are typically shielded from its complexity by higher-level interfaces such as Portals and Problem Solving Environments (PSE).

The core functionality of the Globus toolkit is divided into four major areas – security, data management, execution management, and information services:

- **Security:** One of the key contributions of the Globus toolkit is the Grid Security Infrastructure (GSI) (Foster, Kesselman, Tsudik, Tuecke, 1998). GSI is a public-key system that uses X.509-based certificates signed by trusted third parties called Certificate Authorities (CAs). Typical usage models require that each user be assigned a user credential consisting of a public and private key. Users generate delegated proxy certificates with short life spans that get passed from one component to another and form the basis of authentication and access control.
- **Data management:** The Globus toolkit provides a set of tools for data management. One of the most frequently used services is GridFTP (Allcock, 2001), which provides high-performance, secure, reliable data transfer protocol optimized for high-bandwidth wide-area networks, and based on Grid security mechanisms. A Reliable File Transfer (RFT) service that provides monitoring and management of file transfers via GridFTP is also provided. Other services of interest include ones for data replication and replication location.
- **Execution management:** The key contribution of the Globus toolkit in the area of execution management is the Globus Resource Allocation Manager (GRAM), which provides a single protocol and API for communicating with different batch/cluster job schedulers, using Grid security mechanisms. GRAM uses a scheduler-agnostic Resource Specification Language (RSL) to create jobs, and can be used to interact with most schedulers such as the Sun Grid Engine (SGE), the Portable Batch System (PBS), and Condor (Basney, Livny, Tannenbaum, 1997).
- **Information services:** The Globus toolkit provides a Monitoring and Discovery System (MDS) for providing information about the available resources on the Grid and their status. MDS services provide query and subscription interfaces to arbitrarily detailed resource data and a trigger interface that can be configured to take action when pre-configured trouble conditions are met.

Most of this functionality has been rewritten to be Web services based; however, the older legacy implementations that use custom protocols are also packaged together for backwards compatibility. Running a job via the Globus toolkit typically involves the following steps: a user first retrieves credentials using the security libraries, and then uses the information services to figure out which resource to use to submit a job. The user then uses the data management clients to transfer job inputs to the target resource, and then the execution management clients to launch a job on that particular resource and monitor status. When the job is complete, the user uses the data management clients again to download the results. While this may appear intuitive to a Grid middleware developer, it is too complicated for a scientific application developer or an end-user. We need higher levels of abstractions on top of the standard Grid middleware tools to make it easier for users in the scientific domains to access the Grid infrastructure.

AN ARCHITECTURE FOR SERVICE ORIENTED SCIENCE

As we discussed in the previous section, programming and using distributed Grid resources is quite a challenging task. There is hope that Service-Oriented Architectures (SOA) can address this problem by providing platform-independent, language-neutral service interfaces that hide the complexity of the implementations, while providing a well-defined Qualities of Service (QoS).

In our recent paper (Krishnan, Bhatia, 2007), we claimed that in order to lower the entrance barrier for scientific application developers and end-users to Grid infrastructures, we must do the following:

- **Focus on application-level services:** Until recently, SOAs have only been used to build Grid middleware. For instance, as we discussed above, the Globus toolkit has been redesigned to be consistent with latest Web service technologies. However, from the perspective of a scientist, the services that are most relevant are services that perform a scientific operation and where the semantics of the operations are defined in terms of the domain science. For example, a BLAST service can provide biologists with the capability to compare multiple DNA sequences against each other or against standard publicly available data sets. The particular infrastructure used for its calculation is irrelevant. The tool developers would rather focus on the algorithms and the appropriate interfaces to present to the scientific end-users. In (Foster, 2005), Foster introduces the term Service-Oriented Science. In his words, "Grid technologies can accelerate the development and adoption of service-oriented science by enabling a separation of concerns between discipline-specific content and domain-independent software and hardware infrastructure" enabling access to scientific applications and tools using a service-oriented approach allows the developers of scientific tools to focus on the domain science, and delegate the management of the complex back-end resources to others who are more proficient in Grid middleware.
- **Provide users access to virtualized resources:** Users accessing the cyberinfrastructure should not be provided with the raw resources — rather, the resource access should be virtualized. Resources include data files, computational cycles, or even application services. For instance, providing individual user accounts on computation resources does not scale very well with the number of users, or the number of computational clusters. Instead, individual application services running on the virtualized back-end computational resources may be provided to authenticated users. The services can hide the disparity between the heterogeneous resources in the back-end, and provide a uniform interface to the users. Similarly, data files should be accessed

through generic object identifiers or uniform resource identifiers, and not through direct file paths on hosts. This enables the cyberinfrastructure to perform optimizations such as replication or data migration for better fault tolerance, and higher availability. The use of virtualized resources through a layer of indirection allows the system developers to add or move resources as needed without any interruption. For example, additional clusters can be added, data moved, while the end-user is completely oblivious to such details.

- **Provide multiple user interfaces to scientific users:** There are many classes of users using the cyberinfrastructure. On one end of the spectrum, we have the beginners or casual users who may wish to use simplified interfaces provided by Web-based graphical user interfaces. On the other hand, power users, typically find that graphical interfaces are more of a hindrance, and prefer command-line tools that can be scripted. From a technical standpoint, it is also difficult for Web-based Graphical User Interfaces (GUIs) to represent many of the complex functionality of today's scientific applications. Desktop applications can provide better visualization and interaction than Web-based tools, but they are harder to develop and require platform-specific considerations. For all these reasons, it is more and more difficult to imagine a single user interface satisfying all the different users for any particular cyberinfrastructure. Hence, we advocate application services that can be used by a number of end-user tools, including Web-based portals, rich desktop applications, command-line tools and language-specific libraries.
- **Lower deployment costs:** Deploying a scientific application on a Grid resource should be as simple as possible. The simpler it is, the more likely it is that scientific users will embrace it when the scientific needs are also met. Scientific end-users don't have the time or money to invest in trying to figure out the complexity of the back-end resources.
- **Follow the Web 2.0 paradigm by enabling user participation and contribution** (Clementi et al, 2007): The Web 2.0 paradigm provides several lessons that can be leveraged by the scientific and grid communities. In "What is Web 2.0" (O'Reilly, 2005), O'Reilly lists some design patterns and general recommendations that are a large part of the Web 2.0 paradigm: (1) use the Web as a platform for development, (2) harness the collective intelligence of the community, (3) focus on services, and not prepackaged software, with cost-effective scalability, (4) enable light-weight programming models, and (5) provide a rich user experience. In principle, the Web 2.0 paradigm allows users to contribute their applications, share data, and participate in collaborative activities with ease and keep their focus on the process of creation, instead of maintaining and learning new tools and interfaces. The virtualization trend in computational resources is going to play an important role in enabling Web 2.0 for the masses.

With the above points in mind, we have built the Opal toolkit (Krishnan et al, 2006) for wrapping scientific applications as Web services, and lowering the entrance barrier for scientific users in using Grid resources.

The Opal Toolkit

Figure 1 shows the overall multi-tiered architecture employed by the Opal toolkit. The compute resources are the bottom tier, the Web services layer is the middle tier, and the end-user interfaces are the top tier. The bottom tier is composed of multiple clusters hosted at different sites. Clusters located at different geographical locations run different schedulers of their choice, e.g. Condor, the Sun Grid Engine (SGE), the Portable Batch System (PBS), etc. The Web services in the middle tier provide access to the scientific applications on these clusters through SOAP APIs that are independent of the infrastructure being used at the back-end. At the top tier, the user-interfaces provide application-specific interfaces that can be easily used by the scientific end-user. These include Web portals, Problem Solving Environments such as Continuity (http://www.continuity.ucsd.edu), which is a tool for cardiac biomechanics, and AutoDockTools (ADT) (Sanner, Duncan, Carrillo, Olson, 1999), which is a tool for protein docking, and Kepler (http://kepler-project.org), which is a workflow toolkit.

Scheduling and Cluster Management

Since different sites run different schedulers, it is mandatory that these be accessed in a uniform way for maximum code reuse. Furthermore, these schedulers need to be accessible programmatically by the Web service implementations, and not via their regular command-line interfaces. The Opal toolkit leverages a couple of mechanisms to do so. It can use the Globus Resource Allocation Manager (GRAM) API for submitting and managing Grid jobs, and interact with most schedulers such as SGE, PBS, and Condor. We use the Java CoG Kit (von Laszewski, Foster, Gawor, Lane, 2001) for the client side libraries for GRAM. This makes the Web services code easily portable across sites that use different schedulers. Using a different scheduler is as easy as changing the Uniform Resource Locator (URL) for the Globus gatekeeper, assuming that the Globus GRAM service is properly configured on the Grid resource. Another way to interact with the schedulers is by using the Distributed Resource Management Application API (DRMAA: http://www.drmaa.net). Schedulers such as Condor and SGE provide DRMAA bindings, and the Opal toolkit uses DRMAA Java binding to submit and monitor jobs.

The Web services themselves are hosted inside an Apache Tomcat container, which is responsible for providing desirable qualities of service such as scalability and reliability (with the possible use of load balancing over a set of Tomcat servers).

Data Management and Persistence

When a client invokes an application wrapped as a Web service, a new working directory is created for every run. All inputs are transferred there, and

Figure 1. The end-to-end services oriented architecture

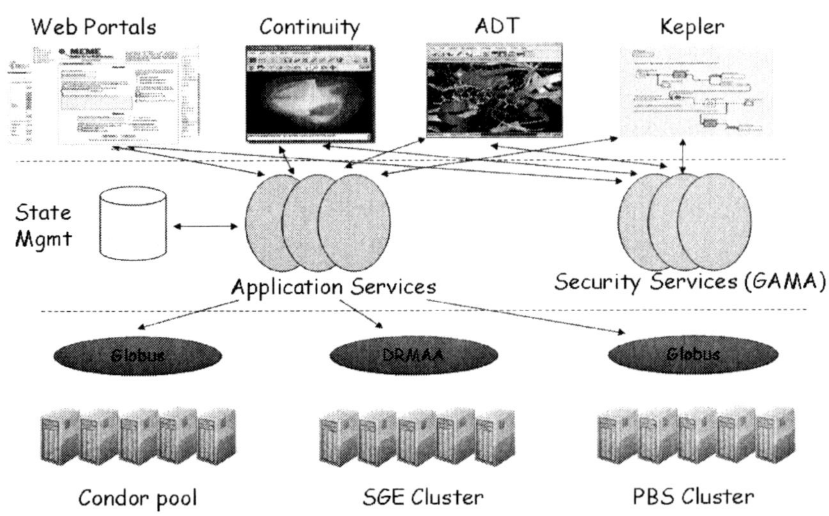

the application is run inside this working directory. The user receives a unique *jobID* that can be used in the future to query for job status and retrieve outputs. This model of detached execution is especially beneficial for long running jobs since the client does not have to block until the job is completed. Furthermore, multiple jobs submitted by different users can be run concurrently since they are executed in separate working directories. This approach makes the Web services stateful. The state of the service consists of the job status and metadata about job inputs and outputs, and also the user information and job history. If a service happens to crash at any point, this state can possibly be lost. However, these services can be optionally configured to store their above-mentioned state into a PostgreSQL (http://www.postgresql.org/) database, accessed via Java Database Connectivity API (JDBC http://java.sun.com/javase/technologies/database/). As discussed later, this information may be easily displayed through a Web interface. Note that we do not provide any ability to checkpoint the applications themselves at this point – this is left to the application developers who may implement checkpointing mechanisms to restart from failures.

Security

GSI-based systems are known to be very complex to administer and use. We use the Grid Account Management Architecture (GAMA) (Bhatia, Chandra, Mueller, 2006) to address this problem for two reasons. First, it provides a simple portal interface for an end-user and site-administrator to create GSI credentials. This interface abstracts out several software packages for building CAs into a single set of services for easy deployment and administration. Second, it provides Web service APIs to securely retrieve proxies from the back-end server, which can then be used by a variety of clients on different platforms, without the installation of any complicated CA software, as shown in Figure 1.

Authentication between clients and the Web services is performed at the transport level for performance reasons. It relies on the creation of a secure point-to-point connection between the client and server, using a GSI-based Secure Sockets Layer (SSL) implementation that can be set up using the security libraries provided by the Java CoG kit. Once the client is authenticated, an Apache Axis Handler is called to perform authorization before the Web service is invoked. Currently, we provide a grid-map (access-control list) based authorization, and a ca-map based authorization where all users from an acceptable CA are authorized to launch jobs. However, Axis Handlers can be written to perform call-outs to more complex authorization packages such as the Community Authorization Service (CAS) (Perlman et al, 2002).

WSDL API

Automatically generated source code and WSDL's can be very specific to the SOAP toolkits being used. With an eye on simplicity and configurability, we use a static WSDL for every service that is deployed using Opal. Legacy scientific applications expose only a command line interface to users, and can only be run with command line arguments. Since they do not expose specific operations that can be run with different parameters (unlike traditional Web services), it suffices to simply expose one single method to execute them — *launchJob*. The input parameter to this operation is an XML data structure containing the argument list as a string (which exactly corresponds to the arguments of the scientific application), and a list of data structures encapsulating the input files. The data structure for an input file consists of a name, and the *content*, which is a Base64 encoded binary representation of the file contents. In the future, the contents could also point to a URL where the inputs can be staged. The *launchJob* operation returns a *jobId* that can be used to query for job status using the *querySta-*

tus operation, and eventually retrieve job outputs using the *getOutputs* operation. The *getOutputs* operation returns a list of URLs from where the output files for a particular run can be retrieved. Furthermore, Opal services also expose a *getMetadata* operation that can be used to retrieve information about the scientific application such as the usage, input and output information, etc. This information is retrieved directly from the application configuration described in the section titled "Deployment of an Application in Opal." Furthermore, jobs can be killed at any time via the destroy operation using the *jobID*.

Since Opal services use static WSDLs, every scientific application deployed as an Opal service has the same WSDL. Two Opal services can be distinguished by their unique URLs, and their associated metadata. This metadata can be published in registries, so that Opal-based scientific applications can be discovered and used dynamically by various clients.

Argument Description and Dynamic Web Interface Generation

In order to further describe application inputs, we added an optional description of command-line arguments to the Opal metadata. We broadly classify the command-line argument types into the following members:

- **Flags:** These parameters are not ordered and are usually represented with a character prefixed with a dash and they activate a functionality in the application (-verbose);
- **Tagged parameters:** These parameters are usually formed by a prefix and input word (-input <filename>), they can appear in any order;
- **Untagged parameter:** These parameters are not prefixed therefore their order is relevant.

Each of the above elements also consists of other meta-data, e.g. the type for tagged or untagged parameters (viz., Integer, Float, Boolean), the default values, a short text description, etc. Moreover we can group parameters that belong together. A group can also be defined as exclusive, which implies that only one parameter can be selected among the ones belonging to that group. The following XML snippet shows an example of a configuration file for the PDB2PQR application (Dolinsky, 2007), which provides a routine to translate between the Protein Data Bank (PDB) and PQR molecule formats, which is a variant of the PDB format. In particular, notice the use of flags, tagged and untagged parameters, and groups.

```
<types xmlns="http://nbcr.sdsc.edu/opal/types">
  <!-- list of flags -->
  <flags>
    <flag>
      <id>nodebump</id>
      <tag>--nodebump</tag>
      <textDesc>
        Do not perform the debumping operation
      </textDesc>
    </flag>
    ...
  </flags>

  <!-- list of tagged parameters -->
  <taggedParams>
    <separator>=</separator>
    <param>
      <id>ffout</id>
      <tag>--ffout</tag>
      <paramType>STRING</paramType>
      <textDesc>Instead of using the standard
        canonical naming scheme for residue and
        atom names, use the names from the given
        forcefield</textDesc>
    </param>
    ...
```

```
 </taggedParams>

 <!-- list of untagged parameters, in order -->
 <untaggedParams>
  <param>
   <id>output-path</id>
   <paramType>FILE</paramType>
   <ioType>OUTPUT</ioType>
    <textDesc>The desired output name of the
PQR
     file to be generated</textDesc>
  </param>
  ...
 </untaggedParams>

 <!— groups of parameters and flags -->
 <groups>
  <group>
   <name>inputParam</name>
   <elements>inFile inId</elements>
   <required>true</required>
   <exclusive>true</exclusive>
   <textDesc>Input file to be used (choose
   one of the two options)</textDesc>
  </group>
  ...
 </groups>
</types>
```

While this model is fairly simplistic, it is capable of describing the syntax of most command-line arguments. In its current form, it is not possible to describe semantic information of the various types, or relations and dependencies between the values of various parameters. This will be addressed as part of our future work.

The above XML description is used for two purposes. Firstly, the Opal Web service can use the command-line description and validate user inputs before launching a job. This is useful because it is conceivable that a submitted job is stuck in a queue for a long period of time, depending on system load. It is not efficient to wait for a job to begin running to determine if its inputs are correct. Secondly, it can be used for automatic interface generation as follows.

We provide an advanced Web submission interface that can fetch the deployment configuration as metadata from the Opal service via its WSDL API at run time, and create a personalized input form based on the command-line argument description. The form is created following these simple rules:

- Flag parameters are rendered as a check boxes
- Tagged and Untagged parameters are rendered as:
    - Check boxes if their type is Boolean
    - File input fields if their type is file and they are input parameters
    - Radio buttons if their type is a list of exclusive values
    - Input text fields for all the other cases

All the groups are displayed using a descriptive heading and then listing the associated parameters. If the group is exclusive, the form ensures that no more than one of the parameters in the group is active. Figure 2 shows a typical scenario where a user requests the submission form for the PDB2PQR service. The Opal GUI dynamically queries the Opal server to fetch the deployment configuration file, and a customized submission form is rendered to the user. Once the submission button is pushed, the Opal GUI launches the PDB2PQR via the Opal Web services API.

While the above demonstrates the use of this approach for Web interfaces, it is general enough to be used in any environment. In short, a PSE that wishes to dynamically generate Opal interfaces must provide an Opal "shell" that can request the application configuration at run-time, and translate it to the presentation logic of the corresponding system. This is possible because,

Figure 2. Service Invocation with Opal GUI

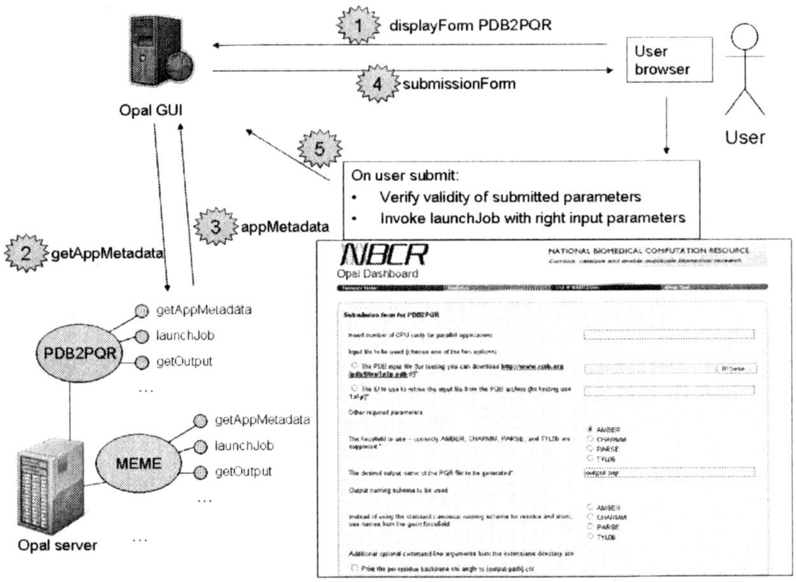

unlike other approaches like Web Services for Remote Portlets (WSRP: http://www.oasis-open.org/committees/wsrp), the Opal services do not push any presentation logic. They only enable access to application metadata in XML form, which is easy to translate into a number of markup languages (even as trivially as via Extensible Stylesheet Language Transformations (XSLT) transformations). We discuss how interfaces are automatically generated for rich desktop clients in the section titled "Automatic Interface Generation for Workflow Toolkits."

Deployment of an Application in Opal

Figure 3 shows the implementation of the Opal services inside the Web services container. Opal services are developed using the commodity Apache Axis toolkit, and are hosted within the Apache Tomcat container. Multiple instances of the Opal services may exist within the same container, each wrapping a different scientific application, and accessible via unique URLs.

Figure 3. Opal implementation details

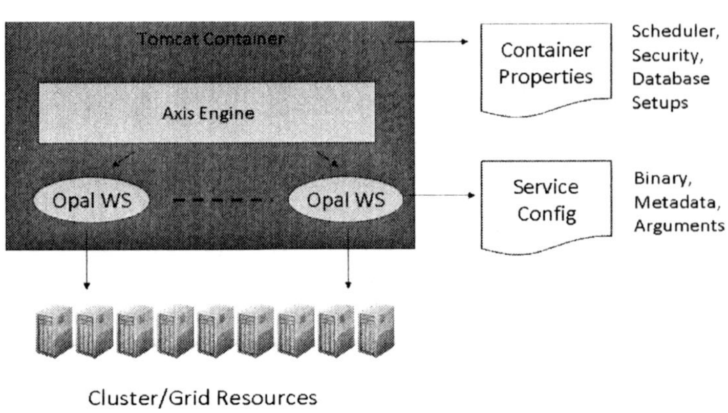

Figure 4. Opal container properties

```
# parallel parameters
num.procs=16
mpi.run=/opt/mpich/gnu/bin/mpirun

# database information
database.use=false
database.url=jdbc:postgresql://localhost/opal_db
database.user=opal_user
database.passwd=opal_passw

# drmma information
drmaa.use=false
drmaa.pe=mpich

# globus information, if drmaa is not being used
globus.use=false
globus.gatekeeper=localhost:2119/jobmanager-sge
globus.service_cert=/etc/certs/app_service.cert.pem
globus.service_privkey=/etc/certs/app_service.privkey
```

The container itself can be configured with a static set of properties. These include properties of the computational infrastructure such as Globus and database setup information, the number of nodes in the compute cluster, etc.

Figure 4 shows an example of the container properties. The number of available processors is specified by the property *num.procs*, and the location of MPI for parallel execution is specified by the property *mpi.run*. If the *database.use* property is set to true, the *database.url*, *database.user*, and *database.passwd* properties are used to persist Web service state into a database (assuming the tables are set up appropriately using the SQL scripts provided). Similarly, if the *globus.use* property is set to true, the *globus.gatekeeper*, *globus.service_cert*, and *globus.service_privkey* properties are used to submit jobs to an appropriate scheduler using Globus GRAM. If the *drmaa.use* property is set to true, the DRMAA API is used to submit jobs instead. If neither Globus nor DRMAA is being used, all jobs are run using simple local process forks.

An application configuration file describes every application. This specifies the location of the binaries (*binaryLocation*), whether it is a parallel application or not (*parallel*), the default arguments (*defaultArgs*), and application metadata (*metadata*). Application metadata includes the usage and other optional information (*info*) specified by the application provider. This can include helpful messages for understanding the various command line arguments, inputs and outputs. The metadata also contains optional command-line argument *type* information, described in the section "Argument Description & Dynamic Web Interface Generation." A sample application configuration, without the optional argument type information, is shown in Figure 5. Note that in its current form, Opal only provides support for two types of applications – serial and parallel. In the future, we plan to support other types of applications such as parameter sweeps (as array jobs).

Service deployment for Opal is extremely straightforward. The Apache Ant toolkit is used for building and deploying our services. Ant is a Java-based build tool, which has become the *de facto* standard for building cross platform Java applications. Ant uses XML-based configuration files, and builds a system by calling out a target tree where various tasks get executed. In our case,

Figure 5. Sample application configuration

```
<appConfig xmlns="http://nbcr.sdsc.edu/opal/types"
           Xmlns:xsd="http://www.w3.org/2001/XMLSchema">
  <metadata>
    <usage><![CDATA[psize.py [opts} <filename>]]></usage>
    <info xsd:type="xsd:string">
    <![CDATA[
      --help              : Display this text
      --CFAC=<value>      : Factor by which to expand mol
                            dims to get coarse grid dims
                            [default = 1.7]
      --FADD=<value>      : Amount to add to mol dims to
                            get find grid dims
                            [default = 20]

      .....
      .....
    ]]>
    </info>
  </metadata>
  <binaryLocation>/bin/psize.py</binaryLocation>
  <defaultArgs>--GMEMCEIL=1000</defaultArgs>
  <parallel>false</parallel>
</appConfig>
```

a scientific application can be deployed using an Apache Ant target named *deploy*. Two system properties need to be set — a *serviceName* for the new service, and an *appConfig* that points to the location of the application configuration file. The Ant task automatically generates the Web Services Deployment Descriptor (WSDD) that is used by Apache Axis for service deployment, and configures the service appropriately. Clients can now access this new service via a unique URL that is created by appending the serviceName to the base URL for the Opal Web Application (*webapp*). Typically, the URL for the service is of the form http://host:port/opal/services/<serviceName>. When a service is invoked, Opal has all the information it needs from the container properties and the application configuration to launch jobs on a cluster, and manage its state.

Opal also provides Ant targets to *undeploy* a service, to upgrade installed versions of Opal, and completely uninstall older versions as well.

Compatibility and Interaction with WSRF

We are working on not only making Opal services available as WSRF services, but also on abilities to access distributed WSRF resources via Opal. In this section, we discuss Opal OP, which is an extension to Opal to make it WSRF compatible, and Opal-CSF4, which uses a WSRF-based Metascheduler to access distributed Grid resources.

Opal OP

Since Opal services are stateful in nature, it is natural to make Opal services available via WSRF mechanisms. Our collaborators from Osaka University have used the Opal toolkit as a building block for wrapping scientific applications as WSRF services. The resulting toolkit is called Opal OP, which integrates the Opal toolkit into the Globus Toolkit 4 (GT4) framework. Opal

OP is being used by several projects, for instance by the Protein Data Bank Japan (PDBj) for its structural navigator, and by Osaka University for Quantum Mechanics & Molecular Dynamics Mechanics (QM/MM) hybrid simulations. More information about Opal OP can be found elsewhere (Ichikawa et al, 2007).

CSF4 Based Access to Distributed Resources

The base version of Opal can only submit jobs to a single scheduler managing a cluster. To achieve greater scalability, it is desirable that we leverage multiple distributed clusters. For this reason, we have developed a plug-in for Opal, which allows it to dispatch jobs to the Community Scheduler Framework (CSF4) (Ding et al, 2001). CSF4 is a WSRF-compliant Metascheduler, which is able to dispatch jobs to several distributed computational resources using the Globus GRAM submission protocol. CSF4 also supports array jobs and data staging using the GridFTP (Allcock et al, 2001) protocol. However, at this point, the Opal CSF4 plug-in only supports the same types of applications supported by the base versions of Opal, serial and parallel.

To provide easy seamless integration with Opal, CSF4 has been extended so that it models applications as resources as well, in addition to traditional cluster resources. The description of an application resource includes necessary parameters such as "name," "path," "clusters deployed on" and optional parameters such as "version," "compiler" and "dependent libraries." To submit a job to CSF4, a user only needs to specify the application name, and CSF4 will select one of the computational resources on which the requested application is installed.

This new feature has allowed us to easily leverage Grid resources such as TeraGrid, and PRAGMA. The typical setup of Opal-CSF4 on distributed Grid resources is shown in Figure 6. When users request Opal to run their applications (possibly using a number of processors), Opal forwards the request to CSF4, which picks a computing resource that can satisfy the user's requirements. It then stages the user inputs to the target resource, and runs the job on that particular resource. When the execution is complete, it stages the results back for the user. From the perspective of end-users, they simply use the Opal Web service APIs, and are completely oblivious to where their jobs are being run and how.

Figure 6. Opal CSF4 architecture

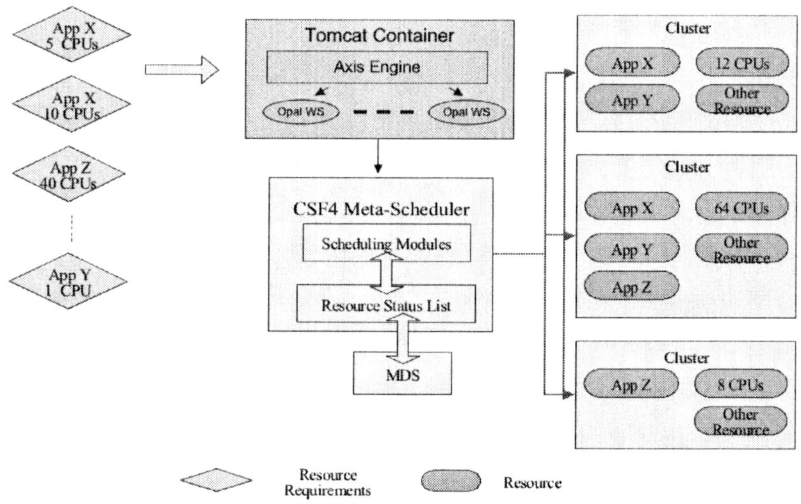

Using Opal and CSF4, we have been able to leverage several TeraGrid clusters to execute a set of NAMD simulations (in particular, DataStar at SDSC, Tungsten at NCSA, and Lonestar at TACC). NAMD (Nanoscale Molecular Dynamics) is a molecular dynamics simulation package, which can scale up to thousands of CPUs. In our simulation of the Avian Influenza surface protein Hemagglutinin binding with avian and human receptors, we have been using up to 256 CPUs. In addition, we have been able to submit AutoDock jobs to perform virtual screening in the avian flu grid project (Ding, 2008).

Although it is desirable to be able to choose the number of processes dynamically depending on the number of resources available, we do not support this feature because there is no generic way to do this while being agnostic to the application implementation. For example, the number of processes may be governed by specific parameters inside file inputs – in the case of APBS; the number of processes depends on the grid parameters specified within the input file. If the number of processors is different from the ones calculated from the grid parameters, the application will fail. Hence we made the decision to let the end-users specify the number of processes for each job – which may be scheduled on any distributed resources using CSF4. In the future, we may consider adding support for dynamically varying the number of processes for certain classes of applications, if the user explicitly specifies that it is ok to do so via the configuration files

INTEGRATED GRID INFRASTRUCTURE USING OPAL

The Opal toolkit has been used in production by several biomedical and bioinformatics applications on a distributed set of resources. In this section, we discuss the types of applications enabled, the back-end infrastructure behind such applications, and the types of clients that are used to access them.

Biomedical and Bioinformatics Applications

A number of popular applications from the National Biomedical Computation Resource (NBCR: Li et al, 2006) and Community Cyberinfrastructure for Advanced Marine Microbial Ecology Research and Analysis projects (CAMERA: http://camera.calit2.net) have been exposed as Web services using the Opal toolkit. Some of the interesting applications are as follows:

1. *APBS*, the Adaptive Poisson-Boltzmann Solver, which is a classical modeling package for electrostatics computations, important in several areas of Biomolecular simulations,
2. *AutoDock* and *AutoGrid*, a suite of automated molecular docking software tools used in applications such as X-ray crystallography and structure-based drug design,
3. *SIESTA*, macromolecular plane wave density functional software that is used for performing electronic structure calculations and ab initio molecular dynamics simulations of molecules and solids,
4. *MolPrep*, a tool enabling manipulation of molecular structures such as rotational conformations,
5. *PDB2PQR*, a utility for converting protein files from the Protein Data Bank (PDB) format to PQR format used by applications such as APBS,
6. *Babel*, a toolkit that enables users to search, convert, analyze, or store data from molecular modeling, chemistry, solid-state materials, biochemistry, or related areas,
7. *MEME* and *MAST*, software tools for discovery and search of motifs (highly conserved regions) in groups of related DNA or protein sequences,
8. *BLAST*, the Basic Local Alignment Search Tool, which enables users to compare a biological sequence against a database of sequences, and identify any resemblance.

Most of these services are running on an NBCR cluster called ws.nbcr.net, and have been used in production since Sept 2007. Figure 7 shows a usage graph of our services running on ws.nbcr.net. As is evident from the graph, some services are more popular than others – in particular, our most heavily used services are MEME and PDB-2PQR. There are other services such as APBS and Continuity, which get fewer hits – however, they can be far more computationally intensive, and hence it is justified to run them on remote clusters, if need be.

Use Cases of Opal Based Services

The principle of "deploy once, access by many" is exemplified by the Opal based services. Opal based services are accessed by portals (Ding et al, 2006), rich desktop clients, lightweight web based clients (Baldridge et al, 2005), and command-line interfaces. Some of the user interfaces being used in practice by our scientific end-users are as follows:

- *Gemstone,* a rich client application (Baldridge et al, 2005) built on top of the Mozilla Firefox platform. Gemstone runs on the user's desktop, and provides an interface for discovering and accessing remote application services. Gemstone can be used by scientific end-users to perform "exploratory workflows", which is defined as the integration of multiple application services in a highly interactive user-directed manner.
- *MGLTools*, developed by The Scripps Research Institute (TSRI), is a set of tools for visualization and analysis of molecular structures. This includes *AutoDockTools* (ADT), which is a graphical front-end for setting up and running *AutoGrid* and *AutoDock*; the *Python Molecular Viewer* (PMV), which provides many pluggable commands ranging from displaying molecular surfaces to running molecular dynamics simulations and energy minimization calculations; and *Vision*, which is a visual-programming environment in which a user can interactively represent novel combinations of computational methods as workflows, and visualize data without actually writing code.
- *Kepler* (http://kepler-project.org), which is a scientific workflow toolkit being developed at the University of California, San Diego (UCSD). Kepler provides a visual interface to create a scientific pipeline, and provides a standard set of tools to interact seamlessly with emerging Grid-based resources.

Figure 7. Opal services usage on the NBCR cluster

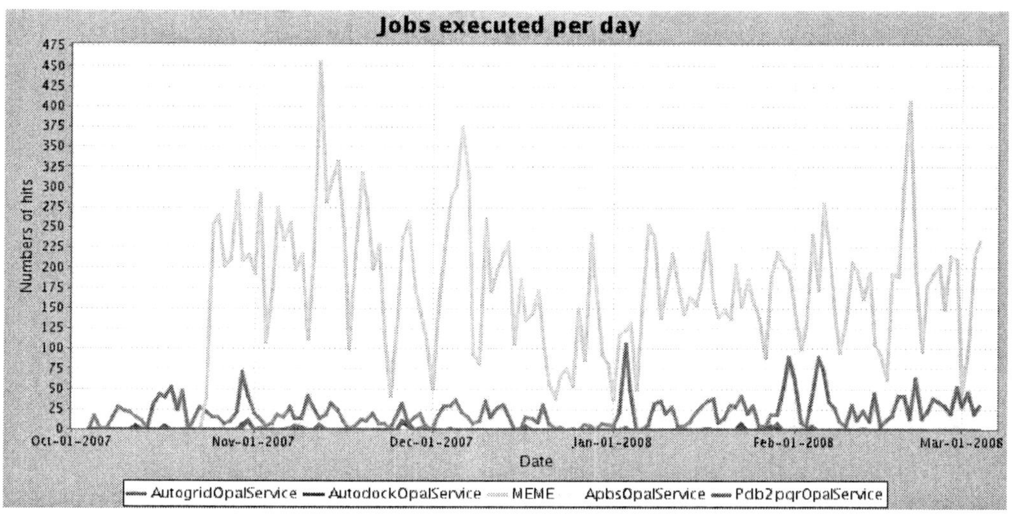

- *My Worksphere* (Ding et al, 2006), based upon the Gridsphere portal framework (Novotny, Russell, Wehrens, 2004), provides a portlet based online environment for job submission, data management and specific application interfaces. The portlet model provides portal developers a standardized model to create pluggable and dynamic application support. The portlets are reusable in JSR168 compliant portal environment such as Open Grid Computing Environment (OGCE http://www.ogce.org/index.php).

Gemstone, Vision, Kepler or similar programming environments allow users to compose workflows based upon not only local software modules but also remotely accessible services, whether data-centric or compute-centric. This allows users to experience rich functionalities all from their desktop environment.

Bioinformatics Workflow Using Opal and Kepler

As mentioned earlier, the MEME and MAST applications have been deployed as Web services using the Opal toolkit on the http://ws.nbcr.net cluster. MEME enables a user to discover motifs (highly conserved regions) in groups of related DNA or protein sequences. MAST enables a user to search sequence databases using motifs. The MEME and MAST services can be coupled into a scientific workflow, where a user first runs MEME to generate motifs for an uploaded sequence dataset, and then searches the sequence databases for these motifs using MAST. We have implemented this scientific workflow using the Kepler framework, as shown in Figure 8.

Kepler has a notion of *actors*, which are components reusable in multiple workflows. Actors read data in via input ports, perform a certain action, and output resulting data via output ports. Kepler provides a standard set of actors that are useful for most scientific workflows, including ours. However, we had to implement a few actors for input generation for the MEME and MAST services. The MEME input generator accepts user inputs, generates a SOAP message, and then makes an invocation on the remote MEME service to launch a MEME job. The service returns an XML message containing a jobID, which is extracted using XML manipulation tools such as XPath/XSLT (since the jobID is contained within the response XML). Using the jobID, the client queries for job status and blocks until the job is complete. Once it is done, the service returns a URL from where the results can be viewed and downloaded. The MEME output is then downloaded and sent to the MAST service inside a SOAP message as a Base64 encoded binary string by the MAST input actor, along with the other MAST parameters. The MAST service also returns a jobID, and the client again queries for the job status and blocks until the job is complete. Once MAST is done, a URL is returned that can be used to retrieve the MAST results.

Thus, with the use of a Web services wrapper toolkit like Opal, and a workflow toolkit like Kepler or Vision, we are able to easily compose workflows involving two or more scientific applications running on complex back-end computational infrastructure. It is worthwhile to note that the Kepler workflow toolkit is completely oblivious to several back-end details — how the applications are deployed, how the jobs are submitted on the Grid, etc. These are exactly the details that we have strived to simplify with the Opal toolkit.

Molecular Science Using Opal and Gemstone

Estimating correct three-dimensional atomic structures of complexes between proteins and ligands is an important component of the drug-design process in the pharmaceutical industry. This process usually involves extracting separate geometries for the protein and the ligand of interest from structural databases, and varying the relative orientation between the protein and the ligand until an optimal orientation is found. A highly accurate quantum chemical model is used for the smaller ligand, employing the computational chemistry package GAMESS (Schmidt et al, 1993). A less accurate electrostatics model, provided by the APBS (Baker et al, 2001) computational package, is used for the larger ligand-protein complex since quantum techniques become prohibitively expensive for molecules of that size. The associated tool, QMView (Baldridge, Greenberg, 1995), is then employed for visualization and analysis.

These computations are wrapped as Web services to fit within our architecture, and are accessed via the Gemstone rich client described earlier. Gemstone provides a highly interactive environment to plug in different services at the back-end, and enables novel exploratory science. It differs from the other traditional workflow toolkits in the sense that the workflows do not need to be built top down, and "pre-wired."

Not all of the above services as Opal-based. The services for Babel, MolPrep, and PDB2PQR are deployed using Opal, while the services for GAMESS and APBS are hand-written for a number of reasons. Firstly, these services were written before the lifetime of the Opal toolkit. Secondly, they provide a sophisticated WSDL API for exchanging and extracting data structures, which is currently not possible with the Opal toolkit. Stronger data typing of this nature is beneficial for easy extraction of data, and translations from one format to another (e.g. with the use of languages such as XSLT). However, Opal-based services and other hand-written Web services can co-exist, as is demonstrated below.

Figure 9 shows the workflow that has been implemented for the above scientific scenario, and accessed via the Gemstone rich desktop client. First, an interesting protein and a ligand are downloaded from the Protein Data Bank (PDB) database. Next, the Opal-based PDB2PQR service is used to convert the protein from the PDB to the PQR format. In parallel, the Opal-based

Figure 8. MEME and MAST workflow using Kepler and Opal

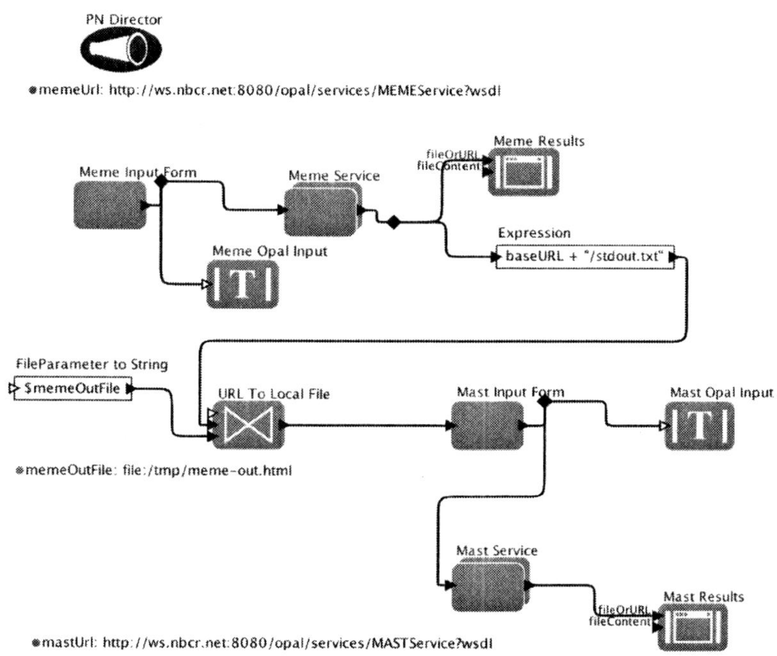

Figure 9. Ligand protein interaction studies using Gemstone

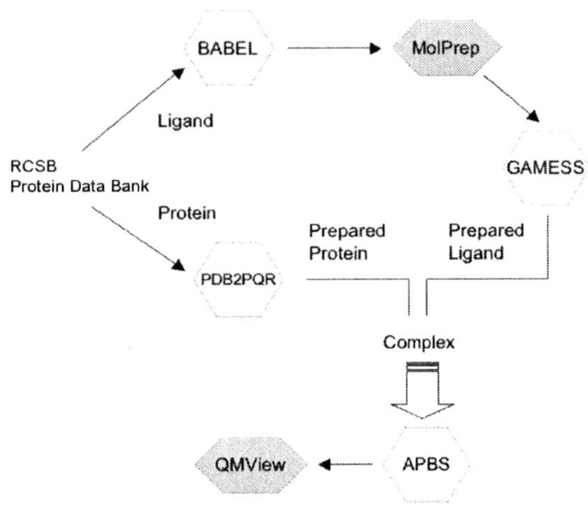

Babel service is used to add hydrogens to the ligand, which is missing from the PDB format. The MolPrep service, also Opal-based, is then used to generate rotational conformations for the ligand. Next, the strongly typed GAMESS service is used to generate accurate charges for the ligand. The strongly typed APBS service is then used to calculate the binding energy for the resulting complex generated by the protein and ligand. The results from APBS are then visualized using the QMView client.

FUTURE TRENDS

Graphical user interface will play an important role in Grid computing in future years because they can simplify the adoption of new scientific applications reducing their learning time. In particular Workflow toolkits can provide to the final user a high level view of the scientific process (or scientific pipeline) without loosing flexibility. Being able to promptly support new submission systems and new parallel application styles is crucial in this rapidly growing field. In this context it is essential to completely shield client applications from these quickly changing back-end technologies. Finally to keep deployment and installation time lows, we are considering virtualization solutions, since they make possible to distribute fully fledge virtual appliances which works out of the box.

In the next subsections we present how we are planning to address these new technological challenges within our SOA framework.

Support for Multiple Workflow Toolkits

We believe that there is no single workflow toolkit that is suitable for all applications, and all scientific users. Several factors influence the end-users, viz. ease of use, support for visualization, the availability of out of the box components for certain scientific tasks, the implementation language, etc. For instance, the Kepler toolkit is written in Java, and provides several useful tools for access to Grid and Web services. It also provides predefined actors for applications in the realm of bio-informatics and geo-sciences. Scientific users within this community, and especially the ones who are familiar with Java, would tend to use a workflow tool such as Kepler. On the other hand, the Vision environment, written in Python, has a foothold in the life sciences community because of its rich support for visualization of molecular systems. Python developers within this community would tend to pick Vision over Kepler. Similarly, Gemstone is based on the Mozilla platform for a more interactive user experience, and uses JavaScript and Web 2.0 technologies. Web 2.0 savvy users would tend to pick Gemstone or similar tools over the others. Our philosophy is, hence, to enable access to scientific applications as Web services so that they can be accessed by a number of tools as the users deem fit.

Automatic Interface Generation

We have been working with the developers of the MGLTools package at TSRI to enable automatic interface generation for the Vision workflow toolkit, and we now have an initial prototype ready. Vision workflows are similar to Kepler workflows – and the term node is used analogous to a Kepler actor. We now have a generic Opal *node* in Vision, which can dynamically retrieve the application metadata from the remote service, and create an input form for the service. Vision is tightly coupled with the other sets of tools that are part of MGLTools, which can be leveraged. In particular, a user can invoke the APBS service remote to calculate the electrostatic potential for a protein using Vision, and then visualize the results using the Python Molecular Viewer (PMV). We are also extending this functionality of interface auto-generation to other workflow tools such as Kepler. We believe that automatic interface generation could be extended to other types of rich client applications as well, to fully utilize the potential of Opal based services.

Opal 2.0

We are current in the implementation phase for the next generation of the Opal toolkit. Opal 2 will be a complete redesign of the current version of Opal for greater flexibility and performance. One of the key benefits of the new version of Opal would be the implementation of a plug-in architecture for the various back-ends. In particular, there will be separate plug-ins for back-ends such as GRAM (Foster, Kesselman, 1997), DRMAA (http://www.drmaa.net), CSF4 (Ding et al, 2001), and GridWay (Huedo, S. Montero, M. Llorente, 2004). As new computational back-ends become available, it will be very easy to integrate it into Opal as this would only require the implementation of an appropriate plug-in. State management will be provided with the help of Hibernate (http://www.hibernate.org), which enables us to use a number of database implementations at the back-end. For improved performance, we will also support other mechanisms for user data upload such as GridFTP, which is missing from the current version of Opal. We are also investigating better validation mechanisms for user inputs, in particular the ones that are file-based. Finally, we will support multiple styles of application execution, in addition to the serial and parallel ones currently provided. This includes modules for parameter-sweep, map-reduce (Dean, Ghemawat, 2004), etc. Users will have to notify Opal about their application types via the configuration files – and Opal will use the appropriate back-end plug-in such as Nimrod (Abramson, 1995) or Apache Hadoop (http://hadoop.apache.org) respectively.

Virtualization

For better packaging of Opal for clusters and virtualization using VMWare or Xen virtual machines, we are building *rocks rolls* (Bruno et al, 2004) that can be easily deployed as Rocks based virtual computing environment. This will allow applications to be distributed with virtual machines as a ready to use virtual appliances. These appliances may contain better support for interoperability with other WSRF services so that we can leverage all the novel functionality provided by GT4 and other WSRF resources, such as those from the caGrid services (Oster et al, 2007).

RELATED WORK

The area of Web service wrappers to scientific applications is not without precedent. The Java CoG Kit (von Laszewski et al, 2001) provided a way to expose legacy applications as Web services as early as 2002. It used a *serviceMap* document to generate source code for the Web service implementation, along with the WSDL's for the same. However, source code generation is not very flexible. It is highly dependent on the version of the Web service software being used, and not very conducive to the addition of new features. Furthermore, the above was only a prototype, and did not support complex features such as Grid scheduling, concurrent and asynchronous job submissions, data management for jobs, authentication and authorization, etc.

The Generic Factory Service (GFac) (Kandaswamy et al, 2006) is a much more sophisticated toolkit for wrapping scientific applications, that relies on the XSUL SOAP libraries (http://www.extreme.indiana.edu/xgws/xsul) for Web services support. It also uses a *serviceMap* to generate new WSDL descriptions for every scientific application. However, it does not rely on source code generation. Instead, it uses an XSUL *Message Processor* to intercept the SOAP calls for a particular Web service and route it to a generic class that invokes the scientific application using the information provided by the serviceMap document. The disadvantage of this implementation is that this approach is very XSUL specific. Most scientific communities such as NBCR, CAMERA, GEON (http://www.geongrid.org), etc. prefer to use commodity SOAP toolkits (e.g. Apache Axis) for Web services development.

SoapLab (http://www.ebi.ac.uk/soaplab) is another advanced toolkit for wrapper generation. They define application configurations using the *ACD* format, and generate the WSDL, and the source code for the Web service implementations from them. As we discussed earlier, source code generation has several disadvantages. Furthermore, they use CORBA (http://www.corba.org) for discovering, starting, and controlling applications. Although the performance and ease of use of CORBA versus Grid software is debatable, the use of CORBA is non-standard in the Grid world – and any additional software that needs to be installed and accessed is traditionally met with a lot of inertia. In our experience, it has proven to be most prudent to use standard commodity software to build our applications, and also to keep our software requirements to the bare minimum. This has been our primary consideration in developing the Opal toolkit.

PISE (Letondal, 2001) is an application to create Web based user interfaces, which allows specification of command line parameters using an XML based configuration file. With a set of PERL scripts, PISE is able to create Web forms and CGI scripts for the execution of applications. Its complex configuration file, which also includes PERL statements, enables definition of advanced rules for command line argument validation. However its complexity can prevent application developers from adopting the same. Another drawback of PISE is its static interface generation engine, which can create submission forms only at compile time, while the Opal GUI is able to dynamically discover application parameters and build submission forms (Clementi et al, 2007).

CONCLUSION

Grid computing provides an enormous opportunity to scientific end-users to use distributed computing and storage resources to enable novel science hitherto impossible using traditional computing mechanisms. However, most Grid middleware tools are complicated to install and too jargon rich for scientists to learn. In order to simplify the use of Grid infrastructures, one has to provide easy to use abstractions to the complex back-end functionality. Towards this end, we have developed the Opal toolkit, which enables wrapping scientific applications running on Grid resources as simple Web services, which can be accessed via a number of user interfaces. The toolkit is beneficial to both end-users as well as scientific application developers. The former can access the scientific applications through Web interfaces or rich desktop clients without having to learn about the infrastructure that the jobs are being run on, while the latter can rapidly expose deployed applications as Web services with the help of Opal toolkit without having to write a single line of code. The automatic interface generation feature of the Opal toolkit further simplifies the creation of human-usable interfaces to the scientific applications, while the use of a metascheduler such as CSF4 enables the use of multiple sets of resources at the back end. We believe that the use of services-oriented science, along with the Web 2.0 style techniques for access, will go a long way towards lowering the entrance barrier for scientific end-users to effectively use Grid middleware.

ACKNOWLEDGMENT

The authors would like to acknowledge support from the NIH P41 RR 08605 to NBCR; TATRC W81XWH-07-2-0014 to PWA, the Gordon and Betty Moore Foundation grant for the CAMERA project, and the NSF grants INT-0314015 and OCI-0627026 for the PRAGMA project. The authors would also like to thank Ilkay Altintas, Peter Arzberger, Kim Baldridge, Nathan Baker, Karan Bhatia, Sargis Dallakyan, Wes Goodman, Fred Lionetti, Jane Ren, Michel Sanner, and Guillaume Vareille for their feedback and help with various pieces of the end-to-end system.

REFERENCES

Abramson, D., Sosic, R., Giddy, J., & Hall, B. (1995) Nimrod: a tool for performing parametrised simulations using distributed workstations. *Fourth IEEE International Symposium on High Performance Distributed Computing* (pp 112-121).

Allcock, W., Bester, J., Bresnahan, J., Chervenak, A., Liming, L., & Tuecke, S. (2001, March). In GridFTP: Protocol Extension to FTP for the Grid. Unpublished Internet-Draft, Grid Forum.

Arzberger, P. W., & Papadopoulos, P. (Feb 2006). PRAGMA: Example of Grass-Roots Grid Promoting Collaborative e-Science Teams. *CTWatch Quarterly, 2* (1).

Atkins, D., Kroegemeier, K., Feldman, S., Garcia-Molina, H., Klein, M., Messerschmitt, D., et al. (2003). Revolutionizing Science and Engineering Through Cyberinfrastructure. In *Report of the National Science Foundation Blue-Ribbon Advisory Panel on Cyberinfrasturcuture.* National Science Foundation: Arlington, VA.

Baker, N. A., Sept, D., Joseph, S., Holst, M. J., & McCammon, J. A. (2001). Electrostatics of nanosystems: application to microtubules and the ribosome. *Proceedings of the National Academy of Sciences of the United States of America 2001,* (Vol. 98), (pp. 10037-41).

Baldridge, K. K., & Greenberg, J. P. (1995). QMView: a computational chemistry three-dimensional visualization tool at the interface between molecules and mankind. *Journal of Molecular Graphics, 13,* 63-66).

Baldridge, K., Bhatia, K., Greenberg, J. P., Stearn, B., Mock, S., Sudholt, W., Krishnan, S., Bowen, A., Amoreira, C., & Potier, Y. (2005). In GEMSTONE: Grid Enabled Molecular Science Through Online Networked Environments. *Life Sciences Grid Workshop, Singapore*. Singapore: World Scientific Press.

Basney, J., Livny, M., & Tannenbaum T. (1997, June). High Throughput Computing with Condor. *HPCU news, 1*(2).

Bhatia, K., Chandra, S., & Mueller, K. (2006). *In GAMA: Grid Account Management Architecture.* 1st IEEE International Conference on e- Science and Grid Computing, Melbourne, Australia.

Bruno, G., Katz, M., Sacerdoti, F. D., & Papadopoulos, P. M. (2004) *In Rolls: Modifying a Standard System Installer to Support User-Customizable Cluster Frontend Appliances.* IEEE International Conference on Cluster Computing, San Diego, CA (pp 421-430).

Clementi, L., Ding, Z., Krishnan, S., Wei, X., Arzberger, P. W., & Li, W. W. (2007). *In Provid-*

ing dynamic virtualized access to grid resources via the web 2.0 paradigm. Grid Computing Environments Workshop (Supercomputing 2007), Reno, NV.

Dean, J., & Ghemawat, S. (2004). *In MapReduce: Simplified Data Processing on Large Clusters*. Sixth Symposium on Operating System Design and Implementation (OSDI'04), San Francisco, CA.

Ding, Z., Luo, Y., Wei, X., Misleh, C., Li, W. W., Arzberger, P. W., & Tatebe, O. (2006). *In My WorkSphere: Integrative Work Environment for Grid-unaware Biomedical Researchers and Applications*. Supercomputing Conference 2006, SC06, 2nd Grid Computing Environment Workshop, Tampa, FL.

Ding, Z., Wei, X., Tatebe, O., Papadopoulos, P. M., Arzberger, P. W., & Li, W. W. (2008). Cyberinfrastructure for Biomedical Applications: Metascheduling as Essential Component for Pervasive Computing. In, J. Cao, (Ed.), *Cyberinfrastructure Technologies and Applications*,. Hauppauge, NY: Nova Science.

Dolinsky, T. J., Czodrowski, P., Li, H., Nielsen, J. E., Jensen, J. H., Klebe, G., & Baker, N. A. (2007). PDB2PQR: expanding and upgrading automated preparation of biomolecular structures for molecular simulations. *Nucleic Acids Res, 35*, 522-525.

Foster, I., Kesselman, C., Nick, J.M., & Tuecke, S. (2002). Grid Services for Distributed Systems Integration. *Computer 35*(6).

Foster, I. (2005). Service-oriented science. *Science, 308*, 814-817.

Foster, I., & Kesselman, C. (1997). Globus: a Metacomputing Infrastructure Toolkit. *The International Journal of Supercomputer Applications and High Performance Computing, 11* (2), 115-128.

Foster, I., & Kesselman, C. (2004). *The Grid 2: Blueprint for a New Computing Infrastructure*, (2 Ed). San Francisco, CA: Morgan Kaufmann Publishers, Inc..

Foster, I., & Kesselman, C. (Ed.) (1999). *The Grid: Blueprint for a New Computing Infrastructure*. San Francisco, CA: Morgan Kaufmann Publishers, Inc..

Foster, I., Kesselman, C., Tsudik, G., & Tuecke, S. (1998). *In A Security Architecture for Computational Grids*. ACM Conference on Computer and Communications Security, San Francisco, CA.

Grethe, J. S., Baru, C., Gupta, A., James, M., Ludaescher, B., Martone, M. E., Papadopoulos, P. M., et al. (2005). Biomedical informatics research network: Building a national collaboratory to hasten the derivation of new understanding and treatment of disease. *Stud Health Technol Inform, 112*, 100-109.

Hey, T., & Trefethen, A. E. (2005). Cyberinfrastructure for e-Science. *Science, 308*, 817-821.

Huedo, E., S. Montero, R., & M. Llorente, I. (2004). A framework for adaptive execution in grids. *Software: Practice and Experience, 34*(7), 631-651).

Ichikawa, K., Date, S., Krishnan, S., Li, W. W., Nakata, K., Yonezawa, Y., Nakamura, H., Shimojo, S. (2007). *In OPAL OP: An Extensible Grid-Enabling Wrapping Tool For Legacy Applications*. 3rd International Workshop on Grid Computing & Applications (GCA2007), Singapore.

Kandaswamy, G., Fang, L., Huang, Y., Shirasuna, S., Marru, S., & Gannon, D. (2006). Building web services for scientific grid applications. *IBM Journal of Research and Development, 50*(2/6), 249-260.

Krishnan, S., Stearn, B., Bhatia, K., Baldridge, K. K., Li, W. W., and Arzberger, P. (2006). Opal: Simple Web Services Wrappers for Scientific

Applications. In *IEEE International Conference on Web Services*, (pp 823-832).

Krishnan, S., & Bhatia, K. (2007). SOAs for Scientific Applications: Experiences and Challenges. In *the 3rd International Conference on e-Science and Grid computing.*

Letondal, C. (2001). A Web interface generator for molecular biology programs in Unix. *Bioinformatics, 17,* 73-82).

Li, W. W., Baker, N. A., Baldridge, K., McCammon, J. A., Ellisman, M. H., Gupta, A., et al. (2006). National Biomedical Computation Resource (NBCR): Devloping End-to-End Cyberinfrastructure for Multiscale Modeling in Biomedical Research. *CTWatch Quarterly, 2,* 6-17.

Li, W. W., Krishnan, S., Mueller, K., Ichikawa, K., Date, S., Dallakyan, S., et al. (2006). *In Building Cyberinfrastructure for Bioinformatics Using Service Oriented Architecture: Cluster Computing and the Grid Workshops.* Sixth IEEE International Symposium on Cluster Computing & the Grid, Singapore.

Novotny, J., Russell, M., & Wehrens, O. (2004). GridSphere: a portal framework for building collaborations. *Concurrency and Computation: Practice and Experience, 16,* 503-513.

O'Reilly, T. (2005). *What is Web 2.0: Design patterns and business models for the next generation of software.* Retrieved on April 10, 2008 from http://www.oreillynet.com/pub/a/oreilly/tim/news/2005/09/30/what-is-web-20.html.

Oster, S., Langella, S., Hastings, S., Ervin, D., Madduri, R., Phillips, J., et al. (2007). caGrid 1.0: An Enterprise Grid Infrastructure for Biomedical Research. *Journal of the American Medical Informatics Association, 15,* 138-149.

Perlman, L., Welch, V., Foster, I., Kesselman, C., & Tuecke, S. (2002). *In A Community Authorization Service for Group Collaboration.* IEEE 3rd International Workshop on Policies for Distributed Systems and Networks.

Sanner, M. F., Duncan, B. S., Carrillo, C. J., & Olson, A. J. (1999). Integrating computation and visualization for biomolecular analysis: an example using python and AVS. *Pacific Symposium in Biocomputing, 99,* 401-412.

Schmidt, M. W., Baldridge, K. K., Boatz, J. A., Elbert, S. T., Gordon, M. S., Jensen, J. H., et al. (1993). General atomic and molecular electronic structure system. *Journal of Computational Chemistry, 14,* 1347-1363.

Siebenlist, F., Magaratnam, N., Welch, V., & Neuman, C. (2004). Security for Virtual Organizations: Federating Trust and Policy Domains. In *The Grid: Blueprint for a New Computing Infrastructure* (2nd Ed.). Amsterdam: Morgan Kaufmann.

Smarr, L., & Catlett, C. E. (June 1992). Metacomputing. *Communications of the ACM, 35*(6), 44-52).

Stevens, R. C. (2006). Trends in Cyberinfrastructure for Bioinformatics and Computational Biology. *CTWatch Quarterly, 2,* 6-17.

von Laszewski, G., Foster, I., Gawor, J., & Lane, P. (2001). A Java Commodity Grid Kit. *Concurrency and Computation: Practice and Experience, 13,* 643-662.

KEY TERMS AND DEFINITIONS

Automatic Interface Generation: Refers to the capability of generating dynamically customized GUI for a particular application simply using an intermediate language.

Cyberinfrastructure: A research environment that support advanced data manipulation capabilities and other computing processing services over the internet.

Grid Computing: In general used to refer to a special type of parallel computing that relies on a distributed set of loosely coupled commodity computers interconnected by a network.

Interoperability: The capability of two or more software components to interact exchanging information.

OPAL: A software framework which allows the deployment of scientific application as a Web Service without writing any code.

Service Oriented Science: A software architectural paradigm that promotes the development of system which providing platform-independent, language-neutral service interfaces that hide the complexity of the implementations.

Web 2.0: A term adopted to describe the use of the World Wide Web platform to enhance creativity, information sharing, and, mainly, collaboration among users.

Web Service: A software system designed to support interoperable machine-to-machine interaction over a network, defined by the W3C consortium.

Chapter XXIX
The LIBI Grid Platform for Bioinformatics

The LIBI Grid Platform Developers, Italy[1]

ABSTRACT

The LIBI project (International Laboratory of BioInformatics), which started in 2005 and will end in 2009, was initiated with the aim of setting up an advanced bioinformatics and computational biology laboratory, focusing on basic and applied research in modern biology and biotechnologies. One of the goals of this project has been the development of a Grid Problem Solving Environment, built on top of EGEE, DEISA and SPACI infrastructures, to allow the submission and monitoring of jobs mapped to complex experiments in bioinformatics. In this work we describe the architecture of this environment and describe several case studies and related results which have been obtained using it.

INTRODUCTION

Modern biology, and particularly the molecular aspects of genomics, proteomics and transcriptomics, require multidisciplinary skills and the use of high-level technologies. Bioinformatics has been developed to meet these needs; it encompasses biological knowledge and expertise in advanced information technology. Public and private interests and investments have definitely boosted this discipline in many European and non-European countries. In Italy in particular many important initiatives have been funded to accelerate the development of bioinformatics. Among them, a project named LIBI (www.libi.it) (International Laboratory of BioInformatics, starting date September 2005), aims at setting up an advanced bioinformatics and computational biology laboratory, focusing on basic and applied research in modern biology and biotechnologies.

LIBI has its headquarters in Bari, Italy. It has been conceived as a bioinformatics laboratory "without walls" and carries out a number of activities, including: i) multidisciplinary collaborations to encourage interactions between researchers in public institutions and industry working in the biomedical and biotechnological field; ii) technological transfer and dissemination to promote training and provision of bioinformatic services; iii) the implementation of a high performance distributed infrastructure supporting the access to large data sets and high computational power; iv) the promotion of bioinformatics research at the (inter)national level maintaining and creating new specialized databases, developing new algorithms, developing and implementing new and more efficient processes and mechanisms for automatic bioinformatics analysis.

Even though this project features only Italian partners the idea is to open the laboratory to other international research groups in order to test and therefore improve the offered services.

To achieve a high level of transparency we designed and developed the LIBI Grid portal which represents the entry point to the LIBI platform, allowing easy and persistent access to distributed and heterogeneous data and software.

The LIBI project will terminate in 2009 and for four years more it will guarantee a virtual working space for the academic and industrial partners belonging to public and private institutions which shall provide their infrastructures, use the platform tools and equipment, and benefit from the same capacity and expertise made available by the other partners. The activities, technological platforms and applications will be integrated within the same environment, aiming at producing and increasing scientific knowledge, providing services and training as well as enhancing international competitiveness.

In this chapter a description of the project will be presented, proposing the architecture of the Grid PSE and the main services grouped in computational and data grids; finally, several case studies implemented with this infrastructure will be discussed as well.

BACKGROUND

A key requirement, considered during the design phase of the LIBI laboratory, has been the consideration that bioinformatics applications are naturally distributed, because experimental data and biological databases are themselves usually distributed. Also many experiments require huge computing power, owing to the large size of data sets and the complexity of processing, and may need to access heterogeneous data, where heterogeneity is multifaceted (data format, access policy, distribution, etc.), and require a secure infrastructure to protect and secure the access to private data owned by different organizations.

The Problem Solving Environment (PSE) (Houstis, 1997) is an approach and a technology that can fulfill such bioinformatics requirements. The PSE can be used for the definition and composition of complex applications, hiding programming and configuration details to the user that can concentrate instead only on the specific biological problem. Moreover, computational grids (Foster, 1999) can be used for building geographically distributed collaborative problem solving environments and grid-aware PSEs (Laszewski, 2001) can search and use dispersed high performance computing, networking and data resources. In this work, the PSE solution has been chosen as the integration platform for bioinformatic tools and data sources.

THE LIBI - INTERNATIONAL LAB FOR BIOINFORMATICS

The main goal of this project is to set up and test an advanced bioinformatics and computational biology laboratory, focusing mainly on the following activities:

- The construction and the maintenance of general genomic, proteomic and transcriptomic databases (e.g. ENSEMBL) as well as specialized databases developed by LIBI partners (e.g. MitoRes, UTRdb, UTRsite, ASPicDB, CSTdb, etc.);
- The design and implementation of new algorithms and software for the analysis of genomes and their expression products and for the prediction of the structure of proteins.

It is a public and private distributed laboratory targeting the following requirements of the bioinformatic community:

- Access to shared tools (analysis, data and displaying tools) in the form of simple or composed (workflow) services and their optimization;
- Distribution of data and their transparent access (database federation);
- Development of a collaborative environment allowing users to solve complex biological problems;
- The possibility to customize the user environment (content management) and manage or share/exchange personal data according to well defined authorization policies.

To satisfy these requirements the LIBI strategy was defined considering the following points: i) choice of high level and interoperable technologies based on open standards; ii) applications and services following the SOA (Service Oriented Architecture) approach; iii) priority of the end-user in defining functionalities (user driven approach); iv) easy access to the platform; v) dissemination and opening to public services.

The Grid Problem Solving Environment (Grid PSE) meets these requirements providing an integrated software platform for solving complex experiments in bioinformatics. Moreover, the Grid offers the computational power needed for solving large scale simulations, through easy-to-use interfaces that hide implementation details.

Each component (application or data) is seen as an atomic service that the user can use "as is" or compose into a graph (workflow).

In the design phase of the laboratory, these applications have been classified considering the computing needs for the following two classes of experiments:

- *High Throughput Computing (HTC)* solves problems that can be split up into many smaller tasks that are essentially independent (coarse-grained problems). They require large amounts of weakly coupled resources;
- *High Performance Computing (HPC)* solves problems that can be split up into small tasks that need to exchange information and often require synchronization (fine-grained problems). They require large amounts of highly coupled resources such as memory, CPU, disk, etc.

According to this classification, a re-engineering of the applications has been needed for their porting in the grid environment. Major details will be given in the Section "Porting Bioinformatic Applications On The Grid" which focuses on several case studies implemented within LIBI.

The partners involved in the laboratory are grouped into 8 research units (RUs), located in Italy (Fig. 1): 4 technological units, whose specific task is to design and implement the technological platform of the Laboratory and 4 scientific-bioinformatic units which, besides dictating the contents and function of the laboratory, are the main users and exploit the platform for developing their research activities.

The technological RUs and their roles in the project are:

- **SPACI, University of Salento:** it offers a HPC/HTC platform and collaborates in the porting of multiple sequence alignment (HTC), protein tertiary structure prediction (HTC) and molecular dynamics simulations (HPC), sequence pattern search (HTC) and Bayesian phylogenies (HPC) applications;

- **INFN (Sections from Bari, Catania, Padova and CNAF-Bologna):** it provides the HTC infrastructure, monitors and gives access to the INFN production GRID and to the EGEE infrastructure (http://www.eu-egee.org/) and collaborates in the porting of applications for sequence similarity search, sequence pattern search and Bayesian phylogenies;
- **IBM Italia S.p.A (IBM Innovation Lab – Bari):** is working on the data federation platform by integrating several structured and unstructured bioinformatic data sources. This RU includes also Exhicon S.r.l which collaborates in the realization of the data federation platform, the University of Bari, Dept. of Computing Science which is involved in providing a number of text analysis services, and the University of Bari, Dept. of Biochemistry and Molecular Biology which is setting-up a specialized a human mitochondrial genomic resource based on variability studies for supporting population genetics and biomedical research;
- **CINECA (the Interuniversity consortium for supercomputing in North-East Italy)** provides an HPC platform and collaborates in the porting of applications for the molecular dynamics simulations and Bayesian phylogenies. As a sub-contractor, CASPUR (the Interuniversity consortium for supercomputing applications for universities and research) offers a HPC/HTC bioinformatic platform for comparative genomic and gene expression studies. It also works with scientific RUs in developing new applications within a Service-Oriented Architecture, where application integration is facilitated by using bioinformatic web services as the building blocks.

The scientific RUs are:

- **CNRBA (CNR Institute for Biomedical Technologies, Bari):** its research activity mainly focuses on genomic and proteomic databases;
- **UNIBO (University of Bologna)** proposes data and services related to the bioinformatics characterization of the proteome. The activities include: the development of new algorithms, software and web resources for the study and characterization of human and other organism proteomes and the development of proteomic databases;
- **UNIMI (University of Milan):** it proposes the development of new algorithms for the functional analysis and annotation of genomes, particularly focused on the identification and characterization of regulatory elements regulating gene expression at the transcriptional and post-transcriptional level;
- **CBMTS (Centre of Molecular Biomedicine, Trieste):** proposes the development of new algorithms for the characterization of human and other organism transcriptomes and contributes to the development of transcriptomic databases.

LIBI Architecture

In practical terms, the LIBI project aims to set-up a virtual bioinformatics research workplace by providing a high performance infrastructure consisting of a grid computing facility that includes a number of data and computational nodes (both HPC and standard nodes) as well as bioinformatic application services and resources. Figure 2 provides a high-level description of the LIBI functional environment.

The architecture of the system (Figure 3) is Service-Oriented (SO), and is built on a network infrastructure for the exchange of information.

Figure 1. LIBI partners

Figure 2. The LIBI environment

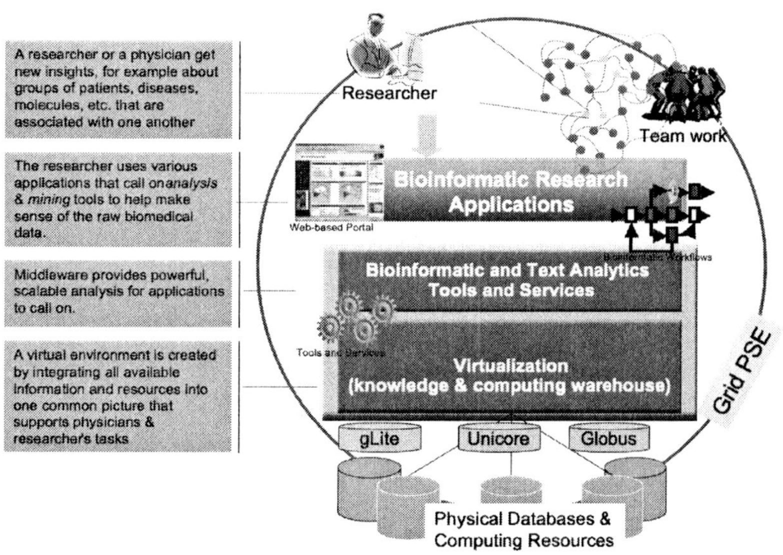

On the top layer there are the application services that can be run separately or composed into a graph by means of a workflow.

Applications in the Laboratory include, for example:

- *BLAST* (Basic Local Alignment and Search Tool) and *PSI-BLAST* (Position-Specific Iterated Blast) (Altschul, 1997): both programs carry out the similarity searches in a data bank, with the latter using a position-specific scoring matrix in the search; these tools are both HTC;
- *PatSearch* (Grillo, 2003) searches complex sequence motifs (i.e. any assortment of regular expression patterns, secondary structures and position weight matrices) and is HTC;
- *DNAFan* (Gisel, 2004) searches specific genomic regions (DNA) defined by a Feature ANalyser where specific patterns can be matched by using PatSearch, hence it is HTC.
- *FT-COMAR* (Vassura, 2007; Vassura, 2008) performs the reconstruction of 3D protein structure from its contact map. It is HTC;
- *AntiHunter* (Lavorgna, 2005): tool for the identification of expressed sequence tag (EST) antisense transcripts from BLAST output. It is HPC;
- *GROMACS* (Lindahl, 2001) /*NAMD* (James, 2005) molecular dynamics software. It is HPC;
- *MrBayes* (Ronquist and Huelsenbeck, 2003) performs Bayesian inference of phylogeny and is HPC/HTC;
- CSTminer (Castrignanò, 2004) identifies conserved tracts in the comparison of two sequences and classifies them as coding (likely belonging to protein coding genes) or non-coding (potential regulatory regions).

These are described in detail in the "Porting bioinformatic applications on the Grid" section in several case studies deployed in the Laboratory.

Access to the applications is obtained via high level Resource Management and Data Management services.

The Resource Management service contains the execution logic of the applications and exploits the so-called "Meta Scheduler" which is based on several highly efficient C libraries implementing functionality for job submission, job monitoring, file transfer and management of users' credentials (Aloisio, 2007).

These include technologies for the development of Web Services based PSE through the use of grid portals that contain the logic for accessing the distributed system by means of easy-to-use interfaces such as the Web. Web Service technology and the PSE allow a secure and ubiquitous access to computing resources and implemented services. The Meta Scheduler is built on top of basic services offered by different grid middlewares such as gLite for HTC applications, Unicore for HPC and Globus for HTC/HPC applications. These grid middlewares are described in the "The LIBI Resources" section.

The functionalities provided by the Resource Management service include scheduling and

Figure 3. The LIBI architecture

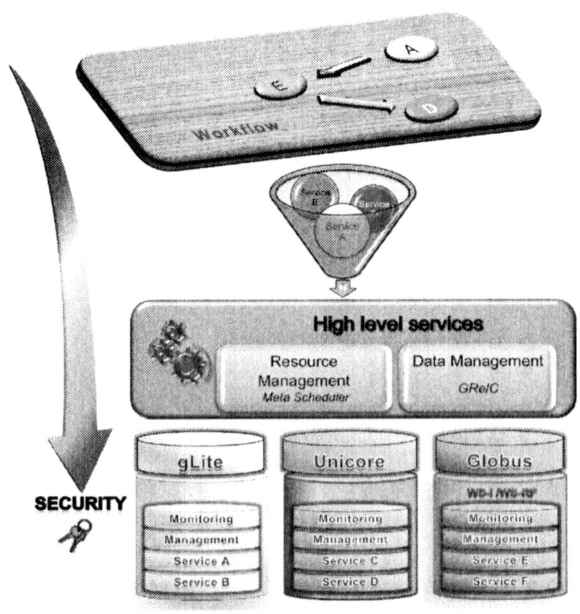

monitoring of the user requests, the optimal allocation of resources in a distributed environment, and so on.

The Data Management service incorporates the grid logic for advanced data management, allowing a transparent and dynamic access to heterogeneous and geographically distributed data. Such functionalities are offered within the LIBI platform by the GRelC Data Access Service (DAS) (Aloisio, 2005a).

This grid service includes in addition secure mechanisms for data transfer on the Grid using advanced protocols to guarantee a high level of efficiency. The Security component handles the management and access to both services and resources within the LIBI distributed environment. It provides mechanisms for authentication, authorization, data protection, etc.

THE GRID IMPLEMENTATION OF THE VIRTUAL LAB

This section presents the Resource and Data Management services implemented for these platforms. For the Data Management service, the implementation of the Federated DB is also described.

Resource Management: Interoperability in the Grid

One of the key aspects in life science applications is the execution of a high number of jobs for large-scale analysis. The Grid is a powerful tool for solving this kind of problem, but in a distributed environment, bioinformatic applications need to be re-engineered in order to provide better performance. In addition, a tool for orchestrating the execution of different kinds of applications and the access to the needed data is fundamental, because different and heterogeneous grid infrastructures have to coexist. The interoperability issue must be addressed analyzing and taking into account the differences between the adopted approaches in the following areas: job submission and monitoring, file transfer, authentication and authorization, and information services.

The solution adopted within the LIBI infrastructure consists of the development of an easily extensible Meta Scheduler able to support different middleware. The development leverages on the GRB technology, developed by the University of Salento, Lecce & the SPACI Consortium, Italy and includes a set of grid libraries, the Meta Scheduler and the LIBI grid portal.

The LIBI grid portal provides an integrated approach to grid resource management, through a user-friendly web GUI and back-end GSI-enabled Meta scheduler Web Service, developed using gSOAP (Van Engelen, 2003) and our GSI plugin for gSOAP (Aloisio, 2003). The Meta Scheduler provides the following features:

- It can be user-centric or VO-centric. Each user is allowed to create his or her own personal grid on-the-fly by adding, removing or updating his or her computer configuration. Alternatively, the Meta Scheduler allows a user to be associated to a GRB managed VO; in this case the user will automatically benefit from all the computational resources belonging to that VO;
- It allows the use of multiple credentials (and thus the usage of multiple machines belonging to different VOs) in the same job submission;
- It provides the automatic, transparent refresh of job credentials to support long running jobs;
- It notifies automatically the user via email each time the status of a job changes;
- It is based on the Open Grid Forum (OGF) Job Submission Description Language, vastly improved to support the addition of VO description, credentials utilized, software installed on remote machines, parameter sweep and workflow job submissions;

- Workflow description based on WF management coalition WorkFlow Description Language (WFDL); the Meta Scheduler is currently partially compliant with WFDL.

The Resource Management service includes a workflow management system for supporting complex "in silico" experiments. It has an editor, implemented in the Java programming language, which allows the definition of an experiment using a graphical formalism for modeling the experiment and describing it in detail.

In a first version we chose UML (Unified Modeling Language) since this provides a visual modeling notation valuable for designing and understanding complex systems. It has been extended for supporting specific grid requirements such as data stage-in or stage-out, parameter sweep jobs and so on. The application components are chosen in the editor and the constructed graph is checked before execution by using rules derived from metadata and ontology (Aloisio, 2005b).

We classified workflow components as data banks, bioinformatic algorithms, graphics tools and input data types. Such a classification is done with a simple ontology written in DAML+OIL.

The workflow engine is embedded in the Meta Scheduler (Cafaro, 2008).

The Meta Scheduler supports batch job submission, supporting X-Windows applications, parameter sweep jobs and workflow jobs. Resource brokering is done using the resource related information gathered from the iGrid Information Service (developed within the GridLab Project), but Meta Scheduler supports also LDAP based information services such as Globus-MDS and gLite BDII.

Within the LIBI project the GRB framework has been extensively extended in order to support the interoperability among different middlewares, particularly Globus, gLite and Unicore. The design of the Meta Scheduler was driven mainly by the definition of a modular architecture which can be extended easily by a plug-in based approach. In order for the Meta Scheduler to support different grid middlewares the specific "adaptor" for the target middleware must be developed and plugged in the Meta Scheduler architecture. The adaptor must include the mechanisms, specific for the target grid middleware, to address job submission and monitoring, authentication and authorization, file transfer and the gathering of resource related information. The Meta Scheduler is then able to use gLite grid resources through its "grb_glite adaptor." Figure 4 illustrates the interchange mechanism between gLite Workload Management

Figure 4. Meta Scheduler support for gLite WMS

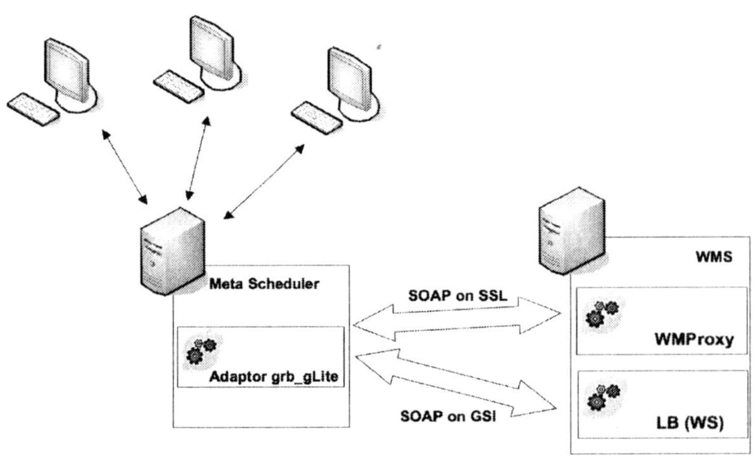

System (WMS) and Meta Scheduler. The Meta Scheduler can access the WMS using the user's credentials with the VOMS extension obtained from the VOMS server. The job is described according to the JDL language and submitted to the WMProxy web service using the GSI enabled web service interface; the job monitoring is performed contacting the Logging and Bookkeeping (LB) web service while the staging in and out of files is explicitly done via GridFTP contacting the appropriate Storage (SE).

In addition to the standard case, for those applications which require the submission to the gLite grid of a large number of jobs, the Meta Scheduler uses a tool developed by the INFN-Bari section for monitoring and bookkeeping called the Job Submission Tool (JST). A description of the tool can be found later in this book.

Within the LIBI project the "grb_unicore adaptor" has also been developed, in order to support a Unicore based grid infrastructure. The Meta Scheduler can contact the Network Job Supervisor (NJS) service, through the Unicore Gateway service, by using the user's credentials through a GSI connection in order to submit batch jobs. The job is submitted to the Gateway using the AJO description file format; the Gateway exposes also functionalities for job monitoring and output retrieval; the file transfer mechanism is performed through a GridFTP connection with the remote NJS.

Data Management: Federated Approaches in the Grid

One of the general challenges in life science research is the heterogeneous nature of both bioinformatic data sources and the nature of the content hosted in them. It is not uncommon to find in every laboratory or even for a specific life science project the adoption of a variety of data source formats, such as relational databases (such as Oracle, DB2, and MySQL) mixed with non-relational sources. Such non-relational sources include XML documents, spreadsheets, ASCII text files and BLAST/FAST/HMMR files. Additionally, independently of the data source format, data sets may include various kinds of structured and unstructured data, such as collections of texts, notes, images, etc. The ability to manage, integrate and analyse related structured and unstructured information, possibly hosted in heterogeneous data source repositories, is fundamental for accelerating bioinformatics research.

Two key elements that must be addressed by the data management framework concern:

1. The need to federate several and heterogeneous data sources within the LIBI context;
2. The need to securely and transparently access the federated LIBI data bank in the grid.

In the following sections we describe the LIBI solution to data management, access and integration of biological data sources.

Data Federation Issues in the Bioinformatics Domain

Within the LIBI platform both bioinformatic databases and algorithms that access database data can coexist. In such an environment an extensive integration of data and information is highly desirable, both from the point of view of information technicians (developers), who would be able to manage a repository with just a single interface, and for final users, who would benefit from this integration to perform, in a much simpler way, powerful queries that combine a number of data sources. According to the LIBI strategy, there are two main approaches to obtain data integration: i) data federation, and ii) a data-warehouse (consolidation) of various data sources.

In the first case, federation is the ability to transparently access relevant information from a variety of sources and platforms as though it were a single resource. A federated approach to both data and information integration provides the ability to synchronize distributed data without requiring that it be moved to a central repository, as shown in Figure 5.

On the other hand, consolidating data into a single physical data store has been, for many problems, the best way to achieve fast, highly available, and integrated access to related information. The drawbacks of this solution include: latency, data aging, and lack of a unique access interface to heterogeneous data.

For the LIBI platform and, in general, for the class of IT bioinformatics problems this platform is devoted to, the solution we designed and experimented is a hybrid approach: in our case it seems to give the best results.

In the following section we provide additional details about the LIBI data federation server we implemented which is based on the IBM WebSphere Information Integrator (Alur, 2005). This kind of data federation uses SQL as the only language to access all data sources. This enables all data sources to be accessed in a standardized format, whether they are an application program, tool or program product. Such architectures can also be efficiently implemented in a grid architecture.

The LIBI Data Federator Server

Figure 6 depicts the realization of the architecture in Figure 5 in the LIBI environment. This picture shows the four wrappers which have been used so far in the LIBI data federation architecture:

1. The Relational Wrapper allows the access of relational databases of some LIBI partners such as MitoRes, UTRsite, UTRdb and HmtDB;
2. The Entrez Wrapper is the gateway for accessing the well-known remote databases hosted by the NCBI: e.g. PubMed, GenBank, OMIM;
3. The EMBL/FASTA Wrapper (Leo, 2008), developed within the LIBI project by using the Information Integrator SDK, allows data sources stored in EMBL format flat-files to be federated. It has been used to access local replicas of the EMBL, UniProt and EMBL CDS data sources;
4. Finally, the Web Services Wrapper has been applied with the aim of integrating the remote database Species 2000.

A Simplified Environment to Data Mash-Up, Query and Discover Data from the Structured World

The basic interface provided by the LIBI federator server is SQL, thus directly targeted by specialized bioinformatic applications. In order

Figure 5. Conceptual operational model for the architecture of a Data Federator Server

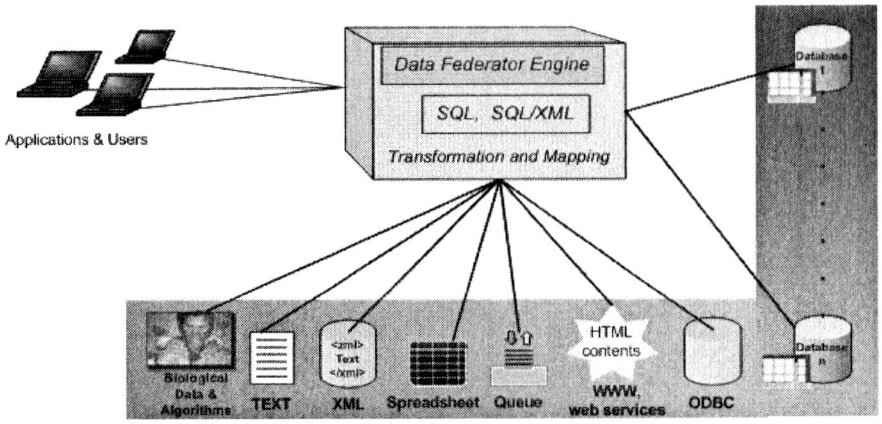

Figure 6. Current component model of the LIBI data federation architecture

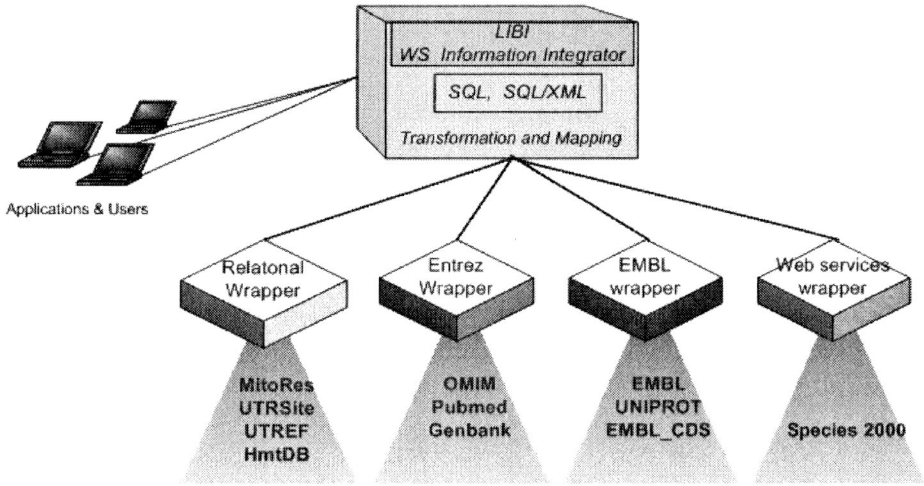

to allow also non-technical people access and directly benefit from the federated database layer a generic system to retrieve information as well as to allow data mash-up is also provided. To do this a new, specialized tool has been introduced in the LIBI platform: the IBM Data Discovery and Query Builder for Healthcare and Life Sciences (DDQB) (DDQB, 2006). It is a powerful search capability tool with a graphical interface that enables users with various levels of expertise to easily configure queries and leverage the full spectrum of information assets.

Figure 7 shows the operation flow of DDQB. The key concepts of the tool are the Data Abstraction Model (DAM) and the Query Abstraction Model (QAM). The former consists of an XML repository that stores all the information concerning the DB physical model, the abstract entities the end-user can handle, and their mapping (this logical to physical mapping is designed just once by the db administrator). Most DDQB behaviour is driven by this component. On the other hand the Query Abstraction Model provides a XML representation of an abstract query in terms of

Figure 7. Data discovery and query builder operation flow

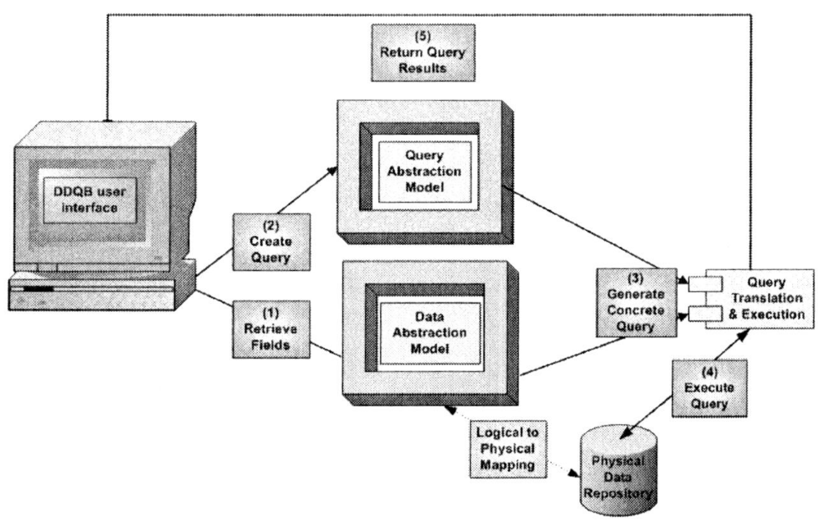

logical fields, and is formulated by considering both the authorization level of the user (with a granularity at the field level) and the selection logic imposed by the query criteria. The *Transformation and Mapping* block, in Figure 6, is the component devoted to concrete query translation and execution.

The end user perspective and a portion of the DDQB interface are shown in Figure 8. End-users construct queries by operating on DAM and QAM in a graphical way by navigating and acting on a generic taxonomy pointing to the LIBI federated database biological concepts. Figure 8 reports a portion of the DAM the user works with (on the left hand side) and a step of the query building phase. The end-user, in this case, is asking with one query for *H. Sapiens* DNA sequences, extracted from GenBank, for which the annotation field includes the **globin** term and the abstracts of the PubMed reference linked to these sequences include the **anemia** term.

Grid Data Access to LIBI Federator Server

In the LIBI project, the grid access to the LIBI Federated DB is carried out through the GRelC Data Access Service (GRelC DAS) (Aloisio, 2005a). The two layers (GRelC DAS and LIBI federator server) are complementary and work at different levels: while the LIBI federator server performs the data federation task (discussed in detail in the previous Section), the GRelC DAS exposes the federated DB in the grid making it available to all the LIBI grid users (see Figure 9).

The GRelC DAS is a GSI/VOMS enabled web service providing extreme performance, interoperability and security. It efficiently and transparently manages databases on the grid across VOs, complying with emerging and consolidated grid standards (W3C and OASIS) and specifications (OGF) as well as production grid middleware (gLite & Globus). It currently provides a uniform grid access interface to a wide range of relational (for instance IBM DB2, on which the LIBI federator server is based, as well as Mysql, Oracle, Postgresql, SQLite, Microsoft SQL Server, etc.) and non-relational data sources (XML DB engines such as eXist, XIndice, etc.).

THE LIBI RESOURCES

In this section, the grid resources involved in the Laboratory are described.

Figure 8. An example of data abstraction model (on the left) and the query building interface (on the right)

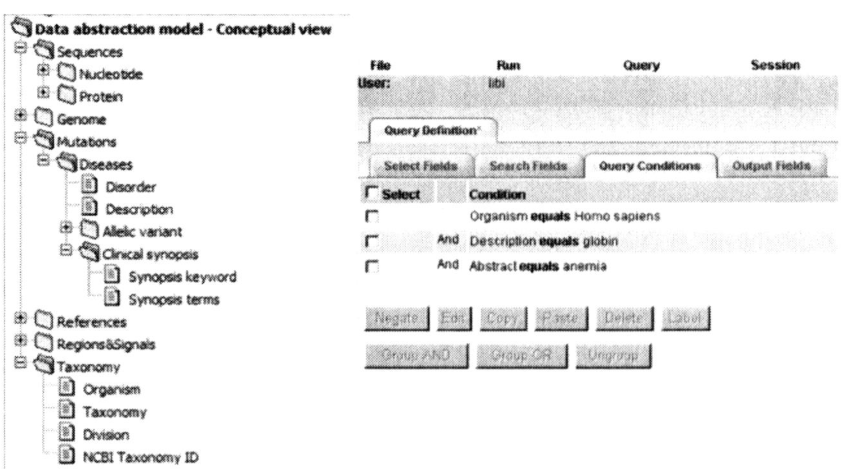

Figure 9. GRelC DAS and DB2 Federated DB

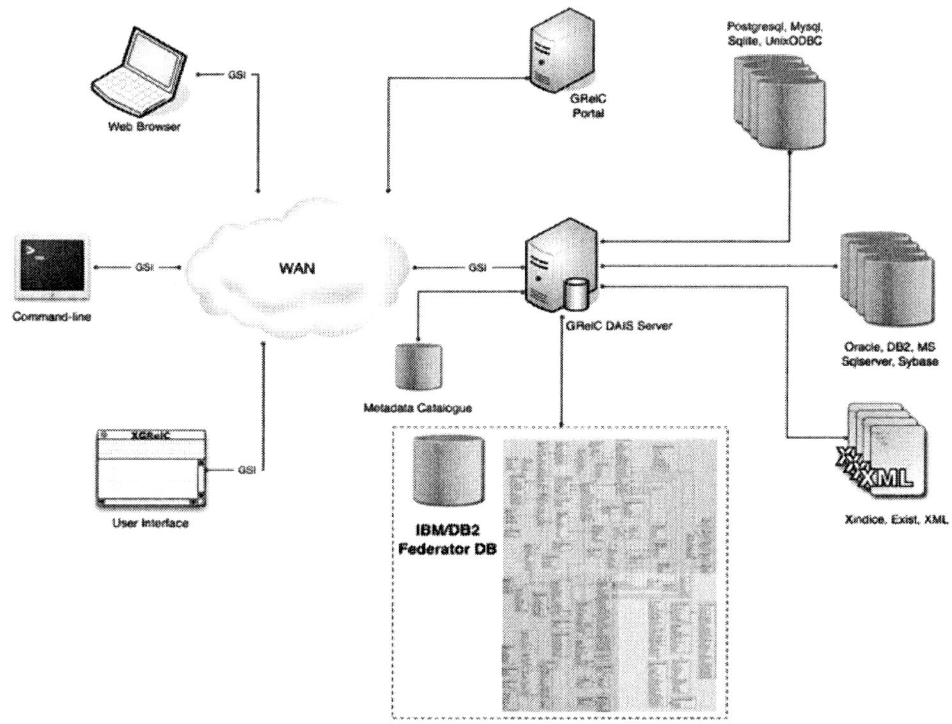

The SPACI Grid

The Italian Southern Partnership for Advanced Computational Infrastructures (SPACI) is a consortium consisting of the University of Salento, Lecce, the University of Calabria, Hewlett- Packard Italy and an ICT company based on spinoff companies belonging to the universities involved in the consortium. The consortium is the follow-up of the SPACI project, a partnership between the University of Calabria, the University of Lecce and the University of Naples "Federico II", based on three geographically distributed High Performance Computing centers located in southern Italy, namely, the MIUR/HPC Center of Excellence of University of Calabria, the CACT (Center for Advanced Computational Technologies) of University of Salento and the Naples/ DMA (Dept. of Mathematics and Applications) of University of Naples "Federico II".

The SPACI Grid comprises more than two hundred dual processor nodes (Itanium2-based HP Integrity) running Linux, two clusters Compaq AlphaServer SC Tru64/Unix (4+16 nodes), plus services and applications.

SPACI is also an associate center of the Euromediterranean Center for Climate Change (CMCC) which provides more than thirty Tflops of aggregated computing power. Deployed middleware includes Globus and gLite and it supports HPC/HTC applications.

The EGEE Grid Infrastructure

The LIBI High Throughput applications, which can be decomposed into many smaller independent tasks, can exploit the unique features of the EGEE Grid infrastructure (http://www.eu-egee.org/). The EGEE grid is in fact the largest multi-disciplinary grid infrastructure in Europe exploiting the gLite middleware (http://glite.web.

cern.ch/glite/), which brings together more than 140 institutions to produce a reliable and scalable computing resource available to the European and global research community. At present, it consists of approximately 300 sites in 50 countries and gives its roughly 10,000 users access to 80,000 CPU cores around-the-clock. Of these resources, more than 17 thousand CPUs are accessible to the bioinformatic community using the "biomed" VO. The LIBI partners contribute to the EGEE infrastructure with a) the INFN Production Grid (http://grid-it.cnaf.infn.it/) including more than 20 sites with a total of 4500 CPUs, of which up to 1,500 CPUs and 5 TB of disk space can be used by the Italian bioinformatic community using the "bio" or "libi" VO; b) the ITB-Bari farm with 20 CPUs and c) the SPACI-LECCE farm with 30 CPUs and 0.5 TB of disk space. GridICE (Aiftimiei, 2007), a grid-monitoring tool developed by INFN, was used to monitor the infrastructure and its usage. GridICE provides a complete monitoring of the grid resources giving users the possibility of retrieving data according to their "Grid role", such as VO managers, site administrators or standard users, providing a secure and authenticated direct access to the data which interest them most.

The INFN Certification Authority (INFN CA) issues the user certificates necessary to access the grid in collaboration with the local Registration Authority (RA). Two of the local RAs, those of ITB-Bari and IBM, have been set up explicitly for the LIBI project.

The DEISA Grid

DEISA is a consortium of leading national supercomputing centres that currently deploys and operates a persistent, production quality, distributed supercomputing environment with continental scope. The purpose of this infrastructure is to enable scientific discovery across a broad spectrum of science and technology, by enhancing and reinforcing European capabilities in the area of high performance computing. This becomes possible through a deep integration of existing national high-end platforms, tightly coupled by a dedicated network and supported by innovative system and grid software.

It is difficult to quote the computing power of this network because the clusters involved are being constantly updated but at the beginning of 2007 the aggregated computing power of the DEISA supercomputers was estimated to be equivalent to approximately 35,000 CPUs and 270 Tflops.

Each DEISA site is interconnected by a dedicated 10 Gb/s network which is used to mount a Global Parallel Filesystem. User data stored on the logical DEISA filesystem can be accessed anywhere in the DEISA network.

DEISA provides users with many facilities for job and data management and also allows the construction and submission of workflows via Unicore. The utilization of the DEISA infrastructure is facilitated by the so-called Common Production Environment which allows the user to deploy software modules to customize the working environment and launch applications in an identical way regardless of site location.

Currently the LIBI project can take advantage of the CINECA resources in DEISA, which amounts to about 30 Tflops of processing power. Further resources are available by applying through CINECA for the DEISA Extreme Computing Initiative (DECI) projects.

A summary of the resources involved in the project is reported in the Table 1.

PORTING BIOINFORMATIC APPLICATIONS ON THE GRID

This section summarises a number of case studies that have been performed with LIBI. They were carried out not only to test the platform design itself but also to experiment the environment in a number of real bioinformatics application scenarios. Involved applications in the case studies

Table 1. Involved resources

Infrastructure	Resources
SPACI	More than two hundred bi-processor nodes (Itanium2-based HP Integrity)
EGEE	More than 17,000 CPUs (biomed VO); more than 1,500 CPUs (bio and libi VOs)
DEISA	Aggregated computing power is given by 35,000 CPUs and 270 TFlop/s

are available, by using the LIBI platform, at the following url: http://www.libi.it/biotools.

CASE STUDY 1: "Genome Wide Identification of Conserved Coding and Non-Coding Regions"

One of the major tasks of genomic analysis is the identification of genes and regulatory elements of gene expression. Among the different methods used to solve this task, comparative analysis has shown to be both efficient and reliable.

CSTminer (Castrignanò, 2004) is a comparative analysis tool that compares two sequences to identify conserved tracts and classifies them as coding (likely belonging to protein coding genes) or non-coding (potential regulatory regions). This classification is based upon the computation of a coding potential score (CPS) that takes into account the evolutionary dynamics of coding sequences.

Unfortunately given the size of the genomes (Giga base pairs – Gbp – in higher eukaryotes) it becomes difficult to run genome scale analysis.

For this reason we exploited the possibility to use a grid based approach performing human-mouse whole genome comparison.

Method

We split genomes in 100Kbp long slices and compared all human slices against the mouse ones. Clearly the number of comparisons needed to complete the analysis is considerable but this kind of problem fits very well the distributed approach offered by the EGEE grid platform. To execute on the grid the huge number of jobs required by this application we used the JST tool described previously.

In order to optimize the ratio between the time needed to exchange data with the work and the time required to perform the analysis we ran groups of 1,000 comparisons (a "task") for each job. This also provides something similar to a checkpoint: whenever a job crashes only a small part of the calculation is lost.

The procedure required to run the analysis is controlled by a shell script that performs the following operations: it downloads the input sequences, decompresses them, starts the executable with the correct parameters, checks if the output was correct and, in case of success, stores the output on a Storage Element randomly chosen from a list of alternatives (granting the possibility to store the output also if some of the nodes are not working at that moment and contributing to reduce the load on each machine).

Because of the high number of jobs needed to complete all the comparisons, it is difficult track every job that must be submitted or that are already completed. Moreover in a distributed environment it is foreseen to experience failures that must be tracked in order to detect and resubmit failed jobs.

To handle this problem we setup a RDBMS procedure to submit each task, and retrieve results. By using this approach it is also possible to save information about the Storage Element used to store the output file and to monitor the status of each tasks. It is also possible to save information related to grid usage (like the job that is running or that has executed that task).

Results

From the CSTminer analysis we obtained a collection of nearly 1,500,000 CSTs with an average length of 190 nt. More than 400,000 CSTs were labelled as coding, 550,000 CSTs as non-coding, while 500,000 CSTs remained undefined as the CPS fell in a twilight zone between the coding and the non-coding thresholds used for the correct classification. Finally nearly 20,000 CSTs displayed more than 95% of identity and were labeled as ultra-conserved (Figure 10).

CASE STUDY 2: "Identification of Regulatory Elements in mRNA Untranslated Regions"

The detection of regulatory elements controlling gene expression at the transcriptional and post-transcriptional level, generally embedded in the non-coding part of the genome, represents a major challenge in post-genomic biology. Transcriptional or post-transcriptional control of gene expression generally involves short DNA or RNA tracts, respectively interacting with transcription factors or RNA-binding proteins.

A large number of oligonucleotide patterns involved in transcriptional and post-transcriptional regulation have been experimentally characterized and this information collected in a number of specialized databases such as UTRsite (Mignone, 2005) for UTR-specific functional motifs. UTRsite is a collection of regulatory elements located in 5' and 3' untranslated sequences (UTRs) whose function and structure have been experimentally determined and annotated in UTRdb (Mignone, 2005), a curated database of 5' and 3' UTRs of eukaryotic mRNAs, derived from several sources of primary data.

Thus, it is of utmost importance to develop specific software tools that are able to identify these functional elements in genomic or cDNA sequences in order to aid their functional characterization.

Method

The PatSearch (Grillo, 2003) software belongs to this family and is currently used to identify the presence of UTRsite patterns in UTRdb entries for their automatic annotation and localisation. It can be used as well for the annotation of newly produced or previously unannotated sequences for UTRsite or other user-defined patterns.

PatSearch is a flexible and fast pattern matcher able to search user submitted sequences for any combination of Position Weight Matrices (PWMs), primary sequence patterns and structural motifs also allowing mismatches and/or mispairings below a user fixed threshold.

Figure 10. Distribution of CSTs among different coding classes assigned by CSTminer. Uc: ultraconserved sequences (identity >95%), COD: coding CSTs, NCOD: non-coding CSTs, UND: CSTs with undefined coding potential.

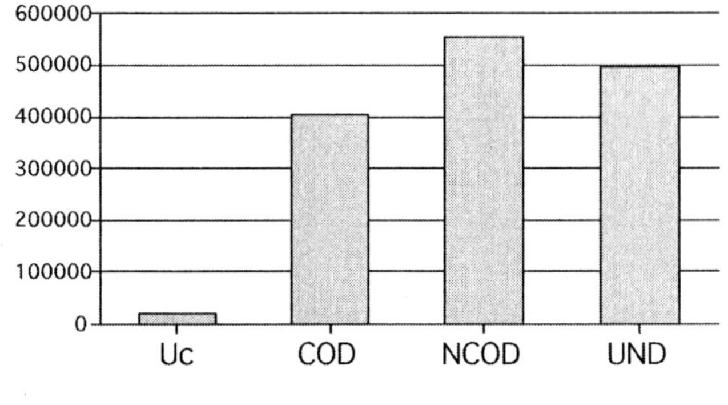

It is also a pattern-matching tool able to find a well-defined pattern against sequences or complete primary or specialized biosequence databases.

The pattern description has been inspired by 'regular expression' rules, although both the syntax and the semantics are different, especially for the inclusion of specific operators for finding PWMs, complementary helices and palindromes.

A pattern is made by a combination of pattern units, defined in syntax rules as: String pattern; Range pattern; Palindrome pattern; Hairpin loop pattern; Position weight matrix (PWM) pattern; Repeat pattern; Length constraints.

The program is also able to do a post-processing of the already matched pattern in order to refine the search. The computational complexity of the PatSearch software is strictly influenced by the structure of pattern composition, since there are some pattern units of syntax which require a back-tracking process and a massive comparison; this affects the whole computation process heavily.

Grid Application

To detect the regulatory motifs in the UTRdb sequence collection, each UTRsite element is defined using the syntax in a format suitable for PatSearch analysis. Generally, the motifs are defined by several and complex pattern units which cannot be separately processed because they are closely inter-dependent.

To accomplish the annotation of these regulatory motifs in the UTRdb database (actually about 1,300,000 sequences) it is necessary to run a PatSearch analysis against each UTR sequence.

A typical high throughput computing approach in a grid infrastructure represents an effective solution to this huge analytical effort. In fact it is advisable to split the UTRdb database into separated subsets in order to run different PatSearch processes against them.

A typical architecture for this task involves a massive data transfer among the main server (that submits the jobs on grid), the data repository and the computational nodes of the grid.

A solution to this bottleneck is solved by using GRelC, a database middleware for the grid. Using this layer is possible to access the data partition directly from the grid node avoiding transfers between main and data servers.

In our test case we have successfully used GRelC with the database federator (IBM Information Integrator) to access the UTRdb.

The porting of this strategy solution on the grid infrastructure has yielded a drastic reduction of the total CPU time through the distribution of processes and database subsets on the computational grid nodes.

CASE STUDY 3: "FT-COMAR: A Large Scale Application for Reconstructing Protein Folding"

To perform their biological function most proteins fold into unique three-dimensional (3D) structures. The prediction of the 3D protein structure from its residue sequence is one of the most important and still unsolved problems in bioinformatics (Lesk, 2006). A promising approach to this challenging problem is based on the prediction of the protein contact map, a reduced in plane 2D representation of the 3D structure, a sort of intermediate step between the primary and tertiary structure. In this representation, two residues are in contact if their distance is lower than a given threshold, usually 1.2 nm in the protein space, when the Calpha trace of the backbone is adopted. A contact map is a binary symmetric matrix containing 1 for each pair of residues in contact, and 0 otherwise. The problem of predicting the protein folding from its residue sequence may be sub-divided into two steps: 1) the prediction of the contact map, 2) the reconstruction of the 3D structure from the contact map. Currently several efforts are addressing the problem of predicting the protein contact map but the methods developed so far are still far from satisfactory. In this large scale effort, we address the second step of the protein folding prediction problem: the reconstruction of the 3D protein structure from its contact map. The general version of this problem has been proved to be NP-hard (Breu et al., 1998). Various heuristic algorithms have been proposed to solve the problem (Galaktionov, 1994; Vendruscolo, 1997; Bohr, 1993; Pollastri, 2006) but most of these methods have been tested only on the relative predictions and none of them are available to the scientific community. We have designed and implemented a new heuristic algorithm, called FT-COMAR, which can reconstruct the 3D structure of a protein from its contact map and which outperforms all existing algorithms (Vassura, 2007 a, b; Vassura, 2008). Grid technology is used to provide the processing power required to cope with the enormous computational effort that results when the effect of noise on the protein reconstruction is analyzed.

The Method: How to Reconstruct the Protein 3D Structure from the Contact Map

In any prediction problem usually a blurred putative solution is given and this is the case when the prediction of the contact map is considered. The algorithm consists of two phases. The first phase allows the prediction of a partially random set of starting coordinates. This initial phase relies essentially on statistical values of inter-residues distances which are randomized and then used to generate a starting set of 3D coordinates through the metric matrix embedding algorithm (Havel, 1998). The second phase of the algorithm iteratively applies a correction/perturbation procedure to the randomly generated set of coordinates. The correction procedure moves each pair of residues whose relative positions are not satisfying the corresponding contact map values according to the contact map value itself. The perturbation procedure slightly moves residues improving their mobility for the next correction. This iteration is performed in order to obtain a new set of coordinates as consistent as possible with the given contact map. The refinement applies until the set of coordinates is consistent with the given contact map or until a stopping condition, based on a control parameter, is reached.

The "in GRID" Extension

FT-COMAR has been previously tested on a data set of 1,760 exact contact maps (Vassura, 2007a; Vassura, 2008) and on a subset of 120 erroneous contact maps (Vassura, 2007b).

However a statistical validation should use a much larger data set. To extend our results to a consistent set of proteins, we downloaded all the protein structures from the Protein Data Bank and we retained only those files with coordinates obtained from X-ray experiments, with resolution <2.5 Å, and without missed internal residues. Using BLAST (Altschul et al., 1997) we removed sequence redundancies. We thus obtained a dataset of 3,613 proteins with sequence similarity lower than 30% and with protein size ranging from 39 to 1,392 residues. For each protein we computed its contact map with a fixed threshold set at 12 Å and, for each percentage of error, we generated 100 random noisy contact maps. Each of these maps was then reconstructed with FT-COMAR. The same experiment was replicated for random errors, random missing contacts, random added contacts, random unknown entries, random unknown entries together with random errors, and random errors with a pre-filtering procedure. The total number of running jobs was 8,467,577 and the experiment was run on the SPACI grid. The parallel time is shown in Figure 11 as a function of the number of processors.

The parallel time (Tp) is given by sequential time (Ts) on processors number (nproc) i.e. Tp =Ts/nproc, so the Tp values are showed in the Table 2.

Figure 11. Estimated computing time in days for 100 runs of FT-COMAR for each type of test on 3,613 proteins as function of the number of used processors

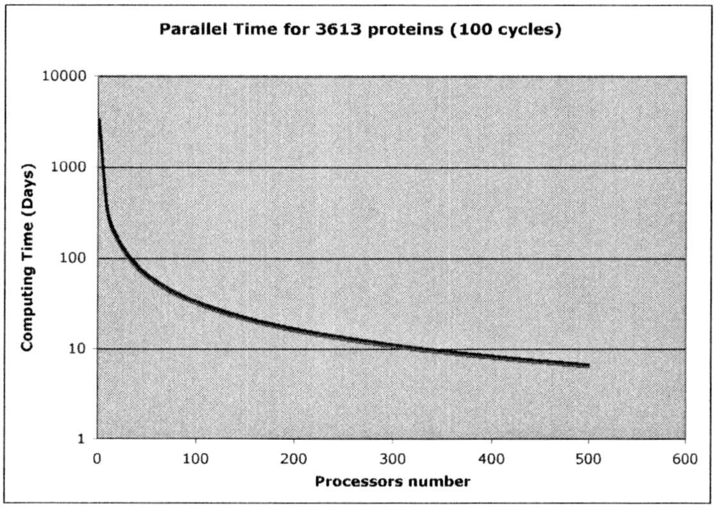

Results

We have shown on a small protein set that FT-COMAR reconstruction is more tolerant to missing contacts than to added contacts in the contact map (Vassura, 2007b). We also introduced a pre-filtering procedure on noisy contact maps based on common neighbors. The filter is based on statistical values of a set of 120 proteins (Vassura, 2007b): the contact map values are not considered if the number of common neighbors between two non-contacting residues is above a given threshold or if the number of common neighbors between two contacting residues is below a given threshold. As shown in Figure 12, when a large testing set is considered (3,613 proteins) the introduction of this filter still improves reconstruction quality for percentages of random errors ranging between 1% and 10%. All together these results indicate that the robustness of FT-COMAR in reconstructing 3D protein structure even in the presence of random errors is good enough to be consistent when a large set of non redundant examples is tested.

Table 2. Parallel Time (in days)

Nproc	Tparallel (Days)
128	26
512	6.50
1024	3.25

CASE STUDY 4: "Large Scale Proteome Comparison"

Large scale genome sequencing projects provide biological data that are available for addressing basic biological problems at the molecular level. An exciting problem that emerges as a consequence of the availability of complete genome sequences is the comparison of all the available putative/annotated proteins in order to highlight basic aspects of life at the molecular level (Marsden, 2007; Bateman, 2006; Lee, 2007). The enormous computational effort in performing such an experiment is clear when the size of the dataset is considered. A total of 599 complete genomes, comprising 551 Prokaryotes and 48 Eukaryotes, can be downloaded from public databases. Data were extracted from NCBI, National Center for Biotechnology Information (http://www.ncbi.

Figure 12. Average Root Mean Square Distance (RMSD) (Å) between reconstructed and native protein 3D structure at increasing percentages of random errors. FT-COMAR refers to FT-COMAR with common neighbors filtering (Vassura, 2007 b).*

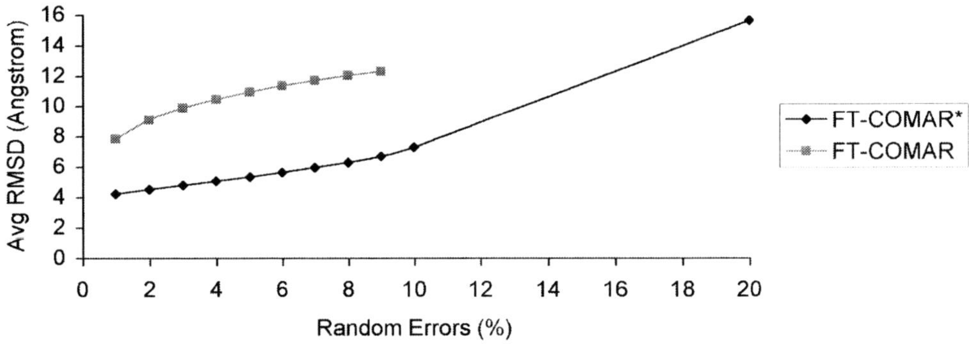

nlm.nih.gov/) and from Ensembl genome browser (http://www.ensembl.org/) updated on 27/07/2007. The overall database consists of 2,624,555 protein sequences in FASTA format with the goal of comparing each of the protein sequences to all the others in the dataset.

Method

Proteome comparison was performed on the EGEE Grid infrastructure through the BLAST program (Altshul, 1990) (http://www.ncbi.nlm.nih.gov/blast/, version 2.2.15) which aligns biological sequences to find regions of similarity; in our case each protein sequence was aligned against all the others. For each sequence, BLAST gives an output file that contains all the sequences in the dataset that have a match with the query and scores these matches on the basis of statistical indices. The E-value accounts for the number of alignments with scores better or equal than the actual expected to occur by chance in a database search. The lower is the E value, the more reliable is the alignment; we fixed a 10^{-10} threshold value. We also fixed the effective length of the database to 100,000 sequences in order to make the BLAST searches independent from the database size. The input and output files were stored and replicated on two different storage devices (Storage Elements, SE) in order to balance the data traffic on the grid and to have a robust fault tolerance. We developed an automated job submission procedure that used 4 different Resource Brokers (RB), distributed all around Europe. The overall production run consisted of approximately 2.6 million pairwise comparisons that were grouped in independent tasks of about 500 comparisons each. Even in this case the jobs were submitted to the grid using the JST tool. The complete comparison of all the protein sequences in the dataset, carried out on the grid, took about 1 week, which should be compared with the about 130 weeks estimated for the run of the same job performed on a single medium power CPU. The duration of the overall task was optimized employing the result of a previous study of the grid job dynamics on a sample of BLAST comparisons (Carota, 2008).

The fast and high-throughput comparison performed through the grid middleware, allowed us to fine tune our analysis and all the 2,624,555 outputs obtained from the BLAST all-against-all comparison were then processed. We searched for clusters with at least one sequence from each genome, to assess the most conserved domain through different species. We developed a clustering algorithm based on a 40% sequence identity threshold. Indeed, several analyses have suggested that a 40% pairwise sequence identity can be used as a significant threshold for grouping sequences in order to obtain common functional and structural biological relevant information. The second constraint used was a 90% coverage. Given two sequences of a BLAST hit, we defined the coverage as: COV=I/U, where I is the length of the intersection of the aligned regions on the two sequences and U is the union is the length of the alignment, namely the length of the two sequences minus their intersection. Both constraints were adopted to detect groups of related protein sequences. BLAST outputs were represented by means of a graph structure. The nodes are the protein sequences and two nodes are connected by an edge if the corresponding proteins share a BLAST hit with the constraints on sequence identity and coverage. We thus computed the connected components of the graph, where every connected component consists of proteins that are connected to each other by a BLAST match in a direct or indirect way.

Results

After clustering, we ended up with 187,594 clusters, containing from 2 up to 15,875 protein sequences (see Figure 13).

Among the clusters, 122,686 contain only prokaryotic organisms while 63,230 are specific for eukaryotes. We focused on the most populated cluster (15,875 sequences) that is also highly specific for prokaryotic organisms, since it has at least one representative sequence for each of them. It turns out that the ABC transporter protein ATP-binding domain (Figure 14) belonging to the ATP-Binding Cassette (ABC) superfamily is the common protein structure. This protein sequence analysis may help in further speculating on possible original functions at the basis of molecular evolution.

Figure 13. Distribution of the number of elements across all the clusters. Only clusters with less than 200 elements are represented: they cover 99% of the overall number of clusters (186,482 over 187,594 clusters). 1,112 clusters contain more than 200 sequences and are not represented in the above distribution.

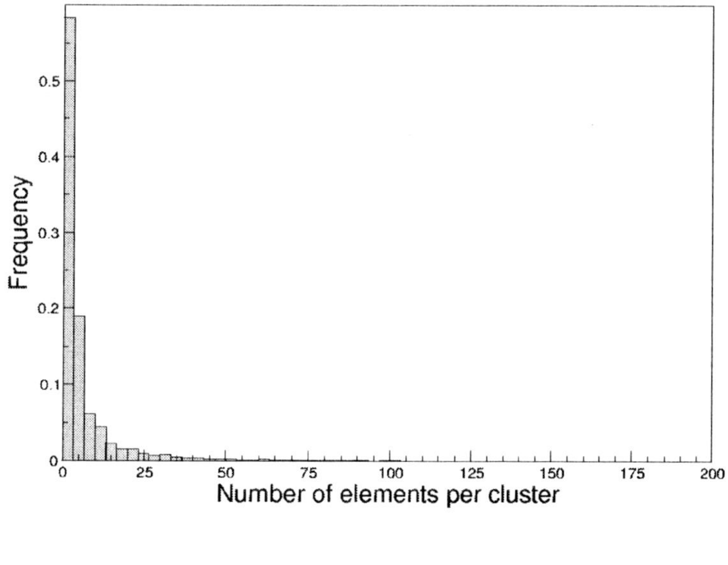

Figure 14. Visualization of an ABC transporter protein structure (PDB ID 1l7v). The darkest chain is the conserved domain represented in the most populated cluster.

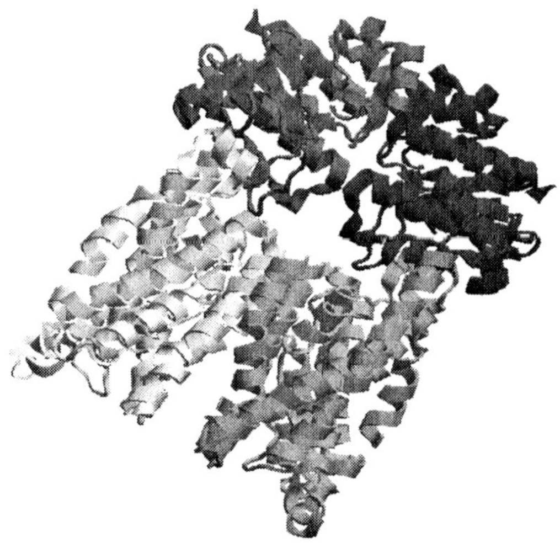

CASE STUDY 5: "Analysis of the Distribution of Introns in mtDNA Genes of Ascomycota"

The case-study presented here highlights how a new query approach developed within the LIBI database federation system integrating public and private molecular and taxonomic databases allows a data retrieval for specific information usually present in the entries of a biological database, but which is hard to extract using the classical query systems. The analysis is based essentially on the use of GenBank (the NIH genetic sequence database), an annotated primary collection of all publicly available DNA sequences (Benson, 2007).

The database query systems commonly used such as SRS (Etzold, 2003) or Entrez (http://www.ncbi.nlm.nih.gov/sites/gquery) appear rather deficient when it is necessary to retrieve selected information contained in particular lines of a database entry. This problem is particularly evident when we wish to extract the data present in the so-called Feature Tables that, in a typical entry of a primary database, contain a detailed description of the various functional portions of the sequence. The described case study demonstrated that the query system developed within the federation of biological databases produced in the LIBI platform can constitute a really effective alternative solution.

The Biological Case

The biological case under analysis is part of an extensive project which aims at the development of an effective and standardized species discrimination molecular system in Ascomycota fungi. Because of the peculiar and favourable features of the mitochondrial genome, mitochondrial DNA has been considered for use as a "species molecular marker" or "barcode". However, a possible problem in the PCR amplification of mitochondrial fungi genes and in their bioinformatic surveys may result from the presence of mobile introns (Hamari et al., 2002; Seifert et al., 2007), non coding sequences which interrupt genes. The aim of our study case is to explore 11 protein coding genes and 2 rRNA coding genes currently available in public databases to find one or more regions free from introns in all Ascomycota and large enough to be proposed as species molecular marker candidates. The most effective and accurate way to obtain the distribution, conceived as the number and position, of the introns in each gene and inside a specific taxonomic class, is to extract this information directly from the features tables of GenBank. An approach based on the LIBI database federation query system seemed immediately desirable to reach this goal. It can effectively avoid a series of alignment and retrieval stages based on the similarity calculation, intrinsic in a Blast approach, which inevitably produce false positives and negatives in the final results.

Results

The results of our study reveal the pervasiveness of the introns across almost all the mtDNA genes, even if the density of distribution and the size can vary remarkably between them. CO1 and CYTB genes show the maximum density of introns suggesting that the COI gene, designated as a core barcode region for animals (Hebert, 2003), has to be reconsidered in fungi. The analysis suggests that at least three genes could be considered the best barcode candidates for their scarcity of introns, namely ND2, ND4 and ND6. In conclusion, the higher efficiency of the query-based system in revealing introns seemed to be evident compared with a Blast-based approach, reducing significantly false positives and negatives.

Technological Aspects

This section reports some performance parameters evaluated during the activities of data retrieval of intron locations from the LIBI federated DB, in particular from its GenBank subsection. It is important to bear in mind that the queries against the GenBank subsection of the LIBI DB are propagated to the NCBI DB in real time and so they experience internet-network latency.

CASE STUDY 6: "The Use of Bayesian Phylogenetic Inference and Complex Phylogenetic Workflow on the LIBI platform, both in a Distributed and a Parallel Environment"

MrBayes is a program for Bayesian inference of molecular evolution parameters and topologies from a set of molecular sequences. Molecular evolution is generally studied only by inference: a pattern (in this case a set of biological sequences) is observed and different possible processes are evaluated to infer what process produced this pattern.

The inference of the values of parameters in the program is Bayesian. The program uses a markovian integration to obtain the sample from the posterior distribution of the parameters. The posterior distribution is the probabilities of the parameter values given the data and assuming that the true process is included in the model of evolution taken under consideration. The markovian integration implemented in MrBayes is a Metropolis-coupled Markov Chains Monte Carlo (MCMCMC) (Geyer, 1991; Altekar, 2004).

The Overview of the Test Cases Test on the MrBayes Parallel Implementation

The MCMCMC implementation of the markovian integration allows MrBayes to be a natural candidate for execution in a parallel environment. The availability of a large number of CPUs on the platform allows us to deal with biological problems that require a great number of chains to adequately explore the parameter space and to achieve convergence.

The test case for this kind of implementation was simply a series of large sequence data sets that represent the nascent barcode reference database for Lepidopterans. The goal was to inform the query system of the database of the phylogenetic structure of the data in order to facilitate the

Performances

Query sent against the federated LIBI DB		
	Number of queries	Approximate response-time
Queries against the GenBank subsection to extract the IDs of the entries to be evaluated	1	9 min
Query against the GenBank subsection to extract features related to the previous IDs	7,089	2 h

species diagnosis of unknown sequences that would query the database. Because of the large number of sequences that typically make up these databases, the query system uses simplified phylogenetic inference as the overall similarity approach, neighbor-joining, or parsimony in order to inform the query (Geiser, 2004; Hibbett, 2005; Ratnasingham, 2007). The LIBI platform allowed us to test the possibility of more realistic inference, that cyclically updated, would inform the query system. The 3,523 sequences were reduced to 2,080 non redundant sequences and were subdivided in 13 groups based on a priori phylogenetic information. The 12 groups ranged from 635 to 1 with 4 groups with more than 200 taxa. All the data set had the approximately the same length. We chose a realistic but generic nucleotide model composed of a GTR substitution matrix (Lanave, 1984) and a gamma distribution to model site variability (Yang, 1994), branch lengths without molecular clock constraints and unconstrained topology.

The execution time of the 13 phylogenies are listed in Table 3 together with the number of taxa (that represent the difficulties of the problem), the number of chains and CPUs used to perform the calculation. For this experiment, the SPACI infrastructure has been employed. The last column shows the temperature used.

Test on the Distributed MrBayes Implementation and Workflow

The use of a distributed version of MrBayes on the LIBI platform is more problematic given the nature of the markovian integration. Typically each MrBayes run, although starting from a fairly small input data, has a quite long execution time. This does not fit with a typical distributed application where generally each job needs to be short. We therefore decided to restrict the distribution of jobs, consisting of single complete analyses, to high performance nodes of the EGEE grid configured to accept MPI jobs. This ensured that each job would run in a sufficiently efficient manner and they would arrive at completion before the maximum run time allowed by the chosen nodes of the EGEE grid.

The case study for the distributed implementation of MrBayes, in reality involves different

Table 3. The comparison between computational effort and execution time for the 13 phylogenetic inferences. The computational effort is function of number of taxa and number of phylogenies evaluated. The number of phylogenies is the result of the product of number of generation, runs and chains.

Data sets	N° of Taxa	N° of phylogenies evaluated	N° of Generation	N° of runs	N° of chains	N° of CPUs	Temperature	Time (h)
Arsenurinae	77	8,000,000	1,000,000	2	4	4	0.20	5.232
Ceratocampinae	150	24,000,000	2,000,000	2	6	6	0.20	38.522
Hemileucinae	215	100,000,000	5,000,000	2	10	20	0.05	86.760
Hesperiinae	208	36,000,000	3,000,000	2	6	6	0.20	35.853
Limenitidinae	29	8,000,000	1,000,000	2	4	4	0.20	1.163
Macroglossinae	515	60,000,000	3,000,000	2	10	20	0.03	109.310
Pyrginae	635	60,000,000	3,000,000	2	10	20	0.05	380.600
Pyrrhopyginae	49	8,000,000	1,000,000	2	4	1	0.20	3.816
Saturniinae	55	8,000,000	1,000,000	2	4	4	0.20	4.095
Sphinginae	119	36,000,000	3,000,000	2	6	6	0.20	25.215
Tortricinae	27	8,000,000	1,000,000	2	4	1	0.20	2.233

capabilities of the LIBI platform from the multiple database querying, blast runs, alignment and phylogenetic inference components. The goal was to evaluate the ability to discriminate among species of the 13 mitochondrial genes in vertebrates using all vertebrate mitochondrial sequences available on EMBL. We were interested to record how often the mitochondrial data were able, using realistic phylogenetic inference, to discriminate between species belonging to the same genus, and if there are differences in discrimination power among the different genes that make up the mitochondrial genome. The goal of the study was to check the validity of the choice of cytochrome oxydase for vertebrates as barcode markers by the CBOL.

The full workflow included at first a database query to collect all vertebrate mitochondrial coding sequences sorted by gene name.

The query system involved first the use of blastP on UniProt and then cross-linked the results with EMBL CDS to obtain the nucleotide sequences. The performances of the queries against the federated LIBI DB are shown in the Table 4.

The data sets divided per gene were then grouped per genera (the systematic grouping that include the more related species). On the LIBI platform each genera data set was aligned as protein in muscle (Edgar, 2004) and the nucleotide alignment was built using protein alignment information as a guide. The platform transformed the fasta alignment in nexus files with a MrBayes command block and submitted them to the distributed version of MrBayes. The model of evolution chosen is the same as before: GTR substitution matrix, gamma distribution for variability among site, and unconstrained branch length and topology.

The complete results were downloaded locally and the burn-in for each analysis was determined using the routine of the coda library within the R environment. Steady state was tested on all parameters of the model and not only on likelihood scores.

The results were then summarized choosing a focal species (the one with more sequences) per genus data set and recording the number of times that the species was reported to be monophyletic in each analysis. The output of the query results and subsequent check of quality are summarized in Table 5.

CASE STUDY 7: "Large Scale Molecular Dynamics on High Performance Platforms"

The structural and dynamic properties of proteins and other large biomolecules can be investigated experimentally by X-ray crystallography or Nuclear Magnetic Resonance. The information available by these methods is however generally limited and experimental difficulties, such as the need to crystallize or dissolve the protein for example, restrict their applicability to a subset of molecules of interest. Therefore, very often researchers use computational approaches such as molecular dynamics (MD) simulation to obtain data which is not readily available from experiment alone. But realistic MD simulations are computationally expensive, so in order to achieve the necessary performance, the program must be able to distribute the workload over many

Table 4. Performances of queries sent against the federated LIBI DB

Query type	Number of queries	Approximate response-time
Queries against the UniProt subsection of the fed. DB	165,384	5 h
Query against the EMBL CDS subsection of the fed. DB	150,620	16 h

Table 5. Query result. The table shows for each gene the total sequences found, the number of correct data sets and the number of shared haplotypes across species that invalidated part of the data set.

	atp6	atp8	coI	Co2	co3	Cytb	nd1	nd2	nd3
Sequences	8,017	6,635	10,904	3,289	3,251	69,384	5,340	17,398	4,293
Clean_Data_set	112	62	376	42	26	46	52	76	34
same_aplotype cross_species	37	48	26	9	3	238	2	2	15

	nd4	nd4L	nd5	nd6	Total	Unknown
Sequences	9,447	2,482	4,128	3,198	139,749	2,662
Clean_Data_set	164	-	54	34	1,044	
same_aplotype cross_species	15	-	10	37		

processors. In this case study, we analyse the parallel performance of two common MD programs, GROMACS (Lindahl, 2001) and NAMD (James, 2005), on two different architectures: the LINUX CLX cluster (Intel Xeon processors and Myrinet interconnects) at the CINECA supercomputer centre and the Itanium XC6000 (Intel Itanium 2 processors and Quadrics interconnects) at the SPACI consortium, University of Salento.

We have performed short simulations of a small protein, bovine β-lactoglobulin, in a solvent (water/urea + NaCl) at 300K and at constant pressure (1bar). Other common features include a time step of 1 fs, the Particle Mesh Ewald method for the electrostatics and SHAKE for constraint dynamics. We note that while it is virtually impossible to do exactly equivalent simulations between any two molecular dynamics programs, we have endeavoured to use equivalent parameters where possible.

Figure 15 shows the performance, measured as the number of picoseconds that can be simulated in 1 day, as a function of the number of processors. We see that the XC6000 is generally faster for both programs: this is probably due to the faster processors and also the faster internal network of this machine. It can also be seen that for a small number of processors, GROMACS is faster than NAMD on both architectures. This is to be expected since GROMACS has been optimised for serial performance, particularly for Intel processors where some sections have been implemented with assembly language instructions. NAMD, on the other hand, has been designed for high parallel scalability and by the time we get to 8 processors, it is performing better than GROMACS on the CLX. But for the XC6000 GROMACS is always faster than NAMD, although the small number of processors used provides only a limited test.

The often-reported superior parallel scalability of NAMD is highlighted by Figure 16 for CLX which shows the speedup, defined as the ratio of the performance for N processors with the performance for 1 processor. On CLX the NAMD parallel scaling is almost linear. GROMACS, however, scales quite poorly and reaches a maximum at 16 processors. After this limit the program actually runs more slowly with the addition of new processors, presumably because communication overheads start to dominate the overall execution time. For the XC6000, however, GROMACS scales well compared to NAMD, presumably because of the faster Quadrics network, although we repeat the number of processors utilised is quite small.

To summarise, in this study we have shown that just based on speed considerations GROMACS

Figure 15. Performance as a function of number of processors

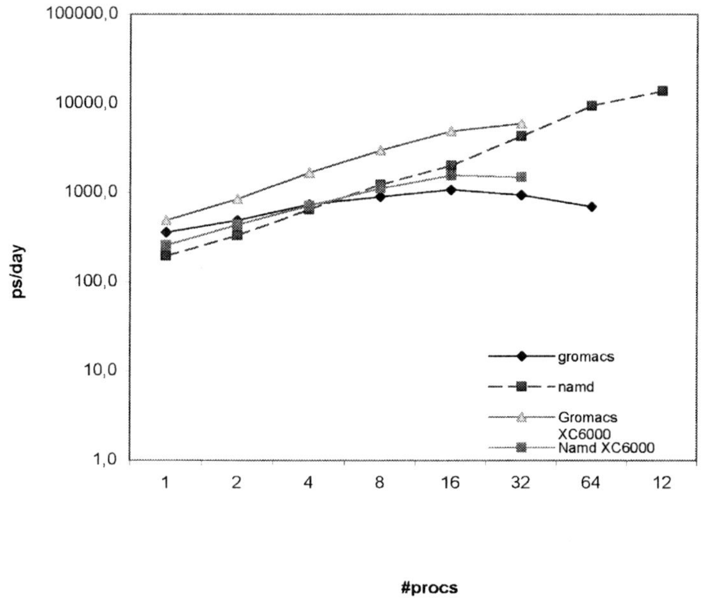

is the program of choice when only four or fewer processors are available, regardless of hardware architecture. For medium-sized clusters (8-32 processors), either could be chosen, depending on the characteristics of the cluster such as the internal network. Based on the CLX results, NAMD is generally more suited for simulations with higher numbers of processors.

CASE STUDY 8: "AntiHunter 3.0: Enhanced Sensitivity in Large Scale Identification of Antisense Transcripts"

The functional role of mRNA antisense (AS) transcripts in prokaryotes and viruses has been documented in several works. Further studies in eukaryotes suggest that AS transcripts are involved in gene regulation including transcriptional interference, genomic imprinting, RNA interference, translational regulation, alternative splicing, X-inactivation, RNA editing and promoter methylation (Wagner, 2002; Lavorgna, 2004).

Methods

AntiHunter's goal is to identify potential AS transcripts within a genomic region of interest (Lavorgna, 2005).

AntiHunter version 3.0 web interface uses the LIBI Meta Scheduler for the submission of jobs on Unicore. The Meta Scheduler manages the needed logic for submitting and monitoring the jobs on Unicore as described in a previous section.

AntiHunter takes in input a genomic sequence and a list of annotated transcripts of the same region and some details or, alternatively, a pre-loaded and annotated genome stored in a resident database. In both cases a filtering of the repeated sequences has to be performed because these can falsify the results of the following BLAST step (Altschul, 1990).

The next step compares the filtered sequences against the whole dbEST database (Boguski, 1993) or a restricted selection of ESTs, using the NCBI BLASTN tool or MEGABLAST (Megablast). The resultant output is the list of the anti-sense ESTs that can interact with the genomic region

Figure 16. Parallel Speedup. The solid line would be for a program with perfect parallel scaling.

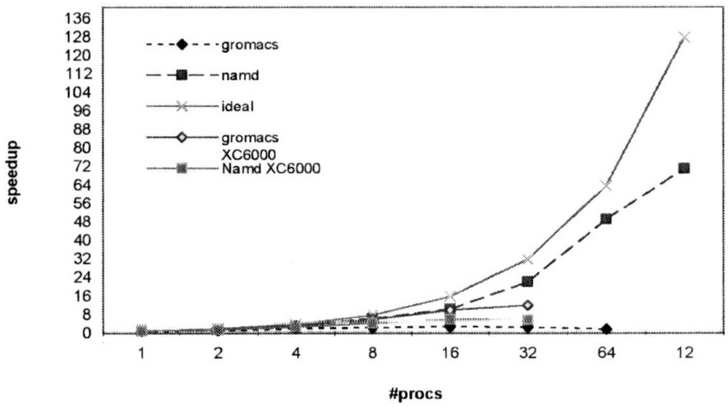

given in the input. Other information, such as the length of the spanned genomic region, the length of the EST, the actual splicing sites plus some flanking sequences, etc. and a quality control check are added to the output as well. The result is available immediately when the job is finished or a link where it can be downloaded is sent by email. In any case the relevant output data are stored for a week.

The accuracy of AntiHunter was tested using genes, which had been shown previously to possess AS transcripts. Fifteen genomic regions, containing overlapping transcriptional units in mammalian genomes described previously in the literature, were used as input to the program.

Results

The program correctly determined the presence of EST AS transcripts in 14 out of 15 cases. In the missing case, given by the human distal-less homeobox protein 1 (DLX1) gene, the presence of AS transcripts was not detected because of the lack of double-checked EST entries in the dbEST.

An excerpt from AntiHunter output is shown on Figure 17. It refers to a 96,700 bp sequence from the human chromosome 4, containing the FGF-2 gene. Several AS ESTs, are detected and the accession numbers are shown in the first column of Figure 17, corresponding to the exons of NUDT6, a known AS transcript to the FGF-2 gene (Li, 2000). Intriguingly, AntiHunter was also able to identify, despite the fact that it was used to query a human sequence, the presence of EST antisense transcripts from species other than human (i.e. *Rattus norvegicus*, *Mus musculus* and *Bos taurus*), unravelling the possible evolutionary conservation of the phenomenon (data not shown).

Figure 17. An AntiHunter output

AI370586	H. sapiens	0e+00	3'	482	7 - 482	0.99	1	n	n	y	P	66333 - 66816	483	FGF2	+	not reported
AI492978	H. sapiens	0e+00	3'	347	13 - 347	0.97	1	n	n	y	P	66333 - 66667	334	FGF2	+	kidney
BQ774001	H. sapiens	0e+00	3'	799	12 - 798	0.98	4	y	y	y	P	66333 - 91288	24955	FGF2	+	Human Chondrosarcoma Cell Line
BU619706	H. sapiens	0e+00	3'	805	12 - 797	0.98	4	y	y	y	P	66333 - 91288	24955	FGF2	+	Cell Line
BM970451	H. sapiens	0e+00	3'	686	12 - 686	0.98	3	y	y	y	P	66333 - 86294	19961	FGF2	+	Lung
BF739878	H. sapiens	0e+00	3'	630	2 - 630	1.00	2	y	y	y	P	66333 - 71358	5025	FGF2	+	kidney
AI803148	H. sapiens	0e+00	3'	559	9 - 547	0.96	1	n	n	y	P	66333 - 66873	540	FGF2	+	not reported
AA628770	H. sapiens	0e+00	3'	553	8 - 553	0.99	1	n	n	y	P	66333 - 66879	546	FGF2	+	not reported
AI337876	H. sapiens	0e+00	3'	604	34 - 604	0.95	1	n	n	y	P	66344 - 66914	570	FGF2	+	uterus
AI074587	H. sapiens	0e+00	3'	558	6 - 558	0.99	2	y	y	y	P	66416 - 71366	4950	FGF2	+	Liver and Spleen
CA392430	H. sapiens	0e+00	5'	716	1 - 716	1.00	4	y	y	y	M	66490 - 91369	24879	FGF2	+	RPE/choroid
BQ434362	H. sapiens	1e-157	5'	933	1 - 909	0.97	5	y	y	y	M	66511 - 96198	29687	FGF2	+	retinoblastoma
BE897694	H. sapiens	1e-152	5'	692	84 - 561	0.69	3	y	y	y	M	66546 - 86312	19766	FGF2	+	melanotic melanoma
BF239070	H. sapiens	1e-106	5'	733	5 - 715	0.97	5	y	y	y	M	66609 - 96584	29975	FGF2	+	from chronic myelogenous leukemia
HSAA14389	H. sapiens	1e-134	5'	448	3 - 419	0.93	3	y	y	y	M	66616 - 86314	19698	FGF2	+	Pooled human melanocyte, fetal heart
BE894167	H. sapiens	1e-107	5'	708	1 - 708	1.00	5	y	y	y	M	66695 - 96196	29501	FGF2	+	melanotic melanoma
BG386740	H. sapiens	1e-135	5'	1048	2 - 777	0.74	5	y	y	y	M	66704 - 96272	29568	FGF2	+	adenocarcinoma cell line
CB128685	H. sapiens	1e-145	5'	616	4 - 616	1.00	5	y	y	y	M	66887 - 96293	29406	FGF2	+	Liver

In conclusion, AntiHunter is a new tool capable of performing an in-silico search for putative EST AS transcripts. It can effectively use relatively raw, but frequently updated, material such as EST sequences and provide useful preliminary results for guiding the design of further experimental analyses.

Furthermore it has the additional advantage of providing the user with an HPC tool capable of tackling complex problems with a friendly web-based user interface.

FUTURE TRENDS

In the last year of the project, the goals of the laboratory will be to integrate the services developed in order to test the entire platform and provide a production infrastructure.

Several improvements are planned in different areas: as regards the resource management, we intend to test the workflow management system in order to compose different applications and run them on different grid middlewares. In addition, we plan to develop a client library for the Web Service applications. For the data management, tools for content-based retrieval, especially those based on visual properties, will be integrated in the platform, in order to provide enhanced analysis for the produced data. Finally, other applications such as the R package will be integrated into the PSE.

BENEFITS OF THE LIBI PLATFORM AND CONCLUSION

This chapter describes the LIBI virtual laboratory for bioinformatics based on a high performance and distributed infrastructure supporting access to large datasets and the execution of single or complex jobs. One of the main goals of the LIBI project has been the development of a Grid Problem Solving Environment, built on top of EGEE, DEISA and SPACI infrastructures, to allow the submission and monitoring of jobs mapped to complex experiments in bioinformatics. At the same time we have developed several case studies and successfully verified the effectiveness of the Grid Problem Solving Environment in supporting experiments in a number of real world bioinformatics application scenarios.

As a summary, in Table 6, we compare the main issues (from the biological and computation perspectives) that characterize each case study and highlight the solutions and benefits that the LIBI platform provided.

Concluding, the LIBI platform involves a large set of resources belonging to three different grid middlewares: gLite, Unicore and Globus. These toolkits provide basic services for managing the resources. Built on top of these services, a set of enhanced and novel services has been implemented related to resource and data management. For the resource management, a Meta Scheduler service has been implemented. It embeds several plug-ins for accessing different grid middlewares. The Meta Scheduler acts as an engine not only for batch, MPI and parameter sweep jobs, but also for complex jobs described by a workflow. Finally, a workflow editor has been implemented in order to compose different application runs on different middlewares. For data management, the LIBI federator server performs the data federation task enabling federated databases to access ten heterogeneous and distributed data sources. It provides a unified data-management interface (both for query and insert statements) based on SQL, whereas the GRelC DAS exposes the federated DB in grid making it available to all the LIBI grid users. It currently provides a uniform grid access interface to a wide range of relational (Mysql, Oracle, Postgresql, SQLite, Microsoft SQL Server, etc.) and non-relational data sources (XML DB engines such as eXist, XIndice, etc.). Finally, several case studies with related results have been presented.

Table 6. summary of case studies

Applications	Issue		Grid/HPC solutions
	Biological	*Computational*	
Case Study 1: CSTMiner	Identification of genes and regulatory elements of gene expression.	Huge size of the genomes (Gbp in higher eukaryotes) makes it difficult to run genome scale analysis. 1,500,00 jobs to be run.	HTC. The genomes have been split in 100Kbp long slices; all human slices are compared against the mouse ones.
Case Study 2: PatSearch	Detection of regulatory elements controlling gene expression at the transcriptional and post-transcriptional level.	Computational complexity is strictly influenced by the structure of pattern composition, since there are some pattern units of syntax which require a backtracking process and a massive comparison; this affects the whole process heavily.	HTC. Data partition directly from the grid node avoiding transfers between main and data servers.
Case Study 3: FT-COMAR	Reconstruction of 3D protein structure from protein contact maps	Large scale fault tolerance tests. Different types of random errors introduced on a data set of 3613 protein contact maps (8,467,577 jobs to be run).	HTC. The protein contact map reconstructions have been split in chunks of jobs.
Case Study 4: BLAST	Comparison of 599 genomes in order to highlight basic aspects of life at the molecular level.	599 complete genomes (551 Prokaryotes and 48 Eukaryotes) were compared for a total of 2,624,555 protein sequences (in the FASTA format).	HTC. ~2,6 million of pairwise comparisons that were grouped in independent tasks of about 500 comparisons each.
Case Study 5: DB Federated	Exploring mtDNA coding sequences of all Ascomycota entries contained in GenBank in order to search for regions free from introns and large enough to be proposed as species molecular marker candidates (barcodes).	Commonly used database query systems (SRS, Entrez, etc.) are rather limited for accessing and retrieving information in some fields of a database entry (e.g. those contained in the so-called Feature Tables). It usually quite difficult for researchers to extract useful information by these systems.	The database federation system integrating public and private molecular and taxonomic databases developed within the LIBI platform constituted a really effective alternative solution to classical query systems for the bioinformatic domain. It allowed the design of a single query able to retrieve all pieces of information needed to locate introns in the Ascomycota mtDNA coding regions referred to GenBank in one go (i.e. in a single result set).
Case Study 6: MrBayes	Bayesian inference of molecular evolution parameters and topologies from sets of molecular sequences.	Biological problems that require a great number of chains to adequately explore the parameter space and reach convergence need the availability of a large number of CPUs.	HPC/HTC. 13 phylogenesis with a huge amount of taxa were run by using at least 8 and no more than 40 CPUs with i between 70 and 620 input sequences.
Case Study 7: Gromacs/NAMD	Molecular dynamics (MD) simulation for determing structural and dynamic properties of proteins.	MD simulations are computationally expensive; there is the need to distribute the workload over many processors.	HPC. Short simulations of a small protein were run on two clusters and the performance as a function of the number of processors was recorded.
Case Study 8: Antihunter	AntiHunter searches for potential AntiSense ESTs within a given genomic sequence.	It performs blast search over many sequences.	HPC. The blast search can run over more than 1 processor.

ACKNOWLEDGMENT

This work was supported by the MIUR (Italian Ministry for Education, University and Research) in the LIBI "International Laboratory of BioInformatics" project, F.I.R.B. 2003, under grant RBLA039M7M.

REFERENCES

Aiftimiei, C., Andreozzi, S., Cuscela, G., Donvito, G., Fantinel, S., Fattibene, E., Misurelli, G., Pierro, A., Dudhalkar, V. & Maggi, G.P. (2007). Recent Evolutions of GridICE: a Monitoring Tool for Grid Systems. In C. Kesselman, J. Dongarra, D. Walker (Eds.), *International Symposium on High Performance Distributed computing (HPDC 2007)* (pp. 1-8). New York: ACM.

Aloisio, G., Cafaro, M., Lezzi, D. & Van Engelen, R. (2003). Secure Web Services with Globus GSI and gSOAP. In *Euro-Par*: (LNCS 2790, pp.421-426). Berlin/Heidelberg: Springer.

Aloisio, G., Cafaro, M., Fiore, S. & Mirto, M. (2005a). The Grid Relational Catalog Project. In L. Grandinetti (Ed.), *Advances in Parallel Computing, "Grid Computing: The New Frontiers of High Performance Computing"* (pp.129-155). Philadelphia: Elsevier.

Aloisio, G., Cafaro, M., Fiore, S. & Mirto, M. (2005b). ProGenGrid a Grid-enabled platform for Bioinformatics In T. Solomonides et al. (Eds.), *Studies in Health Technology and Informatics, From grid to Healthgrid*: *Vol. 112*, (pp. 113-126). Amsterdam: IOS Press.

Aloisio, G., Cafaro, M., Fiore, S., & Mirto, M. (2005c). ProGenGrid: a Workflow Service Infrastructure for Composing and Executing Bioinformatics Grid Services. *Proceedings of the 18th IEEE International Symposium on Computer-Based Medical Systems* (IEEE CBMS 2005), Dublin, June 23-24 2005, Ireland. Washington, DC: IEEE Computer Society.

Aloisio, G., Cafaro, M., Carteni, G., Epicoco, I., Fiore, S., Lezzi, D., Mirto, M., & Mocavero, S. (2007). The Grid Resource Broker Portal. *Concurrency and Computation: Practice and Experience*, *19*(12), 1663-1670.

Altekar, G., Dwarkadas, S., Huelsenbeck, J.P. & Ronquist, F. (2004). Parallel Metropolis coupled Markov chain Monte Carlo for Bayesian phylogenetic inference. *Bioinformatics*, *20* (3), 407–415.

Altschul, S.F., Gish, W., Myers, E.W. & Lipman D.J. (1990). Basic local alignment search tool. *J. Mol. Biol.*, *215*, 403-10.

Altschul, S.F., Madden, T.L., Schaffer, A.A., Zhang, J., Zhang, Z., Miller, W., Lip-man, D.J. (1997). Gapped BLAST and PSI-BLAST: a new generation of protein database search programs. *Nucleic Acids Res.*, *25*(17), 3389-402.

Alur, N., et al. (2005). *Patterns: Information Aggregation and Data Integration with DB2 Information Integrator.* Armonk, NY: IBM Redbook.

Bateman, A. & Valencia, A. (2006). Structural genomics meets computational biology. *Bioinformatics*, *22*, 2319.

Benson, D.A., Karsch-Mizrachi, I., Lipman, D.J., Ostell, J. & Wheeler, D.L. (2007). GenBank. *Nucleic Acids Res.*, *35*(Database issue), D21-5.

Boguski, M.S., Lowe T. M. & Tolstoshev, C.M. (1993). dbEST- database for "expressed sequence tags." *Nat. Genet.*, *4*(4), 332-333.

Bohr, J. et al. (1993). Protein structures from distance inequalities. *J. Mol. Biol.*, *231*, 861-869.

Breu, H. & Kirkpatrick, D.G. (1998). Unit disk graph recognition is NP-hard. *Computational Geometry*, *9*, 3-24.

Cafaro, M., Epicoco, I., Mirto, M., Lezzi, D. & Aloisio, G. (2008). The Grid Resource Broker Workflow Engine. *Concurrency and Computation: Practice and Experience, Special Issue: 2nd International Workshop on Workflow Management and Applications in Grid Environments (WaGe2007)*, *20*(15), 1725-1739.

Carota, L., Bartoli, L., Fariselli, P., Martelli, P.L., Montanucci, L., Maggi G. & Casadio, R. (2008, May). *High Throughput Comparison of Prokaryotic Genomes* (LNCS 4967, pp 1200-1209). Berlin/ Heidelberg: Springer.

Castrignanò, T., Canali, A., Grillo, G., Liuni, S., Mignone, F. & Pesole, G (2004). CSTminer: a web tool for the identification of coding and noncoding conserved sequence tags through cross-species genome comparison. *Nucleic Acids Res.*, *32*(Web Server issue),W624-7.

DDQB: IBM tool designed to help medical providers and researchers perform complex search queries without programming knowledge or IT backgrounds. Retrieved 2006, from http://www-03.ibm.com/systems/deepcomputing/solutions/ddqb.html.

Edgar, R. C. (2004), MUSCLE: multiple sequence alignment with high accuracy and high throughput. *Nucleic Acids Research, 32*(5), 1792-97.

EGEE Grid infrastructure. Retrieved from http://www.eu-egee.org/.

Etzold, T., Harris H. & Beulah, S. (2003). SRS: An integration platform for databanks and analysis tools in bioinformatics. *Bioinformatics Managing Scientific Data,* 35–74.

Foster I. & Kesselman C. (1999). *The Grid: Blueprint for a Future Computing Infrastructure.* San Francisco, CA: Morgan Kaufmann Publishers.

Galaktionov, S.G. & Marshall, G.R. (1994). Properties of intraglobular contacts in proteins: an approach to prediction of tertiary structure. *System Sciences, Vol.V:, Proceedings of the Twenty-Seventh Hawaii International Conference on Biotechnology Computing*: Vol. 5, (pp. 326 – 335). Washington, DC: IEEE Computer Society Press.

Geiser, D.M., Jiménez-Gasco, M., Kang, S., Makalowska, I., Veerarahavan, N., Ward, T.J., Zhang, N., Kuldau, G.A. & O'Donnell, K. (2004). FUSARIUM-ID v.1.0: A DNA sequence database for identifying Fusarium. *European Journal of Plant Pathology, 110*, 473-479.

Geyer, C.J. (1991). Markov chain Monte Carlo maximum likelihood. In Keramidas (ed.), *Computing Science and Statistics: Proceedings of the 23rd Symposium on the Interface* (pp. 156–163). Fairfax Station, VA: Interface Foundation.

gLite middleware. Retrieved from http://glite.web.cern.ch/glite/.

Grillo, G., Licciulli, F., Liuni S., Sbisà, E. & Pesole G. (2003). PatSearch: a program for the detection of patterns and structural motifs in nucleotide sequences. *Nucleic Acids Res., 31*, 3608–3612.

Hamari, Z., Juhász, A. & Kevei, F. (2002). Role of mobile introns in mitochondrial genome diversity of fungi (a mini review). *Acta Microbiol Immunol Hung. 49*(2-3), 331-5.

Havel, T.F. (1998). *Distance Geometry: Theory, Algorithms, and Chemical Applications in the Encyclopedia of Computational Chemistry.* Hoboken, NJ: J. Wiley & Sons.

Hebert, P.D., Ratnasingham, S., deWaard J.R. (2003). Barcoding animal life: cytochrome c oxidase subunit 1 divergences among closely related species. *Proc Biol Sci., 270,* (1), S96-9.

Hibbett, D. S., Nilsson, R. H., Snyder, M., Fonseca, M., Costanzo, J. & Honfeld M. (2005). Automated phylogenetic taxonomy: an example in the homobasidiomycetes (mushroom-forming fungi). *Syst. biol., 54*(4), 660–668.

HmtDB: Human Mitochondrial Genomic Resource Based on Variability Studies. Retrieved from http://www.hmtdb.uniba.it/.

Houstis, E., Gallopoulos, E., Bramley, R. & Rice J. (1997). Problem-Solving Environments for Computational Science. *IEEE Comput. Sci. Eng.*, *4*(3), 18–21.

INFN Production Grid. Retrieved from http://grid-it.cnaf.infn.it/.

Phillips, J. C., Braun, R., Wang, W., Gumbart, J., Tajkhorshid, E., Villa, E., Chipot, C., Skeel R. D., Kale, L. & Schulten, K (2005). Scalable molecular dynamics with NAMD. *Journal of Computational Chemistry*, *26*, 1781-1802.

Lanave, C., Preparata, G., Saccone, C. & Serio, G. (1984). A new method for calculating evolutionary substitution rates. *Journal of Molecular Evolution, 20*, 86-93.

Lavorgna, G., Dahary, D., Lehner, B., Sorek R., Sanderson, C.M. & Casari, G. (2004). In search of antisense. *Trends Biochem. Sci.*, *29*, 88–94.

Lavorgna, G., Triunfo, R., Santoni, F., Orfanelli, U., Noci, S., Bulfone, A., Zanetti, G. & Casari, G. (2005). AntiHunter 2.0: increased speed and sensitivity in searching BLAST output for EST antisense transcripts. *Nucleic Acids Res.*,*1*(33) W665-8.

Lee, D., Redfern, O. & Orengo, C. (2007). Predicting protein function from sequence and structure. *Nat Rev Mol Cell Biol.*, *8*, 995-1005.

Leo, P., Scioscia, G., Pappadà, G. & Quinto, V. (2008). *EMBL Wrapper for WebSphere Information Integrator.* Retrieved from http://www.alphaworks.ibm.com/tech/emblwrapper.

Lesk, A. (2006). *Introduction to Bioinformatics*. New York: Oxford University Press.

Li, A.W. & Murphy, P.R. (2000). Expression of alternatively spliced FGF-2 antisense RNA transcripts in the central nervous system: regulation of FGF-2 mRNA translation. *Mol. Cell. Endocrinol.*, *170*, 233–242.

Lindahl, E., Hess, B. & van der Spoel, D. (2001). Gromacs 3.0: A package for molecular simulation and trajectory analysis. *J. Mol. Mod.*, *7*, 306-317.

Marsden, R.L., Lewis, T.A. & Orengo, C. (2007). Towards a comprehensive structural coverage of completed genomes: a structural genomics viewpoint. *BMC Bioinformatics*. *8*, 86-103.

Mignone, F., Grillo, G., Licciulli, F., Iacono, M., Liuni, S., Kersey, P.J., Duarte, J., Saccone C. & Pesole G. (2005). UTRdb and UTRsite: a collection of sequences and regulatory motifs of the untranslated regions of eukaryotic mRNAs. *Nucleic Acids Res.*, *33*, D141-D146.

MitoRes: bio-sequences resource for mitochondria-related genes, transcripts and proteins. Retrieved from http://www2.ba.itb.cnr.it/MitoRes/.

Pollastri, G., Vullo, A., Fiasconi, P. & Baldi, P. (2006). Modular DAG-RNN Architectures for Assembling Coarse Protein Structures. *J. Comp. Biol.*, *13*(3), 631-650.

Ratnasingham, S. & Hebert, P. (2007). BOLD: The Barcode of Life Data System. *Molecular Ecology Notes, 7*(3), 355-364. Retrieved from www.barcodinglife.org.

Ronquist, F. & Huelsenbeck, J. P. (2003). MRBAYES 3: Bayesian phylogenetic inference under mixed models. *Bioinformatics*, *19*, 1572-1574.

Seifert, K.A., Samson, R.A., Dewaard, J.R, Houbraken, J., Lévesque, C.A., Moncalvo, J.M., Louis-Seize G. & Hebert, P.D. (2007). Prospects for fungus identification using CO1 DNA barcodes, with Penicillium as a test case. *Proc Natl Acad Sci U S A*, *104*(10), 3901-6.

UTREF: untranslated regions database from RefSeq. Retrieved from http://www.ba.itb.cnr.it/BIG/UTRHome/.

UTRsite: collection of functional sequence patterns located in 5' or 3' UTR sequences. Retrieved from http://www2.ba.itb.cnr.it/UTRsite/.

Van Engelen R. et al. (2003). Developing Web Services for C and C++. *IEEE Internet Computing Journal*, 53–61.

Vassura, M., Margara, L., Di Lena, P., Medri, F., Fariselli, P. & Casadio, R. (2007a). Reconstruction of 3D Structures From Protein Contact Maps. In D. I. Mandoiu, A. Zelikovsky (Eds), *Third International Symposium (ISBRA 2007)*: Vol. 4463. *Bioinformatics Research and Applications*, (pp. 578-589)., Berlin, Germany: Springer.

Vassura, M., Margara, L., Di Lena, P., Medri, F., Fariselli, P. & Casadio, R. (2007b). Fault Tolerance for Large Scale Protein 3D Reconstruction from Contact Maps. In R. Giancarlo and S. Hannenhalli (Eds.) *Seventh International Workshop on Algorithms in Bioinformatics (WABI 2007)*, (Lecture Notes in Bioinformatics 4645, pp. 25-37). Berlin, Germany: Springer-Verlag.

Vassura, M., Margara, L., Di Lena, P., Medri, F., Fariselli, P. & Casadio, R. (2008). Reconstruction of 3D Structures From Protein Contact Maps. *Computational Biology and Bioinformatics*, *5*(3), 357-367.

Vendruscolo, M., Kussell, E. & Domany, E. (1997). Recovery of protein structure from contact maps. *Folding and Design*, *2*(5), 295-306.

Von Laszewski, G., Foster, I., Gawor, J., Lane, P., Rehn, N. & Russell, M. (2001). Designing Grid-based Problem Solving Environments and Portals. In *Proceedings of International Conference on System Sciences* (HICSS-34).

Wagner, E.G., Altuvia, S. & Romby, P. (2002). Antisense RNAs in bacteria and their genetic elements. *Adv. Genet.*, *46*, 361-398.

Yang, Z. (1994). Maximum likelihood phylogenetic estimation from DNA sequences with variable rates over sites: Approximate methods. *Journal of Molecular Evolution*, *39*, 306-314.

KEY TERMS AND DEFINITIONS

Computational Grid: By providing scalable, secure, high-performance mechanisms for discovering and negotiating access to remote resources, the Grid promises to make it possible for scientific collaborations to share resources on an unprecedented scale, and for geographically distributed groups to work together in ways that were previously impossible (Foster, 1999).

Data Federation: It refers to platforms able to provide virtualized federation integration of multiple disparate heterogeneous information sources enabling applications to access and integrate diverse data and content sources as if they were a single resource, regardless of where the information resides, while retaining the autonomy and integrity of the data and content sources.

Grid-DBMS: A distributed system which automatically, transparently and dynamically manages Data Resources, according to the Grid state, in order to maintain a desired performance level. It must offer an efficient, robust, intelligent, transparent, uniform access to Grid-Databases by means of a Grid Data Access Service (Grid-DAS) interface.

Grid Data Access Service: The Grid-DBMS virtualized access interface to Grid-Databases.

Grid Problem Solving Environment (Grid-PSE): A PSE is a computer system that provides all the computational facilities needed to solve a target class of problems. These features include advanced solution methods, automatic and semi-automatic selection of solution methods, and ways to easily incorporate novel solution methods. Moreover, PSEs use the language of the target class of problems, so users can run them without specialized knowledge of the underlying computer hardware or software. By exploiting modern technologies such as interactive color graphics, powerful processors, and networks of specialized services, PSEs can track extended problem solving tasks and allow users to review them easily (Houstis, 1997). A Grid PSE is a grid-based PSE that integrates heterogeneous components into an environment providing transparent access to distributed resources, collaborative modeling and simulation, and grid portals.

Grid Resource Management: It refers to all of the actions to be considered in order to efficiently use the available computational resources in a distributed environment. The grid resource management must take into account the heterogeneity of the resource in terms of processors, storage and network performance.

Grid Workflow Management System: A system that defines, creates and manages the execution of workflows through the use of grid computing technologies, running on one or more workflow engines, which is able to interpret the process definition, interact with workflow participants and, where required, integrate distributed resources and legacy software modules.

Interoperability: It is a property referring to the ability of diverse systems and organizations to work together (inter-operate). Mainly referred to the basic grid services like resource management, information management, data management, security.

High Performance Computing (HPC): It uses supercomputers and computer clusters to solve advanced computing problems characterized by intensive computing requirements and consist of tightly-coupled tasks.

High Throughput Computing (HTC): It is a computer science term to describe the use of many computing resources over long periods of time to accomplish a computational problem. A typical HTC problem consists of many loosely-coupled tasks that can be executed in parallel.

ENDNOTE

[1] The members of the LIBI Grid Platform developers are:

Maria Mirto, *University of Salento, Lecce, Italy & SPACI Consortium, Italy*
Italo Epicoco, *University of Salento, Lecce, Italy & SPACI Consortium, Italy*
Sandro Fiore, *University of Salento, Lecce, Italy & SPACI Consortium, Italy*
Massimo Cafaro, *University of Salento, Lecce, Italy & SPACI Consortium, Italy*
Alessandro Negro, *University of Salento, Lecce, Italy & SPACI Consortium, Italy*
Daniele Tartarini, *Università del Salento, Lecce & SPACI Consortium, Italy*
Daniele Lezzi, *Università del Salento, Lecce & SPACI Consortium, Italy*
Osvaldo Marra, *Università del Salento, Lecce & SPACI Consortium, Italy*
Antonio Turi, *Università del Salento, Lecce & SPACI Consortium, Italy, ITB-CNR, Bari, Italy*
Alessandra Ferramosca, *DISTEBA, Università del Salento, Lecce, Italy*
Vincenzo Zara, *DISTEBA, Università del Salento, Lecce, Italy*
Giovanni Aloisio, *Università del Salento, Lecce & SPACI Consortium, Italy*
Giacinto Donvito, *INFN, Sezione di Bari, Italy*
Luciana Carota, *INFN-CNAF, Bologna, Italy*

Guido Cuscela, *INFN, Sezione di Bari, Italy*
Giorgio Maggi, *INFN Bari, Italy & Università e Politecnico di Bari, Italy*
Giuseppe La Rocca, *INFN, Sezione di Catania, Italy*
Mirco Mazzucato, *INFN, Sezione di Padova, Italy*
Salvatore My, *INFN, Sezione di Bari, Italy, Dipartimento Interateneo di Fisica di Bari, Italy*
Giovanna Selvaggi, Dipartimento Interateneo di Fisica di Bari, Italy
Gaetano Scioscia, *IBM Italy S.p.A. - Innovation Lab, Bari, Italy*
Pietro Leo, *IBM Italy S.p.A. - Innovation Lab, Bari, Italy*
Luigi Di Pace, *IBM Italy S.p.A. - Innovation Lab, Bari, Italy*
Graziano Pappada', *Exhicon srl, Bari, Italy*
Vincenzo Quinto, *Exhicon srl, Bari, Italy*
Margherita Berardi, *Exhicon srl, Bari, Italy*
Francesco Falciano, *CINECA, Bologna, Italy*
Andrew Emerson, *CINECA, Bologna, Italy*
Elda Rossi, *CINECA, Bologna, Italy*
Giovanni Lavorgna, *DIBIT-HSR, Milano, Italy*
Andrea Vanni, *CINECA, Bologna, Italy*
Lisa Bartoli, *Biocomputing Group, CIRB/CIG-Department of Biology, University of Bologna, Italy*
Pietro Di Lena, *Bioinformatics Group-Department of Computer Science, University of Bologna, Italy*
Piero Fariselli, *Biocomputing Group, CIRB/CIG-Department of Biology, University of Bologna, Italy*
Raffaele Fronza, *Biocomputing Group, CIRB/CIG-Department of Biology, University of Bologna, Italy*
Luciano Margara, *Bioinformatics Group-Department of Computer Science, University of Bologna, Italy*
Ludovica Montanucci, *Biocomputing Group, CIRB/CIG-Department of Biology, University of Bologna, Italy*
Pier Luigi Martelli, *Biocomputing Group, CIRB/CIG-Department of Biology, University of Bologna, Italy*
Ivan Rossi, *Biocomputing Group, CIRB/CIG-Department of Biology, University of Bologna, Italy*
Marco Vassura, *Bioinformatics Group-Department of Computer Science, University of Bologna, Italy*
Rita Casadio, *Biocomputing Group, CIRB/CIG-Department of Biology, University of Bologna, Italy*
Tiziana Castrignanò, *CASPUR, Roma, Italy*
Domenica D'Elia, *ITB-CNR, Bari, Italy*
Giorgio Grillo, *ITB-CNR, Bari, Italy*
Flavio Licciulli, *ITB-CNR, Bari, Italy*
Sabino Liuni, *ITB-CNR, Bari, Italy*
Andreas Gisel, *Istituto di Tecnologie Biomediche, CNR, Italy*
Monica Santamaria, *ITB-CNR, Bari, Italy*
Saverio Vicario, *ITB-CNR, Bari, Italy*
Cecilia Saccone, *ITB-CNR, Bari, Italy, &"E. Quagliariello", Università di Bari, Italy*
Anna Anselmo, *Dipartimento di Scienze Biomolecolari e Biotecnologie, Università di Milano, Italy*
David Horner, *Dipartimento di Scienze Biomolecolari e Biotecnologie, Università di Milano, Italy*
Flavio Mignone, *Dipartimento di Chimica Strutturale e Stereochimica Inorganica, Università di Milano, Italy*
Giulio Pavesi, *Dipartimento di Scienze Biomolecolari e Biotecnologie, Università di Milano, Italy*
Ernesto Picardi, *"E. Quagliariello", Università di Bari, Italy*

Viviana Piccolo, *Università di Milano, Italy*
Matteo Re, *Dipartimento di Scienze Biomolecolari e Biotecnologie, Università di Milano, Italy*
Federico Zambelli, *Dipartimento di Scienze Biomolecolari e Biotecnologie, Università di Milano, Italy*
Graziano Pesole, *ITB-CNR, Bari, Italy, "E. Quagliariello", Università di Bari, Italy, & Università di Milano, Italy*

Section VII
Building and Deploying HealthGrids

Chapter XXX
UNICORE:
A Middleware for Life Sciences Grids

Piotr Bała
ICM University of Warsaw & N. Copernicus University, Poland

Kim Baldridge
University of Zurich, Switzerland

Emilio Benfenati
Istituto Mario Negri, Italy

Mosè Casalegno
Istituto Mario Negri, Italy

Uko Maran
University of Tartu, Estonia

Łukasz Mirosław
University of Zürich, Switzerland

Vitaliy Ostropytskyy
University of Ulster, UK

Katharina Rasch
Technische Universität Dresden, Germany

Sulev Sild
University of Tartu, Estonia

Robert Schöne
Technische Universität Dresden, Germany

Bernd Schuller
Research Centre Juelich, Germany

Nadya Williams
University of Zürich, Switzerland

ABSTRACT

This chapter provides an overview of Grid middleware and applications related to biomedical and life sciences disciplines. Various technologies, including web-based solutions, are presented. One of the solutions, the UNICORE framework, in its recent version implements key grid standards and specifications. The system architecture and capabilities, such as security, workflow and data management are described. Special attention is given to the idea of a 'gridbean', which expands the UNICORE use for different applications. Examples of gridbeans are provided and the capabilities of UNICORE are illustrated through specific examples built using this grid middleware. In particular, the Chemomentum workbench and its use for in-silico design and modeling in chemistry and life sciences are both described.

INTRODUCTION

The past decade has witnessed tremendous growth in our chemical and biological understanding at the molecular scale that impacts all facets of our daily lives. Inherently driven by information technologies (IT) and massive data acquisition, this growth in understanding requires a stronger coupling between computation and experiment. The expansion of high-throughput technologies-in chemical design, microarray analysis, device technology, physical and chemical analysis processes, combinatorial chemical synthesis, and screening, as well as their computational counterparts, has generated an unprecedented volume and diversity of data that must be abstracted into manageable units. Historically, chemistry has evolved a sophisticated symbolic and iconographic way of projecting a myriad of chemical phenomena onto a model of molecular structure and associated properties. This rapid growth of both experimental and computational data has driven the proliferation of computational tools and the development of first-generation chemoinformatics for data storage, analysis, mining, management, and presentation. However, it is generally accepted that first-generation chemoinformatics tools have not been widely adopted, nor have they met the needs of the researchers. There is a general need for integrating data, information and knowledge that can be accomplished through innovative ways of computational access, resource integration, and creation of transparent toolsets.

The computational science challenge, therefore, is to provide new algorithms and data analysis tools that can exploit more efficiently the existing computational power for more detailed chemical analysis. The corresponding challenges in information technology and visualization include providing richer access to existing repositories of data, enabling transformation of that data into interpretable and predictable scientific information, and leveraging workflows to parallelize the efforts of many researchers for general access.

Over the past few years our team has been developing middleware and end-user interface technologies to enable novice and advanced users to access capabilities of sophisticated scientific applications running on large-scale grid computing resources. In this paper, we present some of these developments, from web-based tools to integrated toolkits such as NIMROD (Abramson, Lewis, Peachy 2001) and BOINC (Boinc 2008), to full-featured grid middleware such as Globus (Foster 2006) and UNICORE (Erwin, Snelling 2002). The main focus is on the latter, which is known from a number of successful deployments in the life sciences area.

BACKGROUND

Web Based Solutions

In the past, the primary access to computational resources was based on batch processing. The input files and command scripts were typically prepared using simple text editors, and the command line tools were used to transfer files from local workstations to remote servers located in the dedicated supercomputer centers, in the campus or departmental networks. The progress in web technology enabled replacement of some command line tools with a web browser. This opened the field for an easy access to distributed resources. A few successful examples of such interface tools are described here. All have differing capabilities, some offer very limited grid functionality yet clearly illustrate a change in the mode of access to the computational resources.

WebMO

WebMo (WebMo 2008) is a web based submission system for quantum chemistry software such as Gaussian (Gaussian03 2004), GAMESS (Schmidt et al. 1993), and MOPAC (Stewart 1990). Users can draw molecular structures in a 3D editor, run calculations, and view results directly in a web browser. WebMO is installed on a single web server that runs the computational chemistry programs or provides access to software through a batch system. This tool is limited to the local batch systems and thus has no true grid capabilities. WebMo is a useful tool for education, and can be considered as representative example of a type of portal access to computational resources, although these advantages are limited (Rauchant 2002).

BioCoRE

BioCoRE is a collaborative work environment for biomedical research, research management, and training, developed by the Theoretical Biophysics Group at the University of Illinois (Bhandarkar et al. 1999). With a standard web browser, users can create a project, store different data and share that data with other project members. The BioCoRE can be used to submit jobs to local or remote sites, carrying out visualization, and create input files for computational chemistry and biochemistry simulations. The files and visualizations can be shared among partners. Users have an access to co-authoring tools for papers and other documents, lab books, project wide bookmarks, chats and notification systems. Sessions are automatically recorded and can be reviewed later by the project leader or other team members.

The BioCoRE Job Management tool provides a simplified interface to supercomputer centers, local computational resources, and even to the workstation on the researchers' own desk. BioCoRE uses Globus (Foster 2006) and MyProxy (Novotny, Tuecke, Welch 2001, Basney, Humphrey, Welch 2005), as well as the ubiquitous SSH protocol, to submit and monitor computations. The monitoring of running simulations and their output is possible from a single web page during the course of the simulation. In addition, the user has an access to the dedicated tools for transfer of files from the BioCoRE file system to the remote system before the job begins. The output files are automatically retrieved upon completion of the job.

Currently, this setup is limited to molecular dynamics simulations performed with NAMD (Kal et al. 1999), and to visualizations using a single package VMD (Humphrey, Dalke, Schulten, 1996). Nevertheless, the set of tools provides an integrated solution for quite a significant research community (Brandon et al. 2005).

NPACI Portal

One early effort to facilitate the submission of jobs to remote platforms are the SDSC portals based on the NPACI GridPort software (Thomas, Boisseau 2003, Thomas et al. 2003). The portal developers provided two main APIs. A Perl API is for establishing secure web portals capable of submitting, controlling, and gathering the output of jobs to remote high performance computing resources. The GridPort API provides data files transfer capabilities that could be used to transfer input and output data using GridFTP (Allcock et al. 2005) or the Storage Resource Broker (SRB) (Moore 2001). Through the web sites built with GridPort, users were shielded from the complexity of the remote platforms used for computations and archival storage. However, building portals is non-trivial and involves the intricacies associated with writing html files and cgi scripts. As a result, any reconfiguration is not trivial. The Baldridge group was involved in early efforts to enable applications via the portals, including, for example, GAMESS, APBS (Baker et al. 2001), AUTODOCK (Morris et al. 1998), etc. To access a GridPort portal such as the GAMESS portal (Baldrige et al. 2002), a user logs onto a website, selects input files and chooses a remote platform for running their jobs. The portal provides the job status information and stores the output on the SRB system. GridPort uses the GSI software provided by the Globus Toolkit to handle authentication, security, and job submission and control. Figure 1 gives an example portal, job, and data flowchart.

Figure 1. Flowchart illustrating the operation of the GAMESS portal. © 2009 Kim K. Baldridge. Used with permission.

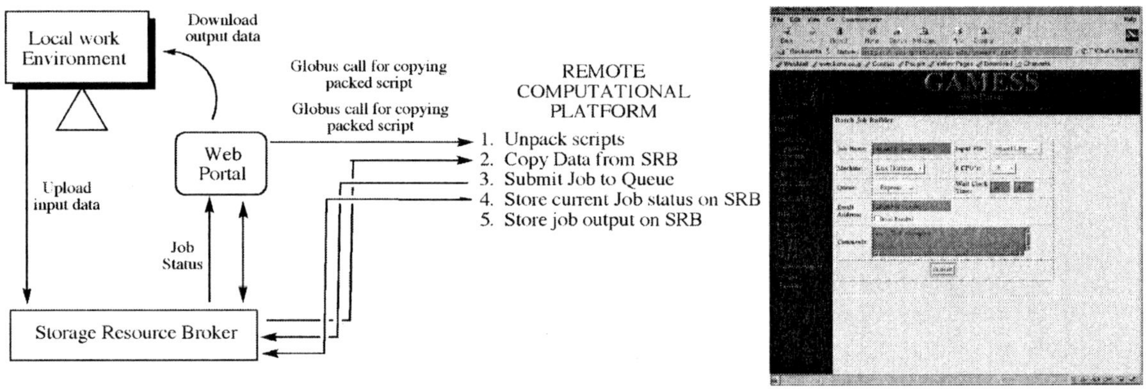

In the initial implementations of the GAMESS portal, the "storage device" used to archive the computational results was an exported NFS file system. In an updated system, all data files were "collections" within the SRB. Access to data stored on the SRB is provided by the web portals, which allowed a user to download and upload files using SRB client tools on Unix-based systems and the "srbBrowser" program on Windows systems, or via the SRB web site.

While GridPort proved to be a useful tool and was widely utilized, it lacked a basis for building complex and dynamically re-configurable workflows. While complexity was hidden from the end user, it was not hidden from the application scientist or portal develop. Many cgi scripts and HTML documents had to be newly created, or copied and modified from the documents for the existing portals. This process was time consuming because it required fine-tuning for each new portal or application. Complications arose when new computational platforms were added, since such additions required implementation of all runtime details for the new platform, including the queuing mechanism, the location of the executable, and the directory structure. What was lacking in this design was an overall architecture that provided the end user and the application programmer with an interface that hides complexity of the available computational resources, while allowing scientific programmers to integrate the software into the system in a simple effective manner.

The next generation of portal technology provided a further layer of abstraction for the applications programmer and exploited the latest grid technologies. A big advantage is that such web and grid technologies can be used to run application jobs, for example, GAMESS, without changing the program nor how it is deployed. To enhance the interoperability with the other programs, and to provide an ability to execute using the same technology, the output needed to be parsed in a form other than special ASCII text. From this point, work has continued towards enabling software such as GAMESS with capabilities to have an output option for structured data in the form of XML documents and JAVA objects based on the associated schemas, thereby facilitating data transfer between programs in actual workflow and providing a basis for database structure.

GEMSTONE: Grid Enabled Molecular Science Though On-Line Networked Environments

GEMSTONE (Gemstone 2008) is a tool that enables molecular science researchers to discover remote grid application services and compose them into a chain of steps, or a workflow, that needs to be executed for a given scientific problem. The interface is a rich desktop application that enables users to browse remote registries for available service instances and to directly access remote applications in innovative ways. In particular, the software provides access to a suite of computational chemistry- and biochemistry-related tools software and associated tools and applications, all wrapped by a service layer and integrated with back-end computational clusters. GEMSTONE supports GSI-based secure web services infrastructure. The current services provide access to classical macromolecular algorithms (APBS), *ab initio* quantum techniques (GAMESS), molecular docking (AUTODOCK), and plane-wave density function theory (SIESTA) (Artacho et al. 2008) amongst others.

A primary focus of GEMSTONE is on creating a "composable service-oriented architecture for grid applications" by leveraging the web services infrastructure. There are three major components in this focus, the Rich-client Desktop Interface, the Workflow Integration, and the Strongly Typed Data Schemas. The first component is built on top of the Mozilla engine (Mozilla 2008), runs on all platforms, and includes dynamic integration of GUI elements. This component provides a "dynamic" user interface to backend computational and visualization services that automatically adds new services as they become available. The portal-like environment enables novice users to invoke the tools with little difficulty. To accomplish this, a local database and state manager, and various remote services, applications and application wrappers are being built. The second component, the Workflow Integration, was already begun in phase one of NMI-middleware developments. Additional work in this phase is necessary to support workflow publication and discovery, as well as brokering and fault tolerance. For the third component, the created web services have strongly typed methods and standardized data types for sending input data and receiving the output results across different applications. Recently, community efforts for incorporating XML- and CML-based interoperability enabled significant progress in bringing more software together into workflow systems, particularly those related to Quantum Chemistry. Since the release of version 1.0 of the GEMSTONE in September of 2006, the software has been used to access these services by the users across 14 countries and 8 US states.

GRID TOOLS FOR LIFE SCIENCES

The main motivation for the development of web-based interfaces is to make access to resources simpler for end users and to overcome aspects of distribution. Many such users must share large amounts of data and run the simulations on diverse computational systems, including large-scale facilities, clusters, workstations, and common PCs. Access to such large and distributed resources requires more sophisticated tools that enable an easy and secure use. Recent advances in Grid tools make them good candidates for development of uniform user interfaces to distributed resources. Computational Grids enable selection, aggregation and sharing of distributed resources across multiple organizations, for the purpose of solving large scale computational and data intensive problems. This includes access to various databases important for life sciences research. Current technology enables a wider range of activity, such as complete process initiation and execution with necessary data location and retrieval. As a result, one can carry out significantly bigger computational tasks that cannot be solved with traditional methods, and can use new capabilities for knowledge generation.

Currently, grid tools such as Globus Toolkit, Avaki (Grimshaw et al. 2003) or UNICORE (Erwin, Snelling 2002), provide a relatively low level of operational control. Grid middleware can be used for the development of tools such as web interfaces, portals, or integrated workflows that will help to automate low level processes and will provide the user interfaces to the applications. The representative solutions are presented below together with the examples of successful deployment of these solutions in the life sciences.

NIMROD Distributed Parametric Modeling Toolkit

Nimrod is a grid computing software toolset, created by David Abramson's research group at Monash University, Australia (Abramson, Lewis, Peachy 2001). Nimrod tools facilitate parametric computational experiments to be carried out, enabling one to perform embarrassingly parallel parameter sweeps (using Nimrod/G (Abramson, Buyya, Giddy 2002)) and optimizations (using Nimrod/O (Peachey, Abramson, Lewis 2006)) that are distributed on a Computational Grid. This specialized system uses a simple declarative parametric modeling language to express a parametric experiment, and provides capability to automate the task of formulating, running, monitoring, and collating the results from the multiple individual experiments. Equally important, Nimrod incorporates a distributed scheduling component that can manage the scheduling of individual experiments to idle computers in a local area network. Together, these features mean that even complex parametric experiments can be defined and run with little programming effort. The user simply specifies a list of available computer resources, and for each computational 'experiment' (i.e., collection of parallel jobs), selects the particular resources to use. A short, declarative 'plan file' describes the computational job parameters and the tasks to be performed for each single job. Nimrod then schedules, allocates, executes, controls, and collects the individual job's results via 'agents' submitted automatically to the involved resources. In many cases, it is possible to establish a new experiment in minutes. Nimrod uses services and features provided by the Globus Toolkit (Foster 2006). The architecture is extensible and can be used with other grid middleware, such as Condor and Netsolve, as well as stand-alone batch schedulers, such as PBS. At the backend of Nimrod there is a PostgreSQL database and Python code. Nimrod has a command-line interface, a web portal (Nimrod Portal), and an Excel spreadsheet interface (Active Sheets) (Abramson et al. 2001). Nimrod is a well established solution for a particular class of computationally intensive problems, and has been applied to Nimrod. In particular, QSAR, docking, and molecular modeling applications are included in our interest.

BOINC

BOINC, Berkeley Open Infrastructure for Network Computing, is an open source grid software toolkit created by David Anderson's research group at UC Berkeley (Boinc 2008). This tool enables applications to be run in the background of private desktop computers connected through the internet. Anybody with an internet access and a computer may sign up to sponsor a research project, volunteering their unused computer resources. Nimrod and BOINC are similar in that they both eanble more efficient processing by distributing computational workloads across a number of resources.

The scheduling policies in the BOINC client enable hosts to successfully contribute to an unlimited number of projects. Most of the time, jobs complete on their deadlines, even when the host is slow, its availability is low or highly variable, or job completion time estimates are consistently erroneous. Support for "account managers" web sites was added so that computer owners can browse and sign up for volunteer computing projects with a single click. The BOINC scheduler is "memory aware", and can suspend the running applications if the associated memory usage exceeds user-configurable limits. This makes it feasible for volunteer computing applications to use large or unpredictable amounts of RAM with respect to the resource usage boundaries set on a given computer.

GAMESS has been enabled within a BOINC environment (BoincGamess 2008). The objective was to enable Nimrod to execute GAMESS quantum chemistry calculations on BOINC desktop resources. The first step was to generate the interface software between BOINC and the Nimrod resource. Ultimately, comparison of results performed on typical Nimrod resources and on desktops confirmed the feasibility of the use of BOINC as a Nimrod resource. A second application project enabled by BOINC middleware is Docking@Home, created by Michela Taufer (Docking@Home 2008). This project enables investigations of potential drug molecules or ligands by computationally simulating the behavior as they are docked into an active protein structure site. Users of Docking@Home can invoke one of several algorithms to adapt to the complexity and characteristics of the proteins under investigation.

Globus

The development of the open source Globus Toolkit started in the 1990's (Foster, Kesselman 1997) and has resulted in a popular technology for the Grid. This toolkit allows users to securely share online compute power, databases, and other tools across corporate, institutional, and geographic boundaries. The first version was made available in 1998, followed by a 2.0 release in 2002. The further development was based on the new Open Grid Services Architecture (OGSA) and resulted in the 4.0 version (Foster 2005, Foster 2006). Essential background of the Globus Toolkit has been provided by its main architects: Ian Foster, Carl Kesselman and Steve Tuecke (Foster, Kesselman, Tuecke 2001, Foster et al. 2002), and has been followed by the work of numerous developers and users.

The Globus Toolkit includes software services and libraries for resource monitoring, discovery, and management, as well as security and file management. The toolkit is packaged as a set of components that can be used either independently or together to develop services. The core services, interfaces and protocols allow users to access remote resources as if they were located locally. Simultaneously, Globus preserves local control over resources, and enables control of who uses them and when.

In addition to being widely adopted by the research community, the Globus Toolkit is a substrate on which significant commercial Grid products are being built. The Globus Toolkit has been used to build large grid infrastructures for the different application domains. One of the best examples of deployment in life science is WISDOM (Jacq et al. 2008), high-throughput computing performed by the Enabling Grids for E-sciencE (EGEE) project (Galiardi 2005). The project performed over 46 million docking simulations to find new inhibitors for a family of proteins produced by *Plasmodium falciparum*. Over 1700 computers distributed in 15 countries were used in about 6 weeks, which is the equivalent of 80 years on a single machine (Breton, Kasam, Jacq 2006). An enhanced version has been used to produce over two millions docking complexes to search for potential drugs for the predicted variants of the *avian flu* virus H5N1 (Lee et al. 2006).

Several applications have been ported to Globus and have been run in production to calculate observables for chemical reactions, simulate molecular dynamics of complex systems, and calculate electronic structure and properties of molecules, molecular aggregates, liquids and solids under the framework of GEMS molecular simulator (Gervasi, Dittamo, Laganà 2005). One should note that deployment of commercial applications important for the life sciences community is not straightforward due to the licensing problem. A notable success was making the Gaussian and Turbomole software packages available via a dedicated virtual organization (Sterzel, Szepieniec 2007).

The virtual screening projects revealed the limitations and bottlenecks of the current EGEE infrastructure. Overall grid efficiency was reported to be about 50 percent on average, because of license server failures, workload management failures and sites failures (Jacq et al. 2006). This means that much still remains to be accomplished in improving availability and performance of Globus-based infrastructures for life science.

PRAGMA

The Pacific Rim Application and Grid Middleware Assembly (PRAGMA),[Pragma 2008] founded in 2002, is an open international organization that focuses on a variety of practical issues of building international scientific collaborations in a number of application areas using grid technologies. With representation initiating in the Pacific Rim and Western USA, the main focus of PRAGMA is on scientific applications. Integration of coordination issues, design, implementation, availability, and performance of a PRAGMA Grid Testbed (Abramson et al. 2006) with associated application spaces and technology collaborations via multidisciplinary teams of researchers in different countries is the driving force to advance the applications and the grid technology.

The PRAGMA test bed itself is a "grass roots" effort to build a long-term working experiment that enables application scientists to work with a variety of applications across international boundaries. Today, close to 29 institutions from four continents and 12 countries have contributed both physical resources and, critical to success, human resources to the test bed. These types of efforts are an instantiation of highly useful, interoperable, and consistently available grid systems that are neither dictated by a single science domain, nor funded by a single national agency. They differ significantly from experiments such as PlanetLab (Bavier 2004, Chun et al. 2006) and TeraGrid (Catlett 2002, Reed 2003) because they do not have uniform infrastructure management. Scientific applications that regularly run on these resources have helped to address important technological and social issues for providing such resources on a routine basis, across a variety of application spaces and underlying grid technologies. The integration of the applications and the middleware is carried out by the close interaction of the working groups, where creation of new components or developing of existing ones is achieved via ongoing scientific investigations and dissemination of knowledge. Although PRAGMA is not focused on the development of software or middleware, it plays an important role in providing sustainable multinational grid infrastructure for life science users.

UNICORE 5

Another production-ready Grid system with integrated security, workflow management, and support for a variety of resources, in particular high performance compute resources, is UNICORE 5. In particular, this toolkit is used to provide access to large infrastructures of D-Grid (Neuroth, Kerzel, Gentzsch 2007) and DEISA (Niederberger, Alessandrini 2004). UNICORE was initiated in 1997 to enable German supercomputer centers to provide their users with a seamless, secure, and intuitive access to heterogeneous computing resources. Two years later the development became international and the UNICORE Forum – a non-profit association to foster the distribution, maintain the specifications and to coordinate further development – was established.

The UNICORE software provides a general framework for running user applications. This includes input preparation, job submission and control, as well as post processing of results. Advanced visualization and access to databases and other distributed sources of information are also considered here. Jobs to be run on the same system can be grouped in job-groups. Jobs can have complicated structures that can be run on different systems with various job subgroups and file transfers ordered with user defined dependency. The user input is mapped to the target

Figure 2. UNICORE 5 architecture

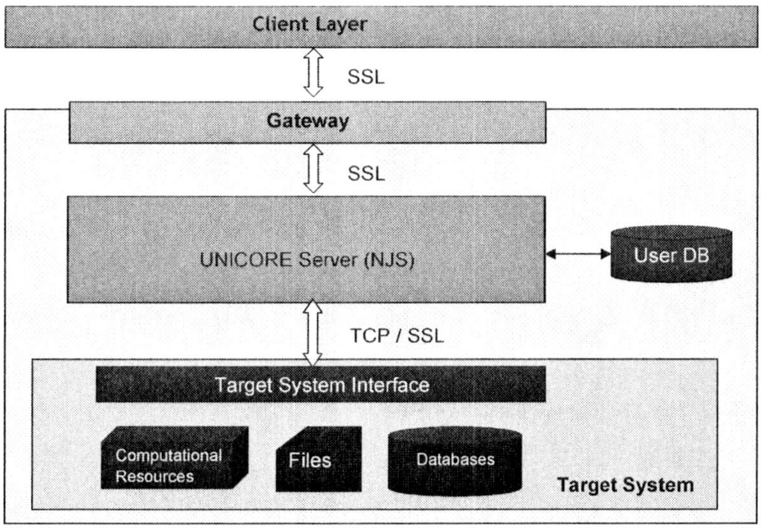

system specific commands and options by the UNICORE infrastructure.

Similar to the other grid middleware, the UNICORE architecture is based on a three-tier model, namely a user, server and target system tier. The user tier consists of the graphical user interface, written as a Java application. This interface offers functions to prepare and control jobs and to set up and maintain the user's security environment. UNICORE security is based on the Secure Socket Layer (SSL) protocol and the X.509 certificates. SSL uses public key cryptography to establish connections between a client and a server. Therefore, each component of the UNICORE infrastructure has a public-private key pair with the public part known by others. The keys are certified by a certification authority (CA).

Within the UNICORE environment, a user can comfortably use distributed computing resources without having to learn site or system specifics. The UNICORE client provides the entry point to a consistent user environment, abstracting Grid resources and services. The job components can be executed asynchronously or independently on different systems at different UNICORE sites. The basic elements of the environment are:

- Script task, to submit job scripts;
- Transfer task, to specify data transfer between different jobs groups;
- Job groups, to build sub-jobs for other systems.

All elements can be added and edited with the graphical interface in the UNICORE client. Users can define target systems for each job group as well as dependencies among job groups and file transfers. Each job group must define required resources, such as the number of processors or nodes, memory allocation, CPU time and disk space. A check is made as to whether the target system can fulfil these requirements. This task is performed during the process of preparing the job, before it is submitted.

Within the UNICORE infrastructure the user program can be run as a script submitted to the target system, analogous to the batch execution mode. The advantage is a single login and transparent file transfer, however the submitted job is similar to the one created in a traditional queuing system. The UNICORE 5 client has built-in capabilities for checking software resource entries on a particular target system, informing the user if the selected program is not available.

This mechanism, however, cannot be used with the simple user script jobs and is accessible only when the user job is prepared with the dedicated plugin for the UNICORE client.

Plugin Concept

The UNICORE plugin mechanism is a very attractive and efficient method for integration of applications with grid middleware. Plugins are loaded into the UNICORE client during initiation or upon request. Once a plugin is available in the client, a user obtains access to the menu that enables, for example, preparation of application specific jobs. Dedicated plugins can also be used for database access or for post-processing. Since plugins are written in Java, they can be easily integrated with external applications or Java applets.

A plugin that enables users to automatically download plugins to the local systems and install them in the proper directory has also been developed. In this way, required components can be easily managed at the target systems and propagated to the end user. This mechanism provides the user with a transparent mechanism of software updates and allows automatic synchronization to modifications of installed software.

UNICORE Plugins for Biomolecular Applications

During the last years, plugins have been developed and made available for leading scientific applications such as Gaussian98, Amber (Pytliński et al. 2002) and Carr-Parinello Molecular Dynamics (CPMD) (Huber 2001, Pytliński et al. 2002). We have also developed plugins for access to remote databases from the UNICORE client (Pytliński et al. 2003). This allows users to prepare queries, for example, to the PDB database, a widely used crystallographic database, as well as to the nucleotides database Entrez. A query is submitted to the databases web interface and the result is returned to the UNICORE client. The search results can be browsed, visualized, or saved for later processing. Since copies of both databases are available at different locations, the plugin checks automatically which mirror database is currently available and which provides the answers in the shortest time. Another plugin has been developed for accessing SQL databases. A user can submit to SQL directly from the UNICORE Client and integrate with other applications as part of the workflow (Wypychowski et al., 2004).

A uniform set of plugins was developed in the framework of the Open Computing Grid for Molecular Science and Engineering (OpenMolGRID), which is a Grid enabled system for molecular design and engineering applications (Slid et al., 2006, Romberg, 2005). The main motivation for the development was *in silico* testing, which has become a crucial part in the molecular design process of new drugs, pesticides, biopolymers, and biomaterials. One of the challenges within OpenMolGRID was data management, addressed by research towards the unified data warehouse concept (Dubitzky et al., 2004). A second challenge was related to providing the ability for geographically distant grid applications to work together in the framework of one virtual organization. Realization of this effort resulted in the XML based simple definition of complex scientific workflows (Schuller et al., 2005, Sild et al., 2005), capable of using grid middleware functionality while assigning required grid resources automatically.

The OpenMolGRID data warehouse component used plugins to provide access to toxicity data for data mining environments from selected public data sources, namely National Toxicology Program (NTP) database and ECOTOX database. Data mining and molecular engineering included plug-ins for the following: fast 2D to 3D conversion with distance geometry approach (Crippen, 1981) implemented in MOLGEO (Gordeeva et al., 1993), semi-empirical quantum chemical package MOPAC, molecular descriptor calculation and QSAR model development from CODESSA Pro software (Codessa, 2008). Numerous smaller plugins were generated eg, to facilitate data conversion. Example uses of the distributed computing technologies include the modeling of inhibition of aspartyl protease enzyme (Maran et al., 2007a), acute toxicity to fish (*Pimephales promelas*) (Maran et al., 2007b) and aqueous solubility (Sild et al., 2005).

UNICORE 6

UNICORE 6 builds upon a number of concepts, such as Service Oriented Architecture (SOA) and messaging. The term "Service Oriented Architecture" implies a certain architectural style for software solutions. As defined by the OASIS SOA Reference Model (Reference Model for Service Oriented Architecture 2006), SOA is a "paradigm for organizing and utilizing distributed capabilities that may be under the control of different ownership domains". In effect, the capabilities of a distributed system are organized into well-defined chunks of functionality, the "services". Other parts of the system may then interact with the services to exploit their capabilities. To this end, both service and service consumer must agree on the service interface, the service semantics, and the real world effects, of a service invocation. To promote loose coupling and reusability of services, consumers rarely bind to services directly, but usually engage in a process of dynamic discovery to 'discover' required services. Consumers can look up required services in a service repository or registry.

The web services technology enables the building of document oriented distributed systems. In addition, web services cater to interoperability, an element that is of crucial importance for realizing SOA. Because web services use XML as their basic message exchange format, services and service consumers need not share the same implementation platform. The Web Services Resource Framework (WSRF) is an application of web service technology to realize access to and management of stateful resources. In grid computing, with its strong focus on heterogeneous resources that have to interoperate, web services have emerged as key technology on which current Grid middleware is built.

Architecture

UNICORE, in its latest version 6, is compliant with the Open Grid Services Architecture (Foster, Kesselman, Nick, Tuecke 2002) and uses standards such as WSRF 1.2 (WSRF 2008) and job submission definition language JSDL 1.0 (JSDL 2008). An overview of the basic UNICORE 6 architecture is provided in Figure 3. The system is layered into three tiers: the client tier, the Gateway tier, and the WSRF service container tier. Communication between these three tiers is carried out through the use of web services.

The Client tier consists of number of diverse clients, varying in design and capabilities. The command line client can be used to submit simple script jobs or workflows described in the XML document according to the JSDL standard. Less advanced users can benefit from the GPE (Grid Programming Environment) Application Client that through gridbeans provides a graphical interface to the installed applications. The most advanced functionality, which includes workflow editing, is available with a dedicated client developed using Eclipse Rich Client Platform. The web client for job or workflow submission has been developed using the UNICORE Basic Client libraries. The same tools are used to introduce grid capabilities to existing graphical applications.

The Gateway authenticates messages and forwards them to the services. This mechanism provides a "web services firewall" and is used to protect several servers.

The services tier consists of a web services resource framework (WSRF) container hosting a set of basic Grid services (atomic services), and additional services such as a registry. To access

Figure 3. UNICORE 6 basic architecture

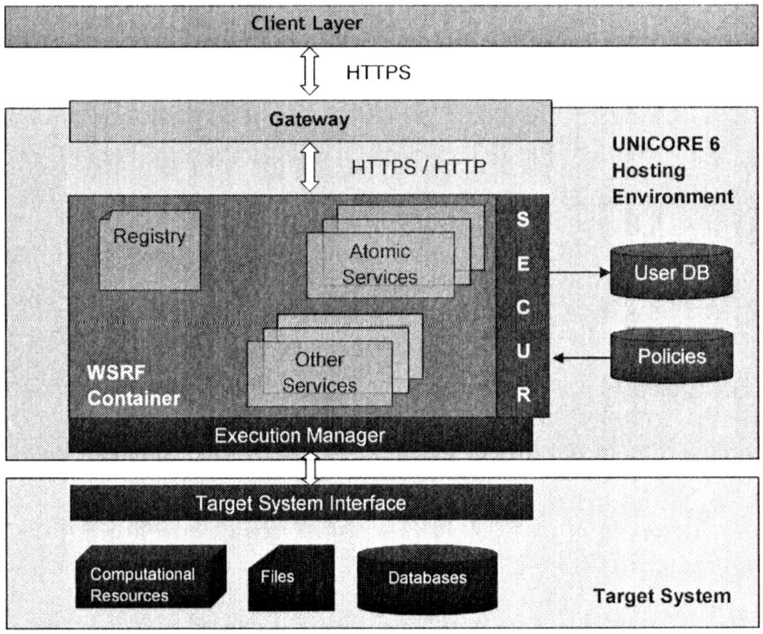

any service instance, clients need the Web Services Addressing End Point Reference (EPR) of that service instance. Via the Execution Management and Target System Interface (TSI) components, the services have access to physical resources such as batch systems or disks.

UNICORE 6 provides a generic web services hosting environment. Services can be deployed into this environment, in order to benefit from its general features: persistence, security infrastructure, and so on. Since all services are hosted in the same environment, a common web service level security infrastructure is achieved.

The basic Grid services are often called "atomic" as they provide small, elementary services that may be composed into higher-level services. The UNICORE 6 provides such basic services, most importantly, a job execution and a file access. All services are WSRF services, so they provide a set of properties (WSRF resource properties), and the usual WSRF lifecycle and lifetime methods. The functions such as "job execution", "file management", "file transfer" and "service registry" are covered by these services. Abstraction and virtualization is used to hide target system specifics such as paths to binaries or storage mount points. In a typical system, many instances of most of these services are running simultaneously.

The current UNICORE 6 security infrastructure offers detailed access control, centralized user and role management and basic transport level security. UNICORE 6 provides basic communication security using secure socket layer and transport layer security (SSL and TLS). Users and server components are identified using a X.509 certificate issued by a trusted certification authority.

Requests are authorized using UNICORE authorization service (XUUDB), and a set of eXtensible Access Control Markup Language (XACML) policies that governs the final decision whether a request is allowed to pass. Authorization policies can contain one or more rules and the XACML standard defines how rules are combined, and how policies can be composed of multiple subpolicies. Example rules expressed in natural language allow access to individual service for users with the role "admin", allow querying the registry by everybody, deny access to job management instances by non-owners, or deny all access for the users based on a specific domain name in their email address. As a result, the flexible and powerful user management is obtained.

The connections between components are protected using a SSL authenticated client. Client authentication means that both parties identify themselves using their certificates, and one party must trust the other party certificate. For the communication between the Client layer and the Gateway, SSL is mandatory, whereas deployed services may communicate with the Gateway and the XUUDB using either SSL or unencrypted HTTP.

Workflow Management System

Workflow processing is at the core of the UNICORE 6 system. The Workflow System is subdivided into two layers of abstraction: the Workflow Engine and Service Orchestration. The Workflow Engine processes a workflow on a logical level, whereas the Service Orchestrator deals with the actual execution and supervision of tasks using different services and resources on the Grid. Thus, the workflow processing logic is cleanly separated from the re-occurring invocations of lower level Grid services.

In order to handle different domain specific workflow languages, the workflow engine translates incoming workflow descriptions into a generic workflow language using pluggable Domain Specific Logic (DSL) modules. The translation can be a complex process. For example, a translation may involve the splitting of single activities into sets of activities that can be executed in parallel within the translated workflow. Splitting and distribution of computational effort is one of the core features of the system. Each activity of the translated workflow results in an atomic unit of work for the Grid, a so-called *Work Assignment (WA)*. Work Assignments (WAs) are abstract, in the sense that they are not bound to specific service endpoints on the Grid. They are individually submitted to the Service Orchestrator for execution. Due to dependencies and conditions in the translated workflow, WAs cannot be executed in an arbitrary order. For instance, one WA may depend

Figure 4. Main components of the Workflow System and their interfaces

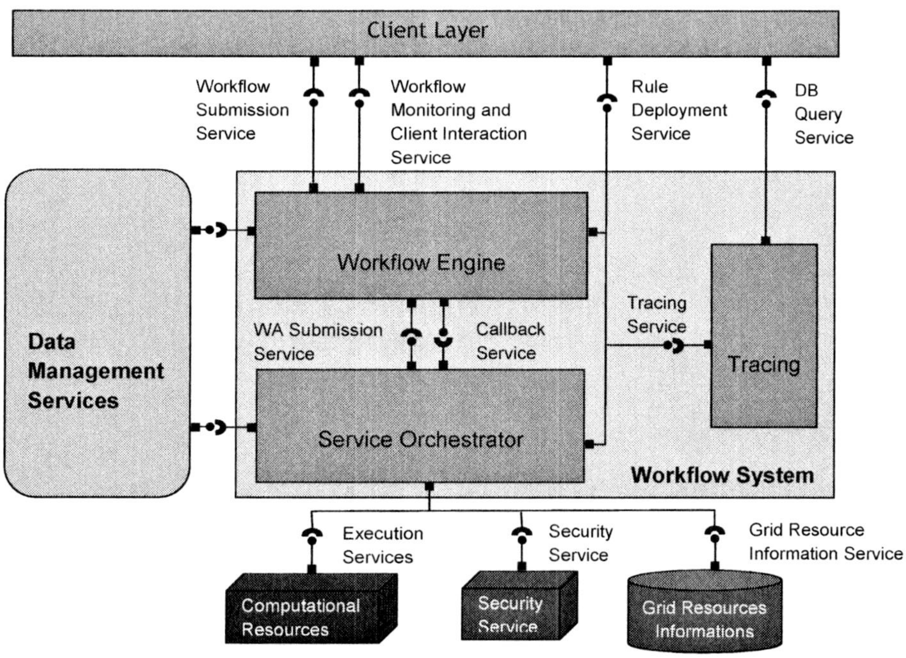

on the output data of another WA. The Workflow Engine keeps track of such dependencies and does not submit WAs with unmet conditions. The Service Orchestrator transforms each incoming WA into a job, given in JSDL. It exchanges the logical names of input files for addresses of physical file locations, submits the job to a computing resource, supervises job execution and informs the workflow engine of job completion/failure.

Data Management System

As a pluggable feature to UNICORE 6, a Data Management system (DMS) has been developed. This system is primarily intended for use within the project Chemomentum, but can be used independently as well. The system provides functionality to store, retrieve, update, and delete data, and gives its global view independent of the location. The DMS establishes a distributed data storage that can be used not only to store data, but also to store and generate augmenting, extendable metadata that thoroughly describe this data. The support for this extendable metadata has been crucial for the ambition of the system to enable a development of Grid technologies applicable in any scientific, economic or other field.

The architecture of the DMS is depicted in Figure 5. The central interface for workflows and clients to the data management system is the Data Management System Access Service (DMSAS). The DMSAS offers high-level access to data management functionality. The system manages the handling of incoming service requests, identifies the underlying services necessary for the processing of the requests, and forwards sub-requests to them, collects and unifies the results, and returns them to the requester. Also, the DMSAS offers a basic transaction mechanism that rolls back changes made to the underlying metadata and data spaces if a service request fails.

The heart of the Data Management System is the Documented Data Space (DDS). The DDS consists of data storages, a location database, and metadata databases. Each of these components is accessed by a specialized service. The data storages contain data in flat files, typically parts of the input and output data produced by single jobs that were run during a workflow. The current transfer protocol to store and retrieve data from these storages is based on the OGSA ByteIO standard (Morgan, Chue-Hong & Drescher 2006), but can be extended to other protocols as well. The location database acts as a global file location directory:

Figure 5. Components and layout of the Data Management system

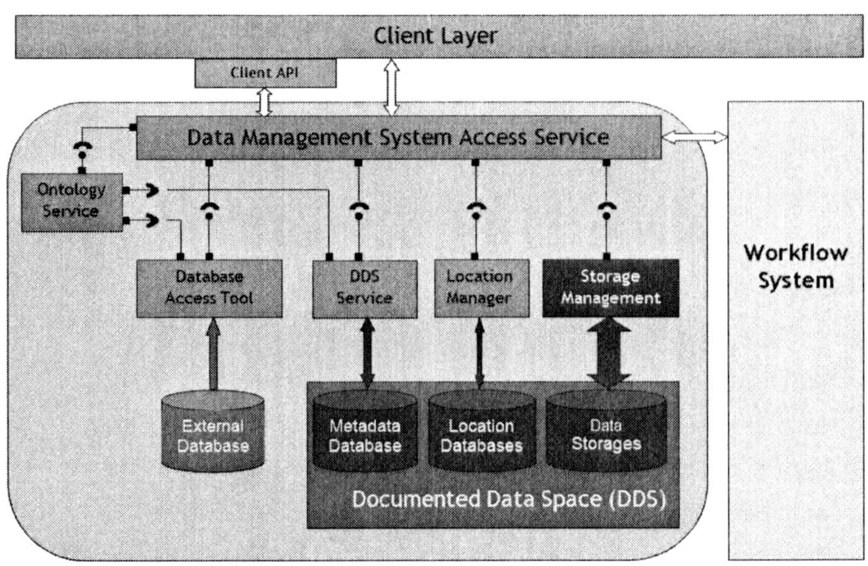

it indexes these files and assigns globally unique logical names to them. The metadata databases contain metadata that describe the files in the data storages, referencing them only by their logical names. Using only the logical names allows an easy migration of files from one file storage to another. To increase the reliability as well as the performance of the whole system, the location database also supports the replication of files. One logical name can be assigned not only to the master copy of a file, but also to an arbitrary number of slave copies. As another layer, the DMS offers the possibility to create and manage easily metadata domains with arbitrary metadata attributes and their properties. This makes it independent of the underlying database system. To allow the use of any database, the access logic provides pluggable database handlers.

The DMS is not limited to access of data produced by the UNICORE 6 system. The Database Access Tool (DBAT) serves as a uniform interface to external data sources. DBAT transforms the original query passed by the DMSAS into the native query language used by the external data source, queries the source, and transforms the result back into a format the client understands. The current prototype provides access to the two Ecotoxicology (Ecotox 2008) databases: Aquire and Terretox. A specialized service – the ontology service – broadens the queries, converts the values between units, provides a list of all possible unit conversions for specific measurements and unifies the results. This is necessary, particularly for the query of external data sources and the subsequent interpretation of results. The ontology service provides access to external ontology sources and additional knowledge, such as ChEBI (Degtyarenko et al., 2008), about names, types, units and vocabulary, enabling interpretation of the data.

A special API is provided for client-side communication with the DMS. This Client API can be used by the UNICORE 6 workflow system, GUIs and other clients. The Client API procures objects and methods, which enable programmers to access the data management system without any knowledge of the underlying web service schemas or security issues. The Client API supports access to the system in various levels of complexity. The API consists of low-level implementations that resembles the web service interface, but also includes high-level methods that enable the client developer to initiate a request with a single line of code.

GridBean Concept

Similar to the previous version, UNICORE 6 provides a unified framework for design and implementation of gridbeans, i.e. special libraries that manage application input and output. In this way, a job is defined on a client side where it is submitted, and the corresponding application is later executed on the server side.

A gridbean is an object responsible for generating a job description for grid services and providing a graphical user interface for handling input and output data. Typically, a gridbean is composed of a job description generation module and a graphical user interface module. The gridbean stores parameters with corresponding values and uses them to generate the job description. The job description is requested by the client application before submission, and is received from the gridbean.

The gridbean determines the names of parameters with their corresponding values and generates a JSDL description. Some fields may also be marked as input or output parameters. These are used to declare input and output files or other arbitrary data obtained from or transferred to the local machine or other jobs. One can also reconstruct field values in the gridbean's GUI from previous tasks. A gridbean can be made to interact with application-related services available on the grid to obtain, for example, a list of possible values for specific parameters.

BLAST Gridbean

One of the first gridbeans developed for life sciences application was the BLAST gridbean (Kluszczyński, 2007). Basic Local Alignment Search Tool (BLAST) is an important and the most widely used tool in the field of bioinformatics, enabling discovery of statistically significant local similarities between user-defined protein or DNA sequence and sequences from databases (Altschul, 1997; Tatusowa, Madden, 1999).

The user interface provided by the BLAST gridbean is organized in a similar way to the NCBI-BLAST website (NCBI BLAST, 2008) to ensure ease of use, particularly by those who are used to the existing interface. In the main panel, a particular sequence can be entered as a text and reread from the local file. The user can choose the type of BLAST program and the database to search for the sequence. The gridbean checks the list of available databases before task preparation begins and the user can see only those databases that are currently available in the system.

The second panel in the BLAST gridbean provides optional choices that affect the presentation of the search results, as well as additional parameters for carrying out the blast search, such as scoring matrix to be used or the level of accuracy for a successful match. The results are presented in a dedicated output panel in textual form, however graphical representation is forthcoming.

NAMD Gridbean

The NAMD gridbean (Kluszczyński, Bała, 2008) provides a user interface to the NAMD molecular dynamics code. With the NAMD gridbean, users can easily set parameter values without worrying about syntax of a configuration file. The user interface is organized into panels that contain groups of parameter components responsible for similar features.

The main panel is dedicated to enter the job name, number of CPUs required, and other mandatory parameters required to initiate the simulation, such as determination of the number of steps for the simulation, specification of necessary files containing molecular system description, and files where the molecule data should be saved after the simulation. A second panel provides parameter fields that determine file formats to be used during loading and/or saving of molecular systems, and a force field description. The frequency of writing out energy and pressure values can be controlled in this panel as well. There is a possibility to save simulated trajectories of a system by selection of appropriate options. A third panel contains components related to details of the actual simulation. For example, the user can enter the timestep value, determine the cutoff distance, set the pair list distance, or other parameters that govern algorithmic procedures, for instance methods of electrostatics calculation.

Additional simulation parameters can be determined in the gridbean's fourth panel. As an example, harmonic constraints, boundary conditions or specification of atoms that should be held fixed during an MD simulation can be controlled there. The user has the possibility to run energy minimizations, perform Langevin dynamics, or control temperature coupling, rescaling and reassignment, or other such processes.

In the last panel, there is a text area where an advanced user can manipulate rarely used parameters using the NAMD configuration file format. This could, for example, apply to activation of locally enhanced sampling (LES) techniques or SMD features.

QSAR Gridbean

The Chemomentum project has developed a gridbean for quantitative structure-activity relationship (QSAR) modeling. The QSAR modeling uses statistical analysis methods to develop predictive models for estimating various biological activities or physicochemical properties of chemicals. The model development process requires a data set of molecules (chemicals) with known experimental activity property data values and theoretically calculated molecular descriptors. This data set is analyzed with various statistical methods to find the best correlation between the experimental data and computed molecular descriptors. The developed QSAR models can be then used for predicting experimental activity data for new molecules.

The QSAR gridbean uses a grid service to integrate the MDA module from the CODESSA Pro software package (Codessa, 2008) to the UNICORE 6 infrastructure. The MDA provides several statistical methods for QSAR/QSPR model development, including multi-linear regression (MLR) and partial least squares (PLS). The MLR methods can be used with various descriptor selection algorithms, including best multi-linear regression (BMLR) and heuristic method. The user interface of the gridbean provides GUI panels for model building jobs and associated visualizing. The input panel enables users to select statistical methods for model building together with tunable parameters. Users may also apply various data pretreatment options, including elimination of insignificant descriptor scales and handling of missing data points. The output panel provides detailed information on developed models, which includes statistical parameters of the models, plots for experimental and calculated values, residual plots, tables with activity and descriptor values, and plots between experimental activities and descriptors.

GAMESS Gridbean

Quantum mechanical methods (QM), in particular, are being incorporated into UNICORE in order to enable the calculation of more accurate molecular descriptors directly from the molecular wave function. The theoretical basis behind the computed properties that result from a high-level quantum calculation is well defined and rich in physical information. This solid theoretical foundation can be used to derive and calculate descriptors to describe chemical reactivity and the nature of intermolecular forces that drive biological activity, often masked by less accurate methods. This is especially true in the case of quantum mechanical treatments of solvation. Hydrophobic descriptors and hydrogen bond estimates have been particularly difficult to determine using empirical based methods.

We have provided a gridbean for quantum mechanics computations based on the GAMESS software. The gridbean enables QM calculations in general, and QSAR modeling with UNICORE in particular. The gridbean allows the user to generate GAMESS input files interactively. Subsequent to running GAMESS, the gridbean is able to fetch results for further analysis.

The GAMESS gridbean is composed of input and output panels, where the user can set up and edit application input parameters, and do a quick input check before the job is submitted. The user is presented with choices such as file type for molecular description, choice of 2D/3D structure representation, GAMESS specific parameters, and more. Currently, key GAMESS parameters are relatively limited and include only the most important settings: the type of computation, choice of the molecular electronic state, choice of molecular charge, basis set specification, and choice of wave function.

Tools have also been created for extraction of specific results from the GAMESS output to be used for calculation of quantum chemical descriptors, to be used in subsequent QSAR analysis. Examples of such descriptors include highest occupied molecular orbital (HOMO) energy, lowest unoccupied molecular orbital (LUMO) energy, electron density matrix elements, dipole moment, energetic frequencies of vibrations, various thermodynamic properties, solvation energies and charges, and optimized coordinates. From these values, one can easily compute QSAR descriptors and use them for biological study investigations.

CHEMOMENTUM WORKBENCH

The focus of the grid middleware is to provide low level tools to access distributed infrastructure. The middleware enables integrated solutions to be built, such as Chemomentum. The Chemomentum framework provides Grid-based solutions for workflow-centric, complex applications, such as risk assessment, toxicity prediction and drug design. The system concentrates tool access to facilitate computational prediction and handling of data and knowledge, in an efficient, reliable, and transparent manner. Intuitive, task-oriented user interfaces enable the users to concentrate on their research, masking Grid associated complexities. The system is based on the most recent version of the UNICORE 6 middleware, which implements Grid Services technology. Advanced workflow management functionality is included, for example creation of subtasks that can be run on different target systems. The UNICORE workflow model allows for branches, loops and conditional execution, thereby spanning many common job scenarios.

A data management system developed for the use in a UNICORE 6 environment offers comprehensive possibilities for storage and retrieval of data and metadata. Such a system can be comfortably coupled with a workflow system, so that extensive metadata can be generated during job execution, and subsequently automatically stored in the data management system. The metadata is not limited to a specific domain, but can be adapted easily to any application scenario. User management in the Chemomentum workbench is carried out using the concept of Virtual Organizations (VO). For this purpose, a dedicated service base of the UNICORE security has been created, providing tools and user-friendly administrative interfaces for the users.

The Chemomentum workbench is general and can be used in diverse application domains. Below we present examples from the life sciences that have motivated development.

QSAR Workflow

One of the most time-consuming steps of QSAR model development is data pre-processing. Before the actual QSAR modeling can be initiated, a number of tasks must be carried out using different software packages to obtain the required input data in a form suitable for model development application. The entire model development process can be described with workflows, and automation of this process can reduce development time and avoid potential human errors that may occur when the process is carried out manually. In the Chemomentum system, we have developed workflows for automating the model development process.

A common QSAR modeling scenario is to develop a model from data available in some chemical database. In most chemical databases, chemicals are represented by two-dimensional structures that, while sufficient for visual depiction on a computer screen, are not directly applicable for QSAR modeling. This is primarily due to the fact that the statistical methods require input data in numerical forms. Typically, additional processing steps are necessary to characterize molecules with theoretically calculated numerical data, often called molecular descriptors. The entire workflow is carried out by the following workflow involving four application services:

1. Generate 3D coordinates for input structures (MOLGEO software.)
2. Perform semi-empirical or full quantum chemical calculation (MOPAC or GAMESS software).
3. Calculate molecular descriptors with the MDC module from CODESSA Pro software.
4. Develop QSAR model with MDA module from CODESSA Pro software.

For each of the above steps, a corresponding application has been integrated into the UNICORE

6 infrastructure and can be used in workflows through the gridbean API. The Chemomentum client, based on the Eclipse Rich Client Platform, includes a workflow editor plugin for the authoring and submission of workflows. The client also contains a tracer plugin that can be used to follow the workflow execution on the grid in real time. The workflow editor fits naturally into the client framework. Users can compose workflows by connecting existing gridbeans into a sequence of actions. The application parameters can be changed by gridbeans, and parameters are stored to the workflow definition. Similarly, the outcome of workflows can be visualized by gridbeans.

Docking GAMESS-APBS

The Chemomentum project offers a unique contribution to the pharmacophore approach for understanding a ligand-receptor system, by improving the theoretical foundation that underlies the molecular based descriptors themselves. Such improvements provide a more accurate perspective on the molecular mechanisms responsible for ligand's selectivity and specificity. However, some levels of molecular dynamics must be included when considering ligand-protein, drug-receptor type interactions. For such investigations, hybrid strategies or purely classical-based docking strategies are typically employed.

Grid-based tools offer a unique opportunity to investigate new strategies for study of large molecular systems, by enabling a smoother coupling between physically appropriate modeling strategies. For example, current research involves a hybrid method invoking quantum chemical techniques (GAMESS) with electrostatic potential prediction (APBS) to estimate the binding energy of the receptor and a variety of ligands in a large number of generated conformations. There are a number of variants of such a scenario, including the use of molecular dynamics, and various strategies for incorporation of the quantum based techniques. A scoring function, based on electrostatic investigations of the ligand-protein complex and estimation of internal energy of the ligand, is typically developed to further rank the ligand conformations within the ligand-protein complex.

In an initial simplified strategy for docking, we have created a command line gridbean that launches executables on the server side with user-

Figure 6. Example QSAR workflow in the workflow editor. © 2009 University of Tartu. Used with permission.

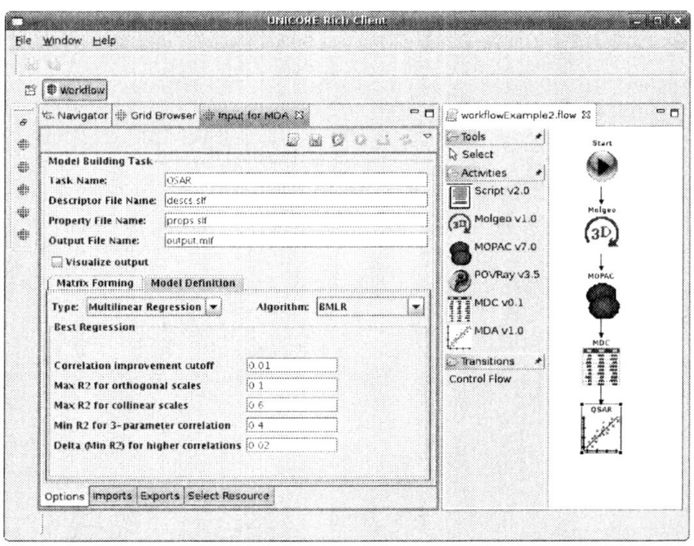

defined input parameters on a client side. The gridbean can be easily integrated with GAMESS-APBS docking. The docking scenario consists of all steps necessary to carry out the QM-APBS hybrid technique. As such, there are several independent algorithms employing different physical-based methods and software components, various molecular structure representations to be done, tools for proper orientation of a ligand docking into active sites, and various visualization and analysis steps. As a whole, the steps describe the hybrid-methodology for the docking procedure. The algorithmic steps can be executed by the command line gridbean sequentially or in parallel, while the UNICORE middleware manages the communication among them.

The workflow begins with a structure from the PDB (Berman et al. 2000). All conformational possibilities are generated according to the specification of torsion angle of the ligand (future work will use a fully flexible method for conformation generation as specified by the user). The generated structures are checked to avoid unrealistic close contact between atoms of the small molecule itself with any nearby residues of the binding site. All remaining structures are considered for computational investigation. Quantum chemical computations are performed on the ligand to estimate charges, optimized hydrogen position, and estimates of internal energy across the various structures. Subsequently, the ligand is docked into the active site, and binding energies are computed using the software APBS. This latter energy requires refinement using a more complex scoring function, taking in account the respective ligand's internal energy calculated with QM techniques (with a ranking procedure), the solvent accessible surface area estimate, and a tension surface term (the latter two terms are computed with APBS tools).

Workflow Overview:

1. **Structural file extraction:** The workflow begins with the extraction from a PDB file containing the structure of the complex of interest. The ligand and the water molecule are deleted from the PDB file. The ligand can be extracted from either the Ligand Depot database in PDB format or from the complex obtained directly from the PDB.
2. **Protonation/water molecule:** A small conversion from PDB to PQR format is required in order to prepare the input for the next step. The Hydrogen atoms, atomic radii and charges are added to the protein using the tool PDB2PQR, which can be invoked as a web service or from a command line. This program takes a file in PDB format and produces a file in PQR format. The Hydrogen atoms on the ligand are added using the Babel program.
3. **Docking program:** The docking program is written in C and is used to generate the different conformations of the ligand, using a chosen metric, in this case one based on torsional motion. Taking a PQR file format of the protein and the ligand as input, this procedure produces a PQR file for each new ligand conformation, places the ligand in the complex and checks for steric interactions or bumping with atoms of the protein.
4. ***Quantum* mechanics:** A tool called PDB2INP, written in Perl, converts the PDB file into an input file for GAMESS. The GAMESS computation optimizes the Hydrogen positions and provides a good estimation of atomic charges. A Fortran program called GAM2PQR takes the GAMESS output file and generates the PQR file for the ligand by extracting the predicted atomic charge and associated atomic radius for each atom.
5. **Classical mechanics:** The APBS input file that defines all parameters required to solve the Poisson-Boltzmann Equation is prepared. Some needed parameters can be estimated automatically by invoking tools included with the APBS package, e.g., de-

termination of the size of the grid to perform the computation. After the computation completes, the binding energy is extracted from the output file and is adjusted by the addition of an apolar contribution of the binding. This apolar contribution is estimated by the Solvent-Accessible Surface Area (SASA) weighted by a tension surface term. The SASA is estimated using the Python tool called ACC, provided with the APBS package.

This process extracts the best candidates for binding to the receptor by applying the full scoring function with the total binding energy weighted by the internal energy of the ligand. A series of small Perl programs and shell scripts were written to launch automatically the tools to extract the data, to store them in a CSV file format, and to load the result into excel sheets directly after the run.

Applications of Chemomentum Workbench

The Chemomentum workbench enables *in silico* design and modeling to be used in many aspects of chemistry and life science related fields. Prominent examples include:

- The European REACH policy – Registration and evaluation of chemicals.
- Research and development in the pharmaceutics – Computational solutions (including Grid technologies) are used more and more for virtual screening of the databases of compounds in order to identify potential leads concerning a given biological activity.
- Research and development in biotechnology – screening and determining technologically relevant new substances.

With the possibility for use in many possible fields, the current emphasis is on the European REACH policy (Registration and evaluation of chemicals). According to the REACH policy substances must be carefully evaluated and assessed for possible risks to human health and the environment, prior to marketing and distribution. This process currently requires massive efforts in terms of time, money, and animal testing. The Chemomentum system has the potential to reduce time and effort in the experimental laboratory and eliminate the need for animal testing, by providing a wide variety of solutions for the computational, *in silico*, testing of chemicals.

There are two primary capabilities offered by Chemomentum:

- Development of QSAR models in a flexible and powerful way;
- Use of already developed QSAR models.

The second capability is a simplified version of the first one, from an information technology point of view. Usage of standard well-developed and accepted QSAR models for a series of conditions with respect to the REACH legislation enables reliable verification and certification possibilities. Indeed, from the theoretical point of view, a large number of QSAR models can be developed for the same endpoint, but only selected models may be suitable for the REACH legislation.

The specific purpose of Chemomentum is to develop QSAR models for efforts such as the REACH legislation, dictating the need to highlight conditions defined by the application specificity. This in turn governs the direction of the QSAR modeling activities. For instance, for regulatory purposes, the environmental and safety authorities do not allow use of a chemical compound that does not meet certain safety standards. Therefore, any predictive QSAR model should be able to distinguish between safe and unsafe chemicals under the desired specification, without the possibility for either false positive (type I) or false negative (type II) predictions. While the latter errors are not so critical, the former are to be avoided.

On the other hand, during the development of a new pharmaceutical compound the industry would like to avoid false positives. This means processing a high number of candidate drugs, which in the end show non-effectiveness in activity. Indeed, in this case the cost for the drug development is too high, due to the time and effort of the experimental processes and workups, which at a certain stage may have to be abandoned.

Even in the case of legislative purposes, there are different conditions that may be applicable depending on the regulation and on conditions identified by the law. For instance, the cosmetic regulation in Europe will ban all experiments that use animals. Thus, only alternative methods should be used. On the other hand, under the current European regulations on pesticides requires experiments to be performed on animals for the parent compounds.

The REACH legislation states that animal experiments should be avoided as much as possible. The use of animal experiments is typically carried out if no other methods can be found to obtain the requested information. A series of alternative allowed methods to obtain regulatory information includes data sharing and computational investigations, including QSAR based methods. Also, depending on the scale that a certain chemical is produced or utilized, values for certain endpoints must be provided. In some of these cases, a classification is sufficient (such as bioaccumulative or not), in other cases a continuous value must be provided.

For many of these reasons, the QSAR developer focusing on regulatory issues must be fully aware of the requirements and conditions defined by law. One risk is the development of an unacceptable model, which is unacceptable not because it is scientifically wrong, but simply because it is not appropriate for the target application.

Any QSAR model starts from a training set, which is a collection of structural and property data for a list of chemical compounds. Obviously, it is highly desirable if the property data has been obtained according to official protocols, rather than from any source. As such, a quality check of a QSAR model for the REACH legislation is necessary for the input of the model (the property and chemical data), for the output of the model (if suitable for the purpose, in terms of uncertainty and use of the output), and for the quality of the modeling procedure. Some official criteria have been described, resulting in a format useful to describe the QSAR model. These criteria are used to characterize the quality of the modeling procedure.

Chemomentum offers a platform enabling a fully reproducible procedure to obtain predicted values on chemicals, and to report and trace the procedure used to obtain the predicted value. Therefore it has major advantages specifically for the purpose of the REACH legislation. We refer here to two fundamental features for regulatory purposes: reproducibility and transparency. It is clear that regulators need reproducibility in the data prediction; that is, the same property value for a specified compound can be produced by the model at any time and by any user. This may seem trivial, but many computational models, in particular those in which manual optimization of a 3D molecular conformation is requested and whereby the user may choose from a palette of many options, could easily provide many different results.

All model components in any QSAR model by necessity must be verifiable. Chemomentum offers such capabilities through uniform usage and disclosure of methodology. In those cases where proprietary software (e.g., industrial software) is incorporated, it ensures access rights and ownership for the model components and assures transparent reproducibility of predictions.

Overall, with an incorporated REACH regulatory layer for QSAR in Chemomentum, the system would have the potential to play a fundamental role for determination of chemical toxicity under specified regulations. Normally, such checks must be carried out manually. With the Chemomentum system capabilities, there is now the possibility to carry out one of the two scenarios. First, one could imagine development of a model specifically governed by a series of regulations, with proven robustness, predictiveness, reproducibility and transparency. Then, with the incorporated workflow capabilities, such a model can be made available as one of a series of accepted REACH QSAR models. Second, certain available QSAR models that already follow regulatory constraints can be offered in a pre-set workflow procedure, which cannot be altered for predictions. The model can as well be used for general unconstrained research investigations that are not so concerned with specific regulatory constraints. In this manner, a user would be able to manipulate many of the available options and control features.

FUTURE TRENDS

The most important are support for Virtual Organizations and security infrastructure suitable for service oriented architecture. The UNICORE will than become solid platform for development of integrated solutions in various applications areas such as computational chemistry, medical image processing and others.

CONCLUSION

The UNICORE middleware has been successfully used as a framework providing uniform access to resources and applications important for life sciences such as computational chemistry and molecular biology. This includes seamless access to computational resources, databases, as well as a framework for application specific interface development. The examples presented in this chapter demonstrate the capabilities of the UNICORE middleware. Modular architecture, based on the plugin/gridbean concept and demonstrated visualization capabilities, offer the possibility for a number of applications in diverse areas.

The overall system architecture and the resulting software are innovative in providing a rich and dynamic environment for the user interface, together with a lightweight framework for application developers. For users, the software offers powerful means to discover and access complex applications running on large-scale computational clusters, and to integrate applications in creative ways. For developers, the software combines support for grid technology, adherence to open standards, and integration with back-end service-oriented architectures. A highlight of the effort is the ability to offer a chemical regulatory layer, specifically geared toward the REACH initiative, that enables consistent and reliable testing of chemical toxicity across diverse needs and usage scenarios. Such a regulatory layer is enabled by key middleware components.

The future of the Chemomentum effort will involve expansion of the current platform with additional features. These include improved APIs and documentation, enhanced stability and robustness of predicted data, enhanced regulatory capabilities, and outreach to educational, research, and developer communities. Additional new developments will continue enhanced software frameworks with additional capabilities for visualization and analyses, integrated relational database and remote data management capabilities, together providing many additional core features.

ACKNOWLEDGMENT

This work has been supported by the Chemomentum project within European Commission contract IST-033437.

REFERENCES

Abramson, D, Buyya, R. & Giddy, J. (2002). A Computational Economy for Grid Computing and its Implementation in the Nimrod-G Resource Broker. *Future Generation Computer Systems, 18*(8) 1061-1074.

Abramson, D, Lewis, A. & Peachy, T. (2001). *Case Studies in Automatic Design Optimisation using the P-BFGS Algorithm.* 2001 High Performance Computing Symposium (HPC'01), Advanced Simulation Technologies Conference, April 22-26, , Seattle, WA (pp.104 – 109). Retrieved from http://www.csse.monash.edu.au/~davida/papers/CaseStudies.pdf

Abramson D., Lynch A., Takemiya H., Tanimura Y., Date S., Nakamura H., Jeong K., Lee H.-C., Arzberger P. A., Baldridge K. K., Li W. W., & Molina T. (2006). Deploying Scientific Applications to the PRAGMA Grid Testbed: Lessons Learned. In *IEEE/CCGRID International Grid Computing, Singapore.*

Abramson D., Roe P., Kotler L., & Mather D. (2001). *ActiveSheets: Super-Computing with Spreadsheets. 2001 High Performance Computing Symposium (HPC'01).* Advanced Simulation Technologies Conference, April 22-26, Seattle, WA (pp. 110–115). Retrieved from http://www.csse.monash.edu.au/~davida/papers/ActiveSheets.pdf.

Allcock W., Bresnahan J., Kettimuthu R., & Link M. (2005). *The Globus Striped GridFTP Framework and Server.* ACM/IEEE SC 2005 Conference (p. 54).

Altschul S. F., Madden T. L., Schaffer A. A., Zhang J., Zhang Z., Miller W., & Lipman D. J. (1997). Gapped BLAST and PSI-BLAST: a new generation of protein database search programs. *Nucleic Acids Research, 25*(17), 3389-3402.

Artacho E., Anglada E., Di´eguez O., Gale J. D, Garc´ıa A., Junquera J., Martin R. M., Ordej´on P., Pruneda J. M., S´anchez-Portal D., & Soler J. M. (2008). The SIESTA method; developments and applicability. *J. Phys.: Condens. Matter, 20,* 064208.

Baker N. A., Sept D., Joseph S., Holst M. J., & McCammon J. A. (2001). *Electrostatics of nanosystems: application to microtubules and the ribosome.* Proc. Natl. Acad. Sci. USA 98, (pp. 10037-10041).

Baldridge K. K., Greenberg J. P., Elbert S. T., Mock, S., & Papadopoulos, P. (2002). *QMView and GAMESS: Integration into the World Wide Computational Grid.* Supercomputing, ACM/IEEE 2002 Conference, 64.

Basney J., Humphrey M., & Welch V. (2005). The MyProxy online credential repository. *Software: Practice and Experience, 35*(9) 801-816.

Bavier M., Bowman M., Chun B., Culler D., Karlin S., Muir S., Peterson L., Roscoe T., Spalink T., & Wawrzoniak M. (2004). *Operating System Support for Planetary-Scale Network Services.* NSDI '04 Proceedings. Retrieved from http://www.cs.princeton.edu/nsg/papers/plos_nsdi_04/nsdi.pdf

Berman F. Fox G., & Hey A. (2003). *Grid Computing: Making the Global.* Hoboken, NJ: John Wiley and Sons.

Berman H. M., Westbrook J., Feng Z., Gilliland G., Bhat T. N., Weissig H., Shindyalov I. N., & Bourne P. E. (2000). The Protein Data Bank. *Nucleic Acids Research, 28,* 235-242.

Breton V., Kasam V., & Jacq N. (2006). *High Throughput Grid Enabled Virtual Screening.* Proceedings of the NETTAB2006: Santa Margherita, 14–18.

Bhandarkar M., Budescu G., Humphrey W. F., Izaguirre J. A., Izrailev S., Kalé L. V., Kosztin D., Molnar F., Phillips J. C., & Schulten K. (1999). BioCoRE: A collaboratory for structural biology. In A. G. Bruzzone, A. Uchrmacher, & E. H. Page

(Eds.), *Proceedings of the SCS International Conference on Web-Based Modeling and Simulation* (pp. 242-251).

Boinc (2008). Retreived March 2008 from http://boinc.berkeley.edu/

BoincGamess (2008). Retreived March 2008 from http://cvs.rocksclusters.org/viewcvs/viewcvs.cgi/gamess/boincgamess-doc/

Brandon D., Vandivort K., Brunner R. & Bach M. (2005). *2005 BioCoRE Survey Report*. Retrieved from http://www.ks.uiuc.edu/Research/biocore/eval/surveyData/BioCoRE_2005_Survey_Report%20.pdf

Chun B., Culler D., Roscoe T., Bavier A., Peterson L., Wawrzoniak M., & Bowman M. (2003). PlanetLab: An Overlay Testbed for Broad-Coverage Services. *ACM Computer Communications Review, 33*(3).

Catlett C. (2002). The philosophy of TeraGrid: Building an open, extensible, distributed TeraScale facility. In *Cluster Computing and the Grid 2nd IEEE/ACM International Symposium CCGRID2002*.

CODESSA PRO (2008). Retrieved March 2008 from http://www.codessa-pro.com/

Crippen, G. M. (1981). *Distance Geometry and Conformational Calculations*. New York: Research Studies.

Degtyarenko K. et al. (2008). ChEBI: a database and ontology for chemical entities of biological interest. *Nucleic Acids Research, 36*. Retrieved from http://www.ebi.ac.uk/chebi/

Docking@Home. Retrieved March 5, 2008 from http://docking.cis.udel.edu/

Dubitzky, D., McCourt, D., Galushka, M., Romberg, M., & Schuller, B. (2004). Grid-enabled Data Warehousing for Molecular Engineering. *Parallel Computing, 30,* 1019-1035.

ECOTOX (ECOTOXicology) database. Retrieved March 4, 2008 from http://www.epa.gov/ecotox/

Erwin D. W, & Snelling D. F. (2001). UNICORE: A Grid Computing Environment. In. J. Darlington et al. (Eds.) *Euro-Par 2001 Parallel Processing*. Berlin, Germany: Springer.

Foster, I. (2006). Globus Toolkit Version 4: Software for Service-Oriented Systems. In *IFIP International Conference on Network and Parallel Computing* (LNCS 3779, pp. 2-13). Berlin, Germany: Springer-Verlag.

Foster, I. (2005). Service-Oriented Science. *Science, 308* (5723) 814-17.

Foster, I., & Kesselman C. (1997). Globus: A metacomputing infrastructure toolkit. *International Journal of Scientific Applications, 11*(2) 115-128.

Foster I., Kesselman C., & Tuecke S. (2001). The Anatomy of the Grid: Enabling Scalable Virtual Organizations. *International J. Supercomputer Applications, 15*(3).

Foster I., Kesselman C., Nick J., & Tuecke S. (2002). *The Physiology of the Grid: An Open Grid Services Architecture for Distributed Systems Integration. Open Grid Service Infrastructure WG, Global Grid Forum.* Retrieved from http://www.globus.org/alliance/publications/papers/ogsa.pdf

Galiardi F. (2005). EGEE – a first results and future plans. In M. Bubak, M. Turała, K. Wiatr (Eds.) Proceedings of Cracow'04 Grid Workshop.

Gaussian 03 (2004). Frisch, M. J.; Trucks, G. W.; Schlegel, H. B.; Scuseria, G. E.; Robb, M. A.; Cheeseman, J. R.; Montgomery, Jr., J. A.; Vreven, T.; Kudin, K. N.; Burant, J. C.; Millam, J. M.; Iyengar, S. S.; Tomasi, J.; Barone, V.; Mennucci, B.; Cossi, M.; Scalmani, G.; Rega, N.; Petersson, G. A.; Nakatsuji, H.; Hada, M.; Ehara, M.;

Toyota, K.; Fukuda, R.; Hasegawa, J.; Ishida, M.; Nakajima, T.; Honda, Y.; Kitao, O.; Nakai, H.; Klene, M.; Li, X.; Knox, J. E.; Hratchian, H. P.; Cross, J. B.; Bakken, V.; Adamo, C.; Jaramillo, J.; Gomperts, R.; Stratmann, R. E.; Yazyev, O.; Austin, A. J.; Cammi, R.; Pomelli, C.; Ochterski, J. W.; Ayala, P. Y.; Morokuma, K.; Voth, G. A.; Salvador, P.; Dannenberg, J. J.; Zakrzewski, V. G.; Dapprich, S.; Daniels, A. D.; Strain, M. C.; Farkas, O.; Malick, D. K.; Rabuck, A. D.; Raghavachari, K.; Foresman, J. B.; Ortiz, J. V.; Cui, Q.; Baboul, A. G.; Clifford, S.; Cioslowski, J.; Stefanov, B. B.; Liu, G.; Liashenko, A.; Piskorz, P.; Komaromi, I.; Martin, R. L.; Fox, D. J.; Keith, T.; Al-Laham, M. A.; Peng, C. Y.; Nanayakkara, A.; Challacombe, M.; Gill, P. M. W.; Johnson, B.; Chen, W.; Wong, M. W.; Gonzalez, C.; & Pople, J. A.; Wallingford, CT: Gaussian, Inc..

Gemstone (2008). Retrieved March 3, 2008 from http://gemstone.mozdev.org/

Gervasi O., Dittamo C., & Laganà A. (2005). A Grid Molecular Simulator for E-Science. In *Advances in Grid Computing - EGC 2005* (LNCS 3470, pp. 16-22). Berlin, Germany: Springer.

Gordeeva E., Katritzky A. R.. Shcherbukhin, V. V., & Zefirov, N. S. (1993). Rapid Conversion of Molecular Graphs to Three-dimensional Representation Using the MOLGEO Program. *J. Chem. Inf. Comput. Sci., 33*, 102-111.

Grimshaw A. S., Natrajan A., Humphrey M. A., Lewis M. J., Nghuyen-Thong A., Karpovich J. F., Morgan M. M., & Ferrari A. (2003). J. From Legion to Avaki. In Berman F. Fox G., Hey A. (Eds.) *Grid Computing: Making the Global*. Hoboken, NJ: John Wiley and Sons.

Huber V. (2001). Supporting Car-Parrinello Molecular Dynamics with Unicore. In *Computational Science ICCS - San Francisco, USA, Proceedings, Pt. 1*, (Vol. 2073), (pp. 560-567). Berlin, Germany: Springer Verlag.

Humphrey W., Dalke A. & Schulten K. (1996). VMD - Visual Molecular Dynamics. *J. Molec. Graphics, 14*, 33-38.

Jacq N., Breton B., Chen H., Ho L., Hofmann M., Lee H., et al. (2006). *Large Scale In Silico Screening on Grid Infrastructures*. Proceedings of the LSGRID2006, Yokohama, Japan (pp.123–136).

Jacq N., Salzemann J., Jacq F., Legré Y., Medernach E., Montagnat J., Maaß A., et al. (2008). Grid-enabled Virtual Screening Against Malaria. *J Grid Computing 6*,29–43.

JSDL 1.0. Job Submission Language Specification (2006). Retrieved March 6, 2008 from http://www.gridforum.org/documents/GFD.56.pdf

Kal L., Skeel R., Bhandarkar M., Brunner R., Gursoy A., Krawetz N., et al. (1999). NAMD2: Greater Scalability for Parallel Molecular Dynamics. *J. Comp. Phys., 151*, 283–312.

Kluszczyński R., & Bała P. (2008). Supporting NAMD Application on the Grid using GPE. To appear in LNCS.

Kluszczyński R., Borcz M., & Bała P. (2007). BLAST Application on the GPE/UnicoreGS grid. In *Euro-Par 2006 Workshops* (pp. 245-253). Berlin, Germany: Springer-Verlag.

Kollman P. (2001). *AMBER (Assisted Model Building with Energy Refinement)*. San Francisco, CA: University of California..

Lee H., Salzemann J., Jacq N., Ho L., Chen H., Breton,V., Merelli L., Milanesi L., Lin S., & Wu Y. (2006). *Grid-enabled High Throughput in-silico Screening Against Influenza A Neuraminidase*. Proceedings of the NETTAB2006, Santa Margherita, 19–25

Maran U., Sild S., Kahn I., & Takkis K. (2007). *Future Gen. Comput. Syst. 23*, 76-83

Maran U., Sild S., Mazzatorta P., Casalegno M., Benfenati E., & Romberg M. (2007). In Dubitzky

W., Schuster A., Sloot P. M. A., Schhroeder M., Romberg M. (Eds.), *Grid Computing in Computational Biology*, (LNCS 4360, pp. 60–74). Berlin Heidelberg: Springer-Verlag.

Moore R. (2001) Data Management Systems for Scientific Applications. In Boisvert R., Tang P. (Eds.) *The Architecture of Scientific Software* (pp. 273-284). Norwell, MA: Kluwer Academic Publishers.

Morgan, M., Chue-Hong, N., & Drescher, M., (2006). *ByteIO Specification 1.0*. Retrieved on July 10, 2008 from https://forge.gridforum.org/sf/docman/do/download/Document/ projects.byteiowg/docman.root.current_documents/doc13719/1

Morris G. M., Goodsell D. S., Halliday R.S., Huey R., Hart W. E., Belew R. K., & Olson A. J. (1998). Automated Docking Using a Lamarckian Genetic Algorithm and and Empirical Binding Free Energy Function. *J. Computational Chemistry, 19*, 1639-1662.

Mozilla (2008). Retrieved on July 10, 2008, from http://www.mozilla.org/newlayout/doc/

National Toxicology Program server at http://ntp-server.niehs.nih.gov/

Niederberger R. & Alessandrini V. (2004). The DEISA Project: Motivations, Strategies, Technologies. In *International Supercomputer Conference*, Heidelberg, Germany.

Neuroth H., Kerzel, M., & Gentzsch, W. (Eds) (2007). *Die D-Grid Initiative*. Goettingen, Germany: Goettingen University Press. Retrieved from http://webdoc.sub.gwdg.de/univerlag/2007/D-Grid_de.pdf

Novotny J., Tuecke S., & Welch V. (2001). *An Online Credential Repository for the Grid: MyProxy*. HPDC, 10th IEEE International Symposium on High Performance Distributed Computing (HPDC-10 '01) (p. 0104).

Peachey T., Abramson D., & Lewis A. (2006). *Model Optimization and Parameter Estimation with Nimrod/O*. The International Conference on Computational Science, May 28-31, University of Reading, UK. Retrieved from: http://www.csse.monash.edu.au/~davida/papers/iccs06.pdf

Pragma 2008, http://www.pragma-grid.net/

Pytliński J., Skorwider Ł., Benedyczak K., Wroński M., Bała P., & Huber V. (2003). Uniform Access to the Distributed Resources for the Computational Chemistry Using UNICORE. In: Sloot M. A et al. (Eds) Computational Science - ICCS 2003 (pp.307-315). Berlin, Germany: Springer-Verlag.

Pytliński J., Skorwider Ł., Huber V., Benedyczak K., & Bała P. (2002). UNICORE - An Uniform Platform for Chemistry on the Grid. *Journal of Computational Methods in Science and Engineering 2* (3s-4s), 369-376.

Rauhut G. (2002). Web Site: Your Web Browser as a Computational Chemistry Frontend. *Angewandte Chemie International Edition, 41*(12), 2205.

Reed D. A. (2003). Grids, the TeraGrid, and Beyond. *Computer, 36*, 62-68. http://portal.acm.org/citation.cfm?id=642202.

Reference Model for Service Oriented Architecture (2006). Committee Draft 1.0. Retrieved from http://www.oasis-open.org/committees/download.php/16587/wd-soa-rm-cd1ED.pdf

Romberg, M. (Ed.) (2005). *OpenMolGRID - Open Computing Grid for Molecular Science and Engineering*, Final Report, NIC Directors, Germany, Jülich, (Vol. 29).

Schmidt M. W., Baldridge K. K., Boatz J. A., Gordon M. S., Elbert S. T., Jensen J. H., et al. (1993). General Atomic and Molecular Electronic Structure System. *J. Comput. Chem., 14*, 1347–1363.

Schuller B., Romberg M., & Kirtchakova L. (2005). Application Driven Grid Developments in the OpenMolGRID project. In P.M.A. Sloot, A.G. Hoekstra, T. Priol, A. Reinefeld, M. Bubak (Eds.), *Advances in Grid Computing* (Vol. 3470) (pp. 23–29). Berlin, Germany: Springer-Verlag.

Sild S., Maran U., Lomaka A., & Karelson M. (2006). Open Computing Grid for Molecular Science and Engineering. *J. Chem. Inf. Model. 46,* 953–959

Sild S., Maran U., Romberg M., Schuller B., & Benfenati E. (2005). OpenMolGRID: Using Automated Workflows in GRID Computing Environment. In P.M.A. Sloot, A.G. Hoekstra, T. Priol, A. Reinefeld, M. Bubak (Eds.), Advances in Grid Computing (Vol. 3470) (pp. 464–473). Berlin, Germany: Springer-Verlag.

Stewart J. J. P. (1990). MOPAC: a Semiempirical Molecular Orbital Program. *J. Comput.-Aided Mol. Des., 4,* 1-45.

Sterzel M., & Szepieniec T. (2007). Enabling commercial chemical software on EGEE Grid – Gaussian VO. In M. Bubak, M. Turała, K. Wiatr (Eds.) *Cracow'06 Grid Workshop. Proceedings* (pp. 499–504).

Tatusova T. A., & Madden T. L. (1999). BLAST 2 Sequences, a new tool for comparing protein and nucleotide sequences. *FEMS Microbiology Letters, 174* (2), 247–250.

Thomas M., & Boisseau J. (2003). Building Grid Computing Portals: The NPACI Grid Portal Toolkit. In F. Berman, G. Fox & T. Hey, (Eds). *Grid Computing: Making the Global Infrastructure a Reality.* Hoboken, NJ: John Wiley & Sons.

Thomas M., Boisseau J., Mock S., Dahan M., Mueller K., & Sutto D. (2003). *The GridPort Toolkit Architecture for Building Grid Portals.* Proceedings of the 10th IEEE Intl. Symp. on High Perf. Dist. Comp.

WebMo (2008). Retrieved from http://www.webmo.net

WSRF Web Services Resource Framework. Retrieved from http://docs.oasis-open.org/wsrf/wsrf-primer-1.2-primer-cd-02.pdf

Wypychowski J., Pytlinski J., Skorwider Ł, Nazaruk M., Benedyczak K., Wroński M., Bała P. (2004). Life Sciences Grid in EUROGRID and GRIP projects. *New Generation Computing, 22* (2), 147-156.

KEY TERMS AND DEFINITIONS

CML (Chemical Markup Language): Domain specific implementation based strictly on XML. It uses XML's portability to help developers and chemists design interoperable documents.

General Atomic and Molecular Electronic Structure System (GAMESS): A general *ab initio* quantum chemistry package.

Grid Enabled Molecular Science Through Online Networked Environments (GEMSTONE): A framework that provides researchers in the molecular sciences with a tool to discover remote grid application services and compose them as appropriate to the chemical and physical nature of the problem at hand.

Job Submission Description Language (JSDL): An extensible XML specification for the description of simple tasks to non-interactive computer execution systems. It describes the submission aspects of a job, and does not attempt to describe the state of running or historic jobs.

Open Grid Services Architecture (OGSA): An architecture for a service-oriented grid computing environment. OGSA is based on several other Web service technologies but it aims to be largely agnostic in relation to the transport-level handling of data.

Pacific Rim Application and Grid Middleware Assembly (PRAGMA): Open organization in which Pacific Rim institutions collaborate to develop grid-enabled applications and deploy the infrastructure throughout the Pacific Region to allow data, computing, and other resource sharing.

Protein Data Bank (PDB): An archive of experimentally-determined, biological macromolecule structures from the Brookhaven National Laboratory.

Quantitative Structure-Activity Relationship (QSAR): The process by which chemical structure is quantitatively correlated with a well defined process, such as biological activity or chemical reactivity.

REACH (Registration, Evaluation, Authorisation and Restriction of Chemical substances): European Community Regulation on chemicals and their safe use (EC 1907/2006). The aim of REACH is to improve the protection of human health and the environment through the better and earlier identification of the intrinsic properties of chemical substances.

Storage Resource Broker (SRB): A client-server middleware that provides a uniform interface for connecting to heterogeneous data resources over a network and accessing replicated data sets.

UNICORE (UNiform Interface to the COmputational REsources): Grid system including client and server software. UNICORE makes distributed computing and data resources available in a seamless and secure way in the internet.

Web Services Resource Framework (WSRF): An application of web service technology to realise access to and management of statefull resources.

XACML (eXtensible Access Control Markup Language): A declarative access control policy language implemented in XML and a processing model, describing how to interpret the policies.

Chapter XXXI
A Grid Paradigm for e-Science Applications

Livia Torterolo
University of Genoa, Italy

Luca Corradi
University of Genoa, Italy

Barbara Canesi
University of Genoa, Italy

Marco Fato
University of Genoa, Italy

Roberto Barbera
University of Catania, Italy & INFN-Catania, Italy

Salvatore Scifo
Consorzio Cometa of Catania, Italy

Antonio Calanducci
INFN-Catania, Italy

Diego Scardaci
INFN (National Institute of Nuclear Physics), Italy

Giordano Scuderi
Unico Informatica s.r.l, Catania, Italy

ABSTRACT

This chapter describes a Grid oriented platform -the Bio Med Portal- as a new tool to promote collaboration and cooperation among scientists and healthcare research groups, enabling the remote use of resources integrated in complex software platform services forming a virtual laboratory. In fact, nowadays many biomedicine studies are dealing with large, distributed, and heterogeneous repositories as well as with computationally demanding analyses, and complex integration techniques are more often required to handle this complexity. The Bio Med Portal is designed to host several medical services and it is able to deploy several analysis algorithms. The scope of this chapter is both to present a Grid application with its own medical use case and to emphasize the benefit that a new Design Paradigm based on Grid could provide to research groups spread in geographically distributed sites.

INTRODUCTION

Nowadays, many clinical and biomedicine studies need large, distributed, and heterogeneous repositories as well as computationally demanding analysis tools and integration among different levels and disciplines is more often required. In order to maximize the results of such experimental scenarios, Grid based approaches may provide a shared, standardized, and reliable solution both for storage of distributed biomedical data and metadata and for access to distributed analysis tools. This is possible thanks both to specialized software layers called frameworks (based on the principle of interchangeable building blocks), and high added value independent tools (developed to cover specific application requirements). Being perfectly integrated with the official middleware, these tools allow to implement complex service platforms according to a modular architecture, scalable enough for accessing remotely large amounts of distributed data as well as for scaling computational performances on terabyte (1012 bytes) datasets. These frameworks are also implemented to hide the complexity of the underlying Grid Infrastructure at two different levels: end users, to make them able to easily access available services and data for life science research and clinical applications; developers, to better design and implement shared research environments and high added value services and tools.

The BM Portal project, a Grid based software solution thought to promote collaboration and cooperation among scientists and healthcare research groups, enables the remote use of electronic medical instruments integrated in a complex software platform services called virtual laboratory. This project is carried out by the University of Genoa, Department of Communication, Computer and System Sciences, BIOLAB laboratory with the cooperation of some Italian SMEs (Aitek SpA, Nice Srl, R&T enginering Srl, Unico informatica SrL) the INFN (National Institute of Nuclear Physics - department of Catania, www.ct.infn.it) as well as the Official Training and Support team of the EGEE middleware and the COMETA Consortium (www.consorzio-cometa.it) as Grid infrastructure provider and integration support.

Background

Research in life sciences is at an important nexus driven by the generation of large volumes of rapidly expanding data and the capabilities of technology to share distributed data and resources. The core of this nexus affords researchers the opportunity to address large and complex problems and model even entire organisms. Grid technology and high-performance computing provide the ability to process, analyse, store and share data at unprecedented scales. Biomedical applications often require high-performance computing and large data handling which exceeds the computing capacity of a single institution (Krishnan, 2004). Sharing of unpublished data is also important in promoting collaborative research among institutions (Li, Byrnes, Hayes, Birnbaum, Reyes, Shahab, Mosley, Pekurovsky, Quinn, Shindyalov, Casanova, Ang, Berman, Arzberger, Miller, & Bourne, 2004), as well as sharing of public databases, biomedical tools and web services (Umetsu, Ohki, Fukuzaki, Konagaya, Shinbara, Saito, Watanabe, Kitagawa, & Hoshino, 2006; Konishi, Yagi, & Konagaya, 2006). Biological knowledge, such as ontology and metadata, also plays an important role in analysis of experimental data and integration of data coming from different fields such as bioinformatics and neuroinformatics (Cannata, Merelli & Altman, 2005). Grid computing has a great potential to become the standard model for cyberinfrastructures for life sciences and it is a promising information technology which meets the above requirements (Arzberger, Farazdel, Konagaya, Ang, Shimojo & Stevens,2004). However, many features of it remain to be improved in terms of availability, performance and security, to name a few.

Computing Grid

Some projects are trying to support large-scale computational applications in life sciences by utilizing distributed high-performance computing, high speed networks, massive data collections and archives as well as the necessary software tools and data integration capabilities. The Swiss Bio Grid (Podvinec, Schwede, Cerutti, Kunszt, Nyffeler, Stockinger, Maffioletti, Thomas, & Turker, 2007) goal is to establish an added value collaborative platform focused on the creation of a dynamic IT infrastructure allowing to use resources across institutions and disciplines, delivering services for compute intensive applications and data sharing. The DEISA project (http://www.deisa.eu/) intends to contribute to the significant enhancement of capabilities and capacities of high-performance computing (HPC) in Europe by the integration of leading national supercomputing infrastructures. To realise this mission, it deploys and operates the distributed multi-terascale European computing platform based on the strong coupling of existing national supercomputers and operates as a virtual European supercomputing centre. At the moment the DEISA research infrastructure is not interfaced with the rest of the European IT infrastructure. By adopting Grid technologies as gLite, it could contribute to the development of an extended, heterogeneous Grid computing environment for HPC in Europe.

Data Grid

Data Grid Technology has been developed to gather and to securely process huge and various datasets and databases. In addition, a variety of databases are used collaboratively and a large amount of computational resources are needed to process data. OGSA provides a general framework for sharing of resources among institutions over firewalls based on the Web Service Resource Framework (WSRF). It enables execution of applications and workflows with remote resources through web services in a secure way. The Japanese BioGrid project designed application metadata and data service metadata to fill the semantic gap among gene-protein databases, interaction databases and compound databases necessary for drug-design using Globus Toolkit 3 and OGSA-DAI for the implementation of a heterogeneous database federation (Tohsato, Kosaka, Date, Shimojo & Matsuda, 2005). This project aims to develop a Computer Grid Technology that can be used as the core technology to realize a future Super Computer Network environment.

The European Union's @neurIST project (Arbona, Benkner, Fingberg, Engelbrecht, Hofmann, Kumpf, Lonsdale, & Woehrer, 2007) is developing a Grid-based IT infrastructure for the management of all processes linked to research, diagnosis, and treatment development for complex and multi-factorial diseases encompassing data repositories, computational analysis services, and information systems handling multi-scale, multi-modal information at distributed sites.

The @neurIST Grid middleware provides advanced compute and data services to support computationally demanding modelling and simulation tasks and to access heterogeneous distributed data sources through semantic integration.

Other programming tools such as SAGA (Simple API for Grid Application) and ByteIO (web service oriented POSIX-like library) are specifications of the OGF (Open Grid Forum, www.ogf.org) which provide standard access mechanisms for files stored on the Grid.

SAGA provides a uniform API to the evolving and changing Grid middleware and greatly simplifies the implementation of Grid applications hiding middleware details from the application through a dynamic adaptor mechanism to bind to multiple middleware, and choose the appropriate one on the fly.

OGSA ByteIO defines a minimal Web Service interface for providing "POSIX-like" file functionality. Essentially, this allows any service which implements the interface to be accessed in a file-like way.

The OGSA Data design team has identified a need for an interface for reading a stream of bytes nobody had addressed before and has finalized its work on the definition of a minimal web service interface for this functionality. These solutions seem quite powerful but oriented to middleware developers more than to software engineers and developers of general purpose applications. They are still vertical, service oriented, and do not solve the basic problems of the fragmentation of the data services (and its consequent rapid growth of the learning curve), due to the SOA design. Moreover, in the specific case of the EGEE gLite middleware, there are neither tools nor APIs nor specifications that bind both data and metadata management tools together.

Many bioinformatics databases are public and freely available but it is often the case that access to the data needs to be strictly controlled in distributed collaborative research. A secure framework is needed to access clinical data that exist across regional, national, and international boundaries for clinical trials and unbiased evaluations of their outcome.

GSAF (Scifo, 2007) is a useful software interface (built on top of several data management APIs), to make developers able to design and implement data access layer for enterprise applications independently both from their own domain and the middleware technological details.

The term "information security" (Foster, Kishimoto, Savva, Berry, Djaoui, Grimshaw, Horn, Maciel, Siebenlist, Subramaniam, Treadwell & Von Reich, 2005) means protecting information and information systems from unauthorized access, use, disclosure, disruption, modification, or destruction in order to provide integrity, confidentiality, and availability. The information integrity can be defined as the protection against improper modification or destruction, and includes authenticity. Confidentiality preserves authorized restrictions on information access and disclosure. Availability ensures timely and reliable access to and use of information. At the present time the EGEE infrastructure grid Middleware is affected by the insider abuse problem since the storage element administrator can access files on the machine directly, bypassing the Grid security infrastructure with its root privileges.

The Secure Storage service (Scardaci & Scuderi, 2007) allows users to manage confidential information/data in a Grid environment ensuring the above described properties (data integrity, confidentiality, and availability), and solving the inside abuse problem. Some other groups are trying to solve the same problems such as the developers of Hydra (https://edms.cern.ch/file/697751/1.0/EGEE-TECH-697751-v1.0.pdf). The following table shows the comparison between the two services.

Table 1. Comparison between EDS and Secure Storage Service

	Hydra	Secure Storage
gLite 3.1 support	No	Yes
File encryption/decryption	Yes	Yes
Data block encryption/decryption (POSIX-like API)	No	Yes
Grid Keystore	Yes	Yes
Strong Encryption	No	Yes
Replica management and automatic key deletion	No	Yes
Shamir's shared secret scheme support	Yes	No

Currently, Secure Storage provides a more complete set of functionality, in particular the Strong Encryption which is the essential part to manage confidential data in a very secure way.

Grid Portals

Regarding data management aspects, some other projects have been working on this area realizing useful Grid portals and powerful APIs. GENIUS (https://genius.ct.infn.it/) and P-Grade (http://www.lpds.sztaki.hu/pgportal/) are custom solutions designed to offer a Portal access to Grid infrastructures. The P-GRADE Portal is a Grid portal solution with such typical characteristics. Using its integrated workflow editor component or its graphical Web interface the users can comfortably organise sequential or parallel jobs into workflows, can execute, monitor and visualize them, and the latest release allows to achieve the aggregation and sharing of knowledge, resources and results. GENIUS, in addition to the support for some workflow system as Triana, offers an easy way for application specific interface development, interactive analysis and web access to personal spooling areas. These portals provide great functionalities to integrate grid-enabled applications and support both Grid job workload and data management systems but they do not deal with the problem of synchronizing data and metadata on the Grid. Many data management tools are involved in the middleware and, due to the general purpose of their nature, operations on data remain decoupled.

Our work tries to solve the above mentioned challenges, only partially faced by other research groups and projects. To summarize, from our experiences in the life sciences domain, we think that the data-centric approach is essential in biomedical contexts where remote and secure data access are strong requirements. Our implementation proposes to provide users with a new and easy way to run their applications in a Grid environment without taking care of the complexity of the infrastructure underneath.

BIO MED PORTAL

Bio Med (BM) Portal provides methods and tools for accessing remote distributed data satisfying security issues, extracting information from those data without moving files on the network, managing related metadata, building and maintaining catalogues, submitting jobs on the Grid together with all the needed parameters, etc.. The implementation of the BM Portal system is based on gLite 3.1, the official Grid middleware for the European Grid infrastructure and covers Computation of Algorithms, Data Organization and Management and Security aspects. The wider base of users and applications (together with the fact of being the outcome of a larger project), makes gLite a reliable middleware providing a large set of resources in a production-quality environment with a complete documentation and an effective technical support. This platform attracts several different applications as well as Genomics and Proteomics, Medical Imaging and Neuroinformatics. BM Portal is designed to host several medical use cases and it is able to deploy several analysis algorithms. A set of independent applications can be published on the portal and either single algorithms or a combination of them might be invoked by the user through a workflow strategy available through a web interface. The services are implemented within an existing Grid computing infrastructure to solve problems concerning both the large datasets storage (data intensive problem) and the implied large computational time (computing intensive problem). From a functional point of view, the adopted portal allows to deploy both server components (visible to users as traditional web pages) and services (Grid or web-services) exposing key components to the public with standard interfaces.

The Portal Architecture

In the traditional multi-layer view of the Grid, BM Portal deploys services at the Service Ware level, and end-users applications will be based on tools and services provided by the Applications layer. This means that all integration and software development work is done on top of existing middleware. The portal is a Java 2 Enterprise Edition (J2EE) web application. It offers Dynamic HTML interfaces to users (such as Google Web Tool Kit widgets) and can interact both with the operating system and the external Grid resources through their command line tools or Application Programmer Interfaces (APIs). This application represents the main component of the whole architecture and it is based on the EnginFrame framework (developed by NICE SrL) which acts as an application container for several applications and provides several utility services (like authentication and authorization) and common infrastructure (like temporary storage, user data areas) to applications and users. While the standard interface is a Web GUI built for human users, a web service interface is provided allowing SOAP based clients to connect to the platform and consume the provided services. Through an homogeneous interface, users can access local data storage areas, Grid Storage Elements, local or remote High Performance Computing facilities (compute clusters) and Grid based computing resources thus allowing user applications to be run on almost any kind of available resources.

Looking at Figure 1 we can see a detailed view of the portal, the main component of the platform, and the resources available to users through it.

Thinking about BM portal as a platform of services we can identify three main components, each including several key modules:

- **The client:** A standard web browser or a standard SOAP client. The client is represented by a user who can access directly the web user interface (B) or interact first with the workflow designer (A) to create its own workflow to submit to the engine. The same workflow designer could be a SOAP client able to call the platform services and orchestrate it in very complex flow diagrams.

Figure 1. The BM Portal architecture

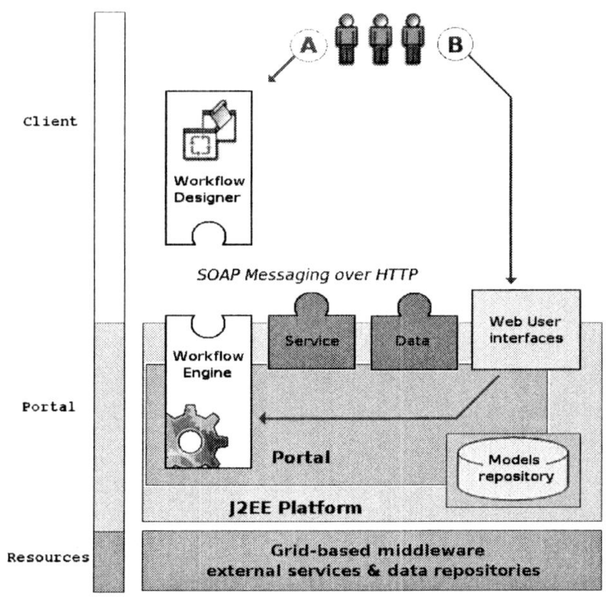

- **The Portal:** It is structured in a portal server and one or more portal remote agents, depending on the distribution level of resources. The server relies on J2EE platform and provides services, tools and data through both web and web services interfaces. The agent is the interface to local and Grid computing resources and to data resources. The portal is supported by a SQL backend database holding configuration and runtime data.
- **Resources:** The system is able to interact with different kind of resources. At the moment the following resources have been selected:
- **Grid infrastructure:** EGEE gLite middleware;
- **Local computing infrastructure:** A local small computer cluster with LSF scheduler for small highly interactive computing tasks;
- **Two Metadata containers:** A Grid metadata container for production data and a local metadata container for temporary data and local jobs.

The platform is flexible enough to manage different kind of data and algorithms. It can be adopted in several biomedical scenarios combining text data for patients, studies, and reports as well as medical imaging acquisition volumes, time series signals and/or genomic information. Data, metadata and computing services are protected and available only to authenticated (and authorized) users. A user can perform a particular action on the portal only if he/she belongs to a valid group. It is not mandatory that portal groups correspond to real life working groups. This allows keeping user authentication credentials on a separate container, while ACLs are managed by the portal itself.

The Use Case

The use case presented here is related to the sharing of medical images among different hospitals for the early diagnosis of Alzheimer's disease (AD), the leading cause of dementia, accounting for more than half of all dementias in elderly people. Clinically, AD is characterized by a progressive loss of cognitive abilities, the memory loss typically being the earliest sign of the disease. The qualitative analysis of medical images does not provide useful suggestions for the diagnosis in AD, and a statistical comparison of PET/SPECT images from suspect AD patients with images from normal cases is a powerful tool for an early diagnosis (Signorini, Paulesu, Friston, Perani, Colleluori, Lucignani, Grassi, Bettinardi, Frackowiak & Fazio, 1999). The use of Statistical Parametric Mapping Analysis (SPM) for the quantification of hypo-metabolic patterns in brain PET/SPECT studies has already been proposed in literature. The SPM software library was originally developed and it is made freely available by the Functional Imaging Lab (FIL) at the Wellcome Department of Imaging Neuroscience (London University College) for activation studies in functional MRI (Friston, 1996). Since then, the use of SPM was extended and, through a specifically defined analysis protocol, SPM routines are presently the standard within the neurological research community as regards a voxel-based analysis of PET/SPECT studies for the early diagnosis of AD. The algorithm performs a statistical analysis in order to compare the perfusion values in the test images against the corresponding values in a set of normal images, taking into account several parameters such as age and average cerebral flow producing in output hypo-metabolic maps and the related views of the brain, directly analysed by clinicians.

Results of a complete SPM analysis for the early diagnosis of AD, including hypo-metabolic maps and the related views of the brain, are shown in Figure 2.

Figure 2. Results of a SPM analysis on a PET study of glucose metabolism in a patient with dementia

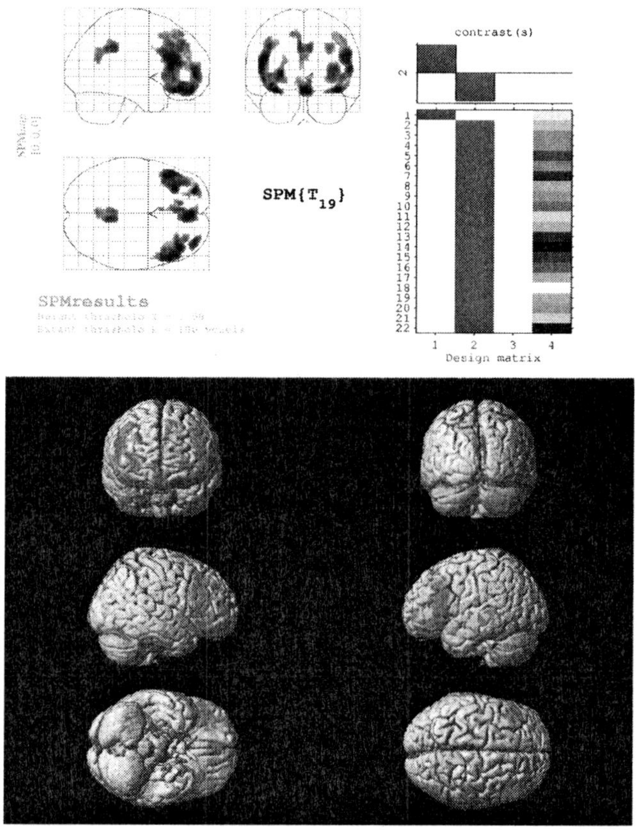

It is worth noting that during the statistical parametric mapping a large set of images of normal patients is required to be used for comparison, because the accuracy of hypo-perfusion maps is strictly related to the number of normal studies compared to the test image. On the other hand, due to ethical issues and to the high costs and the level of risks of neuroimaging nuclear technologies, PET and SPECT studies on normal subjects are very rare. The NEST-DD project, funded by the European Commission, collected a database of more than 100 images in order to make available the first large dataset for these studies. Moreover, the images of normal subjects are covered by privacy and security issues and for this reason they cannot be freely moved on the net or published by the centre that performed the analysis. As a consequence, only doctors working at very large institutions, locally owning large databases of normal images, can usually carry out SPM-based analyses.

Starting from these considerations, the aim of our project has been to enable doctors from small peripheral hospitals to use large sets of normal PET/SPECT images provided by medical research institutes distributed on the network, by remotely extracting the information needed for the statistical analysis from the normal images and collecting it without moving the original image files. Furthermore, the execution time of the analysis must be compatible with an interactive clinical application in a busy medical environment. The time required for the analysis can be reduced, since some aspects of calculation could be parallelized and distributed on the computational resources associated to the remote databases of normal images and the time required for data transfers over the network would be reduced, since the code amounts to just few kB, compared to images sizing up to 100 MB. Due to all these reasons a secure Data Grid approach here is fundamental. In fact, this allows solving the two relevant issues of this use case: first, it allows handling data in a secure way, protecting them and making them available according to real world access rights. On the other hand, the key issues of having data in hospitals and force applications to access them without requiring copies is addressed by the adoption of a remote file access library available on gLite.

The steps needed to perform the complete SPM analysis sequence in our Grid environment are listed below:

1. User accesses the BM Portal website through his/her browser;
2. User logs in the BM Portal and accesses the SPM application;
3. Grid middleware manages authorization and certification procedures;
4. User uploads, through the browser, the test image from a suspect AD patient;
5. User selects the correct reference brain atlas;
6. The test image is sent to the management node together with information about patient age;
7. Thanks to the information in the metadata catalogue, a number of normal images suitable for comparison are selected on the distributed repository nodes;
8. A small piece of software is sent, with a Grid job submission, to the distributed worker nodes together with logical names of files of related normal images to be used for comparison;
9. The job accesses data on the repository nodes through Grid certificates;
10. The code remotely computes on worker nodes, from normal images on repository nodes, data matrices needed for statistical analysis;
11. Data matrices are returned to the management node;
12. Overall SPM statistical analysis is performed on the management node;
13. Results of SPM analysis are shown through the browser on the user node;
14. Depending on user request, results can both be sent to and saved on the user node.

The whole software analysis sequence and the related results have been scientifically and extensively validated. Even small alterations may influence the final results and, therefore, even minor corrections in the original code would require a new validation by the scientific community. This implies that the original SPM routines for data analysis have been used as they are and have not been modified in order to achieve an easier Grid implementation. Finally, doctors prefer not to be forced to interact with technical computing procedures concerning accounting, data transfer, resource sharing, database access, etc. The whole structure of software layers has been hidden to users through the BM Portal interface. In order to start the application, users have just to access the portal with their own browser, to login and to upload their test image. Interactive inputs are performed again on simple web pages and results are visualized on the browser and can be saved on a local repository. This procedure does not require doctors to have any technical skill neither special local computing or storage resources, thus widening the population of potential users. This Alzheimer's disease early diagnosis application well suited the BM Portal as first implemented service since (i) it has been firstly developed and is therefore better defined, (ii) it is being validated by medical doctors and (iii) it has a relevant social impact.

Potential Impact of the Alzheimer's Use Case

Neurodegenerative diseases are sporadic or hereditary conditions that are characterized by progressive nervous system dysfunction due to neuronal and synaptic loss that gives rise to irreversible brain tissue loss (atrophy). The most prevalent neurodegenerative diseases are Alzheimer's and Parkinson's Diseases, but include also others such as Dementia with Lewy Bodies, Frontotemporal Degeneration, Corticobasal Degeneration, Multiple System Atrophy, and many others. Data on the prevalence on Alzheimer's disease and other dementias are available across the EU collected by Alzheimer Europe. However, these estimates must be treated with caution as the type of source varies considerably in scale and accuracy by Member State. An estimated 5.3 million Europeans (in 21 Member States) between 30 and 99 years of age suffered from different types of dementias in 2000 (11.9 per 1,000 inhabitants). In other words, about one in every 20 people over the age of 65 has Alzheimer's disease (AD) The size of the problem is growing rapidly as it is strictly related to ageing. Within the next 20 years there will be a highly significant increase in the proportion of people over 65 years of age, with the fastest growing population in most countries being those who are very old (i.e., aged 80 years and above). In the next 30 years, the proportion of people aged over 80 years (as a share of the over-65 population) will increase, in Europe as a whole from 22% to over 30%. Disease estimation and cost estimation are focused on mild to severe dementia and only estimated figures could be provided merging different EU Member States independent studies. This targets the estimated cost per patient between 9.000€ and 16.000€ per year. There is therefore potentially a really big impact if applications and services such as the ones developed in this project will be able to move from a research context to clinical practice. This is an achievable result if security, reliability, and usability have been kept in mind throughout the whole design and implementation process.

Grid Integration

In general, Grid architecture provides tools and facilities on top of any hardware configuration and network infrastructure that allows data management in several clinical contexts, as the one we are treating here. As we said above, Grid technologies have been adopted to enable doctors from small peripheral hospitals to use large sets of normal PET/SPECT images by remotely extracting the information needed for the statistical analysis and collecting them without moving the original image files. Since such images cannot be exported outside the boundaries of institutions where acquisitions were performed (even if with heavy encryption), privacy and security issues are really critical. This is however true in most health information systems where medical images must be accessed through severe authorization policies. The use of Grid technologies well matches all of the above issues and allows easy access to distributed data as well as to distributed computational resources. In particular, the Data Grid distributed analysis paradigm ("move code rather than data") is well suited for implementing such a use case. Through data-management grid services doctors can access normal PET/SPECT image databases without moving images between hospitals, thus complying with privacy regulations.

Security

The gLite security service is based on X.509 standard certificates and every Grid Services is accessible by users registered in a VOMS (Virtual Organization Membership Service). This is a service able to provide information on the relationship of users with their Virtual Organization (http://glite.web.cern.ch/glite/security/): groups, roles and capabilities. The VO manager can administrate it remotely by using either command line tools or a web interface. As a consequence, there is no need to make changes twice, and several VO users can be mapped onto a single Application Role if needed. This mechanism is fully inherited by the BM Portal in order to provide its users with a complete and reliable Access Control System. From a functional point of view this allow BM Portal administrators to implement in the applications virtual organization and permission schemas already defined in real organizations.

Job Management

From the computing point of view, our strategy has been to study SPM routines, implemented in Matlab, to find the most suitable solution to port SPM on Grid. A detailed analysis revealed that the SPM code is structured in functional blocks, each dedicated to a specific functionality. Following the data Grid paradigm, only the first part related to data access has been modified. The SPM routines concerning access to images of normal patients and extraction of information from such images have been rewritten in C language in order to allow parallelization and remote access to distributed images. In fact, through computational Grid services, statistical information can be extracted from normal images, and the image descriptive data matrix and other information needed for the statistical analysis can be transferred to the management node without moving images. Moreover, this process can be executed in a parallel way on distributed machines and up to a single job for each slice contained into the images, to improve computing performances. The second part of the statistical process, implementing the data analysis, is the same as the original in order not to incur in new algorithm validation process, since the actual Matlab scripts are adopted in the clinical practice "as they are". The voxel-by-voxel based statistical analysis is performed outside the Grid environment through the statistical parametric mapping algorithm. Statistically significant parameters such as the age and sex of the patient and the average cerebral flow are taken into account in order to produce the correct results. These parameters are the only relevant metadata for the analysis of raw data.

Data Management

One of the main benefits of using Grid infrastructure is the possibility to use huge distributed storage space. A medical community might prefer to use storage elements owned/managed by an external organization to outsource the management of these machines and to avoid to buy specialized hardware and to hire specialized staff. The community could rent the storage space as needed, minimizing human and hardware costs. In this way an easy adoption of the Grid technology is allowed also to communities without expertise in Grid computing and computer science in general, overcoming the problems due to a long training process and moving to the "resources as a service" paradigm. The portal is build on top of the gLite Data Management System in order to ensure data access and management. The proposed environment permits users to upload/download their data and results on/from the Grid storage resources providing security (encryption/decryption of data) and reliability (Grid data redundancy). A remarkable metadata management system is provided by the ARDA Metadata Catalogue Project (AMGA, amga.web.cern.ch/amga/) metadata management server recently integrated in the gLite environment. Grid File Catalogue (LCG File Catalogue) allows to register all files stored on the Storage Elements while all logical relationship among files and research teams are kept inside the Metadata Catalogue. Data and metadata input in the system is directly accessible through the web interface where users fill a multi-page web form with required fields. The form is then processed and submitted to the catalogues, as soon as the uploading of data on Grid provides a consistent logical reference to the data themselves. Data and metadata search/browse activities are carried out through a simple tabular interface representing anonymous reference to available normal patients and the relevant metadata. In order to develop a functional distributed management environment, particular attention has been paid to the description of experiments using a standard annotation through the definition of metadata schemas in XML format. These schemas are reflected to AMGA collection implementing the metadata container. This integration, according to the Grid Paradigm, is based on the adoption of two specific tools: GSAF (developed both by IR&T Engineering SrL and the COMETA Consortium) and the Secure Storage Service (developed both by INFN and UNICO SrL).

Grid Storage Access Framework

This section covers some aspects of the Grid data management area looking at the integration, standardization and interoperability among "Data Grid" based applications interfaced to the EGEE gLite Grid Middleware (glite.web.cern.ch/glite). Cultural Heritage applications, Healthcare applications, Earth Sciences applications (just to mention some examples) have their own application logic, presentation styles and also data management rules according to a "vertical dimension" (proprietary customization), indicating the uniqueness of the model of the application itself. However, looking at the layered architecture of each application, there is always a common level where most of required operations are shared. For instance, all application need to create files, browse directories, add files, delete files, rename files, define metadata schema, change metadata values and a lot of other data/metadata management operations. We can also talk of a "horizontal dimension" made by all those common functions. On the other hand, the realization of applications that adopt Grid ad a digital repository is not very easy since several problems affect both design and programming principles. First of all, Grid architecture respects the SOA (Service Oriented Architecture) recommendations. It means that several services (each independent from the others) work in a standalone mode cooperating to provide data and metadata management capabilities (Chervenak, Foster, Kesselman, Salisbury, & Tuecke, 2001). Moreover, within the Grid environment, files are stored inside Storage Elements (the Grid services that take care of data persistency). They can be replicated on several SEs for ubiquity, security, fault-tolerance, and sharing purposes and relationships among locations of file replicas and theirs logical names are kept within a specific File Catalogue Service. For each file it is possible to associate some descriptive attributes (as metadata schema) and they can be valorized (schema instance), modified, quantified and queried through a specific Metadata Catalogue Service. Further, specific modules take care of data transfers from/ to different sites (File Transfer Services).

This scenario, characterized by a considerable fragmentation in terms both of number of services and related APIs leads to several problems:

- APIs fragmentation forces software engineers to implement specialized and decoupled software components according to a vertical architecture;
- Any application must handle by itself the semantic coherence and integrity consistency among sequential operations on the same data set;
- Developers need specific technical knowledge about Grid programming (several APIs means several syntaxes), therefore they are usually busy to learn API usage rather than debug and improve their code;
- The absence of appropriate design patterns produces different codes to implement the same features many times and the maintenance of the code and of a linear architecture become hard.

This is considered a big limitation of the gLite Grid Storage System and our effort has been made in this direction trying to design and implement a software layer that could solve these common problems. According to a 3-layer software architecture of an enterprise application, GSAF can be effectively classified as the Data Access Layer for the Grid middleware. It is an appropriate interface that allows business components to manage data stored on the Data Management System and presentation objects to search and retrieve data from it. Figure 3 shows the logic architecture of the framework.

From the design point of view, GSAF identifies a kind of Grid software engineering pattern offering a standard solution for common problems according to the principle "write once and use many times". Note that GSAF can not remove the vertical design but includes it for all use cases wrapping each gLite data service with a corresponding software module built on top of the related API. The File Manager provides functions to interact with the Storage Element allowing the upload/download of files to/from the SE. The Catalog Manager is related to the File Catalog Service and provides functions to register and de-register files. It also allows browsing and managing virtual directories. The Metadata Manager interfaces the Metadata Catalogue Service wrapping all metadata related functions. The VOMS Manager is responsible to get a valid proxy from the VOMS server. The main module, the GSAF Interface, works as a dispatcher for application requests and as a coordinator of the underlying components. Concluding, we can talk about GSAF as an Object Oriented Framework built on top of the gLite Metadata Service and gLite File Services that exposes classes and related methods for applications located above ("black box" principle).

Framework Core Overview

The core of the application is designed to be a plug-in in order to be easily embedded it many applications. Its design covers several Object Oriented design patterns, such as the singleton method, the strategy method, the factory method, the template method, the iterator and the composite method, that make the module generic enough and adoptable by any applications independently of their specific domain. This also ensures a clean

Figure 3. GSAF software architecture

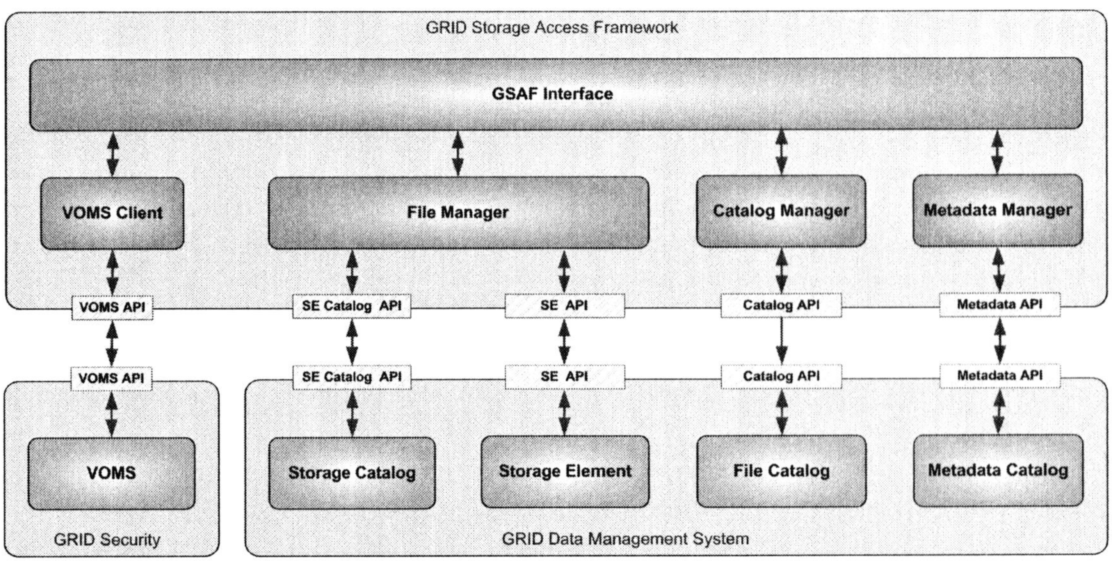

and simple software architecture with an high degree of cohesion and decoupling (best practices in software engineering).

Inside the framework, all logical entities (such as file, directory, collection, entry, attribute, ACL, user, group, etc.), are modeled by Java objects and they represent the basic data structures (data/object mapping principle). One instance for each data object composes the GSAF Context Base object that is used as a kind of execution environment for each GSAF Operation (information hiding and composite pattern). The GSAF Operation interface defines all functions allowed on data set and each function is bound inside a specific class identified by an operation code (command pattern). The functional library is thus build by each GSAF Operation implementation which encapsulates the algorithms to be performed in a given context and on the given data (template pattern). GSAF supports both types of operation: Command, which changes the state of the storage, and Query that performs only searches and returns result sets (iterator pattern). Every operation of the functional library is an elementary building block. The interface GSAF Macro defines a meta-type for operations composted by several elementary operations (composition pattern). It implements the same interface of all its sub components and it is able to perform them sequentially (façade pattern). This mechanism allows developers to easily assemble their own complex functions according to the interface specification of GSAF. A GSAF Manager object which creates and executes operations (strategy pattern) is also defined. Finally, a GSAF Operation Factory object is used to generate on the fly the right operation needed by the application (factory method).

Thanks to GSAF, developers, in order to perform any data management operation, just need to instantiate the context and quantify it with the right parameters (data references and operations on them), according to the business rules of their application and pass it to the manager object.

Secure Storage Service

In the case of applications regarding biomedical research and clinical practice, the communities have strict privacy requirements because they have to manage confidential data (as, for example, patients' medical history). Then, to apply the model described above (confidential data saved in a Storage Element and managed by an external Organization) a mechanism to prevent the administrator of the machine accessing/modifying the sensible data stored on it is required. At the present time, the gLite Middleware provides the same security infrastructure for all grid services. The authentication is performed using a X.509-based PKI infrastructure (http://www.ietf.org/rfc/rfc3280.txt) and the VOMS (http://voms.forge.cnaf.infn.it/home.html) attributes are used to authorize users according to their roles. Moreover, an authorization method based on Access Control Lists (ACL) grants data access only to their owners. However, data are stored in a clear format. The Storage Element administrator can access them bypassing the Grid security infrastructure. This is known as the insider abuse problem (http://www.ots.treas.gov/docs/4/422134.pdf).

In order to solve this so called "insider abuse problem", preventing also the administrators of the Storage Elements to access confidential data, we have designed and developed a new gLite service, the Secure Storage Service, which allows users to manage confidential information/data in a Grid environment ensuring the propertied described below. Secure Storage guarantees data confidentiality and integrity:

- Saving data in an encrypted format;
- Performing the encryption operation on a trusted machine;
- Storing all needed information (e.g., the keys) to be used to encrypt/decrypt the data in a trusted machine and making them accessible only to their owners (or users authorized by them).

Thanks to this service, the complexity of the operations necessary to satisfy these properties is hidden to the end user. Secure Storage provides users with suitable and simple tools to save confidential data in storage elements owned by an external organization in a transparent and secure way. The Secure Storage Service has been designed to be integrated in the gLite middleware. It is made of the following components:

- **Command Line Applications:** Commands integrated in the gLite User Interface to encrypt and upload, decrypt and download files on the storage elements;
- **An Application Program Interface:** The API allows the developer to write programs able to manage confidential data using the Secure Storage service;
- **The Keystore:** A new Grid element used to store and retrieve the users' keys in a secure way;
- **The Secure Storage Framework:** A component of the service, internally used by the other components. It provides encryption/decryption functions and other utility functions. It takes care of interaction with the Grid Data Management System.

Encryption Algorithms

Encryption and decryption operations are performed using symmetric-key algorithms (http://en.wikipedia.org/wiki/Symmetric-key_algorithm). These algorithms are a class of algorithms for cryptography that use the same cryptographic key both for decryption and encryption. We choose this kind of algorithms because they are generally much less computationally intensive than asymmetric key algorithms. In practice, this means that a quality asymmetric key algorithm is hundreds or thousands of times slower than a corresponding quality symmetric key algorithm. Advanced Encryption Standard Algorithm (AES) (http://csrc.nist.gov/CryptoToolkit/aes/rijndael/) is the default encryption algorithm used in the Secure Storage Service with a 256 bits key length. However, the Service allows to support other symmetric algorithms thanks to its modular architecture (Figure 4).

Command Line Applications

The Secure Storage service provides a set of new Command Line Applications on the gLite User Interface. These applications allow users to manage confidential data in a secure way. A list of the main Command Line Applications of the Service is showed in the following section:

- **lcg-scr:** The input parameters of this command are: a local file, a storage element, a Logical File Name (LFN) and a list of users authorized to access the file. The command generates an encryption key, encrypts the input file and uploads it on the storage element, registering its LFN in a LFC file catalogue. Moreover, it stores the key generated

Figure 4. The Secure Storage Service architecture

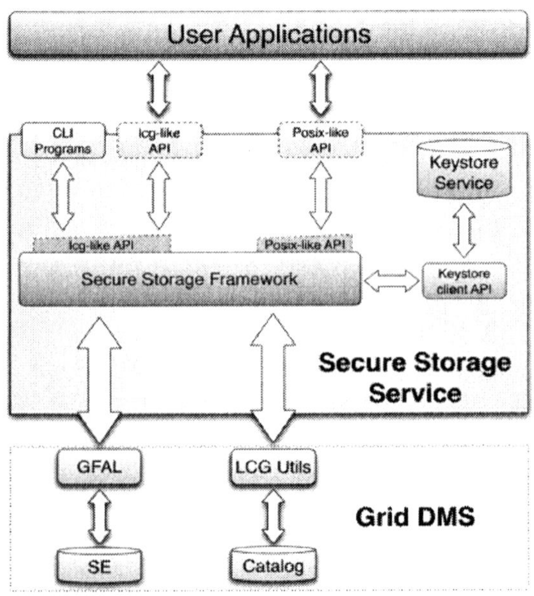

and used to encrypt the file on the keystore. An Access Control List (ACL) will be created and associated to the encryption key on the keystore. This ACL will contain all users authorized to access the file (as a list of distinguished names DNs, http://www.ietf.org/rfc/).

- **lcg-scp:** The input parameter of this command is a LFN. It downloads the encrypted file identified by the input LFN, gets the key to decrypt the file from the keystore, decrypts the file and then stores it on the local file-system. This command successfully executes only if the user is an authorized user (his/her DN is on the ACL associated to the key needed to decrypt the file).

Other commands are available to delete files (lcg-sdel) and to manage the ACL associated to a file (lcg-add-dn and lcg-del-dn).

Data Integrity

The encryption operation guarantees only data privacy. Files saved on the Grid could be maliciously modified breaking the Grid Security Infrastructure. Secure Storage Service ensures also data integrity. The lcg-scr commands can digital sign (http://www.itl.nist.gov/fipspubs/fip186.htm) the confidential data. Optionally, users can require the data digital signature before of the encryption operation, specifying a command option. The data and the digital signature will be saved encrypted on the selected Grid Storage Element. The lcg-scp command automatically verifies data integrity if the data have been signed.

Application Program Interface

Developers can use the Secure Storage API in their Grid Applications to manage confidential data in a secure way. The API is divided in two groups according to their functionalities:

- The first group of functions behaves like Command Line Applications. They are the secure version of the lcg-utils API (http://www.cern.ch/lcg/);
- The second set of functions can be used to manage encrypted remote files as local files. These functions allow developers to read or write encrypted file blocks stored on remote storage elements in a simple way. The functions hide the complexity of working with encrypted remote files. In this way, users can manage remote encrypted files as they were clear local files. They are the secure version of the LCG GFAL API.

Keystore

The Keystore is a new Grid element used to store and retrieve the users' key(s) in a secure way. It is identified by a host X.509 digital certificate and all its Grid transactions are mutually authenticated and encrypted as required by the Grid Security Infrastructure (GSI) model (http://www.globus.org/toolkit/docs/4.0/security/key-index.html). The Keystore is a critical node of the Secure Storage service because the data security depends on its integrity. For this reason the Keystore should be placed in a trusted domain (e.g., inside the local network of the community using the Secure Storage service) and should be appropriately protected by undesired connections. So, a well configured firewall is required.

At the application level, the Keystore is a black box with a single interface towards the external world. This interface accepts only mutually authenticated connections and decides to serve or not the request through an authorization process, once the client is identified. The operations performed during the authorization process are the following:

- The client request is processed only if the client is a member of the list of allowed users and/or it belongs to an enabled Virtual

Organization or to a specific Virtual Organization Group. The request is discarded in any other case;

- If the client wants to retrieve a key (or wants to modify the access permission), the keystore checks if the request is coming from an authorized user inserted in the ACL associated to the request key. It knows the Distinguished Name of the user thanks to the GSI authentication.

The clients of the Keystore are the command line applications and the API functions of the previously described services. For example, the Keystore saves the encryption keys and the associated ACLs received by the lcg-scr commands on its repository and provides the lcg-scp commands with the keys.

FUTURE TRENDS

BM Portal is now a reality and some medical and bioinformatics communities are looking forward to adopt it for their research purposes. The ability of the platform to allow users to manage both algorithms and their input data sets, the capability to host several heterogeneous types of analysis tools, the natural distribution of computing, storage and human resources make BM Portal very attractive as cooperative virtual laboratory. Moreover, BM Portal is not just a portal but it aims at identifying a real paradigm to design, implement and deploy Grid oriented applications in an easy way. In order to enhance the SPM service, some new functionalities regarding spatial normalization and smoothing, as well as other kind of preprocessing on functional images, will be implemented in the BM Portal as soon as possible. With such a complete application it will be possible to include some important research centers into medical trial, and BM Portal will be able to represent the conjunction point between different realities. At present, BM Portal is a very powerful portal concerning the back-end infrastructure, but developers are also working on integration between the framework and different content management systems (as Zope/Plone) that will allow users to access both informative areas and computing/ data services in a trasparent way. Moreover, an upgrade of BM Portal will be performed in order to add new functionalities related to visualization of functional and structural images of the brain and distributed computing on different types of cerebral signals. Its flexible architecture will allow developers to implement other biomedical services, not only in Neuroinformatics field as presented here, but also for genomics, proteomics and bio-informatics needs. From the technical point of view, several interfaces and services to allow database access on the Grid environment are under development (in order to make applications more flexible and specific domain independent), making integration more complex.

In this perspective, GSAF takes advantage from the Object Oriented paradigm offering a high level abstraction and a transparent way to interface applications with Grid middleware. Its design allows an easy and fast encapsulation into existing systems and speeds up the development of Grid applications. The framework is very extensible in order to support several Grid middleware and/or classic or emergent infrastructure for interoperability among hybrid applications. In the next version, GSAF will also provide a Transactional Service to support atomic, consistent, isolate and durable multiple sequential operations on the same data/metadata set.

The Secure Storage Service is going to support Shamir's shared secret scheme in the next release. This algorithm splits the encryption key in N parts and allows to rebuilt the key using K=N-1 parts. Using the Shamir's algorithm, the Secure Storage Service will be able to store the N sub-keys on N different Keystores. In this way data confidentiality will be preserved even if the integrity of K-1 Keystores is compromised, ensuring service availability at the same time.

CONCLUSION

Problems concerning the storage of large datasets, storage safety, and large computational times are solved with a Grid computing infrastructures. Grid portals act as user interfaces both for data storage, metadata management, data analysis and result retrieval. Data access from processing job is done directly on distributed file system, without moving files belonging to datasets to computing nodes' local file system. Furthermore, data annotations are gathered and stored with the use of a metadata schema facilitating the gathering and sharing of information. The proposed environment is being tested by using a specific experimental scenario: an application for Alzheimer's Disease early diagnosis based on normal cases image sharing and processing. Its validation process has recently started and it will be adopted in clinical practice in the next future. The BM Portal is already used for testing of new bioinformatics applications and represents one of the principal Grid Portals for the development of National and International Biomedical projects. According to the Virtual Physiological Human paradigm, a novel approach to knowledge sharing and data analysis which stresses concepts such as virtual experiment and models are required not only on a geographic perspective but also among different levels and disciplines. Large, distributed, and heterogeneous repositories, as well as computationally demanding analysis tools, are more and more involved in biomedical studies and the approach described here exactly addresses these requirements.

ACKNOWLEDGMENT

This project is partially funded by the Italian MIUR-FIRB project LITBIO (www.litbio.org, Laboratory for Interdisciplinary Technologies in Bioinformatics).

This work makes use of results produced by the PI2S2 Project managed by the Consorzio COMETA, a project co-funded by the Italian Ministry of University and Research (MUR) within the Programma Operativo Nazionale "Ricerca Scientifica, Sviluppo Tecnologico, Alta Formazione" (PON 2000-2006). More information is available at http://www.pi2s2.it and http://www.consorzio-cometa.it.

The PET/SPECT images on normal cases are owned by San Raffaele Hospital (Milan, Italy) . Biolab team thanks the Nuclear Medicine Department of HSR for the clinical support during all phases of the work on SPM service.

REFERENCES

Arbona, T., Benkner, S., Fingberg, J., Engelbrecht, G., Hof-mann, M., Kumpf, K., Lonsdale, G., & Woehrer, A. (2007). A Service-oriented Grid Infrastructure for Biomedical Data and Compute Services. *IEEE Transaction on NanoBioscience*, *6*(2), 136-141.

Arzberger, P., Farazdel, A., Konagaya, A., Ang, L., Shimojo, S. & Stevens, R. (2004). Life sciences and cyberinfrastructure: dual and interacting revolutions that will drive future science. *New Generation Computing*, *22* (2), 97-110.

Cannata, N., Merelli, E. & Altman, R., (2005). Time to organize the bioinformatics resourceome. *PLoS Computational Biology*, *1*(7), e76.

Chervenak, A., Foster, I,. Kesselman, C., Salisbury, C., & Tuecke, S. (2001). The Data Grid: Towards an Architecture for the Distributed Management and Analysis of Large Scientific Datasets. *Journal of Network and Computer Applications*, *23*, 187-200.

Foster, I., Kishimoto, H., Savva, A., Berry, D., Djaoui, A., Grimshaw, A., et al. (2005).The Open Grid Services Architecture, Version 1.0. *Informational Document, Global Grid Forum (GGF).*

Friston, K. (1996). Statistical Parametric Mapping and Other Analysis of Functional Imaging Data. In *Brain Mapping The Methods,* (pp. 363-385). New York: Academic Press.

Konishi, F., Yagi, T., & Konagaya, A. (2006). MolWorks+G: Integrated Platform for the Acceleration of Molecular Design by Grid Computing. In *Grid Computing in Life Science* (pp. 134-141). Singapore: World Scientific.

Krishnan, A. (2004). A Survey of life sciences applications on the grid. *New Generation Computing, 22* (2), 111-126.

Li, W., Byrnes, R., Hayes, J., Birnbaum, A., Reyes, V., Shahab, A., et al. (2004). The encyclopaedia of life project: grid software and deployment. *New Generation Computing, 22*, 127-136.

Podvinec, M., Schwede, T., Cerutti, L., Kunszt, P., Nyffeler, B., Stockinger, H., et al (2007). The SwissBioGrid Project: Objectives, Preliminary Results and Lessons Learned. Proceedings of the *2nd IEEE International Conference on e-Science and Grid Computing*, Amsterdam, Netherlands

Scardaci, D. & Scuderi, G. (2007). A Secure Storage Service for the gLite Middleware. In the *proceedings of third International Symposium on Information Assurance and Security* (pp. 261-266), Manchester, UK. Washington, DC: IEEE Computer Society..

Scifo, S. (2007). GSAF - Grid Storage Access Framework. In the *proceeding of 16th IEEE International Workshops on Enabling Technologies: Infrastructure for Collaborative Enterprises* (pp.296-297), Evry, France, June 2007. Washington, DC: IEEE Computer Society.

Signorini, M., Paulesu, E., Friston, K., Perani, D., Colleluori, A., Lucignani, G., et al (1999). Rapid assessment of regional cerebral metabolic abnormalities in single subjects with quantitative and nonquantitative [18F] FDG PET: A clinical validation of statistical parametric mapping. *Neuroimage*, *9*(1), 63-80.

Tohsato, Y., Kosaka, T., Date, S., Shimojo, S. & Matsuda, H. (2005). Heterogeneous Database Federation using Grid Technology for Drug Discovery Process. In *Grid Computing in Life Science* (pp.43-52). New York: Springer.

Umetsu, R., Ohki, S., Fukuzaki, A., Konagaya, A., Shinbara, D., Saito, M., Watanabe, K., Kitagawa, T., & Hoshino, T. (2006). An Architectural Design of Open Genome Services. In *Grid Computing in Life Science* (pp. 87-98). Singapore: World Scientific.

KEY TERMS AND DEFINITIONS

EnginFrame: Grid Portal providing an efficient infrastructure to put Grid-empowered applications on your corporate Intranet/Internet.

Grid Computing: A style of computing that dynamically pools IT resources together for use based on resource need. It allows organizations to provision and scale resources as needs arise, thereby preventing the underutilization of resources (computers, networks, data archives, instruments).

Grid Applications: It shares and reuses application code but uses software technologies like service oriented architectures that facilitate sharing business logic among multiple applications.

Grid Security Infastructure (GSI): The security infrastructure of a gLite grid middleware

Grid Storage Access Framework (GSAF): Data Access Layer for grid middleware. It is interface that allows business components to manage data stored on the Data Management System and presentation objects to search and retrieve data from it.

Positron emission tomography (PET): A nuclear medicine medical imaging technique which produces a three-dimensional image or map of functional processes in the body. The system detects pairs of gamma rays emitted indirectly by a positron-emitting radioisotope, which is introduced into the body on a metabolically active molecule; images of metabolic activity in space are then reconstructed by computer analysis.

Secure Storage Service (SSS): Secure Storage is a grid service that allows users to manage confidential information/data in a Grid environment solving the Insider Abuse problem.

Single Photon Emission Computed Tomography (SPECT): A nuclear medicine tomographic imaging technique using gamma rays. It is very similar to conventional nuclear medicine planar imaging using a gamma camera. However, it is able to provide true 3D information.

Statistical Parametric Mapping (SPM): An academic software toolkit for the analysis of functional imaging data, for users familiar with the underlying statistical, mathematical and image processing concepts. It examines differences in brain activity recorded during functional neuroimaging experiments, using brain imaging technologies such as or PET.

Workflow: A reliably repeatable pattern of activity enabled by a systematic organization of resources, defined roles and mass, energy and information flows, into a work process that can be documented and learned.

Chapter XXXII
gLibrary/DRI:
A Grid-Based Platform to Host Multiple Repositories for Digital Content

Roberto Barbera
University of Catania, Italy & INFN-Catania, Italy

Antonio Calanducci
INFN-Catania, Italy

Juan Manuel González Martín
maat Gknowledge, Spain

Francisco Prieto Castrillo
CETA-CIEMAT, Spain

Raúl Ramos Pollán
CETA-CIEMAT, Spain

Manuel Rubio del Solar
CETA-CIEMAT, Spain

Dorin Tcaci
maat Gknowledge, Spain

ABSTRACT

This chapter presents the gLibrary/DRI (Digital Repositories Infrastructure) platform. The main goal of the platform is to reduce the cost in terms of time and effort that a repository provider spends to get its repository deployed. This is achieved by providing a common infrastructure and a set of mechanisms (APIs and specifications) that repository providers use to define the data model, the access to the content (by navigation trees and filters) and the storage model. It also provides an algorithm launching mechanism. gLibrary/DRI offers a generic way to provide all this functionality; however, providers can add specific behaviours to the default functions for their repositories. The architecture is Grid based. Two use cases are also presented: A mammograms repository example that provides clinicians with a tool that eases diagnostics process and an algorithmic repository based on the Poincare Surface Section.

INTRODUCTION

This chapter describes the gLibrary/Digital Repositories Infrastructure (gLibrary/DRI) developed jointly by Istituto Nazionale di Fisica Nucleare (INFN) at Catania (Italy) and Centro Extremeño de Tecnologías Avanzadas, Centro de Investigaciones Energéticas, Medioambientales y Tecnológicas (CETA-CIEMAT) at Trujillo (Spain). Repositories are digital stores that manage data and metadata providing their access to users. In this sense, gLibrary (gLibrary, 2007) was conceived as a Grid implementation of a digital repository that takes advantage of the Grid features (VO authentication, file catalogues, metadata services, etc) offering an easy-to-use service and a powerful system to handle digital assets. The gLibrary/DRI infrastructure starts from this concept and extends the platform by offering a general multi-repository environment designed to support and manage several repositories. A gLibrary/DRI repository is made of large digital content (as image files, video, etc) and metadata associated with it (annotations, descriptions, etc.) The main goal of gLibrary/DRI is to reduce the cost in terms of time and effort that a repository provider spends in order to get its repository fully deployed.

In this sense, it is possible to distinguish between repository providers and repository users. A repository provider represents his community by defining the nature of the repository that his community will use. A repository user is a member of such community who interacts with the repository (to query, access, update and modify its content, as well as launch algorithms). gLibrary/DRI offers a platform that allows repository providers to (1) define the data structure (the data model) hold by their repository and (2) provide, if needed, specific viewers for their particular type of data. It also allows repository users to manage their data according to the specifications and tools defined by their repository providers along with launching algorithms that act on the repository data.

gLibrary/DRI defines four APIs so that repository providers can describe the characteristics of their repository. First, the data model API specifies how repository providers describe their data. Second, the storage API defines the rules for describing their data to be stored. For instance, a repository for digitalized manuscripts might define their data model to be composed of a set of images representing the digitalization of the pages of a manuscript, and then a series of metadata describing the historical context of the manuscript (year of creation, authors, historical description, etc). Another example, a repository of mammograms, could contain the images representing the mammograms, and metadata describing clinician diagnoses on them. Once this issue is achieved, the repository provider can determine that images will be stored in the Grid storage elements (profiting from replication of large data chunks, etc.) and then eventually decide that the metadata will be stored in a federated database. Third, the user interface API allows repository providers to describe; (1) what tools are needed to view their data (a specific viewer that shows pages of a manuscript on a book-like fashion, or mammograms belonging to a patient in the same physical application window for visual comparison) and, (2) what are the navigational structures which allow users to navigate through the repository, for instance, manuscripts will be browsed through two trees: one by year and another one by content subject. Finally, an algorithm launching API allows repository providers to define which processes will be available for their users to launch selected repository content. For instance, those processes could be able to select a set of manuscript pages and launch an image improvement algorithm on each page, followed by a process to extract the text content of the pages.

In summary, gLibrary/DRI is a multi-repository platform encompassing the most important Grid features such as VO authentication, data federation and replication, high computing power, etc. drastically reducing the effort required by a repository provider to deploy its digital content.

Two repository examples are provided as use case. The first one is a mammograms repository. It is fully compliant with the gLibrary/DRI specification. It is composed of both a repository and a viewer application that represents the patient data. The second one is an experimental plan that describes how gLibrary/DRI can be used not only to host conventional repositories, but also repositories plenty focused on launching a scientific production. This example is based on the Poincaré Surface Section analysis in a given phase space.

BACKGROUND

gLibrary/DRI was born to address the need to make Grid resources easily available to potential users within large scientific communities. Grid middleware currently offers low level tools for federative job submission and sharing storage around the notion of a Virtual Organization (VO). A VO represents a certain community (for instance, the departments of hospitals devoted to breast cancer), regulating what users have access to the resources devoted to the VO. On the other side, VO representatives negotiate with different Grid resource centers how many resources (computing power and storage space) are guaranteed for their VO users. Both in terms of resource availability and user community a VO is a dynamic organization, opened at any time to the possibility of accessing to more Grid sites by providing additional resources to the VO, and by holding more users launching the VO committed resources.

Grid advantages seem appropriate to the notion of digital repositories, as this environment provides federated resource sharing and digital certificate based security contexts for regulating access for VO members. Thus, our main effort is to construct a platform for communities who can easily use Grid for hosting their digital repositories. Then, repository providers must only provide; (1) a description of their data model and, (2) viewer tools to display their repository data. Relying on our platform, repository providers will manage the GRID specific low level mechanisms in a straightforward manner.

Existing Grid based repositories are mostly very specific applications (CERN LHC, health eChild, etc.) and provide little room for reusability of code amongst other problems. Furthermore, Grid middleware complexities have hampered the possibility of organizations to fully engage into efforts to provide more general solutions. Being both INFN and CETA-CIEMAT institutions devoted to promote and enable the use of Grid technologies, to develop such kind of platforms is considered a natural commitment for us.

An interesting initiative within the trend followed by gLibrary/DRI is DSpace (DSpace, 2003). It is a framework conceived to capture any piece of digital data or work and make it distributed over the web. This distribution is performed through an indexing mechanism that eases users to search and share the data items. Its main idea is to publish any research results, and make them available to the worldwide audience. Although there are no restrictions concerning the kind of data managed by gLibrary/DRI at the present moment, this platform lacks the indexing facilities present in DSpace (gLibrary/DRI features strong data-search mechanisms but they are conceived to locate data into the same repository, not at worldwide range, and Google compatible, as DSpace does). On the other hand, DSpace presents a fixed architecture, both for the User Interface part and the Data Persistance Layer. The main advantage which gLibrary/DRI introduces is that repository providers develop their own UI modules and Storage modules and deploy them into the platform, in order to get them working. The only requirement is that these modules have to follow the interfaces provided by gLibrary/DRI.

Cheshire 3 (Cheshire3, 2005) is conceived as a framework with two main objectives. The first one is to offer large scale data persistence mechanisms as gLibrary/DRI or DSpace does. The other one is to base the data mining by semantic mechanisms, processing natural language, so it offers ontological features, document rendering, text retrieval tools. This is the main feature this system presents over gLibrary/DRI.

TMAinspect (TMAinspect, 2008) is not conceived as a framework to store organized data as gLibrary/DRI does, but it is a web-based tool that allows researchers to analyze several tissue samples in a parallel way. This work is addressed in this section because it shares relevant issues with gLibrary/DRI. First of all, the usage of Grid for solving high-throughput problems. Second, the Grid is used as a resource for sharing data and to solve intensive storage and computing problems. Third, it provides a simplified usage of Grid resources. The use case that TMAinspect presents, offering clinicians a collaborative way to work and a powerful problem-solving tool over a Grid environment, fits perfectly with the objectives of gLibrary/DRI.

GridSphere (GridSphere, 2003) shares some concepts with the gLibrary/DRI working mode. Gridsphere is a framework that provides developers quick development for portlets web applications. They are run and managed by the Gridsphere container. The parallel concept to the portlet in gLibrary/DRI is the repository, which is compound of the data model and the modules (UI Module, Storage Module and Job Module) that take advantage of a gLite-based (gLite, 2006) Grid infrastructure.

DIGITAL REPOSITORIES

A digital repository can be defined as a set of annotated digitalized data offered to users in a structured manner. Digitalized data and its annotations can vary greatly both in complexity and nature (such as sets of mammograms for the same patient, session and studies, digitalized ancient manuscripts annotated with historical contexts, etc.) Nevertheless, most digital repositories (of any kind of content) have the following common features:

1. There is a base informational unit of digitalized data: a mammogram, a page of an ancient manuscript, a 3D model, etc.
2. Units of digitalized data are created by domain experts. This task can be achieved either by annotating directly the digitalized data or by creating metadata describing them. For instance, diagnostics are created for mammograms by marking features on them and by describing possible explanations for these features. Additionally, interpretations from manuscripts can be performed by making transcriptions of text, classifying them and setting its historical context, etc.
3. Repository specific algorithms can process the units of digitalized data: generic image enhancement processes, micro calcification detection for mammograms, text recognition for manuscripts, etc.
4. Users can browse through the repositories which can be accessed by them.
5. Users can search the repositories through the annotations (metadata) and possibly through the search algorithms on the digitalized data.
6. Digitalized data and annotations are stored and managed in a federated manner, so that different institutions can contribute to the same repository with their Digital Content preserving the ownership of their contribution and keeping privacy of sensible selected data. In the case of mammograms, a hospital can contribute to the federation of mammograms even when the patient information is intended to remain private. By declaring the corresponding mammograms as anonymous, the hospital ensures that this information will be inaccessible to other members of the federation.

GRID FACILITIES FOR DIGITAL REPOSITORIES

A Grid infrastructure offers interesting features for the deployment, hosting and management of digital repositories. In particular, it allows data federation facilities to both share information and collaborate with different centres related to the repository content. On the other hand, it provides large storage capacity for storing the digital content (an image, a video or another multimedia content supposed to require large storage capacity), high computing power, strong authentication, etc.

Therefore, a Grid-based repository supports federation of sites (e.g. different hospitals in the case of a Mammograms repository, libraries and museums for an Ancient Manuscripts repository, etc.). Each site would hold Grid storage space for large files along with a database for the repository metadata. In particular, Grid technology is used to support the following features:

- Each site owns its data and makes a subset of them available to the federation including related annotations (e.g. a diagnosis for medical repository) keeping private information unshared when necessary. (In medical cases, this is specially interesting, because patient identification has to be anonymous).
- Grid security contexts provide the mechanisms to ensure the confidentiality of private data, making them available only to the appropriate access scopes.
- The sites view the federation data space as a single entity, making no functional distinction on the geographical location of the data.
- User benefits from distributed computer capacity to build and run a defined list of algorithms related to the repository itself (e.g. in case of an Ancient Manuscripts repository, an algorithm could be an optical text recognition over the image).
- Users are able to perform collaborative work over the same piece of digital content, no matter where they are.

The above features and facilities are possible due to the following essential components of the Grid:

- The Virtual Organization (VO) system provides an appropriate security context. Through it, the user can access and control different parts of the repository effectively. This is possible because multiple roles can be defined in a VO.
- Federation and distribution of data is given by the distributed nature of the institutions participating in the VOs.
- The data composed by the repositories persist into the Storage Elements (S.E.) and the federated database. Each site will set up these storage facilities.
- The digitalized data is the source of the algorithms that will be run on the Grid Computing Elements (CE)

The unit of data that builds up a repository is regarded as a "node". A node contains both the metadata (or annotations) and the digital content. In the case of the Mammograms repository example, the metadata could be the patient name, age, etc. and its diagnoses. The digital content for this case would be the mammogram image. Since the metadata has to be stored in a structured way and its relationships have to be maintained, a federated database is used for storing it. As stated above, the digital content is supposed to require high storage capacities, so this part of the node is stored on the SE.

gLIBRARY OVERVIEW

The gLibrary project was the pioneer development in hosting a repository on a Grid-based environment. It was created by the Grid INFN Laboratory for Dissemination Activities (GILDA, 2005) team with the purpose of offering a secure, robust and friendly system to handle digital content and metadata spread throughout the Grid.

gLibrary offers an intuitive interface for browsing the lists of entries (each one referring to a node), and an advanced navigation system to locate any resource. It is based on a hierarchical organization, a set of filters and customizable search types. Another important feature which deserves to be highlighted is a powerful system to describe the file contents.

Thanks to its flexibility, gLibrary can be used for several purposes, e.g. a personal multimedia library to host users' personal data or a community document repository to share presentations or documents (which can be grouped itself into projects).

gLibrary is built on top of the Grid services of the EGEE gLite Middleware: DPM –Disk pool manager- Storage Elements (DPM Storage Element, 2007), AMGA Metadata Catalogue (AMGA, 2006) and VOMS –Virtual Organization Membership Service- authorization (VOMS, 2003) The web interface is the result of various mingled technologies: PHP on the server side, PHP-APIs to interact with AMGA and the SE (Storage Element) and AJAX (AJAX, 2007) and JavaScript technologies for the client side.

The security is based on X509 certificates: the user can access to the system with a .p12 or a .pfx certificate. A Java applet, which acts as a login gateway, creates a Grid proxy certificate on the user's desktop PC. This proxy is transferred over a secure SSL channel to the gLibrary web server. Once on the server, the user's VOMS extensions are added to this proxy in order to allow authentication. This VOMS-extended proxy is also used to directly interact with the rest of the Grid elements (such as LFC -Large File Catalog-, AMGA).

The download and upload processes of the repository files are also done by a Java applet. This applet performs the communication with the Storage Element containing the user requested files and then it transfers them to it. This transfer method, based on GridFTP (GridFTP, 2005) API calls, does not cache files into the web server, but transfers them directly. The download applet receives as input the SURL -Site URL- of the file to get, as well as a copy of the user proxy. With these two parameters, GridFTP perform the secured request and the file is brought. The upload applet works on a similar way; a SURL is generated at the moment of sending the file to the SE.

THE GLIBRARY/DRI PLATFORM

As stated in the introduction, gLibrary/DRI is a multi-repository environment which stems from the original gLibrary basis (i.e. Grid-based repository, annotated content, high storage capacities, structured data, search facilites). It extends this basis in the sense of providing a platform with functions aimed to reducing the cost of implementation and hosting Grid federated repositories for any community or domain.

gLibrary/DRI Main Features

The above functions can be classified in five different groups each one offering a main gLibrary/DRI feature:

Data layer abstraction. Repository providers describe the structure of the repository contents; that is, the way of structuring the data model. This description has to be done by following the DRI Data Model specification, which establishes rules for creating XML files with the definition of the node structure. The DRI Data Model specification supports the following features:

- Arbitrary relational data models. This will allow to organize the model into different relational entities.

- Possibility to specify what parts of the data will be considered digital content and, hence, stored on the SE, and what part will be treated as meta data and stored on the federated database.
- Possibility to mark fields that will be stored on encrypted way.

Once the data model is defined, the repository provider will choose the storage technology to use. gLibrary/DRI publishes an API the provider has to implement in case he needs to use its proper implementation system. On the other hand, gLibrary/DRI has its own storage implementation that can be used simply by calling the gLibrary/DRI storage API. This will be explained with more detail in next sections.

Storage and GUI facilities through DRI API specification. The Storage DRI API Specification provides method definitions for loading and persisting nodes. Through this API we isolate data management from its storage technology. (However we provide an implementation of this API using Grid SE and AMGA technologies.) These methods are transparent to the node complexity and content, and also to the storage system chosen for storing the data.

The GUI Navigation functions are used for providing to the user a quick and effective way of finding any node of the data model into the repository. The navigation system is based on categories trees and a set of filters that reduce the nodes search to the user. The different branches of the tree represent a first level of filtering. Each branch has an associated database query. When the user clicks over a tree branch, its associated query is executed and a first set of nodes that fits with the query is shown in the nodes list.

The filters provide a second level of filtering. They are composed of a list of fields of the data model and their allowed values. When a field is selected its different values are shown and the user can select one of these values. Finally all nodes containing the selected value will be presented on the nodes list.

gLibrary/DRI provides API interfaces for implementing both the Storage and GUI functions, and a XML syntax for defining the data model. These specifications allow decoupling the DRI platform from the repository implementation. Repository providers use the specifications for describing the data model for its repository, as well as the storage policies, the navigation environment, the application viewers, etc.

Default functionality implementation. In order to offer a worthy service in the objective to help the provider to get its repository deployed, the gLibrary/DRI platform offers default implementations for the API modules, so that new repositories can be quickly set up. For example, in the case of the Storage functions used through the Storage API, default implementations are provided for them. Therefore, the provider can use the Load() and Insert() functions by inheriting them from the developed modules. These functions will manage the nodes through the storage layer freeing the provider to implement concrete loading and saving functions. There are also defined default functionality for the GUI navigation functions. More details about the default implementation can be found in the description of the gLibrary/DRI default implementation section.

Algorithms launching system. The platform integrates an API specification for defining jobs with algorithms running upon repository data, and a UI for users to launch and manage those jobs. For each repository, there will be defined a set of algorithms applicable to any piece of data of that repository. For example, in the case of a Mammograms repository, it might be of interest the existence of a set of image treatment algorithms for enhancing or modifying them in order to ease a diagnosis or detect a possible abnormality.

gLibrary/DRI specification contains a Job Definition class that users will extend to implement every algorithm they want to incorporate. These algorithms can be launched over two different drivers; one that sends the Job object to a Grid CE, and other that runs the Jobs locally in the user's machine. Both driver's functionality is also offered by gLibrary/DRI. There will be a listener for reporting continuously the different Job status.

The algorithms are to be applied either to the whole totality of a node, or to any of the entities that structures it. When an algorithm finishes, it will return the data encapsulated into another node in order to ease its posterior treatment. The developer of each algorithm defines what actions will be possible to perform with the returned results, since that actions are fully dependant to the algorithm's nature and objective.

Multiple repository hosting. The gLibrary/DRI platform has the capability to host multiple repositories of different kinds and objectives, such as collections of medical images for working collaboratively on a diagnosis process, catalogues of museum pieces, etc.

Each repository must provide a Java implementation for the different DRI API modules so that the repository provider can describe his data model, create navigation trees, and specify how data is stored and how jobs run on these data. Alternatively, he can use the Java inheritance for the default implementations of every API component mentioned above. This default implementation will suffice for most cases. In both cases (developing a specific implementation for the different modules or inheriting the DRI functionality), each repository implementation is packaged into a set of Java JAR files that are dropped in the proper place within the gLibrary/DRI workspace. The gLibrary/DRI engine will detect all existing JAR files and effectively will host and support the different repositories.

gLibrary/DRI API

The repository provider either develops its own Storage Module and GUI Module implementations or uses those that gLibrary/DRI provides by default. The case for Jobs Module is similar since the provider can use the provided Local and gLite CE drivers or, alternatively, develop a new one in case he wishes to use another technology -such as Condor (Condor, 2001), PBS (PBS, 1995) –Portable Batch System- or different models such as BOINC (BOINC, 2004)–Berkeley Open Infrastructure for Network Computing-. The provider sets up an algorithm list and then chooses the proper running driver for each one.

In the next subsections the interface API for each module is shown, explaining the purpose for each function. The detailed description of the default implementation gLibrary/DRI provided for each module will be addressed in subsequent sections.

Data Model API

Any data item of a repository is represented by a node, following the ideas addressed in JSR170 (JSR170, 2005). The platform supports any data structure that can be represented by a relational model. An abstract Java class, DRIGenericNode, represents any piece of data in a repository. All classes representing data in a repository must inherit from this class. The section referred to the mammograms repository shows an example of data model with a reasonable complexity.

Storage Module API

Typically, parts of the data model that are represented by large chunks of data (such as images or digitalization) are to be stored in Grid SEs, and regular metadata (such as annotations, patient information, or diagnoses, etc) are to be stored in a federated database. In both cases, they are distributed across the sites forming the virtual organization supporting each repository. Thus, storage module functions can be classified in those that interact with the federated database and those that manage the digital content files (large pieces of data such as images or digitalization's that require high storage capacity). For the federated database interaction, an interface named DRIStorageInterface has been created. For the digital content files, a *manager* class interacting with the SE has been developed. No interface has been created for this manager, but its functionality is intended to be used for all implementations. This manager is introduced in the default functionality section.

The DRIStorageInterface implementation works with instances of DRIGenericNode. This class must be inherited by any repository at the moment of defining its data model, so that each entity that structures the data model is an instance of this class. The DRIGenericNode class is designed as a Java bean (Bean) in order to allow introspection (Introspection, 2004) functions that are to be used in the default implementation. The DRIGenericNode also implements the Java Serializable class to allow any instance to be serialized in order to perform network communications. (E.g. A viewer applet that needs to represent the information by certain node will receive it by a servlet in serialized way). Figure 1 shows the abstract class for the DRIGenericNode.

The storage module functions are used to load and save the repository nodes. Figure 2 shows the interface that publishes this module functionality.

The **Load** method returns an instance of a class inherited by DRIGenericNode. The node to be loaded is identified by its Id. This is an example of a Load call used in the Mammograms repository:

```
DRIStorageManager driMgr= new DRIGenericStoragc();
```

(DRIGenericStorage is a class that implements DRIStorageInterface)

```
MammogramsRepPatientNode ptNode = driMgr.Load(ptId);
```

The **Insert** method is used to persist the node instances on the underlying storage technology. It first checks if the node to insert exists on the data base. In that case, it just updates the information. If the node does not exist it will create a new one.

Remove method deletes from the storage layer all the information of the node referred by the Id and all entities that depends of it.

Figure 1. Class for DRIGenericNode

```
public abstract class DRIGenericNode implements DRINodeInterface, java.io.Serializable
{
 int id;
 String NodeType;
 public void setId (int _id)
    {
            id = _id;
    }
public int getId ()
    {
        return id;
    }
 public void setNodeType (String _NodeType)
    {
       NodeType = _NodeType;
    }

 public String getNodeType ()
    {
        return NodeType;
    }
}
```

Figure 2. Storage module interface

```
Public interface DRIStorageInterface
{
public DRIGenericNode Load(String id);
public int Insert(DRIGenericNode node);
public int Remove(String id);
}
```

UI Module API

This module is used to allow the repository provider to set up the information search facilities in the user interface, i.e. those parts that the final user will use in order to find a concrete piece of data. Figure 3 shows the UI Module Interface.

The **getRepositoryTrees** method obtains from the federated database the definitions for the different navigation trees included in the repository. It returns a vector of a "Tree" objects, which contains information about the name and the description of the tree. Once the trees have been shown on screen and the user selects one of them, its branches are retrieved, through the **getTreeHierarchy** method. This function returns a TreeHierarchy object, which is basically a tree entries vector. Each entry corresponds to a branch. A *TreeEntry* object contains information about the branches hierarchy in order to build properly the tree, as well as the query that is run when the branch is selected.

The **getBranchQuery** method obtains the query assigned to a concrete branch. This query will be annexed with those queries returned from the filters. The **getFilterNameInstances** method returns the list of filters defined for a given repository. This information is also stored in the federated database. **getFilterEntries** method obtains the different entries that compose the filter. It works on similar way as the tree branches, except no hierarchy information has to be stored. When a given filter entry is clicked, its query is obtained and the **buildQuery** method is called in order to append the query from the tree branch (and, possibly the query coming from previous filters) to this additional query. By this way, multiple-level filtering is performed. Finally, when a complete query is built by gathering parts coming from the tree branch and the different filters, the results of executing that query are returned through the **getEntriesList** method.

Jobs Module API

The algorithms launching system is based on three premises:

- The input for an algorithm is wrapped into one or more DRIGenericNode instances. This eases the data management and availability. The algorithms output follows the same rule.
- There exist a defined collection of algorithms for each repository. New algorithms for a repository can be added by using the API shown in Figure 4. The **prepare** method

Figure 3. UI module interface

```
Public interface DRIUIInterface
{
 public Vector<Tree> getRepositoryTrees(String repositoryName);
 public TreeHierarchy getTreeHierarchy(String treeName);
 public String getBranchQuery(String queryID);
 public Vector getFilterNameInstances();
 public Vector<FilterEntry> getFilterEntries(String filterName);
 public String buildQuery(String filters,String query);
 public Vector getEntriesList(String fields, String query);
}
```

adapts the input data coming into a DRI-GenericNode instance to the specific needs for the algorithm. The **startExec** method involves two main steps. First, the exec method (to be explained below) is called. Then, the output data is adapted from the exec method to a DRIGenericNode. The **exec** method encapsulates the concrete implementation for the algorithm, and returns the result to the previous startExec method. The concrete implementation for the algorithm can be either Java code directly developed or a external call to an imported library.

- When an algorithm is to be launched, an instance of one specific algorithm class (from the algorithms list) is created. In order to run that instance, a driver has to be chosen. A driver allows to define the basic execution details such as the underlying technology or the notification status system. The interface for the algorithms driver is shown on Figure 5. The constructor method accepts the listener object (used for the notifications) and the algorithms instance. The **runJob** method loads from the storage system the data for the referred node and activates the notification system. Finally, it starts a thread that creates the object that manages the chosen algorithm.

Description of the gLibrary/DRI Default Implementation

Repository providers implement the four APIs described above in order to provide the specifics functions of his repository. Although this provides a clean abstraction and enables us to make a modular architecture for the platform, implementing these APIs implies typically a lot of work. Therefore we provide default implementations of the APIs ready to be used by repository providers. For instance, the default implementations for the Data Model and Storage APIs allows repository providers to simply describe their data model in XML files and manages automatically how instances of this data model are stored and loaded from/to Grid SEs and an AMGA database. This way, the repository deployment process is almost immediate, shielding the provider from the following issues:

- Development of the storage mechanisms.
- Implementation of the navigation GUI system.

Figure 4. Algorithms interface

```
Public interface DRIAlgorithmInterface
{
 public void prepare(DRIGenericNode inputData);
 public void DRIGenericNode startExec();
 public int exec();
}
```

Figure 5. Algorithms interface

```
public interface DRIJobManagerInterface
{
 public DRIGridJobManager(DRIJobListener l, DRIJobNode driJob);
 public void runJob(int NodeRefId);
 public run();
}
```

- Implementation of algorithm execution drivers.

The following subsections describe the different default implementations provided by gLibrary/DRI.

Data Model Default Implementation

Through this default implementation, repository providers can define their repository as a relational model described in a set of XML files. Each XML file describes one entity (or table in the database) and contains:

- The fields list. There exists three field types:
  - **Normal fields.** Used to define regular fields of the metadata (e.g., Patient name, manuscript origin, parameter value, etc) and can be of various data types (String, integer, date, etc.)
  - **Relationship fields.** Used to describe relations between entities. For instance, in the Mammograms repository example, a patient is related to a set of studies. When the gLibrary/DRI engine reads a relationship field, it automatically searches the XML file describing such entity.
  - **LFN fields.** They contain a reference to a Digital Content file in the SE.
- The primary key in order for the node to be referenced and the foreign key in case the node belongs to a hierarchical relationship.

Figure 6 shows the specification that must be followed when defining a data model entity.

Storage Module Default Implementation

The default implementation for the storage module performs the persistence (and retrieval) of both the metadata and the large digital content fields. Regular fields are stored in an AMGA metadata catalogue, which is part of the EGEE gLite Grid middleware, and fields marked as LFN in the XML are stored in a Grid storage element. As any other storage module does, our default implementation follows the interface defined in Figure 2 and behaves in the following way. The insert method updates existing data or inserts new data, depending if the node Id already exists in the database. In both cases our implementation is similar, requiring as input a repository node as a Java Bean object instantiated from a class derived form the DRIGenericNode class. When invoked, it uses Java introspection on the Java Bean to locate the XML file of the repository and reads out how fields are to be stored in AMGA. Once this is discovered, it actually stores the repository node in AMGA by issuing the appropriate AMGA commands. Also, if a file is defined as LFN in the XML, our default implementation stores that specific field in the Grid SE. Both the XML and the Java Bean are inspected recursively so that all related instances are correctly stored.

Figure 6. XML datamodel specification

```
<TableName name=string primaryIdAttr=string foreignAttr=string>
 ............
 <attr name=string>
     <dbAttrName>string</dbAttrName>
     <type>int|string|date|float|LFN</type>
     <dbAttrType>int|varchar|datetime|float</dbAttrType>
 </attr>
 ............
</TableName>
```

The load operation acts likewise but reverting the order of the steps. From a number that identifies the node to load, the function also uses introspection to locate the XML defining the repository node and build a new Java Bean representing it by querying AMGA. Naturally, this method is recursive to retrieve all related instances following the repository relational model.

Note that repository providers simply need to mark the desired piece of data as of type LFN, and the storage API default implementation takes care of interacting with the Grid SEs on a transparent way. When loading a repository node, fields marked as LFN are not retrieved until the application using the module explicitly requests it through a utility class named DRIGridFileManager. Typically, viewers will use this class when the user requests to view the digital content, and it will communicate directly with the storage element. It retrieves the Grid proxy the user created in the server and use it to secure the GridFTP (GridFTP) communication. The methods are based on the GridFTP Globus API (GridFTPAPI, 2005).

DRIGridFileManager reports continuously the transfer status that viewer's developers can use to show users operation progress information.

UI Module Default Implementation

The default implementation for the UI module assumes that navigation trees and filters are defined in AMGA. Under the /config branch, each repository must define a new AMGA collection which, in turn, contains two subcollections, a first one for storing filters and other one for tree descriptions. Figure 7 depicts the structure stored for managing the trees and filters information. The different tree entries correspond to the tree branches that are shown on screen and contain information about the node type to be filtered (it can be any type into the hierarchy) and the associated query. The filter collection contains information about the list of fields that are shown on each filter.

The default implementation of the navigation system retrieves all tree and filter information from AMGA and creates the appropriate Java objects representing them. Then the platform builds the web interfaces with the trees and filter for the user to interact with them.

Default Implementation for Algorithms Execution

In order to run one of the algorithms defined for a given repository, the platform creates an

Figure 7. AMGA database structure for trees and filters

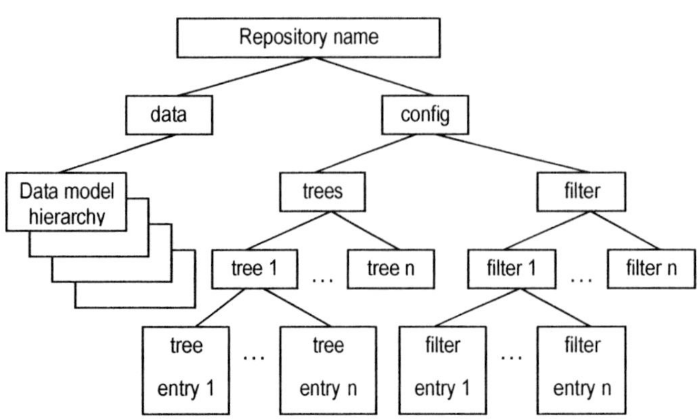

instance of the Job class defining the algorithm and one instance of the driver to use. gLibrary/DRI provides default implementation for two different execution drivers:

- **Local driver.** Runs the algorithm class on the user's machine. This driver is intended for lightweight algorithms that do not require many resources (CPU) for running.
- **gLite CE driver**. (Grid driver). This driver allows running the algorithm as Grid jobs sent to a CE. When a job and the Grid driver have been chosen, the input data for the algorithm is prepared. The input data consists of the job parameters, the algorithm class and the repository node upon which the algorithm will act. When this process finishes the driver creates a Job wrapper which is sent to the Grid CE together with the input data. When the job arrives to a Grid computing, the Job wrapper deserializes the file and runs the algorithm class. The output from the algorithm is also serialized and returned by the wrapper as the result of the Grid job previously submitted.

Architectural Example

As was said in the features section, the gLibrary/DRI platform can host multiple repositories, each one implementing their own modules for node storage, navigation system and job-drivers, or using the default ones as explained above. Figure 8 shows the gLibrary/DRI platform hosting two repositories, one for mammograms, which uses the default implementation for navigation and storage modules, and one for manuscripts that uses its own implementation for these two modules. Both repositories use their specific custom viewer.

Technological Issues

The user interface (the web portal and the login system) is inherited from the original gLibrary implementation by INFN. Since the DRI APIs have been developed in Java we use a PHP-Java bridge (PhpJavaBridge, 2006) to enable PHP code call Java libraries. Figure 9 shows an example of Java calls embedded into PHP code to obtain a list of all repositories that are currently hosted in the platform shown in Figure 9.

Figure 8. Multiple repository hosting

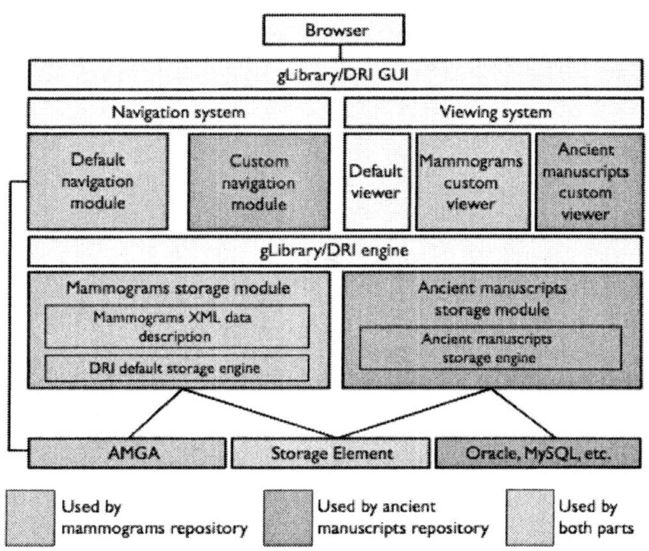

Figure 9. Invoking Java methods into PHP code

```
$obj = new Java("es.cetaciemat.dri.DRIDefaultBootstrap");
$repositoriesList = $obj->getRepositories();
for ($i=0;$i<$repositoriesList->size();$i++)
{
echo "Repository $i name: ".java _ values(($vectorReps[$i]->getName());
}
```

The `DRIDefaultBootStrap` class is instanced as a PHP object (in the `$obj` element). All methods of this class are available from PHP, as we can see on lines 2, 3 and 5.

The web portal is implemented as Java servlets which are later invoked by clients (such as a viewer). The save() operation, requested from the client side, just performs a call to the server-side insert() method. It takes the node to persist and serializes it. Then sends it to a servlet that deserializes the node and invokes the insert() method belonging to the DRI Storage API (see Figure 3) which actually persists the node to the underlying storage technology of the repository. The load() operation acts on a similar way. Digital content files management are performed directly from the client applet, by using the Globus GridFTP API.

REPOSITORY APPLICATION EXAMPLES

This section presents two examples representing two different repositories hosted in the gLibrary/DRI platform. The first one is a Mammograms repository. It contains mamograms and information related to them (patient information, diagnoses, second opinion, etc.). It also uses this information for helping clinicians to make diagnosis in collaborative way. The second use case is an example that demonstrates how the gLibrary/DRI platform can be used to host a Algorithms-oriented repository, so that users can manage the creation of scientific productions through massive calculations. The example we show is based on nonlinear dynamic analysis on the phase space.

Mammograms Repository Example

This repository is used as a tool that helps clinicians to keep track of patient's breast examinations and diagnosis. A VO represents a federation of hospitals sharing data. Since many sites (i.e., hospitals, research centers, etc.) can be added to the VO, clinicians and researchers from these spread sites can share their data and knowledge in order to make a more rich and effective diagnosis.

Data Model and Storage

The data model of this repository is a hierarchical relational structure composed of 5 nodes. This data model is a subset of the DICOM (DICOM, 1997) medical file format applied to mammograms. Since the DICOM standard contains hundreds of fields spread along dozens of entities, we have simplified the data model for this repository just taking the most relevant fields for a single mammograms repository. Figure 10 shows the data model contents.

Italic fields represent relationships between entities. This way, a patient contains one or more studies, which contains one or more series, which in turn contains one or more images which may contain several annotations. The LFN field for this repository (I.e. the one that refers to the actual mammograms) is set in the Images entity. Images also contain some information needed for represent properly the image. Finally, the Annotations

Figure 10. Mammograms repository data model

Patient	Study	Series	Images	Annotations
PatientName	StudyId	SerieID	ImageID	xPos
PatientID	StudyDate	SerieDate	OriginalFileName	yPos
SecondaryID	StudyTime	SerieTime	PixelSize	type
Gender	PathologyId	PathologyId	DisplayPosition	text
AgeAtMenarche	PathologyName	PathologyName	***LFN***	imageI
AgeAtMenopause	Diagnose	BodyPart	SerieID	
DateOfBirth	InstitutionName	StudyID	*AnnotationsEntity*	
StudiesEntity	PhisicianName	*ImagesEntity*		
	PatientID			
	SeriesEntity			

entity contains information needed for supporting graphical annotations, such as circles or polylines over the mammograms.

UI Navigation Module

The mammograms repository uses the default implementation of the UI module. By using the database structure depicted on figure 7, two navigation trees and three filters have been defined. The first navigation tree classifies patients by their alphabetical name, whereas the second one groups patients by the pathologies diagnosed for them. Figure 11 shows the mammograms repository loaded with these two navigation trees where the Pathologies one is selected. In the list of main nodes (for this repository the main node represents a patient) just the patients fulfilling the condition set up by the selected tree branch and the filters appears. These conditions are "Female patients with the benign pathology, any menarche age and a Menopause age range between 50 and 55". For each patient appearing on screen, there is a link that opens the mammograms viewer for such patient.

Mammograms Viewer

This repository has a custom viewer used for representing patient data. It is based on the Tudor (Tudor, 2008) DICOM Viewer, distributed under the LGPL license and it is compatible with a wide range of DICOM images as well as images in other formats (JPEG, TIFF, etc.) It is based mainly on ImageJ (ImageJ, 2004) and dcm4che (dcm4che, 2006) Java libraries, which are also distributed as open source.

We extended the Tudor Viewer with the following features:

Figure 11. User interface for the mammograms repository

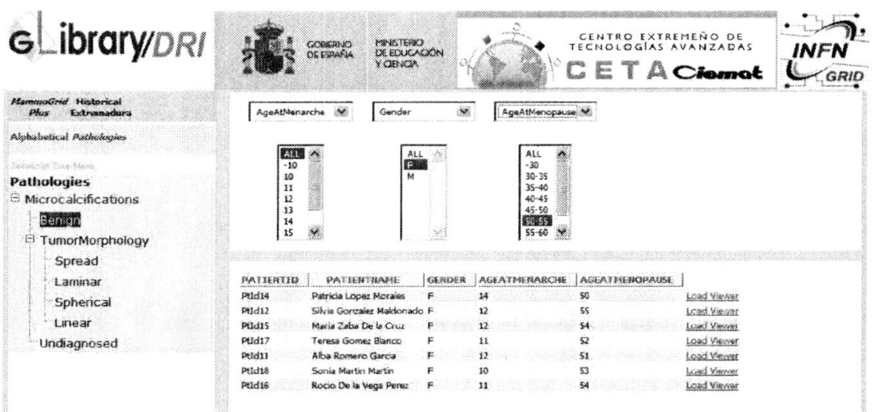

- Compliance to the Mammograms repository data model.
- Patient data can be browsed in a hierarchical tree mode.
- The metadata loaded can be edited/updated/deleted.
- Support for textual and graphic annotations over mammography images to ease the diagnostic. All these changes are saved preserving the original image.
- Multiple related mammogram images can be viewed simultaneously trough configurable screen layouts.
- Data can be serialized and deserialized as well decoding/encoding images from DICOM files is supported, so that it can be sent over network streams.
- The communications to the LFC server are performed directly without passing the Digital Content to the gLibrary/DRI server. This is performed under certificate security way.

Figure 12 shows the viewer representing the information loaded for a patient. It shows a mammogram image that has been retrieved from the SE and a polyline annotation has been drawn over the image. Note that annotations are stored as metadata and do not modify the original image.

Figure 12. Mammograms repository viewer

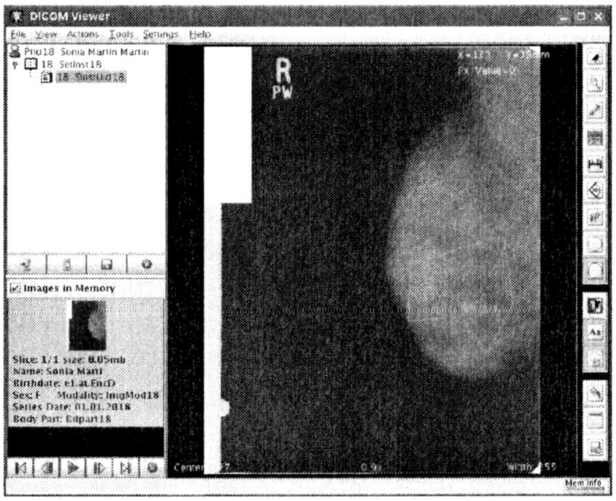

Phase Space, Poincaré Section Example

This example is an experimental use case introduced to demonstrate how a repository focused to perform a scientific production is deployed under gLibrary/DRI.

Problem Statement

Almost every scientific problem can be set in terms of the evolution of some characteristic state. As it is well known, the set of all possible states constitutes the *phase space* (or *state space*). For mechanical systems, the phase space usually consists of all possible values of position and momentum variables.

However, the notion of phase space has been successfully extended from physics to other scientific disciplines such as biology, economy, or the theory of dynamical systems in general.

Under the action of the equations of motion, every point in the phase space is mapped onto another point (also in the phase space). By repeating this process, it is possible to find patterns and structures that are of special interest to researchers (Ott, 1993), (Liberman, 1992).

The goal of a typical nonlinear dynamics analysis is to highlight these structures by an intensive examination of the phase space. The task is performed by an integration of the equations of motion starting from each point of a prescribed region. This region constitutes the initial condition (IC) set. Within this framework, it is clear that larger IC maps lead to a deeper insight into the dynamics.

Depending on the nature of the problem under investigation, the phase space dimension can be large and the type of analysis described becomes soon unfeasible as the number of degrees of freedom grows.

Amongst the phase space analysis techniques, the Poincaré Surface of Section (PSS) has shown to be the standard *by default* method in the everyday calculus, (Ott), (Liberman). Poincaré introduced a dimension reduction technique, namely, the Poincaré Surface of Section, which allows both the visualization and quantification of the dynamics in a subset of the phase space. The PSS is usually implemented as a geometrical condition in the coordinates of the phase space; if the condition is reached, a Poincaré point is obtained. However, other conditions on the variables can also be valid definitions of the PSS.

The time-ordered collection of points obtained by integration of the equations of motion constitutes the orbit (or phase trajectory). In the phase space, every point (and, hence, every trajectory) is independent of the others. This fact has great consequences in relation with the translation of the problem to the GRID infrastructure. In fact, the partition of initial conditions subsets of prescribed size from the IC set is straightforward. This suggest a natural and immediate Grid application, since many independent processes requiring high computing power can be defined. Each process can be identified as a collectionthat can be sent as single job to the GRID.

In a recent study (Ibergrid, 2008) it has been noticed that the Poincaré section technique can benefit of GRID capacities to a high extent.

A typical PSS production process consists in two nested loops; one in the integration of the equations of motion and other for initial conditions. The integration loop is continuously monitored by the PSS condition. If this condition is reached, a Poincaré point is stored. This activity is embedded into another loop in initial conditions continuing until the initial condition (IC) map is exhausted.

It is intended to embed the described analysis into the gLibrary/DRI platform in order to develop a Grid- based repository focused on a scientific production in its most abstract manner.

As has been shown in the case of the mammograms repository, a data model, a storage implementation, the GUI Navigation system, and a repository viewer have to be defined. This repository also has defined a list of algorithms that allow performing the PSS production.

Data Model and Storage System for Phase Space Analysis

The data model for this repository is simpler than the mammograms one, since it is oriented to a massive scientific production, instead of maintaining a large collection of data and metadata. For this reason the data model is composed of one entity. The proposed list of field for this entity can be seen on Figure 13.

Parameters are usually composed by magnitudes lying in two main categories; model parameters and integration arguments. Through the model parameters, the system under study is characterized. On the other hand, integration arguments set up some values needed by the executable (e.g. the integration step size, integration time, accuracy, etc.). Each parameter will be an attribute on the data model that is stored as metadata on the federated database. Both the initial condition set and the input file are stored on the SE.

As stated above, an initial conditions set is a basic element of every phase space analysis. It consists of a set of points in the phase space. Every initial condition along with the parameters (and, optionally, an input file) is used by the executable for an execution.

On the other hand, different IC sets can be used for a calculation. In fact, each IC set con-

Figure 13. PSS repository data model

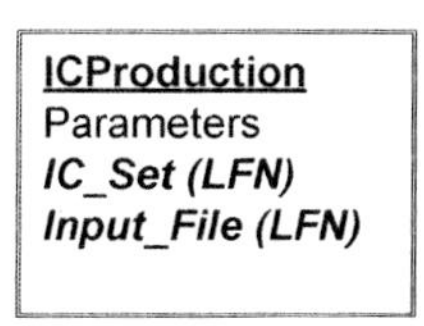

stitutes itself a *strategy* to find out the desired patterns. It is a desirable fact that the user can choose from a repository of IC sets. This allows him to test different strategies in order to match his problem to the best result. For this reason this field appears as a LFN content that the user can refer when designing the production. (I.e. creating the nodes)

Finally, some systems may require the addition of an external file (e.g. the driven force of a pendulum). The input file field is intended to assess this task. It could also be managed as a LFN file so that it can directly accessed whenever the process may require it.

The storage mechanisms are inherited from the gLibrary/DRI default implementation.

GUI Navigation System

The main objective for the navigation system is to highlight a set of nodes (each one representing a PSS computation) in the main nodes list. In a later step, when an algorithm and its driver have been chosen, they are applied to the nodes appearing in the main list. For this reason, the navigation trees and the filters have been configured to allow as many combination of different PSS computations as possible.

Algorithms for the PSS Computations

Three algorithms need to be defined in order to allow a complete operation along the PSS computation. As we said in the previous section, a list of PSS computation nodes is limited through the navigation system and t he available algorithms will run over the selected nodes.

The first algorithm to be run in a logical order is named "Run PSS". Its basic goal is to perform the tasks of the Poincaré section analysis. Each node generates as many jobs as initial conditions of its corresponding IC set. All of these jobs are sent to the proper algorithm driver for their execution. Since the tasks to perform need high computing resources, the "Grid" driver is used.

Next step is to build up the Poincaré section itself which is done by the "Build PSS results" algorithm. When all jobs are finished, their results are gathered to a file, stored into a SE and a node containing the LFN for this file is returned. This way, the user can access the PSS results through the GUI by selecting this algorithm.

Measures in the PSS can be valuable to researchers in some situations. For instance, the Lyapunov exponents (Ott) and the attractor entropy (Liberman) are of special interest. This can be achieved through the "Get measure" algorithm which incorporates some standard techniques for stability quantification.

PSS Repository Viewer

Viewing PSS maps consists of navigating through the set of maps generated by a give IC. Therefore our phase space viewer is simply a method for collecting and plotting all the Poincaré points. Once the PSS is displayed, the user can either save the file to a local systems or keep it as a LFN file in the GRID.

FUTURE TRENDS

The future plans for the gLibrary/DRI project are listed below:

- Extend implementation of the algorithms launching system.
- Development of a generic viewer that accepts any node, shows its data and recognize the LFN-referred digital data for representing it in its proper way (E.g. showing an image, playing a video, show a PDF file, etc). This should avoid to the repository provider to develop a custom viewer in most of the cases.
- Development of an standard for providing the initial data for the GUI navigation system, and a tool that integrates this data into

the default gLibrary/DRI GUI navigation modules.

- To publish all the API implementation through Web Services. This will extend the interoperability for the platform and will separate even more the modules development from the underlying technologies.
- To reach new medical communities for showing and offering the gLibrary/DRI capacities and benefits, as well as extending the usefulness of Grid-based repositories.

It is also intended to align and integrate this platform with other initiatives such as DILIGENT or JSR170 as they consolidate, in order to offer the user community a clean road towards fully using digital repositories into their day to day life.

Future trends in the field of the digital repositories, applied either to medical environments or any other domain play an essential role in the retrieval and search of information. In this regard it deserves to be mentioned the building of a semantic grid, which contains (and analyses) information over the whole repository and offers it through an intelligent/heuristic way as an ontology case.

On the other hand, the gLibrary/DRI project offers an API for developing new repositories and managing data. It also provides a generic viewer which can be customized and transformed to match the needs of the repository provider. This way, any member of the medical community or expert in the domain could develop its own customized repository-based tool covering a concrete set of requirements at a given moment.

Another important feature is the inclusion of advanced streaming capacities for digital stored data, e.g. videos, audio files, large images segmented in multiple parts, etc. This advanced streaming is not only constrained to common PCs or laptops; other kind of devices such as PDAs, mobile phones, etc. could be available. There are also important features for advanced data management, such as OCR functions, complex algorithms (as stated above), or search of information along the digital data (video, images, audio, etc.)

CONCLUSION

Grid infrastructures offer a powerful platform to support and run processes that require high amounts of computational resources, in terms of storage space and computing power. The Grid computing model also offers an appropriate frame for hosting digital repositories due to its high storage capacity, federated resources and strong authentication system.

For this reason INFN and CETA-CIEMAT became involved on the development of the gLibrary/DRI platform, whose main objective is to reduce the effort and cost required by a repository provider to deploy and host its repository.

This is achieved by designing gLibrary/DRI as a modular architecture that exposes APIs for providing data model definitions, storage and GUI navigation functions and algorithms launching systems. It also offers default modules implementing these APIs, ready to use. This way, repository providers can choose either to provide their own implementation of those APIs or use the default modules. In this later case development effort is limited to providing repository specific viewers, and repository providers are simply required to describe their repository data model in XML files.

Two repository examples have been shown. The first one is a fully working mammograms repository that offers a solid structure for storing mammograms and patient data, and a viewer that clinicians can use to make diagnosis parting from this data. The mammograms repository represents a worthy feature for the medical community since it uses the main Grid benefits to offer a collaborative tool, for mammograms federated storage and diagnoses. We extended the TUDOR DICOM open source viewer to include repository specific functionality (such as the capability to annotate mammograms) and integration with the platform.

The second example shows how the gLibrary/DRI platform can also be used for hosting repositories generated by large scientific production over computing resources. In our specific case, the data this repository contains consists of initial conditions for scientific productions of problems formulated through a phase space analysis.

REFERENCES

Abramoff, M., Magalhaes, P., & Ram, S. (2004). Image Processing with ImageJ. *Biophotonics International* (pp 36-42).

Alfieri, R., Cecchini, R., Ciaschini, V., dell' Agnello, L., Frohner, Á,. Gianoly, A., et al. (2003). VOMS, an Authorization System for Virtual Organizations. *1st European Across Grid Conference.*

Allcock, W., Bresnahan, J., Kettimuthu, R., & Link, M. (2005). The Globus Striped GridFTP Framework and Server. (p.54) *Proceedings of the 2005 ACM/IEEE conference on Supercomputing.* Washington, DC: IEEE Computer Society.

Anderson, D. (2006). Boinc: A System for public-resource computing and storage. *Proceedings of the 2006 ACM/IEEE conference on Supercomputing* (a. 126). New York: ACM.

Andronico, G., Ardizzone, V., Barbera, R., Catania, R., Carrieri, A., Falzone, A., et al. (2005). GILDA: the Grid INFN virtual Laboratory for Dissemination Activities. *Proceedings of First International Conference on Testbeds and Research Infrastructures for the Development of NeTworks and COMmunities* (pp. 304-305).

Baker, M., & Lakhoo, R. (2006). Application Reuse Through Portal Frameworks. *Proceedings of the UK e-Science All Hands Conference* (pp. 510-518).

Bidgood, W.D., Horii, S.C., Prior, F.W., & Van Syckle, D.E. (1997). Understanding and Using DICOM, the Data Interchange Standard for Biomedical Imaging. *Journal of the American Medical Informatics Association, 4,* 199-212.

Burke, S., Campana, S., Delgado Peris, A., Donno, F., Mendez Lorenzo, P., Santinelli, R., & Sciaba, A. gLite 3 User Guide. *gLite 3.1 User Guide Manual Series.*

Calanducci, A., Cherubino, C., Ciuffo, L.N., & Scardaci, D. (2007). A Digital Library Management System for the Grid. *Proceedings of the 16th IEEE International Workshops on Enabling Technologies: Infrastructure for Collaborative Enterprises* (pp. 269-272). Washington, DC: IEEE Computer Society.

Englander, R. (1997). *Developing Java Beans.* Sebastopol, CA: O'Reilly & Associates, Inc. Bellia, M., Ochiutto, M.E. (2005). Higher order Programming in Java: Introspection, Subsumption and Extraction. *Concurrency Specification and Programming* (pp. 29-44). Amsterdam, The Netherlands: IOS Press.

Fielding, R.T. (2005). JSR170 Overview. *Stardandizing the Content Repository Interface.*

Frey, J., Tannembaum, T., Foster, I., Livny, M., & Tuecke, S. (2001). Condor-G: A Computation Management Agent for Multi-Institutional Grids. Proceedings of the *Tenth IEEE Symposium on High Performance Distributed Computing* (pp. 237-246). Hingham, MA: Kluwer Academic Publishers.

Galizia, A., Viti, F., Orro, A., D'Agostino,D. Merelli, I., Milanes, L., & Clematis, A. (2008). TMAinspect, an EGEE Framework for Tissue MicroArray Image Handling. *Proceedings of 8th Internal Symposium on Cluster Computing and the Grid* (pp 741-746), IMATI-CNR, Genoa, Italy.

Henderson, R.L. (1995). Job Scheduling Under the Portable Batch System. *Proceedings of the workshop on Job Scheduling Strategies for Parallel Scheduling* (pp. 279.294). London: Springer-Verlag.

Liberman, M.A., & Lichtenberg, A.J. (1992). *Regular and Chaotic Dynamics*. Berlin, Germany: Springer-Verlag.

Novotny, J., Russell, M., & Wherens, O. (2004). Gridsphere: A Portal framework for building collaborations. *Concurrency and Computation: Practice and Experience* (pp. 503-513). Hoboken, NJ: Wiley & Sons.

Ott. E. (1993). *Chaos in Dynamical Systems*. New York: Cambridge University Press.

Prieto, F., Cardenas, M., & Rubio, M. (2008). gPhase: A GRID-based technology for the analysis of phase space structure in nonlinear dynamics. Applications to rocking systems under harmonic loading. Proceedings of the 2nd Iberian Grid Infrastructure Conference.

Santos, N., & Koblitz, B. (2006). Distributed Metadata with the AMGA Metadata Catalog. *In workshop on Next Generation Distributed Data Management*

Smith, M., Barton, M., Bass, M., Branchofsky, M., McClellan, G., Stuve, D., Tansley, R., & Harford J. (2003). An Open Source Dynamic Digital Repository. *D-Lib Magazine, 9.*

Stewart, G., Cameron, D., Cowan, G., & McCance, G. (2007). Storage and Data management in EGEE. *Proceedings of Fifth Australasian Symposium on Grid Computing and e-Research* Australian Computer Society, Inc., Darlinghurst, Australia, (pp. 69-77).

Tudor DICOM Viewer. *Main page*. Retrieved February 12, 2008 from http://www.santec.lu/projects/optimage/dicomtools

Ullman, C. & Dykes, L. (2007). *Beginning Ajax*. Hoboken, NJ: John Wiley & Sons. Retrieved from www.Wrox.com.

Watry, P., & Larson, R. (2005). Implementing support for digital repositories in a data grid environment. *Local to Global data interoperability – Challenges and technologies* (pp. 60-64) LGDI.

Zeilinger, G. (2006). dcm4che, a DICOM implementation in Java and dcm4chee J2EE based PACS. *Main page*. Retrieved February 14, 2008 from http://www.dcm4che.org

KEY TERMS AND DEFINITIONS

Algorithm Driver: Component included in the gLibrary/DRI platform that defines the execution architecture of an algorithm, as well as other parameters.

Computing Element: Grid component that represents a computing resource. Its main functionality is the jobs management.

Data Model: Description of the node structure that defines its entities, fields and relationships.

Default Implementation: Feature of the gLibrary/DRI platform that allows inheritance of functionality previously defined avoiding to the repository providers from the task to implement it.

Digital Repository: Digital stores that manage data and metadata and provide users with access to them.

Grid: Computing infrastructure based on the resource sharing on a non-centralized way. It provides high resources capacities to the sites belonging to a Virtual Organization.

Mammogram: Medical tool doctors have to help them diagnose, evaluate and follow women who have had breast cancer.

Node. Digital repository data unit.

Poincaré Surface Section: Standard method used in the phase space analysis techniques.

Storage Element: Grid component that provides a common access mechanism to disk based and tape based storage.

Section VIII
Selected Readings

Chapter XXXIII
Porting Applications to Grids and Clouds

Wolfgang Gentzsch
Duke University, USA

ABSTRACT

A Grid enables remote, secure access to a set of distributed, networked computing and data resources. Clouds are a natural next step of Grids towards the provisioning of computing as a service. To "Grid-enable" applications, users have to cope with: complexity of grid architectures; different compute and data nodes; wide spectrum of grid middleware tools and services; the e-science application architectures, algorithms and programs. Therefore, the aim of this chapter is to guide users through the important stages of implementing applications on Grid and Cloud infrastructures, together with a discussion of important challenges and their potential solutions. As a case study, we present the DEISA Distributed European Infrastructure for Supercomputing Applications and describe the DEISA Extreme Computing Initiative DECI for porting and running scientific grand challenge applications on the DEISA Grid. This chapter concludes with an outlook on Compute Clouds, and the top ten rules of building a sustainable Grid.

INTRODUCTION

Over the last 40 years, the history of computing is deeply marked by the application developers who continuously port and optimize their application codes to the latest and greatest computing architectures and environments. After the von-Neumann mainframe, there was the vector computer, then the shared-memory parallel computer, the distributed-memory parallel computer, the very-long-instruction word computer, the workstation cluster, the meta-computer, and the Grid (never mind, it continues with SOA, Cloud, Virtualization, Many-core, and so on). There is no easy solution to this, and the real solution would be a separation of concerns between discipline-specific content, domain-independent software and hardware infrastructure. However, this often comes along with a loss of performance stemming from the overhead of the infrastructure layers. Recently, users and developers face another wave of complex computing infrastructures: the Grid.

Let's start with answering the question: What is a Grid? Foster and Kesselman (1998) attempted the following definition: A computational grid is a hardware and software infrastructure that provides dependable, consistent, pervasive, and inexpensive access to high-end computational capabilities. In a subsequent article (Foster, 2002), this definition was changed to include social and policy issues, stating that Grid computing is concerned with coordinated resource sharing and problem solving in dynamic, multi-institutional virtual organizations. The key concept is the ability to negotiate resource-sharing arrangements among a set of participating parties (providers and consumers) and then to use the resulting resource pool for some purpose. Furthermore, the sharing that we are concerned with is not primarily file exchange but rather direct access to computers, software, data, and other resources, as is required by a range of collaborative problem-solving and resource-brokering strategies emerging in industry, science, and engineering. This sharing is, necessarily, highly controlled, with resource providers and consumers defining clearly and carefully just what is shared, who is allowed to share, and the conditions under which sharing occurs. A set of individuals and/or institutions defined by such sharing rules form what we call a virtual organization. This author's concern, from the beginning (Gentzsch, 2002), was that the new definition seemed very ambitious, and as history has proven, many of the Grid projects with a focus on these ambitious objectives did not lead to a sustainable grid production environment, so far. We can only repeat that the simpler the grid infrastructure, the easier it is to use with better chance for success. It is for this reason that currently the so-called Clouds are becoming more and more popular (Amazon, 2007).

Over the last ten years, hundreds of applications in science, industry and enterprises have been ported to Grid infrastructures, mostly prototypes in the early definition of Foster and Kesselman (1998). Each application is unique in that it solves a specific problem, based on modeling, for example, a specific phenomenon in nature (physics, chemistry, biology, etc.), presented as a mathematical formula together with appropriate initial and boundary conditions. Applications could be represented by its discrete analogue using sophisticated numerical methods, translated into a programming language computers can understand, adjusted to the underlying computer architecture, embedded in a workflow, and accessible remotely by the user through a secure, transparent and application-specific portal. These examples summarize the wide spectrum and complexity we face in problem solving on grid infrastructures.

The user (and especially the developer) faces several layers of complexity when porting applications to a computing environment, especially to a compute or data grid of distributed networked nodes ranging from desktops to supercomputers. These nodes, usually, consist of several to many loosely or tightly coupled processors (frequently, these processors contain few to many cores). To run efficiently on such systems, applications have to be adjusted to the different layers, taking into account different levels of granularity, from fine-grain structures deploying multi-core architectures at processor level to the coarse granularity found in application workflows representing for example multi-physics applications. Even, the user has to take into account the specific requirements of the grid, coming from the different components of the grid services architecture, such as security, resource management, information services and data management.

Obviously, in this article, it seems impossible to present and discuss the complete spectrum of applications and their adaptation and implementation on Grids. Therefore, we restrict ourselves in the following to briefly describe the different application classes, present a checklist (or classification) with respect to grouping applications according to their appropriate grid-enabling strategy. Also, for lack of space here, we will not discuss the mental, social, or legal aspects which sometimes might be the knock-out criteria for running applications on a grid. Other show-stoppers such as sensitive data, security concerns, licensing issues, and intellectual property, were discussed in some detail in Gentzsch (2007a).

In the following, we will consider the three main areas of impact on porting applications to grids: infrastructure, data management and application architecture issues. These issues can have an impact on effort and success of porting, on the resulting performance of the grid application, the user-friendly access to the resources, the grid services, the application, the data and the final processing results, among others.

Applications and the Grid Infrastructure

As mentioned before, the successful porting of an application to a grid environment highly depends on the underlying distributed resource infrastructure. The main services components offered by a grid infrastructure are security, resource management, information services and data management. Jacob et al. (2003) suggest that each of these components can affect the application architecture, its design, deployment, and performance. Therefore, the user has to go through the process of matching the application (structure and requirements) with those components of the grid infrastructure, as described here, closely following the description in Jacob at al. (2003).

Applications and Security

The security functions within the grid architecture are responsible for the authentication and authorization of the user, and for the secure communication between the grid resources. Fortunately, these functions are an inherent part of most grid infrastructures and don't usually affect the applications themselves, supposed the user (and thus the user's application) is authorized to use the required resources. Also, security from an application point of view might be taken into account in the case that sensitive data is passed to a resource to be processed by a job and is written to the local disk in a non-encrypted format, and other users or applications might have access to that data.

Applications and Resource Management

The resource management component provides the facilities to allocate a job to a particular resource, provides a means to track the status of the job while it is running and its completion information, and provides the capability to cancel a job or otherwise manage it. In conjunction with Monitoring and Discovery Service (described below) the application must ensure that the appropriate target resource(s) are used. This requires that the application accurately specifies the required environment (operating system, processor, speed, memory, and so on). The more the application developer can do to eliminate specific dependencies, the better the chance that an available resource can be found and that the job will complete. If an application includes multiple jobs, the user must understand (and may be reduced) their interdependencies. Otherwise, logic has to be built to handle items such as inter-process communication, sharing of data, and concurrent job submissions. Finally, the job management provides mechanisms to query the status of the job as well as perform operations such as canceling the job. The application may need to utilize these capabilities to provide feedback to the user or to clean up or free up resources when required. For instance, if one job within an application fails, other jobs that may be dependent on it may need to be cancelled before needlessly consuming resources that could be used by other jobs.

Applications and Resource Information Services

An important part of the process of grid-enabling an application is to identify the appropriate (if not optimal) resources needed to run the application, i.e. to submit the respective job to. The service which maintains and provides the knowledge about the grid resources is the Grid Information Service (GIS), also known as the Monitoring and Discovery Service (e.g. MDS in Globus [xx]). MDS provides access to static and dynamic information of resources. Basically, it contains the following components:

- Grid Resource Information Service (GRIS), the repository of local resource information derived from information providers.
- Grid Index Information Service (GIIS), the repository that contains indexes of resource information registered by the GRIS and other GIISs.
- Information providers, translate the properties and status of local resources to the format defined in the schema and configuration files.
- MDS client which initially performs a search for information about resources in the grid environment.

Resource information is obtained by the information provider and it is passed to GRIS. GRIS registers its local information with the GIIS, which can optionally also register with another GIIS, and so on. MDS clients can query the resource information directly from GRIS (for local resources) and/or a GIIS (for grid-wide resources).

It is important to fully understand the requirements for a specific job so that the MDS query can be correctly formatted to return resources that are appropriate. The user has to ensure that the proper information is in MDS. There is a large amount of data about the resources within the grid that is available by default within the MDS. However, if the application requires special resources or information that is not there by default, the user may need to write her own information providers and add the appropriate fields to the schema. This may allow the application or broker to query for the existence of the particular resource/requirement.

Applications and Data Management

Data management is concerned with collectively maximizing the use of the limited storage space, networking bandwidth, and computing resources. Within the application, data requirements have been built in, which determine how data will be moved around the infrastructure or otherwise accessed in a secure and efficient manner. Standardizing on a set of grid protocols will allow to communicate between any data source that is available within the software design. Especially data intensive applications often have a federated database to create a virtual data store or other options including Storage Area Networks, network file systems and dedicated storage servers. Middleware like the Globus Toolkit provide GridFTP and Global Access to Secondary Storage data transfer utilities in the grid environment. The GridFTP facility (extending the FTP File Transfer Protocol) provides secure and reliable data transfer between grid hosts.

Developers and users face a few important data management issues that need to be considered in application design and implementation. For large datasets, for example, it is not practical and may be impossible to move the data to the system where the job will actually run. Using data replication or otherwise copying a subset of the entire dataset to the target system may provide a solution. If the grid resources are geographically distributed with limited network connection speeds, design considerations around slow or limited data access must be taken into account. Security, reliability and performance become an issue when moving data across the Internet. When the data access may be slow or prevented one has to build the required logic to handle this situation. To assure that the data is available at the appropriate location by the time the job requires it, the user should schedule the data transfer in advance. One should also be aware of the number and size of any concurrent transfers to or from any one resource at the same time.

Beside the above described main requirements for applications for running efficiently on a grid infrastructure, there are a few more issues which are discussed in Jacob (2003), such as scheduling, load balancing, grid broker, inter-process communication, and portals for easy access, and non-functional requirements such as performance, reliability, topology aspects, and consideration of mixed platform environments.

The Simple API for Grid Applications (SAGA)

Among the many efforts in the grid community to develop tools and standards which simplify the porting of applications to Grids by enabling the application to make easy use of the Grid middleware services as described above, one of the more predominant ones is SAGA, a high-level Application Programmers Interface (API), or programming abstraction, defined by the Open Grid Forum (OGF, 2008), an international committee that coordinates standardization of Grid middleware and architectures. SAGA intends to simplify the development of grid-enabled applications, even for scientists without any background in computer science or grid computing. Historically, SAGA was influenced by the work on the GAT Grid Application Toolkit, a C-based API developed in the EU-funded project GridLab (GAT, 2005). The purpose of SAGA is two-fold:

1. Provide a simple API that can be used with much less effort compared to the interfaces of existing grid middleware.
2. Provide a standardized, portable, common interface for the various grid middleware systems.

According to Goodale (2008), SAGA facilitates rapid prototyping of new grid applications by allowing developers a means to concisely state very complex goals using a minimum amount of code.

SAGA provides a simple, POSIX-style API to the most common Grid functions at a sufficiently high-level of abstraction so as to be able to be independent of the diverse and dynamic Grid environments. The SAGA specification defines interfaces for the most common Grid-programming functions grouped as a set of functional packages. Version 1.0 (Goodale, 2008) defines the following packages:

- File package - provides methods for accessing local and remote file systems, browsing directories, moving, copying, and deleting files, setting access permissions, as well as zero-copy reading and writing
- Replica package - provides methods for replica management such as browsing logical file systems, moving, copying, deleting logical entries, adding and removing physical files from a logical file entry, and search logical files based on attribute sets.
- Job package - provides methods for describing, submitting, monitoring, and controlling local and remote jobs. Many parts of this package were derived from the largely adopted DRMAA [11] specification.
- Stream package - provides methods for authenticated local and remote socket connections with hooks to support authorization and encryption schemes.

- RPC package - is an implementation of the OGF GridRPC API definition and provides methods for unified remote procedure calls.

The two critical aspects of SAGA are its simplicity of use and the fact that it is well on the road to becoming a community standard. It is important to note, that these two properties provided the added value of using SAGA for Grid application development. Simplicity arises from being able to limit the scope to only the most common and important grid-functionality required by applications. There are major advantages arising from its simplicity and imminent standardization. Standardization represents the fact that the interface is derived from a wide-range of applications using a collaborative approach and the output of which is endorsed by the broader community.

More information about the SAGA C++ Reference Implementation (developed at the Center for Computation and Technology at the Louisiana State University) and various aspects of grid enabling toolkits is available on the SAGA implementation home page (SAGA, 2006). It also provides additional information with regard to different aspects of grid enabling toolkits.

Grid Applications and Data

Any e-science application at its core has to deal with data, from input data (e.g. in the form of output data from sensors, or as initial or boundary data), to processing data and storing of intermediate results, to producing final results (e.g. data used for visualization). Data has a strong influence on many aspects of the design and deployment of an application and determines whether a grid application can be successfully ported to the grid. Therefore, in the following, we present a brief overview of the main data management related aspects, tasks and issues which might affect the process of grid-enabling an application, such as data types and size, shared data access, temporary data spaces, network bandwidth, time-sensitive data, location of data, data volume and scalability, encrypted data, shared file systems, databases, replication and caching. For a more in-depth discussion of data management related tasks, issues, and techniques, we refer to Bart Jacob's tutorial on application enabling with Globus (Jacob, 2003).

Shared Data Access

Sharing data access can occur with concurrent jobs and other processes within the network. Access to data input and the data output of the jobs can be of various kinds. During the planning and design of the grid application, potential restrictions on the access of databases, files, or other data stores for either read or write have to be considered. The installed policies need to be observed and sufficient access rights have to be granted to the jobs. Concerning the availability of data in shared resources, it must be assured that at run-time of the individual jobs the required data sources are available in the appropriate form and at the expected service level. Potential data access conflicts need to be identified up front and planned for. Individual jobs should not try to update the same record at the same time, nor dead lock each other. Care has to be taken for situations of concurrent access and resolution policies imposed.

The use of federated databases may be useful in data grids where jobs must handle large amounts of data in various different data stores. They offer a single interface to the application and are capable of accessing data in large heterogeneous environments. Federated database systems contain information about location (node, database, table, record) and access methods (SQL, VSAM, privately defined methods) of connected data sources. Therefore, a simplified interface to the user (a grid job or other client) requires that the essential information for a request should not include the data source, but rather use a discovery service to determine the relevant data source and access method.

Data Topology

Issues about the size of the data, network bandwidth, and time sensitivity of data determine the location of data for a grid application. The total amount of data within the grid application may exceed the amount of data input and output of the grid application, as there can be a series of sub-jobs that produce data for other sub-jobs. For permanent storage, the grid user needs to be able to locate where the required storage space is available in the grid. Other temporary data sets that may need to be copied from or to the client also need to be considered.

The amount of data that has to be transported over the network is restricted by available bandwidth. Less bandwidth requires careful planning of the data traffic among the distributed components of a grid application at runtime. Compression and decompression techniques are useful to reduce the data amount to be transported over the network. But in turn, it raises the issue of consistent techniques on all involved nodes. This may exclude the utilization of scavenging for a grid, if there are no agreed standards universally available.

Another issue in this context is time-sensitive data. Some data may have a certain lifetime, meaning its values are only valid during a defined time period. The jobs in a grid application have to reflect this in order to operate with valid data when executing. Especially when using data caching or other replication techniques, it has to be assured that the data used by the jobs is up-to-date, at any given point in time. The order of data processing by the individual jobs, especially the production of input data for subsequent jobs, has to be carefully observed.

Depending on the job, the authors Jacob at al. (2003) recommend to consider the following data-related questions which refer to input as well as output data of the jobs within the grid application:

- Is it reasonable that each job or set of jobs accesses the data via the network?
- Does it make sense to transport a job or set of jobs to the data location?
- Is there any data access server (for example, implemented as a federated database) that allows access by a job locally or remotely via the network?
- Are there time constraints for data transport over the network, for example, to avoid busy hours and transport the data to the jobs in a batch job during off-peak hours?
- Is there a caching system available on the network to be exploited for serving the same data to several consuming jobs?
- Is the data only available in a unique location for access, or are there replicas that are closer to the executable within the grid?

Data Volume

The ability for a grid job to access the data it needs will affect the performance of the application. When the data involved is either a large amount of data or a subset of a very large data set, then moving the data set to the execution node is not always feasible. Some of the considerations as to what is feasible include the volume of the data to be handled, the bandwidth of the network, and logical interdependences on the data between multiple jobs.

Data volume issues: In a grid application, transparent access to its input and output data is required. In most cases the relevant data is permanently located on remote locations and the jobs are likely to process local copies. This access to the data results in a network cost and it must be carefully quantified. Data volume and network bandwidth play an important role in determining the scalability of a grid application.

Data splitting and separation: Data topology considerations may require the splitting, extraction, or replication of data from data sources involved. There are two general approaches that are suitable for higher scalability in a grid application: Independent tasks per job and a static input file for all jobs. In the case of independent tasks, the application can be split into several jobs that are able to work independently on a disjoint subset of the input data. Each job produces its own output data and the gathering of all of the results of the jobs provides the output result by itself. The scalability of such a solution depends on the time required to transfer input data, and on the processing time to prepare input data and generate the final data result. In this case the input data may be transported to the individual nodes on which its corresponding job is to be run. Preloading of the data might be possible depending on other criteria like timeliness of data or amount of the separated data subsets in relation to the network bandwidth. In the case of static input files, each job repeatedly works on the same static input data, but with different parameters, over a long period of time. The job can work on the same static input data several times but with different parameters, for which it generates differing results. A major improvement for the performance of the grid application may be derived by transferring the input data ahead of time as close as possible to the compute nodes.

Other cases of data separation: More unfavorable cases may appear when jobs have dependencies on each other. The application flow may be carefully checked in order to determine the level of parallelism to be reached. The number of jobs that can be run simultaneously without dependences is important in this context. For independent jobs, there needs to be synchronization mechanisms in place to handle the concurrent access to the data.

Synchronizing access to one output file: Here all jobs work with common input data and generate their output to be stored in a common data store. The output data generation implies that software is needed to provide synchronization between the jobs. Another way to process this case is to let each job generate individual output files, and then to run a post-processing program to merge all these output files into the final result. A similar case is that each job has its individual input data set, which it can consume. All jobs then produce output data to be stored in a common data set. As described above, the synchronization of the output for the final result can be done through software designed for the task.

Hence, thorough evaluation of the input and output data for jobs in the grid application is needed to properly handle it. Also, one should weigh the available data tools, such as federated databases, a data joiner, and related products and technologies, in case the grid application is highly data oriented or the data shows a complex structure.

Porting and Programming Grid Applications

Besides taking into account the underlying grid resources and the application's data handling, as discussed in the previous two paragraphs, another challenge is the porting of the application program itself. In this context, developers and users are facing mainly two different approaches when implementing their application on a grid. Some developers port an existing application code on a set of distributed grid resources. Often, in the past, the application previously has been developed and optimized with a specific computer architecture in mind, for example, mainframes or servers, single- or multiple-CPU vector computers, shared- or distributed-memory parallel computers, or loosely coupled distributed systems like workstation clusters, for example. Other developers start from the scratch and design and develop a new application program with the grid in mind, often such application architectures with its inherent numerical algorithms are optimally mapped onto the best-suited (set of) resources in a grid.

In both scenarios, the effort of implementing an application can be huge. Therefore, it is important to perform a careful analysis beforehand on: the user requirements for running the application on a grid (e.g. cost, time); on application type (e.g. compute or data intensive); application architecture and algorithms (e.g. explicit, or implicit) and application components and how they interact (e.g. loosely or tightly coupled, or workflows); what is the best way to map the application onto a grid; and which is the best suited grid architecture to run the application in an optimally performing way. Therefore, in the following, we summarize the most popular strategies for porting an existing application to a grid, and for designing and developing a new grid application.

Many scientific papers and books deal with the issues of designing, programming, and porting grid applications, and it is difficult to recommend the best suited among them. Here, we mainly follow the books from Ian Foster and Carl Kesselman (1999 & 2004), the IBM Redbook (Jacob, 2003), the SURA Grid Technology Cookbook (SURA, 2007), several research papers on programming models and environments, e.g. Soh (2006), Badia (2003), Karonis (2002), Seymour (2002), Buyya (2000), Venugopal (2004), Luther (2005), Altintas (2004) and Frey (2005), and our own experience at Sun Microsystems and MCNC (Gentzsch, 2004), RENCI (Gentzsch, 2007), D-Grid (2008), Neuroth (2007), and currently in DEISA-2 (2008).

Grid Programming Models and Environments

Our own experience in porting applications to distributed resource environments is very similar to the one from Soh et al. (2006) who present a useful discussion on grid programming models and environments which we briefly summarize in the following. In their paper, they start with differentiating application porting into resource composition and program composition. Resource composition, i.e. matching the application to the grid resources needed, has already been discussed in paragraphs 2 and 3 above.

Concerning program composition, there is a wide spectrum of strategies of distributing an application onto the available grid resources. This spectrum ranges from the ideal situation of simply distributing a list of, say, n parameters together with n identical copies of that application program onto the Grid, to the other end of the spectrum where one has to compose or parallelize the program into chunks or components that can be distributed to the grid resources for execution. In the latter case, Soh (2006) differentiates between implicit parallelism, where programs are automatically parallelized by the environment, and explicit parallelism which requires the programmer to be responsible for most of the parallelization effort such as task decomposition, mapping tasks to processors and inter-task communication. However, implicit approaches often lead to non-scalable parallel performance, while explicit approaches often are complex and work- and time-consuming. In the following, we summarize the approaches and methods discussed in detail in Soh (2006):

Superscalar, sequential applications composed of tasks of a certain granularity are automatically converted into a parallel application where the tasks are executed in different servers of a computational Grid. An example is Grid Superscalar (Badia, 2003).

Explicit Communication, such as Message Passing and Remote Procedure Call (RPC). A message passing example is MPICH-G2 (Karonis, 2002), a Grid-enabled implementation of the Message Passing Interface (MPI) which defines standard functions for communication between processes and groups of processes, extended by the Globus Toolkit. An RPC example is GridRPC, an API for Grids (Seymour, 2002), which offers a convenient, high-level abstraction whereby many interactions with a Grid environment can be hidden.

Bag of Tasks, which can be easily distributed on grid resources. An example is the Nimrod-G Broker (Buyya, 2000) which is a Grid-aware version of Nimrod, a specialized parametric modeling system. Nimrod uses a simple declarative parametric modeling language and automates the task of formulating, running, monitoring, and aggregating results. Another example is the Gridbus Broker (Venugopal, 2004) that permits users access to heterogeneous Grid resources transparently.

Distributed Objects, as in ProActive (2005), a Java based library that provides an API for the creation, execution and management of distributed active objects. Proactive is composed of only standard Java classes and requires no changes to the Java Virtual Machine (JVM) allowing Grid applications to be developed using standard Java code.

Distributed Threads, for example Alchemi (Luther, 2005), a Microsoft .NET Grid computing framework, consisting of service-oriented middleware and an application program interface (API). Alchemi features a simple and familiar multithreaded programming model.

Grid Workflows. Many Workflow Environments have been developed in recent years for Grids, such as Triana, Taverna, Simdat, P-Grade, and Kepler. Kepler, for example, is a scientific workflow management system along with a set of Application Program Interfaces (APIs) for heterogeneous hierarchical modeling (Altintas, 2004). Kepler provides a modular, activity oriented programming environment, with an intuitive GUI to build complex scientific workflows.

Grid Services. An example is the Open Grid Services Architecture (OGSA) (Frey, 2005) which is an ongoing project that aims to enable interoperability between heterogeneous resources by aligning Grid technologies with established Web services technology. The concept of a Grid service is introduced as a Web service that provides a set of well defined interfaces that follow specific conventions. These grid services can be composed into more sophisticated services to meet the needs of users.

Grid-Enabling Application Programs and Numerical Algorithms

In many cases, restructuring (grid-enabling, decomposing, parallelizing) the core algorithm(s) within a single application program doesn't make sense, especially in the case of a more powerful higher-level grid-enabling strategy. For example, in the case of parameter jobs (see below), many identical copies of the application program together with different data-sets can easily be distributed onto many grid nodes, or where the application program components can be mapped onto a workflow, or where applications (granularity, run time, special dimension, etc.) simply are too small to efficiently run on a grid, and the grid latencies and management overhead become too dominant. In other cases, however, where e.g. just one long run has to be performed, grid-enabling the application program itself can lead to dramatic performance improvements and, thus, time savings. In an effort to better guide the reader through this complex field, in the following, we will briefly present a few popular application codes and their algorithmic structure and provide recommendations for some meaningful grid-enabling strategies.

General Approach. First, we have to make sure that we gain an important benefit from running our application on a grid. We should start asking a few more general questions, top-down. Has this code been developed in-house, or is it a third-party code, developed elsewhere? Will I submit many jobs (as e.g. in a parameter study), or is the overall application structure a workflow, or is it a single monolithic application code? In case of the latter, are the core algorithms within the application program of explicit or of implicit nature? In many cases, grid-enabling those kinds of applications can be based on experience made in the past with parallelizing them for the moderately or massively parallel systems, see e.g. Fox et al. (1994) and Dongarra et al. (2003).

In-house Codes. In case of an application code developed in-house, the source code of this application is often still available, and ideally the code developers are still around. Then, we have the possibility to analyze the structure of the code, its components (subroutines), dependencies, data handling, core algorithms, etc. With older codes, sometimes, this analysis has already been done before, especially for the vector and parallel computer architectures of the eighties and nineties. Indeed, some of this knowledge can be re-used now for the grid-enabling process, and often only minor adjustments are needed to port such a code to the grid.

Third-Party Codes. Licensed from so-called Independent Software Vendors (ISVs) cannot be grid-enabled without the support from these ISVs. Therefore, in this case, we recommend to contact the ISV. In case the ISV receives similar requests from other customers as well, there might be a real chance that the ISV will either provide a grid-enabled code or completely change its sales strategy and sell its software as a service, or develops its own application portal to provide access to the application and the computing resources. Obviously, this requires patience and is thus not a solution if you are under a time constraint.

Parameter Jobs. In science and engineering, often, the application has to run many times: same code, different data. Only a few parameters have to be modified for each individual job, and at the end of the many job runs, the results are analyzed with statistical or stochastic methods, to find a certain optimum. For example, during the design of a new car model, many crash simulations have to be performed, with the aim to find the best-suited material and geometry for a specific part of the wire-frame model of the car.

Application Workflows. It is very common in so-called Problem Solving Environments that the application program consists of a set of components or modules which interact with each other. This can be modeled in grid workflow environments which support the design and the execution of the workflow representing the application program. Usually, these grid workflow environments contain a middleware layer which maps the application modules onto the different resources in the grid. Many Workflow Environments have been developed in recent years for Grids, such as Triana (2003), Taverna (2008), Simdat (2008), P-Grade (2003), and Kepler (Altintas, 2004). One application area which is well suited for such a workflow is climate simulation. Today's climate codes consist of modules for simulating the weather on the continent with mesoscale meteorology models, and include other modules for taking into account the influence from ocean and ocean currents, snow and ice, sea ice, wind, clouds and precipitation, solar and terrestrial radiation, absorption, emission, and reflection, land surface processes, volcanic gases and particles, and human influences. Interactions happen between all these components, e.g. air-ocean, air-ice, ice-ocean, ocean-land, etc. resulting in a quite complex workflow which can be mapped onto the underlying grid infrastructure.

Highly Parallel Applications. Amdahl's Law states that the scalar portion of a parallel program becomes a dominant factor as processor number increases, leading to a loss in application scalability with growing number of processors. Gustafson (1988) proved that this holds only for fixed problem size, and that in practice, with increasing number of processors, the user increases problem size as well, always trying to solve the largest possible problem on any given number of CPUs. Gustafson demonstrated this on a 1028-processor parallel system, for several applications. For example, he was able to achieve a speed-up factor of over 1000 for a Computational Fluid Dynamics application with 1028 parallel processes on the 1028-processor system. Porting these highly parallel applications to a grid, however, has shown that many of them degrade in performance simply because overhead of communication for message-passing operations (e.g. send and receive) drops from a few microseconds on a tightly-coupled parallel system to a few milliseconds on a (loosely-coupled) workstation cluster or grid. In this case, therefore, we recommend to implement a coarse-grain Domain Decomposition approach, i.e. to dynamically partition the overall computational domain into sub-domains (each consisting of as many parallel processes, volumes, finite elements, as possible), such that each sub-domain completely fits onto the available processors of the corresponding parallel system in the grid. Thus, only moderate performance degradation from the reduced number of inter-system communication can be expected. A prerequisite for this to work successfully is that the subset of selected parallel systems is of homogeneous nature, i.e. architecture and operating system of these parallel systems should be identical. One Grid infrastructure which offers this feature is the Distributed European Infrastructure for Supercomputing Applications (DEISA, 2008), which (among others) provides a homogeneous cluster of parallel AIX machines distributed over several of the 11 European supercomputing centers which are part of DEISA (see also Section 5 in this article).

Moderately Parallel Applications. These applications, which have been parallelized in the past, often using Message Passing MPI library functions for the inter-process communication on workstation clusters or on small parallel systems, are well-suited for parallel systems with perhaps a few dozen to a few hundreds of processors, but they won't scale easily to a large number of parallel processes (and processors). Reasons are a significant scalar portion of the code which can't run in parallel and/or the relatively high ratio of inter-process communication to computation, resulting in relatively high idle times of the CPUs waiting for the data. Many commercial codes fall in this category, for example finite-element codes such as Abaqus, Nastran or Pamcrash. Here we recommend to check if the main goal is to analyze many similar scenarios with one and the same code but on different data sets, and run as many codes in parallel as possible, on as many moderately parallel sub-systems as possible (this could be virtualized sub-systems on one large supercomputer, for example).

Explicit versus Implicit Algorithms. Discrete Analogues of systems of partial differential equations, stemming from numerical methods such as finite difference, finite volume, or finite element discretizations, often result in large sets of explicit or implicit algebraic equations for the unknown discrete variables (e.g. velocity vectors, pressure, temperature). The explicit methods are usually slower (in convergence to the exact solution vector of the algebraic system) than the implicit ones but they are also inherently parallel, because there is no dependence of the solution variables among each other, and therefore there are no recursive algorithms. In case of the more accurate implicit methods, however, solution variables are highly inter-dependent leading to recursive sparse-matrix systems of algebraic equations which cannot easily split (parallelized) into smaller systems. Again, here, we recommend to introduce a Domain Decomposition approach as described in the above section on Highly Parallel Algorithms, and solve an implicit sparse-matrix system within each domain, and bundle sets of 'neighboring' domains into super-sets to submit to the (homogeneous) grid.

Domain Decomposition. This has been discussed in the paragraphs on Highly Parallel Applications and on Explicit versus Implicit Algorithms.

Job Mix. Last but not least, one of the most trivial but most widely used scenarios often found in university and research computer centers is the general job mix, stemming from hundreds or thousands of daily users, with hundreds or even thousands of different applications, with varying requirements for computer architecture, data handling, memory and disc space, timing, priority, etc. This scenario is ideal for a grid which is managed by an intelligent Distributed Resource Manager (DRM), for example GridWay (2008) for a global grid, Sun Grid Engine Enterprise Edition (Chaubal, 2003) for an enterprise grid, or the open source Grid Engine (2001) for a departmental grid or a simple cluster. These DRMs are able to equally balance the overall job load across the distributed resource environment and submit the jobs always to the best suited and least loaded resources. This can result in overall resource utilization of 90% and higher.

Applications and Grid Portals

Grid portals are an important part of the process of grid-enabling, composing, manipulating, running, and monitoring applications. The lower layers of the grid-enabling process have been described in the previous paragraphs, but often, the user is still exposed to the many details of the grid services and even has to take care of configuring, composing, provisioning, etc. the application and the services "by hand". This however can be drastically simplified and mostly hidden from the user through a Grid portal, which is a Web-based portal able to expose Grid services and resources through a browser to allow users remote, ubiquitous, transparent and secure access to grid services (computers, storage, data, applications, etc). The main goal of a Grid portal is to hide the details and complexity of the underlying Grid infrastructure from the user in order to improve usability and utilization of the Grid, especially by greatly simplifying the use of Grid-enabled applications through a user-friendly interface.

Grid portals have become popular in research and the industry communities. Using Grid portals, computational and data-intensive applications such as genomics, financial modeling, crash test analysis, oil and gas exploration, and many more, can be provided over the Web as traditional services. Examples of existing scientific application portals are the GEONgrid (2008) and CHRONOS (2004) portals that provide a platform for the Earth Science community to study and understand the complex dynamics of Earth systems; the NEESGrid project (2008) focuses on earthquake engineering research; the BIRN portal (2008) targets biomedical informatics researchers; and the MyGrid portal (2008) provides access to bioinformatics tools running on a back-end Grid infrastructure. As it turns out, scientific portals are usually being developed inside specific research projects. As a result, they are specialized for specific applications and services satisfying project requirements for that particular research application area.

In order to rapidly build customized Grid portals in a flexible and modular way, several more generic toolkits and frameworks have been developed. These frameworks are designed to meet the diverse needs and usage models arising from both research and industry. One of these frameworks is EnginFrame, which simplifies development of highly functional Grid portals exposing computing services that run on a broad range of different computational Grid systems. EnginFrame (Beltrame, 2006) has been adopted by many industrial companies, and by organizations in research and education.

Example: The EnginFrame Portal Environment

EnginFrame (2008) is a Web-based portal technology that enables the access and the exploitation of grid-enabled applications and infrastructures. It allows organizations to provide application-oriented computing and data services to both users (via Web browsers) and in-house or ISV applications (via SOAP/WSDL based Web services), thus hiding the complexity of the underlying Grid infrastructure. Within a company or department, an enterprise portal aggregates and consolidates the services and exposes them to the users, through the Web. EnginFrame can be integrated as Web application in a J2EE standard application server or as a portlet in a JSR168 compliant portlet container.

As a Grid portal framework, EnginFrame offers a wide range of functionalities to IT developers facing the task to provide application-oriented services to the end users. EnginFrame's plug-in mechanism allows to easily and dynamically extend its set of functionalities and services. A plug-in is a self-contained software bundle that encapsulates XML Extensible Markup Language service descriptions, custom layout or XSL Extensible Stylesheet Language and the scripts or executables involved with the services actions. A flexible authentication delegation offers a wide set of pre-configured authentication mechanisms: OS/NIS/PAM, LDAP, Microsoft Active Directory, MyProxy, Globus, etc. It can also be extended throughout the plug-in mechanism.

Besides authentication, EnginFrame provides an authorization framework that allows defining groups of users and Access Control Lists (ACLs), and to bind ACLs to resources, services, service parameters and service results. The Web interface of the services provided by the portal can be authorized and thus tailored to the specific users' roles and access rights.

EnginFrame supports a wide variety of compute Grid middleware like LSF, PBS, Sun Grid Engine, Globus, gLite and others. An XML virtualization layer invokes specific middleware commands and translates results, jobs and Grid resource descriptions into a portable XML format called GridML that abstracts from the actual underlying Grid technology. For the GridML, as for the service description XML, the framework provides pre-built XSLs to translate GridML into HTML. EnginFrame data management allows for browsing and handling data on the client side or remotely archived in the Grid and then to host a service working environment in file system areas called spoolers.

The EnginFrame architecture is structured into three tiers, Client, Resource, Server. The Client Tier normally consists of the user's Web browser and provides an easy-to-use interface based on established Web standards like XHTML and JavaScript, and it is independent from the specific software and hardware environment used by the end user. When needed, the client tier also provides integration with desktop virtualization technologies like Citrix Metaframe (ICA), VNC, X, and Nomachine NX. The Resource Tier consists of one or more Agents deployed on the back-end Grid infrastructure whose role is to control and provide distributed access to the actual computing resources. The Server Tier consists of a server component that provides resource brokering to manage resource activities in the back-end.

The EnginFrame server authenticates and authorizes incoming requests from the Web, and asks an Agent to execute the required actions. Agents can perform different kind of actions that range from the execution of a simple command on the underlying Operating System, to the submission of a job to the Grid. The results of the executed action are gathered by the agent and sent back to the Server which applies post processing transformations, filters the output according to ACLs and transforms the results into a suitable format according to the nature of the client: HTML for Web browsers and XML in a SOAP message for Web services client applications.

Case Study: Applications on the DEISA Infrastructure

As one example, in the following, we will discuss the DEISA Distributed European Infrastructure for Supercomputing Applications. DEISA (2008) is different from many other Grid initiatives which aim at building a general purpose grid infrastructure and therefore have to cope with many (almost) insurmountable barriers such as complexity, resource sharing, crossing administrative (and even national) domains, handling IP and legal issues, dealing with sensitive data, working on interoperability, and facing the issue to expose every little detail of the underlying infrastructure services to the grid application user. DEISA avoids most of these barriers by staying very focused: The main focus of DEISA is to provide the European supercomputer user with a flexible, dynamic, user-friendly supercomputing ecosystem (one could say Supercomputing Cloud, see next paragraph) for easy handling, submitting, and monitoring long-running jobs on the best-suited and least-loaded supercomputer(s) in Europe, trying to avoid the just mentioned barriers. For a similar European funded initiative for enterprise applications, we refer the reader to the BEinGRID project (2008), which consists of 18 so-called business experiments each dealing with a pilot application that addresses a concrete business case, and is represented by an end-user, a service provider, and a Grid service integrator. Experiments come from key business sectors such as multimedia, financial, engineering, chemistry, gaming, environmental science, and logistics and so on, based on different Grid middleware solutions, see (BEinGRID, 2008).

The DEISA Project

DEISA is the Distributed European Initiative for Supercomputing Applications, funded by the EU in Framework Program 6 (DEISA1, 2004 – 2008) and Framework Program 7 (DEISA2, 2008 – 2011). The DEISA Consortium consists of 11 partners, MPG-RZG (Germany, consortium lead), BSC (Spain), CINECA (Italy), CSC (Finland), ECMWF (UK), EPCC (UK), FZJ (Germany), LRS (Germany), IDRIS (France), LRZ (Germany), and SARA (Netherlands), and 3 asociated partners KTH (Sweden), CSCS (Switzerland), and JSCC (Russia).

DEISA develops and supports a distributed high performance computing infrastructure and a collaborative environment for capability computing and data management. The resulting infrastructure enables the operation of a powerful Supercomputing Grid built on top of national supercomputing services, facilitating Europe's ability to undertake world-leading computational science research. DEISA is certainly instrumental for advancing computational sciences in scientific and industrial disciplines within Europe and is paving the way towards the deployment of a cooperative European HPC ecosystem. The existing infrastructure is based on the coupling of eleven leading national supercomputing centers, using dedicated network interconnections (currently 10 GBs) of GÉANT2 and the NRENs.

DEISA2 develops activities and services relevant for applications enabling, operation, and technologies, as these are indispensable for the effective support of computational sciences in the area of supercomputing. The service provisioning model is extended from one that supports a single project (in DEISA1) to one supporting Virtual European Communities (now in DEISA2). Collaborative activities will be carried out with new European and other international initiatives. Of strategic importance is the cooperation with the PRACE (2008) initiative which is preparing for the installation of a limited number of leadership-class Tier-0 supercomputers in Europe.

The DEISA Infrastructure Services

The essential services to operate the infrastructure and support its efficient usage are organized in the three Service Activities Operations, Technologies, and Applications:

Operations refer to operating the infrastructure including all existing services, adopting approved new services from the Technologies activity, and advancing the operation of the DEISA HPC infrastructure to a turnkey solution for the future European HPC ecosystem by improving the operational model and integrating new sites.

Technologies cover monitoring of technologies in use in the project, identifying and selecting technologies of relevance for the project, evaluating technologies for pre-production deployment, and planning and designing specific sub-infrastructures to upgrade existing services or deliver new ones based on approved technologies. User-friendly access to the DEISA Supercomputing Grid is provided by DEISA Services for Heterogeneous management Layer (DESHL, 2008) and the UNiforme Interface for COmputing Resources (UNICORE, 2008).

Applications cover the areas applications enabling and extreme computing projects, environment and user related application support, and benchmarking. Applications enabling focuses on enhancing scientific applications from the DEISA Extreme Computing Initiative (DECI), Virtual Communities and EU projects. Environment and user related application support addresses the maintenance and improvement of the DEISA application environment and interfaces, and DEISA-wide user support in the applications area. Benchmarking refers to the provision and maintenance of a European Benchmark Suite for supercomputers.

In DEISA2, two *Joint Research Activities (JRA) complement the portfolio of service activities.* JRA1 (Integrated DEISA Development Environment) aims at an integrated environment for scientific application development, based on a software infrastructure for tools integration, which provides a common user interface across multiple computing platforms. JRA2 (Enhancing Scalability) aims at the enabling of supercomputer applications for the efficient exploitation of current and future supercomputers, to cope with a production infrastructure characterized by an aggressive parallelism on heterogeneous HPC architectures at a European scale.

DECI – DEISA Extreme Computing Initiative for Supercomputing Applications

The DEISA Extreme Computing Initiative (DECI, 2008) was launched in May 2005 by the DEISA Consortium, as a way to enhance its impact on science and technology. The main purpose of this initiative is to enable a number of "grand challenge" applications in all areas of science and technology. These leading, ground breaking applications must deal with complex, demanding and innovative simulations that would not be possible without the DEISA infrastructure, and which benefit from the exceptional resources provided by the Consortium. The DEISA applications are expected to have requirements that cannot be fulfilled by the national services alone.

In DEISA2, the single-project oriented activities (DECI) will be qualitatively extended towards persistent support of Virtual Science Communities. This extended initiative will benefit from and build on the experiences of the DEISA scientific Joint Research Activities where selected computing needs of various scientific communities and a pilot industry partner were addressed. Examples of structured science communities with which close relationships are planned to be established are EFDA and the European climate community. DEISA2 will provide a computational platform for them, offering integration via distributed services and web applications, as well as managing data repositories.

Applications Adapted to the DEISA Grid Infrastructure

In the following, we describe examples of application profiles and use cases that are well-suited for the DEISA supercomputing Grid, and that can benefit from the computational resources made available by the DECI Extreme Computing Initiative.

International collaboration involving scientific teams that access the nodes of the AIX super-cluster in different countries, can benefit from a common data repository and a unique, integrated programming and production environment (via common global file systems). Imagine, for example, that team A in France and team B in Germany dispose of allocated resources at IDRIS in Paris and FZJ in Juelich, respectively. They can benefit from a shared directory in the distributed super-cluster, and for all practical purposes it looks as if they were accessing a single supercomputer.

Extreme computing demands of a challenging project requiring a dominant fraction of a single supercomputer. Rather than spreading a huge, tightly coupled parallel application on two or more supercomputers, DEISA can organize the management of its distributed resource pool such that it is possible to allocate a substantial fraction of a single supercomputer to this project which is obviously more efficient than splitting the application and distributing it over several supercomputers.

Workflow applications involving at least two different HPC platforms. Workflow applications are simulations where several independent codes act successively on a stream of data, the output of one code being the input of the next one in the chain. Often, this chain of computations is more efficient if each code runs on the best-suited HPC platform (e.g. scalar, vector, or parallel supercomputers) where it develops the best performance. Support of these applications via UNICORE (2008) which allows treating the whole simulation chain as a single job is one of the strengths of the DEISA Grid.

Coupled applications involving more than one platform. In some cases, it does make sense to spread a complex application over several computing platforms. This is the case of multi-physics, multi-scale application codes involving several computing modules each dealing with one particular physical phenomenon, and which only need to exchange a moderate amount of data in real time. DEISA has already developed a few applications of this kind, and is ready to consider new ones, providing substantial support to their development. This activity is more prospective, because systematic production runs of coupled applications require a co-allocation service which is currently being implemented.

Applications in the Cloud

With increasing demand for higher performance, efficiency, productivity, agility and lower cost, Information Communication Technologies (ICT) are dramatically changing from static silos with manually managing resources and applications, towards dynamic virtual environments with automated and shared services, i.e. from silo-oriented to service-oriented architectures.

With sciences and businesses turning global and competitive, applications, products and services becoming more complex, and research and development teams being distributed, ICT is in transition again. Global challenges require global approaches: on the horizon, so-called virtual organizations and partner grids will provide the necessary communication and collaboration platform, with grid portals for secure access to resources, applications, data, and collaboratories.

One component which will certainly foster this next-generation scenario is Cloud Computing, as recently offered by companies like Sun (2006) Network.com, IBM (2008), Amazon (2007), Elastic Compute Cloud, Google (2008) App Engine, Google Group (2008), and CloudCamp (2008), and many more in the near future. Clouds will become important dynamic components of research and enterprise grids, adding a new 'external' dimension of flexibility to them by enhancing their 'home' resource capacity whenever needed, on demand. Existing businesses will use them for their peak demands and for new projects, service providers will host their applications on them and provide Software as a Service (SaaS), start-ups will integrate them in their offerings without the need to buy resources upfront, and setting up new Web 2.0 communities will become very easy.

To Cloud-enable applications will follow similar strategies as with grid-enabling, as discussed in the previous paragraphs. Similarly challenging as with Grids, though, are the cultural, mental, legal, and political aspects in the Cloud context.

Building trust and reputation among the users and the providers will help in some scenarios. But it is currently difficult to imagine that users may easily entrust their corporate core assets and sensitive data to Cloud service providers. In 2008, the status of Clouds seems to be similar to the status of Grids in the early 2000s: a few simple and well-suited application scenarios run on Clouds, but by far most of the more complex and demanding applications in research and enterprises will face many barriers on Clouds which still have to be removed, one by one.

One example of an early innovative Cloud system came from Sun when it truly built its SunGrid (2005) from scratch, based on the vision that the network is the computer. As with other early technologies in the past, Sun paid a high price for being first and doing all the experiments and the evangelization, but their reputation as an innovator is here to stay. Its successor, Sun Network.com (2008), is very popular among its few die-hard clients. This is because of an easy-to use technology (Grid Engine, Jini, JavaSpaces), but it's especially because of their innovative early users, such as CDO2 (2008), and because of the instant support users get from the Sun team.

A similar promising example in the future might be the DEISA Distributed European Infrastructure for Supercomputing Applications, with its DECI – DEISA Extreme Computing Initiative. Why is DECI currently so successful in offering millions of supercomputing cycles to the European e-Science community and helping scientists gain new scientific insights? Several reasons abound: DEISA has a very targeted focus on specific (long-running) supercomputing applications and most of the applications just run on one – best-suited - system; its user-friendly access - through technology like DESHL (2008) and UNICORE (2008); its staying away from those more ambitious general-purpose Grid efforts; its coordinating function which leaves the consortium partners (the European supercomputer centers) fully independent. It is equally similar to Sun Network.com because of ATASKF (DECI (2008) and the application task force. The application experts also help the users with porting their applications to the DEISA infrastructure. If all this is here to stay and the (currently funded) activities are taken over by the individual supercomputer centers, DEISA will have a good chance to exist for a long time, even after the funding dries up. We might end up with a DEISA Cloud which will become an (external) HPC node within a Grid application workflow.

With this sea-change ahead of us, there will be a continuous strategic importance for sciences and businesses to support the work of the Open Grid Forum (OGF, 2008). Because only standards will enable building e-infrastructures and grid-enabled applications easily from different components and to transition towards an agile platform for federated services. Standards, developed in OGF, guarantee upfront an interoperation of components best suited for applications, and thus reducing dependency from proprietary building blocks and keeping cost under control with increasing research and business flexibility.

CONCLUSION

Sustainable grid-enabled applications require sustainable grid infrastructures. It doesn't make any sense, for example, in a three-year funded Grid project, to develop or port a complex application to the Grid which will shut down after the project ends. We have to make sure that we are able to build sustainable grid infrastructures which will last for a long time. The author offers 10 rules for building a sustainable grid, available also from the OGF Thought Leadership (2008). These rules are derived from mainly four sources: author's research on major grid projects published in a RENCI report (Gentzsch, 2007a), the e-IRG Workshop on "A Sustainable Grid Infrastructure for Europe" in (Gentzsch, 2007b), the 2nd International Workshop on Campus and Community

Grids at OGF20 in Manchester (McGinnis, 2007) and author's experience with coordinating the German D-Grid Initiative (D-Grid, 2008). The rules presented here are mainly non-technical, because the author believes most of the challenges in building and operating a grid are in the form of cultural, legal and regulatory barriers.

Rule 1: Identify your specific benefits. The first thought should be about users and organizations. What's in it for them? Identify the benefits which fit best: transparent access to and better utilization of resources; almost infinite compute and storage capacity; flexibility, adaptability and automation through dynamic and concerted interoperation of networked resources; cost reduction through utility model; shorter time-to-market because of more simulations at the same time on the grid. Grid technology helps to adjust an enterprise's IT architecture to real business requirements (and not vice versa). For example, global companies will be able to decompose their highly complex processes into modular components of a workflow which can be distributed around the globe such that on-demand availability and access to suitable workforce and resources are assured, productivity increased, and cost reduced. Application of grid technology in these processes, guarantees seamless integration of and communication among all distributed components and provides transparent and secure access to sensitive company information and other proprietary assets, world-wide. Grid computing is especially of great benefit for those research and business groups which cannot afford expensive IT resources . It enables engineers to remotely access any IT resource as a utility, to simulate any process and any product (and product life cycle) before it is built, resulting in higher quality, increased functionality, and cost and risk reduction.

Rule 2: Evangelize decision makers first. Decision makers give the money and authority for

the grid project. The more they know about the project and the more they believe in it, the more money and time will be available, and the easier the task to lead and motivate the team and to get things done. Present a business case (current deficiencies, specific benefits of the grid (see Rule #1), how much will it cost and how much will it return, etc. They might also have to modify existing policies, top down, to make it easier for users (and providers) to cope with the challenges of and to accept and use the new services. For example, why would a researcher (or a department in an enterprise) stop buying computers when money continues to be allocated for buying it? This policy should be changed to support a utility model instead of an ownership model. If the building of a national grid, for example, convince the government to modify its research funding model is a tough task.

Rule 3: Don't re-invent the wheels. In the early grid days, many grid projects tried to develop the whole software stack themselves: from the middleware layer to the software tools, to grid-enabling the applications, to the portal and Web layer...and got troubled by the next technology change. Today, so many grid technologies, products and projects exist that one can examine, select the ones that fit best one's needs, and perhaps 'copy' for a prototype. These systems can be optimized to fit the requirements of users. Consider, however, that all grids are different. For example, research grids are mainly about sharing (e.g. sharing resources, knowledge, data), commercial enterprise grids are about cost and revenue (e.g. TCO, ROI, productivity). Therefore, if one is in the academic community, search for academic use cases, while if commercial, check for commercial use cases.

Rule 4: KISS (Keep It Simple and Stupid). It took users years to get acquainted with their current working environment and tools. Try hard to stick with what is at hand and how things are done. Plan for an incremental approach and spend time interacting. Social effects dominate in grids. Join forces with the system people to change/modify mainly the lower layers of the architecture. Users are your customers, thus king. Differentiate between two groups of users: the end users who are designing and developing the products (or the research results) which account for all the earnings of the company (or reputation and therefore funding for your research institute), and the system experts who are eager to support the end users with the best possible services. Therefore, one can only succeed if one demonstrates a handful of clear benefits to these two user groups.

Rule 5: Evolution, not revolution. As the saying goes: never change a running system. We all hate changes in our daily lives, except when we are sure that things will drastically improve. Users and their applications deeply depend on a reliable infrastructure. So, whenever changing the user layer, only change it in small steps and in large time cycles. Start with enhancing existing service models moderately, and test suitable utility models first as pilots. Part of business plan has to be an excellent training and communications strategy.

Rule 6: Establish a governance structure. Define clear responsibilities and dependencies for specific tasks, duties and people during and after the project. An advisory board should include representatives of the end-users as well as application and system experts. In case of more complex projects, e.g. consisting of an integration project and several application or community projects, an efficient management board should lead and steer coordination and collaboration among the projects and the working groups. The management board (Steering Committee) should consist of leaders of the sub-projects. Regular face-to–face meetings are very important.

Rule 7: Money, money, money. Don't have unrealistic expectations that grid computing will save money initially. In the early stage, grid projects needed enough funding to get over the early-adopter phase into a mature state with a rock-solid grid infrastructure such that other user communities can join easily. In research grids, for example, we estimate this funding phase currently to be in the order of 3-5 years, with more funding in the beginning for the grid infrastructure, and later more funding for the application communities. In larger (e.g. global) research grids, funding must cover Teams or Centers of Excellence, for building, managing and operating the grid infrastructure, and for middleware tools, application support, and training. Today's funding models in research and education are often project based and thus not ready for a utilitarian approach where resource usage is based on a pay-as-you-go approach. Old funding models first have to be adjusted accordingly before a utility model can be introduced successfully. For example, today's existing government funding models are often counter-productive when establishing new and efficient forms of utility services (see Rule #2). In the long run, grid computing will save you money through a much more efficient, flexible and productive infrastructure.

Rule 8: Secure some funding for after the end of the project. Continuity especially for maintenance and support are extremely important for the sustainability of your grid infrastructure. Make sure at the beginning of your project that additional funding will be available after the end of the project, to guarantee service and support and continuous improvement and adjustment of the infrastructure.

Rule 9: Try not to grid-enable your applications in the first place. Adjusting application to changing technologies costs a lot of effort and money, and takes a lot of precious time. Did one macro-assemble, vectorize, multi-task, parallelize, or multithread application in the past? Then, grid-enabling that code is relatively easy, as we have seen in this article. But doing this from scratch is not what the user should do. Better to use the money to buy (lease, rent, subscribe to) software as a service or to hire a few consultants who can grid-enable application and/or (even better) help enable grid architecture to dynamically cope with the applications and user requirements (instead of vice versa). Today, in grids, we are looking more at chunks of independent jobs, (or chunks of transactions), but let schedulers and brokers decide how to distribute these chunks onto the best-suited and least-loaded servers in the grid, or let the servers decide themselves to share the chunks with their neighbors automatically whenever they become overloaded.

Role 10: Adopt a 'human' business model. Don't invent new business models. This usually increases the risk for failure. Learn from the business models we have with other service infrastructures: water, gas, telephony, electricity, mass transportation, the Internet, and the World Wide Web. Despite this wide variety of areas, there is only a handful of successful business models: on one end of the spectrum, one pays the total price for the service or part service that is used. Sometimes, however, there are hidden or secondary applications. For example, electrical power alone doesn't help. It's only useful if it generates something, e.g. light, or heat, or cold, etc. The infrastructure is what creates a whole new industry of new appliances: light bulbs, heaters, refrigerators, etc. Back to grids: providing the right (transparent) infrastructure (services) and the right (simple) business model will most certainly create a new set of services which most probably will improve the quality of life in the future.

REFERENCES

Altintas, I., Berkley, C., Jaeger, E., Jones, M., Ludascher, B., & Mock, S. (2004). Kepler: an extensible system for design and execution of scientific workflows. Proceedings of the 16th International Conference on Scientific and Statistical Database Management (SSDBM), Santorini Island, Greece. Kepler website at http://kepler-project.org

Amazon Elastic Compute Cloud (2007). Retrieved from www.amazon.com/ec2

Badia, R.M., Labarta, J.S., Sirvent, R.L., Perez, J.M., Cela, J.M., & Grima, R. (2003). Programming Grid Applications with GRID Superscalar. Journal of Grid Computing, vol. 1, pp. 151-170.

Baker, S. (2007). Google and the Wisdom of Clouds. Business Week, Dec. 13. Retrieved from www.businessweek.com/magazine/content/07_52/b4064048925836.htm

BEinGRID (2008). Business Experiments in Grids. Retrieved from www.beingrid.com.

Beltrame, F., Maggi, P., Melato, M., Molinari, E., Sisto, R., & Torterolo, L. (2006). SRB Data Grid and Compute Grid Integration via the EnginFrame Grid Portal. Proceedings of the 1st SRB Workshop, 2-3 February 2006, San Diego, USA. PDF retrieved from www.sdsc.edu/srb/Workshop/SRB-handout-v2.pdf

BIRN (2008). Biomedical Informatics Research Network. Retrieved from www.nbirn.net/index.shtm

Buyya, R., Abramson, D., & Giddy, J. (2000). Nimrod/G: an architecture for a resource management and scheduling system in a global computational grid", Proceedings of the 4th International Conference on High Performance Computing in the Asia-Pacific Region. www.csse.monash.edu.au/~davida/nimrod/nimrodg.htm

Chaubal, Ch. (2003). Sun Grid Engine, Enterprise Edition—Software Configuration Guidelines andUse Cases. Sun Blueprints. www.sun.com/blueprints/0703/817-3179.pdf

CHRONOS Portal (2004). Retrieved from http://portal.chronos.org/gridsphere/gridsphere

CloudCamp (2008). Retrived from http://www.cloudcamp.com/

CDO2 (2008). CDOSheet for Pricing and Risk Analysis. Retrieved from www.cdo2.com

DECI (2008). DEISA Extreme Computing Initiative. Retrieved from www.deisa.eu/science/deci

DEISA. (2008). Distributed European Infrastructure for Supercomputing Applications. Retrieved from www.deisa.eu

DESHL. (2008). DEISA Services for Heterogeneous management Layer. http://forge.nesc.ac.uk/projects/deisa-jra7/

D-Grid (2008). Retrieved from www.d-grid.de/index.php?id=1&L=1

Dongarra, J., Foster, I., Fox, G., Gropp, W., Kennedy, K., Torczon, L., & White, A. (2003). Sourcebook of Parallel Computing. Morgan Kaufmann Publishers.

EnginFrame (2008). Grid and Cloud Portal. Retrieved from www.nice-italy.com

Foster, I., & Kesselman, C. (Eds.). (1999). The Grid: Blueprint for a New Computing Infrastructure. Morgan Kaufmann Publishers.

Foster, I. (2000). Internet Computing and the Emerging Grid. Nature. Retrieved from www.nature.com/nature/webmatters/grid/grid.html

Foster, I. (2002). What is the Grid? A Three Point Checklist. Retrieved from http://www-fp.mcs.anl.gov/~foster/Articles/WhatIsTheGrid.pdf

Foster, I., Kesselman, & C., Tuecke, S. (2002), The Anatomy of the Grid: Enabling Scalable Virtual Organizations. Retrieved from www.globus.org/alliance/publications/papers/anatomy.pdf

Foster, I., & Kesselman, C. (Eds.). (2004). The Grid 2: Blueprint for a New Computing Infrastructure. Morgan Kaufmann Publishers.

Fox, G., Williams, R., & Messina, P. (1994). Parallel Computing Works! Morgan Kaufmann Publishers.

Frey, J., Mori, T., Nick, J., Smith, C., Snelling, D., Srinivasan, L., & Unger, J. (2005). The Open Grid Services Architecture, Version 1.0. www.ggf.org/ggf_areas_architecture.htm

GAT. (2005). Grid Application Toolkit. www.gridlab.org/WorkPackages/wp-1/

Gentzsch, W. (2002). Response to Ian Foster's "What is the Grid?" GRIDtoday. August 5. www.gridtoday.com/02/0805/100191.html

Gentzsch, W. (2004). Grid Computing Adoption in Research and Industry. In A. Abbas (Ed.), Grid Computing: A Practical Guide to Technology and Applications (pp. 309 – 340). Charles River Media Publishers.

Gentzsch, W. (2004). Enterprise Resource Management : Applications in Research and Industry. In I. Foster & C. Kesselman (Eds.). The Grid 2: Blueprint for a New Computing Infrastructure (pp. 157 – 166). Morgan Kaufmann Publishers.

Gentzsch, W. (2007a). Grid Initiatives: Lessons Learned and Recommendations, RENCI Report retrieved from www.renci.org/publications/reports.php

Gentzsch, W. (Ed.) (2007b). A Sustainable Grid Infrastructure for Europe, Executive Summary of the e-IRG Open Workshop on e-Infrastructures,

Heidelberg, Germany. Retrieved from the e-IRG Website www.e-irg.org/meetings/2007-DE/workshop.html

Gentzsch (2008). Top 10 Rules for Building a Sustainable Grid. In: Grid Thought Leadership Series. Open Grid Forum website at www.ogf.org/TLS/?id=1

GEONgrid (2008). Retrieved from www.geongrid.org

Goodale, T., Jha, S., Kaiser, H., Kielmann, T., Kleijer, P., Merzky, A., Shalf, J., Smith, Ch. (2008). A Simple API for Grid Applications (SAGA). Grid Forum Document GFD.90. Open Grid Forum. Retrieved from www.ogf.org/documents/GFD.90.pdf

Google (2008). Google App Engine. Retrieved from http://code.google.com/appengine/

Google Groups (2008). Cloud Computing. Retrieved from http://groups.google.ca/group/cloud-computing

GridWay (2008). Metascheduling Technologies for the Grid. Retrieved from www.gridway.org/

Grid Engine (2001). Open Source Project. Retrieved from http://gridengine.sunsource.net/

GridSphere (2008). Retrieved from www.gridsphere.org/gridsphere/gridsphere

Gustafson, J. (1987). Reevaluating Amdahl's Law, CACM, vol. 31, pp. 532-533.

Jacob, B., Ferreira, L., Bieberstein, N., Gilzean, C., Girard, J.-Y., Strachowski, R., & Yu, S. (2003). Enabling applications for Grid computing with Globus. IBM Redbook, retrieved from www.redbooks.ibm.com/abstracts/sg246936.html?Open

Jha, S., Kaiser, H., El Khamra, Y., & Weidner, O. (2007). Design and Implementation of Network Performance Aware Applications Using SAGA and Cactus. 3rd IEEE Conference on eScience and Grid Computing, Bangalore, India, 10-13 Dec, pp. 143 – 150.

Karonis, N.T., Toonen, B., & Foster, I. (2002). MPICH-G2: A Grid-Enabled Implementation of the Message Passing Interface. Journal of Parallel and Distributed Computing (JPDC), vol. 63, pp. 551-563.

Lee, C. (2003). Grid Programming Models: Current Tools, Issues and Directions. *Grid Computing*. In G. F. Fran Berman, Tony Hey, Ed., pp. 555–578, Wiley Press, USA.

Luther, A., Buyya, R., Ranjan, R., & Venugopal, S. (2005). Peer-to-Peer Grid Computing and a .NET-based Alchemi Framework. High Performance Computing: Paradigm and Infrastructure. In Guo, M. (Ed.). Wiley Press, USA. www.alchemi.net

McGinnis, L., Wallom, D., & Gentzsch, W. (Eds.) (2007). 2nd International Workshop on Campus and Community Grids. Report retrieved from http://forge.gridforum.org/sf/go/doc14617?nav=1

MyGrid (2008). Retrieved from www.mygrid.org.uk

NEESgrid (2008). Retrieved from www.nees.org/

Neuroth, H., Kerzel, M., & Gentzsch, W. (Eds.). (2007). German Grid Initiative D-Grid. Universitätsverlag Göttingen Publishers. Retrieved from www.d-grid.de/index.php?id=4&L=1

OGF (2008). Open Grid Forum. Retrieved from www.ogf.org

P-GRADE (2003). Parallel Grid Run-time and Application Development Environment. Retrieved from www.lpds.sztaki.hu/pgrade/

PRACE (2008). Partnership for Advanced Computing in Europe. Retrieved from www.prace-project.eu/

Proactive. (2005). "Proactive Manual REVISED 2.2", Proactive, INRIA. Retrieved from http://www-sop.inria.fr/oasis/Proactive/

Saara Väärtö, S. (Ed.). (2008). Advancing Science in Europe. DEISA – Distributed European Infrastructure for Supercomputing Applications. EU FP6 Project. Retrieved from www.deisa.eu/press/DEISA-AdvancingScienceInEurope.pdf

SAGA (2006). SAGA implementation home page. http://fortytwo.cct.lsu.edu:8000/SAGA

Seymour, K., Nakada, H., Matsuoka, S., Dongarra, J., Lee, C., & Casanova, H. (2002). Overview of GridRPC: A Remote Procedure Call API for Grid Computing. Proceedings of the Third International Workshop on Grid Computing, Lecture Notes in Computer Science, Springer, ed 2536, pp. 274–278, Baltimore, USA.

SIMDAT (2008). Grids for Industrial Product Development. Retrieved from www.scai.fraunhofer.de/about_simdat.html

Soh, H., Shazia Haque, S., Liao, W., & Buyya, R. (2006). Grid Programming Models and Environments. In Yuan-Shun Dai, et. al. Advanced Parallel and Distributed Computing (pp. 141–173). Nova Science Publishers.

SunGrid (2005). Sun Utility Computing. Retrieved from www.sun.com/service/sungrid/

Sun Network.com (2008). retrieved from www.network.com/

SURA -Southeastern Universities Research Association. (2007). The Grid Technology Cookbook. Programming Concepts and Challenges. Retrieved from www.sura.org/cookbook/gtcb/

TAVERNA (2008). The Taverna Workbench 1.7. Retrieved from http://taverna.sourceforge.net/

TRIANA (2003). The Triana Project. Retrieved from www.trianacode.org/

UNICORE (2008). UNiform Interface to COmputing Resources. Retrieved from www.unicore.eu/

Venugopal, S., Buyya, R., & Winton, L. (2004). A grid service broker for scheduling distributed data-oriented applications on global grids. Proceedings of the 2nd workshop on Middleware for grid computing, pp. 75–80, Toronto, Canada. www.Gridbus.org/broker

This work was previously published in International Journal of Grid and High Performance Computing, Vol. 1, Issue 1, edited by E. Udoh, pp. 55-77, copyright 2009 by IGI Publishing (an imprint of IGI Global).

Chapter XXXIV
Evaluating a Bio-Inspired Approach for the Design of a Grid Information System: The SO-Grid Portal

Agostino Forestiero
Institute of High Performance Computing and Networking CNR-ICAR, Italy

Carlo Mastroianni
Institute of High Performance Computing and Networking CNR-ICAR, Italy

Fausto Pupo
Institute of High Performance Computing and Networking CNR-ICAR, Italy

Giandomenico Spezzano
Institute of High Performance Computing and Networking CNR-ICAR, Italy

ABSTRACT

This chapter proposes a bio-inspired approach for the construction of a self-organizing Grid information system. A dissemination protocol exploits the activity of ant-inspired mobile agents to replicate and reorganize metadata information on the basis of the characteristics of the related Grid resources. Resource reorganization emerges from simple operations of a large number of agents, in a "swarm intelligence" fashion. Moreover, a discovery protocol allows Grid clients to locate useful resources on the Grid through a semi-informed approach. This chapter also describes the SO-Grid Portal, a simulation portal through which registered users can simulate and analyze the ant-based protocols. This portal can be used by researchers to perform "parameter sweep" studies, as it allows for the graphical comparison of results obtained in previous sessions. We believe that the deployment of the SO-Grid portal, along with the definition and discussion of the protocols presented in this chapter, can foster the understanding and use of swarm intelligence, multi-agent and bio-inspired paradigms in the field of distributed computing.

INTRODUCTION

To support the design and execution of complex applications, modern distributed systems must provide enhanced services such as the retrieval and access to content, the creation and management of content, and the placement of content at appropriate locations. In a Grid, these services are offered by a pillar component of Grid frameworks, the *information system*. This chapter discusses a novel approach for the construction of a Grid information system which allows for an efficient management and discovery of information. The approach, proposed in (Forestiero et al., 2005) in its basic version, exploits the features of (1) epidemic mechanisms tailored to the dissemination of information in distributed systems (Peterson et al., 1997, Eugster & al., 2004) and (2) self organizing systems in which "swarm intelligence" emerges from the behavior of a large number of agents which interact with the environment (Bonabeau & al., 1999, Dasgupta, 2004).

The proposed ARMAP protocol (*Ant-based Replication and MApping Protocol*) disseminates Grid resource *descriptors* (i.e., metadata documents) in a controlled way, by spatially mapping these descriptors according to their semantic classification, so to achieve a logical reorganization of resources. A resource descriptor can be composed of a syntactical description of a Grid service (e.g. a Web Services Description Language - WSDL - document) and/or a semantic description of the capabilities of the service.

Descriptor reorganization results from pick and drop operations performed by a large number of agents. Each ARMAP agent travels the Grid through P2P interconnections among Grid hosts, and uses simple probability functions to decide whether or not to *pick* descriptors from or *drop* descriptors into the current Grid host. This approach is inspired by the activity of some species of ants and termites that cluster and map items within their environment (Bonabeau & al., 1999).

Furthermore, a self-organization approach based on ants' pheromone (Van Dyke & al., 2005) enables each agent to regulate its activity, i.e. its operation *mode*, only on the basis of local information. Indeed, each agent initially works in the *copy* mode: it can generate new descriptors and disseminate them on the Grid. However, when it realizes from its own past activity that a sufficient number of replicas have been generated, it switches to the *move* mode: it only moves descriptors from one host to another without generating new replicas. This switch is performed when the level of a pheromone variable, which depends on agent's activity, exceeds a given threshold.

The ARMAP protocol can effectively be used to build a Grid information system in which (1) resource descriptors are properly replicated and (2) the overall entropy is reduced. A balance between these two features is achieved by regulating the pheromone threshold, i.e., by shortening or extending the time interval in which agents operate under the *copy* mode.

A semi-informed discovery protocol exploits the logical resource organization achieved by ARMAP. Indeed, whenever a large number of descriptors of a specific class are accumulated in a restricted region of the Grid, it becomes convenient to drive query messages (issued by users to locate descriptors of this class) towards this region, in order to maximize the number of discovered descriptors and minimize the response time. While this chapter focuses on enhancements and performance of the ARMAP protocol, the discovery protocol, namely ARDIP (*Ant-Based Resource Discovery Protocol*) is here shortly discussed, whereas an extensive analysis can be found in (Forestiero & al., 2007).

This chapter also introduces the SO-Grid (*Self Organizing Grid*) portal, a Web portal which gives remote users access to an event-based simulator written in Java. This portal, available at the URL *http://so-grid.icar.cnr.it*, allows for the experimental reproduction of simulation results and can be used by researchers to perform "parameter sweep" studies. Through the portal front-end, a user can (1) register to the portal and create a private space in the file system that will contain the results of all the performed simulations; (2) run a new simulation, after setting network and protocol parameters, and graphically monitor the performance indices and (3) choose a subset of simulation data to graphically compare the related performance results.

We hope that the SO-Grid portal will help foster the understanding and use of swarm intelligence, multi-agent and bio-inspired paradigms in the field of distributed computing.

BACKGROUND

The main purpose of the protocol presented in this chapter is the dissemination of metadata documents and their intelligent reorganization. These two objectives are correlated, since an intelligent dissemination of information can increase efficiency management and facilitate discovery, as discussed in (Forestiero & al., 2007).

Information dissemination is a fundamental and frequently occurring problem in large, dynamic and distributed systems whose main purpose is the management and delivery of content. In (Iamnitchi & Foster, 2005) it is proposed to disseminate information selectively to groups of users with common interests, so that data is sent only to where it is wanted. In our chapter, instead of classifying users, the proposal is to exploit the classification of resources: resource descriptors are replicated and disseminated with the purpose of creating regions of the network that are specialized in specific classes of resources. In (Aktas & al., 2007) information dissemination is combined with the issue of effective replica placement, since the main interest is to place replicas in the proximity of requesting clients by taking into account changing demand patterns. Specifically, a metadata document is replicated if its demand is higher than a defined threshold and each replica is placed according to a multicast mechanism that aims to discover the data server which is the closest to demanding clients.

The ARMAP protocol, proposed in this chapter, is inspired by biological mechanisms, and in particular exploits the self-organizing and swarm intelligence characteristics of several biological systems, ranging from ant colonies to wasp swarms and bird flocks. In these systems, a number of small and autonomous entities perform very simple operations driven by local information, but from the combination of such operations a complex and intelligent behavior emerges (Dasgupta, 2005): for example, ants are able to establish the shortest path towards a food source; birds travel in large flocks and rapidly adapt their movements to the ever changing characteristics of the environment.

These biological systems can be quite naturally emulated in a Grid through the multi-agent paradigm (Sycara, 1998): the behavior of insects and birds is imitated by mobile agents which travel through the hosts of a Grid and perform their operations. Multi-agent computer systems can inherit interesting and highly beneficial properties from their biological counterparts, namely: (1) self-organization, since decisions are based on local information, i.e., without any central coordinator; (2) adaptivity, since agents can flexibly react to the ever-changing environment; (3) stigmergy awareness (Grassè, 1959), since agents are able to interact and cooperate through the modifications of the environment that are induced by their operations.

The ARMAP protocol is specifically inspired to ant algorithms, a class of agent systems which aim to solve very complex problems by imitating the behavior of some species of ants (Bonabeau & al., 1999). A technique based on ant pheromone is exploited to tune the behavior of ARMAP agents, making them able to autonomously switch from the *copy* to the *move* mode. This kind of approach is discussed in (Van Dyke & al., 2005), where a decentralized scheme, inspired by insect pheromone, is used to control the activity of a single agent and better accomplish the system goal.

The ARMAP pick and drop probability functions are inspired by probability functions that were introduced in (Deneubourg & al., 1990), and later elaborated and discussed in (Bonabeau & al., 1999), to emulate the behavior of some species of ants that build cemeteries by aggregating corpses in clusters. In ARMAP, such functions have been adapted through the following main modifications: (i) descriptors are not only moved and aggregated, as are corpses in the mentioned papers, but also *replicated*; (ii) descriptors are reorganized according to the class to which they belong, so as to reduce the overall entropy of the system.

DISSEMINATION OF RESOURCE DESCRIPTORS

The aim of the ARMAP protocol (Forestiero & al., 2005) is to achieve a logical organization information on the Grid by spatially mapping resource descriptors according to the semantic classification of related resources. It is assumed that the resources have been previously classified into a number of classes *Nc*, according to their semantics and functionalities (Crespo & Garcia Molina, 2002).

The ARMAP protocol has been analyzed in a Grid in which hosts are connected in a P2P fashion and each host is connected to at most 8 neighbor peers. The Grid has a dynamic nature, and hosts can disconnect and rejoin the network. When connecting to the Grid, a host generates a number of agents given by a discrete Gamma stochastic function, with average *Ngen*, and sets the life time of these agents to *PlifeTime*, which is the average connection time of the host, calculated on the basis of the host's past activity. This mechanism allows for controlling the number of agents that operate on the Grid: indeed, the number of agents is maintained to a value which is about *Ngen* times the number of hosts.

Periodically each ARMAP agent sets off from the current host and performs a number of hops through the P2P links that interconnect the Grid hosts. Then the agent uses random *pick* and *drop* functions in order to replicate and move descriptors from one peer to another. More specifically, at each host an agent must decide whether or not to *pick* the descriptors of a given class, and then carry them in its successive movements, or to *drop* the descriptors that it has previously picked from another host.

As a consequence of pick and drop operations, each host can maintain descriptors of local resources, which are never removed, as well as descriptors of resources published by other hosts, which can be picked up and discarded by agents. In the following, when distinction is relevant, such descriptors will respectively be referred to as *local* and *remote* descriptors.

Pick and drop probability functions are discussed in the following.

Pick Operation

Whenever an ARMAP agent hops to a Grid host, it must decide, for each resource class, whether or not to *pick* the descriptors of this class which are managed by the current host, unless it already carries some descriptors of the same class. In order to achieve replication and mapping functionalities, a *pick* random function is defined with the intention that the probability of picking the descriptors of a class decreases as the local region of the Grid accumulates these descriptors and vice versa. This assures that as soon as the equilibrium condition is broken (i.e., descriptors belonging to different classes begin to be accumulated in different regions), a further reorganization of descriptors is favored.

The basic *Ppick* probability function is shown in formula (1). The agent evaluates this function, generates a real number between 0 and 1, from a uniform random distribution, and then it performs the pick operation if the generated number is lower than the value of the *Ppick* function. This function is the product of two factors, which take into account, respectively, the *absolute* accumulation of descriptors of a given class and their *relative* accumulation (i.e., with respect to other classes). While the absolute factor is inspired by the pick probability defined in (Bonabeau & al., 1999), the relative factor was introduced here to achieve the spatial separation of descriptors.

$$P_{pick} = \left(\frac{fa}{k1 + fa} \right)^2 \cdot \left(\frac{k2}{k2 + fr} \right)^2 \qquad (1)$$

The *fa* fraction is computed as the number of *local* descriptors of the class of interest, maintained by the hosts located in the *visibility region*, out of the overall number of descriptors of the class of interest (including both *local* and *remote*) that are maintained by the same hosts. The value of *fa* is comprised between 0 and 1, as well as the value of *fr*, defined below. The *visibility region* includes all the hosts that are reachable from the current host with a given number of hops, i.e. within the *visibility radius*, which is an algorithm parameter. As more *remote* descriptors of a class are accumulated in the visibility region, *fa* decreases, the first factor of the pick function decreases as well, and the probability that the agent will picks the descriptors becomes lower. This facilitates the formation of accumulation regions. Conversely, the pick probability is large if a small number of *remote* descriptors are located in the local region of the Grid.

The *fr* fraction is computed as the number of descriptors of the class of interest, accumulated in the hosts located in the visibility region, divided by the overall number of descriptors, of all classes, that are accumulated in the same region. As the local region accumulates more descriptors of a class, with respect to other classes, *fr* increases and accordingly the value of the pick probability for this class decreases, and vice versa. This facilitates the spatial separation of descriptors of different classes. The constants *k1, k2* are set to the value 0.1, as in analogous formulas defined in (Bonabeau et al., 1999).

The pick operation can be performed with two different *modes*. If the *copy* mode is used, the agent, when executing a pick operation, leaves the descriptors on the current host, generates a replica of them, and carries such replicas until it will drop them in another host. Conversely, with the *move* mode, as an agent picks the descriptors, it removes them from the current host (except for the *local* descriptors, which cannot be removed), thus preventing an excessive proliferation of replicas.

Drop Operation

As well as the pick function, the drop function is first used to break the initial equilibrium and then to strengthen the mapping of descriptors of different classes in different Grid regions. Whenever an agent gets to a new Grid host, it must decide, if it is carrying some descriptors of a given class, whether or not to *drop* such descriptors in the current host. As opposed to the pick operation, the drop probability function *Pdrop*, shown in formula (2), is directly proportional to the relative accumulation of descriptors of the class of interest in the visibility region. Therefore, as accumulation regions begin to emerge, the drop operations are and more and more favored in these regions. In (2) the threshold constant *k3* is set to 0.3, as in (Bonabeau & al., 1999).

$$P_{drop} = \left(\frac{fr}{k3 + fr} \right)^2 \qquad (2)$$

A Pheromone Mechanism for Reduction of System Entropy

A spatial entropy function, based on the well known Shannon's formula for the calculation of information content, is defined to evaluate the effectiveness of the ARMAP protocol. For each peer *p*, the local entropy *Ep* gives an estimation of the extent to which the descriptors have already been mapped within the visibility region centered in *p*. *Ep* has been normalized, so that its value is comprised between 0 and 1. As shown in formula (3), the overall entropy *E* is defined as the average of the entropy values *Ep* computed at all the Grid hosts. In (3), *fr(i)* is the fraction of descriptors of class *Ci* that are located in the visibility region with respect to the overall number of descriptors located in the same region.

$$Ep = \frac{\sum_{i=1..Nc} fr(i) \cdot \log_2 \frac{1}{fr(i)}}{\log_2 Nc}, \quad E = \frac{\sum_{p \varepsilon Grid} Ep}{Np} \qquad (3)$$

In (Forestiero et al., 2005) it was shown that the overall spatial entropy can be minimized if each agent exploits both the ARMAP modes, i.e. *copy* and *move*. In the first period of its life, each agent *copies* the descriptors that it picks from a Grid host, but when it realizes from its own activeness that the mapping process is at an advanced stage, it begins simply to *move* descriptors from one host to another, without creating new replicas.

In fact, agents cannot always operate under the *copy* mode, since eventually every host would be assigned a very large number of descriptors of all classes, thus weakening the efficacy of descriptor mapping. The protocol is effective only if each agent, after replicating a number of descriptors, switches from *copy* to *move*. A self-organization approach based on ants' pheromone mechanism enables an agent to perform this mode switch only on the basis of local information. This approach is inspired by the observation that agents perform more operations when the system entropy is high, whereas operation frequency gradually decreases as descriptors are properly reorganized. In particular, at given time intervals, i.e. every 2,000 seconds, each agent counts up the number of times that it has evaluated the pick and drop probability functions, and the number of times that it has actually performed *pick* and *drop* operations, as the generated random number is lower than the value of the probability function. At the end of each time interval, the agent makes a deposit into its pheromone base, by adding a pheromone amount equal to the ratio between the number of "unsuccessful" (not actually performed) operations and the overall number of operation attempts. At the end of the i-th time interval, the pheromone level Φi is computed with formula (4).

$$\Phi i = Ev \cdot \Phi i - 1 + \varphi i \qquad (4)$$

In this formula, φi is the fraction of unsuccessful operations performed in the last time interval. An evaporation mechanism is used to give a higher weigh to recent behavior of the agent, and the evaporation rate *Ev* is set to 0.9. With these settings, the value of Φi is always comprised between 0 and 10. As soon as the pheromone level exceeds a pheromone threshold *Tf* (whose value is also in the range [0,10]), the agent realizes that the frequency of *pick* and *drop* operations has remarkably reduced, so it switches its protocol mode from *copy* to *move*. The value of *Tf* can be used to tune the number of agents that work in *copy* mode and are therefore able to create new descriptor replicas, as discussed in the next section.

PERFORMANCE EVALUATION

The performance of the ARMAP protocol has been evaluated with an event-based simulator written in Java, which will be described in the chapter section related to the SO-Grid Portal.

Simulation runs have been performed with the following setting of network and protocol parameters. The number of peers *Np*, or Grid size, is set to 2500. The average connection time of a specific peer, *Plifetime*, is generated according to a Gamma distribution function, with an average value set to 100,000 seconds. The use of the Gamma function assures that the Grid contains very dynamic hosts, that frequently disconnect and rejoin the network, as well as much more stable hosts. Every time a peer disconnects from the Grid, it loses all the *remote* descriptors previously deposited by agents, thus contributing to the removal of obsolete information. The average number of Grid resources owned and published by a single peer is set to 15. Grid resources are classified in a number of classes *Nc*, which is equal to 5. The mean number of agents that travel the Grid is set to *Np*/2: this is accomplished, as explained in the previous section, by setting the mean number of agents generated by a peer, *Ngen*, to 0.5. The average time *Tmov* between two successive agent movements (i.e., between two successive evaluations of *pick* and *drop* functions) is 60 s. The maximum number of P2P hops that are performed within a single agent movement, *Hmax*, is set to 3. The visibility radius *Rv*, defined in the previous section and used for the evaluation of pick and drop functions, is set to 1. Finally, the pheromone threshold *Tf* ranges from 3 to 10.

It is worth noting that the So-Grid replication algorithm is very robust with respect to variations of the above mentioned parameters. The values of algorithm parameters, e.g., the number of resources per peer, the number of resource classes, the average connection time of a peer, etc., can affect the rapidity of the process, and in some cases can slightly influence the steady values of performance indices, but the qualitative behavior is always preserved. This can be verified by performing "parameter sweep" simulation through the SO-Grid Portal, as discussed in the next section.

The following performance indices are used. The overall entropy *E*, discussed in the precious section, is used to estimate the effectiveness of the ARMAP protocol in the reorganization of descriptors. The *Nrpr* index is defined as the mean number of descriptors that are generated for each resource. Since new descriptors are only generated by ARMAP agents that work in the *copy* mode, the number of such agents, *Ncopy*, is another interesting performance index.

Prior to the numerical analysis, a graphical description of the behavior of the replication algorithm is given in Figure 1. For the sake of clearness, here the number of classes is set to 3, and the peers are arranged in a 2-dimensional mesh. The figure shows only a portion of the Grid, though the simulation was performed on a network with 2500 hosts.

In this figure, different symbols are associated with the three classes. Each peer is visualized by the symbol that corresponds to the class to which the largest number of descriptors, maintained by this peer, belong. Furthermore, the symbol size is

Figure 1. Accumulation and reorganization of resource descriptors, belonging to 3 resources classes, in a region of the Grid

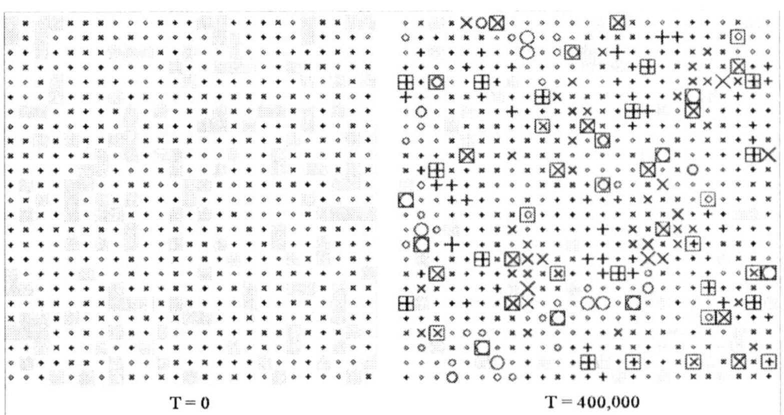

proportional to the number of descriptors of the dominant class. Two snapshots of the network are depicted: the first is taken when the ARMAP process is initiated, at time 0, the second is taken 400,000 seconds later, in a quite steady situation. This figure shows that descriptors are initially distributed in a completely random fashion, but subsequently they are accumulated and reorganized by agents in separate regions of the Grid, according to their class. The peers with a marked border are *representative* peers that work as attractors for query messages, as briefly described in the next section.

A set of simulation runs have been performed to evaluate the performance of the basic pick and drop probability functions and the effectiveness of the pheromone mechanism that drives the mode switch of agents. These simulations were performed with a Grid size *Np* equal to 2500 and a number of classes *Nc* equal to 5. Figure 2 reports the number of agents that work in *copy* mode, *Ncopy* (also called *copy agents* in the following), versus time, for different values of the pheromone threshold *Tf*. When ARMAP is initiated, all the agents (about 1250, half the number of peers) are generated in the *copy* mode, but subsequently several agents switch to *move*, as soon as their pheromone value exceeds the threshold *Tf*. This corresponds to the sudden drop of curves that can be observed in Figure 2. This drop does not occur if *Tf* is equal to 10 because this value can never be reached by the pheromone (see formula (4)); therefore, with *Tf*=10 all agents remain in *copy* along all their lives.

After the first phase of the ARMAP process, an equilibrium is reached because the number of new agents which are generated by hosts (such agents always set off in *copy* mode) and the number of agents that switch from *copy* to *move* get balanced. Moreover, if the pheromone threshold *Tf* is increased, the average interval of time in which an agent works in *copy* becomes longer, because the pheromone level takes more time to reach this threshold; therefore the average value of *Ncopy* at the equilibrium becomes larger. In conclusion, the setting of *Tf* is a simple and efficient method to tune the number of *copy* agents, and therefore the rate of generation of new replicas.

Figure 3 shows the mean number of descriptor replicas per resource and confirms that descriptor dissemination is more intense if the pheromone threshold is increased, because a larger number of *copy* agents operate on the network. However, a more intense dissemination is not always as-

Figure 2. Number of agents in copy mode for different values of the pheromone threshold Tf

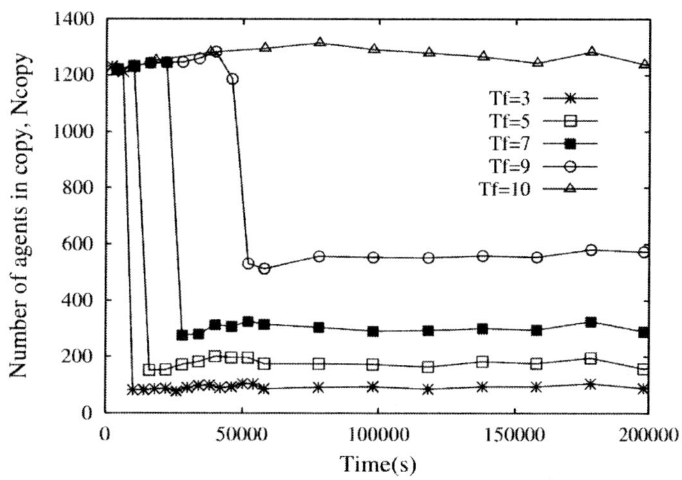

sociated to a better reorganization, i.e. to a more effective spatial separation of descriptors belonging to different classes.

Figure 4 shows that when the number of *copy* agents is increased, which is achieved by increasing the threshold *Tf*, the reduction of the overall entropy is lower. For example, with *Tf*=3, the value of the overall entropy decreases from the initial value of about 1 (maximum disorder) to less than 0.72. With *Tf*=9, however, the entropy is only reduced to about 0.87. As an extreme case, virtually no entropy decrease is observed if all the agents operate in *copy* (*Tf*=10), which confirms that the presence of *move* agents is necessary to perform an effective descriptor reorganization.

It can be concluded that *copy* agents are useful to replicate and disseminate descriptors but it is the *move* agents that actually perform the descriptor

Figure 3. Mean number of replicas per resource for different values of the pheromone threshold Tf

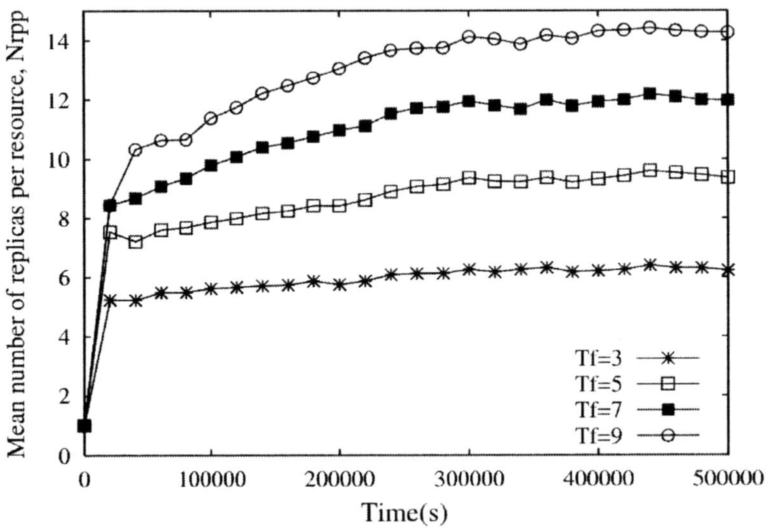

Figure 4. Overall system entropy for different values of the pheromone threshold Tf

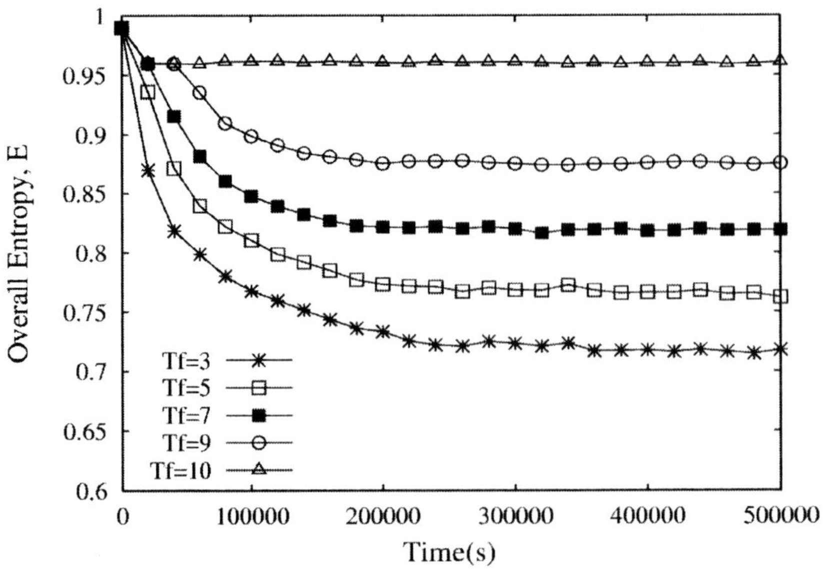

reorganization and are able to create Grid regions specialized in specific classes of resources. A balance between the two features (replication and reorganization) can be performed by appropriately tuning the pheromone threshold.

As opposed to the indices described so far, the processing load *L*, i.e., the average number of agents per second that are processed by a peer, does not depend on the pheromone threshold, but on the number of agents and on the frequency of their movements across the Grid. Recalling that the number of agents *Na* is approximately equal to the number of peers *Np* times the mean number of agents generated by a peer, *Ngen*, *L* can be obtained as follows:

$$L = \frac{Na}{Tmov \cdot Np} \approx \frac{Ngen}{Tmov}$$

With the parameter setting discussed before, each peer receives and processes about one agent every 120 seconds, which can be considered an acceptable load.

RESOURCE DISCOVERY

The logical resource organization achieved by AR-MAP can be exploited through a semi-informed discovery protocol, in order to maximize the number of discovered resources and minimize the response time. Such a discovery protocol, namely ARDIP (Ant-Based Resource Discovery Protocol), is based on the notion of *representative peers*. As descriptors of a class are accumulated in a Grid region, the peer that, within this region, collects the largest number of descriptors is elected as a representative peer for this class. The objective of a discovery operation is to direct a query message to a representative peer as fast as possible, since this peer likely maintains a large number of useful descriptors.

The key features of the ARDIP protocol are discussed in the following, while a detailed discussion and evaluation can be found in (Forestiero & al., 2007). With ARDIP, a discovery operation is performed in two phases, one *blind* and one *informed*. In the *blind* phase, the *random walks* technique is adopted (Lv & al., 2002): a number of query messages are issued by the requesting peer and travel the Grid in a *blind* (random) fashion. Whenever a query gets close enough to a representative peer, the search becomes *informed* and the query is driven towards this representative peer. When the query arrives at the representative peer, a queryHit message is generated and gets back to the requesting peer by following the same path in the opposite direction. The proximity of a representative peer is detected through another kind of pheromone mechanism, which is inspired by the behavior of ants that search for a food source. When a query message first discovers a representative peer in its random path, the queryHit message leaves an amount of pheromone in the first peers of its return journey, and this pheromone can be later recognized by other peers that casually approach this representative peer.

The semi-informed protocol aims to combine the benefits of both blind and informed resource discovery approaches which are currently used in P2P networks (Tsoumakos & Roussopoulos, 2003). In fact, a pure blind approach (e.g. using flooding or random walks techniques) is simple to implement but has limited performance and can cause an excessive network load, whereas a pure informed approach generally requires a very structured resource organization which is impractical in a large, heterogeneous and dynamic Grid.

THE SO-GRID PORTAL

The ICAR research institute of the Italian CNR recently developed the So-Grid (Self Organizing Grid) Portal, available at the URL *http://so-grid.icar.cnr.it*, in order to allow remote users to experience and evaluate the bio-inspired protocols described in this chapter. The portal, designed according to the Java J2EE technology and built upon the JBoss application server, provide users with a Web interface through which it is possible to run a remote simulation, after setting its parameters, graphically monitor the simulation execution at run time and compare the results of several simulations in order to perform "parameter sweep" analysis.

The overall architecture of the portal is depicted in Figure 5. The portal is composed of three main tiers: (1) the portal front-end, which gives access to the Web interface; (2) the portal back-end, which includes the So-Grid controller, a database and a file repository and (3) the So-Grid event-based simulator.

In the following, the portal components will be briefly described, in a right to left order, starting from the simulator.

The SO-Grid Simulator

The performances of the ARMAP and ARDIP protocols have been evaluated through an event-based simulator written in Java. Simulation *objects* are used to emulate Grid hosts and bio-inspired agents. Each object reacts to external *events* according to a finite state automaton and responds by performing specific operations and/or by generating new messages/events to be delivered to other objects.

For example, a peer visited by an agent gives it information about the descriptors that this peer maintains; afterwards the agent uses the probability functions define in formulas (1) and (2) to decide whether or not to *pick* descriptors from or *drop* descriptors into the peer. Finally, the agent sets the simulation time in which it will perform its next movement on the Grid and creates a related event that will be delivered at the specified time to the peer to which the agent will move. Events are ordered in a general queue by a simulation *manager* component, and they are delivered to corresponding destination objects according to their expiration time, so that peers and agents can operate concurrently along the simulation. Moreover, the simulation environment exploits the visual facilities and libraries offered by the Swarm environment (Minar & al., 1996).

The simulator was integrated in the SO-Grid portal through interactions with the SO-Grid Controller and the file repository. In particular, the Controller prepares the simulation configuration file which contains the parameter values set by the user, sends this file to the simulator and starts the simulation. As the simulation proceeds, the partial results are stored in the file repository, so as to let the Controller access them. When the simulation terminates, the simulator informs the Control-

Figure 5. SO-Grid portal: Overall architecture

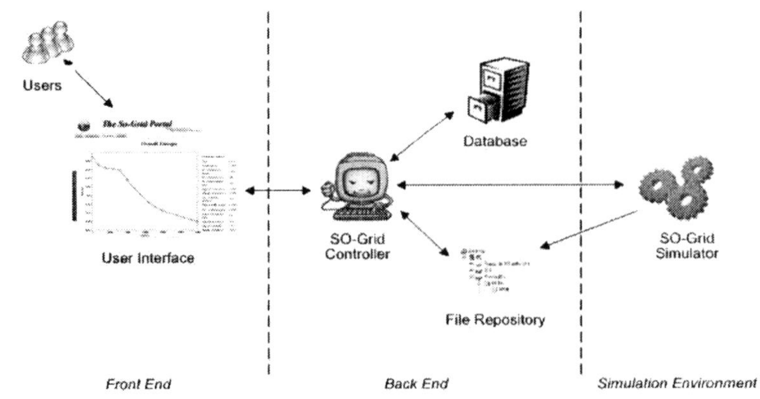

ler, which in turn updates database information pertaining to the running simulations.

File Repository and Database

Both the file repository and the database contain information used by the Controller to manage registered users, running simulations and simulation results.

Specifically, the file repository is organized in folders. Each folder is assigned to a registered user, and essentially contains configuration files and results files. A configuration file contains a set of parameter-value associations and is given as input to the simulator. Parameters can be related either to the simulated network (number of nodes, average connection time of nodes, number of resources, etc.) or to the bio-inspired protocols (number of agents, values of probability function constants, etc.). Each parameter has a default value that can be modified by the user: the resulting configuration file is saved in the user folder, in order to maintain all information about simulation settings, for future use.

Moreover, a user folder stores the files which contain the results of the simulations performed by the user. The name of a results file is chosen by the user before starting the simulation. A results file contains the values of performance indices (system entropy, number of replicas, discovery response times, etc.) as they are periodically computed by the simulator. A results file is organized in rows and columns so as it can be given as input to Gnuplot, a freeware software that offers batch plotting utilities and is able to automatically build graphical plots from data files. Gnuplot is used both to visualize simulation results at run time and compare results related to several simulations performed by the same user.

The database stores two kinds of information. The first is the account data (name, password, e-mail address etc.) of registered users, which is checked when a user requests to log on the portal. Furthermore, the database maintains information about the running simulations; in particular the database associates the process id of each running simulation to the user which has requested it and to the name of the file that will contain the results. This helps the management of simulations: for example, after the log in procedure of a user, the portal checks if there is a simulation started by this user which is still running, and in this case the user is redirected to the portal form which shows the results in real time. Moreover, the database is used to control the number of ongoing simulations: indeed, to maintain the performance of the simulator at an acceptable level, each user can execute only one simulation at a time and no more than 5 simulations can be executed concurrently by different users.

SO-GRID CONTROLLER

The SO-Grid Controller is the component that receives, through the graphical interface, user requests and interacts with other components to satisfy them. It is built upon the J2EE technology and its main functionalities are performed by Java Servlets. Some of the tasks that the Controller performs are the following:

- It manages the user log on procedure, by checking account data of users stored on the database;
- It manages the user choice of simulation parameters, and creates the correspondent configuration file, which is passed as input to the simulator;
- It interacts with the simulator to start and stop simulations on behalf of users, and updates related information on the database;
- It accesses the file repository to retrieve simulation results and show them to users, according to their preferences, through the graphical interface;
- When requested by a user, it builds a Gnuplot file for the comparison of results related to different simulation runs, and passes it to the graphical interface.

SO-Grid GUI

The portal front-end provides the user GUI through which users can operate on the portal, execute simulations and visualize results. In this section, we show some of the GUI Web forms, in order to illustrate how the portal can be actually used.

Figure 6 depicts the login form that is presented to users when they want to enter the simulation area. Notice that the other areas, i.e. the "Home" area and the areas that give access to information concerning documents and involved people, is available to all users and does not need a log on procedure.

Once in the simulation area, the user can choose whether to run a new simulation or visualize the results of one or more simulations executed previously. If a simulation started in a previous session is still running, the user can also monitor its execution.

Before running a new simulation, the user accesses a Web form in which he/she can set the duration of the simulation and the values of network and simulation parameters, and can choose the name of the results file, as shown in Figure 7. For each parameter, the user can accept the default value or decide to change and set it within a given range. For example, the default number of classes is 5, but values between 3 and 12 are admitted.

While the simulation proceeds, the user can visualize, at run time, the values related to the different performance indices, by means of plots built by the Gnuplot software. As an example, Figure 8 shows a snapshot of the SO-Grid interface, in which the user can observe the trend of the overall entropy, plotted versus time. Whenever he/she wants, the user can decide to stop a running simulation, otherwise this will terminate when the simulation time expires; in both cases the results are automatically saved in the file whose name was specified in the parameter setting phase.

The user can also configure and visualize a plot that compares results related to different simulations. Through the form shown in Figure 9, he/she can check the results files corresponding to past simulations (in this case results obtained with different numbers of classes), and the performance index of interest. After clicking on the "Create graphic" button, the user will be shown a plot that compares the values of the chosen performance index obtained with the selected simulations. This is a very user friendly and powerful tool that allows users to carry out "parameter sweep" analysis of the bio-inspired protocols presented in this chapter. For example, Figure 10 depicts the values of the overall entropy obtained in three

Figure 6. SO-Grid graphical interface: the login page

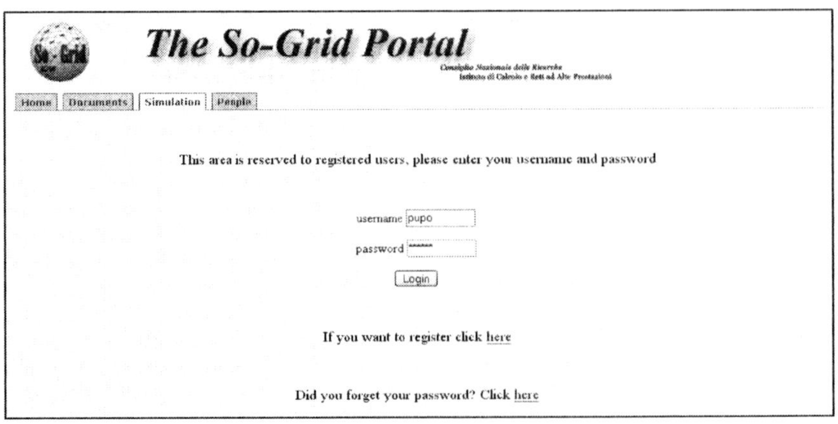

Figure 7. SO-Grid graphical interface: Web form for parameter setting

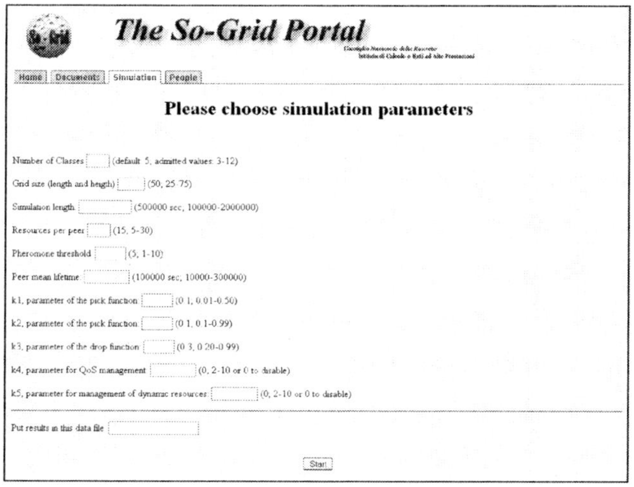

Figure 8. SO-Grid graphical interface: Web page that shows simulation results at run time, in this case the system overall entropy

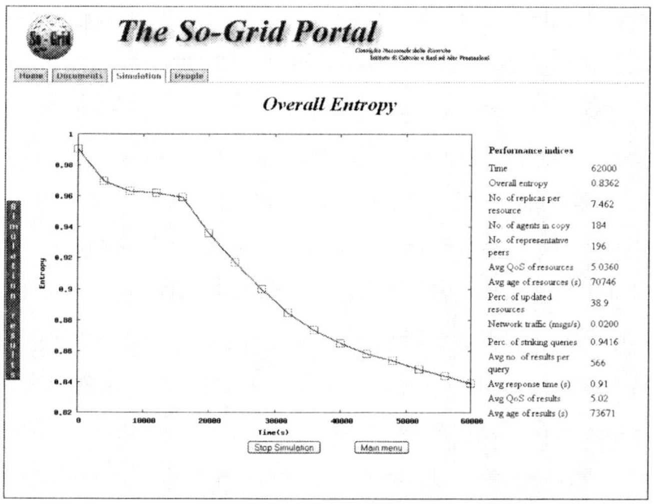

Figure 9. SO-Grid graphical interface: Web form that allows the user to compare results related to different simulations

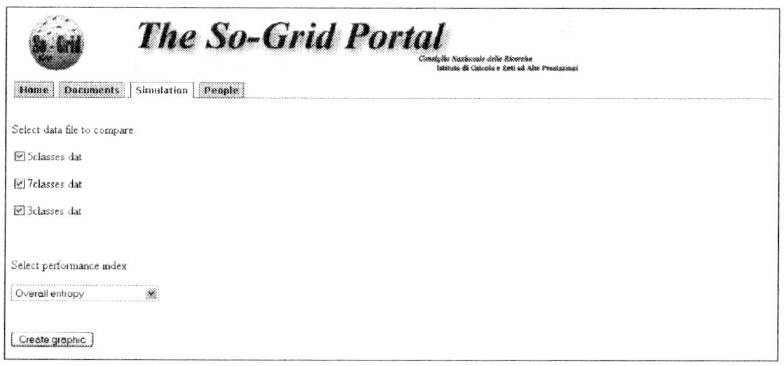

Figure 10. SO-Grid graphical interface: comparison of overall entropy values obtained when the number of classes is set to 3, 5 and 7

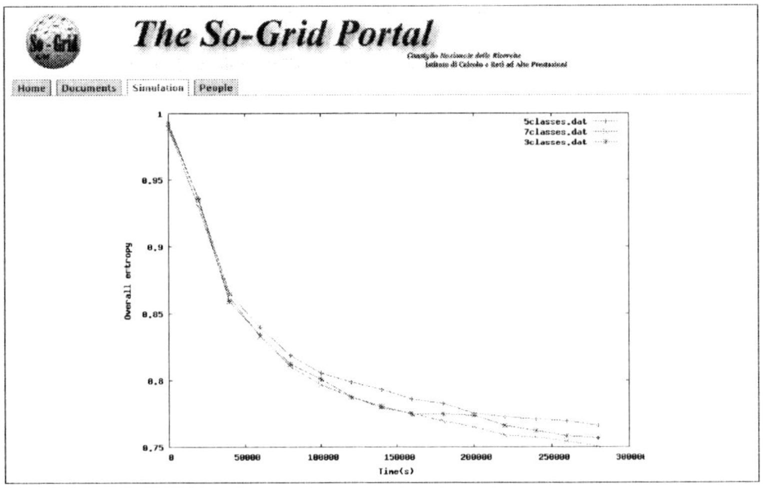

simulations in which the number of classes is set to 3, 5 and 7.

CONCLUSION AND FUTURE WORK

This chapter proposes a bio-inspired approach for the construction of a Grid information system. The ant-inspired ARMAP protocol allows for the dissemination and reorganization of resource descriptors according to the characteristics of related resources. This in turn enables the use of another ant-inspired protocol, ARDIP, which copes with the discovery of resources on behalf of users. ARMAP is executed by a number of ant-like mobile agents that travel the Grid through P2P interconnections among hosts. Agents disseminate metadata documents on the Grid and aggregate information related to similar resources in neighbor Grid nodes, so contributing to decrease the overall system entropy. Resource replication and reorganization can be tuned by appropriately setting a pheromone threshold in order to foster or reduce the activeness of agents. Simulation results show that the ARMAP protocol is able to achieve the mentioned objectives, and is inherently scalable, as agent operations are driven by self-organization and fully decentralized mechanisms, and no information is required about the global state of the system.

To foster the understanding and use of bio-inspired algorithms in distributed systems, this chapter also illustrates the SO-Grid Portal, a simulation portal through which registered users can simulate and experience the behaviour of the discussed protocols. This portal is specifically designed to carry out "parameter sweep" analysis, since a user can graphically compare the results of different simulations executed in previous sessions.

The exploitation of bio-inspired algorithms in distributed systems surely needs more analysis and evaluation in future years. An important issue is that results are always obtained empirically and tuned by experience: a very important step forward would be the construction of an analytical model that relates the operations of simple agents to the overall results obtained by their collective work.

A set of research avenues could also be explored to enhance the efficiency of the discussed protocols. For example, in a future version proposed in this chapter, agents could replicate and disseminate descriptors on the basis of the quality of service (QoS) of the corresponding resources. If the descriptors pertaining to high quality and recently updated resources were disseminated more rapidly than other descriptors, the users would be likely given access to more useful and appropriate resources.

Other improvements could be obtained by better driving the movements of agents: for example, an agent that has reached an accumulation region on the Grid could give other agents this information so as to make their work more effective. This kind of interaction could be inspired by another class of bio-inspired mechanisms, which try to emulate the movements of birds in a flock. Another possible enhancement could be obtained by reorganizing the P2P overlay of Grid hosts according to the "small-world" paradigm (Kleinberg, 2000). A small number of short distance P2P connections should be substituted by long-distance connections, that would directly link hosts located in distant regions of the Grids. It is conceivable that this approach, without significantly altering the load of the system, would allow agents to move on the network more rapidly and efficiently, which could fasten the dissemination and reorganization of descriptors.

REFERENCES

Aktas, M. S., Fox, G. C., & Pierce, M. (2007). Fault tolerant high performance Information Services for dynamic collections of Grid and Web services. *Future Generation Computer Systems*, 23(3), 317-337.

Bonabeau, E., Dorigo, M., & Theraulaz, G. (Eds.). (1999). *Swarm Intelligence: From Natural to Artificial Systems*, New York, USA: Oxford University Press.

Crespo, A., & Garcia-Molina, H. (2002). *Routing indices for peer-to-peer systems.* Paper presented at the 22nd International Conference on Distributed Computing Systems (ICDCS 2002), Wien, Austria.

Dasgupta, P. (2004). *Intelligent Agent Enabled P2P Search Using Ant Algorithms*, Paper presented at the 8th International Conference on Artificial Intelligence, Las Vegas, USA.

Deneubourg, J. L., Goss, S., Franks, N., Sendova-Franks, A., Detrain, C., & Chretien, L. (1990) *The dynamics of collective sorting robot-like ants and ant-like robots.* Paper presented at the 1st International Conference on Simulation of Adaptive Behavior on From Animals to Animats, Cambridge, USA.

Eugster, P., Guerraoui, R., Kermarres, A.M., & Massoulieacute, L. (2004). *Epidemic Information Dissemination in Distributed System*, IEEE Computer, 37(5), 60-67

Forestiero, A., Mastroianni, C., & Spezzano, G. (2005). *A Multi Agent Approach for the Construction of a Peer-to-Peer Information System in Grids*, Paper presented at the International Conference on Self-Organization and Adaptation of Multi-agent and Grid Systems (SOAS 2005), Glasgow, United Kingdom.

Forestiero, A., Mastroianni, C., & Spezzano, G. (2007). *A Decentralized Ant-Inspired Approach for Resource Management and Discovery in Grids*, Multiagent and Grid Systems - An International Journal, 3(1), 43-63.

Grassè, P. (1959). *La reconstruction du nid et les coordinations inter-individuelles chez belicositermes natalensis et cubitermes sp. la theorie de la stigmergie: Essai d'interprtation du comportement des termites constructeurs*, Insectes Sociaux 6, 41–84.

Iamnitchi, A., & Foster, I. (2005). *Interest-aware information dissemination in small-world communities*, Paper presented at the 14th IEEE International Symposium on High Performance Distributed Computing (HPDC 2005), Research Triangle Park, USA.

Kleinberg, J. (2000). Navigation in a Small World , Nature 406(2000), 845.

Lv, C., Cao, P. Cohen, E. Li, L., & Shenker, S. (2002). Search and *replication in unstructured peer-to-peer networks*, ACM, Sigmetrics.

Minar, N., Burkhart, R., Langton, C., & Askenazi, M. (1996). *The swarm simulation system, a toolkit for building multi-agent simulations*, from http://www.santafe.edu.

Petersen, K., Spreitzer, M., Terry, D., Theimer, M., & Demers, A. (1997). *Flexible Update Propagation for Weakly Consistent Replication*, Paper presented at the 16th ACM Symposium on Operating System Principles, Saint-Malo, France.

Sycara, K. (1998). *Multiagent systems*, Artificial Intelligence Magazine, Association for the Advancement of Artificial Intelligence, 10(2), 79–93.

Tsoumakos, D., & Roussopoulos, N. (2003). *A Comparison of Peer-to-Peer Search Methods*, Paper presented at the 6th International Workshop on the Web and Databases (WebDB), San Diego, USA.

Van Dyke Parunak, H., Brueckner, S. A., Matthews, R., & Sauter, J. (2005). *Pheromone Learning for Self-Organizing Agents*, IEEE Transactions on Systems, Man, and Cybernetics, Part A: Systems and Humans, 35(3).

ADDITIONAL READING

To better understand the concepts concerning bio-inspired techniques for the management of resources in distributed systems, the reader can refer to "Swarm Intelligence: From Natural to Artificial Systems" authored by Bonabeau, Dorigo and Theraulaz, pioneers of so called *ant optimization* and the simulation of social insects. With their book , they provide an overview of the state of the art in swarm intelligence. They go further by outlining future directions and areas of research where the application of biologically inspired algorithms seems to be promising. The book describes several phenomenona in social insects that can be observed in nature and explained by models. The book also provides examples of the latter being transferred successfully to algorithms, multi-agent systems and robots, or at least describes promising approaches for doing this.

Another interesting book is "Ant Colony Optimization" authored by Dorigo and Stützle. This book presents an overview of this rapidly growing field, from its theoretical inception to practical applications, including descriptions of many available ACO algorithms and their uses. The book first describes the translation of observed ant behavior into working optimization algorithms. The ant colony metaheuristic is then introduced and viewed in the general context of combinatorial optimization. This is followed by a detailed description and guide to all major ACO algorithms and a report on current theoretical findings. The book surveys ACO applications now in use, including routing, assignment, scheduling, subset, machine learning, and bioinformatics problems.

Other suggested works in the field of swarm intelligence are the following:

Camazine, S., Franks, N. R., Sneyd, J., Bonabeau, E., Deneubourg, J.-L., and Theraula, G. (2001). Self-Organization in Biological Systems. Princeton University Press, Princeton, NJ, USA.

Deneubourg, J. L., Goss, S., Franks, N., Sendova-Franks, A., Detrain, C., & Chretien, L. (1990) *The dynamics of collective sorting robot-like ants and ant-like robots*. Paper presented at the 1st International Conference on Simulation of Adaptive Behavior on From Animals to Animats, Cambridge, USA.

Grassè, P. (1959). *La reconstruction du nid et les coordinations inter-individuelles chez belicositermes natalensis et cubitermes sp. la theorie de la stigmergie: Essai d'interprtation du comportement des termites constructeurs*, Insectes Sociaux 6, 41–84.

Martin, M., Chopard, B., and Albuquerque, P. (2002). Formation of an ant cemetery: swarm intelligence or statistical accident? Future Generation Computer Systems 18, 7, 951–959.

Parunak, H. V. D., Brueckner, S., Matthews, R. S., and Sauter, J. A. (2005). Pheromone learning for self-organizing agents. IEEE Transactions on Systems, Man, and Cybernetics, Part A 35, 3, 316–326.

Van Dyke Parunak, H., Brueckner, S. A., Matthews, R., & Sauter, J. (2005). *Pheromone Learning for Self-Organizing Agents*, IEEE Transactions on Systems, Man, and Cybernetics, Part A: Systems and Humans, 35(3).

Additional readings concerning the management of resources both in Grid environment and in peer to peer networks, are:

Anand, S., Padmanabhan, G., Shaowen, R. and Wang, B. (2005). A self-organized grouping (sog) method for efficient grid resource discovery. In Proc. of the 6th IEEE/ACM International Workshop on Grid Computing. Seattle, Washington, USA. ACM Journal Name, Vol. V, No. N, April 2007.

Chakravarti, A. J., Baumgartner, G., and Lauria, M. (2005). The organic grid: self-organizing computation on a peer-to-peer network. IEEE Transactions on Systems, Man, and Cybernetics, Part A 35, 3, 373–384.

Cheema, A. S., Muhammad, M., and Gupta, I. (2005). Peer-to-peer discovery of computational resources for grid applications. In Proc. of the 6th IEEE/ACM International Workshop on Grid Computing. Seattle, WA, USA, 179–185.

Cohen, E. and Shenker, S. 2002. Replication strategies in unstructured peer-to-peer networks. In Proc. of the Special Interest Group on Data Communication ACM SIGCOMM'02. Pittsburgh, Pennsylvania, USA.

Foster, I. and Kesselman, C. (2003). The Grid 2: Blueprint for a New Computing Infrastructure. Morgan Kaufmann Publishers Inc., San Francisco, CA, USA.

Iamnitchi, A., Foster, I., Weglarz, J., Nabrzyski, J., Schopf, J., and Stroinski, M. (2003). A peer-to-peer approach to resource location in grid environments. In Grid Resource Management. Kluwer Publishing.

Petersen, K., Spreitzer, M. J., Terry, D. B., Theimer, M. M., and Demers, A. J. (1997). Flexible update propagation for weakly consistent replication. In Proc. of the Sixteenth ACM symposium on Operating systems principles SOSP '97. ACM Press, New York, NY, USA, 288–301.

Ran, S. (2003). A model for web services discovery with qos. ACM SIGecom Exchanges 4, 1, 1–10.

Sharma, P., Estrin, D., Floyd, S., and Jacobson, V. (1997). Scalable timers for soft state protocols. In Proc. of the 16th Annual Joint Conference of the IEEE Computer and Communications Societies, INFOCOM'97. Vol. 1. IEEE Computer Society, Washington, DC, USA,222–229.

Stoica, I., Morris, R., Karger, D., Kaashoek, M. F., and Balakrishnan, H. (2001). Chord: A scalable peer-to-peer lookup service for internet applications. In Proc. of the Conference on Applications, technologies, architectures, and protocols for computer communications SIGCOMM'01. ACM Press, New York, NY, USA, 149–160.

TheGlobusAlliance. (2006). The web services resource framework. http://www.globus.org/wsrf/.

Trunfio, P., Talia, D., Fragopoulou, P., Papadakis, C., Mordacchini, M., Pennanen, M., Popov, K., Vlassov, V., and Haridi, S. (2006). Peer-to-peer models for resource discovery on grids. Tech. Rep. TR-0028, Institute on System Architecture, CoreGRID Network of Excellence. March. For an extended version, see "Peer-to-Peer resource discovery in Grids: Models and systems", Future Generation Computer Systems, 2007.

Tsoumakos, D. and Roussopoulos, N. (2003a). Adaptive probabilistic search for peer-to-peer networks. In Proc. of the Third IEEE International Conference on P2P Computing P2P'03. 102–109.

Vu, L. H., Hauswirth, M., and Aberer, K. (2005). QoS-based Service Selection and Ranking with Trust and Reputation Management. In Proc. of the International Conference on Cooperative Information Systems CoopIS'05, 31 Oct - 4 Nov 2005, Agia Napa, Cyprus. Vol. 3760. 446–483.

Finally, we propose the following readings that discuss bio-inspired approaches for the dissemination and discovery of resources in Grid environment:

Erdil, D. C., Lewis, M. J., and Abu-Ghazaleh, N. (2005). An adaptive approach to information dissemination in self-organizing grids. In Proc. of the International Conference on Autonomic and Autonomous Systems ICAS'06. Silicon Valley, CA, USA.

Eugster, P., Guerraoui, R., Kermarres, A.M., & Massoulieacute, L. (2004). *Epidemic Information Dissemination in Distributed System*, IEEE Computer, 37(5), 60-67

Forestiero, A., Mastroianni, C., & Spezzano, G. (2005). *A Multi Agent Approach for the Construction of a Peer-to-Peer Information System in Grids*, Paper presented at the International Conference on Self-Organization and Adaptation of Multi-agent and Grid Systems (SOAS 2005), Glasgow, United Kingdom.

Forestiero, A., Mastroianni, C., & Spezzano, G. (2007). *A Decentralized Ant-Inspired Approach for Resource Management and Discovery in Grids*, Multiagent and Grid Systems - An International Journal, 3(1), 43-63.

Forestiero, A., Mastroianni, C., and Spezzano, G. (2006). An agent based semi-informed protocol for resource discovery in grids. In Proc. of the International Conference on Computational Science(4) ICCS'06. 1047–1054.

This work was previously published in Advancing Artificial Intelligence through Biological Process Applications, edited by A. Pazos, A. Sierra, and W. Buceta, pp. 291-310, copyright 2009 by Medical Information Science Reference (an imprint of IGI Global).

Chapter XXXV
Large-Scale Co-Phylogenetic Analysis on the Grid

Heinz Stockinger
Swiss Institute of Bioinformatics, Switzerland

Alexander F. Auch
University of Tübingen, Germany

Markus Göker
University of Tübingen, Germany

Jan Meier-Kolthoff
University of Tübingen, Germany

Alexandros Stamatakis
Ludwig-Maximilians-University Munich, Germany

ABSTRACT

Phylogenetic data analysis represents an extremely compute-intensive area of Bioinformatics and thus requires high-performance technologies. Another compute- and memory-intensive problem is that of host-parasite co-phylogenetic analysis: given two phylogenetic trees, one for the hosts (e.g., mammals) and one for their respective parasites (e.g., lice) the question arises whether host and parasite trees are more similar to each other than expected by chance alone. CopyCat is an easy-to-use tool that allows biologists to conduct such co-phylogenetic studies within an elaborate statistical framework based on the highly optimized sequential and parallel AxParafit program. We have developed enhanced versions of these tools that efficiently exploit a Grid environment and therefore facilitate large-scale data analyses. Furthermore, we developed a freely accessible client tool that provides co-phylogenetic analysis capabilities. Since the computational bulk of the problem is embarrassingly parallel, it fits well to a computational Grid and reduces the response time of large scale analyses.

INTRODUCTION

The generation of novel insights in many scientific domains such as biology, physics, or chemistry increasingly relies on compute-intensive applications that require high-performance or large-scale, distributed high-throughput computing technology and infrastructure. In the discipline of bioinformatics, biological insight is typically generated via data analysis pipelines that use a plethora of distinct and highly specialized tools. Most commonly, bioinformaticians and biologists collaborate to analyze data extracted from large databases containing DNA and/or protein data in order to study, e.g., the function of living beings, the effect and influence of diseases and defects, or their evolutionary history. Early "classic" bioinformatics tools, such as CLUSTALW (Thompson et al., 1994) or BLAST (Altschul et al., 1997) that have been ported to Grid computing environments deal with biological sequence search, analysis, and comparison. Typically, these programs are embarrassingly parallel and therefore represent ideal candidate applications for Grid computing environments (Stockinger et al., 2006).

The study of the genome represents a way to obtain new insight and extract novel knowledge about living beings. In particular, stand-alone phylogenetic analyses have many important applications in biological and medical research. Applications range from predicting the development of emerging infectious diseases (Salzberg et al., 2007), over the study of Papillomavirus evolution that is associated with cervical cancer (Gottschling et al., 2007), to the determination of the common origin of Caribbean frogs (Heinicke et al., 2007).

Recent years have witnessed significant progress in the field of stand-alone phylogeny reconstruction algorithms, which represent an NP-complete optimization problem (Chor and Tuller, 2005), with the release of programs such as TNT (Goloboff, 1999), RAxML (Stamatakis, 2006), MrBayes (Ronquist and Huelsenbeck, 2003) or GARLI (Zwickl, 2006). Because of the continuous explosive accumulation and availability of molecular sequence data coupled with advances in phylogeny reconstruction methods, it has now become feasible to reconstruct and fully analyze large phylogenetic trees comprising hundreds or even thousands of sequences (organisms). However, current meta-analysis methods for phylogenetic trees such as programs that conduct co-phylogenetic tests can currently not handle such large datasets.

To alleviate this bottleneck in the meta-analysis pipeline, we recently parallelized, and released the highly optimized co-phylogenetic analysis program AxParafit (Axelerated Parafit - Stamatakis et al., 2007) that implements an elaborate statistical test of congruence between host and parasite trees (Legendre et al., 2002). AxParafit is a typical stand-alone Linux/Unix command line program. AxParafit has been integrated and can be invoked via a user-friendly graphical interface for co-phylogenetic analyses called CopyCat (Meier-Kolthoff et al., 2007). In this article, we present an enhanced version of this tool suite (henceforth denoted as CopyCat(AxParafit)) for co-phylogenetic analyses, that is packaged into a client tool which makes use of a world-wide Grid environment and thereby allows for large-scale data analysis. In the current version, the underlying Grid middleware is gLite (Laure et al., 2006) that is coupled with an efficient submission and execution model called Run Time Sensitive (RTS) scheduling and execution (Stockinger et al., 2006).

The remainder of this article is organized as follows: initially, we provide a brief introduction to the field of phylogenetic inference, co-phylogenetic analyses, and related software packages in Section 2. Next, we discuss the implementation and architecture of our new approach for efficient adaptation of the CopyCat(AxParafit) tool-suite to a Grid environment. Finally, we provide detailed performance results on the EGEE (Enabling Grids for E-SciencE, http://www.eu-egee.org) Grid infrastructure (where the gLite middleware is deployed in production mode) and demonstrate the performance as well as scalability of our proposed bioinformatics tool.

BACKGROUND

Phylogenetic (evolutionary) trees are used to represent the evolutionary history of a set of *s* currently living organisms, roughly comparable to a genealogical tree of species rather than individual organisms. Phylogenetic trees or simply phylogenies are typically unrooted binary trees. The *s* organisms, which are represented by their DNA or AA (Amino Acid/Protein) sequences that are used as input data for the computation, are located at the leave nodes (tips) of the tree while the inner nodes of the topology represent common extinct ancestors. There exist various methods and models to reconstruct such trees which differ in their computational complexity and also in the accuracy of the final results, i.e., there exists a "classic" trade-off between speed and accuracy. As already mentioned in the introduction, phylogenetic analysis has many important applications in medical and biological research. In Figure 1, we provide a simple example for the phylogenetic tree of monkeys.

In the context of this article, however, we will not address stand-alone phylogenetic inference, but consider the problem of co-phylogenetic analysis. Given two phylogenetic trees that represent the evolutionary histories of hosts and their respective parasites, the "classic" example being mammals and lice, and given the extant associations between the former and the latter, we want to determine whether the parasite phylogeny is more similar to the phylogeny of the respective hosts than expected by chance alone. The main interpretation of such a congruence between the trees is that parasites have been associated with respect to their evolutionary history and mostly speciated in parallel (co-speciated) with their hosts (Page, 2002). Given a parasite tree with *n* organisms and a host tree with *m* organisms (sequences), their associations can be represented as a *n* times *m* binary matrix, that contains information of the type: does parasite *x* *(x=1...n)* occur or live on host *y* *(y=1...m)*? In addition to

Figure 1. Phylogenetic tree of monkeys

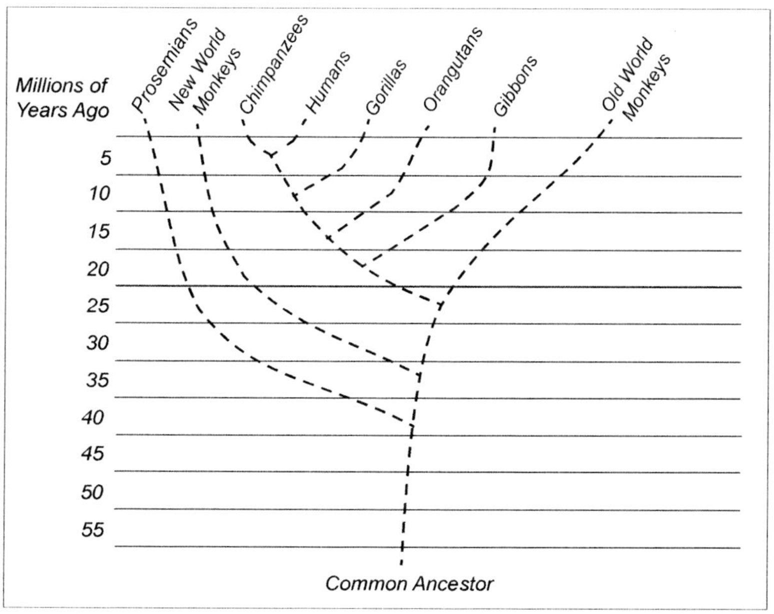

the question of global congruence, one may also be interested in whether individual associations significantly increase the agreement between the phylogenies. Such associations can be interpreted as being caused mainly by co-speciation.

As previously mentioned, recent advances in stand-alone phylogenetic inference methods in combination with the increasing availability of appropriate sequence data, allow for large-scale phylogenetic analyses with several hundred or thousand sequences (Stamatakis, 2006). Thus, large-scale co-phylogenetic studies have, in principle, become feasible. However, most common co-phylogenetic tools or methods such as BPA, Component, TreeMap, TreeFitter (cf. review in Charleston, 2006) or Tarzan (Merkle, 2006) are not able to handle datasets with a large number of organisms or have not been tested in this regard with respect to their statistical properties and scalability. Faster methods based on topological distances between trees, like, e.g., I_{cong} (de Vienne, 2007) are even limited to the analysis of bijective associations only. In this context bijectivity means that each parasite can only be associated to one single host, and vice versa. Therefore, there is a performance and scalability gap between tools for phylogenetic analysis and meta-analysis. The capability to analyze large datasets is important to infer "deep co-phylogenetic" relationships which can otherwise not be assessed (Meier-Kolthoff et al., 2007; Stamatakis et al., 2007). Deep relationships are relationships that determine the extant associations between parasite and host organisms at a high taxonomic level, such as, e.g., families and orders.

Parafit (Legendre, 2002) and the analogous highly optimized AxParafit (Stamatakis et al., 2007) program implement a statistical test to assess hypotheses of global congruence between trees as well as the impact of individual associations. This test is based on the permutation of the entries in the association matrix. The null hypothesis is that the global similarity between the trees, or the respective impact of an individual local association on the similarity, is not larger than expected by pure chance. Extensive simulations have shown that the Parafit test is statistically well-behaved and yields acceptable error rates. The method has been successfully applied in a number of biological studies (Hansen et al., 2003; Ricklefs et al., 2004; Meinilä et al., 2004).

In addition, the type-II statistical error of Parafit decreases with the size of the dataset (see Legendre, 2002), i.e., this approach scales well on large phylogenies of hosts and parasites in terms of accuracy. The AxParafit program is a highly optimized version of Parafit which yields exactly the same results. The sequential version of AxParafit is up to 67 times faster than the original Parafit implementation, while the speedup increases with increasing input size, caused by higher cache efficiency. The speedup of AxParafit has been achieved via low-level optimizations in C, re-design of the algorithm, omission of redundant code, reduction of memory footprint, and integration of highly optimized BLAS (Basic Linear Algebra Subroutines, http://www.netlib.org/blas/) routines.

Earlier work describes these optimizations together with a respective performance study. Moreover, the program was used to conduct the largest co-phylogenetic analysis on real-world data to date. The underlying data were smut fungi and their respective host plants (Stamatakis et al., 2007). Smut fungi are parasitic mushrooms that cause plant diseases. For economically important hosts, such as barley and other cereals, smut fungi can for instance cause considerable yield losses (Thomas and Menzies, 1997).

Workflow of a Co-Phylogenetic Analysis with CopyCat and AxParafit

In this section, we provide an outline of the work-flow for a full co-phylogenetic analysis using CopyCat(AxParafit). The input for a co-phylogenetic analysis with CopyCat(AxParafit) are the host and parasite phylogenies, that might have branch lengths, depending on which method/model was used to calculate the trees. The aforementioned associations are represented as a plain text file containing a list of sequence (organism) name pairs of hosts and parasites, i.e., an adjacency list. This input data representation is henceforth also referred to as list of host-parasite associations. Initially, these files are parsed and transformed into the appropriate file format by CopyCat. In a first step, a principal coordinate analysis is conducted on the respective tree-based distance matrices induced by the host and parasite trees. This analysis is carried out by the AxPcoords (Axelerated Principal Coordinates) program (Stamatakis et al., 2007), which is an optimized version of the analogous DistPCoA program (Legendre and Anderson, 1998). The output of AxPcoords for the host and parasite trees is then parsed and appropriately prepared for the AxParafit analysis which takes the two principal coordinates matrices and the binary matrix with the associations as input. The output of this computation is a list of probabilities for the individual null hypotheses that a certain association does not improve the fit between host and associate phylogenies. In addition, a probability for the global null hypothesis of the absence of congruence between host and parasite trees is computed. Upon termination of AxParafit the output files are read by the CopyCat tool and presented in a human-readable format. It is important to note that the computations with AxParafit represent the by far largest part (over 95%) of the computational effort required to conduct such a co-phylogenetic analysis. Therefore, the AxPcoords and CopyCat parts of the workflow can be handled sequentially and executed locally. We will, thus, mainly focus on the parallel and gridified versions of AxParafit in the next sections. The basic workflow is outlined in Figure 2 (at the end of the article).

Parallel AxParafit

The most compute-intensive operation (95% of execution time) conducted by AxParafit to compute the statistics is a dense matrix-matrix multiplication of double precision floating point numbers. This is the rationale for integration of highly optimized BLAS routines. In the remainder of this article, we thus always refer to the BLAS-based version of AxParafit.

Initially, the program will compute the statistics for the global congruence of the complete list of host-parasite associations. This part of the computation is significantly less expensive than the individual tests for each host-parasite association, which take *nz* times longer, where *nz* is the number of non-zero entries in the binary association matrix, i.e., number of entries in the original host-parasite association list. For large datasets that require parallel and distributed computing resources as well as a sufficient amount of memory typically $nz >> 1$. The statistics computed during the global test of congruence are required as input data for the individual tests of host-parasite associations, hence there is a sequential dependency: global test → *nz* local tests. Thus, in the MPI-based parallel implementation we only parallelized these *nz* local tests which can be computed independently of each other via a straight-forward master-worker scheme. The master simply distributes the *nz* individual host-parasite association tests to the worker processes.

The potential bottleneck induced by the sequential part of the computations can be alleviated by using, e.g., the respective shared-memory implementations of BLAS. With respect to a gridification, this sequential dependency actually has advantages. Since the inference time as well as memory footprint of the global test of congruence are nearly identical (same type of operation, identical matrix sizes, permuted input data) to each of the individual *nz* tests, the information on run-times and memory requirements collected during the global tests can be used for scheduling

Figure 2. Detailed work- and dataflow for co-phylogenetic analysis on the Grid

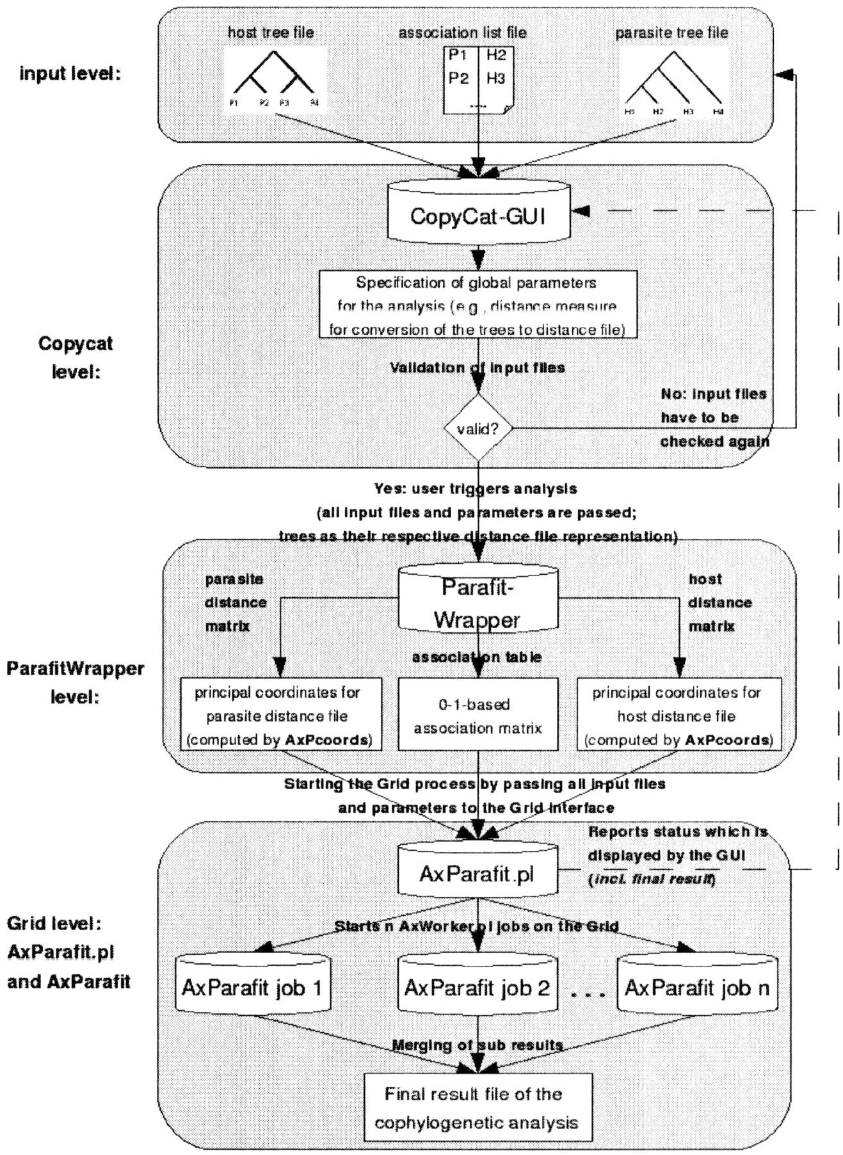

decisions, as well as to determine an optimal level of granularity and to assess respective resource requirements.

FIT FOR THE GRID

In the following section, we describe how CopyCat(AxParafit) has been adapted and modified for use in a Grid environment. The overall architecture of the client tool will be explained as well as the integration with an existing middleware toolkit.

An important design goal of the Grid-based system for co-phylogenetic analyses was to re-use the current graphical user interface of CopyCat such that the deployment of Grid resources is hidden from the end-user. One fundamental difference between the standard and Grid-enabled versions of CopyCat (AxParafit) is that specific

Grid credentials are required (an X.509 user certificate) since Grid jobs can only be submitted by authenticated and authorized users.

Overall Architecture

The basic workflow of a co-phylogenetic study using CopyCat and the AxPcoords/AxParafit programs has already been outlined in the above section. Here, we will describe the architecture and workflow for a gridified analysis in greater detail. The input data consists of three files: a host tree file, a parasite tree file, and a host-parasite association list. In the reminder of this article, the following terminology is used:

- **Individual test (job):** an individual test is the minimal "work unit" or processing entity that has to be conducted by AxParafit to calculate a single host-parasite association. In the context of AxParafit this is also referred to as *job*. In total, *nz* individual tests have to be computed to achieve the final result.
- **Task:** a task consists of a fraction (subset) of the *nz* individual tests that have to be conducted by AxParafit.
- **Grid job:** a Grid job is an executable that is scheduled by the Grid middleware to be executed on a Worker Node of a Grid computing resource (also referred to as Computing Element). In our model, a single Grid job can execute one or several such tasks.

The overall workflow is depicted in Figure 2. The most important Grid-enhancement is the interface to the Grid (represented by the Perl program AxParafit.pl in Figure 2). Once the input files are validated, CopyCat uses AxParafit.pl to determine a specific set of tasks (to be registered with a Task Server as indicated in Figure 3) and Grid jobs which are then submitted to Grid computing resources using the gLite middleware. Each individual Grid job then requests tasks from the Task Server, processes them, and stores the result on a Grid Storage Element.

Figure 3. Interaction of AxParafit.pl with the gLite Grid middleware, a Task Server and a Storage Element. Each submitted Grid job will execute on a Grid Worker Node.

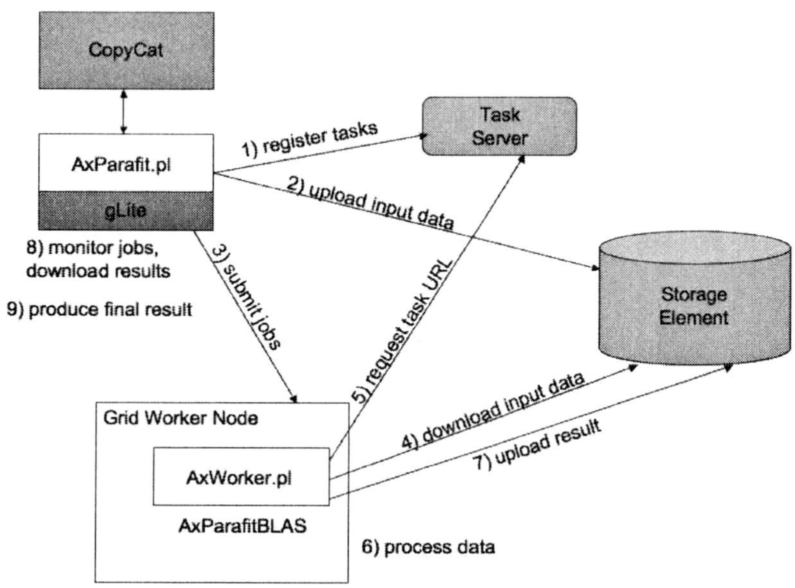

AxParafit.pl will constantly monitor the overall Grid job status and presents intermediate results in a CopyCat control window. Once all results are obtained and merged, CopyCat indicates where the final result can be obtained. Further details about AxParafit.pl, AxWorker.pl etc. will be given in the other section.

Implementation Details

In the following section we describe the necessary modifications and adaptations of the existing CopyCat and AxParafit tools as well as additional components that were necessary to implement the system outlined in Section 3.1.

CopyCat

Previous versions of CopyCat already provided straight-forward GUI-based functionality for the preparation and analysis of co-phylogenetic datasets. The CopyCat GUI is implemented in Java using the Standard Widget Toolkit (SWT). Upon startup, the user can load the host and parasite trees (represented in the standard Newick tree format: http://evolution.genetics.washington.edu/phylip/newicktree.html), together with a host-parasite association list in a simple plain-text format that contains one host-parasite association per line.

When starting an analysis, the user can now utilize a new Grid interface that connects CopyCat to the gridified program AxParafit. Instead of directly calling the AxParafit executable, the interface invokes a Perl script (AxParafit.pl) which hides the Grid-related parts from the user and CopyCat. By delegating the invocation process to a script, dependencies between the user front-end and the Grid software are minimized. Thus, future modifications like the development of a Web interface for job submissions (see Conclusion) or the usage of a different middleware system are possible.

The AxParafit.pl script entirely manages the execution of AxParafit on the Grid and provides status updates to the standard output stream at the same time. As CopyCat is listening to the output stream of the external programs it invokes, it also receives the status updates generated by the aforementioned Perl-script and writes them to the CopyCat log-message window, thus keeping the user informed about the progress of Grid jobs. Upon termination of the script, the output of the Grid jobs (individual tests of host-parasite associations), as well as the global significance test results, are read by CopyCat. The results can then be displayed and further analyzed via the CopyCat evaluation window.

Within the context of an automated Grid-driven simultaneous analysis of several distinct datasets (and other potential script-based applications, based on CopyCat), the program has been extended by a command-line interface. As a side-effect, this enables CopyCat users to speed-up certain analyses by simply executing a specific command-line call with a defined set of parameters (please refer to the CopyCat manual for detailed information on the command-line options).

Application-Side Modifications of AxParafit

As outlined in other section, the parallel MPI implementation of AxParafit uses a simple master-worker scheme. In order to devise a distributed version of AxParafit we modified the code as follows: initially, we appropriately modified the global test of congruence in AxParafit to write an additional file called "gridData.RUN-ID" where RUN-ID is the output file name appendix for a specific analysis that is passed to AxParafit via a command line parameter (for details see the AxParafit Manual at http://icwww.epfl.ch/~stamatak/). This file contains the necessary data to make scheduling decisions for the distributed computation of the n individual tests of host-parasite associations, i.e., the number of jobs nz, e.g. Jobs=2000, and the approximate execution time per job in seconds, e.g., ComputeTime=10 . This data can then be used to determine the level of granularity for individual Grid tasks since in the current example the scheduling overhead induced by distributing 2,000 jobs of 10 seconds each, along with the comparatively large input datasets on the Grid, would be immense. We have, thus, extended the implementation of the individual host-parasite association tests in AxParafit by two additional command line parameters -l (lower limit) and -u (upper limit). These parameters allow for computation of several host-parasite associations in one single program run. The lower and upper limits just refer to the order of the *nz* non-zero entries in the binary association matrix. Thus, in the present example, we can schedule larger, in terms of execution times, Grid jobs by only distributing two Grid jobs with -l 0 -u 1000 and -l 1000 -u 2000 that would require approximately 10,000 seconds of execution time each, i.e., Grid job 0 would compute statistics for the first 1,000 host-parasite associations and Grid job 1 for the remaining 1,000 associations. The result files of these distributed Grid jobs only need to be recovered and concatenated in the order of the associations they computed, and the respective result file can then be read and visualized by CopyCat.

Grid-Side Adaptation

Parafit.pl provides the actual link between CopyCat and the gridified version of AxParafit. First, it reads the file "gridData.RUN-ID" to determine the number of tasks to be created (registered) for execution on the Grid (Step 1 in Figure 3). As outlined in Section 3.2.1, the basic idea consists of combining appropriate fractions (subsets) of the *nz* individual tests into a single *task*, i.e., a set of individual tests $k < nz$ are executed by a Grid job. In order to make efficient use of the Grid and to reduce scheduling overhead, a task contains a minimum of *k* individual tests, such that the respective job requires at least 30 seconds on an average CPU. After the number of tasks has been determined, a certain number of Grid jobs (approximately *nz/k*) needs to be submitted (Step 3 in Figure 3) which then ask for tasks to be executed, i.e., issue work requests. An individual Grid job can request and execute several tasks, as long as the Task Server can provide more work (Steps 5 and 6 in Figure 3). The protocol used for the Task Server is HTTP which allows for fast communication between the client and the server. For additional background and fault tolerance features of this processing model with a Task Server please refer to Stockinger et al. (2006).

Box 1.

```
Executable = "AxWorker.pl";
Arguments = "-j ax-May1319-41-28 -p 100 -1 2048 -2 2048 -3 2025 -4 2031 \
–A gsiftp://example.org/dpm/home/biomed/heinz/selection_2048.mat-ax-May1319-41-28 \
-B gsiftp://example.org/dpm/home/biomed/heinz/selection_2048_P.pco-ax-May1319-41-28 \
-C gsiftp://example.org/dpm/home/biomed/heinz/selection_2048_H.tra-ax-May1319-41-28 -i
1";
Stdoutput = "output.txt";
InputSandbox = {"/home/stockinger/AxWorker.pl", "/home/stockinger/AxParafitBLAS"};
OutputSandbox = {"output.txt"}
```

Before Grid jobs can be submitted, AxParafit.pl creates the Grid job specification, i.e. the job description file to decide which files (data and/or executables) to send to Grid computing resources. A typical job description file looks as follows:(See Box 1).

The wrapper code (identified as "Executable" in the JDL file above) is AxWorker.pl using the command line arguments specified by "Arguments". Once AxParafit.pl is running on a Grid Worker Node, it is responsible for requesting tasks from a Task Server and executing AxParafitBLAS. The two programs (AxWorker.pl and AxParafitBLAS) are transferred to the Grid Worker Node as specified in the InputSandbox in the example above, i.e. gLite provides the means to transfer data from the client machine to the actual computing resource.

In parallel to the execution of Grid jobs, the script AxParafit.pl monitors the status and is responsible for providing and assembling the final result (Steps 8 and 9 in Figure 3).

In particular, when tasks have been processed successfully, they are downloaded from the Storage Element and transferred to the client. Note that an alternative implementation option is to transfer the output of individual tasks via the gLite middleware (using the OutputSandbox). However, because of performance and reliability considerations, it has turned out to be more efficient to store files at an external Storage Element and retrieve them from there: one reason is that the actual job output can only be retrieved if gLite indicates that a job has been finished. However, because of update latencies in the Grid-wide information and motoring system, jobs might have finished already several dozens of seconds or even a few minutes ago while the job status is still indicated as pending or running.

As a final remark: since the gLite services can only be accessed by authorized users, the execution of the AxParafit.pl script requires the usage of a valid X.509 proxy certificate.

EXPERIMENTAL RESULTS

The main goal of the gridified version of CopyCat(AxParafit) is to accelerate and facilitate large-scale analyses. We present two experiments with large computational demands and study their performance on the Grid. The performance improvement is outlined with respect to running the application sequentially on a single machine. Moreover, we conduct a performance comparison between a dedicated compute cluster and the Grid.

Test Environment

The Grid platform that is supported by our application is gLite 3. Tests are conducted using gLite on the EGEE production infrastructure. In particular, we use the Virtual Organization (VO) that is dedicated to biomedical applications: "biomed". Members of this VO have access to about 50 Computing Elements (acting as front-ends to computing clusters), each having between 2 and a few hundred processing cores. The exact number of processing cores available to a single user at a given time cannot be easily obtained since it depends on the current system load as well as the general availability of a Computing Element at a certain point in time. Currently, gLite does not support resource reservation nor job priorities, which means that experimental results can not be fully reproduced. However, once one is correctly registered with the Virtual Organization, one can use it any time of the day.

On the client side, we used gLite on GNU/Linux on an AMD Opteron machine (2 GB RAM, 2.2 GHz CPU) located in Lyon, France – previous tests (in particular with the installation of CopyCat and the Grid interface have been conducted on a machine located in Lausanne, Switzerland). The gLite components used are the workload management system (for job submission and status monitoring) as well as data management clients for file transfer. Additionally, we deployed and used a Task Server that is located in Lausanne, using resources provided by the Vital-IT group of the Swiss Institute of Bioinformatics. In the second experiment, we used a dedicated compute cluster with 128 CPUs. In contrast to the Grid, the cluster had to be reserved in advance.

Experiment with Real-world Data

In the first experiment we are interested in the raw performance (response time) of AxParafit.pl, i.e., how long does it take to fully process a set of tasks on the Grid. In this experiment, we do not include CopyCat but directly invoke AxParafit.pl as follows:(See Box 2).

The parameters -1, -2, -3 and -4 specify the number of rows and columns in the association matrix as well as the number of rows and columns in the parasite and host matrices; -p represents the number of permutations conducted by the statistical test; -A, -B, and -C are used to read the plain-text input files; -n specifies a run ID that is appended to all output files (for details on the AxParafit program parameters please refer to the AxParafit manual at http://icwww.epfl.ch/~stamatak/). The dataset we used is the aforementioned (Section 2) dataset for the study of smut-fungi, that was used to demonstrate performance of the stand-alone AxParafit code by Stamatakis et al. (2007). As already mentioned, this dataset represents the largest *real-world* co-phylogenetic study conducted to date. While the sequential execution time for this dataset still appears to be acceptable, such studies were previously not feasible with Parafit which is between 1-2 orders of magnitude slower than AxParafit. Since the host-parasite association list contains *nz*=2,362 entries, 2,362 individual tests need to be performed. The execution of AxParafit to compute global congruence of the trees returned an estimated run time of 3 seconds per job, i.e., an overall expected run time of almost two hours (2,362 x 3 seconds). The main goal of the first test is therefore to minimize the expected response time. We also executed the full test, as specified above, on a single machine and observed that the estimated run time of about 2 hours (7,000 seconds) is almost identical to the measured run time (7,200 seconds). Therefore, we deduce that the run time prediction mechanism is sufficiently accurate for our application. In our experiments, we varied the number of tasks (in the range between 60 and 162) as well as the number of parallel Grid jobs (in the range between 24 to 124) to experimentally determine the minimal response time. However, because of varying response times of the Grid (i.e. the various Computing Elements and their job queues etc.) it was not possible to determine an optimal number of Grid jobs and tasks. Finally, in the experiment we used 124 Grid jobs and 150 tasks which have been proposed by the work distribution algorithm outlined in aforementioned section. The overall response time to produce the final output was 11 minutes and 15 seconds (cf. Figure 4). Consequently, we observe a clear runtime improvement with respect to a single, sequential run. Note that the AxParafit.pl program had to be adapted to allow for parallel downloads of the individual results: originally, results were downloaded sequentially, which

Box 2.

```
AxParafit.pl -p 10 -1 413 -2 1400 -3 1390 -4 411 \
    -A smuts010907.mat -B smuts010907_P.pco
-C smuts010907_H.tra -n RUN_1
```

Figure 4. Comparison of smut-fungi dataset on a single CPU and on a Grid using 124 Grid jobs and 150 tasks. Note that there is a rather high redundancy in Grid jobs and not all 124 jobs really participate in the overall calculation because of start-up latencies. In fact, a few Grid jobs (AxWorker.pl) started, requested tasks and found out that there were no more tasks available and gracefully finished.

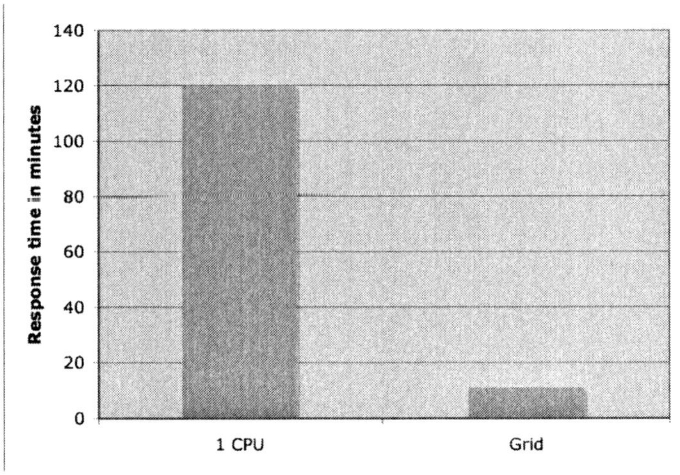

increased the overall response time by several minutes. By overlapping communication with computation, this problem was resolved.

Experiment with Synthetic Data

In another experiment, we used a larger (*synthetic*) test dataset that had been extracted from a larger empirical dataset to test scalability of AxParafit and compared the runtime of the Grid with the infiniband cluster at the Technical University of Munich equipped with 128 AMD Opteron 2.4 GHz CPUs. In the association list, there were nz=2,048 non-zero entries (equivalent to 2,048 tasks) and we used 100 permutations. The expected runtime of a single task was 568 seconds, i.e., about 10 minutes. As a result, the expected sequential response time to finish all 2,048 tasks is about 13.4 days. We used the wrapper as follows: (See Box 3).

Note that the input files are bigger than in the previous experiment: they cannot be directly submitted with the Grid job but they are uploaded to a Storage Element and then dynamically downloaded by Grid jobs when needed.

A direct performance comparison between the cluster and a Grid is not feasible since the cluster we used had several favorable features that a multi-institutional Grid does not have: a shared file system between all processing nodes which minimizes the data transfer time; homogeneous hardware infrastructure; pre-defined number of CPUs that are available which does not require an automatic task assignment, no overhead for job submission etc. However, the cluster needed to be reserved in advance (larger slots can only be obtained overnight) which means that it was only available at a specific time, whereas Grid resources are available on demand at any time. Intuitively, one expects a cluster to provide a better response time to a large size application but it has a considerable "reservation latency", a fact that should not be underestimated.

The final performance results of the experiments are depicted in Figure 5. For the Grid execution, we used between 90 and 175 parallel jobs (the number varied during the overall execution time). Given the number of parallel jobs used in the Grid, the cluster performed better. However, if the number of jobs is increased on the Grid,

Figure 5. Performance comparison of a 128 CPU cluster with a Grid using between 90 and 175 parallel jobs. Note that the number of Grid job varied and was never constant.

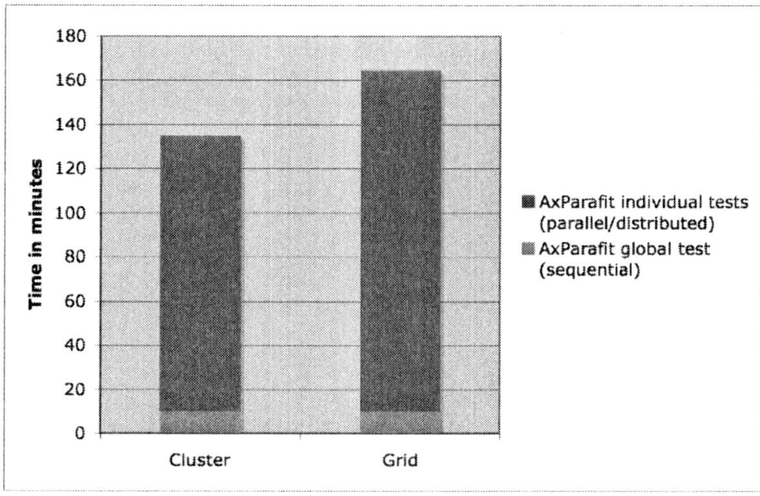

Box 3.

AxParafit.pl -p 100 -1 2048 -2 2048 -3 2025 -4 2031 -A selection_2048.mat \
-B selection_2048_P.pco -C selection_2048_H.tra -n RUN_2

the cluster can actually be out-performed. Note that, the Grid response time comprises the sequential run time that is necessary to determine the number of tasks and jobs and compute the test for global congruence that is then used as input data for the *nz* individual tests. This initial part of the analysis also needs to be executed sequentially on the cluster. In addition, the Grid response time also includes the job submission overhead that is imposed by the gLite workload management systems. In order to avoid congestion problems at the submission server, only a certain number of jobs are submitted at a given time by AxParafit.pl. The actual processing time of the 2,048 AxParafit tests can then be better compared to the cluster performance. Another interesting observation is the average processing time of 13.3 min per single task on the Grid compared to the local execution time of 11 min on the Grid client machines. This indicates that distinct Computing Elements have CPUs with rather different CPU speeds and latencies.

Overall, our Grid-based approach requires computing times that are in the same order of magnitude as those of a dedicated cluster. Consequently, the gridified version provides an easier to use alternative to a compute cluster with comparable performance.

CONCLUSION

We have demonstrated how a compute-intensive application for a statistical test of congruence between host and parasite phylogenies can efficiently be distributed on the Grid. The proposed Grid-based implementation can greatly contribute to the reduction of response times for large-scale analyses and to the computation of a larger number of test permutations, which in turn improve upon accuracy. Moreover, we have integrated the access to Grid resources into an easy-to-use Graphical User Interface (CopyCat) which entirely hides the technical details related to the exploitation of Grid resources from the user. Note that in particular for non-expert users, easy accessibility and usability of HPC resources represents a major criterion for the selection of software and systems. We thus believe that the proposed architecture will greatly facilitate access to HPC resources for real-world biological studies on host-parasite evolution. Nonetheless, the requirement to obtain access and accreditation to use Grid resources (valid X.509 proxy certificate) will possibly hinder a large amount of potential users to exploit these new possibilities offered by the Grid. Based on previous experience with the development of the freely accessible RAxML Web servers for phylogenetic reconstruction (Stamatakis et al., 2008, over 8,000 job submissions in the first 8 months of operation) that are however scheduling jobs to dedicated clusters instead of the Grid, we believe that a freely accessible Web server for this Grid-enabled system for co-phylogenetic analyses can contribute to the generation of biological insights, by further simplifying the access to HPC resources. Thus, future work will concentrate on the development of such a Web server, as well as the integration with the aforementioned RAxML servers such as to provide a comprehensive phylogenetic and co-phylogenetic analysis pipeline.

ACKNOWLEDGMENT

This work was funded in part by the EU project EMBRACE Grid which is funded by the European Commission within its FP6 Program, under the thematic area "Life sciences, genomics and biotechnology for health", contract number LUNG-CT-2004-512092. The Exelixis lab (AS) is funded under the auspices of the Emmy-Noether program by the German Science Foundation (DFG).

REFERENCES

Altschul, S.F., Madden, T.L., Schaffer, A.A., et al. (1997). Gapped BLAST and PSI-BLAST: a new generation of protein database search programs. *Nucleic Acids Research*, 25(17):3389-3402.

Chor, B., & Tuller, T. (2005). Maximum likelihood of evolutionary trees: hardness and approximation. *Bioinformatics*, 21(1):97-106.

Goloboff, P. (1999). Analyzing Large Data Sets in Reasonable Times: Solutions for Composite Optima. *Cladistics* 15(4): 415-428.

Charleston, M.A., & Perkins L. (2006). Traversing the tangle: Algorithms and applications for cophylogenetic studies. *Journal of Biomedical Informatics*, 39 (2006):62-71.

Gottschling, M., Stamatakis, A., Nindl, I., et al. (2007). Multiple Evolutionary Mechanisms Drive Papillomavirus Diversification. *Molecular Biology and Evolution*, 24(5):1242-1258.

Hansen, H., Bachmann, L., & Bakke, T.A. (2003). Mitochondrial DNA variation of *Gyrodactylus* spp. *Monogenea, Gyrodactylidae* populations infecting Atlantic salmon, grayling, and rainbow trout in Norway and Sweden. *International Journal of Parasitology*, 33(13): 1471-1478.

Heinicke, M.P., Duellman, W.E., & Hedges, S.B. (2007). From the Cover: Major Caribbean and Central American frog faunas originated by ancient oceanic dispersal. *Proceedings of the National Academy of Sci*

Laure, E., Fisher, S., Frohner, A., et al. (2006). Programming the Grid with gLite. *Computational Methods in Science and Technology*, 12(1):33-45.

Legendre, P., Anderson, M.J. (1998). DistPCOA program description, source code, executables, and documentation: http://www.bio.umontreal.ca/Casgrain/en/labo/distpcoa.html

Legendre, P., Desdevises, Y., & Bazin, E. (2002). A Statistical Test for Host-Parasite Coevolution. *Systematic Biology*, 51(2):217-234.

Meier-Kolthoff, J.P, Auch, A.F., Huson, D.H., & Göker, M.(2007). COPYCAT: Co-phylogenetic Analysis tool. *Bioinformatics*, 23(7):898-900.

Meinilä, M., Kuusela, J., Zietara, M.S., & Lumme, J. (2004). Initial steps of speciation by geographic isolation and host switch in salmonid pathogen *Gyrodactylus salaris (Monogenea: Gyrodactylidae). International Journal of Parasitology*, 34(4):515-526

Merkle, D., & Middendorf. M. (2005). Reconstruction of the cophylogenetic history of related phylogenetic trees with divergence timing information. *Theory in Biosciences*, 123(4):277-299.

Ricklefs, R.E., Fallon, S.M., & Birmingham, E. (2004). Evolutionary relationships, cospeciation, and host switching in avian malaria parasites. *Systematic Biology*, 53(1):111-119.

Ronquist, F., & Huelsenbeck, J. (2003). MrBayes 3: Bayesian phylogenetic inference under mixed models. *Bioinformatics*, 19(12): 1572-1574.

Salzberg, S.L., Kingsford, C., Cattoli, G., et al. (2007). Genome analysis linking recent European and African influenza (H5N1) viruses. *Emerg Infect Dis*,. 13(5):713-8.

Stamatakis, A. (2006). RAxML-VI-HPC: maximum likelihood-based phylogenetic analyses with thousands of taxa and mixed models. *Bioinformatics*, 22(21): 2688-2690.

Stamatakis, A., Hoover, P., & Rougemont, J. (2008). A Rapid Bootstrapping Algorithm for the RAxML Web Servers. *Systematic Biology*, in press.

Stamatakis, A., Auch, A.F., Meier-Kolthoff, J., & Göker, M. (2007). *AxPcoords & parallel AxParafit: statistical co-phylogenetic analyses on thousands of taxa. BMC Bioinformatics* 2007, **8:**405.

Stockinger, H., Pagni, M., Cerutti, L., & Falquet, L. (2006). Grid Approach to Embarrassingly Parallel CPU-Intensive Bioinformatics Problems. *2nd IEEE International Conference on e-Science and Grid Computing (e-Science 2006)*, IEEE Computer Society Press, Amsterdam, The Netherlands.

Thomas, P.L., & Menzies, J.G. (1997). Cereal smuts in Manitoba and Saskatchewan, 1989-95. *Canadian Journal of Plant Pathology*, 19(2):161-165.

Thompson, J.D., Higgins, D.G.,, & Gibson, T.J. (1994). CLUSTAL W: improving the sensitivity of progressive multiple sequence alignment through sequence weighting, position-specific gap penalties and weight matrix choice. *Nucleic Acids Research*, 22(22):4673-4680.

de Vienne, D.M., Giraud, T., & Martin, O.C. (2007). A congruence index for testing topological similarity between trees. *Bioinformatics*, 23(23):3119-3124.

Zwickl, D. (2006). Genetic algorithm approaches for the phylogenetic analysis of large biological sequence datasets under the maximum likelihood criterion. *PhD Thesis*, The University of Texas at Austin.

This work was previously published in International Journal of Grid and High Performance Computing, Vol. 1, Issue 1, edited by E. Udoh, pp. 39-54, copyright 2009 by IGI Publishing (an imprint of IGI Global).

Compilation of References

Aalst, W., Hofstede, A., Kiepuszewski, B., & Barros, A. (2003). Workflow Patterns. *Distributed and Parallel Databases, 14*(3), 5-51.

Abraham, A., & Nath, B. (2000). Hybrid heuristics for optimal design of artificial neural networks. In R. John & R. Birkenhead (Eds.), *Advances in Soft Computing Techniques and Applications* (pp. 15-22). Berlin, Germany: Springer-Verlag.

Abraham, A., Grosan, C., & Ramos, V. (Eds.). (2006). *Swarm Intelligence in Data Mining, Studies in Computational Intelligence*. Berlin, Germany: Springer-Verlag.

Abramoff, M., Magalhaes, P., & Ram, S. (2004). Image Processing with ImageJ. *Biophotonics International* (pp 36-42).

Abramson D., Lynch A., Takemiya H., Tanimura Y., Date S., Nakamura H., Jeong K., Lee H.-C., Arzberger P. A., Baldridge K. K., Li W. W., & Molina T. (2006). Deploying Scientific Applications to the PRAGMA Grid Testbed: Lessons Learned. In *IEEE/CCGRID International Grid Computing, Singapore*.

Abramson D., Roe P., Kotler L., & Mather D. (2001). *ActiveSheets: Super-Computing with Spreadsheets. 2001 High Performance Computing Symposium (HPC'01).* Advanced Simulation Technologies Conference, April 22-26, Seattle, WA (pp. 110–115). Retrieved from http://www.csse.monash.edu.au/~davida/papers/ActiveSheets.pdf.

Abramson, D, Buyya, R. & Giddy, J. (2002). A Computational Economy for Grid Computing and its Implementation in the Nimrod-G Resource Broker. *Future Generation Computer Systems, 18*(8) 1061-1074.

Abramson, D, Lewis, A. & Peachy, T. (2001). *Case Studies in Automatic Design Optimisation using the P-BFGS Algorithm.* 2001 High Performance Computing Symposium (HPC'01), Advanced Simulation Technologies Conference, April 22-26, , Seattle, WA (pp.104 – 109). Retrieved from http://www.csse.monash.edu.au/~davida/papers/CaseStudies.pdf

Abramson, D., Sosic, R., Giddy, J., & Hall, B. (1995). Nimrod: a tool for performing parametrised simulations using distributed workstations. *Fourth IEEE International Symposium on High Performance Distributed Computing* (pp 112-121).

Achlioptas, D. (2001). Database Friendly Random Projections. *Proc. Symposium on Principals of Database Systems (PODS01)*, (pp. 274-281).

Achtman, M., & Wagner, M. (2008). Microbial Diversity and genetic nature of microbial species. *Nat Rev Microbiology, 6*, 431-440.

Adai, A., Johnson, C., Mlotshwa, S., Archer-Evans, S., Manocha,V., Vance, V. & Sundaresan, V. (2005). Computational prediction of miRNAs in Arabidopsis thaliana. *Genome Res., 15*, 78-91.

Adam, B.L Qu, Y.. Davis, J.W. Ward, M.D Clements, M.A Cazares, et al. (2002). Serum Protein Fingerprinting Coupled with a Pattern-matching Algorithm Distinguishes Prostate Cancer from Benign Prostate Hyperplasia and Healthy Men. *Cancer Research, 62*(13), 3609–3614.

Adcock, S. A., & McCammon, J. A. (2006). Molecular dynamics: Survey of methods for simulating the activity of proteins. *Chemical Review, 106*, 1589-1615.

Aebersold, R., & Mann, M., (2003). Mass spectrometry-based proteomics. *Nature, 422*(1), 198–207.

Aflori, C., & Craus, M. (2007). Grid implementation of the Apriori algorithm. *Advanced Engineering Software, 38*(5), 295-300.

AG. *The Access Grid.* Retrieved from http://www.accessgrid.org/

Agrawal, R., Mannila, H., Srikant, R., Toivonen, H., & Verkamo, A. I. (1996). Fast discovery of association rules. In , U. Fayyad et al. (Eds.), *Advances in Knowledge Discovery and Data Mining* (pp. 307-328). Menlo Park, CA: AAAI Press.

Ahrens, J., Law, C., Schroeder, W., Martin, K. & Papka, M. (2000). *A Parallel Approach for Visualizing Extremely Large, Time-Varying Datasets*(Tech. Rep. LAUR-00-1620). Los Alamos, NM: Los Alamos National Laboratory. Retrieved from http://www.vtk.org/pdf/pvtk.pdf

Aiftimiei, C., Andreozzi, S., Cuscela, G., Donvito, G., Fantinel, S., Fattibene, E., Misurelli, G., Pierro, A., Dudhalkar, V. & Maggi, G.P. (2007). Recent Evolutions of GridICE: a Monitoring Tool for Grid Systems. In C. Kesselman, J. Dongarra, D. Walker (Eds.), *International Symposium on High Performance Distributed computing (HPDC 2007)* (pp. 1-8). New York: ACM.

Ajayi, O., Sinnott, R. O., Stell, A. J. & Young, A. (2008). *Towards a Virtual Anonymisation Grid for Unified Access to Remote Clinical Data.* Proceedings of 6th International HealthGrid conference, Chicago.

Ajayi, O., Sinnott, R. O., Watt, J., Jiang, J. & Stell, A. J. (2006). *Single-Sign on and Authorization for Dynamic Virtual Organizations.* Proceedings of International Conference on Virtual Enterprises, (PRO-VE'06), Helsinki, Finland.

Aktas, M. S., Fox, G. C., & Pierce, M. (2007). Fault tolerant high performance Information Services for dynamic collections of Grid and Web services. *Future Generation Computer Systems*, 23(3), 317-337.

Alfieri, R., Cecchini, R., Ciaschini, V., dell' Agnello, L., Frohner, Á,. Gianoly, A., et al. (2003). VOMS, an Authorization System for Virtual Organizations. *1st European Across Grid Conference.*

Alfieri, R., Cecchini, R., Ciaschini, V., dell'Agnello, L., Frohner, A., Gianoli, A., et al. (2004). VOMS, an Authorization System for Virtual Organizations. In F. Fernández Rivera (Ed.), *Grid computing: first European Across Grids Conference, Santiago de Compostela, Spain, February 13-14, 2003 : revised papers. Lecture notes in computer science, 2970* (pp. xi, 328 p.). New York: Springer.

Alfieri, R., et al (2003). *VOMS: an authorization system for virtual organizations.* Proceedings of 1st European across Grids conference, Santiago de Compostela, Spain.

Ali, A. S., & Taylor, I. J. (2005). Web services composition for distributed data mining. In *The 2005 IEEE International Conference on Parallel Processing Workshops* (pp. 11-18).

Allaert, F. A. (2002). *Security standards for healthcare information systems: a perspective from the EU ISIS MEDSEC Project.* Amsterdam, the Netherlands: IOS Press.

Allcock W., Bresnahan J., Kettimuthu R., & Link M. (2005). *The Globus Striped GridFTP Framework and Server.* ACM/IEEE SC 2005 Conference (p. 54).

Allcock, B., Bester, J., Bresnahan, J., Chervenack, A. L., Foster, I., Kesselman, C., et al. (2002). Data management and transfer in high-performance computational grid environments. *Parallel Computing, 28*(5), 749-771.

Allcock, B., Bester, J., Bresnahan, J., Chervenak, A.L., Foster, I., Kesselman, C., et al. (2002). Data Management, Transfer in High Performance Computational Grid Environments. *Parallel Computing Journal, 28* (5), 749-771.

Allcock, W., Bester, J., Bresnahan, J., Chervenak, A., Liming, L., & Tuecke, S. (2002, April). *GridFTP: Protocol Extensions to FTP for the Grid.* Argonne, IL: Argonne National Laboratory.

Allcock, W., Bester, J., Bresnahan, J., Chervenak, A., Liming, L., & Tuecke, S. (2001, March). In GridFTP: Protocol Extension to FTP for the Grid. Unpublished Internet-Draft, Grid Forum.

Allcock, W., Bresnahan, J., Kettimuthu, R., & Link, M. (2005). The Globus Striped GridFTP Framework and Server. (p.54) *Proceedings of the 2005 ACM/IEEE conference on Supercomputing.* Washington, DC: IEEE Computer Society.

Almeida, J., et al (2003). EURISWEB – Web-based epidemiological surveillance of antibiotic-resistant pneumococci in Day Care Centers. *BMC Medical Informatics and Decision Making, 3*(9).

Almond, J. & Snelling, D. (1999). UNICORE: uniform access to supercomputing as an element of electronic commerce. *Future Generation Computer Systems, 15*(5-6), 539–548.

Aloisio G., Cafaro M., Fiore S., & Mirto M. (2005). ProGenGrid: A Workflow Service Infrastructure for Composing and Executing Bioinformatics Grid Services. In *18th IEEE Symposium on Computer-Based Medical Systems* (pp. 555-560). Dublin, Ireland: Trinity College..

Aloisio, G., Breton, V., Mirto, M., Murli, A. & Solomonides, T. (2008). Special section: Life science Grids for biomedicine and bioinformatics. *Future Generation Computer Systems, 23*(3), 367-370

Aloisio, G., Cafaro, M., Cartenì, G., Epicoco, I., Fiore, S., Lezzi, D., Mirto, M., & Mocavero, S. (2007). The Grid Resource Broker Portal. *Concurrency and Computation: Practice and Experience, 19*(12), 1663-1670.

Aloisio, G., Cafaro, M., Cartenì, G., Epicoco, I., Fiore, S., Lezzi, D., Mirto, M., & Mocavero, S. (2007). The Grid Resource Broker Portal. *Concurrency and Computation: Practice and Experience, 19*(12), 1663-1670.

Aloisio, G., Cafaro, M., Epicoco, I., Fiore, S. Lezzi, D., Mirto, M., & Mocavero, S. (2005). iGrid, a Novel Grid Information Service. In P.M.A. Sloot et al. (Eds.), *Proceedings of the First European Grid Conference (EGC) 2005,* (LNCS 3470)(pp. 506-515). Berlin, Germany: Springer-Verlag,.

Aloisio, G., Cafaro, M., Fiore, S. & Mirto, M. (2005a). ProGenGrid a Grid-enabled platform for Bioinformatics, Studies in Health Technology and Informatics, From grid to Healthgrid. In T. Solomonides et al. (Eds.), *Proceedings of HealthGrid 2005,* (Vol. 112) (pp. 113-126). Amsterdam, the Netherlands: IOS Press..

Aloisio, G., Cafaro, M., Fiore, S. & Mirto, M. (2005a). The Grid Relational Catalog Project. In L. Grandinetti (Ed.), *Advances in Parallel Computing, "Grid Computing: The New Frontiers of High Performance Computing"* (pp.129-155). Philadelphia: Elsevier.

Aloisio, G., Cafaro, M., Fiore, S. & Mirto, M. (2005). ProGenGrid a Grid-enabled platform for Bioinformatics In T. Solomonides et al. (Eds.), *Studies in Health Technology and Informatics, From grid to Healthgrid: Vol. 112,* (pp. 113-126). Amsterdam: IOS Press.

Aloisio, G., Cafaro, M., Fiore, S. & Mirto, M. (2005). The Grid Relational Catalog Project. In L. Grandinetti (Ed.), *Advances in Parallel Computing, "Grid Computing: The New Frontiers of High Performance Computing"* (pp.129-155). Philadelphia: Elsevier.

Aloisio, G., Cafaro, M., Fiore, S., & Mirto, M. (2005). ProGenGrid: a Workflow Service Infrastructure for Composing and Executing Bioinformatics Grid Services. *Proceedings of the 18th IEEE International Symposium on Computer-Based Medical Systems* (IEEE CBMS 2005), Dublin, June 23-24 2005, Ireland. Washington, DC: IEEE Computer Society.

Aloisio, G., Cafaro, M., Fiore, S., & Mirto, M. (2005, June 23-24). ProGenGrid: a Workflow Service Infrastructure for Composing and Executing Bioinformatics Grid Services. *Proceedings of the 18th IEEE International Symposium on Computer-Based Medical Systems* (IEEE CBMS 2005), Dublin, Ireland.

Aloisio, G., Cafaro, M., Fiore, S., Mirto, M. (2007). A Grid System for the Ingestion of Biological Data into a Relational DBMS. In *Proceedings of IEEE International Symposium on Bioinformatics and Life Science Computing (BLSC-07),* (pp. 713-718), Niagara Falls, Canada.

Aloisio, G., Cafaro, M., Lezzi, D. & Van Engelen, R. (2003). Secure Web Services with Globus GSI and gSOAP. In *Euro-Par*: (LNCS 2790, pp.421-426). Berlin/Heidelberg: Springer.

Aloisio, G., Cafaro, M., Lezzi, D. & Van Engelen, R. (2005). The GSI plug-in for gSOAP: Enhanced Security, Performance, and Reliability. In *Proceedings of Information Technology Coding and Computing (ITCC 2005),* (Vol. I), (pp. 304–309). New York: IEEE Press.

Aloiso, G., Benkner, S., Bilofsky, H., Blanquer Espert, I., Breton, V., Cannataro, M., et al. (2005). *HealthGrid White Paper.* Retrieved from http://whitepaper.healthgrid.org.

Alsairafi, S., Emmanouil, F. S., Ghanem, M., Giannadakis, N., Guo, Y., Kalaitzopoulos, D., et al. (2003). The design of discoverynet: Towards open grid services for knowledge discovery. *International Journal of High Performance Computing and Applications, 17*(3), 297-315.

Altekar, G., Dwarkadas, S., Huelsenbeck, J.P. & Ronquist, F. (2004). Parallel Metropolis coupled Markov chain Monte Carlo for Bayesian phylogenetic inference. *Bioinformatics, 20* (3), 407–415.

Altintas, I., Barney, O., Cheng, Z., Critchlow, T., Ludaescher, B., Parker, S., Shoshani, A., & Vouk, M. (2006). Accelerating the scientific exploration process with scientific workflows. *J. Phys.: Conf. Ser., 46*(1), 468+.

Altintas, I., Berkley, C., Jaeger, E., Jones, M., Ludascher, B., & Mock, S. (2004). Kepler: an extensible system for design and execution of scientific workflows. Proceedings of the 16th International Conference on Scientific and Statistical Database Management (SSDBM), Santorini Island, Greece. Kepler website at http://kepler-project.org

Altschul S. F., Madden T. L., Schaffer A. A., Zhang J., Zhang Z., Miller W., & Lipman D. J. (1997). Gapped BLAST and PSI-BLAST: a new generation of protein database search programs. *Nucleic Acids Research, 25*(17), 3389-3402.

Altschul, S.F., Gish, W., Miller, W., Myers, E.W. & Lipman, D.J. (1990). Basic Local Alignment Search Tool. *J. Mol. Biol., 215*, 403-410.

Altschul, S.F., Madden, T.L., Schaffer, A.A., et al. (1997). Gapped BLAST and PSI-BLAST: a new generation of protein database search programs. *Nucleic Acids Research*, 25(17):3389-3402.

Altschul, S.F., Madden, T.L., Schaffer, A.A., Zhang, J., Zhang, Z., Miller, W., Lip-man, D.J. (1997). Gapped BLAST and PSI-BLAST: a new generation of protein database search programs. *Nucleic Acids Res., 25*(17), 3389-402.

Altschul, S.F., Madden, T.L., Schffer, A.A., Zhang, J., Zhang, Z., Miller, W. & Lipman D.J. (1997). Gapped BLAST and PSI-BLAST: a new generation of protein database search programs. *Nucleic Acids Res., 17*(25), 3389-3402. Retrieved from http://www.ncbi.nlm.nih.gov/cgi-bin/BLAST/nph-psi.

Alur, N., et al. (2005). *Patterns: Information Aggregation and Data Integration with DB2 Information Integrator.* Armonk, NY: IBM Redbook.

Alvarez, J. (2004). High-throughput docking as a source of novel drug leads. *Curr Opin Chem Biol, 8*, 365-370.

Alves, D., Hass, V. & Caliri, A. (2003). The predictive power of R0 in an epidemic probabilistic model. *Journal of Biological Physics, 29*(1), 63.

Amazon Elastic Compute Cloud (2007). Retrieved from www.amazon.com/ec2

Ammar, R.A., Demurjian, S.A., Greenshields, I.R., Pattipatti, K. & Rajasekaran, S. (2005). Analysis of Heterogeneous Data in Ultrahigh Dimensions. In R. Popp and J. Yen (Eds.), *Emergent Information Technologies and Enabling Policies for Counter-Terrorism*, (pp. 105-124). Hoboken, NJ: Wiley Interscience.

Andersen, A. H., & Kak, C. A. (1984). Simultaneous algebraic reconstruction technique (SART): a superior implementation of the art algorithm. *Ultrasonic imaging, 6*(1), 81-94.

Anderson, D. (2006). Boinc: A System for public-resource computing and storage. *Proceedings of the 2006 ACM/IEEE conference on Supercomputing* (a. 126). New York: ACM.

Anderson, D., & Fedak, G. (2006). The Computational and Storage Potential of Volunteer Computing, In *Proceedings of the 6th IEEE/ACM International Symposium on Cluster Computing and the Grid*, (pp. 73-80. New York: IEEE Computer Society.

Anderson, D.P. (2004). BOINC: A System for Public-Resource Computing and Storage. In: *Proceedings of the 5th IEEE/ACM International Workshop on Grid Computing*, (pp. 4-10) New York: IEEE Computer Society.

Anderson, D.P., Cobb, J., Korpela, E., Lebofsky, M., & Werthimer, D. (2002). SETI@home: an experiment in public resource computing. *Communications of the ACM, 45*(11), 56-61.

Andronico, G., Ardizzone, V., Barbera, R., Catania, R., Carrieri, A., Falzone, A., et al. (2005). GILDA: the Grid INFN virtual Laboratory for Dissemination Activities. *Proceedings of First International Conference on Test-*

beds and Research Infrastructures for the Development of NeTworks and COMmunities (pp. 304-305).

Angele, J., & Weiten, M. (2005). Semantic Technologies in the SIMDAT Grid Project. In C. Goble, C. Kesselman, and Y. Sure, (Ed.), *Semantic Grid: The Convergence of Technologies, number 05271 in Dagstuhl Seminar Proceedings*. Schloss Dagstuhl, Germany: Internationales Begegnungs und Forschungszentrum fuer Informatik (IBFI).

Anjomshoaa A., Brisard F., Drescher M., Fellows D., Ly A., McGough S., Pulsipher D., Savva A. (2005). *GFD-R-P.056 Job Submission Description Language (JSDL) Specification v1.0*. Retrieved from http://forge.gridforum.org/projects/jsdl-wg.

Antonioletti, M., Atkinson, M., Baxter, R., Borley, A., Hong, N. P. C., Collins, B., et al. (2005). The design and implementation of Grid database services in OGSA-DAI. *Concurrency and Computation-Practice & Experience, 17*(2-4), 357-376.

Antonioletti, M., Atkinson, M., Baxter, R., Borley, A., Hong, N.P.C., Collins, B., et al. (2005). The design and implementation of grid database services in OGSA-DAI. *Concurrency Computation Practice and Experience*, 17(2-4), 357-376.

Arabia, M., & Akutsu, T. (1984). A New Test Circulatory System for Research in Cardiovascular Engineering. *Annals of Biomedical Engineering, 12*, 29-48.

Arbona, T., Benkner, S., Fingberg, J., Engelbrecht, G., Hof-mann, M., Kumpf, K., Lonsdale, G., & Woehrer, A. (2007). A Service-oriented Grid Infrastructure for Biomedical Data and Compute Services. *IEEE Transaction on NanoBioscience, 6*(2), 136-141.

Arnold K., Schwede, T., & Podvinec, M. (2007). SWISS-TANDEM: A Web-based Workspace for MS/MS Protein Identification on PC Grids. In Dominique Zosso, (Ed.) *20th IEEE International Symposium on Computer-Based Medical Systems* (Vol. 1) (pp 406-411). New York: IEEE press.

Arsanjani, A. (2004). *Service-oriented modeling and architecture*, (IBM technical article). Retrieved October 20, 2008, from http://www.ibm.com/developerworks/library/ws-soa-design1/.

Artacho E., Anglada E., Di´eguez O., Gale J. D, Garc´ıa A., Junquera J., Martin R. M., Ordej´on P., Pruneda J. M., S´anchez-Portal D., & Soler J. M. (2008). The SIESTA method; developments and applicability. *J. Phys.: Condens. Matter, 20*, 064208.

Arzberger, P. W., & Papadopoulos, P. (Feb 2006). PRAGMA: Example of Grass-Roots Grid Promoting Collaborative e-Science Teams. *CTWatch Quarterly, 2* (1).

Arzberger, P., Farazdel, A., Konagaya, A., Ang, L., Shimojo, S. & Stevens, R. (2004). Life sciences and cyberinfrastructure: dual and interacting revolutions that will drive future science. *New Generation Computing, 22* (2), 97-110.

Ashburner, M., Ball, C.A., et al. (2000). Gene Ontology: tool for the unification of biology. The Gene Ontology Consortium. *Nat Genet 25*(1): 25-9.

Assel, M., & Kalyoncu, O. (2008). Dynamic Access Control Management for Distributed Biomedical Data Resources. In Cunningham, P. & Cunningham, M. (Eds.), *Collaboration and the Knowledge Economy: Issues, Applications, Case Studies* (pp. 1593-1599). IOS Press.

Astakhov, V., Gupta, A., Grethe, J., Ross, E., Little, D., Yilmaz, A., Qian, X., Santini, S., Martone, M., Ellisman, M. (2006). Semantically Based Data Integration Environment for Biomedical Research. In *Proceedings of the 19th IEEE Symposium on Computer-Based Medical Systems (CBMS'06)*. New York: IEEE Computer Society Press.

Atkins, D. (2003). *Report of Blue-Ribbon Advisory Panel on Cyberinfrastructure*. Retrieved from http://www.nsf.gov/publications

Atkins, D., Kroegemeier, K., Feldman, S., Garcia-Molina, H., Klein, M., Messerschmitt, D., et al. (2003). Revolutionizing Science and Engineering Through Cyberinfrastructure. In *Report of the National Science Foundation Blue-Ribbon Advisory Panel on Cyberinfrasturcuture*. National Science Foundation: Arlington, VA.

Atkins, D.E., Droegemeier, K.K., Feldman, S.I., Garcia-Molina, H. Klein, M.L., Messerschmitt, D.G., et al. (n.d.). *Revolutionizing Science and Engineering Through Cyberinfrastructure: Report to the National Science Foundation Blue-Ribbon Advisory Panel on*

Cyberinfrastructure. Retrieved on January 15, 2008, from http://www.communitytechnology.org/nsf_ci_report/report.pdf.

Attwood, T.K., Flower, D.R., Lewis, A. P., Mabey, J.E., Morgan, S.R., Scordis, P, Selley, J., & Wright, W. (2000). PRINTS-S: the database formerly known as PRINTS. *Nucleic Acids Research, 27,* 220-225.

Austin, J., Jackson, T., Fletcher, M., Jessop, M., Cowley, P., & Lobner, P. (2004). Predictive maintenance: distributed aircraft engine diagnostics. In Elsevier (Ed.), *The grid 2* (2nd Ed., pp. 69-79). San Francisco, CA: Morgan Kaufmann.

Axelson, O. (1999). The character and the public health implications of ecological analyses. In A. Lawson, A. Biggeri, D. Böhning, E. Lesaffre, J.-F. Viel & R. Bertollini (Ed.), *Disease Mapping and Risk Assessment for Public Health,* (pp. 301-309). New York: Wiley and Sons.

Azencott, R. (1988). Simulated Annealing. *Seminaire Bourbaki 40eme annee, 1987-88, No. 697,* (pp. 223-237).

Azencott, R.(1992). *Simulated Annealing: Parallelization Techniques.* Hoboken, NJ: Wiley.

Badia, R.M., Labarta, J.S., Sirvent, R.L., Perez, J.M., Cela, J.M., & Grima, R. (2003). Programming Grid Applications with GRID Superscalar. Journal of Grid Computing, vol. 1, pp. 151-170.

Bagnasco, A., Poggi, A., Parodi, G., & Scapolla, A.M. (2006). Exposing Measurement Instruments As Grid Services. In *Distributed Cooperative Laboratories: Networking, Instrumentation, and Measurements* (pp. 321-329). New York: Springer.

Bailey, T. C., Gatrell, A. C. (1995). *Interactive Spatial Data Analysis.* Essex, UK: Longman Scientific and Technical.

Bairoch, A., Apweiler, R., Wu, C.H., Barker, W.C., Boeckmann, B., Ferro, S. & et al (2005). The Universal Protein Resource (UniProt). *Nucleic Acids Res.*, (33), 154–159. Retrieved from http://www.uniprot.org.

Bajaj, C. and S. Cutchin (1999). Web Based Collaborative Visualization of Distributed and Parallel Simulation. In *Proc. of the 1999 Symposium on Parallel Visualization and Graphics, San Francisco,* (pp. 47-54).

Baker N. A., Sept D., Joseph S., Holst M. J., & McCammon J. A. (2001). *Electrostatics of nanosystems: application to microtubules and the ribosome.* Proc. Natl. Acad. Sci. USA 98, (pp. 10037-10041).

Baker, M., & Lakhoo, R. (2006). Application Reuse Through Portal Frameworks. *Proceedings of the UK e-Science All Hands Conference* (pp. 510-518).

Baker, N. A., Sept, D., Joseph, S., Holst, M. J., & McCammon, J. A. (2001). Electrostatics of nanosystems: application to microtubules and the ribosome. *Proceedings of the National Academy of Sciences of the United States of America 2001,* (Vol. 98), (pp. 10037-41).

Baker, S. (2007). Google and the Wisdom of Clouds. Business Week, Dec. 13. Retrieved from www.businessweek.com/magazine/content/07_52/b4064048925836.htm

Bala, P., Pytlinski, J., Nazaruk, M., Alessandrini, V., Girou, D., Erwin, D.W., et al. (2002). BioGRID: An European Grid for Molecular Biology. *11th IEEE International Symposium on High Performance Distributed Computing* (p. 412)., Edinburgh, UK: IEEE Computer Society.

Baldridge K. K., Greenberg J. P., Elbert S. T., Mock, S., & Papadopoulos, P. (2002). *QMView and GAMESS: Integration into the World Wide Computational Grid.* Supercomputing, ACM/IEEE 2002 Conference, 64.

Baldridge, K. K., & Greenberg, J. P. (1995). QMView: a computational chemistry three-dimensional visualization tool at the interface between molecules and mankind. *Journal of Molecular Graphics, 13,* 63-66).

Baldridge, K., Bhatia, K., Greenberg, J. P., Stearn, B., Mock, S., Sudholt, W., Krishnan, S., Bowen, A., Amoreira, C., & Potier, Y. (2005). In GEMSTONE: Grid Enabled Molecular Science Through Online Networked Environments. *Life Sciences Grid Workshop, Singapore.* Singapore: World Scientific Press.

Balis, B., Bubak, M., & Wach, J. (2007) User-Oriented Querying over Repositories of Data and Provenance. In G. Fox, K. Chiu, & R. Buyya, (Eds.), *Third IEEE International Conference on e-Science and Grid Computing, e-Science 2007, Bangalore, India, 10-13 December 2007,* (pp.77-84). New York: IEEE Computer Society.

Balis, B., Bubak, M., & Wach, J. (2008). Provenance Tracking in the ViroLab Virtual Laboratory. In *Proceedings of PPAM 2007, Seventh International Conference on Parallel Processing and Applied Mathematics, Lecture Notes in Computer Science*, (Vol. 4967), (pp. 381-390). New York: Springer.

Balis, B., Bubak, M., & Wach, K. (2007). User-Oriented Querying over Repositories of Data and Provenance. In *Proceedings of the 3rd IEEE International Conference on e-Science and Grid Computing, e-Science 2007*, Bangalore, India, Dec. 2007, (pp. 187-194). New York: IEEE Computer Society.

Balis, B., Bubak, M., & Wach, K.(2008). Provenance Tracking in the ViroLab Virtual Laboratory. In R. Wyrzykowski, (Ed.), *Proceedings of the Seventh International Conference on Parallel Processing and Applied Mathematics*, LNCS, Gdansk, Poland, Sept. 2007. Berlin, Germany: Springer.

Baloa, L., Boston, J., & Antaki, J. (2001). Elastance-Based Control of a Mock Circulatory System. *Annals of Biomedical Engineering, 29*, 244-251.

Banks, D. (2001). *Game Theory and. Risk Analysis for. Counterterrorism*. Retrieved November 2008 from http://www.galaxy.gmu.edu/interface/I03/I2003Proceedings/BanksDavid/Banks David.presentation.pdf

Barker, W.C., Pfeiffer, F., & George, D.G. (1996). Superfamily classification in FIR-international protein sequence database. *Methods in Enzymology, 266*, 59-71.

Barla, G. Jurman, S. Riccadonna, M. Chierici, S. Merler, & Furlanello,C. (2008). Machine learning methods for predictive proteomics. *Briefings in Bioinformatics, 9*(2), 119-128.

Barrett C. L., Eubank S. G., and Smith J. P. (2005). If Smallpox Strikes Portland. *Scientific American Magazine*. Retrieved July 2008 from http://www.sciam.com/article.cfm?id=if-smallpox-strikes-portl

Baru, C., Moore, R., Rajasekar, A., & Wan, M. (1998) The SDSC Storage Resource Broker. In *Proceedings of the 1998 Conference of the Centre for Advanced Studies on Collaborative Research* (p. 5). Toronto, Canada: IBM Press.

Baruah, S., Cohen, N., Plaxton, C. & Varvel, D. (1996). Proportionate Progress: A Notion of Fairness in Resource Allocation. *Algorithmica, 15*(6), 600-625.

Basney J., Humphrey M., & Welch V. (2005). The MyProxy online credential repository. *Software: Practice and Experience, 35*(9) 801-816.

Basney, J., Livny, M., & Tannenbaum T. (1997, June). High Throughput Computing with Condor. *HPCU news, 1*(2).

Basu, S., Talwar, V., Agarwalla, B., & Kuma, R. (2003). *Interactive Grid Architecture for Application Service Providers* (HP Technical Report HPL-2003-84R1).

Bata, P., Alessandrini, V., Girou, D., MacLaren, J., Brooke, J., Pytlinski, J., et al. (2002). BIOGRID-A European grid for molecular biology. In *Proceedings of the 11th IEEE International Symposium on High Performance Distributed Computing (HPDC-11)*, Edinburgh, UK.

Bateman, A. & Valencia, A. (2006). Structural genomics meets computational biology. *Bioinformatics, 22*, 2319.

Bateman, A., Birney, E., Durbin, R., Eddy, S.R., Finn, R.D., & Sonnhammer, E.L. (2000). The Pfam protein families database. *Nucleic Acids Research, 28*, 263-266.

Bates, D. G. A. (2003). Improving safety with information technology. *New England Journal of Medicine, 348*, 2526-2534.

Bavier M., Bowman M., Chun B., Culler D., Karlin S., Muir S., Peterson L., Roscoe T., Spalink T., & Wawrzoniak M. (2004). *Operating System Support for Planetary-Scale Network Services*. NSDI '04 Proceedings. Retrieved from http://www.cs.princeton.edu/nsg/papers/plos_nsdi_ 04/nsdi.pdf

Bechhofer, S., Carr, L., Goble, C., Kampa, S., & Miles-Board, T. (2002). The semantics of semantic annotation. In Meersman, R. & Tari, Z., (Ed.), *On the Move to Meaningful Internet Systems 2002: CoopIS, DOA, and ODBASE*, , London, (pp. 1152-1167). Berlin, Germany: Springer-Verlag.

Beck M., Moore T., & Plank J.S. (2002 July). An End-To-End Approach to Globally Scalable Network Storage. In *SIGCOMM '02: Proceedings of the 2002 conference on*

Applications, technologies, architectures, and protocols for computer communications, (pp. 339-346).

BEinGRID (2008). Business Experiments in Grids. Retrieved from www.beingrid.com.

Belhajjame, K., Embury, S.M., Paton, N. (2008). Automatic annotations of web services based on workflow definitions. *ACM Transactions on the Web, 2* (2), 1-34.

Bell, W.H., Cameron, D.G., Capozza, L., Millar A.P., Stockinger, K., & Zini, F. (2002 November). Simulation of Dynamic Grid Replication Strategies in OptorSim. *In Proceedings of the 3rd Int'l. IEEE Workshop on Grid Computing (Grid'2002)*, Lecture Notes in Computer Science. Berlin, Germany: Springer Verlag.

Bellotti, R., Cerello, P., Tangaro, S., Bevilacqua, V., Castellano, M., Mastronardi, G., et al. (2007). Distributed medical images analysis on a Grid infrastructure. *Future Gener. Comput. Syst., 23* (3), 475-484.

Beltrame, F., Maggi, P., Melato, M., Molinari, E., Sisto, R., & Torterolo, L. (2006). SRB Data Grid and Compute Grid Integration via the EnginFrame Grid Portal. Proceedings of the 1st SRB Workshop, 2-3 February 2006, San Diego, USA. PDF retrieved from www.sdsc.edu/srb/Workshop/SRB-handout-v2.pdf

Benson, D.A., Karsch-Mizrachi, I., Lipman, D.J., Ostell, J. & Wheeler, D.L. (2007). GenBank. *Nucleic Acids Res., 35*(Database issue), D21-5.

Bergamaschi, S., Castano, S., Vincini, M., & Beneventano, D. (2001). Semantic integration of heterogeneous information sources. *Data and Knowledge Engineering, 36*(3), 215-249.

Bergmann, J., Bott, O. J., Hoffmann, I., & Pretschner, D. P. (2005). An eConsent-based System Architecture Supporting Cooperation in Integrated Healthcare Networks. *Stud Health Technol Inform, 116*, 961-966.

Berman F. Fox G., & Hey A. (2003). *Grid Computing: Making the Global*. Hoboken, NJ: John Wiley and Sons.

Berman H. M., Westbrook J., Feng Z., Gilliland G., Bhat T. N., Weissig H., Shindyalov I. N., & Bourne P. E. (2000). The Protein Data Bank. *Nucleic Acids Research, 28*, 235-242.

Berman, F. (2001). Viewpoint: From teragrid to knowledge grid. *Commun. ACM, 44*(11), 27-28.

Berman, F. Wolski, R. & Casanova, H. (2003). Adaptive Computing on the Grid Using AppLeS. *IEEE Transactions on Parallel and Distribributed Systems, 14*(4), 369–382.

Berman, F., Fox, G. &, Hey, A.J.G. (2003). *Grid Computing: Making the Global Infrastructure a Reality.* Hoboken, NJ: John Wiley.

Berman, F., Fox, G., & Hey, T. (Ed.). (2003). *Grid computing: Making the global infrastructure a reality.* Chichester, UK: Wiley.

Berman, F., Hey, A. J. G & Fox, G. C. (Eds.), (2003). *Grid Computing: Making the Global Infrastructure a Reality.* New York: John Wiley.

Berman, H.M., Henrick, K., & Nakamura, H. (2003). Announcing the worldwide Protein Data Bank. *Nature Structural Biology, 10*(12), 980.

Berman, H.M., Westbrook, J., Feng, Z., Gilliland, G., Bhat, T.N. & et al. (2000). The Protein Data Bank. *Nucleic Acids Res.*, (28), 235–242. Retrieved from www.rcsb.org/pdb/.

Berman, H.M., Westbrook, J., Feng, Z., Gilliland, G., Bhat, T.N., Weissig, H., Shindayalov, I.N. & P.E. Bourne (2000). The Protein Data Bank. *Nucleic Acids Research, 28*, 235-242.

Berrar, D., Granzow, M., & Dubitzky, W. (2007). Introduction to Genomic and Proteomic Data Analysis. In D. Berrar, M. Granzow, & W. Dubitzky (Eds.), *Fundamentals of Data Mining in Genomics and Proteomics*, (pp. 1-37). Berlin, German: Springer Science.

Berry, D., Germain-Renaud, C., Hill, D., Pieper, F. & Saltz, J. (2004). Report on the Workshop IMAGE'03: Images, medical analysis and grid environments (TR UKeS-2004-02). UK National e-Science Centre, Glasgow, UK.

Bhandarkar M., Budescu G., Humphrey W. F., Izaguirre J. A., Izrailev S., Kalé L. V., Kosztin D., Molnar F., Phillips J. C., & Schulten K. (1999). BioCoRE: A collaboratory for structural biology. In A. G. Bruzzone, A. Uchrmacher, & E. H. Page (Eds.), *Proceedings of the*

SCS International Conference on Web-Based Modeling and Simulation (pp. 242-251).

Bhatia, K., Chandra, S., & Mueller, K. (2006). *In GAMA: Grid Account Management Architecture*. 1st IEEE International Conference on e- Science and Grid Computing, Melbourne, Australia.

Bhuiya, A. K. & Stanley, E. (1963). The refinement of atomic parameters by direct calculation of the minimum residual, *Acta Crystallogr, 16*, 981-984.

Bidgood, W.D., Horii, S.C., Prior, F.W., & Van Syckle, D.E. (1997). Understanding and Using DICOM, the Data Interchange Standard for Biomedical Imaging. *Journal of the American Medical Informatics Association, 4*, 199-212.

Bilbao-Castro, J. R., Carazo, J. M., García, I., & Fernandez, J. J. (2006). Parallelization of Reconstruction Algorithms in Three-Dimensional Electron Microscopy. *Applied Mathematical Modelling, 30*(8), 688-701.

Bilbao-Castro, J. R., García, I., & Fernández, J. J. (2007). EGEETomo: a user-friendly, fault-tolerant and grid-enabled application for 3D reconstruction in electron tomography. *Bioinformatics, 23*(24), 3391-3393.

Bilbao-Castro, J. R., Merino, A. J., García, I., Carazo, J. M., & Fernández, J. J. (2007b). Parameter optimization in 3D reconstruction on a large scale grid. *Parallel Computing, 33*(4-5), 250-263.

Bilgin, A., Zweig, G., & Marcellin, M.W. (2000). Three-dimensional image compression with integer wavelet transforms. *Applied Optics, 39*, 1799-1814.

Bilykh, I., Bychkov, Y., Dahlem, D., Jahnke, J.H., McCallum, G., Obry, C., et al. (2003). Can grid services provide answers to the challenges of national health information sharing? *In Proceedings of the 2003 conference of the Centre for Advanced Studies on Collaborative research (CASCON '03)* (pp. 39-53). Armonk, NY: IBM Press.

BioWBI (2004). *Bioinformatic Workflow Builder Interface*. Retrieved from http://www.alphaworks.ibm.com/tech/biowbi

BIRN (2008). Biomedical Informatics Research Network. Retrieved from www.nbirn.net/index.shtm

BIRN. (2000). *Biomedical Informatics Research Network*. Retrieved November 2008 from http://www.nbirn.net/

Biton, O., Cohen-Boulakia, S. & Davidson, S. (2007). Zoom* userviews: Querying relevant provenance in workflow systems. In *VLDB* (pp. 1366–1369).

Biton, O., Cohen-Boulakia, S., Davidson, S. & Hara, C. (2008) Querying and managing provenance through user-views in scientific workflows. In *ICDE*. Los Alamitos, CA: IEEE.

Black, D., Morris S. N., Smith, C., Townsend, P. (1982). The pattern of present health inqualities. In P. Townsend & N. Davidson (orgs.), *Inequalities in health: the black report* (pp. 51-64). New York: Penguin Books.

Blackledge, T.A., Gillespie, R.G. (2004). Convergent evolution of behavior in an adaptive radiation of Hawaiian web-building spiders. *Proc Natl Acad Sci USA, 101*(46):16228-33.

Blanchet, C., Lefort, V., Combet, C. & Deléage, G. (2006). GPS@ bioinformatics portal: from network to EGEE grid. *Stud Health Technol Inform, 120*, 187-193.

Blanquer, I., Hernandez, V., & Segrelles, D. (2006). TRENCADIS – A Grid Architecture for Creating Virtual Repositories of DICOM Objects in an OGSA-Based Ontological Framework. In *Biological and Medical Data Analysis* (pp. 183-194).

Blessing, R. H. & Smith, G. D. (1999). Difference Structure Factor Normalization for Determining Heavy-Atom or Anomalous Scattering Substructures, *J. Appl. Cryst. 32*, 664-670.

Bobashev, G. V., Goedecke, D. M., Yu, F., Epstein, J. M. (2007). A Hybrid Epidemic Model: Combining the Advances of Agent-Based and Equation-Based Approaches. In *Proceedings of the 2007 Winter Simulation Conference*. New York: IEEE Press.

Boccia, V., Guarracino, M.R., D'Amore, L. & Laccetti, G. (2005). A grid enabled PSE for medical imaging: experiences on MedIGrid. In *Proceedings of the 18th IEEE Symposium on computer-based medical systems* (pp. 529-536). Washington, DC: IEEE Computer Society.

Bodenreider, O. (2004) The Unified Medical Language System (UMLS): integrating biomedical terminology. *Nucleic Acids Research, 32*, D267-D270.

Boguski, M.S., Lowe T. M. & Tolstoshev, C.M. (1993). dbEST- database for "expressed sequence tags." *Nat. Genet.*, *4*(4), 332-333.

Bohr, J. et al. (1993). Protein structures from distance inequalities. *J. Mol. Biol.*, *231*, 861-869.

Boinc (2008). Retreived March 2008 from http://boinc.berkeley.edu/

BoincGamess (2008). Retreived March 2008 from http://cvs.rocksclusters.org/viewcvs/viewcvs.cgi/gamess/boincgamess-doc/

Bolze, R. Cappello, F., Caron, E., Daydé, R., Desprez, F., & Jeannot, E., et al. (2006) Grid'5000: a large scale and highly reconfigurable experimental Grid testbed. *International Journal of High Performance Computing Applications*, *20*(4), 481-494.

Bolze, R., Cappello, F., Caron, E., Daydé, M., Desprez, F., Jeannot, E., et al. (2006). Grid'5000: A Large Scale and Highly Reconfigurable Experimental Grid Testbed. *International Journal of High Performance Computing Applications*, *20*(4) 481-494.

Bonabeau, E., Dorigo, M., & Theraulaz, G. (Eds.). (1999). *Swarm Intelligence: From Natural to Artificial Systems*, New York, USA: Oxford University Press.

Bond, P. J., & Sansom, M. S. P. (2006). Insertion and assembly of membrane proteins via simulation. *Journal of the American Chemical Society, 128*, 2697-2704.

Bond, P. J., Derrick, J. P., & Sansom, M. S. P. (2007). Membrane simulations of OpcA: Gating in the loops? *Biophysical Journal, 92*, L23-L25.

Borking, J. J. (2005, October 3, 2005). *Privacy Standards for Trust.* Paper presented at the 27th International Conference on Privacy and Personal Data Protection, Montreux, Switzerland.

Boser, B. E., Guyon, I. M., & Vapnik, V. N. (1992). A training algorithm for optimal margin classifiers. *In Proceedings of the fifth annual workshop on Computational learning theory* (pp. 144-152). New York: ACM.

Bowers, S., McPhillips, T. M., Ludascher, B., Cohen, S., & Davidson, S. B. (2006). A Model for User-Oriented Data Provenance in Pipelined Scientific Workflows. In L. Moreau & I. T. Foster, (Eds.), *Provenance and Annotation of Data, International Provenance and Annotation Workshop, IPAW 2006, Lecture Notes in Computer Science*, (Vol. 4145) (pp. 133-147). Chicago, IL: Springer.

Brady, J. M., Gavaghan, D. J., Simpson, A. C., Parada, M. M. & Highnam, R. P. (2003). eDiaMoND: A grid-enabled federated database of annotated mammograms. In Berman, F., Fox, G. C. & Hey, A. J. C. (Eds.). *Grid computing: making the global infrastructure a reality*, (3rd Ed.) (pp. 923-943). Hoboken, NJ: Wiley.

Brandon D., Vandivort K., Brunner R. & Bach M. (2005). *2005 BioCoRE Survey Report.* Retrieved from http://www.ks.uiuc.edu/Research/biocore/eval/surveyData/BioCoRE_2005_Survey_Report%20.pdf

Braun, R.C., Pedretti, K.T., Casavant, T.L., Scheetz T.E., Birkett, C.L., & Roberts, C.A. (2001). Parallelization of Local BLAST Service on Workstation Clusters. *Future Genereration Computers Systems, 17*(6), 745-754.

Brenner, S., Chothia, C., & Hubbard, T. (1998). Assessing sequence comparison methods with reliable structurally identified distant evolutionary relationships. *Biochemistry, 95*(11) 6073-6078.

Breton V., Kasam V., & Jacq N. (2006). *High Throughput Grid Enabled Virtual Screening*. Proceedings of the NETTAB2006: Santa Margherita, 14–18.

Breton, V. (2007). A perspective on biomedical applications on Grids. In *eScience Conference 2007.*

Breton, V., Blanchet, C., Legré, Y., Maigne, L. and Montagnat, J. (2005). Grid Technology for Biomedical Applications. In M. Daydé et al. (Eds.), *VECPAR 2004*, (LNCS 3402), (pp. 204–218).

Breton, V., Dean, K. & Solomonides, T. (2005). The HealthGrid White Paper. In *Proceedings of Third HealthGrid Conference: Studies in Health Technology and Informatics,* Vol. 112. Amsterdam, The Netherlands: IOS Press (pp 249–321). Retrieved from http://www.healthgrid.org

Breton, V., Dean, K. & Solomonides, T. (2005). The HealthGrid white paper. *Studies in Health Technology and Informatics, 112,* 249-321.

Breu, H. & Kirkpatrick, D.G. (1998). Unit disk graph recognition is NP-hard. *Computational Geometry, 9*, 3-24.

Brevik, J.,Nurmi, D. & Wolski, R. (2006, March*). Predicting bounds on queuing delay for batch-scheduled parallel machines.* In *11th ACM SIGPLAN symposium on principles and practice of parallel programming* (pp 110–118). New York, NY: ACM Press.

Brezany, P., Hofer, J., Tjoa, A., & Wohrer, A. (2003). Gridminer: An infrastructure for data mining on computational grids. *In Data Mining on Computational Grids APAC'03.*

Brin, S., Motwani, R., Ullman, J. D., & Tsur, S. (1997). Dynamic itemset counting and implication rules for market basket data. In J. Peckham (Ed.), *International Conference on Management of Data* (pp. 255-264). New York: ACM Press.

Brooks, B. R., Bruccoleri, R. E., Olafson, B. D., States, D. J., Swaminathan, S., & Karplus, M. (1983). CHARMM: a program for macromolecular energy, minimization, and dynamics calculations. *Journal of Computational Chemistry, 4*, 187–217.

Brooks, D. R., & McLennan D. A. (2002). *The Nature of diversity: An evolutionary voyage of discovery.* Chicago, IL: University of Chicago Press.

Brownstein, J.S., Freifeld, C.C. (2007). HealthMap: the development of automated real-time internet surveillance for epidemic intelligence. *Euro Surveill 12: E071129 071125.* Retrieved February 28, 2008 from http://www.eurosurveillance.org/ew/2007/071129.asp#5.

Bruno, G., Katz, M., Sacerdoti, F. D., & Papadopoulos, P. M. (2004) *In Rolls: Modifying a Standard System Installer to Support User-Customizable Cluster Frontend Appliances.* IEEE International Conference on Cluster Computing, San Diego, CA (pp 421-430).

Bubak, M., Gubala, T., Kapalka, M., Malawski, M., and Rycerz, K. (2005). Workflow composer and service registry for grid applications. *Future Generation Computer Systems, 21*(1), 79-86.

Bubak, M., Gubala, T., Kasztelnik, M., Malawski, M., Nowakowski, P. & Sloot, P. M. A. (2007). Collaborative virtual laboratory for e-health. In *Expanding the Knowledge Economy: Issues, Applications, Case Studies, eChallenges e-2007 Conference Proceedings* (pp. 537-544). Amsterdam, the Netherlands: IOS Press.

Budura, A., Cudré-Mauroux, P., & Aberer, K. (2007). From bioinformatic web portals to semantically integrated Data Grid networks. *Future Gener. Comput. Syst., 23* (3), 485-496.

Buetow, K.H. (2003). Cyberinfrastructure: Empowering a "Third Way" in Biomedical Research. *Science*, 44, 417.

Bulger, E. & Nathens, A. (2002). Management of severe head injury: institutional variations in care and effect on outcome. *Critical Care Med, 30*, 1870-1876.

Bundesministerium für Bildung und Forschung (BMBF). (2008). *3. D-Grid call: BMBF-Förderbekanntmachung „Grid-Dienste für Wirtschaft und Wissenschaft".* Retrieved May 26, 2008, from http://www.bmbf.de/foerderungen/12378.php

Buneman, P., Chapman, A. & Cheney, J. (2006) Provenance management in curated databases. In Chaudhuri, et al.

Burge, C. & Karlin, S. (1997). Prediction of complete gene structures in human genomic DNA. *J. Mol. Biol., 268*, 78-94.

Burke, S. Campana, S., Mendez Lorenzo, P., Nater C., Santinelli, R., Sciabà, A. (2008). gLite 3.1 User Guide. Retrieved June 6, 2008, from https://edms.cern.ch/file/722398//gLite-3-UserGuide.pdf

Burke, S., Campana, S., Delgado Peris, A., Donno, F., Mendez Lorenzo, P., Santinelli, R., & Sciaba, A. gLite 3 User Guide. *gLite 3.1 User Guide Manual Series.*

Burla, M.C., Camalli, M., Cascarano, G., Giacovazzo, C., Polidori, G., Spagna, R. & Viterbo, D. (1989). SIR88 - a direct-methods program for the automatic solution of crystal structures. *J. Appl. Cryst. 22*, 389-393.

Burnum, J.F. (1989). The misinformation era; the fall of the medical record. *Annals of Internal Medicine, 110*(6), 482-484.

Burrage, K., Hood, L. & Ragan, M. A. (2006). Advanced computing for systems biology. *Brief Bioinform, 7*(4), 390-398.

Butler, D. (2006) Disease surveillance needs a revolution. *Nature, 440*(March 2), 6-7.

Butte, A. J., & Kohane, I. S. (2006). Creation and implications of a phenome-genome network. *Nat Biotechnol, 24*(1), 55-62.

Buyya, R., Abramson, D., & Giddy, J. (2000). Nimrod/G: an architecture for a resource management and scheduling system in a global computational grid", Proceedings of the 4th International Conference on High Performance Computing in the Asia-Pacific Region. www.csse.monash.edu.au/~davida/nimrod/nimrodg.htm

Caas, D. Lpez-Ferrer, A. Ramos-Fernndez, E. Camafeita, & E. Calvo. (2006). Mass spectrometry technologies for Proteomics. *Briefing in Functional Genomics and Proteomics, 4*(4), 295–320.

Cafaro, M., Epicoco, I., Mirto, M., Lezzi, D. & Aloisio, G. (2008). The Grid Resource Broker Workflow Engine. *Concurrency and Computation: Practice and Experience, Special Issue: 2nd International Workshop on Workflow Management and Applications in Grid Environments (WaGe2007), 20*(15), 1725-1739.

Calanducci, A., Cherubino, C., Ciuffo, L.N., & Scardaci, D. (2007). A Digital Library Management System for the Grid. *Proceedings of the 16th IEEE International Workshops on Enabling Technologies: Infrastructure for Collaborative Enterprises* (pp. 269-272). Washington, DC: IEEE Computer Society.

Callahan, S.P., Freire, J., Santos, E., Eduardo Scheidegger, C., Silva, C.T. & Vo, H.T. (2006) Vistrails: visualization meets data management. In Chaudhuri, et al. (pp. 745–747).

Cameron G., (2003). Bioinformatics and Genomics Grid for European Research (BIGGER). In *Proceedings of the First European HealthGrid Conference (HealthGrid'03)*, Lyon, France.

Camon, E., Barrell, D., et al. (2004). The Gene Ontology Annotation (GOA) Database--an integrated resource of GO annotations to the UniProt Knowledgebase. *In Silico Biol 4*(1): 5-6.

Camon, E., Magrane, M., et al. (2004). The Gene Ontology Annotation (GOA) Database: sharing knowledge in Uniprot with Gene Ontology. *Nucleic Acids Res, 32*, D262-6.

Cannata, N., Merelli, E. & Altman, R., (2005). Time to organize the bioinformatics resourceome. *PLoS Computational Biology, 1*(7), e76.

Cannataro, M. (2008). Computational proteomics: management and analysis of proteomics data. *Briefings in Bioinformatics, 9*(2), 9-10.

Cannataro, M Barla, A. Flor, R. Jurman, G. Merler, S. Paoli, S. Tradigo, G. Veltri, P & Furlanello, C (2007). A grid environment for high-throughput proteomics. *IEEE Transaction on NanoBiosciences, 6*(2),117-123.

Cannataro, M & Veltri, P. (2007). MS-Analyzer: Composing and Executing Preprocessing and Data Mining Services for Proteomics Applications on the Grid. *Concurrency and Computation: Practice and Experience, 19*(15), 2047-2066,

Cannataro, M Tradigo, G. & Veltri, P. (2007). Sharing Mass Spectrometry Data in a Grid-based Distributed Proteomics Laboratory. *Information Processing & Management, 43*(3), 577-591

Cannataro, M. & Talia, D. (2003). The knowledge grid: An architecture for distributed knowledge discovery. *Commun. ACM, 46*(1), 89-93.

Cannataro, M. (2006). Next-generation Grids: requirements and knowledge-based services: Research Articles, *Concurrency and Computation: Practice & Experience, 18*(8), 887-898.

Cannataro, M., Barla, A., Flor, R., Jurman, G., Merler, S., Paoli, S., Tradigo, G., Veltri, P. & Furlanello, C. (2007). A grid environment for high-throughput proteomics. *IEEE Trans Nanobioscience, 6*(2),117-123.

Cannataro, M., Comito, C., Lo Schiavo, F., & Veltri, P. (2004). *Proteus. A Grid Based Problem Solving for Bioinformatics: Architecture and Experiments.* IEEE Intelligent Informatics Bulletin, *3*(1), 7-18.

Cannataro, M., Comito, C., Lo Schiavo, F., & Veltri, P. (2004, February). Proteus, a grid based problem solving environment for bioinformatics: Architecture and experiments. *IEEE Computational Intelligence Bulletin*, 3(1), 7–18.

Cannataro, M., Cuda, G., Gaspari, M., Greco, S., Tradigo, G., & Veltri, P. (2007). The EIPeptiDi Tool: Enhancing

Peptides Discovering in ICAT-based LC MS/MS Experiments. *BMC Bioinformatics, 8*(255).

Cannataro, M., Guzzi, P. H., Mazza, T., Tradigo, G., and Veltri, P. (2007). Using ontologies for preprocessing and mining spectra data on the grid. *Future Generation Computer Systems, 23*(1), 55-60.

Cannataro, M., Guzzi, P.H., Mazza, T., Tradigo, G., & Veltri, P. (2006). Managing Ontologies for Grid Computing, Multiagent and Grid Systems, *2*(1), pp. 29-44.

Cannataro, M., Guzzi, P.H., Mazza, T., Tradigo, G., & Veltri, P. (2007). Using Ontologies for Preprocessing and Mining Spectra Data on the Grid. *Future Generation Computer Systems, 23*(1), 55-60

Capriotti, E., Fariselli, P. & Casadio, R. (2005). I-Mutant2.0: predicting stability changes upon mutation from the protein sequence or structure. *Nucleic Acids Research,* (33), 303–305.

Cardoso, J. (2007). The semantic web vision: Where are we? *IEEE Intelligent Systems, 22*(5), 84-88.

Carey, L., Darling, A. E., & Chun Feng, W. (2003). The Design, Implementation, and Evaluation of mpiBLAST. In *Proceedings of ClusterWorld 2003*. San Jose, CA: IEEE.

Caron, E., & Desprez, F. (2006) DIET: A Scalable Toolbox to Build Network Enabled Servers on the Grid. *International Journal of High Performance Computing Applications, 20*(3), 335-352.

Caron, E., Chis, A., Desprez, F., & Su, A. (2008, January). Design of Plug-In Schedulers for a GridRPC Environment. *Future Generation Computer Systems, 24*(1), 46-57.

Carota, L., Bartoli, L., Fariselli, P., Martelli, P.L., Montanucci, L., Maggi G. & Casadio, R. (2008, May). *High Throughput Comparison of Prokaryotic Genomes* (LNCS 4967, pp 1200-1209). Berlin/Heidelberg: Springer.

Carrasco, R. C., & J. Oncina, J. (1994). Learning stochastic regular grammar by means of state merging method. In *Proceedings of the 2nd International Colloquium on Grammatical Inference (ICGI '94)*, Alicante, Spain. (Lecture Notes in Artificial Intelligence LNAI 862), (pp. 139 – 152). Berlin, Germany: Springer - Verlag.

Carvajal-Schiaffino, R., Cameron, D.G., Millar, A.P., Nicholson, C., Stockinger, K. & Zini F. (2003, November). Evaluating Scheduling and Replica Optimisation Strategies in OptorSim. In *4th International Workshop on Grid Computing (Grid2003)*. New York: IEEE Computer Society Press.

Castrignanò, T., Canali, A., Grillo, G., Liuni, S., Mignone, F. & Pesole, G (2004). CSTminer: a web tool for the identification of coding and noncoding conserved sequence tags through cross-species genome comparison. *Nucleic Acids Res., 32*(Web Server issue),W624-7.

Catlett C. (2002). The philosophy of TeraGrid: Building an open, extensible, distributed TeraScale facility. In *Cluster Computing and the Grid 2nd IEEE/ACM International Symposium CCGRID2002.*

Catlett, C. et al (2007). TeraGrid: Analysis of Organization, System Architecture, and Middleware Enabling New Types of Applications. In L. Grandinetti (Ed.), *Advances in Parallel Computing*. Amsterdam, the Netherlands: IOS Press.

CDC (2007). *Center for Disease Control of Atlanta.* Retrieved November 2008 from http://www.cdc.gov/

CDI (2007). *Cyber-Enabled Discovery and Innovation (CDI), Computer and Information Science and Engineering Directorate, National Science Foundation.* Paper presented at the National Science Foundation.

CDO2 (2008). CDOSheet for Pricing and Risk Analysis. Retrieved from www.cdo2.com

Chadwick, D. W., & Otenko, A. (2002). The PERMIS X.509 role based privilege management infrastructure. *Proceedings of the seventh ACM symposium on Access control models and technologies*. Monterey, CA: ACM.

Chadwick, D. W., Otenko, O. (2002). The PERMIS X.509 Role Based Privilege Management Infrastructure. *Future Generation Computer Systems, 936*, 1–13.

Chadwick, D., & Otenko, O. (2003). *A Comparison of the Akenti and PERMIS Authorization Infrastructures in Ensuring Security in IT Infrastructures.* Paper presented at the Proceedings of the ITI First International Conference on Information and Communications Technology (ICICT 2003) Cairo University, Egypt.

Chakrabarti, A. (2007). *Grid computing security*. New York: Springer.

Chakrabarti, A., Dheepak, R.A., & Sengupta S. (July 2004). *Integration of Scheduling and Replication in Data Grids* (Tech. Rep. TR-0407-001). New York: Infosys Tech. Ltd.

Chang, C.-S., DeTitta, G., Hauptman, H., Jones, R., Miller, R., Thuman, P., & Weeks, C. (1993). Solving the phase problem of x-ray crystallography on parallel machines. In R.F. Sincovec, D.E. Keyes, M.R. Leuze, L.R. Petzold, and D.A. Reed, (Eds.) *Proceedings of The Sixth SIAM Conference on Parallel Processing for Scientific Computing*, (pp. 304-307).

Chang, C.-S., DeTitta, G.T., Miller, R. & Weeks, C.M. (1994). On the application of parallel genetic algorithms in x-ray crystallography. *Proceedings of the 1994 Scalable High Performance Computing Conference*, (pp. 796-802). New York: IEEE Computer Society Press.

Channabasavaiah, K., Holley, K., & Tuggle, E. (2003). Migrating to a service-oriented architecture. *IBM DeveloperWorks*. Retrieved from http://www-128.ibm.com/developerworks/library/ws-migratesoa/

Charleston, M.A., & Perkins L. (2006). Traversing the tangle: Algorithms and applications for cophylogenetic studies. *Journal of Biomedical Informatics*, 39 (2006):62-71.

Charnock, S.J., Brown, I.E., Turkenburg, J.P., Black, G.W., Davies, G.J. (2002). Convergent evolution sheds light on the anti-beta-elimination mechanism common to family 1 and 10 polysaccharide lyases. *Proc Natl Acad Sci USA, 99*(19),12067-72.

Chaubal, Ch. (2003). Sun Grid Engine, Enterprise Edition—Software Configuration Guidelines and Use Cases. Sun Blueprints. www.sun.com/blueprints/0703/817-3179.pdf

Chaudhuri, S., Hristidis, V. & Polyzotis, N. (2006) *Proceedings of the ACM SIGMOD International Conference on Management of Data, Chicago, Illinois, USA, June 27-29, 2006*. New York: ACM.

Cheeseman, P., & Stutz, J. (1996). Bayesian classification (autoclass): Theory and results. *In Advances in Knowledge Discovery and Data Mining* (pp. 153-180).

Chen, C., Schmidt, B., Liu, W., & Mueller-Wittig, K.W. (2008). GPU-MEME: Using Graphics Hardware to Accelerate Motif Finding in DNA Sequences. *Lecture Notes in Bioinformatics, 5295*, 448-459.

Chen, L., Reddy, K., & Agrawal, G. (2004). Gates: A grid-based middleware for processing distributed data streams. *In the 13th International Symposium on High Performance Distributed Computing* (pp. 192-201). New York: IEEE Computer Society.

Cheng, Y., Zak, O., Aisen, P., Harrison, S., & Walz, T. (2004). Structure of the human transferrin receptor-transferrin complex. *Cell, 116*(1), 565-576.

Chervenak, A., Foster, I,. Kesselman, C., Salisbury, C., & Tuecke, S. (2001). The Data Grid: Towards an Architecture for the Distributed Management and Analysis of Large Scientific Datasets. *Journal of Network and Computer Applications, 23*, 187-200.

Cheung, D. & Xiao, Y. (1998). Effect of data skewness in parallel mining of association rules. *In the Second Pacific-Asia Conference on Research and Development in Knowledge Discovery and Data Mining* (pp. 48–60). Berlin: Springer-Verlag.

Childers, L., Disz, T., Olson, R., Papka, M.E., Stevens, R. and T. Udeshi (2000). Access Grid: Immersive Group-to-Group Collaborative Visualization. In *Proc. 4'th International Projection Technology Workshop*. Retrieved from http://www-unix.mcs.anl.gov/fl/publications/childers00.pdf

Cho, S., Kim, D., & Pearlman, W.A. (2004). Lossless compression of volumetric medical images with improved three-dimensional SPIHT algorithm. *Journal of Digital Imaging, 17*, 57-63.

Cho, V., & Wüthrich, B. (2002). Distributed mining of classification rules. *Knowledge Information Systems, 4*(1), 1-30.

Chor, B., & Tuller, T. (2005). Maximum likelihood of evolutionary trees: hardness and approximation. *Bioinformatics*, 21(1):97-106.

CHRONOS Portal (2004). Retrieved from http://portal.chronos.org/gridsphere/gridsphere

Chun B., Culler D., Roscoe T., Bavier A., Peterson L., Wawrzoniak M., & Bowman M. (2003). PlanetLab: An

Overlay Testbed for Broad-Coverage Services. *ACM Computer Communications Review, 33*(3).

Churches, D., Gombas, G., Harrison, A., Maassen, J., Robinson, C., Shields, M., Taylor, I., & Wang, I. (2005). Programming scientific and distributed workflow with triana services. *Concurrency and Computation: Practice and Experience, 18*(10), 1021-1037.

Ciepiela, E., Kocot, J., Gubala, T., Malawski, M., Kasztelnik, M., & Bubak, M. (2007). Virtual laboratory engine – gridspace engine. In Bubak, M., Turala, M., and Wiatr, K., (Eds.), *Proceedings of Cracow Grid Workshop - CGW'07,* Krakow, Poland, (pp. 53-58). ACC CYFRONET AGH.

Cirne, W., Paranhos, D., Costa, L., Santos-Neto, E., Brasileiro, F., Sauve, J., Silva, F.A.B., Barros, C. O. & Silveira, C. (2003) Running Bag-of-Tasks Applications on Computational Grids: The MyGrid Approach. In *Proc. International Conference on Parallel Processing (ICPP'03),* pp.407-416.

Clark, T. (2003). Editorial: Identity and interoperability in bioinformatics. *Briefings in Bioinformatics, 4*(1),4–6.

Clementi, L., Ding, Z., Krishnan, S., Wei, X., Arzberger, P. W., & Li, W. W. (2007). *In Providing dynamic virtualized access to grid resources via the web 2.0 paradigm.* Grid Computing Environments Workshop (Supercomputing 2007), Reno, NV.

CloudCamp (2008). Retrived from http://www.cloudcamp.com/

CNRS – IBCP (2006). *Grid Genomic Web Portal.* Retrieved November 2008 from gpsa.ibcp.fr.

CODESSA PRO (2008). Retrieved March 2008 from http://www.codessa-pro.com/

Cohen, W. W. (1995). Fast effective rule induction. In A. Prieditis and S. Russell (Eds.), *12th International Conference on Machine Learning,* (pp 115-123). San Francisco, CA: Morgan Kaufmann.

Coiera E. (2003). *Guide to health informatics.* London: Hodder Arnold.

Colacino, F.M., Arabia, M., Danieli, G.A., Moscato, F., Nicosia, S., Piedimonte, F., et al. (2005). Hybrid test bench for evaluation of any device related to mechanical cardiac assistance. *The International Journal of Artificial Organs, 28*(8), 817-826.

Colacino, F.M., Moscato, F., Piedimonte, F., Danieli, G., Nicosia, S., & Arabia, M. (2008). A modified elastance model to control mock ventricles in real-time: numerical and experimental validation. *Asaio J, 54*(6), 563-573.

Colinge, J., Masselot A., Giron, M., Dessingy, T., & Magnin, J. (2003). OLAV: towards high-throughput tandem mass spectrometry data identification. *Proteomics, 3*(8), 1454-1463.

Collobert, R. , Bengio, S., & Bengio, Y. (2002). A parallel mixture of svms for very large scale problems. In *Advances in Neural Information Processing Systems.* Cambridge, MA: MIT Press.

Congiusta, A., Talia, D., & Trunfio, P. (2007). Distributed data mining services leveraging WSRF. *Future Generation Computing Systems, 23*(1), 34-41.

Corcho, O., Alper, P., Kotsiopoulos, I., Missier, P., Bechhofer, S., & Goble, C. (2006). An overview of s-ogsa: A reference semantic grid architecture. *Web Semantics: Science, Services and Agents on the World Wide Web, 4*(2), 102-115.

Corpet, F., Gouzy, J., & Kahn, D. (2000). ProDom and ProDom-CG: tools for protein domain analysis and whole genome comparisons. *Nucleic Acids Research, 28*, 267-269.

Council of Europe (1997). *Recommendation No. R (97) 5 on the Protection of Medical Data of Feb. 13, 1997.*

Covitz, P.A., Hartel, F., Schaefer, C., De Coronado, S., Fragoso, G., Sahni H., Gustafson, S., & Buetow K.H. (2003). *caCORE: A Common Infrastructure for Cancer Informatics.*

Craig, R. & Beavis, R. C. (2004). TANDEM: matching proteins with tandem mass spectra. *Bioinformatics, 20*(9), 1466-1467.

Craig, R., Cortens, J.P., & . Beavis, R.C, (2004). Open source system for analyzing, validating, and storing protein identification data. *Journal of Proteome Research 3*(6) 1234–1242.

Crawfis, R., Max, N. and B. Becker (1994). Vector Field Visualization. *IEEE Computer Graphics and Applications, 14* (5), 50-56.

Crespo, A., & Garcia-Molina, H. (2002). *Routing indices for peer-to-peer systems*. Paper presented at the 22nd International Conference on Distributed Computing Systems (ICDCS 2002), Wien, Austria.

Crippen, G. M. (1981). *Distance Geometry and Conformational Calculations*. New York: Research Studies.

Crompton, S. Y., Matthews, B. M., Gray, W. A., Jones, A. C., White, R. J. & Pahwa, J. S. (2006). OGSA-DAI and bioinformatics grids: Challenges, experience and strategies. In *Proceedings of the 6th IEEE International Symposium on Cluster Computing and the Grid (CCGRID 2006)*, (pp. 193-200). Washington, DC: IEEE Computer Society.

CrossGrid.org. (2008). *CrossGrid project homepage*. Retrieved March 8th, 2008, from http://www.crossgrid.org/main.html

Cui, Y. & Widom, J. (2003). Lineage tracing for general data warehouse transformations. *VLDB J, 12*(1),41–58.

Culler, G.J. & Fried, B.D. (1963). An On-Line Computing Center for Scientific Problems. In *Proc. 1963 Pacific Computer Conf.*, (pp. 221–242). , Piscataway, NJ: IEEE.

Cunha, J. C. (2003). *Future Trends in Distributed Applications and PSEs*. Talk held at the Euresco Conference on Advanced Environments and Tools for High Performance Computing, Albufeira, Portugal.

CVIMO (2007). Retrieved July 2008 from http://www.grycap.upv.es/cvimo

Dasgupta, P. (2004). *Intelligent Agent Enabled P2P Search Using Ant Algorithms*, Paper presented at the 8th International Conference on Artificial Intelligence, Las Vegas, USA.

Data, A. (2006). Multiple Sequence Alignment in Parallel on a Cluster of Workstations. In A. Zomaya (Ed.), *Parallel Computing for Bioinformatics and Computational Biology*, (pp. 193-210). Hoboken, NJ: Wiley.

DATAGRID. The European DataGrid Project. Retrieved from http://www.eu-datagrid.org

DataTag (2008). *DataTag project description*. Retrieved March 8th 2008, from http://datatag.web.cern.ch/datatag/project.html

Date, C. J. (2000). *An Introduction to database systems*. Reading, MA: Addison-Wesley.

DDQB: IBM tool designed to help medical providers and researchers perform complex search queries without programming knowledge or IT backgrounds. Retrieved 2006, from http://www-03.ibm.com/systems/deepcomputing/solutions/ddqb.html.

De Beenhouwer, J., Staelens, S., Vandenberghe, S., & Lemahieu, I. (2008). Acceleration of GATE SPECT simulations. *Med Phys, 35*(4), 1476-1485.

De Carvalho, R.L., & Lifschitz, S. (2003). Database Allocation Strategies for Parallel BLAST Evaluation on Clusters. *Distributed and Parallel Databases, 13*, 99-127.

de Groot, B. L., van Aalten, D. M. F., Scheek, R. M., Amadei, A., Vriend, G., & Berendsen, H. J. C. (1997). Prediction of protein conformational freedom from distance constraints. *Proteins, 29*, 240-251.

de la Calle, G., Maojo, V., & Benito, M. (2006). The INFOBIOMED Network of Excellence: facilitating training and mobility in biomedical informatics in Europe. *Stud Health Technol Inform, 124*, 893-898.

De Roure, D., & Frey, J. (2007). Three Perspectives on Collaborative Knowledge Acquisition in e-Science. In *Proceedings of Workshop on Semantic Web for Collaborative Knowledge Acquisition* (SWeCKa) 2007, Hyderabad, India.

De Roure, D., Jennings, N.R., & Shadbolt, N. (2003): The Semantic Grid: A future e-Science infrastructure. In *Grid Computing - Making the Global Infrastructure a Reality*, (pp. 437-470). Hoboken, NJ: John Wiley and Sons Ltd.

De Roure, D., Jennings, N.R., & Shadbolt, N.R. (2005). The Semantic Grid: Past, Present, and Future. In *Proceedings of the IEEE, 93*(3), 669-681.

de Vienne, D.M., Giraud, T., & Martin, O.C. (2007). A congruence index for testing topological similarity between trees. *Bioinformatics*, 23(23):3119-3124.

Deacon, A. M., Ni, Y. S., Coleman, W. G. Jr. & Ealick, S. E. (2000). The crystal structure of ADP-L-glycero-D-mannoheptose 6-epimerase: catalysis with a twist. *Structure, 8*, 453-462.

Dean, J., & Ghemawat, S. (2004). *In MapReduce: Simplified Data Processing on Large Clusters*. Sixth Symposium on Operating System Design and Implementation (OSDI'04), San Francisco, CA.

Debaerdemaeker, T., Tate, C. & Woolfson, M.M. (1985). On the application of phase relationships to complex structures (XXIV: The Sayre tangent formula). *Acta Cryst.*A41), 286-290.

DECI (2008). DEISA Extreme Computing Initiative. Retrieved from www.deisa.eu/science/deci

Deelman, E., Blythe, J., Gil, Y., & Kesselman, C. (2004). *Workflow management in GriPhyN. In Grid resource management: state of the art and future trends*. Norwell, MA: Kluwer Academic Publishers.

Deelman, E., Blythe, J., Gil, Y., Kesselman, C., Mehta, G., Patil, S., Su, M. H., Vahi, K. & Livny, M. (2004). Pegasus: Mapping scientific workflows onto the Grid. In Grid *Computing: Second European AcrossGrids Conference, AxGrids,* (LNCS 3165, pp 11-20). Berlin, Germany: Springer.

Deelman, E., Kesselman, C., Mehta, G., Meshkat, L., Pearlman, L., Blackburn, K., et al. (2002). Griphyn and ligo, building a virtual data grid for gravitational wave scientists. *In HPDC* (pp. 225-234). New York: IEEE Computer Society.

Deelman, E., Singh, G., Su, M., Blythe, J., Gil, Y. Kesselman, C., Mehta, G., Vahi, K., Berriman, G.B., Foster, I., & Kesselman, C. (1997). Globus: A Metacomputing Infrastructure Toolkit. *Intl J. Supercomputer Applications*, 11(2), 115–128.

Degtyarenko K. et al. (2008). ChEBI: a database and ontology for chemical entities of biological interest. *Nucleic Acids Research, 36*. Retrieved from http://www.ebi.ac.uk/chebi/

DEISA. (2008). Distributed European Infrastructure for Supercomputing Applications. Retrieved from www.deisa.eu

Delcher, A.L., Harmon, D., Kasif, S., White, O. & Salzberg, S.L. (1999). Improved microbial gene identification with GLIMMER. *Nucl. Acids Res., 27*(23), 4636-4641.

Delhay, B., Clarysse, P., Pera, C. & Magnin, I. E. (2006). A spatio-temporal deformation model for dense motion estimation in periodic cardiac image sequence. In *From Statistical Atlases to Personalized Models: Understanding Complex Diseases in Populations and Individuals, Satellite Workshop MICCAI 2006*. Copenhagen, Denmark.

Delhay, B., Clarysse, P., Pera, C. & Magnin, I. E. (2007, June). Locally adapted spatio-temporal deformation model for dense motion estimation in periodic cardiac image sequences. In *Functional Imaging and Modeling of the Heart*. Salt Lake City, UT, (LNCS 4466, pp. 393-402). Berlin, Germany: Springer-Verlag.

Deneubourg, J. L., Goss, S., Franks, N., Sendova-Franks, A., Detrain, C., & Chretien, L. (1990) *The dynamics of collective sorting robot-like ants and ant-like robots*. Paper presented at the 1st International Conference on Simulation of Adaptive Behavior on From Animals to Animats, Cambridge, USA.

DESHL. (2008). DEISA Services for Heterogeneous management Layer. http://forge.nesc.ac.uk/projects/deisa-jra7/

Desiere, F., Deutsch, E., Nesvizhskii, A., Mallick, P., King, N., Eng, J., et al. (2004). Integration with the human genome of peptide sequences obtained by high-throughput mass spectrometry. *Genome Biology, 6*(1).

Desprez, F. & Vernois, A. (2007, September). Semi-Static Algorithms for Data Replication and Scheduling Over the Grid. In *IEEE 3rd International Conference on Intelligent Computer Communication and Processing, workshop on Grid Computing*, Cluj-Napoca, Romania.

Desprez, F., & Vernois, A. (2006, March). Simultaneous Scheduling of Replication and Computation for Data-Intensive Applications on the Grid. *Journal Of Grid Computing, 4*(1), 19-31.

DeTitta, G. T., Weeks, C. M., Thuman, P., Miller, R. & Hauptman, H. A. (1994). Structure solution by minimal function phase refinement and Fourier filtering, (I. Theoretical basis). *Acta Crystallogr.* A50, 203-210.

DeTitta, G., Hauptman, H., Miller, R., Pagels, M. Sabin, T., Thuman, P., & Weeks, C. (1991). Parallel solutions to the phase problem in x-ray crystallography: an update. In

Proceedings of The Sixth Distributed Memory Computing Conference, (pp. 587-594). New York: IEEE Computer Society Press.

D-Grid (2008). Retrieved from www.d-grid.de/index.php?id=1&L=1

D-Grid. (2008). *D-Grid*. Retrieved July 21, 2008, from www.d-grid.de

Dhillon, I. S., & Modha, D. S. (2000). A data-clustering algorithm on distributed memory multiprocessors. In *Large-Scale Parallel Data Mining* (pp. 245-260). Lecture Notes in Artificial Intelligence.

Dietterich, T. G. (2000). An experimental comparison of three methods for constructing ensembles of decision trees: Bagging, boosting, and randomization. *Machine Learning, 40*, 139-157.

DIMACS (2002) *Special Focus on Computational and Mathematical Epidemiology*. Retrieved July from http://dimacs.rutgers.edu/SpecialYears/2002_Epid/

Dinda, P. (2006). Design, Implementation, and Performance of an Extensible Toolkit for Resource Prediction In Distributed Systems. *IEEE Transactions on Parallel and Distributed Systems, 17*(2), 160-173.

Ding, Z., Luo, Y., Wei, X., Misleh, C., Li, W. W., Arzberger, P. W., & Tatebe, O. (2006). *In My WorkSphere: Integrative Work Environment for Grid-unaware Biomedical Researchers and Applications*. Supercomputing Conference 2006, SC06, 2nd Grid Computing Environment Workshop, Tampa, FL.

Ding, Z., Wei, X., Tatebe, O., Papadopoulos, P. M., Arzberger, P. W., & Li, W. W. (2008). Cyberinfrastructure for Biomedical Applications: Metascheduling as Essential Component for Pervasive Computing. In, J. Cao, (Ed.), *Cyberinfrastructure Technologies and Applications*,. Hauppauge, NY: Nova Science.

Dixit, S. B., Beveridge, D. L., Case, D. A., Cheatham III, T. E., Guidice, E., Lankas, F., Lavery, R., Maddocks, J. H., Osman, R., Sklenar, H., Thayer, K. M., & Varnai, P. (2005). Molecular dynamics simulations of the 136 unique tetranucleotide sequences of DNA oligonucleotides. II: Sequence context effects on the dynamical structures of the 10 unique dinucleotide steps, *Biophysical Journal, 89*, 3721-3740.

Docking@Home. Retrieved March 5, 2008 from http://docking.cis.udel.edu/

Dodson, C.T.J. and T. Poston (1991). *Tensor Geometry: The Geometric Viewpoint and its Uses*. Berlin, Germany: Springer-Verlag.

Dolin, R. H., Alschuler, L., Beebe, C., Biron, P. V., Boyer, S. L., Essin, D., et al. (2001). The HL7 Clinical Document Architecture. *Journal of the American Medical Informatics Association, 8*(6), 552-569.

Dolin, R. H., Alschuler, L., Boyer, S., Beebe, C., Behlen, F. M., Biron, P. V., et al. (2006). HL7 Clinical Document Architecture, Release 2. *J Am Med Inform Assoc, 13*(1), 30-39.

Dolinsky, T. J., Czodrowski, P., Li, H., Nielsen, J. E., Jensen, J. H., Klebe, G., & Baker, N. A. (2007). PDB-2PQR: expanding and upgrading automated preparation of biomolecular structures for molecular simulations. *Nucleic Acids Res, 35*, 522-525.

Dongarra, J., Foster, I., Fox, G., Gropp, W., Kennedy, K., Torczon, L., & White, A. (2003). Sourcebook of Parallel Computing. Morgan Kaufmann Publishers.

Donnely, C. (2002). FMD 2001: Using Statistics and Mathematics for Outbreak Control and Eradication. *International Conference on Computational and Mathematical Epidemiology*. Retrieved July 2008 from http://dimacs.rutgers.edu

Donovan, F.J. (1975). Design of a hydraulic analog of the circulatory system fo evaluating artificial Hearts. *Biomaterials, Medical Devices, and Artificial Organs, 3*(4), 439-449.

Dowsey, A. W., Dunn, M. J. & Yang, G. (2004). ProteomeGRID: towards a high-throughput proteomics pipeline through opportunistic cluster image computing for two-dimensional gel electrophoresis. *Proteomics, 4*(12), 3800-3812.

Droegemeier, K. K., Gannon, D., Reed, D. A., Plale, B., Alameda, J. Baltzer, T., et al. (2005). Service-oriented Environments in Research and Education for Dynamically Interacting with Mesoscale Weather. *Computing in Science and Engineering* 7(6)12-29.

Droegemeier, K. K., Gannon, D., Reed, D., Plale, B., Alameda, J., Baltzer, T., et al. (2005). Service-oriented environments for dynamically interacting with mesoscale weather. *Computing in Science and Engineering, 7*(6), 12-29.

Dubitzky, D., McCourt, D., Galushka, M., Romberg, M., & Schuller, B. (2004). Grid-enabled Data Warehousing for Molecular Engineering. *Parallel Computing, 30*, 1019-1035.

Dublin Core Metadata Initiative (2008). Retrieved July 20, 2008 from http://www.dublincore.org

Dubochet, J., Adrian, M., Chang, J. J., Home, J. C., Lepault, J., McDowall, A. W., et al. (1988). Cryo-electron microscopy of vitrified specimens. *Quarterly Review of Biophysics, 21*(1), 129-228.

Dwyer III, S.J. (2000). A personalized view of the history of PACS in the USA. In G. J. Blaine & E. L. Siegel, *Proceedings of the SPIE, "Medical Imaging 2000: PACS Design and Evaluation: Engineering and Clinical Issues"*, (Vol. 3980), (pp. 2-9).

Dzenowagis, J. & Kernen, G. (2005). *Connecting for Health: global vision, local insight* (Report for the World Summit on Information Society). Tunis, Tunisia: World Health Organization Press

ECOTOX (ECOTOXicology) database. Retrieved March 4, 2008 from http://www.epa.gov/ecotox/

Edgar, R. C. (2004), MUSCLE: multiple sequence alignment with high accuracy and high throughput. *Nucleic Acids Research, 32*(5), 1792-97.

EGEE Grid infrastructure. Retrieved from http://www.eu-egee.org/.

EGEE SDJ WG (2006). *Report of the Short Deadline Jobs Working Group*. Retrieved from http://egee-na4.ct.infn.it/wiki/index.php/ShortJobs.

EGEE, Enabling the Grid for E-sciencE 2007 Conference, Live Demo Competition (2007). Retrieved October 20, 2008, from http://indico.cern.ch/contributionDisplay.py?contribId=174&sessionId=23&confId=18714.

EGEE-SEE ROC (2007). *Web Authentication and Authorization using MyProxy APIs and Grid Portals*. Retrieved July 2008 from http://wiki.egee-see.org/index.php/Web_Authentication_and_Authorization_using_MyProxy_APIs_and_Grid_Portals

EGI Consortium (2007). *European Grid Initiative*. Retrieved July 2008 from http://www.eu-egi.org

Eilemann, S. & Pajorola, R. (2007). Direct Send Compositing for Parallel Sort-Last Rendering. *Eurographics Symposium on Parallel Graphics and Visualization*, (pp. 29-36).

Ekman A., Klint A., Dickman P.W. , Adami H.O., Litton J. (2007). Optimizing the design of web-based questionnaires - experience from a population-based study among 50.000 women. *European Journal of Epidemiology 22*, 293-300.

Ekman A., Litton J. (2007). New times, new needs; e-epidemiology. *European Journal of Epidemiology, 22*, 285-292.

El Sayed, G. and I. Greenshields (2008). Low FM: A Robust Non-Blocking Low-Frequency Messaging Checkpoint for Parallel/Grid Applications. In *Proc. 23rd International Conference on Computers and Their Applications, Cancun, Mexico, April 9-11.*

Ellisman, M. & Peltier, S. (2004). Medical data federation: the biomedical informatics research network. In Foster, I. & Kesselman, C. (Eds.), *The Grid: Blueprint for a New Computing Infrastructure* (2nd Ed.). San Francisco, CA: Morgan Kaufmann.

Embrace. Embrace website. Retrieved from http://www.embracegrid.info

Emmen, A. (2008). National Digital Mammography Archive announces installation at University of Pennsylvania Health System. *EnterTheGrid - PrimeurMonthly* Retrieved March 8, 2008 from *http://www.hoise.com/primeur/05/articles/monthly/AE-PR-02-05-1.html* .

Enabling Grids for E-Science (n.d.). Retrieved March 17, 2008, from http://www.eu-egee.org

Eng, J. K., McCormack, A. L. & Yates III, J. R. (1994). An approach to correlate tandem mass spectral data of peptides with amino acid sequences in a protein database. *Journal of the American Society for Mass Spectrometry, 5*(11), 976-989.

EnginFrame (2008). Grid and Cloud Portal. Retrieved from www.nice-italy.com

Englander, R. (1997). *Developing Java Beans*. Sebastopol, CA: O'Reilly & Associates, Inc. Bellia, M., Ochiutto, M.E. (2005). Higher order Programming in Java: Introspection, Subsumption and Extraction. *Concurrency Specification and Programming* (pp. 29-44). Amsterdam, The Netherlands: IOS Press.

Ennis, D.B., Kindlmann, G., Rodriguez, I., Helm, P.A. & McVeigh, E.R. (2005).Visualization of Tensor Fields Using Superquadric Glyphs. *Magn. Resonance Med., 53*(1) 169-176.

Epstein, J. M., Cummings, D. A., Chakravarty, S., Singa, R. M., Burke, D. S. (2002, December). *Toward a Containment Strategy for Smallpox Bioterror: An Individual-Based Computational Approach. Brookings Institution, Center on Social and Economic Dynamics* (Working Paper No. 31).

EQU. (n.d.). *Equalizer*. Retrieved from http://www.equalizergraphics.com/

Erberich, S., Silverstein, J., Chervenak, A., Schuler, R., Nelson, M. & Kesselman, K. (2007). Globus MEDICUS – Federation of DICOM Medical Imaging Devices into Healthcare Grids. *Studies in Health Technology and Informatics, 126*, 269-278.

Erberich, S., Silverstein, J., Kratz M., Chervenak, A., Schuler, R., Nelson, M. & Kesselman, K. (2007, April). *Globus MEDICUS – Standards Based Enterprise for Medical Image Publication, Discovery, and Archiving in HealthGrids*. Paper presented at the Open Source Solutions for Multi-Center Information Management. Saint Louis, MO.

Eres, M.H., Pound, G.E., Jiao, Z., Wason, J.L., Xu, F., Keane, A.J. & Cox, S.J. (2005) Implementation and utilisation of a Grid-enabled problem solving environment in Matlab. *Future Generation Computer Systems, 21* (6) 920-929.

Erickson, B. (2002). Irreversible Compression of Medical Images. *Journal of Digital imaging, 15*(1), 5-14.

Erwin D. W, & Snelling D. F. (2001). UNICORE: A Grid Computing Environment. In. J. Darlington et al. (Eds.) *Euro-Par 2001 Parallel Processing*. Berlin, Germany: Springer.

Estrella, F., Hauer, T., McClatchey, R., Odeh, M., Rogulin, D. & Solomonides, T. (2007). Experiences of engineering Grid-based medical software. *International Journal of Medical Informatics, 76*(8), 621-32.

Eswar, N., Marti-Renom, M. A., Webb, B., Madhusudhan, M. S., Eramian, D., Shen, M., Pieper, U. & Sali, A. (2000). Comparative Protein Structure Modeling With MODELLER. *Current Protocols in Bioinformatics* (Supplement 15, 5.6.1-5.6.30). Hoboken, NJ: John Wiley & Sons, Inc..

Etzold, T., Harris H. & Beulah, S. (2003). SRS: An integration platform for databanks and analysis tools in bioinformatics. *Bioinformatics Managing Scientific Data*, 35–74.

Eubank, S. (2002). *Scalable, Efficient, Epidemiological Simulation*. New York: ACM. Retrieved March 1, 2008 from http://www.ccs.lanl.gov/ccs5/documents/pdf/2001/LA-UR-01-5513.pdf.

Eubank, S., Guclu, H.,Kumar, V. S. A., Marathe, M. V., Srinivasan, A., Toroczkai, Z.& Wang, N. (2004). Modeling disease outbreaks in realistic urban social networks. *Nature*, 429(May 13), 180.

Eugster, P., Guerraoui, R., Kermarres, A.M., & Massoulieacute, L. (2004). *Epidemic Information Dissemination in Distributed System*, IEEE Computer, 37(5), 60-67

EUROCARE Consortium. (2008). *EUROCARE*. Retrieved July 2008 from http://www.eurocare.it/

European Parliament. (1995). *Official Journal of the European Communities 281*, 31.

European Parliament. (2002). *Official Journal 201*, 0037 - 0047.

European Union (2002). *Resource Book of IST Projects Relating to Health (1998-2002)*. Retrieved from http://cordis.europa.eu/ist/ka1/health/projectbooklet/index.htm.

EUROREC (2002). Retrieved July 2008 from http://www.eurorec.org/

EU-SHARE. (2006). *SHARE - Roadmap for a European Health Grid*. Retrieved 28.06.2006, from http://www.eu-share.org/

Falkner, J., & Sax, U. (2008). *Service Certificates and Portal Security - Longterm requirements towards security mechanisms (German).* Paper presented at the D-Grid Security Workshop 2008, Göttingen.

Fan, W., Stolfo, S. J., & Zhang, J. (1999). The application of adaboost for distributed, scalable and on-line learning. In *The fifth ACM SIGKDD international conference on Knowledge discovery and data mining* (pp. 362-366). New York: ACM.

FDA. (2004, April 1). *FDA 21 CFR Part 11*. Retrieved September 21, 2005, from http://www.accessdata.fda.gov/scripts/cdrh/cfdocs/cfcfr/CFRsearch.cfm?CFRPart=11

Federrath, H. (2005, 22-26 August). *Privacy Enhanced Technologies: Methods – Markets – Misuse* Paper presented at the Trust, Privacy and Security in Digital Business Copenhagen, Denmark.

Feig, M., Abdullah, M., Johnsson, L., & Pettitt, B. M. (1999). Large scale distributed data repository: Design of a molecular dynamics trajectory database. *Future Generation Computer Systems, 16,* 101-110.

Fensel, D. (2003). *Ontologies: Silver Bullet for Knowledge Management and Electronic Commerce (*2nd Ed.). Berin, Germany: Springer.

Ferguson N. M., Cummings, D. A.T., Cauchemez, S., Fraser, C., Riley, S., Meeyai, A., Iamsirithaworn, S., Burke, D. S. (2005). Strategies for containing an emerging influenza pandemic in Southeast Asia. *Nature 437,* 209 - 214.

Fernández, J. J., Bilbao-Castro, J. R., Marabini, R., Carazo, J. M., & García, I. (2005). On the Suitability of Biological Structure Determination by Electron Microscopy to Grid Computing. *New Generation Comput, 23*(1), 101-112.

Fernández, J. J., García, I., Carazo, J. M., & Marabini, R. (2007). Electron tomography of complex biological specimens on the Grid. *Future Generation Computing Systems, 23*(3), 435-446.

Fernández, J. J., Lawrence, A. F., Roca, J., García, I., Ellisman, M. H., & Carazo, J. M. (2002). High-performance electron tomography of complex biological specimens. *Journal of Structural Biology, 138*(1-2), 6-20.

Ferrari, G., Kozarski, M., De Lazzari, C., Gorczynska, K., Tosti, G., & Darowski, M. (2005). Development of a hybrid (numerical-hydraulic) circulatory model: Prototype testing and its response to IABP assistance. *International Journal of Artificial Organs, 28*(7), 750-759.

Fielding, R.T. (2005). JSR170 Overview. *Stardandizing the Content Repository Interface.*

Finn, R.D., Tate, J., et al. (2008). The Pfam protein families database. *Nucleic Acids Res, 36,* D281-8.

Fiolet, V., Olejnik, R., Lefait, G., & Toursel, B. (2006). Optimal grid exploitation algorithms for data mining. *In the Fifth International Symposium on Parallel and Distributed Computing* (pp. 246-252). New York: IEEE Computer Society.

Fiore, S., Negro, A., Vadacca, S., Cafaro, M., Mirto, M. & Aloisio, G. (2007). Advanced Grid DataBase Management with the GRelC Data Access Service. *Proceedings of ISPA 2007 - August 29-31, 2007 - Niagara Falls, Ontario, Canada,* (LNCS 4742) (pp. 683-694).

Foley, J.D., van Dam, A., Feiner, S.K., Hughes, J.F. & Philips, R.L. (1993). *Introduction to Computer Graphics.* Upper Saddle River, NJ: Addison-Wesley.

Forestiero, A., Mastroianni, C., & Spezzano, G. (2005). *A Multi Agent Approach for the Construction of a Peer-to-Peer Information System in Grids*, Paper presented at the International Conference on Self-Organization and Adaptation of Multi-agent and Grid Systems (SOAS 2005), Glasgow, United Kingdom.

Forestiero, A., Mastroianni, C., & Spezzano, G. (2007). *A Decentralized Ant-Inspired Approach for Resource Management and Discovery in Grids*, Multiagent and Grid Systems - An International Journal, 3(1), 43-63.

Forge, S., & Blackman, C., (2006). *Commercial Exploitation of Grid Technologies and Services. Drivers and Barriers, Business Models and Impacts of Using Free and Open Source Licensing Schemes,* (Final Report by SCF Associates for European Commission DG Information Society and Media). Retrieved March 9, 2008 from IST website http://cordis.europa.eu/ist/grids/pub-report.htm

Foster I. & Kesselman C. (1999). *The Grid: Blueprint for a Future Computing Infrastructure*. San Francisco, CA: Morgan Kaufmann Publishers.

Foster I., Kesselman C., & Tuecke S. (2001). The Anatomy of the Grid: Enabling Scalable Virtual Organizations. *International J. Supercomputer Applications, 15*(3).

Foster I., Kesselman C., Nick J., & Tuecke S. (2002). *The Physiology of the Grid: An Open Grid Services Architecture for Distributed Systems Integration. Open Grid Service Infrastructure WG, Global Grid Forum.* Retrieved from http://www.globus.org/alliance/publications/papers/ogsa.pdf

Foster, I. & Kesselman, C. (2005). Globus Toolkit Version 4: Software for Service-Oriented Systems. In IFIP *International Conference on Network and Parallel Computing*, (LNCS 3779) (pp. 2–13). Berlin, Germany: Springer-Verlag.

Foster, I. & Kesselman, C. (2006). Scaling system-level science: Scientific exploration and it implications. *Computer, 39*(11), 31-39.

Foster, I. & Kesselman, K. (Eds.), (2003). *The Grid 2: Blueprint for a New Computing Infrastructure.* San Francisco, CA: Morgan Kaufmann Publishers.

Foster, I. & Kesselmann, C. (Eds.) (1999). *The Grid: Blueprint for a New Computing Intrastructure.* San Francisco, CA: Morgan Kaufmann Publishers.

Foster, I. (2000). Internet Computing and the Emerging Grid. Nature. Retrieved from www.nature.com/nature/webmatters/grid/grid.html

Foster, I. (2002) What is the Grid? A Three Point Checklist. *GRID Today, 1*(6). Retrieved February 13, 2008 from http://www.gridtoday.com/02/0722/100136.html.

Foster, I. (2002). *The Grid: Enabling Resource Sharing With Virtual Organizations.* Keynote speech at WWW2002, Honolulu, HI. Retrieved from http://www2002.org/foster.pdf

Foster, I. (2005). Globus Toolkit Version 4: Software for Service-Oriented Systems. *IFIP International Conference on Network and Parallel Computing*, (LNCS 3779, pp. 2-13). Berlin, Germany: Springer-Verlag.

Foster, I. (2005). Service-oriented science. *Science, 308,* 814-817.

Foster, I. (2006). *Globus Toolkit Version 4: Software for Service-Oriented Systems.* Paper presented at the IFIP International Conference on Network and Parallel Computing.

Foster, I. (2006). Globus Toolkit Version 4: Software for Service-Oriented Systems. In *IFIP International Conference on Network and Parallel Computing* (LNCS 3779, pp. 2-13). Berlin, Germany: Springer-Verlag.

Foster, I. (2006, October). Globus Toolkit Version 4: Software for Service-Oriented Systems. *IFIP International Conference on Network and Parallel Computing,* Tokyo, Japan (LNCS 3779, pp 2-13). Berlin, Germany: Springer-Verlag.

Foster, I. (Ed.), Kesselman, C. (Ed.). (1998). *The Grid: Blueprint for a New Computing Infrastructure.* San Francisco, CA: Morgan Kaufmann Publishers.

Foster, I., & Kesselman C. (1997). Globus: A metacomputing infrastructure toolkit. *International Journal of Scientific Applications, 11*(2) 115-128.

Foster, I., & Kesselman, C. (2004). *The Grid 2: Blueprint for a New Computing Infrastructure*, (2 Ed). San Francisco, CA: Morgan Kaufmann Publishers, Inc..

Foster, I., & Kesselman, C. (Ed.) (1999). *The Grid: Blueprint for a New Computing Infrastructure.* San Francisco, CA: Morgan Kaufmann Publishers, Inc..

Foster, I., & Kesselman, C. (Eds.). (2004). The Grid 2: Blueprint for a New Computing Infrastructure. Morgan Kaufmann Publishers.

Foster, I., Czajkowski, K., Ferguson, D. E., Frey, J., Graham, S., & Maguire, T., et al. (2005). Modeling and managing state in distributed systems: The role of OGSI and WSRF. *Proceedings of the IEEE, 93*(3), 604-612.

Foster, I., Kesselman, & C., Tuecke, S. (2002), The Anatomy of the Grid: Enabling Scalable Virtual Organizations. Retrieved from www.globus.org/alliance/publications/papers/anatomy.pdf

Foster, I., Kesselman, C. & Tuecke, S. (1999). *The Grid.* San Francisco: Morgan Kaufmann Publishers.

Foster, I., Kesselman, C. & Tuecke, S. (2001). The Anatomy of the Grid. Retrieved june 6, 2008, from http://www.globus.org/alliance/publications/papers/anatomy.pdf

Foster, I., Kesselman, C. & Tuecke, S. (2004). *The Grid 2*. San Francisco: Morgan Kaufmann Publishers.

Foster, I., Kesselman, C. & Tuecke, S. (n.d.). *The Anatomy of the Grid: Enabling Scalable Virtual Organizations*. Retrieved from http://www.globus.org/alliance/publications/papers/anatomy.pdf

Foster, I., Kesselman, C. (2006). Scaling system-level science: Scientific exploration and its implications. *IEEE Computer, 39* (11), 31-39.

Foster, I., Kesselman, C., & Tuecke, S. (2001). The anatomy of the grid: enabling scalable virtual organizations. *International Journal of High Performance Computing Applications*, *15*(3), 200-222.

Foster, I., Kesselman, C., & Tuecke, S. (2001). The Anatomy of the Grid: Enabling Scalable Virtual Organizations. *International Journal of High Performance Computing Applications*, *15*(3), 200–222.

Foster, I., Kesselman, C., Nick, J. & Tuecke, S. (2002). Grid Services for Distributed System Integration. *Computer*, *35*(6), 37-46.

Foster, I., Kesselman, C., Nick, J., Tuecke, S. (2002). The Physiology of the Grid: An Open Grid Services Architecture for Distributed Systems Integration. Retrieved February 14, 2008 from http://www.globus.org/alliance/publications/papers/ogsa.pdf.

Foster, I., Kesselman, C., Nick, J.M., & Tuecke, S. (2002). Grid Services for Distributed Systems Integration. *Computer 35*(6).

Foster, I., Kesselman, C., Tsudik, G., & Tuecke, S. (1998). *In A Security Architecture for Computational Grids*. ACM Conference on Computer and Communications Security, San Francisco, CA.

Foster, I., Kesselman, K. & Tuecke, S. (2001). The Anatomy of the Grid: Enabling Scalable Virtual Organizations. *International Journal Supercomputer Applications*, *15*(3), 200-222.

Foster, I., Kishimoto, H., Savva, A., Berry, D., Djaoui, A., Grimshaw, A., et al. (2005).The Open Grid Services Architecture, Version 1.0. *Informational Document, Global Grid Forum (GGF).*

Foster. I. (2006). Globus Toolkit Version 4: Software for Service-Oriented Systems. *LNCS*, 3779, 2-13.

Foti, D., Lipari, D., Pizzuti, C., & Talia, D. (2000). Scalable parallel clustering for data mining on multicomputers. *In the 15th Workshops on Parallel and Distributed Processing* (pp. 390-398). Berlin, Germany: Springer-Verlag.

Fowler, P., Khalid, S., Mokrab, Y., Sansom, M. S. P., Stansfeld, P., & Tai, K. (2008). KDB: A potassium channel model and simulation database. *Biophysical Journal, 94,* 2038.

Fox, G. C., Guha, R., Mcmullen, D. F., Mustacoglu, A. F., Pierce, M. E., Topcu, A. E., & Wild, D. J. (in print, 2007). Web 2.0 for grids and e-science. In *Proceedings of INGRID 2007 - Instrumenting the Grid 2nd International Workshop on Distributed Cooperative Laboratories*, S. Margherita Ligure Portofino, Italy.

Fox, G., Williams, R., & Messina, P. (1994). Parallel Computing Works! Morgan Kaufmann Publishers.

Fox, P., McGuinness, D., Raskin, R., and Sinha, K. (2007). A volcano erupts: semantically mediated integration of heterogeneous volcanic and atmospheric data. In Mitra, P., Giles, L. C., & Carr, L., (Eds), *CIMS '07: Proceedings of the ACM first workshop on CyberInfrastructure: information management in eScience*, (pp. 1-6). New York: ACM.

Frank, J. (1992). *Electron Tomography: Three-Dimensional Imaging whith the Transmission Electron Microscope.* New York: Plenum Press.

Frank, J. (2006). *Three-Dimensional Electron Microscopy of Macromolecular Assemblies: Visualization of Biological Molecules in Their Native State*. Oxford, UK: Oxford University Press.

Fraser, C.M., Eisen, J., Fleischmann, R.D., Ketchum, & K.A., Peterson, S. (2000). Comparative Genomics and Understanding of Microbial Biology. *Emerging Infectious Diseases*, *6*(5), 505-512.

Frazão, C., Sieker, L., Sheldrick, G. M., Lamzin, V., LeGall, J. & Carrondo, M. A. (1999). Ab initio structure solution of a dimeric cytochrome c3 from Desulfovibrio gigas containing disulfide bridges. *J. Biol. Inorg. Chem. 4*, 162-165.

Freitas, A. A. (1998). A survey of parallel data mining. In *2nd International Conference on the Practical Applications of Knowledge Discovery and Data Mining* (pp. 287-300). London: The Practical Application Company.

Freitas, A. A. (Ed.). (2002). *Data Mining & Knowledge Discovery with Evolutionary Algorithms.* Berlin, Germany: Springer-Verlag.

Freund, J., Comaniciu, D., Ioannis, Y., Liu, P., McClatchey, R., Morley-Fletcher, E., et al. (2006). Health-e-child: an integrated biomedical platform for grid-based paediatric applications. *Stud Health Technol Inform, 120,* 259-270.

Freund, Y. & Schapire, R. E. (1997). A decision-theoretic generalization of on-line learning and an application to boosting. *Journal of Computing System Science, 55*(1), 119-139.

Frey, J., Mori, T., Nick, J., Smith, C., Snelling, D., Srinivasan, L., & Unger, J. (2005). The Open Grid Services Architecture, Version 1.0. www.ggf.org/ggf_areas_architecture.htm

Frey, J., Tannembaum, T., Foster, I., Livny, M., & Tuecke, S. (2001). Condor-G: A Computation Management Agent for Multi-Institutional Grids. Proceedings of the *Tenth IEEE Symposium on High Performance Distributed Computing* (pp. 237-246). Hingham, MA: Kluwer Academic Publishers.

Frey, J., Tannenbaum, T., Livny, M., Foster, I., & Tuecke, S. (2002). Condor-G: A Computation Management Agent for Multi-Institutional Grids. *Cluster Computing, 5*(3), 237-246.

Frist, W. (2005). Health care in the 21st century. *New England Journal of Medicine, 352,* 267-272.

Friston, K. (1996). Statistical Parametric Mapping and Other Analysis of Functional Imaging Data. In *Brain Mapping The Methods,* (pp. 363-385). New York: Academic Press.

Frizziero, E., Gulmini, M., Lelli, F., Maron, G., Oh, A., Orlando, S., et al. (2006). *Instrument Element: a new Grid component that enables the control of remote instrumentation.* Paper presented at the Proceedings of the Sixth IEEE International Symposium on Cluster Computing and the Grid Workshops (CCGRIDW'06).

Frohner, A., Jouvenot, D., Kunszt, P., Montagnat, J., Pera, C., Koblitz, B., Santos, N, Loomis, C. (2007). A Secure Grid Medical Data Manager Interfaced to the gLite Middleware. *Journal of Grid Computing, 6*(1), 45-59.

Funika, W. & Pegiel, P. (2007). GScript Editor as Part of the ViroLab Presentation Layer. In *Proceedings of Cracow Grid Workshop 2006 (CGW06),* (pp. 320-327). Krakow, Poland: ACC CYFRONET AGH.

Furey, W. & Swaminathan, S. (1997). PHASES-95: A Program Package for the Processing and Analysis of Diffraction Data from Macromolecules. *Meth. Enzymol. 277,* 590-620.

G., Cafaro, M., Fiore, S., & Mirto, M. (2005). ProGenGrid: A Workflow Service Infrastructure for Composing and Executing Bioinformatics Grid Services. In A. Tsymbal & P. Cunningham (Eds.), *18th IEEE Symposium on Computer Based Medical Systems* (Vol. 1) (pp. 555-560). New York: IEEE Press

Gagliardi, F., Jones, B., Grey, F., Bégin, M. E., Heikkurinen, M. (2005). Building an infrastructure for scientific grid computing: status and goals of the EGEE project. *Phylosophical Transactions: Mathematica, Physical and Engineering Sciences, 363*(1833), 1729-1742.

Gagliardi, H. F., Silva, F. A. B., Alves, D. (2006). Automata Network Simulator Applied to the Epidemiology of Urban Dengue Fever. *ICCS 2006, Part III,* (LNCS 3993, pp. 297 – 304).

Galaktionov, S.G. & Marshall, G.R. (1994). Properties of intraglobular contacts in proteins: an approach to prediction of tertiary structure. *System Sciences, Vol.V:, Proceedings of the Twenty-Seventh Hawaii International Conference on Biotechnology Computing*: Vol. 5, (pp. 326 – 335). Washington, DC: IEEE Computer Society Press.

Galiardi F. (2005). EGEE – a first results and future plans. In M. Bubak, M. Turała, K. Wiatr (Eds.) Proceedings of Cracow'04 Grid Workshop.

Galizia, A., Viti, F., Orro, A., D'Agostino,D. Merelli, I., Milanes, L., & Clematis, A. (2008). TMAinspect, an EGEE Framework for Tissue MicroArray Image Handling. *Proceedings of 8th Internal Symposium on Cluster Computing and the Grid* (pp 741-746), IMATI-CNR, Genoa, Italy.

Gallo, E., Gagliardi, H.F., Neto, V. C., Alves, D., Silva, F.A.B. (2007) GISE: A Data Access and Integration Service for a Grid-Based Epidemic Surveillance System. *Proceedings of the 40th Annual Simulation Symposium*. New York: IEEE Computer Society Press.

Gao, H., Sengupta, J., Valle, M., Korostelev, A., Eswar, N., Stagg, S. M., et al. (2003). Study of the structural dynamics of the E. coli 70S ribosome using real-space refinement. *Cell, 113*(1), 789-801.

Gao, J., Opiteck, G. J., Friedrichs, M. S. & Haefta, S.. (2003). Changes in the Protein Expression of Yeast as a Function of Carbon Source. *Journal of Proteome Research*, 643-649.

Garcia, J. L., Dikaiakos, M. D., Kyprianou, T., Bilas, A., & Marazakis, M. (2008). Data Privacy considerations in Intesive Care Grids. *HealthGrid Conference 2008*. Chicago, IL.

Gardarin, G.; Sha, F.; Ngoc, T. (1999) XML-based Components for Federating Multiple Heterogeneous Data Sources. In *Proc. of International Conference on Conceptual Modeling*, Paris, France.

Gärdén, P., Alm, R., & Häkkinen, J. (2005). PROTEIOS: an open source proteomics initiative. *Bioinformatics, 21*(9), 2085-2087.

Gardner, S. P. (2005). Ontologies and semantic data integration. *Drug Discovery Today, 10*(14), 1001-1007.

Garduño, E., Gabor, G. T., & Herman, G. T. (2004). Optimization of Basis Functions for Both Reconstruction and Visualization. *Discrete Applied Mathematics, 139* (1-3), 95-111.

Gasteiger, E., Jung, E., et al. (2001). SWISS-PROT: connecting biomolecular knowledge via a protein database. *Curr Issues Mol Biol*, 3(3), 47-55.

GAT. (2005). Grid Application Toolkit. www.gridlab.org/WorkPackages/wp-1/

Gaudeau, Y., & Moureaux, J.M. (2007, November). *Lossy compression of volumetric medical images with 3D dead zone lattice vector quantization*. 26th Picture Coding Symposium, Lisboa, Portugal.

Gauss, C., Kalkum, M., Lowe, M., Lehrach, H., & Klose, J. (1999). Analysis of the mouse proteome. (I) Brain proteins: separation by two-dimensional electrophoresis and identification by mass spectrometry and genetic variation. *Electrophoresis, 20*, 575-600.

Gaussian 03 (2004). Frisch, M. J.; Trucks, G. W.; Schlegel, H. B.; Scuseria, G. E.; Robb, M. A.; Cheeseman, J. R. et al. Wallingford, CT: Gaussian, Inc.

GDMP. Retrieved from http://project-gdmp.web.cern.ch/project-gdmp

g-Eclipse project (2008). Retrieved October 20, 2008, from http://www.geclipse.eu/

Geddes, J., Lloyd, S., Simpson, A., Rossor, M., Fox, N., Hill, D., Et al. (2005). NeuroGrid: using grid technology to advance neuroscience. In *Proceedings of the 18th IEEE symposium on computer-based medical systems*, (pp. 570-572). Washington, DC: IEEE Computer Society.

Geiser, D.M., Jiménez-Gasco, M., Kang, S., Makalowska, I., Veerarahavan, N., Ward, T.J., Zhang, N., Kuldau, G.A. & O'Donnell, K. (2004). FUSARIUM-ID v.1.0: A DNA sequence database for identifying Fusarium. *European Journal of Plant Pathology, 110*, 473-479.

Geman, S. and D. Geman (1984). Stochastic Relaxation, Gibbs Distributions, and the Bayesian Restoration of Images. *IEEE Trans. PAMI, 6*, 721-741.

Gemstone (2008). Retrieved March 3, 2008 from http://gemstone.mozdev.org/

Gene Ontology Consortium (2001). Creating the Gene Ontology resource: design and implementation. *Genome Res, 11*(8), 1425-33.

Gene Ontology Consortium (2008). The Gene Ontology project in 2008. *Nucleic Acids Res, 36*, D440-4.

Gentzsch (2008). Top 10 Rules for Building a Sustainable Grid. In: Grid Thought Leadership Series. Open Grid Forum website at www.ogf.org/TLS/?id=1

Gentzsch, W. (2002). Response to Ian Foster's "What is the Grid?" GRIDtoday. August 5. www.gridtoday.com/02/0805/100191.html

Gentzsch, W. (2004). Enterprise Resource Management : Applications in Research and Industry. In I. Foster & C. Kesselman (Eds.). The Grid 2: Blueprint for a New Computing Infrastructure (pp. 157 – 166). Morgan Kaufmann Publishers.

Gentzsch, W. (2004). Grid Computing Adoption in Research and Industry. In A. Abbas (Ed.), Grid Computing: A Practical Guide to Technology and Applications (pp. 309 – 340). Charles River Media Publishers.

Gentzsch, W. (2007a). Grid Initiatives: Lessons Learned and Recommendations, RENCI Report retrieved from www.renci.org/publications/reports.php

Gentzsch, W. (Ed.) (2007b). A Sustainable Grid Infrastructure for Europe, Executive Summary of the e-IRG Open Workshop on e-Infrastructures, Heidelberg, Germany. Retrieved from the e-IRG Website www.e-irg.org/meetings/2007-DE/workshop.html

GEONgrid (2008). Retrieved from www.geongrid.org

GeoServer website (2008). Retrieved July 20, 2008 from http://geoserver.org

Germain, G. & Woolfson, M. M. (1968). On the application of phase relationships to complex structures, *Acta Crystallogr. B24*, 91-96.

Germain-Renaud, C., Loomis, C., Moscicki, JT., & Texier. R. (2008). Scheduling for responsive grids. *Journal of Grid Computing, 6*(1), 15-27.

Germain-Renaud, C., Perez, J., Kégl, B. & Loomis, C. (2008, May). Grid Differentiated Services: a Reinforcement Learning Approach. *8th IEEE Symposium on Cluster Computing and the Grid.* Lyon, France: IEEE Computer Society.

Gervasi O., Dittamo C., & Laganà A. (2005). A Grid Molecular Simulator for E-Science. In *Advances in Grid Computing - EGC 2005* (LNCS 3470, pp. 16-22). Berlin, Germany: Springer.

Geyer, C.J. (1991). Markov chain Monte Carlo maximum likelihood. In Keramidas (ed.), *Computing Science and Statistics: Proceedings of the 23rd Symposium on the Interface* (pp. 156–163). Fairfax Station, VA: Interface Foundation.

Giannadakis, N., Rowe, A., Ghanem, M., & Guo, Y. (2003). Infogrid: providing information integration for knowledge discovery. *Information Sciences 155*, 199-226.

Gil, Y., Deelman, E., Ellisman, M., Fahringer, T., Fox, G. Gannon, D., Goble, C., Livny, M., Moreau, L., & Myers J.(2007). Examining the Challenges of Scientific Workflows. *IEEE Computer, 40*(12), 24-32

Gil, Y., Deelman, E., Ellisman, M., Fahringer, T., Fox, G., Gannon, D., Goble, C., Livny, M., Moreau, L., & Myers, J. (2007). Examining the challenges of scientific workflows. *IEEE Computer, 40*(12), 24-32.

Gkekas, C., Psomopoulos, F., & Mitkas, P. (2008). *A Parallel Data Mining Application for Gene Ontology Term Prediction, 3rd EGEE User Forum*, Clermont-Ferrand, France.

Glatard, T., Montagnat, J., Emsellem, D. & Lingrand, D. (in press). A Service-Oriented Architecture enabling dynamic services grouping for optimizing distributed workflows execution. In *Future Generation Computer Systems.*

Glatard, T., Montagnat, J., Erberich S., Carmeli B. & Kaufman, J. (2007, November). *Cross Enterprise Document and Image Exchange (XDS/-I) in HealthGrids.* Paper presented at the Radiology Society of North America (RSNA'07), Chicago, IL.

Glatard, T., Montagnat, J., Lingrand, D. & Pennec, X. (in press). Flexible and efficient workflow deployement of data-intensive applications on grids with MOTEUR. *International Journal of High Performance Computing and Applications.*

Glatard, T., Pennec, X. & Montagnat, J. (2006). Performance evaluation of grid-enabled registration algorithms using bronze-standards. In *Proc. of MICCAI'06, Part II*, (LNCS 4191, pp. 152-160).

Glich, G.L., & Vachet, R.W. (2003). The basics of mass spectrometry in the twenty first century. *Nature Reviews Drug Discovery, 2*(2), 140-150.

gLite middleware. Retrieved from http://glite.web.cern.ch/glite/.

glite, A lightweight middleware for grid computing (2008). Retrieved October 20, 2008, from http://glite.web.cern.ch/glite/.

GLOBUS (n.d.). *The Globus Alliance.* Retrieved from http://www.globus.org/

Globus Alliance (2008). Retrieved July 20, 2008 from http://www.globus.org

Globus. Retrieved from http://www.globus.org

Goad, W.B., & Kanehisa, M.I. (1982). Pattern Recognition in Nucleic Acid Sequences: A General Method for Finding Local Homologies and Symmetries. *Nucleic Acids Research, 10*(1), 247-263.

Goble, C. A., & De Roure, D. (2004). The Semantic Grid: Myth Busting and Bridge Building. In *Proceedings of the 16th Eureopean Conference on Artificial Intelligence, ECAI'2004, including Prestigious Applicants of Intelligent Systems, PAIS 2004, Valencia, Spain, August 22-27, 2004* (pp. 1129-1135). Amsterdam, the Netherlands: IOS Press.

Goble, C., Corcho, O., Alper, P. & De Roure, D. (2006). E-science and the semantic web: A symbiotic relationship. In *Discovery Science 2006, Barcelona, Spain, LNAI* (Vol. 4265)(pp. 1-12). New York: Springer.

Goble, C.A., Pettifer, S., Stevens R. & Greenhalgh, C. (2003). Knowledge Integration: In silico Experiments in Bioinformatics. In I. Foster, C. Kesselman (Eds.), *The Grid: Blueprint for a New Computing Infrastructure* (pp. 121-124). San Francisco, CA: Morgan Kaufmann Publishers Inc.

Goloboff, P. (1999). Analyzing Large Data Sets in Reasonable Times: Solutions for Composite Optima. *Cladistics,* 15(4): 415-428.

Gonzales, J., Pomel, S., Breton, V., Clot, B., Gutknecht, J.L., Irrthum, B. & Legré, Y. (2005) Empowering humanitarian medical development using grid technology. *Methods of Information in Medicine, 44,* 186-189.

Good, J., Laity, A., Jacob, J.C., & Katz, D.S. (2005). Pegasus: a framework for mapping complex scientific workflows onto distributed systems. *Scientific Programming, 13*(3), 219-237.

Goodale, T., Jha, S., Kaiser, H., Kielmann, T., Kleijer, P., Merzky, A., Shalf, J., Smith, Ch. (2008). A Simple API for Grid Applications (SAGA). Grid Forum Document GFD.90. Open Grid Forum. Retrieved from www.ogf.org/documents/GFD.90.pdf

Goodhew, P. J. (1984). *Specimen Preparation for Transmission Electron Microscopy of Materials.* Oxford, UK: Oxford University Press.

Google (2008). Google App Engine. Retrieved from http://code.google.com/appengine/

Google Groups (2008). Cloud Computing. Retrieved from http://groups.google.ca/group/cloud-computing

Gordeeva E., Katritzky A. R.. Shcherbukhin, V. V., & Zefirov, N. S. (1993). Rapid Conversion of Molecular Graphs to Three-dimensional Representation Using the MOLGEO Program. *J. Chem. Inf. Comput. Sci., 33,* 102-111.

Gordon, P. M., Trinh, Q., & Sensen, C. W. (2007). Semantic web service provision: a realistic framework for bioinformatics programmers. *Bioinformatics, 23*(9), 1178-1180.

Gordon, R., Bender, R., & Herman, G. T. (1970). Algebraic reconstruction techniques (ART) for three-dimensional electron microscopy and x-ray photography. *Journal of Theoretical Biology, 29*(3), 471-481.

Gottschling, M., Stamatakis, A., Nindl, I., et al. (2007). Multiple Evolutionary Mechanisms Drive Papillomavirus Diversification. *Molecular Biology and Evolution,* 24(5):1242-1258.

Govindarajan, S., Goldstein, R.A. (1996). Why are some proteins structures so common? *Proc Natl Acad Sci USA, 93*(8), 3341-5.

Gracy, J., & Argos, P. (1998). Automated protein sequence database classification. *Bioinformatics, 14,* 164-187.

Grassè, P. (1959). *La reconstruction du nid et les coordinations inter-individuelles chez belicositermes natalensis et cubitermes sp. la theorie de la stigmergie: Essai d'interprtation du comportement des termites constructeurs,* Insectes Sociaux 6, 41–84.

Green, M. L. & Miller, R. (2003). Grid computing in Buffalo. *Annals of the European Academy of Sciences,* 191-218.

Green, M. L. & Miller, R. (2004). Molecular structure determination on a computational & data grid. *Parallel Computing, 30,* 1001-1017.

Green, M. L. & Miller, R. (2004). Evolutionary molecular structure determination using grid-enabled data mining. *Parallel Computing, 30,* 1057-1071.

Green, M. L. & Miller, R. (2004). A client-server prototype for application grid-enabling template design. *Parallel Processing Letters, 14*, 241-253.

Grethe, J. S., Baru, C., Gupta, A., James, M., Ludaescher, B., Martone, M. E., Papadopoulos, P. M., et al. (2005). Biomedical informatics research network: Building a national collaboratory to hasten the derivation of new understanding and treatment of disease. *Stud Health Technol Inform, 112*, 100-109.

Grethe, J.S., Baru, C., Gupta, A., James, M., Ludaescher, B., Martone, M.E., Papadopoulos, P.M., Peltier, S.P., Rajasekar, A., Santini, S., et al (2005). Biomedical Informatics Research Network: Building a National Collaboratory to Hasten the Derivation of New Understanding and Treatment of Disease. *Stud Health Technol Inform, 112*, 100-109.

Grid Engine (2001). Open Source Project. Retrieved from http://gridengine.sunsource.net/

GridSphere (2008). Retrieved from www.gridsphere.org/gridsphere/gridsphere

GridWay (2008). Metascheduling Technologies for the Grid. Retrieved from www.gridway.org/

Grillo, G., Licciulli, F., Liuni S., Sbisà, E. & Pesole G. (2003). PatSearch: a program for the detection of patterns and structural motifs in nucleotide sequences. *Nucleic Acids Res., 31*, 3608 3612.

Grimshaw A. S., Natrajan A., Humphrey M. A., Lewis M. J., Nghuyen-Thong A., Karpovich J. F., Morgan M. M., & Ferrari A. (2003). J. From Legion to Avaki. In Berman F. Fox G., Hey A. (Eds.) *Grid Computing: Making the Global*. Hoboken, NJ: John Wiley and Sons.

GriPhyN (n.d.). *The Grid Physics Network*. Retrieved from http://www.griphyn.org

Groth, P., Jiang, S. Miles, S. Munroe, S., Tan, V., Tsasakou, S., & Moreau, L. (2006). *An Architecture for Provenance Systems* (Tech. Rep.). Southampton, UK: University of Southampton..

GRyCAP (2007). *Grid Biocomputation Portal*. Retrived July 2008 from http://www.grycap.upv.es/bio

Guan, J. W., & Bell, D. A. (1991). *Evidence Theory and Its Application*. Amsterdam, the Netherlands: Elsevier Science.

Guardian (2007). *Family doctors shun national database of patients' records, 20 Nov 2007*. Retrieved from www.guardian.co.uk/society/2007/nov/20/nhs.health

Guarracino, M. R., Cifarelli, C., Seref, O., & Pardalos, P. M. (2006). A Parallel Classification Method for Genomic and Proteomic Problems. *In 20th international Conference on Advanced information Networking and Applications* (Vol. 02) (pp. 588-592). AINA. Washington, DC: IEEE Computer Society,.

Gubala, T. & Hoheisel, A. (2006). Highly dynamic workflow orchestration for scientific applications. In *CoreGRID Integation Workshop 2006 (CIW06)*, (pp. 309-320). Krakow, Poland: ACC CYFRONET AGH.

Gubala, T., Balis, B., Malawski, M., Kasztelnik, M, Nowakowski, P., Assel, M., et al. (2008). ViroLab Virtual Laboratory. In *Proceedings of Cracow Grid Workshop '07, ACC CYFRONET AGH, Krakow*.

Gubala, T., Balis, B., Malawski, M., Kasztelnik, M., Nowakowski, P., Assel, M., et al. (2007). ViroLab Virtual Laboratory. In Bubak, M., Turala, M., and Wiatr, K., (Eds.), *Proceedings of Cracow Grid Workshop - CGW'07*, (pp. 35-40). Krakow, Poland: ACC CYFRONET AGH

Gubala, T., Kasztelnik, M., Malawski, M., & Bubak, M. (2007). Development and execution of collaborative application on the virolab virtual laboratory. In Bubak, M., Turala, M., and Wiatr, K., editors, *Proceedings of Cracow Grid Workshop - CGW'07*, (pp. 41-46). Krakow, Poland: ACC CYFRONET AGH.

Gustafson, J. (1987). Reevaluating Amdahl's Law, CACM, vol. 31, pp. 532-533.

Guyton, A.C., Jones, C.E., & Coleman, T.G. (1973). *Circulatory Physiology: Cardiac Output and its Regulation* (2nd Ed.). Philadelphia: W.B. Saunders.

Gwak. K.-W., Paden, B.E., Noh, M.D., & Antaki, J.F. (2006). Fluidic Operational Amplifier for Mock Circulatory Systems. *IEEE Transactions of Control Systems Technology, 14*(4), 602-612.

Gygi, S.P., Rist, B., Gerber, S.A., Turecek, F., Gelb, M.H., & Aebersold, R. (1999). Quantitative analysis of complex protein mixtures using isotope-coded affinity tags. *Nature Biotechnology, 17*(10), 994–999

H. F. Gagliardi, F. A.B. da Silva, & D. Alves. (2006) Automata Network Simulator Applied to the Epidemiology of Urban Dengue Fever. In *Proceedings of the ICCS'06, Lecture Notes in Computer Science, 3993* (pp. 297-304). Berlin, Germany: Springer

Hall, L. O., Chawla, N., & Bowyer, K. W. (1998). Combining decision trees learned in parallel. In *Working Notes of the KDD-97 Workshop on Distributed Data Mining* (pp. 10-15).

Hamari, Z., Juhász, A. & Kevei, F. (2002). Role of mobile introns in mitochondrial genome diversity of fungi (a mini review). *Acta Microbiol Immunol Hung. 49*(2-3), 331-5.

Han, D.K., Eng, J., Zhou. H. & Aebersold, R. (2001). Quantitative profiling of differentiation-induced microsomal proteins using isotope- coded affinity tags and mass spectrometry. *Nature Biotechnology, 19*, 946-51.

Handschuh, S., Staab, S., & Ciravegna, F. (2002). S-CREAM - semi-automatic creation of metadata. In Gómez-Pérez, A. and Benjamins, R. V., (Eds.), *EKAW '02: Proceedings of the 13th International Conference on Knowledge Engineering and Knowledge Management. Ontologies and the Semantic Web*, London, (pp. 358-372). Berlin, Germany: Springer-Verlag.

Hansen, C.D. & C.R. Johnson (2005). *The Visualization Handbook*. Burlington, MA: Elsevier Butterworth-Heineman..

Hansen, H., Bachmann, L., & Bakke, T.A. (2003). Mitochondrial DNA variation of *Gyrodactylus* spp. *Monogenea, Gyrodactylidae* populations infecting Atlantic salmon, grayling, and rainbow trout in Norway and Sweden. *International Journal of Parasitology*, 33(13): 1471-1478.

Harris, M. A., Clark, J., et al. (2004). The Gene Ontology (GO) database and informatics resource. *Nucleic Acids Res, 32*, D258-61.

Hastie, T., Tibshirani, R., & Friedman, J. (2002). *The Elements of Statistical Learning*. New York: Springer.

Havel, T.F. (1998). *Distance Geometry: Theory, Algorithms, and Chemical Applications in the Encyclopedia of Computational Chemistry*. Hoboken, NJ: J. Wiley & Sons.

Hawkins, P., & Skillman, G. (2006). Ligand-Based Design Workflow.

Health Level 7 (1997). Retrieved July 2008 from http://www.hl7.org/

Health Level Seven Inc. (2008). *Health Level Seven homepage*. Retrieved March 8, 2008 from http://www.hl7.org/

Health-e-Child, The EU FP6 Information Societies Technology Project (2008). Retrieved October 20, 2008, from http://www.health-e-child.org/.

Heath, T. & Motta, E. (in press). Ease of interaction plus ease of integration: Combining web2.0 and the semantic web in a reviewing site. *Journal of Web Semantics*, (Special Issue on Web2.0 and the Semantic Web).

Hebert, P.D., Ratnasingham, S., deWaard J.R. (2003). Barcoding animal life: cytochrome c oxidase subunit 1 divergences among closely related species. *Proc Biol Sci., 270*, (1), S96-9.

Heinicke, M.P., Duellman, W.E., & Hedges, S.B. (2007). From the Cover: Major Caribbean and Central American frog faunas originated by ancient oceanic dispersal. *Proceedings of the National Academy of Sci*

Heinrichs, W.L. (1999). The Critical Path from Tissue Slices to Surgical Simulation: What do Surgeons Want? In *Proc. 12'th IEEE Symposium on Computer-Based Medical Systems, June 18-20, Stamford, CT*, (pp. 118-123).

Heinzlreiter, P. & Kranzlmueller, D. (2003). Visualization Services on the Grid: The Grid Visualization Kernel. *Parallel Processing Letters, 13*(2), 135–148.

Henderson, R. (1995). The potential and limitations of neutrons, electrons and x-rays for atomic resolution microscopy of unstained biological molecules. *Quarterly Review of Biophysics, 28*(1), 171-193.

Henderson, R. (2004). Realising the potential of electron cryomicroscopy. *Quarterly Review of Biophysics, 37*(1), 3-13.

Henderson, R.L. (1995). Job Scheduling Under the Portable Batch System. *Proceedings of the workshop on Job Scheduling Strategies for Parallel Scheduling* (pp. 279.294). London: Springer-Verlag.

Hendrix, E. M. T., Ortigosa, P. M., & García, I. (2001). On success rates for controlled random search. *Journal of Global Optimization, 21*(3), 239-263.

Henikoff, J.G., Henikoff, S., & Pietrokovski, S. (2000). Increased coverage of protein families with the Blocks Database servers. *Nucleic Acids Research, 28*, 228-230.

Henikoff, S., & Henikoff, J.G. (1992). Amino Acid Substitution Matrices from Protein Blocks. *Proceedings of the National Academy of Sciences of the United States of America, 89*(22), 10915-10919.

Hennessy, J. L., & Patterson, D. A., (2006). *Computer Architecture (4th Ed.): A Quantitative Approach*. San Francisco, CA: Morgan Kaufmann.

Hernandez, P., Gras, R., Frey, J., & Appel, R. D. (2003). Popitam: towards new heuristic strategies to improve protein identification from tandem mass spectrometry data. *Proteomics, 3*(6), 870-878.

Hernandez, P., Müller, M. & Appel, R. D. (2006). Automated protein identification by tandem mass spectrometry: issues and strategies. *Mass Spectrom Rev, 25*(2) 235-254.

Herveg, J. A. M., & Poullet, Y. (2003, January 16-172003). *Directive 95/46 and the use of GRID technologies in the heathcare sector: selected legal issues.* Paper presented at the Healthgrid 2003, Lyon, France.

Hes, R., & Borking, J. J. (1998). *Privacy-enhancing technologies: the path to anonymity* (Rev. Rd.). The Hague, the Netherlands: Registratiekamer.

Hey, A. J. G., & Trefethen, A. E. (2002). The UK e-Science Core Programme and the Grid. *Future Generation Computer Systems 18*(8), 1017-1031.

Hey, T., & Trefethen, A. E. (2005). Cyberinfrastructure for e-Science. *Science, 308*, 817-821.

Hibbett, D. S., Nilsson, R. H., Snyder, M., Fonseca, M., Costanzo, J. & Honfeld M. (2005). Automated phylogenetic taxonomy: an example in the homobasidiomycetes (mushroom-forming fungi). *Syst. biol., 54*(4), 660–668.

HIPPA (2006). Retrieved July 2008 from http://www.hhs.gov/ocr/hipaa/

HL7. (2005). HL7 Receives ANSI Approval of Three Version 3 Specifications Including CDA, Release 2.

Hluchy, L., Habala, O., Babik, M., Laclavik, M., Balogh, Z., & Gatial, E. (2007). Knowledge-based platform for environmental risk management. In Volkert, J., Kranzlmüller, D., & Schreiner, W., (Eds.), *ISPDC '07: Proceedings of the Sixth International Symposium on Parallel and Distributed Computing*, Washington, DC, (p.1). New York: IEEE Computer Society.

HmtDB: Human Mitochondrial Genomic Resource Based on Variability Studies. Retrieved from http://www.hmtdb.uniba.it/.

Hofer, J. (2004). *Distributed decision tree induction within the grid data mining framework gridminer-core* (Tech. Rep. No. 2004-04). Vienna, Austria: University of Vienna, Institute for Software Science.

Hoffmann, K., Bucher, P., Falquet, L., & Bairoch, A. (1999). The PROSITE database, its status in 1999. *Nucleic Acids Research, 27*, 215-219.

Hohne, K.H. & Bernstein, R. (1986). Shading 3D-Images from CT Using Gray-Level Gradients. *IEEE Transactions on Medical Imaging, 5*(1), 45-47.

Honore, B., Ostergaard, M., & Vorum, H. (2004). Functional genomics studied by proteomics. *Bioessays, 26*(8), 901-915.

Hoogland, C., Sanchez, J.C., Tonella, L., Bairoch, A., Hochstrasser, D.F., & Appel, R.D.(1998). Current status of the SWISS-2DPAGE database, Nucleic. *Acids Research 26*(1), 332–333

Housley, R. & Polk, T. (2001). *Planning for PKI: Best Practices Guide for Deploying Public Key Infrastructures.* Hoboken, NJ: Wiley Computer Publishing.

Houstis, E., Gallopoulos, E., Bramley, R. & Rice J. (1997). Problem-Solving Environments for Computational Science. *IEEE Comput. Sci. Eng., 4*(3), 18–21.

Houstis, E., Gallopoulos, E., Bramley, R., & Rice, J. (1997). Problem-Solving Environments for Computational Science. *IEEE Comput. Sci. Eng., 4*(3), 18–21.

Huang, X., Huang, L. & Li, M. (2006). Grid-enabled medical image processing application system based on OGSA-DAI techniques: In Proceedings of the International workshops on advanced Web, network technologies and Applications, pp. 460-464. In Shen, H. T., Li, J., Li, M., Ni, J. & Wang, W. (Eds.), *Lecture notes in Computer Science,* (Vol. 3842). Aachen, Germany: Springer-Verlag.

Huber V. (2001). Supporting Car-Parrinello Molecular Dynamics with Unicore. In *Computational Science ICCS - San Francisco, USA, Proceedings, Pt. I,* (Vol. 2073), (pp. 560-567). Berlin, Germany: Springer Verlag.

Huedo, E., S. Montero, R., & M. Llorente, I. (2004). A framework for adaptive execution in grids. *Software: Practice and Experience*, *34*(7), 631-651).

Huhns, M. N. & Singh, M. P. (2005). Service-oriented computing: Key concepts and principles. *IEEE Internet Computing*, *9*(1), 75-81

Humphrey, W., Dalke, A., & Schulten, K. (1996). VMD: Visual molecular dynamics. *Journal of Molecular Graphics, 14,* 33-38.

Humphreys, G., Houston, M., Ng, R., Frank, R., Ahern, Kirchner, P.D. & J.T. Klosowski (2002). Chromium: A Stream-Processing Framework for Interactive Rendering on Clusters. *ACM Transactions on Graphics*, *21*(3), 693-702.

Hwang, S., & Kesselman, C. (2004). A Flexible Framework for Fault Tolerance on the Grid. *Journal of Grid Computing, 1*(3), 251-272.

I. Andoulsi, J. Herveg, V. Stroetmann, K. Stroetmann, A. Dobrev, C. Van Doosselaere, P. Wilson (2007). *Bottlenecks & Challenges and RTD Responses for Legal, Ethical, Social, and Economic Aspects of Healthgrids – Roadmap II.* Retrieved July 2008 from http://eu-share.org/about-share/deliverables-and-documents.html

I. Blanquer, V. Hernandez, D. Segrelles (2007). TRENCADIS – Secure Architecture to Share and Manage DICOM Objects in a Ontological Framework Based on OGSA. *From Genes to Personalized HealthCare: Grid Solutions for the Life Sciences, Studies in Health Technology and Informatics, 126,* 115-124.

Iakovidis, I. (2005). *HealthGrid – 3 sided concept: ICT for Health, ISTAG WG.* Retrieved August 30, 2007, from http://www.who.int/classifications/terminology/iakovidis.pdf

Iamnitchi, A., & Foster, I. (2005). *Interest-aware information dissemination in small-world communities*, Paper presented at the 14th IEEE International Symposium on High Performance Distributed Computing (HPDC 2005), Research Triangle Park, USA.

Iavindrasana, J., Loiacono, L., Müller, H., Periz, I., Summers, P., & Wright, J. (2008). Access to Clinical Information Systems for Research in the Life Sciences - Security and Privacy Considerations, *HealthGrid Conference 2008.* Chicago, IL.

Ichikawa, K., Date, S., Krishnan, S., Li, W. W., Nakata, K., Yonezawa, Y., Nakamura, H., Shimojo, S. (2007). *In OPAL OP: An Extensible Grid-Enabling Wrapping Tool For Legacy Applications.* 3rd International Workshop on Grid Computing & Applications (GCA2007), Singapore.

INFN Production Grid. Retrieved from http://grid-it.cnaf.infn.it/.

INSPIRE Project (2008). Retrieved July 12, 2008 from http://www.inspire-geoportal.eu

Institut Pasteurand the HKU-Pasteur Research Centre (2005). *Information on the international outbreak of the illness known as severe acute respiratory syndrome (SARS).* Retrieved July 2008 from http://www.pasteur.fr/recherche/unites/REG/SARS.html

Integrating the Healthcare Enterprise (2008). Retrieved July from 2008 from http://www.ihe.net/

International Organization for Standardization (2007). *Codes for the representation of names of countries and their subdivisions – Part 2: Country subdivision code.* Geneva, Switzerland: International Organization for Standardization.

Internet2 Shibboleth, (2006). *Shibboleth Architecture Protocols and Profiles.* Retrieved from http://shibboleth.internet2.edu/docs/draft-mace-shibboleth-arch-protocols-latest.pdf

Internet2 Shibboleth, (2006). *Shibboleth Architecture Technical Overview.* Retrieved from http://shibboleth.internet2.edu/docs/draft-mace-shibboleth-tech-overview-latest.pdf

Itoh, M., Akutsu, T., & Kanehisa, M. (2004) Clustering of database sequences for fast homology search using upper bounds on alignment score. *Genome Informatics, 15*(1), 93-104.

Itoh, M., Goto, S., Akutsu, T., & Kanehisa, M. (2005). Fast and accurate database homology search using upper bounds of local alignment scores. *Bioinformatics, 21*(7), 912-921.

ITU-T Recommendation X.509 (2001) | ISO/IEC 9594-8: 2001, Information technology – Open Systems Interconnection – Public-Key and Attribute Certificate Frameworks.

ITU-T Recommendation X.812 (1995). | ISO/IEC 10181-3:1995, Security Frameworks for open systems: Access control framework.

J. Roboz, (1968). *Mass spectrometry instrumentation and techniques.* New York: Intersience Publisher.

Jaakkola, T., et al. (2000). A discriminative framework for detecting remote protein homologies. *Journal of Molecular Biology, 7,* 95-114.

Jacob, B., Ferreira, L., Bieberstein, N., Gilzean, C., Girard, J.-Y., Strachowski, R., & Yu, S. (2003). Enabling applications for Grid computing with Globus. IBM Redbook, retrieved from www.redbooks.ibm.com/abstracts/sg246936.html?Open

Jacq N., Breton B., Chen H., Ho L., Hofmann M., Lee H., et al. (2006). *Large Scale In Silico Screening on Grid Infrastructures.* Proceedings of the LSGRID2006, Yokohama, Japan (pp.123–136).

Jacq N., Salzemann J., Jacq F., Legré Y., Medernach E., Montagnat J., Maaß A., et al. (2008). Grid-enabled Virtual Screening Against Malaria. *J Grid Computing 6,*29–43.

Jacq, N., Blanchet, C., Combet, C., Cornillot, E., Duret, L., Kurata, K., Nakamura, H., Silvestre, T., & Breton, V. (2004). Grid as a Bioinformatic Tool. *Parallel Computing, 30*(9-10), 1093-1107.

Jacq, N., Blanchet, C., Combet, C., Cornillot, E., Duret, L., Kurata, K., Nakamura, H., Silvestre, T., Breton, V. (2004). Grid as a bioinformatics tool. *Parallel Computing, Special Issue: High-performance parallel biocomputing, 30.*

Jacq, N., Salzemann, J., Legre, Y., Reichstadt, M., Jacq, F., Zimmermann, M., et al. (2006). Demonstration of in silico docking at a large scale on Grid infrastructure. *Stud Health Technol Inform, 120,* 155-157.

JaJa, J., Varshney, A. & Shi, Q. (2008). Parallel Algorithms for Volumetric Surface Construction. In S. Rajasekaran & J. Reif (eds.), *Handbook of Parallel Computing,* (pp. 27-17). Boca Raton, FL: Chapman and Hall.

Jankun-Kelly, T.J. & K. Meta (2006). Superellipsoid-based, real Symmetric Traceless Tensor Glyphs Motivated by Nematic Liquid Crystal Alignment Visualization. *IEEE Transactions on Visualization and Computer Graphics,* (pp. 1197-1204).

Januzaj, E., Kriegel, H. P., & Pfeifle, M. (2004). Scalable density-based distributed clustering. *In the 8th European Conference on Principles and Practice of Knowledge Discovery in Databases* (pp. 231-244). Berlin, Germany: Springer-Verlag.

Jeannot, E. (2005, April), Improving middleware performance with AdOC: an Adaptive Online Compression library for data transfer. *IEEE International Parallel and Distributed Processing Symposium* (p. 70). Denver, CO. New York: IEEE Computer Society.

Jeffrey, S., Morris, M., Kevin, R., Coombes, J., Koomen, K., Baggerly, K., & Kobayashi, R. (2005). Feature extraction and quantification for mass spectrometry in biomedical applications using the mean spectrum. *Bioinformatics, 21*(9) 1764–1775.

Jeffries, N. (2005). Algorithms for alignment of mass spectrometry proteomic data. *Bioinformatics, 21*(14), 3066–3073.

Jensen, ED., Locke, CD. & Tokuda, H. (1985, December). *A time driven scheduling model for real-time operating systems* (pp. 112–122). IEEE Real-Time Systems Symposium, San Diego, CA. New York: IEEE Computer Society.

Jha, S., Kaiser, H., El Khamra, Y., & Weidner, O. (2007). Design and Implementation of Network Performance Aware Applications Using SAGA and Cactus. 3rd IEEE Conference on eScience and Grid Computing, Bangalore, India, 10-13 Dec, pp. 143 – 150.

Jithesh, P. V., Kelly, N., Wasnik, S., Donachy, P., Harmer, T., Perrott, R., et al. (2005). Bioinformatics application integration in genegrid. *In the UK e-Science All Hands Meeting.*

Johnson, G. (1998). Collaborative Visualization 101. *ACM SIGGRAPH, 32*(2), 8-11.

Jones, D. M., Fenner, J. W., Berti, G., Kruggel, F., Mehrem, R. A., Backfrieder, W., Moore, R. and Geltmeier, A. (2004). The GEMSS Grid: An evolving HPC environment for medical applications. In *Proceedings of HealthGrid 2004.* Retrieved March 8, 2008, from http://www.ccrl-nece.de/gemss/Reports/Jones-healthgrid2004.pdf

Jones, K. E., Pattel, N. G., Levy, M. A., Storeygard, A., Balk, D., Gittleman, J. L., Daszak, P. (2008). Global Trends in Emerging Infectious Diseases. *Nature, 451*(February 21), 990-994.

Jong, M.D., & Hien, T.T. (2006). Avian influenza A (H5N1). *J Clin Virol, 35*(1), 2-13.

Joshi, A., Karypis, G., & Kumar, V. (1998). ScalparC: A new scalable and efficient parallel classification algorithm for mining large datasets. In *11th Inter national Parallel Processing Symposium.* New York: IEEE Computer Society Press.

Jourdan L., Dhaenens C., Talbi E.-G. (2004). A Parallel adaptive GA for linkage disequilibrium in genomics. NIDISC'2004 Workshop. *IEEE International Parallel & Distributed processing symposium (IPDPS 2004),* Santa Fe, NM.

JSDL 1.0. Job Submission Language Specification (2006). Retrieved March 6, 2008 from http://www.gridforum.org/documents/GFD.56.pdf

Kabsch, W. & Sander, C. (1983). Dictionary of protein secondary structure: pattern recognition of hydrogen-bonded and geometrical features. *Biopolymers, 22,* 2577–2637.

Kakazu, K., Cheung, L., & Lynne, W. (2004). The Cancer Biomedical Informatics Grid (caBIG): pioneering an expansive network of information and tools for collaborative cancer research. *Hawaii Med J, 63*(9), 273-275.

Kal L., Skeel R., Bhandarkar M., Brunner R., Gursoy A., Krawetz N., et al. (1999). NAMD2: Greater Scalability for Parallel Molecular Dynamics. *J. Comp. Phys., 151,* 283–312.

Kalé, L., Skeel, R., Bhandarkar, M., Brunner, R., Gursoy, A., Krawetz, N., Phillips, J., Shinozaki, A., Varadarajan, K., & Schulten, K. (1999). NAMD2: Greater scalability for parallel molecular dynamics. *Journal of Computational Physics, 151,* 283-312.

Kalfoglou, Y. & Schorlemmer, M. (2003). Ontology mapping: the state of the art. *The Knowledge Engineering Review, 18*(1), 1-31.

Kandaswamy, G., Fang, L., Huang, Y., Shirasuna, S., Marru, S., & Gannon, D. (2006). Building web services for scientific grid applications. *IBM Journal of Research and Development, 50*(2/6), 249-260.

Kapp, E. A., Schütz, F., Connolly, L. M., Chakel, J. A., Meza, J. E., Miller, C. A., et al. (2005). 'An evaluation, comparison, and accurate benchmarking of several publicly available MS/MS search algorithms: sensitivity and specificity analysis. *Proteomics, 5*(13), 3475-3490.

Karasavvas, K., Antonioletti, M., Atkinson, M., Hong, N. C., Sugden, T., Hume, A., et al. (2005). Introduction to OGSA-DAI Services. In *Scientific Applications of Grid Computing, First International Workshop* (Vol. 3458)(pp. 1-12). Beijing: Springer Berlin / Heidelberg.

Kargupta, H., & Chan, P. (Eds.) (2000). *Advances in Distributed and Parallel Knowledge Discovery.* Cambridge, MA: MIT Press.

Karle, J. & Hauptman, H. A. (1956). A theory of phase determination for the four types of non-centrosymmetric space groups $1P222$, $2P22$, $3P_12$, $3P_22$. *Acta Crystallogr., 9,* 635-651.

Karonis, N.T., Toonen, B., & Foster, I. (2002). MPICH-G2: A Grid-Enabled Implementation of the Message Passing Interface. Journal of Parallel and Distributed Computing (JPDC), vol. 63, pp. 551-563.

Kasztelnik, M., Bubak, M., Gorka, C., Malawski, M., & Gubala, T. (2007). Fault tolerant grid registry. In Kacsuk, P., Fahringer, T., & Nemeth, Z. (Eds.), *Distributed and Parallel Systems,* Distributed and Parallel Systems, (pp. 135-144). Berlin, Germany: Springer.

Katz, D.S., Anagnostou, N., Berriman, G.B., Deelman, E., Good, J., Jacob, J.C., Kesselman, C., et al. (2006). Astronomical Image Mosaicking on a Grid: Initial Experiences. In Martino, B.D., Dongarra, J., Hoisie, A., Yang, L., & Zima, H. (Ed.), *Engineering the Grid-Status and Perspective*. Valencia, CA: American Scientific Publishers.

Kazutaka, K., Kazuharu, M., Keichi, K. & Takashi, M. (2002). MAFFT: A novel method for rapid multiple sequence alignment based on fast Fourier transform. *Nucl. Acids Res.*, *30*(14), 3059–3066.

Keahey, K., Foster, I., Freeman, T. & Zhang, X. (2006). Virtual Workspaces: Achieving Quality of Service and Quality of Life in the Grid. *Scientific Programming, 13*(4), 265-275.

Keller, A., Eng. J., Zhang, N., Li. X., & Aebersold, R. (2005). A uniform proteomics MS/MS analysis platform utilizing open XML file formats. *Mol Syst Biol, 1*, 2005.0017

Keller, A., Nesvizhskii, A. I., Kolker, E., & Aebersold R., (2002). Empirical statistical model to estimate the accuracy of peptide identifications made by MS/MS and database search. *Analytical. Chemistry, 74*(20), 5383-92

Kelly, N., Jithesh, P.V., Wasnik, S., McLaughlin, R., Fragoso, F., Donachy, P., et al. (2005). The GeneGrid Portal: A User Interface for a Virtual Bioinformatics Laboratory. In *UK e-Science All Hands Meeting (AHM 2005).*

Ken Lin, Booz Allen Hamilton, Gary Daemer. (2006) *caBIG Security Technology Evaluation.*(White Paper). Retrieved July 2008 from https://cabig.nci.nih.gov/workspaces/Architecture/Security_Tech_Eval_White_Paper/.

Kesselman, C., Foster, I., & Prudhomme, T. (2004). Distributed Telepresenze:The NESSgrid Earthquake Engineering Collaboratory. In Elsevier (Ed.), *The grid 2* (2nd Ed, pp. 81-93). San Francisco, CA: Morgan Kaufmann.

Khoussainov, R., Zuo, X., & Kushmerick, N. (2004). Grid-enabled weka: A toolkit for machine learning on the grid. *ERCIM News, 59.*

Kikinis, R. Warfield, S. & Westin, C. (1998). *High Performance Computing (HPC) in Medical Image Analysis (MIA) at the Surgical Planning Laboratory (SPL)*. Paper presented at the 3rd High Performance Computing Asia Conference & Exhibition, Singapore, Malaysia.

Kim, J., Deelman, E., Gil, Y., Mehta, G., & Ratnakar, V. (2007). Provenance trails in the Wings/Pegasus system. In *Concurrency and Computation: Practice and Experience*, San Francisco, CA: Wiley InterScience, *20*(5), 587-597.

Kim, J., Kim, H., & Han, D. (2006). Performance Evaluation of BLAST on SMP Machines. *Lecture Notes in Computer Sciences*, 4331, 668.

Kindlmann, G. (2004). Superquadric Tensor Glyphs, Joint Eurographics. In *IEEE TCVG Symposium on Visualization*, (pp. 147-154).

Kirkpatrick, S., Gellatt Jr., C.D. & Vecchi, M.P. (1982). *Optimization by simulated annealing.* Yorktown Heights, NY: IBM T.J. Watson Research Center.

Kiryakov, A., Popov, B., Terziev, I., Manov, D., & Ognyanoff, D. (2004). Semantic annotation, indexing, and retrieval. *Web Semantics: Science, Services and Agents on the World Wide Web, 2*(1), 49-79.

Kleinberg, J. (2000). Navigation in a Small World , Nature 406(2000), 845.

Klose, J. (1999). Genotypes and phenotypes. *Electrophoresis, 20*, 643-652.

Kluszczyński R., & Bała P. (2008). Supporting NAMD Application on the Grid using GPE. To appear in LNCS.

Kluszczyński R., Borcz M., & Bała P. (2007). BLAST Application on the GPE/UnicoreGS grid. In *Euro-Par 2006 Workshops* (pp. 245-253). Berlin, Germany: Springer-Verlag.

Knierbein, B., Reul, H., Eilers, R., Lange, M., Kaufmann, R., & Rau, G. (1992). Compact mock loops of the systemic and pulmonary circulation for blood pump testing. *The International Journal of Artificial Organs, 15*(1), 40-48.

Kohlbacher, O., Reinert, K., Gröpl, C., Lange, E., Pfeifer, N., Schulz-Trieglaff, O., & Sturm, M. (2007). TOPP - The OpenMS Proteomics Pipeline. *Bioinformatics, 23*(1), 191-197.

Kolff ,W.J. (1959). Mock circulation to test pumps designed for permanent replacement of damaged hearts. *Cleve Clin Q, 26*, 223-226.

Kollman P. (2001). *AMBER (Assisted Model Building with Energy Refinement)*. San Francisco, CA: University of California..

Konagaya, A. (2006). Trends in life science grid: from computing grid to knowledge grid. *BMC Bioinformatics, 7 Suppl 5*, S10.

Kondylakis, H., Analyti, A., & Plexousakis, D. (2007). *Quete: Ontology-based query system for distributed sources*. (*LNCS* 4690, pp. 359-375).

Konishi, F., Yagi, T., & Konagaya, A. (2006). MolWorks+G: Integrated Platform for the Acceleration of Molecular Design by Grid Computing. In *Grid Computing in Life Science* (pp. 134-141). Singapore: World Scientific.

Korf. I., Yandell, M., & Bedell, J., (2003). BLAST. Sebastopol, CA: O'Reilly.

Korhonen, J. (2006). *Finpro, Post workshop report, Fifth EGEE industry day concentrated on Challenges in Bioscience*, November 2006, Helsinki, Finland.

Kosar, T. & Livny M. (2004). Stork: Making Data Placement a First Class Citizen in the Grid. In *Proceedings of 24th IEEE International Conference on Distributed Computing Systems (ICDCS2004)*, Tokyo, Japan.

Krafzig, D., Banke, K., & Slama, D. (2004). *Enterprise SOA: Service-Oriented Architecture Best Practices (The Coad Series)*. Upper Saddle River, NJ: Prentice Hall PTR.

Kranzlmueller, D. (n.d.). *The GVK/GVid Approach: Visualization on the Grid.* Retrieved from http://www.ogf.org/OGF19/materials/562/kranzlmueller.pdf

Kranzlmueller, D., Heinzlreiter, P. and J. Volkert (2004). *Grid-Enabled Visualization with GVK, in Grid Computing* (LNCS Vol. 2970) (pp. 139-146). Berlin, Germany: Springer-Verlag.

Krause, A., & Vingron, M. (1998). A set-theoretic approach to database searching and clustering. *Bioinformatics, 14*, 430-438.

Krefting, D., Bart, J., Beronov, K., Falkner, J., Hartung, M., Hoheisel, A., et al. (2008). MediGRID: A Userfriendly Secured Grid Infrastructure. *Future Generation Computer Systems Journal - International Journal of Grid Computing.*

Krishnan, A. (2004). A Survey of life sciences applications on the grid. *New Generation Computing, 22*(2), 111-126.

Krishnan, A. (2005). GridBLAST: a Globus-based high-throughput implementation of BLAST in a Grid Computing framework. *Concurrency and Computation: Practice and Experience, 17*(13), 1607-1623.

Krishnan, S., & Bhatia, K. (2007). SOAs for Scientific Applications: Experiences and Challenges. In *the 3rd International Conference on e-Science and Grid computing.*

Krishnan, S., Stearn, B., Bhatia, K., Baldridge, K. K., Li, W. W., and Arzberger, P. (2006). Opal: Simple Web Services Wrappers for Scientific Applications. In *IEEE International Conference on Web Services*, (pp 823-832).

Krogh, A., et al. (1994). Hidden Markov models in computational biology: applications to protein modeling. *Journal of Molecular Biology, 235*, 1501-1531.

Kryza, B., Dutka, L., Slota, R., & Kitowski, J. (2008). Supporting management of dynamic virtual organizations in the grid through contracts. In Bubak, M., Turala, M., and Wiatr, K., (Eds.), *Proceedings of Cracow'07 Grid Workshop*, (pp. 140-147). Krakow, Poland: ACC CYFRONET AGH.

Kuperman, G. J. (1999). Improving response to critical laboratory values with automation: results of a randomised controlled trial. *J Am Med Inform Assoc, 6*, 512-522.

L. Moreau, et al. (2007). The first provenance challenge. *Concurrency and Computation: Practice and Experience, 20*(5), 409-418. San Francisco, CA: Wiley InterScience.

Laidlaw, D.H., Ahrens, E.T., Kremers, D., Avalos, M.J., Jacobs, R.E. & Readhead, C. (1998). Visualizing Diffusion Tensor Images of the Mouse Spinal Cord. In *IEEE Visualization '98*, (pp. 127-134).

Lamehamedi, H., Szymanski, B., Shentu, Z., & Deelman, E. (2002). Data replication strategies in grid environments. In *Proceedings Of the Fifth International Conference on Algorithms and Architectures for Parallel Processing*, Beijing, China, (pp. 378-383). New York: IEEE Computer Society Press.

Lamehamedi, H., Szymanski, B., Zujun Shentu, & Deelman E. (2002, October). Data Replication Strategies in Grid Environments. In *Proceedings of 5th International Conference on Algorithms and Architecture for Parallel Processing, ICA3PP'2002*, (378-383). New York: IEEE Computer Science Press.

Lanave, C., Preparata, G., Saccone, C. & Serio, G. (1984). A new method for calculating evolutionary substitution rates. *Journal of Molecular Evolution, 20*, 86-93.

Lane, C. S. (2005). Mass spectrometry-based proteomics in the life sciences. *Cell Mol Life Sci, 62*(7-8), 848-869.

Langerhans, R.B., Knouft, J.H., Losos, J.B. (2006). Shared and unique features of diversification in Greater Antillean Anolis ecomorphs. *Evolution, 60*(2), 362-9.

Laskowski, R.A., MacArthur, M.W., Moss, D.S. & Thornton, J.M. (1993). PROCHECK: a program to check the stereochemical quality of protein structures. *J. Appl. Cryst.* (26), 283–291.

Lathers, A., Su, M.H., Kulungowski, A., Lin, A.W., Mehta, G., Peltier, S.T., Deelman, E., & Ellisman, M.H. (2006). Enabling Parallel Scientific Applications with Workflow Tools. *Proceedings of Challenges of Large Applications in Distributed Environments* (pp. 55-60). New York: IEEE Computer Society.

Laure, E. et al. (2004, September). *Middleware for the next generation Grid infrastructure*. Paper presented at the 2004 Conference for Computing in High Energy and Nuclear Physics, Interlaken, Switzerland.

Laure, E., Fisher, S. M., Frohner, A., Grandi, C., Kunszt, P., Krenek, A., et al. (2006). Programming the Grid with gLite. *Computational Methods in Science and Technology, 12*(1), 33-45.

Laure, E., Hemmer F. et al., (2004). Middleware for the Next Generation Grid Infrastructure. *Computing in High Energy and Nuclear Physics* (CHEP), Interlaken, Switzerland.

Lavenier, D., Leroy, H., Mac Wing, M., Andonov, R., Hurfin, M., Raipin-Parvedy, P., Mouchard, L., & Guinand, F. (2003, January). GénoGRID: An Experimental Grid for Genomic Applications. In *HealthGrid 2003* Lyon.

Lavorgna, G., Dahary, D., Lehner, B., Sorek R., Sanderson, C.M. & Casari, G. (2004). In search of antisense. *Trends Biochem. Sci., 29*, 88–94.

Lavorgna, G., Triunfo, R., Santoni, F., Orfanelli, U., Noci, S., Bulfone, A., Zanetti, G. & Casari, G. (2005). AntiHunter 2.0: increased speed and sensitivity in searching BLAST output for EST antisense transcripts. *Nucleic Acids Res., 1*(33)W665-8.

Lawford-davies, J., Jenkins, J. (2005). Legal Issues to Address when Managing Clinical Information across Europe: The ECIT Case Study. *Healthgrid 2005, Studies in Health Technology and Informatics, IOS Press; 120*: 210-221.

Lawn, R.M., Schwartz, K., Patthy, L.(1997). Convergent evolution of apolipoprotein(a) in primates and hedgehog. *Proc Natl Acad Sci USA*, 94(22), 11992-7.

Lawrence, P. (1997). *Workflow handbook*. New York: John Wiley & Sons.

Lazarevic, A., & Obradovic, Z. (2002). Boosting algorithms for parallel and distributed learning. *Distributed and Parallel Databases, 11*(2), 203-229.

Lazarevic, A., Pokrajac, D., & Obradovic, Z. (2000). Distributed clustering and local regression for knowledge discovery in multiple spatial databases. *In 8th European Symposium on Artificial Neural Networks* (pp. 129-134).

Lee H., Salzemann J., Jacq N., Ho L., Chen H., Breton,V., Merelli L., Milanesi L., Lin S., & Wu Y. (2006). *Grid-enabled High Throughput in-silico Screening Against Influenza A Neuraminidase.* Proceedings of the NETTAB2006, Santa Margherita, 19–25

Lee, C. (2003). Grid Programming Models: Current Tools, Issues and Directions. *Grid Computing.* In G. F. Fran Berman, Tony Hey, Ed., pp. 555–578, Wiley Press, USA.

Lee, D., Redfern, O. & Orengo, C. (2007). Predicting protein function from sequence and structure. *Nat Rev Mol Cell Biol., 8*, 995-1005.

Lee, S., Wang, T. D., Hashmi, N., & Cummings, M. P. (2007). Bio-STEER: A semantic web workflow tool for Grid computing in the life sciences. *Future Gener. Comput. Syst., 23*(3), 497-509.

Leenaars, M., Heikkurinen, M., Louridas, M., & Karayannis, F. (2005). e-infrastructures roadmap. Espoo, Finland: e-IRGSG. Retrieved from http://www.e-irg.org/roadmap/eIRG-roadmap.pdf

Legendre, P., Anderson, M.J. (1998). DistPCOA program description, source code, executables, and documentation: http://www.bio.umontreal.ca/Casgrain/en/labo/distpcoa.html

Legendre, P., Desdevises, Y., & Bazin, E. (2002). A Statistical Test for Host-Parasite Coevolution. *Systematic Biology,* 51(2):217-234.

Legré, Y., Y., M., Viezens, F., Rienhoff, O., & Sax, U. (2008). HealthGRID/ SHARE: Grids for Health - International Interoperability. In *eHealth Conference 2007 - Experts' Special Interest Sessions* (Vol. 58) (pp. 81-99). Bonn, Germany: nanos Verlag oHG.

Lelli, F., Frizziero, E., Gulmini, M., Maron, G., Orlando, S., Petrucci, A., et al. (2007). The many Faces of the Integration of Instruments and the Grid. *International Journal of Web and Grid Services, 3*(3), 239-266.

Leo, P., Scioscia, G., Pappadà, G. & Quinto, V. (2008). *EMBL Wrapper for WebSphere Information Integrator.* Retrieved from http://www.alphaworks.ibm.com/tech/emblwrapper.

Lesk, A. (2006). *Introduction to Bioinformatics*. New York: Oxford University Press.

Leslie, C., Eskin, E., & Noble W.S. (2002). The Spectrum Kernel: A string kernel for SVM protein classification. *Pacific Symposium on Biocomputing, 7*, 566-575.

Leslie, C., et al. (2002). Mismatch string kernels for SVM protein classification. *Advanced Neural Information Processing Systems, 15*, 1441-1448.

Letondal, C. (2001). A Web interface generator for molecular biology programs in Unix. *Bioinformatics, 17,* 73-82).

Lewitt, R. M. (1992). Alternatives to voxels for image representation in iterative reconstruction algorithms. *Physics in Medicine and Biology, 3*(12), 705-716.

Li, A.W. & Murphy, P.R. (2000). Expression of alternatively spliced FGF-2 antisense RNA transcripts in the central nervous system: regulation of FGF-2 mRNA translation. *Mol. Cell. Endocrinol., 170*, 233–242.

Li, W. W., Baker, N. A., Baldridge, K., McCammon, J. A., Ellisman, M. H., Gupta, A., et al. (2006). National Biomedical Computation Resource (NBCR): Devloping End-to-End Cyberinfrastructure for Multiscale Modeling in Biomedical Research. *CTWatch Quarterly, 2,* 6-17.

Li, W. W., Krishnan, S., Mueller, K., Ichikawa, K., Date, S., Dallakyan, S., et al. (2006). *In Building Cyberinfrastructure for Bioinformatics Using Service Oriented Architecture: Cluster Computing and the Grid Workshops.* Sixth IEEE International Symposium on Cluster Computing & the Grid, Singapore.

Li, W., Byrnes, R., Hayes, J., Birnbaum, A., Reyes, V., Shahab, A., et al. (2004). The encyclopaedia of life project: grid software and deployment. *New Generation Computing, 22*, 127-136.

Li, X.-J. Zhang, H. Ranish, J.R. & Aebersold, R. (2003). Automated statistical analysis of protein abundance ratios from data generated by stable isotope dilution and tandem mass spectrometry. *Analytical Chemistry, 75*(23), 6648-57

Liberman, M.A., & Lichtenberg, A.J. (1992). *Regular and Chaotic Dynamics.* Berlin, Germany: Springer-Verlag.

Libin P., Deforche K., Van Laethem K., Camacho R. & Vandamme A-M. (2007). RegaDB: An Open Source, Community-Driven HIV Data and Analysis Management Environment. In *Reviews in Antiretroviral Therapy* (Vol. 2007-2). Fifth European HIV Drug Resistance Workshop, Cascais, Portugal, March 2007.

Lickerman, E., Bry, L., Herring, N., Pandelidis, Y., Afsheen, N., & Sharif, A. (2006). *Grid-Enabled Bio-Sample Management Application for Data-Intensive Biomarker Analysis.* Paper presented at the Proceedings of the 19th IEEE Symposium on Computer-Based Medical Systems.

Lin, B. & Dinda, PA. (2005, November). *VSched: Mixing Batch And Interactive Virtual Machines Using Periodic Real-time Scheduling.* ACM/IEEE conference on Supercomputing. New York: IEEE Computer Society

Lindahl, E., Hess, B., & van der Spoel, D. (2001). GROMACS 3.0: a package for molecular simulation and trajectory analysis. *Journal of Molecular Modeling, 7,* 306-317.

Lipinski, C.A., Lombardo, F., Dominy, B.W., & Feeney, P.J. (1997). Experimental and computational approaches to estimate solubility and permeability in drug discovery and development settings. *Adv. Drug Delivery Rev, 23,* 3-25.

Lippmann, R. P. (1987). An introduction to computing with neural nets. *IEEE ASSP Magazine, 4*(22), 7-25.

Listgarten, J. & Emili, A. (2005). Statistical and computational methods for comparative proteomic profiling using liquid chromatography-tandem mass spectrometry. *Molecular Cellular Proteomics, 4,* 419–434.

Liu, W., Schmidt, B., Voss, and G., Mueller-Wittig, K.W. (2008). Accelerating Molecular Dynamics Simulations Using Graphics Processing Units with CUDA. *Computer Physics Communications, 179*(9), 634-641.

Liu, W., Schmidt, B., Voss, G., Schroder, A., & Muller-Wittig, W. (2006). Bio-Sequence Database Scanning on a GPU. *20th IEEE International Parallel & Distributed Processing Symposium (IPDPS 2006) (HICOMB Workshop),* Rhode Island, Greece.

Lord, P. W., Stevens, R.D., et al. (2003). Investigating semantic similarity measures across the Gene Ontology: the relationship between sequence and annotation. *Bioinformatics,* 19(10), 1275-83.

Lorensen, W.E. & Cline, H. (1987). Marching cubes: A High Resolution 3D Surface Construction Algorithm. *ACM SIGGRAPH, 21*(4), 163-169.

Lučić, V., Förster, F., & Baumeister, W. (2005). Structural studies by electron tomography: From cells to molecules. *Annual Review of Biochemistry, 74*(1), 833-865.

LUCY (n.d.). *Stanford Visible Female.* Retrieved from http://lucy.stanford.edu/home.html

Ludäscher, B., Altintas, I., Berkley, C., Higgins, D., Jaeger, E., Jones, M., et al. (2006). Scientific workflow management and the kepler system. *Concurrency and Computation: Practice and Experience, 18*(10), 1039–1065.

Luengo, F., Cofino, A. S., & Gutierrez, J. M. (2004). Grid oriented implementation of self-organizing maps for data mining in meteorology. *Lecture Notes in Computer Science, 2970,* 163-171.

Luo, C., Pereira, A. L., & Chung, S. M. (2006). Distributed mining of maximal frequent itemsets on a data grid system. *Journal of Supercomputing, 37*(1), 71-90.

Luther, A., Buyya, R., Ranjan, R., & Venugopal, S. (2005). Peer-to-Peer Grid Computing and a .NET-based Alchemi Framework. High Performance Computing: Paradigm and Infrastructure. In Guo, M. (Ed.). Wiley Press, USA. www.alchemi.net

Luz, P.M., Codeço, C.T., Massad, E., Struchiner C.J. (2003, October). Uncertainties Regarding Dengue Modeling. *Mem Inst Oswaldo Cruz, 98*(7), 871-8.

Lv, C., Cao, P. Cohen, E. Li, L., & Shenker, S. (2002). Search and *replication in unstructured peer-to-peer networks,* ACM, Sigmetrics.

Macfarlane, A.; Mccann, Liddell, H. (1996). A Common Data Model For Meta-Data Interoperable Environments. In *Proceedings of II International Baltic Workshop on Database and Information Systems.*

Madoff, L., Woodall, J. (2005). The Internet and the Global Monitoring of emerging Diseases: Lessons from the First 10 Years of ProMED-mail. *Archives of Medical Research, 36,* 724-730.

Main, P., Fiske, S.J., Hull, S.E., Lessinger, L., Germain, G., Declercq, J.P. & Woolfson, M.M. (1980). *MULTAN80: a system of computer programs for the automatic solution of crystal structures from x-ray diffraction data* (Universities of York and Louvain).

Malawski, M., Bartynski, T., & Bubak, M. (2008). A tool for building collaborative applications by invocation of Grid operations. In Bubak, M., van Albada, D. G., Dongarra, J., and Sloot, P. M. A., (Eds.), *Computational science - ICCS 2008, 8th International Conference, Kraków, Poland, June 23-25, 2008, Proceedings, part III,* (LNCS 5103 pp. 243-252). Berlin, Germany: Springer.

Malawski, M., Bubak, M., Placek, M., Kurzyniec, D. & Sunderam, V. (2006). Experiments with distributed component computing across Grid boundaries. In *Proceedings of the HPC-GECO/CompFrame workshop in conjunction with HPDC 2006*, (pp. 109-116), Paris, France.

Malawski, M., Gubala, T., Kasztelnik, M., Bartynski, T., Bubak, M., Baude, F., & Henrio, L. (2008). High-level scripting approach for building component-based applications on the grid. In *CoreGRID Workshop on Grid Programming Model Grid and P2P Systems Architecture Grid Systems, Tools and Environments*, Heraklion, Crete. Berlin, Germany: Springer.

MammoGrid (2008). *European federated mammogram database implemented on a grid structure.* Retrieved March 8, 2008 from http://mammogrid.vitamib.com

Manavski, S.A., & Valle, G. (2008). CUDA compatible GPU cards as efficient hardware accelerators for Smith-Waterman sequence alignment. *BMC Bioinformatics, 9*(S2), S10.

Manocha, D. (2005). General Purpose computations using graphics processors. *IEEE Computer, 20*(7), 85-88.

Manssour, I.H. & Dal Sasso Freitas, C.M. (2000). Collaborative Visualization in Medicine. In *Proc. 8'th International Conference in Central Europe on Computer Graphics, Visualization and Interactive Digital Media (WSCG2000), Plzen*, (pp. 266-273).

Maojo V, de la Calle G, Martín-Sánchez F, Díaz C, Sanz F. (2005). INFOBIOMED: European Network of Excellence on Biomedical Informatics to support individualised healthcare. *AMIA Annual Symposium Proceedings:* 1041.

Marabini, R., Herman, G. T., & Carazo, J. M. (1998). 3D reconstruction in electron microscopy using ART with smooth spherically-symmetric volume elements (blobs). *Ultramicroscopy, 72*(1), 53-65.

Maran U., Sild S., Kahn I., & Takkis K. (2007). *Future Gen. Comput. Syst. 23*, 76-83

Maran U., Sild S., Mazzatorta P., Casalegno M., Benfenati E., & Romberg M. (2007). In Dubitzky W., Schuster A., Sloot P. M. A., Schhroeder M., Romberg M. (Eds.), *Grid Computing in Computational Biology*, (LNCS 4360, pp. 60–74). Berlin Heidelberg: Springer-Verlag.

Maran, U., Sild, S., Mazzatorta, P., Casalegno, M., Benfenati, E., & Romberg, M. (2007). Grid Computing for the Estimation of Toxicity: Acute Toxicity on Fathead Minnow (Pimephales promelas). In *Distributed, High-Performance and Grid Computing in Computational Biology* (pp. 60-74).

Marinescu, D. C., & Ji, Y. (2003). A computational framework for the 3D structure determination of viruses with unknown symmetry. *Parallel and Sitributed Computing, 63*(1), 738-758.

Markus, W. & Manfred, J. (2007). ProSA-web: interactive web service for the recognition of errors in three-dimensional structures of proteins. *Nucleic Acids Res.* (35), 407–410.

Marrink, S. J., de Vries, A. H., & Mark, A. E. (2004). Coarse grained model for semiquantitative lipid simulations. *Journal of Physical Chemistry B, 108,* 750-760.

Marsden, R.L., Lewis, T.A. & Orengo, C. (2007). Towards a comprehensive structural coverage of completed genomes: a structural genomics viewpoint. *BMC Bioinformatics. 8*, 86-103.

Martelli, P.L., Fariselli, P. & Casadio, R. (2003). An ENSEMBLE machine learning approach for the prediction of all-alpha membrane proteins. *Bioinformatics 19*(1), 205–211.

Martens, L., Hermjakob, H. Jones, P. Adamski, M. Taylor, C. States, D. Gevaert, K. Vandekerckhove, J. & Apweiler, R. (2005). Pride: the proteomics identifications database. *Proteomics, 5*(13), 3537-3545

Martin, D., Brun, C., Remy, E., Mouren, P., Thieffry, D., Jacq, B. (2004). GOTool-Box: functional analysis of gene datasets based on Gene Ontology. *Genome Biol, 5*, R101.

Martino, R. L. (2006). Parallel Computing in the Analysis of Gene Expression Relationships. In Zomaya A. (Ed.), *Parallel Computing for Bioinformatics and Computational Biology*, (pp. 265-284). Berlin, Germany: Wiley.

Martonne, M. (2005). The BIRNLex: Principles and practices of community ontology development. *Re-use or*

Re-invention - a Roadmap for Data Integration (Schizophrenia as a Test Case). NeSC Events. Retrieved July 2008 from http://www.nesc.ac.uk/action/esi/download.cfm?index=3359

Mastroianni, C., Talia, D. & Trunfio, P. (2004). Metadata for managing grid resources in data mining applications. *Journal of Grid Computing, 2*(1), 85-102.

Mattevi, A., Vanoni, M.A., Tosone, F., Rizzi, M., Teplyakov, A., Coda, A., Bolognesi, M., Curti, B. (1996). Crystal structure of D-amino acid oxidase: a case of active site mirror-image convergent evolution with flavocytochrome b2. *Proc Natl Acad Sci USA, 93*(15), 7496-501.

Matthiesen, R. (2007). Methods, algorithms and tools in computational proteomics: A practical point of view. *Proteomics, 7*, 2815-2832.

Mawudeku A., Blench M. (2006). Global Public Health Intelligence Network (GPHIN). In *Proceedings of the 7th Conference of the Association for Machine Translation in the Americas*. Retrieved February 11, 2008 from www.mt-archive.info/MTS-2005-Mawudeku.pdf

McClatchey, R., Manset, D., & Solomonides, T. (2006). Lessons learned from MammoGrid for integrated biomedical solutions. In *19th IEEE Int. Symposium on Computer-Based Medical Systems, CBMS 2006.*

McGinnis, L., Wallom, D., & Gentzsch, W. (Eds.)(2007). 2nd International Workshop on Campus and Community Grids. Report retrieved from http://forge.gridforum.org/sf/go/doc14617?nav=1

McIlraith, S.A., Son, T.C. & Zeng, H. (2001). Semantic web services. *IEEE Intelligent Systems, 16*(2), (46–53.

McMullen, D.F., Devadithya, T., & Chiu, K. (2005). *Integrating Instruments and Sensors into the Grid with CIMA Web Services.* Paper presented at the Proceedings of the Third APAC Conference on Advanced Computing, Grid Applications and e-Research.

Medical Dictionary for Regulatory Activities (2008). Retrieved March 8, 2008, from http://www.meddramsso.com/MSSOWeb/index.htm

MediGRID. (2005). *www.medigrid.de*. Retrieved March 14, 2007.

Meier-Kolthoff, J.P, Auch, A.F., Huson, D.H., & Göker, M.(2007). COPYCAT: Co-phylogenetic Analysis tool. *Bioinformatics*, 23(7):898-900.

Meinilä, M., Kuusela, J., Zietara, M.S., & Lumme, J. (2004). Initial steps of speciation by geographic isolation and host switch in salmonid pathogen *Gyrodactylus salaris (Monogenea: Gyrodactylidae). International Journal of Parasitology,* 34(4):515-526

Melville, J., Harmon, L.J., Losos, J.B. (2006). Intercontinental community convergence of ecology and morphology in desert lizards. *Proc Biol Sci., 273*(1586), 557-63.

Merelli, I. & Milanesi, L. (2005) *High performance implementation of BLAST using GRID technology.* Paper presented at the meeting of the Italian Bioinformatics Society, Milan, Italy.

Merkle, D., & Middendorf. M. (2005). Reconstruction of the cophylogenetic history of related phylogenetic trees with divergence timing information. *Theory in Biosciences,* 123(4):277-299.

Metropolis, N., Rosenblath, A.W., Rosenblath, M.N., Teller A.H., & Teller, E. (1953). Equations of State Calculations by Fast Computing Machines. *J. Chem. Physics, 21*, 1087-1091.

Mignone, F., Grillo, G., Licciulli, F., Iacono, M., Liuni, S., Kersey, P.J., Duarte, J., Saccone C. & Pesole G. (2005). UTRdb and UTRsite: a collection of sequences and regulatory motifs of the untranslated regions of eukaryotic mRNAs. *Nucleic Acids Res., 33*, D141-D146.

Mikler, S. Venkatachalam, A.& (2005). Towards computational epidemiology: Using stochastic cellular automata in modeling spread of diseases. *Proceedings. of the 4th Annual International Conference on Statistics.* Retrieved July 2008 from http://www.cerl.unt.edu/publications/2005/slides/ppt/Hawaii.ppt

Milanesi, L., & Merelli, I. (2006). High Performance GRID Based Implementation for Genomics and Protein Analysis. *Studies in health technology and informatics, 120*, 374-380.

Miles, S., Wong, S.C., Fang, W., Groth, P.T., Zauner, K.P. & Moreau, L. (2007) Provenance-based validation of e-science experiments. *J. Web Sem., 5*(1), 28–38.

Miller, R., Bednasz, J.J., Chiu, K., Gallo, S.M., Govindaraju, M., Lewis, M., Ruby, C., & Weeks, C.M. (2008). Grid-based research, development, and deployment in New York State, In *Proceedings of the International Parallel and Distributed Processing Symposium*, in press.

Miller, R., DeTitta, G. T., Jones, R., Langs, D. A., Weeks, C. M. & Hauptman, H. A. (1993). On the application of the minimal principle to solve unknown structures, *Science*, 259, 1430–1433.

Miller, R., Gallo, S. M., Khalak, H. G. & Weeks, C. M. (1994). *SnB*: Crystal structure determination via *Shake-and-Bake*. *J. Appl. Cryst.*. *27*, 613–621.

Minar, N., Burkhart, R., Langton, C., & Askenazi, M. (1996). *The swarm simulation system, a toolkit for building multi-agent simulations*, from http://www.santafe.edu.

Mirto, M., Cafaro, M., Fiore, S., Tartarini, D. & Aloisio, G. (2007, June). A Grid-enabled Protein Secondary Structure Predictor. *IEEE Transactions on NanoBioscience*, *6* (2),124-130.

Mirto, M., Ferramosca, A., Tartarini, D., Romano S., Negro A., Tasco, G., Fiore, S., Zara, V., Casadio, R., & Aloisio, G. (2008). A Protein Structure Prediction Service in the ProGenGrid System. *Global HealthGrid: e-Science Meets Biomedical Informatics Proceedings of HealthGrid 2008*, (pp 135-146), Chicago, June 2-4.

Mirto, M., Fiore, S., Epicoco, I., Cafaro, M., Mocavero, S., Blasi, E. & Aloisio G. (2008). A Bioinfomatics Grid Alignment Toolkit. *Future Generation Computer Systems*, *24*(7), 752-762. Philadelphia: Elsevier

Mirto, M., Rossi, I., Epicoco, I., Fiore, S., Fariselli, P., Casadio, R. & Aloisio, G. (2007). High Throughput Protein Similarity Searches in the LIBI Grid Problem Solving Environment. *Proceedings of the 5th International Symposium on Parallel and Distributed Processing and Applications (ISPA07)*, (pp. 414-423), Niagara Falls, Canada,.

Missier, P., Belhajjame, K, Zhao, J. & Goble, C.A. (2008). Data lineage model for Taverna workflows with lightweight annotation requirements. *IPAW 2008*. New York: Springer.

Missier, P., Embury, S.M. & Stapenhurst, R. (2008). Exploiting provenance to make sense of automated data acceptance decisions in scientific workflows. *IPAW 2008*. Springer.

MitoRes: bio-sequences resource for mitochondria-related genes, transcripts and proteins. Retrieved from http://www2.ba.itb.cnr.it/MitoRes/.

Mohamed, H.H., & Epema, D.H.J. (2004). An Evaluation of the Close-to-Files Processor and Data Co-Allocation Policy in Multiclusters. In *Cluster*, (pp. 287-298). New York: IEEE Computer Society Press.

Mohammed, Y. (2006). Erweiterte Sicherheit und Datenschutz Techniken. In U. Sax, Y. Mohammed, F. Viezens & O. Rienhoff (Eds.), *Grid-Computing in der Biomedizinischen Forschung – Datenschutz und Datensicherheit* (Vol. 90, pp. 70-77). München, Germany: Urban &Vogel.

Mohammed, Y., Sax, U., Viezens, F., & Rienhoff, O. (2007). Shortcomings of Current Grid Middlewares Regarding Privacy in HealthGrids. *Stud Health Technol Inform*, *126*, 322-329.

Montagnat J., Frohner A., Jouvenot D., Pera C., Kunszt P., Koblitz B., et al. (2008). A Secure Grid Medical Data Manager Interfaced to the gLite Middleware. *Journal of Grid Computing (JGC)*, *6* (1): 45-59.

Montagnat, J., Bellet, F., Benoit-Cattin, H., Breton, V., Brunie, L., Duque, H., et al. (2004). Medical images simulation, storage, and processing on the European datagrid testbed. *Journal of Grid Computing*, *2* (4), 387-400.

Montagnat, J., Breton, V., & Magnin, I. (2003). Using Grid technologies to face medical image analysis challenges. In *Proceedings of IEEE/ACM CCGrid'03*, (pp. 588-593). New York: IEEE Computer Society.

Montagnat, J., Frohner, Á., Jouvenot, D., Pera, C., Kunszt, P., Koblitz, B., et al. (2008). A Secure Grid Medical Data Manager Interfaced to the gLite Middleware. *Journal of Grid Computing*, *6*(1), 45-59.

Moore R. (2001). Data Management Systems for Scientific Applications. In Boisvert R., Tang P. (Eds.) *The Architecture of Scientific Software* (pp. 273-284). Norwell, MA: Kluwer Academic Publishers.

Moore, GA. (1991). *Crossing the Chasm: Marketing and Selling High-Tech Products to Mainstream Customers.* New York: Harper Business.

Morgan, M., Chue-Hong, N., & Drescher, M., (2006). *ByteIO Specification 1.0.* Retrieved on July 10, 2008 from https://forge.gridforum.org/sf/docman/do/download/Document/ projects.byteiowg/docman.root.current_documents/doc13719/1

Mori, S. (2007). *Introduction to Diffusion Tensor Imaging.* Amsterdam, the Netherlands: Elsevier.

Morris G. M., Goodsell D. S., Halliday R.S., Huey R., Hart W. E., Belew R. K., & Olson A. J. (1998). Automated Docking Using a Lamarckian Genetic Algorithm and and Empirical Binding Free Energy Function. *J. Computational Chemistry, 19*, 1639-1662.

Moscicki, J.T., Lee, H.C., Guatelli, S., Lin, S.C. & Pia, M.G. (2004). *A Biomedical Applications on the GRID: Efficient Management of Parallel Jobs.* Paper presented at the Nuclear Science Symposium, Geneve, Switzerland.

Mozilla (2008). Retrieved on July 10, 2008, from http://www.mozilla.org/newlayout/doc/

Mulder, N. J., Kersey, P., et al. (2008). In Silico Characterization of Proteins: UniProt, InterPro and Integr8. *Mol Biotechnol, 38*(2), 165-77.

Murdock, S. E., Tai, K., Ng, M. H., Johnston, S., Wu, B., Fangohr, H., Laughton, C. A., Essex, J. W., & Sansom, M. S. P. (2006). Quality assurance for biomolecular simulations. *Journal of Chemical Theory and Computation, 2,* 1477-1481.

MyGrid (2008). Retrieved from www.mygrid.org.uk

Myhre, S., Tveit, H., et al. (2006). Additional gene ontology structure for improved biological reasoning. *Bioinformatics, 22*(16), 2020-7.

Nagesh, H., Goil, S., & Choudhary, A. (1999). *Mafia: Efficient and scalable subspace clustering for very large data sets* (Tech. Rep. No. 9906-010), Center for Parallel and Distributed Computing.

Narayan, R. M. (2002). Clinical Trials in Head Injury. *Journal of Neurotrauma, 19*, 503-557.

Naseer, A. & Stergioulas, L. K. (2006). Resource discovery in grids and other distributed environments: States of the art. *Multiagent and Grid Systems - An International Journal, 2*(2), 163-182.

Naseer, A. & Stergioulas, L. K. (2006). "A Taxonomy of HealthGrids and HealthGrid Resources". *11th International Symposium on Health Information Management Research* (iSHIMR2006), Halifax, Canada (pp. 93-99).

Naseer, A. and Stergioulas, L. K. (2006). Discovering HealthGrid services. In *Proceedings of the IEEE International Conference on Services Computing (SCC'06), Chicago, IL*, (pp.301-306).

Naseer, A. and Stergioulas, L. K. (2007). Combining web services and grid services: practical approaches and implications to resource discovery. In Periorellis, P. (Ed.), *Securing Web Services: Practical Usage of Standards and Specifications,* (pp. 298-317). Hershey, PA: IGI Global..

National Center for Biotechnology Information. Retrieved from http://www.ncbi.nlm.nih.gov/BLAST/.

National Institutes of Health (2000). *Working definition of bioinformatics and computational biology.* Retrieved March 17, 2008, from http://www.bisti.nih.gov/CompuBioDef.pdf

National Toxicology Program server at http://ntp-server.niehs.nih.gov/

Nature (2007). Editorial: The database revolution. *Nature, 445,* 229-230.

NCBI Blast (2007). Retrieved July 2008 from http://www.ncbi.nlm.nih.gov/blast/Blast.cgi.

NCBI. Ncbi website. Retrieved from http://www.ncbi.nih.gov/

NCSA (2007). *MyProxy Credentials Management Service.* Retrieved July 2008 from http://grid.ncsa.uiuc.edu/myproxy/

Needleman, S., Wunsch, C. (1970). A general method applicable to the search for similarities in the amino acid sequence of two proteins. *J Mol Biol, 48*(3), 443-53.

NEESgrid (2008). Retrieved from www.nees.org/

NEMA – ACR (2008). *DICOM Standard*. Retrieved July 2008 from http://medical.nema.org/

Nesvizhskii, A. I., Keller, A., Kolker, E. & Aebersold, R. (2003). A statistical model for identifying proteins by tandem mass spectrometry. *Analytical Chemistry, 75*(17): 4646-4658.

Nesvizhskii, A. I., Vitek, O. & Aebersold, R. (2007). Analysis and validation of proteomic data generated by tandem mass spectrometry. *Nat Methods, 4*(10). 787-797.

Neto V. C., Gagliardi H. F., Rezende A., Gonçalves E. S. , Gallo E. , Barbosa F. A., Torres I., Alves D. (2006). Data Access Service in a Computational Grid Platform Applied to the Monitoring and Control of Epidemics on Georeferenced Dynamic Maps. *Proceedings of the e-Science conference*. (p. 142).

Neto, V.C., Gagliardi, H. F., Rezende, A., Gonçalves, E.S., Gallo, E., Silva, F.A.B., Pisa, I.T., Alves, D. (2006). Data Access Service in a Computational Grid Platform Applied to the Monitoring and Control of Epidemics on Georeferenced Dynamic Maps. In *Proceedings of the 2nd IEEE Conference on e-Science and Grid Computing*. New York: IEEE Computer Society Press.

NeuGrid Project (2008). Retrieved October 20, 2008, from http://www.neugrid.eu/.

Neumann, U. (1994). Communication Costs for Parallel Volume-Rendering Algorithms. *IEEE Computer Graphics and Applications, 15*(4), 49-58.

Neuroth H., Kerzel, M., & Gentzsch, W. (Eds) (2007). *Die D-Grid Initiative*. Goettingen, Germany: Goettingen University Press. Retrieved from http://webdoc.sub.gwdg.de/univerlag/2007/D-Grid_de.pdf

Neuroth, H., Kerzel, M., & Gentzsch, W. (Eds.). (2007). German Grid Initiative D-Grid. Universitätsverlag Göttingen Publishers. Retrieved from www.d-grid.de/index.php?id=4&L=1

Nevill-Manning, C.G., Wu, T.D., & Brutlag, D.L. (1998). Highly Specific Protein Sequence Motifs for Genome. *Proceedings of the National Academy of Sciences USA, 95*, 5865-5871.

Ng, M. H., Johnston, S., Wu, B., Murdock, S. E., Tai, K., Fangohr, H., Cox, S. J., Essex, J. W., Sansom, M. S. P., & Jeffreys, P. (2006). BioSimGrid: Grid-enabled biomolecular simulation data storage and analysis. *Future Generation Computer Systems, 22*, 657-664.

Niederberger R. & Alessandrini V. (2004). The DEISA Project: Motivations, Strategies, Technologies. In *International Supercomputer Conference*, Heidelberg, Germany.

Nielsen, S. O., Lopez, C. F., Srinivas, G., & Klein, M. L. (2004). Coarse grain models and the computer simulation of soft materials. *Journal of Physics: Condensed Matter, 16*, R481-R512.

NIH. (2008). *Models of Infectious Disease Agent Study*. Retrieved July 2008 from http://www.nigms.nih.gov/Initiatives/MIDAS

NLM (n.d.). *Visible Human Project*. Bethesda, MD: National Library of Medicine., Retrieved from http://www.nlm.nih.gov/research/visible/visible_human.html

Notredame, C., Higgins, D.G. & Heringa, J. (2000). T-coffee: a novel method for fast and accurate multiple sequence alignment. *J. Mol. Biol., 302*, 205–217.

Novotny J., Tuecke S., & Welch V. (2001). *An Online Credential Repository for the Grid: MyProxy*. HPDC, 10th IEEE International Symposium on High Performance Distributed Computing (HPDC-10 '01) (p. 0104).

Novotny, J. (2003). The grid portal development kit. In F. Berman, G. Fox, & A. Hey, *Grid Computing: Making the Global Infrastructure a Reality* (pp. 657-674). Hoboken, NJ: John Wiley and Sons.

Novotny, J., Russell, M., & Wehrens, O. (2004). GridSphere: a portal framework for building collaborations. *Concurrency and Computation: Practice and Experience, 16*, 503-513.

Novotny, J., Russell, M., & Wherens, O. (2004). Gridsphere: A Portal framework for building collaborations. *Concurrency and Computation: Practice and Experience* (pp. 503-513). Hoboken, NJ: Wiley & Sons.

Noy, N. F. (2004). Semantic integration: A survey of ontology-based approaches. *SIGMOD Record, 33*(4), 65-70.

NRCAM (n.d.). *The National Resource For Cell Analysis and Modeling*. Farmington, CT: University of Connecticut Health Center. Retrieved from http://www.nrcam.uchc.edu

NVIDIA Compute Unified Device Architecture-Programming Guide (2007). Retrieved from http://developer.download.nvidia.com/compute/cuda/1_0/NVIDIA_CUDA_Programming_Guide_1.0.pdf.

O'Farell, P.H. (1975). High-resolution two-dimensional gel electrophoresis of proteins. *Journal of Biolical Chemistry, 250*(10), 4007-4021.

O'Reilly, T. (2005). *What is Web 2.0: Design patterns and business models for the next generation of software*. Retrieved on April 10, 2008 from http://www.oreillynet.com/pub/a/oreilly/tim/news/2005/09/30/what-is-web-20.html.

OASIS (2003). *Assertions and Protocol for the OASIS Security Assertion Markup Language (SAML) v1.1.* Retrieved from http://www.oasis-open.org/committees/security/

OASIS Consortium (2008). Retrieved July 20, 2008, from http://www.oasis-open.org.

OGF - Open Grid Forum (2008). Retrieved July 20, 2008, from http://www.ogf.org.

OGF (2008). Open Grid Forum. Retrieved from www.ogf.org

OGSA-DAI (2008) Open Grid Service Architecture – OGSA-DAI Documentation. Retrieved February 20, 2008, from http://www.ogsadai.org.uk/documentation/.

OGSA-DAI project (2007). *OGSA-DAI.* 2007. Retrieved July 2008 from http://www.ogsadai.org.uk/

Oinn, T., Addis, M., Ferris, J., Marvin, D., Senger, M., Greenwood, M., et al. (2004). Taverna: A tool for the composition and enactment of bioinformatics workflows. *Bioinformatics, 20*(17), 3045-3054.

Oinn, T., Greenwood, M., Addis, M., Nedim Alpdemir, M., Ferris, J., et al. (2006). Taverna: lessons in creating a workflow environment for the life sciences. *Concurrency and Computation: Practice and Experience, 18*(10) 1067-1100.

Oinn, T.M, Greenwood, R.M., Addis, M., Alpdemir, N., Ferris, J., Glover, K., et al. (2006). Taverna: lessons in creating a workflow environment for the life sciences. *Concurrency and Computation: Practice and Experience, 18*(10) 1067–1100.

Oinn, T.M., Addis, M. Ferris, J., Marvin, D. Greenwood, R.M. Carver, T. Pocock, M.R. Wipat, A., & Li P.(2004) Taverna: A tool for the composition and enactment of bioinformatics workflows. *Bioinformatics, 20*(17), 3045–3054

Olesen, J. L. M. (2003). The burden of brain disease in Europe. *European Journal of Neurology, 10*, 471-477.

Oliveira, M. C., Cirne, W., & de Azevedo Marques, P. M. (2007). Towards applying content-based image retrieval in the clinical routine. *Future Gener. Comput. Syst., 23* (3), 466-474.

Oliver, T., Schmidt, B., & Maskell, D. (2005). Reconfigurable Architectures for Bio-sequence Database Scanning on FPGAs. *IEEE Transactions on Circuits and Systems II, 52*(12), 851-855.

Open Grid Services Architecture - Data Access and Integration Services. Retrieved March 13, Accessed 2006, from www.ogsadai.org.uk/

OpenDX. Retrieved from http://www.opendx.org/

OpenEHR (2007). Retrieved July 2008 from http://www.openehr.org/home.html

Operation and maintenance manual for the mock circulatory system. (1965). Rockville, MD: Hydrospace Research Corporation.

Orchard, S., & Hermjakob, H., (2008) The HUPO proteomics standards initiative easing communication and minimizing data loss in a changing world. *Briefing in Bioinformatics, 9*(2), 166-173

Orchard, S., Hermjakob, H., Binz, P.A., Hoogland, C., Taylor, C.F., Zhu, W., Julian, R.K., & Apweiler R., (2005) Further steps towards data standardization: The Proteomic standard initiative. *Proteomics, 5*(2), 337-339

Oren, E., Haller, A., Hauswirth, M., Heitmann, B., Decker, S., & Mesnage, C. (2007). A flexible integration framework for semantic web 2.0 applications. *IEEE Software, 24*(5), 64-71.

Osorio A., Valette PJ., Mihalcea A., Atif J. & Ripoche X. (2002) *A new PC based software to perform semi-automatic hepatic segmentation using CT and MR images.* Presented at InfoRAD-RSNA, Chicago, IL.

Oster, S., Langella, S., Hastings, S., Ervin, D., Madduri, R., Phillips, J., et al. (2007). caGrid 1.0: An Enterprise Grid Infrastructure for Biomedical Research. *Journal of the American Medical Informatics Association, 15,* 138-149.

Ott. E. (1993). *Chaos in Dynamical Systems.* New York: Cambridge University Press.

Ourselin, S., Roche, A., Prima, S. & Ayache, N. (2000). Block Matching: A General Framework to Improve Robustness of Rigid Registration of Medical Images. In proc. of *MICCAI 2000*, , Pittsburgh, PA, (LNCS 1935, pp. 557-566). Berlin, Germany: Springer.

Owens, J., Luebke, D., Govindaraju, N., Harris, M., Krüger, J., Lefohn, A.E., & Purcell, T.J. (2007). A survey of general-purpose computation on graphics hardware. *Computer Graphics Forum, 26*(1), 80-113.

Palmieri, F. (2004). The mitochondrial transporter family (SLC25): physiological and pathological implications. *Pflugers Arch., 447,* 689–709.

Palmieri, L., Palmieri, F., Runswick, M. & Walker, J. (1996). Identification by bacterial expression and functional reconstitution of the yeast genomic sequence encoding the mitochondrial dicarboxylate carrier protein. *FEBS Lett.* (399), 299–302.

Pang, A., Arinaminpathy, Y., Sansom, M. S. P., & Biggin, P. C. (2005). Comparative molecular dynamics – similar folds and similar motions? *Proteins: Structure, Function, and Bioinformatics, 61,* 809-822.

Pantalos, G., Koenig, S., Gillars, K., Giridharan, G., & Ewert, D. (2004). Characterization of an Adult Mock Circulation for Testing Cardiac Support Devices. *ASAIO Journal, 50*(1), 37-46.

Papazoglou, M.P. (2003). Service-oriented computing: concepts, characteristics and directions. In *Proceedings of the 4th International Conference on Web Information Systems Engineering, WISE 2003,* Roma, Italy.

Pappin, D. J., Hojrup, P. & Bleasby, A. J. (1993). 'Rapid identification of proteins by peptide-mass fingerprinting. *Curr Biol 3*(6), 327-332.

Park, J., et al. (1998). Sequence comparisons using multiple sequences detect twice as many remote homologues as pairwise methods. *Journal of Molecular Biology, 284,* 1201-1210.

Parker, J. (2007). A Flexible, Large-Scale, Distributed Agent Based Epidemic Model. *Proceedings of the 2007 Winter Simulation Conference.* New York: IEEE Press.

Paterson, S., Viney M. E. (2000). The interface between epidemiology and population genetics. *Parasitology Today, 16*(12), 528.

Patil, A.A., Oundhakar, S.A., Sheth, A.P. & Verma; K. (2004) Meteor-s web service annotation framework. In S. I. Feldman, M. Uretsky, M. Najork, & C. E. Wills (Eds.), *WWW,* (pp. 553–562). New York; ACM.

Pautasso, C., Bausch, W., & Alonso, G. (2006). Autonomic Computing for Virtual Laboratories. In J. Kohlas, B. Meyer, A. Schiper (Eds.), *Dependable Systems: Software, Computing, Networks,* (LNCS 4028). Berlin, Germany: Springer Verlag.

Peachey T., Abramson D., & Lewis A. (2006). *Model Optimization and Parameter Estimation with Nimrod/O.* The International Conference on Computational Science, May 28-31, University of Reading, UK. Retrieved from: http://www.csse.monash.edu.au/~davida/papers/iccs06.pdf

Pearlman, D. A., Case, D. A., Caldwell, J. W., Ross, W. S., Cheatham III, T. E., DeBolt, S., Ferguson, D., Seibel G., & Kollman, P. (1995). AMBER, a computer program for applying molecular mechanics, normal mode analysis, molecular dynamics and free energy calculations to elucidate the structures and energies of molecules. *Computer Physics Communications, 91,* 1-41.

Pearlman, L. W., V. Foster, I. Kesselman, C. Tuecke, S. (2002). A Community Authorization Service for Group Collaboration. In IEEE Computer Society, TC on Distributed Processing, Naval Postgraduate School (U.S.) & United States, Office of Naval Research. (Eds.), *Third International Workshop on Policies for Distrib-*

uted Systems and Networks, June 5-7, 2002, Monterey, California, USA: proceedings (pp. 50-59). Los Alamitos, CA.: IEEE Computer Society.

Pearson, W.R. (2000). Flexible sequence similarity searching with the FASTA3 program package. *Methods in Molecular Biology, 132*, 185-219.

Pearson, W.R., & Lipman, D.J. (1988). Improved Tools for Biological Sequence Comparison. *Proceedings of the National Academy of Sciences of the United States of America, 85*(8), 2444-2448.

Pebay-Peyroula, E., Dahout-Gonzalez, C., Kahn, R., Trezeguet, V., Lauquin, G. & Brandolin, G. (2003). Structure of mitochondrial ADP/ATP carrier in complex with carboxyatractyloside r. *Nature* (426), 39–44.

Pedretti, K.T., Casavant, T.L., Braun, R.C., Scheetz, T.E., Birkett, C.L., & Roberts, C.A. (1999). Three Complementary Approaches to Parallelization of Local BLAST Service on Workstation Clusters. In *PaCT '999: Proceedings of the 5th International Conference on Parallel Computing Technologies*, (pp. 271-282). London: Springer-Verlag.

Pedrioli, P. G. A., Eng, J. K., Hubley, R., Vogelzang, M., Deutsch, E. W., Raught, B., et al. (2004). A common open representation of mass spectrometry data and its application to proteomics research. *Nature Biotechnology, 22*(11), 1459+.

Peltier, ST. & Ellisman MH. (2003). The Biomedical Informatics Research Network (BIRN). In: I Foster and C Kesselman (Eds.), *The Grid: Blueprint for a New Computing Infrastructure* (pp 109-120). San Francisco, CA: Morgan Kaufmann Publishers.

Penedo, M., Souto, M., Tahoces, P.G., Carreira, J.M., Villalón, J., Porto, G., et al. (2005). Free-response receiver operating characteristic evaluation of lossy JPEG2000 and object-based set partitioning in hierarchical trees compression of digitized mammograms. *Radiology, 237*, 450-457.

Pennec, X & Thirion, J-P. (1997). A Framework for Uncertainty and Validation of 3D Registration Methods based on Points and Frames. *Int. Journal of Computer Vision, 25*(3) 203-229.

Pennec, X., Ayache, N. & Thirion, J.P. (2000). Landmark-based registration using features identified through differential geometry. In I. Bankman, (Ed.), *Handbook of Medical Imaging*, (pp. 499-513). Academic Press.

Pereira-Leal, J. B., Audit, B., Peregrin-Alvarez, J. M., & Ouzounis C. A. (2005). An Exponential Core in the Heart of the Yeast Protein Interaction Network. *Molecular Biology and Evolution, 22*(3), 421-425.

Perelmutov, T., Petravick, D., Gu, J., Barring, O., Baud, J.-P., De Witt, S., et al. (2002, May). *SRM Interface Specification v2.2* (FNAL technical report). Retrieved from http://sdm.lbl.gov/srm-wg/doc/SRM.v2.2.html

Pérez, M. S., Sànchez, A., Herrero, P., Robles, V., & Pena, J. M. (2005). Adapting the weka data mining toolkit to a grid based environment. *In 3rd Atlantic Web Intelligence Conference* (pp. 492-497).

Perkins, D. N., Pappin, D. J., Creasy, D. M., & Cottrell, J. S. (1999). Probability-based protein identification by searching sequence databases using mass spectrometry data. *Electrophoresis*, 20(18), 3551-3567.

Perlman, L., Welch, V., Foster, I., Kesselman, C., & Tuecke, S. (2002). *In A Community Authorization Service for Group Collaboration*. IEEE 3rd International Workshop on Policies for Distributed Systems and Networks.

Pesole, G., Liuni, S. & et al (1999). UTRdb: a specialized database of 5' and 3' untranslated regions of eukaryotic mRNAs. *Nucleic Acids Res.*, (27):188–191. http://www.ba.itb.cnr.it/UTR/.

Peters, T.J., Greenshields, I.R. & Dorney, S. (1992). Topological Fidelity in Surface Reconstruction, Curves and Surfaces in Computer Vision and Graphics III. In *Proc. SPIE* (Vol. 0277-786X), (pp. 221-225).

Petersen, K., Spreitzer, M., Terry, D., Theimer, M., & Demers, A. (1997). *Flexible Update Propagation for Weakly Consistent Replication*, Paper presented at the 16th ACM Symposium on Operating System Principles, Saint-Malo, France.

Petricoin, E. F., Ardekani, A. M., Hitt, B. A., Levine, P. J., Fusaro, V. A., Steinberg, S. M., et al. (2002). Use of proteomic patterns in serum to identify ovarian cancer. *Lancet, 359*(9306), 572-577

P-GRADE (2003). Parallel Grid Run-time and Application Development Environment. Retrieved from www.lpds.sztaki.hu/pgrade/

Phillips, J. C., Braun, R., Wang, W., Gumbart, J., Tajkhorshid, E., Villa, E., Chipot, C., Skeel R. D., Kale, L. & Schulten, K (2005). Scalable molecular dynamics with NAMD. *Journal of Computational Chemistry, 26,* 1781-1802.

Piggott, D., Teljeur, C. & Kelly, A. (2004). Exploring the potential for using the grid to support health impact assessment modelling. *Parallel Computing, 30*(9-10), 1073-1091.

Piper, I., Citerio, G., & the BrainIT Group (2003). The BrainIT Group: Concept and Core Dataset Definition. *Acta Neurochirurgica, 145,* 615-629

Podvinec, M., Maffioletti, S., Kunszt, P., Arnold, K., Cerutti, L., Nyffeler, B., et al. (2006). The SwissBioGrid Project: Objectives, Preliminary Results and Lessons Learned. *2nd IEEE International Conference on e-Science and Grid Computing (e-Science 2006) - Workshop on Production Grids.*

Podvinec, M., Schwede, T., Cerutti, L., Kunszt, P., Nyffeler, B., Stockinger, H., et al (2007). The SwissBioGrid Project: Objectives, Preliminary Results and Lessons Learned. Proceedings of the *2nd IEEE International Conference on e-Science and Grid Computing,* Amsterdam, Netherlands

Politi L, Groppi, A. & Polettini, A. (2005). Review: Applications of liquid chromatography- mass spectrometry in doping control. *Journal of Analytical Toxicology, 29,* (1), 1–14

Pollastri, G., Vullo, A., Fiasconi, P. & Baldi, P. (2006). Modular DAG-RNN Architectures for Assembling Coarse Protein Structures. *J. Comp. Biol., 13*(3), 631-650.

Polychroniadou, H., Psomopoulos, F., & Miktas, P. (2006). G-Class: A divide and conquer application for protein classification. In *Proceedings of the 2nd ADBIS Workshop on Data Mining and Knowledge Discovery,* Thessaloniki, Greece, (pp. 121-132).

Pommerening, K. (2006). Grundlagen des sicheren Grid-Computing. In *Grid-Computing in der biomedizinischen Forschung* (Vol. 90, pp. 63-69). München, Germany: Urban & Vogel.

Ponting C.P., & Russell R.R. (2002). The natural history of protein domains. *Annu Rev Biophys Biomol Struct. 31,* 45-71.

Ponting, C.P., Schultz, J., Milpetz, F., & Bork, P. (1999). SMART: identification and annotation of domains from signalling and extracellular protein sequences. *Nucleic Acids Research, 27,* 229-232.

PostGIS website (2008). Retrieved July 20, 2008 from http://postgis.refractions.net.

Potter, C. S., Zhu, Y., & Carragher, B. (2004). Automated particle selection for cryo-electron microscopy. *Journal of Structural Biology, 145*(1), 1-180.

Power, D. J., Politou, E. A., Slaymaker, M. A. & Simpson, A. C. (2006). Securing web services for deployment in HealthGrids. *Future Generation Computer Systems, 22*(5), 547-570.

PRACE (2008). Partnership for Advanced Computing in Europe. Retrieved from www.prace-project.eu/

Pragma 2008, http://www.pragma-grid.net/

Preparata, F.P. & Shamos, M.I. (1990). *Computational Geometry.* Berlin, Germany: Springer-Verlag.

Prieto, F., Cardenas, M., & Rubio, M. (2008). gPhase: A GRID-based technology for the analysis of phase space structure in nonlinear dynamics. Applications to rocking systems under harmonic loading. *Proceedings of the 2nd Iberian Grid Infrastructure Conference.*

Proactive. (2005). "Proactive Manual REVISED 2.2", Proactive, INRIA. Retrieved from http://www-sop.inria.fr/oasis/Proactive/

Provost, F. P., & Hennessy, D. N. (1996). Scaling up: Distributed machine learning with cooperation. *In the Thirteenth National Conference on Artificial Intelligence* (74-79). Menlo Park, CA: AAAI Press.

Psomopoulos, F., & Mitkas, P. (2005). A protein classification engine based on stochastic finite state automata. *Lecture Series on Computer and Computational Sciences VSP/Brill, 4B,* 1371-1374.

Psomopoulos, F., Diplaris, S., & Mitkas, P. (2004). A finite state automata based technique for protein classification rules induction. In *the Proceedings of the Second European Conference on Data Mining and Text Mining in Bioinformatcs*, ECML/PKDD, Piza, Italy, (pp. 54-60).

PsyGrid Consortium (2008). *PsyGrid*. Retrieved July 2008 from http://www.psygrid.org/

Pujol, S, Kikinis, R & Gollub, R. (2008). Lowering the barriers inherent in translating advances in neuroimage analysis to clinical research applications. *Acad Radiology, 15*(1), 114-118.

Pytliński J., Skorwider Ł., Benedyczak K., Wroński M., Bała P., & Huber V. (2003). Uniform Access to the Distributed Resources for the Computational Chemistry Using UNICORE. In: Sloot M. A et al. (Eds.) Computational Science - ICCS 2003 (pp.307-315). Berlin, Germany: Springer-Verlag.

Pytliński J., Skorwider Ł., Huber V., Benedyczak K., & Bała P. (2002). UNICORE - An Uniform Platform for Chemistry on the Grid. *Journal of Computational Methods in Science and Engineering 2* (3s-4s), 369-376.

Quandt, A., Hernandez, P., Kunzst, P., Pautasso, C., Tuloup, M., Hernandez, C., & Appel, R. (2007). Grid-based Analysis of Tandem Mass Spectrometry Data in Clinical Proteomics. In (Eds.) N. Jacq, N. Müller, I. Blanquer, Y. Legré, V. Breton, D. Hausser, V. Hernández, T. Solomonides, M. Apitius, *From Genes to Personalized HealthCare: Grid Solutions for the Life Sciences - Proceedings of HealthGrid 2007*, (126), (pp.13-22).

Quandt, A., Hernandez, P., Masselot, A., Hernandez, C., Maffioletti, S., Pautasso, C., et al. (2008). swissPIT: A novel approach for pipelined analysis of mass spectrometry data. *Bioinformatics*, 24(11), 1416-1417.

Quinlan, J. R. (Ed.). (1993). *C4.5: Programs for Machine Learning*. San Francisco, CA: Morgan Kaufmann.

Radwan1, A., Younis, A., Luykx, P. & Khuri, S. (2008). Prediction and analysis of nucleosome exclusion regions in the human genome. *BMC Genomic, 9*, 186.

Rajasekar, A., Wan, M., Moore, R., Schroeder, W., Kremenek, G., Jagatheesan, A., et al. (2003.). Storage Resource Broker - Managing Distributed Data in a Grid. *Computer Society of India Journal, 33*(4), 42-54

Ram, J. D., Sreenivas, T.H., & Subramaniam, K. G. (1996). Parallel Simulated Annealing Algorithms. *Jnl. Parallel and Distributed Computing, 37*, 207-212.

Ranganathan, K., & Foster, I. (2003). Simulation Studies of Computation and Data Scheduling Algorithms for Data Grids. *Journal of Grid Computing, 1*(1), 53-62.

Rangwala, H., Lantz, E., Musselman, R., Pinnow, K., Smith, B., & Wallenfelt, B. (2005). Massively Parallel BLAST for the Blue Gene/L. *High Availability and Performance Workshop*.

Rappleye, J., Innus, M., Weeks, C. M. & Miller, R. (2002). *SnB* v2.2: An Example of Crystallographic Multiprocessing. *J. Appl. Cryst., 35*, 374-376.

Rarey, M., Kramer, B., & Lengauer, T. (1996). A fast flexible docking method using an incremental construction algorithm. *J Mol Biol, 261*, 470-489.

Ratnasingham, S. & Hebert, P. (2007). BOLD: The Barcode of Life Data System. *Molecular Ecology Notes, 7*(3), 355-364. Retrieved from www.barcodinglife.org.

Rauhut G. (2002). Web Site: Your Web Browser as a Computational Chemistry Frontend. *Angewandte Chemie International Edition, 41*(12), 2205.

Reed D. A. (2003). Grids, the TeraGrid, and Beyond. *Computer, 36*, 62-68. http://portal.acm.org/citation.cfm?id=642202.

Reed, D. A., Mendes, C. L., Lu, C., Foster, I., & Kesselman, C. (Eds.). (2003). *The Grid 2: Blueprint for a New Computing Infrastructure - Application Tuning and Adaptation*. San Francisco, CA: Morgan Kaufman.

Reference Model for Service Oriented Architecture (2006). Committee Draft 1.0. Retrieved from http://www.oasis-open.org/committees/download.php/16587/wd-soa-rm-cd1ED.pdf

Ren, Y., Zhang, Q., Wan, J., Huang, X., Xie, P., & Yang, G. (2006). Implementation of CCNUGrid-Based Drug Virtual Screening Applications Using Workflow Techniques. *Proceedings of the Fifth International Conference on Grid and Cooperative Computing Workshops* (pp. 371-374). New York: IEEE Computer Society.

Resnik, P. (1999). Semantic similarity in a taxonomy: an information based measure and its application to problems of ambiguity in natural language. *J Artif Intelligence Res, 11*, 95–130.

Richards, W.G. (2002). Virtual Screening Using Grid Computing: the Screensaver Project. *Nat Rev Drug Discov, 1*, 551-555.

Ricklefs, R.E., Fallon, S.M., & Birmingham, E. (2004). Evolutionary relationships, cospeciation, and host switching in avian malaria parasites. *Systematic Biology,* 53(1):111-119.

Roche A., Pennec X., Malandain G. & Ayache N. (2001). Rigid Registration of 3D Ultrasound with MR Images: a New Approach Combining Intensity and Gradient Information. *IEEE Transactions on Medical Imaging, 20*(10), 1038-1049.

Romberg, M. (Ed.) (2005). *OpenMolGRID - Open Computing Grid for Molecular Science and Engineering,* Final Report, NIC Directors, Germany, Jülich, (Vol. 29).

Romei, A., Ruggieri, S., & Turini, F. (2006). Kddml: a middleware language and system for knowledge discovery in databases. *Data Knowledge Engineering, 57*(2), 179-220.

Romei, A., Sciolla, M., Turini, F., & Valentini, M. (2007). Kddml-G: a grid-enabled knowledge discovery system. *Concurr. Comput. : Pract. Exper., 19*(13), 1785-1809.

Ronquist, F. & Huelsenbeck, J. P. (2003). MRBAYES 3: Bayesian phylogenetic inference under mixed models. *Bioinformatics, 19*, 1572-1574.

Rosenberg, G., Phillips, W., Landis, D., & Pierce, W. (1981). Design and Evaluation of the Pennsylvania State University Mock Circulation System. *ASAIO Journal, 4*(1), 41-49.

Rosenfield, B. A. & Dorman, T. (2000). Intensive care unit telemedicine; alternative paradigm for providing continuous intensive care. *Crit. Care Med, 28*, 3925-3931.

Rosenthal, A., Seligman, L., & Renner, S. (2004). From semantic integration to semantics management: Case studies and a way forward. *SIGMOD Record, 33*(4), 44-50.

Rushing, J., Ramachandran, R., Nair, U., Graves, S., Welch, R., & Lin, H. (2005). ADaM: a data mining toolkit for scientists and engineers. *Computers and Geosciences, 31*, 607-618.

Rutten, M., de Hart, J., de Mol, B., van de Molengraft, R., & de Vosse, F. (2007). Analysis of LVADs in a controlled mock systemic circulation. *ASAIO Journal, 53*(2), 26A.

Rycerz, K., Bubak, M., Sloot, P., Getov, V. (2008). Problem solving environment for distributed interactive simulations. In Gorlatch, S., Bubak, M., Priol, T., (Eds.), *Achievements in European Reseach on Grid Systems. CoreGRID Integration Workshop 2006 (Selected Papers),* (pp. 55—66). Berlin, Germany: Springer.

Saara Väärtö, S. (Ed.). (2008). Advancing Science in Europe. DEISA – Distributed European Infrastructure for Supercomputing Applications. EU FP6 Project. Retrieved from www.deisa.eu/press/DEISA-AdvancingScienceInEurope.pdf

SAGA (2006). SAGA implementation home page. http://fortytwo.cct.lsu.edu:8000/SAGA

Šali, A., & Blundell, T. L. (1993).Comparative protein modelling by satisfaction of spatial restraints. *Journal of Molecular Biology, 234*, 779-815.

Saltz, J., Oster, S., Hastings, S., Langella, S., Kurc, T., Sanchez, W., et al. (2006). *caGrid:* Design and Implementation of the Core Architecture of the Cancer Biomedical Informatics Grid. *Bioinformatics, 22*(15), 1910.

Salzberg, S.L., Kingsford, C., Cattoli, G., et al. (2007). Genome analysis linking recent European and African influenza (H5N1) viruses. *Emerg Infect Dis,.* 13(5):713-8.

Salzemann, J., Jacq, N., & Breton, V. (2007, June). Replication and Update of Molecular Biology Databases. *IEEE Transactions on Nanobioscience, 6*(2).

Samatova, N. F., Ostrouchov, G., Geist, A., & Melechko, A. (2002). RACHET: An Efficient Cover-Based Merging of Clustering Hierarchies from Distributed Datasets. *Distributed and Parallel Databases, 11*(2), 157-180.

Sandhu, R., Coyne, E. J., Feinstein, H. L. & Youman, C.E. (1996). Role-Based Access Control Models. *IEEE Computer 29* (2), 38-47.

Sanner, M. F., Duncan, B. S., Carrillo, C. J., & Olson, A. J. (1999). Integrating computation and visualization for biomolecular analysis: an example using python and AVS. *Pacific Symposium in Biocomputing, 99,* 401-412.

Sansom, M. S. P., Scott, K. A., & Bond, P. J. (2008). Coarse-grained simulation: A high-throughput computational approach to membrane proteins. *Biochemistry Society Transactions, 36,* 27-32.

Santos, N. & Koblitz, B. (2006, June). *Distributed Metadata with the AMGA Metadata Catalog.* Paper presented at the workshop on Next-Generation Distributed Data Management, Paris, France. Retrieved from http://project-arda-dev.web.cern.ch/project-arda-dev/metadata

Santos, N., & Koblitz, B. (2006). Distributed Metadata with the AMGA Metadata Catalog. *In workshop on Next Generation Distributed Data Management*

Sax, U., Mohammed, Y., Viezens, F., & Rienhoff, O. (2006). Grid-Computing in der biomedizinischen Forschung. In K. Überla, O. Rienhoff & N. Victor (Eds.), *Medizinische Informatik, Biometrie und Epidemiologie* (Vol. 90). München, Germany: Urban & Vogel.

Scardaci, D. & Scuderi, G. (2007). A Secure Storage Service for the gLite Middleware. In the *proceedings of third International Symposium on Information Assurance and Security* (pp. 261-266), Manchester, UK. Washington, DC: IEEE Computer Society..

Scheres, S. H. W., Merino, A. J., Sorzano, C. O. S., & Carazo, J. M. (2005). Grid Computing in 3D-EM Image Processing Using Xmipp. In *Proceedings of the 18th IEEE Symposium on Computer-Based Medical Systems,* Washington, D.C., (pp. 561-563).New York: IEEE Computer Society.

Scheres, S. H. W., Valle, M., Nuñez, R., Sorzano, C. O. S., Marabini, R., Herman, G. T., et al. (2005). Maximum-likelihood Multi-reference Refinement for Electron Microscopy Images. *Journal of Molecular Biology, 348*(1), 139-149.

Schiffers, M. (2008). VO Management, *IVOM Workshop.* Regionales Rechenzentrum für Niedersachen, Hannover, Germany.

Schmidt, M. W., Baldridge, K. K., Boatz, J. A., Elbert, S. T., Gordon, M. S., Jensen, J. H., et al. (1993). General atomic and molecular electronic structure system. *Journal of Computational Chemistry, 14,* 1347-1363.

Schneider, T. R. & Sheldrick, G. M. (2002). Substructure solution with *SHELXD. Acta Crystallogr., D58,* 1772-1779.

Schroeder, W., Martin, K. & B. Lorensen (2006). *Visualization Toolkit: An Object-Oriented Approach to 3D Graphics.* Clifton Park, NY: Kitware.

Schuchardt, K., Didier, B., & Black, G. (2002). Ecce - a problem-solving environment's evolution toward Grid services and a Web architecture. *Concurrency and Computation: Practice and Experience, 14,*1221–1239.

Schuller B., Romberg M., & Kirtchakova L. (2005). Application Driven Grid Developments in the OpenMolGRID project. In P.M.A. Sloot, A.G. Hoekstra, T. Priol, A. Reinefeld, M. Bubak (Eds.), *Advances in Grid Computing* (Vol. 3470) (pp. 23–29). Berlin, Germany: Springer-Verlag.

Schulze, W & Mann,M (2003) A novel proteomic screen for peptide-protein interactions. *Journal of Biological Chemistry, 279*(11), 10756-64

Schuster, A., & Wolff, R. (2001). Communication-efficient distributed mining of association rules. *In ACM SIGMOD international conference on Management of data* (pp. 473-484). New York: ACM.

Schuster, A.,Wolff, R., & Trock, D. (2003). A High-Performance Distributed Algorithm for Mining Association Rules. In *Third IEEE International Conference on Data Mining.*

Scifo, S. (2007). GSAF - Grid Storage Access Framework. In the *proceeding of 16th IEEE International Workshops on Enabling Technologies: Infrastructure for Collaborative Enterprises* (pp.296-297), Evry, France, June 2007. Washington, DC: IEEE Computer Society.

Scott, K. A., Alonso, D. O. V., Sato, S., Fersht, A. R., & Daggett, V. (2007). Conformational entropy of alanine versus glycine in protein denatured states. *Proceedings of the National Academy of Sciences USA, 104,* 2661-2666.

Sederberg, T. W., & Parry, S. R. (1986). Free-form deformation of solid geometric models. *ACM Computer Graphic, 20,* 151-160.

Seifert, K.A., Samson, R.A., Dewaard, J.R, Houbraken, J., Lévesque, C.A., Moncalvo, J.M., Louis-Seize G. & Hebert, P.D. (2007). Prospects for fungus identification using COI DNA barcodes, with Penicillium as a test case. *Proc Natl Acad Sci U S A, 104*(10), 3901-6.

Seitz, L., Montagnat, J., Pierson, J. M., Oriol, D., & Lingrand, D. (2005). Authentication and Authorisation Prototype on the μgrid for Medical Data Management. In T. Solomonides, R. McClatchey, V. Breton, Y. Legré & S. Nørager (Eds.), *From Grid to Healthgrid - Proceedings of Healthgrid 2005* (pp. 222 - 233). Washington, DC: IOS Press.

Seo, J. and Lee, K.J. (2004). Post-translational modifications and their biological functions: proteomic analysis and systematic approaches. *J Biochem Mol Biol, 37*(1), 35-44.

Seymour, K., Lee, C., Desprez, F., Nakada, H., & Tanaka, Y. (2004, September). The End-User and Middleware APIs for GridRPC. In *Workshop on Grid Application Programming Interfaces, In conjunction with GGF12*, Brussels, Belgium.

Seymour, K., Nakada, H., Matsuoka, S., Dongarra, J., Lee, C., & Casanova, H. (2002). Overview of GridRPC: A Remote Procedure Call API for Grid Computing. Proceedings of the Third International Workshop on Grid Computing, Lecture Notes in Computer Science, Springer, ed 2536, pp. 274–278, Baltimore, USA.

Shadforth, I., Crowther, D. & Bessant, C. (2005). Protein and peptide identification algorithms using MS for use in high-throughput, automated pipelines. *Proteomics, 5*(16), 4082-4095.

Shafer, J. C., Agrawal, R., & Mehta, M. (1996). SPRINT: A scalable parallel classifier for data mining. In T. M. Vijayaraman, A. P. Buchmann, C. Mohan, & N. L. Sarda (Eds.), *22nd Int. Conf. Very Large Databases* (pp. 544-555). San Francisco, CA: Morgan Kaufmann.

Shah, S.P., He, D.Y.M., Sawkins, J.N., Druce, J.C., Quon, G., Lett, D., Zheng, G.X.Y., Xu, T., Quellette, B.F.F. (2004). Pegasys: Software for executing and integrating analyses of biological sequences. *BMC Bioinformatics, 5*, 40.

Shamir, A. (1979). How to share a secret. *Communications of the ACM, 22*, 612-613.

SHARE Consortium (2006). *Structuring Healthgrid Activities and Research in Europe: Developing a Roadmap.* Retrieved July 2008 from http://www.eu-share.org

Shaw, M. & Piper, I. (2008). The Brain Monitoring with Information Technology (BrainIT) collaborative network: Data Validation Results. *Acta Neurochirurgica, ICP Symposium Journal.*

Sheldrick, G.M. (1990). Phase annealing in SHELX-90: direct methods for larger structures. *Acta Cryst. A46*, 467–473.

Shen, J., Xu, X., & Cheng, F. (2003). Virtual screening on natural products for discovering active compounds and target clues. *Curr Med Chem, 10*, 2327-2342.

Shibboleth Project (2006). Retrieved July 2008 from http://shibboleth.internet2.edu/

Shindyalov, I.N. & Bourne, P.E. (1983). Protein structure alignment by incremental combinatorial extension (CE) of the optimal path. *Protein Engineering 11*(9), 739–747.

Shoichet, B. (2004). Virtual screening of chemical libraries. *Nature, 432*, 862-865.

Siebenlist, F., Magaratnam, N., Welch, V., & Neuman, C. (2004). Security for Virtual Organizations: Federating Trust and Policy Domains. In *The Grid: Blueprint for a New Computing Infrastructure* (2nd Ed.). Amsterdam: Morgan Kaufmann.

Signorini, M., Paulesu, E., Friston, K., Perani, D., Colleluori, A., Lucignani, G., et al (1999). Rapid assessment of regional cerebral metabolic abnormalities in single subjects with quantitative and nonquantitative [18F] FDG PET: A clinical validation of statistical parametric mapping. *Neuroimage, 9*(1), 63-80.

Sild S., Maran U., Lomaka A., & Karelson M. (2006). Open Computing Grid for Molecular Science and Engineering. *J. Chem. Inf. Model. 46*, 953–959

Sild S., Maran U., Romberg M., Schuller B., & Benfenati E. (2005). OpenMolGRID: Using Automated Workflows in GRID Computing Environment. In P.M.A. Sloot, A.G. Hoekstra, T. Priol, A. Reinefeld, M. Bubak (Eds.),

Advances in Grid Computing (Vol. 3470) (pp. 464–473). Berlin, Germany: Springer-Verlag.

Silva, F.A.B., Gagliardi, H. F., Gallo, E., Madope, A.M. Neto, V.C. , Pisa, I.T., Alves, D.(2007) IntegraEPI: a Grid-based Epidemic Surveillance System. In N. Jacq, Y. Legré, H. Muller, I. Blanquer, V. Breton, D. Hausser, V. Hernández, T. Solomonides and M. Hofman-Apitius (Eds.), *From Genes to Personalized HealthCare: Grid Solutions for the Life Sciences - Proceedings of HealthGrid 2007.* Amsterdam, the Netherlands: IOS Press.

Sim, A. & Shoshani, A. (2007) *The Storage Resource Manager Interface Specification Version 2.2.* Retrieved from https://forge.gridforum.org/ sf/projects/gsm-wg

SIMDAT (2008). Grids for Industrial Product Development. Retrieved from www.scai.fraunhofer.de/about_simdat.html

Simmhan, Y. L., & Plale, B., & Gannon, D. (2007). Query capabilities of the Karma provenance framework. In *Concurrency and Computation: Practice and Experience, 20*(5) 441-451. San Francisco, CA:Wiley InterScience

Simmhan, Y. L., Plale, B., & Gannon, D. (2005). A survey of data provenance in e-science. *SIGMOD Rec. 34* (3), 31-36.

Simmhan, Y., Plale, B. & Gannon, D. (2005) A survey of data provenance in e-science. *SIGMOD Record, 34*(3) 31–36.

Simmhan, Y., Plale, B., Gannon, D. & Marru, S. (2006) Performance evaluation of the karma provenance framework for scientific workflows. In L. Moreau and I. T. Foster (Eds.), *IPAW* (pp. 222–236). New York: Springer.

Sims, L.D., et al. (2005). Origin and evolution of highly pathogenic H5N1 avian influenza in Asia. *Vet Rec. 157*(6), 159-64.

Sinnott, R. O., Jiang, J., Stewart, G., Stell, A. J., Martin, D., Doherty, T., Su, L. & Watt, J. (2008). *Advanced Security for Virtual Organizations: Exploring the Pros and Cons of Centralized vs Decentralized Security Models.* Proceedings of 8th IEEE International Symposium on Cluster Computing and the Grid (CCGrid 2008), Lyon, France.

Sinnott, R. O., Stell, A. J., Chadwick, D. W., & Otenko, O. (2005). Experiences of Applying Advanced Grid Authorisation Infrastructures. In P.M.A. Sloot, et al, (eds.), *Proceedings of European Grid Conference (EGC)*, Amsterdam, the Netherlands (pp. 265-275).

Sinnott, R. O., Stell, A. J., Chadwick, D. W., Otenko, O., Watt, J., Koetsier, J. & Nguyen, T. A. (2006). *Supporting Decentralized, Security focused Dynamic Virtual Organizations across the Grid.* Proceedings of 2nd IEEE International Conference on e-Science and Grid Computing, Amsterdam, Holland.

Sinnott, R. O., Watt, J., & Jiang, J. (2007). *The GLASS Project: Supporting Secure Shibboleth-based Single Sign-On to Campus Resources.* Proceedings of UK e-Science All Hands Meeting, Nottingham, UK.

Sinnott, R. O., Watt, J., Ajayi, O. & Jiang, J. (2006). *Shibboleth-based Access to and Usage of Grid Resources.* Proceedings of IEEE International Conference on Grid Computing, Barcelona, Spain.

Sinnott, R., Atkinson, M., Bayer, M., Berry, D., Dominiczak, A., Ferrier, M., Gilbert, D., Hanlon, N., Houghton, D., Hunt, E., & White, D. (2004). Grid services supporting the usage of secure federated, distributed biomedical data. *In the UK e-Science All Hands Conference. Engineering and Physical Sciences Research Council.*

Sinnott, R.O. Ajayi,O. Stell,A.J. (2006). Development of Grid Frameworks for Clinical Trials and Epidemiological Studies. *HealthGrid 2006 Conference, Studies in Health Technology and Informatics, 120,* 117-130.

Skillicorn, D. B. (2002). The case for datacentric grids. *In the 16th International Parallel and Distributed Processing Symposium* (p. 313). New York: IEEE Computer Society.

Skolnick, J. and Fetrow, J.S. (2000). From genes to protein structure and function: novel applications of computational approaches in the genomic era. *Trends Biotechnol, 18*(1), 34-9.

Slaughter, L. A., Soergel, D. & Rindflesch, T. C. (2006). Semantic representation of consumer questions and physician answers. *International Journal of Medical Informatics, 75*(7), 513-529.

Sloot, P. M. A., Ramos, A. T., Altintas, I., Bubak, M., & Boucher, C. (2006). From molecule to man: Decision support in individualized e-health. *Computer, 39*(11), 40-46.

Smarr, L., & Catlett, C. E. (June 1992). Metacomputing. *Communications of the ACM, 35*(6), 44-52).

Smith, C. A., Want, E. J., O'Maille, G., Abagyan, R., & Siuzdak, G. (2006). Xcms: processing mass spectrometry data for metabolite profiling using nonlinear peak alignment, matching, and identification. *Analytical Chemistry, 78*(3), 779-787.

Smith, M., Barton, M., Bass, M., Branchofsky, M., McClellan, G., Stuve, D., Tansley, R., & Harford J. (2003). An Open Source Dynamic Digital Repository. *D-Lib Magazine, 9.*

Smith, T., & Waterman, M. (1981). Identification of common molecular subsequences. *Journal of Molecular Biology, 147,* 195-197.

Smith, T.F., & Waterman, M.S. (1981). Identification of Common Molecular Subsequences. *Journal of Molecular Biology, 147,* 195-197.

Snell, Q., Clement, MJ., Jackson DB. & Gregory, C. (2000, May). *The Performance Impact of Advance Reservation Meta-scheduling.* 6th workshop on Job Scheduling Strategies for Parallel Processing, Cancun, Mexico (LNCS 1911, pp 137–153). Berlin, Germany: Springer-Verlag.

Soh, H., Shazia Haque, S., Liao, W., & Buyya, R. (2006). Grid Programming Models and Environments. In Yuan-Shun Dai, et. al. Advanced Parallel and Distributed Computing (pp. 141–173). Nova Science Publishers.

Sorzano, C. O. S., Marabini, R., Herman, G. T., & Carazo, J. M. (2005b). Multi-objective algorithm parameter optimization using multi-variate statistics in three-dimensional electron microscopy reconstruction. *Pattern Recognition, 38*(12), 2587-2601.

Sprinzl, M., & Vassilenko, K.S. (2005). Compilation of tRNA sequences and sequences of tRNA genes. *Nucleic Acids Research, 33,* D139–D140.

SRB – The SDSC Storage Resource Broker. (2008). Retrieved October 20, 2008, from http://www.sdsc.edu/srb/index.php/Main_Page.

Srivastava, A., Han, E., Kumar, V., & Singh, V., (1999). Parallel formulations of decision-tree classification algorithms. *Data Mining and Knowledge Discovery, 3,* 237-261.

Stallman, R. M. (2002). *Free software, free society.* Boston, MA: GNU Press.

Stamatakis, A. (2006). Parallel and Distributed Computation of Large Phylogenetic Trees. In Zomaya A. (Ed.), *Parallel Computing for Bioinformatics and Computational Biology,* (pp. 327-346). Hoboken, NJ: Wiley.

Stamatakis, A. (2006). RAxML-VI-HPC: maximum likelihood-based phylogenetic analyses with thousands of taxa and mixed models. *Bioinformatics,* 22(21): 2688-2690.

Stamatakis, A., Auch, A.F., Meier-Kolthoff, J., & Göker, M. (2007). *AxPcoords & parallel AxParafit: statistical co-phylogenetic analyses on thousands of taxa. BMC Bioinformatics*, 8, 405.

Stamatakis, A., Hoover, P., & Rougemont, J. (2008). A Rapid Bootstrapping Algorithm for the RAxML Web Servers. *Systematic Biology,* in press.

Stankovski, V., Swain, M., Kravtsov, V., Niessen, T., Wegener, D., Kindermann, J., & Dubitzky, J. (2007). Grid-enabling data mining applications with datamining-grid: An architectural perspective. *Future Generation Computing Systems, 24*(4), 259-279.

Stankovski, V., Swain, M., Kravtsov, V., Niessen, T., Wegener, D., Kindermann, J., & Dubitzky, W. (2008). Grid-enabling data mining applications with DataMiningGrid: An architectural perpective. *Future Generation Computer Systems, 24,* 259-279.

Stark, C., Breitkreutz, B.-J., Reguly, T., Boucher, L., Breitkreutz, A., & Tyers, M. (2006). BioGRID: A General Repository for Interaction Datasets. *Nucleic Acids Research, 34,* 535-539.

Starling, E. (1918). *The linacre lecture on the law of the heart.* London: Green and Co.

Steadm, D.A., Paton, N.W., Missier, P., Embury, S.M. Hedeler, C. Jin, B. Brown, A.J., & Preece, A. (2008) Information quality in proteomic *Briefing in Bioinformatics,* 9(2), 174-188.

Stein, L. D. (2008). Towards a cyber-infrastructure for the biological sciences: progress, visions and challenges. *Nature Rev. Genet.*, *9*, 678-688.

Stell A. J., Sinnott R. O., Watt J. P. (2005). Comparison of advanced authorisation infrastructures for grid computing. *International Symposium on High Performance Computing Systems and Applications 2005 (HPCS2005)*, (pp. 195-201).

Stell, A. J., Sinnott, R. O. & Ajayi, O. (2008). *Supporting Nationwide e-Clinical Trials and Studies*. Proceedings of 15th Mardi Gras Conference, Baton Rouge, LA.

Stell, A. J., Sinnott, R. O. & Ajayi, O. (2006). *Secure, Reliable and Dynamic Access to Distributed Clinical Data*. Proceedings of Life Science Grid Conference, Yokohama, Japan.

Stell, A. J., Sinnott, R. O., & Watt, J. (2005). *Comparison of Advanced Authorisation Infrastructures for Grid Computing*. Proceedings of International Conference on High Performance Computing Systems and Applications, Guelph, Canada.

STEP (2002). *Seeding the EuroPhysiome: A Roadmap to the Virtual Physiological Human*. Retrieved from http://www.biomedtown.org/biomed_town/STEP/Reception/step_presentations/RoadMap/vph_roadmap_printed_3.pdf

Sterzel M., & Szepieniec T. (2007). Enabling commercial chemical software on EGEE Grid – Gaussian VO. In M. Bubak, M. Turała, K. Wiatr (Eds.) *Cracow'06 Grid Workshop. Proceedings* (pp. 499–504).

Stevens, R, Robinson, A., & Goble, C. (2003). myGrid: personalised bioinformatics on the information Grid. *Bioinformatics*, *19*(Suppl. 1), i302-i304.

Stevens, R. C. (2006). Trends in Cyberinfrastructure for Bioinformatics and Computational Biology. *CTWatch Quarterly*, *2*, 6-17.

Stevens, R., Goble, C., Paton, N.W., Bechhofer, S., Ng, G., Baker, P., and Brass A. (2003. May). Complex Query Formulation Over Diverse Information Sources in TAMBIS. In Z. Lacroix & T. Critchlow, (Eds.), *Bioinformatics: Managing Scientific Data*. San Fransisco, CA : Morgan Kaufmann.

Stevens, R., McEntire, R., Goble, C., Greenwood, M., Zhao, J., Wipat, A. & Li, P. (2004). myGrid and the drug discovery process. *Drug Discovery Today: BIOSILICO*, *2*(4), 140-148.

Stevens, R.D., Robinson, A. J. & Goble C. A. (2004). myGrid: Personalised Bioinformatics on the Information Grid. *Bioinformatics*, *19*(1), 302–302. Retrieved from http://www.mygrid.org.uk/.

Stevens, R.D., Robinson, A.J., & Goble, C.A. (2003). *myGrid: Personalised Bioinformatics on the Information Grid.*

Stevens, RD, Robinson, AJ, & Goble CA (2004). myGrid: Personalised bioinformatics on the information Grid. *Bioinformatics*, 19(1), 302–302C.

Stewart J. J. P. (1990). MOPAC: a Semiempirical Molecular Orbital Program. *J. Comput.-Aided Mol. Des.*, *4*, 1-45.

Stewart, C. A. (2004). Introduction to the edition. *Communications of the ACM*, *47*(11), 30-33.

Stewart, G., Cameron, D., Cowan, G., & McCance, G. (2007). Storage and Data management in EGEE. *Proceedings of Fifth Australasian Symposium on Grid Computing and e-Research* Australian Computer Society, Inc., Darlinghurst, Australia, (pp. 69-77).

Stockinger, H., Pagni, M., Cerutti, L., & Falquet, L. (2006). Grid Approach to Embarrassingly Parallel CPU-Intensive Bioinformatics Problems. *2nd IEEE International Conference on e-Science and Grid Computing (e-Science 2006)*, IEEE Computer Society Press, Amsterdam, The Netherlands.

Straatsma, T. P., Philippopoulos, M., & McCammon, J. (2000). NWChem: Exploiting parallelism in molecular simulations. *Computer Physics Communications*, *128*, 377-385.

Sun Network.com (2008). retrieved from www.network.com/

SunGrid (2005). Sun Utility Computing. Retrieved from www.sun.com/service/sungrid/

SURA -Southeastern Universities Research Association. (2007). The Grid Technology Cookbook. Programming

Concepts and Challenges. Retrieved from www.sura.org/cookbook/gtcb/

Swain, M., Hunniford, T., Mandel, J., Palfreyman, N., & Dubitzky, W., (2005). Reverse-engineering gene-regulatory networks using evolutionary algorithms and grid computing, *Journal of Clinical Monitoring and Computing, 19*(4-5), 329-337.

Switula, D. (2000). Principles of good clinical practice (GCP) in clinical research. *Sci Eng Ethics, 6*(1), 71-77.

Swofford, D., Olsen, G., Waddell, P. & Hillis, D.M. (1996). Phylogenetic inference. In D.N. Hillis, C. Moritz, & B.K. Mable, (Eds.), *Molecular Systematics,* (pp. 407-514). Sunderland, MA: Sinauer Associated.

Sycara, K. (1998). *Multiagent systems,* Artificial Intelligence Magazine, Association for the Advancement of Artificial Intelligence, 10(2), 79–93.

Tai, K., Baaden, M., Murdock, S., Wu, B., Ng, M. H., Johnston, S., Boardman, R., Fangohr, H., Cox, K., Essex, J. W., & Sansom M. S. P. (2007). Three hydrolases and a transferase: Comparative analysis of active-site dynamics via the BioSimGrid database. *Journal of Molecular Graphics and Modelling, 25,* 896-902.

Tai, K., Murdock, S., Wu, B., Ng, M. H., Johnston, S., Fangohr, H., Cox, S. J., Jeffreys, P., Essex, J. W., & Sansom, M. S. P. (2004). BioSimGrid: Towards a worldwide repository for biomolecular simulations. *Organic and Biomolecular Chemistry, 2,* 3219-3221.

Talia, D., Trunfio, P., & Verta, O. (2005). Weka4WS: A WSRF-Enabled Weka Toolkit for Distributed Data Mining on Grids. *In the 9th European Conference on Principles and Practice of Knowledge Discovery in Databases, 3721,* 309-320.

Tan, H., Chen, X. & Gu, J. (2007). Semanteme-based processing mechanism under data grid environment. *Journal of Huazhong University of Science and Technology (Natural Science Edition), 35*(4), 22-25.

Tan, W. C. (2007). Provenance in Databases: Past, Current, and Future. In *IEEE Data Eng. Bull. 30*(4),3-12.

Tanner, S., Shu, H., Frank, A., Wang, L., Zandi, E., Mumby, M., et al. (2005), InsPecT: identification of post-translationally modified peptides from tandem mass spectra. *Analytical Chemistry* 7(14), 4626-4639.

Tatusov, R.L., Eugene, V.K., & David, J.L. (1997). A genomic perspective on protein families. *Science, 278,* 631-637.

Tatusova T. A., & Madden T. L. (1999). BLAST 2 Sequences, a new tool for comparing protein and nucleotide sequences. *FEMS Microbiology Letters, 174* (2), 247–250.

Taubenberger, J.K., et al. (2005). Characterization of the 1918 influenza virus polymerase genes. *Nature, 437*(7060), 889-893.

Taufer, M., An, C., Kerstens, A., & Brooks, C.L. (2006). Predictor@Home: Protein Structure Prediction Supercomputer Based on Global Computing. *IEEE Transactions on Parallel and Distributed Computing, 17*(8), 786-796.

TAVERNA (2008). The Taverna Workbench 1.7. Retrieved from http://taverna.sourceforge.net/

Taylor, C. F., Hermjakob, H., Julian, R. K., Garavelli, J. S., Aebersold, R., & Apweiler, R. (2006). The work of the human proteome organisation's proteomics standards initiative (hupo psi). *OMICS, 10*(2), 145-151.

Taylor, I. (2006) Triana generations. In *e-Science*, (pp. 143). Los Alamitos, CA: IEEE Computer Society.

Taylor, I., Shields, M., Wang, I., & Harrison, A. (2005). Visual Grid workflow in Triana. *Journal of Grid Computing, 3*(3-4), 153-169.

Tesauro, GJ. & Kephart, J. O. (2004, May). *Utility functions in autonomic systems.* 1st International Conference on Autonomic Computing (pp. 70–77). New York: IEEE Computer Society.

Texier, R (2007) *Systèmes de globalisation des ressources données-calculs. Application à l'imagerie médicale.* Unpublished doctoral dissertation, Université Paris-Sud, France.

The Gene Ontology Consortium (2000). Gene Ontology: tool for the unification of biology, *Nature Genetics, 25,* 25-29.

The Globus Security Team. (2005). *Globus Toolkit Version 4 Grid Security Infrastructure: A Standards Perspective.*

Thomas M., & Boisseau J. (2003). Building Grid Computing Portals: The NPACI Grid Portal Toolkit. In F. Berman, G. Fox & T. Hey, (Eds). *Grid Computing: Making the Global Infrastructure a Reality*. Hoboken, NJ: John Wiley & Sons.

Thomas M., Boisseau J., Mock S., Dahan M., Mueller K., & Sutto D. (2003). *The GridPort Toolkit Architecture for Building Grid Portals*. Proceedings of the 10th IEEE Intl. Symp. on High Perf. Dist. Comp.

Thomas, P.L., & Menzies, J.G. (1997). Cereal smuts in Manitoba and Saskatchewan, 1989-95. *Canadian Journal of Plant Pathology*, 19(2):161-165.

Thompson, J.D., Higgins, D.G. & Gibson, T.J. (1994). CLUSTAL W: improving the sensitivity of progressive multiple sequence alignment through sequence weighting, position-specific gap penalties and weight matrix choice. *Nucl. Acids Res.*, *22*(22), 4673-4680.

Tian C., Ding W., Cao R. and Jiang S. (2007). Extensive Epidemic Spreading Model Based on Multi-agent System Framework (LNCS 4490, pp. 129-133).

Ting, K. M., & Witten, I. H. (1997). Stacked generalizations: When does it work? *In International Joint Conference on Artificial Intelligence* (pp. 866-873).

Tishkoff S.A., Varkonyi R., Cahinhinan N., Abbes S., Argyropoulos G., Destro-Bisol G., et al. (2001). Haplotype diversity and linkage disequilibrium at human G6PD: recent origin of alleles that confer malarial resistance. *Science, 293*.

TMF. (2006). *AG Datenschutz*. Retrieved March 16, 2006, from www.tmf-ev.de/site/DE/int/AG/DS/container_ag_ds.php

Tohsato, Y., Kosaka, T., Date, S., Shimojo, S. & Matsuda, H. (2005). Heterogeneous database federation using grid technology for drug discovery process,. In *Lecture notes in Computer Science*, (Vol. 3370) (pp. 43-52). Aachen,Germany: Springer-Verlag.

Toivonen, H. (1996). Sampling large databases for association rules. In T. M. Vijayaraman, A. P. Buchmann, C. Mohan, and N. L. Sarda (Eds.), *International Conference on Very Large Data Bases*. San Francisco, CA: Morgan Kaufman.

Torres E., De Alfonso C., Blanquer I., Hernandez V. (2006). Privacy Protection in HealthGrid: Distributing Encryption Management over the VO. *Studies in Health Technology and Informatics, 120,* 131-41.

Traxer, O., Osorio A., Pasqui, F., Sebe, P., Merran, S., Ripoche, X., et al. (2005, March). *New PC based software using an Augmented Reality system to simulate PCNL*. 20th European Association of Urology (EAU) Congress, Istanbul, Turkey.

TRIANA (2003). The Triana Project. Retrieved from www.trianacode.org/

Tsaregorodtsev, A., Garonne, V. & Stokes-Rees, I. (2004). DIRAC: A Scalable Lightweight Architecture for High Throughput Computing. In *5th IEEE/ACM International Workshop on Grid Computing*, (pp 19-25). New York: IEEE Computer Society.

Tsiknakis, M., Kafetzopoulos, D., Potamias, G., Analyti, A., Marias, K., & Manganas, A. (2006). Building a European biomedical grid on cancer: the ACGT Integrated Project. *Stud Health Technol Inform, 120*, 247-258.

Tsoumakas, G. & Vlahavas, I. (2002). Distributed data mining of large classifier ensembles. *In the Second Hellenic Conference on Artificial Intelligence* (pp. 249-256).

Tsoumakos, D., & Roussopoulos, N. (2003). *A Comparison of Peer-to-Peer Search Methods*, Paper presented at the 6th International Workshop on the Web and Databases (WebDB), San Diego, USA.

Tsymbal, A., Huber, M., Zillner, S., Hauer, T., & Zhou, K. (2007a). Visualizing patient similarity in clinical decision support. In *Lernen - Wissensentdeckung - Adaptivität, LWA 2007, Joint Workshops of the German Society for Informatics (GI), Workshop on Knowledge & Experience Management, FGWM'07.*

Tsymbal, A., Zillner, S., & Huber, M. (2007b). Ontology - supported machine learning and decision support in biomedicine, In *Data Integration in the Life Sciences, 4th International Workshop, DILS 2007, Philadelphia, USA*, (LNCS 4544, pp. 156-171). Berlin-Heidelberg: Springer.

Tudor DICOM Viewer. *Main page*. Retrieved February 12, 2008 from http://www.santec.lu/projects/optimage/dicomtools

Tuecke, S., Czajkowski, K., Foster, I., Frey, J., Graham, S., & Kesselman, C., et al. (2003). Open grid services infrastructure (OGSI) (No. Ver. 1.0)., *Global Grid Forum (GGF)*. Retrieved March 8, 2008 from http://www.globus.org/alliance/publications/papers.php#GSSpec

Tulipano, A., Donvito, G., Licciulli, F., Maggi, G., Gisel, A.(2007). Gene analogue finder: a GRID solution for finding functionally analogous gene products. *BMC Bioinformatics, 8*, 329.

Tuloup, M., Hernandez, C., Coro, I., Hoogland, C., & Binz, P.,-A,. (2003) Aldente and BioGraph: An improved peptide mass fingerprinting protein identification environment. In Ducret A, Eberle AN, Faupel MD, Hofsteenge J, Jenö P, Langen H, et al. (Eds.), *Proceedings of the Swiss Proteomics Proceedings of the Swiss Proteomics Society, 2003 Congress, Understanding biological systems* (pp. 174–176). Lausanne, Switzerland: FontisMedia Publishers.

Tveit, A., & Engum, H. (2003). Parallelization of the Incremental Proximal Support Vector Machine Classifier using a Heap-based Tree Topology. In *Parallel & Distributed computing for Machine Learning*.

Udupa, J.K. & Hermann, G.T. (1991). 3D Imaging in Medicine. Boca Raton: CRC Press..

Udupa, J.K. (1991). Computer Aspects of 3D Imaging in Medicine: A Tutorial. In J.K. Udupa and G.T. Hermann (Eds.), *3D imaging in Medicine*, (pp. 1-70).

Ullman, C. & Dykes, L. (2007). *Beginning Ajax*. Hoboken, NJ: John Wiley & Sons. Retrieved from www.Wrox.com.

Umetsu, R., Ohki, S., Fukuzaki, A., Konagaya, A., Shinbara, D., Saito, M., Watanabe, K., Kitagawa, T., & Hoshino, T. (2006). An Architectural Design of Open Genome Services. In *Grid Computing in Life Science* (pp. 87-98). Singapore: World Scientific.

UNICORE (2008). UNiform Interface to COmputing Resources. Retrieved from www.unicore.eu/

UNICORE (Uniform Interface to Computing Resources) (2008). Retrieved October 20, 2008, from http://www.unicore.eu/.

United States. Dept. of Defense. (1999). *Common criteria & methodology for information technology security evaluation*

Unlu, M., Morgan, M.E., & Minden, J.S. (1997). Difference gel electrophoresis: A single gel method for detecting changes in protein extracts. *Electrophoresis, 18*(11), 2071-2077.

Uren, V., Cimiano, P., Iria, J., Handschuh, S., Vargas-Vera, M., Motta, E., and Ciravegna, F. (2006). Semantic annotation for knowledge management: Requirements and a survey of the state of the art. *Web Semantics: Science, Services and Agents on the World Wide Web, 4*(1), 14-28.

UTREF: untranslated regions database from RefSeq. Retrieved from http://www.ba.itb.cnr.it/BIG/UTRHome/.

UTRsite: collection of functional sequence patterns located in 5' or 3' UTR sequences. Retrieved from http://www2.ba.itb.cnr.it/UTRsite/.

Valle, M., Zavialov, A., Li, W., Stagg, S. M., Sengupta, J., Nielsen, R. C., et al. (2003). Incorporation of aminoacyl-tRNA into the ribosome as seen by cryoelectron microscopy. *Nature Structural Biology, 10*(1), 899-906.

Van Dyke Parunak, H., Brueckner, S. A., Matthews, R., & Sauter, J. (2005). *Pheromone Learning for Self-Organizing Agents*, IEEE Transactions on Systems, Man, and Cybernetics, Part A: Systems and Humans, 35(3).

Van Engelen R. et al. (2003). Developing Web Services for C and C++. *IEEE Internet Computing Journal*, 53–61.

van Heel, M. (1987). Similarity measures between images. *Ultramicroscopy, 21*, 95-100.

van Heel, M., & Schatz, M. (2005). Fourier Shell Correlation threshold criteria. *Journal of Structural Biology, 151*(3), 250-262.

van Rossum, G., & Drake, Jr, F. L. (2003). *An Introduction to Python*. Bristol, UK: Network Theory.

Vandenberghe, S., Segers, P., Meyns, B., & Verdonck, P. (2001). Hydrodynamic Characterization of Ventricular Assist Devices. *International Journal of Artificial Organs, 24*(7), 470-477.

Vannier, M.W., Hildebolt, C.F., Gayou, D.E., & Marsh, J.L. (1991). Introduction to 3D Imaging. In J.K. Udupa and G.T. Hermann (Eds.), *3D imaging in Medicine*, (pp. 71-88).

Vassura, M., Margara, L., Di Lena, P., Medri, F., Fariselli, P. & Casadio, R. (2007). Reconstruction of 3D Structures From Protein Contact Maps. In D. I. Mandoiu, A. Zelikovsky (Eds), *Third International Symposium (ISBRA 2007)*: Vol. 4463. *Bioinformatics Research and Applications*, (pp. 578-589)., Berlin, Germany: Springer.

Vassura, M., Margara, L., Di Lena, P., Medri, F., Fariselli, P. & Casadio, R. (2007). Fault Tolerance for Large Scale Protein 3D Reconstruction from Contact Maps. In R. Giancarlo and S. Hannenhalli (Eds.) *Seventh International Workshop on Algorithms in Bioinformatics (WABI 2007)*, (Lecture Notes in Bioinformatics 4645, pp. 25-37). Berlin, Germany: Springer-Verlag.

Vassura, M., Margara, L., Di Lena, P., Medri, F., Fariselli, P. & Casadio, R. (2008). Reconstruction of 3D Structures From Protein Contact Maps. *Computational Biology and Bioinformatics*, *5*(3), 357-367.

Vendruscolo, M., Kussell, E. & Domany, E. (1997). Recovery of protein structure from contact maps. *Folding and Design*, *2*(5), 295-306.

Ventrone, V.; Heiler, S., (1991, December). Semantic Heterogeneity as a Result of Domain Evolution. *ACM SIGMOD Record, 20*(4), 16-20.

Venugopal, S., Buyya, R., & Ramamohanarao, K. (2006). A Taxonomy of Data Grids for Distributed Data Sharing, Management, And Processing. *ACM Computing Surveys, 38*(1), 3.

Venugopal, S., Buyya, R., & Winton, L. (2004). A grid service broker for scheduling distributed data-oriented applications on global grids. Proceedings of the 2nd workshop on Middleware for grid computing, pp. 75–80, Toronto, Canada. www.Gridbus.org/broker

Verschelde, J.L., Santos, M.C.D., Deray, T., Smith, B., & Ceusters, W. (2004) Ontology-Assisted Database Integration to Support Natural Language Processing and Biomedical Data-mining. *Journal of Integrative Bio-informatics*.

Von Delft, F., Inoue, T., Saldanha, S. A., Ottenhof, H. H., Schmitzberger, F., Birch, L. M., et al. (2003). Structure of E. coli Ketopantoate Hydroxymethyl Transferase Complexed with Ketopantoate and Mg2+, Solved by Locating 160 Selenomethionine Sites. *Structure, 11*, 985-996.

von Laszewski, G., Foster, I., Gawor, J., & Lane, P. (2001). A Java Commodity Grid Kit. *Concurrency and Computation: Practice and Experience, 13*, 643-662.

Von Laszewski, G., Foster, I., Gawor, J., Lane, P., Rehn, N. & Russell, M. (2001). Designing Grid-based Problem Solving Environments and Portals. In *Proceedings of International Conference on System Sciences* (HICSS-34).

VOTES project. (2007). *Virtual Organisations for Trials and Epidemiological Studies*. Retrieved July 2008 from http://labserv.nesc.gla.ac.uk/projects/votes/index.html

VTK, (n.d.). *The Visualization Toolkit*. Clifton Park, NY: Kitware. Retrieved from http://www.vtk.org/

W3C (2001). *Web Services Description Language (WSDL)*. Retrieved from http://www.w3.org/TR/wsdl.
Warren, R. et al (2007a). A Prototype Distributed Mammographic Database for Europe 2007. *Clinical Radiology 62*(11) pp. 1044-51. Amsterdam, The Netherlands: Elsevier.

W3C (2004). *OWL Web Ontology Language*. Retrieved July 2008 from http://www.w3.org/TR/owl-features/

W3C (2007). *RDF Resource Description Framework*. Retrieved July 2008 from http://www.w3.org/RDF/

W3C. *XML Protocol Working Group*. Retrieved from http://www.w3.org/2000/xp/Group/

Wache, H., Vogele, T., Visser, U., Stuckenschmidt, H., Schuster, G., Neumann, H., and Hubner, S. (2001). Ontology-based information integration: A survey of existing approaches. In Gomez-Perez, A., Gruninger, M., Uschold, M., & Stuckenschmidt, H., (Eds.), *International Joint Conference on Artificial Intelligence; Workshop: Ontologies and Information Sharing*, (pp. 108-117). New York: ACM.

Wagner, E.G., Altuvia, S. & Romby, P. (2002). Antisense RNAs in bacteria and their genetic elements. *Adv. Genet.*, *46*, 361-398.

Wagstaff, A. (2000). Socioeconomic inequalities in child mortality: comparisons across nine developing countries. *Bull World Health Organ 2000, 78*,19-29.

Walker, D., Rana, O. F., Li, M., Shields, M. S., & Huang Y. (2000). The Software Architecture of a Distributed Problem-Solving Environment. *Concurrency: Practice and Experience, 12*(15),1455–1480

Wang, G., & Dunbrack, R.L. (2003). PISCES: improvements to a protein sequence culling server. *Bioinformatics, 19*,1589-1591. http://dunbrack.fccc.edu/Guoli/pisces_download.php

Warfield S.K., Zou, K.H, & Wells, W.M. (2004). Simultaneous Truth and Performance Level Estimation (STAPLE): An Algorithm for the Validation of Image Segmentation. *IEEE Transactions on Medical Imaging, 23*(7) 903-921

Warren R., Solomonides A.E., del Frate C., Warsi I., Ding J., Odeh M., et al. (2007). MammoGrid — a prototype distributed mammographic database for Europe. *Clinical Radiology, 62* (11), 1044-1051.

Warren, R. et al (2007b). A Comparison of Some Anthropometric Parameters Between an Italian and a UK Population: "proof of principle" of a European project using MammoGrid. *Clinical Radiology 62*(11), pp. 1052-60. Amsterdam, The Netherlands: Elsevier

Watry, P., & Larson, R. (2005). Implementing support for digital repositories in a data grid environment. *Local to Global data interoperability – Challenges and technologies* (pp. 60-64) LGDI.

Watt, J., Sinnott, R. O., Ajayi, O., Jiang, J. & Stell, A. J. (2006). *User Oriented Access to Secure Biomedical Resources through the Grid.* Proceedings of Life Science Grid Conference, Yokohama, Japan.

Watt, J., Sinnott, R. O., Jiang, J., Stewart, G., Stell, A. J., Martin, D. & Doherty, T. (2007). Federated Authentication and Authorisation for e-Science. In *Proceedings of APAC conference*, Perth, Australia.

WebMo (2008). Retrieved from http://www.webmo.net

Weeks, C. M. & Miller, R. (1999). The design and implementation of *SnB* v2.0. *J. Appl. Cryst., 32*, 120-124.

Weeks, C. M., Blessing, R. H., Miller, R., Mungee, R., Potter, S. A., Rappleye, J., Smith, G. D., Xu, H. & Furey, W. (2002). Towards automated protein structure determination: *BnP*, the *SnB*-PHASES interface. *Z. Kristallogr., 217*, 686-693.

Weeks, C. M., DeTitta, G. T., Hauptman, H. A., Thuman, P. & Miller, R. (1994). Structure solution by minimal function phase refinement and Fourier filtering (II. Implementation and applications). *Acta Crystallogr., A50*, 210-220.

Weeks, C.M., DeTitta, G.T., Miller, R. & Hauptman, H.A. (1993). Applications of the minimal principle to peptide structures. *Acta Cryst., D49*, 179-181.

Weicker, R. (1984). Dhrystone: A Synthetic Systems Programming Benchmark. *Communications of the ACM, 27*(10), 1013-1030.

Welch, V. (2005). *Globus Toolkit Version 4 Grid Security Infrastructure: A Standards Perspective*. Retrieved from http://www.globus.org/toolkit/docs/4.0/security/GT4-GSI-Overview.pdf.

Welch, V., Siebenlist, F., Foster, I., Bresnahan, J., Czajkowski, K., Gawor, J., et al. (2003). Security for Grid services. *Proceedings of the 12th IEEE International Symposium on High Performance Distributed Computing, 2003*, (pp. 48-57).

Westerhof, N., Elzinga, G., & Sipkema, P. (1971). An Artificial Arterial System for Pumping Hearts. *Journal of Applied Physiology, 31*(5), 776-781.

Weston, J., Leslie, C., Ie, E., Zhou, D., Elisseeff, A., & Noble William S. (2005). Semi-supervised protein classification using cluster kernels. *Bioinformatics, 21*(15): 3241-3247.

Whiddett, R., Hunter, I., Engelbrecht, J. & Handy, J. (2006). Patients' attitudes towards sharing their health information. *International Journal of Medical Informatics, 75*(7), 530-541.

White, S.A. (2004). *Process Modeling Notations and Workflow Patterns*. Retrieved from http://BPTrends.com

WHO (2007). *The world health report 2007 - A safer future*. Geneva, Switzerland: WHO publications. Retrieved July from http://www.who.int/whr/2007/en/index.html

WHO (2008). *International Classification of Diseases*. Retrieved July 2008 from. http://www.who.int/classifications/icd/en/.

Wiens, J.J., Brandley, M.C., Reeder, T.W.(2006). Why does a trait evolve multiple times within a clade? Repeated evolution of snakelike body form in squamate reptiles. *Evolution*, 60(1),123-41.

Wikipedia (2007). *Ontology*. Retrieved July 2008 from http://en.wikipedia.org/wiki/Ontology_%28computer_science%29

Wikipedia (2008). *Epidemiology*. Retrived July 2008 from http://en.wikipedia.org/wiki/Epidemiology

Williams, D. B., & Carter, C. B. (2004). *Transmission Electron Microscopy: A Textbook for Materials Science*. Berlin, Germany: Springer.

Winkler, G. (1995). *Random Fields and Dynamic Monte Carlo Methods*. Berlin, Germany: Springer-Verlag.

WISDOM (2006). *Wide In-Silico Docking On Malaria*. Retrieved from http://wisdom.eu-egee.fr/

Witten, I. H., & Frank, E. (2005). *Data mining: Practical machine learning tools and techniques*. San Francisco: Morgan Kaufmann.

Witten, I. H., & Frank, E. (Eds.). (2005). *Data mining: Practical machine learning tools and techniques*. San Francisco, CA: Morgan Kaufmann.

Witten, I., Frank, E. (2005). *Practical Machine Learning Tools and Techniques*, (2nd Ed.). San Francisco, CA: Morgan Kaufmann.

Wittkopp, P.J., Williams, B.L., Selegue, J.E., Carroll, S.B.(2003). Drosophila pigmentation evolution: divergent genotypes underlying convergent phenotypes. *Proc Natl Acad Sci USA*, *100*(4), 1808-13.

Witzig, C. (2008). Interoperability Shibboleth - gLite, *IVOM Workshop*. Regionales Rechenzentrum für Niedersachen, Hannover, Germany.

Wolf, J., Gerber, A.P., & Keller, W. (2002). tadA, an essential tRNA-specific adenosine deaminase from Escherichia coli. *The EMBO Journal*, *21*, 3841–3851.

Wolniewicz, P., Meyer, N., Stroiński, M., Stuempert, M., Kornmayer, H., Polak, M., et al. (2007). Accessing Grid Computing Resources with g-Eclipse Platform. *Computational Methods in Science and Technology*, *13*(2), 131-141.

Wolpert, D. H. (1992). Stacked generalization. *Neural Networks*, *5*, 241-259.

Wolpert, D., & Macready, W. (1997). No free lunch theorems for optimization. *IEEE Transactions on Evolutionary Computing*, *1*(1), 67-82.

Wong, J. Cagney, G, & Cartwright, H.M (2005). Specalign—processing and alignment of mass spectra datasets. *Bioinformatics*, *21* (9), 2088–2090.

Wong, S.C., Miles, S., Fang, W., Groth, P.T. & Moreau, L. (2005) Provenance-based validation of e-science experiments. In Y. Gil, E. Motta, V. R. Benjamins, & M. A. Musen (Eds.), *International Semantic Web Conference: Lecture Notes in Computer Science*, (Vol. 3729), (pp. 801–815). New York: Springer.

Wood T.E., Burke J.M., & Rieseberg L.H. (2005). Parallel genotypic adaptation: when evolution repeats itself. *Genetica*, *123*(1-2), 157-70.

Wood, J., Wright, H. & K. Brodie (1997). Collaborative Visualization. *Proc. Visualization '97*, (pp. 253-259).

Wozniak, J. M., Brenner, P., Thain, D., Striegel, A., & Izaguirre, J. A. (2005). Generosity and gluttony in GEMS: Grid Enabled Molecular Simulations. In *Proceedings of the 14th IEEE International Symposium on High Performance Distributed Computing*. Los Alamitos, CA: IEEE Computer Society Press.

Wroe, C., Stevens, R., Goble, C., Roberts, A., & Greenwood, M. (2003). A suite of DAML+OIL Ontologies to Describe Bioinformatics Web Services and Data. *International Journal of Cooperative Information Systems*, *12*(2), 441-445.

WSRF Web Services Resource Framework. Retrieved from http://docs.oasis-open.org/wsrf/wsrf-primer-1.2-primer-cd-02.pdf

Wu, X., & Tseng, C.W. (2006). Searching Sequence Databases Using High-Performance BLASTs. In A.Y. Zomaya, (Ed.), *Parallel Computing for Bioinformatics and Computational Biology*, (pp. 211-232). Hoboken, NJ: John Wiley & Sons, Inc.

Wüthrich, B. (1997). Discovering probabilistic decision rules. *International Journal of Intelligent Systems in Accounting, Finance and Management*, *6*(4), 269-277.

Wypychowski J., Pytlinski J., Skorwider Ł, Nazaruk M., Benedyczak K., Wroński M., Bała P. (2004). Life Sciences Grid in EUROGRID and GRIP projects. *New Generation Computing, 22* (2), 147-156.

Xiong, Z., Wu, X., Cheng, S., & Hau, J. (2003). Lossy-to-Lossless compression of medical volumetric data using three-dimensional integer wavelet transform. *IEEE Transactions on Medical Imaging*, *22*, 459-470.

Yafchak, M.F. & Trauner, M., (Eds.) (2008). *The Grid Technology Cookbook.* Retrieved January 15, 2008, from http://www.sura.org/cookbook/gtcb/index.php?topic=12&mlevel=1&parent=0

Yang, C. T., Tsai, S. T., & Li, K. C. (2005). Decision tree construction for data mining on grid computing environments. *In the 19th International Conference on Advanced Information Networking and Applications* (pp. 421-42). New York: IEEE Computer Society.

Yang, C.T., Luo, T.-L., & Lai, C.-L. (April 2005). Designing Computing Platform for BioGrid. *International Journal of Computer Applications in Technology, 22*, pp. 3-13.

Yang, Z. & Greenshields, I. (2001). *Ensemble Classification of the VF Dataset with Limited Merging.* In *Proc. 14'th IEEE Symposium on Computer-Based Medical Systems*, Bethesda, MD, (pp. 117-122).

Yang, Z. (1994). Maximum likelihood phylogenetic estimation from DNA sequences with variable rates over sites: Approximate methods. *Journal of Molecular Evolution, 39*, 306-314.

Yasui Y., Pepe, M., Thompson, M.L., Adam, B.L., Wright, G.L., Qu, Y., et al. (2003). A data-analytic strategy for protein biomarker discovery: profiling of high-dimensional proteomic data for cancer detection. *Biostatistics, 4*(3), 449–463.

Yona, G., Linial, N., & Linial, M. (2000). ProtoMap: automatic classification of protein sequences and hierarchy of protein families. *Nucleic Acids Research, 28*(1), 49-55.

Yoon, S., Smellie, A., Hartsough, D., & Filikov, A. (2005). Surrogate Docking: Structure-based Virtual Screening at High Throughput Speed. *Jouranl of Computer-Aided Molecular Design, 19*, 483-497.

Yu, J. and Buyya, R. (2005). A taxonomy of workflow management systems for grid computing. *Journal of Grid Computing*, *3*(3-4), 171-200.

Yu, J., Buyya, R., *Ong, SE., Blagoev, B., Kratchmarova, I., Kristensen, DB., et al. (2002).* Stable isotope labeling by amino acids in cell culture, SILAC, as a simple and accurate approach to expression proteomics. *Molecular & Cell Proteomics, 1, 376–86.*

Zaki, M. J. (1999). Parallel and Distributed Association Mining: A Survey. *IEEE Concurrency.*

Zaki, M. J., & Ho, C. T. (Eds.). (2000). *Large-Scale Parallel Data Mining.* Lecture Notes in Computer Science. Berlin, Germany: Springer.

Zara, V., Ferramosca, A., Capobianco, L., Baltz, K.M., Randel, O., Rassow, J., Palmieri, F. & Papatheodorou, P. (2007). Biogenesis of yeast dicarboxylate carrier: The carrier signature facilitates translocation across the mitochondrial outer membrane. *J. Cell Science* (120), 4099–4106.

Zdobnov, E.M., & Apweiler, R. (2001). InterProScan: protein domains identifier. *Bioinformatics*, *17*, 847-848.

Zeilinger, G. (2006). dcm4che, a DICOM implementation in Java and dcm4chee J2EE based PACS. *Main page*. Retrieved February 14, 2008 from http://www.dcm4che.org

Zhang, J. (2008). *Visualization for information retrieval.* Berlin-Heidelberg: Springer.

Zhao, J., Goble, C., Stevens, R., & Turi, D. (2007). Mining Taverna's Semantic Web of Provenance. In *Concurrency and Computation: Practice and Experience*, *20*(5), 463-472.

Zhao, J., Wroe, C., Goble, C. A., Stevens, R., Quan, D., Greenwood, R. M. (2004). Using Semantic Web Technologies for Representing E-science Provenance. In *International Semantic Web Conference, LNCS* (Vol. 3298), (pp. 92–106). New York: Springer.

Zhao, Y., Wilde, M., & Foster, I. T. (2006). Applying the virtual data provenance model. In L. Moreau and I. T. Foster (Eds.) *Provenance and Annotation of Data, International Provenance and Annotation Workshop, IPAW 2006, Chicago, IL, USA, May 3-5, 2006, Revised Selected Papers, Lecture Notes in Computer Science,* (Vol. 4145) (pp. 148–161). New York: Springer.

Zmasek, C.M. (n.d.). *Forester: software libraries for evolutionary biology research.* Retrieved June 6, 2008, from http://www.phylosoft.org/forester.

Zomaya, A.Y. (Ed.) (2006). *Parallel Computing for Bioinformatics and Computational Biology.* Hoboken, NJ: John Wiley & Sons, Inc.

Zosso, D., Arnold, K., Schwede, T., Podvinec, M., (2007). SWISS-TANDEM: A Web-Based Workspace for MS/MS Protein Identification on PC Grids. *CMBS,* 406-411.

Zwickl, D. (2006). Genetic algorithm approaches for the phylogenetic analysis of large biological sequence datasets under the maximum likelihood criterion. *PhD Thesis*, The University of Texas at Austin.

About the Contributors

Mario Cannataro is an associate professor of Computer Engineering at University Magna Græcia, Catanzaro, Italy, since 2002. He received his Laurea Degree Magna cum Laude in Computer Engineering from the University of Calabria, Rende, Italy, in 1993. He was a researcher at the CRAI Research Consortium and a senior researcher at the Italian National Research Council (ICAR-CNR), Rende, Italy, where he worked on parallel computing, massively parallel architectures, and parallel implementation of logic programs and cellular automata. He is currently the responsible of the Bioinformatics Laboratory at University Magna Græcia of Catanzaro. His current research interests are on grid computing, bioinformatics, computational genomics and proteomics, and ontologies. He authored a book on parallel implementation of logic programs and over 150 scientific papers published on international journals and conference proceedings. He is a member of ACM, IEEE Computer Society, BITS (Italian Bioinformatics Society), and HealthGrid. He is a co-founder and a member of Exeura (www.exeura.com). Mario Cannataro can be reached at cannataro@unicz.it.

* * *

Giovanni Aloisio is full professor of Information Processing Systems at the Engineering Faculty of the University of Salento, Lecce. His research interests are in the area of High Performance Computing, Distributed and Grid Computing. He has founded SPACI Consortium (www.spaci.it). He is also the Vice-director of the NanoSciences and Grid Computing Section of the "Scuola Superiore ISUFI" and CEO of the SPACI Consortium. He is also director of the Division "ICT and Operations" of the Euromediterranean Center for Climate Chances (CMCC). He is the author of more than 100 papers in refereed journals on parallel & grid computing.

Domingos Alves obtained his doctoral degree on Biologic Physics in São Paulo University, USP in 1999. He is currently a professor of the Social Medicine Department at São Paulo University, Brazil. He has been working in the area of Theoretical Biology, Modeling and Simulation. His current research focuses on the transmission dynamics and within-host population biology of infectious disease. It combines viral quasispecies modeling with epidemic modeling to address questions in these areas.

Isabelle Andoulsi is a lawyer at the Brussels Bar, working within the law firm Exelia, as well as a researcher and at doctoral candidate at the Institute of European Studies of the Brussels University, Belgium (www.iee-ulb.eu). She currently works on the assumption of a use or a transposition of the assets as regards data protection within the framework of the first pillar in the third pillar of the European Union Treaty. She also undertook research concerning the possibility of an innovating use of information technologies in order to allow a broad access of patients to effective, profitable and of high-quality health care services. She is thus particularly interested in personal data processing and in personal data transfers, but also in the questions of interoperability of personal databases, of shared responsibilities within the framework of telematics and telemedicine networks, and in intellectual property rights and competition law.

Ashiq Anjum has recently completed a PhD on data-intensive and network-aware Grid scheduling and previously held a MSc in Computer Science.He has been influential in the Grids community and has responsibility for UWE's involvement in the Grid aspects of the Health-e-Child project. in addition Anjum has led research into the provision of generic, reusable services in the context of a service-oriented architecture for the FP7 neuGRID project. Dr Anjum has a growing publications list with over 30 papers in international journals or conferences. His current research is in Grid-based medical systems and the role of SOAs and Cloud computing in applied computer science.

Anna Anselmo obtained her Master's degree in Functional Genomics and Bioinformatics in 2005 carrying out a thesis on alternative splicing prediction at the University of Milan, Italy. She is now concluding her PhD in "Molecular and Cellular Biology" at the University of Milan. The main topic of her PhD project has been the in silico characterization of cancer specific isoforms. Her research interests include also the study of coding and non-coding conserved sequence tags in human and mouse genomes.

Maurizio Arabia received the Laurea Degree in Electrical Engineering in 1957 and specialized in Nuclear Engineering in 1959 from the University of Rome (Italy). He worked in the bioengineering field and particularly in adaptive control applied to cardiovascular system, cardiac mechanics, total artificial heart, ventricular assist devices, and mock circulatory systems. Together with Prof. Westerhof (Institute for Cardiovascular Research, Free University of Amsterdam, Netherland) he launched the European project HEART. He collaborated with the Univ. of Rome Italy (Department of Applied Mechanics Dept.), ENEA Agency, University of Mississippi Medical Center (Jackson, Mississippi, USA), Texas Heart Institute (Houston, USA), National Cardiovascular Center (Osaka, Japan), and University of Calabria (Department of Mechanical Engineering, Rende, Italy). He is now working on volumetric continuous flow-high efficiency pumping devices for industrial applications.

Matthias Assel received his diploma degree in computer science from the Technische Universität München in 2005 and has joined the High Performance Computing Center of University Stuttgart (HLRS) in 2006 in order to write his PhD thesis in the field of (bio)medical IT solutions. He is currently working on Grid-related European projects on research and development level, examining and making use of emerging technologies. His actual research focus rests on secure environments for distributed data access, distributed data and text mining in biomedicine, but also intelligent infrastructures including e.g. the automatic reconfiguration of relevant services.

Piotr Bala is professor at ICM Uniwersity od Warsaw and at Faculty of Mathematics and Computer Science N. Copernicus University, director of Department of Parallel and Distributed Processing. He obtained PhD in Physics in 1993 in Institute of Physics, N. Copernicus University and in 2000 habilitation in physics for development and application of new methods to study enzymatic reactions in biological systems. He has published over 60 papers. His main research interest is in development of parallel and grid tools for scientific applications. He has been involved in number of European grid projects, currently acts as coordinator of Chemomentum project.

Kim Baldridge, professor of Theoretical Chemistry at the University of Zürich. She graduated with a double BA in chemistry and mathematics, received MA in mathematics and a PhD in theoretical chemistry in 1988 in the USA. She is a theoretical and computational chemist with expertise in the design, development, implementation, and application of computational chemistry methodology. She is the coauthor, developer, and supporter of the computational quantum chemistry software, GAMESS, in addition to several key associated tools for Grid computing, analysis and visualization. She has over 100 publications that cover topics in these areas and directs a programme in computational Grid applications and computing.

Bartosz Balis obtained his MSc degree in Computer Science from the AGH University of Science and Technology, Krakow, Poland. He is employed as a research and teaching assistant at the Institute of Computer Science, AGH. Co-author of about 50 international publications. His research interests include monitoring and performance analysis of applications, Grid computing, scientific workflows, adapting legacy codes to modern environments, and provenance tracking. Participant of several international research projects including EU-IST CrossGrid, CoreGRID, K-Wf Grid, ViroLab, Gredia. Member of Program Committee for ICCS conference and a reviewer at journals: Future Generation Computer Systems, Scientific Programming, and conferences: ICCS, EuroPVM/MPI, and e-Science 2006.

Roberto Barbera graduated in Physics "cum laude" at the University of Catania in 1986 and since 1990 he holds a PhD in Physics from the same University. He is currently Associated professor at the Department of Physics and Astronomy. His main research activity has been done in the realm of Experimental Nuclear and Particle Physics. Since 1997 he is involved in the NA57 Experiment at CERN SPS and in the ALICE Experiment at CERN LHC. Since late 1999 he is interested in Grid Computing. He is a member of the Technical Board of the Italian INFN Grid Project.

Tomasz Bartynski is a computer scientist at the Academic Computer Center CYFRONET AGH in Krakow. He received his MSc degree in computer science from the AGH University of Science and Technology in Krakow. Since 2006 he has been working for ACC Cyfronet AGH in Krakow. Currently he is involved in EU-funded projects: ViroLab and GREDIA as a designer and developer. His scientific interests involve distributed systems, parallel computing, Web technologies and programming paradigms.

Lisa Bartoli has a degree in Physics. Actually she is a PhD student in Physics at the Biocomputing Group, University of Bologna. Her main research interests concern Computational Biology, in particular protein structure prediction, prediction of protein-protein interactions, study of protein contact maps with machine learning methods.

Khalid Belhajjame is a researcher at the school of computer science of the University of Manchester. He obtained his PhD from the University of Grenoble in 2004. His research interests include semantic annotations of web services, scientific workflows, dataspaces, and data mapping and integration.

Emilio Benfenati has been researcher at Stanford University, CA, in 1983-84. In that period he also had a collaborative research at the Berkeley University, CA. He has gained a considerable experience in the management of international programs, coordinating 11 European projects and participating to 16 others. Many of them are on toxicity and environmental modelling. He has been Member of the National Commission for the evaluation of pesticides, Italian Ministry of Health, 1997-1999. His research activities include: computer chemistry, QSAR, toxicity prediction, risk assessment and many others. He is author or co-author of about 200 papers in international journals and books.

Margherita Berardi is a computer science researcher in the field of data mining. She received the PhD Degree (Curriculum of Artificial Intelligence) from the University of Bari by discussing a work entitled "Towards Semantic Indexing of documents: a Data Mining perspective" under the supervision of Prof. D. Malerba. She worked as a research consultant for Exhicon s.r.l. and the University of Bari within projects for IBM Innovation Centre focusing on unstructured data management for Bioinformatics.

José Román Bilbao Castro received the MSc and PhD degrees in computer science from the University of Almería, Spain, in 2001 and 2006, respectively. As a PhD student he worked at the Bio-Computing Unit of the National Center for Biotechnology, Spain. Since 2007 he holds a postdoctoral position at the University of Almería under the "Juan de la Cierva" contract. He is a member of the Supercomputing-Algorithms group at the Department of Computer Architecture and Electronics and his research interests include parallel computing in 3D electron microscopy, grid computing and image processing algorithms.

Ignacio Blanquer (PhD) is assistant professor of the Computer System Department (DSIC) of the Polytechnic University of Valencia (UPV) since 1999. He has been involved in Parallel Computation and Medical Image processing since 12 years ago participating in 17 national and European Research Projects. He is also a research fellow of the Institute for the Applications of Advanced Information and Comunication Technologies (ITACA) and Network Centre for Biomedical Engineering (CRIB) and member of the board of directors of HealthGrid association.

Vincent Breton received his engineer degree from Ecole Centrale de Paris in 1985 and his PhD in Nuclear Physics from the University of Paris XI- Orsay in 1990. Since 1990, he has been a research associate at the French National Centre for Scientific Research (CNRS). In 2001, he founded a research group (http://clrpcsv.in2p3.fr) on the application to biomedical sciences of the IT technologies and tools used in high energy physics. He is also investigating the relevance of grid technology for biomedical applications. He co-founded of the Healthgrid initiative, chaired the first European conferences on grids for health in January 2003 and January 2004. He is leading application support and identification in the EGEE european project (FP6). He is involved in various European initiatives in the biomedical area.

Marian Bubak, PhD, is an adjunct at the Institute of Computer Science AGH, a staff member at the ACC CYFRONET AGH, and the professor at the Informatics Institute of the Universiteit van Amsterdam. His research interests include distributed and grid systems for scientific simulations. He co-authored about 230 papers. He lead the architecture team of the EU IST CrossGrid Project, he was the Scientific Coordinator of K-WfGrid Project and the member of the Integration Monitoring Committee of CoreGRID. He served as a program committee member, chairman and organizer of several international conferences (HPCN, Physics Computing, EuroPVM/MPI, SupEur, HiPer, ICCS, HPCC, e-Science'2006); he is co-editor of 17 proceedings of international conferences.

Massimo Cafaro is an assistant professor at the Department of Innovation Engineering of the University of Salento. His research covers High Performance, Distributed and Grid Computing. He is the principal investigator and developer of the Grid Resource Broker, a web portal to computational grids and of the GSI plug-in for gSOAP. He received a PhD in Computer Science from the University of Bari, is a member of ACM, IEEE and IEEE Computer Society, and holds a visitor position at CACR Caltech. He authored more than 70 refereed papers on parallel, distributed and Grid Computing. He co-authored and holds a patent on distributed database technologies.

Antonio Calanducci graduated in Computer Science "cum laude" at the University of Catania in 2004 and his final dissertation was carried out at NYU in the field of BioInformatics. Since his graduation, he has been involved in the realm of grid computing: starting as a collaborator at INAF, since 2005 he has been working at the INFN Catania within EGEE projects. His main activities are related to the dissemination and training of Grid computing, focusing on gLite. He also has carried out research activities on Grid-based Digital Libraries, and has a particular expertise on data and metadata management in grids.

Barbara Canesi received the master degree cum laude in Biomedical Engineering from the University of Genoa, Italy, in 2004. In 2004-2005 she collaborated to many biomedical projects, aimed to support diagnosis of neurodegenative diseases. Since 2006, she has been a PhD student at DIST, University of Genoa, Italy. In 2007-2008 she was visiting student at University of Maastricht, The Netherlands. Her research activity is focused on neuroimaging and integration of different kinds of biomedical signals and data.

Eddy Caron is an assistant professor at Ecole Normale Supérieure de Lyon and holds a position with the LIP laboratory (ENS Lyon, France). He is a member of GRAAL project and technical manager for the DIET software package. He received his PhD in C.S. from University de Picardie Jules Verne in 2000. His research interests include parallel libraries for scientific computing on parallel distributed memory machines, problem solving environments, and grid computing. He is co-chair of the GridRPC working group in OGF. He is the coordinator of LEGO a french ANR project (ANR-CICG05-11). See http://graal.ens-lyon.fr/~ecaron for further information.

Luciana Carota is working in grid tecnology from 2004 at the National Institute of Nuclear Physics (INFN), Bologna, Italy. She received a PhD in Physics in January 2004 from L'Aquila University, Italy, working in collaboration with INFN, Rome, on neural dynamics and learning in VLSI recurrent neural networks. She received a master degree in Physics in 2000 and a degree in Electronic Engineering in 2006.

Rita Casadio is Full professor of Biochemistry/Biophysics at the University of Bologna and Group Leader of the Biocomputing Group (www.biocomp.unibo.it). She has a degree in Physics and has been working mainly in the fields of membrane and protein Biophysics both experimentally and theoretically. Presently she is interested in computer modelling of relevant biological processes and her researches are devoted to different aspects of protein modelling, including prediction of secondary and tertiary structures with neural networks, hidden Markov models, genetic algorithms, molecular docking, protein reaction mechanisms and de novo design of helical peptides. RC is the author of about 200 scientific papers and presented her work at several (over 300) national and international meetings.

Mosè Casalegno is senior research fellow at Mario Negri Institute for Pharmacological Research. In 2002 he obtained a PhD degree in theoretical and computational chemistry from the University of Milan. As a post-doctoral fellow he has been working on quantum Monte Carlo methods. In 2003 he joined the department of Health and Environment at Mario Negri. His main research interest is the development of algorithms and models in the field of chemical information and QSAR. He has been involved in different European funded projects in the area of computational chemistry, toxicity prediction and Grid computing.

Tiziana Castrignanò is a bioinformatics senior technologist at CASPUR (www.caspur.it). She received her PhD in Biophysics in 1999 from the University of Rome "La Sapienza". She has developed bioinformatics web tools and databases largely used by the scientific community. Her primary activity is the development of high performance bioinformatics services and databases and leads the technological support activity to many national and international bioinformatics research projects.

Chen Chen received his Bachelor from the Fudan University, China in 2003, and the PhD degree from the Nanyang Technological University (NTU), Singapore, in 2008. His research interests include 3D modeling and reconstruction, computer animation, physical simulation and general purpose GPU computing.

Eryk Ciepiela received his MSc title in Computer Science from the AGH University of Science and Technology in Kraków, Poland for the thesis "Monitoring of component-based applications". In 2005 - 2006 he was a memeber of Technologies Group in Reseach and Development departament in ComArch, Kraków. Since 2006 he has worked for CYFRONET supercomputing centre in Kraków where he has been involved in EU-funded project ViroLab and Gredia working on GridSpace Engine. He's also a founder of IT enterprise it-xpander.pl. His area of interest ranges from enterprise systems, web technologies to distributed systems, monitoring, programming paradigms, models and methodologies.

Patrick Clarysse received a MSc degree in 1987 and a PhD in 1991 from the Scientific and Technological University of Lille, France. Since 1992, Patrick Clarysse is with the French National Center for Scientific Research (CNRS) at CREATIS-LRMN, CNRS UMR #5220, Inserm U630, Lyon, France. His primary research interests are in the bioengineering and medical image analysis fields, and include multimodal image segmentation and registration, motion estimation and deformable models with applications to the analysis of the heart functions and the motion of thoracic structures.

Luca Clementi received his MSc in Software Engineering in 2005 from the University of Bologna, Italy. He worked form 2005 till 2007 as a researcher at CINECA where he was involved in several EU founded projects on Grid computing (DEISA and A-WARE). Since 2007, he is working at the University of California, San Diego in the National Biomedical Computation Resource project where he is in charge of the development of the SOA infrastructure which integrates biomedical applications with high end computational resources.

Francesco Maria Colacino received the Laurea Degree (summa cum laude) in Mechanical Engineering in 2000 from the University of Calabria (Rende, Italy), and the PhD in 2004 from the same University with a thesis entitled "An Elastance-Based Artificial Ventricle for Testing any Cardiac Assist Device". From 2005 to 2008 he was Research Fellow at the University of Calabria as well as at the Department of Surgery, University of Kentucky Chandler Medical Center (Lexington, KY - USA). He is currently Research Fellow at the Institute of Sound and Vibration Research (ISVR), University of Southampton (UK). His present research interests deal with the applications of bio-feedback to skeletal muscle models.

Antonio Congiusta received his PhD in systems and computer engineering from University of Calabria, Italy, in 2006. From 2002 to 2003 he was research fellow at ICAR-CNR, Institute of the Italian National Research Council, working on Grid-based data mining systems. From 2004 to 2006 he was research fellow at University of Calabria, working on the Knowledge Grid project. Currently he is research fellow at University of Salerno, Italy. His research interests are focused on Grid programming environments, workflow-based Grid systems, and virtual organizations.

Luca Corradi received his master degree in Biomedical Engineering at the University of Genoa, Italy, in 2006 and gained admission into the engineering profession, in 2007. Since 2007, he has been a PhD student at DIST, University of Genoa, Italy. His research activity is focused on data and application integration applied to bioinformatics and neuroinformatics fields.

Paolo Cozzi received the Laurea degree in Bioinformatics from the university of "Milano – Bicocca", Italy, in 2007 with a thesis on the development of databases for the correlation between protein functions and molecular surfaces. His research activity concerns the development of software for functional analysis in genomics and proteomics. Moreover, he has implemented HPC solutions for grid computing, with a particular interest in searching for surface similarities and molecular visualization. At the moment he works at Institute of Biomedical technology supported by a fellowship of the PhD School of Molecular Medicine at the University of Milan.

Guido Cuscela has a software engineer degree from the "Politecnico di Bari", Italy. He is a member of EGEE I and EGEE II project since March 2006 as SA1 and JRA1. He has worked for GridICE project (INFN Grid monitoring tool) deploying specific sensors for jobs and batch systems monitoring. He is now a member of LIBI project .

Bertrand Delhay received an engineering degree and a PhD degree in Image Processing from the National Institute of Applied Sciences (INSA) of Lyon, France in 2002 and 2006, respectively. He is currently a postdoctoral research fellow at the CREATIS-LRMN lab. His research interests are medical image processing, image registration, spatio-temporal motion estimation, object tracking and intensive computing.

Domenica D'Elia received a summa cum laude degree in Biological Sciences in 1991 and a PhD in Biochemistry and Molecular Biology in 1996 from the University of Bari (IT). She presently works as a researcher at the Institute of Biomedical Technologies of the National Research Council (CNR-IT). Her research interests include bioinformatics (database and analysis tools development) applied to the study of mitochondrial biogenesis and to genome expression studies. She is node manager of the European Molecular Biology Network node (http://www.embnet.org/) in Italy, member of the editorial board of EMBner.news journal and of EMBnet Publicity and Public Relations Project Committee.

Giulia De Sario is a researcher of the Italian CNR (National Research Council). She graduated in Physics from the University of Bari in 2003. She works on GRID infrastructure applied on bioinformatics, in particular her research field is the implementation of bioinformatic tools and databases on grid.

Frédéric Desprez is a director of research at INRIA and holds a position at LIP laboratory (ENS Lyon, France). He received is PhD in C.S. from the Institut National Polytechnique de Grenoble in 1994 and his MS in C.S. from the ENS Lyon in 1990. His research interests include scheduling for large scale distributed platforms, SOA architectures, data management, and grid computing. See http://graal.ens-lyon.fr/~desprez for further information.

Luigi Di Pace is the director of the IBM GBS Innovation Center. He began his career at the IBM Scientific Center based in Rome in 1986. Since 1991 he was a researcher involved in complex AI projects for large customers. In 1995 he was the funder of the Innovation Center located in Bari starting a team of 16 researchers working on Object-Oriented and Internet technologies. He currently heads, directly and/or receive his functional direction, an organization of more than 150 researchers and specialists providing innovative services projects to all relevant IBM customers for all main industrial sectors and business areas.

Frank Dickmann is research assistant in the department of Medical Informatics, Georg-August-University Göttingen, Germany. In 2005 he finished his diploma degree in Information Systems at the Georg-August-University Göttingen. Since 2006 he is a PhD-student at the Georg-August-University Göttingen. His main scope of study is strategic IT management of universities. From 2006 to 2007 he worked as academic assistant in the department of IT Strategy at the University Medical Center Göttingen. His main field of work is the development of economic sustainability for biomedical Grid Computing.

Zhaohui Ding was born in 1979. He received the BS and MS degrees in Computer Science from Jilin University, in 2001 and 2006, respectively. Currently, he is a PhD student of College of Computer Science and Technology of Jilin University, P.R.C. His research fields include grid computing, meta-scheduling and distributed system.

Alexander Dobrev holds a Bachelor of Arts Honours Degree in Philosophy, Politics and Economics from the University of Oxford (2005). He has an extensive record of experience in international work and education environments, including international projects and conferences across Europe. He joined the *empirica* eHealth team as a full time consultant in 2005. Alexander is focusing on the socio-economic and commercial aspects of healthcare and eHealth. Recent EU studies he has been involved in include Scenarios4Health, a foresight study on scenarios for ICT-enabled new models of healthcare; eHealth ERA, a study analysing priorities and strategies for eHealth in 32 European countries; eHealth Indicators, a representative survey of European GPs. Being responsible for Financing eHealth – a study on investment decision making in eHealth, Alexander investigated decision making processes and success factors for eHealth investments. Alexander has gained extensive experience and expertise with good practice case identification, selection, description and presentation through his substantial contribution to a number of studies, including eHealth IMPACT, a Study on the Economic Impact of eHealth and Good eHealth, Study of best practice across Europe in providing innovative health-related services, both for European Commission, DG INFSO. Through his contributions to SHARE, he has also explored the socio-economic and business issues concerning HealthGrids. Currently, he is leading the EHR IMPACT study on the economic impact of interoperable electronic health records and ePrescribing systems and is responsible for the socio-economic impact evaluations in several large EU research projects.

Giacinto Donvito is currently working as technical coordinator for LIBI (International Laboratory of Bioinformatics) a National project for supporting bioinformatics application over gLite grid. As system administrator of the Bari Grid farm since 2004, has gained experience in deploying and using several storage software solutions and database grid access software tools.

Celine Van Doosselaere has a background in liberal arts and in development management. Since 2000, she has worked in Brussels on European policies impacting health - including the fascinating field of eHealth. At the time of her collaboration with the SHARE team, Celine worked for the European Health Management Association (EHMA), a non-profit membership association bringing together academics & researchers, policy-makers, trainers and managers to improve health management with a view to raising standards in care.

Gamal El-Sayed is an assistant professor in the Department of Electrical Engineering at Assiut University in Egypt. He obtained his PhD in computer science at the University of Connecticut, where he also completed his post-doctoral training. His research interests include general grid computing with specific interests in dynamic task migration over the Grid.

Andrew Emerson received his PhD in computational chemistry in 1991 from the University of Southampton (UK). In 2000 he took up a post at the CINECA supercomputer centre (Bologna) where apart from performing research, he also provides support and gives courses to researchers using the centre's facilities. His current interests include the molecular dynamics of biomolecular systems, systems biology and grid computing.

Italo Epicoco is an assistant professor at the University of Salento, Italy. He received PhD in "Materials and Innovative Technologies" in June 2003 at University of Salento. He is staff member of the Center for Advanced Computational Technology (CACT) of the National Nanotechnology Laboratory (NNL CNR-INFM). He is responsible of the High Performance Computing group within the Scientific Computing and Operation division of the Euro-Mediterranean Center for Climate Changes (CMCC). His research interests include high performance computing, distributed and grid computing. He is author of about 40 papers in conference proceedings, journals and book chapters.

Francesco Falciano obtained his master degree in Biology with a specialization in Bioinformatics from the university of Milano Bicocca. After taking up a position at the CINECA supercomputing center he obtained a PhD in Biochemistry and Molecular Biology from the University of Ferrara. At CINECA he works in the high performance computing group with interests in computational biology and bioinformatics.

Juergen Falkner works in the business unit Software Technology at the Fraunhofer Institute for Industrial Engineering (IAO). He has been a key contributer to the development and setup of the Fraunhofer Resource Grid. He has been working within several Grid-related projects on European as well as the German national level (D-Grid) with focus on Grid user interfaces such as Grid portals. Within MediGrid he gathered experience on a variety of biomedical applications. Juergen Falkner has also been working in the field of IT security and incident response with a focus on VO and user management in Grids, authentication mechanisms and tools and the procedural aspects of IT security.

Piero Fariselli has a PhD in Biophysics and a Laurea in Physics. Author of over 80 publications he is a permanent researcher at the Biocomputing Group, University of Bologna. His main research interests concern Computational Biology and Machine Learning. PF main achievements are in the field of bioinformatics, where he developed machine learning codes, such as neural networks and hidden Markov models, and other algorithms for the prediction of structural and functional features of biomolecules.

Marco Fato is associate professor in Biomedical Engineering at School of Engineering of the University of Genoa since 2006. He has a career covering 20 years as a Research Collaborator at DIST and other national and international laboratories. He held a research position at DIST from 2001 to 2005, during which time he authored more than research 50 papers.

José-Jesús Fernández received the MSc and PhD degrees in computer science from the University of Granada, Spain, in 1992 and 1997, respectively. He was a PhD student at the BioComputing Unit of the National Center for Biotechnology (CNB), Spanish Research Council (CSIC), Madrid, Spain. He became an assistant professor in October 1997 and since 2000 he is an Associate professor of Computer Architecture at the University of Almería, Almería, Spain. He is a member of the Supercomputing-Algorithms Research Group at the University of Almería. His current research interests include high performance computing, image processing, and tomographic reconstruction.

Alessandra Ferramosca is researcher of Biochemistry at the University of Salento (Lecce). Her research includes the following topics: study of the import pathway of some proteins, in particular the metabolite carriers, into isolated mitochondria; purification of carrier proteins of the inner mitochondrial membrane and characterization of their structural and functional properties; effect of different nutritional states on the hepatic fatty acid biosynthesis; energetic metabolism of spermatozoa. She is the author of 43 scientific articles, most of them published in specialized international journals. She has participated as a speaker at various national conferences.

Sandro Luigi Fiore was born in Galatina, Italy (1976). He received a summa cum laude Laurea degree in Computer Engineering from the University of Lecce (Italy), (2001). He received a PhD degree in Informatic Engineering from the ISUFI-University of Lecce, Italy, (2004). Since 2001 he is the Principal Investigator of the GRelC project. He was involved in EGEE and EGEE-II projects and currently in EGEE-III. Since 2006, leads the Data Grid Group at the Euro-Mediterranean Centre for Climate Change (CMCC). He is author/co-author of about 45 papers on parallel & grid computing and he got 1 patent on data management.

Raffaele Fronza has a Degree in Molecular Biology. He worked for 5 years on cancer research. Now is Phd student at the University of Bologna and is involved in a project of applied genomics aimed to identify the basic biological process affecting the etiology and pathogenesis of complex congenital disorders.

Gionata Fragomeni graduated in Engineering at the University of Calabria in 1997. He received the PhD in Bioengineering from the University of Bologna in 2003. Gionata Fragomeni was a temporary chair in Cardiovascular Mechanics and Mechanics for Biomedical Applications at Calabria University and Magna Graecia University. He is an assistant professor at the Faculty of Medicine at Magna Graecia University of Catanzaro. He was the coordinator of 2 national research projects financed by the Italian Ministry of Research, and has authored over 30 publications on journals and international conference proceedings. His research interests include Biomechanics, Virtual Reality Applied to Orthopaedic, Bio-Fluid Dynamics, and Cardiovascular Mechanics.

Wlodzimierz Funika holds MSc degree in Mathematics and PhD in Computer Science. From 1978 to 1993 he was with Moldavian Academy of Sciences. At present he is with the Institute of Computer Science AGH. His research interests include parallel and distributed computing, development tool construction, performance monitoring, semantic-oriented systems, visualisation techniques. He has authored and co-authored ca. 100 papers. He participated in EU funded and national projects, e.g. CrossGrid, K-Wf Grid, CoreGRID. Currently he is involved in the EU IST ViroLab project where he is responsible for the presentation layer.

Henrique Fabricio Gagliardi has a master degree in Computer Science from Universidade Católica de Santos, Brazil. His recent work is related to grid computing and health care applications applied to the modeling of biological epidemic systems, design and development of grid based simulation services for epidemic forecasting tools and development of health based applications for PalmOS. His current work-in-progresses consists on the development of a peer-to-peer network-based communication service for Tidia-AE, a Brazilian e-learning project and Grid to Peer-to-Peer integration for use in health care applications.

Eduardo Gallo has a master degree in Computer Science from Universidade Católica de Santos (2006). He has also a specialist degree in Information Technology applied to Disabled People (Universidade Federal do Mato Grosso do Sul, 2000) and has a bachelor degree in Computer Science (UNIBAN, 2000). Currently he is System Manager at APRAESPI. He is interested in the following areas: grid computing, distributed systems, web development, database administration, database integration and epidemiological surveillance systems.

Inmaculada García Fernández received the BSc degree in physics in 1977 from the Complutense University of Madrid, Spain, and the PhD degree in 1986 from the University of Santiago de Compostela, Spain. From 1977 to 1987 she was an assistant professor at the University of Granada, from 1987 to 1996 an Associated professor at the University of Almería, and since 1997 she is a Full professor at the University of Almería, Spain. She is head of the Department of Computer Architecture and Electronics and the Supercomputing-Algorithms Research Group. Her research interest lies in the field of parallel algorithms for irregular problems related to image processing, global optimization, and matrix computation.

Yann Gaudeau received the engineering degree from Université Technologique de Troyes,France, in 2003, and, in 2006, the PhD degree in Automatic Control and Signal Processing from Université Henri Poincaré, Nancy, France. He is currently a postdoctoral fellow with the Centre de Recherche en Automatique de Nancy (CRAN, Nancy Université, UMR CNRS 7039). His research interests are image coding, lattice vector quantization and medical image compression.

Cécile Germain-Renaud received the agrégation de mathématiques in 1979, her PhD in Computer Science in 1989 and her Habilitation à Diriger les Recherches in 2005. She is a professor at Paris-Sud University (Laboratoire de Recherche en Informatique). Her research interests are in models and optimisation of high performance computing systems, parallel architectures and grids.

Andreas Gisel is a researcher of the Italian CNR (National Research Council). He graduated in molecular biology from the Swiss Federal Institute of Technology Zuerich in 1990 and obtained a PhD in 1995 from the same institution. His research field is bioinformatics with emphasize on biological ontologies, bio-sequence analysis, bio-databases and grid infrastructure applied to bioinformatics. He is involved in various European and National projects on those topics.

Tristan Glatard obtained a master degree from Ecole Centrale de Lyon (France) in 2004 and a PhD from the University of Nice - Sophia Antipolis in 2007. He is a researcher at CNRS (French National Scientific Research Center). Tristan Glatard is working on data-intensive workflows. He is the main developer of the MOTEUR grid-enabled workflow manager.

Carole Goble is a full professor in the the School of Computer Science in the University of Manchester, UK, where she has co-led the Information Management Group since 1997. She has an international reputation in the Semantic Web, e-Science and Grid communities and has led the application of Semantic Web technologies to both the Grid and e-Science, a fusion dubbed the Semantic Grid.

Ian Greenshields is an associate professor of Computer Science and Engineering at the University of Connecticut. He obtained his PhD in Electrical Engineering from Brunel University in the United Kingdom. His current research interests include general computational medicine, medical image processing, scheduling on the Grid and medical visualization.

Giorgio Grillo is a researcher at the Institute Biomedical Technologies CNR of Bari, Italy. He obtained a degree in Computer Science at University of Bari and since 1994 he has been involved in Bioinformatics and particularly in the development of software for genomics and proteomics. He takes part in many national and international research projects for his expertise in the analysis of biosequences and their functional characterization, in the linguistic and computational analysis in nucleotide sequences. Efforts include design, development and implementation of bioinformatic databases largely used by international researchers and in the management of query systems.

Tomasz Gubala received his MSc title in Computer Science from the AGH University of Science and Technology in Kraków, Poland. Since then he worked for CYFRONET supercomputing centre in Kraków as a researcher and for the Section of Computational Science (SCS) at the University of Amsterdam, the Netherlands as a scientific programmer and computer science research assistant. Currently he is involved in a major EU-funded project ViroLab as a chief designer of the virtual laboratory for infectious diseases. He is an external PhD student at the SCS (University of Amsterdam). His main scientific interests are Grid and distributed (cloud) computing, problem-solving environments and services for eScience.

Vicente Hernández García: Full professor in Computer Science and Artificial Intelligence and leader of the Grid and High Performance Computing Group (GRyCAP). He has large experience in Parallel and Grid Computing and Numerical Methods. He has managed and participated in more than 25 European projects, from the III to the VI Framework Programme, and national projects. He has been the Vice-chancellor of Research, Development and Innovation of the Polytechnic University of Valencia (UPV) during the period 2000-2005. He is currently the Scientific Coordinator of the Spanish e-Science Network (NGI).

Pietro Hiram Guzzi received his PhD in Biomedical Engineering in 2008, from Magna Græcia University of Catanzaro. He received his Laurea degree in Computer Engineering in 2004 from the University of Calabria, Rende, Italy. Currently, he is an assistant professor of the University of Catanzaro. His research interests comprise bioinformatics, the analysis of proteomics data, and the analysis of protein interaction networks. Pietro H. Guzzi can be reached at hguzzi@unicz.it.

Moez Ben Hadj Hamida received the Master degree in computer science from the University of Tunis El Manar, Tunisia, in 2005. He is currently pursuing the PhD in large computer sciences at the same university. He is currently lecturer assistant at the Department of Computer Science of Faculty of Sciences of Tunis, Tunisia. His current research interests include parallel, distributed and Grid-based data mining applications.

Daniel Harezlak is a computer scientist at the Academic Computing Center CYFRONET in Krakow. He has graduated from the AGH University of Science and Technology ang got MSc degree in computer science. He has been working on EU-funded projects such as K-WfGrid, ViroLab and GREDIA in the areas of workflow management and web computing, which resulted in many publications. He is also interested in building enterprise applications and agile programming.

Jean Herveg is a member of the Bar of Brussels. Since 2000, he has been with the Research Center on IT & Law (www.crid.be) at Facultés Universitaires de Notre-Dame de la Paix, Namur, Belgium (FUNDP). He has been a lecturer at FUNDP since September 2005, teaching a course on "Medical Law and ICT". He is also teaching a course on "Data Protection" in the post-graduate program "INFOSAFE". His research activity deals with Data Protection and ehealth products and services. His full biography is at: http://www.fundp.ac.be/universite/personnes/page_view/01004558/cv.html).

David Horner received a PhD in Biochemistry from the University of London in 1995. He worked for 7 years at the Natural History Museum in London on molecular sequence evolution and is currently a researcher at the Department of Molecular Biosciences and Biotechnology at the University of Milan. His research interests include comparative genomics and bioinformatics of non-coding RNAs.

Joshua Zhexue Huang is an assistant director at the E-Business Technology Institute (ETI), Honorary professor at the Department of Mathematics of the University of Hong Kong (HKU). Dr. Huang received his PhD degree from the Royal Institute of Technology in Sweden. Dr. Huang is acquired many years of experience in databases and data mining research and has contributed to the development of a series of k-means type algorithms in data mining, including k-modes, fuzzy k-modes and k-prototypes which are being widely used in research and real world applications. Dr. Huang has written more than 90 research papers published in international conferences and journals.

Martin Huber studied at the University of Karlsruhe, Germany and received a PhD degree in computer science in 1999. Since 1996 he has been with Siemens Corporate Technology and Siemens Medical Solutions including an assignment as scientific collaboration manager for computed tomography in the U.S.A. Currently, he is heading an interdisciplinary team in Erlangen, Germany consisting of specialists in imaging and machine learning focusing on information-driven software in healthcare. Further, he is technical coordinator of the EU funded project Health-e-Child as well as coordinator of THESEUS-MEDICO funded by the German ministy of economy. His research interests are in medical imaging and semantic data integration.

Nicolas Jacq is a consultant in industry in grid computing and data center automation. He received his Engineer degree from CUST, Clermont-Ferrand, France, in 2000. He achieved his PhD in Bioinformatics in 2006 in the University of Clermont-Ferrand, France, in collaboration with an industrial partner Communication & System. As research engineer, he deployed a genomic platform on a grid infrastructure within the framework of the DataGrid project (2000-2003). He was involved in the RUGBI project (2004-2006), suported by the French ministry of research, which aimed to create a set of grid services for the bioinformatics. In 2005 and 2006, he was in charge of the first and the second biomedical datachallenges, called WISDOM, in the EGEE and EGEE-II project about the drug discovery against malaria and avian flu. These experiences were the first large scale productions of biological results on several grid infrastructures. He was the technical coordinator of the European SHARE project for the HealthGrid association in 2007.

Emmanuel Jeannot is a researcher at INRIA (Institut National de Recherche en Informatique et en Automatique), LORIA laboratory. He got his PhD and master degree of computer science (resp. in 1996 and 1999) both from Ecole Normale supérieure de Lyon. His main research interests are scheduling for heterogeneous and large- scale environments, data redistribution, experimental tools, adaptive online compression and programming models.

Marek Kasztelnik graduated from the AGH University of Science and Technology in Kraków, Poland and received his MSc in Computer Science. He joined Technologies Group in Research and development department in ComArch in 2005 where he is responsible for creating IT solutions for telecommunication management. He has been also involved in EU-funded projects ViroLab, Gredia and CoreGRID in ACC CYFRONET AGH supercomputing centre in Kraków since 2005. He is applying his scientific knowledge there by developing the concept of resource discovery mechanisms. His scientific interests include enterprise systems, resource discovery, distributed computing, solutions for e-science and telecommunication applications.

Joanna Kocot received her MSc degree in computer science from the AGH University of Science and Technology in Krakow. Since 2006 she has been working for ACC Cyfronet AGH in Krakow. Currently she is involved in EU-funded projects: ViroLab and GREDIA as a designer and developer. Her scientific interests involve distributed computing, Web technologies, programming paradigms and e-science/e-health applications.

Dagmar Krefting is researcher and lecturer at the Institute of Medical Informatics of the Charité - Universitätsmedizin Berlin, Germany since 2004. She obtained her Ph.D in physics at the University of Göttingen in 2003 and worked as postdoctoral fellow at the Fritz-Haber-Institute, Germany. Her current research focus is on image processing of medical ultrasound data and she has taken special research interest in integration of grid based medical image processing applications. She heads of the medical image processing module of the german MediGrid project.

Sriram Krishnan is a Senior Distributed Systems Researcher at the San Diego Supercomputer Center. He received his Master's and PhD degrees in Computer Science from Indiana University in 2001 and 2004 respectively. He is currently the Project Manager for the NEES Cyberinfrastructure Center (NEESit) at SDSC, which is a service-focused organization created to deliver information technology tools to enable earthquake engineers to remotely participate in experiments, perform simulations, organize and share data, and collaborate with colleagues. He is also the lead architect for the Services Oriented Architecture for the National Biomedical Computation Resource (NBCR), which provides transparent access to emerging cyberinfrastructures by "grid-enabling" biomedical codes.

Dariusz Krol is a student at the Institute of Computer Science, AGH University of Science and Technology in Krakow. Since 2007 he has been working for ACC Cyfronet AGH in Krakow where he has been involved in EU-funded projects: ViroLab and Gredia as a scientific developer. He is particularly interested in Web technologies, distributed computing and semantic based monitoring.

Peter Kunszt has a PhD in theoretical physics (lattice QCD) from the University of Bern. He holds a builder status in the Sloan Digital Sky Survey SDSS project, where he was one of the developers of its science archive database at the Johns Hopkins University in Baltimore. After 2001, he has contributed to the EU DataGrid and EGEE projects at CERN, Geneva, where he was leading the development efforts for the data management middleware in both projects. Currently, he is responsible for several projects at the Swiss National Supercomputing Centre CSCS in Manno, an institution of the ETH Zurich.

Giuseppe La Rocca is doctor degree in Software Engineer from the University of Catania, Italy. Since 2004 he joined the Italian INFN of Catania as a scientific collaborator. He has been from 2004 to 2006 scientific collaborator for NA4 work package in the context of the Enabling Grid for E-sciencE (EGEE) Project. Afterward, he has been involved from 2006 to 2007 as a scientific collaborator in the WP6 of the BIOINFOGRID Project. Up to April 2008 he was involved in LIBI Project. Starting from May 2208 he's involved in EGEE-III project. He is author of 7 proceedings of international conferences.

Giovanni Lavorgna is research scientis at DIBIT- Ospedale San Raffaele and professor of Bioinformatics at the Vita-Salute University in Milan. After graduating, he has spent several years at the National Institutes of Health (NIH) in Bethesda, MD, working on transcriptional regulation. After moving back to

Italy, Giovanni has developed several innovative algorithms for the analysis of nuclei acid and proteins. He is the author of several bioinformatics papers and is reviewer for several journals in the field.

Yannick Legré is one of the co-founders and current president of the International HealthGrid Association. He works in the domain of Grid technologies (network grids for healthcare applications and contexts, application integration, service-oriented architectures and software engineering research and development projects involving health service providers and the software industry) since August 2000, in particular within the European Project named DataGrid and now within EGEE, as well as several others. Since early June 2008, he's employed as Consultant and International Relations Manager by Maat-G, a major Grid player (www.maat-g.com)

Pietro Di Lena is research assistant of Computer Science at the University of Bologna, Italy. He received the Laurea degree in Informatica (2003) and the Dottorato di Ricerca in Informatica (PhD in Computer Science, 2007) from the University of Bologna. His research supervisors are Prof. Luciano Margara and Prof. Rita Casadio. His research interests are in Combinatorial Optimization, Cellular Automata, and recently in Bioinformatics.

Pietro Leo is an executive IT architect at IBM GBS Innovation Center in Bari, Italy. He is a permanent member of the IBM Italy TEC, affiliated with the IBM Academy of Technology. His areas of expertise include data, information and application integration, unstructured information management, mining and semantic/conceptual indexing and search, bioinformatics, healthcare, and wireless solutions. He was the technical responsible in a number of multi-millions complex projects for IBM customers and for Industrial Research projects. He is the author or co-author of more than 45 scientific or technical publications, and is the co-author of two books edited by IBM Redbooks.

Daniele Lezzi gained a PhD in Information Technology Engineering in the Center for Advanced Computing Technologies division of the National Nanotechnology Laboratory of the University of Salento. His research covers High Performance, Distributed and Grid Computing and he is also interested in computer security. He is team member of the Grid Resource Broker, of the iGrid Information Service and of the GSI plugin for gSOAP projects. He has been a researcher in the Euro-Mediterranean Centre for Climate Changes involved in the design of the computational infrastructure and it is currently researcher in the Computer Science department of the Barcelona Supercomputing Center.

Wilfred W. Li received his PhD in Biochemistry and Molecular Biology from the University of Southern California (USC), and his postdoctoral training in signal transduction at the University of California, San Diego (UCSD). He is currently the Executive director of the National Biomedical Computation Resource (NBCR at http://nbcr.net), senior fellow in Bioinformatics at the San Diego Supercomputer Center, UCSD, and adjunct professor at the College of Computer Science and Technology, Jilin University, PRC. His current research interests include cluster and grid computing, cyberinfrastructure, high throughput proteome annotation, data mining, virtual screening, structure based drug discovery, multiscale modeling and visualization.

Vito Flavio Licciulli obtained a degree in Computer Science at University of Bari in 1992. Since 1994 he is involved in Bioinformatics with the Bari BioInformatics Group and he is a researcher at the

C.N.R.- Istitute for Biomedical Technologies (ITB). His main research activity are design, development, implementation and maintenance of biological databases, development of bioinformatics software and WEB interfaces, and he has published over 20 scientific publications in these fields. He is responsible for the database and operating systems management at ITB-CNR-Bari institute.

Thomas Lingner received the Diploma degree in computer science in 2005 from the University of Bielefeld, Germany. In 2005 he became a member of the MediGRID project and a group member of the Bioinformatics Department at the University of Göttingen, Germany. He is currently working towards his PhD degree with focus on the application of machine learning techniques for biological sequence analysis. His research interests also include Grid computing and high performance computing.

Frédérique Lisacek received a PhD in Computer Science (Artificial Intelligence) from the University Pierre & Marie Curie, Paris, France in 1984. From then on she has held research positions in bioinformatics in France, Japan and Australia working on knowledge representation and sequence analysis in various fields of biology. She has been involved in proteomics projects since 1999 through work with companies Proteome Systems (Australia) and GeneBio where she acted as CSO until 2005. She currently manages the Proteome Informatics Group of the SIB while being on leave from a tenured position in bioinformatics at the University of Versailles-Saint-Quentin, France.

Weiguo Liu received his bachelor and master degree from the Xi'an JiaoTong University, China in 1998 and 2002, and the PhD degree from the Nanyang Technological University (NTU), Singapore, in 2006. He is currently a research fellow with the Center for Advanced Media Technology at NTU. His research interests include computational biology, parallel algorithms and architectures, high-performance computing, and data mining.

Sabino Liuni has a degree in Biological Sciences at the University of Bari and is researcher at Institute for Biomedical Technologies in Bari (Italy). His interest has been focused mainly on the design and implementation of specialized databases of nucleotide sequences and on development of algorithms for the analysis of biosequence information. He is the coordinator of some research lines and member of research projects in the field of Bioinformatics. In particular his is the responsible of research unit into regional project for development of new bioinformatics and biotechnological tools to analyze gene profile expression of cancer using microarray data.

Maria Antónia Alberto Madope is presently a member of Ford Foundation Alumni Association - Mozambique, and a Computer Science Advisor for Transportes Públicos de Maputo Company. She has a master degree in Computer Science (2007) from Universidade Católica de Santos UNISANTOS, Brazil with a thesis about the IntegraEPI Project titled IntegrAnalysis - an Epidemic Analysis Service on the Grid. She has a bachelor degree in Computer Science (2000) from Universidade Eduardo Mondlane, Mozambique. Her areas of interest are health informatics, distributed systems and informatics systems analysis.

Sergio Maffiolettti has been working as scientist at the Swiss National Supercomputing Centre (CSCS) since June 2004. Previously, he worked in the Pervasive and Artificial Intelligence group at the University of Fribourg where he obtained his PhD in 2004. Currently, he is responsible for the infrastructure management and operation in national and international Grid related projects. He is also co-responsible for the design, planning and operation of the computing infrastructure for the WLCG project as Tier 2 center in Switzerland. Sergio is a member of the Scientific Council and active member of the Swiss National Grid Association (SwiNG).

Giorgio Maggi is professor at the Politecnico di Bari. He graduated in physics at the University of Bari (Italy) in 1972. His research field is in particle physics in close connection with the INFN research program. He has been involved in several high energy physic experiments both at the Fermilab and CERN laboratories, in particular in the Aleph experiment and, now days, in CMS. He is also collaborating to the EGEE european projects to the establishment of a Grid infrastructure and its exploitation both in high energy physics and bioinformatics.

Gaël Le Mahec is a PhD student in computer science at University of Clermont-Ferrand, France. He is a member of the PCSV team (IN2P3 Clermont-Ferrand) since 2005 and his research interests include data management and scheduling in parallel and distributed environments. He received his M.S. from Amiens University, France, in 2004.

Maciej Malawski MSc in Computer Science (AGH, Krakow) and in Physics (Jagiellonian University, Krakow). Since 2001 he is the research and teaching assistant at the Institute of Computer Science AGH. He was involved in EU-funded projects such as CrossGrid, K-WfGrid and CoreGRID at ACC CYFRONET AGH. In 2004-2005 he was a research fellow at Distributed Computing Laboratory at Emory University, Atlanta. Currently he is involved in EU IST ViroLab project, where he is responsible for component-based middleware. His scientific interests include parallel computing, Grid systems, distributed service- and component-based systems, and e-science applications.

David Manset, born in 1978, Mph in Computer Sciences. David Manset is CEO of MAAT France and director of Biomedical Applications for MAAT Gknowledge industrial group. During the last 7 years, he has contributed to European and International projects focussing on e-health with a special emphasis on the application of Grid technologies. His work and team have been awarded several times in major events (such as Salon International des Inventions de Genève 2007, EGEE 2007 Conference, EGEE 2008 User Forum, ICT 2008 etc). David Manset is specialized in the application of formal Model Driven Engineering techniques to Grid system architecture design and development.

Uko Maran is senior research fellow of the Molecular Technology of University of Tartu. He holds a PhD (since 1997) in theoretical and computational chemistry from University of Tartu. His research interests include molecular design and modelling of bio- and macromolecular systems, quantum chemistry, statistical analysis, development of molecular descriptors, distributed computing, etc. He has published more than 50 papers in international pre-reviewed journals related to these areas. He has gained experience as participant and/or site principal investigator in number of EU projects such as IMAGETOX, OpenMolGRID, NANOQUANT, Cardioworkbench and Chemomentum.

Luciano Margara is full professor of Computer Science at the University of Bologna, Italy. He received the Laurea degree in Scienze dell'Informazione (1991) and the Dottorato di Ricerca in Informatica (PhD in Computer Science, 1995) from the University of Pisa. He Joined the University of Bologna in

1995 (Research Associate from 1995 to 2000 and Associate professor from 2000 to 2005). He has been a visiting scientist at the International Computer Science Institute (Berkeley) and a visiting professor at the Cornell University (Department of Computer Science). His research interests are in Discrete Time Dynamical Systems, Optical Networks, Computational Complexity, and recently in Bioinformatics.

Osvaldo Marra received a Laurea Degree in Computer Science from the University of Bari, Italy, in 1994. From December 1994 to January 2002, he was the IT Systems Engineer of the Networks Lab at the University of Lecce. From February 2002 to date, he is the senior IT Systems Engineer of the High Performance Computing Lab at the University of Salento. Recently he has been involved in the SPACI Project (Southern Partnership for Advanced Computational Infrastructures), and he is currently leading the IT technical staff of the Lecce site of the SPACI Consortium. He has been also involved, as a site-manager and a IA64 supporter, in the European Project EGEE (Enabling Grids for E-Science in Europe), and now he is continuing his collaboration with the third phase of the project (EGEE-II).

Pier Luigi Martelli is researcher in Biochemistry at the University of Bologna. He has a degree in Physics and a PhD degree in Physics. His research is mainly focused on the design and implementation of systems based on neural networks, hidden Markov models and support vector machines for the prediction of different features in proteins. PLM is author of more than 30 paper on international peer-reviewed journals and presented the results of his research to more than 50 national and international meetings. He was awarded for the best paper at the ISMB (Intelligent Systems for Molecular Biology) 2002 meeting.

Juan Manuel Gonzalez Martin studied Computing at the University of Castilla La Mancha, now he finishes his studies at UNED. He worked as a programmer in Maat G Knowledge since 2003, until 2006 he was working at diverse proyects of this company, some related to grid tecnology . On March 2006 he joined CETA-CIEMAT, he employed at UAA(Unit of Applications Architecture). He is currently working in projects related to Grid Digital Repositories.

Mirco Mazzucato: Current Activities: Present position: INFN director of Research. CNAF (INFN Advanced Computing Centre) director, since 2004. INFN Grid Project Project Manager since 2000. INFN delegate and member of the Management Board in the (W)LCG, World Computing Grid for LHC Computing Grid Project at CERN since 2002 (~ 6000 physicists in Europe, USA and Asia). Member of the Project Management Board and co-ordinator of the Italian Federation in the FP6 European project EGEE (and its successor EGEE II, EGEE-III) MIUR Italian delegate in the e-Infrastructure Reflection Group.

Richard McClatchey is director of the Centre for Complex Co-operative Systems at the University of the West of England, Bristol and a CERN Associate. He is a Fellow of the Institute of Engineering and Technology (FIET) and of the British Computer Society (FBCS) as well as a Chartered Information Technology Practitioner and a Chartered Engineer. McClatchey established his Research Centre in 1997 and was given a Chair in Applied Computer Science in 2000. His specialities are in the management of distributed data and processes, Grids Computing and the design of data models for the capture of workflow information. He has published over 120 articles in journals or in international conferences and has been on programme committees of numerous conference series during his 18 years at UWE.

McClatchey holds a PhD in High Energy Physics (from University of Sheffield in 1982) and was awarded a DPhil by UWE in 2000 acknowledging his contribution to the field of Computer Science. He has successfully supervised ten PhD students and 12 Masters students and currently supervises six Doctoral students. He has considerable experience in large European projects (e.g. ARTEMIS, CICERO, CRISTAL, MAMMOGRID, STAFF, Health-e-Child and neuGRID) and has several years consultancy experience including ORACLE. McClatchey has held research awards from the Royal Society, the Royal Academy of Engineering and CERN and has been a Visiting professor at the University of the Savoie, France and the National University of Science & Technology at Rawalpindi, Pakistan.

Ivan Merelli received the Laurea degree in Biomedical Engineering from the Polytechnic University, Milan, Italy, in 2003 with a thesis about molecular surface modelling and analysis. His research activity concerns the development of software for sequence based genomics and for structural proteomics researches, with particular interest in protein-protein interactions. He works actively on the high performance implementation of bioinformatics components using parallel programming and distributed platforms. Currently, he is researcher at the Institute of Biomedical Technologies of National Research Council and works at the FIRB - MIUR projects "Italian Bioinformatics Network - ITALBIONET" and "Laboratory of Interdisciplinary Bioinformatics Technologies - LITBIO".

Flavio Mignone is assistant professor of Computer Science at the University of Milan, Italy. He received a PhD in "Molecular and Cell Biology" at the University of Milan in 2004 working on the in-silico characterization of untranslated sequences. He received a master degree in Biological Sciences in 1998 at the University of Milan and a degree in Information Engineering in 2008 at the Politecnico of Turin, Italy.

Luciano Milanesi is currently researcher of the Italian National Research Council (CNR). Since 1988 he is head of the Bioinformatics Division of CNR-ITB. He focused on developing tools for the genome sequence analysis and prediction of gene structure in different organisms. More recently his work has shifted towards studying the functions of genes, promoter prediction, gene expression analysis and the development of databases integrated with sequence analysis tools. He has been principle investigator for the European Projects: TRADAT "TRAnscription Database and Analysis Tools", ORIEL "an Online Research Information Environment for the Life Sciences" and Bioinfogrid "Grid for Bioinformatics applications". He has published contributions in books and scientific publications in bioinformatics.

Russ Miller is UB Distinguished professor of Computer Science and Engineering and senior scientist at the Hauptman-Woodward Medical Research Institute. He has co-authored seminal papers in the fields of parallel algorithms for the mesh, pyramid, hypercube, and reconfigurable architectures in domains that include computational geometry, image analysis, and fundamental data movement operations. He has published nearly 200 papers, chapters, and abstracts of presentations at national and international meetings and has co-authored two textbooks on parallel and sequential algorithms. His work in theoretical and applied parallel algorithms led to his work in molecular structure determination, which was recognized by the IEEE poster "Top Algorithms of the 20th Century." His current research interests are in cyberinfrastructure and grid computing.

Lukasz Miroslaw holds a post-doc position at the University of Zurich. He received a PhD (since 2006) in automation and robotics from Wroclaw University of Technology. His research interests include grid computing, image processing and statistical pattern recognition with expertise in the software design and development. As a grant holder of Max Planck Society, Herbert Quandt Stiftung and Ministry of Saxony he performed several research projects at Max Planck Institute of Cell Biology and Genetics in Dresden, Germany (with Dr. Frank Buchholz) and at Dresden University of Technology (with Prof. Siegfried Fuchs). Author and co-author of several papers in peer reviewed journals including Nature Cell Biology.

Maria Mirto received her Laurea Degree in Computer Engineering from the University of Lecce, Italy in 2002. She received the PhD degree in Computer Engineering from the ISUFI-University of Lecce, in 2006. Since 2006, she has a postdoc position in the LIBI project for the University of Salento. Since 2003 she is the Principal Investigator of the ProGenGrid project. Her research interests include High Performance, Distributed and Grid Computing, Bioinformatics and Workflow. She received the best paper award at the ITCC 2003 conference. She is co-author of more than 45 papers on parallel, grid computing and Bioinformatics.

Paolo Missier Since 2004 Paolo Missier has been a Research Associate at the School of Computer Science, University of Manchester, UK, where he is now a Resarch Fellow. Prior to joining academia, he was a Research Scientist at Telcordia Technologies (formerly Bellcore), NJ, USA from 1994 through 2001, where he gained research and industrial experience in the area of information management and software architectures. He has also been a lecturer in databases at Universita' of Milano Bicocca, in Italy; an independent IT consultant for the Italian public sector; a course instructor in Software Engineering for a number of private companies, and has contributed to research projects in Europe in the area of Information Quality management and information extraction from Web sources. He holds a PhD in Computer Science from the University of Manchester, UK, a MSc in Computer Science from University of Houston, Tx., USA and a BSc and MSc in Computer Science from Universita' di Udine, Italy.

Wayne Mitchell is group leader, Informatics at the ASTAR Experimental Therapeutics Centre, Biopolis, Singapore and Adjunct Associate professor, Division of Information Sciences, School of Computer Engineering, Nanyang Technological University, Singapore. He holds an BA from Harvard University, a BS from the University of Massachusetts, Boston and a PhD from the University of California, Berkeley. He was scientific founder and Chief Genomics Officer at Tao Biosciences, and has published in genomics, bioinformatics, and knowledge engineering.

Pericles A. Mitkas is a professor of Electrical and Computer Engineering at the Aristotle University of Thessaloniki (AUTH) in Greece. Between 1990 and 2000 he was a faculty member with the Department of Electrical and Computer Engineering at Colorado State University in USA. Dr. Mitkas' research interests include databases and knowledge bases, data mining, software agents, enviromatics and bioinformatics. His work has been published in over 150 papers, book chapters, and conference publications. Dr Mitkas received his Diploma of Electrical Engineering from AUTH in 1985 and an MSc and PhD in Computer Engineering from Syracuse University, USA, in 1987 and 1990, respectively.

Yassene Mohammed is research assistant in the department of medical informatics at the Georg-August-University Göttingen, Germany. Studied 1997-2002 biomedical engineering at the University of Damaskus, Licencié in September 2002. Studied 2002-2005 Photonics Engineering at the Faculty of Sciences and Technology University for Applied Science and Art, Göttingen, MSc, January 2005. 2002-2005 research assistant in the medical image processing group of the Department of Medical Informatics, University Göttingen. Since 2005 PhD-thesis in Medical Informatics, Georg-August University, Göttingen, Germany and research assistant in the project MediGRID.

Johan Montagnat is a researcher at CNRS (French National Scientific Research Center). He graduated from the Ecole Normale Supérieure de Cachan in 1996. He obtained a PhD in 1999 and an habilitation thesis in 2006, both from the University of Nice - Sophia Antipolis. His main research interests are medical imaging and grid infrastructures applied to biomedicine. He is involved in various European and national projects on these topics.

Ludovica Montanucci has a PhD in Physics and a Laurea in Physics. Actually she is post-doc reasearcher at the Biocomputing Group, University of Bologna. Her main research interests concern Computational Biology, in particular cross genome comparison by means of machine learning methods and multivariate data analysis techniques.

Jean-Marie Moureaux received his PhD degree in Electrical Engineering from the University of Nice-Sophia Antipolis, France in 1994. Since 1995, he is with the Centre de Recherche en Automatique de Nancy (CRAN, Nancy Université and CNRS). His research interests are image coding, watermarking, lattice vector quantization, multimedia transport over the Internet and medical image compression.

Salvatore My is researcher at Facoltà di Ingegneria of the Technical University of Bari since July 1st 2002 (scientific sector: Experimental Physics). He is member of the Bari's Physics Department. The main activity of Dr. S. My have been in the building of micro-strip silicon modules for the CMS tracker and he has especially been involved in the test of these modules. He is collaborating to the LIBI project, in particular to the porting of bioinformatics applications to grid.

Aisha Naseer is a research fellow working on an EPSRC-funded project RIGHT (Research into Global Healthcare Tools) at the School of Information Systems, Computing and Mathematics, Brunel University, UK. She received her PhD (2007) in Computer Science from Brunel University, UK, specialising in Grid-Based Semantic Integration of Heterogeneous Data Resources on HealthGrids. She has taught various Computer Science and Information Systems courses at both the undergraduate and postgraduate level. She has published several papers in journals and international conferences and also has chapters published in several books. She has chaired various conference sessions and reviewed many papers and book chapters. Her research interests include HealthGrids, semantic interoperability, intelligent data management, HealthGrid applications, health informatics, Grid computing, artificial intelligence and simulation & modelling of healthcare systems.

Alessandro Negro was born in S. Pietro Vernotico (BR) in 1981. He received a summa cum laude bachelor degree in Computer Engineering from the University of Salento, Lecce, Italy in February 2004 and a summa cum laude master degree in Computer Engineering from the same University in April

2006. His research interests include GUI Development, Data Management, Distributed, Grid Computing and Web Services. He is also interested in web and pattern-oriented design. Since 2003 he is a Team Member of the GRelC Project. Since September 2006 he holds a contract position in the LIBI project for the University of Salento.

Virgilio Cavicchioli Neto has a bachelor degree in computer engineering, and is currently pursuing a master degree in information technology applied to public health since 2006 at Federal University of Sao Paulo – UNIFESP, in the Health Informatics department. His thesis will be about georeferencing epidemic cases of dengue fever, tuberculoses and leprosies using open source tools and developing a service which make georreferenced data available to users.

Piotr Nowakowski holds an MSc degree from the AGH University of Science and Technology and is currently employed as a research associate at the Academic Computing Center CYFRONET AGH in Krakow. He has been involved in a number of Grid and distributed computing projects, including CrossGrid, K-WfGrid, ViroLab and GREDIA. He has authored or co-authored approximately 20 publications in the field of Grid and distributed computing. His interests revolve around problems of distributed data access and resource sharing.

Mark Olive is a doctoral candidate in Information Systems at the Bristol Institute of Technology. He graduated from the University of the West of England, Bristol, with a BA in Systems Analysis and was engaged by the SHARE project after a period of work in industry and independent technical consultancy. His subject is knowledge management in healthgrids, with particular emphasis on the interpretation of Integrated Care Pathways as a vehicle for knowledge representation and evolutionary learning.

Alessandro Orro received his PhD in Electronics and Computer Science Engineering in February 2005, after a three-year course at the University of Cagliari, Italy. He currently works at the Institute of Biomedical Technologies of the National Research Council in Segrate, Italy. His research interests are in the field of artificial intelligence (mainly soft computing and multi-agent systems) applied to problems in computational biology and medicine, in particular protein multiple alignment, protein structure prediction and protein function annotation, genetic and clinical data analysis. Presently, he collaborates in the BioinfoGRID project.

Angel Osorio is physician and physicist. He obtained his PhD in 1971. After working as an Associated professor in Madrid (Spain), San Francisco (USA) and London (UK) he joined the CNRS (French National Research Center) in Paris (France) in September 1973 and still working there as a Research director. Since 1995, his research activity is focusing on 3D medical imaging treatment. He is the head of many software packages about automatic 3D treatment of DICOM images. He directed 27 PhD and is currently directing 3 ones. He is the author of 151 published articles and scientific communications. He obtained 8 international awards for his scientific work.

Vitaliy Ostropytskyy is a research associate at the School of Biomedical Sciences at the University of Ulster. He holds a Degree in Computer Sciences, and a PhD (since 2000) in Progressive Information Technologies from Ukraine. He has published over 40 papers. His main research interests are database management, data warehousing and Internet technologies. He has been involved in different international projects.

Graziano Pappadà is a bioinformatic scientist in Italy responsible of the Exhicon s.r.l Bioinformatics unit. He graduated in biology from the Faculty of Science of the University of Bari in 2002 and since then has worked on several bioinformatics projects concerning biological database development, data integration and workflow design in collaboration with the CNR, University of Bari and IBM Innovation Centre.

Marco Passante is PhD student in Information Technology Engineering at the University of Salento, Lecce, Italy. He received a summa cum laude master degree in Computer Engineering from the same University, in October 2008. Since January 2009, he has a contract position in the LIBI International laboratory of BioInformatics, project FIRB 2003, at the University of Salento. His interests concern the area of Biological Data Management, Grid Problem Solving and Grid Workflow Management System.

Cesare Pautasso is assistant professor of Informatics at the University of Lugano, Switzerland. He was a researcher at the IBM Zurich Research Lab and a senior researcher at ETH Zurich. His research focuses on building experimental systems to explore the intersection of model-driven software composition techniques, business process modeling languages, and autonomic/Grid computing. He has developed an interest in Web 2.0 Mashups and Architectural Decision Modeling. He is the lead architect of JOpera, a Grid workflow management tool for Eclipse. His teaching and training activities both in academia and in industry cover Web Development, SOA and emerging Web services technologies.

Giulio Pavesi is assistant professor of Computer Science at the Department of Biomolecular Science and Biotechnology of the University of Milano, where he had obtained both his M.S. and PhD degrees in Computer Science and currently works in the Bioinformatics and Comparative Genomics lab.

Michal Pelczar holds an MSc degree from the AGH University of Science and Technology. He is currently involved in EU-funded ViroLab project. He has co-authored several publications in the field of collecting and mining information about virtual experiments. He is mostly interested in representation and processing of knowledge in Semantic Web.

Xavier Pennec is a research director at the ASCLEPIOS team at INRIA Sophia-Antipolis (France). He holds an Engineering degree from the French Ecole Polytechnique in 1992 and a PhD degree in Computer Science from the same institution in 1996. He was a post-doctoral associate at MIT AI Lab (Mass. USA) in 1997, before joining INRIA in 1998. His main research interests are about statistics on geometric data, in particular for medical image analysis, biomedical image registration and computational anatomy. He is associate editor of the Medical Image Analysis Journal (Elsevier) and chairs the International Workshop on Mathematical Foundations of Computational Anatomy since its inception in 2006.

Graziano Pesole is full professor of Molecular Biology at the University of Bari and leads a research team in Bioinformatics and Comparative Genomics at the Institute of Biomedical Technology (National Research Council). His research interests include bioinformatics, development of tools for genome annotation, comparative genomics and molecular evolution. He is a member of the Editorial Board of international journals (BMC Bioinformatics, BMC Genomics, Computational Biology and Chemistry,

Briefings in Bioinformatics; Comparative and Functional Genomics), author of over 130 papers published in international journals and co-author of books on Bioinformatics and Genomics published by italian (Zanichelli, Gnocchi) and international (Wiley) editors.

Ernesto Picardi is assistant professor at the University of Bari, Italy. He received a PhD in Plant biology in January 2006 at University of Calabria. His research interests are molecular evolution, comparative genomics and bioinformatics. He is the author of 8 papers in international peer-reviewd journals.

Viviana Piccolo is a PhD student of Molecular and Cellular Biology at the University of Milan. She obtained her Master's degree in Biology at the University ROMA TRE (Rome) and her post lauream Master in Bioinformatics at the University of BICOCCA (Milan). She is working in the functional and comparative genomics of non coding RNAs with particular interest in micro RNAs, small interfering RNAs and targets prediction in plants and animals. Her research interests include also the development of tools focused on Support Vector Machine classification for the ab initio discrimination of micro RNA precursors from other genomic hairpin structures.

Ian Piper was awarded a BSc in Biological Sciences (University of Edinburgh) with Honours in Physiology in 1979, an Electronics Diploma (Devry Institute of Technology: Toronto, Canada) in 1984 and a PhD (University of Edinburgh) Dept. Clinical Neurosciences, Faculty of Medicine. Ian has worked within the field of traumatic brain injury research since 1979 first at the National Hospital for Nervous Diseases, Queen Square, London; in Spinal Cord injury research at the University of Toronto, and for the last 20 years first with professor Douglas Miller in Edinburgh and latterly with professor Sir Graham Teasdale in Glasgow, Scotland. During this time he has been responsible for key developments in the field of computerised detection and quantification of adverse events and the measurement of intracranial compliance in patients with traumatic brain injury. He is the coordinator of the Brain Monitoring with information Technology (BrainIT) research group.

Ivan Torres Pisa is currently an associate professor at the Health Informatics Department, Federal University of São Paulo (UNIFESP) and vice-president of the department's graduate program. His areas of interest are Health Informatics, Telemedicine, Medical Decision Support Systems, PACS, and Distributed Systems. He has a PhD in Health Informatics, University of São Paulo (USP), 2003, with a thesis about the MIDster Project. Ivan Pisa also has a MSc degree in Health Informatics, USP, 1999, and a bachelor degree in Mathematics from São Paulo State University (UNESP), Rio Claro Campus, 1997.

Raul Ramos Pollan graduated as Computer Engineer from University of A Coruña, Spain in 1995 and obtained his MSc in Artificial Intelligence in 1997. He is since 2007 director of CETA-CIEMAT in Trujillo, Spain, a grid based computer centre of CIEMAT. He worked during 5 years at CERN (European Laboratory for Particle Physics) IT Division as a software engineer and lecturer. Between 2000 and 2004 he worked for Sun Microsystems as member of their Java Centre providing consultancy in software engineering, as responsible of software development processes and their methodology team. In 2004 he created an SME specialized in the field of aeronautics and aerospace and was in charge of their systems department. In 2006 he joined CETA-CIEMAT as Assistance director until 2007 when he took over his current possition.

Frederic Pourraz After five years of scientific studies, Frédéric Pourraz has joined the LISTIC Laboratory at the University of Savoie and has participated to the ArchWare European Project (IST-2001-32360). During this project, Frédéric has made a PHD in the domain of SOAs. He has then joined the maatG. Knowledge company and work as an SOA expert in the context of the Health-e-Child European Project (IST-2004-027749). Since 2008, Frédéric is an assistant professor of the University of Savoie affiliated to the Software an Systems group of the LISTIC Laboratory.

Francisco Prieto graduated at Universidad Autonoma de Madrid (UAM) in Theoretical Physics. He was granted by the European Commission for the Safety Assessment for Earthquake Risk Reductions (SAFERR) Project. Within SAFERR, he joined outstanding scientific institutions such as Joint Research Centre (JRC) at Ispra (Italy). He obtained a PhD in Physics at UAM in the field of complex systems and chaos theory. He has relevant contributions in prestigious scientific journals in the fields of physics and engineering. Recently leads the Science and Technology Unit at the Extremadura Research Centre for Advanced Technologies.

Fotis E. Psomopoulos is a PhD candidate with the Electrical and Computer Engineering Department of the Aristotle University of Thessaloniki, Greece, where he received his Diploma in 2004. He is a research associate of the Informatics and Telematics Institute of CERTH, where he is a member of the Intelligent Systems and Software Engineering Lab. His research interests include bioinformatics, data mining and grid computing. He has worked on various National and European projects, including "eTHMMY" and "ASSIST" among others. Also, he is a member of the IEEE Computer Society and the Technical Chamber of Greece.

Andreas Quandt received a PhD in bioinformatics from the University of Geneva. He studied Biotechnology in Berlin and was involved in several research projects focusing on tissue engineering (TU Hamburg-Harburg) and stem cells proliferation (Charite Berlin) before he moved more towards the data analysis aspect in life sciences. He continued studying Bioinformatics in Berlin (MSc) and worked in different research centers such as the Max Delbrück Centrum Berlin-Buch, the Max Planck Institute for Molecular Genetics, and the Karolinska Institute Stockholm. His current research is focusing on analysis workflows for mass-spectrometry data in combination with large-scale computing resources.

Vincenzo Quinto is an IT specialist with extensive experience in Java technologies and bioinformatic applications. He worked as an IT consultant for Exhicon s.r.l. within projects for IBM Innovation Centre focusing on concerns related to biological data access and integration. He has developed particular experience in application development, client/server architectures and web services.

Hanene Boussi Rahmouni is a doctoral candidate in Computer Science at the Bristol Institute of Technology. She gained a first class honours degree in Computer Science from the University of the West of England, Bristol, and joined the SHARE team on graduation. She has worked on issues of privacy, confidentiality and security, specialising in the automatic interpretation of policies into permissions and obligations at the operational level in healthgrids.

Alfredo Tirado-Ramos is a researcher at the University of Amsterdam's Informatics Institute, working on software architectures for distributed biomedical applications. He chairs committees in international conferences in topics ranging from Grids for healthcare to teaching computational science and electronic commerce. He is currently a member of scientific committees on health Grids, high performance medicine and biology, learning technologies, and computational science. Alfredo worked previously at Philips Research North America on medical informatics interoperability, and as modeling and methodology facilitator for the DICOM/HL7 joint committee. He has co-authored 3 patents pending and more than 30 peer-reviewed publications in refereed journals and conferences.

Katharina Rasch is a research fellow at the Center for Information Services and High Performance Computing at the Technische Universität Dresden. She holds a degree in Computer Science with a specialization in data management and communication systems. Before joining the Chemomentum project she has been researching music database systems and automatic transcription of music.

Matteo Rè, PhD in Molecular Biology, is a post doctoral fellow at the department of Information Science of University of Milano. He currently works in the Machine learning and Bioinformatics lab.

Jérôme Revillard, born in 1978, Doctor in Computer Sciences from the University of Savoie, in France. Before joining the company MAAT France, he has been working for MAAT Gknowledge for which he was responsible of the developments achieved in various bio-informatics European projects such as the EU FP6 Health-e-Child. Currently, R&D director of MAAT France, he is responsible for all the technical decisions and developments orientations of the company.

Alfonso Rios, born in 1972, electronic engineer and PhD in Physics at CERN (2000), he is the CTO of MAAT Gknowledge. He has been working in several R&D groups in Databases, Data Mining, supercomputers, semantic Web and their applications to health, anthropology, High Energy Physics etc. He is the main author of MAAT's core technology "G".

Elda Rossi has a degree in Chemistry from the University of Modena . Siince 1984 she has been working at CINECA where she is responsible for computational applications in fields related to Chemistry and Life Science. Her expertise is in support to the sceintific research using advanced computing tools, in addition to writing and integrating scientific programs in the ICT field. She has been the principal investigator of several projects funded by the European Community and she was the coordinator of three of them. The last projects (ImmunoGrid) started in February 2006 and is about the simulation of the human immune system.

Ivan Rossi holds a PhD in Computational Chemistry. He has been Adjunct professor of Computational Biology at the University of Bologna and authored around 40 peer-reviewed publications. His main interest in the field of bioinformatics is the application of bioinformatics methods to investigate problems in biology and biochemistry.

Cecilia Saccone has been full professor of Molecular Biology, University of Bari, Italy since 1973. She was President of the CNR Research Area of Bari (1987-2000). She has been and/or is a member of many scientific committees among which: EMBO (European Molecular Biology Organization), European Commission for Nucleotide Sequence Databases, Task Force Biodiversity Committee of the OCSE Forum of Megasciences, CBOL Scientific Advisory Board (Consortium for the Barcoding of Life). Her main research areas of interest are: Functional and Comparative Genomics, Molecular Evolution and Biodiversity, Bioinformatics and Biocomputing.

Mark S. P. Sansom is in the Department of Biochemistry at the University of Oxford, where he is head of the Structural Bioinformatics and Computational Biochemistry unit. He received his degree from Oxford in 1983, and worked for seven years in the University of Nottingham, before returning to Oxford. He has interests in many aspects of computational studies of membrane proteins with special interest in ion channels, bacterial outer-membrane proteins, and the development of e-science and high-end computing for simulation and modelling of complex membrane systems.

Monica Santamaria received a summa cum laude degree in Biological Sciences in 2000 and a PhD in Genetics and Molecular Evolution in 2004 from the University of Bari (IT). At present she works as a researcher at the Institute of Biomedical Technologies of the National Research Council (CNR-IT), where she is involved in Molecular Biodiversity studies and in particular in Fungi DNA barcode surveys. Furthermore she collaborates to human molecular variability studies focused on the pyrosequencing applications on mitochondrial genome.

Ulrich Sax is assistant professor for Medical Informatics at the Georg-August-University in Göttingen, Germany. He finished his PhD-thesis in Medical Informatics 2002 at the Georg-August University, Göttingen, Germany and was Postdoctoral Research Fellow in Children's HST Informatics Program, Harvard-MIT Division of Health Sciences & Technology, Harvard Medical School, Boston 2003-2005. He is head of the CIOffice Medical Research Networks, Göttingen, Germany, vice head of the BMBF-funded project MediGRID within D-Grid. His research is focused on Privacy and security aspects in Medical informatics, Electronic Patient Records, Personal Health Records and how to deal with genomic data in research and health care.

Diego Scardaci was born in Catania, Italy, the 11th September, 1975. He got the degree in Computer Science Engineering on 21th July, 2000. Since 2006 he is working as Researcher for the INFN (Italian National Institute of Nuclear Physics) involved in Grid Computing activities. He worked on Grid interoperability, Grid Security and Grid API development. Moreover, he has been involved in the Dissemination of Advanced Grid knowledge activities, organizing many induction courses. Since 2002-2005 he has worked as Researcher for Telecom Italia Lab and he has designed and developed innovative platforms and services for the mobile telephony operator TIM.

Joël Schaerer received an Engineering degree in Telecommunications from INSA Lyon, France in 2004, and a MSc in Signal and Image Processing also in 2004 from the same university. He is currently a PhD student at CREATIS-LRMN, Lyon, France. His research interests include medical image segmentation, deformable elastic models, and cardiac tractography.

Bertil Schmidt is an Associate professor at the School of Computer Engineering at NTU, Singapore. He received his Master degree in Computer Science from the Kiel University, Germany, in 1995 and his PhD degree from the Loughborough University, UK, in 1999. He has worked with the company ISATEC in the area of embedded parallel systems. His research interests include parallel algorithms and architectures, high-performance computing, and bioinformatics.

Robert Schöne is a research fellow at the Center for Information Services and High Performance Computing at the Technische Universität Dresden. He holds a degree in Computer Science from the Technische Universität Dresden. Before joining the Chemomentum project he was involved in analyzing the performance of parallel programs and in developing performance analysis tools.

Bernd Schuller is core UNICORE developer team and co-head of the open source UNICORE project at SourceForge, as well as a member of the Technical Board of the UNICORE Forum (www.unicore.org). He has extensive experience with development of Grid middleware, deployment and application integration. He has been involved in number of EU projects such as OpenMolGRID and Chemomentum.

Salvatore Scifo is graduated in Computer Science at the University of Catania in 2001. Since 2006 he was involved as researcher in the Grid R&D and Application Support fields with both National Institute of Nuclear Physics of Catania (2006-2007) and Cometa Consortium of Catania (since 2007). From 2000 to 2005 he was employed with several Italian SMEs: as a "Software Designing Engineer " (2000-2002) and as a "Senior Software Engineer" (2002-2005) on both TLC and ITC areas. He has also carried out research activities on Grid-based Digital Libraries and has a particular expertise on data and metadata management in grids.

Gaetano Scioscia is an IT architect at IBM GBS Innovation Centre in Bari, Italy. He graduated in 1995 in physics and earned a PhD in theoretical physics in 1999 in Bari. In 1998, Dr. Scioscia joined IBM, where he worked as an IT specialist and then an IT architect in such fields as data and application integration, and information and knowledge management. During recent years, his work has been focused on bioinformatics and currently he is the IBM Innovation Lab bioinformatic team technical leader.

Giordano Scuderi is graduating in Computer Science at the University of Catania, Italy. Since 1998 he has collaborated with several private companies in projects related to software development, security and IT Infrastructure design. Since 2006 he has been involved in the design and the development of the Secure Storage Service, in the context of Trigrid VL Project.

Giovanna Selvaggi is professor at the University of Bari. She graduated in physics at the University of Bari (Italy) in 1972. She has been involved in several high energy physic experiments both at the Fermilab and CERN laboratories, in particular in the Aleph experiment and, now days, in CMS. She is now collaborating to the LIBI project in particular to the porting of bioinformatics applications to grid.

Sulev Sild is a senior research fellow at the University of Tartu. He holds a PhD (since 2001) in theoretical and computational chemistry from University of Tartu. He has been participating in several international research projects (with prof. Katritzky at University of Florida, with prof. Diercksen at Max Planck Institute in Munich, Germany, and in the OpenMolGRID project funded by European Union). His research interests include molecular design and modelling, application of artificial neural networks, computational chemistry, and Grid computing.

Fabricio Alves Barbosa da Silva obtained his doctoral degree in Computer Science from Université Paris VI (Pierre et Marie Curie) in 2000. He is currently an assistant professor of the Informatics Department at Universidade de Lisboa, Portugal. His research interests are scientific computing, modeling and simulation, distributed computing and grid computing. Recently he has been working on the application of these technologies to the epidemiology and public health areas.

Richard O. Sinnott was awarded a BSc (Honours) in Theoretical Physics from the University of East Anglia, Norwich, England in 1988; an MSc in Software Engineering from the University of Stirling, Scotland in 1983 and a PhD also at Stirling in 1997 in the area of formal modeling of distributed systems. He has worked across a wide range of computing science research areas both in academia and in industry. He has over 100 publications across a range of computing science and application oriented fields, most recently in the area of Grid security and usability especially in the life sciences. He has edited several ITU-T/ISO international standards.

Konstantin Skaburskas, born in 1972, MSc in Biomedical Engineering and Physics from University of Tartu, Estonia. Before joining IT Department at CERN in February 2006, he has been working in the area of computational biomedical engineering, cluster and distributed computing and grids for about seven years. Currently, holding a Fellowship at CERN IT Grid Deployment group, he is involved in two projects: EGEE/WLCG by providing Grid middleware distributed monitoring solutions and in Health-e-Child by providing Grid middleware expertise to the project.

Peter Sloot, PhD, is the group leader of the Section Computational Science and director of the Computer Science Institute of the University of Amsterdam. He is an expert in computational science, modeling and simulation and Grid computing. He has over 200 publications in these fields, has laid the foundations for the conference series of computational science (http://www.iccs-meeting.org/), organized the first European Grid Conference, in February 2005 in Amsterdam (http://genias.biz/egc2005/) and the IEEE eScience Conference in 2006. Moreover, he is editor in chief of a new established international Elsevier-Science journal on grid computing (http://www.elsevier.com/locate/future).

Manuel Rubio del Solar graduated in Computing at the University of Extremadura in 2003. He worked as programmer in e-learning projects and also in the computing service at the University of Extremadura. On 2003 he began his Phd, currently he works on his thesis. In 2005 he began working as project manager in the computing service on the Agriculture department of the Extremadura regional government. On March 2006 he joined CETA-CIEMAT, where leads the Unit of Application Architecture, managing projects related to Grid Digital Repositories and Boinc middleware development. He is participating on the EELA2 and EDGeS projects.

Tony Solomonides is a Reader in Computer Science and Medical Informatics at the Bristol Institute of Technology, University of the West of England, Bristol. He switched from Mathematics to Computer Science in the 1980's and worked in clinical medical informatics with particular reference to diabetes and inflammatory bowel disease for a decade before beginning work on healthgrids. He was a coinvestigator (and briefly at the end of the project Scientific Coordinator) of MammoGrid, the principal investigator for the integrated road map in SHARE, and has continued to contribute to other healthgrid projects, notably Health-e-Child and neuGRID. He is also leading a non-framework European project on pre-implantation genetic diagnosis (PGD) in assisted reproduction. He recently completed a UK-funded project on knowledge sharing in the National Health Service.

Lampros K. Stergioulas is a Reader and director of Undergraduate Studies in the Department of Information Systems and Computing at Brunel University, UK. He is a qualified Chartered Engineer. He has published over 150 papers in journals and international conferences. He has supervised and examined numerous PhD dissertations in the areas of information systems, information processing, and computer science in general. He has held many National and European Grants in educational computing, human-centred communications and computing, information processing, modelling and simulation, and intelligent systems. He has chaired several International Conferences and has been the coordinator of interdisciplinary, international projects. His research interests include human-centred information systems, educational computing, intelligent data processing, information management, modelling and simulation of business processes, and healthcare information systems.

Karl A. Stroetmann, MBA PhD FRSM, is senior associate with *empirica* Communication & Technology Research in Bonn, Germany. Dr. Stroetmann received an MBA from the Free University in Berlin, Germany, and a PhD in Business Administration and Finance from the University of British Columbia in Vancouver, Canada. He has been a consultant to the European Commission, OECD and global industrial players. He was / is project coordinator or principal investigator in European projects on eHealth policy and strategic issues, on market validation and business development (various eTEN studies), economic impact assessment (eHealth IMPACT and EHR IMPACT socio-economic studies), and in health RTD studies (patient safety; interoperability; semantics; business case development for patient data mining applications to improve clinical use of antibiotics as well as of Virtual Physiological Human [VPH] simulation applications for osteoporosis diagnosis and treatment). Recent work also concerned a foresight study on scenarios for ICT-enabled new models of healthcare, research on financing models, and a representative survey of European GPs. In Nov. 2002, he and his wife, Dr. Veli Stroetmann, received an award for: „Telemedicine - Human Applications of IT in Medicine" (Integrata Foundation). They regularly publish in peer-reviewed journals and present at international conferences, including the European Union High Level Ministerial eHealth conferences, national eHealth symposia, and European and global scientific and professional meetings.

Velitchka N. Stroetmann, MD PhD, studied at the Academy for Medical Sciences, Sofia/Bulgaria, and obtained an M.D. degree (1984). After working as a family doctor she continued studies in Informatics and received a PhD in Medical Informatics (1992). After four years as a doctor and researcher in Medical and Health Information Science in Sofia, she came to Germany and joined *empirica* in 1992. There she has been involved in various telemedicine pilot and validation projects, research on older or disabled people, on eHealth applications and information policy issues. She was the chief editor of the highly acclaimed expert report on "Telematic Applications in the Health Sector", and principal investigator for the SATS "Study on the use of advanced telecommunications services by health care establishments in Europe". She was actively involved in the eHealth IMPACT and eHealth ERA projects. Presently, she coordinates a project for the European Commission to improve semantic interoperability among Member State health systems (Semantic Interoperability Deployment and Research Roadmap) and is involved in longer-term eHealth RTD studies (economic issues and business case development of patient data mining applications to improve clinical use of antibiotics [DebugIT] as well as of Virtual Physiological Human [VPH] simulation applications for osteoporosis diagnosis and treatment [VPHOP]). In Nov. 2002, she received an award for: „Telemedicine - Human Applications of IT in Medicine" (Integrata Foundation).

Heinz Stockinger has been working in Grid projects in Europe (SIB, CERN, etc.) and in the USA (Stanford Linear Accelerator Center) in various technical, scientific and management functions. Heinz is affiliated with the Swiss Institute of Bioinformatics where he works on diverse Grid subjects. He has been appointed "Privatdozent" at the University of Vienna - leading the Research Lab for Computational Technologies and Applications in 2005. Currently, he is also a lecturer at the Swiss Federal Institute of Technology in Lausanne (EPFL). Heinz holds a PhD degree in Computer Science and Business Administration from the University of Vienna, Austria.

Kaihsu Tai is a research associate in the Department of Biochemistry, University of Oxford. He led the team to develop BioSimGrid. He received a B.Sc. (Hon.) from California Institute of Technology (1998) and PhD in chemistry from the University of California, San Diego (2002) with Prof. J. Andrew McCammon. His research interests include simulations of biomolecular systems. He is a Chatered Chemist, a Member of the Royal Society of Chemistry, and an advisory member of the Green Economics Institute.

Daniele Tartarini is PhD student in "Interdisciplinary sciences and technologies" for the Nanoscience and Grid computing section of the "Scuola Superiore ISUFI", University of Salento, Lecce, Italy. He received a summa cum laude master degree in Computer Engineering from the University of Lecce, Italy, in July 2006. Since September 2006, he has a contract position in the LIBI International laboratory of BioInformatics, project FIRB 2003, at the University of Salento. His research interests concern the area of High Performance Computing, Distributed and Grid Computing, Neural networks, BioInformatics and NanoBiosciences. He is coauthor of some papers in refereed journals on Grid Computing & NanoBioscience.

Dorin Tcaci graduated in Mechanics at Moldova Technical University (Chişinău, Moldova) in 1998. He worked in Moldova's companies as software engineer on many projects related to geoinformational systems, e-commerce and e-learning services. From 2006 he started to work at "MAAT-G knowledge" and has been involved in projects like Health-e-Child and MammoGrid+. On 2006 he began his Phd at University Jaume I, Castellon. On January 2007, as a result of an egreement between "MAAT-G knowledge" and "CETA-CIEMAT", he started to work at CETA on Grid Digital Repositories especially on the DRI & MammoGrid+ projects.

Romain Texier completed his PhD in 2007 at University of Paris Sud, which focused on quality of service adapted to responsiveness for EGEE computing resources. He has been the main developer of the gPTM3D software. He joined the I3S laboratory as CNRS engineer since 2007 and currently contributes to the Medical Data Manager. Antonio Turi is working in bioinformatic from 2003 at the Institute for Biomedical Technologies, Bari, Italy. From 2007 he is interesting in grid technology at the University of Salento, Lecce & SPACI Consortium, Italy. He received a master degree in Computer Science in 2002 and, at present, he is last year PhD student in knowledge discovery in databases (KDD) at University of Bari, Italy.

Livia Torterolo received her master degree cum laude in Biomedical Engineering from the University of Genoa, Italy, in 2004. After her admission into the engineering profession in 2004, she started collaborating with several private companies related to grid computing technologies. Since 2006, she has been a PhD student at DIST, University of Genoa, Italy. In 2007-2008 she was visiting student at Trinity College in Dublin, Ireland. Her research activity is focused on grid portal technology and grid-enabled biomedical applications.

Giuseppe Tradigo received his Laurea degree in Computer Engineering from the University of Calabria, Rende, Italy, in 2003. Currently he is working with the bioinformatics group of University "Magna Græcia" of Catanzaro. His research interests are in geographical databases and GIS, grid computing, software engineering, protein structure prediction, and grid environments for bioinformatics applications. Giuseppe Tradigo can be reached at gtradigo@si.deis.unical.it.

Gabriele Trombetti received the PhD degree in Computer Science Engineering in 2008 at the University of Bologna, Italy with a thesis on Enabling Bioinformatics Applications on the Grid Platform. Previously he had received the Laurea (MS) degree with full marks (100/100 cum Laude) in Computer Science Engineering in 2004 at the same university with a thesis on Modelling and Validation Tools for QoS Enabled Distributed Systems and Applications. The thesis was developed at the DOC group at Vanderbilt University, Nashville, TN where he also worked as a Research assistant in 2004 for researches on the same topics. In 1996 he was graduated at the High school Liceo Scientifico Augusto Righi in Bologna, Italy with full marks (60/60). Since 2008 he is a researcher at the Institute for Biomedical Technologies, National Research Council, Italy.

Alexey Tsymbal is a research scientist at Siemens AG in Erlangen, Germany. He holds a university diploma with distinction in software engineering gained from National Tech University, Ukraine in 1997. He received his PhD in computer science in 2002 from the University of Jyväskylä, Finland. In 2003-2006 he also took part in research projects at Trinity College Dublin, Ireland. His research interests include machine learning and data mining, knowledge-based systems, and applications of AI techniques to biomedical domains. He has more than 50 peer-reviewed publications in these areas".

Angelica Tulipano graduated in Physics at the University of Bari in 2003 and obtained a PhD in 2008 from the same institution working on microarray data analysis using clustering methods developed in mechanical statistics. She was involved in an european project on grid applications to Bioinformatics.

Andrea Vanni gained a degree in Computer Science from Bologna University. He joined CINECA in 2002 where he works in Grid computing. He is involved in various European projects such: DEISA2, eDeisa, HPC-Europa, Unigrids and PRACE. He has a wide experience in grid solutions developed for Unicore, Globus and Gridsphere. Since 2006 he has been the co-chair of the DEISA2 WP4 and eDEISA Service Activity 3.

Marco Vassura is research assistant of Computer Science at the University of Bologna, Italy. He received the Laurea degree in Informatica (2001) and the Dottorato di Ricerca in Informatica (PhD in Computer Science, 2005) from the University of Bologna. His research supervisors are Prof. Luciano Margara and Prof. Rita Casadio. His research interests are in Combinatorial Optimization, Optical Networks, and recently in Bioinformatics.

Pierangelo Veltri is an assistant professor of Computer Engineering at University "Magna Græcia" of Catanzaro, Italy, since 2002. He received his PhD in computer science in 2002, from University of Orsay at INRIA Rocquencourt, France. From 1998 to 2000 he was member of the Verso group and he participated to the Chorochronous project (1998-1999) and at the Xyleme project (1999-2001). From 2002 he partecipate to several projects, whose main topics are pure computer science (such as Grid computing, geographical and temporal databases, XML views) or more recently bioinformatics topics (spectra data analysis for cancer prevention, proteomics, prediction of protein structures). His main research interests are: XML and semi-structured database, data modelling, spatial and multidimensional databases, protein matching structure, proteins structure prediction, data management on grid computing. Pierangelo Veltri can be reached at veltri@unicz.it

Saverio Vicario is a researcher the Institute of Biomedical Technologies of the National Research Council (CNR-IT), where he works on Molecular Biodiversity within the LIBI project. In 1999 he graduated in Biology at University of Rome "TorVergata" and got a PhD in 2006 from the Department of Ecology and Evolutionary Biology at Yale University. His research interests are molecular evolution and molecular systematics both in basic research, as evolution of codon usage, and applied research, as the barcode initiative.

Fred Viezens is research assistant in the department of medical informatics at the Georg-August-University Göttingen, Germany. Studied 1988-1993 Informatics at the Technical University "Otto-von-Guericke", Magdeburg. 1994-1997 Lecture at the DEKRA Academy GmbH. 1998-2001 Freelance Activity in an Engineer's Office. 2001-2002 research assistant at the Otto-von-Guericke University Magdeburg, Institute for Biometry and Medical Informatics. 2003 assistant in the MBR Computing Centre GmbH, Magdeburg. 2004-2005 research assistant at the Otto-von-Guericke University Magdeburg, Faculty for Mechanical Engineering, Institute for Material Engineering and Material Testing, Institute for Logistical Systems. Since 2006 research assistant at the University Medical Center Göttingen.

Jakub Wach holds and MSc degree from the AGH University of Science and Technology. He is currently employed as a Technical Consultant at the Volantis Systems, leader in mobile solutions middleware. He has also experience from other companies, including ACK Cyfronet AGH and GridwiseTech. His main fields of interest are large distributed systems and leading-edge novel solutions for e-Science. He has been invloved in EU-IST K-Wf Grid and ViroLab projects, primarily as an architect and leading engineer of provenance systems. He is a co-author of eight scientific papers.

Qiang Wang is now a PhD candidate student at the School of Computer Science of Harbin Institute of Technology (HIT). Mr. Wang's research interests include data mining and bioinformatics, especially the study and development of novel clustering algorithms for microarray data analysis. He is now taking active part in the research and development of China Drug Discovery Grid (DDG) project.

Charles Weeks is a senior research scientist in the Department of Structural Biology at the Hauptman-Woodward Institute. His research interests lie in computational crystallography, in particular the development of algorithms and software for automating the process of protein structure determination. He is a coauthor of the widely used programs *SnB* and *BnP*.

Nadya Williams is a software architect at the Institute of Organic Chemistry at the University of Zurich. She has a Master of Science degree in Oceanography from Far-East State University, Vladivostok, Russia, and a Master of Science in Computer Science from University of California, San Diego. She has worked at the University of California, San Diego for various projects involving oceanographic research, and dealing with all aspects of software and hardware installation and maintenance, and grid computing.

Petra Wilson is a director in the Healthcare team of Cisco's Internet Business Solutions Group (IBSG), focusing on helping healthcare providers and deciders develop their strategies and implementation plans for "Connected Health" information management and technology. Previously, she was deputy director of the European Health Management Association (EHMA), a membership organisation committed to improving healthcare in Europe by raising standards of managerial performance in the health sector. At EHMA she had responsibility for the association's Brussels' office and, in particular, EHMA's relationship with the European Commission. In that role, Petra advised a wide number of European healthcare organisations on EU-level health policy and assisted them in engaging with policy-makers in the Commission. Prior to EHMA, Petra worked for the European Commission in its eHealth directorate. She worked specifically on the European eHealth Action Plan, which commits all member states of the European Union to adopt road maps for the implementation of Connected Health solutions. Petra also advised the Commission on legal and ethical issues surrounding use of information technologies in healthcare. Alongside her role at Cisco Petra currently holds the chairmanship of the European Policy Workgroup of the Continua Health Alliance, as well as the chair of the Governing Council of Health Information and Management Systems Society Europe. She is also acts as an advisor to the Health on the Net Foundation. Originally trained in law, Petra spent several years as a lecturer and Senior Research Fellow at Nottingham University, UK, specializing in healthcare law and European law. She has a PhD in public health law from Oxford University, and has published widely on the legal aspects of using information technologies in healthcare. Originally from the UK, she is fluent in English, German, and French.

Yunming Ye is a vice director at the E-Business & Intelligent Enterprise Computing Research Center (ICES), Associate professor and PhD Supervisor at the School of Computer Science of Harbin Institute of Technology (HIT). Dr. Ye received his PhD degree form the Shanghai Jiao Tong University, visiting scholar at E-Business Technology Institute (ETI) of Hong Kong University. Dr. Ye has many years of experience in business intelligence, data mining and web searching research and has written more than 40 research papers published in international conferences and journals. Dr. Ye also is now responsible for 3 national projects.

Kunqian Yu is an associate professor of Shanghai Institute of Materia Medica (SIMM), CAS. He graduated from the State Key Lab of Institute of Theoretical Chemistry of Jilin University July 2002, and then worked as a post-doctor in Shanghai Institutes for Life Sciences (SIBS), CAS. Dr. Yu did much research work on Molecular Dynamics (MD) simulations for polymer, studied the recognition and interaction of ligands with proteins by computer simulation, and also preformed virtual drug screening and computer aided drug design for tens of important protein targets. He now focuses on the drug development for avian influenza virus, and also is responsible for the construction of China drug discovery grid (DDG).

Federico Zambelli is a PhD student of Molecular Biology at the Department of Biomolecular Science and Biotechnology of the University of Milano, where he currently works in the Bioinformatics and Comparative Genomics lab.

Vincenzo Zara is full professor of Biochemistry at the University of Salento. His research has included the following topics: study of the import pathway of metabolite carriers into isolated mitochondria; study of the assembly of the hydrophobic proteins in the inner mitochondrial membrane; purification and structural and functional characterization of mitochondrial carriers; effect of different hormonal and nutritional states on the hepatic fatty acid biosynthesis; study of the biogenesis of the yeast cytochrome bc_1 complex. He is the author of 157 scientific articles, most of them published in specialized international journals. He is also a reviewer for several international journals.

Index

C

D

H

I

J

K

L

M

N

O

P

Q

R

S

T

U

V

W

X

Z